Prepare for Class "Read t[...]

Feature	Description		[...]ge
Every chapter begins with....			
Chapter Opening Article & Project	Each chapter begins with a current article and ends with a related project.	The Article describes a real situation. The Project lets you apply what you learned to solve a related problem.	103, 203
Every section begins with....			
Learning Objectives ✓1	Each section begins with a list of objectives. Objectives also appear in the text where the objective is covered.	These focus your studying by emphasizing what's most important and where to find it.	104
Most sections contain...			
PREPARING FOR THIS SECTION	Most sections begin with a list of key concepts to review with page numbers.	Ever forget what you've learned? This feature highlights previously learned material to be used in this section. Review it, and you'll always be prepared to move forward.	104
Now Work the 'Are You Prepared?' Problems	Special problems that support the Preparing for This Section feature.	Not sure you need the Preparing for This Section review? Work the 'Are You Prepared?' problems. If you get one wrong, you'll know exactly what you need to review and where to review it!	104, 113
"Now Work" PROBLEMS	These follow most examples and direct you to a related exercise.	We learn best by doing. You'll solidify your understanding of examples if you try a similar problem right away, to be sure you understand what you've just read.	110
WARNING	Warnings are provided in the text.	These point out common mistakes and help you to avoid them.	119
Explorations and **Seeing the Concept**	These represent graphing utility activities to foreshadow a concept or solidify a concept just presented.	You will obtain a deeper and more intuitive understanding of theorems and definitions.	69, 165
In Words	These provide alternative descriptions of select definitions and theorems.	Does math ever look foreign to you? This feature translates math into plain English.	152
Calculus Icon △	These appear next to information essential for the study of calculus.	Pay attention – if you spend extra time now, you'll do better later!	53
Showcase New! EXAMPLES	These examples provide "how-to" instruction by offering a guided, step-by-step approach to solving a problem.	With each step presented on the left and the mathematics displayed on the right, students can immediately see how each step is employed.	168–169
Model It! Examples and Problems 🌑	Marked with 🌑 . These are examples and problems that require you to build a mathematical model from either a verbal description or data.	It is rare for a problem to come in the form, "Solve the following equation". Rather, the equation must be developed based on an explanation of the problem. These problems require you to develop models that will allow you to describe the problem mathematically and suggest a solution to the problem.	127

Practice "Work the Problems"

Feature	Description	Benefit	Page
"Assess Your Understanding" contains a variety of problems at the end of each section.			
'Are You Prepared?' Problems	These assess your retention of the prerequisite material you'll need. Answers are given at the end of the section exercises. This feature is related to the Preparing for This Section feature.	Do you always remember what you've learned? Working these problems is the best way to find out. If you get one wrong, you'll know exactly what you need to review and where to review it!	161, 172
Concepts and Vocabulary	These short-answer questions, mainly Fill-in-the-Blank and True/False items, assess your understanding of key definitions and concepts in the current section.	Learning math is more than memorization – it's about discovering connections. These problems help you understand the 'big ideas' before diving into skill building.	172
Skill Building	Correlated to section examples, these problems provide straightforward practice.	It's important to dig in and develop your skills. These problems provide you with ample practice to do so.	172–175
Mixed Practice	These problems offer comprehensive assessment of the skills learned in the section by asking problems that relate to more than one concept or objective. These problems may also require you to utilize skills learned in previous sections.	Learning mathematics is a building process. Many concepts are interrelated. These problems help you see how mathematics builds on itself and also see how the concepts tie together.	175
Applications and Extensions	These problems allow you to apply your skills to real-world problems. These problems also allow you to extend concepts learned in the section.	You will see that the material learned within the section has many uses in everyday life.	175–176
Discussion and Writing	"Discussion and Writing" problems are colored red. These support class discussion, verbalization of mathematical ideas, and writing and research projects.	To verbalize an idea, or to describe it clearly in writing, shows real understanding. These problems nurture that understanding. Many are challenging but you'll get out what you put in.	176
"Now Work" PROBLEMS	Many examples refer you to a related homework problem. These related problems are marked by a pencil and yellow numbers.	If you get stuck while working problems, look for the closest Now Work problem and refer back to the related example to see if it helps.	163

Review "Study for Quizzes and Tests"

Feature	Description	Benefit	Page
Chapter Reviews at the end of each chapter contain...			
"Things to Know"	A detailed list of important theorems, formulas, and definitions from the chapter.	Review these and you'll know the most important material in the chapter!	196–197
"You Should Be Able To..."	Contains a complete list of objectives by section, examples that illustrate the objective, and practice exercises that test your understanding of the objective.	Do the recommended exercises and you'll have mastery over the key material. If you get something wrong, review the suggested page numbers and try again.	198
Review Exercises	These provide comprehensive review and practice of key skills, matched to the Learning Objectives for each section.	Practice makes perfect. These problems combine exercises from all sections, giving you a comprehensive review in one place.	199–201
CHAPTER TEST	About 15–20 problems that can be taken as a Chapter Test. Be sure to take the Chapter Test under test conditions—no notes!	Be prepared. Take the sample practice test. This will get you ready for your instructor's test. If you get a problem wrong, watch the Chapter Test Prep video.	202
CUMULATIVE REVIEW	These problem sets appear at the end of each chapter, beginning with Chapter 2. They combine problems from previous chapters, providing an ongoing cumulative review.	These are really important. They will ensure that you are not forgetting anything as you go. These will go a long way toward keeping you constantly primed for quizzes and tests.	202
CHAPTER PROJECTS	The Chapter Project applies what you've learned in the chapter. Additional projects are available on the Instructor's Resource Center (IRC).	The Project gives you an opportunity to apply what you've learned in the chapter to solve a problem related to the opening article. If your instructor allows, these make excellent opportunities to work in a group, which is often the best way of learning math.	203

Fifth Edition

Trigonometry
A Right Triangle Approach

Michael Sullivan
Chicago State University

Michael Sullivan, III
Joliet Junior College

PEARSON

Prentice
Hall

Upper Saddle River, New Jersey 07458

000007

Library of Congress Cataloging-in-Publication Data

Sullivan, Michael/Sullivan, Michael, III
  Trigonometry: A Right Triangle Approach, 5e
    p. cm.
  ISBN 13: 978-0-13-602896-3; ISBN 10: 0-13-602896-9

Complete CIP data is on file at the Library of Congress

Senior Editor: *Adam Jaworski*
Editorial Director, Mathematics: *Christine Hoag*
Sponsoring Editor: *Dawn Murrin*
Executive Marketing Manager: *Kate Valentine*
Editorial Assistant/Print Supplements Editor: *Joseph Colella*
Project Manager, Production: *Bob Walters, Prepress Management, Inc.*
Senior Managing Editor: *Linda Mihatov Behrens*
Senior Operations Supervisor: *Diane M. Peirano*
Executive Marketing Manager: *Becky Anderson*
Marketing Assistant: *Kathleen DeChavez*
Art Director: *Heather Scott*
Interior Designer: *Tamara Newnam*
Cover Designer: *Tamara Newnam*
A V Project Manager: *Thomas Benfatti*
TestGen: *Mary Durnwald*
MathXL: *Phil Oslin*
Director, Image Resource Center: *Melinda Patelli*
Manager, Rights and Permissions: *Zina Arabia*
Manager, Visual Research: *Beth Brenzel*
Image Permission Coordinator: *Joanne Dippel*
Photo Researcher: *Kathy Ringrose*
Art Studios: *Laserwords/Artworks*
    Senior Managing Editor, Art Production and Management: *Patricia Burns*
    Manager, Production Technologies: *Matthew Haas*
    Manager, Illustration Production: *Sean Hogan*
    Assistant Manager, Illustration Production: *Ronda C. Whitson*
    Illustrators: *Stacy Smith, Mark Landis, and Audrey Simonetti*
    Quality Assurance: *Pamela Taylor, Timothy Nguyen, Trinell Hartranft*
Compositor: *ICC Macmillan Inc.*

© 2009, 2006, 2003, 2000, 1996 Pearson Education, Inc.
Pearson Prentice Hall
Pearson Education, Inc.
Upper Saddle River, New Jersey 07458

Pearson Prentice Hall™ is a trademark of Pearson Education, Inc.

Printed in the United States of America
10 9 8 7 6 5 4 3 2

ISBN: 978-0-13-602896-3
ISBN: 0-13-602896-9

Pearson Education LTD., *London*
Pearson Education Australia PTY, Limited, *Sydney*
Pearson Education Singapore, Pte. Ltd
Pearson Education North Asia Ltd. *Hong Kong*
Pearson Education Canada, Ltd., *Toronto*
Pearson Educación de Mexico, S.A. de C.V.
Pearson Education – Japan, *Tokyo*
Pearson Education Malaysia, Pte. Ltd

For Our Students
Past and Present

Contents

Three Distinct Series

Students have different goals, learning styles, and levels of preparation. Instructors have different teaching philosophies, styles, and techniques. Rather than write one series to fit all, the Sullivans have written three distinct series. All share the same goal—to develop a high level of mathematical understanding and an appreciation for the way mathematics can describe the world around us. The manner of reaching that goal, however, differs from series to series.

Contemporary Series, Eighth Edition

The Contemporary Series is the most traditional in approach, yet modern in its treatment of precalculus mathematics. Graphing utility coverage is optional and can be included or excluded at the discretion of the instructor: *College Algebra, College Algebra Essentials, Algebra & Trigonometry, Trigonometry, Precalculus.*

Enhanced with Graphing Utilities Series, Fifth Edition

This series provides a more thorough integration of graphing utilities into topics, allowing students to explore mathematical concepts and foreshadow ideas usually studied in later courses. Using technology, the approach to solving certain problems differs from the Contemporary Series, while the emphasis on understanding concepts and building strong skills does not: *College Algebra, Algebra & Trigonometry, Trigonometry: A Right Triangle Approach, Precalculus.*

Concepts through Functions Series, First Edition

This series differs from the others, utilizing a functions approach that serves as the organizing principle tying concepts together. Functions are introduced early in various formats. This approach supports the Rule of Four, which states that functions are represented symbolically, numerically, graphically, and verbally. Each chapter introduces a new type of function and then develops all concepts pertaining to that particular function. The solutions of equations and inequalities, instead of being developed as stand-alone topics, are developed in the context of the underlying functions. Graphing utility coverage is optional and can be included or excluded at the discretion of the instructor: *College Algebra; Precalculus, with a Unit Circle Approach to Trigonometry; Precalculus, with a Right Triangle Approach to Trigonometry.*

To the Student

As you begin, you may feel anxious about the number of theorems, definitions, procedures, and equations. You may wonder if you can learn it all in time. Don't worry, your concerns are normal. This textbook was written with you in mind. If you attend class, work hard, and read and study this book, you will build the knowledge and skills you need to be successful. Here's how you can use the book to your benefit.

Read Carefully

When you get busy, it's easy to skip reading and go right to the problems. Don't. . . the book has a large number of examples and clear explanations to help you break down the mathematics into easy-to-understand steps. Reading will provide you with a clearer understanding, beyond simple memorization. Read before class (not after) so you can ask questions about anything you didn't understand. You'll be amazed at how much more you'll get out of class if you do this.

Use the Features

We use many different methods in the classroom to communicate. Those methods, when incorporated into the book, are called "features." The features serve many purposes, from providing timely review of material you learned before (just when you need it), to providing organized review sessions to help you prepare for quizzes and tests. Take advantage of the features and you will master the material.

To make this easier, we've provided a brief guide to getting the most from this book. Refer to the 'Prepare for Class,' 'Practice,' and 'Review' pages on the inside front cover of this book. Spend fifteen minutes reviewing the guide and familiarizing yourself with the features by flipping to the page numbers provided. Then, as you read, use them. This is the best way to make the most of your textbook.

Please do not hesitate to contact us, through Pearson Education, with any questions, suggestions, or comments that would improve this text. We look forward to hearing from you, and good luck with all of your studies.

Best Wishes!

Michael Sullivan

Michael Sullivan, III

Preface to the Instructor

As professors at both an urban university and a community college, Michael Sullivan and Michael Sullivan, III, are aware of the varied needs of Trigonometry students, ranging from those who have little mathematical background and a fear of mathematics courses, to those having a strong mathematical education and a high level of motivation. For some of your students, this will be their last course in mathematics, whereas others will further their mathematical education. This text is written for both groups.

As a teacher, and as an author of precalculus, engineering calculus, finite mathematics, and business calculus texts, Michael Sullivan understands what students must know if they are to be focused and successful in upper-level math courses. However, as a father of four, he also understands the realities of college life. As an author of a developmental mathematics series, Michael's co-author and son, Michael Sullivan, III, understands the trepidations and skills students bring to the Trigonometry course. Michael, III also believes in the value of technology as a tool for learning that enhances understanding without sacrificing math skills. Together, both authors have taken great pains to ensure that the text contains solid, student-friendly examples and problems, as well as a clear and seamless writing style.

A tremendous benefit of authoring a successful series is the broad-based feedback we receive from teachers and students. We are sincerely grateful for their support. Virtually every change in this edition is the result of their thoughtful comments and suggestions. We are sincerely grateful for this support and hope that we have been able to take these ideas and, building upon a successful fourth edition, make this series an even better tool for learning and teaching. We continue to encourage you to share with us your experiences teaching from this text.

Features in the Fifth Edition

Rather than provide a list of new features here, that information can be found on the endpapers in the front of this book. This places the new features in their proper context, as building blocks of an overall learning system that has been carefully crafted over the years to help students get the most out of the time they put into studying. Please take the time to review the features listed in the front of this book and to discuss them with your students at the beginning of your course. Our experience has been that when students utilize these features, they are more successful in the course.

New to the Fifth Edition

- **Showcase Examples** are used to present examples in a guided, step-by-step format. Students can immediately see how each of the steps in a problem are employed.

The "How To" examples have a two-column format in which the left column describes the step in solving the problem and the right column displays the algebra complete with annotations.

- **Model It** examples and exercises are clearly marked with a ● icon. These examples and exercises are meant to develop the student's ability to build models from both verbal descriptions and data. Many of the problems involving data require the students to first determine the appropriate model (linear, quadratic, and so on) to fit to the data and justify their choice.

- **Exercise Sets** at the end of each section have been classified according to purpose. Where appropriate, more problems to challenge the better student have been added. In addition, **Mixed Practice** exercises have been added where appropriate so that students may synthesize skills from a variety of sections.

- **Applied Problems** have been updated and many new problems involving sourced information as well as data sets have been added to bring relevance and timeliness to these exercises.

- **New Content**
  - Section A.2, Geometry Essentials, now contains a discussion of congruent and similar triangles.
  - Examples and exercises that involve a more in-depth analysis of graphing and exponential and logarithmic functions are provided.
  - Examples and exercises that involve graphing a wider variety of inverse trigonometric functions are provided.

Using the Fifth Edition Effectively with Your Syllabus

To meet the varied needs of diverse syllabi, this book contains more content than is likely to be covered in a Trigonometry course. As the chart illustrates, this book has been organized with flexibility of use in mind. Within a given chapter, certain sections are optional (see the detail following the flowchart) and can be omitted without loss of continuity.

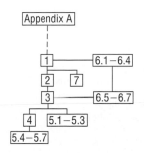

Chapter 1 Graphs and Functions

This chapter lays the foundation for the entire course.

Chapter 2 Trigonometric Functions

Section 2.8 may be omitted in a brief course.

Chapter 3 Analytic Trigonometry

Sections 3.2, 3.6, and 3.8 may be omitted in a brief course.

Chapter 4 Applications of Trigonometric Functions

Sections 4.4 and 4.5 may be omitted in a brief course.

Chapter 5 Polar Coordinates; Vectors

Sections 5.1–5.3 and Sections 5.4–5.7 are independent and may be covered separately.

Chapter 6 Analytic Geometry

Sections 6.1–6.4 follow in sequence. Sections 6.5, 6.6, and 6.7 are independent of each other, but each requires Sections 6.1–6.4.

Chapter 7 Exponential and Logarithmic Functions

Sections 7.1–7.4 follow in sequence. Sections 7.5, 7.6, and 7.7 are optional.

Appendix Review

This chapter consists of review material. It may be used as the first part of the course or later as a just-in-time review when the content is required. Specific references to this chapter occur throughout the book to assist in the review process.

Acknowledgments

Textbooks are written by authors but evolve from an idea to final form through the efforts of many people. It was Don Dellen who first suggested this book and series. Don is remembered for his extensive contributions to publishing and mathematics.

Thanks are due the following people for their assistance and encouragement in the preparation of this edition:

- From Pearson Education: Adam Jaworski for his substantial contributions, ideas, and enthusiasm; Becky Anderson, for her creativity and energy; Kate Valentine, who remains a huge fan and supporter; Dawn Murrin, for her unmatched talent at getting the details right; Bob Walters for his superb organizational skills in directing production; Chris Hoag for her continued support and genuine interest; Greg Tobin for his leadership and commitment to excellence; and the Pearson Math and Science Sales team, for their continued confidence and personal support of our books.

- Accuracy checkers: Kevin Bodden and Randy Gallaher, for creating the Solutions Manuals and accuracy checking answers; LaurelTech's editors, Teri Lovelace, Harold Whipple, Jill McClain-Wardynski and Peggy Irish who read the entire manuscript and checked all the answers.

- Reviewers: Debra Kopcso, *Louisiana State University;* Lynn Marecek, *Santa Ana College;* Kathleen Miranda, *SUNY at Old Westbury;* Karla Neal, *Louisiana State University;* Leticia Oropesa, *University of Miami;* Laura Pyzdrowski, *West Virginia University;* Mike Rosenthal, *Florida International University;* Phoebe Rouse, *Louisiana State University;* Brenda Santistevan, *Salt Lake Community College;* Larissa Williamson, *University of Florida.*

Finally, we offer our grateful thanks to the dedicated users and reviewers of our books, whose collective insights form the backbone of each textbook revision.

Our list of indebtedness just grows and grows. And, if we've forgotten anyone, please accept our apology. Thank you all.

James Africh, College of DuPage
Steve Agronsky, Cal Poly State University
Grant Alexander, Joliet Junior College
Dave Anderson, South Suburban College
Joby Milo Anthony, University of Central Florida
James E. Arnold, University of Wisconsin-Milwaukee
Adel Arshaghi, Center for Educational Merit
Carolyn Autray, University of West Georgia
Agnes Azzolino, Middlesex County College
Wilson P Banks, Illinois State University
Sudeshna Basu, Howard University
Dale R. Bedgood, East Texas State University
Beth Beno, South Suburban College
Carolyn Bernath, Tallahassee Community College
William H. Beyer, University of Akron
Annette Blackwelder, Florida State University
Richelle Blair, Lakeland Community College
Kevin Bodden, Lewis and Clark College
Larry Bouldin, Roane State Community College
Bob Bradshaw, Ohlone College
Trudy Bratten, Grossmont College
Tim Bremer, Broome Community College
Tim Britt, Jackson State Community College
Joanne Brunner, Joliet Junior College

Warren Burch, Brevard Community College
Mary Butler, Lincoln Public Schools
Melanie Butler, West Virginia University
Jim Butterbach, Joliet Junior College
William J. Cable, University of Wisconsin-Stevens Point
Lois Calamia, Brookdale Community College
Jim Campbell, Lincoln Public Schools
Roger Carlsen, Moraine Valley Community College
Elena Catoiu, Joliet Junior College
Mathews Chakkanakuzhi, Palomar College
John Collado, South Suburban College
Nelson Collins, Joliet Junior College
Jim Cooper, Joliet Junior College
Denise Corbett, East Carolina University
Carlos C. Corona, San Antonio College
Theodore C. Coskey, South Seattle Community College
Donna Costello, Plano Senior High School
Paul Crittenden, University of Nebraska at Lincoln
John Davenport, East Texas State University
Faye Dang, Joliet Junior College
Antonio David, Del Mar College
Duane E. Deal, Ball State University
Timothy Deis, University of Wisconsin-Platteville

Vivian Dennis, Eastfield College
Deborah Dillon, R. L. Turner High School
Guesna Dohrman, Tallahassee Community College
Cheryl Doolittle, Iowa State University
Karen R. Dougan, University of Florida
Louise Dyson, Clark College
Paul D. East, Lexington Community College
Don Edmondson, University of Texas-Austin
Erica Egizio, Joliet Junior College
Laura Egner, Joliet Junior College
Jason Eltrevoog, Joliet Junior College
Christopher Ennis, University of Minnesota
Kathy Eppler, Salt Lake Community College
Ralph Esparza, Jr., Richland College
Garret J. Etgen, University of Houston
Scott Fallstrom, Shoreline Community College
Pete Falzone, Pensacola Junior College
W.A. Ferguson, University of Illinois-Urbana/Champaign
Iris B. Fetta, Clemson University
Mason Flake, student at Edison Community College
Timothy W. Flood, Pittsburg State University
Merle Friel, Humboldt State University
Richard A. Fritz, Moraine Valley Community College
Dewey Furness, Ricke College

Randy Gallaher, Lewis and Clark College
Tina Garn, University of Arizona
Dawit Getachew, Chicago State University
Wayne Gibson, Rancho Santiago College
Robert Gill, University of Minnesota Duluth
Sudhir Kumar Goel, Valdosta State University
Joan Goliday, Sante Fe Community College
Frederic Gooding, Goucher College
Donald Goral, Northern Virginia Community College
Sue Graupner, Lincoln Public Schools
Jennifer L. Grimsley, University of Charleston
Ken Gurganus, University of North Carolina
James E. Hall, University of Wisconsin-Madison
Judy Hall, West Virginia University
Edward R. Hancock, DeVry Institute of Technology
Julia Hassett, DeVry Institute-Dupage
Christopher Hay-Jahans, University of South Dakota
Michah Heibel, Lincoln Public Schools
LaRae Helliwell, San Jose City College
Celeste Hernandez, Richland College
Brother Herron, Brother Rice High School
Robert Hoburg, Western Connecticut State University
Lynda Hollingsworth, Northwest Missouri State University
Charla Holzbog, Denison High School
Lee Hruby, Naperville North High School
Miles Hubbard, St. Cloud State University
Kim Hughes, California State College-San Bernardino
Ron Jamison, Brigham Young University
Richard A. Jensen, Manatee Community College
Sandra G. Johnson, St. Cloud State University
Tuesday Johnson, New Mexico State University
Moana H. Karsteter, Tallahassee Community College
Donna Katula, Joliet Junior College
Arthur Kaufman, College of Staten Island
Thomas Kearns, North Kentucky University
Jack Keating, Massasoit Community College
Shelia Kellenbarger, Lincoln Public Schools
Rachael Kenney, North Carolina State University
Debra Kopcso, Louisiana State University
Lynne Kowski, Raritan Valley Community College
Keith Kuchar, Manatee Community College
Tor Kwembe, Chicago State University
Linda J. Kyle, Tarrant Country Jr. College
H.E. Lacey, Texas A & M University
Harriet Lamm, Coastal Bend College
James Lapp, Fort Lewis College
Matt Larson, Lincoln Public Schools
Christopher Lattin, Oakton Community College
Julia Ledet, Lousiana State University
Adele LeGere, Oakton Community College
Kevin Leith, University of Houston
JoAnn Lewin, Edison College
Jeff Lewis, Johnson County Community College
Janice C. Lyon, Tallahassee Community College
Jean McArthur, Joliet Junior College
Virginia McCarthy, Iowa State University

Karla McCavit, Albion College
Michael McClendon, University of Central Oklahoma
Tom McCollow, DeVry Institute of Technology
Marilyn McCollum, North Carolina State University
Jill McGowan, Howard University
Dave McGuire, Joliet Junior College
Angela McNulty, Joliet Junior College
Laurence Maher, North Texas State University
Jay A. Malmstrom, Oklahoma City Community College
Rebecca Mann, Apollo High School
Lynn Marecek, Santa Ana College
Sherry Martina, Naperville North High School
Alec Matheson, Lamar University
Nancy Matthews, University of Oklahoma
James Maxwell, Oklahoma State University-Stillwater
Marsha May, Midwestern State University
Judy Meckley, Joliet Junior College
David Meel, Bowling Green State University
Carolyn Meitler, Concordia University
Samia Metwali, Erie Community College
Rich Meyers, Joliet Junior College
Eldon Miller, University of Mississippi
James Miller, West Virginia University
Michael Miller, Iowa State University
Kathleen Miranda, SUNY at Old Westbury
Thomas Monaghan, Naperville North High School
Maria Montoya, Our Lady of the Lake University
Craig Morse, Naperville North High School
Samad Mortabit, Metropolitan State University
Pat Mower, Washburn University
A. Muhundan, Manatee Community College
Jane Murphy, Middlesex Community College
Richard Nadel, Florida International University
Gabriel Nagy, Kansas State University
Bill Naegele, South Suburban College
Karla Neal, Lousiana State University
Lawrence E. Newman, Holyoke Community College
Dwight Newsome, Pasco-Hernando Community College
James Nymann, University of Texas-El Paso
Mark Omodt, Anoka-Ramsey Community College
Seth F. Oppenheimer, Mississippi State University
Leticia Oropesa, University of Miami
Linda Padilla, Joliet Junior College
E. James Peake, Iowa State University
Kelly Pearson, Murray State University
Philip Pina, Florida Atlantic University
Michael Prophet, University of Northern Iowa
Laura Pyzdrowski, West Virginia University
Neal C. Raber, University of Akron
Thomas Radin, San Joaquin Delta College
Ken A. Rager, Metropolitan State College
Kenneth D. Reeves, San Antonio College
Elsi Reinhardt, Truckee Meadows Community College
Jane Ringwald, Iowa State University

Stephen Rodi, Austin Community College
William Rogge, Lincoln Northeast High School
Howard L. Rolf, Baylor University
Mike Rosenthal, Florida International University
Phoebe Rouse, Lousiana State University
Edward Rozema, University of Tennessee at Chattanooga
Dennis C. Runde, Manatee Community College
Alan Saleski, Loyola University of Chicago
Susan Sandmeyer, Jamestown Community College
Brenda Santistevan, Salt Lake Community College
Linda Schmidt, Greenville Technical College
A.K. Shamma, University of West Florida
Martin Sherry, Lower Columbia College
Tatrana Shubin, San Jose State University
Anita Sikes, Delgado Community College
Timothy Sipka, Alma College
Lori Smellegar, Manatee Community College
Gayle Smith, Loyola Blakefield
Vicki Smith, Nicholls State University
John Spellman, Southwest Texas State University
Karen Spike, University of North Carolina, Wilmington
Rajalakshmi Sriram, Okaloosa-Walton Community College
Becky Stamper, Western Kentucky University
Judy Staver, Florida Community College-South
Neil Stephens, Hinsdale South High School
Sonya Stephens, Florida A&M Univeristy
Patrick Stevens, Joliet Junior College
Matthew TenHuisen, University of North Carolina, Wilmington
Christopher Terry, Augusta State University
Diane Tesar, South Suburban College
Tommy Thompson, Brookhaven College
Martha K. Tietze, Shawnee Mission Northwest High School
Richard J. Tondra, Iowa State University
Suzanne Topp, Salt Lake Community College
Marvel Townsend, University of Florida
Jim Trudnowski, Carroll College
Robert Tuskey, Joliet Junior College
Richard G. Vinson, University of South Alabama
Mary Voxman, University of Idaho
Jennifer Walsh, Daytona Beach Community College
Donna Wandke, Naperville North High School
Kathryn Wetzel, Amarillo College
Darlene Whitkenack, Northern Illinois University
Suzanne Williams, Central Piedmont Community College
Larissa Williamson, University of Florida
Christine Wilson, West Virginia University
Brad Wind, Florida International University
Mary Wolyniak, Broome Community College
Canton Woods, Auburn University
Tamara S. Worner, Wayne State College
Terri Wright, New Hampshire Community Technical College, Manchester
George Zazi, Chicago State University
Steve Zuro, Joliet Junior College

Michael Sullivan
Chicago State University

Michael Sullivan, III
Joliet Junior College

RESOURCES

Available to students are the following supplements:

- *Student Solutions Manual (ISBN 10: 0136029418; ISBN 13: 9780136029410)*

  Fully worked solutions to odd-numbered exercises.

- *Algebra Review (ISBN 10: 0131480065; ISBN 13: 9780131480063)*

  Four chapters of Intermediate Algebra review. Perfect for a slower-paced course or for individual review.

- *Instructor Solutions Manual (ISBN 10: 013602968X; ISBN 13: 9780136029687)*

  Fully worked solutions to all textbook exercises and chapter projects

Chapter Test Prep Video CD

Packaged at the back of the text, this video CD provides students with step-by-step solutions for each of the exercises in the book's chapter tests.

MyMathLab®

MyMathLab® is a series of text-specific, easily customizable online courses for Pearson Education's textbooks in mathematics and statistics. Powered by CourseCompass™ (our online teaching and learning environment) and MathXL® (our online homework, tutorial, and assessment system), MyMathLab gives you the tools you need to deliver all or a portion of your course online, whether your students are in a lab setting or working from home. MyMathLab provides a rich and flexible set of course materials, featuring free-response exercises that are algorithmically generated for unlimited practice and mastery. Students can also use online tools, such as video lectures, animations, and a multimedia textbook, to independently improve their understanding and performance. Instructors can use MyMathLab's homework and test managers to select and assign online exercises correlated directly to the textbook, and they can also create and assign their own online exercises and import TestGen tests for added flexibility. MyMathLab's online gradebook—designed specifically for mathematics and statistics—automatically tracks students' homework and test results and gives the instructor control over how to calculate final grades. Instructors can also add offline (paper-and-pencil) grades to the gradebook. MyMathLab is available to qualified adopters. For more information, visit our website at **www.mymathlab.com** or contact your sales representative.

MathXL®

MathXL® is a powerful online homework, tutorial, and assessment system that accompanies Pearson Education's textbooks in mathematics or statistics. With MathXL, instructors can create, edit, and assign online homework and tests using algorithmically generated exercises correlated at the objective level to the textbook. They can also create and assign their own online exercises and import TestGen tests for added flexibility. All student work is tracked in MathXL's online gradebook. Students can take chapter tests in MathXL and receive personalized study plans based on their test results. The study plan diagnoses weaknesses and links students directly to tutorial exercises for the objectives they need to study and retest. Students can also access supplemental animations and video clips directly from selected exercises. MathXL is available to qualified adopters. For more information, visit our website at **www.mathxl.com**, or contact your sales representative.

InterAct Math Tutorial Website: www.interactmath.com

Get practice and tutorial help online! This interactive tutorial website provides algorithmically generated practice exercises that correlate directly to the exercises in the textbook. Students can retry an exercise as many times as they like with new values each time for unlimited practice and mastery. Every exercise is accompanied by an interactive guided solution that provides helpful feedback for incorrect answers, and students can also view a worked-out sample problem that steps them through an exercise similar to the one they're working on.

List of Applications

Photo Credits

Trigonometry

Graphs and Functions

1

U.S. Consumers Make Wireless Top Choice

July 11, 2005—WASHINGTON, D.C., CTIA—The Wireless Association President and CEO Steve Largent highlights the Federal Communications Commission (FCC) report on local telephone competition, which indicates for the first time there are more wireless subscribers in the United States than wireline service lines. According to the FCC report, as of December 31, 2004, there were 181.1 million wireless subscribers in the U.S., compared to a combined 178 million incumbent local exchange carrier switched access lines and competitive local exchange carrier switched access lines. Also according to the report, there was a 15% increase in wireless subscribers for the 12-month period ending December 31, 2004.

"This significant milestone reflects consumers' approval of a vibrant, ultracompetitive industry that lets us communicate how we want, where we want and with whom we want," said Largent.

"Wireless is so popular because it lets people communicate on their terms. And it's not just about talking anymore. You can browse the Web, take pictures and video, download music, play games or conduct business. Wireless can satisfy a lot of different communication needs and desires and that's extremely popular to millions of consumers."

Source: "U.S. Consumers Make Wireless Top Choice," July 11, 2005 http://ctia.org/news_media/press/body.cfm?record_id=1533. Used with the permission of CTIA—The Wireless Association®.

—See Chapter Project I—

A Look Back

In the Appendix we review Algebra Essentials, Geometry Essentials, and equations in one variable.

A Look Ahead

Here we connect algebra and geometry using the rectangular coordinate system and use it to graph equations in two variables. We will discuss a special type of equation involving two variables called a *function*. This chapter explains what a function is, how to graph functions, properties of functions, one-to-one functions, and inverse functions. The word *function* apparently was introduced by René Descartes in 1637. For him, a function simply meant any positive integral power of a variable *x*. Gottfried Wilhelm Leibniz (1646–1716), who always emphasized the geometric side of mathematics, used the word *function* to denote any quantity associated with a curve, such as the coordinates of a point on the curve. Leonhard Euler (1707–1783) employed the word to mean any equation or formula involving variables and constants. His idea of a function is similar to the one most often seen in courses that precede calculus. Later, the use of functions in investigating heat flow equations led to a very broad definition, due to Lejeune Dirichlet (1805–1859), which describes a function as a rule or correspondence between two sets. It is his definition that we use here.

Outline

1.1 Rectangular Coordinates; Introduction to Graphing Equations

PREPARING FOR THIS SECTION *Before getting started, review the following:*

- Algebra Essentials (Appendix, Section A.1, pp. A1–A10)
- Geometry Essentials (Appendix, Section A.2, pp. A14–A19)

Now Work the 'Are You Prepared?' problems on page 13.

OBJECTIVES 1 Use the Distance Formula (p. 4)
2 Use the Midpoint Formula (p. 7)
3 Graph Equations by Hand by Plotting Points (p. 7)
4 Graph Equations Using a Graphing Utility (p. 9)
5 Use a Graphing Utility to Create Tables (p. 11)
6 Find Intercepts from a Graph (p. 12)
7 Use a Graphing Utility to Approximate Intercepts (p. 12)

Rectangular Coordinates

Figure 1

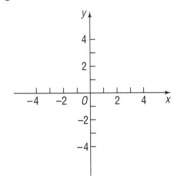

We locate a point on the real number line by assigning it a single real number, called the *coordinate of the point*. For work in a two-dimensional plane, we locate points by using two numbers.

We begin with two real number lines located in the same plane: one horizontal and the other vertical. We call the horizontal line the **x-axis,** the vertical line the **y-axis,** and the point of intersection the **origin O.** See Figure 1. We assign coordinates to every point on these number lines using a convenient scale. Recall that the scale of a number line is the distance between 0 and 1. In mathematics, we usually use the same scale on each axis; in applications, a different scale is often used on each axis.

The origin O has a value of 0 on both the x-axis and y-axis. Points on the x-axis to the right of O are associated with positive real numbers, and those to the left of O are associated with negative real numbers. Points on the y-axis above O are associated with positive real numbers, and those below O are associated with negative real numbers. In Figure 1, the x-axis and y-axis are labeled as x and y, respectively, and we have used an arrow at the end of each axis to denote the positive direction.

The coordinate system described here is called a **rectangular** or **Cartesian* coordinate system.** The plane formed by the x-axis and y-axis is sometimes called the **xy-plane,** and the x-axis and y-axis are referred to as the **coordinate axes.**

Any point P in the xy-plane can then be located by using an **ordered pair** (x, y) of real numbers. Let x denote the signed distance of P from the y-axis (*signed* means that, if P is to the right of the y-axis, then $x > 0$, and if P is to the left of the y-axis, then $x < 0$); and let y denote the signed distance of P from the x-axis. The ordered pair (x, y), also called the **coordinates** of P, then gives us enough information to locate the point P in the plane.

Figure 2

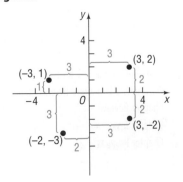

For example, to locate the point whose coordinates are $(-3, 1)$, go 3 units along the x-axis to the left of O and then go straight up 1 unit. We **plot** this point by placing a dot at this location. See Figure 2, in which the points with coordinates $(-3, 1), (-2, -3), (3, -2)$, and $(3, 2)$ are plotted.

The origin has coordinates $(0, 0)$. Any point on the x-axis has coordinates of the form $(x, 0)$, and any point on the y-axis has coordinates of the form $(0, y)$.

If (x, y) are the coordinates of a point P, then x is called the **x-coordinate,** or **abscissa,** of P and y is the **y-coordinate,** or **ordinate,** of P. We identify the point P by its coordinates (x, y) by writing $P = (x, y)$. Usually, we will simply say "the point (x, y)" rather than "the point whose coordinates are (x, y)."

The coordinate axes divide the xy-plane into four sections called **quadrants,** as shown in Figure 3. In quadrant I, both the x-coordinate and the y-coordinate of all points are positive; in quadrant II, x is negative and y is positive; in quadrant III,

*Named after René Descartes (1596–1650), a French mathematician, philosopher, and theologian.

Figure 3

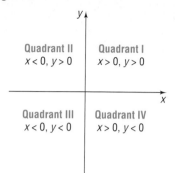

both x and y are negative; and in quadrant IV, x is positive and y is negative. Points on the coordinate axes belong to no quadrant.

➤ **Now Work** PROBLEM 11

Graphing Utilities

All graphing utilities, that is, all graphing calculators and all computer software graphing packages, graph equations by plotting points on a screen. The screen itself actually consists of small rectangles, called **pixels.** The more pixels the screen has, the better the resolution. Most graphing calculators have 48 pixels per square inch; most computer screens have 32 to 108 pixels per square inch. When a point to be plotted lies inside a pixel, the pixel is turned on (lights up). The graph of an equation is a collection of lighted pixels. Figure 4 shows how the graph of $y = 2x$ looks on a TI-84 Plus graphing calculator.

Figure 4
$Y_1 = 2x$

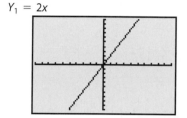

The screen of a graphing utility will display the coordinate axes of a rectangular coordinate system. However, you must set the scale on each axis. You must also include the smallest and largest values of x and y that you want included in the graph. This is called **setting the viewing rectangle** or **viewing window.** Figure 5 illustrates a typical viewing window.

Figure 5

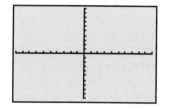

To select the viewing window, we must give values to the following expressions:

Xmin: the smallest value of x shown on the viewing window

Xmax: the largest value of x shown on the viewing window

Xscl: the number of units per tick mark on the x-axis

Ymin: the smallest value of y shown on the viewing window

Ymax: the largest value of y shown on the viewing window

Yscl: the number of units per tick mark on the y-axis

Figure 6 illustrates these settings and their relation to the Cartesian coordinate system.

Figure 6

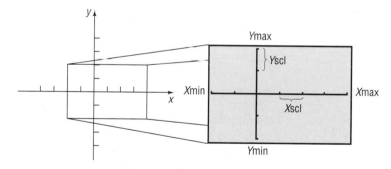

If the scale used on each axis is known, we can determine the minimum and maximum values of x and y shown on the screen by counting the tick marks. Look again at Figure 5. For a scale of 1 on each axis, the minimum and maximum values of x are -10 and 10, respectively; the minimum and maximum values of y are also -10 and 10. If the scale is 2 on each axis, then the minimum and maximum values of x are -20 and 20, respectively; and the minimum and maximum values of y are -20 and 20, respectively.

Conversely, if we know the minimum and maximum values of x and y, we can determine the scales being used by counting the tick marks displayed. We shall follow the practice of showing the minimum and maximum values of x and y in our illustrations so that you will know how the window was set. See Figure 7.

Figure 7

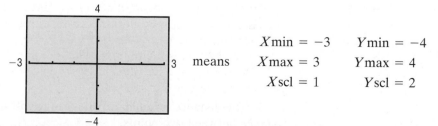

EXAMPLE 1 | **Finding the Coordinates of a Point Shown on a Graphing Utility Screen**

Find the coordinates of the point shown in Figure 8. Assume the coordinates are integers.

Figure 8

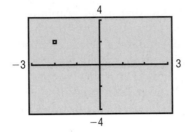

Solution | First we note that the viewing window used in Figure 8 is

$$X\text{min} = -3 \qquad Y\text{min} = -4$$
$$X\text{max} = 3 \qquad Y\text{max} = 4$$
$$X\text{scl} = 1 \qquad Y\text{scl} = 2$$

The point shown is 2 tick units to the left on the horizontal axis (scale = 1) and 1 tick up on the vertical scale (scale = 2). The coordinates of the point shown are $(-2, 2)$.

━━━━━ **Now Work** PROBLEMS **15** AND **25**

1 Use the Distance Formula

If the same units of measurement, such as inches, centimeters, and so on, are used for both the x-axis and y-axis, then all distances in the xy-plane can be measured using this unit of measurement.

EXAMPLE 2 | **Finding the Distance between Two Points**

Find the distance d between the points $(1, 3)$ and $(5, 6)$.

Solution | First we plot the points $(1, 3)$ and $(5, 6)$ and connect them with a straight line. See Figure 9(a). We are looking for length d. We begin by drawing a horizontal line from $(1, 3)$ to $(5, 3)$ and a vertical line from $(5, 3)$ to $(5, 6)$, forming a right triangle, as in Figure 9(b). One leg of the triangle is of length 4 (since $|5 - 1| = 4$) and the other is of length 3 (since $|6 - 3| = 3$). By the Pythagorean Theorem, the square of the distance d that we seek is

$$d^2 = 4^2 + 3^2 = 16 + 9 = 25$$
$$d = \sqrt{25} = 5$$

Figure 9

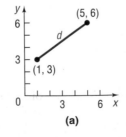 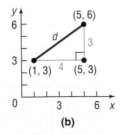

(a)   (b)

The **distance formula** provides a straightforward method for computing the distance between two points.

THEOREM

Distance Formula

The distance between two points $P_1 = (x_1, y_1)$ and $P_2 = (x_2, y_2)$, denoted by $d(P_1, P_2)$, is

$$d(P_1, P_2) = \sqrt{(x_2 - x_1)^2 + (y_2 - y_1)^2} \qquad \textbf{(1)}$$

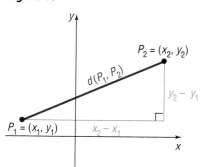

Figure 10

> **In Words**
>
> To compute the distance between two points, find the difference of the *x*-coordinates, square it, and add this to the square of the difference of the *y*-coordinates. The square root of this sum is the distance.

Figure 10 illustrates the theorem.

Proof of the Distance Formula   Let (x_1, y_1) denote the coordinates of point P_1, and let (x_2, y_2) denote the coordinates of point P_2. Assume that the line joining P_1 and P_2 is neither horizontal nor vertical. Refer to Figure 11(a). The coordinates of P_3 are (x_2, y_1). The horizontal distance from P_1 to P_3 is the absolute value of the difference of the *x*-coordinates, $|x_2 - x_1|$. The vertical distance from P_3 to P_2 is the absolute value of the difference of the *y*-coordinates, $|y_2 - y_1|$. See Figure 11(b). The distance $d(P_1, P_2)$ that we seek is the length of the hypotenuse of the right triangle, so, by the Pythagorean Theorem, it follows that

$$[d(P_1, P_2)]^2 = |x_2 - x_1|^2 + |y_2 - y_1|^2$$
$$= (x_2 - x_1)^2 + (y_2 - y_1)^2$$
$$d(P_1, P_2) = \sqrt{(x_2 - x_1)^2 + (y_2 - y_1)^2}$$

Figure 11

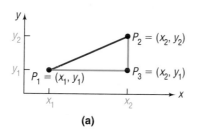

(a)

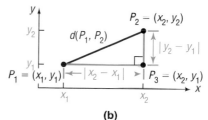

(b)

Now, if the line joining P_1 and P_2 is horizontal, then the *y*-coordinate of P_1 equals the *y*-coordinate of P_2; that is, $y_1 = y_2$. Refer to Figure 12(a). In this case, the distance formula (1) still works, because, for $y_1 = y_2$, it reduces to

$$d(P_1, P_2) = \sqrt{(x_2 - x_1)^2 + 0^2} = \sqrt{(x_2 - x_1)^2} = |x_2 - x_1|$$

Figure 12

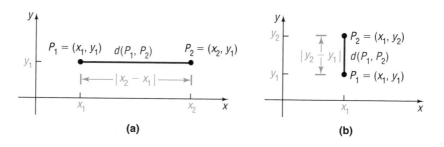

(a)　　　　　　　(b)

A similar argument holds if the line joining P_1 and P_2 is vertical. See Figure 12(b). The distance formula is valid in all cases. ∎

EXAMPLE 3   Finding the Length of a Line Segment

Find the length of the line segment shown in Figure 13 on page 6.

Solution

Figure 13

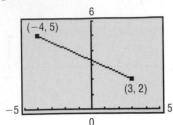

The length of the line segment is the distance between the points $P_1 = (x_1, y_1) = (-4, 5)$ and $P_2 = (x_2, y_2) = (3, 2)$. Using the distance formula (1) with $x_1 = -4$, $y_1 = 5$, $x_2 = 3$, and $y_2 = 2$, the length d is

$$d = \sqrt{(x_2 - x_1)^2 + (y_2 - y_1)^2} = \sqrt{[3 - (-4)]^2 + (2 - 5)^2}$$
$$= \sqrt{7^2 + (-3)^2} = \sqrt{49 + 9} = \sqrt{58} \approx 7.62$$

▬▬ **Now Work** PROBLEM 31

The distance between two points $P_1 = (x_1, y_1)$ and $P_2 = (x_2, y_2)$ is never a negative number. Furthermore, the distance between two points is 0 only when the points are identical, that is, when $x_1 = x_2$ and $y_1 = y_2$. Also, because $(x_2 - x_1)^2 = (x_1 - x_2)^2$ and $(y_2 - y_1)^2 = (y_1 - y_2)^2$, it makes no difference whether the distance is computed from P_1 to P_2 or from P_2 to P_1; that is, $d(P_1, P_2) = d(P_2, P_1)$.

The introduction to this chapter mentioned that rectangular coordinates enable us to translate geometry problems into algebra problems, and vice versa. The next example shows how algebra (the distance formula) can be used to solve geometry problems.

EXAMPLE 4 | **Using Algebra to Solve Geometry Problems**

Consider the three points $A = (-2, 1)$, $B = (2, 3)$, and $C = (3, 1)$.

(a) Plot each point and form the triangle ABC.
(b) Find the length of each side of the triangle.
(c) Verify that the triangle is a right triangle.
(d) Find the area of the triangle.

Solution

Figure 14

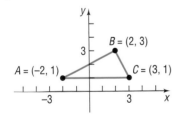

(a) Points A, B, C and triangle ABC are plotted in Figure 14.

(b) $d(A, B) = \sqrt{[2 - (-2)]^2 + (3 - 1)^2} = \sqrt{16 + 4} = \sqrt{20} = 2\sqrt{5}$
$\quad d(B, C) = \sqrt{(3 - 2)^2 + (1 - 3)^2} = \sqrt{1 + 4} = \sqrt{5}$
$\quad d(A, C) = \sqrt{[3 - (-2)]^2 + (1 - 1)^2} = \sqrt{25 + 0} = 5$

(c) To show that the triangle is a right triangle, we need to show that the sum of the squares of the lengths of two of the sides equals the square of the length of the third side. (Why is this sufficient?) Looking at Figure 14, it seems reasonable to conjecture that the right angle is at vertex B. We shall check to see whether

$$[d(A, B)]^2 + [d(B, C)]^2 = [d(A, C)]^2$$

We find that

$$[d(A, B)]^2 + [d(B, C)]^2 = (2\sqrt{5})^2 + (\sqrt{5})^2$$
$$= 20 + 5 = 25 = [d(A, C)]^2$$

so it follows from the converse of the Pythagorean Theorem that triangle ABC is a right triangle.

(d) Because the right angle is at vertex B, the sides AB and BC form the base and height of the triangle. Its area is

$$\text{Area} = \frac{1}{2}(\text{Base})(\text{Height}) = \frac{1}{2}(2\sqrt{5})(\sqrt{5}) = 5 \text{ square units}$$

▬▬ **Now Work** PROBLEM 49

2 Use the Midpoint Formula

Figure 15

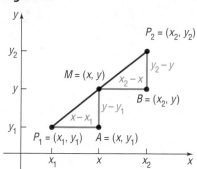

We now derive a formula for the coordinates of the **midpoint of a line segment.** Let $P_1 = (x_1, y_1)$ and $P_2 = (x_2, y_2)$ be the endpoints of a line segment, and let $M = (x, y)$ be the point on the line segment that is the same distance from P_1 as it is from P_2. See Figure 15. The triangles P_1AM and MBP_2 are congruent. [Do you see why? Angle AP_1M = angle BMP_2,* angle P_1MA = angle MP_2B, and $d(P_1, M) = d(M, P_2)$ is given. So, we have angle–side–angle.] Hence, corresponding sides are equal in length. That is,

$$x - x_1 = x_2 - x \qquad \text{and} \qquad y - y_1 = y_2 - y$$
$$2x = x_1 + x_2 \qquad\qquad\qquad 2y = y_1 + y_2$$
$$x = \frac{x_1 + x_2}{2} \qquad\qquad\qquad y = \frac{y_1 + y_2}{2}$$

THEOREM

Midpoint Formula

The midpoint $M = (x, y)$ of the line segment from $P_1 = (x_1, y_1)$ to $P_2 = (x_2, y_2)$ is

$$M = (x, y) = \left(\frac{x_1 + x_2}{2}, \frac{y_1 + y_2}{2} \right) \qquad\qquad \textbf{(2)}$$

In Words

To find the midpoint of a line segment, average the x-coordinates and average the y-coordinates of the endpoints.

EXAMPLE 5 | Finding the Midpoint of a Line Segment

Find the midpoint of a line segment from $P_1 = (-5, 5)$ to $P_2 = (3, 1)$. Plot the points P_1 and P_2 and their midpoint.

Solution

We apply the midpoint formula (2) using $x_1 = -5$, $y_1 = 5$, $x_2 = 3$, and $y_2 = 1$. Then the coordinates (x, y) of the midpoint M are

$$x = \frac{x_1 + x_2}{2} = \frac{-5 + 3}{2} = -1 \quad \text{and} \quad y = \frac{y_1 + y_2}{2} = \frac{5 + 1}{2} = 3$$

That is, $M = (-1, 3)$. See Figure 16.

Figure 16

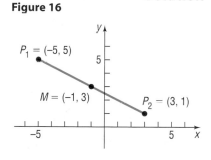

Now Work PROBLEM 55

3 Graph Equations by Hand by Plotting Points

An **equation in two variables,** say x and y, is a statement in which two expressions involving x and y are equal. The expressions are called the **sides** of the equation. Since an equation is a statement, it may be true or false, depending on the value of the variables. Any values of x and y that result in a true statement are said to **satisfy** the equation.

For example, the following are all equations in two variables x and y:

$$x^2 + y^2 = 5 \qquad 2x - y = 6 \qquad y = 2x + 5 \qquad x^2 = y$$

The first of these, $x^2 + y^2 = 5$, is satisfied for $x = 1$, $y = 2$, since $1^2 + 2^2 = 1 + 4 = 5$. Other choices of x and y also satisfy this equation. It is not satisfied for $x = 2$ and $y = 3$, since $2^2 + 3^2 = 4 + 9 = 13 \neq 5$.

---

*A postulate from geometry states that the transversal $\overline{P_1P_2}$ forms congruent corresponding angles with the parallel line segments $\overline{P_1A}$ and $\overline{MB}$.

The **graph of an equation in two variables** x and y consists of the set of points in the xy-plane whose coordinates (x, y) satisfy the equation.

| EXAMPLE 6 | Determining Whether a Point Is on the Graph of an Equation |

Determine if the following points are on the graph of the equation $2x - y = 6$.

(a) $(2, 3)$ (b) $(2, -2)$

Solution

(a) For the point $(2, 3)$, we check to see if $x = 2$, $y = 3$ satisfies the equation $2x - y = 6$.

$$2x - y = 2(2) - 3 = 4 - 3 = 1 \neq 6$$

The equation is not satisfied, so the point $(2, 3)$ is not on the graph.

(b) For the point $(2, -2)$, we have

$$2x - y = 2(2) - (-2) = 4 + 2 = 6$$

The equation is satisfied, so the point $(2, -2)$ is on the graph.

■

- **Now Work** PROBLEM 65

| EXAMPLE 7 | How to Graph an Equation by Hand by Plotting Points |

Graph the equation: $y = -2x + 3$

Step-by-Step Solution

STEP 1 Find all points (x, y) that satisfy the equation. To determine these points, choose values of x and use the equation to find the corresponding values for y. See Table 1.

Table 1

x	$y = -2x + 3$	(x, y)
-2	$-2(-2) + 3 = 7$	$(-2, 7)$
-1	$-2(-1) + 3 = 5$	$(-1, 5)$
0	$-2(0) + 3 = 3$	$(0, 3)$
1	$-2(1) + 3 = 1$	$(1, 1)$
2	$-2(2) + 3 = -1$	$(2, -1)$

STEP 2 Plot the points listed in the table as shown in Figure 17(a). Now connect the points to obtain the graph of the equation (a *line*), as shown in Figure 17(b).

Figure 17

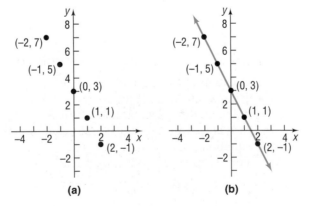

(a)   (b)

■

| EXAMPLE 8 | Graphing an Equation by Hand by Plotting Points |

Graph the equation: $y = x^2$

Solution Table 2 provides several points on the graph. In Figure 18 we plot these points and connect them with a smooth curve to obtain the graph (a *parabola*).

Table 2

x	($y = x^2$)	(x, y)
-4	16	(-4, 16)
-3	9	(-3, 9)
-2	4	(-2, 4)
-1	1	(-1, 1)
0	0	(0, 0)
1	1	(1, 1)
2	4	(2, 4)
3	9	(3, 9)
4	16	(4, 16)

Figure 18

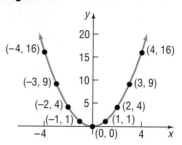

The graphs of the equations shown in Figures 17(b) and 18 do not show all the points that are on the graph. For example, in Figure 17(b), the point $(20, -37)$ is a part of the graph of $y = -2x + 3$, but it is not shown. Since the graph of $y = -2x + 3$ could be extended out as far as we please, we use arrows to indicate that the pattern shown continues. It is important when illustrating a graph to present enough of the graph so that any viewer of the illustration will "see" the rest of it as an obvious continuation of what is actually there. This is referred to as a **complete graph.**

One way to obtain a complete graph of an equation is to plot a sufficient number of points on the graph until a pattern becomes evident. Then these points are connected with a smooth curve following the suggested pattern. But how many points are sufficient? Sometimes knowledge about the equation tells us. For example, if an equation is of the form $y = mx + b$, then its graph is a line. In this case, two points would suffice to obtain the graph.

One purpose of this book is to investigate the properties of equations in order to decide whether a graph is complete. Sometimes we shall graph equations by plotting a sufficient number of points on the graph until a pattern becomes evident and then connect these points with a smooth curve, following the suggested pattern. (Shortly, we shall investigate various techniques that will enable us to graph an equation without plotting so many points.) Other times we shall graph equations using a graphing utility.

4 Graph Equations Using a Graphing Utility

From Examples 7 and 8, we see that a graph can be obtained by plotting points in a rectangular coordinate system and connecting them. Graphing utilities perform these same steps when graphing an equation. For example, the TI-84 Plus determines 95 evenly spaced input values,* uses the equation to determine the output values, plots these points on the screen, and finally (if in the connected mode) draws a line between consecutive points.

To graph an equation in two variables x and y using a graphing utility requires that the equation be written in the form $y = \{$expression in $x\}$. If the original equation is not in this form, replace it by equivalent equations until the form $y = \{$expression in $x\}$ is obtained. See the Appendix, Section A.4, page A26 for procedures that result in equivalent equations.

EXAMPLE 9 | Expressing an Equation in the Form $y = \{$expression in $x\}$

Solve for y: $2y + 3x - 5 = 4$

---

*These input values depend on the values of Xmin and Xmax. For example, if Xmin $= -10$ and Xmax $= 10$, then the first input value will be -10 and the next input value will be $-10 + (10 - (-10))/94 = -9.7872$, and so on.

Solution We replace the original equation by a succession of equivalent equations.

$$2y + 3x - 5 = 4$$
$$2y + 3x - 5 + 5 = 4 + 5 \quad \text{Add 5 to both sides.}$$
$$2y + 3x = 9 \quad \text{Simplify.}$$
$$2y + 3x - 3x = 9 - 3x \quad \text{Subtract 3x from both sides.}$$
$$2y = 9 - 3x \quad \text{Simplify.}$$
$$\frac{2y}{2} = \frac{9 - 3x}{2} \quad \text{Divide both sides by 2.}$$
$$y = \frac{9 - 3x}{2} \quad \text{Simplify.}$$

Now we are ready to graph equations using a graphing utility.

EXAMPLE 10	**How to Graph an Equation Using a Graphing Utility**

Use a graphing utility to graph the equation: $6x^2 + 3y = 36$

Step-by-Step Solution

STEP 1 Solve the equation for y in terms of x.

$$6x^2 + 3y = 36$$
$$3y = -6x^2 + 36 \quad \text{Subtract } 6x^2 \text{ from both sides of the equation.}$$
$$y = -2x^2 + 12 \quad \text{Divide both sides of the equation by 3 and simplify.}$$

STEP 2 Enter the equation to be graphed into your graphing utility. Figure 19 shows the equation to be graphed entered on a TI-84 Plus.

Figure 19

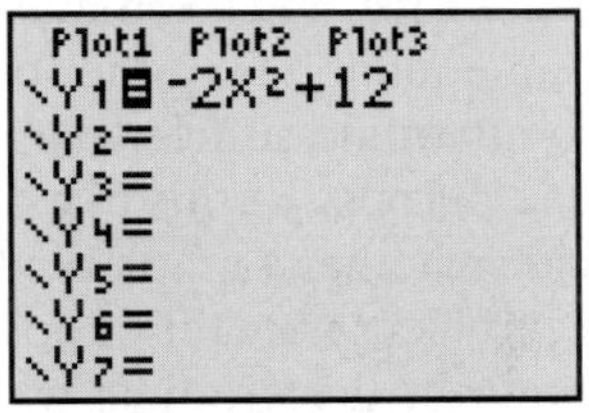

```
Plot1  Plot2  Plot3
\Y1■-2X²+12
\Y2=
\Y3=
\Y4=
\Y5=
\Y6=
\Y7=
```

STEP 3 Choose an initial viewing window. Without any knowledge about the behavior of the graph, it is common to choose the standard viewing window as the initial viewing window. The standard viewing window is

$$Xmin = -10 \quad Ymin = -10$$
$$Xmax = 10 \quad Ymax = 10$$
$$Xscl = 1 \quad Yscl = 1$$

See Figure 20.

Figure 20

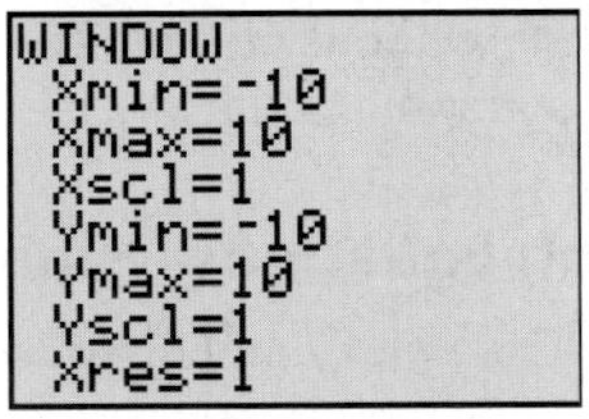

```
WINDOW
 Xmin=-10
 Xmax=10
 Xscl=1
 Ymin=-10
 Ymax=10
 Yscl=1
 Xres=1
```

STEP 4 Graph the equation. See Figure 21.

Figure 21

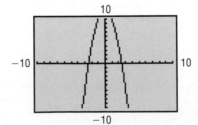

STEP 5 Adjust the viewing window until a complete graph is obtained.

The graph of $y = -2x^2 + 12$ is not complete. The value of $Ymax$ must be increased so that the top portion of the graph is visible. After increasing the value of $Ymax$ to 12, we obtain the graph in Figure 22. The graph is now complete.

Figure 22

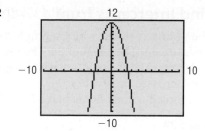

NOTE Some graphing utilities have a ZOOM-STANDARD feature that automatically sets the viewing window to the standard viewing window. In addition, some graphing utilities have a ZOOM-FIT feature that determines the appropriate Ymin and Ymax for a given Xmin and Xmax. Consult your owner's manual for the appropriate keystrokes. ■

━ **Now Work** PROBLEM 83

5 Use a Graphing Utility to Create Tables

In addition to graphing equations, graphing utilities can also be used to create a table of values that satisfy the equation. This feature is especially useful in determining an appropriate viewing window when graphing an equation.

EXAMPLE 11	**How to Create a Table Using a Graphing Utility**

Create a table that displays the points on the graph of $6x^2 + 3y = 36$ for $x = -3, -2, -1, 0, 1, 2,$ and 3.

Step-by-Step Solution

STEP 1 Solve the equation for y in terms of x.

We solved the equation for y in terms of x in Example 10 and obtained $y = -2x^2 + 12$.

STEP 2 Enter the expression in x following the $Y =$ prompt of the graphing utility.

See Figure 19 on page 10.

STEP 3 Set up the table. Graphing utilities typically have two modes for creating tables. In the AUTO mode, the user determines a starting point for the table (TblStart) and ΔTbl (pronounced "delta table"). The ΔTbl feature determines the increment for x in the table. The ASK mode requires the user to enter values of x, and then the utility determines the corresponding value of y.

We will create a table using AUTO mode. The table we wish to create starts at -3, so TblStart $= -3$. The increment in x is 1, so ΔTbl $= 1$. See Figure 23.

Figure 23

```
TABLE SETUP
 TblStart=-3
 △Tbl=1
Indpnt: Auto Ask
Depend: Auto Ask
```

STEP 4 Create the table. See Table 3.

Table 3

The user can scroll within the table if it is created in AUTO mode.

In looking at Table 3, we notice that $y = 12$ when $x = 0$. This information could have been used to help to create the initial viewing window by letting us know that Ymax needs to be at least 12 in order to get a complete graph.

6 Find Intercepts from a Graph

The points, if any, at which a graph crosses or touches the coordinate axes are called the **intercepts.** See Figure 24. The *x*-coordinate of a point at which the graph crosses or touches the *x*-axis is an *x*-intercept, and the *y*-coordinate of a point at which the graph crosses or touches the *y*-axis is a **y-intercept.** For a graph to be complete, all its intercepts must be displayed.

Figure 24

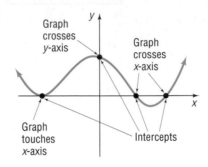

EXAMPLE 12 | **Finding Intercepts from a Graph**

Find the intercepts of the graph in Figure 25. What are its *x*-intercepts? What are its *y*-intercepts?

Solution The intercepts of the graph are the points

$$(-3, 0), \quad (0, 3), \quad \left(\frac{3}{2}, 0\right), \quad \left(0, -\frac{4}{3}\right), \quad (0, -3.5), \quad (4.5, 0)$$

The *x*-intercepts are -3, $\dfrac{3}{2}$, and 4.5; the *y*-intercepts are -3.5, $-\dfrac{4}{3}$, and 3.

Figure 25

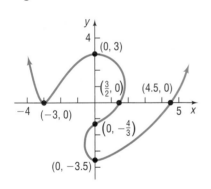

In Example 12, you should notice the following usage: If we do not specify the type of intercept (*x*- versus *y*-), then we report the intercept as an ordered pair. However, if we specify the type of intercept, then we only report the coordinate of the intercept. For *x*-intercepts, we report the *x*-coordinate of the intercept; for *y*-intercepts, we report the *y*-coordinate of the intercept.

━━━━ **Now Work** PROBLEM 71

7 Use a Graphing Utility to Approximate Intercepts

We can use a graphing utility to approximate the intercepts of the graph of an equation.

EXAMPLE 13 | **Approximating Intercepts Using a Graphing Utility**

Use a graphing utility to approximate the intercepts of the equation $y = x^3 - 16$.

Solution Figure 26(a) shows the graph of $y = x^3 - 16$.

The eVALUEate feature of a TI-84 Plus graphing calculator accepts as input a value of *x* and determines the value of *y*. If we let $x = 0$, we find that the *y*-intercept is -16. See Figure 26(b).

The ZERO feature of a TI-84 Plus is used to find the *x*-intercept(s). See Figure 26(c). Rounded to two decimal places, the *x*-intercept is 2.52.

Figure 26

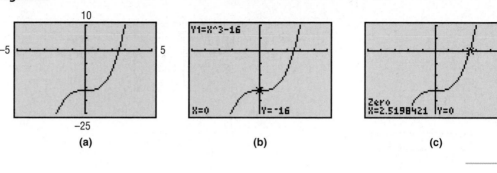

(a)   (b)   (c)

— **Now Work** PROBLEM 93

We discuss finding intercepts algebraically in the next section.

1.1 Assess Your Understanding

'Are You Prepared?' *Answers are given at the end of these exercises. If you get a wrong answer, read the pages listed in red.*

1. If -3 and 5 are the coordinates of two points on the real number line, the distance between these points is _____. (pp. A5–A6)

2. If 3 and 4 are the legs of a right triangle, the hypotenuse is _____. (p. A14)

3. Use the converse of the Pythagorean Theorem to show that a triangle whose sides are of lengths 11, 60, and 61 is a right triangle. (pp. A14–A15)

4. The _____-_____-_____ case states that two triangles are congruent if two of the angles are congruent and the lengths of the corresponding sides between the two angles are equal. (p. A17)

Concepts and Vocabulary

5. If (x, y) are the coordinates of a point P in the xy-plane, then x is called the _____ of P and y is the _____ of P.

6. The coordinate axes divide the xy-plane into four sections called _____.

7. If three distinct points P, Q, and R all lie on a line and if $d(P, Q) = d(Q, R)$, then Q is called the _____ of the line segment from P to R.

8. **True or False** The distance between two points is sometimes a negative number.

9. **True or False** The point $(-1, 4)$ lies in quadrant IV of the Cartesian plane.

10. **True or False** The midpoint of a line segment is found by averaging the x-coordinates and averaging the y-coordinates of the endpoints.

Skill Building

In Problems 11 and 12, plot each point in the xy-plane. Tell in which quadrant or on what coordinate axis each point lies.

11. (a) $A = (-3, 2)$ (d) $D = (6, 5)$
    (b) $B = (6, 0)$ (e) $E = (0, -3)$
    (c) $C = (-2, -2)$ (f) $F = (6, -3)$

12. (a) $A = (1, 4)$ (d) $D = (4, 1)$
    (b) $B = (-3, -4)$ (e) $E = (0, 1)$
    (c) $C = (-3, 4)$ (f) $F = (-3, 0)$

13. Plot the points $(2, 0)$, $(2, -3)$, $(2, 4)$, $(2, 1)$, and $(2, -1)$. Describe the set of all points of the form $(2, y)$, where y is a real number.

14. Plot the points $(0, 3)$, $(1, 3)$, $(-2, 3)$, $(5, 3)$, and $(-4, 3)$. Describe the set of all points of the form $(x, 3)$, where x is a real number.

In Problems 15–18, determine the coordinates of the points shown. Tell in which quadrant each point lies. Assume the coordinates are integers.

15.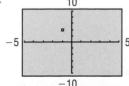

16.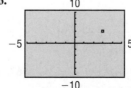

17.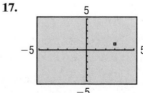

18.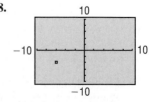

In Problems 19–24, select a setting so that each given point will lie within the viewing window.

19. $(-10, 5)$, $(3, -2)$, $(4, -1)$

20. $(5, 0)$, $(6, 8)$, $(-2, -3)$

21. $(40, 20)$, $(-20, -80)$, $(10, 40)$

22. $(-80, 60)$, $(20, -30)$, $(-20, -40)$

23. $(0, 0)$, $(100, 5)$, $(5, 150)$

24. $(0, -1)$, $(100, 50)$, $(-10, 30)$

In Problems 25–30, determine the viewing window used.

25.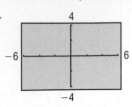

26.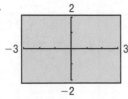

27.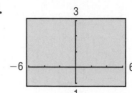

28.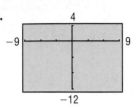

29.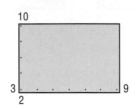

30.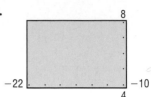

In Problems 31–44, find the distance $d(P_1, P_2)$ between the points P_1 and P_2.

31.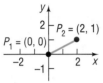

32.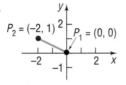

33.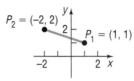

34.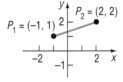

35. $P_1 = (3, -4)$; $P_2 = (5, 4)$

36. $P_1 = (-1, 0)$; $P_2 = (2, 4)$

37. $P_1 = (-5, -3)$; $P_2 = (11, 9)$

38. $P_1 = (2, -3)$; $P_2 = (10, 3)$

39. $P_1 = (4, -3)$; $P_2 = (6, 4)$

40. $P_1 = (-4, -3)$; $P_2 = (6, 2)$

41. $P_1 = (-0.2, 0.3)$; $P_2 = (1.3, 2.1)$

42. $P_1 = (1.2, 2.3)$; $P_2 = (-0.5, 1.6)$

43. $P_1 = (a, b)$; $P_2 = (0, 0)$

44. $P_1 = (a, a)$; $P_2 = (0, 0)$

In Problems 45–48, find the length of the line segment. Assume that the endpoints of each line segment have integer coordinates.

45.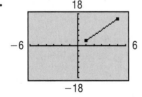

46.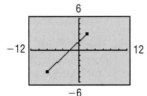

47.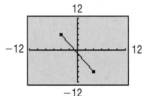

48.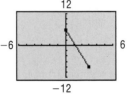

In Problems 49–54, plot each point and form the triangle ABC. Verify that the triangle is a right triangle. Find its area.

49. $A = (-2, 5)$; $B = (1, 3)$; $C = (-1, 0)$

50. $A = (-2, 5)$; $B = (12, 3)$; $C = (10, -11)$

51. $A = (-5, 3)$; $B = (6, 0)$; $C = (5, 5)$

52. $A = (-6, 3)$; $B = (3, -5)$; $C = (-1, 5)$

53. $A = (4, -3)$; $B = (0, -3)$; $C = (4, 2)$

54. $A = (4, -3)$; $B = (4, 1)$; $C = (2, 1)$

In Problems 55–64, find the midpoint of the line segment joining the points P_1 and P_2.

55. $P_1 = (3, -4)$; $P_2 = (5, 4)$

56. $P_1 = (-2, 0)$; $P_2 = (2, 4)$

57. $P_1 = (-5, -3)$; $P_2 = (11, 9)$

58. $P_1 = (2, -3)$; $P_2 = (10, 3)$

59. $P_1 = (4, -3)$; $P_2 = (6, 1)$

60. $P_1 = (-4, -3)$; $P_2 = (2, 2)$

61. $P_1 = (-0.2, 0.3)$; $P_2 = (1.3, 2.1)$

62. $P_1 = (1.2, 2.3)$; $P_2 = (-0.5, 1.6)$

63. $P_1 = (a, b)$; $P_2 = (0, 0)$

64. $P_1 = (a, a)$; $P_2 = (0, 0)$

In Problems 65–70, tell whether the given points are on the graph of the equation.

65. Equation: $y = x^4 - \sqrt{x}$
Points: $(0, 0); (1, 1); (-1, 0)$

66. Equation: $y = x^3 - 2\sqrt{x}$
Points: $(0, 0); (1, 1); (1, -1)$

67. Equation: $y^2 = x^2 + 9$
Points: $(0, 3); (3, 0); (-3, 0)$

68. Equation: $y^3 = x + 1$
Points: $(1, 2); (0, 1); (-1, 0)$

69. Equation: $x^2 + y^2 = 4$
Points: $(0, 2); (-2, 2); \left(\sqrt{2}, \sqrt{2}\right)$

70. Equation: $x^2 + 4y^2 = 4$
Points: $(0, 1); (2, 0); \left(2, \dfrac{1}{2}\right)$

In Problems 71–78, the graph of an equation is given. List the intercepts of the graph.

71.
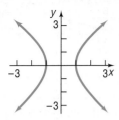

72.
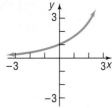

73.
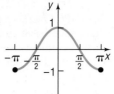

74.
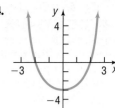

75.
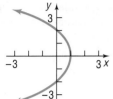

76.
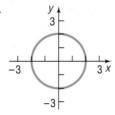

77.
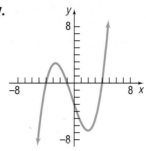

78.
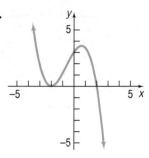

In Problems 79–90, graph each equation by hand by plotting points. Verify your results using a graphing utility.

79. $y = x + 2$ **80.** $y = x - 6$ **81.** $y = 2x + 8$ **82.** $y = 3x - 9$

83. $y = x^2 - 1$ **84.** $y = x^2 - 9$ **85.** $y = -x^2 + 4$ **86.** $y = -x^2 + 1$

87. $2x + 3y = 6$ **88.** $5x + 2y = 10$ **89.** $9x^2 + 4y = 36$ **90.** $4x^2 + y = 4$

In Problems 91–98, graph each equation using a graphing utility. Use a graphing utility to approximate the intercepts rounded to two decimal places. Use the TABLE feature to help to establish the viewing window.

91. $y = 2x - 13$ **92.** $y = -3x + 14$ **93.** $y = 2x^2 - 15$ **94.** $y = -3x^2 + 19$

95. $3x - 2y = 43$ **96.** $4x + 5y = 82$ **97.** $5x^2 + 3y = 37$ **98.** $2x^2 - 3y = 35$

Applications and Extensions

99. The **medians** of a triangle are the line segments from each vertex to the midpoint of the opposite side (see the figure). Find the lengths of the medians of the triangle with vertices at $A = (0, 0)$, $B = (6, 0)$, and $C = (4, 4)$.

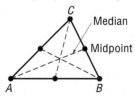

100. An **equilateral triangle** is one in which all three sides are of equal length. If two vertices of an equilateral triangle are $(0, 4)$ and $(0, 0)$, find the third vertex. How many of these triangles are possible?

*In Problems 101–104, find the length of each side of the triangle determined by the three points P_1, P_2, and P_3. State whether the triangle is an isosceles triangle, a right triangle, neither of these, or both. (An **isosceles triangle** is one in which at least two of the sides are of equal length.)*

101. $P_1 = (2, 1)$; $P_2 = (-4, 1)$; $P_3 = (-4, -3)$

102. $P_1 = (-1, 4)$; $P_2 = (6, 2)$; $P_3 = (4, -5)$

103. $P_1 = (-2, -1)$; $P_2 = (0, 7)$; $P_3 = (3, 2)$

104. $P_1 = (7, 2)$; $P_2 = (-4, 0)$; $P_3 = (4, 6)$

105. Baseball A major league baseball "diamond" is actually a square, 90 feet on a side (see the figure). What is the distance directly from home plate to second base (the diagonal of the square)?

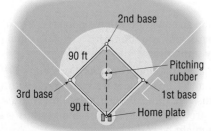

106. Little League Baseball The layout of a Little League playing field is a square, 60 feet on a side. How far is it directly from home plate to second base (the diagonal of the square)?

Source: Little League Baseball, Official Regulations and Playing Rules, 2007.

107. Baseball Refer to Problem 105. Overlay a rectangular coordinate system on a major league baseball diamond so that the origin is at home plate, the positive *x*-axis lies in the direction from home plate to first base, and the positive *y*-axis lies in the direction from home plate to third base.

(a) What are the coordinates of first base, second base, and third base? Use feet as the unit of measurement.

(b) If the right fielder is located at $(310, 15)$, how far is it from there to second base?

(c) If the center fielder is located at $(300, 300)$, how far is it from there to third base?

108. Little League Baseball Refer to Problem 106. Overlay a rectangular coordinate system on a Little League baseball diamond so that the origin is at home plate, the positive x-axis lies in the direction from home plate to first base, and the positive y-axis lies in the direction from home plate to third base.

(a) What are the coordinates of first base, second base, and third base? Use feet as the unit of measurement.
(b) If the right fielder is located at $(180, 20)$, how far is it from there to second base?
(c) If the center fielder is located at $(220, 220)$, how far is it from there to third base?

109. Distance between Moving Objects A Dodge Neon and a Mack truck leave an intersection at the same time. The Neon heads east at an average speed of 30 miles per hour, while the truck heads south at an average speed of 40 miles per hour. Find an expression for their distance apart d (in miles) at the end of t hours.

110. Distance of a Moving Object from a Fixed Point A hot-air balloon, headed due east at an average speed of 15 miles per hour at a constant altitude of 100 feet, passes over an intersection (see the figure). Find an expression for its distance d (measured in feet) from the intersection t seconds later.

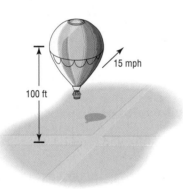

111. Drafting Error When a draftsman draws three lines that are to intersect at one point, the lines may not intersect as intended and subsequently will form an **error triangle.** If this error triangle is long and thin, one estimate for the location of the desired point is the midpoint of the shortest side. The figure shows one such error triangle.

(a) Find an estimate for the desired intersection point.
(b) Find the length of the median for the midpoint found in part (a). See Problem 99.

Source: www.uwgb.edu/dutchs/STRUCTGE/s100.htm

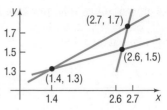

112. Net Sales The figures illustrates how net sales of Wal-Mart Stores, Inc., have grown from 2002 through 2006. Use the midpoint formula to estimate the net sales of Wal-Mart Stores, Inc., in 2004. How does your result compare to the reported value of $256 billion?

Source: Wal-Mart Stores, Inc., 2006 Annual Report

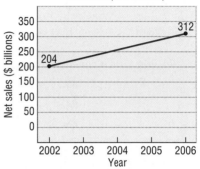

Wal-Mart Stores, Inc.
Net sales (in $ billions)

113. Poverty Threshold Poverty thresholds are determined by the U.S. Census Bureau. A poverty threshold represents the minimum annual household income for a family not to be considered poor. In 1995, the poverty threshold for a family of four with two children under the age of 18 years was $15,455. In 2005, the poverty threshold for a family of four with two children under the age of 18 years was $19,806. Assuming poverty thresholds increase in a straight-line fashion, use the midpoint formula to estimate the poverty threshold of a family of four with two children under the age of 18 in 2000. How does your result compare to the actual poverty threshold in 2000 of $17,463?

Source: U.S. Census Bureau

114. Horizontal and Vertical Shifts Suppose that $A = (2, 5)$ are the coordinates of a point in the xy-plane.

(a) Find the coordinates of the point if A is shifted 3 units to the right and 2 units down.
(b) Find the coordinates of the point if A is shifted 2 units to the left and 8 units up.

115. Completing a Line Segment Plot the points $A = (-1, 8)$ and $M = (2, 3)$ in the xy-plane. If M is the midpoint of a line segment AB, find the coordinates of B.

Discussion and Writing

In Problem 116, you may use a graphing utility, but it is not required.

116. (a) Graph $y = \sqrt{x^2}$, $y = x$, $y = |x|$, and $y = (\sqrt{x})^2$, noting which graphs are the same.
(b) Explain why the graphs of $y = \sqrt{x^2}$ and $y = |x|$ are the same.
(c) Explain why the graphs of $y = x$ and $y = (\sqrt{x})^2$ are not the same.
(d) Explain why the graphs of $y = \sqrt{x^2}$ and $y = x$ are not the same.

117. Make up an equation with the points $(2, 0)$, $(4, 0)$, and $(0, 1)$. Compare your equation with a friend's equation. Comment on any similarities.

118. Draw a graph that contains the points $(-2, -1)$, $(0, 1)$, $(1, 3)$, and $(3, 5)$. Compare your graph with those of other students. Are most of the graphs almost straight lines? How many are "curved"? Discuss the various ways that these points might be connected.

119. Explain what is meant by a complete graph.

120. Write a paragraph that describes a Cartesian plane. Then write a second paragraph that describes how to plot points in the Cartesian plane. Your paragraphs should include the terms "coordinate axes," "ordered pair," "coordinates," "plot," "x-coordinate," and "y-coordinate."

'Are You Prepared?' Answers

1. 8   **2.** 5   **3.** $11^2 + 60^2 = 61^2$ **4.** angle–side–angle

---

1.2 Intercepts; Symmetry; Graphing Key Equations; Circles

PREPARING FOR THIS SECTION   *Before getting started, review the following:*

- Completing the Square (Appendix, Section A.4, p. A31)
- Square Root Method (Appendix, Section A.4, pp. A30–A31)

Now Work the 'Are You Prepared?' problems on page 26.

OBJECTIVES **1** Find Intercepts Algebraically from an Equation (p. 17)
    2 Test an Equation for Symmetry (p. 18)
    3 Know How to Graph Key Equations (p. 19)
    4 Write the Standard Form of the Equation of a Circle (p. 22)
    5 Graph a Circle by Hand and by Using a Graphing Utility (p. 23)
    6 Work with the General Form of the Equation of a Circle (p. 24)

1 Find Intercepts Algebraically from an Equation

Figure 27

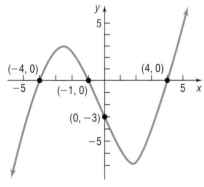

In Section 1.1, we discussed how to find intercepts from a graph and how to approximate intercepts from an equation using a graphing utility. Now we discuss how to find intercepts from an equation algebraically. To help understand the procedure, we present Figure 27. From the graph, we can see that the intercepts are $(-4, 0)$, $(-1, 0)$, $(4, 0)$, and $(0, -3)$. The x-intercepts are -4, -1, and 4. The y-intercept is -3. Notice that x-intercepts have y-coordinates that equal 0; y-intercepts have x-coordinates that equal 0. This leads to the following procedure for finding intercepts.

Procedure for Finding Intercepts

1. To find the x-intercept(s), if any, of the graph of an equation, let $y = 0$ in the equation and solve for x, where x is a real number.

2. To find the y-intercept(s), if any, of the graph of an equation, let $x = 0$ in the equation and solve for y, where y is a real number.

EXAMPLE 1   **Finding Intercepts from an Equation**

Find the x-intercept(s) and the y-intercept(s) of the graph of $y = x^2 - 4$. Then graph $y = x^2 - 4$ by plotting points.

Solution   To find the x-intercept(s), we let $y = 0$ and obtain the equation

$$x^2 - 4 = 0 \quad \text{\textit{y = x}}^2 \text{\textit{ − 4 with y = 0}}$$
$$(x + 2)(x - 2) = 0 \quad \text{\textit{Factor.}}$$
$$x + 2 = 0 \quad \text{or} \quad x - 2 = 0 \quad \text{\textit{Zero-Product Property}}$$
$$x = -2 \quad \text{or} \quad x = 2 \quad \text{\textit{Solve.}}$$

The equation has two solutions, -2 and 2. The x-intercepts are -2 and 2.

To find the y-intercept(s), we let $x = 0$ in the equation.

$$y = x^2 - 4$$
$$= 0^2 - 4 = -4$$

The y-intercept is -4.

Since $x^2 \geq 0$ for all x, we deduce from the equation $y = x^2 - 4$ that $y \geq -4$ for all x. This information, the intercepts, and the points from Table 4 enable us to graph $y = x^2 - 4$ by hand. See Figure 28.

Table 4

x	$y = x^2 - 4$	(x, y)
-3	$(-3)^2 - 4 = 5$	$(-3, 5)$
-1	-3	$(-1, -3)$
1	-3	$(1, -3)$
3	5	$(3, 5)$

Figure 28

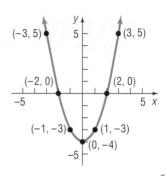

━━━ **Now Work** PROBLEM 1 5

2 Test an Equation for Symmetry

Another helpful tool for graphing equations by hand involves *symmetry,* particularly symmetry with respect to the x-axis, the y-axis, and the origin.

Symmetry often occurs in nature. Consider the picture of the butterfly. Do you see the symmetry?

DEFINITION

A graph is said to be **symmetric with respect to the x-axis** if, for every point (x, y) on the graph, the point $(x, -y)$ is also on the graph.

A graph is said to be **symmetric with respect to the y-axis** if, for every point (x, y) on the graph, the point $(-x, y)$ is also on the graph.

A graph is said to be **symmetric with respect to the origin** if, for every point (x, y) on the graph, the point $(-x, -y)$ is also on the graph.

Figure 29 illustrates the definition. Notice that, when a graph is symmetric with respect to the x-axis, the part of the graph above the x-axis is a reflection or mirror image of the part below it, and vice versa. When a graph is symmetric with respect to the y-axis, the part of the graph to the right of the y-axis is a reflection of the part to the left of it, and vice versa. Symmetry with respect to the origin may be viewed in two ways:

1. As a reflection about the y-axis, followed by a reflection about the x-axis
2. As a projection along a line through the origin so that the distances from the origin are equal

Figure 29

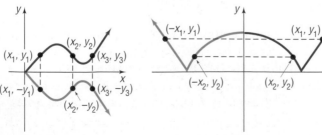

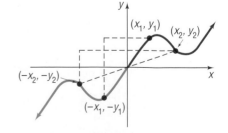

Symmetry with respect to the x-axis

Symmetry with respect to the y-axis

Symmetry with respect to the origin

<table>
<tr><td>EXAMPLE 2</td><td>Symmetric Points</td></tr>
</table>

(a) If a graph is symmetric with respect to the x-axis and the point $(4, 2)$ is on the graph, then the point $(4, -2)$ is also on the graph.

(b) If a graph is symmetric with respect to the y-axis and the point $(4, 2)$ is on the graph, then the point $(-4, 2)$ is also on the graph.

(c) If a graph is symmetric with respect to the origin and the point $(4, 2)$ is on the graph, then the point $(-4, -2)$ is also on the graph.

_____ ■

━━━━━ **Now Work** PROBLEM 23

When the graph of an equation is symmetric with respect to the x-axis, the y-axis, or the origin, the number of points that you need to plot in order to see the pattern is reduced. For example, if the graph of an equation is symmetric with respect to the y-axis, then, once points to the right of the y-axis are plotted, an equal number of points on the graph can be obtained by reflecting them about the y-axis. Because of this, before we graph an equation, we first want to determine whether it has any symmetry. The following tests are used for this purpose.

Tests for Symmetry

To test the graph of an equation for symmetry with respect to the

x-Axis   Replace y by $-y$ in the equation. If an equivalent equation results, the graph of the equation is symmetric with respect to the x-axis.

y-Axis   Replace x by $-x$ in the equation. If an equivalent equation results, the graph of the equation is symmetric with respect to the y-axis.

Origin   Replace x by $-x$ and y by $-y$ in the equation. If an equivalent equation results, the graph of the equation is symmetric with respect to the origin.

━━━━━ **Now Work** PROBLEM 53

3 Know How to Graph Key Equations

There are certain equations whose graphs we should be able to easily visualize in our mind's eye. We have already discussed the graph of $y = x^2$ in Example 8 in Section 1.1. The next three examples use intercepts, symmetry, and point plotting to obtain the graphs of additional key equations. It is important to know the graphs of these key equations because we use them later.

<table>
<tr><td>EXAMPLE 3</td><td>Graphing the Equation $y = x^3$ by Finding Intercepts and Checking for Symmetry</td></tr>
</table>

Graph the equation $y = x^3$ by hand by plotting points. Find any intercepts and check for symmetry first.

Solution

First, we seek the intercepts of $y = x^3$. When $x = 0$, then $y = 0$; and when $y = 0$, then $x = 0$. The origin $(0, 0)$ is the only intercept. Now we test $y = x^3$ for symmetry.

x-Axis: Replace y by $-y$. Since $-y = x^3$ is not equivalent to $y = x^3$, the graph is not symmetric with respect to the x-axis.

y-Axis: Replace x by $-x$. Since $y = (-x)^3 = -x^3$ is not equivalent to $y = x^3$, the graph is not symmetric with respect to the y-axis.

Origin: Replace x by $-x$ and y by $-y$. Since $-y = (-x)^3 = -x^3$ is equivalent to $y = x^3$ (multiply both sides by -1), the graph is symmetric with respect to the origin.

Figure 30

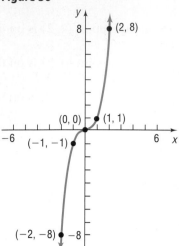

To graph by hand, we use the equation to obtain several points on the graph. Because of the symmetry with respect to the origin, we only need to locate points on the graph for which $x \geq 0$. See Table 5 below. Points on the graph could also be obtained using the TABLE feature on a graphing utility. See Table 6. Do you see the symmetry with respect to the origin from the table? Figure 30 shows the graph.

Table 5

x	y = x³	(x, y)
0	0	(0, 0)
1	1	(1, 1)
2	8	(2, 8)
3	27	(3, 27)

Table 6

X	Y1
-3	-27
-2	-8
-1	-1
0	0
1	1
2	8
3	27

Y1 ▊X^3

EXAMPLE 4   **Graphing the Equation $x = y^2$**

Graph the equation $x = y^2$. Find any intercepts and check for symmetry first.

Solution

The lone intercept is $(0, 0)$. The graph is symmetric with respect to the x-axis since $x = (-y)^2$ is equivalent to $x = y^2$. The graph is not symmetric with respect to the y-axis or the origin.

To graph $x = y^2$ by hand, we use the equation to obtain several points on the graph. Because the equation is solved for x, it is easier to assign values to y and use the equation to determine the corresponding values of x. Because of the symmetry, we can restrict ourselves to points whose y-coordinates are positive. We then use the symmetry to find additional points on the graph. See Table 7. For example, since $(1, 1)$ is on the graph, so is $(1, -1)$. Since $(4, 2)$ is on the graph, so is $(4, -2)$, and so on. We plot these points and connect them with a smooth curve to obtain Figure 31.

Table 7

y	x = y²	(x, y)
0	0	(0, 0)
1	1	(1, 1)
2	4	(4, 2)
3	9	(9, 3)

Figure 31

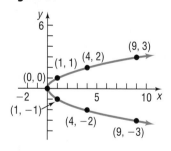

To graph the equation $x = y^2$ using a graphing utility, we must write the equation in the form $y = \{\text{expression in } x\}$. We proceed to solve for y.

$$x = y^2$$
$$y^2 = x$$
$$y = \pm\sqrt{x} \quad \text{Square Root Method}$$

To graph $x = y^2$, we need to graph both $Y_1 = \sqrt{x}$ and $Y_2 = -\sqrt{x}$ on the same screen. Figure 32 shows the result. Table 8 shows various values of y for a given value of x when $Y_1 = \sqrt{x}$ and $Y_2 = -\sqrt{x}$. Notice that when $x < 0$ we get an error. Can you explain why?

Figure 32

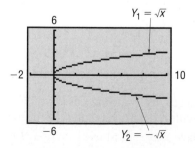

Table 8

X	Y1	Y2
-1	ERROR	ERROR
0	0	0
1	1	-1
2	1.4142	-1.414
3	1.7321	-1.732
4	2	-2
5	2.2361	-2.236

Y1 ▊√(X)

Look again at either Figure 31 or 32. If we restrict y so that $y \geq 0$, the equation $x = y^2$, $y \geq 0$, may be written as $y = \sqrt{x}$. The portion of the graph of $x = y^2$ in quadrant I plus the origin is the graph of $y = \sqrt{x}$. See Figure 33.

Figure 33

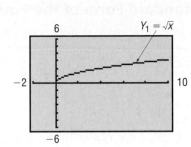
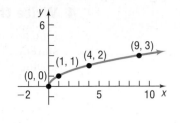

EXAMPLE 5

Graphing the Equation $y = \dfrac{1}{x}$

Graph the equation $y = \dfrac{1}{x}$. Find any intercepts and check for symmetry first.

Solution

We check for intercepts first. If we let $x = 0$, we obtain a 0 in the denominator, which is not defined. We conclude that there is no y-intercept. If we let $y = 0$, we get the equation $\dfrac{1}{x} = 0$, which has no solution. We conclude that there is no x-intercept. The graph of $y = \dfrac{1}{x}$ does not cross or touch the coordinate axes.

Next we check for symmetry.

x-Axis: Replacing y by $-y$ yields $-y = \dfrac{1}{x}$, which is not equivalent to $y = \dfrac{1}{x}$.

y-Axis: Replacing x by $-x$ yields $y = \dfrac{1}{-x} = -\dfrac{1}{x}$, which is not equivalent to $y = \dfrac{1}{x}$.

Origin: Replacing x by $-x$ and y by $-y$ yields $-y = -\dfrac{1}{x}$, which is equivalent to $y = \dfrac{1}{x}$. The graph is symmetric only with respect to the origin.

We can use the equation to form Table 9 and obtain some points on the graph. Because of symmetry, we only find points (x, y) for which x is positive. From Table 9 we infer that, if x is a large and positive number, then $y = \dfrac{1}{x}$ is a positive number close to 0. We also infer that if x is a positive number close to 0 then $y = \dfrac{1}{x}$ is a large and positive number. Armed with this information, we can graph the equation. Figure 34 illustrates some of these points and the graph of $y = \dfrac{1}{x}$. Observe how the absence of intercepts and the existence of symmetry with respect to the origin were utilized. Figure 35 confirms our algebraic analysis using a TI-84 Plus.

Table 9

x	$y = \dfrac{1}{x}$	(x, y)
$\dfrac{1}{10}$	10	$\left(\dfrac{1}{10}, 10\right)$
$\dfrac{1}{3}$	3	$\left(\dfrac{1}{3}, 3\right)$
$\dfrac{1}{2}$	2	$\left(\dfrac{1}{2}, 2\right)$
1	1	$(1, 1)$
2	$\dfrac{1}{2}$	$\left(2, \dfrac{1}{2}\right)$
3	$\dfrac{1}{3}$	$\left(3, \dfrac{1}{3}\right)$
10	$\dfrac{1}{10}$	$\left(10, \dfrac{1}{10}\right)$

Figure 34

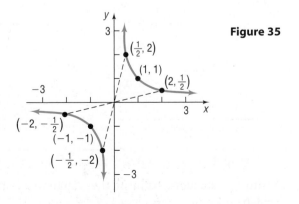

Figure 35

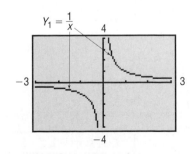

4 Write the Standard Form of the Equation of a Circle

One advantage of a coordinate system is that it enables us to translate a geometric statement into an algebraic statement, and vice versa. Consider, for example, the following geometric statement that defines a circle.

DEFINITION
A **circle** is a set of points in the xy-plane that are a fixed distance r from a fixed point (h, k). The fixed distance r is called the **radius,** and the fixed point (h, k) is called the **center** of the circle.

Figure 36

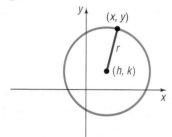

Figure 36 shows the graph of a circle. To find the equation, we let (x, y) represent the coordinates of any point on a circle with radius r and center (h, k). Then the distance between the points (x, y) and (h, k) must always equal r. That is, by the distance formula

$$\sqrt{(x - h)^2 + (y - k)^2} = r$$

After squaring both sides, we obtain

$$(x - h)^2 + (y - k)^2 = r^2$$

DEFINITION
The **standard form of an equation of a circle** with radius r and center (h, k) is

$$(x - h)^2 + (y - k)^2 = r^2 \qquad \textbf{(1)}$$

THEOREM
The standard form of an equation of a circle of radius r with center at the origin $(0, 0)$ is

$$x^2 + y^2 = r^2$$

DEFINITION
If the radius $r = 1$, the circle whose center is at the origin is called the **unit circle** and has the equation

$$x^2 + y^2 = 1$$

See Figure 37.

Figure 37
Unit circle $x^2 + y^2 = 1$

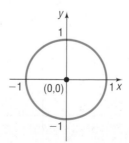

EXAMPLE 6 Writing the Standard Form of the Equation of a Circle

Write the standard form of the equation of the circle with radius 5 and center $(-3, 6)$.

Solution Using the form of equation (1) and substituting the values $r = 5, h = -3$, and $k = 6$, we have

$$(x - h)^2 + (y - k)^2 = r^2$$
$$(x - (-3))^2 + (y - 6)^2 = 5^2$$
$$(x + 3)^2 + (y - 6)^2 = 25$$

━━━▶ **Now Work** PROBLEM 65

5 Graph a Circle by Hand and by Using a Graphing Utility

The graph of any equation of the form $(x - h)^2 + (y - k)^2 = r^2$ is that of a circle with radius r and center (h, k).

EXAMPLE 7	Graphing a Circle by Hand and by Using a Graphing Utility

Graph the equation: $(x + 3)^2 + (y - 2)^2 = 16$

Solution Since the equation is in the form of equation (1), its graph is a circle. To graph the equation by hand, we first compare the given equation to the standard form of the equation of a circle. The comparison yields information about the circle.

$$(x + 3)^2 + (y - 2)^2 = 16$$
$$(x - (-3))^2 + (y - 2)^2 = 4^2$$
$$\uparrow \qquad\qquad \uparrow \qquad\quad \uparrow$$
$$(x - h)^2 + (y - k)^2 = r^2$$

We see that $h = -3$, $k = 2$, and $r = 4$. The circle has center $(-3, 2)$ and a radius of 4 units. To graph this circle, we first plot the center $(-3, 2)$. Since the radius is 4, we can locate four points on the circle by plotting points 4 units to the left, to the right, up, and down from the center. These four points can then be used as guides to obtain the graph. See Figure 38(a).

To graph a circle on a graphing utility, we must write the equation in the form $y = \{\text{expression involving } x\}.*$ We must solve for y in the equation

$$(x + 3)^2 + (y - 2)^2 = 16$$
$$(y - 2)^2 = 16 - (x + 3)^2 \qquad \text{Subtract } (x + 3)^2 \text{ from both sides.}$$
$$y - 2 = \pm\sqrt{16 - (x + 3)^2} \qquad \text{Use the Square Root Method.}$$
$$y = 2 \pm \sqrt{16 - (x + 3)^2} \qquad \text{Add 2 to both sides.}$$

To graph the circle, we graph the top half

$$Y_1 = 2 + \sqrt{16 - (x + 3)^2}$$

and the bottom half

$$Y_2 = 2 - \sqrt{16 - (x + 3)^2}$$

Also, be sure to use a square screen. Otherwise, the circle will appear distorted. Figure 38(b) shows the graph on a TI-84 Plus.

Figure 38

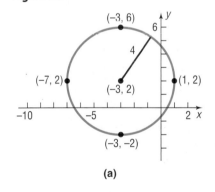

(a)

(b)

━━━▶ **Now Work** PROBLEMS 81(a) AND (b)

*Some graphing utilities (e.g., TI-83, TI-84, and TI-86) have a CIRCLE function that allows the user to enter only the coordinates of the center of the circle and its radius to graph the circle.

EXAMPLE 8	Finding the Intercepts of a Circle

For the circle $(x + 3)^2 + (y - 2)^2 = 16$, find the intercepts, if any, of its graph.

Solution   This is the equation discussed and graphed in Example 7. To find the x-intercepts, if any, let $y = 0$ and solve for x. Then

$$(x + 3)^2 + (y - 2)^2 = 16$$
$$(x + 3)^2 + (0 - 2)^2 = 16 \qquad \text{\small $y = 0$}$$
$$(x + 3)^2 + 4 = 16 \qquad \text{\small Simplify.}$$
$$(x + 3)^2 = 12 \qquad \text{\small Subtract 4 from both sides.}$$
$$x + 3 = \pm\sqrt{12} \qquad \text{\small Apply the Square Root Method.}$$
$$x = -3 \pm 2\sqrt{3} \qquad \text{\small Solve for x.}$$

The x-intercepts are $-3 - 2\sqrt{3} \approx -6.46$ and $-3 + 2\sqrt{3} \approx 0.46$.
To find the y-intercepts, if any, we let $x = 0$ and solve for y. Then

$$(x + 3)^2 + (y - 2)^2 = 16$$
$$(0 + 3)^2 + (y - 2)^2 = 16 \qquad \text{\small $x = 0$}$$
$$9 + (y - 2)^2 = 16$$
$$(y - 2)^2 = 7$$
$$y - 2 = \pm\sqrt{7} \qquad \text{\small Apply the Square Root Method.}$$
$$y = 2 \pm \sqrt{7} \qquad \text{\small Solve for y.}$$

The y-intercepts are $2 - \sqrt{7} \approx -0.65$ and $2 + \sqrt{7} \approx 4.65$.
Look back at Figure 38 to verify the approximate locations of the intercepts.

 **Now Work**   PROBLEM   8 1 (c)

6 Work with the General Form of the Equation of a Circle

If we eliminate the parentheses from the standard form of the equation of the circle given in Example 8, we get

$$(x + 3)^2 + (y - 2)^2 = 16$$
$$x^2 + 6x + 9 + y^2 - 4y + 4 = 16$$

which we find, upon simplifying, is equivalent to

$$x^2 + y^2 + 6x - 4y - 3 = 0$$

It can be shown that any equation of the form

$$x^2 + y^2 + ax + by + c = 0$$

has a graph that is a circle, or a point, or has no graph at all. For example, the graph of the equation $x^2 + y^2 = 0$ is the single point $(0, 0)$. The equation $x^2 + y^2 + 5 = 0$, or $x^2 + y^2 = -5$, has no graph, because sums of squares of real numbers are never negative.

DEFINITION   When its graph is a circle, the equation

$$x^2 + y^2 + ax + by + c = 0$$

is referred to as the **general form of the equation of a circle.**

If an equation of a circle is in the general form, we use the method of completing the square to put the equation in standard form so that we can identify its center and radius.

| EXAMPLE 9 | Graphing a Circle Whose Equation Is in General Form |

Graph the equation $x^2 + y^2 + 4x - 6y + 12 = 0$

Solution   We complete the square in both x and y to put the equation in standard form. Group the expression involving x, group the expression involving y, and put the constant on the right side of the equation. The result is

$$(x^2 + 4x) + (y^2 - 6y) = -12$$

Next, complete the square of each expression in parentheses. Remember that any number added on the left side of the equation must be added on the right.

$$(x^2 + 4x + 4) + (y^2 - 6y + 9) = -12 + 4 + 9$$

$$\left(\frac{4}{2}\right)^2 = 4 \qquad \left(\frac{-6}{2}\right)^2 = 9$$

$$(x + 2)^2 + (y - 3)^2 = 1 \quad \text{Factor}$$

We recognize this equation as the standard form of the equation of a circle with radius 1 and center $(-2, 3)$. To graph the equation by hand, use the center $(-2, 3)$ and the radius 1. See Figure 39(a).

To graph the equation using a graphing utility, we need to solve for y.

$$(y - 3)^2 = 1 - (x + 2)^2$$

$$y - 3 = \pm\sqrt{1 - (x + 2)^2} \qquad \text{Use the Square Root Method.}$$

$$y = 3 \pm \sqrt{1 - (x + 2)^2} \qquad \text{Add 3 to both sides.}$$

Figure 39(b) illustrates the graph on a TI-84 Plus graphing calculator.

Figure 39

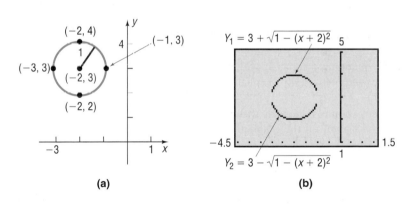

(a)   (b)

── **Now Work** PROBLEM 85

| EXAMPLE 10 | Finding the General Equation of a Circle |

Find the general equation of the circle whose center is $(1, -2)$ and whose graph contains the point $(4, -2)$.

Figure 40

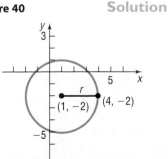

Solution   To find the equation of a circle, we need to know its center and its radius. Here, we know that the center is $(1, -2)$. Since the point $(4, -2)$ is on the graph, the radius r will equal the distance from $(4, -2)$ to the center $(1, -2)$. See Figure 40. Thus,

$$r = \sqrt{(4 - 1)^2 + [-2 - (-2)]^2}$$

$$= \sqrt{9} = 3$$

The standard form of the equation of the circle is

$$(x - 1)^2 + (y + 2)^2 = 9$$

Eliminating the parentheses and rearranging terms, we get the general equation

$$x^2 + y^2 - 2x + 4y - 4 = 0$$

 Now Work PROBLEM 71

Overview

The discussion here and in the Appendix, Section A.9 about lines deal with two main types of problems that can be generalized as follows:

1. Given an equation, classify it and graph it.
2. Given a graph, or information about a graph, find its equation.

This text deals with both types of problems. We shall study various equations, classify them, and graph them. The second type of problem is usually more difficult to solve than the first. In many instances a graphing utility can be used to solve problems when information about the problem (such as data) is given.

1.2 Assess Your Understanding

'Are You Prepared?' *Answers are given at the end of these exercises. If you get a wrong answer, read the pages listed in red.*

1. To complete the square of $x^2 + 10x$, you would (*add/ subtract*) the number _____. (p. A31)

2. Use the Square Root Method to solve the equation $(x - 2)^2 = 9$. (pp. A30–A31)

Concepts and Vocabulary

3. The *x*-intercepts of the graph of an equation are those *x*-values for which _____.

4. If for every point (x, y) on the graph of an equation the point $(-x, y)$ is also on the graph, then the graph is symmetric with respect to the _____.

5. If the graph of an equation is symmetric with respect to the *y*-axis and −4 is an *x*-intercept of this graph, then _____ is also an *x*-intercept.

6. If the graph of an equation is symmetric with respect to the origin and $(3, -4)$ is a point on the graph, then _____ is also a point on the graph.

7. *True or False* To find the *y*-intercepts of the graph of an equation, let $x = 0$ and solve for *y*.

8. *True or False* If a graph is symmetric with respect to the *x*-axis, then it cannot be symmetric with respect to the *y*-axis.

9. *True or False* Every equation of the form
$$x^2 + y^2 + ax + by + c = 0$$
has a circle as its graph.

10. For a circle, the _____ is the distance from the center to any point on the circle.

Skill Building

In Problems 11–22, find the intercepts and graph each equation by plotting points. Be sure to label the intercepts.

11. $y = x + 2$ 12. $y = x - 6$ 13. $y = 2x + 8$ 14. $y = 3x - 9$

15. $y = x^2 - 1$ 16. $y = x^2 - 9$ 17. $y = -x^2 + 4$ 18. $y = -x^2 + 1$

19. $2x + 3y = 6$ 20. $5x + 2y = 10$ 21. $9x^2 + 4y = 36$ 22. $4x^2 + y = 4$

In Problems 23–32, plot each point. Then plot the point that is symmetric to it with respect to (a) the x-axis; (b) the y-axis; (c) the origin.

23. $(3, 4)$ 24. $(5, 3)$ 25. $(-2, 1)$ 26. $(4, -2)$ 27. $(5, -2)$

28. $(-1, -1)$ 29. $(-3, -4)$ 30. $(4, 0)$ 31. $(0, -3)$ 32. $(-3, 0)$

In Problems 33–40, the graph of an equation is given. Indicate whether the graph is symmetric with respect to the x-axis, the y-axis, or the origin.

33.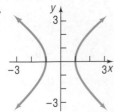

34.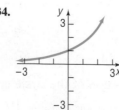

35.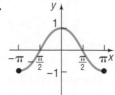

36.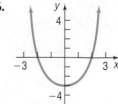

37.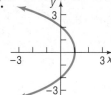

38.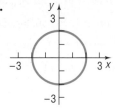

39.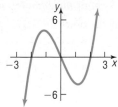

40.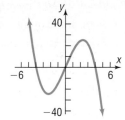

In Problems 41–44, draw a complete graph so that it has the type of symmetry indicated.

41. *y*-axis

42. *x*-axis

43. Origin

44. *y*-axis

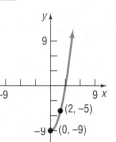

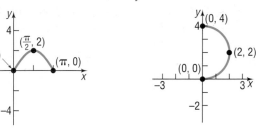

In Problems 45–48, draw a quick sketch of the equation by hand. Be sure to label all intercepts.

45. $y = x^3$

46. $x = y^2$

47. $y = \sqrt{x}$

48. $y = \dfrac{1}{x}$

In Problems 49–64, list the intercepts and test for symmetry.

49. $y^2 = x + 4$

50. $y^2 = x + 9$

51. $y = \sqrt[3]{x}$

52. $y = \sqrt[5]{x}$

53. $y = x^4 - 8x^2 - 9$

54. $y = x^4 - 2x^2 - 8$

55. $9x^2 + 4y^2 = 36$

56. $4x^2 + y^2 = 4$

57. $y = x^3 + x^2 - 9x - 9$

58. $y = x^3 + 2x^2 - 4x - 8$

59. $y = |x| - 4$

60. $y = |x| - 2$

61. $y = \dfrac{3x}{x^2 + 9}$

62. $y - \dfrac{x^2 - 4}{2x}$

63. $y = \dfrac{-x^3}{x^2 - 9}$

64. $y = \dfrac{x^4 + 1}{2x^5}$

In Problems 65–68, find the center and radius of each circle. Write the standard form of the equation.

65.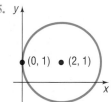

66.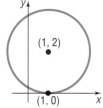

67.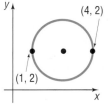

68.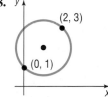

In Problems 69–78, write the standard form of the equation and the general form of the equation of each circle of radius r and center (h, k). *By hand, graph each circle.*

69. $r = 2$; $(h, k) = (0, 0)$

70. $r = 3$; $(h, k) = (0, 0)$

71. $r = 2$; $(h, k) = (0, 2)$

72. $r = 3$; $(h, k) = (1, 0)$

73. $r = 5$; $(h, k) = (4, -3)$

74. $r = 4$; $(h, k) = (2, -3)$

75. $r = 4$; $(h, k) = (-2, 1)$

76. $r = 7$; $(h, k) = (-5, -2)$

77. $r = \dfrac{1}{2}$; $(h, k) = \left(\dfrac{1}{2}, 0\right)$

78. $r = \dfrac{1}{2}$; $(h, k) = \left(0, -\dfrac{1}{2}\right)$

In Problems 79–92, (a) find the center (h, k) *and radius r of each circle; (b) by hand, graph each circle; (c) find the intercepts, if any.*

79. $x^2 + y^2 = 4$

80. $x^2 + (y - 1)^2 = 1$

81. $2(x - 3)^2 + 2y^2 = 8$

82. $3(x + 1)^2 + 3(y - 1)^2 = 6$

83. $x^2 + y^2 - 2x - 4y - 4 = 0$

84. $x^2 + y^2 + 4x + 2y - 20 = 0$

85. $x^2 + y^2 + 4x - 4y - 1 = 0$

86. $x^2 + y^2 - 6x + 2y + 9 = 0$

87. $x^2 + y^2 - x + 2y + 1 = 0$

88. $x^2 + y^2 + x + y - \dfrac{1}{2} = 0$

89. $2x^2 + 2y^2 - 12x + 8y - 24 = 0$

90. $2x^2 + 2y^2 + 8x + 7 = 0$

91. $2x^2 + 8x + 2y^2 = 0$

92. $3x^2 + 3y^2 - 12y = 0$

In Problems 93–100, find the standard form of the equation of each circle.

93. Center at the origin and containing the point $(-2, 3)$

94. Center $(1, 0)$ and containing the point $(-3, 2)$

95. Center $(2, 3)$ and tangent to the x-axis

96. Center $(-3, 1)$ and tangent to the y-axis

97. With endpoints of a diameter at $(1, 4)$ and $(-3, 2)$

98. With endpoints of a diameter at $(4, 3)$ and $(0, 1)$

99. Center $(-1, 3)$ and tangent to the line $y = 2$

100. Center $(4, -2)$ and tangent to the line $x = 1$

Mixed Practice

In Problems 101–108, (a) find the intercepts of each equation, (b) test each equation for symmetry with respect to the x-axis, the y-axis, and the origin, and (c) graph each equation by hand by plotting points. Be sure to label the intercepts on the graph and use any symmetry to assist in drawing the graph. Verify your results using a graphing utility.

101. $y = x^2 - 5$

102. $y = x^2 - 8$

103. $x - y^2 = -9$

104. $x + y^2 = 4$

105. $x^2 + y^2 = 9$

106. $x^2 + y^2 = 16$

107. $y = x^3 - 4x$

108. $y = x^3 - x$

Applications and Extensions

109. Given that the point $(1, 2)$ is on the graph of an equation that is symmetric with respect to the origin, what other point is on the graph?

110. If the graph of an equation is symmetric with respect to the y-axis and 6 is an x-intercept of this graph, name another x-intercept.

111. If the graph of an equation is symmetric with respect to the origin and -4 is an x-intercept of this graph, name another x-intercept.

112. If the graph of an equation is symmetric with respect to the x-axis and 2 is a y-intercept, name another y-intercept.

113. **Microphones**   In studios and on stages, cardioid microphones are often preferred for the richness they add to voices and for their ability to reduce the level of sound from the sides and rear of the microphone. Suppose one such cardioid pattern is given by the equation $(x^2 + y^2 - x)^2 = x^2 + y^2$.
(a) Find the intercepts of the graph of the equation.
(b) Test for symmetry with respect to the x-axis, y-axis, and origin.
Source: www.notaviva.com

114. **Solar Energy**   The solar electric generating systems at Kramer Junction, California, use parabolic troughs to heat a heat-transfer fluid to a high temperature. This fluid is used to generate steam that drives a power conversion system to produce electricity. For troughs 7.5 feet wide, an equation for the cross section is $16y^2 = 120x - 225$.

(a) Find the intercepts of the graph of the equation.
(b) Test for symmetry with respect to the x-axis, y-axis, and origin.
Source: U.S. Department of Energy

115. Find the area of the square in the figure.

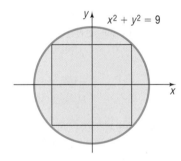

116. Find the area of the blue shaded region in the figure, assuming the quadrilateral inside the circle is a square.

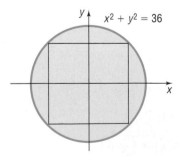

117. **Ferris Wheel**   The original Ferris wheel was built in 1893 by Pittsburg, Pennsylvania bridge builder George W. Ferris. The Ferris wheel was originally built for the 1893 World's Fair in Chicago, but was also later reconstructed for the 1904 World's Fair in St. Louis. It had a maximum height of 264 feet and a wheel diameter of 250 feet. Find an equation for the wheel if the center of the wheel is on the y-axis.
Source: inventors.about.com

118. Ferris Wheel In 2006, the Star of Nanchang (in the Jiangxi province) opened as the world's largest Ferris wheel. It has a maximum height of 160 meters and a diameter of 153 meters, with one full rotation taking approximately 30 minutes. Find an equation for the wheel if the center of the wheel is on the y-axis.

Source: AsiaOne Travel

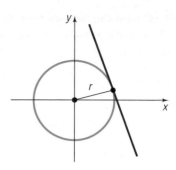

Note: Two other even larger Ferris wheels are reportedly to be completed in Asia by 2008 in time for the 2008 summer Olympics.

119. Weather Satellites Earth is represented on a map of a portion of the solar system so that its surface is the circle with equation $x^2 + y^2 + 2x + 4y - 4091 = 0$. A weather satellite circles 0.6 unit above Earth with the center of its circular orbit at the center of Earth. Find the equation for the orbit of the satellite on this map.

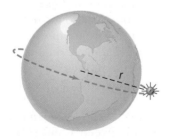

120. The **tangent line** to a circle may be defined as the line that intersects the circle in a single point, called the **point of tangency.** See the figure.

If the equation of the circle is $x^2 + y^2 = r^2$ and the equation of the tangent line is $y = mx + b$, show that:

(a) $r^2(1 + m^2) = b^2$

[Hint: The quadratic equation $x^2 + (mx + b)^2 = r^2$ has exactly one solution.]

(b) The point of tangency is $\left(\dfrac{-r^2 m}{b}, \dfrac{r^2}{b}\right)$.

(c) The tangent line is perpendicular to the line containing the center of the circle and the point of tangency.

121. The Greek Method The Greek method for finding the equation of the tangent line to a circle uses the fact that at any point on a circle the lines containing the center and the tangent line are perpendicular (see Problem 120). Use this method to find an equation of the tangent line to the circle $x^2 + y^2 = 9$ at the point $(1, 2\sqrt{2})$.

122. Use the Greek method described in Problem 121 to find an equation of the tangent line to the circle $x^2 + y^2 - 4x + 6y + 4 = 0$ at the point $(3, 2\sqrt{2} - 3)$.

123. Refer to Problem 120. The line $x - 2y + 4 = 0$ is tangent to a circle at $(0, 2)$. The line $y = 2x - 7$ is tangent to the same circle at $(3, -1)$. Find the center of the circle.

124. Find an equation of the line containing the centers of the two circles

$$x^2 + y^2 - 4x + 6y + 4 = 0$$

and

$$x^2 + y^2 + 6x + 4y + 9 = 0$$

125. If a circle of radius 2 is made to roll along the x-axis, what is an equation for the path of the center of the circle?

126. If the circumference of a circle is 6π, what is its radius?

Discussion and Writing

127. An equation is being tested for symmetry with respect to the x-axis, the y-axis, and the origin. Explain why, if two of these symmetries are present, the remaining one must also be present.

128. Draw a graph that contains the points $(-2, 5)$, $(-1, 3)$, and $(0, 2)$ that is symmetric with respect to the y-axis. Compare your graph with those of other students; comment on any similarities. Can a graph contain these points and be symmetric with respect to the x-axis? the origin? Why or why not?

129. Which of the following equations might have the graph shown? (More than one answer is possible.)

(a) $(x - 2)^2 + (y + 3)^2 = 13$
(b) $(x - 2)^2 + (y - 2)^2 = 8$
(c) $(x - 2)^2 + (y - 3)^2 = 13$
(d) $(x + 2)^2 + (y - 2)^2 = 8$
(e) $x^2 + y^2 - 4x - 9y = 0$
(f) $x^2 + y^2 + 4x - 2y = 0$
(g) $x^2 + y^2 - 9x - 4y = 0$
(h) $x^2 + y^2 - 4x - 4y = 4$

130. Which of the following equations might have the graph shown? (More than one answer is possible.)

(a) $(x - 2)^2 + y^2 = 3$
(b) $(x + 2)^2 + y^2 = 3$
(c) $x^2 + (y - 2)^2 = 3$
(d) $(x + 2)^2 + y^2 = 4$
(e) $x^2 + y^2 + 10x + 16 = 0$
(f) $x^2 + y^2 + 10x - 2y = 1$
(g) $x^2 + y^2 + 9x + 10 = 0$
(h) $x^2 + y^2 - 9x - 10 = 0$

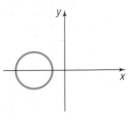

131. Explain how the center and radius of a circle can be used to graph a circle.

132. **What Went Wrong?** A student stated that the center and radius of the graph whose equation is $(x + 3)^2 + (y - 2)^2 = 16$ are $(3, -2)$ and 4, respectively. Why is this incorrect?

'Are You Prepared?' Answers

1. add; 25

2. $\{-1, 5\}$

1.3 Functions

PREPARING FOR THIS SECTION *Before getting started, review the following:*

- Intervals (Appendix, Section A.7, pp. A51–A52)
- Evaluating Algebraic Expressions, Domain of a Variable (Appendix, Section A.1, pp. A6–A7)
- Solving Inequalities (Appendix, Section A.7, pp. A54–A57)

Now Work the 'Are You Prepared?' problems on page 39.

OBJECTIVES **1** Determine Whether a Relation Represents a Function (p. 30)
  2 Find the Value of a Function (p. 34)
  3 Find the Domain of a Function Defined by an Equation (p. 37)

1 Determine Whether a Relation Represents a Function

We often see situations where one variable is somehow linked to the value of another variable. For example, an individual's level of education is linked to annual income. Engine size is linked to gas mileage. When the value of one variable is related to the value of a second variable, we have a *relation*. A **relation** is a correspondence between two sets. If x and y are two elements in these sets and if a relation exists between x and y, then we say that x **corresponds** to y or that y **depends on** x, and we write $x \rightarrow y$.

We have a number of ways to express relations between two sets. For example, the equation $y = 3x - 1$ shows a relation between x and y. It says that if we take some number x, multiply it by 3, and then subtract 1 we obtain the corresponding value of y. In this sense, x serves as the **input** to the relation and y is the **output** of the relation. We can also express this relation as a graph, as shown in Figure 41.

Not only can a relation be expressed through an equation or graph, but we can also express a relation through a technique called *mapping*. A **map** illustrates a relation by using a set of inputs and drawing arrows to the corresponding element in the set of outputs. Ordered pairs can also be used to represent $x \rightarrow y$ as (x, y). We illustrate these two concepts in Example 1.

Figure 41

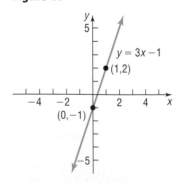

EXAMPLE 1 Maps and Ordered Pairs as Relations

Figure 42 shows a relation between states and the number of representatives each state has in the House of Representatives. The relation might be named "number of representatives."

Figure 42

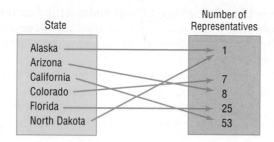

State   Number of Representatives

In this relation, Alaska corresponds to 1, Arizona corresponds to 8, and so on. Using ordered pairs, this relation would be expressed as

{(Alaska, 1), (Arizona, 8), (California, 53), (Colorado, 7), (Florida, 25), (North Dakota, 1)}

We now present one of the most important concepts in algebra—the *function*. A function is a special type of relation. To understand the idea behind a function, let's revisit the relation presented in Example 1. If we were to ask, "How many representatives does Alaska have?," you would respond 1. In fact, each input *state* corresponds to a single output *number of representatives*.

Let's consider a second relation where we have a correspondence between the weight of a brain (in grams) and IQ. See Figure 43. Notice the brain that weighs 1342 grams corresponds to two different IQs: 104 and 111. So, if someone asked, "What is the IQ of an individual whose brain weighs 1342 grams?," you cannot provide a single answer.

Figure 43

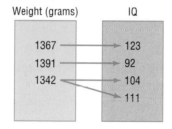

Weight (grams)   IQ

Let's look at one more relation. Figure 44 is a relation that shows a correspondence between *animals* and *life expectancy*. If asked to determine the life expectancy of a dog, we would all respond "11 years." If asked to determine the life expectancy of a rabbit, we would all respond "7 years."

Figure 44

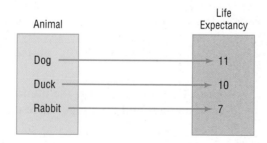

Animal   Life Expectancy

Notice that the relations presented in Figures 42 and 44 have something in common. What is it? The common link between these two relations is that each input corresponds to exactly one output. This leads to the definition of a *function*.

DEFINITION

Let X and Y be two nonempty sets.* A **function** from X into Y is a relation that associates with each element of X exactly one element of Y.

*The sets X and Y will usually be sets of real numbers, in which case a (real) function results. The two sets can also be sets of complex numbers, and then we have defined a complex function. In the broad definition (due to Lejeune Dirichlet), X and Y can be any two sets.

Figure 45

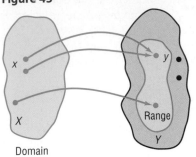

Domain

The set X is called the **domain** of the function. For each element x in X, the corresponding element y in Y is called the **value** of the function at x, or the **image** of x. The set of all images of the elements in the domain is called the **range** of the function. See Figure 45.

Since there may be some elements in Y that are not the image of some x in X, it follows that the range of a function may be a subset of Y, as shown in Figure 45.

Not all relations between two sets are functions. The next example shows how to determine whether a relation is a function.

| EXAMPLE 2 | Determining Whether a Relation Represents a Function

Determine which of the following relations represent a function. If the relation is a function, then state its domain and range.

(a) See Figure 46. For this relation, the domain represents the level of education and the range represents the unemployment rate.

Figure 46

Source: Statistical Abstract of the United States, 2006

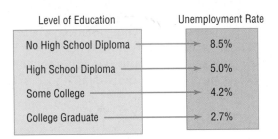

(b) See Figure 47. For this relation, the domain represents the number of calories in a sandwich from a fast-food restaurant and the range represents the fat content (in grams).

Figure 47

Source: Each company's Web site

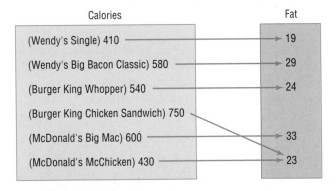

(c) See Figure 48. For this relation, the domain represents the weight (in carats) of pear-cut diamonds and the range represents the price (in dollars).

Figure 48

Source: diamonds.com

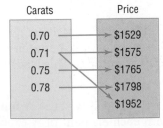

Solution (a) The relation in Figure 46 is a function because each element in the domain corresponds to exactly one element in the range. The domain of the relation is {No High School Diploma, High School Diploma, Some College, College Graduate}, and the range of the relation is {8.5%, 5.0%, 4.2%, 2.7%}.

(b) The relation in Figure 47 is a function because each element in the domain corresponds to exactly one element in the range. The domain of the relation is {410, 580, 540, 750, 600, 430}. The range of the relation is {19, 29, 24, 33, 23}. Notice that it is okay for more than one element in the domain to correspond to the same element in the range (McDonald's and Burger King's chicken sandwich both have 23 grams of fat).

(c) The relation in Figure 48 is not a function because each element in the domain does not correspond to exactly one element in the range. If a 0.71-carat diamond is chosen from the domain, a single price cannot be assigned to it.

— **Now Work** PROBLEM 15

> **In Words**
> For a function, no input has more than one output.

The idea behind a function is its predictability. If the input is known, we can use the function to determine the output. With "nonfunctions," we don't have this predictability. Look back at Figure 47. The inputs are {410, 580, 540, 750, 600, 430}. The correspondence is *number of fat grams*, and the outputs are {19, 29, 24, 33, 23}. If asked, "How many grams of fat are in a 410-calorie sandwich?," we can use the correspondence to answer "19." Now consider Figure 48. If asked, "What is the price of a 0.71-carat diamond?," we could not give a single response because two outputs result from the single input "0.71." For this reason, the relation in Figure 48 is not a function.

> **In Words**
> For a function, the domain is the set of inputs, and the range is the set of outputs.

We may also think of a function as a set of ordered pairs (x, y) in which no ordered pairs have the same first element and different second elements. The set of all first elements x is the domain of the function, and the set of all second elements y is its range. Each element x in the domain corresponds to exactly one element y in the range.

EXAMPLE 3   **Determining Whether a Relation Represents a Function**

Determine whether each relation represents a function. If it is a function, state the domain and range.

(a) {(1, 4), (2, 5), (3, 6), (4, 7)}
(b) {(1, 4), (2, 4), (3, 5), (6, 10)}
(c) {(−3, 9), (−2, 4), (0, 0), (1, 1), (−3, 8)}

Solution   (a) This relation is a function because there are no ordered pairs with the same first element and different second elements. The domain of this function is {1, 2, 3, 4}, and its range is {4, 5, 6, 7}.

(b) This relation is a function because there are no ordered pairs with the same first element and different second elements. The domain of this function is {1, 2, 3, 6}, and its range is {4, 5, 10}.

(c) This relation is not a function because there are two ordered pairs, (−3, 9) and (−3, 8), that have the same first element and different second elements.

In Example 3(b), notice that 1 and 2 in the domain each have the same image in the range. This does not violate the definition of a function; two different first elements can have the same second element. A violation of the definition occurs when two ordered pairs have the same first element and different second elements, as in Example 3(c).

— **Now Work** PROBLEM 19

Up to now we have shown how to identify when a relation is a function for relations defined by mappings (Example 2) and ordered pairs (Example 3). We know that relations can also be expressed as equations. We discuss next the circumstances under which equations are functions.

To determine whether an equation, where y depends on x, is a function, it is often easiest to solve the equation for y. If any value of x in the domain corresponds to more than one y, the equation does not define a function; otherwise, it does define a function.

EXAMPLE 4	**Determining Whether an Equation Is a Function**

Determine if the equation $y = 2x - 5$ defines y as a function of x.

Solution   The equation tells us to take an input x, multiply it by 2, and then subtract 5. For any input x, these operations yield only one output y. For example, if $x = 1$, then $y = 2(1) - 5 = -3$. If $x = 3$, then $y = 2(3) - 5 = 1$. For this reason, the equation is a function.

EXAMPLE 5	**Determining Whether an Equation Is a Function**

Determine if the equation $x^2 + y^2 = 1$ defines y as a function of x.

Solution   To determine whether the equation $x^2 + y^2 = 1$, which defines the unit circle, is a function, we need to solve the equation for y.

$$x^2 + y^2 = 1$$
$$y^2 = 1 - x^2$$
$$y = \pm\sqrt{1 - x^2}$$

For values of x between -1 and 1, two values of y result. For example, if $x = 0$, then $y = \pm 1$, so two different outputs result from the same input. This means that the equation $x^2 + y^2 = 1$ does not define a function.

Now Work PROBLEM 33

2 Find the Value of a Function

Functions are often denoted by letters such as $f, F, g, G,$ and others. If f is a function, then for each number x in its domain the corresponding image in the range is designated by the symbol $f(x)$, read as "f of x" or as "f at x." We refer to $f(x)$ as the **value of f at the number x;** $f(x)$ is the number that results when x is given and the function f is applied; $f(x)$ is the output corresponding to x or the image of x; $f(x)$ does *not* mean "f times x." For example, the function given in Example 4 may be written as $y = f(x) = 2x - 5$. Then $f\left(\dfrac{3}{2}\right) = -2$.

Figure 49 illustrates some other functions. Notice that, in every function, for each x in the domain there is one value in the range.

Figure 49

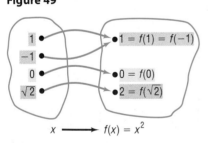

(a) $f(x) = x^2$

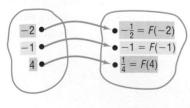

(b) $F(x) = \dfrac{1}{x}$

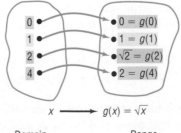

(c) $g(x) = \sqrt{x}$

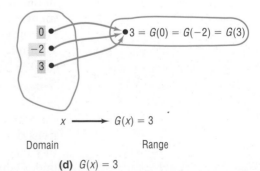

(d) $G(x) = 3$

Figure 50

Input x

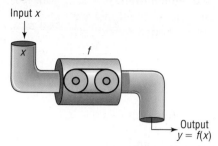

Output
y = f(x)

Sometimes it is helpful to think of a function f as a machine that receives as input a number from the domain, manipulates it, and outputs the value. See Figure 50.

The restrictions on this input/output machine are as follows:

1. It only accepts numbers from the domain of the function.

2. For each input, there is exactly one output (which may be repeated for different inputs).

For a function $y = f(x)$, the variable x is called the **independent variable,** because it can be assigned any of the numbers from the domain. The variable y is called the **dependent variable,** because its value depends on x.

Any symbol can be used to represent the independent and dependent variables. For example, if f is the *cube function,* then f can be given by $f(x) = x^3$ or $f(t) = t^3$ or $f(z) = z^3$. All three functions are the same. Each tells us to cube the independent variable to get the output. In practice, the symbols used for the independent and dependent variables are based on common usage, such as using C for cost in business.

The independent variable is also called the **argument** of the function. Thinking of the independent variable as an argument can sometimes make it easier to find the value of a function. For example, if f is the function defined by $f(x) = x^3$, then f tells us to cube the argument. Thus, $f(2)$ means to cube 2, $f(a)$ means to cube the number a, and $f(x + h)$ means to cube the quantity $x + h$.

EXAMPLE 6 | **Illustrating Language Used with Functions**

Consider the function $f(p) = p^2$. The name of the function is f. The *independent variable* or *argument* of f is p. If $q = f(p)$, then q is the *dependent variable.* Since $f(3) = 9$, we say that 9 is the *value* of f at 3 or 9 is the *image* of 3. Because we can square any real number, the *domain* of the function is the set of all real numbers, or $(-\infty, \infty)$ using interval notation. Because the square of any real number is greater than or equal to zero, the *range* of f is $[0, \infty)$. In words, the function f tells us to square the argument.

EXAMPLE 7 | **Finding Values of a Function**

For the function f defined by $f(x) = 2x^2 - 3x$, evaluate

(a) $f(3)$ (b) $f(x) + f(3)$ (c) $3f(x)$

(d) $f(-x)$ (e) $-f(x)$ (f) $f(3x)$

(g) $f(x + 3)$ (h) $\dfrac{f(x + h) - f(x)}{h}$ $h \neq 0$

Solution   (a) We substitute 3 for x in the equation for f, $f(x) = 2x^2 - 3x$, to get

$$f(3) = 2(3)^2 - 3(3) = 18 - 9 = 9$$

The image of 3 is 9.

(b) $f(x) + f(3) = (2x^2 - 3x) + (9) = 2x^2 - 3x + 9$

(c) We multiply the equation for f by 3.

$$3f(x) = 3(2x^2 - 3x) = 6x^2 - 9x$$

(d) We substitute $-x$ for x in the equation for f and simplify.

$$f(-x) = 2(-x)^2 - 3(-x) = 2x^2 + 3x$$

(e) $-f(x) = -(2x^2 - 3x) = -2x^2 + 3x$

(f) We substitute $3x$ for x in the equation for f and simplify.

$$f(3x) = 2(3x)^2 - 3(3x) = 2(9x^2) - 9x = 18x^2 - 9x$$

(g) We substitute $x + 3$ for x in the equation for f and simplify.

$$f(x + 3) = 2(x + 3)^2 - 3(x + 3) \qquad \text{Notice the use of parentheses here.}$$

$$= 2(x^2 + 6x + 9) - 3x - 9$$

$$= 2x^2 + 12x + 18 - 3x - 9$$

$$= 2x^2 + 9x + 9$$

(h) $\dfrac{f(x + h) - f(x)}{h} \underset{\uparrow}{=} \dfrac{[2(x + h)^2 - 3(x + h)] - [2x^2 - 3x]}{h}$

$f(x + h) = 2(x + h)^2 - 3(x + h)$

$$= \frac{2(x^2 + 2xh + h^2) - 3x - 3h - 2x^2 + 3x}{h} \qquad \text{Simplify.}$$

$$= \frac{2x^2 + 4xh + 2h^2 - 3h - 2x^2}{h} \qquad \begin{array}{l}\text{Distribute and combine}\\ \text{like terms.}\end{array}$$

$$= \frac{4xh + 2h^2 - 3h}{h} \qquad \text{Combine like terms.}$$

$$= \frac{h(4x + 2h - 3)}{h} \qquad \text{Factor out } h.$$

$$= 4x + 2h - 3 \qquad \text{Divide out the } h\text{'s.}$$

Notice in Example 7 that $f(x + 3) \neq f(x) + f(3)$, $f(-x) \neq -f(x)$, and $3f(x) \neq f(3x)$.

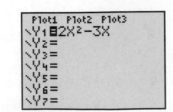 The expression in part (h) is called the **difference quotient** of f, an important expression in calculus.

Now Work PROBLEM 39

Most calculators have special keys that allow you to find the value of certain commonly used functions. For example, you should be able to find the square function $f(x) = x^2$, the square root function $f(x) = \sqrt{x}$, the reciprocal function $f(x) = \dfrac{1}{x} = x^{-1}$, and many others that will be discussed later in this book (such as $\ln x$ and $\log x$). Verify the results of Example 8, which follows, on your calculator.

EXAMPLE 8 | Finding Values of a Function on a Calculator

(a) $f(x) = x^2$ $\qquad f(1.234) = 1.234^2 = 1.522756$

(b) $F(x) = \dfrac{1}{x}$ $\qquad F(1.234) = \dfrac{1}{1.234} \approx 0.8103727715$

(c) $g(x) = \sqrt{x}$ $\qquad g(1.234) = \sqrt{1.234} \approx 1.110855526$

COMMENT Graphing calculators can be used to evaluate any function that you wish. Figure 51 shows the result obtained in Example 7(a) on a TI-84 Plus graphing calculator with the function to be evaluated, $f(x) = 2x^2 - 3x$, in Y_1.

Figure 51

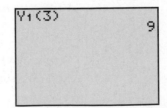

Implicit Form of a Function

In general, when a function f is defined by an equation in x and y, we say that the function f is given **implicitly.** If it is possible to solve the equation for y in terms of x, then we write $y = f(x)$ and say that the function is given **explicitly.** For example,

Implicit Form	**Explicit Form**
$3x + y = 5$	$y = f(x) = -3x + 5$
$x^2 - y = 6$	$y = f(x) = x^2 - 6$
$xy = 4$	$y = f(x) = \dfrac{4}{x}$

COMMENT The explicit form of a function is the form required by a graphing calculator. ∎

We list next a summary of some important facts to remember about a function f.

SUMMARY Important Facts about Functions

(a) For each x in a domain of a function f, there is exactly one image $f(x)$ in the range; however, an element in the range can result from more than one x in the domain.

(b) f is the symbol that we use to denote the function. It is symbolic of the equation that we use to get from an x in the domain to $f(x)$ in the range.

(c) If $y = f(x)$, then x is called the independent variable or argument of f, and y is called the dependent variable or the value of f at x.

---

3 Find the Domain of a Function Defined by an Equation

Often the domain of a function f is not specified; instead, only the equation defining the function is given. In such cases, we agree that the **domain of f** is the largest set of real numbers for which the value $f(x)$ is a real number. The domain of a function f is the same as the domain of the variable x in the expression $f(x)$.

EXAMPLE 9   Finding the Domain of a Function

Find the domain of each of the following functions:

(a) $f(x) = x^2 + 5x$ (b) $g(x) = \dfrac{3x}{x^2 - 4}$ (c) $h(t) = \sqrt{4 - 3t}$

Solution (a) The function tells us to square a number and then add five times the number. Since these operations can be performed on any real number, we conclude that the domain of f is the set of all real numbers.

(b) The function g tells us to divide $3x$ by $x^2 - 4$. Since division by 0 is not defined, the denominator $x^2 - 4$ can never be 0, so x can never equal -2 or 2. The domain of the function g is $\{x | x \neq -2, x \neq 2\}$.

(c) The function h tells us to take the square root of $4 - 3t$. But only nonnegative numbers have real square roots, so the expression under the square root (the radicand) must be nonnegative (greater than or equal to zero). This requires that

$$4 - 3t \geq 0$$
$$-3t \geq -4$$
$$t \leq \frac{4}{3}$$

In Words
The domain of g found in Example 9(b) is $\{x | x \neq -2, x \neq 2\}$. This notation is read, "The domain of the function g is the set of all real numbers x such that x does not equal -2 and x does not equal 2."

The domain of h is $\left\{t | t \leq \dfrac{4}{3}\right\}$ or the interval $\left(-\infty, \dfrac{4}{3}\right]$.

For the functions that we will encounter in this book, the following steps may prove helpful for finding the domain of a function that is defined by an equation whose domain is a subset of real numbers.

> ### Finding the Domain of a Function Defined by an Equation
>
> **1.** Start with the domain as the set of real numbers.
> **2.** If the equation has a denominator, exclude any numbers that give a zero denominator.
> **3.** If the equation has a radical of even index, exclude any numbers that cause the expression inside the radical to be negative.

━━━━━━ **Now Work** PROBLEM 51

If x is in the domain of a function f, we shall say that **f is defined at x,** or **$f(x)$ exists.** If x is not in the domain of f, we say that **f is not defined at x,** or **$f(x)$ does not exist.** For example, if $f(x) = \dfrac{x}{x^2 - 1}$, then $f(0)$ exists, but $f(1)$ and $f(-1)$ do not exist. (Do you see why?)

We have not said much about finding the range of a function. The reason is that when a function is defined by an equation it is often difficult to find the range.* Therefore, we shall usually be content to find just the domain of a function when the function is defined by an equation. We shall express the domain of a function using inequalities, interval notation, set notation, or words, whichever is most convenient.

When we use functions in applications, the domain may be restricted by physical or geometric considerations. For example, the domain of the function f defined by $f(x) = x^2$ is the set of all real numbers. However, if f is used to obtain the area of a square when the length x of a side is known, then we must restrict the domain of f to the positive real numbers, since the length of a side can never be 0 or negative.

| EXAMPLE 10 | Finding the Domain in an Application |

Express the area of a circle as a function of its radius. Find the domain.

Solution

See Figure 52. We know that the formula for the area A of a circle of radius r is $A = \pi r^2$. If we use r to represent the independent variable and A to represent the dependent variable, the function expressing this relationship is

Figure 52

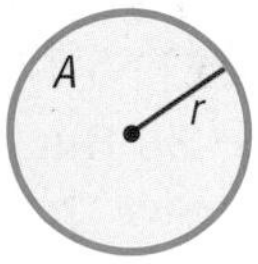

$$A(r) = \pi r^2$$

In this setting, the domain is $\{r \mid r > 0\}$. (Do you see why?)

Observe in the solution to Example 10 that we used the symbol A in two ways: It is used to name the function, and it is used to symbolize the dependent variable. This double use is common in applications and should not cause any difficulty.

━━━━━━ **Now Work** PROBLEM 67

---

*In Section 1.8, we discuss a way to find the range for a special class of functions.

SUMMARY

We list here some of the important vocabulary introduced in this section, with a brief description of each term.

Function	A relation between two sets so that each element x in the first set, the domain, has corresponding to it exactly one element y in the second set
	A set of ordered pairs (x, y) or $(x, f(x))$ in which no first element is paired with two different second elements
	The range is the set of y values of the function that are the images of the x values in the domain.
	A function f may be defined implicitly by an equation involving x and y or explicitly by writing $y = f(x)$.
Unspecified domain	If a function f is defined by an equation and no domain is specified, then the domain will be taken to be the largest set of real numbers for which the equation defines a real number.
Function notation	$y = f(x)$
	f is a symbol for the function.
	x is the independent variable or argument.
	y is the dependent variable.
	$f(x)$ is the value of the function at x, or the image of x.

1.3 Assess Your Understanding

'Are You Prepared?' *Answers are given at the end of these exercises. If you get a wrong answer, read the pages listed in* red.

1. The inequality $-1 < x < 3$ can be written in interval notation as _____. (pp. A51–A52)

2. If $x = -2$, the value of the expression $3x^2 - 5x + \dfrac{1}{x}$ is _____. (pp. A6–A7)

3. The domain of the variable in the expression $\dfrac{x - 3}{x + 4}$ is _____. (p. A7)

4. Solve the inequality: $3 - 2x > 5$. Graph the solution set. (pp. A54–A55)

Concepts and Vocabulary

5. If f is a function defined by the equation $y = f(x)$, then x is called the _____ variable and y is the _____ variable.

6. The set of all images of the elements in the domain of a function is called the _____.

7. If the domain of f is all real numbers in the interval $[0, 7]$ and the domain of g is all real numbers in the interval $[-2, 5]$, the domain of $f + g$ is all real numbers in the interval _____.

8. If $f(5) = 17$, then 17 is the _____ of 5.

9. If $f(x) = x + 1$ and $g(x) = x^3$, then _____ $= x^3 - (x + 1)$.

10. *True or False* Every relation is a function.

11. *True or False* The domain of $(f \cdot g)(x)$ consists of the numbers x that are in the domains of both f and g.

12. *True or False* The independent variable is sometimes referred to as the argument of the function.

13. *True or False* If no domain is specified for a function f, then the domain of f is taken to be the set of real numbers.

14. *True or False* The domain of the function $f(x) = \dfrac{x^2 - 4}{x}$ is $\{x | x \neq \pm 2\}$.

Skill Building

In Problems 15–26, determine whether each relation represents a function. For each function, state the domain and range.

15.
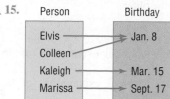

16.
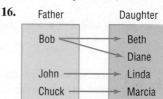

17.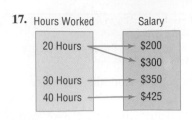

Hours Worked   Salary

18.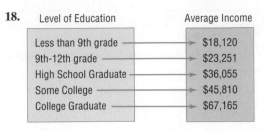

Level of Education   Average Income

19. $\{(2, 6), (-3, 6), (4, 9), (2, 10)\}$ **20.** $\{(-2, 5), (-1, 3), (3, 7), (4, 12)\}$ **21.** $\{(1, 3), (2, 3), (3, 3), (4, 3)\}$

22. $\{(0, -2), (1, 3), (2, 3), (3, 7)\}$ **23.** $\{(-2, 4), (-2, 6), (0, 3), (3, 7)\}$ **24.** $\{(-4, 4), (-3, 3), (-2, 2), (-1, 1), (-4, 0)\}$

25. $\{(-2, 4), (-1, 1), (0, 0), (1, 1)\}$ **26.** $\{(-2, 16), (-1, 4), (0, 3), (1, 4)\}$

In Problems 27–38, determine whether the equation defines y as a function of x.

27. $y = x^2$ **28.** $y = x^3$ **29.** $y = \dfrac{1}{x}$ **30.** $y = |x|$

31. $y = \pm\sqrt{4 - x^2}$ **32.** $y = \pm\sqrt{1 - 2x}$ **33.** $x = y^2$ **34.** $x + y^2 = 1$

35. $y = 2x^2 - 3x + 4$ **36.** $y = \dfrac{3x - 1}{x + 2}$ **37.** $2x^2 + 3y^2 = 1$ **38.** $x^2 - 4y^2 = 1$

In Problems 39–46, find the following for each function:

  (a) $f(0)$ *(b)* $f(1)$ *(c)* $f(-1)$ *(d)* $f(-x)$ *(e)* $-f(x)$ *(f)* $f(x + 1)$ *(g)* $f(2x)$ *(h)* $f(x + h)$

39. $f(x) = 3x^2 + 2x - 4$ **40.** $f(x) = -2x^2 + x - 1$ **41.** $f(x) = \dfrac{x}{x^2 + 1}$ **42.** $f(x) = \dfrac{x^2 - 1}{x + 4}$

43. $f(x) = |x| + 4$ **44.** $f(x) = \sqrt{x^2 + x}$ **45.** $f(x) = \dfrac{2x + 1}{3x - 5}$ **46.** $f(x) = 1 - \dfrac{1}{(x + 2)^2}$

In Problems 47–60, find the domain of each function.

47. $f(x) = -5x + 4$ **48.** $f(x) = x^2 + 2$ **49.** $f(x) = \dfrac{x}{x^2 + 1}$ **50.** $f(x) = \dfrac{x^2}{x^2 + 1}$

51. $g(x) = \dfrac{x}{x^2 - 16}$ **52.** $h(x) = \dfrac{2x}{x^2 - 4}$ **53.** $F(x) = \dfrac{x - 2}{x^3 + x}$ **54.** $G(x) = \dfrac{x + 4}{x^3 - 4x}$

55. $h(x) = \sqrt{3x - 12}$ **56.** $G(x) = \sqrt{1 - x}$ **57.** $f(x) = \dfrac{4}{\sqrt{x - 9}}$ **58.** $f(x) = \dfrac{x}{\sqrt{x - 4}}$

59. $p(x) = \sqrt{\dfrac{2}{x - 1}}$ **60.** $q(x) = \sqrt{-x - 2}$

Applications and Extensions

61. If $f(x) = 2x^3 + Ax^2 + 4x - 5$ and $f(2) = 5$, what is the value of A?

62. If $f(x) = 3x^2 - Bx + 4$ and $f(-1) = 12$, what is the value of B?

63. If $f(x) = \dfrac{3x + 8}{2x - A}$ and $f(0) = 2$, what is the value of A?

64. If $f(x) = \dfrac{2x - B}{3x + 4}$ and $f(2) = \dfrac{1}{2}$, what is the value of B?

65. If $f(x) = \dfrac{2x - A}{x - 3}$ and $f(4) = 0$, what is the value of A? Where is f not defined?

66. If $f(x) = \dfrac{x - B}{x - A}$, $f(2) = 0$ and $f(1)$ is undefined, what are the values of A and B?

67. Geometry Express the area A of a rectangle as a function of the length x if the length of the rectangle is twice its width.

68. Geometry Express the area A of an isosceles right triangle as a function of the length x of one of the two equal sides.

69. Constructing Functions Express the gross salary G of a person who earns \$10 per hour as a function of the number x of hours worked.

70. Constructing Functions Tiffany, a commissioned salesperson, earns \$100 base pay plus \$10 per item sold. Express her gross salary G as a function of the number x of items sold.

71. Population as a Function of Age The function

$$P(a) = 0.015a^2 - 4.962a + 290.580$$

represents the population P (in millions) of Americans in 2005 that are a years of age or older.

Source: U.S. Census Bureau

(a) Identify the dependent and independent variable.

(b) Evaluate $P(20)$. Provide a verbal explanation of the meaning of $P(20)$.

(c) Evaluate $P(0)$. Provide a verbal explanation of the meaning of $P(0)$.

72. Number of Rooms   The function

$$N(r) = -1.44r^2 + 14.52r - 14.96$$

represents the number N of housing units (in millions) in 2005 that have r rooms, where r is an integer and $2 \leq r \leq 9$.

Source: U.S. Census Bureau

(a) Identify the dependent and independent variable.

(b) Evaluate $N(3)$. Provide a verbal explanation of the meaning of $N(3)$.

73. Effect of Gravity on Earth   If a rock falls from a height of 20 meters on Earth, the height H (in meters) after x seconds is approximately

$$H(x) = 20 - 4.9x^2$$

(a) What is the height of the rock when $x = 1$ second? $x = 1.1$ seconds? $x = 1.2$ seconds? $x = 1.3$ seconds?

(b) When is the height of the rock 15 meters? When is it 10 meters? When is it 5 meters?

(c) When does the rock strike the ground?

74. Effect of Gravity on Jupiter   If a rock falls from a height of 20 meters on the planet Jupiter, its height H (in meters) after x seconds is approximately

$$H(x) = 20 - 13x^2$$

(a) What is the height of the rock when $x = 1$ second? $x = 1.1$ seconds? $x = 1.2$ seconds?

(b) When is the height of the rock 15 meters? When is it 10 meters? When is it 5 meters?

(c) When does the rock strike the ground?

75. Cost of Trans-Atlantic Travel   A Boeing 747 crosses the Atlantic Ocean (3000 miles) with an airspeed of 500 miles per hour. The cost C (in dollars) per passenger is given by

$$C(x) = 100 + \frac{x}{10} + \frac{36,000}{x}$$

where x is the ground speed (airspeed $\pm$ wind).

(a) What is the cost per passenger for quiescent (no wind) conditions?

(b) What is the cost per passenger with a head wind of 50 miles per hour?

(c) What is the cost per passenger with a tail wind of 100 miles per hour?

(d) What is the cost per passenger with a head wind of 100 miles per hour?

76. Cross-sectional Area   The cross-sectional area of a beam cut from a log with radius 1 foot is given by the function $A(x) = 4x\sqrt{1 - x^2}$, where x represents the length, in feet, of half the base of the beam. See the figure. Determine the cross-sectional area of the beam if the length of half the base of the beam is as follows:

(a) One-third of a foot

(b) One-half of a foot

(c) Two-thirds of a foot

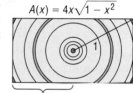

Discussion and Writing

77. Are the functions $f(x) = x - 1$ and $g(x) = \dfrac{x^2 - 1}{x + 1}$ the same? Explain.

78. Investigate when, historically, the use of the function notation $y = f(x)$ first appeared.

79. Write the equation of the following function f. The square root of three times a number x increased by 5.

80. Write the following function in words: $g(t) = t^2 - 2$.

'Are You Prepared?' Answers

1. $(-1, 3)$ **2.** 21.5   **3.** $\{x \,|\, x \neq -4\}$ **4.** $\{x \,|\, x < -1\}$

1.4 The Graph of a Function

PREPARING FOR THIS SECTION   *Before getting started, review the following:*

- Graphs of Equations (Section 1.1, pp. 7–11)   • Intercepts (Section 1.2, pp. 17–18)

Now Work the 'Are You Prepared?' problems on page 46.

OBJECTIVES   **1** Identify the Graph of a Function (p. 42)

2 Obtain Information from or about the Graph of a Function (p. 43)

In applications, a graph often demonstrates more clearly the relationship between two variables than, say, an equation or table would. For example, Table 10 shows the average price of gasoline in California adjusted for inflation (based on 2005 dollars) for the years 1979–2006. If we plot these data and then connect the points, we obtain Figure 53.

Table 10

Year	Price	Year	Price	Year	Price	Year	Price
1979	1.9829	1986	1.3459	1993	1.5189	2000	1.8249
1980	2.4929	1987	1.3274	1994	1.4709	2001	1.7536
1981	2.4977	1988	1.3111	1995	1.4669	2002	1.5955
1982	2.1795	1989	1.3589	1996	1.5397	2003	1.8950
1983	1.8782	1990	1.4656	1997	1.5329	2004	2.1521
1984	1.8310	1991	1.4973	1998	1.3246	2005	2.4730
1985	1.7540	1992	1.3969	1999	1.5270	2006	2.8919

Source: *Statistical Abstract of the United States*

Figure 53

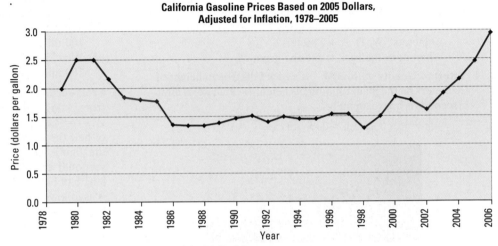

California Gasoline Prices Based on 2005 Dollars, Adjusted for Inflation, 1978–2005

Source: Statistical Abstact of the United States.

We can see from the graph that the price of gasoline (adjusted for inflation) was falling from 2000 to 2002 and has been rising rapidly since 2002. The graph also shows that the lowest price occurred in 1998. To learn information such as this from an equation requires that some calculations be made.

Look again at Figure 53. The graph shows that for each date on the horizontal axis there is only one price on the vertical axis. The graph represents a function, although the exact rule for getting from date to price is not given.

When a function is defined by an equation in x and y, the **graph of the function** is the graph of the equation, that is, the set of points (x, y) in the xy-plane that satisfies the equation.

1 Identify the Graph of a Function

Not every collection of points in the xy-plane represents the graph of a function. Remember, for a function, each number x in the domain has exactly one image y in the range. This means that the graph of a function cannot contain two points with the same x-coordinate and different y-coordinates. Therefore, the graph of a function must satisfy the following **vertical-line test.**

THEOREM

Vertical-line Test

A set of points in the xy-plane is the graph of a function if and only if every vertical line intersects the graph in at most one point.

In other words, if any vertical line intersects a graph at more than one point, the graph is not the graph of a function.

EXAMPLE 1   Identifying the Graph of a Function

Which of the graphs in Figure 54 are graphs of functions?

Figure 54

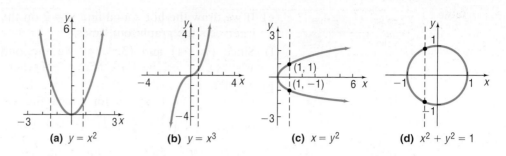

(a) $y = x^2$ **(b)** $y = x^3$ **(c)** $x = y^2$ **(d)** $x^2 + y^2 = 1$

Solution   The graphs in Figures 54(a) and 54(b) are graphs of functions, because every vertical line intersects each graph in at most one point. The graphs in Figures 54(c) and 54(d) are not graphs of functions, because there is a vertical line that intersects each graph in more than one point. Notice in Figure 54(c) that the input 1 corresponds to two outputs, -1 and 1. This is why the graph does not represent a function.

━━━━━━ **Now Work** PROBLEM 15

2 Obtain Information from or about the Graph of a Function

If (x, y) is a point on the graph of a function f, then y is the value of f at x. That is, if $y = f(x)$, then (x, y) is a point on the graph of f. The next example illustrates how to obtain information about a function if its graph is given.

EXAMPLE 2   **Obtaining Information from the Graph of a Function**

Let f be the function whose graph is given in Figure 55. (The graph of f might represent the distance that the bob of a pendulum is from its *at-rest* position. Negative values of y mean that the pendulum is to the left of the at-rest position, and positive values of y mean that the pendulum is to the right of the at-rest position.)

Figure 55

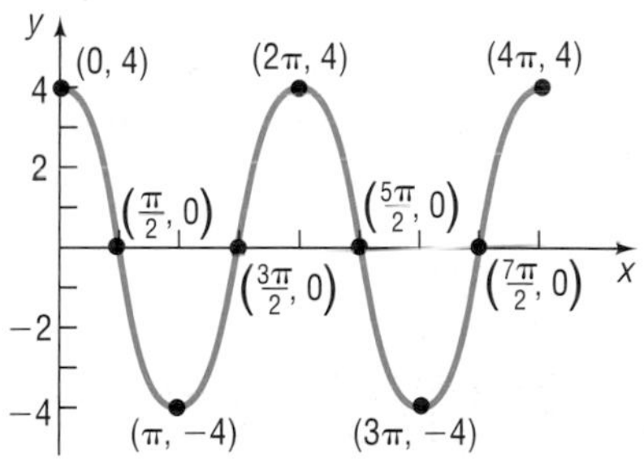

(a) What are $f(0)$, $f\left(\dfrac{3\pi}{2}\right)$, and $f(3\pi)$?

(b) What is the domain of f?

(c) What is the range of f?

(d) List the intercepts. (Recall that these are the points, if any, where the graph crosses or touches the coordinate axes.)

(e) How often does the line $y = 2$ intersect the graph?

(f) For what values of x does $f(x) = -4$?

(g) For what values of x is $f(x) > 0$?

Solution   (a) Since $(0, 4)$ is on the graph of f, the y-coordinate 4 is the value of f at the x-coordinate 0; that is, $f(0) = 4$. In a similar way, we find that when $x = \dfrac{3\pi}{2}$ then $y = 0$, so $f\left(\dfrac{3\pi}{2}\right) = 0$. When $x = 3\pi$, then $y = -4$, so $f(3\pi) = -4$.

(b) To determine the domain of f, we notice that the points on the graph of f have x-coordinates between 0 and 4π, inclusive; and for each number x between 0 and 4π, there is a point $(x, f(x))$ on the graph. The domain of f is $\{x | 0 \le x \le 4\pi\}$ or the interval $[0, 4\pi]$.

(c) The points on the graph all have y-coordinates between -4 and 4, inclusive; and for each such number y, there is at least one number x in the domain. The range of f is $\{y | -4 \le y \le 4\}$ or the interval $[-4, 4]$.

(d) The intercepts are

$$(0, 4), \left(\frac{\pi}{2}, 0\right), \left(\frac{3\pi}{2}, 0\right), \left(\frac{5\pi}{2}, 0\right), \quad \text{and} \quad \left(\frac{7\pi}{2}, 0\right)$$

(e) If we draw the horizontal line $y = 2$ on the graph in Figure 55, we find that it intersects the graph four times.

(f) Since $(\pi, -4)$ and $(3\pi, -4)$ are the only points on the graph for which $y = f(x) = -4$, we have $f(x) = -4$ when $x = \pi$ and $x = 3\pi$.

(g) To determine where $f(x) > 0$, we look at Figure 55 and determine the x-values from 0 to 4π for which the y-coordinate is positive. This occurs on $\left[0, \dfrac{\pi}{2}\right) \cup \left(\dfrac{3\pi}{2}, \dfrac{5\pi}{2}\right) \cup \left(\dfrac{7\pi}{2}, 4\pi\right]$. Using inequality notation, $f(x) > 0$ for

$$0 \le x < \frac{\pi}{2} \text{ or } \frac{3\pi}{2} < x < \frac{5\pi}{2} \text{ or } \frac{7\pi}{2} < x \le 4\pi.$$

When the graph of a function is given, its domain may be viewed as the shadow created by the graph on the x-axis by vertical beams of light. Its range can be viewed as the shadow created by the graph on the y-axis by horizontal beams of light. Try this technique with the graph given in Figure 55.

Now Work PROBLEMS 9 AND 13

EXAMPLE 3	**Obtaining Information about the Graph of a Function**

Consider the function: $f(x) = \dfrac{x + 1}{x + 2}$

(a) Is the point $\left(1, \dfrac{1}{2}\right)$ on the graph of f?

(b) If $x = 2$, what is $f(x)$? What point is on the graph of f?

(c) If $f(x) = 2$, what is x? What point is on the graph of f?

(d) What are the x-intercepts of the graph of f (if any)? What point(s) are on the graph of f?

Solution

(a) When $x = 1$, then

$$f(x) = \frac{x + 1}{x + 2}$$

$$f(1) = \frac{1 + 1}{1 + 2} = \frac{2}{3}$$

The point $\left(1, \dfrac{2}{3}\right)$ is on the graph of f; the point $\left(1, \dfrac{1}{2}\right)$ is not.

(b) If $x = 2$, then

$$f(x) = \frac{x + 1}{x + 2}$$

$$f(2) = \frac{2 + 1}{2 + 2} = \frac{3}{4}$$

The point $\left(2, \dfrac{3}{4}\right)$ is on the graph of f.

(c) If $f(x) = 2$, then

$$f(x) = 2$$

$$\frac{x + 1}{x + 2} = 2$$

$$x + 1 = 2(x + 2) \qquad \text{Multiply both sides by } x + 2.$$

$$x + 1 = 2x + 4 \qquad \text{Remove parentheses.}$$

$$x = -3 \qquad \text{Solve for x.}$$

If $f(x) = 2$, then $x = -3$. The point $(-3, 2)$ is on the graph of f.

(d) The *x*-intercepts of the graph of *f* are the real solutions of the equation $f(x) = 0$. The only real solution of the equation $f(x) = \dfrac{x+1}{x+2} = 0$ is $x = -1$, so -1 is the only *x*-intercept. Since $f(-1) = 0$, the point $(-1, 0)$ is on the graph of *f*.

Now Work PROBLEM 25

| EXAMPLE 4 | **Average Cost Function** |

The average cost $\overline{C}$ of manufacturing *x* computers per day is given by the function

$$\overline{C}(x) = 0.56x^2 - 34.39x + 1212.57 + \frac{20{,}000}{x}$$

Determine the average cost of manufacturing:

(a) 30 computers per day
(b) 40 computers per day
(c) 50 computers per day
(d) Graph the function $\overline{C} = \overline{C}(x), 0 < x \leq 80$.
(e) Create a TABLE with TblStart $= 1$ and ΔTbl $= 1$. Which value of *x* minimizes the average cost?

Solution

(a) The average cost of manufacturing $x = 30$ computers is

$$\overline{C}(30) = 0.56(30)^2 - 34.39(30) + 1212.57 + \frac{20{,}000}{30} = \$1351.54$$

(b) The average cost of manufacturing $x = 40$ computers is

$$\overline{C}(40) = 0.56(40)^2 - 34.39(40) + 1212.57 + \frac{20{,}000}{40} = \$1232.97$$

(c) The average cost of manufacturing $x = 50$ computers is

$$\overline{C}(50) = 0.56(50)^2 - 34.39(50) + 1212.57 + \frac{20{,}000}{50} = \$1293.07$$

(d) See Figure 56 for the graph of $\overline{C} = \overline{C}(x)$.
(e) With the function $\overline{C} = \overline{C}(x)$ in Y_1, we create Table 11. We scroll down until we find a value of *x* for which Y_1 is smallest. Table 12 shows that manufacturing $x = 41$ computers minimizes the average cost at \$1231.74 per computer.

Figure 56

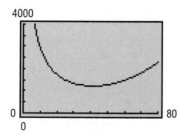

Table 11

X	Y1
1	21179
2	11146
3	7781.1
4	6084
5	5054.6
6	4359.7
7	3856.4

Y1■.56X²−34.39X...

Table 12

X	Y1
38	1240.7
39	1235.9
40	1233
41	1231.7
42	1232.2
43	1234.4
44	1238.1

Y1=1231.74487805

Now Work PROBLEM 31

SUMMARY

Graph of a function The collection of points (x, y) that satisfies the equation $y = f(x)$.

Vertical-line test A collection of points is the graph of a function provided that every vertical line intersects the graph in at most one point.

1.4 Assess Your Understanding

'Are You Prepared?' *Answers are given at the end of these exercises. If you get a wrong answer, read the pages listed in red.*

1. The intercepts of the equation $x^2 + 4y^2 = 16$ are _____. (pp. 17–18)

2. **True or False** The point $(-2, -6)$ is on the graph of the equation $x = 2y - 2$. (pp. 7–8)

Concepts and Vocabulary

3. A set of points in the *xy*-plane is the graph of a function if and only if every _____ line intersects the graph in at most one point.

4. If the point $(5, -3)$ is a point on the graph of f, then $f(\_\_\_) = \_\_\_$.

5. If $g(-2) = 7$, then the point _____ is on the graph of g.

6. **True or False** A function can have more than one *y*-intercept.

7. **True or False** The graph of a function $y = f(x)$ always crosses the *y*-axis.

8. **True or False** The *y*-intercept of the graph of the function $y = f(x)$, whose domain is all real numbers, is $f(0)$.

Skill Building

9. Use the graph below of the function f to answer parts (a)–(n).

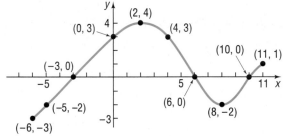

(a) Find $f(0)$ and $f(-6)$.
(b) Find $f(6)$ and $f(11)$.
(c) Is $f(3)$ positive or negative?
(d) Is $f(-4)$ positive or negative?
(e) For what values of x is $f(x) = 0$?
(f) For what values of x is $f(x) > 0$?
(g) What is the domain of f?
(h) What is the range of f?
(i) What are the *x*-intercepts?
(j) What is the *y*-intercept?

(k) How often does the line $y = \dfrac{1}{2}$ intersect the graph?

(l) How often does the line $x = 5$ intersect the graph?
(m) For what values of x does $f(x) = 3$?
(n) For what values of x does $f(x) = -2$?

10. Use the graph below of the function f to answer parts (a)–(n).

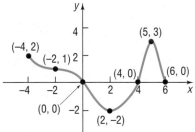

(a) Find $f(0)$ and $f(6)$.
(b) Find $f(2)$ and $f(-2)$.
(c) Is $f(3)$ positive or negative?
(d) Is $f(-1)$ positive or negative?
(e) For what values of x is $f(x) = 0$?
(f) For what values of x is $f(x) < 0$?
(g) What is the domain of f?
(h) What is the range of f?
(i) What are the *x*-intercepts?
(j) What is the *y*-intercept?
(k) How often does the line $y = -1$ intersect the graph?
(l) How often does the line $x = 1$ intersect the graph?
(m) For what value of x does $f(x) = 3$?
(n) For what value of x does $f(x) = -2$?

In Problems 11–22, determine whether the graph is that of a function by using the vertical-line test. If it is, use the graph to find:

  (a) The domain and range
  (b) The intercepts, if any
  (c) Any symmetry with respect to the x-axis, the y-axis, or the origin

11.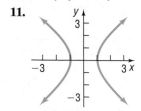

12.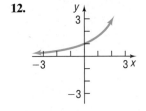

13.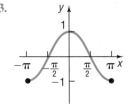

14.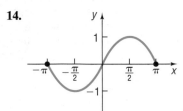

15.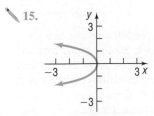

16.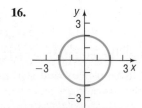

17.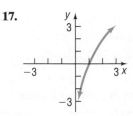

18.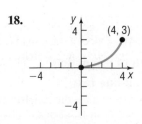

19.

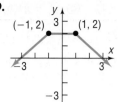

20.

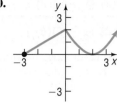

21.

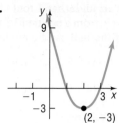

22.

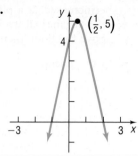

In Problems 23–28, answer the questions about the given function.

23. $f(x) = 2x^2 - x - 1$
(a) Is the point $(-1, 2)$ on the graph of f?
(b) If $x = -2$, what is $f(x)$? What point is on the graph of f?
(c) If $f(x) = -1$, what is x? What point(s) are on the graph of f?
(d) What is the domain of f?
(e) List the x-intercepts, if any, of the graph of f.
(f) List the y-intercept, if there is one, of the graph of f.

24. $f(x) = -3x^2 + 5x$
(a) Is the point $(-1, 2)$ on the graph of f?
(b) If $x = -2$, what is $f(x)$? What point is on the graph of f?
(c) If $f(x) = -2$, what is x? What point(s) are on the graph of f?
(d) What is the domain of f?
(e) List the x-intercepts, if any, of the graph of f.
(f) List the y-intercept, if there is one, of the graph of f.

25. $f(x) = \dfrac{x + 2}{x - 6}$
(a) Is the point $(3, 14)$ on the graph of f?
(b) If $x = 4$, what is $f(x)$? What point is on the graph of f?
(c) If $f(x) = 2$, what is x? What point(s) are on the graph of f?
(d) What is the domain of f?
(e) List the x-intercepts, if any, of the graph of f.
(f) List the y-intercept, if there is one, of the graph of f.

26. $f(x) = \dfrac{x^2 + 2}{x + 4}$
(a) Is the point $\left(1, \dfrac{3}{5}\right)$ on the graph of f?

(b) If $x = 0$, what is $f(x)$? What point is on the graph of f?
(c) If $f(x) = \dfrac{1}{2}$, what is x? What point(s) are on the graph of f?
(d) What is the domain of f?
(e) List the x-intercepts, if any, of the graph of f.
(f) List the y-intercept, if there is one, of the graph of f.

27. $f(x) = \dfrac{2x^2}{x^4 + 1}$
(a) Is the point $(-1, 1)$ on the graph of f?
(b) If $x = 2$, what is $f(x)$? What point is on the graph of f?
(c) If $f(x) = 1$, what is x? What point(s) are on the graph of f?
(d) What is the domain of f?
(e) List the x-intercepts, if any, of the graph of f.
(f) List the y-intercept, if there is one, of the graph of f.

28. $f(x) = \dfrac{2x}{x - 2}$
(a) Is the point $\left(\dfrac{1}{2}, -\dfrac{2}{3}\right)$ on the graph of f?
(b) If $x = 4$, what is $f(x)$? What point is on the graph of f?
(c) If $f(x) = 1$, what is x? What point(s) are on the graph of f?
(d) What is the domain of f?
(e) List the x-intercepts, if any, of the graph of f.
(f) List the y-intercept, if there is one, of the graph of f.

Applications and Extensions

29. Free-throw Shots According to physicist Peter Brancazio, the key to a successful foul shot in basketball lies in the arc of the shot. Brancazio determined the optimal angle of the arc from the free-throw line to be 45 degrees. The arc also depends on the velocity with which the ball is shot. If a player shoots a foul shot, releasing the ball at a 45-degree angle from a position 6 feet above the floor, then the path of the ball can be modeled by the function

$$h(x) = -\frac{44x^2}{v^2} + x + 6$$

where h is the height of the ball above the floor, x is the forward distance of the ball in front of the foul line, and v is the initial velocity with which the ball is shot in feet per second. Suppose a player shoots a ball with an initial velocity of 28 feet per second.
(a) Determine the height of the ball after it has traveled 8 feet in front of the foul line.

(b) Determine the height of the ball after it has traveled 12 feet in front of the foul line.
(c) Find additional points and graph the path of the basketball by hand.
(d) The center of the hoop is 10 feet above the floor and 15 feet in front of the foul line. Will the ball go through the hoop? Why or why not? If not, with what initial velocity must the ball be shot in order for the ball to go through the hoop?

Source: The Physics of Foul Shots, Discover, *Vol. 21, No. 10, October 2000*

30. Granny Shots The last player in the NBA to use an underhand foul shot (a "granny" shot) was Hall of Fame forward Rick Barry who retired in 1980. Barry believes that current NBA players could increase their free-throw percentage if they were to use an underhand shot. Since underhand shots are released from a lower position, the angle of the shot must be increased. If a player shoots an underhand foul shot, releasing the ball at a 70-degree angle from a position 3.5 feet above the floor, then the path of the ball can be modeled by the function $h(x) = -\dfrac{136x^2}{v^2} + 2.7x + 3.5$, where h is the height of the ball above the floor, x is the forward distance of the ball in front of the foul line, and v is the initial velocity with which the ball is shot in feet per second.

(a) The center of the hoop is 10 feet above the floor and 15 feet in front of the foul line. Determine the initial velocity with which the ball must be shot in order for the ball to go through the hoop.

(b) Write the function for the path of the ball using the velocity found in part (a).

(c) Determine the height of the ball after it has traveled 9 feet in front of the foul line.

(d) Find additional points and graph the path of the basketball by hand.

Source: The Physics of Foul Shots, Discover, *Vol. 21, No. 10, October 2000*

31. **Cost of Trans-Atlantic Travel**   A Boeing 747 crosses the Atlantic Ocean (3000 miles) with an airspeed of 500 miles per hour. The cost C (in dollars) per passenger is given by

$$C(x) = 100 + \frac{x}{10} + \frac{36,000}{x}$$

where x is the ground speed (airspeed ± wind).

(a) Use a graphing utility to graph the function $C = C(x)$.

(b) Create a TABLE with TblStart = 0 and ΔTbl = 50.

(c) To the nearest 50 miles per hour, what ground speed minimizes the cost per passenger?

32. **Cross-sectional Area**   The cross-sectional area of a beam cut from a log with radius 1 foot is given by the function $A(x) = 4x\sqrt{1 - x^2}$, where x represents the length, in feet, of half the base of the beam. See the figure.

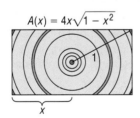

$A(x) = 4x\sqrt{1 - x^2}$

(a) Find the domain of A.

(b) Use a graphing utility to graph the function $A = A(x)$.

(c) Create a TABLE with TblStart = 0 and ΔTbl = 0.1 for $0 \le x \le 1$. Which value of x maximizes the cross-sectional area? What should be the length of the base of the beam to maximize the cross-sectional area?

33. The graphs of two functions, f and g, are illustrated. Use the graphs to answer parts (a)–(f).

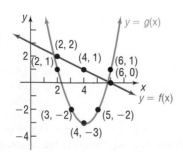

(a) $(f + g)(2)$ (b) $(f + g)(4)$ (c) $(f - g)(6)$

(d) $(g - f)(6)$ (e) $(f \cdot g)(2)$ (f) $\left(\dfrac{f}{g}\right)(4)$

34. **Reading and Interpreting Graphs**   Let C be the function whose graph is given below. This graph represents the cost C of manufacturing q computers in a day.

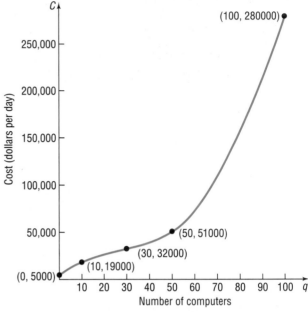

(a) Determine $C(0)$. Interpret this value.

(b) Determine $C(10)$. Interpret this value.

(c) Determine $C(50)$. Interpret this value.

(d) What is the domain of C? What does this domain imply in terms of daily production?

(e) Describe the shape of the graph.

(f) The point $(30, 32000)$ is called an *inflection point.* Describe the behavior of the graph around the inflection point.

35. **Reading and Interpreting Graphs**   Let C be the function whose graph is given below. This graph represents the cost C of using m anytime cell phone minutes in a month for a five-person family plan.

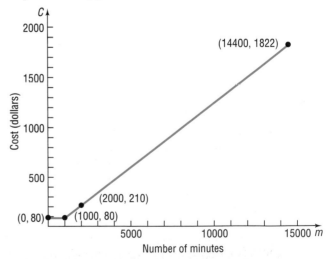

(a) Determine $C(0)$. Interpret this value.

(b) Determine $C(1000)$. Interpret this value.

(c) Determine $C(2000)$. Interpret this value.

(d) What is the domain of C? What does this domain imply in terms of the number of anytime minutes?

(e) Describe the shape of the graph.

Discussion and Writing

36. Describe how you would proceed to find the domain and range of a function if you were given its graph. How would your strategy change if you were given the equation defining the function instead of its graph?

37. How many *x*-intercepts can the graph of a function have? How many *y*-intercepts can the graph of a function have?

38. Is a graph that consists of a single point the graph of a function? Can you write the equation of such a function?

39. Match each of the following functions with the graph that best describes the situation.
   (a) The cost of building a house as a function of its square footage
   (b) The height of an egg dropped from a 300-foot building as a function of time
   (c) The height of a human as a function of time
   (d) The demand for Big Macs as a function of price
   (e) The height of a child on a swing as a function of time

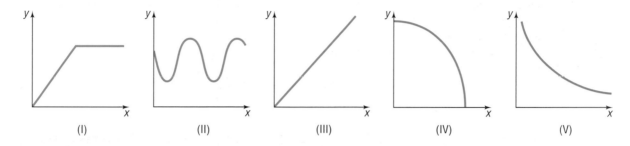

(I) (II) (III) (IV) (V)

40. Match each of the following functions with the graph that best describes the situation.
   (a) The temperature of a bowl of soup as a function of time
   (b) The number of hours of daylight per day over a 2-year period
   (c) The population of Florida as a function of time
   (d) The distance traveled by a car going at a constant velocity as a function of time
   (e) The height of a golf ball hit with a 7-iron as a function of time

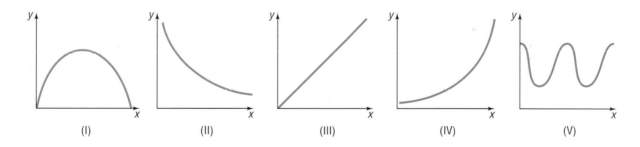

(I) (II) (III) (IV) (V)

41. Consider the following scenario: Barbara decides to take a walk. She leaves home, walks 2 blocks in 5 minutes at a constant speed, and realizes that she forgot to lock the door. So Barbara runs home in 1 minute. While at her doorstep, it takes her 1 minute to find her keys and lock the door. Barbara walks 5 blocks in 15 minutes and then decides to jog home. It takes her 7 minutes to get home. Draw a graph of Barbara's distance from home (in blocks) as a function of time.

42. Consider the following scenario: Jayne enjoys riding her bicycle through the woods. At the forest preserve, she gets on her bicycle and rides up a 2000-foot incline in 10 minutes. She then travels down the incline in 3 minutes. The next 5000 feet are level terrain and she covers the distance in 20 minutes. She rests for 15 minutes. Jayne then travels 10,000 feet in 30 minutes. Draw a graph of Jayne's distance traveled (in feet) as a function of time.

43. The sketch to the right represents the distance *d* (in miles) that Kevin is from home as a function of time *t* (in hours). Answer the questions based on the graph. In parts (a)–(g), how many hours elapsed and how far was Kevin from home during this time?

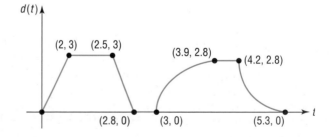

(a) From $t = 0$ to $t = 2$
(b) From $t = 2$ to $t = 2.5$
(c) From $t = 2.5$ to $t = 2.8$
(d) From $t = 2.8$ to $t = 3$
(e) From $t = 3$ to $t = 3.9$
(f) From $t = 3.9$ to $t = 4.2$
(g) From $t = 4.2$ to $t = 5.3$
(h) What is the farthest distance that Kevin is from home?
(i) How many times did Kevin return home?

44. The following sketch represents the speed v (in miles per hour) of Michael's car as a function of time t (in minutes).

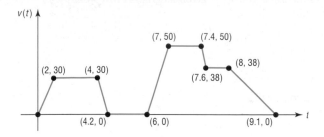

(a) Over what interval of time is Michael traveling fastest?
(b) Over what interval(s) of time is Michael's speed zero?

(c) What is Michael's speed between 0 and 2 minutes?
(d) What is Michael's speed between 4.2 and 6 minutes?
(e) What is Michael's speed between 7 and 7.4 minutes?
(f) When is Michael's speed constant?

45. Draw the graph of a function whose domain is $\{x \mid -3 \leq x \leq 8, \ x \neq 5\}$ and whose range is $\{y \mid -1 \leq y \leq 2, \ y \neq 0\}$. What point(s) in the rectangle $-3 \leq x \leq 8, -1 \leq y \leq 2$ cannot be on the graph? Compare your graph with those of other students. What differences do you see?

46. Is there a function whose graph is symmetric with respect to the x-axis? Explain.

47. Explain why the vertical-line test works.

'Are You Prepared?' Answers

1. $(-4, 0), (4, 0), (0, -2), (0, 2)$ **2.** False

1.5 Properties of Functions

PREPARING FOR THIS SECTION   *Before getting started, review the following:*

- Intervals (Appendix, Section A.7, pp. A51–A52)
- Intercepts (Section 1.2, pp. 17–18)
- Slope of a Line (Appendix, Section A.9, pp. A69–A70)
- Point–Slope Form of a Line (Appendix, Section A.9, p. A73)
- Symmetry (Section 1.2, pp. 18–19)

Now Work the 'Are You Prepared?' problems on page 56.

OBJECTIVES   **1** Determine Even and Odd Functions from a Graph (p. 50)

2 Identify Even and Odd Functions from the Equation (p. 51)

3 Use a Graph to Determine Where a Function Is Increasing, Decreasing, or Constant (p. 52)

4 Use a Graph to Locate Local Maxima and Local Minima (p. 53)

5 Use a Graphing Utility to Approximate Local Maxima and Local Minima and to Determine Where a Function Is Increasing or Decreasing (p. 54)

6 Find the Average Rate of Change of a Function (p. 55)

To obtain the graph of a function $y = f(x)$, it is often helpful to know certain properties that the function has and the impact of these properties on the way that the graph will look.

1 Determine Even and Odd Functions from a Graph

The words *even* and *odd,* when applied to a function f, describe the symmetry that exists for the graph of the function.

A function f is even, if and only if, whenever the point (x, y) is on the graph of f then the point $(-x, y)$ is also on the graph. Using function notation, we define an even function as follows:

DEFINITION   A function f is **even** if, for every number x in its domain, the number $-x$ is also in the domain and

$$f(-x) = f(x)$$

A function f is odd, if and only if, whenever the point (x, y) is on the graph of f then the point $(-x, -y)$ is also on the graph. Using function notation, we define an odd function as follows:

DEFINITION   A function f is **odd** if, for every number x in its domain, the number $-x$ is also in the domain and

$$f(-x) = -f(x)$$

Refer to page 19, where the tests for symmetry are listed. The following results are then evident.

THEOREM   A function is even if and only if its graph is symmetric with respect to the y-axis. A function is odd if and only if its graph is symmetric with respect to the origin.

EXAMPLE 1   **Determining Even and Odd Functions from the Graph**

Determine whether each graph given in Figure 57 is the graph of an even function, an odd function, or a function that is neither even nor odd.

Figure 57

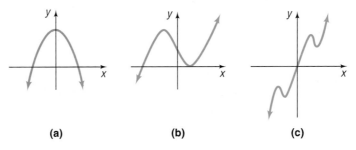

(a)   (b)   (c)

Solution   (a) The graph in Figure 57(a) is that of an even function, because the graph is symmetric with respect to the y-axis.

(b) The function whose graph is given in Figure 57(b) is neither even nor odd, because the graph is neither symmetric with respect to the y-axis nor symmetric with respect to the origin.

(c) The function whose graph is given in Figure 57(c) is odd, because its graph is symmetric with respect to the origin.

Now Work   PROBLEMS 21(a), (b), AND (d)

2 Identify Even and Odd Functions from the Equation

A graphing utility can be used to conjecture whether a function is even, odd, or neither. Remember that, when the graph of an even function contains the point (x, y), it must also contain the point $(-x, y)$. Therefore, if the graph shows evidence of symmetry with respect to the y-axis, we would conjecture that the function is even. In addition, if the graph shows evidence of symmetry with respect to the origin, we would conjecture that the function is odd.

In the next example, we use a graphing utility to conjecture whether a function is even, odd, or neither. Then we verify our conjecture algebraically.

EXAMPLE 2   **Identifying Even and Odd Functions**

Use a graphing utility to conjecture whether each of the following functions is even, odd, or neither. Verify the conjecture algebraically. Then state whether the graph is symmetric with respect to the y-axis or with respect to the origin.

(a) $f(x) = x^2 - 5$ (b) $g(x) = x^3 - 1$ (c) $h(x) = 5x^3 - x$

Solution

Figure 58

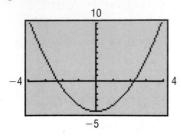

Figure 59

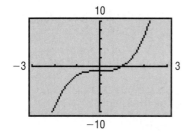

Figure 60

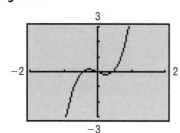

(a) Graph the function. See Figure 58. It appears that the graph is symmetric with respect to the y-axis. We conjecture that the function is even.

To algebraically verify the conjecture, we replace x by $-x$ in $f(x) = x^2 - 5$. Then

$$f(-x) = (-x)^2 - 5 = x^2 - 5 = f(x)$$

Since $f(-x) = f(x)$, we conclude that f is an even function, and the graph is symmetric with respect to the y-axis.

(b) Graph the function. See Figure 59. It appears that there is no symmetry. We conjecture that the function is neither even nor odd.

To algebraically verify that the function is not even, we find $g(-x)$ and compare the result with $g(x)$.

$$g(-x) = (-x)^3 - 1 = -x^3 - 1; \quad g(x) = x^3 - 1$$

Since $g(-x) \neq g(x)$, the function is not even.

To algebraically verify that the function is not odd, we find $-g(x)$ and compare the result with $g(-x)$.

$$-g(x) = -(x^3 - 1) = -x^3 + 1; \quad g(-x) = -x^3 - 1$$

Since $g(-x) \neq -g(x)$, the function is not odd. The graph is not symmetric with respect to the y-axis nor is it symmetric with respect to the origin.

(c) Graph the function. See Figure 60. It appears that there is symmetry with respect to the origin. We conjecture that the function is odd.

To algebraically verify the conjecture, we replace x by $-x$ in $h(x) = 5x^3 - x$. Then

$$h(-x) = 5(-x)^3 - (-x) = -5x^3 + x = -(5x^3 - x) = -h(x)$$

Since $h(-x) = -h(x)$, h is an odd function, and the graph of h is symmetric with respect to the origin.

 Now Work PROBLEM 33

3 Use a Graph to Determine Where a Function Is Increasing, Decreasing, or Constant

Consider the graph given in Figure 61. If you look from left to right along the graph of the function, you will notice that parts of the graph are rising, parts are falling, and parts are horizontal. In such cases, the function is described as *increasing, decreasing,* or *constant,* respectively.

Figure 61

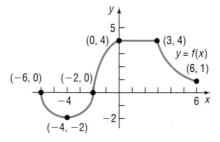

EXAMPLE 3

Determining Where a Function Is Increasing, Decreasing, or Constant from Its Graph

Where is the function in Figure 61 increasing? Where is it decreasing? Where is it constant?

Solution

To answer the question of where a function is increasing, where it is decreasing, and where it is constant, we use strict inequalities involving the independent variable x, or we use open intervals* of x-coordinates. The graph in Figure 61 is rising (increasing) from the point $(-4, -2)$ to the point $(0, 4)$, so we conclude that it is

*The open interval (a, b) consists of all real numbers x for which $a < x < b$.

WARNING We describe the behavior of a graph in terms of its x-values. Do not say the graph in Figure 61 is increasing from the point $(-4, -2)$ to $(0, 4)$. Rather, say it is increasing on the interval $(-4, 0)$. ∎

increasing on the open interval $(-4, 0)$ or for $-4 < x < 0$. The graph is falling (decreasing) from the point $(-6, 0)$ to the point $(-4, -2)$ and from the point $(3, 4)$ to the point $(6, 1)$. We conclude that the graph is decreasing on the open intervals $(-6, -4)$ and $(3, 6)$ or for $-6 < x < -4$ and $3 < x < 6$. The graph is constant on the open interval $(0, 3)$ or for $0 < x < 3$.

More precise definitions follow:

DEFINITIONS

In Words

If a function is decreasing, then, as the values of x get bigger, the values of the function get smaller. If a function is increasing, then, as the values of x get bigger, the values of the function also get bigger. If a function is constant, then, as the values of x get bigger, the values of the function remain unchanged.

A function f is **increasing** on an open interval I if, for any choice of x_1 and x_2 in I, with $x_1 < x_2$, we have $f(x_1) < f(x_2)$.

A function f is **decreasing** on an open interval I if, for any choice of x_1 and x_2 in I, with $x_1 < x_2$, we have $f(x_1) > f(x_2)$.

A function f is **constant** on an open interval I if, for all choices of x in I, the values $f(x)$ are equal.

Figure 62 illustrates the definitions. The graph of an increasing function goes up from left to right, the graph of a decreasing function goes down from left to right, and the graph of a constant function remains at a fixed height.

Figure 62

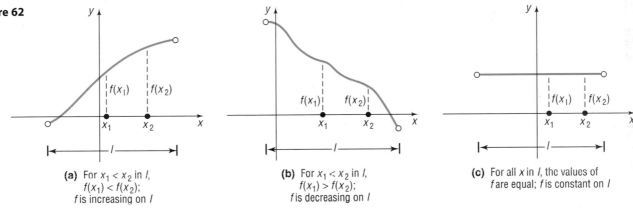

(a) For $x_1 < x_2$ in I, $f(x_1) < f(x_2)$; f is increasing on I

(b) For $x_1 < x_2$ in I, $f(x_1) > f(x_2)$; f is decreasing on I

(c) For all x in I, the values of f are equal; f is constant on I

Now Work PROBLEMS **11, 13, 15,** AND **21(c)**

4 Use a Graph to Locate Local Maxima and Local Minima

When the graph of a function is increasing to the left of $x = c$ and decreasing to the right of $x = c$, then at c the value of f is largest. This value is called a *local maximum* of f.* See Figure 63(a).

Figure 63

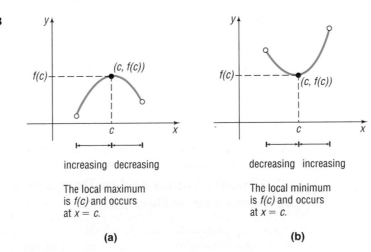

increasing decreasing

The local maximum is $f(c)$ and occurs at $x = c$.

(a)

decreasing increasing

The local minimum is $f(c)$ and occurs at $x = c$.

(b)

*Some texts will use the term *relative* instead of *local*.

When the graph of a function is decreasing to the left of $x = c$ and is increasing to the right of $x = c$, then at c the value of f is the smallest. This value is called a *local minimum* of f. See Figure 63(b).

DEFINITIONS

A function f has a **local maximum** at c if there is an open interval I containing c so that, for all x not equal to c in I, $f(x) < f(c)$. We call $f(c)$ a **local maximum of f.**

A function f has a **local minimum** at c if there is an open interval I containing c so that, for all x not equal to c in I, $f(x) > f(c)$. We call $f(c)$ a **local minimum of f.**

If f has a local maximum at c, then the value of f at c is greater than the values of f near c. If f has a local minimum at c, then the value of f at c is less than the values of f near c. The word *local* is used to suggest that it is only near c that the value $f(c)$ is largest or smallest.

EXAMPLE 4

Finding Local Maxima and Local Minima from the Graph of a Function and Determining Where the Function Is Increasing, Decreasing, or Constant

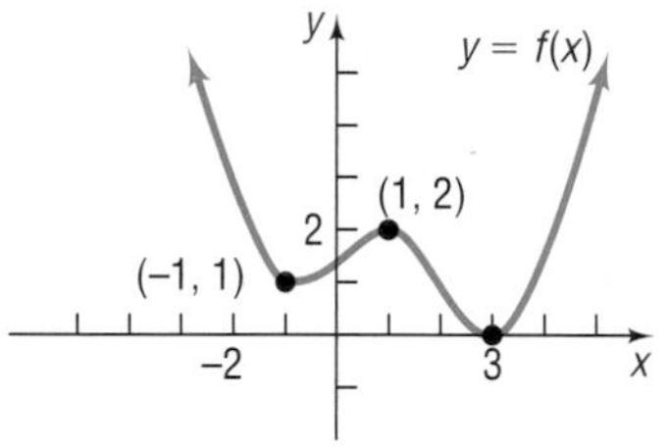

Figure 64 shows the graph of a function f.

Figure 64

(a) At what value(s) of x, if any, does f have a local maximum?
(b) What are the local maxima?
(c) At what value(s) of x, if any, does f have a local minimum?
(d) What are the local minima?
(e) List the intervals on which f is increasing. List the intervals on which f is decreasing.

Solution

The domain of f is the set of real numbers.

WARNING The y-value is the local maximum or local minimum and it occurs at some x-value. For Figure 64, we say the local maximum is 2 and that the local maximum occurs at $x = 1$. ■

(a) f has a local maximum at 1, since for all x close to 1, $x \neq 1$, we have $f(x) < f(1)$.
(b) The local maximum is $f(1) = 2$.
(c) f has a local minimum at -1 and at 3.
(d) The local minima are $f(-1) = 1$ and $f(3) = 0$.
(e) The function whose graph is given in Figure 64 is increasing for all values of x between -1 and 1 and for all values of x greater than 3. That is, the function is increasing on the intervals $(-1, 1)$ and $(3, \infty)$ or for $-1 < x < 1$ and $x > 3$. The function is decreasing for all values of x less than -1 and for all values of x between 1 and 3. That is, the function is decreasing on the intervals $(-\infty, -1)$ and $(1, 3)$ or for $x < -1$ and $1 < x < 3$.

■

Now Work PROBLEMS **17** AND **19**

5 Use a Graphing Utility to Approximate Local Maxima and Local Minima and to Determine Where a Function Is Increasing or Decreasing

To locate the exact value at which a function f has a local maximum or a local minimum usually requires calculus. However, a graphing utility may be used to approximate these values by using the MAXIMUM and MINIMUM features.

EXAMPLE 5	Using a Graphing Utility to Approximate Local Maxima and Minima and to Determine Where a Function Is Increasing or Decreasing

(a) Use a graphing utility to graph $f(x) = 6x^3 - 12x + 5$ for $-2 < x < 2$. Approximate where f has a local maximum and where f has a local minimum.

(b) Determine where f is increasing and where it is decreasing.

Solution

(a) Graphing utilities have a feature that finds the maximum or minimum point of a graph within a given interval. Graph the function f for $-2 < x < 2$. The MAXIMUM and MINIMUM commands require us to first determine the open interval I. The graphing utility will then approximate the maximum or minimum value in the interval. Using MAXIMUM we find that the local maximum is 11.53 and it occurs at $x = -0.82$, rounded to two decimal places. See Figure 65(a). Using MINIMUM, we find that the local minimum is -1.53 and it occurs at $x = 0.82$, rounded to two decimal places. See Figure 65(b).

Figure 65

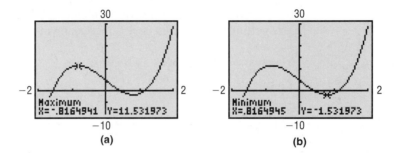

(a)   (b)

(b) Looking at Figures 65(a) and (b), we see that the graph of f is increasing from $x = -2$ to $x = -0.82$ and from $x = 0.82$ to $x = 2$, so f is increasing on the intervals $(-2, -0.82)$ and $(0.82, 2)$ or for $-2 < x < -0.82$ and $0.82 < x < 2$. The graph is decreasing from $x = -0.82$ to $x = 0.82$, so f is decreasing on the interval $(-0.82, 0.82)$ or for $-0.82 < x < 0.82$.

◼

➡️ **Now Work** PROBLEM 45

6 Find the Average Rate of Change of a Function

In the Appendix, Section A.9, we said that the slope of a line could be interpreted as the average rate of change of the line. To find the average rate of change of a function between any two points on its graph, we calculate the slope of the line containing the two points.

DEFINITION

If a and b, $a \neq b$, are in the domain of a function $y = f(x)$, the **average rate of change of f** from a to b is defined as

$$\text{Average rate of change} = \frac{\Delta y}{\Delta x} = \frac{f(b) - f(a)}{b - a} \qquad a \neq b \qquad \textbf{(1)}$$

The symbol Δy in (1) is the "change in y," and Δx is the "change in x." The average rate of change of f is the change in y divided by the change in x.

EXAMPLE 6	Finding the Average Rate of Change

Find the average rate of change of $f(x) = 3x^2$:

(a) From 1 to 3   (b) From 1 to 5   (c) From 1 to 7

Solution

(a) The average rate of change of $f(x) = 3x^2$ from 1 to 3 is

$$\frac{\Delta y}{\Delta x} = \frac{f(3) - f(1)}{3 - 1} = \frac{27 - 3}{3 - 1} = \frac{24}{2} = 12$$

(b) The average rate of change of $f(x) = 3x^2$ from 1 to 5 is

$$\frac{\Delta y}{\Delta x} = \frac{f(5) - f(1)}{5 - 1} = \frac{75 - 3}{5 - 1} = \frac{72}{4} = 18$$

(c) The average rate of change of $f(x) = 3x^2$ from 1 to 7 is

$$\frac{\Delta y}{\Delta x} = \frac{f(7) - f(1)}{7 - 1}$$

$$= \frac{147 - 3}{7 - 1} = \frac{144}{6} = 24$$

Figure 66

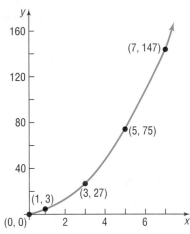

See Figure 66 for a graph of $f(x) = 3x^2$. The function f is increasing for $x > 0$. The fact that the average rates of change are getting larger indicates that the graph is getting steeper; that is, it is increasing at an increasing rate.

Now Work PROBLEM 53

1.5 Assess Your Understanding

'Are You Prepared?' *Answers are given at the end of these exercises. If you get a wrong answer, read the pages listed in red.*

1. The interval $(2, 5)$ can be written as the inequality _____. (pp. A51–A52)

2. The slope of the line containing the points $(-2, 3)$ and $(3, 8)$ is _____. (pp. A69–A70)

3. Test the equation $y = 5x^2 - 1$ for symmetry with respect to the x-axis, the y-axis, and the origin. (pp. 18–19)

4. Write the point–slope form of the line with slope 5 containing the point $(3, -2)$. (p. A73)

5. The intercepts of the equation $y = x^2 - 9$ are _____. (pp. 17–18)

Concepts and Vocabulary

6. A function f is _____ on an open interval I if, for any choice of x_1 and x_2 in I, with $x_1 < x_2$, we have $f(x_1) < f(x_2)$.

7. An _____ function f is one for which $f(-x) = f(x)$ for every x in the domain of f; an _____ function f is one for which $f(-x) = -f(x)$ for every x in the domain of f.

8. ***True or False*** A function f is decreasing on an open interval I if, for any choice of x_1 and x_2 in I, with $x_1 < x_2$, we have $f(x_1) > f(x_2)$.

9. ***True or False*** A function f has a local maximum at c if there is an open interval I containing c so that, for all x not equal to c in I, $f(x) < f(c)$.

10. ***True or False*** Even functions have graphs that are symmetric with respect to the origin.

Skill Building

In Problems 11–20, use the given graph of the function f.

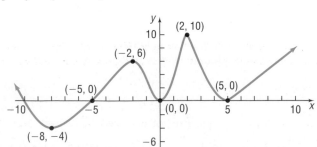

11. Is f increasing on the interval $(-8, -2)$?

12. Is f decreasing on the interval $(-8, -4)$?

13. Is f increasing on the interval $(2, 10)$?

14. Is f decreasing on the interval $(2, 5)$?

15. List the interval(s) on which f is increasing.

16. List the interval(s) on which f is decreasing.

17. Is there a local maximum at 2? If yes, what is it?

18. Is there a local maximum at 5? If yes, what is it?

19. List the numbers at which f has a local maximum. What are these local maxima?

20. List the numbers at which f has a local minimum. What are these local minima?

In Problems 21–28, the graph of a function is given. Use the graph to find:
 (a) *The intercepts, if any*
 (b) *The domain and range*
 (c) *The intervals on which it is increasing, decreasing, or constant*
 (d) *Whether it is even, odd, or neither*

21.
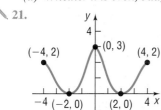

22.
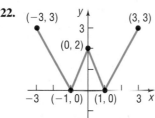

23.
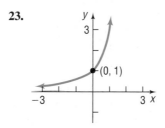

24.
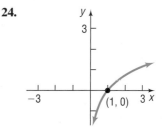

25.
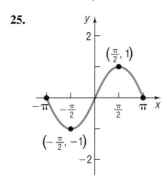

26.
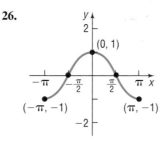

27.
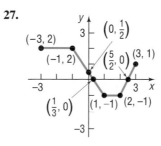

28.
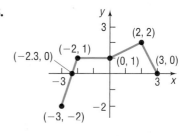

In Problems 29–32, the graph of a function f is given. Use the graph to find:
 (a) *The values, if any, at which f has a local maximum. What are these local maxima?*
 (b) *The values, if any, at which f has a local minimum. What are these local minima?*

29.
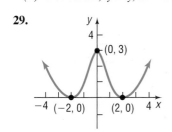

30.
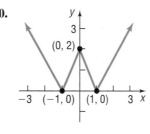

31.
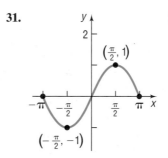

32.
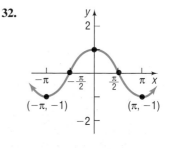

In Problems 33–44, determine algebraically whether each function is even, odd, or neither.

33. $F(x) = 4x^3 + x$

34. $f(x) = 2x^4 - x^2$

35. $g(x) = -3x^2 - 5$

36. $h(x) = 3x^5 + 5x$

37. $F(x) = \sqrt[3]{x}$

38. $G(x) = \sqrt{x}$

39. $f(x) = x + |x|$

40. $f(x) = \sqrt[3]{2x^2 + 1}$

41. $g(x) = \dfrac{1}{x^2}$

42. $h(x) = \dfrac{x}{x^2 - 1}$

43. $h(x) = \dfrac{-x^3}{3x^2 - 9}$

44. $F(x) = \dfrac{2x}{|x|}$

In Problems 45–52, use a graphing utility to graph each function over the indicated interval and approximate any local maxima and local minima. Determine where the function is increasing and where it is decreasing. Round answers to two decimal places.

45. $f(x) = x^3 - 3x + 2$ $(-2, 2)$

46. $f(x) = x^3 - 3x^2 + 5$ $(-1, 3)$

47. $f(x) = x^5 - x^3$ $(-2, 2)$

48. $f(x) = x^4 - x^2$ $(-2, 2)$

49. $f(x) = -0.2x^3 - 0.6x^2 + 4x - 6$ $(-6, 4)$

50. $f(x) = -0.4x^3 + 0.6x^2 + 3x - 2$ $(-4, 5)$

51. $f(x) = 0.25x^4 + 0.3x^3 - 0.9x^2 + 3$ $(-3, 2)$

52. $f(x) = -0.4x^4 - 0.5x^3 + 0.8x^2 - 2$ $(-3, 2)$

53. Find the average rate of change of $f(x) = -2x^2 + 4$

   (a) From 0 to 2
   (b) From 1 to 3
   (c) From 1 to 4

54. Find the average rate of change of $f(x) = -x^3 + 1$

   (a) From 0 to 2
   (b) From 1 to 3
   (c) From -1 to 1

55. Find the average rate of change of $g(x) = x^3 - 2x + 1$

   (a) From -3 to -2
   (b) From -1 to 1
   (c) From 1 to 3

56. Find the average rate of change of $h(x) = x^2 - 2x + 3$

   (a) From -1 to 1
   (b) From 0 to 2
   (c) From 2 to 5

Mixed Practice

57. $g(x) = x^3 - 4x$

   (a) Determine whether g is even, odd, or neither.
   (b) Using a graphing utility, graph g.
   (c) There is a local minimum in quadrant IV. Approximate the local minimum.
   (d) Using the results from part (a), determine the local maximum in quadrant II without using your graphing utility. Verify your result using your graphing utility.

58. $f(x) = -x^3 + 6x$

   (a) Determine whether f is even, odd, or neither.
   (b) Using a graphing utility, graph f.
   (c) There is a local maximum in quadrant I. Approximate the local maximum.
   (d) Using the results from part (a), determine the local minimum in quadrant III without using your graphing utility. Verify your result using your graphing utility.

59. $F(x) = -x^4 + 8x^2 + 8$

   (a) Determine whether F is even, odd, or neither.
   (b) Using a graphing utility, graph F.
   (c) There is a local maximum in quadrant I. Approximate the local maximum.

   (d) Using the results from part (a), determine the local maximum in quadrant II without using your graphing utility.

   (e) Suppose the area under the graph of F between $x = 0$ and $x = 3$ that is bounded below by the x-axis is 47.4 square units. Using the results from part (a), determine the area under the graph of F between $x = -3$ and $x = 0$ bounded below by the x-axis.

60. $G(x) = -x^4 + 32x^2 + 144$

   (a) Determine whether G is even, odd, or neither.
   (b) Using a graphing utility, graph G.
   (c) There is a local maximum in quadrant I. Approximate the local maximum.
   (d) Using the results from part (a), determine the local maximum in quadrant II without using your graphing utility.

   (e) Suppose the area under the graph of G between $x = 0$ and $x = 6$ that is bounded below by the x-axis is 1612.8 square units. Using the results from part (a), determine the area under the graph of F between $x = -6$ and $x = 0$ bounded below by the x-axis.

Applications and Extensions

61. Minimum Average Cost The average cost per hour in dollars, $\overline{C}$, of producing x riding lawn mowers can be modeled by the function

$$\overline{C}(x) = 0.3x^2 + 21x - 251 + \frac{2500}{x}$$

   (a) Use a graphing utility to graph $\overline{C} = \overline{C}(x)$.
   (b) Determine the number of riding lawn mowers to produce in order to minimize average cost.
   (c) What is the minimum average cost?

62. Medicine Concentration The concentration C of a medication in the bloodstream t hours after being administered is modeled by the function

$$C(t) = -0.002x^4 + 0.039t^3 - 0.285t^2 + 0.766t + 0.085$$

   (a) After how many hours will the concentration be highest?
   (b) A woman nursing a child must wait until the concentration is below 0.5 before she can feed him. After taking the medication, how long must she wait before feeding her child?

63. E-coli Growth A strain of E-coli Beu 397-recA441 is placed into a nutrient broth at 30° Celsius and allowed to grow. The data shown in the table are collected. The population is measured in grams and the time in hours. Since population P depends on time t and each input corresponds to exactly one output, we can say that population is a function of time; so $P(t)$ represents the population at time t.

   (a) Find the average rate of change of the population from 0 to 2.5 hours.
   (b) Find the average rate of change of the population from 4.5 to 6 hours.
   (c) What is happening to the average rate of change as time passes?

Time (hours), t	Population (grams), P
0	0.09
2.5	0.18
3.5	0.26
4.5	0.35
6	0.50

64. e-Filing Tax Returns The Internal Revenue Service Restructuring and Reform Act (RRA) was signed into law by President Bill Clinton in 1998. A major objective of the RRA was to promote electronic filing of tax returns. The data in the table show the percentage of individual income tax returns filed electronically for filing years 1998–2006. Since the percentage P of returns filed electronically depends on the filing year y and each input corresponds to exactly one output, the percentage of returns filed electronically is a function of the filing year; so $P(y)$ represents the percentage of returns filed electronically for filing year y.

Year	Percentage of returns e-filed
1998	20.7
1999	23.5
2000	27.6
2001	30.7
2002	35.6
2003	40.2
2004	46.5
2005	51.1
2006	57.1

SOURCE: Internal Revenue Service

(a) Find the average rate of change of the percentage of e-filed returns from 1998 to 2000.
(b) Find the average rate of change of the percentage of e-filed returns from 2001 to 2003.

(c) Find the average rate of change of the percentage of e-filed returns from 2004 to 2006.
(d) What is happening to the average rate of change as time passes?

65. For the function $f(x) = x^2$, compute each average rate of change:

(a) From 0 to 1
(b) From 0 to 0.5
(c) From 0 to 0.1
(d) From 0 to 0.01
(e) From 0 to 0.001
(f) What is happening to the average rate of change as x approaches 0? Is there some number that they are getting closer to? What is that number?

66. For the function $f(x) = x^2$, compute each average rate of change:

(a) From 1 to 2
(b) From 1 to 1.5
(c) From 1 to 1.1
(d) From 1 to 1.01
(e) From 1 to 1.001
(f) What is happening to the average rate of change as x approaches 1? Is there some number that they are getting closer to? What is that number?

Discussion and Writing

67. Draw the graph of a function that has the following properties: domain: all real numbers; range: all real numbers; intercepts: $(0, -3)$ and $(3, 0)$; a local maximum of -2 is at -1; a local minimum of -6 is at 2. Compare your graph with those of others. Comment on any differences.

68. Redo Problem 67 with the following additional information: increasing on $(-\infty, -1)$, $(2, \infty)$; decreasing on $(-1, 2)$. Again compare your graph with others and comment on any differences.

69. How many x-intercepts can a function defined on an interval have if it is increasing on that interval? Explain.

70. Suppose that a friend of yours does not understand the idea of increasing and decreasing functions. Provide an explanation, complete with graphs, that clarifies the idea.

71. Can a function be both even and odd? Explain.

72. Using a graphing utility, graph $y = 5$ on the interval $(-3, 3)$. Use MAXIMUM to find the local maxima on $(-3, 3)$. Comment on the result provided by the calculator.

'Are You Prepared?' Answers

1. $2 < x < 5$ **2.** 1   **3.** symmetric with respect to the y-axis   **4.** $y + 2 = 5(x - 3)$ **5.** $(-3, 0), (3, 0), (0, -9)$

1.6 Library of Functions; Piecewise-defined Functions

PREPARING FOR THIS SECTION   *Before getting started, review the following:*

- Intercepts (Section 1.2, pp. 17–18)

- Graphs of Key Equations (Section 1.1: Example 8, p. 8; Section 1.2: Example 3, p. 19; Example 4, p. 20; Example 5, p. 21)

Now Work the 'Are You Prepared?' problems on page 65.

OBJECTIVES **1** Graph the Functions Listed in the Library of Functions (p. 59)
2 Graph Piecewise-defined Functions (p. 64)

1 Graph the Functions Listed in the Library of Functions

We now introduce a few more functions to add to our list of important functions. We begin with the *square root function*.

Figure 67

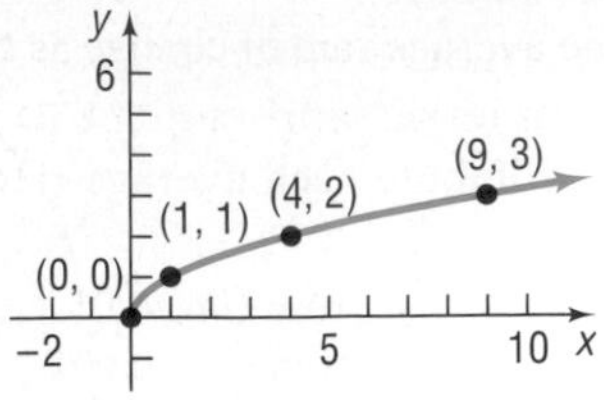

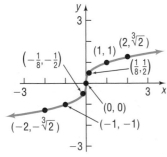

On pages 20–21, we graphed the equation $y = \sqrt{x}$. Figure 67 shows a graph of the function $f(x) = \sqrt{x}$. Based on the graph, we have the following properties:

Properties of $f(x) = \sqrt{x}$

1. The domain and the range are the set of nonnegative real numbers.
2. The x-intercept of the graph of $f(x) = \sqrt{x}$ is 0. The y-intercept of the graph of $f(x) = \sqrt{x}$ is also 0.
3. The function is neither even nor odd.
4. It is increasing on the interval $(0, \infty)$.
5. It has a minimum value of 0 at $x = 0$.

| EXAMPLE 1 | Graphing the Cube Root Function |

(a) Determine whether $f(x) = \sqrt[3]{x}$ is even, odd, or neither. State whether the graph of f is symmetric with respect to the y-axis or symmetric with respect to the origin.
(b) Determine the intercepts, if any, of the graph of $f(x) = \sqrt[3]{x}$.
(c) Graph $f(x) = \sqrt[3]{x}$.

Solution
(a) Because

$$f(-x) = \sqrt[3]{-x} = -\sqrt[3]{x} = -f(x)$$

the function is odd. The graph of f is symmetric with respect to the origin.

(b) The y-intercept is $f(0) = \sqrt[3]{0} = 0$. The x-intercept is found by solving the equation $f(x) = 0$.

$$f(x) = 0$$
$$\sqrt[3]{x} = 0 \quad \text{f(x)} = \sqrt[3]{x}$$
$$x = 0 \quad \text{Cube both sides of the equation.}$$

The x-intercept is also 0.

(c) We use the function to form Table 13 and obtain some points on the graph. Because of the symmetry with respect to the origin, we need to find only points (x, y) for which $x \geq 0$. Figure 68 shows the graph of $f(x) = \sqrt[3]{x}$.

Figure 68

Table 13

x	$y = f(x) = \sqrt[3]{x}$	(x, y)
0	0	$(0, 0)$
$\dfrac{1}{8}$	$\dfrac{1}{2}$	$\left(\dfrac{1}{8}, \dfrac{1}{2}\right)$
1	1	$(1, 1)$
2	$\sqrt[3]{2} \approx 1.26$	$(2, \sqrt[3]{2})$
8	2	$(8, 2)$

From the results of Example 1 and Figure 68, we have the following properties of the cube root function.

Properties of $f(x) = \sqrt[3]{x}$

1. The domain and the range are the set of all real numbers.
2. The x-intercept of the graph of $f(x) = \sqrt[3]{x}$ is 0. The y-intercept of the graph of $f(x) = \sqrt[3]{x}$ is also 0.
3. The graph is symmetric with respect to the origin. The function is odd.
4. It is increasing on the interval $(-\infty, \infty)$.
5. It does not have a local minimum or a local maximum.

EXAMPLE 2	**Graphing the Absolute Value Function**

(a) Determine whether $f(x) = |x|$ is even, odd, or neither. State whether the graph of f is symmetric with respect to the y-axis or symmetric with respect to the origin.
(b) Determine the intercepts, if any, of the graph of $f(x) = |x|$.
(c) Graph $f(x) = |x|$.

Solution (a) Because

$$f(-x) = |-x|$$
$$= |x| = f(x)$$

the function is even. The graph of f is symmetric with respect to the y-axis.

(b) The y-intercept is $f(0) = |0| = 0$. The x-intercept is found by solving the equation $f(x) = 0$ or $|x| = 0$. So the x-intercept is 0.

(c) We use the function to form Table 14 and obtain some points on the graph. Because of the symmetry with respect to the y-axis, we need to find only points (x, y) for which $x \geq 0$. Figure 69 shows the graph of $f(x) = |x|$.

Table 14

x	$y = f(x) = |x|$	(x, y)
0	0	(0, 0)
1	1	(1, 1)
2	2	(2, 2)
3	3	(3, 3)

Figure 69

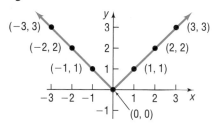

From the results of Example 2 and Figure 69, we have the following properties of the absolute value function.

> **Properties of $f(x) = |x|$**
>
> 1. The domain is the set of all real numbers. The range of f is $\{y \mid y \geq 0\}$.
> 2. The x-intercept of the graph of $f(x) = |x|$ is 0. The y-intercept of the graph of $f(x) = |x|$ is also 0.
> 3. The graph is symmetric with respect to the y-axis. The function is even.
> 4. It is decreasing on the interval $(-\infty, 0)$. It is increasing on the interval $(0, \infty)$.
> 5. It has a local minimum of 0 at $x = 0$.

Seeing the Concept

Graph $y = |x|$ on a square screen and compare what you see with Figure 69. Note that some graphing calculators use abs(x) for absolute value.

We now provide a summary of the key functions that we have encountered. In going through this list, pay special attention to the properties of each function, particularly to the shape of each graph. Knowing these graphs along with key points on the graph will lay the foundation for later graphing techniques.

Figure 70
Constant Function

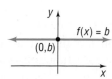

Constant Function

$f(x) = b$ $\quad$ b is a real number

See Figure 70.

Figure 71
Identity Function

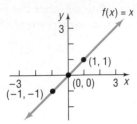

The domain of a **constant function** is the set of all real numbers; its range is the set consisting of a single number b. Its graph is a horizontal line whose y-intercept is b. The constant function is an even function whose graph is a horizontal line over its domain.

Identity Function

$$f(x) = x$$

See Figure 71.

The domain and the range of the **identity function** are the set of all real numbers. Its graph is a line whose slope is $m = 1$ and whose y-intercept is 0. The line consists of all points for which the x-coordinate equals the y-coordinate. The identity function is an odd function that is increasing over its domain. Note that the graph bisects quadrants I and III.

Figure 72
Square Function

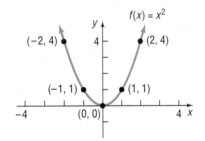

Square Function

$$f(x) = x^2$$

See Figure 72.

The domain of the **square function** f is the set of all real numbers; its range is the set of nonnegative real numbers. The graph of this function is a parabola whose intercept is at $(0, 0)$. The square function is an even function that is decreasing on the interval $(-\infty, 0)$ and increasing on the interval $(0, \infty)$.

Figure 73
Cube Function

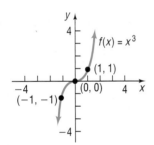

Cube Function

$$f(x) = x^3$$

See Figure 73.

The domain and the range of the **cube function** are the set of all real numbers. The intercept of the graph is at $(0, 0)$. The cube function is odd and is increasing on the interval $(-\infty, \infty)$.

Figure 74
Square Root Function

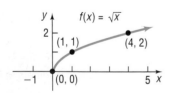

Square Root Function

$$f(x) = \sqrt{x}$$

See Figure 74.

The domain and the range of the **square root function** are the set of nonnegative real numbers. The intercept of the graph is at $(0, 0)$. The square root function is neither even nor odd and is increasing on the interval $(0, \infty)$.

Figure 75
Cube Root Function

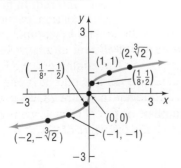

Cube Root Function

$$f(x) = \sqrt[3]{x}$$

See Figure 75.

The domain and the range of the **cube root function** are the set of all real numbers. The intercept of the graph is at $(0, 0)$. The cube root function is an odd function that is increasing on the interval $(-\infty, \infty)$.

Reciprocal Function

$$f(x) = \frac{1}{x}$$

Figure 76
Reciprocal Function

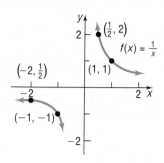

Refer to Example 5, page 21, for a discussion of the equation $y = \dfrac{1}{x}$. See Figure 76.

The domain and the range of the **reciprocal function** are the set of all nonzero real numbers. The graph has no intercepts. The reciprocal function is decreasing on the intervals $(-\infty, 0)$ and $(0, \infty)$ and is an odd function.

Absolute Value Function

$$f(x) = |x|$$

Figure 77
Absolute Value Function

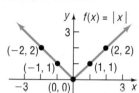

See Figure 77.

The domain of the **absolute value function** is the set of all real numbers; its range is the set of nonnegative real numbers. The intercept of the graph is at $(0, 0)$. If $x \geq 0$, then $f(x) = x$, and the graph of f is part of the line $y = x$; if $x < 0$, then $f(x) = -x$, and the graph of f is part of the line $y = -x$. The absolute value function is an even function; it is decreasing on the interval $(-\infty, 0)$ and increasing on the interval $(0, \infty)$.

The notation $\text{int}(x)$ stands for the largest integer less than or equal to x. For example,

$$\text{int}(1) = 1, \quad \text{int}(2.5) = 2, \quad \text{int}\left(\frac{1}{2}\right) = 0, \quad \text{int}\left(-\frac{3}{4}\right) = -1, \quad \text{int}(\pi) = 3$$

This type of correspondence occurs frequently enough in mathematics that we give it a name.

DEFINITION

Greatest Integer Function

$$f(x) = \text{int}(x)^* = \text{greatest integer less than or equal to } x$$

Table 15

x	$y = f(x)$ $= \text{int}(x)$	(x, y)
-1	-1	$(-1, -1)$
$-\dfrac{1}{2}$	-1	$\left(-\dfrac{1}{2}, -1\right)$
$-\dfrac{1}{4}$	-1	$\left(-\dfrac{1}{4}, -1\right)$
0	0	$(0, 0)$
$\dfrac{1}{4}$	0	$\left(\dfrac{1}{4}, 0\right)$
$\dfrac{1}{2}$	0	$\left(\dfrac{1}{2}, 0\right)$
$\dfrac{3}{4}$	0	$\left(\dfrac{3}{4}, 0\right)$

We obtain the graph of $f(x) = \text{int}(x)$ by plotting several points. See Table 15. For values of x, $-1 \leq x < 0$, the value of $f(x) = \text{int}(x)$ is -1; for values of x, $0 \leq x < 1$, the value of f is 0. See Figure 78 for the graph.

Figure 78
Greatest Integer Function

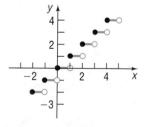

The domain of the **greatest integer function** is the set of all real numbers; its range is the set of integers. The y-intercept of the graph is 0. The x-intercepts lie in the interval $[0, 1)$. The greatest integer function is neither even nor odd. It is constant on every interval of the form $[k, k + 1)$, for k an integer. In Figure 78, we use a solid dot to indicate, for example, that at $x = 1$ the value of f is $f(1) = 1$; we use an open circle to illustrate that the function does not assume the value of 0 at $x = 1$.

*Some books use the notation $f(x) = [x]$ instead of $\text{int}(x)$.

Although a precise definition requires the idea of a limit, discussed in calculus, in a rough sense, a function is said to be **continuous** if its graph has no gaps or holes and can be drawn without lifting a pencil from the paper on which the graph is drawn. We contrast this with a *discontinuous* function. A function is **discontinuous** if its graph has gaps or holes so that its graph cannot be drawn without lifting a pencil from the paper.

From the graph of the greatest integer function, we can see why it is also called a **step function.** At $x = 0$, $x = \pm1$, $x = \pm2$, and so on, this function is discontinuous because, at integer values, the graph suddenly "steps" from one value to another without taking on any of the intermediate values. For example, to the immediate left of $x = 3$, the y-coordinates of the points on the graph are 2, and to the immediate right of $x = 3$, the y-coordinates of the points on the graph are 3. So, the graph has gaps in it.

COMMENT When graphing a function using a graphing utility, you can choose either the **connected mode,** in which points plotted on the screen are connected, making the graph appear without any breaks, or the **dot mode,** in which only the points plotted appear. When graphing the greatest integer function with a graphing utility, it is necessary to be in the dot mode. This is to prevent the utility from "connecting the dots" when $f(x)$ changes from one integer value to the next. See Figure 79.

Figure 79
$f(x) = \text{int}(x)$

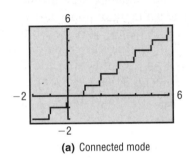

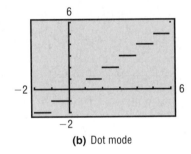

(a) Connected mode   (b) Dot mode

The functions that we have discussed so far are basic. Whenever you encounter one of them, you should see a mental picture of its graph. For example, if you encounter the function $f(x) = x^2$, you should see in your mind's eye a picture like Figure 72.

― **Now Work** PROBLEMS 9 THROUGH 16

2 Graph Piecewise-defined Functions

Sometimes a function is defined differently on different parts of its domain. For example, the absolute value function $f(x) = |x|$ is actually defined by two equations: $f(x) = x$ if $x \geq 0$ and $f(x) = -x$ if $x < 0$. For convenience, we generally combine these equations into one expression as

$$f(x) = |x| = \begin{cases} x & \text{if } x \geq 0 \\ -x & \text{if } x < 0 \end{cases}$$

When functions are defined by more than one equation, they are called **piecewise-defined** functions.

Let's look at another example of a piecewise-defined function.

EXAMPLE 3   Analyzing a Piecewise-defined Function

The function f is defined as

$$f(x) = \begin{cases} -x + 1 & \text{if } -3 \leq x < 1 \\ 2 & \text{if } x = 1 \\ x^2 & \text{if } x > 1 \end{cases}$$

(a) Find $f(0)$, $f(1)$, and $f(2)$.   (b) Determine the domain of f.

(c) Graph f.

(d) Use the graph to find the range of f.

(e) Is f continuous on its domain?

Solution

(a) To find $f(0)$, we observe that when $x = 0$ the equation for f is given by $f(x) = -x + 1$. So we have

$$f(0) = -0 + 1 = 1$$

When $x = 1$, the equation for f is $f(x) = 2$. Thus,

$$f(1) = 2$$

When $x = 2$, the equation for f is $f(x) = x^2$. So

$$f(2) = 2^2 = 4$$

Figure 80

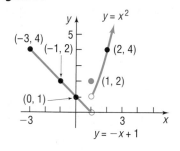

(b) To find the domain of f, we look at its definition. Since f is defined for all x greater than or equal to -3, the domain of f is $\{x \mid x \geq -3\}$, or the interval $[-3, \infty)$.

(c) To graph f, we graph "each piece." First we graph the line $y = -x + 1$ and keep only the part for which $-3 \leq x < 1$. Then we plot the point $(1, 2)$ because, when $x = 1$, $f(x) = 2$. Finally, we graph the parabola $y = x^2$ and keep only the part for which $x > 1$. See Figure 80.

(d) From the graph, we conclude that the range of f is $\{y \mid y > 0\}$, or the interval $(0, \infty)$.

(e) The function f is not continuous because there is a "jump" in the graph at $x = 1$.

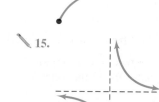 **Now Work** PROBLEM 29

1.6 Assess Your Understanding

'Are You Prepared?' *Answers are given at the end of these exercises. If you get a wrong answer, read the pages listed in red.*

1. Sketch the graph of $y = \sqrt{x}$. (pp. 20–21)

2. Sketch the graph of $y = \dfrac{1}{x}$. (p. 21)

3. List the intercepts of the equation $y = x^3 - 8$. (pp. 17–18)

Concepts and Vocabulary

4. The function $f(x) = x^2$ is decreasing on the interval _____.

5. When functions are defined by more than one equation, they are called _____ functions.

6. **True or False** The cube function is odd and is increasing on the interval $(-\infty, \infty)$.

7. **True or False** The cube root function is odd and is decreasing on the interval $(-\infty, \infty)$.

8. **True or False** The domain and the range of the reciprocal function are the set of all real numbers.

Skill Building

In Problems 9–16, match each graph to its function.

A. *Constant function*

B. *Identity function*

C. *Square function*

D. *Cube function*

E. *Square root function*

F. *Reciprocal function*

G. *Absolute value function*

H. *Cube root function*

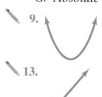

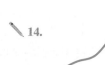

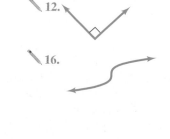

In Problems 17–24, sketch the graph of each function. Be sure to label three points on the graph.

17. $f(x) = x$ **18.** $f(x) = x^2$ **19.** $f(x) = x^3$ **20.** $f(x) = \sqrt{x}$

21. $f(x) = \dfrac{1}{x}$ **22.** $f(x) = |x|$ **23.** $f(x) = \sqrt[3]{x}$ **24.** $f(x) = 3$

25. If $f(x) = \begin{cases} x^2 & \text{if } x < 0 \\ 2 & \text{if } x = 0 \\ 2x + 1 & \text{if } x > 0 \end{cases}$

    find: (a) $f(-2)$ (b) $f(0)$ (c) $f(2)$

26. If $f(x) = \begin{cases} -3x & \text{if } x < -1 \\ 0 & \text{if } x = -1 \\ 2x^2 + 1 & \text{if } x > -1 \end{cases}$

    find: (a) $f(-2)$ (b) $f(-1)$ (c) $f(0)$

27. If $f(x) = \begin{cases} 2x - 4 & \text{if } -1 \le x \le 2 \\ x^3 - 2 & \text{if } 2 < x \le 3 \end{cases}$

    find: (a) $f(0)$ (b) $f(1)$ (c) $f(2)$ (d) $f(3)$

28. If $f(x) = \begin{cases} x^3 & \text{if } -2 \le x < 1 \\ 3x + 2 & \text{if } 1 \le x \le 4 \end{cases}$

    find: (a) $f(-1)$ (b) $f(0)$ (c) $f(1)$ (d) $f(3)$

In Problems 29–40:

  (a) *Find the domain of each function.*   (b) *Locate any intercepts.*   (c) *Graph each function.*

  (d) *Based on the graph, find the range.*   (e) *Is f continuous on its domain?*

29. $f(x) = \begin{cases} 2x & \text{if } x \ne 0 \\ 1 & \text{if } x = 0 \end{cases}$

30. $f(x) = \begin{cases} 3x & \text{if } x \ne 0 \\ 4 & \text{if } x = 0 \end{cases}$

31. $f(x) = \begin{cases} -2x + 3 & x < 1 \\ 3x - 2 & x \ge 1 \end{cases}$

32. $f(x) = \begin{cases} x + 3 & \text{if } x < -2 \\ -2x - 3 & \text{if } x \ge -2 \end{cases}$

33. $f(x) = \begin{cases} x + 3 & \text{if } -2 \le x < 1 \\ 5 & \text{if } x = 1 \\ -x + 2 & \text{if } x > 1 \end{cases}$

34. $f(x) = \begin{cases} 2x + 5 & \text{if } -3 \le x < 0 \\ -3 & \text{if } x = 0 \\ -5x & \text{if } x > 0 \end{cases}$

35. $f(x) = \begin{cases} 1 + x & \text{if } x < 0 \\ x^2 & \text{if } x \ge 0 \end{cases}$

36. $f(x) = \begin{cases} \dfrac{1}{x} & \text{if } x < 0 \\ \sqrt[3]{x} & \text{if } x \ge 0 \end{cases}$

37. $f(x) = \begin{cases} |x| & \text{if } -2 \le x < 0 \\ x^3 & \text{if } x > 0 \end{cases}$

38. $f(x) = \begin{cases} 2 - x & \text{if } -3 \le x < 1 \\ \sqrt{x} & \text{if } x > 1 \end{cases}$

39. $f(x) = 2\,\text{int}(x)$

40. $f(x) = \text{int}(2x)$

In Problems 41–44, the graph of a piecewise-defined function is given. Write a definition for each function.

41.

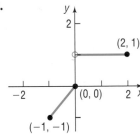

42.

43.

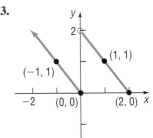

44.

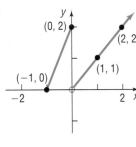

45. If $f(x) = \text{int}(2x)$, find   (a) $f(1.2)$ (b) $f(1.6)$ (c) $f(-1.8)$

46. If $f(x) = \text{int}\left(\dfrac{x}{2}\right)$, find   (a) $f(1.2)$ (b) $f(1.6)$ (c) $f(-1.8)$

Applications and Extensions

47. Cell Phone Service Sprint PCS offers a monthly cellular phone plan for $39.99. It includes 450 anytime minutes and charges $0.45 per minute for additional minutes. The following function is used to compute the monthly cost for a subscriber:

$$C(x) = \begin{cases} 39.99 & \text{if } 0 < x \le 450 \\ 0.45x - 162.51 & \text{if } x > 450 \end{cases}$$

where x is the number of anytime minutes used. Compute the monthly cost of the cellular phone for use of the following anytime minutes:

(a) 200   (b) 465   (c) 451

Source: Sprint PCS

48. Parking at O'Hare International Airport The short-term (no more than 24 hours) parking fee F (in dollars) for parking x hours at O'Hare International Airport's main parking garage can be modeled by the function

$$F(x) = \begin{cases} 3 & \text{if } 0 < x \le 3 \\ 5\,\text{int}(x + 1) + 1 & \text{if } 3 < x < 9 \\ 50 & \text{if } 9 \le x \le 24 \end{cases}$$

Determine the fee for parking in the short-term parking garage for

(a) 2 hours   (b) 7 hours   (c) 15 hours

(d) 8 hours and 24 minutes

Source: O'Hare International Airport

49. Cost of Natural Gas In February 2007, Peoples Energy had the following rate schedule for natural gas usage in single-family residences:

Monthly service charge	$9.45
Per therm service charge	
1st 50 therms	$0.36375/therm
Over 50 therms	$0.11445/therm
Gas charge	$0.8738/therm

(a) What is the charge for using 50 therms in a month?
(b) What is the charge for using 500 therms in a month?
(c) Develop a model that relates the monthly charge C for x therms of gas.
(d) Graph the function found in part (c).

Source: Peoples Energy, Chicago, Illinois, 2007

50. Cost of Natural Gas In February 2007, Nicor Gas had the following rate schedule for natural gas usage in single-family residences:

Monthly customer charge	$8.85
Distribution charge	
1st 20 therms	$0.1473/therm
Next 30 therms	$0.0579/therm
Over 50 therms	$0.0519/therm
Gas supply charge	$0.67/therm

(a) What is the charge for using 40 therms in a month?
(b) What is the charge for using 202 therms in a month?
(c) Develop a model that gives the monthly charge C for x therms of gas.
(d) Graph the function found in part (c).

Source: Nicor Gas, Aurora, Illinois, 2007

Discussion and Writing

In Problems 51–58, use a graphing utility.

51. Exploration Graph $y = x^2$. Then on the same screen graph $y = x^2 + 2$, followed by $y = x^2 + 4$, followed by $y = x^2 - 2$. What pattern do you observe? Can you predict the graph of $y = x^2 - 4$? Of $y = x^2 + 5$?

52. Exploration Graph $y = x^2$. Then on the same screen graph $y = (x - 2)^2$, followed by $y = (x - 4)^2$, followed by $y = (x + 2)^2$. What pattern do you observe? Can you predict the graph of $y = (x + 4)^2$? Of $y = (x - 5)^2$?

53. Exploration Graph $y = |x|$. Then on the same screen graph $y = 2|x|$, followed by $y = 4|x|$, followed by $y = \frac{1}{2}|x|$. What pattern do you observe? Can you predict the graph of $y = \frac{1}{4}|x|$? Of $y = 5|x|$?

54. Exploration Graph $y = x^2$. Then on the same screen graph $y = -x^2$. What pattern do you observe? Now try $y = |x|$ and $y = -|x|$. What do you conclude?

55. Exploration Graph $y = \sqrt{x}$. Then on the same screen graph $y = \sqrt{-x}$. What pattern do you observe? Now try $y = 2x + 1$ and $y = 2(-x) + 1$. What do you conclude?

56. Exploration Graph $y = x^3$. Then on the same screen graph $y = (x - 1)^3 + 2$. Could you have predicted the result?

57. Exploration Graph $y = x^2$, $y = x^4$, and $y = x^6$ on the same screen. What do you notice is the same about each graph? What do you notice that is different?

58. Exploration Graph $y = x^3$, $y = x^5$, and $y = x^7$ on the same screen. What do you notice is the same about each graph? What do you notice that is different?

59. Consider the equation

$$y = \begin{cases} 1 & \text{if } x \text{ is rational} \\ 0 & \text{if } x \text{ is irrational} \end{cases}$$

Is this a function? What is its domain? What is its range? What is its y-intercept, if any? What are its x-intercepts, if any? Is it even, odd, or neither? How would you describe its graph?

60. Define some functions that pass through $(0, 0)$ and $(1, 1)$ and are increasing for $x \geq 0$. Begin your list with $y = \sqrt{x}$, $y = x$, and $y = x^2$. Can you propose a general result about such functions?

'Are You Prepared?' Answers

1.

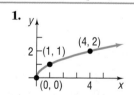

2.

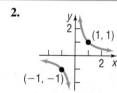

3. $(0, -8), (2, 0)$

1.7 Graphing Techniques: Transformations

OBJECTIVES
1 Graph Functions Using Vertical and Horizontal Shifts (p. 68)
2 Graph Functions Using Compressions and Stretches (p. 70)
3 Graph Functions Using Reflections about the x-Axis or y-Axis (p. 72)

At this stage, if you were asked to graph any of the functions defined by $y = x$, $y = x^2$, $y = x^3$, $y = \sqrt{x}$, $y = \sqrt[3]{x}$, $y = |x|$, or $y = \frac{1}{x}$, your response should be, "Yes, I recognize these functions and know the general shapes of their graphs." (If this is not your answer, review the previous section, Figures 71 through 77.)

Sometimes we are asked to graph a function that is "almost" like one that we already know how to graph. In this section, we look at some of these functions and develop techniques for graphing them. Collectively, these techniques are referred to as **transformations.** We introduce the method of transformations because it is a more efficient method of graphing than point-plotting.

1 Graph Functions Using Vertical and Horizontal Shifts

Exploration

On the same screen, graph each of the following functions:

$$Y_1 = x^2$$
$$Y_2 = x^2 + 2$$
$$Y_3 = x^2 - 2$$

What do you observe? Now create a table of values for $Y_1, Y_2,$ and Y_3. What do you observe?

Figure 81

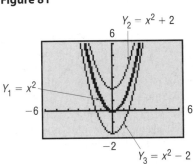

$Y_2 = x^2 + 2$

$Y_1 = x^2$

$Y_3 = x^2 - 2$

Result   Figure 81 illustrates the graphs. You should have observed a general pattern. With $Y_1 = x^2$ on the screen, the graph of $Y_2 = x^2 + 2$ is identical to that of $Y_1 = x^2$, except that it is shifted vertically up 2 units. The graph of $Y_3 = x^2 - 2$ is identical to that of $Y_1 = x^2$, except that it is shifted vertically down 2 units. From Table 16(a), we see that the y-coordinates on $Y_2 = x^2 + 2$ are 2 units larger than the y-coordinates on $Y_1 = x^2$ for any given x-coordinate. From Table 16(b), we see that the y-coordinates on $Y_3 = x^2 - 2$ are 2 units smaller than the y-coordinates on $Y_1 = x^2$ for any given x-coordinate.

Table 16

X	Y1	Y2
-2	4	6
-1	1	3
0	0	2
1	1	3
2	4	6
3	9	11
4	16	18

Y2☐X²+2

X	Y1	Y3
-2	4	2
-1	1	-1
0	0	-2
1	1	-1
2	4	2
3	9	7
4	16	14

Y3☐X²−2

(a)   (b)

We are led to the following conclusions:

If a positive real number k is added to the outputs of a function $y = f(x)$, the graph of the new function $y = f(x) + k$ is the graph of f **shifted vertically up** k units.

If a positive real number k is subtracted from the outputs of a function $y = f(x)$, the graph of the new function $y = f(x) - k$ is the graph of f **shifted vertically down** k units.

EXAMPLE 1 | **Vertical Shift Down**

Use the graph of $f(x) = x^2$ to obtain the graph of $h(x) = x^2 - 4$.

Solution

Table 17 lists some points on the graphs of $Y_1 = f(x) = x^2$ and $Y_2 = h(x) = f(x) - 4 = x^2 - 4$. Notice that each y-coordinate of h is 4 units less than the corresponding y-coordinate of f.

Table 17

X	Y1	Y2
-3	9	5
-2	4	0
-1	1	-3
0	0	-4
1	1	-3
2	4	0
3	9	5

Y2☐X²−4

To obtain the graph of h from the graph of f, subtract 4 from each y-coordinate on the graph of f. The graph of h is identical to that of f, except that it is shifted down 4 units. See Figure 82.

Figure 82

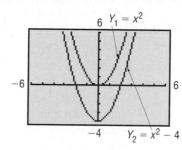
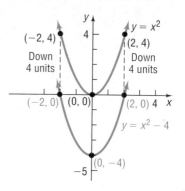

Now Work PROBLEM 39

Exploration

On the same screen, graph each of the following functions:

$$Y_1 = x^2$$
$$Y_2 = (x - 3)^2$$
$$Y_3 = (x + 2)^2$$

What do you observe?

Figure 83

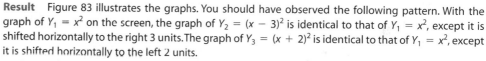

$Y_1 = x^2$

$Y_3 = (x + 2)^2$ $Y_2 = (x - 3)^2$

Result Figure 83 illustrates the graphs. You should have observed the following pattern. With the graph of $Y_1 = x^2$ on the screen, the graph of $Y_2 = (x - 3)^2$ is identical to that of $Y_1 = x^2$, except it is shifted horizontally to the right 3 units. The graph of $Y_3 = (x + 2)^2$ is identical to that of $Y_1 = x^2$, except it is shifted horizontally to the left 2 units.

From Table 18(a), we see the x-coordinates on $Y_2 = (x - 3)^2$ are 3 units larger than they are for $Y_1 = x^2$ for any given y-coordinate. For example, when $Y_1 = 0$, then $x = 0$, and when $Y_2 = 0$, then $x = 3$. Also, when $Y_1 = 1$, then $x = -1$ or 1, and when $Y_2 = 1$, then $x = 2$ or 4. From Table 18(b), we see the x-coordinates on $Y_3 = (x + 2)^2$ are 2 units smaller than they are for $Y_1 = x^2$ for any given y-coordinate. For example, when $Y_1 = 0$, then $x = 0$, and when $Y_3 = 0$, then $x = -2$. Also, when $Y_1 = 4$, then $x = -2$ or 2, and when $Y_3 = 4$, then $x = -4$ or 0.

Table 18

X	Y1	Y2
-1	1	16
0	0	9
1	1	4
2	4	1
3	9	0
4	16	1
5	25	4

Y2日(X-3)²

(a)

X	Y1	Y3
-4	16	4
-3	9	1
-2	4	0
-1	1	1
0	0	4
1	1	9
2	4	16

Y3日(X+2)²

(b)

We are led to the following conclusions:

If the argument x of a function f is replaced by $x - h, h > 0$, the graph of the new function $y = f(x - h)$ is the graph of f **shifted horizontally right** h units. If the argument x of a function f is replaced by $x + h, h > 0$, the graph of the new function $y = f(x + h)$ is the graph of f **shifted horizontally left** h units.

Now Work PROBLEM 43

NOTE Vertical shifts result when adding or subtracting a real number k after performing the operation suggested by the basic function, while horizontal shifts result when adding or subtracting a real number h to or from x before performing the operation suggested by the basic function. For example, the graph of $f(x) = \sqrt{x} + 3$ is obtained by shifting the graph of $y = \sqrt{x}$ up 3 units, because we evaluate the square root function first and then add 3. The graph of $g(x) = \sqrt{x + 3}$ is obtained by shifting the graph of $y = \sqrt{x}$ left 3 units, because we first add 3 to x before we evaluate the square root function. ∎

Vertical and horizontal shifts are sometimes combined.

| EXAMPLE 2 | Combining Vertical and Horizontal Shifts |

Graph the function $f(x) = (x + 3)^2 - 5$.

Solution We graph f in steps. First, we note that the rule for f is basically a square function, so we begin with the graph of $y = x^2$ as shown in Figure 84(a). To get the graph of $y = (x + 3)^2$, we shift the graph of $y = x^2$ horizontally 3 units to the left. See Figure 84(b). Finally, to get the graph of $y = (x + 3)^2 - 5$, we shift the graph of $y = (x + 3)^2$ vertically down 5 units. See Figure 84(c). Note the points plotted on each graph. Using key points can be helpful in keeping track of the transformation that has taken place.

Figure 84

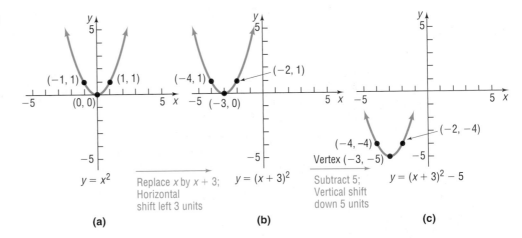

(a)   (b)   (c)

✓ **Check:** Graph $Y_1 = f(x) = (x + 3)^2 - 5$ and compare the graph to Figure 84(c).

In Example 2, if the vertical shift had been done first, followed by the horizontal shift, the final graph would have been the same. (Try it for yourself.)

➤ **Now Work** PROBLEM 45

2 Graph Functions Using Compressions and Stretches

Exploration

On the same screen, graph each of the following functions:

$$Y_1 = |x|$$

$$Y_2 = 2|x|$$

$$Y_3 = \frac{1}{2}|x|$$

Figure 85

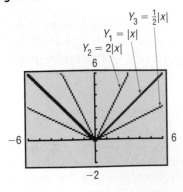

Then create a table of values and compare the y-coordinates for any given x-coordinate.

Result Figure 85 illustrates the graphs.

Now look at Table 19, where $Y_1 = |x|$ and $Y_2 = 2|x|$. Notice that the values for Y_2 are two times the values of Y_1 for each x-value. This means that the graph of $Y_2 = 2|x|$ can be obtained from the graph of $Y_1 = |x|$ by multiplying each y-coordinate of $Y_1 = |x|$ by 2. Therefore, the graph of Y_2 will be vertically *stretched* by a factor of 2.

Look at Table 20 where $Y_1 = |x|$ and $Y_3 = \frac{1}{2}|x|$. The values of Y_3 are half the values of Y_1 for each x-value. So, the graph of $Y_3 = \frac{1}{2}|x|$ can be obtained from the graph of $Y_1 = |x|$ by multiplying each y-coordinate by $\frac{1}{2}$. Therefore, the graph of Y_3 will be vertically *compressed* by a factor of $\frac{1}{2}$.

Table 19

X	Y₁	**Y₂**
-2	2	4
-1	1	2
0	0	0
1	1	2
2	2	4
3	3	6
4	4	8

Y₂⬛2abs(X)

Table 20

X	Y₁	**Y₃**
-2	2	1
-1	1	.5
0	0	0
1	1	.5
2	2	1
3	3	1.5
4	4	2

Y₃⬛.5abs(X)

Based on the Exploration, we have the following result:

When the right side of a function $y = f(x)$ is multiplied by a positive number a, the graph of the new function $y = af(x)$ is obtained by multiplying each y-coordinate on the graph of $y = f(x)$ by a. The new graph is a **vertically compressed** (if $0 < a < 1$) or a **vertically stretched** (if $a > 1$) version of the graph of $y = f(x)$.

━━━▶ **Now Work**   PROBLEM 47

What happens if the argument x of a function $y = f(x)$ is multiplied by a positive number a, creating a new function $y = f(ax)$? To find the answer, we look at the following Exploration.

Exploration

On the same screen, graph each of the following functions:

$$Y_1 = f(x) = \sqrt{x} \qquad Y_2 = f(2x) = \sqrt{2x} \qquad Y_3 = f\left(\frac{1}{2}x\right) = \sqrt{\frac{1}{2}x} = \sqrt{\frac{x}{2}}$$

Create a table of values to explore the relation between the x- and y-coordinates of each function.

Result   You should have obtained the graphs in Figure 86. Look at Table 21(a). Notice that $(1, 1)$, $(4, 2)$, and $(9, 3)$ are points on the graph of $Y_1 = \sqrt{x}$. Also, $(0.5, 1)$, $(2, 2)$, and $(4.5, 3)$ are points on the graph of $Y_2 = \sqrt{2x}$. For a given y-coordinate, the x-coordinate on the graph of Y_2 is $\frac{1}{2}$ of the x-coordinate on Y_1. We conclude that the graph of $Y_2 = \sqrt{2x}$ is obtained by multiplying the x-coordinate of each point on the graph of $Y_1 = \sqrt{x}$ by $\frac{1}{2}$. The graph of $Y_2 = \sqrt{2x}$ is the graph of $Y_1 = \sqrt{x}$ *compressed* horizontally.

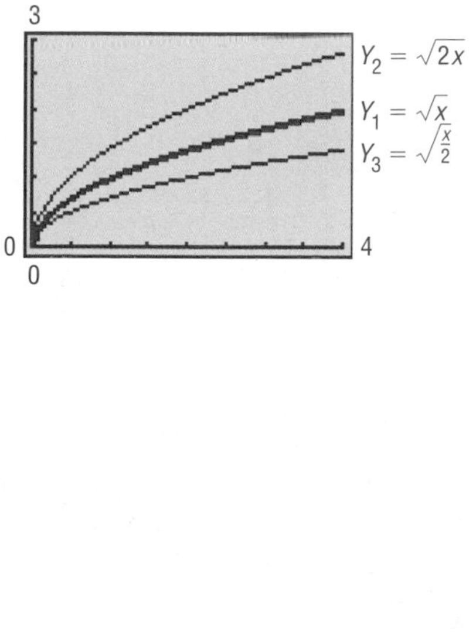

Look at Table 21(b). Notice that $(1, 1)$, $(4, 2)$, and $(9, 3)$ are points on the graph of $Y_1 = \sqrt{x}$. Also notice that $(2, 1)$, $(8, 2)$, and $(18, 3)$ are points on the graph of $Y_3 = \sqrt{\frac{x}{2}}$. For a given y-coordinate, the x-coordinate on the graph of Y_3 is 2 times the x-coordinate on Y_1. We conclude that the graph of $Y_3 = \sqrt{\frac{x}{2}}$ is obtained by multiplying the x-coordinate of each point on the graph of $Y_1 = \sqrt{x}$ by 2. The graph of $Y_3 = \sqrt{\frac{x}{2}}$ is the graph of $Y_1 = \sqrt{x}$ *stretched* horizontally.

Figure 86

$Y_2 = \sqrt{2x}$
$Y_1 = \sqrt{x}$
$Y_3 = \sqrt{\frac{x}{2}}$

Table 21

X	Y₁	**Y₂**
0	0	0
.5	.70711	1
1	1	1.4142
2	1.4142	2
4	2	2.8284
4.5	2.1213	3
9	3	4.2426

Y₂⬛√(2X)

(a)

X	Y₁	**Y₃**
0	0	0
1	1	.70711
2	1.4142	1
4	2	1.4142
8	2.8284	2
9	3	2.1213
18	4.2426	3

Y₃⬛√(X/2)

(b)

Based on the results of the Exploration, we have the following result:

If the argument x of a function $y = f(x)$ is multiplied by a positive number a, the graph of the new function $y = f(ax)$ is obtained by multiplying each x-coordinate of $y = f(x)$ by $\dfrac{1}{a}$. A **horizontal compression** results if $a > 1$, and a **horizontal stretch** occurs if $0 < a < 1$.

Let's look at an example.

EXAMPLE 3 | **Graphing Using Stretches and Compressions**

The graph of $y = f(x)$ is given in Figure 87. Use this graph to find the graphs of

(a) $y = 2f(x)$ (b) $y = f(3x)$

Figure 87

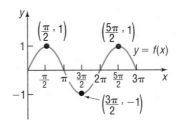

Solution   (a) The graph of $y = 2f(x)$ is obtained by multiplying each y-coordinate of $y = f(x)$ by 2. See Figure 88.

(b) The graph of $y = f(3x)$ is obtained from the graph of $y = f(x)$ by multiplying each x-coordinate of $y = f(x)$ by $\dfrac{1}{3}$. See Figure 89.

Figure 88

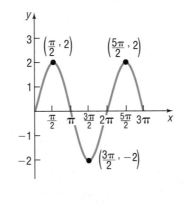

$y = 2f(x)$

Figure 89

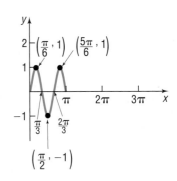

$y = f(3x)$

━━━ **Now Work** PROBLEMS **69(e)** AND **(g)**

3 Graph Functions Using Reflections about the *x*-Axis or *y*-Axis

Exploration

Reflection about the *x*-axis:

(a) Graph and create a table of $Y_1 = x^2$ and $Y_2 = -x^2$.

(b) Graph and create a table of $Y_1 = |x|$ and $Y_2 = -|x|$.

(c) Graph and create a table of $Y_1 = x^2 - 4$ and $Y_2 = -(x^2 - 4) = -x^2 + 4$.

Result See Tables 22(a), (b), and (c) and Figures 90(a), (b), and (c). For each point (x, y) on the graph of Y_1, the point $(x, -y)$ is on the graph of Y_2. Put another way, Y_2 is the reflection about the x-axis of Y_1.

Table 22

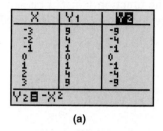 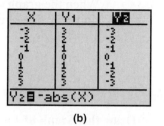 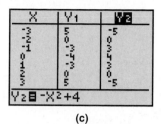

(a)   (b)   (c)

Figure 90

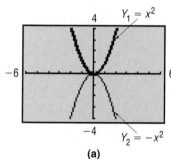 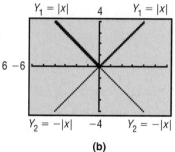 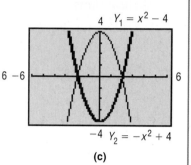

(a)   (b)   (c)

The results of the previous Exploration lead to the following result.

> When the right side of the function $y = f(x)$ is multiplied by -1, the graph of the new function $y = -f(x)$ is the **reflection about the x-axis** of the graph of the function $y = f(x)$.

Now Work PROBLEM 51

Exploration

Reflection about the y-axis:

(a) Graph $Y_1 = \sqrt{x}$, followed by $Y_2 = \sqrt{-x}$.
(b) Graph $Y_1 = x + 1$, followed by $Y_2 = -x + 1$.
(c) Graph $Y_1 = x^4 + x$, followed by $Y_2 = (-x)^4 + (-x) = x^4 - x$.

Result See Tables 23(a), (b), and (c) and Figures 91(a), (b), and (c). For each point (x, y) on the graph of Y_1, the point $(-x, y)$ is on the graph of Y_2. Put another way, Y_2 is the reflection about the y-axis of Y_1.

Table 23

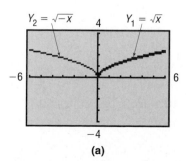 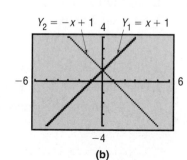 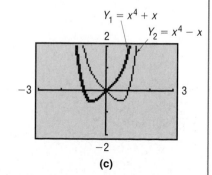

(a)   (b)   (c)

Figure 91

(a)   (b)   (c)

The results of the previous Exploration lead to the following result.

> When the graph of the function $y = f(x)$ is known, the graph of the new function $y = f(-x)$ is the **reflection about the y-axis** of the graph of the function $y = f(x)$.

SUMMARY OF GRAPHING TECHNIQUES

To Graph:	Draw the Graph of f and:	Functional Change to $f(x)$
Vertical shifts		
$y = f(x) + k, \quad k > 0$	Raise the graph of f by k units.	Add k to $f(x)$.
$y = f(x) - k, \quad k > 0$	Lower the graph of f by k units.	Subtract k from $f(x)$.
Horizontal shifts		
$y = f(x + h), \quad h > 0$	Shift the graph of f to the left h units.	Replace x by $x + h$.
$y = f(x - h), \quad h > 0$	Shift the graph of f to the right h units.	Replace x by $x - h$.
Compressing or stretching		
$y = af(x), \quad a > 0$	Multiply each y-coordinate of $y = f(x)$ by a. Stretch the graph of f vertically if $a > 1$. Compress the graph of f vertically if $0 < a < 1$.	Multiply $f(x)$ by a.
$y = f(ax), \quad a > 0$	Multiply each x-coordinate of $y = f(x)$ by $\dfrac{1}{a}$. Stretch the graph of f horizontally if $0 < a < 1$. Compress the graph of f horizontally if $a > 1$.	Replace x by ax.
Reflection about the x-axis		
$y = -f(x)$	Reflect the graph of f about the x-axis.	Multiply $f(x)$ by -1.
Reflection about the y-axis		
$y = f(-x)$	Reflect the graph of f about the y-axis.	Replace x by $-x$.

The examples that follow combine some of the procedures outlined in this section to get the required graph.

EXAMPLE 4   **Determining the Function Obtained from a Series of Transformations**

Find the function that is finally graphed after the following three transformations are applied to the graph of $y = |x|$.

1. Shift left 2 units.
2. Shift up 3 units.
3. Reflect about the y-axis.

Solution   1. Shift left 2 units: Replace x by $x + 2$. $y = |x + 2|$

2. Shift up 3 units: Add 3. $y = |x + 2| + 3$

3. Reflect about the y-axis: Replace x by $-x$. $y = |-x + 2| + 3$

- Now Work PROBLEM 27

| EXAMPLE 5 | **Combining Graphing Procedures** |

Graph the function $f(x) = \dfrac{3}{x-2} + 1$. Find the domain and the range of f.

Solution It is helpful to write f as $f(x) = 3\left(\dfrac{1}{x-2}\right) + 1$. Now we use the following steps to obtain the graph of f:

STEP 1: $y = \dfrac{1}{x}$ Reciprocal function

STEP 2: $y = 3 \cdot \left(\dfrac{1}{x}\right) = \dfrac{3}{x}$ Multiply by 3; vertical stretch of the graph of $y = \dfrac{1}{x}$ by a factor of 3.

STEP 3: $y = \dfrac{3}{x-2}$ Replace x by x − 2; horizontal shift to the right 2 units.

STEP 4: $y = \dfrac{3}{x-2} + 1$ Add 1; vertical shift up 1 unit.

See Figure 92.

Figure 92

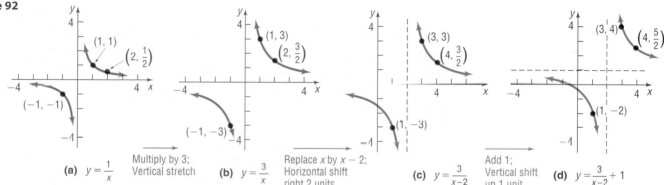

(a) $y = \dfrac{1}{x}$ Multiply by 3; Vertical stretch

(b) $y = \dfrac{3}{x}$ Replace x by x − 2; Horizontal shift right 2 units

(c) $y = \dfrac{3}{x-2}$ Add 1; Vertical shift up 1 unit

(d) $y = \dfrac{3}{x-2} + 1$

The domain of $y = \dfrac{1}{x}$ is $\{x \mid x \neq 0\}$ and its range is $\{y \mid y \neq 0\}$. Because we shifted right 2 units and up 1 unit to obtain f, the domain of f is $\{x \mid x \neq 2\}$ and its range is $\{y \mid y \neq 1\}$.

Other ordering of the steps shown in Example 5 would also result in the graph of f. For example, try this one:

HINT Although the order in which transformations are performed can be altered, you may consider using the following order for consistency:
1. Reflections
2. Compressions and stretches
3. Shifts ∎

STEP 1: $y = \dfrac{1}{x}$ Reciprocal function

STEP 2: $y = \dfrac{1}{x-2}$ Replace x by x − 2; horizontal shift to the right 2 units.

STEP 3: $y = \dfrac{3}{x-2}$ Multiply by 3; vertical stretch of the graph of $y = \dfrac{1}{x-2}$ by a factor of 3.

STEP 4: $y = \dfrac{3}{x-2} + 1$ Add 1; vertical shift up 1 unit.

| EXAMPLE 6 | **Combining Graphing Procedures** |

Graph the function $f(x) = \sqrt{1-x} + 2$. Find the domain and the range of f.

Solution Because horizontal shifts require the form $x - h$, we begin by rewriting $f(x)$ as $f(x) = \sqrt{1-x} + 2 = \sqrt{-(x-1)} + 2$. Now use the following steps:

STEP 1: $y = \sqrt{x}$ Square root function

STEP 2: $y = \sqrt{-x}$ Replace x by −x; reflect about the y-axis.

Step 3: $y = \sqrt{-(x-1)} = \sqrt{1-x}$ Replace x by x − 1; horizontal shift to the right 1 unit.

Step 4: $y = \sqrt{1-x} + 2$ Add 2; vertical shift up 2 units.

See Figure 93.

Figure 93

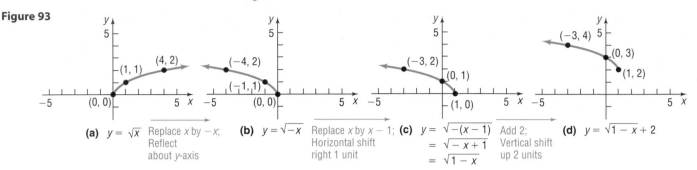

(a) $y = \sqrt{x}$ Replace x by −x; Reflect about y-axis

(b) $y = \sqrt{-x}$ Replace x by x − 1; Horizontal shift right 1 unit

(c) $y = \sqrt{-(x-1)}$ Add 2; $= \sqrt{-x+1}$ Vertical shift $= \sqrt{1-x}$ up 2 units

(d) $y = \sqrt{1-x} + 2$

The domain of f is $(-\infty, 1]$ and its range is $[2, \infty)$.

———— **Now Work** PROBLEM 61

Asymptotes

Look back at Figure 92(d). Notice that as the values of x become more negative, that is, as x becomes **unbounded in the negative direction** ($x \to -\infty$, read as "x **approaches negative infinity**"), the values $f(x)$ approach 1. Also, as the values of x become **unbounded in the positive direction** ($x \to \infty$, read as "x **approaches infinity**"), the values of $f(x)$ also approach 1. That is,

As $x \to -\infty$, the values $f(x)$ approach 1. $[\lim_{x \to -\infty} f(x) = 1]$

As $x \to \infty$, the values $f(x)$ approach 1. $[\lim_{x \to \infty} f(x) = 1]$

This behavior of the graph is depicted by the horizontal line $y = 1$, called *a horizontal asymptote* of the graph.

DEFINITION   Let R denote a function. If, as $x \to -\infty$ or as $x \to \infty$, the values of $R(x)$ approach some fixed number, L, then the line $y = L$ is a **horizontal asymptote** of the graph of R.

Now look again at Figure 92(d). Notice that as x gets closer to 2, but remains less than 2, values of f are becoming unbounded in the negative direction; that is, $f(x) \to -\infty$. And as x gets closer to 2, but remains larger than 2, the values of f are becoming unbounded in the positive direction; that is $f(x) \to \infty$. We conclude that

As $x \to 2$, the values of $|f(x)| \to \infty$ $[\lim_{x \to 2} |f(x)| = \infty]$

This behavior of the graph is depicted by the vertical line $x = 2$, which is called a *vertical asymptote* of the graph.

DEFINITION   If, as x approaches some number c, the values $|R(x)| \to \infty$, then the line $x = c$ is a **vertical asymptote** of the graph of R. The graph of a function never intersects a vertical asymptote.

Even though the asymptotes of a function are not part of the graph of the function, they provide information about how the graph looks. Figure 94 illustrates some of the possibilities.

A horizontal asymptote, when it occurs, describes the end behavior of the graph as $x \to \infty$ or as $x \to -\infty$. **The graph of a function may intersect a horizontal asymptote.**

Figure 94

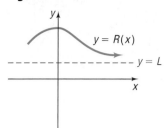

$y = R(x)$
$y = L$

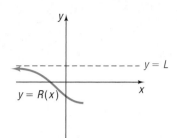

$y = L$
$y = R(x)$

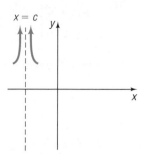

$x = c$

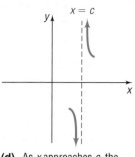

$x = c$

(a) End behavior:
As $x \to \infty$, the values of $R(x)$ approach L
[symbolized by $\lim\limits_{x \to \infty} R(x) = L$].
That is, the points on the graph of R are getting closer to the line $y = L$; $y = L$ is a horizontal asymptote.

(b) End behavior:
As $x \to -\infty$, the values of $R(x)$ approach L
[symbolized by $\lim\limits_{x \to -\infty} R(x) = L$].
That is, the points on the graph of R are getting closer to the line $y = L$; $y = L$ is a horizontal asymptote.

(c) As x approaches c, the values of $R(x) \to \infty$
[for $x < c$, this is symbolized by $\lim\limits_{x \to c^-} R(x) = \infty$; for $x > c$, this is symbolized by $\lim\limits_{x \to c^+} R(x) = \infty$]. That is, the points on the graph of R are getting closer to the line $x = c$; $x = c$ is a vertical asymptote.

(d) As x approaches c, the values of $|R(x)| \to \infty$
[for $x < c$, this is symbolized by $\lim\limits_{x \to c^-} R(x) = -\infty$; for $x > c$, this is symbolized by $\lim\limits_{x \to c^+} R(x) = \infty$]. That is, the points on the graph of R are getting closer to the line $x = c$; $x = c$ is a vertical asymptote.

A vertical asymptote, when it occurs, describes the behavior of the graph when x is close to some number c. **The graph of a function will never intersect a vertical asymptote.**

— **Now Work** PROBLEM 73

1.7 Assess Your Understanding

Concepts and Vocabulary

1. Suppose that the graph of a function f is known. Then the graph of $y = f(x - 2)$ may be obtained by a(n) _____ shift of the graph of f to the _____ a distance of 2 units.

2. Suppose that the graph of a function f is known. Then the graph of $y = f(-x)$ may be obtained by a reflection about the _____-axis of the graph of the function $y - f(x)$.

3. Suppose that the x-intercepts of the graph of $y = f(x)$ are $-2, 1$, and 5. The x-intercepts of $y = f(x + 3)$ are _____, _____, and _____.

4. *True or False* The graph of $y = -f(x)$ is the reflection about the x-axis of the graph of $y = f(x)$.

5. *True or False* To obtain the graph of $y = f(x + 2) - 3$, shift the graph of $y = f(x)$ horizontally to the right 2 units and vertically down 3 units.

6. *True or False* Suppose that the x-intercepts of the graph of $y = f(x)$ are -3 and 2. Then the x-intercepts of the graph of $y = 2f(x)$ are also -3 and 2.

Skill Building

In Problems 7–18, match each graph to one of the following functions:

A. $y = x^2 + 2$ B. $y = -x^2 + 2$ C. $y = |x| + 2$ D. $y = -|x| + 2$
E. $y = (x - 2)^2$ F. $y = -(x + 2)^2$ G. $y = |x - 2|$ H. $y = -|x + 2|$
I. $y = 2x^2$ J. $y = -2x^2$ K. $y = 2|x|$ L. $y = -2|x|$

7.

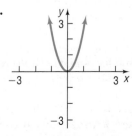

8.

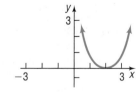

9.

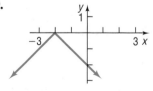

10.

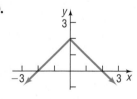

11.

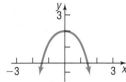

12.

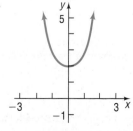

13.

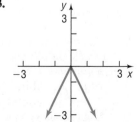

14.

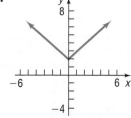

15.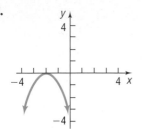
16.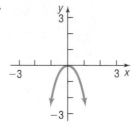
17.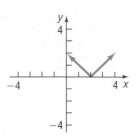
18.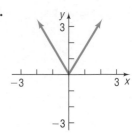

In Problems 19–26, write the function whose graph is the graph of $y = x^3$, but is:

19. Shifted to the right 4 units

20. Shifted to the left 4 units

21. Shifted up 4 units

22. Shifted down 4 units

23. Reflected about the y-axis

24. Reflected about the x-axis

25. Vertically stretched by a factor of 4

26. Horizontally stretched by a factor of 4

In Problems 27–30, find the function that is finally graphed after the following transformations are applied to the graph of $y = \sqrt{x}$.

27. (1) Shift up 2 units
　　(2) Reflect about the x-axis
　　(3) Reflect about the y-axis

28. (1) Reflect about the x-axis
　　(2) Shift right 3 units
　　(3) Shift down 2 units

29. (1) Reflect about the x-axis
　　(2) Shift up 2 units
　　(3) Shift left 3 units

30. (1) Shift up 2 units
　　(2) Reflect about the y-axis
　　(3) Shift left 3 units

31. If $(3, 6)$ is a point on the graph of $y = f(x)$, which of the following points must be on the graph of $y = -f(x)$?
　　(a) $(6, 3)$
　　(b) $(6, -3)$
　　(c) $(3, -6)$
　　(d) $(-3, 6)$

32. If $(3, 6)$ is a point on the graph of $y = f(x)$, which of the following points must be on the graph of $y = f(-x)$?
　　(a) $(6, 3)$
　　(b) $(6, -3)$
　　(c) $(3, 6)$
　　(d) $(-3, 6)$

33. If $(1, 3)$ is a point on the graph of $y = f(x)$, which of the following points must be on the graph of $y = 2f(x)$?
　　(a) $(1, 3)$
　　(b) $(2, 3)$
　　(c) $(1, 6)$
　　(d) $\left(\dfrac{1}{2}, 3\right)$

34. If $(4, 2)$ is a point on the graph of $y = f(x)$, which of the following points must be on the graph of $y = f(2x)$?
　　(a) $(4, 1)$
　　(b) $(8, 2)$
　　(c) $(2, 2)$
　　(d) $(4, 4)$

35. Suppose that the x-intercepts of the graph of $y = f(x)$ are -5 and 3.
　　(a) What are the x-intercepts of the graph of $y = f(x + 2)$?
　　(b) What are the x-intercepts of the graph of $y = f(x - 2)$?
　　(c) What are the x-intercepts of the graph of $y = 4f(x)$?
　　(d) What are the x-intercepts of the graph of $y = f(-x)$?

36. Suppose that the x-intercepts of the graph of $y = f(x)$ are -8 and 1.
　　(a) What are the x-intercepts of the graph of $y = f(x + 4)$?
　　(b) What are the x-intercepts of the graph of $y = f(x - 3)$?
　　(c) What are the x-intercepts of the graph of $y = 2f(x)$?
　　(d) What are the x-intercepts of the graph of $y = f(-x)$?

37. Suppose that the function $y = f(x)$ is increasing on the interval $(-1, 5)$.
　　(a) Over what interval is the graph of $y = f(x + 2)$ increasing?
　　(b) Over what interval is the graph of $y = f(x - 5)$ increasing?
　　(c) What can be said about the graph of $y = -f(x)$?
　　(d) What can be said about the graph of $y = f(-x)$?

38. Suppose that the function $y = f(x)$ is decreasing on the interval $(-2, 7)$.
　　(a) Over what interval is the graph of $y = f(x + 2)$ decreasing?
　　(b) Over what interval is the graph of $y = f(x - 5)$ decreasing?
　　(c) What can be said about the graph of $y = -f(x)$?
　　(d) What can be said about the graph of $y = f(-x)$?

In Problems 39–68, graph each function using the techniques of shifting, compressing, stretching, and/or reflecting. Start with the graph of the basic function (for example, $y = x^2$) and show all stages. Be sure to show at least three key points. Find the domain and the range of each function. Verify your results using a graphing utility.

39. $f(x) = x^2 - 1$

40. $f(x) = x^2 + 4$

41. $g(x) = x^3 + 1$

42. $g(x) = x^3 - 1$

43. $h(x) = \sqrt{x - 2}$

44. $h(x) = \sqrt{x + 1}$

45. $f(x) = (x - 1)^3 + 2$

46. $f(x) = (x + 2)^3 - 3$

47. $g(x) = 4\sqrt{x}$

48. $g(x) = \dfrac{1}{2}\sqrt{x}$

49. $h(x) = \dfrac{1}{2x}$

50. $h(x) = \sqrt[3]{2x}$

51. $f(x) = -\sqrt[3]{x}$

52. $f(x) = -\sqrt{x}$

53. $g(x) = \sqrt[3]{-x}$

54 $g(x) = \dfrac{1}{-x}$

55. $h(x) = -x^3 + 2$

56. $h(x) = \dfrac{1}{-x} + 2$

57. $f(x) = 2(x + 1)^2 - 3$

58. $f(x) = 3(x - 2)^2 + 1$

59. $g(x) = 2\sqrt{x - 2} + 1$

60. $g(x) = 3|x + 1| - 3$

61. $h(x) = \sqrt{-x} - 2$

62. $h(x) = \dfrac{4}{x} + 2$

63. $f(x) = -(x + 1)^3 - 1$

64. $f(x) = -4\sqrt{x - 1}$

65. $g(x) = 2|1 - x|$

66. $g(x) = 4\sqrt{2 - x}$

67. $h(x) = 2\,\text{int}(x - 1)$

68. $h(x) = \text{int}(-x)$

In Problems 69–72, the graph of a function f is illustrated. Use the graph of f as the first step toward graphing each of the following functions:

(a) $F(x) = f(x) + 3$

(b) $G(x) = f(x + 2)$

(c) $P(x) = -f(x)$

(d) $H(x) = f(x + 1) - 2$

(e) $Q(x) = \dfrac{1}{2}f(x)$

(f) $g(x) = f(-x)$

(g) $h(x) = f(2x)$

69.

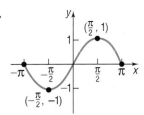

70.

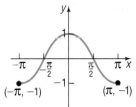

71.

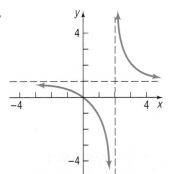

72.

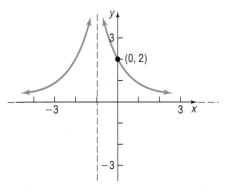

In Problems 73–76, use the graph to find the vertical and horizontal asymptotes.

73.

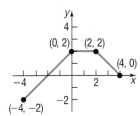

74.

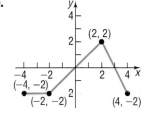

75.

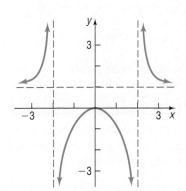

76.

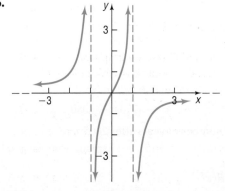

Mixed Practice

77. (a) Using a graphing utility, graph $f(x) = x^3 - 9x$ for $-4 < x < 4$.
  (b) Find the x-intercepts of the graph of f.
  (c) Approximate any local maxima and local minima.
  (d) Determine where f is increasing and where it is decreasing.
  (e) Without using a graphing utility, repeat parts (b)–(d) for $y = f(x + 2)$.
  (f) Without using a graphing utility, repeat parts (b)–(d) for $y = 2f(x)$.
  (g) Without using a graphing utility, repeat parts (b)–(d) for $y = f(-x)$.

78. (a) Using a graphing utility, graph $f(x) = x^3 - 4x$ for $-3 < x < 3$.
  (b) Find the x-intercepts of the graph of f.
  (c) Approximate any local maxima and local minima.
  (d) Determine where f is increasing and where it is decreasing.
  (e) Without using a graphing utility, repeat parts (b)–(d) for $y = f(x - 4)$.
  (f) Without using a graphing utility, repeat parts (b)–(d) for $y = f(2x)$.
  (g) Without using a graphing utility, repeat parts (b)–(d) for $y = -f(x)$.

Applications and Extensions

79. The equation $y = (x - c)^2$ defines a *family of parabolas,* one parabola for each value of c. On one set of coordinate axes, graph the members of the family for $c = 0$, $c = 3$, and $c = -2$.

80. Repeat Problem 79 for the family of parabolas $y = x^2 + c$.

81. Thermostat Control Energy conservation experts estimate that homeowners can save 5% to 10% on winter heating bills by programming their thermostats 5 to 10 degrees lower while sleeping. In the given graph, the temperature T (in degrees Fahrenheit) of a home is given as a function of time t (in hours after midnight) over a 24-hour period.

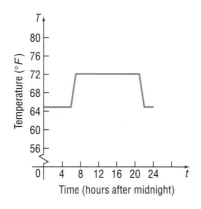

(a) At what temperature is the thermostat set during daytime hours? At what temperature is the thermostat set overnight?
(b) The homeowner reprograms the thermostat to $y = T(t) - 2$. Explain how this affects the temperature in the house. Graph this new function.
(c) The homeowner reprograms the thermostat to $y = T(t + 1)$. Explain how this affects the temperature in the house. Graph this new function.

Source: Roger Albright, *547 Ways to be Fuel Smart,* 2000

82. Digital Music Revenues The total projected worldwide digital music revenues R, in millions of dollars, for the years 2005 through 2010 can be estimated by the function

$$R(x) = 170.7x^2 + 1373x + 1080$$

where x is the number of years after 2005.
(a) Find $R(0)$, $R(3)$, and $R(5)$ and explain what each value represents.
(b) Find $r = R(x - 5)$.

(c) Find $r(5)$, $r(8)$, and $r(10)$ and explain what each value represents.
(d) In the model r, what does x represent?
(e) Would there be an advantage in using the model r when estimating the projected revenues for a given year instead of the model R?

Source: eMarketer.com, May 2006

83. Temperature Measurements The relationship between the Celsius (°C) and Fahrenheit (°F) scales for measuring temperature is given by the equation

$$F = \frac{9}{5}C + 32$$

The relationship between the Celsius (°C) and Kelvin (K) scales is $K = C + 273$. Graph the equation $F = \frac{9}{5}C + 32$ using degrees Fahrenheit on the y-axis and degrees Celsius on the x-axis. Use the techniques introduced in this section to obtain the graph showing the relationship between Kelvin and Fahrenheit temperatures.

84. Period of a Pendulum The period T (in seconds) of a simple pendulum is a function of its length l (in feet) defined by the equation

$$T = 2\pi\sqrt{\frac{l}{g}}$$

where $g \approx 32.2$ feet per second per second is the acceleration of gravity.

(a) Use a graphing utility to graph the function $T = T(l)$.
(b) Now graph the functions $T = T(l + 1)$, $T = T(l + 2)$, and $T = T(l + 3)$.
(c) Discuss how adding to the length l changes the period T.

(d) Now graph the functions $T = T(2l), T = T(3l)$, and $T = T(4l)$.

(e) Discuss how multiplying the length l by factors of 2, 3, and 4 changes the period T.

85. Cigar Company Profits The daily profits of a cigar company from selling x cigars are given by

$$p(x) = -0.05x^2 + 100x - 2000$$

The government wishes to impose a tax on cigars (sometimes called a *sin tax*) that gives the company the option of either paying a flat tax of $10,000 per day or a tax of 10% on profits. As chief financial officer (CFO) of the company, you need to decide which tax is the better option for the company.

(a) On the same screen, graph $Y_1 = p(x) - 10,000$ and $Y_2 = (1 - 0.10)p(x)$.

(b) Based on the graph, which option would you select? Why?

(c) Using the terminology learned in this section, describe each graph in terms of the graph of $p(x)$.

(d) Suppose that the government offered the options of a flat tax of $4800 or a tax of 10% on profits. Which would you select? Why?

86. The graph of a function f is illustrated in the figure.

(a) Draw the graph of $y = |f(x)|$.

(b) Draw the graph of $y - f(|x|)$.

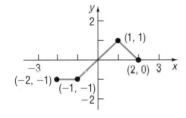

87. The graph of a function f is illustrated in the figure.

(a) Draw the graph of $y = |f(x)|$.

(b) Draw the graph of $y = f(|x|)$.

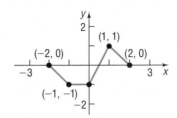

88. Suppose $(1, 3)$ is a point on the graph of $y = f(x)$.

(a) What point is on the graph of $y = f(x + 3) - 5$?

(b) What point is on the graph of $y = -2f(x - 2) + 1$?

(c) What point is on the graph of $y = f(2x + 3)$?

89. Suppose $(-3, 5)$ is a point on the graph of $y = f(x)$.

(a) What point is on the graph of $y = f(x + 1) - 3$?

(b) What point is on the graph of $y = -3f(x - 4) + 3$?

(c) What point is on the graph of $y = f(3x + 9)$?

Discussion and Writing

90. Suppose that the graph of a function f is known. Explain how the graph of $y = 4f(x)$ differs from the graph of $y = f(4x)$.

91. The area under the curve $y = \sqrt{x}$ bounded below by the x-axis, on the left by $x = 0$, and on the right by $x = 4$ is $\dfrac{16}{3}$

square units. Using the ideas presented in this section, what do you think is the area under the curve of $y = \sqrt{-x}$ bounded below by the x-axis, on the left by $x - -4$, and on the right by $x = 0$? Justify your answer.

1.8 One-to-One Functions; Inverse Functions

PREPARING FOR THIS SECTION *Before getting started, review the following:*

- Functions (Section 1.3, pp. 30–39)
- Increasing/Decreasing Functions (Section 1.5, pp. 52–53)

Now Work the 'Are You Prepared?' problems on page 91.

OBJECTIVES **1** Determine Whether a Function Is One-to-One (p. 82)

2 Determine the Inverse of a Function Defined by a Map or a Set of Ordered Pairs (p. 84)

3 Obtain the Graph of the Inverse Function from the Graph of the Function (p. 86)

4 Find the Inverse of a Function Defined by an Equation (p. 87)

1 Determine Whether a Function Is One-to-One

In Section 1.3, we presented four different ways to represent a function: as (1) a map, (2) a set of ordered pairs, (3) a graph, and (4) an equation. For example, Figures 95 and 96 illustrate two different functions represented as mappings. The function in Figure 95 shows the correspondence between states and their population (in millions). The function in Figure 96 shows a correspondence between animals and life expectancy (in years).

Figure 95

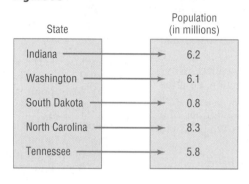

Figure 96

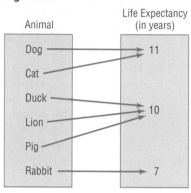

Suppose we asked a group of people to name the state that has a population of 0.8 million based on the function in Figure 95. Everyone in the group would respond South Dakota. Now, if we asked the same group of people to name the animal whose life expectancy is 11 years based on the function in Figure 96, some would respond dog, while others would respond cat. What is the difference between the functions in Figures 95 and 96? In Figure 95, we can see that no two elements in the domain corresponds to the same element in the range. In Figure 96, this is not the case: two different elements in the domain correspond to the same element in the range. We give functions such as the one in Figure 95 a special name.

DEFINITION

A function is **one-to-one** if any two different inputs in the domain correspond to two different outputs in the range. That is, if x_1 and x_2 are two different inputs of a function f, then f is one-to-one if $f(x_1) \neq f(x_2)$.

In Words

A function is not one-to-one if two different inputs correspond to the same output.

Put another way, a function f is one-to-one if no y in the range is the image of more than one x in the domain. A function is not one-to-one if two different elements in the domain correspond to the same element in the range. So the function in Figure 96 is not one-to-one because two different elements in the domain, *dog* and *cat,* both correspond to 11. Figure 97 illustrates the distinction of one-to-one functions, functions that are not one-to-one, and nonfunctions.

Figure 97

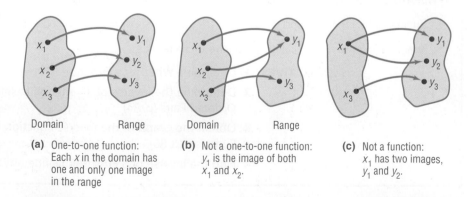

(a) One-to-one function: Each x in the domain has one and only one image in the range

(b) Not a one-to-one function: y_1 is the image of both x_1 and x_2.

(c) Not a function: x_1 has two images, y_1 and y_2.

EXAMPLE 1	Determining Whether a Function Is One-to-One

Determine whether the following functions are one-to-one.

(a) For the following function, the domain represents the age of five males and the range represents their HDL (good) cholesterol (mg/dL).

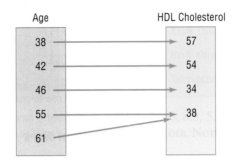

(b) $\{(-2, 6), (-1, 3), (0, 2), (1, 5), (2, 8)\}$

Solution
(a) The function is not one-to-one because there are two different inputs, 55 and 61, that correspond to the same output, 38.

(b) The function is one-to-one because there are no two distinct inputs that correspond to the same output.

 **Now Work** PROBLEMS **9** AND **13**

For functions defined by an equation $y = f(x)$ and for which the graph of f is known, there is a simple test, called the **horizontal-line test,** to determine whether f is one-to-one.

THEOREM	Horizontal-line Test

Figure 98
$f(x_1) = f(x_2) = h$ and $x_1 \neq x_2$; f is not a one-to-one function.

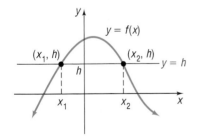

If every horizontal line intersects the graph of a function f in at most one point, then f is one-to-one.

The reason that this test works can be seen in Figure 98, where the horizontal line $y = h$ intersects the graph at two distinct points, (x_1, h) and (x_2, h). Since h is the image of both x_1 and x_2 and $x_1 \neq x_2$, f is not one-to-one. Based on Figure 98, we can state the horizontal-line test in another way: If the graph of any horizontal line intersects the graph of a function f at more than one point, then f is not one-to-one.

EXAMPLE 2	Using the Horizontal-line Test

For each function, use its graph to determine whether the function is one-to-one.

(a) $f(x) = x^2$ (b) $g(x) = x^3$

Solution
(a) Figure 99(a) on page 84 illustrates the horizontal-line test for $f(x) = x^2$. The horizontal line $y = 1$ intersects the graph of f twice, at $(1, 1)$ and at $(-1, 1)$, so f is not one-to-one.

(b) Figure 99(b) illustrates the horizontal-line test for $g(x) = x^3$. Because every horizontal line intersects the graph of g exactly once, it follows that g is one-to-one.

Consider the function $f(x) = 2x$, which multiplies the argument x by 2. The inverse function f^{-1} undoes whatever f does. So the inverse function of f is $f^{-1}(x) = \dfrac{1}{2}x$, which divides the argument by 2. For example, $f(3) = 2(3) = 6$ and $f^{-1}(6) = \dfrac{1}{2}(6) = 3$, so f^{-1} undoes what f did. We can verify this by showing that

$$f^{-1}(f(x)) = f^{-1}(2x) = \frac{1}{2}(2x) = x \quad \text{and} \quad f(f^{-1}(x)) = f\left(\frac{1}{2}x\right) = 2\left(\frac{1}{2}x\right) = x$$

See Figure 101.

Figure 101

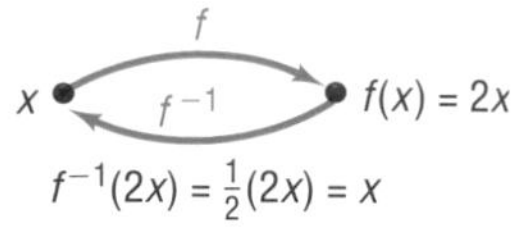

$x \bullet \underset{\overset{\curvearrowleft}{f^{-1}}}{\overset{f}{\curvearrowright}} \bullet f(x) = 2x$

$f^{-1}(2x) = \frac{1}{2}(2x) = x$

| EXAMPLE 5 | Verifying Inverse Functions |

Exploration

Simultaneously graph $Y_1 = x$, $Y_2 = x^3$, and $Y_3 = \sqrt[3]{x}$ on a square screen with $-3 \le x \le 3$. What do you observe about the graphs of $Y_2 = x^3$, its inverse $Y_3 = \sqrt[3]{x}$, and the line $Y_1 = x$?

Repeat this experiment by simultaneously graphing $Y_1 = x$, $Y_2 = 2x + 3$, and $Y_3 = \dfrac{1}{2}(x - 3)$ on a square screen with $-6 \le x \le 3$. Do you see the symmetry of the graph of Y_2 and its inverse Y_3 with respect to the line $Y_1 = x$?

(a) We verify that the inverse of $g(x) = x^3$ is $g^{-1}(x) = \sqrt[3]{x}$ by showing that

$$g^{-1}(g(x)) = g^{-1}(x^3) = \sqrt[3]{x^3} = x \quad \text{for all } x \text{ in the domain of } g$$
$$g(g^{-1}(x)) = g(\sqrt[3]{x}) = (\sqrt[3]{x})^3 = x \quad \text{for all } x \text{ in the domain of } g^{-1}$$

(b) We verify that the inverse of $f(x) = 2x + 3$ is $f^{-1}(x) = \dfrac{1}{2}(x - 3)$ by showing that

$$f^{-1}(f(x)) = f^{-1}(2x + 3) = \frac{1}{2}[(2x + 3) - 3] = \frac{1}{2}(2x) = x \quad \begin{array}{l}\text{for all } x \text{ in the} \\ \text{domain of } f\end{array}$$

$$f(f^{-1}(x)) = f\left(\frac{1}{2}(x - 3)\right) = 2\left[\frac{1}{2}(x - 3)\right] + 3 = (x - 3) + 3 = x \quad \begin{array}{l}\text{for all } x \text{ in the} \\ \text{domain of } f^{-1}\end{array}$$

| EXAMPLE 6 | Verifying Inverse Functions |

Verify that the inverse of $f(x) = \dfrac{1}{x - 1}$ is $f^{-1}(x) = \dfrac{1}{x} + 1$. For what values of x is $f^{-1}(f(x)) = x$? For what values of x is $f(f^{-1}(x)) = x$?

Solution The domain of f is $\{x | x \ne 1\}$ and the domain of f^{-1} is $\{x | x \ne 0\}$. Now

$$f^{-1}(f(x)) = f^{-1}\left(\frac{1}{x - 1}\right) = \frac{1}{\dfrac{1}{x - 1}} + 1 = x - 1 + 1 = x \qquad \text{provided } x \ne 1$$

$$f(f^{-1}(x)) = f\left(\frac{1}{x} + 1\right) = \frac{1}{\dfrac{1}{x} + 1 - 1} = \frac{1}{\dfrac{1}{x}} = x \qquad \text{provided } x \ne 0$$

Now Work PROBLEM 31

3 Obtain the Graph of the Inverse Function from the Graph of the Function

For the functions in Example 5(b), we list points on the graph of $f = Y_1$ and on the graph of $f^{-1} = Y_2$ in Table 24. We notice that whenever (a, b) is on the graph of f then (b, a) is on the graph of f^{-1}. Figure 102 shows these points plotted. Also shown is the graph of $y = x$, which you should observe is a line of symmetry of the points.

Table 24

Figure 102

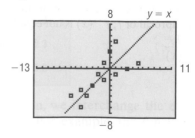

Figure 103

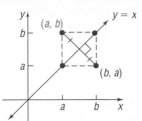

Suppose that (a, b) is a point on the graph of a one-to-one function f defined by $y = f(x)$. Then $b = f(a)$. This means that $a = f^{-1}(b)$, so (b, a) is a point on the graph of the inverse function f^{-1}. The relationship between the point (a, b) on f and the point (b, a) on f^{-1} is shown in Figure 103. The line segment containing (a, b) and (b, a) is perpendicular to the line $y = x$ and is bisected by the line $y = x$. (Do you see why?) It follows that the point (b, a) on f^{-1} is the reflection about the line $y = x$ of the point (a, b) on f.

THEOREM

The graph of a function f and the graph of its inverse f^{-1} are symmetric with respect to the line $y = x$.

Figure 104 illustrates this result. Notice that, once the graph of f is known, the graph of f^{-1} may be obtained by reflecting the graph of f about the line $y = x$.

Figure 104

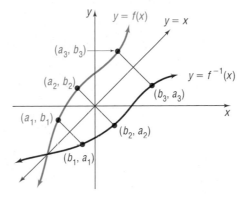

EXAMPLE 7 | **Graphing the Inverse Function**

The graph in Figure 105(a) is that of a one-to-one function $y = f(x)$. Draw the graph of its inverse.

Solution

We begin by adding the graph of $y = x$ to Figure 105(a). Since the points $(-2, -1), (-1, 0)$, and $(2, 1)$ are on the graph of f, we know that the points $(-1, -2), (0, -1)$, and $(1, 2)$ must be on the graph of f^{-1}. Keeping in mind that the graph of f^{-1} is the reflection about the line $y = x$ of the graph of f, we can draw f^{-1}. See Figure 105(b).

Figure 105

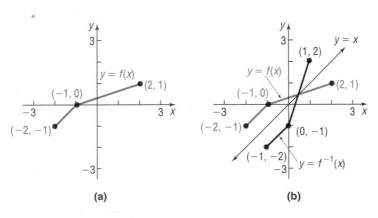

(a)   (b)

Now Work PROBLEM 41

4 Find the Inverse of a Function Defined by an Equation

The fact that the graphs of a one-to-one function f and its inverse function f^{-1} are symmetric with respect to the line $y = x$ tells us more. It says that we can obtain f^{-1}

by interchanging the roles of x and y in f. If f is defined by the equation

$$y = f(x)$$

then f^{-1} is defined by the equation

$$x = f(y)$$

The equation $x = f(y)$ defines f^{-1} *implicitly*. If we can solve this equation for y, we will have the *explicit* form of f^{-1}, that is,

$$y = f^{-1}(x)$$

Let's use this procedure to find the inverse of $f(x) = 2x + 3$. (Since f is a linear function and is increasing, we know that f is one-to-one and so has an inverse function.)

| EXAMPLE 8 | **How to Find the Inverse Function** |

Find the inverse of $f(x) = 2x + 3$. Graph f and f^{-1} on the same coordinate axes.

Step-by-Step Solution

STEP 1 Replace f(x) with y. In $y = f(x)$, interchange the variables x and y to obtain $x = f(y)$. This equation defines the inverse function f^{-1} implicitly.

We replace $f(x)$ with y in $f(x) = 2x + 3$ and obtain $y = 2x + 3$. Now we interchange the variables x and y to obtain

$$x = 2y + 3$$

This equation defines the inverse f^{-1} implicitly.

STEP 2 If possible, solve the implicit equation for y in terms of x to obtain the explicit form of f^{-1}, $y = f^{-1}(x)$.

To find the explicit form of the inverse, we solve $x = 2y + 3$ for y.

$$x = 2y + 3$$
$$2y + 3 = x \qquad \text{Reflexive Property, If } a = b, \text{ then } b = a.$$
$$2y = x - 3 \qquad \text{Subtract 3 from both sides.}$$
$$y = \frac{1}{2}(x - 3) \qquad \text{Divide both sides by 2.}$$

The explicit form of the inverse f^{-1} is therefore

$$f^{-1}(x) = \frac{1}{2}(x - 3)$$

STEP 3 Check the result by showing that $f^{-1}(f(x)) = x$ and $f(f^{-1}(x)) = x$.

We verified that f and f^{-1} are inverses in Example 5(b) on page 86.

The graphs of $f(x) = 2x + 3$ and its inverse $f^{-1}(x) = \frac{1}{2}(x - 3)$ are shown in Figure 106. Note the symmetry of the graphs with respect to the line $y = x$.

Figure 106

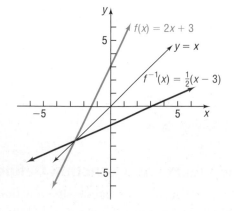

We now list the steps to follow for finding the inverse of a one-to-one function.

> **Procedure for Finding the Inverse of a One-to-One Function**
>
> **STEP 1:** In $y = f(x)$, interchange the variables x and y to obtain
> $$x = f(y)$$
> This equation defines the inverse function f^{-1} implicitly.
>
> **STEP 2:** If possible, solve the implicit equation for y in terms of x to obtain the explicit form of f^{-1}:
> $$y = f^{-1}(x)$$
>
> **STEP 3:** Check the result by showing that
> $$f^{-1}(f(x)) = x \quad \text{and} \quad f(f^{-1}(x)) = x$$

EXAMPLE 9 **Finding the Inverse Function**

The function
$$f(x) = \frac{2x + 1}{x - 1} \quad x \neq 1$$

is one-to-one. Find its inverse and check the result.

Solution **STEP 1:** Replace $f(x)$ with y and interchange the variables x and y in
$$y = \frac{2x + 1}{x - 1}$$

to obtain
$$x = \frac{2y + 1}{y - 1}$$

STEP 2: Solve for y.

$$x = \frac{2y + 1}{y - 1}$$
$$x(y - 1) = 2y + 1 \quad \text{Multiply both sides by } y - 1.$$
$$xy - x = 2y + 1 \quad \text{Apply the Distributive Property.}$$
$$xy - 2y = x + 1 \quad \text{Subtract } 2y \text{ from both sides; add } x \text{ to both sides.}$$
$$(x - 2)y = x + 1 \quad \text{Factor.}$$
$$y = \frac{x + 1}{x - 2} \quad \text{Divide by } x - 2.$$

The inverse is
$$f^{-1}(x) = \frac{x + 1}{x - 2} \quad x \neq 2 \quad \text{Replace } y \text{ by } f^{-1}(x).$$

STEP 3: ✓ Check:

$$f^{-1}(f(x)) = f^{-1}\left(\frac{2x + 1}{x - 1}\right) = \frac{\dfrac{2x + 1}{x - 1} + 1}{\dfrac{2x + 1}{x - 1} - 2} = \frac{2x + 1 + x - 1}{2x + 1 - 2(x - 1)} = \frac{3x}{3} = x \quad x \neq 1$$

$$f(f^{-1}(x)) = f\left(\frac{x + 1}{x - 2}\right) = \frac{2\left(\dfrac{x + 1}{x - 2}\right) + 1}{\dfrac{x + 1}{x - 2} - 1} = \frac{2(x + 1) + x - 2}{x + 1 - (x - 2)} = \frac{3x}{3} = x \quad x \neq 2$$

Exploration

In Example 9, we found that, if $f(x) = \dfrac{2x + 1}{x - 1}$, then $f^{-1}(x) = \dfrac{x + 1}{x - 2}$. Compare the vertical and horizontal asymptotes of f and f^{-1}.

Result The vertical asymptote of f is $x = 1$, and the horizontal asymptote is $y = 2$. The vertical asymptote of f^{-1} is $x = 2$, and the horizontal asymptote is $y = 1$.

━━━━━ **Now Work** PROBLEM 49

Earlier in the chapter, we said that finding the range of a function f is not easy. However, if f is one-to-one, we can find its range by finding the domain of the inverse function f^{-1}.

EXAMPLE 10 | **Finding the Range of a Function**

Find the domain and the range of

$$f(x) = \frac{2x + 1}{x - 1}$$

Solution The domain of f is $\{x \mid x \neq 1\}$. To find the range of f, we use the fact that the domain of f^{-1} equals the range of f. Based on Example 9, we have

$$f^{-1}(x) = \frac{x + 1}{x - 2}$$

The domain of f^{-1} is $\{x \mid x \neq 2\}$, so the range of f is $\{y \mid y \neq 2\}$. Also, because the domain of f is $\{x \mid x \neq 1\}$, the range of f^{-1} is $\{y \mid y \neq 1\}$.

■

━━━━━ **Now Work** PROBLEM 63

If a function is not one-to-one, it has no inverse function. Sometimes, though, an appropriate restriction on the domain of such a function will yield a new function that is one-to-one. Then the function defined on the restricted domain has an inverse function. Let's look at an example of this common practice.

EXAMPLE 11 | **Finding the Inverse of a Domain-restricted Function**

Find the inverse of $y = f(x) = x^2$ if $x \geq 0$.

Solution The function $y = x^2$ is not one-to-one. [Refer to Example 2(a).] However, if we restrict the domain of this function to $x \geq 0$, as indicated, we have a new function that is increasing and therefore is one-to-one. As a result, the function defined by $y = f(x) = x^2$, $x \geq 0$, has an inverse function, f^{-1}.

We follow the steps given previously to find f^{-1}.

STEP 1: In the equation $y = x^2$, $x \geq 0$, interchange the variables x and y. The result is

$$x = y^2 \qquad y \geq 0$$

This equation defines (implicitly) the inverse function.

Figure 107

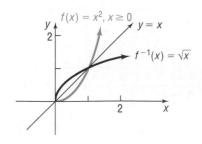

STEP 2: We solve for y to get the explicit form of the inverse. Since $y \geq 0$, only one solution for y is obtained: $y = \sqrt{x}$. So $f^{-1}(x) = \sqrt{x}$.

STEP 3: ✓ Check: $f^{-1}(f(x)) = f^{-1}(x^2) = \sqrt{x^2} = |x| = x$ since $x \geq 0$

$$f(f^{-1}(x)) = f(\sqrt{x}) = (\sqrt{x})^2 = x$$

■

Figure 107 illustrates the graphs of $f(x) = x^2$, $x \geq 0$, and $f^{-1}(x) = \sqrt{x}$.

SUMMARY

1. If a function f is one-to-one, then it has an inverse function f^{-1}.
2. Domain of f = Range of f^{-1}; Range of f = Domain of f^{-1}.
3. To verify that f^{-1} is the inverse of f, show that $f^{-1}(f(x)) = x$ for every x in the domain of f and $f(f^{-1}(x)) = x$ for every x in the domain of f^{-1}.
4. The graphs of f and f^{-1} are symmetric with respect to the line $y = x$.
5. To find the range of a one-to-one function f, find the domain of its inverse function f^{-1}.

1.8 Assess Your Understanding

'Are You Prepared?' *Answers are given at the end of these exercises. If you get a wrong answer, read the pages listed in red.*

1. Is the set of ordered pairs $\{(1, 3), (2, 3), (-1, 2)\}$ a function? Why or why not? (pp. 30–33)
2. Where is the function $f(x) = x^2$ increasing? Where is it decreasing? (pp. 52–53)
3. What is the domain of $f(x) = \dfrac{x + 5}{x^2 + 3x - 18}$? (pp. 37–38)

Concepts and Vocabulary

4. If every horizontal line intersects the graph of a function f at no more than one point, f is a(n) _____ function.
5. If f^{-1} denotes the inverse of a function f, then the graphs of f and f^{-1} are symmetric with respect to the line _____.
6. If the domain of a one-to-one function f is $[4, \infty)$, the range of its inverse, f^{-1}, is _____.
7. *True or False* If f and g are inverse functions, the domain of f is the same as the domain of g.
8. *True or False* If f and g are inverse functions, their graphs are symmetric with respect to the line $y = x$.

Skill Building

In Problems 9–16, determine whether the function is one-to-one.

9.
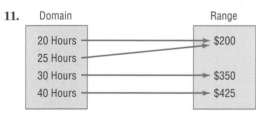

10.
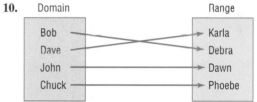

11.
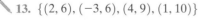

12.

13. $\{(2, 6), (-3, 6), (4, 9), (1, 10)\}$

14. $\{(-2, 5), (-1, 3), (3, 7), (4, 12)\}$

15. $\{(0, 0), (1, 1), (2, 16), (3, 81)\}$

16. $\{(1, 2), (2, 8), (3, 18), (4, 32)\}$

In Problems 17–22, the graph of a function f is given. Use the horizontal-line test to determine whether f is one-to-one.

17.
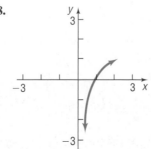

18.

19.
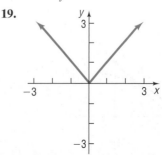

20.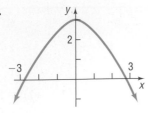

21.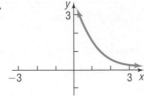

22.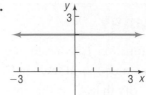

In Problems 23–30, find the inverse of each one-to-one function. State the domain and the range of each inverse function.

23.

Location	Annual Rainfall (inches)
Mt Waialeale, Hawaii	460.00
Monrovia, Liberia	202.01
Pago Pago, American Samoa	196.46
Moulmein, Burma	191.02
Lae, Papua New Guinea	182.87

Source: Information Please Almanac

24.

Title	Domestic Gross (in millions)
Star Wars	$461
Star Wars: Episode One – The Phantom Menace	$431
E.T. the Extra Terrestrial	$400
Jurassic Park	$357
Forrest Gump	$330

Source: Information Please Almanac

25.

Age	Monthly Cost of Life Insurance
30	$7.09
40	$8.40
45	$11.29

Source: eterm.com

26.

State	Unemployment Rate
Virginia	11%
Nevada	5.5%
Tennessee	5.1%
Texas	6.3%

Source: United States Statistical Abstract

27. $\{(-3, 5), (-2, 9), (-1, 2), (0, 11), (1, -5)\}$

28. $\{(-2, 2), (-1, 6), (0, 8), (1, -3), (2, 9)\}$

29. $\{(-2, 1), (-3, 2), (-10, 0), (1, 9), (2, 4)\}$

30. $\{(-2, -8), (-1, -1), (0, 0), (1, 1), (2, 8)\}$

In Problems 31–40, verify that the functions f and g are inverses of each other by showing that $f(g(x)) = x$ and $g(f(x)) = x$. Give any values of x that need to be excluded from the domain.

31. $f(x) = 3x + 4; \quad g(x) = \dfrac{1}{3}(x - 4)$

32. $f(x) = 3 - 2x; \quad g(x) = -\dfrac{1}{2}(x - 3)$

33. $f(x) = 4x - 8; \quad g(x) = \dfrac{x}{4} + 2$

34. $f(x) = 2x + 6; \quad g(x) = \dfrac{1}{2}x - 3$

35. $f(x) = x^3 - 8; \quad g(x) = \sqrt[3]{x + 8}$

36. $f(x) = (x - 2)^2, x \geq 2; \quad g(x) = \sqrt{x} + 2$

37. $f(x) = \dfrac{1}{x}; \quad g(x) = \dfrac{1}{x}$

38. $f(x) = x; \quad g(x) = x$

39. $f(x) = \dfrac{2x + 3}{x + 4}; \quad g(x) = \dfrac{4x - 3}{2 - x}$

40. $f(x) = \dfrac{x - 5}{2x + 3}; \quad g(x) = \dfrac{3x + 5}{1 - 2x}$

In Problems 41–46, the graph of a one-to-one function f is given. Draw the graph of the inverse function f^{-1}. For convenience (and as a hint), the graph of $y = x$ is also given.

41.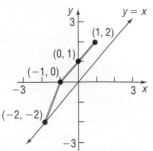

42.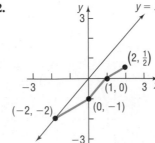

43.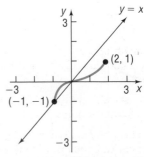

44.

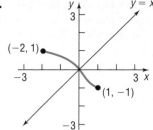

45.

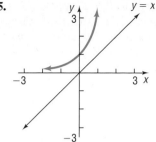

46.

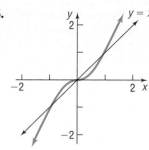

In Problems 47–58, the function f is one-to-one. Find its inverse and check your answer. Graph f, f^{-1}, and y = x on the same coordinate axes.

47. $f(x) = 3x$

48. $f(x) = -4x$

49. $f(x) = 4x + 2$

50. $f(x) = 1 - 3x$

51. $f(x) = x^3 - 1$

52. $f(x) = x^3 + 1$

53. $f(x) = x^2 + 4 \quad x \geq 0$

54. $f(x) = x^2 + 9 \quad x \geq 0$

55. $f(x) = \dfrac{4}{x}$

56. $f(x) = -\dfrac{3}{x}$

57. $f(x) = \dfrac{1}{x - 2}$

58. $f(x) = \dfrac{4}{x + 2}$

In Problems 59–70, the function f is one-to-one. Find its inverse and check your answer. State the domain and the range of f and f^{-1}.

59. $f(x) = \dfrac{2}{3 + x}$

60. $f(x) = \dfrac{4}{2 - x}$

61. $f(x) = \dfrac{3x}{x + 2}$

62. $f(x) = -\dfrac{2x}{x - 1}$

63. $f(x) = \dfrac{2x}{3x - 1}$

64. $f(x) = -\dfrac{3x + 1}{x}$

65. $f(x) = \dfrac{3x + 4}{2x - 3}$

66. $f(x) = \dfrac{2x - 3}{x + 4}$

67. $f(x) = \dfrac{2x + 3}{x + 2}$

68. $f(x) = \dfrac{-3x - 4}{x - 2}$

69. $f(x) = \dfrac{x^2 - 4}{2x^2} \quad x > 0$

70. $f(x) = \dfrac{x^2 + 3}{3x^2} \quad x > 0$

Applications and Extensions

71. Use the graph of $y = f(x)$ given in Problem 41 to evaluate the following:
(a) $f(-1)$ (b) $f(1)$ (c) $f^{-1}(1)$ (d) $f^{-1}(2)$

72. Use the graph of $y = f(x)$ given in Problem 42 to evaluate the following:
(a) $f(2)$ (b) $f(1)$ (c) $f^{-1}(0)$ (d) $f^{-1}(-1)$

73. If $f(7) = 13$ and f is one-to-one, what is $f^{-1}(13)$?

74. If $g(-5) = 3$ and g is one-to-one, what is $g^{-1}(3)$?

75. The domain of a one-to-one function f is $[5, \infty)$, and its range is $[-2, \infty)$. State the domain and the range of f^{-1}.

76. The domain of a one-to-one function f is $[0, \infty)$, and its range is $[5, \infty)$. State the domain and the range of f^{-1}.

77. The domain of a one-to-one function g is the set of all real numbers, and its range is $[0, \infty)$. State the domain and the range of g^{-1}.

78. The domain of a one-to-one function g is $[0, 15]$, and its range is $(0, 8)$. State the domain and the range of g^{-1}.

79. A function $y = f(x)$ is increasing on the interval $(0, 5)$. What conclusions can you draw about the graph of $y = f^{-1}(x)$?

80. A function $y = f(x)$ is decreasing on the interval $(0, 5)$. What conclusions can you draw about the graph of $y = f^{-1}(x)$?

81. Find the inverse of the linear function
$$f(x) = mx + b \quad m \neq 0$$

82. Find the inverse of the function
$$f(x) = \sqrt{r^2 - x^2} \quad 0 \leq x \leq r$$

83. A function f has an inverse function. If the graph of f lies in quadrant I, in which quadrant does the graph of f^{-1} lie?

84. A function f has an inverse function. If the graph of f lies in quadrant II, in which quadrant does the graph of f^{-1} lie?

85. The function $f(x) = |x|$ is not one-to-one. Find a suitable restriction on the domain of f so that the new function that results is one-to-one. Then find the inverse of f.

86. The function $f(x) = x^4$ is not one-to-one. Find a suitable restriction on the domain of f so that the new function that results is one-to-one. Then find the inverse of f.

In applications, the symbols used for the independent and dependent variables are often based on common usage. So, rather than using $y = f(x)$ to represent a function, an applied problem might use $C = C(q)$ to represent the cost C of manufacturing q units of a good since, in economics, q is used for output. Because of this, the inverse notation f^{-1} used in a pure mathematics problem is not used when finding inverses of applied problems. Rather, the inverse of a function such as $C = C(q)$ will be $q = q(C)$. So $C = C(q)$ is a function that represents the cost C as a function of the output q, while $q = q(C)$ is a function that represents the output q as a function of the cost C. Problems 87–90 illustrate this idea.

87. Vehicle Stopping Distance Taking into account reaction time, the distance d (in feet) that a car requires to come to a complete stop while traveling r miles per hour is given by the function

$$d(r) = 6.97r - 90.39$$

(a) Express the speed r at which the car is traveling as a function of the distance d required to come to a complete stop.
(b) Verify that $r = r(d)$ is the inverse of $d = d(r)$ by showing that $r(d(r)) = r$ and $d(r(d)) = d$.
(c) Predict the speed that a car was traveling if the distance required to stop was 300 feet.

88. Height and Head Circumference The head circumference C of a child is related to the height H of the child (both in inches) through the function

$$H(C) = 2.15C - 10.53$$

(a) Express the head circumference C as a function of height H.
(b) Verify that $C = C(H)$ is the inverse of $H = H(C)$ by showing that $H(C(H)) = H$ and $C(H(C)) = C$.
(c) Predict the head circumference of a child who is 26 inches tall.

89. Ideal Body Weight One model for the ideal body weight W for men (in kilograms) as a function of height h (in inches) is given by the function

$$W(h) = 50 + 2.3(h - 60)$$

(a) What is the ideal weight of a 6-foot male?
(b) Express the height h as a function of weight W.
(c) Verify that $h = h(W)$ is the inverse of $W = W(h)$ by showing that $h(W(h)) = h$ and $W(h(W)) = W$.
(d) What is the height of a male who is at his ideal weight of 80 kilograms?

[**Note:** The ideal body weight W for women (in kilograms) as a function of height h (in inches) is given by $W(h) = 45.5 + 2.3(h - 60)$.]

90. Temperature Conversion The function $F(C) = \dfrac{9}{5}C + 32$ converts a temperature from C degrees Celsius to F degrees Fahrenheit.

(a) Express the temperature in degrees Celsius C as a function of the temperature in degrees Fahrenheit F.
(b) Verify that $C = C(F)$ is the inverse of $F = F(C)$ by showing that $C(F(C)) = C$ and $F(C(F)) = F$.
(c) What is the temperature in degrees Celsius if it is 70 degrees Fahrenheit?

91. Income Taxes The function

$$T(g) = 4386.25 + 0.25(g - 31,850)$$

represents the 2007 federal income tax T (in dollars) due for a "single" filer whose modified adjusted gross income is g dollars, where $31,850 \le g \le 77,100$.
(a) What is the domain of the function T?
(b) Given that the tax due T is an increasing linear function of modified adjusted gross income g, find the range of the function T.
(c) Find adjusted gross income g as a function of federal income tax T. What are the domain and the range of this function?

92. Income Taxes The function

$$T(g) = 1565 + 0.15(g - 15,650)$$

represents the 2007 federal income tax T (in dollars) due for a "married filing jointly" filer whose modified adjusted gross income is g dollars, where $15,650 \le g \le 63,700$.
(a) What is the domain of the function T?
(b) Given that the tax due T is an increasing linear function of modified adjusted gross income g, find the range of the function T.
(c) Find adjusted gross income g as a function of federal income tax T. What are the domain and the range of this function?

93. Gravity on Earth If a rock falls from a height of 100 meters on Earth, the height H (in meters) after t seconds is approximately

$$H(t) = 100 - 4.9t^2$$

(a) In general, quadratic functions are not one-to-one. However, the function $H(t)$ is one-to-one. Why?
(b) Find the inverse of H and verify your result.
(c) How long will it take a rock to fall 80 meters?

94. Period of a Pendulum The period T (in seconds) of a simple pendulum as a function of its length l (in feet) is given by

$$T(l) = 2\pi\sqrt{\dfrac{l}{32.2}}$$

(a) Express the length l as a function of the period T.
(b) How long is a pendulum whose period is 3 seconds?

95. Given

$$f(x) = \dfrac{ax + b}{cx + d}$$

find $f^{-1}(x)$. If $c \ne 0$, under what conditions on $a, b, c,$ and d is $f = f^{-1}$?

Discussion and Writing

96. Can a one-to-one function and its inverse be equal? What must be true about the graph of f for this to happen? Give some examples to support your conclusion.

97. Draw the graph of a one-to-one function that contains the points $(-2, -3)$, $(0, 0)$, and $(1, 5)$. Now draw the graph of its inverse. Compare your graph to those of other students. Discuss any similarities. What differences do you see?

98. Give an example of a function whose domain is the set of real numbers and that is neither increasing nor decreasing on its domain, but is one-to-one.

  [**Hint:** Use a piecewise-defined function.]

99. Is every odd function one-to-one? Explain.

100. Suppose $C(g)$ represents the cost C, in dollars, of manufacturing g cars. Explain what $C^{-1}(800{,}000)$ represents.

'Are You Prepared?' Answers

1. Yes; for each input x there is one output y.

2. Increasing on $(0, \infty)$; decreasing on $(-\infty, 0)$

3. $\{x \mid x \neq -6, x \neq 3\}$

CHAPTER REVIEW

Library of Functions

Constant function (p. 61)

$$f(x) = b$$

The graph is a horizontal line with y-intercept b.

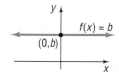

Identity function (p. 62)

$$f(x) = x$$

The graph is a line with slope 1 and y-intercept 0.

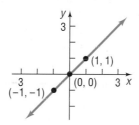

Square function (p. 62)

$$f(x) = x^2$$

The graph is a parabola with intercept at $(0, 0)$.

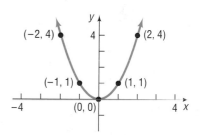

Cube function (p. 62)

$$f(x) = x^3$$

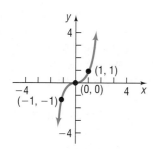

Square root function (p. 62)

$$f(x) = \sqrt{x}$$

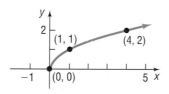

Cube root function (p. 62)

$$f(x) = \sqrt[3]{x}$$

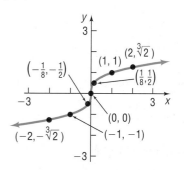

Reciprocal function (p. 63)

$$f(x) = \frac{1}{x}$$

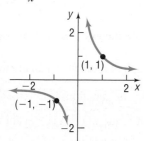

Absolute value function (p. 63)

$$f(x) = |x|$$

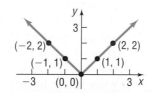

Things to Know

Formulas

Distance formula (p. 5)

$$d = \sqrt{(x_2 - x_1)^2 + (y_2 - y_1)^2}$$

Midpoint formula (p. 7)

$$(x, y) = \left(\frac{x_1 + x_2}{2}, \frac{y_1 + y_2}{2} \right)$$

Equations of Circles

Standard form of an equation of a circle (p. 22)

$(x - h)^2 + (y - k)^2 = r^2$; r is the radius of the circle, (h, k) is the center of the circle

Equation of the unit circle (p. 22)

$x^2 + y^2 = 1$

General form of the equation of a circle (p. 24)

$x^2 + y^2 + ax + by + c = 0$, with restrictions on a, b, and c

Function (pp. 30–34)

A relation between two sets so that each element x in the first set, the domain, has corresponding to it exactly one element y in the second set. The range is the set of y values of the function for the x values in the domain.

A function can also be characterized as a set of ordered pairs (x, y) in which no first element is paired with two different second elements.

Function notation (pp. 34–37)

$y = f(x)$

f is a symbol for the function.

x is the argument, or independent variable.

y is the dependent variable.

$f(x)$ is the value of the function at x, or the image of x.

A function f may be defined implicitly by an equation involving x and y or explicitly by writing $y = f(x)$.

Domain (pp. 37–38)

If unspecified, the domain of a function f is the largest set of real numbers for which $f(x)$ is a real number.

Vertical-line test (p. 42)

A set of points in the plane is the graph of a function if and only if every vertical line intersects the graph in at most one point.

Even function (p. 50)

$f(-x) = f(x)$ for every x in the domain ($-x$ must also be in the domain).

Odd function (p. 51)

$f(-x) = -f(x)$ for every x in the domain ($-x$ must also be in the domain).

Increasing function (p. 53)

A function f is increasing on an open interval I if, for any choice of x_1 and x_2 in I, with $x_1 < x_2$, we have $f(x_1) < f(x_2)$.

Decreasing function (p. 53)

A function f is decreasing on an open interval I if, for any choice of x_1 and x_2 in I, with $x_1 < x_2$, we have $f(x_1) > f(x_2)$.

Constant function (p. 53)

A function f is constant on an open interval I if, for all choices of x in I, the values of $f(x)$ are equal.

Local maximum (p. 54)

A function f has a local maximum at c if there is an open interval I containing c so that, for all x not equal to c in I, $f(x) < f(c)$.

Local minimum (p. 54)

A function f has a local minimum at c if there is an open interval I containing c so that, for all x not equal to c in I, $f(x) > f(c)$.

Average rate of change of a function (p. 55)

The average rate of change of f from a to b is

$$\frac{\Delta y}{\Delta x} = \frac{f(b) - f(a)}{b - a} \qquad a \neq b$$

One-to-one function f (p. 82)

A function for which any two different inputs in the domain correspond to two different outputs in the range.

For any choice of elements x_1, x_2 in the domain of f, if $x_1 \neq x_2$, then $f(x_1) \neq f(x_2)$.

Horizontal-line test (p. 83)

If every horizontal line intersects the graph of a function f in at most one point, then f is one-to-one.

Inverse function f^{-1} of f (pp. 84–87)

Domain of f = range of f^{-1}; range of f = domain of f^{-1}

$f^{-1}(f(x)) = x$ for all x in the domain of f, and $f(f^{-1}(x)) = x$ for all x in the domain of f^{-1}.

The graphs of f and f^{-1} are symmetric with respect to the line $y = x$.

Objectives

Section		You should be able to	Examples	Review Exercises
1.1	1	Use the distance formula (p. 4)	2, 3, 4	1(a)–6(a)
	2	Use the midpoint formula (p. 7)	5	1(b)–6(b)
	3	Graph equations by hand by plotting points (p. 7)	6, 7, 8	9–14
	4	Graph equations using a graphing utility (p. 9)	9, 10	8, 9–14
	5	Use a graphing utility to create tables (p. 11)	11	8
	6	Find intercepts from a graph (p. 12)	12	7
	7	Use a graphing utility to approximate intercepts (p. 12)	13	8
1.2	1	Find intercepts algebraically from an equation (p. 17)	1	9–14, 29–32
	2	Test an equation for symmetry (p. 18)	2–5	15–22
	3	Know how to graph key equations (p. 19)	3–5	23, 24
	4	Write the standard form of the equation of a circle (p. 22)	6	25–28
	5	Graph a circle by hand and by using a graphing utility (p. 23)	7, 8	29–32
	6	Work with the general form of the equation of a circle (p. 24)	9, 10	29–32, 35
1.3	1	Determine whether a relation represents a function (p. 30)	1–5	37, 38
	2	Find the value of a function (p. 34)	6–8	39–44, 99, 100
	3	Find the domain of a function defined by an equation (p. 37)	9, 10	45–52
1.4	1	Identify the graph of a function (p. 42)	1	75–78
	2	Obtain information from or about the graph of a function (p. 43)	2–4	53(a)–(e), 54(a)–(e), 55(a), 55(d), 55(f), 56(a), 56(d), 56(f)
1.5	1	Determine even and odd functions from a graph (p. 50)	1	55(e), 56(e)
	2	Identify even and odd functions from the equation (p. 51)	2	57–64
	3	Use a graph to determine where a function is increasing, decreasing, or constant (p. 52)	3	55(b), 56(b)
	4	Use a graph to locate local maxima and local minima (p. 53)	4	55(c), 56(c)
	5	Use a graphing utility to approximate local maxima and local minima and to determine where a function is increasing or decreasing (p. 54)	5	65–68
	6	Find the average rate of change of a function (p. 55)	6	69–74
1.6	1	Graph the functions listed in the library of functions (p. 59)	1, 2	79–82
	2	Graph piecewise-defined functions (p. 64)	3	95–98
1.7	1	Graph functions using vertical and horizontal shifts (p. 68)	1, 2	53(f), 54(f), 54(g) 83, 84, 87–94
	2	Graph functions using compressions and stretches (p. 70)	3	53(g), 54(h), 85, 86, 93, 94
	3	Graph functions using reflections about the x-axis or y-axis (p. 72)	4–6	53(h), 85, 89, 90, 94
1.8	1	Determine whether a function is one-to-one (p. 82)	1, 2	101(a), 102(a), 103, 104
	2	Determine the inverse of a function defined by a map or a set of ordered pairs (p. 84)	3, 4	101(b), 102(b)
	3	Obtain the graph of the inverse function from the graph of the function (p. 86)	7	103, 104
	4	Find the inverse of a function defined by an equation (p. 87)	8, 9, 11	105–110

Review Exercises

In Problems 1–6, find the following for each pair of points:
  (a) The distance between the points.
  (b) The midpoint of the line segment connecting the points.

1. $(0, 0)$; $(4, 2)$ **2.** $(0, 0)$; $(-4, 6)$ **3.** $(1, -1)$; $(-2, 3)$

4. $(-2, 2)$; $(1, 4)$ **5.** $(4, -4)$; $(4, 8)$ **6.** $(-3, 4)$; $(2, 4)$

7. List the intercepts of the following graph.

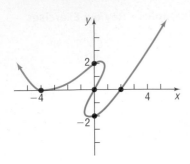

8. Graph $y = -x^2 + 15$ using a graphing utility. Create a table of values to determine a good initial viewing window. Use a graphing utility to approximate the intercepts.

In Problems 9–14, determine the intercepts and graph each equation by hand by plotting points. Verify your results using a graphing utility. Label the intercepts on the graph.

9. $2x - 3y = 6$

10. $x + 2y = 4$

11. $y = x^2 - 9$

12. $y = x^2 + 4$

13. $x^2 + 2y = 16$

14. $2x^2 - 4y = 24$

In Problems 15–22, test each equation for symmetry with respect to the x-axis, the y-axis, and the origin.

15. $2x = 3y^2$

16. $y = 5x$

17. $x^2 + 4y^2 = 16$

18. $9x^2 - y^2 = 9$

19. $y = x^4 + 2x^2 + 1$

20. $y = x^3 - x$

21. $x^2 + x + y^2 + 2y = 0$

22. $x^2 + 4x + y^2 - 2y = 0$

23. Sketch a graph of $y = x^3$.

24. Sketch a graph of $y = \sqrt{x}$.

In Problems 25–28, find the standard form of the equation of the circle whose center and radius are given.

25. $(h, k) = (-2, 3); r = 4$

26. $(h, k) = (3, 4); r = 4$

27. $(h, k) = (-1, -2); r = 1$

28. $(h, k) = (2, -4); r = 3$

In Problems 29–32, find the center and radius of each circle. Graph each circle by hand. Determine the intercepts of the graph of each circle.

29. $x^2 + y^2 - 2x + 4y - 4 = 0$ **30.** $x^2 + y^2 + 4x - 4y - 1 = 0$ **31.** $3x^2 + 3y^2 - 6x + 12y = 0$ **32.** $2x^2 + 2y^2 - 4x = 0$

---

33. Show that the points $A = (3, 4), B = (1, 1),$ and $C = (-2, 3)$ are the vertices of an isosceles triangle.

34. Show that the points $A = (1, 5), B = (2, 4),$ and $C = (-3, 5)$ lie on a circle with center $(-1, 2)$. What is the radius of this circle?

35. The endpoints of the diameter of a circle are $(-3, 2)$ and $(5, -6)$. Find the center and radius of the circle. Write the general equation of this circle.

36. Find two numbers y such that the distance from $(-3, 2)$ to $(5, y)$ is 10.

In Problems 37 and 38, determine whether each relation represents a function. For each function, state the domain and the range.

37. $\{(-1, 0), (2, 3), (4, 0)\}$

38. $\{(4, -1), (2, 1), (4, 2)\}$

In Problems 39–44, find the following for each function:

(a) $f(2)$ (b) $f(-2)$ (c) $f(-x)$ (d) $-f(x)$ (e) $f(x - 2)$ (f) $f(2x)$

39. $f(x) = \dfrac{3x}{x^2 - 1}$

40. $f(x) = \dfrac{x^2}{x + 1}$

41. $f(x) = \sqrt{x^2 - 4}$

42. $f(x) = |x^2 - 4|$

43. $f(x) = \dfrac{x^2 - 4}{x^2}$

44. $f(x) = \dfrac{x^3}{x^2 - 9}$

In Problems 45–52, find the domain of each function.

45. $f(x) = \dfrac{x}{x^2 - 9}$

46. $f(x) = \dfrac{3x^2}{x - 2}$

47. $f(x) = \sqrt{2 - x}$

48. $f(x) = \sqrt{x + 2}$

49. $h(x) = \dfrac{\sqrt{x}}{|x|}$

50. $g(x) = \dfrac{|x|}{x}$

51. $f(x) = \dfrac{x}{x^2 + 2x - 3}$

52. $F(x) = \dfrac{1}{x^2 - 3x - 4}$

53. Using the graph of the function f shown:

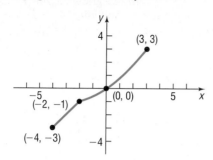

(a) Find the domain and the range of f.
(b) List the intercepts.
(c) Find $f(-2)$.
(d) For what value of x does $f(x) = -3$?
(e) Solve $f(x) > 0$.
(f) Graph $y = f(x - 3)$.

(g) Graph $y = f\left(\dfrac{1}{2}x\right)$.

(h) Graph $y = -f(x)$.

54. Using the graph of the function g shown:

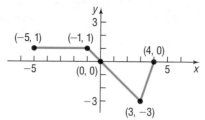

(a) Find the domain and the range of g.
(b) Find $g(-1)$.
(c) List the intercepts.
(d) For what value of x does $g(x) = -3$?
(e) Solve $g(x) > 0$.
(f) Graph $y = g(x - 2)$.
(g) Graph $y = g(x) + 1$.
(h) Graph $y = 2g(x)$.

In Problems 55 and 56, use the graph of the function f to find:
(a) The domain and the range of f.
(b) The intervals on which f is increasing, decreasing, or constant.
(c) The local minima and local maxima.
(d) Whether the graph is symmetric with respect to the x-axis, the y-axis, or the origin.
(e) Whether the function is even, odd, or neither.
(f) The intercepts, if any.

55.

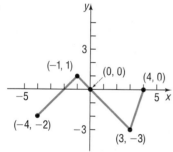

56.

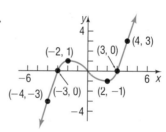

In Problems 57–64, determine (algebraically) whether the given function is even, odd, or neither.

57. $f(x) = x^3 - 4x$

58. $g(x) = \dfrac{4 + x^2}{1 + x^4}$

59. $h(x) = \dfrac{1}{x^4} + \dfrac{1}{x^2} + 1$

60. $F(x) = \sqrt{1 - x^3}$

61. $G(x) = 1 - x + x^3$

62. $H(x) = 1 + x + x^2$

63. $f(x) = \dfrac{x}{1 + x^2}$

64. $g(x) = \dfrac{1 + x^2}{x^3}$

In Problems 65–68, use a graphing utility to graph each function over the indicated interval. Approximate any local maxima and local minima. Determine where the function is increasing and where it is decreasing.

65. $f(x) = 2x^3 - 5x + 1 \quad (-3, 3)$

66. $f(x) = -x^3 + 3x - 5 \quad (-3, 3)$

67. $f(x) = 2x^4 - 5x^3 + 2x + 1 \quad (-2, 3)$

68. $f(x) = -x^4 + 3x^3 - 4x + 3 \quad (-2, 3)$

In Problems 69 and 70, find the average rate of change of f:
(a) From 1 to 2 *(b) From 0 to 1* *(c) From 2 to 4*

69. $f(x) = 8x^2 - x$

70. $f(x) = 2x^3 + x$

In Problems 71–74, find the average rate of change from 2 to 3 for each function f. Be sure to simplify.

71. $f(x) = 2 - 5x$

72. $f(x) = 2x^2 + 7$

73. $f(x) = 3x - 4x^2$

74. $f(x) = x^2 - 3x + 2$

In Problems 75–78, is the graph shown the graph of a function?

75.

76.

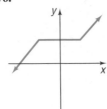

77.

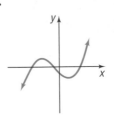

78.

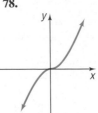

In Problems 79–82, sketch the graph of each function. Be sure to label at least three points.

79. $f(x) = |x|$

80. $f(x) = \sqrt[3]{x}$

81. $f(x) = \sqrt{x}$

82. $f(x) = \dfrac{1}{x}$

In Problems 83–94, graph each function using the techniques of shifting, compressing or stretching, and reflections. Identify any intercepts on the graph. State the domain and, based on the graph, find the range.

83. $F(x) = |x| - 4$

84. $f(x) = |x| + 4$

85. $g(x) = -2|x|$

86. $g(x) = \dfrac{1}{2}|x|$

87. $h(x) = \sqrt{x - 1}$

88. $h(x) = \sqrt{x} - 1$

89. $f(x) = \sqrt{1 - x}$

90. $f(x) = -\sqrt{x + 3}$

91. $h(x) = (x - 1)^2 + 2$

92. $h(x) = (x + 2)^2 - 3$

93. $g(x) = 3(x - 1)^3 + 1$

94. $g(x) = -2(x + 2)^3 - 8$

In Problems 95–98,

(a) Find the domain of each function.

(b) Locate any intercepts.

(c) Graph each function.

(d) Based on the graph, find the range.

(e) Is f continuous on its domain?

95. $f(x) = \begin{cases} 3x & \text{if } -2 < x \le 1 \\ x + 1 & \text{if } x > 1 \end{cases}$

96. $f(x) = \begin{cases} x - 1 & \text{if } -3 < x < 0 \\ 3x - 1 & \text{if } x \ge 0 \end{cases}$

97. $f(x) = \begin{cases} x & \text{if } -4 \le x < 0 \\ 1 & \text{if } x = 0 \\ 3x & \text{if } x > 0 \end{cases}$

98. $f(x) = \begin{cases} x^2 & \text{if } -2 \le x \le 2 \\ 2x - 1 & \text{if } x > 2 \end{cases}$

99. A function f is defined by

$$f(x) = \frac{Ax + 5}{6x - 2}$$

If $f(1) = 4$, find A.

100. A function g is defined by

$$g(x) = \frac{A}{x} + \frac{8}{x^2}$$

If $g(-1) = 0$, find A.

In Problems 101 and 102, (a) verify that the function is one-to-one, and (b) find the inverse of the given function.

101. $\{(1, 2), (3, 5), (5, 8), (6, 10)\}$

102. $\{(-1, 4), (0, 2), (1, 5), (3, 7)\}$

In Problems 103 and 104, state why the graph of the function is one-to-one. Then draw the graph of the inverse function f^{-1}. For convenience (and as a hint), the graph of $y = x$ is also given.

103.

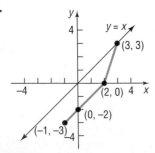

104.

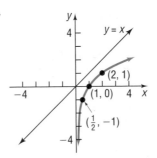

In Problems 105–110, the function f is one-to-one. Find the inverse of each function and check your answer. Find the domain and the range of f and f^{-1}.

105. $f(x) = \dfrac{2x + 3}{5x - 2}$

106. $f(x) = \dfrac{2 - x}{3 + x}$

107. $f(x) = \dfrac{1}{x - 1}$

108. $f(x) = \sqrt{x - 2}$

109. $f(x) = \dfrac{3}{x^{1/3}}$

110. $f(x) = x^{1/3} + 1$

CHAPTER TEST

In Problems 1–2, use $P_1 = (-1, 3)$ and $P_2 = (5, -1)$.

1. Find the distance from P_1 to P_2.

2. Find the midpoint of the line segment joining P_1 and P_2.

3. Graph $y = x^2 - 9$ by plotting points.

4. Sketch the graph of $y^2 = x$.

5. List the intercepts and test for symmetry: $x^2 + y = 9$.

6. Write the standard form of the circle with center $(4, -3)$ and radius 5.

7. Find the center and radius of the circle
$x^2 + y^2 + 4x - 2y - 4 = 0$. Graph this circle.

8. Determine whether each relation represents a function. For each function, state the domain and the range.
   (a) $\{(2, 5), (4, 6), (6, 7), (8, 8)\}$
   (b) $\{(1, 3), (4, -2), (-3, 5), (1, 7)\}$
   (c)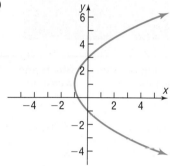
   (d)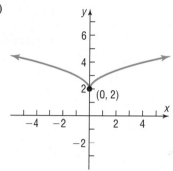

In Problems 9–11, find the domain of each function and evaluate each function at $x = -1$.

9. $f(x) = \sqrt{4 - 5x}$

10. $g(x) = \dfrac{x + 2}{|x + 2|}$

11. $h(x) = \dfrac{x - 4}{x^2 + 5x - 36}$

12. Using the graph of the function f:

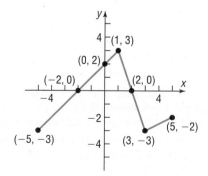

   (a) Find the domain and the range of f.
   (b) List the intercepts.
   (c) Find $f(1)$.
   (d) For what value(s) of x does $f(x) = -3$?
   (e) Solve $f(x) < 0$.

13. Use a graphing utility to graph the function $f(x) = -x^4 + 2x^3 + 4x^2 - 2$ on the interval $(-5, 5)$. Approximate any local maxima and local minima rounded to two decimal places. Determine where the function is increasing and where it is decreasing.

14. Consider the function $g(x) = \begin{cases} 2x + 1 & \text{if } x < -1 \\ x - 4 & \text{if } x \geq -1 \end{cases}$

   (a) Graph the function.
   (b) List the intercepts.
   (c) Find $g(-5)$.
   (d) Find $g(2)$.

15. For the function $f(x) = 3x^2 - 2x + 4$, find the average rate of change of f from 3 to 4.

16. Graph each function using the techniques of shifting, compressing or stretching, and reflections. Start with the graph of the basic function and show all stages.
   (a) $h(x) = -2(x + 1)^3 + 3$
   (b) $g(x) = |x + 4| + 2$

17. Find the inverse of $f(x) = \dfrac{2}{3x - 5}$ and check your answer. State the domain and the range of f and f^{-1}.

18. If the point $(3, -5)$ is on the graph of a one-to-one function f, what point must be on the graph of f^{-1}?

CHAPTER PROJECTS

I. Cell Phone Service This month, as you were paying your bills, you noticed that your cell phone service contract had expired. Since cell phone numbers are now portable, you decided to investigate different companies. Since you regularly travel outside your local calling area, you decide to look only at plans with no roaming charges. Here are your findings:

	Anytime Minutes Included	Charge for Each Extra Minute	Mobile-to-Mobile Minutes	National Long Distance	Nights and Weekends
Company A:					
Plan A1: $39.99	450 with rollover	$0.45	Unlimited	Included	5000 (9 PM–6 AM)
Plan A2: $59.99	900 with rollover	$0.40	Unlimited	Included	Unlimited (9 PM–6 AM)
Company B:					
Plan B1: $39.99	600	$0.40	None	Included	Unlimited (9 PM–7 AM)
Plan B2: $49.99	1000	$0.40	None	Included	Unlimited (9 PM–7 AM)
Company C:					
Plan C1: $59.99	550	551–1050, $5 for each 50 minutes; Above 1050, $0.10 each minute	Unlimited	Included	Unlimited (7 PM–7 AM)
Plan C2: $69.99	800	801–1300, $5 for each 50 minutes; Above 1300, $0.10 each minute	Unlimited	Included	Unlimited (7 PM–7 AM)

Source: Based on rates from the Web sites of the companies Cingular, T Mobile, and Sprint PCS for the zip code 76201 on May 20, 2006. (*www.cingular.com, www.t-mobile.com, www.sprintpcs.com*)

Each plan requires a 2-year contract.

1. Determine the total cost of each plan for the life of the contract, assuming that you stay within the allotted anytime minutes provided by each contract.

2. If you expect to use 400 anytime, 200 mobile-to-mobile, and 4500 night and weekend minutes per month, which plan provides the best deal? If you expect to use 400 anytime, 200 mobile-to-mobile, and 5500 night and weekend minutes per month, which plan provides the best deal? If you expect to use 500 anytime, and 1000 mobile-to-mobile, and 2000 night and weekend minutes per month, which plan provides the best deal?

3. Ignoring any mobile-to-mobile and night and weekend usage, if you expect to use 850 anytime minutes each month, which option provides the best deal? What if you use 1050 anytime minutes per month?

4. Each monthly charge includes a specific number of peak time minutes in the monthly fee. Write a function for each option, where C is the monthly cost and x is the number of anytime minutes used.

5. Graph each function from part 4.

6. For each of the companies A, B, and C, determine the average price per minute for each plan, based on no extra minutes used. For each company, which plan is better?

7. Looking at the three plans that you found to be the best for companies A, B, and C, in part 6, which of these three seems to be the best deal?

8. Based on your own cell phone usage, which plan would be the best for you?

The following projects are available on the Instructor's Resource Center (IRC):

II. Project at Motorola: *Wireless Internet Service* Use functions and their graphs to analyze the total cost of various wireless Internet service plans.

III. Cost of Cable When government regulations and customer preference influence the path of a new cable line, the Pythagorean Theorem can be used to assess the cost of installation.

IV. Oil Spill Functions are used to analyze the size and spread of an oil spill from a leaking tanker.

Trigonometric Functions

2

Surf's Up: Using Models To Predict Huge Waves

GALVESTON, Feb. 15, 2005—If you are a ship captain and there might be 50-foot waves headed your way, you would appreciate some information about them, right?

That's the idea behind a wave model system a Texas A&M University at Galveston professor has developed. His detailed wave prediction system is currently in use in the Gulf of Mexico and the Gulf of Maine.

Vijay Panchang, head of the Department of Maritime Systems Engineering, doesn't make waves; he predicts what they will do, when they will do it, and how high they will get.

Using data provided daily from NOAA [National Oceanic and Atmospheric Administration] and his own complex mathematical models, Panchang and research engineer Doncheng Li provide daily wave model predictions for much of the Texas coast, the Gulf of Mexico, and the Gulf of Maine. Their simulations, updated every 12 hours, provide a forecast for two days ahead.

Because the models use wind data, tsunamis that are created by undersea earthquakes cannot be predicted. But that is not to say his modeling system does not come up with some big waves.

His wave model predicted big waves in November 2003 in the Gulf of Maine, and it was accurate; waves as high as 30 feet were recorded during one storm even in coastal regions.

Last summer during Hurricane Ivan, a buoy located 60 miles south of the Alabama coast recorded a whopping 60-foot wave. "There may have been higher waves because right after recording the 60-foot wave, the buoy snapped and stopped functioning," he says.

Source: Science Daily www.sciencedaily.com/releases/2005/02/050222193810.htm. Posted February 23, 2005.

—See Chapter Project I—

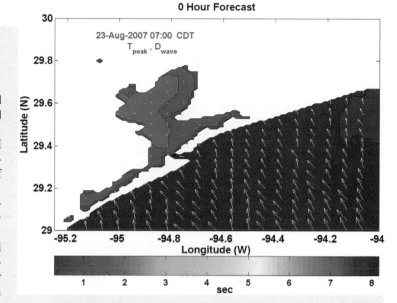

A Look Back

In Chapter 1, we began our discussion of functions. We defined domain and range and independent and dependent variables; we found the value of a function and graphed functions. We continued our study of functions by listing properties that a function might have, like being even or odd, and we created a library of functions, naming key functions and listing their properties, including the graph.

A Look Ahead

In this chapter we define the trigonometric functions, six functions that have wide application. We shall talk about their domain and range, see how to find values, graph them, and develop a list of their properties.

There are two widely accepted approaches to the development of the trigonometric functions: one uses right triangles; the other uses circles, especially the unit circle. In this book, we develop the trigonometric functions using right triangles. In Section 2.5, we introduce trigonometric functions using the unit circle and show that this approach leads to the definition using right triangles.

Outline

2.1 Angles and Their Measure

PREPARING FOR THIS SECTION *Before getting started, review the following:*

- Circumference and Area of a Circle (Appendix, Section A.2, p. A16)

Now Work the 'Are You Prepared?' problems on page 113.

OBJECTIVES 1 Convert between Decimals and Degrees, Minutes, Seconds Forms for Angles (p. 106)
2 Find the Arc Length of a Circle (p. 108)
3 Convert from Degrees to Radians and from Radians to Degrees (p. 108)
4 Find the Area of a Sector of a Circle (p. 111)
5 Find the Linear Speed of an Object Traveling in Circular Motion (p. 112)

Figure 1

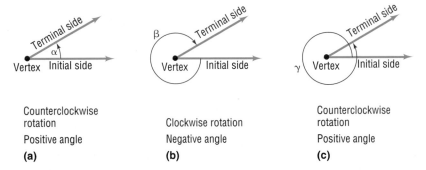

A **ray,** or **half-line,** is that portion of a line that starts at a point V on the line and extends indefinitely in one direction. The starting point V of a ray is called its **vertex.** See Figure 1.

If two rays are drawn with a common vertex, they form an **angle.** We call one ray of an angle the **initial side** and the other the **terminal side.** The angle formed is identified by showing the direction and amount of rotation from the initial side to the terminal side. If the rotation is in the counterclockwise direction, the angle is **positive;** if the rotation is clockwise, the angle is **negative.** See Figure 2.

Figure 2

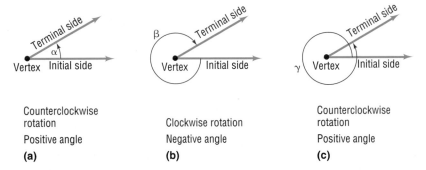

Lowercase Greek letters, such as α (alpha), β (beta), γ (gamma), and θ (theta), will often be used to denote angles. Notice in Figure 2(a) that the angle α is positive because the direction of the rotation from the initial side to the terminal side is counterclockwise. The angle β in Figure 2(b) is negative because the rotation is clockwise. The angle γ in Figure 2(c) is positive. Notice that the angle α in Figure 2(a) and the angle γ in Figure 2(c) have the same initial side and the same terminal side. However, α and γ are unequal, because the amount of rotation required to go from the initial side to the terminal side is greater for angle γ than for angle α.

An angle θ is said to be in **standard position** if its vertex is at the origin of a rectangular coordinate system and its initial side coincides with the positive x-axis. See Figure 3.

Figure 3

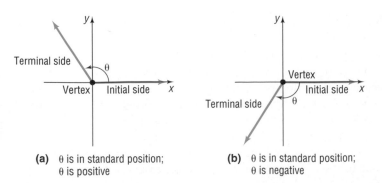

When an angle θ is in standard position, the terminal side will lie either in a quadrant, in which case we say that θ **lies in that quadrant,** or the terminal side will lie on the x-axis or the y-axis, in which case we say that θ is a **quadrantal angle.** For example, the angle θ in Figure 4(a) lies in quadrant II, the angle θ in Figure 4(b) lies in quadrant IV, and the angle θ in Figure 4(c) is a quadrantal angle.

Figure 4

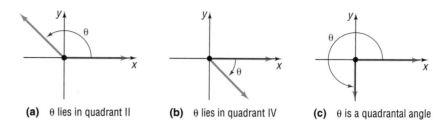

(a) θ lies in quadrant II   **(b)** θ lies in quadrant IV   **(c)** θ is a quadrantal angle

We measure angles by determining the amount of rotation needed for the initial side to become coincident with the terminal side. The two commonly used measures for angles are *degrees* and *radians*.

Degrees

HISTORICAL NOTE One counterclockwise rotation is 360° due to the Babylonian year, which had 360 days in it. ■

The angle formed by rotating the initial side exactly once in the counterclockwise direction until it coincides with itself (1 revolution) is said to measure 360 degrees, abbreviated 360°. **One degree, 1°,** is $\dfrac{1}{360}$ revolution. A **right angle** is an angle that measures 90°, or $\dfrac{1}{4}$ revolution; a **straight angle** is an angle that measures 180°, or $\dfrac{1}{2}$ revolution. See Figure 5. As Figure 5(b) shows, it is customary to indicate a right angle by using the symbol ∟.

Figure 5

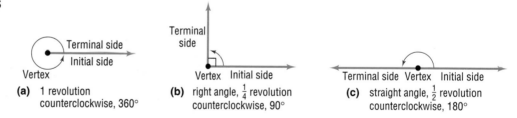

(a) 1 revolution counterclockwise, 360°   **(b)** right angle, $\frac{1}{4}$ revolution counterclockwise, 90°   **(c)** straight angle, $\frac{1}{2}$ revolution counterclockwise, 180°

It is also customary to refer to an angle that measures θ degrees as an angle *of* θ degrees.

EXAMPLE 1   **Drawing an Angle**

Draw each angle.

(a) 45°   (b) −90°   (c) 225°   (d) 405°

Solution   (a) An angle of 45° is $\dfrac{1}{2}$ of a right angle.   (b) An angle of −90° is $\dfrac{1}{4}$ revolution in See Figure 6.   the clockwise direction. See Figure 7.

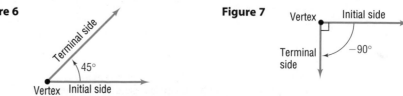

Figure 6   **Figure 7**

(c) An angle of 225° consists of a rotation through 180° followed by a rotation through 45°. See Figure 8.

(d) An angle of 405° consists of 1 revolution (360°) followed by a rotation through 45°. See Figure 9.

Figure 8

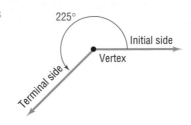

Figure 9

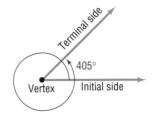

── **Now Work** PROBLEM 11

1 Convert between Decimals and Degrees, Minutes, Seconds Forms for Angles

Although subdivisions of a degree may be obtained by using decimals, we also may use the notion of *minutes* and *seconds*. **One minute,** denoted by **1′**, is defined as $\frac{1}{60}$ degree. **One second,** denoted by **1″**, is defined as $\frac{1}{60}$ minute, or equivalently, $\frac{1}{3600}$ degree. An angle of, say, 30 degrees, 40 minutes, 10 seconds is written compactly as 30°40′10″. To summarize:

$$\begin{array}{c} 1 \text{ counterclockwise revolution} = 360° \\ 1° = 60' \qquad 1' = 60'' \end{array} \tag{1}$$

It is sometimes necessary to convert from the degree, minute, second notation (D°M′S″) to a decimal form, and vice versa. Check your calculator; it should be capable of doing the conversion for you.

Before using your calculator, though, you must set the mode to degrees because there are two common ways to measure angles: degree mode and radian mode. (We will define radians shortly.) Usually, a menu is used to change from one mode to another. Check your owner's manual to find out how your particular calculator works.

Now let's see how to convert from the degree, minute, second notation (D°M′S″) to a decimal form, and vice versa, by looking at some examples:

$$15°30' = 15.5° \quad \text{because} \quad 30' = 30 \cdot 1' = 30 \cdot \left(\frac{1}{60}\right)° = 0.5°$$

$$1' = \left(\frac{1}{60}\right)°$$

$$32.25° = 32°15' \quad \text{because} \quad 0.25° = \left(\frac{1}{4}\right)° = \frac{1}{4} \cdot 1° = \frac{1}{4}(60') = 15'$$

$$1° = 60'$$

| EXAMPLE 2 | Converting between Degrees, Minutes, Seconds, and Decimal Forms |

(a) Convert 50°6′21″ to a decimal in degrees. Round the answer to four decimal places.

(b) Convert 21.256° to the D°M′S″ form. Round the answer to the nearest second.

Algebraic Solution

(a) Because $1' = \left(\dfrac{1}{60}\right)^{\circ}$ and $1'' = \left(\dfrac{1}{60}\right)' = \left(\dfrac{1}{60} \cdot \dfrac{1}{60}\right)^{\circ}$, we convert as follows:

$$
\begin{aligned}
50°6'21'' &= 50° + 6' + 21'' \\
&= 50° + 6 \cdot \left(\dfrac{1}{60}\right)^{\circ} + 21 \cdot \left(\dfrac{1}{60} \cdot \dfrac{1}{60}\right)^{\circ} \\
&\approx 50° + 0.1° + 0.0058° \\
&= 50.1058°
\end{aligned}
$$

(b) We proceed as follows:

$$
\begin{aligned}
21.256° &= 21° + 0.256° \\
&= 21° + (0.256)(60') \qquad \text{\small Convert fraction of degree to} \\
&\qquad\qquad\qquad\qquad \text{\small minutes; } 1° = 60'. \\
&= 21° + 15.36' \\
&= 21° + 15' + 0.36' \\
&= 21° + 15' + (0.36)(60'') \qquad \text{\small Convert fraction of minute to} \\
&\qquad\qquad\qquad\qquad\qquad \text{\small seconds; } 1' = 60''. \\
&= 21° + 15' + 21.6'' \\
&\approx 21°15'22'' \qquad\qquad\qquad \text{\small Round to the nearest second.}
\end{aligned}
$$

Graphing Solution

(a) Figure 10 shows the solution using a TI-84 Plus graphing calculator.

Figure 10

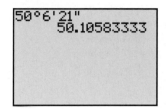

So $50°6'21'' = 51.1058°$ rounded to four decimal places.

(b) Figure 11 shows the solution using a T1-84 Plus graphing calculator.

Figure 11

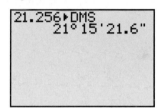

So $21.256° = 21°15'22''$ rounded to the nearest second.

Now Work PROBLEMS **23** AND **29**

 In many applications, such as describing the exact location of a star or the precise position of a ship at sea, angles measured in degrees, minutes, and even seconds are used. For calculation purposes, these are transformed to decimal form. In other applications, especially those in calculus, angles are measured using *radians*.

Radians

A **central angle** is a positive angle whose vertex is at the center of a circle. The rays of a central angle subtend (intersect) an arc on the circle. If the radius of the circle is r and the length of the arc subtended by the central angle is also r, then the measure of the angle is **1 radian.** See Figure 12(a).

For a circle of radius 1, the rays of a central angle with measure 1 radian subtend an arc of length 1. For a circle of radius 3, the rays of a central angle with measure 1 radian subtend an arc of length 3. See Figure 12(b).

Figure 12

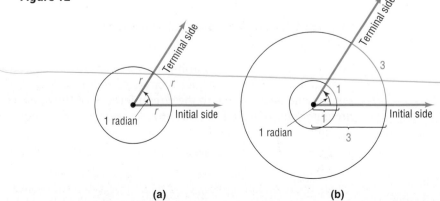

(a)　　　　　　　　(b)

2 Find the Arc Length of a Circle

Now consider a circle of radius r and two central angles, θ and θ_1, measured in radians. Suppose that these central angles subtend arcs of lengths s and s_1, respectively, as shown in Figure 13. From geometry, we know that the ratio of the measures of the angles equals the ratio of the corresponding lengths of the arcs subtended by these angles; that is,

Figure 13

$\dfrac{\theta}{\theta_1} = \dfrac{s}{s_1}$

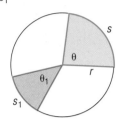

$$\frac{\theta}{\theta_1} = \frac{s}{s_1} \tag{2}$$

Suppose that $\theta_1 = 1$ radian. Refer again to Figure 12(a). The length s_1 of the arc subtended by the central angle $\theta_1 = 1$ radian equals the radius r of the circle. Then $s_1 = r$, so equation (2) reduces to

$$\frac{\theta}{1} = \frac{s}{r} \quad \text{or} \quad s = r\theta \tag{3}$$

THEOREM

Arc Length

For a circle of radius r, a central angle of θ radians subtends an arc whose length s is

$$s = r\theta \tag{4}$$

NOTE Formulas must be consistent with regard to the units used. In equation (4), we write
$$s = r\theta$$
To see the units, however, we must go back to equation (3) and write
$$\frac{\theta \text{ radians}}{1 \text{ radian}} = \frac{s \text{ length units}}{r \text{ length units}}$$
$$s \text{ length units} = r \text{ length units} \frac{\theta \text{ radians}}{1 \text{ radian}}$$
Since the radians cancel, we are left with
$$s \text{ length units} = (r \text{ length units})\theta \quad s = r\theta$$
where θ appears to be "dimensionless" but, in fact, is measured in radians. So, in using the formula $s = r\theta$, the dimension for θ is radians, and any convenient unit of length (such as inches or meters) may be used for s and r. ∎

EXAMPLE 3   Finding the Length of an Arc of a Circle

Find the length of the arc of a circle of radius 2 meters subtended by a central angle of 0.25 radian.

Solution   We use equation (4) with $r = 2$ meters and $\theta = 0.25$. The length s of the arc is
$$s = r\theta = 2(0.25) = 0.5 \text{ meter}$$

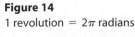 **Now Work** PROBLEM 71

3 Convert from Degrees to Radians and from Radians to Degrees

Figure 14

1 revolution $= 2\pi$ radians

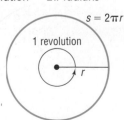

Next we discuss the relationship between angles measured in degrees and angles measured in radians. Consider a circle of radius r. A central angle of 1 revolution will subtend an arc equal to the circumference of the circle (Figure 14). Because the circumference of a circle equals $2\pi r$, we use $s = 2\pi r$ in equation (4) to find that, for an angle θ of 1 revolution,

$$s = r\theta$$
$$2\pi r = r\theta \qquad \theta = 1 \text{ revolution; } s = 2\pi r$$
$$\theta = 2\pi \text{ radians} \qquad \text{Solve for } \theta.$$

From this we have,

$$1 \text{ revolution} = 2\pi \text{ radians} \qquad (5)$$

Since 1 revolution = 360°, we have

$$360° = 2\pi \text{ radians}$$

Dividing both sides by 2 yields

$$180° = \pi \text{ radians} \qquad (6)$$

Divide both sides of equation (6) by 180. Then

$$1 \text{ degree} = \frac{\pi}{180} \text{ radian}$$

Divide both sides of (6) by π. Then

$$\frac{180}{\pi} \text{ degrees} = 1 \text{ radian}$$

We have the following two conversion formulas:

$$1 \text{ degree} = \frac{\pi}{180} \text{ radian} \qquad 1 \text{ radian} = \frac{180}{\pi} \text{ degrees} \qquad (7)$$

| EXAMPLE 4 | **Converting from Degrees to Radians** |

Convert each angle in degrees to radians.
(a) 60°   (b) 150°   (c) −45°   (d) 90°   (e) 107°

Solution (a) $60° = 60 \cdot 1 \text{ degree} = 60 \cdot \frac{\pi}{180} \text{ radian} = \frac{\pi}{3} \text{ radians}$

(b) $150° = 150 \cdot 1° = 150 \cdot \frac{\pi}{180} \text{ radian} = \frac{5\pi}{6} \text{ radians}$

(c) $-45° = -45 \cdot \frac{\pi}{180} \text{ radian} = -\frac{\pi}{4} \text{ radian}$

(d) $90° = 90 \cdot \frac{\pi}{180} \text{ radian} = \frac{\pi}{2} \text{ radians}$

(e) $107° = 107 \cdot \frac{\pi}{180} \text{ radian} \approx 1.868 \text{ radians}$

Example 4, parts (a)–(d), illustrates that angles that are "nice" fractions of a revolution are expressed in radian measure as fractional multiples of π, rather than as decimals. For example, a right angle, as in Example 4(d), is left in the form $\frac{\pi}{2}$ radians, which is exact, rather than using the approximation $\frac{\pi}{2} \approx \frac{3.1416}{2} = 1.5708$ radians. When the fractions are not "nice," we use the decimal approximation of the angle, as in Example 4(e).

Now Work PROBLEMS 35 AND 61

EXAMPLE 5   **Converting Radians to Degrees**

Convert each angle in radians to degrees.

(a) $\dfrac{\pi}{6}$ radian   (b) $\dfrac{3\pi}{2}$ radians   (c) $-\dfrac{3\pi}{4}$ radians

(d) $\dfrac{7\pi}{3}$ radians   (e) 3 radians

Solution   (a) $\dfrac{\pi}{6}$ radian $= \dfrac{\pi}{6} \cdot 1$ radian $= \dfrac{\pi}{6} \cdot \dfrac{180}{\pi}$ degrees $= 30°$

(b) $\dfrac{3\pi}{2}$ radians $= \dfrac{3\pi}{2} \cdot \dfrac{180}{\pi}$ degrees $= 270°$

(c) $-\dfrac{3\pi}{4}$ radians $= -\dfrac{3\pi}{4} \cdot \dfrac{180}{\pi}$ degrees $= -135°$

(d) $\dfrac{7\pi}{3}$ radians $= \dfrac{7\pi}{3} \cdot \dfrac{180}{\pi}$ degrees $= 420°$

(e) 3 radians $= 3 \cdot \dfrac{180}{\pi}$ degrees $\approx 171.89°$

Now Work   PROBLEM 47

Table 1 lists the degree and radian measures of some commonly encountered angles. You should learn to feel equally comfortable using degree or radian measure for these angles.

Table 1

Degrees	0°	30°	45°	60°	90°	120°	135°	150°	180°
Radians	0	$\dfrac{\pi}{6}$	$\dfrac{\pi}{4}$	$\dfrac{\pi}{3}$	$\dfrac{\pi}{2}$	$\dfrac{2\pi}{3}$	$\dfrac{3\pi}{4}$	$\dfrac{5\pi}{6}$	π
Degrees		210°	225°	240°	270°	300°	315°	330°	360°
Radians		$\dfrac{7\pi}{6}$	$\dfrac{5\pi}{4}$	$\dfrac{4\pi}{3}$	$\dfrac{3\pi}{2}$	$\dfrac{5\pi}{3}$	$\dfrac{7\pi}{4}$	$\dfrac{11\pi}{6}$	2π

EXAMPLE 6   **Finding the Distance between Two Cities**

 See Figure 15(a). The latitude of a location L is the angle formed by a ray drawn from the center of Earth to the Equator and a ray drawn from the center of Earth to L. See Figure 15(b). Glasgow, Montana, is due north of Albuquerque, New Mexico. Find the distance between Glasgow (48°9′ north latitude) and Albuquerque (35°5′ north latitude). Assume that the radius of Earth is 3960 miles.

Figure 15

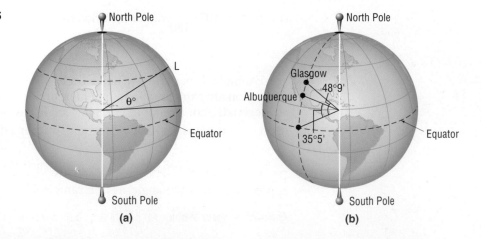

(a)   (b)

Solution   The measure of the central angle between the two cities is $48°9' - 35°5' = 13°4'$. We use equation (4), $s = r\theta$, but first we must convert the angle of $13°4'$ to radians.

$$\theta = 13°4' \approx 13.0667° = 13.0667 \cdot \frac{\pi}{180} \text{ radian} \approx 0.228 \text{ radian}$$

$$4' = \left(\frac{4}{60}\right)°$$

We use $\theta = 0.228$ radian and $r = 3960$ miles in equation (4). The distance between the two cities is

$$s = r\theta = 3960 \cdot 0.228 \approx 903 \text{ miles}$$

NOTE If the measure of an angle is given as 5, it is understood to mean 5 radians; if the measure of an angle is given as 5°, it means 5 degrees. ∎

When an angle is measured in degrees, the degree symbol will always be shown. However, when an angle is measured in radians, we will follow the usual practice and omit the word *radians*. So, if the measure of an angle is given as $\frac{\pi}{6}$, it is understood to mean $\frac{\pi}{6}$ radian.

━━━━ **Now Work**   PROBLEM 101

Figure 16

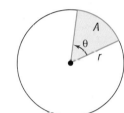

4 Find the Area of a Sector of a Circle

Consider a circle of radius r. Suppose that θ, measured in radians, is a central angle of this circle. See Figure 16. We seek a formula for the area A of the sector (shown in blue) formed by the angle θ.

Now consider a circle of radius r and two central angles θ and θ_1, both measured in radians. See Figure 17. From geometry, we know the ratio of the measures of the angles equals the ratio of the corresponding areas of the sectors formed by these angles. That is,

Figure 17
$$\frac{\theta}{\theta_1} = \frac{A}{A_1}$$

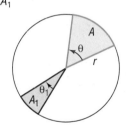

$$\frac{\theta}{\theta_1} = \frac{A}{A_1}$$

Suppose that $\theta_1 = 2\pi$ radians. Then $A_1 = $ area of the circle $= \pi r^2$. Solving for A, we find

$$A = A_1 \frac{\theta}{\theta_1} = \pi r^2 \frac{\theta}{2\pi} = \frac{1}{2} r^2 \theta$$

$$A_1 = \pi r^2$$
$$\theta_1 = 2\pi$$

THEOREM   **Area of a Sector**

The area A of the sector of a circle of radius r formed by a central angle of θ radians is

$$A = \frac{1}{2} r^2 \theta \qquad\qquad \textbf{(8)}$$

EXAMPLE 7   **Finding the Area of a Sector of a Circle**

Find the area of the sector of a circle of radius 2 feet formed by an angle of 30°. Round the answer to two decimal places.

Solution   We use equation (8) with $r = 2$ feet and $\theta = 30° = \frac{\pi}{6}$ radian. [Remember, in equation (8), θ must be in radians.]

$$A = \frac{1}{2} r^2 \theta = \frac{1}{2} (2)^2 \frac{\pi}{6} = \frac{\pi}{3}$$

The area A of the sector is 1.05 square feet, rounded to two decimal places.

━━━━ **Now Work**   PROBLEM 79

5 Find the Linear Speed of an Object Traveling in Circular Motion

We have already defined the average speed of an object as the distance traveled divided by the elapsed time.

DEFINITION

Suppose that an object moves around a circle of radius r at a constant speed. If s is the distance traveled in time t around this circle, then the **linear speed** v of the object is defined as

$$v = \frac{s}{t} \qquad (9)$$

Figure 18
$v = \dfrac{s}{t}$

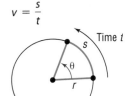

As this object travels around the circle, suppose that θ (measured in radians) is the central angle swept out in time t. See Figure 18.

DEFINITION

The **angular speed** ω (the Greek letter omega) of this object is the angle θ (measured in radians) swept out, divided by the elapsed time t; that is,

$$\omega = \frac{\theta}{t} \qquad (10)$$

Angular speed is the way the turning rate of an engine is described. For example, an engine idling at 900 rpm (revolutions per minute) is one that rotates at an angular speed of

$$900 \frac{\text{revolutions}}{\text{minute}} = 900 \frac{\text{revolutions}}{\text{minute}} \cdot 2\pi \frac{\text{radians}}{\text{revolution}} = 1800\pi \frac{\text{radians}}{\text{minute}}$$

There is an important relationship between linear speed and angular speed:

$$\text{linear speed} = v = \frac{s}{t} = \frac{r\theta}{t} = r\left(\frac{\theta}{t}\right) = r \cdot \omega$$

$$\uparrow \qquad \uparrow \qquad\qquad\qquad \uparrow$$
$$(9) \quad s = r\theta \qquad\qquad (10)$$

So

$$v = r\omega \qquad (11)$$

where ω is measured in radians per unit time.

When using equation (11), remember that $v = \dfrac{s}{t}$ (the linear speed) has the dimensions of length per unit of time (such as feet per second or miles per hour), r (the radius of the circular motion) has the same length dimension as s, and ω (the angular speed) has the dimensions of radians per unit of time. If the angular speed is given in terms of *revolutions* per unit of time (as is often the case), be sure to convert it to *radians* per unit of time using the fact that 1 revolution $= 2\pi$ radians before attempting to use equation (11).

EXAMPLE 8 Finding Linear Speed

A child is spinning a rock at the end of a 2-foot rope at the rate of 180 revolutions per minute (rpm). Find the linear speed of the rock when it is released.

Solution

Look at Figure 19. The rock is moving around a circle of radius $r = 2$ feet. The angular speed ω of the rock is

$$\omega = 180 \frac{\text{revolutions}}{\text{minute}} = 180 \frac{\text{revolutions}}{\text{minute}} \cdot 2\pi \frac{\text{radians}}{\text{revolution}} = 360\pi \frac{\text{radians}}{\text{minute}}$$

Figure 19

From equation (11), the linear speed v of the rock is

$$v = r\omega = 2 \text{ feet} \cdot 360\pi \frac{\text{radians}}{\text{minute}} = 720\pi \frac{\text{feet}}{\text{minute}} \approx 2262 \frac{\text{feet}}{\text{minute}}$$

The linear speed of the rock when it is released is 2262 ft/min $\approx$ 25.7 mi/hr.

Now Work PROBLEM 97

Historical Feature

Trigonometry was developed by Greek astronomers, who regarded the sky as the inside of a sphere, so it was natural that triangles on a sphere were investigated early (by Menelaus of Alexandria about AD 100) and that triangles in the plane were studied much later. The first book containing a systematic treatment of plane and spherical trigonometry was written by the Persian astronomer Nasir Eddin (about AD 1250).

Regiomontanus (1436–1476) is the person most responsible for moving trigonometry from astronomy into mathematics. His work was improved by Copernicus (1473–1543) and Copernicus's student

Rhaeticus (1514–1576). Rhaeticus's book was the first to define the six trigonometric functions as ratios of sides of triangles, although he did not give the functions their present names. Credit for this is due to Thomas Finck (1583), but Finck's notation was by no means universally accepted at the time. The notation was finally stabilized by the textbooks of Leonhard Euler (1707–1783).

Trigonometry has since evolved from its use by surveyors, navigators, and engineers to present applications involving ocean tides, the rise and fall of food supplies in certain ecologies, brain wave patterns, and many other phenomena.

2.1 Assess Your Understanding

'Are You Prepared?' *Answers are given at the end of these exercises. If you get a wrong answer, read the pages listed in red.*

1. What is the formula for the circumference C of a circle of radius r? (p. A16)

2. What is the formula for the area A of a circle of radius r? (p. A16)

Concepts and Vocabulary

3. An angle θ is in _____ _____ if its vertex is at the origin of a rectangular coordinate system and its initial side coincides with the positive x-axis.

4. On a circle of radius r, a central angle of θ radians subtends an arc of length $s =$ _____; the area of the sector formed by this angle θ is $A =$ _____.

5. An object travels around a circle of radius r with constant speed. If s is the distance traveled in time t around the circle and θ is the central angle (in radians) swept out in time t, then the linear speed of the object is $v =$ _____ and the angular speed of the object is $\omega =$ _____.

6. *True or False* $\pi = 180$.

7. *True or False* $180° = \pi$ radians.

8. *True or False* On the unit circle, if s is the length of the arc subtended by a central angle θ, measured in radians, then $s = \theta$.

9. *True or False* The area A of the sector of a circle of radius r formed by a central angle of θ degrees is $A = \frac{1}{2}r^2\theta$.

10. *True or False* For circular motion on a circle of radius r, linear speed equals angular speed divided by r.

Skill Building

In Problems 11–22, draw each angle.

11. $30°$ 12. $60°$ 13. $135°$ 14. $-120°$ 15. $450°$ 16. $540°$

17. $\dfrac{3\pi}{4}$ 18. $\dfrac{4\pi}{3}$ 19. $-\dfrac{\pi}{6}$ 20. $-\dfrac{2\pi}{3}$ 21. $\dfrac{16\pi}{3}$ 22. $\dfrac{21\pi}{4}$

In Problems 23–28, convert each angle to a decimal in degrees. Round your answer to two decimal places.

23. $40°10'25''$ 24. $61°42'21''$ 25. $1°2'3''$ 26. $73°40'40''$ 27. $9°9'9''$ 28. $98°22'45''$

In Problems 29–34, convert each angle to $D°M'S''$ form. Round your answer to the nearest second.

29. $40.32°$ 30. $61.24°$ 31. $18.255°$ 32. $29.411°$ 33. $19.99°$ 34. $44.01°$

In Problems 35–46, convert each angle in degrees to radians. Express your answer as a multiple of π.

35. 30°   **36.** 120°   **37.** 240°   **38.** 330°   **39.** −60°   **40.** −30°

41. 180°   **42.** 270°   **43.** −135°   **44.** −225°   **45.** −90°   **46.** −180°

In Problems 47–58, convert each angle in radians to degrees.

47. $\dfrac{\pi}{3}$ **48.** $\dfrac{5\pi}{6}$ **49.** $-\dfrac{5\pi}{4}$ **50.** $-\dfrac{2\pi}{3}$ **51.** $\dfrac{\pi}{2}$ **52.** 4π

53. $\dfrac{\pi}{12}$ **54.** $\dfrac{5\pi}{12}$ **55.** $-\dfrac{\pi}{2}$ **56.** $-\pi$ **57.** $-\dfrac{\pi}{6}$ **58.** $-\dfrac{3\pi}{4}$

In Problems 59–64, convert each angle in degrees to radians. Express your answer in decimal form, rounded to two decimal places.

59. 17°   **60.** 73°   **61.** −40°   **62.** −51°   **63.** 125°   **64.** 350°

In Problems 65–70, convert each angle in radians to degrees. Express your answer in decimal form, rounded to two decimal places.

65. 3.14   **66.** 0.75   **67.** 2   **68.** 3   **69.** 6.32   **70.** $\sqrt{2}$

In Problems 71–78, s denotes the length of the arc of a circle of radius r subtended by the central angle θ. Find the missing quantity. Round answers to three decimal places.

71. $r = 10$ meters, $\theta = \dfrac{1}{2}$ radian, $s = ?$ **72.** $r = 6$ feet, $\theta = 2$ radians, $s = ?$

73. $\theta = \dfrac{1}{3}$ radian, $s = 2$ feet, $r = ?$ **74.** $\theta = \dfrac{1}{4}$ radian, $s = 6$ centimeters, $r = ?$

75. $r = 5$ miles, $s = 3$ miles, $\theta = ?$ **76.** $r = 6$ meters, $s = 8$ meters, $\theta = ?$

77. $r = 2$ inches, $\theta = 30°$, $s = ?$ **78.** $r = 3$ meters, $\theta = 120°$, $s = ?$

In Problems 79–86, A denotes the area of the sector of a circle of radius r formed by the central angle θ. Find the missing quantity. Round answers to three decimal places.

79. $r = 10$ meters, $\theta = \dfrac{1}{2}$ radian, $A = ?$ **80.** $r = 6$ feet, $\theta = 2$ radians, $A = ?$

81. $\theta = \dfrac{1}{3}$ radian, $A = 2$ square feet, $r = ?$ **82.** $\theta = \dfrac{1}{4}$ radian, $A = 6$ square centimeters, $r = ?$

83. $r = 5$ miles, $A = 3$ square miles, $\theta = ?$ **84.** $r = 6$ meters, $A = 8$ square meters, $\theta = ?$

85. $r = 2$ inches, $\theta = 30°$, $A = ?$ **86.** $r = 3$ meters, $\theta = 120°$, $A = ?$

In Problems 87–90, find the length s and area A. Round answers to three decimal places.

87.

88.

89.

90.

Applications and Extensions

91. Movement of a Minute Hand The minute hand of a clock is 6 inches long. How far does the tip of the minute hand move in 15 minutes? How far does it move in 25 minutes? Round answers to two decimal places.

92. Movement of a Pendulum A pendulum swings through an angle of 20° each second. If the pendulum is 40 inches long, how far does its tip move each second? Round answers to two decimal places.

93. Area of a Sector Find the area of the sector of a circle of radius 4 meters formed by an angle of 45°. Round the answer to two decimal places.

94. Area of a Sector Find the area of the sector of a circle of radius 3 centimeters formed by an angle of 60°. Round the answer to two decimal places.

95. Watering a Lawn   A water sprinkler sprays water over a distance of 30 feet while rotating through an angle of 135°. What area of lawn receives water?

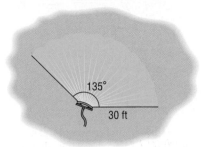

135°
30 ft

96. Designing a Water Sprinkler   An engineer is asked to design a water sprinkler that will cover a field of 100 square yards that is in the shape of a sector of a circle of radius 15 yards. Through what angle should the sprinkler rotate?

97. Motion on a Circle   An object is traveling around a circle with a radius of 5 centimeters. If in 20 seconds a central angle of $\frac{1}{3}$ radian is swept out, what is the angular speed of the object? What is its linear speed?

98. Motion on a Circle   An object is traveling around a circle with a radius of 2 meters. If in 20 seconds the object travels 5 meters, what is its angular speed? What is its linear speed?

99. Bicycle Wheels   The diameter of each wheel of a bicycle is 26 inches. If you are traveling at a speed of 35 miles per hour on this bicycle, through how many revolutions per minute are the wheels turning?

100. Car Wheels   The radius of each wheel of a car is 15 inches. If the wheels are turning at the rate of 3 revolutions per second, how fast is the car moving? Express your answer in inches per second and in miles per hour.

In Problems 101–104, the latitude of a location L is the angle formed by a ray drawn from the center of Earth to the Equator and a ray drawn from the center of Earth to L. See the figure.

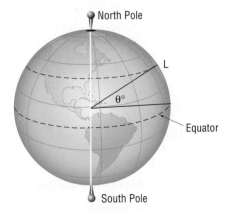

North Pole

L

θ°

Equator

South Pole

101. Distance between Cities   Memphis, Tennessee, is due north of New Orleans, Louisiana. Find the distance between Memphis (35°9′ north latitude) and New Orleans (29°57′ north latitude). Assume that the radius of Earth is 3960 miles.

102. Distance between Cities   Charleston, West Virginia, is due north of Jacksonville, Florida. Find the distance between Charleston (38°21′ north latitude) and Jacksonville (30°20′ north latitude). Assume that the radius of Earth is 3960 miles.

103. Linear Speed on Earth   Earth rotates on an axis through its poles. The distance from the axis to a location on Earth 30° north latitude is about 3429.5 miles. Therefore, a location on Earth at 30° north latitude is spinning on a circle of radius 3429.5 miles. Compute the linear speed on the surface of Earth at 30° north latitude.

104. Linear Speed on Earth   Earth rotates on an axis through its poles. The distance from the axis to a location on Earth 40° north latitude is about 3033.5 miles. Therefore, a location on Earth at 40° north latitude is spinning on a circle of radius 3033.5 miles. Compute the linear speed on the surface of Earth at 40° north latitude.

105. Speed of the Moon   The mean distance of the moon from Earth is 2.39×10^5 miles. Assuming that the orbit of the moon around Earth is circular and that 1 revolution takes 27.3 days, find the linear speed of the moon. Express your answer in miles per hour.

106. Speed of Earth   The mean distance of Earth from the Sun is 9.29×10^7 miles. Assuming that the orbit of Earth around the Sun is circular and that 1 revolution takes 365 days, find the linear speed of Earth. Express your answer in miles per hour.

107. Pulleys   Two pulleys, one with radius 2 inches and the other with radius 8 inches, are connected by a belt. (See the figure.) If the 2-inch pulley is caused to rotate at 3 revolutions per minute, determine the revolutions per minute of the 8-inch pulley.

[**Hint:** The linear speeds of the pulleys are the same; both equal the speed of the belt.]

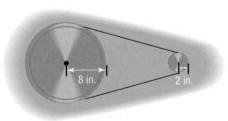

8 in.   2 in.

108. Ferris Wheels   A neighborhood carnival has a Ferris wheel whose radius is 30 feet. You measure the time it takes for one revolution to be 70 seconds. What is the linear speed (in feet per second) of this Ferris wheel? What is the angular speed in radians per second?

109. Computing the Speed of a River Current   To approximate the speed of the current of a river, a circular paddle wheel with radius 4 feet is lowered into the water. If the current causes the wheel to rotate at a speed of 10 revolutions per minute, what is the speed of the current? Express your answer in miles per hour.

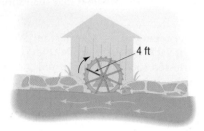

110. **Spin Balancing Tires** A spin balancer rotates the wheel of a car at 480 revolutions per minute. If the diameter of the wheel is 26 inches, what road speed is being tested? Express your answer in miles per hour. At how many revolutions per minute should the balancer be set to test a road speed of 80 miles per hour?

111. **The Cable Cars of San Francisco** At the Cable Car Museum you can see the four cable lines that are used to pull cable cars up and down the hills of San Francisco. Each cable travels at a speed of 9.55 miles per hour, caused by a rotating wheel whose diameter is 8.5 feet. How fast is the wheel rotating? Express your answer in revolutions per minute.

112. **Difference in Time of Sunrise** Naples, Florida, is approximately 90 miles due west of Ft. Lauderdale. How much sooner would a person in Ft. Lauderdale first see the rising Sun than a person in Naples? See the hint.

[**Hint:** Consult the figure. When a person at Q sees the first rays of the Sun, a person at P is still in the dark. The person at P sees the first rays after Earth has rotated so that P is at the location Q. Now use the fact that at the latitude of Ft. Lauderdale in 24 hours an arc of length $2\pi(3559)$ miles is subtended.]

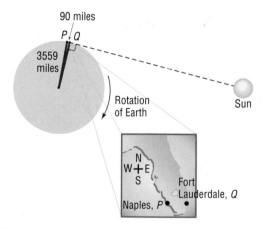

113. **Keeping Up with the Sun** How fast would you have to travel on the surface of Earth at the equator to keep up with the Sun (that is, so that the Sun would appear to remain in the same position in the sky)?

114. **Nautical Miles** A **nautical mile** equals the length of arc subtended by a central angle of 1 minute on a great circle* on the surface of Earth. (See the figure on the top, right of the page.) If the radius of Earth is taken as 3960 miles, express 1 nautical mile in terms of ordinary, or **statute,** miles.

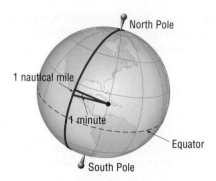

115. **Approximating the Circumference of Earth** Eratosthenes of Cyrene (276–194 BC) was a Greek scholar who lived and worked in Cyrene and Alexandria. One day while visiting in Syene he noticed that the Sun's rays shone directly down a well. On this date 1 year later, in Alexandria, which is 500 miles due north of Syene he measured the angle of the Sun to be about 7.2 degrees. See the figure. Use this information to approximate the radius and circumference of Earth.

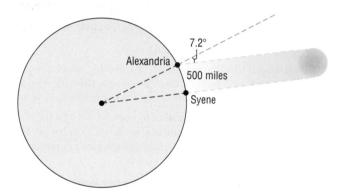

116. **Designing a Little League Field** For a 60-foot Little League Baseball field, the distance from home base to the nearest fence (or other obstruction) on fair territory should be a minimum of 200 feet. The commissioner of parks and recreation is making plans for a new 60-foot field. Because of limited ground availability, he will use the minimum required distance to the outfield fence. To increase safety, however, he plans to include a 10-foot wide warning track on the inside of the fence. To further increase safety, the fence and warning track will extend both directions into foul territory. In total the arc formed by the outfield fence (including the extensions into the foul territories) will be subtended by a central angle at home plate measuring 96°, as illustrated.
(a) Determine the length of the outfield fence.
(b) Determine the area of the warning track.

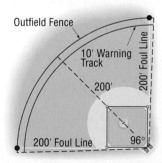

Source: www.littleleague.org

*Any circle drawn on the surface of Earth that divides Earth into two equal hemispheres.

[**Note:** There is a 90° angle between the two foul lines. Then there are two 3° angles between the foul lines and the dotted lines shown. The angle between the two dotted lines outside the 200-foot foul lines is 96°.]

117. Pulleys Two pulleys, one with radius r_1 and the other with radius r_2, are connected by a belt. The pulley with radius r_1 rotates at ω_1 revolutions per minute, whereas the pulley with radius r_2 rotates at ω_2 revolutions per minute. Show that

$$\frac{r_1}{r_2} = \frac{\omega_2}{\omega_1}.$$

Discussion and Writing

118. Do you prefer to measure angles using degrees or radians? Provide justification and a rationale for your choice.

119. What is 1 radian? What is 1°?

120. Which angle has the larger measure: 1 degree or 1 radian? Or are they equal?

121. Explain the difference between linear speed and angular speed.

122. For a circle of radius r, a central angle of θ degrees subtends an arc whose length s is $s = \dfrac{\pi}{180} r\theta$. Discuss whether this is a true or false statement. Give reasons to defend your position.

123. Discuss why ships and airplanes use nautical miles to measure distance. Explain the difference between a nautical mile and a statute mile.

124. Investigate the way that speed bicycles work. In particular, explain the differences and similarities between 5-speed and 9-speed derailleurs. Be sure to include a discussion of linear speed and angular speed.

125. In Example 6, we found that the distance between Albuquerque, New Mexico, and Glasgow, Montana, is approximately 903 miles. According to *mapquest.com*, the distance is approximately 1300 miles. What might account for the difference?

'Are You Prepared?' Answers

1. $C = 2\pi r$ **2.** $A = \pi r^2$

2.2 Right Triangle Trigonometry

PREPARING FOR THIS SECTION   *Before getting started, review the following:*

- Geometry Essentials (Appendix, Section A.2, pp. A14–A19)
- Functions (Section 1.3, pp. 30–39)

Now Work the 'Are You Prepared?' problems on page 125.

OBJECTIVES   **1**   Find the Values of Trigonometric Functions of Acute Angles (p. 117)
               2   Use the Fundamental Identities (p. 119)
               3   Find the Values of the Remaining Trigonometric Functions, Given the Value of One of Them (p. 121)
               4   Use the Complementary Angle Theorem (p. 123)

1 Find the Values of Trigonometric Functions of Acute Angles

A triangle in which one angle is a right angle (90°) is called a **right triangle.** Recall that the side opposite the right angle is called the **hypotenuse,** and the remaining two sides are called the **legs** of the triangle. In Figure 20 we have labeled the hypotenuse as c to indicate that its length is c units, and, in a like manner, we have labeled the legs as a and b. Because the triangle is a right triangle, the Pythagorean Theorem tells us that

$$c^2 = a^2 + b^2$$

Figure 20

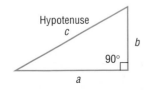

Now, suppose that θ is an **acute angle;** that is, $0° < \theta < 90°$ (if θ is measured in degrees) and $0 < \theta < \dfrac{\pi}{2}$ (if θ is measured in radians). See Figure 21(a) on page 118. Using this acute angle θ, we can form a right triangle, like the one illustrated in

Figure 21(b), with hypotenuse of length c and legs of lengths a and b. Using the three sides of this triangle, we can form exactly six ratios:

$$\frac{b}{c}, \quad \frac{a}{c}, \quad \frac{b}{a}, \quad \frac{c}{b}, \quad \frac{c}{a}, \quad \frac{a}{b}$$

Figure 21

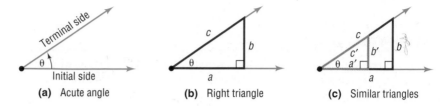

(a) Acute angle   **(b)** Right triangle   **(c)** Similar triangles

In fact, these ratios depend only on the size of the angle θ and not on the triangle formed. To see why, look at Figure 21(c). Any two right triangles formed using the angle θ will be similar and, hence, corresponding ratios will be equal. As a result,

$$\frac{b}{c} = \frac{b'}{c'} \quad \frac{a}{c} = \frac{a'}{c'} \quad \frac{b}{a} = \frac{b'}{a'} \quad \frac{c}{b} = \frac{c'}{b'} \quad \frac{c}{a} = \frac{c'}{a'} \quad \frac{a}{b} = \frac{a'}{b'}$$

Because the ratios depend only on the angle θ and not on the triangle itself, we give each ratio a name that involves θ: sine of θ, cosine of θ, tangent of θ, cosecant of θ, secant of θ, and cotangent of θ.

DEFINITION

The six ratios of a right triangle are called **trigonometric functions of acute angles** and are defined as follows:

Function Name	Abbreviation	Value	Function Name	Abbreviation	Value
sine of θ	$\sin\theta$	$\dfrac{b}{c}$	cosecant of θ	$\csc\theta$	$\dfrac{c}{b}$
cosine of θ	$\cos\theta$	$\dfrac{a}{c}$	secant of θ	$\sec\theta$	$\dfrac{c}{a}$
tangent of θ	$\tan\theta$	$\dfrac{b}{a}$	cotangent of θ	$\cot\theta$	$\dfrac{a}{b}$

Figure 22

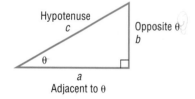

As an aid to remembering these definitions, it may be helpful to refer to the lengths of the sides of the triangle by the names *hypotenuse (c), opposite (b),* and *adjacent (a)*. See Figure 22. In terms of these names, we have the following ratios:

$$\sin\theta = \frac{\text{opposite}}{\text{hypotenuse}} = \frac{b}{c} \qquad \cos\theta = \frac{\text{adjacent}}{\text{hypotenuse}} = \frac{a}{c} \qquad \tan\theta = \frac{\text{opposite}}{\text{adjacent}} = \frac{b}{a}$$

$$\csc\theta = \frac{\text{hypotenuse}}{\text{opposite}} = \frac{c}{b} \qquad \sec\theta = \frac{\text{hypotenuse}}{\text{adjacent}} = \frac{c}{a} \qquad \cot\theta = \frac{\text{adjacent}}{\text{opposite}} = \frac{a}{b} \qquad \textbf{(1)}$$

Since a, b, and c are positive, each of the trigonometric functions of an acute angle θ is positive.

EXAMPLE 1   **Finding the Value of Trigonometric Functions**

Find the value of each of the six trigonometric functions of the angle θ in Figure 23.

Solution   We see in Figure 23 that the two given sides of the triangle are

$$c = \text{hypotenuse} = 5 \qquad a = \text{adjacent} = 3$$

Figure 23

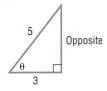

To find the length of the opposite side, we use the Pythagorean Theorem.

$$(\text{adjacent})^2 + (\text{opposite})^2 = (\text{hypotenuse})^2$$

$$3^2 + (\text{opposite})^2 = 5^2$$

$$(\text{opposite})^2 = 25 - 9 = 16$$

$$\text{opposite} = 4$$

WARNING When writing the values of the trigonometric functions, do not forget the argument of the function

$$\sin \theta = \frac{4}{5} \quad correct$$

$$\sin = \frac{4}{5} \quad incorrect \qquad \blacksquare$$

Now that we know the lengths of the three sides, we use the ratios in (1) to find the value of each of the six trigonometric functions:

$$\sin \theta = \frac{\text{opposite}}{\text{hypotenuse}} = \frac{4}{5} \qquad \cos \theta = \frac{\text{adjacent}}{\text{hypotenuse}} = \frac{3}{5} \qquad \tan \theta = \frac{\text{opposite}}{\text{adjacent}} = \frac{4}{3}$$

$$\csc \theta = \frac{\text{hypotenuse}}{\text{opposite}} = \frac{5}{4} \qquad \sec \theta = \frac{\text{hypotenuse}}{\text{adjacent}} = \frac{5}{3} \qquad \cot \theta = \frac{\text{adjacent}}{\text{opposite}} = \frac{3}{4}$$

➤ **Now Work** PROBLEM 11

2 Use the Fundamental Identities

You may have observed some relationships that exist among the six trigonometric functions of acute angles. For example, the **reciprocal identities** are

Reciprocal Identities

$$\csc \theta = \frac{1}{\sin \theta} \qquad \sec \theta = \frac{1}{\cos \theta} \qquad \cot \theta = \frac{1}{\tan \theta} \qquad \textbf{(2)}$$

Two other fundamental identities that are easy to see are the **quotient identities.**

Quotient Identities

$$\tan \theta = \frac{\sin \theta}{\cos \theta} \qquad \cot \theta = \frac{\cos \theta}{\sin \theta} \qquad \textbf{(3)}$$

If $\sin \theta$ and $\cos \theta$ are known, formulas (2) and (3) make it easy to find the values of the remaining trigonometric functions.

EXAMPLE 2 **Finding the Values of the Remaining Trigonometric Functions, Given $\sin \theta$ and $\cos \theta$**

Given $\sin \theta = \dfrac{\sqrt{5}}{5}$ and $\cos \theta = \dfrac{2\sqrt{5}}{5}$, find the value of each of the four remaining trigonometric functions of θ.

Solution Based on formula (3), we have

$$\tan \theta = \frac{\sin \theta}{\cos \theta} = \frac{\dfrac{\sqrt{5}}{5}}{\dfrac{2\sqrt{5}}{5}} = \frac{1}{2}$$

Then we use the reciprocal identities from formula (2) to get

$$\csc\theta = \frac{1}{\sin\theta} = \frac{1}{\frac{\sqrt5}{5}} = \frac{5}{\sqrt5} = \sqrt5 \qquad \sec\theta = \frac{1}{\cos\theta} = \frac{1}{\frac{2\sqrt5}{5}} = \frac{5}{2\sqrt5} = \frac{\sqrt5}{2} \qquad \cot\theta = \frac{1}{\tan\theta} = \frac{1}{\frac12} = 2$$

■

Now Work PROBLEM **21**

Figure 24

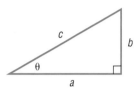

Refer now to the right triangle in Figure 24. The Pythagorean Theorem states that $a^2 + b^2 = c^2$, which we can write as

$$b^2 + a^2 = c^2$$

Dividing each side by c^2, we get

$$\frac{b^2}{c^2} + \frac{a^2}{c^2} = 1 \quad \text{or} \quad \left(\frac{b}{c}\right)^2 + \left(\frac{a}{c}\right)^2 = 1$$

In terms of trigonometric functions of the angle θ, this equation states that

$$(\sin\theta)^2 + (\cos\theta)^2 = 1 \qquad\qquad\textbf{(4)}$$

Equation (4) is an identity, since the equation is true for any acute angle θ.

It is customary to write $\sin^2\theta$ instead of $(\sin\theta)^2$, $\cos^2\theta$ instead of $(\cos\theta)^2$, and so on. With this notation, we can rewrite equation (4) as

$$\boxed{\sin^2\theta + \cos^2\theta = 1 \qquad\qquad\textbf{(5)}}$$

Another identity can be obtained from equation (5) by dividing each side by $\cos^2\theta$.

$$\frac{\sin^2\theta}{\cos^2\theta} + 1 = \frac{1}{\cos^2\theta}$$

Now use formulas (2) and (3) to get

$$\boxed{\tan^2\theta + 1 = \sec^2\theta \qquad\qquad\textbf{(6)}}$$

Similarly, by dividing each side of equation (5) by $\sin^2\theta$, we get $1 + \cot^2\theta = \csc^2\theta$, which we write as

$$\boxed{\cot^2\theta + 1 = \csc^2\theta \qquad\qquad\textbf{(7)}}$$

Collectively, the identities in equations (5), (6), and (7) are referred to as the **Pythagorean Identities.**

Let's pause here to summarize the fundamental identities.

Fundamental Identities

$$\tan\theta = \frac{\sin\theta}{\cos\theta} \qquad \cot\theta = \frac{\cos\theta}{\sin\theta}$$

$$\csc\theta = \frac{1}{\sin\theta} \qquad \sec\theta = \frac{1}{\cos\theta} \qquad \cot\theta = \frac{1}{\tan\theta}$$

$$\sin^2\theta + \cos^2\theta = 1 \qquad \tan^2\theta + 1 = \sec^2\theta \qquad \cot^2\theta + 1 = \csc^2\theta$$

EXAMPLE 3	Finding the Exact Value of a Trigonometric Expression Using Identities

Find the exact value of each expression. Do not use a calculator.

(a) $\tan 20° - \dfrac{\sin 20°}{\cos 20°}$

(b) $\sin^2 \dfrac{\pi}{12} + \dfrac{1}{\sec^2 \dfrac{\pi}{12}}$

Solution   (a) $\tan 20° - \underset{\uparrow}{\dfrac{\sin 20°}{\cos 20°}} = \tan 20° - \tan 20° = 0$

$$\dfrac{\sin \theta}{\cos \theta} = \tan \theta$$

(b) $\sin^2 \dfrac{\pi}{12} + \dfrac{1}{\underset{\uparrow}{\sec^2 \dfrac{\pi}{12}}} = \sin^2 \dfrac{\pi}{12} + \underset{\uparrow}{\cos^2 \dfrac{\pi}{12}} = 1$

$$\cos \theta = \dfrac{1}{\sec \theta} \qquad \sin^2 \theta + \cos^2 \theta = 1$$

━━━━━━ **Now Work**   PROBLEM **39**

3 Find the Values of the Remaining Trigonometric Functions, Given the Value of One of Them

Once the value of one trigonometric function is known, it is possible to find the value of each of the remaining five trigonometric functions.

EXAMPLE 4	Finding the Values of the Remaining Trigonometric Functions, Given sin θ, θ Acute

Given that $\sin \theta = \dfrac{1}{3}$ and θ is an acute angle, find the exact value of each of the remaining five trigonometric functions of θ.

Solution   We solve this problem in two ways: The first way uses the definition of the trigonometric functions; the second method uses the fundamental identities.

Solution 1
Using the Definition
We draw a right triangle with acute angle θ, opposite side of length $b = 1$, and hypotenuse of length $c = 3$ $\left(\text{because } \sin \theta = \dfrac{1}{3} = \dfrac{b}{c}\right)$. See Figure 25. The adjacent side a can be found by using the Pythagorean Theorem.

Figure 25

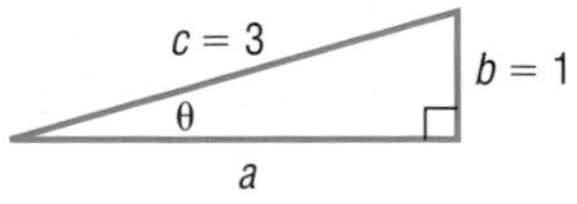

$$a^2 + 1^2 = 3^2 \qquad a^2 + b^2 = c^2; b = 1, c = 3$$
$$a^2 + 1 = 9$$
$$a^2 = 8$$
$$a = 2\sqrt{2}$$

Now the definitions given in equation (1) can be used to find the value of each of the remaining five trigonometric functions. (Refer back to the method used in Example 1.) Using $a = 2\sqrt{2}, b = 1$, and $c = 3$, we have

$$\cos \theta = \dfrac{a}{c} = \dfrac{2\sqrt{2}}{3} \qquad\qquad \tan \theta = \dfrac{b}{a} = \dfrac{1}{2\sqrt{2}} = \dfrac{\sqrt{2}}{4}$$

$$\csc \theta = \dfrac{c}{b} = \dfrac{3}{1} = 3 \quad \sec \theta = \dfrac{c}{a} = \dfrac{3}{2\sqrt{2}} = \dfrac{3\sqrt{2}}{4} \quad \cot \theta = \dfrac{a}{b} = \dfrac{2\sqrt{2}}{1} = 2\sqrt{2}$$

Solution 2
Using Identities

We begin by seeking $\cos \theta$, which can be found by using the Pythagorean Identity from equation (5).

$$\sin^2 \theta + \cos^2 \theta = 1 \qquad \text{Formula (5)}$$

$$\frac{1}{9} + \cos^2 \theta = 1 \qquad \sin \theta = \frac{1}{3}$$

$$\cos^2 \theta = 1 - \frac{1}{9} = \frac{8}{9}$$

Recall that the trigonometric functions of an acute angle are positive. In particular, $\cos \theta > 0$ for an acute angle θ, so we have

$$\cos \theta = \sqrt{\frac{8}{9}} = \frac{2\sqrt{2}}{3}$$

Now we know that $\sin \theta = \frac{1}{3}$ and $\cos \theta = \frac{2\sqrt{2}}{3}$, so we can proceed as we did in Example 2.

$$\tan \theta = \frac{\sin \theta}{\cos \theta} = \frac{\frac{1}{3}}{\frac{2\sqrt{2}}{3}} = \frac{1}{2\sqrt{2}} = \frac{\sqrt{2}}{4} \qquad \cot \theta = \frac{1}{\tan \theta} = \frac{1}{\frac{\sqrt{2}}{4}} = \frac{4}{\sqrt{2}} = 2\sqrt{2}$$

$$\sec \theta = \frac{1}{\cos \theta} = \frac{1}{\frac{2\sqrt{2}}{3}} = \frac{3}{2\sqrt{2}} = \frac{3\sqrt{2}}{4} \qquad \csc \theta = \frac{1}{\sin \theta} = \frac{1}{\frac{1}{3}} = 3$$

Finding the Values of the Trigonometric Functions When One Is Known

Given the value of one trigonometric function of an acute angle θ, the exact value of each of the remaining five trigonometric functions of θ can be found in either of two ways.

Method 1 Using the Definition

STEP 1: Draw a right triangle showing the acute angle θ.

STEP 2: Two of the sides can then be assigned values based on the value of the given trigonometric function.

STEP 3: Find the length of the third side by using the Pythagorean Theorem.

STEP 4: Use the definitions in equation (1) to find the value of each of the remaining trigonometric functions.

Method 2 Using Identities

Use appropriately selected identities to find the value of each of the remaining trigonometric functions.

EXAMPLE 5

Given One Value of a Trigonometric Function, Find the Values of the Remaining Ones

Given $\tan \theta = \frac{1}{2}$, θ an acute angle, find the exact value of each of the remaining five trigonometric functions of θ.

Solution 1
Using the Definition

Figure 26 shows a right triangle with acute angle θ, where

$$\tan \theta = \frac{1}{2} = \frac{\text{opposite}}{\text{adjacent}} = \frac{b}{a}$$

Figure 26

$$\tan \theta = \frac{1}{2}$$

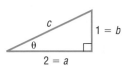

With $b = 1$ and $a = 2$, the hypotenuse c can be found by using the Pythagorean Theorem.

$$c^2 = a^2 + b^2 = 2^2 + 1^2 = 5$$
$$c = \sqrt{5}$$

Now apply the definitions using $a = 2$, $b = 1$, and $c = \sqrt{5}$.

$$\sin \theta = \frac{b}{c} = \frac{1}{\sqrt{5}} = \frac{\sqrt{5}}{5} \qquad \cos \theta = \frac{a}{c} = \frac{2}{\sqrt{5}} = \frac{2\sqrt{5}}{5}$$

$$\csc \theta = \frac{c}{b} = \frac{\sqrt{5}}{1} = \sqrt{5} \qquad \sec \theta = \frac{c}{a} = \frac{\sqrt{5}}{2} \qquad \cot \theta = \frac{a}{b} = \frac{2}{1} = 2$$

Solution 2
Using Identities

Because we know the value of $\tan \theta$, we use the Pythagorean Identity that involves $\tan \theta$:

$$\tan^2 \theta + 1 = \sec^2 \theta \qquad \text{Formula (6)}$$

$$\left(\frac{1}{2}\right)^2 + 1 = \sec^2 \theta \qquad \tan \theta = \frac{1}{2}$$

$$\sec^2 \theta = \frac{1}{4} + 1 = \frac{5}{4} \qquad \text{Proceed to solve for sec } \theta.$$

$$\sec \theta = \frac{\sqrt{5}}{2} \qquad \sec \theta > 0 \quad \text{since } \theta \text{ is acute.}$$

Now we know $\tan \theta = \frac{1}{2}$ and $\sec \theta = \frac{\sqrt{5}}{2}$. Using reciprocal identities, we find

$$\cos \theta = \frac{1}{\sec \theta} = \frac{1}{\dfrac{\sqrt{5}}{2}} = \frac{2}{\sqrt{5}} = \frac{2\sqrt{5}}{5}$$

$$\cot \theta = \frac{1}{\tan \theta} = \frac{1}{\dfrac{1}{2}} = 2$$

To find $\sin \theta$, we use the following reasoning:

$$\tan \theta = \frac{\sin \theta}{\cos \theta}, \quad \text{so} \quad \sin \theta = (\tan \theta)(\cos \theta) = \frac{1}{2} \cdot \frac{2\sqrt{5}}{5} = \frac{\sqrt{5}}{5}$$

$$\csc \theta = \frac{1}{\sin \theta} = \frac{1}{\dfrac{\sqrt{5}}{5}} = \sqrt{5}$$

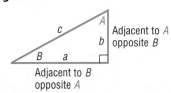 **Now Work** PROBLEM **25**

4 Use the Complementary Angle Theorem

Two acute angles are called **complementary** if their sum is a right angle, or 90°. Because the sum of the angles of any triangle is 180°, it follows that, for a right triangle, the two acute angles are complementary, so the sum of the acute angles in a right triangle is 90°.

Refer now to Figure 27; we have labeled the angle opposite side b as B and the angle opposite side a as A. Notice that side a is adjacent to angle B and is opposite angle A. Similarly, side b is opposite angle B and is adjacent to angle A. As a result,

Figure 27

Adjacent to A
opposite B

Adjacent to B
opposite A

$$\sin B = \frac{b}{c} = \cos A \qquad \cos B = \frac{a}{c} = \sin A \qquad \tan B = \frac{b}{a} = \cot A$$

$$\csc B = \frac{c}{b} = \sec A \qquad \sec B = \frac{c}{a} = \csc A \qquad \cot B = \frac{a}{b} = \tan A$$

(8)

Because of these relationships, the functions sine and cosine, tangent and cotangent, and secant and cosecant are called **cofunctions** of each other. The identities (8) may be expressed in words as follows:

THEOREM

Complementary Angle Theorem

Cofunctions of complementary angles are equal.

Here are examples of this theorem.

Complementary angles
$$\sin 30° = \cos 60°$$
Cofunctions

Complementary angles
$$\tan 40° = \cot 50°$$
Cofunctions

Complementary angles
$$\sec 80° = \csc 10°$$
Cofunctions

If an angle θ is measured in degrees, we will use the degree symbol when writing a trigonometric function of θ, as, for example, in $\sin 30°$ and $\tan 45°$. If an angle θ is measured in radians, then no symbol is used when writing a trigonometric function of θ, as, for example, in $\cos \pi$ and $\sec \dfrac{\pi}{3}$.

If θ is an acute angle measured in degrees, the angle $90° - \theta$ (or $\dfrac{\pi}{2} - \theta$, if θ is in radians) is the angle complementary to θ. Table 2 restates the preceding theorem on cofunctions.

Table 2

θ (Degrees)	θ (Radians)
$\sin \theta = \cos(90° - \theta)$	$\sin \theta = \cos\left(\dfrac{\pi}{2} - \theta\right)$
$\cos \theta = \sin(90° - \theta)$	$\cos \theta = \sin\left(\dfrac{\pi}{2} - \theta\right)$
$\tan \theta = \cot(90° - \theta)$	$\tan \theta = \cot\left(\dfrac{\pi}{2} - \theta\right)$
$\csc \theta = \sec(90° - \theta)$	$\csc \theta = \sec\left(\dfrac{\pi}{2} - \theta\right)$
$\sec \theta = \csc(90° - \theta)$	$\sec \theta = \csc\left(\dfrac{\pi}{2} - \theta\right)$
$\cot \theta = \tan(90° - \theta)$	$\cot \theta = \tan\left(\dfrac{\pi}{2} - \theta\right)$

The angle θ in Table 2 is acute. We will see later (Section 3.4) that these results are valid for any angle θ.

EXAMPLE 6 | Using the Complementary Angle Theorem

(a) $\sin 62° = \cos(90° - 62°) = \cos 28°$

(b) $\tan \dfrac{\pi}{12} = \cot\left(\dfrac{\pi}{2} - \dfrac{\pi}{12}\right) = \cot \dfrac{5\pi}{12}$

(c) $\cos \dfrac{\pi}{4} = \sin\left(\dfrac{\pi}{2} - \dfrac{\pi}{4}\right) = \sin \dfrac{\pi}{4}$

(d) $\csc \dfrac{\pi}{6} = \sec\left(\dfrac{\pi}{2} - \dfrac{\pi}{6}\right) = \sec \dfrac{\pi}{3}$

EXAMPLE 7   **Using the Complementary Angle Theorem**

Find the exact value of each expression. Do not use a calculator.

(a) $\sec 28° - \csc 62°$

(b) $\dfrac{\sin 35°}{\cos 55°}$

Solution   (a) $\sec 28° - \csc 62° = \csc(90° - 28°) - \csc 62°$

$$= \csc 62° - \csc 62° = 0$$

(b) $\dfrac{\sin 35°}{\cos 55°} = \dfrac{\cos(90° - 35°)}{\cos 55°} = \dfrac{\cos 55°}{\cos 55°} = 1$

══════ **Now Work** PROBLEM 43

Historical Feature

The name *sine* for the sine function is due to a medieval confusion. The name comes from the Sanskrit word *jīva*, (meaning *chord*), first used in India by Araybhata the Elder (AD 510). He really meant half-chord, but abbreviated it. This was brought into Arabic as *jība*, which was meaningless. Because the proper Arabic word *jaib* would be written the same way (short vowels are not written out in Arabic), *jība*, was pronounced as *jaib*, which meant bosom or hollow, and *jaib* remains as the Arabic word for sine to this day. Scholars translating the Arabic works into Latin found that the word *sinus* also meant bosom or hollow, and from *sinus* we get the word *sine*.

The name *tangent*, due to Thomas Finck (1583), can be understood by looking at Figure 28. The line segment $\overline{DC}$ is tangent to the circle at C. If $d(O, B) = d(O, C) = 1$, then the length of the line segment $\overline{DC}$ is

$$d(D, C) = \frac{d(D, C)}{1} = \frac{d(D, C)}{d(O, C)} = \tan \alpha$$

The old name for the tangent is *umbra versa* (meaning turned shadow), referring to the use of the tangent in solving height problems with shadows.

Figure 28

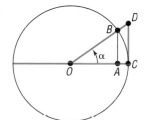

The names of the cofunctions came about as follows. If A and B are complementary angles, then $\cos A = \sin B$. Because B is the complement of A, it was natural to write the cosine of A as *sin co A*. Probably for reasons involving ease of pronunciation, the *co* migrated to the front, and then cosine received a three-letter abbreviation to match sin, sec, and tan. The two other cofunctions were similarly treated, except that the long forms *cotan* and *cosec* survive to this day in some countries.

2.2 Assess Your Understanding

'Are You Prepared?' *Answers are given at the end of these exercises. If you get a wrong answer, read the pages listed in red.*

1. In a right triangle with legs $a = 6$ and $b = 10$, the Pythagorean Theorem tells us that the hypotenuse $c =$ _____. (p. A14)

2. The value of the function $f(x) = 3x - 7$ at 5 is _____. (pp. 34–36)

Concepts and Vocabulary

3. Two acute angles whose sum is a right angle are called _____.

4. The sine and _____ functions are cofunctions.

5. $\tan 28° = \cot$ _____.

6. For any angle θ, $\sin^2 \theta + \cos^2 \theta =$ _____,

7. *True or False* $\tan \theta = \dfrac{\sin \theta}{\cos \theta}$.

8. *True or False* $1 + \tan^2 \theta = \csc^2 \theta$.

9. *True or False* If θ is an acute angle and $\sec \theta = 3$, then $\cos \theta = \dfrac{1}{3}$.

10. *True or False* $\tan \dfrac{\pi}{5} = \cot \dfrac{4\pi}{5}$.

Skill Building

In Problems 11–20, find the value of the six trigonometric functions of the angle θ in each figure.

11.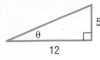

12.

13.

14.

15.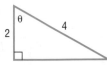

16.

17.

18.

19.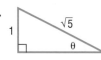

20.

In Problems 21–24, use identities to find the exact value of each of the four remaining trigonometric functions of the acute angle θ.

21. $\sin \theta = \dfrac{1}{2}, \quad \cos \theta = \dfrac{\sqrt{3}}{2}$ **22.** $\sin \theta = \dfrac{\sqrt{3}}{2}, \quad \cos \theta = \dfrac{1}{2}$ **23.** $\sin \theta = \dfrac{2}{3}, \quad \cos \theta = \dfrac{\sqrt{5}}{3}$ **24.** $\sin \theta = \dfrac{1}{3}, \quad \cos \theta = \dfrac{2\sqrt{2}}{3}$

In Problems 25–36, use the definition or identities to find the exact value of each of the remaining five trigonometric functions of the acute angle θ.

25. $\sin \theta = \dfrac{\sqrt{2}}{2}$ **26.** $\cos \theta = \dfrac{\sqrt{2}}{2}$ **27.** $\cos \theta = \dfrac{1}{3}$ **28.** $\sin \theta = \dfrac{\sqrt{3}}{4}$

29. $\tan \theta = \dfrac{1}{2}$ **30.** $\cot \theta = \dfrac{1}{2}$ **31.** $\sec \theta = 3$ **32.** $\csc \theta = 5$

33. $\tan \theta = \sqrt{2}$ **34.** $\sec \theta = \dfrac{5}{3}$ **35.** $\csc \theta = 2$ **36.** $\cot \theta = 2$

In Problems 37–54, use Fundamental Identities and/or the Complementary Angle Theorem to find the exact value of each expression. Do not use a calculator.

37. $\sin^2 20° + \cos^2 20°$ **38.** $\sec^2 28° - \tan^2 28°$ **39.** $\sin 80° \csc 80°$ **40.** $\tan 10° \cot 10°$

41. $\tan 50° - \dfrac{\sin 50°}{\cos 50°}$ **42.** $\cot 25° - \dfrac{\cos 25°}{\sin 25°}$ **43.** $\sin 38° - \cos 52°$ **44.** $\tan 12° - \cot 78°$

45. $\dfrac{\cos 10°}{\sin 80°}$ **46.** $\dfrac{\cos 40°}{\sin 50°}$ **47.** $1 - \cos^2 20° - \cos^2 70°$ **48.** $1 + \tan^2 5° - \csc^2 85°$

49. $\tan 20° - \dfrac{\cos 70°}{\cos 20°}$ **50.** $\cot 40° - \dfrac{\sin 50°}{\sin 40°}$ **51.** $\tan 35° \cdot \sec 55° \cdot \cos 35°$ **52.** $\cot 25° \cdot \csc 65° \cdot \sin 25°$

53. $\cos 35° \sin 55° + \cos 55° \sin 35°$ **54.** $\sec 35° \csc 55° - \tan 35° \cot 55°$

---

55. Given $\sin 30° = \dfrac{1}{2}$, use trigonometric identities to find the exact value of
  (a) $\cos 60°$ (b) $\cos^2 30°$
  (c) $\csc \dfrac{\pi}{6}$ (d) $\sec \dfrac{\pi}{3}$

56. Given $\sin 60° = \dfrac{\sqrt{3}}{2}$, use trigonometric identities to find the exact value of
  (a) $\cos 30°$ (b) $\cos^2 60°$
  (c) $\sec \dfrac{\pi}{6}$ (d) $\csc \dfrac{\pi}{3}$

57. Given $\tan \theta = 4$, use trigonometric identities to find the exact value of
  (a) $\sec^2 \theta$ (b) $\cot \theta$
  (c) $\cot\left(\dfrac{\pi}{2} - \theta\right)$ (d) $\csc^2 \theta$

58. Given $\sec \theta = 3$, use trigonometric identities to find the exact value of
  (a) $\cos \theta$ (b) $\tan^2 \theta$ (c) $\csc(90° - \theta)$ (d) $\sin^2 \theta$

59. Given $\csc \theta = 4$, use trigonometric identities to find the exact value of
  (a) $\sin \theta$ (b) $\cot^2 \theta$ (c) $\sec(90° - \theta)$ (d) $\sec^2 \theta$

60. Given $\cot \theta = 2$, use trigonometric identities to find the exact value of
  (a) $\tan \theta$ (b) $\csc^2 \theta$
  (c) $\tan\left(\dfrac{\pi}{2} - \theta\right)$ (d) $\sec^2 \theta$

61. Given the approximation $\sin 38° \approx 0.62$, use trigonometric identities to find the approximate value of
  (a) $\cos 38°$ (b) $\tan 38°$
  (c) $\cot 38°$ (d) $\sec 38°$
  (e) $\csc 38°$ (f) $\sin 52°$
  (g) $\cos 52°$ (h) $\tan 52°$

62. Given the approximation $\cos 21° \approx 0.93$, use trigonometric identities to find the approximate value of
(a) $\sin 21°$
(b) $\tan 21°$
(c) $\cot 21°$
(d) $\sec 21°$
(e) $\csc 21°$
(f) $\sin 69°$
(g) $\cos 69°$
(h) $\tan 69°$

63. If $\sin \theta = 0.3$, find the exact value of $\sin \theta + \cos\left(\dfrac{\pi}{2} - \theta\right)$.

64. If $\tan \theta = 4$, find the exact value of $\tan \theta + \tan\left(\dfrac{\pi}{2} - \theta\right)$.

65. Find an acute angle θ that satisfies the equation
$$\sin \theta = \cos(2\theta + 30°)$$

66. Find an acute angle θ that satisfies the equation
$$\tan \theta = \cot(\theta + 45°)$$

Applications and Extensions

67. Calculating the Time of a Trip From a parking lot you want to walk to a house on the ocean. The house is located 1500 feet down a paved path that parallels the beach, which is 500 feet wide. Along the path you can walk 300 feet per minute, but in the sand on the beach you can only walk 100 feet per minute. See the illustration.

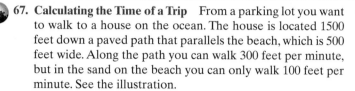

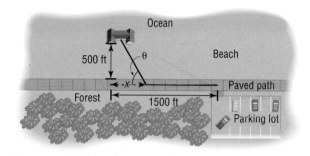

(a) Calculate the time T if you walk 1500 feet along the paved path and then 500 feet in the sand to the house.
(b) Calculate the time T if you walk in the sand directly toward the ocean for 500 feet and then turn left and walk along the beach for 1500 feet to the house.
(c) Express the time T to get from the parking lot to the beachhouse as a function of the angle θ shown in the illustration.
(d) Calculate the time T if you walk directly from the parking lot to the house.
$$\left[\textbf{Hint: } \tan \theta = \frac{500}{1500}\right]$$
(e) Calculate the time T if you walk 1000 feet along the paved path and then walk directly to the house.
(f) Graph $T = T(\theta)$. For what angle θ is T least? What is x for this angle? What is the minimum time?
(g) Explain why $\tan \theta = \dfrac{1}{3}$ gives the smallest angle θ that is possible.

68. Carrying a Ladder around a Corner Two hallways, one of width 3 feet, the other of width 4 feet, meet at a right angle. See the illustration.
(a) Express the length L of the line segment shown as a function of the angle θ.
(b) Discuss why the length of the longest ladder that can be carried around the corner is equal to the smallest value of L.

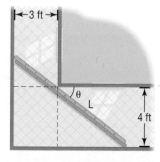

69. Electrical Engineering A resistor and an inductor connected in a series network impede the flow of an alternating current. This impedance Z is determined by the reactance X of the inductor and the resistance R of the resistor. The three quantities, all measured in ohms, can be represented by the sides of a right triangle as illustrated, so $Z^2 = X^2 + R^2$. The angle ϕ is called the **phase angle.** Suppose a series network has an inductive reactance of $X = 400$ ohms and a resistance of $R = 600$ ohms.
(a) Find the impedance Z.
(b) Find the values of the six trigonometric functions of the phase angle ϕ.

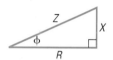

70. Electrical Engineering Refer to Problem 69. A series network has a resistance of $R = 588$ ohms. The phase angle ϕ is such that $\tan \phi = \dfrac{5}{12}$.
(a) Determine the inductive reactance X and the impedance Z.
(b) Determine the values of the remaining five trigonometric functions of the phase angle ϕ.

71. Geometry Suppose that the angle θ is a central angle of a circle of radius 1 (see the figure). Show that
(a) Angle $OAC = \dfrac{\theta}{2}$
(b) $|CD| = \sin \theta$ and $|OD| = \cos \theta$
(c) $\tan \dfrac{\theta}{2} = \dfrac{\sin \theta}{1 + \cos \theta}$

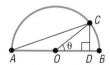

72. Geometry Show that the area A of an isosceles triangle is $A = a^2 \sin \theta \cos \theta$, where a is the length of one of the two equal sides and θ is the measure of one of the two equal angles (see the figure).

73. Geometry Let $n \geq 1$ be any real number and let θ be any angle for which $0 < n\theta < \dfrac{\pi}{2}$. Then we can draw a triangle with the angles θ and $n\theta$ and included side of length 1 (do you see why?) and place it on the unit circle as illustrated.

Now, drop the perpendicular from C to $D = (x, 0)$ and show that

$$x = \frac{\tan(n\theta)}{\tan \theta + \tan(n\theta)}$$

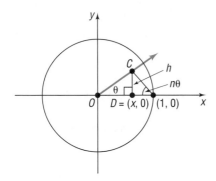

74. Geometry Refer to the figure. The smaller circle, whose radius is a, is tangent to the larger circle, whose radius is b. The ray $\overrightarrow{OA}$ contains a diameter of each circle, and the ray $\overrightarrow{OB}$ is tangent to each circle. Show that

$$\cos \theta = \frac{\sqrt{ab}}{\dfrac{a + b}{2}}$$

(This shows that $\cos \theta$ equals the ratio of the geometric mean of a and b to the arithmetic mean of a and b.)
[Hint: First show that $\sin \theta = (b - a)/(b + a)$.]

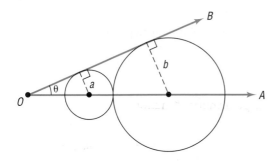

75. Geometry Refer to the figure. If $|OA| = 1$, show that

(a) Area $\triangle OAC = \dfrac{1}{2}\sin \alpha \cos \alpha$

(b) Area $\triangle OCB = \dfrac{1}{2}|OB|^2 \sin \beta \cos \beta$

(c) Area $\triangle OAB = \dfrac{1}{2}|OB| \sin(\alpha + \beta)$

(d) $|OB| = \dfrac{\cos \alpha}{\cos \beta}$

(e) $\sin(\alpha + \beta) = \sin \alpha \cos \beta + \cos \alpha \sin \beta$
[Hint: Area $\triangle OAB = $ Area $\triangle OAC + $ Area $\triangle OCB$]

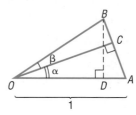

△ **76. Geometry** Refer to the figure, where a unit circle is drawn. The line segment $\overline{DB}$ is tangent to the circle.

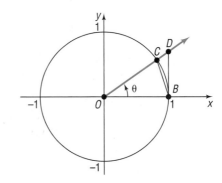

(a) Express the area of $\triangle OBC$ in terms of $\sin \theta$ and $\cos \theta$.
[Hint: Use the altitude from C to the base $\overline{OB} = 1$.]

(b) Express the area of $\triangle OBD$ in terms of $\sin \theta$ and $\cos \theta$.

(c) The area of the sector OBC of the circle is $\dfrac{1}{2}\theta$, where θ is measured in radians. Use the results of parts (a) and (b) and the fact that

$$\text{Area } \triangle OBC < \text{Area of sector } OBC < \text{Area } \triangle OBD$$

to show that

$$1 < \frac{\theta}{\sin \theta} < \frac{1}{\cos \theta}$$

77. If $\cos \alpha = \tan \beta$ and $\cos \beta = \tan \alpha$, where α and β are acute angles, show that

$$\sin \alpha = \sin \beta = \sqrt{\frac{3 - \sqrt{5}}{2}}$$

78. If θ is an acute angle and $\tan \theta = x$, $x \neq 0$, express the remaining five trigonometric functions in terms of x.

Discussion and Writing

79. If θ is an acute angle, explain why $\sec \theta > 1$.

80. If θ is an acute angle, explain why $0 < \sin \theta < 1$.

81. How would you explain the meaning of the sine function to a fellow student who has just completed college algebra?

82. Look back at Example 5. Which of the two solutions do you prefer? Explain your reasoning.

'Are You Prepared?' Answers

1. $2\sqrt{34}$ **2.** $f(5) = 8$

2.3 Computing the Values of Trigonometric Functions of Acute Angles

OBJECTIVES 1 Find the Exact Values of the Trigonometric Functions of $\frac{\pi}{4} = 45°$ (p. 129)

2 Find the Exact Values of the Trigonometric Functions of $\frac{\pi}{6} = 30°$ and $\frac{\pi}{3} = 60°$ (p. 130)

3 Use a Calculator to Approximate the Values of the Trigonometric Functions of Acute Angles (p. 132)

4 Model and Solve Applied Problems Involving Right Triangles (p. 132)

In the previous section, we developed ways to find the value of each trigonometric function of an acute angle when one of the functions is known. In this section, we discuss the problem of finding the value of each trigonometric function of an acute angle when the angle is given.

For three special acute angles, we can use some results from plane geometry to find the exact value of each of the six trigonometric functions.

1 Find the Exact Values of the Trigonometric Functions of $\frac{\pi}{4} = 45°$

EXAMPLE 1 **Finding the Exact Values of the Trigonometric Functions of $\frac{\pi}{4} = 45°$**

Find the exact values of the six trigonometric functions of $\frac{\pi}{4} = 45°$.

Solution Using the right triangle in Figure 29(a), in which one of the angles is $\frac{\pi}{4} = 45°$, it follows that the other acute angle is also $\frac{\pi}{4} = 45°$, and hence the triangle is isosceles.

As a result, side a and side b are equal in length. Since the values of the trigonometric functions of an angle depend only on the angle and not on the size of the triangle, we may assign any values to a and b for which $a = b > 0$. We decide to use the triangle for which

$$a = b = 1$$

Then, by the Pythagorean Theorem,

$$c^2 = a^2 + b^2 = 1 + 1 = 2$$
$$c = \sqrt{2}$$

As a result, we have the triangle in Figure 29(b), from which we find

Figure 29

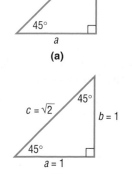

(a)

(b)

$$\sin \frac{\pi}{4} = \sin 45° = \frac{b}{c} = \frac{1}{\sqrt{2}} = \frac{\sqrt{2}}{2} \qquad \cos \frac{\pi}{4} = \cos 45° = \frac{a}{c} = \frac{1}{\sqrt{2}} = \frac{\sqrt{2}}{2}$$

Using Quotient and Reciprocal Identities, we find

$$\tan\frac{\pi}{4} = \tan 45° = \frac{\sin 45°}{\cos 45°} = \frac{\frac{\sqrt{2}}{2}}{\frac{\sqrt{2}}{2}} = 1 \qquad \cot\frac{\pi}{4} = \cot 45° = \frac{1}{\tan 45°} = \frac{1}{1} = 1$$

$$\sec\frac{\pi}{4} = \sec 45° = \frac{1}{\cos 45°} = \frac{1}{\frac{1}{\sqrt{2}}} = \sqrt{2} \qquad \csc\frac{\pi}{4} = \csc 45° = \frac{1}{\sin 45°} = \frac{1}{\frac{1}{\sqrt{2}}} = \sqrt{2}$$

EXAMPLE 2   **Finding the Exact Value of a Trigonometric Expression**

Find the exact value of each expression.

(a) $(\sin 45°)(\tan 45°)$

(b) $\left(\sec\frac{\pi}{4}\right)\left(\cot\frac{\pi}{4}\right)$

Solution   We use the results obtained in Example 1.

(a) $(\sin 45°)(\tan 45°) = \frac{\sqrt{2}}{2} \cdot 1 = \frac{\sqrt{2}}{2}$

(b) $\left(\sec\frac{\pi}{4}\right)\left(\cot\frac{\pi}{4}\right) = \sqrt{2} \cdot 1 = \sqrt{2}$

═══ **Now Work**   PROBLEMS 5 AND 17

2 Find the Exact Values of the Trigonometric Functions of $\frac{\pi}{6} = 30°$ and $\frac{\pi}{3} = 60°$

EXAMPLE 3   **Finding the Exact Values of the Trigonometric Functions of $\frac{\pi}{6} = 30°$ and $\frac{\pi}{3} = 60°$**

Find the exact values of the six trigonometric functions of $\frac{\pi}{6} = 30°$ and $\frac{\pi}{3} = 60°$.

Solution   Form a right triangle in which one of the angles is $\frac{\pi}{6} = 30°$. It then follows that the third angle is $\frac{\pi}{3} = 60°$. Figure 30(a) illustrates such a triangle with hypotenuse of length 2. Our problem is to determine a and b.

We begin by placing next to the triangle in Figure 30(a) another triangle congruent to the first, as shown in Figure 30(b). Notice that we now have a triangle whose angles are each 60°. This triangle is therefore equilateral, so each side is of length 2. In particular, the base is $2a = 2$, so $a = 1$. By the Pythagorean Theorem, b satisfies the equation $a^2 + b^2 = c^2$, so we have

$$a^2 + b^2 = c^2$$
$$1^2 + b^2 = 2^2 \qquad a = 1, c = 2$$
$$b^2 = 4 - 1 = 3$$
$$b = \sqrt{3}$$

Figure 30

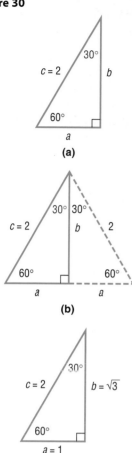

(a)

(b)

(c)

Using the triangle in Figure 30(c) and the fact that $\dfrac{\pi}{6} = 30°$ and $\dfrac{\pi}{3} = 60°$ are complementary angles, we find

$$\sin \frac{\pi}{6} = \sin 30° = \frac{\text{opposite}}{\text{hypotenuse}} = \frac{1}{2} \qquad\qquad \cos \frac{\pi}{3} = \cos 60° = \frac{1}{2}$$

$$\cos \frac{\pi}{6} = \cos 30° = \frac{\text{adjacent}}{\text{hypotenuse}} = \frac{\sqrt{3}}{2} \qquad\qquad \sin \frac{\pi}{3} = \sin 60° = \frac{\sqrt{3}}{2}$$

$$\tan \frac{\pi}{6} = \tan 30° = \frac{\sin 30°}{\cos 30°} = \frac{\dfrac{1}{2}}{\dfrac{\sqrt{3}}{2}} = \frac{1}{\sqrt{3}} = \frac{\sqrt{3}}{3} \qquad\qquad \cot \frac{\pi}{3} = \cot 60° = \frac{\sqrt{3}}{3}$$

$$\csc \frac{\pi}{6} = \csc 30° = \frac{1}{\sin 30°} = \frac{1}{\dfrac{1}{2}} = 2 \qquad\qquad \sec \frac{\pi}{3} = \sec 60° = 2$$

$$\sec \frac{\pi}{6} = \sec 30° = \frac{1}{\cos 30°} = \frac{1}{\dfrac{\sqrt{3}}{2}} = \frac{2}{\sqrt{3}} = \frac{2\sqrt{3}}{3} \qquad\qquad \csc \frac{\pi}{3} = \csc 60° = \frac{2\sqrt{3}}{3}$$

$$\cot \frac{\pi}{6} = \cot 30° = \frac{1}{\tan 30°} = \frac{1}{\dfrac{\sqrt{3}}{3}} = \frac{3}{\sqrt{3}} = \sqrt{3} \qquad\qquad \tan \frac{\pi}{3} = \tan 60° = \sqrt{3}$$

Table 3 summarizes the information just derived for the angles $\dfrac{\pi}{6} = 30°$, $\dfrac{\pi}{4} = 45°$, and $\dfrac{\pi}{3} = 60°$. Rather than memorize the entries in Table 3, you can draw the appropriate triangle to determine the values given in the table.

Table 3

θ (Radians)	θ (Degrees)	$\sin \theta$	$\cos \theta$	$\tan \theta$	$\csc \theta$	$\sec \theta$	$\cot \theta$
$\dfrac{\pi}{6}$	30°	$\dfrac{1}{2}$	$\dfrac{\sqrt{3}}{2}$	$\dfrac{\sqrt{3}}{3}$	2	$\dfrac{2\sqrt{3}}{3}$	$\sqrt{3}$
$\dfrac{\pi}{4}$	45°	$\dfrac{\sqrt{2}}{2}$	$\dfrac{\sqrt{2}}{2}$	1	$\sqrt{2}$	$\sqrt{2}$	1
$\dfrac{\pi}{3}$	60°	$\dfrac{\sqrt{3}}{2}$	$\dfrac{1}{2}$	$\sqrt{3}$	$\dfrac{2\sqrt{3}}{3}$	2	$\dfrac{\sqrt{3}}{3}$

EXAMPLE 4 **Finding the Exact Value of a Trigonometric Expression**

Find the exact value of each expression.

(a) $\sin 45° \cos 30°$ (b) $\tan \dfrac{\pi}{4} - \sin \dfrac{\pi}{3}$ (c) $\tan^2 \dfrac{\pi}{6} + \sin^2 \dfrac{\pi}{4}$

Solution (a) $\sin 45° \cos 30° = \dfrac{\sqrt{2}}{2} \cdot \dfrac{\sqrt{3}}{2} = \dfrac{\sqrt{6}}{4}$

(b) $\tan \dfrac{\pi}{4} - \sin \dfrac{\pi}{3} = 1 - \dfrac{\sqrt{3}}{2} = \dfrac{2 - \sqrt{3}}{2}$

(c) $\tan^2 \dfrac{\pi}{6} + \sin^2 \dfrac{\pi}{4} = \left(\dfrac{\sqrt{3}}{3}\right)^2 + \left(\dfrac{\sqrt{2}}{2}\right)^2 = \dfrac{1}{3} + \dfrac{1}{2} = \dfrac{5}{6}$

Now Work PROBLEMS 9 AND 19

3 Use a Calculator to Approximate the Values of Trigonometric Functions of Acute Angles

Before getting started, you must first decide whether to enter the angle in the calculator using radians or degrees and then set the calculator to the correct MODE. (Check your instruction manual to find out how your calculator handles degrees and radians.) Your calculator has the keys marked $\boxed{\sin}$, $\boxed{\cos}$, and $\boxed{\tan}$. To find the values of the remaining three trigonometric functions (secant, cosecant, and cotangent), we use the reciprocal identities.

$$\sec \theta = \frac{1}{\cos \theta} \qquad \csc \theta = \frac{1}{\sin \theta} \qquad \cot \theta = \frac{1}{\tan \theta}$$

EXAMPLE 5	Using a Calculator to Approximate the Value of Trigonometric Functions

Use a calculator to find the approximate value of:

(a) $\cos 48°$ (b) $\csc 21°$ (c) $\tan \dfrac{\pi}{12}$

Express your answer rounded to two decimal places.

Solution (a) First, we set the MODE to receive degrees. See Figure 31(a). Figure 31(b) shows the solution using a TI-84 Plus graphing calculator.

Figure 31

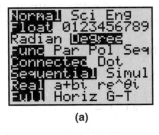
(a)

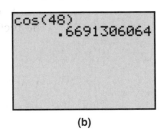
(b)

Figure 32

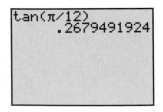

Rounded to two decimal places,

$$\cos 48° = 0.67$$

(b) Most calculators do not have a csc key. The manufacturers assume the user knows some trigonometry. To find the value of csc 21°, we use the fact that $\csc 21° = \dfrac{1}{\sin 21°}$. Figure 32 shows the solution using a TI-84 Plus graphing calculator. Rounded to two decimal places,

$$\csc 21° = 2.79$$

Figure 33

(c) Set the MODE to receive radians. Figure 33 shows the solution using a TI-84 Plus graphing calculator. Rounded to two decimal places,

$$\tan \dfrac{\pi}{12} = 0.27$$

 Now Work PROBLEM 29

4 Model and Solve Applied Problems Involving Right Triangles

Right triangles can be used to model many types of situations, such as the optimal design of a rain gutter.*

---

*In applied problems, it is important that answers be reported with both justifiable accuracy and appropriate significant figures. Throughout the text, we shall assume that angles are measured to the nearest tenth and sides are measured to the nearest hundredth, resulting in sides being rounded to two decimal places and angles being rounded to one decimal place.

| EXAMPLE 6 | Constructing a Rain Gutter |

Figure 34

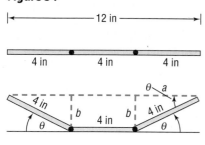

A rain gutter is to be constructed of aluminum sheets 12 inches wide. After marking off a length of 4 inches from each edge, this length is bent up at an angle θ. See Figure 34.

(a) Express the area A of the opening as a function of θ.

　　[**Hint:** Let b denote the vertical height of the bend.]

(b) Find the area A of the opening for $\theta = 30°, \theta = 45°, \theta = 60°$, and $\theta = 75°$.

(c) Graph $A = A(\theta)$. Find the angle θ that makes A largest. (This bend will allow the most water to flow through the gutter.)

Solution

(a) Look again at Figure 34. The area A of the opening is the sum of the areas of two congruent right triangles and one rectangle. Look at Figure 35, showing one of the triangles in Figure 34 redrawn. We see that

Figure 35

$$\cos\theta = \frac{a}{4}, \quad \text{so} \quad a = 4\cos\theta \qquad \sin\theta = \frac{b}{4}, \quad \text{so} \quad b = 4\sin\theta$$

The area of the triangle is

$$\text{area of triangle} = \frac{1}{2}(\text{base})(\text{height}) = \frac{1}{2}ab = \frac{1}{2}(4\cos\theta)(4\sin\theta) = 8\sin\theta\cos\theta$$

So the area of the two congruent triangles is $16\sin\theta\cos\theta$.

The rectangle has length 4 and height b, so its area is

$$\text{area of rectangle} = 4b = 4(4\sin\theta) = 16\sin\theta$$

The area A of the opening is

$$A = \text{area of the two triangles} + \text{area of the rectangle}$$

$$A(\theta) = 16\sin\theta\cos\theta + 16\sin\theta = 16\sin\theta(\cos\theta + 1)$$

(b) 　For $\theta = 30°$: 　$A(30°) = 16\sin 30°(\cos 30° + 1)$

$$= 16\left(\frac{1}{2}\right)\left(\frac{\sqrt{3}}{2} + 1\right) = 4\sqrt{3} + 8 \approx 14.93$$

The area of the opening for $\theta = 30°$ is about 14.93 square inches.

For $\theta = 45°$: 　$A(45°) = 16\sin 45°(\cos 45° + 1)$

$$= 16\left(\frac{\sqrt{2}}{2}\right)\left(\frac{\sqrt{2}}{2} + 1\right) = 8 + 8\sqrt{2} \approx 19.31$$

The area of the opening for $\theta = 45°$ is about 19.31 square inches.

For $\theta = 60°$: 　$A(60°) = 16\sin 60°(\cos 60° + 1)$

$$= 16\left(\frac{\sqrt{3}}{2}\right)\left(\frac{1}{2} + 1\right) = 12\sqrt{3} \approx 20.78$$

Figure 36

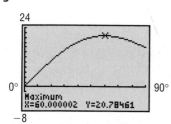

The area of the opening for $\theta = 60°$ is about 20.78 square inches.

For $\theta = 75°$: 　$A(75°) = 16\sin 75°(\cos 75° + 1) \approx 19.45$

The area of the opening for $\theta = 75°$ is about 19.45 square inches.

(c) Figure 36 shows the graph of $A = A(\theta)$. Using MAXIMUM, the angle θ that makes A largest is 60°.

Now Work PROBLEM 61

In addition to developing models using right triangles, we can use right triangle trigonometry to measure heights and distances that are either awkward or impossible to measure by ordinary means. When using right triangles to solve these problems, pay attention to the known measures. This will indicate what trigonometric function to use. For example, if we know the measure of an angle and the length of the side adjacent to the angle, and we wish to find the length of the opposite side, we would use the tangent function. Do you know why?

EXAMPLE 7 **Finding the Width of a River**

A surveyor can measure the width of a river by setting up a transit at a point C on one side of the river and taking a sighting of a point A on the other side. Refer to Figure 37. After turning through an angle of 90° at C, the surveyor walks a distance of 200 meters to point B. Using the transit at B, the angle θ is measured and found to be 20°. What is the width of the river rounded to the nearest meter?

Figure 37

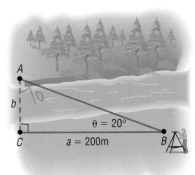

Solution We seek the length of side b. We know a and θ. So we use the fact that b is opposite θ and a is adjacent to θ and write

$$\tan \theta = \frac{b}{a}$$

which leads to

$$\tan 20° = \frac{b}{200}$$
$$b = 200 \tan 20° \approx 72.79 \text{ meters}$$

The width of the river is 73 meters, rounded to the nearest meter.

Now Work PROBLEM 69

Vertical heights can sometimes be measured using either the *angle of elevation* or the *angle of depression*. If a person is looking up at an object, the acute angle measured from the horizontal to a line of sight to the object is called the **angle of elevation.** See Figure 38(a).

Figure 38

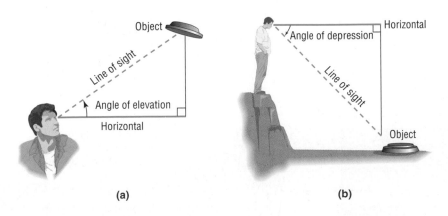

(a) (b)

If a person is standing on a cliff looking down at an object, the acute angle made by the line of sight to the object and the horizontal is called the **angle of depression.** See Figure 38(b).

EXAMPLE 8   Finding the Height of a Cloud

Meteorologists find the height of a cloud using an instrument called a **ceilometer.**
A ceilometer consists of a **light projector** that directs a vertical light beam up to
the cloud base and a **light detector** that scans the cloud to detect the light beam.
See Figure 39(a). On December 1, 2006, at Midway Airport in Chicago, a ceilometer was employed to find the height of the cloud cover. It was set up with its light
detector 300 feet from its light projector. If the angle of elevation from the light detector to the base of the cloud is 75°, what is the height of the cloud cover?

Figure 39

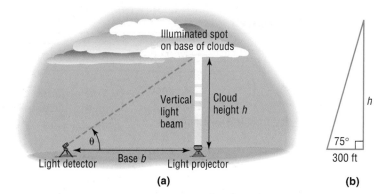

(a)   (b)

Solution   Figure 39(b) illustrates the situation. To find the height h, we use the fact that
$\tan 75° = \dfrac{h}{300}$, so

$$h = 300 \tan 75° \approx 1120 \text{ feet}$$

The ceiling (height to the base of the cloud cover) is approximately 1120 feet.

━━━━ **Now Work** PROBLEM 71

The idea behind Example 8 can also be used to find the height of an object with
a base that is not accessible to the horizontal.

EXAMPLE 9   Finding the Height of a Statue on a Building

Adorning the top of the Board of Trade building in Chicago is a statue of Ceres, the
Roman goddess of wheat. From street level, two observations are taken 400 feet
from the center of the building. The angle of elevation to the base of the statue is
found to be 55.1°, and the angle of elevation to the top of the statue is 56.5°. See
Figure 40(a). What is the height of the statue?

Figure 40

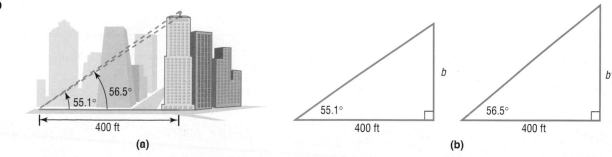

(a)   (b)

Solution   Figure 40(b) shows two triangles that replicate Figure 40(a). The height of the statue of Ceres will be $b' - b$. To find b and b', we refer to Figure 40(b).

$$\tan 55.1° = \frac{b}{400} \qquad\qquad \tan 56.5° = \frac{b'}{400}$$

$$b = 400 \tan 55.1° \approx 573.39 \qquad b' = 400 \tan 56.5° \approx 604.33$$

The height of the statue is approximately $604.33 - 573.39 = 30.94$ feet ≈ 31 feet.

━━━━➤ **Now Work** PROBLEM 77

2.3 Assess Your Understanding

Concepts and Vocabulary

1. $\tan \dfrac{\pi}{4} + \sin 30° =$ _____.

2. Using a calculator, $\sin 2 =$ _____, rounded to two decimal places.

3. *True or False* Exact values can be found for the trigonometric functions of 60°.

4. *True or False* Exact values can be found for the sine of any angle.

Skill Building

5. Write down the exact value of each of the six trigonometric functions of 45°.

6. Write down the exact value of each of the six trigonometric functions of 30° and of 60°.

In Problems 7–16, $f(\theta) = \sin\theta$ and $g(\theta) = \cos\theta$. Find the exact value of each expression if $\theta = 60°$. Do not use a calculator.

7. $f(\theta)$

8. $g(\theta)$

9. $f\left(\dfrac{\theta}{2}\right)$

10. $g\left(\dfrac{\theta}{2}\right)$

11. $[f(\theta)]^2$

12. $[g(\theta)]^2$

13. $2f(\theta)$

14. $2g(\theta)$

15. $\dfrac{f(\theta)}{2}$

16. $\dfrac{g(\theta)}{2}$

In Problems 17–28, find the exact value of each expression. Do not use a calculator.

17. $4 \cos 45° - 2 \sin 45°$

18. $2 \sin 45° + 4 \cos 30°$

19. $6 \tan 45° - 8 \cos 60°$

20. $\sin 30° \cdot \tan 60°$

21. $\sec \dfrac{\pi}{4} + 2 \csc \dfrac{\pi}{3}$

22. $\tan \dfrac{\pi}{4} + \cot \dfrac{\pi}{4}$

23. $\sec^2 \dfrac{\pi}{6} - 4$

24. $4 + \tan^2 \dfrac{\pi}{3}$

25. $\sin^2 30° + \cos^2 60°$

26. $\sec^2 60° - \tan^2 45°$

27. $1 - \cos^2 30° - \cos^2 60°$

28. $1 + \tan^2 30° - \csc^2 45°$

In Problems 29–46, use a calculator to find the approximate value of each expression. Round the answer to two decimal places.

29. $\sin 28°$

30. $\cos 14°$

31. $\tan 21°$

32. $\cot 70°$

33. $\sec 41°$

34. $\csc 55°$

35. $\sin \dfrac{\pi}{10}$

36. $\cos \dfrac{\pi}{8}$

37. $\tan \dfrac{5\pi}{12}$

38. $\cot \dfrac{\pi}{18}$

39. $\sec \dfrac{\pi}{12}$

40. $\csc \dfrac{5\pi}{13}$

41. $\sin 1$

42. $\tan 1$

43. $\sin 1°$

44. $\tan 1°$

45. $\tan 0.3$

46. $\tan 0.1$

Mixed Practice

In Problems 47–56, $f(x) = \sin x$, $g(x) = \cos x$, $h(x) = 2x$, and $p(x) = \dfrac{x}{2}$. Find the value of each of the following:

47. $(f + g)(30°)$

48. $(f - g)(60°)$

49. $(f \cdot g)\left(\dfrac{\pi}{4}\right)$

50. $(f \cdot g)\left(\dfrac{\pi}{3}\right)$

51. $(f \circ h)\left(\dfrac{\pi}{6}\right)$

52. $(g \circ p)(60°)$

53. $(p \circ g)(45°)$

54. $(h \circ f)\left(\dfrac{\pi}{6}\right)$

55. (a) Find $f\left(\dfrac{\pi}{4}\right)$. What point is on the graph of f?

  (b) Using the result of part (a), what point is on the graph of f^{-1}?

  (c) What point is on the graph of $y = f\left(x + \dfrac{\pi}{4}\right) - 3$ if $x = \dfrac{\pi}{4}$?

56. (a) Find $g\left(\dfrac{\pi}{6}\right)$. What point is on the graph of g?

  (b) Using the result of part (a), what point is on the graph of g^{-1}?

  (c) What point is on the graph of $y = 2g\left(x - \dfrac{\pi}{6}\right)$ if $x = \dfrac{\pi}{6}$?

Applications and Extensions

Problems 57–60 require the following discussion.

Projectile Motion   The path of a projectile fired at an inclination θ to the horizontal with initial speed v_0 is a parabola. See the figure. The **range R** of the projectile, that is, the horizontal distance that the projectile travels, is found by using the function

$$R(\theta) = \frac{2v_0^2 \sin \theta \cos \theta}{g}$$

where $g \approx 32.2$ feet per second per second ≈ 9.8 meters per second per second is the acceleration due to gravity. The maximum height H of the projectile is given by the function

$$H(\theta) = \frac{v_0^2 \sin^2 \theta}{2g}$$

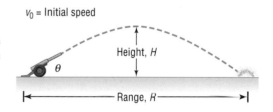

In Problems 57– 60, find the range R and maximum height H of the projectile. Round answers to two decimal places.

57. The projectile is fired at an angle of $45°$ to the horizontal with an initial speed of 100 feet per second.

58. The projectile is fired at an angle of $30°$ to the horizontal with an initial speed of 150 meters per second.

59. The projectile is fired at an angle of $25°$ to the horizontal with an initial speed of 500 meters per second.

60. The projectile is fired at an angle of $50°$ to the horizontal with an initial speed of 200 feet per second.

61. Inclined Plane   See the illustration. If friction is ignored, the time t (in seconds) required for a block to slide down an inclined plane is modeled by the function

$$t(\theta) = \sqrt{\frac{2a}{g \sin \theta \cos \theta}}$$

where a is the length (in feet) of the base and $g \approx 32$ feet per second per second is the acceleration due to gravity. How long does it take a block to slide down an inclined plane with base $a = 10$ feet when

(a) $\theta = 30°$?   (b) $\theta = 45°$?   (c) $\theta = 60°$?

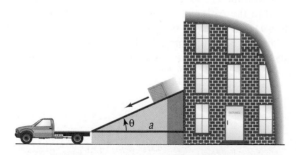

62. Piston Engines   See the illustration. In a certain piston engine, the distance x (in inches) from the center of the drive shaft to the head of the piston is modeled by the function

$$x(\theta) = \cos \theta + \sqrt{16 + 0.5(2 \cos^2 \theta - 1)}$$

where θ is the angle between the crank and the path of the piston head. Find x when $\theta = 30°$ and when $\theta = 45°$.

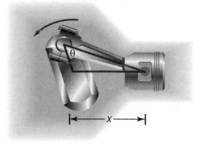

63. Calculating the Time of a Trip   Two oceanfront homes are located 8 miles apart on a straight stretch of beach, each a distance of 1 mile from a paved road that parallels the ocean. Sally can jog 8 miles per hour along the paved road, but only 3 miles per hour in the sand on the beach. Because a river flows between the two houses, it is necessary to jog on the sand to the road, continue on the road, and then jog on the sand to get from one house to the other. See the illustration.

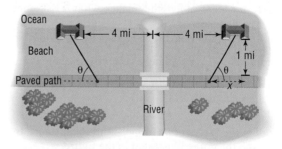

(a) Express the time T to get from one house to the other as a function of the angle θ shown in the illustration.

(b) Calculate the time T for $\theta = 30°$. How long is Sally on the paved road?

(c) Calculate the time T for $\theta = 45°$. How long is Sally on the paved road?

(d) Calculate the time T for $\theta = 60°$. How long is Sally on the paved road?

(e) Calculate the time T for $\theta = 90°$. Describe the path taken.

(f) Calculate the time T for $\tan \theta = \dfrac{1}{4}$. Describe the path taken. Explain why θ must be larger than $14°$.

(g) Graph $T = T(\theta)$. What angle θ results in the least time? What is the least time? How long is Sally on the paved road?

64. **Designing Fine Decorative Pieces** A designer of decorative art plans to market solid gold spheres encased in clear crystal cones. Each sphere is of fixed radius R and will be enclosed in a cone of height h and radius r. See the illustration. Many cones can be used to enclose the sphere, each having a different slant angle θ.

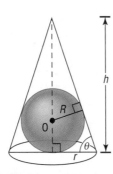

(a) Express the volume V of the cone as a function of the slant angle θ of the cone.

   [**Hint:** The volume V of a cone of height h and radius r is $V = \dfrac{1}{3}\pi r^2 h$.]

(b) What volume V is required to enclose a sphere of radius 2 centimeters in a cone whose slant angle θ is $30°$? $45°$? $60°$?

(c) What slant angle θ should be used for the volume V of the cone to be a minimum? (This choice minimizes the amount of crystal required and gives maximum emphasis to the gold sphere.)

65. **Geometry** A right triangle has a hypotenuse of length 8 inches. If one angle is $35°$, find the length of each leg.

66. **Geometry** A right triangle has a hypotenuse of length 10 centimeters. If one angle is $40°$, find the length of each leg.

67. **Geometry** A right triangle contains a $25°$ angle.

(a) If one leg is of length 5 inches, what is the length of the hypotenuse?

(b) There are two answers. How is this possible?

68. **Geometry** A right triangle contains an angle of $\dfrac{\pi}{8}$ radian.

(a) If one leg is of length 3 meters, what is the length of the hypotenuse?

(b) There are two answers. How is this possible?

69. **Finding the Width of a Gorge** Find the distance from A to C across the gorge illustrated in the figure.

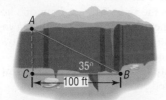

70. **Finding the Distance across a Pond** Find the distance from A to C across the pond illustrated in the figure.

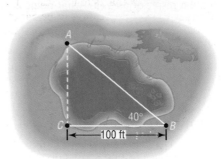

71. **The Eiffel Tower** The tallest tower built before the era of television masts, the Eiffel Tower was completed on March 31, 1889. Find the height of the Eiffel Tower (before a television mast was added to the top) using the information given in the illustration.

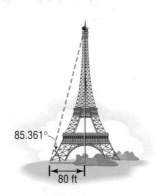

72. **Finding the Distance of a Ship from Shore** A person in a small boat, offshore from a vertical cliff known to be 100 feet in height, takes a sighting of the top of the cliff. If the angle of elevation is found to be $30°$, how far offshore is the ship?

73. **Finding the Distance to a Plateau** Suppose that you are headed toward a plateau 50 meters high. If the angle of elevation to the top of the plateau is $60°$, how far are you from the base of the plateau?

74. **Finding the Reach of a Ladder** A 22-foot extension ladder leaning against a building makes a $70°$ angle with the ground. How far up the building does the ladder touch?

75. **Finding the Distance between Two Objects** A blimp flying at an altitude of 500 feet, lies directly over a line from Soldier Field to the Adler Planetarium on Lake Michigan (see the figure). If the angle of depression from the blimp to the stadium is $32°$ and from the blimp to the planetarium is $23°$, find the distance between Soldier Field and the Adler Planetarium.

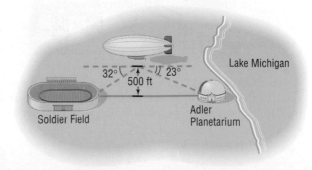

76. Hot-air Balloon While taking a ride in a hot-air balloon in Napa Valley, Francisco wonders how high he is. To find out, he chooses a landmark that is to the east of the balloon and measures the angle of depression to be 54°. A few minutes later, after traveling 100 feet east, the angle of depression to the same landmark is determined to be 61°. Use this information to determine the height of the balloon.

77. Mt. Rushmore To measure the height of Lincoln's caricature on Mt. Rushmore, two sightings 800 feet from the base of the mountain are taken. If the angle of elevation to the bottom of Lincoln's face is 32° and the angle of elevation to the top is 35°, what is the height of Lincoln's face?

78. The CN Tower The CN Tower, located in Toronto, Canada, is the tallest structure in the world. While visiting Toronto, a tourist wondered what the height of the tower above the top of the Sky Pod is. While standing 4000 feet from the tower, she measured the angle to the top of the Sky Pod to be 20.1°. At this same distance, the angle of elevation to the top of the tower was found to be 24.4°. Use this information to determine the height of the tower above the Sky Pod.

79. Finding the Length of a Guy Wire A radio transmission tower is 200 feet high. How long should a guy wire be if it is to be attached to the tower 10 feet from the top and is to make an angle of 45° with the ground?

80. Finding the Height of a Tower A guy wire 80 feet long is attached to the top of a radio transmission tower, making an angle of 45° with the ground. How high is the tower?

81. Washington Monument The angle of elevation of the Sun is 35.1° at the instant the shadow cast by the Washington Monument is 789 feet long. Use this information to calculate the height of the monument.

82. Finding the Length of a Mountain Trail A straight trail with an inclination of 17° leads from a hotel at an elevation of 9000 feet to a mountain lake at an elevation of 11,200 feet. What is the length of the trail?

83. Constructing a Highway A highway whose primary directions are north–south is being constructed along the west

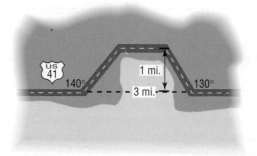

coast of Florida. Near Naples, a bay obstructs the straight path of the road. Since the cost of a bridge is prohibitive, engineers decide to go around the bay. The illustration shows the path that they decide on and the measurements taken. What is the length of highway needed to go around the bay?

84. Photography A camera is mounted on a tripod 4 feet high at a distance of 10 feet from George, who is 6 feet tall. See the illustration. If the camera lens has angles of depression and elevation of 20°, will George's feet and head be seen by the lens? If not, how far back will the camera need to be moved to include George's feet and head?

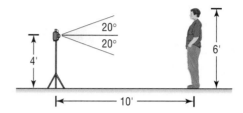

85. Calculating Pool Shots A pool player located at X wants to shoot the white ball off the top cushion and hit the red ball dead center. He knows from physics that the white ball will come off a cushion at the same angle as it hits a cushion. Where on the top cushion should he hit the white ball?

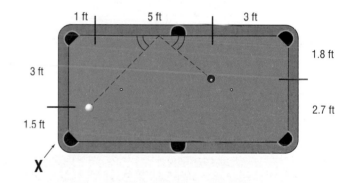

86. The Freedom Tower The Freedom Tower is to be the centerpiece of the rebuilding of the World Trade Center in New York City. The tower will be 1776 feet tall (not including a broadcast antenna). The angle of elevation from the base of an office building to the top of the tower is 34°. The angle of elevation from the helipad on the roof of the office building to the top of the tower is 20°.

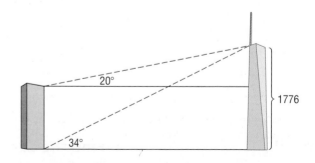

(a) How far away is the office building from the Freedom Tower? Assume the side of the tower is vertical. Round to the nearest foot.
(b) How tall is the office building? Round to the nearest foot.

△ **87.** Use a calculator set in radian mode to complete the following table. What can you conclude about the value of $f(\theta) = \dfrac{\sin\theta}{\theta}$ as θ approaches 0?

θ	0.5	0.4	0.2	0.1	0.01	0.001	0.0001	0.00001
$f(\theta) = \dfrac{\sin\theta}{\theta}$								

△ **88.** Use a calculator set in radian mode to complete the following table. What can you conclude about the value of $g(\theta) = \dfrac{\cos\theta - 1}{\theta}$ as θ approaches 0?

θ	0.5	0.4	0.2	0.1	0.01	0.001	0.0001	0.00001
$g(\theta) = \dfrac{\cos\theta - 1}{\theta}$								

89. Find the exact value of $\tan 1° \cdot \tan 2° \cdot \tan 3° \cdot \ldots \cdot \tan 89°$.

90. Find the exact value of $\cot 1° \cdot \cot 2° \cdot \cot 3° \cdot \ldots \cdot \cot 89°$.

91. Find the exact value of $\cos 1° \cdot \cos 2° \cdot \ldots \cdot \cos 45° \cdot \csc 46° \cdot \ldots \cdot \csc 89°$.

92. Find the exact value of $\sin 1° \cdot \sin 2° \cdot \ldots \cdot \sin 45° \cdot \sec 46° \cdot \ldots \cdot \sec 89°$.

Discussion and Writing

93. Write a brief paragraph that explains how to quickly compute the trigonometric functions of $30°, 45°,$ and $60°$.

94. Explain how you would measure the width of the Grand Canyon from a point on its ridge.

95. Explain how you would measure the height of a TV tower that is on the roof of a tall building.

2.4 Trigonometric Functions of General Angles

OBJECTIVES 1 Find the Exact Values of the Trigonometric Functions for General Angles (p. 140)

2 Use Coterminal Angles to Find the Exact Value of a Trigonometric Function (p. 143)

3 Determine the Signs of the Trigonometric Functions of an Angle in a Given Quadrant (p. 144)

4 Find the Reference Angle of a General Angle (p. 145)

5 Use a Reference Angle to Find the Exact Value of a Trigonometric Function (p. 146)

6 Find the Exact Values of Trigonometric Functions of an Angle, Given Information about the Functions (p. 147)

Figure 41

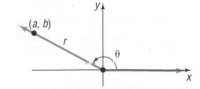

1 Find the Exact Values of the Trigonometric Functions for General Angles

To extend the definition of the trigonometric functions to include angles that are not acute, we employ a rectangular coordinate system and place the angle in the standard position so that its vertex is at the origin and its initial side is along the positive x-axis. See Figure 41.

DEFINITION

Let θ be any angle in standard position, and let (a, b) denote the coordinates of any point, except the origin $(0, 0)$, on the terminal side of θ. If $r = \sqrt{a^2 + b^2}$ denotes the distance from $(0, 0)$ to (a, b), then the **six trigonometric functions of θ** are defined as the ratios

$$\sin \theta = \frac{b}{r} \qquad \cos \theta = \frac{a}{r} \qquad \tan \theta = \frac{b}{a}$$

$$\csc \theta = \frac{r}{b} \qquad \sec \theta = \frac{r}{a} \qquad \cot \theta = \frac{a}{b}$$

provided no denominator equals 0. If a denominator equals 0, that trigonometric function of the angle θ is not defined.

Notice in the preceding definitions that if $a = 0$, that is, if the point (a, b) is on the y-axis, then the tangent function and the secant function are undefined. Also, if $b = 0$, that is, if the point (a, b) is on the x-axis, then the cosecant function and the cotangent function are undefined.

By constructing similar triangles, you should be convinced that the values of the six trigonometric functions of an angle θ do not depend on the selection of the point (a, b) on the terminal side of θ, but rather depend only on the angle θ itself. See Figure 42 for an illustration of this when θ lies in quadrant II. Since the triangles are similar, the ratio $\dfrac{b}{r}$ equals the ratio $\dfrac{b'}{r'}$, which equals $\sin \theta$. Also, the ratio $\dfrac{|a|}{r}$ equals the ratio $\dfrac{|a'|}{r'}$, so $\dfrac{a}{r} = \dfrac{a'}{r'}$, which equals $\cos \theta$. And so on.

Also, observe that if θ is acute these definitions reduce to the right triangle definitions given in Section 2.2, as illustrated in Figure 43.

Finally, from the definition of the six trigonometric functions of a general angle, we see that the Quotient and Reciprocal Identities hold. Also, using $r^2 = a^2 + b^2$ and dividing each side by r^2, we can derive the Pythagorean Identities for general angles.

Figure 42

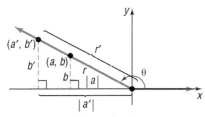

Figure 43

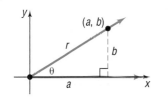

EXAMPLE 1

Finding the Exact Values of the Six Trigonometric Functions of θ, Given a Point on the Terminal Side

Find the exact value of each of the six trigonometric functions of a positive angle θ if $(4, -3)$ is a point on its terminal side.

Solution

Figure 44 illustrates the situation. For the point $(a, b) = (4, -3)$, we have $a = 4$ and $b = -3$. Then $r = \sqrt{a^2 + b^2} = \sqrt{16 + 9} = 5$, so

Figure 44

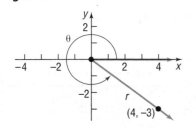

$$\sin \theta = \frac{b}{r} = -\frac{3}{5} \qquad \cos \theta = \frac{a}{r} = \frac{4}{5} \qquad \tan \theta = \frac{b}{a} = -\frac{3}{4}$$

$$\csc \theta = \frac{r}{b} = -\frac{5}{3} \qquad \sec \theta = \frac{r}{a} = \frac{5}{4} \qquad \cot \theta = \frac{a}{b} = -\frac{4}{3}$$

Now Work PROBLEM 11

In the next example, we find the exact value of each of the six trigonometric functions at the quadrantal angles $0, \dfrac{\pi}{2}, \pi,$ and $\dfrac{3\pi}{2}$.

| EXAMPLE 2 | **Finding the Exact Values of the Six Trigonometric Functions of Quadrantal Angles** |

Find the exact values of each of the six trigonometric functions of

(a) $\theta = 0 = 0°$ (b) $\theta = \dfrac{\pi}{2} = 90°$ (c) $\theta = \pi = 180°$ (d) $\theta = \dfrac{3\pi}{2} = 270°$

Solution

(a) We can choose any point on the terminal side of $\theta = 0 = 0°$. For convenience, we choose the point $P = (1, 0) = (a, b)$, which is a distance of $r = 1$ unit from the origin. See Figure 45. Then

Figure 45
$\theta = 0 = 0°$

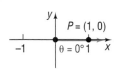

$$\sin 0 = \sin 0° = \frac{b}{r} = \frac{0}{1} = 0 \qquad \cos 0 = \cos 0° = \frac{a}{r} = \frac{1}{1} = 1$$

$$\tan 0 = \tan 0° = \frac{b}{a} = \frac{0}{1} = 0 \qquad \sec 0 = \sec 0° = \frac{r}{a} = \frac{1}{1} = 1$$

Since the y-coordinate of P is 0, $\csc 0$ and $\cot 0$ are not defined.

(b) The point $P = (0, 1) = (a, b)$ is on the terminal side of $\theta = \dfrac{\pi}{2} = 90°$ and is a distance of $r = 1$ unit from the origin. See Figure 46. Then

Figure 46
$\theta = \dfrac{\pi}{2} = 90°$

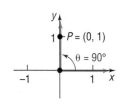

$$\sin \frac{\pi}{2} = \sin 90° = \frac{b}{r} = \frac{1}{1} = 1 \qquad \cos \frac{\pi}{2} = \cos 90° = \frac{a}{r} = \frac{0}{1} = 0$$

$$\csc \frac{\pi}{2} = \csc 90° = \frac{r}{b} = \frac{1}{1} = 1 \qquad \cot \frac{\pi}{2} = \cot 90° = \frac{a}{b} = \frac{0}{1} = 0$$

Since the x-coordinate of P is 0, $\tan \dfrac{\pi}{2}$ and $\sec \dfrac{\pi}{2}$ are not defined.

(c) The point $P = (-1, 0)$ is on the terminal side of $\theta = \pi = 180°$ and is a distance of $r = 1$ unit from the origin. See Figure 47. Then

Figure 47
$\theta = \pi = 180°$

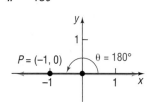

$$\sin \pi = \sin 180° = \frac{0}{1} = 0 \qquad \cos \pi = \cos 180° = \frac{-1}{1} = -1$$

$$\tan \pi = \tan 180° = \frac{0}{-1} = 0 \qquad \sec \pi = \sec 180° = \frac{1}{-1} = -1$$

Since the y-coordinate of P is 0, $\csc \pi$ and $\cot \pi$ are not defined.

(d) The point $P = (0, -1)$ is on the terminal side of $\theta = \dfrac{3\pi}{2} = 270°$ and is a distance of $r = 1$ unit from the origin. See Figure 48. Then

Figure 48
$\theta = \dfrac{3\pi}{2} = 270°$

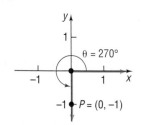

$$\sin \frac{3\pi}{2} = \sin 270° = \frac{-1}{1} = -1 \qquad \cos \frac{3\pi}{2} = \cos 270° = \frac{0}{1} = 0$$

$$\csc \frac{3\pi}{2} = \csc 270° = \frac{1}{-1} = -1 \qquad \cot \frac{3\pi}{2} = \cot 270° = \frac{0}{-1} = 0$$

Since the x-coordinate of P is 0, $\tan \dfrac{3\pi}{2}$ and $\sec \dfrac{3\pi}{2}$ are not defined.

Table 4 summarizes the values of the trigonometric functions found in Example 2.

Table 4

θ (Radians)	θ (Degrees)	$\sin \theta$	$\cos \theta$	$\tan \theta$	$\csc \theta$	$\sec \theta$	$\cot \theta$
0	0°	0	1	0	Not defined	1	Not defined
$\dfrac{\pi}{2}$	90°	1	0	Not defined	1	Not defined	0
π	180°	0	-1	0	Not defined	-1	Not defined
$\dfrac{3\pi}{2}$	270°	-1	0	Not defined	-1	Not defined	0

Solution (a) Refer to Figure 63. The reference angle for 135° is 45° and $\sin 45° = \dfrac{\sqrt{2}}{2}$. The angle 135° is in quadrant II, where the sine function is positive, so

$$\sin 135° = \sin 45° = \frac{\sqrt{2}}{2}$$

Figure 63

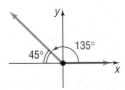

(b) Refer to Figure 64. The reference angle for 600° is 60° and $\cos 60° = \dfrac{1}{2}$. The angle 600° is in quadrant III, where the cosine function is negative, so

$$\cos 600° = -\cos 60° = -\frac{1}{2}$$

Figure 64

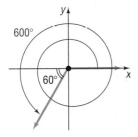

(c) Refer to Figure 65. The reference angle for $\dfrac{17\pi}{6}$ is $\dfrac{\pi}{6}$ and $\cos\dfrac{\pi}{6} = \dfrac{\sqrt{3}}{2}$. The angle $\dfrac{17\pi}{6}$ is in quadrant II, where the cosine function is negative, so

$$\cos\frac{17\pi}{6} = -\cos\frac{\pi}{6} = -\frac{\sqrt{3}}{2}$$

Figure 65

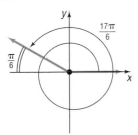

(d) Refer to Figure 66. The reference angle for $-\dfrac{\pi}{3}$ is $\dfrac{\pi}{3}$ and $\tan\dfrac{\pi}{3} = \sqrt{3}$. The angle $-\dfrac{\pi}{3}$ is in quadrant IV, where the tangent function is negative, so

$$\tan\left(-\frac{\pi}{3}\right) = -\tan\frac{\pi}{3} = -\sqrt{3}$$

Figure 66

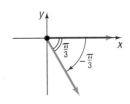

Finding the Values of the Trigonometric Functions of a General Angle

- If the angle θ is a quadrantal angle, draw the angle, pick a point on its terminal side, and apply the definition of the trigonometric functions.
- If the angle θ lies in a quadrant:

  1. Find the reference angle α of θ.
  2. Find the value of the trigonometric function at α.
  3. Adjust the sign (+ or −) of the value of the trigonometric function based on the quadrant in which θ lies.

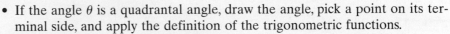

Now Work PROBLEMS 59 AND 61

6 Find the Exact Values of Trigonometric Functions of an Angle, Given Information about the Functions

EXAMPLE 7 Finding the Exact Values of Trigonometric Functions

Given that $\cos\theta = -\dfrac{2}{3}, \dfrac{\pi}{2} < \theta < \pi$, find the exact value of each of the remaining trigonometric functions.

Solution The angle θ lies in quadrant II, so we know that $\sin\theta$ and $\csc\theta$ are positive and the other four trigonometric functions are negative. If α is the reference angle for θ, then $\cos\alpha = \dfrac{2}{3} = \dfrac{\text{adjacent}}{\text{hypotenuse}}$. The values of the remaining trigonometric functions

Figure 67

$\cos \alpha = \dfrac{2}{3}$

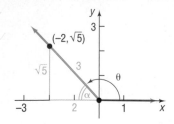

of the reference angle α can be found by drawing the appropriate triangle and using the Pythagorean Theorem to determine that the side opposite α is $\sqrt{5}$. We use Figure 67 to obtain

$$\sin \alpha = \frac{\sqrt{5}}{3} \qquad \cos \alpha = \frac{2}{3} \qquad \tan \alpha = \frac{\sqrt{5}}{2}$$

$$\csc \alpha = \frac{3}{\sqrt{5}} = \frac{3\sqrt{5}}{5} \qquad \sec \alpha = \frac{3}{2} \qquad \cot \alpha = \frac{2}{\sqrt{5}} = \frac{2\sqrt{5}}{5}$$

Now we assign the appropriate signs to each of these values to find the values of the trigonometric functions of θ.

$$\sin \theta = \frac{\sqrt{5}}{3} \qquad \cos \theta = -\frac{2}{3} \qquad \tan \theta = -\frac{\sqrt{5}}{2}$$

$$\csc \theta = \frac{3\sqrt{5}}{5} \qquad \sec \theta = -\frac{3}{2} \qquad \cot \theta = -\frac{2\sqrt{5}}{5}$$

Now Work PROBLEM 83

EXAMPLE 8 | **Finding the Exact Values of Trigonometric Functions**

If $\tan \theta = -4$ and $\sin \theta < 0$, find the exact value of each of the remaining trigonometric functions of θ.

Solution Since $\tan \theta = -4 < 0$ and $\sin \theta < 0$, it follows that θ lies in quadrant IV. If α is the reference angle for θ, then $\tan \alpha = 4 = \dfrac{4}{1} = \dfrac{b}{a}$. With $a = 1$ and $b = 4$, we find $r = \sqrt{1^2 + 4^2} = \sqrt{17}$. See Figure 68. Then

$$\sin \alpha = \frac{4}{\sqrt{17}} = \frac{4\sqrt{17}}{17} \qquad \cos \alpha = \frac{1}{\sqrt{17}} = \frac{\sqrt{17}}{17} \qquad \tan \alpha = \frac{4}{1} = 4$$

Figure 68

$\tan \alpha = 4$

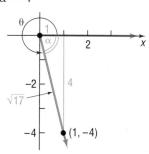

$$\csc \alpha = \frac{\sqrt{17}}{4} \qquad \sec \alpha = \frac{\sqrt{17}}{1} = \sqrt{17} \qquad \cot \alpha = \frac{1}{4}$$

We assign the appropriate sign to each of these to obtain the values of the trigonometric functions of θ.

$$\sin \theta = -\frac{4\sqrt{17}}{17} \qquad \cos \theta = \frac{\sqrt{17}}{17} \qquad \tan \theta = -4$$

$$\csc \theta = -\frac{\sqrt{17}}{4} \qquad \sec \theta = \sqrt{17} \qquad \cot \theta = -\frac{1}{4}$$

Now Work PROBLEM 93

2.4 Assess Your Understanding

Concepts and Vocabulary

1. For an angle θ that lies in quadrant III, the trigonometric functions _____ and _____ are positive.

2. Two angles in standard position that have the same terminal side are _____.

3. The reference angle of 240° is _____.

4. **True or False** $\sin 182° = \cos 2°$.

5. **True or False** $\tan \dfrac{\pi}{2}$ is not defined.

6. **True or False** The reference angle is always an acute angle.

7. What is the reference angle of 600°?

8. In which quadrants is the cosine function positive?

9. If $0 \le \theta < 2\pi$, for what angles θ, if any, is $\tan \theta$ undefined?

10. What is the reference angle of $\dfrac{13\pi}{3}$?

Skill Building

In Problems 11–20, a point on the terminal side of an angle θ is given. Find the exact value of each of the six trigonometric functions of θ.

11. $(-3, 4)$ **12.** $(5, -12)$ **13.** $(2, -3)$ **14.** $(-1, -2)$ **15.** $(-3, -3)$

16. $(2, -2)$ **17.** $\left(\dfrac{\sqrt{3}}{2}, \dfrac{1}{2}\right)$ **18.** $\left(-\dfrac{1}{2}, \dfrac{\sqrt{3}}{2}\right)$ **19.** $\left(\dfrac{\sqrt{2}}{2}, -\dfrac{\sqrt{2}}{2}\right)$ **20.** $\left(-\dfrac{\sqrt{2}}{2}, -\dfrac{\sqrt{2}}{2}\right)$

In Problems 21–32, use a coterminal angle to find the exact value of each expression. Do not use a calculator.

21. $\sin 405°$ **22.** $\cos 420°$ **23.** $\tan 405°$ **24.** $\sin 390°$ **25.** $\csc 450°$ **26.** $\sec 540°$

27. $\cot 390°$ **28.** $\sec 420°$ **29.** $\cos \dfrac{33\pi}{4}$ **30.** $\sin \dfrac{9\pi}{4}$ **31.** $\tan(21\pi)$ **32.** $\csc \dfrac{9\pi}{2}$

In Problems 33–40, name the quadrant in which the angle θ lies.

33. $\sin \theta > 0, \quad \cos \theta < 0$ **34.** $\sin \theta < 0, \quad \cos \theta > 0$ **35.** $\sin \theta < 0, \quad \tan \theta < 0$

36. $\cos \theta > 0, \quad \tan \theta > 0$ **37.** $\cos \theta > 0, \quad \cot \theta < 0$ **38.** $\sin \theta < 0, \quad \cot \theta > 0$

39. $\sec \theta < 0, \quad \tan \theta > 0$ **40.** $\csc \theta > 0, \quad \cot \theta < 0$

In Problems 41–58, find the reference angle of each angle.

41. $-30°$ **42.** $-60°$ **43.** $120°$ **44.** $210°$ **45.** $300°$ **46.** $330°$

47. $\dfrac{5\pi}{4}$ **48.** $\dfrac{5\pi}{6}$ **49.** $\dfrac{8\pi}{3}$ **50.** $\dfrac{7\pi}{4}$ **51.** $-135°$ **52.** $-240°$

53. $-\dfrac{2\pi}{3}$ **54.** $-\dfrac{7\pi}{6}$ **55.** $440°$ **56.** $490°$ **57.** $\dfrac{15\pi}{4}$ **58.** $\dfrac{19\pi}{6}$

In Problems 59–82, use the reference angle to find the exact value of each expression. Do not use a calculator.

59. $\sin 150°$ **60.** $\cos 210°$ **61.** $\sin 510°$ **62.** $\cos 600°$ **63.** $\cos(-45°)$ **64.** $\sin(-240°)$

65. $\sec 240°$ **66.** $\csc 300°$ **67.** $\cot 330°$ **68.** $\tan 225°$ **69.** $\sin \dfrac{3\pi}{4}$ **70.** $\cos \dfrac{2\pi}{3}$

71. $\cos \dfrac{13\pi}{4}$ **72.** $\tan \dfrac{8\pi}{3}$ **73.** $\sin\left(-\dfrac{2\pi}{3}\right)$ **74.** $\cot\left(-\dfrac{\pi}{6}\right)$ **75.** $\tan \dfrac{14\pi}{3}$ **76.** $\sec \dfrac{11\pi}{4}$

77. $\sin(8\pi)$ **78.** $\cos(-2\pi)$ **79.** $\tan(7\pi)$ **80.** $\cot(5\pi)$ **81.** $\sec(-3\pi)$ **82.** $\csc\left(-\dfrac{5\pi}{2}\right)$

In Problems 83–100, find the exact value of each of the remaining trigonometric functions of θ.

83. $\sin \theta = \dfrac{12}{13}, \quad \theta$ in quadrant II   **84.** $\cos \theta = \dfrac{3}{5}, \quad \theta$ in quadrant IV   **85.** $\cos \theta = -\dfrac{4}{5}, \quad \theta$ in quadrant III

86. $\sin \theta = -\dfrac{5}{13}, \quad \theta$ in quadrant III   **87.** $\sin \theta = \dfrac{5}{13}, \quad 90° < \theta < 180°$ **88.** $\cos \theta = \dfrac{4}{5}, \quad 270° < \theta < 360°$

89. $\cos \theta = -\dfrac{1}{3}, \quad 180° < \theta < 270°$ **90.** $\sin \theta = -\dfrac{2}{3}, \quad 180° < \theta < 270°$ **91.** $\sin \theta = \dfrac{2}{3}, \quad \tan \theta < 0$

92. $\cos \theta = -\dfrac{1}{4}, \quad \tan \theta > 0$ **93.** $\sec \theta = 2, \quad \sin \theta < 0$ **94.** $\csc \theta = 3, \quad \cot \theta < 0$

95. $\tan \theta = \dfrac{3}{4}, \quad \sin \theta < 0$ **96.** $\cot \theta = \dfrac{4}{3}, \quad \cos \theta < 0$ **97.** $\tan \theta = -\dfrac{1}{3}, \quad \sin \theta > 0$

98. $\sec \theta = -2, \quad \tan \theta > 0$ **99.** $\csc \theta = -2, \quad \tan \theta > 0$ **100.** $\cot \theta = -2, \quad \sec \theta > 0$

101. Find the exact value of $\sin 40° + \sin 130° + \sin 220° + \sin 310°$.

102. Find the exact value of $\tan 40° + \tan 140°$.

Mixed Practice

In Problems 103–106, $f(x) = \sin x, g(x) = \cos x, h(x) = \tan x, F(x) = \csc x, G(x) = \sec x,$ and $H(x) = \cot x.$

103. (a) Find $f(315°)$. What point is on the graph of f?

(b) Find $G(315°)$. What point is on the graph of G?

(c) Find $h(315°)$. What point is on the graph of h?

104. (a) Find $g(120°)$. What point is on the graph of g?

(b) Find $F(120°)$. What point is on the graph of F?

(c) Find $H(120°)$. What point is on the graph of H?

105. (a) Find $g\left(\dfrac{7\pi}{6}\right)$. What point is on the graph of g?

(b) Find $F\left(\dfrac{7\pi}{6}\right)$. What point is on the graph of F?

(c) Find $H(-315°)$. What point is on the graph of H?

106. (a) Find $f\left(\dfrac{7\pi}{4}\right)$. What point is on the graph of f?

(b) Find $G\left(\dfrac{7\pi}{4}\right)$. What point is on the graph of G?

(c) Find $F(-225°)$. What point is on the graph of F?

Applications and Extensions

107. If $f(\theta) = \sin \theta = 0.2$, find $f(\theta + \pi)$.

108. If $g(\theta) = \cos \theta = 0.4$, find $g(\theta + \pi)$.

109. If $F(\theta) = \tan \theta = 3$, find $F(\theta + \pi)$.

110. If $G(\theta) = \cot \theta = -2$, find $G(\theta + \pi)$.

111. If $\sin \theta = \dfrac{1}{5}$, find $\csc(\theta + \pi)$.

112. If $\cos \theta = \dfrac{2}{3}$, find $\sec(\theta + \pi)$.

113. Find the exact value of
$\sin 1° + \sin 2° + \sin 3° + \cdots + \sin 358° + \sin 359°$

114. Find the exact value of
$\cos 1° + \cos 2° + \cos 3° + \cdots + \cos 358° + \cos 359°$

115. Projectile Motion An object is propelled upward at an angle θ, $45° < \theta < 90°$, to the horizontal with an initial velocity of v_0 feet per second from the base of a plane that makes an angle of $45°$ with the horizontal. See the illustration. If air resistance is ignored, the distance R that it travels up the inclined plane is given by the function

$$R(\theta) = \frac{v_0^2 \sqrt{2}}{32}[\sin(2\theta) - \cos(2\theta) - 1]$$

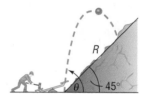

(a) Find the distance R that the object travels along the inclined plane if the initial velocity is 32 feet per second and $\theta = 60°$.

(b) Graph $R = R(\theta)$ if the initial velocity is 32 feet per second.

(c) What value of θ makes R largest?

Discussion and Writing

116. Give three examples that demonstrate how to use the theorem on reference angles.

117. Write a brief paragraph that explains how to quickly compute the value of the trigonometric functions of $0°, 90°, 180°,$ and $270°$.

118. Explain what a reference angle is. What role does it play in finding the value of a trigonometric function?

2.5 Unit Circle Approach; Properties of the Trigonometric Functions

PREPARING FOR THIS SECTION *Before getting started, review the following:*

- Unit Circle (Section 1.2, p. 22)
- Functions (Section 1.3, pp. 30–39)
- Even and Odd Functions (Section 1.5, pp. 50–52)

Now Work the 'Are You Prepared?' problems on page 159.

OBJECTIVES **1** Find the Exact Values of the Trigonometric Functions Using the Unit Circle (p. 151)

2 Know the Domain and Range of the Trigonometric Functions (p. 155)

3 Use the Periodic Properties to Find the Exact Values of the Trigonometric Functions (p. 156)

4 Use Even–Odd Properties to Find the Exact Values of the Trigonometric Functions (p. 158)

In this section, we develop important properties of the trigonometric functions. We begin by introducing the trigonometric functions using the unit circle. This approach will lead to the definition given earlier of the trigonometric functions of a general angle.

1 Find the Exact Values of the Trigonometric Functions Using the Unit Circle

Recall that the unit circle is a circle whose radius is 1 and whose center is at the origin of a rectangular coordinate system. Also recall that any circle of radius r has circumference of length $2\pi r$. Therefore, the unit circle (radius $= 1$) has a circumference of length 2π. In other words, for 1 revolution around the unit circle the length of the arc is 2π units.

The following discussion sets the stage for defining the trigonometric functions using the unit circle.

Let $t \geq 0$ be any real number and let s be the distance from the origin to t on the real number line. See the red portion of Figure 69(a). Now look at the unit circle in Figure 69(a). Beginning at the point $(1, 0)$ on the unit circle, travel $s = t$ units in the counterclockwise direction along the circle to arrive at the point $P = (a, b)$. In this sense, the length $s = t$ units is being **wrapped** around the unit circle.

If $t < 0$, we begin at the point $(1, 0)$ on the unit circle and travel $s = |t|$ units in the clockwise direction to arrive at the point $P = (a, b)$. See Figure 69(b).

If $t > 2\pi$ or if $t < -2\pi$, it will be necessary to travel around the unit circle more than once before arriving at point P. Do you see why?

Figure 69

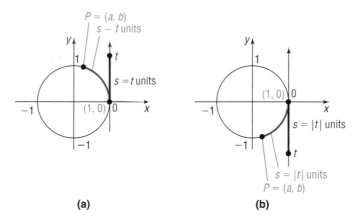

(a)   (b)

Let's describe this process another way. Picture a string of length $s = |t|$ units being wrapped around a circle of radius 1 unit. We start wrapping the string around the circle at the point $(1, 0)$. If $t \geq 0$, we wrap the string in the counterclockwise direction; if $t < 0$, we wrap the string in the clockwise direction. The point $P = (a, b)$ is the point where the string ends.

This discussion tells us that, for any real number t, we can locate a unique point $P = (a, b)$ on the unit circle. We call this point **the point P on the unit circle that corresponds to t.** This is the important idea here. No matter what real number t is chosen, there is a unique point P on the unit circle corresponding to it. We use the coordinates of the point $P = (a, b)$ on the unit circle corresponding to the real number t to define the **six trigonometric functions of t.** Be sure to consult Figures 69(a) and (b) as you read these definitions.

DEFINITION

Let t be a real number and let $P = (a, b)$ be the point on the unit circle that corresponds to t.

The **sine function** associates with t the y-coordinate of P and is denoted by

$$\sin t = b$$

The **cosine function** associates with t the x-coordinate of P and is denoted by

$$\cos t = a$$

If $a \neq 0$, the **tangent function** associates with t the ratio of the y-coordinate to the x-coordinate of P and is denoted by

$$\tan t = \frac{b}{a}$$

If $b \neq 0$, the **cosecant function** is defined as

$$\csc t = \frac{1}{b}$$

If $a \neq 0$, the **secant function** is defined as

$$\sec t = \frac{1}{a}$$

If $b \neq 0$, the **cotangent function** is defined as

$$\cot t = \frac{a}{b}$$

Once again, notice in these definitions that if $a = 0$ (that is, if the point P is on the y-axis) the tangent function and the secant function are undefined. Also, if $b = 0$ (that is, if the point P is on the x-axis), the cosecant function and the cotangent function are undefined.

Because we use the unit circle in these definitions of the trigonometric functions, they are also sometimes referred to as **circular functions.**

EXAMPLE 1   **Finding the Values of the Trigonometric Functions Using a Point on the Unit Circle**

Find the values of $\sin t$, $\cos t$, $\tan t$, $\csc t$, $\sec t$, and $\cot t$ if $P = \left(-\frac{1}{2}, \frac{\sqrt{3}}{2}\right)$ is the point on the unit circle that corresponds to the real number t.

Solution   See Figure 70. We follow the definition of the six trigonometric functions using

Figure 70

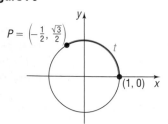

$$P = \left(-\frac{1}{2}, \frac{\sqrt{3}}{2}\right) = (a, b). \text{ Then, with } a = -\frac{1}{2} \text{ and } b = \frac{\sqrt{3}}{2}, \text{ we have}$$

$$\sin t = b = \frac{\sqrt{3}}{2} \qquad \cos t = a = -\frac{1}{2} \qquad \tan t = \frac{b}{a} = \frac{\frac{\sqrt{3}}{2}}{-\frac{1}{2}} = -\sqrt{3}$$

$$\csc t = \frac{1}{b} = \frac{1}{\frac{\sqrt{3}}{2}} = \frac{2}{\sqrt{3}} = \frac{2\sqrt{3}}{3} \qquad \sec t = \frac{1}{a} = \frac{1}{-\frac{1}{2}} = -2 \qquad \cot t = \frac{a}{b} = \frac{-\frac{1}{2}}{\frac{\sqrt{3}}{2}} = -\frac{1}{\sqrt{3}} = -\frac{\sqrt{3}}{3}$$

Now Work PROBLEM 9

Trigonometric Functions of Angles

Let $P = (a, b)$ be the point on the unit circle corresponding to the real number t. See Figure 71(a). Let θ be the angle in standard position, measured in radians, whose terminal side is the ray from the origin through P. See Figure 71(b). Since the unit circle has radius 1 unit, if $s = |t|$ units, then from the arc length formula $s = r\theta$, we have $\theta = t$ radians. See Figures 71(c) and (d).

Figure 71

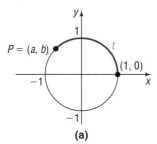 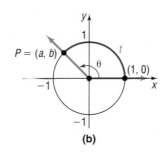 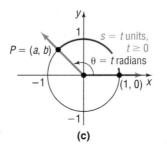 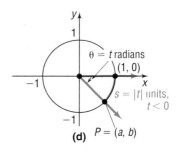

    (a)   (b)   (c)   (d)

The point $P = (a, b)$ on the unit circle that corresponds to the real number t is the point P on the terminal side of the angle $\theta = t$ radians. As a result, we can say that

$$\sin t = \sin \theta$$
$$\quad \uparrow \qquad\qquad \uparrow$$
$$\text{Real number} \quad \theta = t \text{ radians}$$

and so on. We can now define the trigonometric functions of the angle θ.

DEFINITION

If $\theta = t$ radians, the six **trigonometric functions of the angle θ** are defined as

$\sin \theta = \sin t$	$\cos \theta = \cos t$	$\tan \theta = \tan t$
$\csc \theta = \csc t$	$\sec \theta = \sec t$	$\cot \theta = \cot t$

Even though the trigonometric functions can be viewed both as functions of real numbers and as functions of angles, it is customary to refer to trigonometric functions of real numbers and trigonometric functions of angles collectively as *the trigonometric functions*. We will follow this practice from now on.

Since the values of the trigonometric functions of an angle θ are determined by the coordinates of the point $P = (a, b)$ on the unit circle corresponding to θ, the units used to measure the angle θ are irrelevant. For example, it does not matter whether we write $\theta = \dfrac{\pi}{2}$ radians or $\theta = 90°$. In either case, the point on the unit circle corresponding to this angle is $P = (0, 1)$. As a result,

$$\sin \frac{\pi}{2} = \sin 90° = 1 \quad \text{and} \quad \cos \frac{\pi}{2} = \cos 90° = 0$$

The discussion based on Figure 71 implies the following: To find the exact value of a trigonometric function of an angle θ requires that we locate the corresponding point $P^* = (a^*, b^*)$ on the unit circle. In fact, though, any circle whose center is at the origin can be used.

Let θ be any nonquadrantal angle placed in standard position. Let $P = (a, b)$ be the point on the circle $x^2 + y^2 = r^2$ that corresponds to θ, and let $P^* = (a^*, b^*)$ be the point on the unit circle that corresponds to θ. See Figure 72.

Notice that the triangles OA^*P^* and OAP are similar, so ratios of corresponding sides are equal.

$$\frac{b^*}{1} = \frac{b}{r} \qquad \frac{a^*}{1} = \frac{a}{r} \qquad \frac{b^*}{a^*} = \frac{b}{a}$$

$$\frac{1}{b^*} = \frac{r}{b} \qquad \frac{1}{a^*} = \frac{r}{a} \qquad \frac{a^*}{b^*} = \frac{a}{b}$$

Figure 72

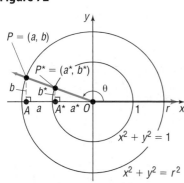

These results lead us to formulate the following theorem:

THEOREM

For an angle θ in standard position, let $P = (a, b)$ be any point on the terminal side of θ that is also on the circle $x^2 + y^2 = r^2$. Then

$$\sin \theta = \frac{b}{r} \qquad \cos \theta = \frac{a}{r} \qquad \tan \theta = \frac{b}{a} \quad a \neq 0$$

$$\csc \theta = \frac{r}{b} \quad b \neq 0 \qquad \sec \theta = \frac{r}{a} \quad a \neq 0 \qquad \cot \theta = \frac{a}{b} \quad b \neq 0$$

This result coincides with the definition given in Section 2.4 for the six trigonometric functions of a general angle θ.

Consider Figure 73, which shows a circle of radius $\sqrt{2}$. Inside the circle we have drawn a right triangle whose acute angles both measure $45°$. The lengths of the legs are then 1 unit each. The right triangle is Figure 29(b) on page 129. Therefore, the point on the circle that corresponds to an angle of $45°$ is $(1, 1) = (a, b)$. From this,

$$\sin 45° = \frac{b}{c} = \frac{b}{r} = \frac{1}{\sqrt{2}} = \frac{\sqrt{2}}{2}, \ \cos 45° = \frac{a}{c} = \frac{a}{r} = \frac{1}{\sqrt{2}} = \frac{\sqrt{2}}{2}, \text{and so on.}$$

Now consider Figure 74(a), which shows a circle of radius 2. Inside the circle we have drawn a right triangle whose acute angles measure $30°$ and $60°$. Refer to Figure 30(c) on page 131. If we place the $30°$ angle at the origin, we find that the point on the circle that corresponds to an angle of $30°$ is $\left(\sqrt{3}, 1\right)$. From this,

$$\sin 30° = \frac{b}{c} = \frac{b}{r} = \frac{1}{2}, \ \cos 30° = \frac{a}{c} = \frac{a}{r} = \frac{\sqrt{3}}{2}, \text{and so on. Figure 74(b) also shows}$$

a circle of radius 2, but the angle at the origin is $60°$. Now we can see that

$$\sin 60° = \frac{b}{c} = \frac{b}{r} = \frac{\sqrt{3}}{2}, \text{and so on.}$$

Figure 73

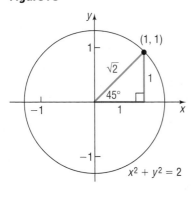

Figure 74

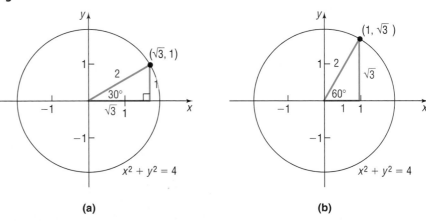

(a)   (b)

EXAMPLE 2   **Finding the Exact Values of the Six Trigonometric Functions**

Find the exact values of each of the six trigonometric functions of an angle θ if $(4, -3)$ is a point on its terminal side.

Figure 75

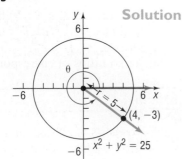

Solution   See Figure 75. The point $(4, -3)$ is on a circle of radius $r = \sqrt{4^2 + (-3)^2} = \sqrt{16 + 9} = \sqrt{25} = 5$ with the center at the origin.

For the point $(a, b) = (4, -3)$, we have $a = 4$ and $b = -3$. Since $r = 5$, we find

$$\sin \theta = \frac{b}{r} = -\frac{3}{5} \qquad \cos \theta = \frac{a}{r} = \frac{4}{5} \qquad \tan \theta = \frac{b}{a} = -\frac{3}{4}$$

$$\csc \theta = \frac{r}{b} = -\frac{5}{3} \qquad \sec \theta = \frac{r}{a} = \frac{5}{4} \qquad \cot \theta = \frac{a}{b} = -\frac{4}{3}$$

$\blacksquare$

━━━ **Now Work** PROBLEM 15

2 Know the Domain and Range of the Trigonometric Functions

Let θ be an angle in standard position, and let $P = (a, b)$ be the point on the unit circle that corresponds to θ. See Figure 76. Then, by the definition given earlier:

Figure 76

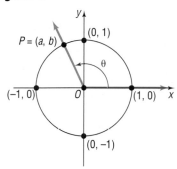

$$\sin \theta = b \qquad \cos \theta = a \qquad \tan \theta = \frac{b}{a} \quad a \neq 0$$

$$\csc \theta = \frac{1}{b} \quad b \neq 0 \qquad \sec \theta = \frac{1}{a} \quad a \neq 0 \qquad \cot \theta = \frac{a}{b} \quad b \neq 0$$

For $\sin \theta$ and $\cos \theta$, θ can be any angle, so it follows that the domain of the sine function and cosine function is the set of all real numbers.

> The domain of the sine function is the set of all real numbers.
>
> The domain of the cosine function is the set of all real numbers.

If $a = 0$, then the tangent function and the secant function are not defined. That is, for the tangent function and secant function, the x-coordinate of $P = (a, b)$ cannot be 0. See Figure 76. On the unit circle, there are two such points, $(0, 1)$ and $(0, -1)$. These two points correspond to the angles $\frac{\pi}{2} (90°)$ and $\frac{3\pi}{2} (270°)$ or, more generally, to any angle that is an odd integer multiple of $\frac{\pi}{2} (90°)$, such as $\pm \frac{\pi}{2} (\pm 90°)$, $\pm \frac{3\pi}{2} (\pm 270°)$, and $\pm \frac{5\pi}{2} (\pm 450°)$. Such angles must therefore be excluded from the domain of the tangent function and secant function.

> The domain of the tangent function is the set of all real numbers, except odd integer multiples of $\frac{\pi}{2} (90°)$.
>
> The domain of the secant function is the set of all real numbers, except odd integer multiples of $\frac{\pi}{2} (90°)$.

If $b = 0$, then the cotangent function and the cosecant function are not defined. For the cotangent function and cosecant function, the y-coordinate of $P = (a, b)$ cannot be 0. See Figure 76. On the unit circle, there are two such points, $(1, 0)$ and $(-1, 0)$. These two points correspond to the angles $0 (0°)$ and $\pi (180°)$ or, more generally, to any angle that is an integer multiple of $\pi (180°)$, such as $0(0°)$, $\pm \pi (\pm 180°)$, $\pm 2\pi (\pm 360°)$, and $\pm 3\pi (\pm 540°)$. Such angles must be excluded from the domain of the cotangent function and cosecant function.

> The domain of the cotangent function is the set of all real numbers, except integer multiples of $\pi (180°)$.
>
> The domain of the cosecant function is the set of all real numbers, except integer multiples of $\pi (180°)$.

Next, we determine the range of each of the six trigonometric functions. Refer again to Figure 76. Let $P = (a, b)$ be the point on the unit circle that corresponds to the angle θ. It follows that $-1 \leq a \leq 1$ and $-1 \leq b \leq 1$. Since $\sin \theta = b$ and $\cos \theta = a$, we have

$$-1 \leq \sin \theta \leq 1 \quad \text{and} \quad -1 \leq \cos \theta \leq 1$$

The range of both the sine function and the cosine function consists of all real numbers between -1 and 1, inclusive. Using absolute value notation, we have $|\sin \theta| \leq 1$ and $|\cos \theta| \leq 1$.

If θ is not an integer multiple of $\pi(180°)$, then $\csc\theta = \dfrac{1}{b}$. Since $b = \sin\theta$ and $|b| = |\sin\theta| \le 1$, it follows that $|\csc\theta| = \dfrac{1}{|\sin\theta|} = \dfrac{1}{|b|} \ge 1$. The range of the cosecant function consists of all real numbers less than or equal to -1 or greater than or equal to 1. That is,

$$\csc\theta \le -1 \quad \text{or} \quad \csc\theta \ge 1$$

If θ is not an odd integer multiple of $\dfrac{\pi}{2}(90°)$, then $\sec\theta = \dfrac{1}{a}$. Since $a = \cos\theta$ and $|a| = |\cos\theta| \le 1$, it follows that $|\sec\theta| = \dfrac{1}{|\cos\theta|} = \dfrac{1}{|a|} \ge 1$. The range of the secant function consists of all real numbers less than or equal to -1 or greater than or equal to 1. That is,

$$\sec\theta \le -1 \quad \text{or} \quad \sec\theta \ge 1$$

The range of both the tangent function and the cotangent function consists of all real numbers. That is,

$$-\infty < \tan\theta < \infty \quad \text{and} \quad -\infty < \cot\theta < \infty$$

You are asked to prove this in Problems 91 and 92.
Table 6 summarizes these results.

Table 6

Function	Symbol	Domain	Range
sine	$f(\theta) = \sin\theta$	All real numbers	All real numbers from -1 to 1, inclusive
cosine	$f(\theta) = \cos\theta$	All real numbers	All real numbers from -1 to 1, inclusive
tangent	$f(\theta) = \tan\theta$	All real numbers, except odd integer multiples of $\dfrac{\pi}{2}(90°)$	All real numbers
cosecant	$f(\theta) = \csc\theta$	All real numbers, except integer multiples of $\pi(180°)$	All real numbers greater than or equal to 1 or less than or equal to -1
secant	$f(\theta) = \sec\theta$	All real numbers, except odd integer multiples of $\dfrac{\pi}{2}(90°)$	All real numbers greater than or equal to 1 or less than or equal to -1
cotangent	$f(\theta) = \cot\theta$	All real numbers, except integer multiples of $\pi(180°)$	All real numbers

━━━━━ **Now Work** PROBLEMS 61 AND 65

3 Use the Periodic Properties to Find the Exact Values of the Trigonometric Functions

Look at Figure 77. This figure shows that for an angle of $\dfrac{\pi}{3}$ radians the corresponding point P on the unit circle is $\left(\dfrac{1}{2}, \dfrac{\sqrt{3}}{2}\right)$. Notice that for an angle of $\dfrac{\pi}{3} + 2\pi$ radians the corresponding point P on the unit circle is also $\left(\dfrac{1}{2}, \dfrac{\sqrt{3}}{2}\right)$. As a result,

$$\sin\frac{\pi}{3} = \frac{\sqrt{3}}{2} \quad \text{and} \quad \sin\left(\frac{\pi}{3} + 2\pi\right) = \frac{\sqrt{3}}{2}$$

$$\cos\frac{\pi}{3} = \frac{1}{2} \quad \text{and} \quad \cos\left(\frac{\pi}{3} + 2\pi\right) = \frac{1}{2}$$

Figure 77

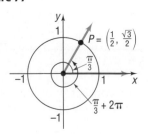

Figure 78

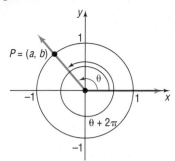

This example illustrates a more general situation. For a given angle θ, measured in radians, suppose that we know the corresponding point $P = (a, b)$ on the unit circle. Now add 2π to θ. The point on the unit circle corresponding to $\theta + 2\pi$ is identical to the point P on the unit circle corresponding to θ. See Figure 78. The values of the trigonometric functions of $\theta + 2\pi$ are equal to the values of the corresponding trigonometric functions of θ.

If we add (or subtract) integer multiples of 2π to θ, the values of the sine and cosine function remain unchanged. That is, for all θ

$$\sin(\theta + 2\pi k) = \sin \theta \qquad \cos(\theta + 2\pi k) = \cos \theta$$
$$\text{where } k \text{ is any integer} \qquad\qquad\qquad \textbf{(1)}$$

Functions that exhibit this kind of behavior are called *periodic functions*.

DEFINITION

A function f is called **periodic** if there is a positive number p such that, whenever θ is in the domain of f, so is $\theta + p$, and

$$f(\theta + p) = f(\theta)$$

If there is a smallest such number p, this smallest value is called the **(fundamental) period** of f.

Figure 79

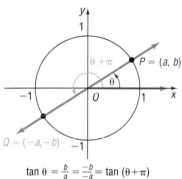

$$\tan \theta = \frac{b}{a} = \frac{-b}{-a} = \tan(\theta + \pi)$$

Based on equation (1), the sine and cosine functions are periodic. In fact, the sine, cosine, secant, and cosecant functions have period 2π. You are asked to prove this in Problems 93 through 96.

The tangent and cotangent functions are periodic with period π. See Figure 79 for a partial justification. You are asked to prove this statement in Problems 97 and 98.

These facts are summarized as follows:

Periodic Properties

$$\sin(\theta + 2\pi) = \sin \theta \qquad \cos(\theta + 2\pi) = \cos \theta \qquad \tan(\theta + \pi) = \tan \theta$$
$$\csc(\theta + 2\pi) = \csc \theta \qquad \sec(\theta + 2\pi) = \sec \theta \qquad \cot(\theta + \pi) = \cot \theta$$

In Words
Tangent and cotangent have period π; the others have period 2π.

Because the sine, cosine, secant, and cosecant functions have period 2π, once we know their values for $0 \le \theta < 2\pi$, we know all their values; similarly, since the tangent and cotangent functions have period π, once we know their values for $0 \le \theta < \pi$, we know all their values.

EXAMPLE 3

Using Periodic Properties to Find Exact Values

Find the exact value of:

(a) $\sin 420°$ 　　　　(b) $\tan \dfrac{5\pi}{4}$ 　　　　(c) $\cos \dfrac{11\pi}{4}$

Solution

(a) $\sin 420° = \sin(60° + 360°) = \sin 60° = \dfrac{\sqrt{3}}{2}$

(b) $\tan \dfrac{5\pi}{4} = \tan\left(\dfrac{\pi}{4} + \pi\right) = \tan \dfrac{\pi}{4} = 1$

(c) $\cos \dfrac{11\pi}{4} = \cos\left(\dfrac{3\pi}{4} + \dfrac{8\pi}{4}\right) = \cos\left(\dfrac{3\pi}{4} + 2\pi\right) = \cos \dfrac{3\pi}{4} = -\dfrac{\sqrt{2}}{2}$

The periodic properties of the trigonometric functions will be very helpful to us when we study their graphs in the next section.

— **Now Work** PROBLEMS 21 AND 79

4 Use Even–Odd Properties to Find the Exact Values of the Trigonometric Functions

Recall that a function f is even if $f(-\theta) = f(\theta)$ for all θ in the domain of f; a function f is odd if $f(-\theta) = -f(\theta)$ for all θ in the domain of f. We will now show that the trigonometric functions sine, tangent, cotangent, and cosecant are odd functions and the functions cosine and secant are even functions.

THEOREM

Even–Odd Properties

$$\sin(-\theta) = -\sin\theta \qquad \cos(-\theta) = \cos\theta \qquad \tan(-\theta) = -\tan\theta$$

$$\csc(-\theta) = -\csc\theta \qquad \sec(-\theta) = \sec\theta \qquad \cot(-\theta) = -\cot\theta$$

Figure 80

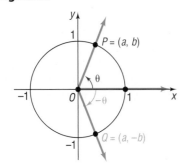

Proof Let $P = (a, b)$ be the point on the unit circle that corresponds to the angle θ. See Figure 80. The point Q on the unit circle that corresponds to the angle $-\theta$ will have coordinates $(a, -b)$. Using the definition for the trigonometric functions, we have

$$\sin\theta = b \qquad \sin(-\theta) = -b \qquad \cos\theta = a \qquad \cos(-\theta) = a$$

so

$$\sin(-\theta) = -\sin\theta \qquad\qquad \cos(-\theta) = \cos\theta$$

Now, using these results and some of the Fundamental Identities, we have

$$\tan(-\theta) = \frac{\sin(-\theta)}{\cos(-\theta)} = \frac{-\sin\theta}{\cos\theta} = -\tan\theta$$

$$\cot(-\theta) = \frac{1}{\tan(-\theta)} = \frac{1}{-\tan\theta} = -\cot\theta$$

$$\sec(-\theta) = \frac{1}{\cos(-\theta)} = \frac{1}{\cos\theta} = \sec\theta$$

$$\csc(-\theta) = \frac{1}{\sin(-\theta)} = \frac{1}{-\sin\theta} = -\csc\theta \qquad\blacksquare$$

In Words
Cosine and secant are even functions; the others are odd functions.

EXAMPLE 4 | **Finding Exact Values Using Even–Odd Properties**

Find the exact value of:

(a) $\sin(-45°)$ (b) $\cos(-\pi)$ (c) $\cot\left(-\dfrac{3\pi}{2}\right)$ (d) $\tan\left(-\dfrac{37\pi}{4}\right)$

Solution

(a) $\sin(-45°) = -\sin 45° = -\dfrac{\sqrt{2}}{2}$

       ↑

     Odd function

(b) $\cos(-\pi) = \cos\pi = -1$

    ↑

  Even function

(c) $\cot\left(-\dfrac{3\pi}{2}\right) = -\cot\dfrac{3\pi}{2} = 0$
  ↑
  Odd function

(d) $\tan\left(-\dfrac{37\pi}{4}\right) = -\tan\dfrac{37\pi}{4} = -\tan\left(\dfrac{\pi}{4} + 9\pi\right) = -\tan\dfrac{\pi}{4} = -1$
  ↑   ↑
  Odd function   *Period is π.*

━━━━▸ **Now Work** P R O B L E M S **37 A N D 73**

2.5 Assess Your Understanding

'Are You Prepared?' *Answers are given at the end of these exercises. If you get a wrong answer, read the pages listed in* red.

1. What is the equation of the unit circle? (p. 22)

2. The domain of the function $f(x) = \dfrac{3x - 6}{x - 4}$ is _____.
   (pp. 37–38)

3. A function for which $f(x) = f(-x)$ for all x in the domain of f is called a(n) _____ function. (p. 50)

Concepts and Vocabulary

4. The sine, cosine, cosecant, and secant functions have period _____; the tangent and cotangent functions have period _____.

5. The domain of the tangent function is _____.

6. The range of the sine function is _____.

7. If $\sin\theta = 0.2$, then $\sin(-\theta) = $ _____ and $\sin(\theta + 2\pi) = $ _____.

8. *True or False* The only even trigonometric functions are the cosine and secant functions.

Skill Building

In Problems 9–14, the point P on the unit circle that corresponds to a real number t is given. Find $\sin t, \cos t, \tan t, \csc t, \sec t,$ *and* $\cot t$.

9. $\left(\dfrac{\sqrt{3}}{2}, -\dfrac{1}{2}\right)$
10. $\left(-\dfrac{\sqrt{3}}{2}, -\dfrac{1}{2}\right)$
11. $\left(-\dfrac{\sqrt{2}}{2}, -\dfrac{\sqrt{2}}{2}\right)$
12. $\left(\dfrac{\sqrt{2}}{2}, -\dfrac{\sqrt{2}}{2}\right)$
13. $\left(\dfrac{\sqrt{5}}{3}, \dfrac{2}{3}\right)$
14. $\left(-\dfrac{\sqrt{5}}{5}, \dfrac{2\sqrt{5}}{5}\right)$

In Problems 15–20, the point P on the circle $x^2 + y^2 = r^2$ that is also on the terminal side of an angle θ in standard position is given. Find $\sin\theta, \cos\theta, \tan\theta, \csc\theta, \sec\theta,$ and $\cot\theta$.

15. $(3, -4)$
16. $(4, -3)$
17. $(-2, 3)$
18. $(2, -4)$
19. $(-1, -1)$
20. $(-3, 1)$

In Problems 21–36, use the fact that the trigonometric functions are periodic to find the exact value of each expression. Do not use a calculator.

21. $\sin 405°$
22. $\cos 420°$
23. $\tan 405°$
24. $\sin 390°$

25. $\csc 450°$
26. $\sec 540°$
27. $\cot 390°$
28. $\sec 420°$

29. $\cos\dfrac{33\pi}{4}$
30. $\sin\dfrac{9\pi}{4}$
31. $\tan(21\pi)$
32. $\csc\dfrac{9\pi}{2}$

33. $\sec\dfrac{17\pi}{4}$
34. $\cot\dfrac{17\pi}{4}$
35. $\tan\dfrac{19\pi}{6}$
36. $\sec\dfrac{25\pi}{6}$

In Problems 37–54, use the even–odd properties to find the exact value of each expression. Do not use a calculator.

37. $\sin(-60°)$
38. $\cos(-30°)$
39. $\tan(-30°)$
40. $\sin(-135°)$
41. $\sec(-60°)$

42. $\csc(-30°)$
43. $\sin(-90°)$
44. $\cos(-270°)$
45. $\tan\left(-\dfrac{\pi}{4}\right)$
46. $\sin(-\pi)$

47. $\cos\left(-\dfrac{\pi}{4}\right)$
48. $\sin\left(-\dfrac{\pi}{3}\right)$
49. $\tan(-\pi)$
50. $\sin\left(-\dfrac{3\pi}{2}\right)$

51. $\csc\left(-\dfrac{\pi}{4}\right)$
52. $\sec(-\pi)$
53. $\sec\left(\dfrac{\pi}{6}\right)$
54. $\csc\left(-\dfrac{\pi}{3}\right)$

In Problems 55–60, find the exact value of each expression. Do not use a calculator.

55. $\sin(-\pi) + \cos(5\pi)$

56. $\tan\left(-\dfrac{5\pi}{6}\right) - \cot\dfrac{7\pi}{2}$

57. $\sec(-\pi) + \csc\left(-\dfrac{\pi}{2}\right)$

58. $\tan(-6\pi) + \cos\dfrac{9\pi}{4}$

59. $\sin\left(-\dfrac{9\pi}{4}\right) - \tan\left(-\dfrac{9\pi}{4}\right)$

60. $\cos\left(-\dfrac{17\pi}{4}\right) - \sin\left(-\dfrac{3\pi}{2}\right)$

61. What is the domain of the sine function?

62. What is the domain of the cosine function?

63. For what numbers θ is $f(\theta) = \tan\theta$ not defined?

64. For what numbers θ is $f(\theta) = \cot\theta$ not defined?

65. For what numbers θ is $f(\theta) = \sec\theta$ not defined?

66. For what numbers θ is $f(\theta) = \csc\theta$ not defined?

67. What is the range of the sine function?

68. What is the range of the cosine function?

69. What is the range of the tangent function?

70. What is the range of the cotangent function?

71. What is the range of the secant function?

72. What is the range of the cosecant function?

73. Is the sine function even, odd, or neither? Is its graph symmetric? With respect to what?

74. Is the cosine function even, odd, or neither? Is its graph symmetric? With respect to what?

75. Is the tangent function even, odd, or neither? Is its graph symmetric? With respect to what?

76. Is the cotangent function even, odd, or neither? Is its graph symmetric? With respect to what?

77. Is the secant function even, odd, or neither? Is its graph symmetric? With respect to what?

78. Is the cosecant function even, odd, or neither? Is its graph symmetric? With respect to what?

79. If $\sin\theta = 0.3$, find the value of:
$$\sin\theta + \sin(\theta + 2\pi) + \sin(\theta + 4\pi)$$

80. If $\cos\theta = 0.2$, find the value of:
$$\cos\theta + \cos(\theta + 2\pi) + \cos(\theta + 4\pi)$$

81. If $\tan\theta = 3$, find the value of:
$$\tan\theta + \tan(\theta + \pi) + \tan(\theta + 2\pi)$$

82. If $\cot\theta = -2$, find the value of:
$$\cot\theta + \cot(\theta - \pi) + \cot(\theta - 2\pi)$$

In Problems 83–88, use the periodic and even–odd properties.

83. If $f(x) = \sin x$ and $f(a) = \dfrac{1}{3}$, find the exact value of:
   (a) $f(-a)$ (b) $f(a) + f(a + 2\pi) + f(a + 4\pi)$

84. If $f(x) = \cos x$ and $f(a) = \dfrac{1}{4}$, find the exact value of:
   (a) $f(-a)$ (b) $f(a) + f(a + 2\pi) + f(a - 2\pi)$

85. If $f(x) = \tan x$ and $f(a) = 2$, find the exact value of:
   (a) $f(-a)$ (b) $f(a) + f(a + \pi) + f(a + 2\pi)$

86. If $f(x) = \cot x$ and $f(a) = -3$, find the exact value of:
   (a) $f(-a)$ (b) $f(a) + f(a + \pi) + f(a + 4\pi)$

87. If $f(x) = \sec x$ and $f(a) = -4$, find the exact value of:
   (a) $f(-a)$ (b) $f(a) + f(a + 2\pi) + f(a + 4\pi)$

88. If $f(x) = \csc x$ and $f(a) = 2$, find the exact value of:
   (a) $f(-a)$ (b) $f(a) + f(a + 2\pi) + f(a + 4\pi)$

Applications and Extensions

In Problems 89–90, use the figure to approximate the value of the six trigonometric functions at t to the nearest tenth. Then use a calculator to approximate each of the six trigonometric functions at t.

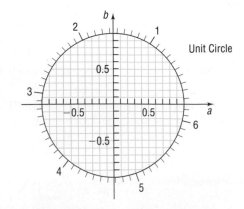

89. (a) $t = 1$ (b) $t = 5.1$

90. (a) $t = 2$ (b) $t = 4$

91. Show that the range of the tangent function is the set of all real numbers.

92. Show that the range of the cotangent function is the set of all real numbers.

93. Show that the period of $f(\theta) = \sin\theta$ is 2π.
   [**Hint:** Assume that $0 < p < 2\pi$ exists so that $\sin(\theta + p) = \sin\theta$ for all θ. Let $\theta = 0$ to find p. Then let $\theta = \dfrac{\pi}{2}$ to obtain a contradiction.]

94. Show that the period of $f(\theta) = \cos\theta$ is 2π.

95. Show that the period of $f(\theta) = \sec\theta$ is 2π.

96. Show that the period of $f(\theta) = \csc\theta$ is 2π.

97. Show that the period of $f(\theta) = \tan\theta$ is π.

98. Show that the period of $f(\theta) = \cot\theta$ is π.

99. If θ, $0 < \theta < \pi$, is the angle between a horizontal ray directed to the right (say, the positive x-axis) and a nonhorizontal, nonvertical line L, show that the slope m of L equals $\tan\theta$. The angle θ is called the **inclination** of L.

[**Hint:** See the illustration, where we have drawn the line M parallel to L and passing through the origin. Use the fact that M intersects the unit circle at the point $(\cos\theta, \sin\theta)$.]

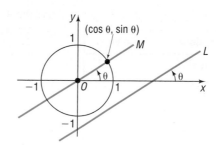

Discussion and Writing

100. Explain how you would find the value of $\sin 390°$ using periodic properties.

101. Explain how you would find the value of $\cos(-45°)$ using even–odd properties.

102. Write down five properties of the tangent function. Explain the meaning of each.

103. Describe your understanding of the meaning of a periodic function.

'Are You Prepared?' Answers

1. $x^2 + y^2 = 1$　　**2.** $\{x \mid x \neq 4\}$　　**3.** even

2.6 Graphs of the Sine and Cosine Functions*

PREPARING FOR THIS SECTION　*Before getting started, review the following:*

- Graphing Techniques: Transformations (Section 1.7, pp. 67–76)

　Now Work the 'Are You Prepared?' problems on page 172.

OBJECTIVES　**1** Graph Functions of the Form $y = A\sin(\omega x)$ Using Transformations (p. 163)
　2 Graph Functions of the Form $y = A\cos(\omega x)$ Using Transformations (p. 164)
　3 Determine the Amplitude and Period of Sinusoidal Functions (p. 166)
　4 Graph Sinusoidal Functions Using Key Points (p. 167)
　5 Find an Equation for a Sinusoidal Graph (p. 171)

Since we want to graph the trigonometric functions in the xy-plane, we shall use the traditional symbols x for the independent variable (or argument) and y for the dependent variable (or value at x) for each function. So we write the six trigonometric functions as

$y = f(x) = \sin x$	$y = f(x) = \cos x$	$y = f(x) = \tan x$
$y = f(x) = \csc x$	$y = f(x) = \sec x$	$y = f(x) = \cot x$

 Here the independent variable x represents an angle, measured in radians. In calculus, x will usually be treated as a real number. As we said earlier, these are equivalent ways of viewing x.

The Graph of the Sine Function $y = \sin x$

Since the sine function has period 2π, we need to graph $y = \sin x$ only on the interval $[0, 2\pi]$. The remainder of the graph will consist of repetitions of this portion of the graph.

*For those who wish to include phase shifts here, Section 2.8 can be covered immediately after Section 2.6 without loss of continuity.

Table 7

x	y = sin x	(x, y)
0	0	(0, 0)
$\dfrac{\pi}{6}$	$\dfrac{1}{2}$	$\left(\dfrac{\pi}{6}, \dfrac{1}{2}\right)$
$\dfrac{\pi}{2}$	1	$\left(\dfrac{\pi}{2}, 1\right)$
$\dfrac{5\pi}{6}$	$\dfrac{1}{2}$	$\left(\dfrac{5\pi}{6}, \dfrac{1}{2}\right)$
π	0	$(\pi, 0)$
$\dfrac{7\pi}{6}$	$-\dfrac{1}{2}$	$\left(\dfrac{7\pi}{6}, -\dfrac{1}{2}\right)$
$\dfrac{3\pi}{2}$	-1	$\left(\dfrac{3\pi}{2}, -1\right)$
$\dfrac{11\pi}{6}$	$-\dfrac{1}{2}$	$\left(\dfrac{11\pi}{6}, -\dfrac{1}{2}\right)$
2π	0	$(2\pi, 0)$

We begin by constructing Table 7, which lists some points on the graph of $y = \sin x$, $0 \le x \le 2\pi$. As the table shows, the graph of $y = \sin x$, $0 \le x \le 2\pi$, begins at the origin. As x increases from 0 to $\dfrac{\pi}{2}$, the value of $y = \sin x$ increases from 0 to 1; as x increases from $\dfrac{\pi}{2}$ to π to $\dfrac{3\pi}{2}$, the value of y decreases from 1 to 0 to -1; as x increases from $\dfrac{3\pi}{2}$ to 2π, the value of y increases from -1 to 0. If we plot the points listed in Table 7 and connect them with a smooth curve, we obtain the graph shown in Figure 81.

Figure 81
$y = \sin x$, $0 \le x \le 2\pi$

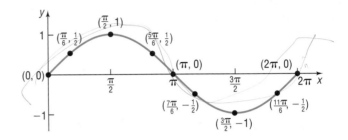

The graph in Figure 81 is one period, or **cycle,** of the graph of $y = \sin x$. To obtain a more complete graph of $y = \sin x$, we continue the graph in each direction, as shown in Figure 82(a). Figure 82(b) shows the graph on a TI-84 Plus graphing calculator.

Figure 82
$y = \sin x$, $-\infty < x < \infty$

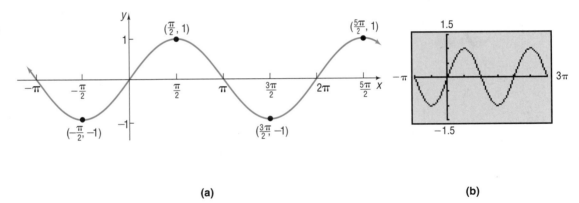

(a)   (b)

The graph of $y = \sin x$ illustrates some of the facts that we already know about the sine function.

Properties of the Sine Function $y = \sin x$

1. The domain is the set of all real numbers.
2. The range consists of all real numbers from -1 to 1, inclusive.
3. The sine function is an odd function, as the symmetry of the graph with respect to the origin indicates.
4. The sine function is periodic, with period 2π.
5. The x-intercepts are $\ldots, -2\pi, -\pi, 0, \pi, 2\pi, 3\pi, \ldots$; the y-intercept is 0.
6. The maximum value is 1 and occurs at $x = \ldots, -\dfrac{3\pi}{2}, \dfrac{\pi}{2}, \dfrac{5\pi}{2}, \dfrac{9\pi}{2}, \ldots$;

   the minimum value is -1 and occurs at $x = \ldots, -\dfrac{\pi}{2}, \dfrac{3\pi}{2}, \dfrac{7\pi}{2}, \dfrac{11\pi}{2}, \ldots$.

Now Work PROBLEMS **9**, **11**, AND **13**

1 Graph Functions of the Form $y = A \sin(\omega x)$ Using Transformations

EXAMPLE 1	**Graphing Functions of the Form $y = A \sin(\omega x)$ Using Transformations**

Graph $y = 3 \sin x$ using transformations. Use the graph to determine the domain and the range of $y = 3 \sin x$.

Solution   Figure 83 illustrates the steps.

Figure 83

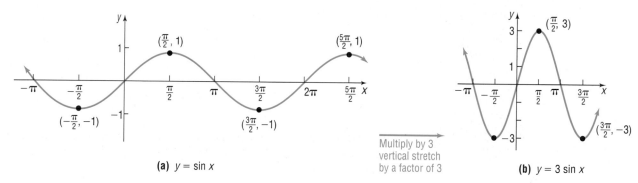

(a) $y = \sin x$

Multiply by 3
vertical stretch
by a factor of 3

(b) $y = 3 \sin x$

The domain of $y = 3 \sin x$ is the set of all real numbers or $(-\infty, \infty)$. The range of $y = 3 \sin x$ is $\{y \mid -3 \le y \le 3\}$ or $[-3, 3]$.

✓ Check: Graph $Y_1 = 3 \sin x$ to verify the graph shown in Figure 83(b).

EXAMPLE 2	**Graphing Functions of the Form $y = A \sin(\omega x)$ Using Transformations**

Graph $y = -\sin(2x)$ using transformations. Use the graph to determine the domain and the range of $y = -\sin(2x)$.

Solution   Figure 84 illustrates the steps.

Figure 84

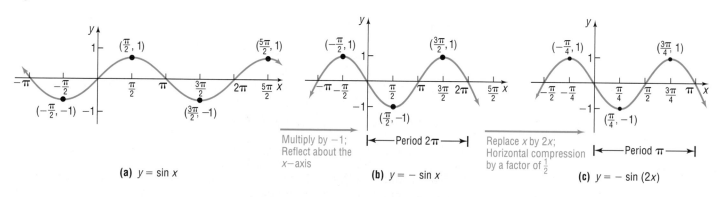

(a) $y = \sin x$

Multiply by -1;
Reflect about the
x-axis

(b) $y = -\sin x$

|←——Period 2π——→|

Replace x by $2x$;
Horizontal compression
by a factor of $\frac{1}{2}$

|←——Period π——→|

(c) $y = -\sin(2x)$

The domain of $y = -\sin(2x)$ is the set of all real numbers. The range of $y = -\sin(2x)$ is $\{y \mid -1 \le y \le 1\}$ or $[-1, 1]$.

Notice in Figure 84(c) that the period of the function $y = -\sin(2x)$ is π due to the horizontal compression of the original period 2π by a factor of $\dfrac{1}{2}$.

✓ Check: Graph $Y_1 = -\sin(2x)$ to verify the graph shown in Figure 84(c).

Now Work PROBLEM 45 USING TRANSFORMATIONS

Table 8

x	$y = \cos x$	(x, y)
0	1	$(0, 1)$
$\dfrac{\pi}{3}$	$\dfrac{1}{2}$	$\left(\dfrac{\pi}{3}, \dfrac{1}{2}\right)$
$\dfrac{\pi}{2}$	0	$\left(\dfrac{\pi}{2}, 0\right)$
$\dfrac{2\pi}{3}$	$-\dfrac{1}{2}$	$\left(\dfrac{2\pi}{3}, -\dfrac{1}{2}\right)$
π	-1	$(\pi, -1)$
$\dfrac{4\pi}{3}$	$-\dfrac{1}{2}$	$\left(\dfrac{4\pi}{3}, -\dfrac{1}{2}\right)$
$\dfrac{3\pi}{2}$	0	$\left(\dfrac{3\pi}{2}, 0\right)$
$\dfrac{5\pi}{3}$	$\dfrac{1}{2}$	$\left(\dfrac{5\pi}{3}, \dfrac{1}{2}\right)$
2π	1	$(2\pi, 1)$

The Graph of the Cosine Function

The cosine function also has period 2π. We proceed as we did with the sine function by constructing Table 8, which lists some points on the graph of $y = \cos x$, $0 \le x \le 2\pi$. As the table shows, the graph of $y = \cos x, 0 \le x \le 2\pi$, begins at the point $(0, 1)$. As x increases from 0 to $\dfrac{\pi}{2}$ to π, the value of y decreases from 1 to 0 to -1; as x increases from π to $\dfrac{3\pi}{2}$ to 2π, the value of y increases from -1 to 0 to 1.

As before, we plot the points in Table 8 to get one period or cycle of the graph. See Figure 85.

Figure 85
$y = \cos x, 0 \le x \le 2\pi$

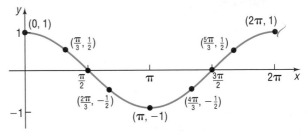

A more complete graph of $y = \cos x$ is obtained by continuing the graph in each direction, as shown in Figure 86(a). Figure 86(b) shows the graph on a TI-84 Plus graphing calculator.

Figure 86
$y = \cos x, -\infty < x < \infty$

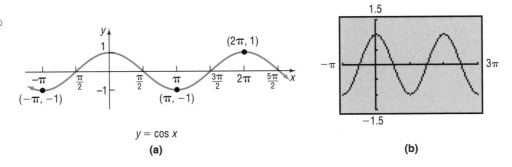

The graph of $y = \cos x$ illustrates some of the facts that we already know about the cosine function.

Properties of the Cosine Function

1. The domain is the set of all real numbers.
2. The range consists of all real numbers from -1 to 1, inclusive.
3. The cosine function is an even function, as the symmetry of the graph with respect to the y-axis indicates.
4. The cosine function is periodic, with period 2π.
5. The x-intercepts are $\ldots, -\dfrac{3\pi}{2}, -\dfrac{\pi}{2}, \dfrac{\pi}{2}, \dfrac{3\pi}{2}, \dfrac{5\pi}{2}, \ldots$; the y-intercept is 1.
6. The maximum value is 1 and occurs at $x = \ldots, -2\pi, 0, 2\pi, 4\pi, 6\pi, \ldots$; the minimum value is -1 and occurs at $x = \ldots, -\pi, \pi, 3\pi, 5\pi, \ldots$.

2 Graph Functions of the Form $y = A \cos(\omega x)$ Using Transformations

EXAMPLE 3 Graphing Functions of the Form $y = A \cos(\omega x)$ Using Transformations

Graph $y = 2\cos(3x)$ using transformations. Use the graph to determine the domain and the range of $y = 2\cos(3x)$.

Solution   Figure 87 shows the steps.

Figure 87

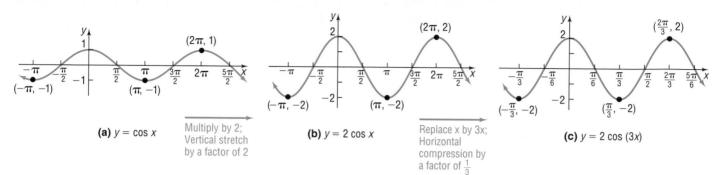

(a) $y = \cos x$ — Multiply by 2; Vertical stretch by a factor of 2 → **(b)** $y = 2 \cos x$ — Replace x by 3x; Horizontal compression by a factor of $\frac{1}{3}$ → **(c)** $y = 2 \cos (3x)$

The domain of $y = 2\cos(3x)$ is the set of all real numbers or $(-\infty, \infty)$. The range of $y = 2\cos(3x)$ is $\{y \mid -2 \le y \le 2\}$ or $[-2, 2]$.

✓ **Check:** Graph $Y_1 = 2\cos(3x)$ to verify the graph shown in Figure 87(c). ∎

Notice in Figure 87(c) that the period of the function $y = 2\cos(3x)$ is $\dfrac{2\pi}{3}$ due to the compression of the original period 2π by a factor of $\dfrac{1}{3}$.

Now Work   PROBLEM 53 USING TRANSFORMATIONS

Sinusoidal Graphs

Shift the graph of $y = \cos x$ to the right $\dfrac{\pi}{2}$ units to obtain the graph of $y = \cos\left(x - \dfrac{\pi}{2}\right)$. See Figure 88(a). Now look at the graph of $y = \sin x$ in Figure 88(b). We see that the graph of $y = \sin x$ is the same as the graph of $y = \cos\left(x - \dfrac{\pi}{2}\right)$.

Figure 88

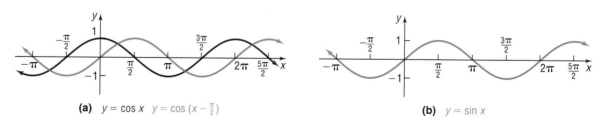

(a) $y = \cos x$ $y = \cos\left(x - \frac{\pi}{2}\right)$ **(b)** $y = \sin x$

Based on Figure 88, we conjecture that

$$\sin x = \cos\left(x - \frac{\pi}{2}\right)$$

Seeing the Concept

Graph $Y_1 = \sin x$ and $Y_2 = \cos\left(x - \dfrac{\pi}{2}\right)$.

How many graphs do you see?

(We shall prove this fact in Chapter 3.) Because of this relationship, the graphs of functions of the form $y = A\sin(\omega x)$ or $y = A\cos(\omega x)$ are referred to as **sinusoidal graphs.**

Let's look at some general properties of sinusoidal graphs.

3 Determine the Amplitude and Period of Sinusoidal Functions

In Figure 89(b) we show the graph of $y = 2 \cos x$. Notice that the values of $y = 2 \cos x$ lie between -2 and 2, inclusive.

Figure 89

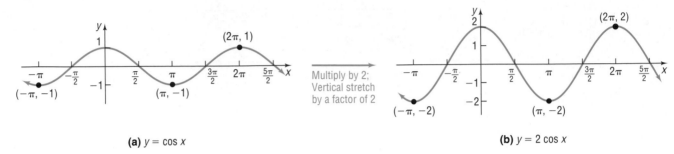

(a) $y = \cos x$ Multiply by 2; Vertical stretch by a factor of 2   **(b)** $y = 2 \cos x$

In general, the values of the functions $y = A \sin x$ and $y = A \cos x$, where $A \neq 0$, will always satisfy the inequalities

$$-|A| \leq A \sin x \leq |A| \quad \text{and} \quad -|A| \leq A \cos x \leq |A|$$

respectively. The number $|A|$ is called the **amplitude** of $y = A \sin x$ or $y = A \cos x$. See Figure 90.

Figure 90

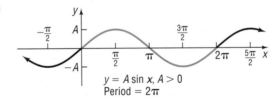

$y = A \sin x, A > 0$
Period $= 2\pi$

In Figure 91(b), we show the graph of $y = \cos(3x)$. Notice that the period of this function is $\dfrac{2\pi}{3}$, due to the horizontal compression of the original period 2π by a factor of $\dfrac{1}{3}$.

Figure 91

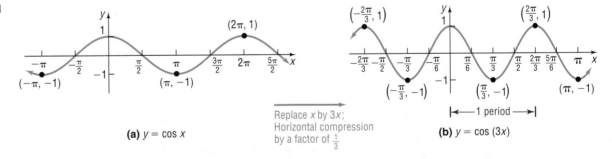

(a) $y = \cos x$ Replace x by $3x$; Horizontal compression by a factor of $\frac{1}{3}$ **(b)** $y = \cos(3x)$

In general, if $\omega > 0$, the functions $y = \sin(\omega x)$ and $y = \cos(\omega x)$ will have period $T = \dfrac{2\pi}{\omega}$. To see why, recall that the graph of $y = \sin(\omega x)$ is obtained from the graph of $y = \sin x$ by performing a horizontal compression or stretch by a factor $\dfrac{1}{\omega}$. This horizontal compression replaces the interval $[0, 2\pi]$, which contains one period of the graph of $y = \sin x$, by the interval $\left[0, \dfrac{2\pi}{\omega}\right]$, which contains one period of the graph of $y = \sin(\omega x)$.

For example, for the function $y = \cos(3x)$, graphed in Figure 91(b), $\omega = 3$, so the period is $\dfrac{2\pi}{\omega} = \dfrac{2\pi}{3}$.

One period of the graph of $y = A\sin(\omega x)$ or $y = A\cos(\omega x)$ is called a **cycle.** Figure 92 illustrates the general situation. The blue portion of the graph is one cycle.

Figure 92

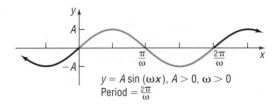

$y = A\sin(\omega x),\ A > 0,\ \omega > 0$
Period $= \frac{2\pi}{\omega}$

NOTE A function f is even if $f(-x) = f(x)$; a function is odd if $f(-x) = -f(x)$. Since the sine function is odd, $\sin(-\theta) = -\sin\theta$; since the cosine function is even, $\cos(-\theta) = \cos\theta$. ∎

If $\omega < 0$ in $y = A\sin(\omega x)$ or $y = A\cos(\omega x)$, we use the Even–Odd Properties of the sine and cosine functions as follows:

$$A\sin(-\omega x) = -A\sin(\omega x) \quad \text{and} \quad A\cos(-\omega x) = A\cos(\omega x)$$

This gives us an equivalent form in which the coefficient of x in the argument is positive. For example,

$$\sin(-2x) = -\sin(2x) \quad \text{and} \quad \cos(-\pi x) = \cos(\pi x)$$

Because of this, we can assume $\omega > 0$.

THEOREM

If $\omega > 0$, the amplitude and period of $y = A\sin(\omega x)$ and $y = A\cos(\omega x)$ are given by

$$\text{Amplitude} = |A| \qquad \text{Period} = T = \frac{2\pi}{\omega} \tag{1}$$

EXAMPLE 4 **Finding the Amplitude and Period of a Sinusoidal Function**

Determine the amplitude and period of $y = 5\sin(-4x)$.

Solution First, we use the fact that the sine function is odd and write $y = 5\sin(-4x)$ as $y = -5\sin(4x)$. Comparing $y = -5\sin(4x)$ to $y = A\sin(\omega x)$, we find that $A = -5$ and $\omega = 4$. From equation (1),

$$\text{Amplitude} = |A| = |-5| = 5 \qquad \text{Period} = T = \frac{2\pi}{\omega} = \frac{2\pi}{4} = \frac{\pi}{2}$$

✏️ **Now Work** PROBLEM 23

4 Graph Sinusoidal Functions Using Key Points

So far, we have graphed functions of the form $y = A\sin(\omega x)$ or $y = A\cos(\omega x)$ using transformations. We now introduce another method that can be used to graph these functions.

Figure 93 on page 168 shows one cycle of the graphs of $y = \sin x$ and $y = \cos x$ on the interval $[0, 2\pi]$. Notice that each graph consists of four parts corresponding to the four subintervals:

$$\left[0, \frac{\pi}{2}\right] \qquad \left[\frac{\pi}{2}, \pi\right] \qquad \left[\pi, \frac{3\pi}{2}\right] \qquad \left[\frac{3\pi}{2}, 2\pi\right]$$

Each subinterval is of length $\dfrac{\pi}{2}$ (the period 2π divided by 4, the number of parts), and the endpoints of these intervals, $x = 0$, $x = \dfrac{\pi}{2}$, $x = \pi$, $x = \dfrac{3\pi}{2}$, $x = 2\pi$, give rise to five key points on each graph:

For $y = \sin x$: $\quad (0, 0), \left(\dfrac{\pi}{2}, 1\right), (\pi, 0), \left(\dfrac{3\pi}{2}, -1\right), (2\pi, 0)$

For $y = \cos x$: $\quad (0, 1), \left(\dfrac{\pi}{2}, 0\right), (\pi, -1), \left(\dfrac{3\pi}{2}, 0\right), (2\pi, 1)$

Look again at Figure 93.

Figure 93

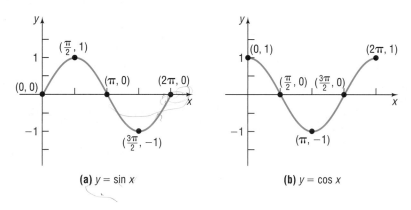

(a) $y = \sin x$ $\qquad\qquad$ **(b)** $y = \cos x$

Let's look at an example that illustrates how to graph a sinusoidal function using key points.

| **EXAMPLE 5** | **How to Graph a Sinusoidal Function Using Key Points** |

Graph $y = 3\sin(4x)$ using key points.

Step-by-Step Solution

STEP 1 Determine the amplitude and period of the sinusoidal function.

Comparing $y = 3\sin(4x)$ to $y = A\sin(\omega x)$, we see that $A = 3$ and $\omega = 4$, so the amplitude is $|A| = 3$ and the period is $\dfrac{2\pi}{\omega} = \dfrac{2\pi}{4} = \dfrac{\pi}{2}$. Because the amplitude is 3, the graph of $y = 3\sin(4x)$ will lie between -3 and 3 on the y-axis. Because the period is $\dfrac{\pi}{2}$, one cycle will begin at $x = 0$ and end at $x = \dfrac{\pi}{2}$.

STEP 2 Divide the interval $\left[0, \dfrac{2\pi}{\omega}\right]$ into four subintervals of the same length.

We divide the interval $\left[0, \dfrac{\pi}{2}\right]$ into four subintervals, each of length $\dfrac{\pi}{2} \div 4 = \dfrac{\pi}{8}$, as follows:

$$\left[0, \frac{\pi}{8}\right] \quad \left[\frac{\pi}{8}, \frac{\pi}{8} + \frac{\pi}{8}\right] = \left[\frac{\pi}{8}, \frac{\pi}{4}\right] \quad \left[\frac{\pi}{4}, \frac{\pi}{4} + \frac{\pi}{8}\right] = \left[\frac{\pi}{4}, \frac{3\pi}{8}\right] \quad \left[\frac{3\pi}{8}, \frac{3\pi}{8} + \frac{\pi}{8}\right] = \left[\frac{3\pi}{8}, \frac{\pi}{2}\right]$$

The endpoints of the subintervals are $0, \dfrac{\pi}{8}, \dfrac{\pi}{4}, \dfrac{3\pi}{8}, \dfrac{\pi}{2}$. These values represent the x-coordinates of the five key points on the graph.

STEP 3 Use the endpoints of these subintervals to obtain five key points on the graph.

To obtain the y-coordinates of the five key points of $y = 3\sin(4x)$, we multiply the y-coordinates of the five key points for $y = \sin x$ in Figure 93(a) by $A = 3$. The five key points are

$$(0, 0) \quad \left(\frac{\pi}{8}, 3\right) \quad \left(\frac{\pi}{4}, 0\right) \quad \left(\frac{3\pi}{8}, -3\right) \quad \left(\frac{\pi}{2}, 0\right)$$

STEP 4 Plot the five key points and draw a sinusoidal graph to obtain the graph of one cycle. Extend the graph in each direction to make it complete.

We plot the five key points obtained in Step 3 and fill in the graph of the sine curve as shown in Figure 94(a). We extend the graph in each direction to obtain the complete graph shown in Figure 94(b). Notice that additional key points appear every $\frac{\pi}{8}$ radian.

Figure 94

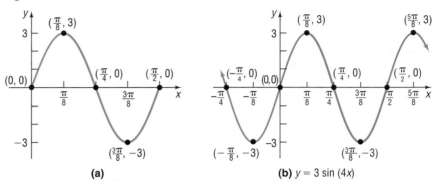

(a)

(b) $y = 3 \sin(4x)$

NOTE In Step 3 of Example 5, we could also obtain the five key points by evaluating $y = 3 \sin(4x)$ at each value of x. ∎

✓ **Check:** Graph $y = 3 \sin(4x)$ using transformations. Which graphing method do you prefer?

✏️ **Now Work** PROBLEM 45 USING KEY POINTS

SUMMARY **Steps for Graphing a Sinusoidal Function of the Form $y = A \sin(\omega x)$ or $y = A \cos(\omega x)$ Using Key Points**

STEP 1: Determine the amplitude and period of the sinusoidal function.

STEP 2: Divide the interval $\left[0, \dfrac{2\pi}{\omega}\right]$ into four subintervals of the same length.

STEP 3: Use the endpoints of these subintervals to obtain five key points on the graph.

STEP 4: Plot the five key points with a sinusoidal graph to obtain the graph of one cycle. Extend the graph in each direction to make it complete.

EXAMPLE 6 **Graphing a Sinusoidal Function Using Key Points**

Graph $y = 2 \sin\left(-\dfrac{\pi}{2}x\right)$ using key points.

Solution Since the sine function is odd, we can use the equivalent form:

$$y = -2 \sin\left(\frac{\pi}{2}x\right)$$

STEP 1: Comparing $y = -2 \sin\left(\dfrac{\pi}{2}x\right)$ to $y = A \sin(\omega x)$, we find that $A = -2$ and $\omega = \dfrac{\pi}{2}$. The amplitude is $|A| = |-2| = 2$, and the period is $T = \dfrac{2\pi}{\omega} = \dfrac{2\pi}{\dfrac{\pi}{2}} = 4$.

The graph of $y = -2 \sin\left(\dfrac{\pi}{2}x\right)$ will lie between -2 and 2 on the y-axis. One cycle will begin at $x = 0$ and end at $x = 4$.

STEP 2: We divide the interval $[0, 4]$ into four subintervals, each of length $4 \div 4 = 1$. The x-coordinates of the five key points are:

$$\underset{\text{1st x-coordinate}}{0} \qquad \underset{\text{2nd x-coordinate}}{0 + 1 = 1} \qquad \underset{\text{3rd x-coordinate}}{1 + 1 = 2} \qquad \underset{\text{4th x-coordinate}}{2 + 1 = 3} \qquad \underset{\text{5th x-coordinate}}{3 + 1 = 4}$$

STEP 3: Since $y = -2 \sin\left(\dfrac{\pi}{2}x\right)$, we multiply the y-coordinates of the five key points in Figure 93(a) by $A = -2$. The five key points on the graph are

$$(0, 0) \quad (1, -2) \quad (2, 0) \quad (3, 2) \quad (4, 0)$$

COMMENT To graph a sinusoidal function of the form $y = A \sin(\omega x)$ or $y = A \cos(\omega x)$ using a graphing utility, we use the amplitude to set $Y_{\min}$ and $Y_{\max}$ and use the period to set $X_{\min}$ and $X_{\max}$. ∎

STEP 4: We plot these five points and fill in the graph of the sine function as shown in Figure 95(a). Extending the graph in each direction, we obtain Figure 95(b).

Figure 95(c) shows the graph using a graphing utility.

Figure 95

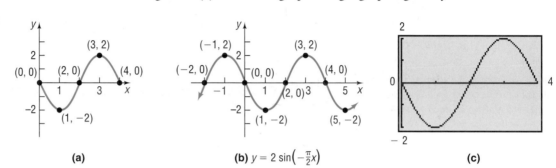

(a)   **(b)** $y = 2 \sin\left(-\frac{\pi}{2}x\right)$ **(c)**

✓ Check: Graph $y = 2 \sin\left(-\dfrac{\pi}{2}x\right)$ using transformations. Which graphing method do you prefer?

─────── **Now Work** PROBLEM 49 USING KEY POINTS

If the function to be graphed is of the form $y = A \sin(\omega x) + B$ [or $y = A \cos(\omega x) + B$], first graph $y = A \sin(\omega x)$ [or $y = A \cos(\omega x)$] and then use a vertical shift.

EXAMPLE 7 | **Graphing a Sinusoidal Function Using Key Points**

Graph $y = -4 \cos(\pi x) - 2$ using key points. Use the graph to determine the domain and the range of $y = -4 \cos(\pi x) - 2$.

Solution   We begin by graphing the function $y = -4 \cos(\pi x)$. Comparing $y = -4 \cos(\pi x)$ with $y = A \cos(\omega x)$, we find that $A = -4$ and $\omega = \pi$. The amplitude is $|A| = |-4| = 4$, and the period is $T = \dfrac{2\pi}{\omega} = \dfrac{2\pi}{\pi} = 2$.

The graph of $y = -4 \cos(\pi x)$ will lie between -4 and 4 on the y-axis. One cycle will begin at $x = 0$ and end at $x = 2$.

We divide the interval $[0, 2]$ into four subintervals, each of length $2 \div 4 = \dfrac{1}{2}$. The x-coordinates of the five key points are:

$$\underset{\text{1st x-coordinate}}{0} \qquad \underset{\text{2nd x-coordinate}}{0 + \frac{1}{2} = \frac{1}{2}} \qquad \underset{\text{3rd x-coordinate}}{\frac{1}{2} + \frac{1}{2} = 1} \qquad \underset{\text{4th x-coordinate}}{1 + \frac{1}{2} = \frac{3}{2}} \qquad \underset{\text{5th x-coordinate}}{\frac{3}{2} + \frac{1}{2} = 2}$$

Since $y = -4 \cos(\pi x)$, we multiply the y-coordinates of the five key points of $y = \cos x$ shown in Figure 93(b) by $A = -4$ to obtain the five key points on the graph of $y = -4 \cos(\pi x)$:

$$(0, -4) \quad \left(\frac{1}{2}, 0\right) \quad (1, 4) \quad \left(\frac{3}{2}, 0\right) \quad (2, -4)$$

We plot these five points and fill in the graph of the cosine function as shown in Figure 96(a). Extending the graph in each direction, we obtain Figure 96(b), the graph of $y = -4\cos(\pi x)$.

A vertical shift down 2 units gives the graph of $y = -4\cos(\pi x) - 2$, as shown in Figure 96(c).

Figure 96

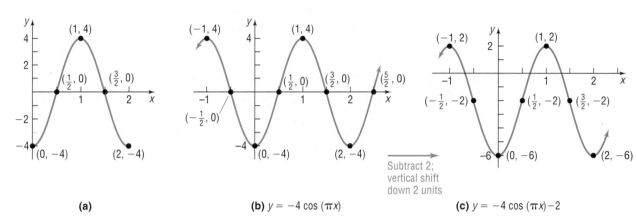

(a)

(b) $y = -4\cos(\pi x)$

Subtract 2;
vertical shift
down 2 units

(c) $y = -4\cos(\pi x) - 2$

The domain of $y = -4\cos(\pi x) - 2$ is the set of all real numbers or $(-\infty, \infty)$. The range of $y = -4\cos(\pi x) - 2$ is $\{y \,|\, -6 \le y \le 2\}$ or $[-6, 2]$.

Now Work PROBLEM 59

5 Find an Equation for a Sinusoidal Graph

We can also use the ideas of amplitude and period to identify a sinusoidal function when its graph is given.

| EXAMPLE 8 | Finding an Equation for a Sinusoidal Graph |

Find an equation for the graph shown in Figure 97.

Figure 97

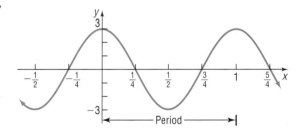

Period

Solution   The graph has the characteristics of a cosine function. Do you see why? The maximum value occurs at $x = 0$. So we view the equation as a cosine function $y = A\cos(\omega x)$ with $A = 3$ and period $T = 1$. Then $\dfrac{2\pi}{\omega} = 1$, so $\omega = 2\pi$. The cosine function whose graph is given in Figure 97 is

$$y = A\cos(\omega x) = 3\cos(2\pi x)$$

✓ Check: Graph $Y_1 = 3\cos(2\pi x)$ and compare the result with Figure 97.

| EXAMPLE 9 | Finding an Equation for a Sinusoidal Graph |

Find an equation for the graph shown in Figure 98 on the following page.

Figure 98

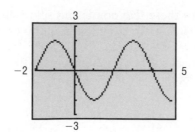

Solution   The graph is sinusoidal, with amplitude $|A| = 2$. The period is 4, so $\dfrac{2\pi}{\omega} = 4$ or $\omega = \dfrac{\pi}{2}$. Since the graph passes through the origin, it is easiest to view the equation as a sine function,* but notice that the graph is actually the reflection of a sine function about the x-axis (since the graph is decreasing near the origin). This requires that $A = -2$. The sine function whose graph is given in Figure 98 is

$$y = A\sin(\omega x) = -2\sin\left(\frac{\pi}{2}x\right)$$

✓ **Check:**   Graph $Y_1 = -2\sin\left(\dfrac{\pi}{2}x\right)$ and compare the result with Figure 98.

Now Work   PROBLEMS 67 AND 71

---

*The equation could also be viewed as a cosine function with a horizontal shift, but viewing it as a sine function is easier.

2.6 Assess Your Understanding

'Are You Prepared?'   *Answers are given at the end of these exercises. If you get a wrong answer, read the pages listed in red.*

1. Use transformations to graph $y = 3x^2$ (pp. 70–72)

2. Use transformations to graph $y = \sqrt{2x}$. (pp. 70–72)

Concepts and Vocabulary

3. The maximum value of $y = \sin x, 0 \le x \le 2\pi$, is _____ and occurs at $x =$ _____.

4. The function $y = A\sin(\omega x)$, $A > 0$, has amplitude 3 and period 2; then $A =$ _____ and $\omega =$ _____.

5. The function $y = 3\cos(6x)$ has amplitude _____ and period _____.

6. *True or False*   The graphs of $y = \sin x$ and $y = \cos x$ are identical except for a horizontal shift.

7. *True or False*   For $y = 2\sin(\pi x)$, the amplitude is 2 and the period is $\dfrac{\pi}{2}$.

8. *True or False*   The graph of the sine function has infinitely many x-intercepts.

Skill Building

In Problems 9–18, if necessary, refer to a graph to answer each question.

9. What is the y-intercept of $y = \sin x$?

10. What is the y-intercept of $y = \cos x$?

11. For what numbers x, $-\pi \le x \le \pi$, is the graph of $y = \sin x$ increasing?

12. For what numbers x, $-\pi \le x \le \pi$, is the graph of $y = \cos x$ decreasing?

13. What is the largest value of $y = \sin x$?

14. What is the smallest value of $y = \cos x$?

15. For what numbers $x, 0 \le x \le 2\pi$, does $\sin x = 0$?

16. For what numbers $x, 0 \le x \le 2\pi$, does $\cos x = 0$?

17. For what numbers $x, -2\pi \le x \le 2\pi$, does $\sin x = 1$? Where does $\sin x = -1$?

18. For what numbers $x, -2\pi \le x \le 2\pi$, does $\cos x = 1$? Where does $\cos x = -1$?

In Problems 19–28, determine the amplitude and period of each function without graphing.

19. $y = 2 \sin x$

20. $y = 3 \cos x$

21. $y = -4 \cos(2x)$

22. $y = -\sin\left(\dfrac{1}{2}x\right)$

23. $y = 6 \sin(\pi x)$

24. $y = -3 \cos(3x)$

25. $y = -\dfrac{1}{2}\cos\left(\dfrac{3}{2}x\right)$

26. $y = \dfrac{4}{3}\sin\left(\dfrac{2}{3}x\right)$

27. $y = \dfrac{5}{3}\sin\left(-\dfrac{2\pi}{3}x\right)$

28. $y = \dfrac{9}{5}\cos\left(-\dfrac{3\pi}{2}x\right)$

In Problems 29–38, match the given function to one of the graphs (A)–(J).

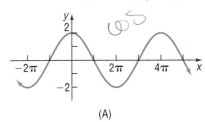

(A)

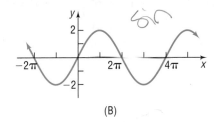

(B)

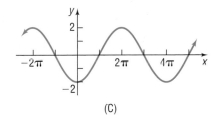

(C)

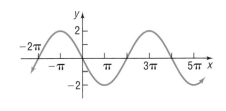

(D)

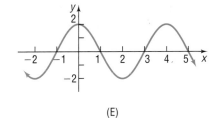

(E)

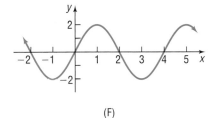

(F)

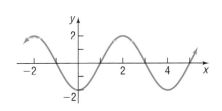

(G)

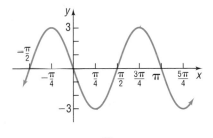

(H)

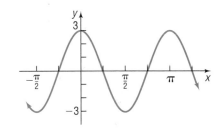

(I)

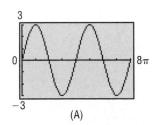

(J)

29. $y = 2 \sin\left(\dfrac{\pi}{2}x\right)$

30. $y = 2 \cos\left(\dfrac{\pi}{2}x\right)$

31. $y = 2 \cos\left(\dfrac{1}{2}x\right)$

32. $y = 3 \cos(2x)$

33. $y = -3 \sin(2x)$

34. $y = 2 \sin\left(\dfrac{1}{2}x\right)$

35. $y = -2 \cos\left(\dfrac{1}{2}x\right)$

36. $y = -2 \cos\left(\dfrac{\pi}{2}x\right)$

37. $y = 3 \sin(2x)$

38. $y = -2 \sin\left(\dfrac{1}{2}x\right)$

In Problems 39–42, match the given function to one of the graphs (A)–(D).

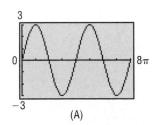

(A)

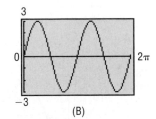

(B)

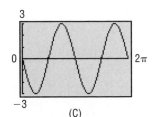

(C)

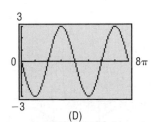

(D)

39. $y = 3 \sin\left(\dfrac{1}{2}x\right)$

40. $y = -3 \sin(2x)$

41. $y = 3 \sin(2x)$

42. $y = -3 \sin\left(\dfrac{1}{2}x\right)$

In Problems 43–66, graph each function. Be sure to label key points and show at least two cycles. Verify the graph using a graphing utility. Use the graph to determine the domain and the range of each function.

43. $y = 4 \cos x$

44. $y = 3 \sin x$

45. $y = -4 \sin x$

46. $y = -3 \cos x$

47. $y = \cos(4x)$

48. $y = \sin(3x)$

49. $y = \sin(-2x)$

50. $y = \cos(-2x)$

51. $y = 2 \sin\left(\dfrac{1}{2} x\right)$

52. $y = 2 \cos\left(\dfrac{1}{4} x\right)$

53. $y = -\dfrac{1}{2} \cos(2x)$

54. $y = -4 \sin\left(\dfrac{1}{8} x\right)$

55. $y = 2 \sin x + 3$

56. $y = 3 \cos x + 2$

57. $y = 5 \cos(\pi x) - 3$

58. $y = 4 \sin\left(\dfrac{\pi}{2} x\right) - 2$

59. $y = -6 \sin\left(\dfrac{\pi}{3} x\right) + 4$

60. $y = -3 \cos\left(\dfrac{\pi}{4} x\right) + 2$

61. $y = 5 - 3 \sin(2x)$

62. $y = 2 - 4 \cos(3x)$

63. $y = \dfrac{5}{3} \sin\left(-\dfrac{2\pi}{3} x\right)$

64. $y = \dfrac{9}{5} \cos\left(-\dfrac{3\pi}{2} x\right)$

65. $y = -\dfrac{3}{2} \cos\left(\dfrac{\pi}{4} x\right) + \dfrac{1}{2}$

66. $y = -\dfrac{1}{2} \sin\left(\dfrac{\pi}{8} x\right) + \dfrac{3}{2}$

In Problems 67–70, write the equation of a sine function that has the given characteristics.

67. Amplitude: 3
Period: π

68. Amplitude: 2
Period: 4π

69. Amplitude: 3
Period: 2

70. Amplitude: 4
Period: 1

In Problems 71–84, find an equation for each graph.

71.

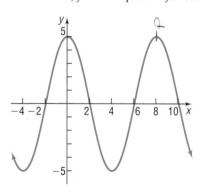

72.

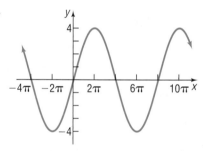

73.

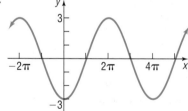

74.

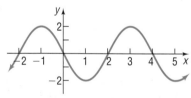

75.

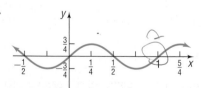

76.

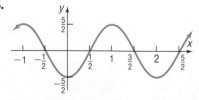

77.

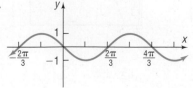

78.

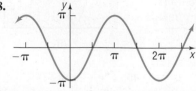

2.7 Graphs of the Tangent, Cotangent, Cosecant, and Secant Functions

PREPARING FOR THIS SECTION *Before getting started, review the following:*

- Vertical Asymptotes (Section 1.7, pp. 76–77)

Now Work the 'Are You Prepared?' problems on page 183.

OBJECTIVES 1 Graph Functions of the Form $y = A\tan(\omega x) + B$
and $y = A\cot(\omega x) + B$ (p. 179)

2 Graph Functions of the Form $y = A\csc(\omega x) + B$
and $y = A\sec(\omega x) + B$ (p. 181)

The Graph of the Tangent Function

Because the tangent function has period π, we only need to determine the graph over some interval of length π. The rest of the graph will consist of repetitions of that graph. Because the tangent function is not defined at $\ldots, -\dfrac{3\pi}{2}, -\dfrac{\pi}{2}, \dfrac{\pi}{2}, \dfrac{3\pi}{2}, \ldots$, we will concentrate on the interval $\left(-\dfrac{\pi}{2}, \dfrac{\pi}{2}\right)$, of length π, and construct Table 9, which lists some points on the graph of $y = \tan x, -\dfrac{\pi}{2} < x < \dfrac{\pi}{2}$. We plot the points in the table and connect them with a smooth curve. See Figure 99 for a partial graph of $y = \tan x$, where $-\dfrac{\pi}{3} \le x \le \dfrac{\pi}{3}$.

Table 9

x	$y - \tan x$	(x, y)
$-\dfrac{\pi}{3}$	$-\sqrt{3} \approx -1.73$	$\left(-\dfrac{\pi}{3}, -\sqrt{3}\right)$
$-\dfrac{\pi}{4}$	-1	$\left(-\dfrac{\pi}{4}, -1\right)$
$-\dfrac{\pi}{6}$	$-\dfrac{\sqrt{3}}{3} \approx -0.58$	$\left(-\dfrac{\pi}{6}, -\dfrac{\sqrt{3}}{3}\right)$
0	0	$(0, 0)$
$\dfrac{\pi}{6}$	$\dfrac{\sqrt{3}}{3} \approx 0.58$	$\left(\dfrac{\pi}{6}, \dfrac{\sqrt{3}}{3}\right)$
$\dfrac{\pi}{4}$	1	$\left(\dfrac{\pi}{4}, 1\right)$
$\dfrac{\pi}{3}$	$\sqrt{3} \approx 1.73$	$\left(\dfrac{\pi}{3}, \sqrt{3}\right)$

Figure 99

$y = \tan x, \; -\dfrac{\pi}{3} \le x \le \dfrac{\pi}{3}$

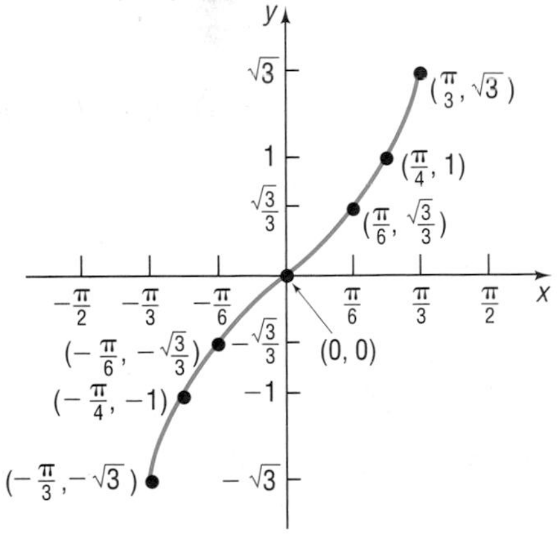

To complete one period of the graph of $y = \tan x$, we need to investigate the behavior of the function as x approaches $-\dfrac{\pi}{2}$ and $\dfrac{\pi}{2}$. We must be careful, though, because $y = \tan x$ is not defined at these numbers. To determine this behavior, we use the identity

$$\tan x = \frac{\sin x}{\cos x}$$

See Table 10. If x is close to $\dfrac{\pi}{2} \approx 1.5708$, but remains less than $\dfrac{\pi}{2}$, then $\sin x$ will be close to 1 and $\cos x$ will be positive and close to 0. (To see this, refer back to the graphs of the sine function and the cosine function.) So the ratio $\dfrac{\sin x}{\cos x}$ will be positive and large. In fact, the closer x gets to $\dfrac{\pi}{2}$, the closer $\sin x$ gets to 1 and $\cos x$ gets to 0, so $\tan x$ approaches $\infty \left(\lim\limits_{x \to \frac{\pi}{2}^-} \tan x = \infty \right)$. In other words, the vertical line $x = \dfrac{\pi}{2}$ is a vertical asymptote to the graph of $y = \tan x$.

Table 10

x	sin x	cos x	y = tan x
$\dfrac{\pi}{3} \approx 1.05$	$\dfrac{\sqrt{3}}{2}$	$\dfrac{1}{2}$	$\sqrt{3} \approx 1.73$
1.5	0.9975	0.0707	14.1
1.57	0.9999	7.96E−4	1255.8
1.5707	0.9999	9.6E−5	10,381
$\dfrac{\pi}{2} \approx 1.5708$	1	0	Undefined

If x is close to $-\dfrac{\pi}{2}$, but remains greater than $-\dfrac{\pi}{2}$, then $\sin x$ will be close to -1 and $\cos x$ will be positive and close to 0. The ratio $\dfrac{\sin x}{\cos x}$ approaches $-\infty \left(\lim\limits_{x \to -\frac{\pi}{2}^+} \tan x = -\infty \right)$. In other words, the vertical line $x = -\dfrac{\pi}{2}$ is also a vertical asymptote to the graph.

With these observations, we can complete one period of the graph. We obtain the complete graph of $y = \tan x$ by repeating this period, as shown in Figure 100.

Figure 101 shows the graph of $y = \tan x$, $-\infty < x < \infty$, using a graphing utility. Notice we used dot mode when graphing $y = \tan x$. Do you know why?

Figure 100
$y = \tan x$, $-\infty < x < \infty$, x not equal to odd multiples of $\dfrac{\pi}{2}$

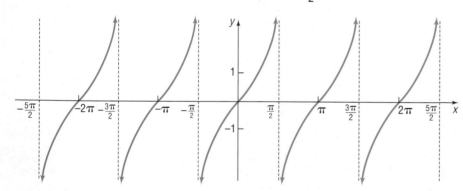

Figure 101

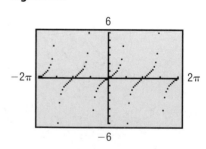

The graph of $y = \tan x$ in Figure 100 illustrates the following properties.

Properties of the Tangent Function

1. The domain is the set of all real numbers, except odd multiples of $\dfrac{\pi}{2}$.
2. The range is the set of all real numbers.
3. The tangent function is an odd function, as the symmetry of the graph with respect to the origin indicates.

4. The tangent function is periodic, with period π.

5. The x-intercepts are $\ldots, -2\pi, -\pi, 0, \pi, 2\pi, 3\pi, \ldots$; the y-intercept is 0.

6. Vertical asymptotes occur at $x = \ldots, -\dfrac{3\pi}{2}, -\dfrac{\pi}{2}, \dfrac{\pi}{2}, \dfrac{3\pi}{2}, \ldots$.

 **Now Work** PROBLEMS 7 AND 15

1 Graph Functions of the Form $y = A \tan(\omega x) + B$

For tangent functions, there is no concept of amplitude since the range of the tangent function is $(-\infty, \infty)$. The role of A in $y = A \tan(\omega x) + B$ is to provide the magnitude of the vertical stretch. The period of $y = \tan x$ is π, so the period of $y = A \tan(\omega x) + B$ is $\dfrac{\pi}{\omega}$, caused by the horizontal compression of the graph by a factor of $\dfrac{1}{\omega}$. Finally, the presence of B indicates that a vertical shift is required.

EXAMPLE 1	**Graphing Functions of the Form $y = A \tan(\omega x) + B$**

Graph $y = 2 \tan x - 1$. Use the graph to determine the domain and the range of $y = 2 \tan x - 1$.

Solution   Figure 102 shows the steps using transformations.

Figure 102

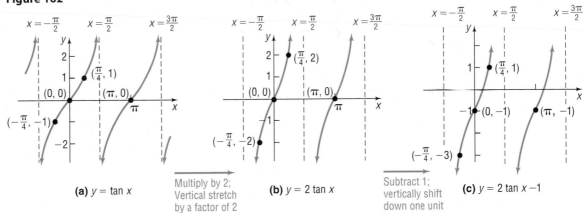

(a) $y = \tan x$

Multiply by 2;
Vertical stretch
by a factor of 2

(b) $y = 2 \tan x$

Subtract 1;
vertically shift
down one unit

(c) $y = 2 \tan x - 1$

The domain of $y = 2 \tan x - 1$ is $\left\{ x \mid x \neq \dfrac{k\pi}{2}, k \text{ is an odd integer} \right\}$ and the range is the set of all real numbers or $(-\infty, \infty)$.

✓**Check:** Graph $Y_1 = 2 \tan x - 1$ to verify the graph shown in Figure 102(c).

EXAMPLE 2	**Graphing Functions of the Form $y = A \tan(\omega x) + B$**

Graph $y = 3 \tan(2x)$. Use the graph to determine the domain and the range of $y = 3 \tan(2x)$.

Solution   Figure 103 shows the steps using transformations.

Figure 103

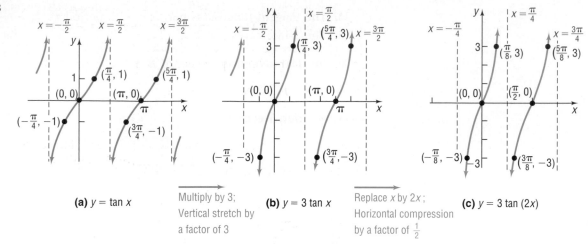

(a) $y = \tan x$

Multiply by 3;
Vertical stretch by
a factor of 3

(b) $y = 3 \tan x$

Replace x by $2x$;
Horizontal compression
by a factor of $\frac{1}{2}$

(c) $y = 3 \tan (2x)$

The domain of $y = 3 \tan (2x)$ is $\{x \mid x \ne \dfrac{k\pi}{4}, k \text{ is an odd integer}\}$ and the range is the set of all real numbers or $(-\infty, \infty)$.

✓ **Check:** Graph $Y_1 = 3 \tan(2x)$ to verify the graph in Figure 103(c).

Notice in Figure 103(c) that the period of $y = 3 \tan(2x)$ is $\dfrac{\pi}{2}$ due to the compression of the original period π by a factor of $\dfrac{1}{2}$. Notice that the asymptotes are $x = -\dfrac{\pi}{4}$, $x = \dfrac{\pi}{4}$, $x = \dfrac{3\pi}{4}$, and so on, also due to the compression.

✏ **Now Work** PROBLEM **21**

The Graph of the Cotangent Function

We obtain the graph of $y = \cot x$ as we did the graph of $y = \tan x$. The period of $y = \cot x$ is π. Because the cotangent function is not defined for integer multiples of π, we will concentrate on the interval $(0, \pi)$. Table 11 lists some points on the graph of $y = \cot x$, $0 < x < \pi$. As x approaches 0, but remains greater than 0, the value of $\cos x$ will be close to 1 and the value of $\sin x$ will be positive and close to 0. Hence, the ratio $\dfrac{\cos x}{\sin x} = \cot x$ will be positive and large; so as x approaches 0, with $x > 0$, $\cot x$ approaches ∞ ($\lim\limits_{x \to 0^+} \cot x = \infty$). Similarly, as x approaches π, but remains less than π, the value of $\cos x$ will be close to -1, and the value of $\sin x$ will be positive and close to 0. So the ratio $\dfrac{\cos x}{\sin x} = \cot x$ will be negative and will approach $-\infty$ as x approaches π ($\lim\limits_{x \to \pi^-} \cot x = -\infty$). Figure 104 shows the graph.

Table 11

x	$y = \cot x$	(x, y)
$\dfrac{\pi}{6}$	$\sqrt{3}$	$\left(\dfrac{\pi}{6}, \sqrt{3}\right)$
$\dfrac{\pi}{4}$	1	$\left(\dfrac{\pi}{4}, 1\right)$
$\dfrac{\pi}{3}$	$\dfrac{\sqrt{3}}{3}$	$\left(\dfrac{\pi}{3}, \dfrac{\sqrt{3}}{3}\right)$
$\dfrac{\pi}{2}$	0	$\left(\dfrac{\pi}{2}, 0\right)$
$\dfrac{2\pi}{3}$	$-\dfrac{\sqrt{3}}{3}$	$\left(\dfrac{2\pi}{3}, -\dfrac{\sqrt{3}}{3}\right)$
$\dfrac{3\pi}{4}$	-1	$\left(\dfrac{3\pi}{4}, -1\right)$
$\dfrac{5\pi}{6}$	$-\sqrt{3}$	$\left(\dfrac{5\pi}{6}, -\sqrt{3}\right)$

Figure 104
$y = \cot x$, $-\infty < x < \infty$, x not equal to integer multiples of π, $-\infty < y < \infty$

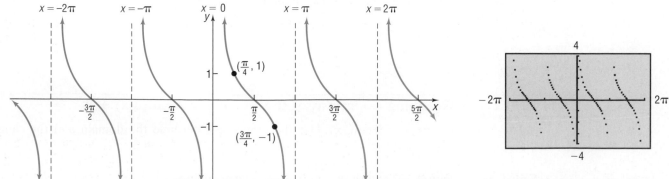

The graph of $y = A\cot(\omega x) + B$ has similar characteristics to those of the tangent function. The cotangent function $y = A\cot(\omega x) + B$ has period $\dfrac{\pi}{\omega}$. The cotangent function has no amplitude. The role of A is to provide the magnitude of the vertical stretch; the presence of B indicates a vertical shift is required.

Now Work PROBLEM 23

The Graph of the Cosecant Function and the Secant Function

The cosecant and secant functions, sometimes referred to as **reciprocal functions,** are graphed by making use of the reciprocal identities

$$\csc x = \frac{1}{\sin x} \quad \text{and} \quad \sec x = \frac{1}{\cos x}$$

For example, the value of the cosecant function $y = \csc x$ at a given number x equals the reciprocal of the corresponding value of the sine function, provided that the value of the sine function is not 0. If the value of $\sin x$ is 0, then x is an integer multiple of π. At such numbers, the cosecant function is not defined. In fact, the graph of the cosecant function has vertical asymptotes at integer multiples of π. Figure 105 shows the graph.

Figure 105

$y = \csc x,\ -\infty < x < \infty,\ x$ not equal to integer multiples of π, $|y| \geq 1$

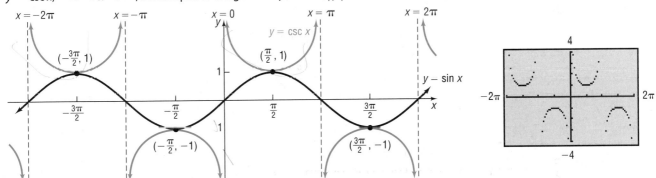

Using the idea of reciprocals, we can similarly obtain the graph of $y = \sec x$. See Figure 106.

Figure 106

$y = \sec x,\ -\infty < x < \infty,\ x$ not equal to odd multiples of $\dfrac{\pi}{2}$, $|y| \geq 1$

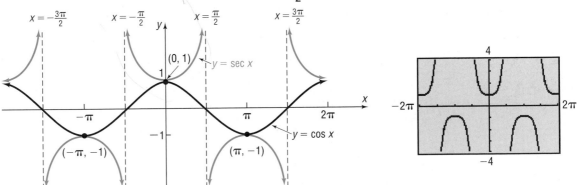

2 Graph Functions of the Form $y = A\csc(\omega x) + B$

The role of A in these functions is to set the range. The range of $y = \csc x$ is $\{y \mid y \leq -1 \text{ or } y \geq 1\}$ or $\{y \mid |y| \geq 1\}$; the range of $y = A\csc x$ is $\{y \mid |y| \geq |A|\}$, due to the vertical stretch of the graph by a factor of $|A|$. Just as with the sine and cosine functions, the period of $y = \csc(\omega x)$ and $y = \sec(\omega x)$ becomes $\dfrac{2\pi}{\omega}$, due to the horizontal compression of the graph by a factor of $\dfrac{1}{\omega}$. The presence of B indicates a vertical shift is required.

We shall graph these functions in two ways: using transformations and using the reciprocal function.

EXAMPLE 3   **Graphing Functions of the Form $y = A \csc(\omega x) + B$**

Graph $y = 2 \csc x - 1$. Use the graph to determine the domain and the range of $y = 2 \csc x - 1$.

Solution
Using Transformations

Figure 107 shows the required steps.

Figure 107

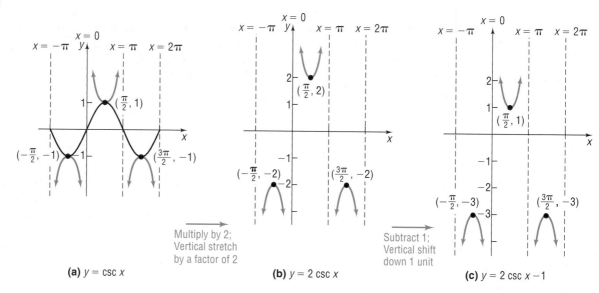

(a) $y = \csc x$ (b) $y = 2 \csc x$ (c) $y = 2 \csc x - 1$

Multiply by 2;
Vertical stretch
by a factor of 2

Subtract 1;
Vertical shift
down 1 unit

Solution
Using the Reciprocal
Function

We graph $y = 2 \csc x - 1$ by first graphing the reciprocal function $y = 2 \sin x - 1$ and then filling in the graph of $y = 2 \csc x - 1$, using the idea of reciprocals. See Figure 108.

Figure 108

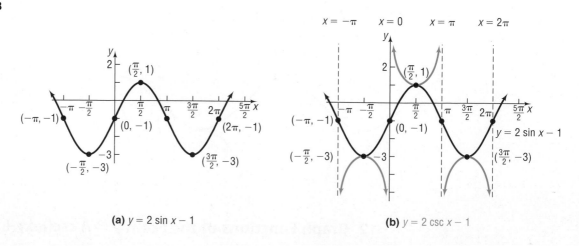

(a) $y = 2 \sin x - 1$ (b) $y = 2 \csc x - 1$

The domain of $y = 2 \csc x - 1$ is $\{x \mid x \neq k\pi, k \text{ is an integer}\}$ and the range is $\{y \mid y \leq -3 \text{ or } y \geq 1\}$, or, using interval notation $(-\infty, -3] \cup [1, \infty)$.

✓ **Check:** Graph $Y_1 = 2 \csc x - 1$ to verify the graph shown in Figure 107 or 108.

━━━━━━ **Now Work** PROBLEM **29**

2.7 Assess Your Understanding

'Are You Prepared?' *Answers are given at the end of these exercises. If you get a wrong answer, read the pages listed in red.*

1. The graph of $y = \dfrac{3x - 6}{x - 4}$ has a vertical asymptote. What is it?
   (pp. 76–77)

2. **True or False** If $x = 3$ is a vertical asymptote of a function R, then $\lim\limits_{x \to 3} |R(x)| = \infty$. (pp. 76–77)

Concepts and Vocabulary

3. The graph of $y = \tan x$ is symmetric with respect to the _____ and has vertical asymptotes at _____.

4. The graph of $y = \sec x$ is symmetric with respect to the _____ and has vertical asymptotes at _____.

5. It is easiest to graph $y = \sec x$ by first sketching the graph of _____.

6. **True or False** The graphs of $y = \tan x$, $y = \cot x$, $y = \sec x$, and $y = \csc x$ each have infinitely many vertical asymptotes.

Skill Building

In Problems 7–16, if necessary, refer to the graphs to answer each question.

7. What is the y-intercept of $y = \tan x$?

8. What is the y-intercept of $y = \cot x$?

9. What is the y-intercept of $y = \sec x$?

10. What is the y-intercept of $y = \csc x$?

11. For what numbers x, $-2\pi \le x \le 2\pi$, does $\sec x = 1$? For what numbers x does $\sec x = -1$?

12. For what numbers x, $-2\pi \le x \le 2\pi$, does $\csc x = 1$? For what numbers x does $\csc x = -1$?

13. For what numbers x, $-2\pi \le x \le 2\pi$, does the graph of $y = \sec x$ have vertical asymptotes?

14. For what numbers x, $-2\pi \le x \le 2\pi$, does the graph of $y = \csc x$ have vertical asymptotes?

15. For what numbers x, $-2\pi \le x \le 2\pi$, does the graph of $y = \tan x$ have vertical asymptotes?

16. For what numbers x, $-2\pi \le x \le 2\pi$, does the graph of $y = \cot x$ have vertical asymptotes?

In Problems 17–40, graph each function. Be sure to label key points and show at least two cycles. Verify the graph using a graphing utility. Use the graph to determine the domain and the range of each function.

17. $y = 3 \tan x$

18. $y = -2 \tan x$

19. $y = 4 \cot x$

20. $y = -3 \cot x$

21. $y = \tan\left(\dfrac{\pi}{2}x\right)$

22. $y = \tan\left(\dfrac{1}{2}x\right)$

23. $y = \cot\left(\dfrac{1}{4}x\right)$

24. $y = \cot\left(\dfrac{\pi}{4}x\right)$

25. $y = 2 \sec x$

26. $y = \dfrac{1}{2} \csc x$

27. $y = -3 \csc x$

28. $y = -4 \sec x$

29. $y = 4 \sec\left(\dfrac{1}{2}x\right)$

30. $y = \dfrac{1}{2} \csc(2x)$

31. $y = -2 \csc(\pi x)$

32. $y = -3 \sec\left(\dfrac{\pi}{2}x\right)$

33. $y = \tan\left(\dfrac{1}{4}x\right) + 1$

34. $y = 2 \cot x - 1$

35. $y = \sec\left(\dfrac{2\pi}{3}x\right) + 2$

36. $y = \csc\left(\dfrac{3\pi}{2}x\right)$

37. $y = \dfrac{1}{2} \tan\left(\dfrac{1}{4}x\right) - 2$

38. $y = 3 \cot\left(\dfrac{1}{2}x\right) - 2$

39. $y = 2 \csc\left(\dfrac{1}{3}x\right) - 1$

40. $y = 3 \sec\left(\dfrac{1}{4}x\right) + 1$

Mixed Practice

In Problems 41–44, find the average rate of change of f from 0 to $\dfrac{\pi}{6}$.

41. $f(x) = \tan x$ **42.** $f(x) = \sec x$ **43.** $f(x) = \tan(2x)$ **44.** $f(x) = \sec(2x)$

In Problems 45–48, find $(f \circ g)(x)$ and $(g \circ f)(x)$ and graph each of these functions.

45. $f(x) = \tan x$
   $g(x) = 4x$

46. $f(x) = 2 \sec x$
   $g(x) = \dfrac{1}{2} x$

47. $f(x) = -2x$
   $g(x) = \cot x$

48. $f(x) = \dfrac{1}{2} x$
   $g(x) = 2 \csc x$

Applications and Extensions

49. Carrying a Ladder around a Corner   Two hallways, one of width 3 feet, the other of width 4 feet, meet at a right angle. See the illustration.

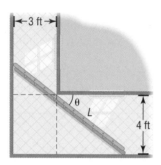

(a) Show that the length L of the line segment shown as a function of the angle θ is

$$L(\theta) = 3 \sec \theta + 4 \csc \theta$$

(b) Graph $L = L(\theta), 0 < \theta < \dfrac{\pi}{2}$.

(c) For what value of θ is L the least?

(d) What is the length of the longest ladder that can be carried around the corner? Why is this also the least value of L?

50. A Rotating Beacon   Suppose that a fire truck is parked in front of a building as shown in the figure.

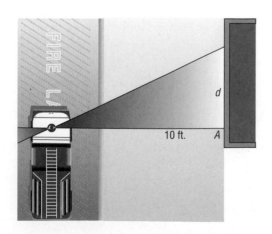

The beacon light on top of the fire truck is located 10 feet from the wall and has a light on each side. If the beacon light rotates 1 revolution every 2 seconds, then a model for determining the distance d, in feet, that the beacon of light is from point A on the wall after t seconds is given by

$$d(t) = |10 \tan(\pi t)|$$

(a) Graph $d(t) = |10 \tan(\pi t)|$ for $0 \le t \le 2$.

(b) For what values of t is the function undefined? Explain what this means in terms of the beam of light on the wall.

(c) Fill in the following table.

t	0	0.1	0.2	0.3	0.4
$d(t) = 10 \tan(\pi t)$					

(d) Compute $\dfrac{d(0.1) - d(0)}{0.1 - 0}, \dfrac{d(0.2) - d(0.1)}{0.2 - 0.1}$, and so on, for each consecutive value of t. These are called **first differences**.

(e) Interpret the first differences found in part (d). What is happening to the speed of the beam of light as d increases?

51. Exploration   Graph

$$y = \tan x \quad \text{and} \quad y = -\cot\left(x + \dfrac{\pi}{2}\right)$$

Do you think that $\tan x = -\cot\left(x + \dfrac{\pi}{2}\right)$?

'Are You Prepared?' Answers

1. $x = 4$ **2.** True

2.8 Phase Shift; Sinusoidal Curve Fitting

OBJECTIVES   **1** Graph Sinusoidal Functions of the Form $y = A \sin(\omega x - \phi) + B$
      (p. 185)
  2 Build Sinusoidal Models from Data (p. 189)

1 Graph Sinusoidal Functions of the Form $y = A \sin(\omega x - \phi) + B$

Figure 109
One cycle of
$y = A\sin(\omega x), A > 0, \omega > 0$

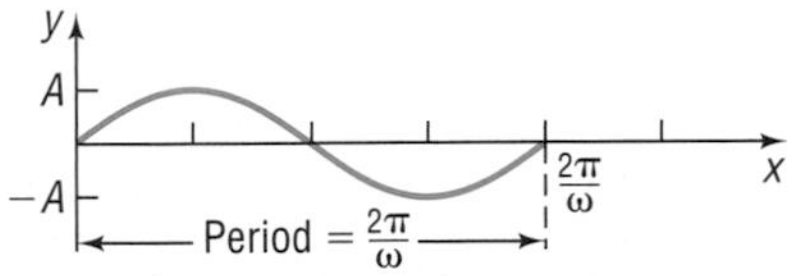

We have seen that the graph of $y = A \sin(\omega x)$, $\omega > 0$, has amplitude $|A|$ and period $T = \dfrac{2\pi}{\omega}$. One cycle can be drawn as x varies from 0 to $\dfrac{2\pi}{\omega}$ or, equivalently, as ωx varies from 0 to 2π. See Figure 109.

We now want to discuss the graph of

$$y = A \sin(\omega x - \phi)$$

which may also be written as

$$y = A \sin\left[\omega\left(x - \frac{\phi}{\omega}\right)\right]$$

where $\omega > 0$ and ϕ (the Greek letter phi) are real numbers. The graph will be a sine curve with amplitude $|A|$. As $\omega x - \phi$ varies from 0 to 2π, one period will be traced out. This period will begin when

$$\omega x - \phi = 0 \quad \text{or} \quad x = \frac{\phi}{\omega}$$

and will end when

$$\omega x - \phi = 2\pi \quad \text{or} \quad x = \frac{\phi}{\omega} + \frac{2\pi}{\omega}$$

See Figure 110.

NOTE We can also find the beginning and end of the period by solving the inequality:

$$0 \le \omega x - \phi \le 2\pi$$
$$\phi \le \omega x \le 2\pi + \phi$$
$$\frac{\phi}{\omega} \le x \le \frac{2\pi}{\omega} + \frac{\phi}{\omega} \qquad ■$$

Figure 110
One cycle of
$y = A\sin(\omega x - \phi), A > 0, \omega > 0, \phi > 0$

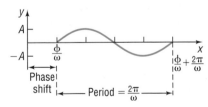

We see that the graph of $y = A \sin(\omega x - \phi) = A \sin\left[\omega\left(x - \dfrac{\phi}{\omega}\right)\right]$ is the same as the graph of $y = A \sin(\omega x)$, except that it has been shifted $\dfrac{\phi}{\omega}$ units (to the right if $\phi > 0$ and to the left if $\phi < 0$). This number $\dfrac{\phi}{\omega}$ is called the **phase shift** of the graph of $y = A \sin(\omega x - \phi)$.

For the graphs of $y = A \sin(\omega x - \phi)$ or $y = A \cos(\omega x - \phi)$, $\omega > 0$,

$$\text{Amplitude} = |A| \qquad \text{Period} = T = \frac{2\pi}{\omega} \qquad \text{Phase shift} = \frac{\phi}{\omega}$$

The phase shift is to the left if $\phi < 0$ and to the right if $\phi > 0$.

EXAMPLE 1 **Finding the Amplitude, Period, and Phase Shift of a Sinusoidal Function and Graphing It**

Find the amplitude, period, and phase shift of $y = 3\sin(2x - \pi)$ and graph the function.

Solution We will use the same four steps used to graph sinusoidal functions of the form $y = A\sin(\omega x)$ or $y = A\cos(\omega x)$ given on page 169.

STEP 1: Comparing $\quad y = 3\sin(2x - \pi) = 3\sin\left[2\left(x - \dfrac{\pi}{2}\right)\right]$

to

$$y = A\sin(\omega x - \phi) = A\sin\left[\omega\left(x - \dfrac{\phi}{\omega}\right)\right]$$

we find that $A = 3$, $\omega = 2$, and $\phi = \pi$. The graph is a sine curve with amplitude $|A| = 3$, period $T = \dfrac{2\pi}{\omega} = \dfrac{2\pi}{2} = \pi$, and phase shift $= \dfrac{\phi}{\omega} = \dfrac{\pi}{2}$.

NOTE We also can find the interval defining one cycle by solving the inequality

$$0 \le 2x - \pi \le 2\pi$$

Then

$$\pi \le 2x \le 3\pi$$

$$\frac{\pi}{2} \le x \le \frac{3\pi}{2} \quad\blacksquare$$

STEP 2: The graph of $y = 3\sin(2x - \pi)$ will lie between -3 and 3 on the y-axis. One cycle will begin at $x = \dfrac{\phi}{\omega} = \dfrac{\pi}{2}$ and end at $x = \dfrac{\phi}{\omega} + \dfrac{2\pi}{\omega} = \dfrac{\pi}{2} + \pi = \dfrac{3\pi}{2}$.

To find the five key points, we divide the interval $\left[\dfrac{\pi}{2}, \dfrac{3\pi}{2}\right]$ into four subintervals, each of length $\pi \div 4 = \dfrac{\pi}{4}$, by finding the following values of x:

$$\underset{\text{1st x-coordinate}}{\dfrac{\pi}{2}} \quad \underset{\text{2nd x-coordinate}}{\dfrac{\pi}{2} + \dfrac{\pi}{4} = \dfrac{3\pi}{4}} \quad \underset{\text{3rd x-coordinate}}{\dfrac{3\pi}{4} + \dfrac{\pi}{4} = \pi} \quad \underset{\text{4th x-coordinate}}{\pi + \dfrac{\pi}{4} = \dfrac{5\pi}{4}} \quad \underset{\text{5th x-coordinate}}{\dfrac{5\pi}{4} + \dfrac{\pi}{4} = \dfrac{3\pi}{2}}$$

STEP 3: Use these values of x to determine the five key points on the graph:

$$\left(\dfrac{\pi}{2}, 0\right) \quad \left(\dfrac{3\pi}{4}, 3\right) \quad (\pi, 0) \quad \left(\dfrac{5\pi}{4}, -3\right) \quad \left(\dfrac{3\pi}{2}, 0\right)$$

STEP 4: We plot these five points and fill in the graph of the sine function as shown in Figure 111(a). Extending the graph in each direction, we obtain Figure 111(b).

Figure 111

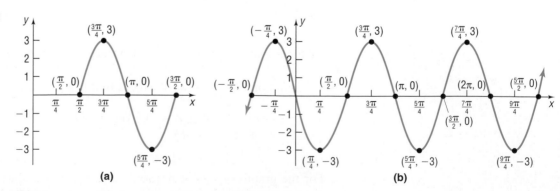

(a)   (b)

The graph of $y = 3\sin(2x - \pi) = 3\sin\left[2\left(x - \dfrac{\pi}{2}\right)\right]$ may also be obtained using transformations. See Figure 112.

Figure 112

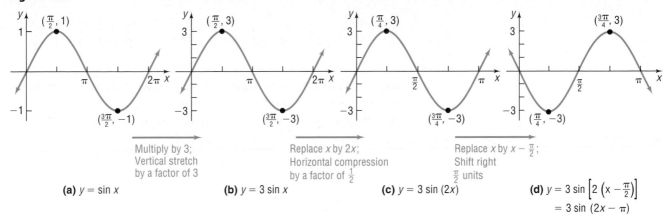

(a) $y = \sin x$

Multiply by 3;
Vertical stretch
by a factor of 3

(b) $y = 3 \sin x$

Replace x by $2x$;
Horizontal compression
by a factor of $\frac{1}{2}$

(c) $y = 3 \sin (2x)$

Replace x by $x - \frac{\pi}{2}$;
Shift right
$\frac{\pi}{2}$ units

(d) $y = 3 \sin \left[2 \left(x - \frac{\pi}{2}\right)\right]$
$= 3 \sin (2x - \pi)$

Figure 113

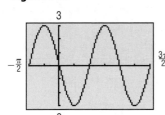

✓ **Check:** Figure 113 shows the graph of $Y_1 = 3 \sin(2x - \pi)$ using a graphing utility.

■

To graph a sinusoidal function of the form $y = A \sin(\omega x - \phi) + B$, we first graph the function $y = A \sin(\omega x - \phi)$ and then apply a vertical shift.

EXAMPLE 2	**Finding the Amplitude, Period, and Phase Shift of a Sinusoidal Function and Graphing It**

Find the amplitude, period, and phase shift of $y = 2 \cos(4x + 3\pi) + 1$ and graph the function.

Solution   **STEP 1:** We begin by graphing $y = 2 \cos(4x + 3\pi)$. Comparing

$$y = 2 \cos(4x + 3\pi) = 2 \cos\left[4\left(x + \frac{3\pi}{4}\right)\right]$$

to

$$y = A \cos(\omega x - \phi) = A \cos\left[\omega\left(x - \frac{\phi}{\omega}\right)\right]$$

we see that $A = 2$, $\omega = 4$, and $\phi = -3\pi$. The graph is a cosine curve with amplitude $|A| = 2$, period $T = \dfrac{2\pi}{\omega} = \dfrac{2\pi}{4} = \dfrac{\pi}{2}$, and phase shift $= \dfrac{\phi}{\omega} = -\dfrac{3\pi}{4}$.

NOTE We can also find the interval defining one cycle by solving the inequality

$$0 \le 4x + 3\pi \le 2\pi$$

Then

$$-3\pi \le 4x \le -\pi$$

$$-\frac{3\pi}{4} \le x \le -\frac{\pi}{4}$$ ■

STEP 2: The graph of $y = 2 \cos(4x + 3\pi)$ will lie between -2 and 2 on the y-axis. One cycle will begin at $x = \dfrac{\phi}{\omega} = -\dfrac{3\pi}{4}$ and end at $x = \dfrac{\phi}{\omega} + \dfrac{2\pi}{\omega} = -\dfrac{3\pi}{4} + \dfrac{\pi}{2} = -\dfrac{\pi}{4}$. To find the five key points, we divide the interval $\left[-\dfrac{3\pi}{4}, -\dfrac{\pi}{4}\right]$ into four subintervals, each of the length $\dfrac{\pi}{2} \div 4 = \dfrac{\pi}{8}$, by finding the following values.

$$-\frac{3\pi}{4} \qquad -\frac{3\pi}{4} + \frac{\pi}{8} = -\frac{5\pi}{8} \qquad -\frac{5\pi}{8} + \frac{\pi}{8} = -\frac{\pi}{2} \qquad -\frac{\pi}{2} + \frac{\pi}{8} = -\frac{3\pi}{8} \qquad -\frac{3\pi}{8} + \frac{\pi}{8} = -\frac{\pi}{4}$$

1st x-coordinate   2nd x-coordinate   3rd x-coordinate   4th x-coordinate   5th x-coordinate

STEP 3: The five key points on the graph of $y = 2\cos(4x + 3\pi)$ are

$$\left(-\frac{3\pi}{4}, 2\right) \quad \left(-\frac{5\pi}{8}, 0\right) \quad \left(-\frac{\pi}{2}, -2\right) \quad \left(-\frac{3\pi}{8}, 0\right) \quad \left(-\frac{\pi}{4}, 2\right)$$

STEP 4: We plot these five points and fill in the graph of the cosine function as shown in Figure 114(a). Extending the graph in each direction, we obtain Figure 114(b), the graph of $y = 2\cos(4x + 3\pi)$.

STEP 5: A vertical shift up 1 unit gives the final graph. See Figure 114(c).

Figure 114

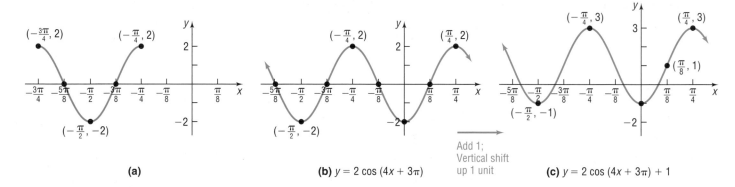

(a)

(b) $y = 2\cos(4x + 3\pi)$

Add 1;
Vertical shift
up 1 unit

(c) $y = 2\cos(4x + 3\pi) + 1$

The graph of $y = 2\cos(4x + 3\pi) + 1 = 2\cos\left[4\left(x + \frac{3\pi}{4}\right)\right] + 1$ may also be obtained using transformations. See Figure 115.

Figure 115

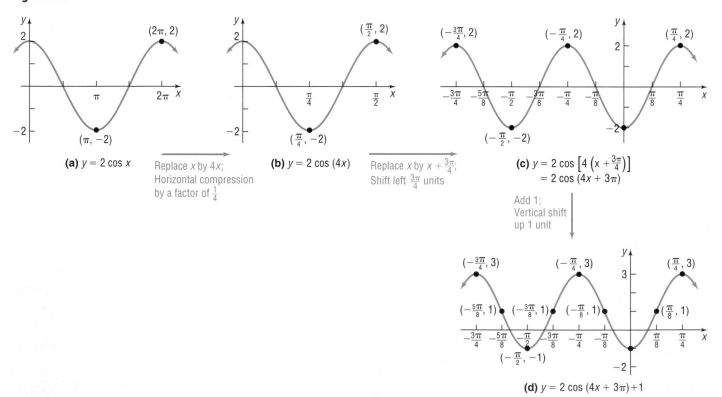

(a) $y = 2\cos x$

Replace x by $4x$;
Horizontal compression
by a factor of $\frac{1}{4}$

(b) $y = 2\cos(4x)$

Replace x by $x + \frac{3\pi}{4}$;
Shift left $\frac{3\pi}{4}$ units

(c) $y = 2\cos\left[4\left(x + \frac{3\pi}{4}\right)\right]$
$= 2\cos(4x + 3\pi)$

Add 1;
Vertical shift
up 1 unit

(d) $y = 2\cos(4x + 3\pi) + 1$

✓ **Check:** Graph $Y_1 = 2\cos(4x + 3\pi) + 1$ to verify the graph in Figure 114(c) or Figure 115(d).

Now Work PROBLEM 3

> **SUMMARY** Steps for Graphing Sinusoidal Functions $y = A \sin(\omega x - \phi) + B$
> or $y = A \cos(\omega x - \phi) + B$
>
> **STEP 1:** Determine the amplitude $|A|$ and period $T = \dfrac{2\pi}{\omega}$.
>
> **STEP 2:** Determine the starting point of one cycle of the graph, $\dfrac{\phi}{\omega}$. Determine the ending point of one cycle of
> the graph, $\dfrac{\phi}{\omega} + \dfrac{2\pi}{\omega}$. Divide the interval $\left[\dfrac{\phi}{\omega}, \dfrac{\phi}{\omega} + \dfrac{2\pi}{\omega}\right]$ into four subintervals, each of length $\dfrac{2\pi}{\omega} \div 4$.
>
> **STEP 3:** Use the endpoints of the subintervals to find the five key points on the graph.
>
> **STEP 4:** Plot the five key points with a sinusoidal graph to obtain one cycle of the graph. Extend the graph in each direction to make it complete.
>
> **STEP 5:** If $B \neq 0$, apply a vertical shift.

2 Build Sinusoidal Models from Data

Scatter diagrams of data sometimes take the form of a sinusoidal function. Let's look at an example.

The data given in Table 12 represent the average monthly temperatures in Denver, Colorado. Since the data represent *average* monthly temperatures collected over many years, the data will not vary much from year to year and so will essentially repeat each year. In other words, the data are periodic. Figure 116 shows the scatter diagram of these data repeated over 2 years, where $x = 1$ represents January, $x = 2$ represents February, and so on.

Table 12

Month, x	Average Monthly Temperature, °F
January, 1	29.7
February, 2	33.4
March, 3	39.0
April, 4	48.2
May, 5	57.2
June, 6	66.9
July, 7	73.5
August, 8	71.4
September, 9	62.3
October, 10	51.4
November, 11	39.0
December, 12	31.0

Source: U.S. National Oceanic and Atmospheric Administration

Figure 116

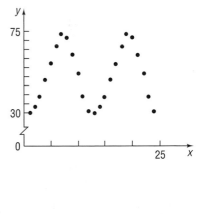

Notice that the scatter diagram looks like the graph of a sinusoidal function. We choose to fit the data to a sine function of the form

$$y = A \sin(\omega x - \phi) + B$$

where A, B, ω, and ϕ are constants.

| EXAMPLE 3 | **Finding a Sinusoidal Function from Temperature Data** |

Find a sine function that models the data in Table 12.

Solution We begin with a scatter diagram of the data for 1 year. See Figure 117. The data will be fitted to a sine function of the form

$$y = A \sin(\omega x - \phi) + B$$

Figure 117

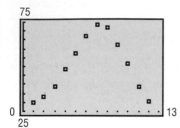

STEP 1: To find the amplitude A, we compute

$$\text{Amplitude} = \frac{\text{largest data value} - \text{smallest data value}}{2}$$

$$= \frac{73.5 - 29.7}{2} = 21.9$$

To see the remaining steps in this process, we superimpose the graph of the function $y = 21.9 \sin x$, where x represents months, on the scatter diagram.

Figure 118 shows the two graphs. To fit the data, the graph needs to be shifted vertically, shifted horizontally, and stretched horizontally.

Figure 118

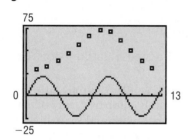

STEP 2: We determine the vertical shift by finding the average of the highest and lowest data values.

$$\text{Vertical shift} = \frac{73.5 + 29.7}{2} = 51.6$$

Now we superimpose the graph of $y = 21.9 \sin x + 51.6$ on the scatter diagram. See Figure 119.

We see that the graph needs to be shifted horizontally and stretched horizontally.

Figure 119

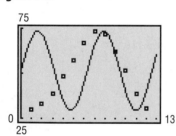

STEP 3: It is easier to find the horizontal stretch factor first. Since the temperatures repeat every 12 months, the period of the function is $T = 12$. Since $T = \dfrac{2\pi}{\omega} = 12$, we find

$$\omega = \frac{2\pi}{12} = \frac{\pi}{6}$$

Figure 120

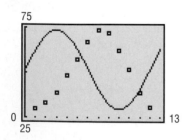

Now we superimpose the graph of $y = 21.9 \sin\left(\dfrac{\pi}{6}x\right) + 51.6$ on the scatter diagram. See Figure 120. We see that the graph still needs to be shifted horizontally.

STEP 4: To determine the horizontal shift, we use the period $T = 12$ and divide the interval $[0, 12]$ into four subintervals of length $12 \div 4 = 3$:

$$[0, 3] \quad [3, 6] \quad [6, 9] \quad [9, 12]$$

The sine curve is increasing on the interval $(0, 3)$ and is decreasing on the interval $(3, 9)$, so a local maximum occurs at $x = 3$. The data indicate that a maximum occurs at $x = 7$ (corresponding to July's temperature), so we must shift the graph of the function 4 units to the right by replacing x by $x - 4$. Doing this, we obtain

$$y = 21.9 \sin\left(\frac{\pi}{6}(x - 4)\right) + 51.6$$

Multiplying out, we find that a sine function of the form $y = A \sin(\omega x - \phi) + B$ that fits the data is

$$y = 21.9 \sin\left(\frac{\pi}{6}x - \frac{2\pi}{3}\right) + 51.6$$

Figure 121

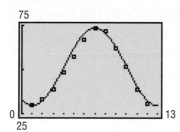

The graph of $y = 21.9 \sin\left(\dfrac{\pi}{6}x - \dfrac{2\pi}{3}\right) + 51.6$ and the scatter diagram of the data are shown in Figure 121.

The steps to fit a sine function

$$y = A \sin(\omega x - \phi) + B$$

to sinusoidal data follow:

Steps for Fitting a Sine Function $y = A \sin(\omega x - \phi) + B$ to Data

STEP 1: Determine A, the amplitude of the function.

$$\text{Amplitude} = \frac{\text{largest data value} - \text{smallest data value}}{2}$$

STEP 2: Determine B, the vertical shift of the function.

$$\text{Vertical shift} = \frac{\text{largest data value} + \text{smallest data value}}{2}$$

STEP 3: Determine ω. Since the period T, the time it takes for the data to repeat, is $T = \dfrac{2\pi}{\omega}$, we have

$$\omega = \frac{2\pi}{T}$$

STEP 4: Determine the horizontal shift of the function by using the period of the data. Divide the period into four subintervals of equal length. Determine the x-coordinate for the maximum of the sine function and the x-coordinate for the maximum value of the data. Use this information to determine the value of the phase shift, $\dfrac{\phi}{\omega}$.

Now Work PROBLEM 29(a)–(c)

Let's look at another example. Since the number of hours of sunlight in a day cycles annually, the number of hours of sunlight in a day for a given location can be modeled by a sinusoidal function.

The longest day of the year (in terms of hours of sunlight) occurs on the day of the summer solstice. For locations in the northern hemisphere, the summer solstice is the time when the sun is farthest north. In 2007, the summer solstice occurred on June 21 (the 172nd day of the year) at 2:06 PM EDT. The shortest day of the year occurs on the day of the winter solstice. The winter solstice is the time when the Sun is farthest south (again, for locations in the northern hemisphere). In 2007, the winter solstice occurred on December 22 (the 356th day of the year) at 1:08 AM (EST).

EXAMPLE 4   **Finding a Sinusoidal Function for Hours of Daylight**

According to the *Old Farmer's Almanac,* the number of hours of sunlight in Boston on the summer solstice is 15.30 and the number of hours of sunlight on the winter solstice is 9.08.

(a) Find a sinusoidal function of the form $y = A \sin(\omega x - \phi) + B$ that models the data.

(b) Use the function found in part (a) to predict the number of hours of sunlight on April 1, the 91st day of the year.

(c) Draw a graph of the function found in part (a).

(d) Look up the number of hours of sunlight for April 1 in the *Old Farmer's Almanac* and compare it to the results found in part (b).

Source: The *Old Farmer's Almanac,* www.almanac.com/rise

Solution

(a) **STEP 1:** Amplitude $= \dfrac{\text{largest data value} - \text{smallest data value}}{2}$

$$= \dfrac{15.30 - 9.08}{2} = 3.11$$

STEP 2: Vertical shift $= \dfrac{\text{largest data value} + \text{smallest data value}}{2}$

$$= \dfrac{15.30 + 9.08}{2} = 12.19$$

STEP 3: The data repeat every 365 days. Since $T = \dfrac{2\pi}{\omega} = 365$, we find

$$\omega = \dfrac{2\pi}{365}$$

So far we have $y = 3.11 \sin\left(\dfrac{2\pi}{365}x - \phi\right) + 12.19$.

STEP 4: To determine the horizontal shift, we use the period $T = 365$ and divide the interval $[0, 365]$ into four subintervals of length $365 \div 4 = 91.25$:

$$[0, 91.25] \quad [91.25, 182.5] \quad [182.5, 273.75] \quad [273.75, 365]$$

The sine curve is increasing on the interval $(0, 91.25)$ and is decreasing on the interval $(91.25, 273.75)$, so a local maximum occurs at $x = 91.25$. Since the maximum occurs on the summer solstice at $x = 172$, we must shift the graph of the function $172 - 91.25 = 80.75$ units to the right by replacing x by $x - 80.75$. Doing this, we obtain

$$y = 3.11 \sin\left(\dfrac{2\pi}{365}(x - 80.75)\right) + 12.19$$

Multiplying out, we find that a sine function of the form $y = A \sin(\omega x - \phi) + B$ that models the data is

$$y = 3.11 \sin\left(\dfrac{2\pi}{365}x - \dfrac{323\pi}{730}\right) + 12.19$$

(b) To predict the number of hours of daylight on April 1, we let $x = 91$ in the function found in part (a) and obtain

$$y = 3.11 \sin\left(\dfrac{2\pi}{365}\cdot 91 - \dfrac{323}{730}\pi\right) + 12.19$$

$$\approx 12.74$$

So we predict that there will be about 12.74 hours = 12 hours, 44 minutes of sunlight on April 1 in Boston.

(c) The graph of the function found in part (a) is given in Figure 122.

(d) According to the *Old Farmer's Almanac,* there will be 12 hours 43 minutes of sunlight on April 1 in Boston.

Figure 122

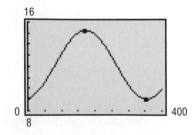

──── **Now Work** PROBLEM 35

Certain graphing utilities (such as a TI-83, TI-84 Plus, and TI-86) have the capability of finding the sine function of best fit for sinusoidal data. At least four data points are required for this process.

EXAMPLE 5	Finding the Sine Function of Best Fit

Use a graphing utility to find the sine function of best fit that models the data in Table 12. Graph this function with the scatter diagram of the data.

Solution Enter the data from Table 12 and execute the SINe REGression program. The result is shown in Figure 123.

The output that the utility provides shows the equation

$$y = a \sin(bx + c) + d$$

The sinusoidal function of best fit is

$$y = 21.15 \sin(0.55x - 2.35) + 51.19$$

where x represents the month and y represents the average temperature.

Figure 124 shows the graph of the sinusoidal function of best fit on the scatter diagram.

Figure 123

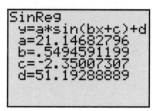

Figure 124

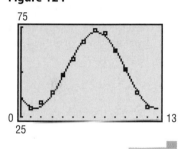

Now Work PROBLEMS 29(d) AND (e)

2.8 Assess Your Understanding

Concepts and Vocabulary

1. For the graph of $y = A \sin(\omega x - \phi)$, the number $\dfrac{\phi}{\omega}$ is called the _____ .

2. *True or False* Only two data points are required by a graphing utility to find the sine function of best fit.

Skill Building

In Problems 3–14, find the amplitude, period, and phase shift of each function. Graph each function. Be sure to label key points. Show at least two periods. Verify your graph using a graphing utility.

3. $y = 4 \sin(2x - \pi)$

4. $y = 3 \sin(3x - \pi)$

5. $y = 2 \cos\left(3x + \dfrac{\pi}{2}\right)$

6. $y = 3 \cos(2x + \pi)$

7. $y = -3 \sin\left(2x + \dfrac{\pi}{2}\right)$

8. $y = -2 \cos\left(2x - \dfrac{\pi}{2}\right)$

9. $y = 4 \sin(\pi x + 2) - 5$

10. $y = 2 \cos(2\pi x + 4) + 4$

11. $y = 3 \cos(\pi x - 2) + 5$

12. $y = 2 \cos(2\pi x - 4) - 1$

13. $y = -3 \sin\left(-2x + \dfrac{\pi}{2}\right)$

14. $y = -3 \cos\left(-2x + \dfrac{\pi}{2}\right)$

In Problems 15–18, write the equation of a sine function that has the given characteristics.

15. Amplitude: 2
Period: π
Phase shift: $\dfrac{1}{2}$

16. Amplitude: 3
Period: $\dfrac{\pi}{2}$
Phase shift: 2

17. Amplitude: 3
Period: 3π
Phase shift: $-\dfrac{1}{3}$

18. Amplitude: 2
Period: π
Phase shift: -2

Mixed Practice

In Problems 19–26, apply the methods of this and the previous section to graph each function. Be sure to label key points and show at least two periods. Verify your graph using a graphing utility.

19. $y = 2\tan(4x - \pi)$ **20.** $y = \dfrac{1}{2}\cot(2x - \pi)$ **21.** $y = 3\csc\left(2x - \dfrac{\pi}{4}\right)$ **22.** $y = \dfrac{1}{2}\sec(3x - \pi)$

23. $y = -\cot\left(2x + \dfrac{\pi}{2}\right)$ **24.** $y = -\tan\left(3x + \dfrac{\pi}{2}\right)$ **25.** $y = -\sec(2\pi x + \pi)$ **26.** $y = -\csc\left(-\dfrac{1}{2}\pi x + \dfrac{\pi}{4}\right)$

Applications and Extensions

27. Alternating Current (ac) Circuits   The current I, in amperes, flowing through an ac (alternating current) circuit at time t, in seconds, is

$$I(t) = 120\sin\left(30\pi t - \dfrac{\pi}{3}\right) \qquad t \geq 0$$

What is the period? What is the amplitude? What is the phase shift? Graph this function over two periods.

28. Alternating Current (ac) Circuits   The current I, in amperes, flowing through an ac (alternating current) circuit at time t, in seconds, is

$$I(t) = 220\sin\left(60\pi t - \dfrac{\pi}{6}\right) \qquad t \geq 0$$

What is the period? What is the amplitude? What is the phase shift? Graph this function over two periods.

29. Monthly Temperature   The following data represent the average monthly temperatures for Juneau, Alaska.

Month, x	Average Monthly Temperature, °F
January, 1	24.2
February, 2	28.4
March, 3	32.7
April, 4	39.7
May, 5	47.0
June, 6	53.0
July, 7	56.0
August, 8	55.0
September, 9	49.4
October, 10	42.2
November, 11	32.0
December, 12	27.1

Source: U.S. National Oceanic and Atmospheric Administration

(a) Draw a scatter diagram of the data for one period.
(b) Find a sinusoidal function of the form
    $y = A\sin(\omega x - \phi) + B$ that models the data.
(c) Draw the sinusoidal function found in part (b) on the scatter diagram.
(d) Use a graphing utility to find the sinusoidal function of best fit.
(e) Draw the sinusoidal function of best fit on a scatter diagram of the data.

30. Monthly Temperature   The following data represent the average monthly temperatures for Washington, D.C.
(a) Draw a scatter diagram of the data for one period.

(b) Find a sinusoidal function of the form
    $y = A\sin(\omega x - \phi) + B$ that models the data.
(c) Draw the sinusoidal function found in part (b) on the scatter diagram.
(d) Use a graphing utility to find the sinusoidal function of best fit.
(e) Graph the sinusoidal function of best fit on a scatter diagram of the data.

Month, x	Average Monthly Temperature, °F
January, 1	34.6
February, 2	37.5
March, 3	47.2
April, 4	56.5
May, 5	66.4
June, 6	75.6
July, 7	80.0
August, 8	78.5
September, 9	71.3
October, 10	59.7
November, 11	49.8
December, 12	39.4

Source: U.S. National Oceanic and Atmospheric Administration

31. Monthly Temperature   The following data represent the average monthly temperatures for Indianapolis, Indiana.

Month, x	Average Monthly Temperature, °F
January, 1	25.5
February, 2	29.6
March, 3	41.4
April, 4	52.4
May, 5	62.8
June, 6	71.9
July, 7	75.4
August, 8	73.2
September, 9	66.6
October, 10	54.7
November, 11	43.0
December, 12	30.9

Source: U.S. National Oceanic and Atmospheric Administration

(a) Draw a scatter diagram of the data for one period.
(b) Find a sinusoidal function of the form
$y = A \sin(\omega x - \phi) + B$ that models the data.
(c) Draw the sinusoidal function found in part (b) on the scatter diagram.
(d) Use a graphing utility to find the sinusoidal function of best fit.
(e) Graph the sinusoidal function of best fit on a scatter diagram of the data.

32. **Monthly Temperature** The following data represent the average monthly temperatures for Baltimore, Maryland.
(a) Draw a scatter diagram of the data for one period.
(b) Find a sinusoidal function of the form
$y = A \sin(\omega x - \phi) + B$ that models the data.
(c) Draw the sinusoidal function found in part (b) on the scatter diagram.
(d) Use a graphing utility to find the sinusoidal function of best fit.
(e) Graph the sinusoidal function of best fit on a scatter diagram of the data.

Month, x	Average Monthly Temperature, °F
January, 1	31.8
February, 2	34.8
March, 3	44.1
April, 4	53.4
May, 5	63.4
June, 6	72.5
July, 7	77.0
August, 8	75.6
September, 9	68.5
October, 10	56.6
November, 11	46.8
December, 12	36.7

Source: U.S. National Oceanic and Atmospheric Administration

33. **Tides** Suppose that the length of time between consecutive high tides is approximately 12.5 hours. According to the National Oceanic and Atmospheric Administration, on Saturday, August 7, 2007, in Savannah, Georgia, high tide occurred at 3:38 AM (3.6333 hours) and low tide occurred at 10:08 AM (10.1333 hours). Water heights are measured as the amounts above or below the mean lower low water. The height of the water at high tide was 8.2 feet, and the height of the water at low tide was −0.6 foot.
(a) Approximately when will the next high tide occur?
(b) Find a sinusoidal function of the form
$y = A \sin(\omega x - \phi) + B$ that models the data.
(c) Draw a graph of the function found in part (b).
(d) Use the function found in part (b) to predict the height of the water at the next high tide.

34. **Tides** Suppose that the length of time between consecutive high tides is approximately 12.5 hours. According to the National Oceanic and Atmospheric Administration, on Saturday, August 7, 2007, in Juneau, Alaska, high tide occurred at 8:11 AM (8.1833 hours) and low tide occurred at 2:14 PM (14.2333 hours). Water heights are measured as the amounts above or below the mean lower low water. The height of the water at high tide was 13.2 feet, and the height of the water at low tide was 2.2 feet.
(a) Approximately when will the next high tide occur?
(b) Find a sinusoidal function of the form
$y = A \sin(\omega x - \phi) + B$ that models the data.
(c) Draw a graph of the function found in part (b).
(d) Use the function found in part (b) to predict the height of the water at the next high tide.

35. **Hours of Daylight** According to the *Old Farmer's Almanac*, in Miami, Florida, the number of hours of sunlight on the summer solstice of 2007 was 13.75, and the number of hours of sunlight on the winter solstice was 10.55.
(a) Find a sinusoidal function of the form
$y = A \sin(\omega x - \phi) + B$ that models the data.
(b) Use the function found in part (a) to predict the number of hours of sunlight on April 1, the 91st day of the year.
(c) Draw a graph of the function found in part (a).
(d) Look up the number of hours of sunlight for April 1 in the *Old Farmer's Almanac*, and compare the actual hours of daylight to the results found in part (c).

36. **Hours of Daylight** According to the *Old Farmer's Almanac*, in Detroit, Michigan, the number of hours of sunlight on the summer solstice of 2007 was 15.30, and the number of hours of sunlight on the winter solstice was 9.10.
(a) Find a sinusoidal function of the form
$y = A \sin(\omega x - \phi) + B$ that models the data.
(b) Use the function found in part (a) to predict the number of hours of sunlight on April 1, the 91st day of the year.
(c) Draw a graph of the function found in part (a).
(d) Look up the number of hours of sunlight for April 1 in the *Old Farmer's Almanac*, and compare the actual hours of daylight to the results found in part (c).

37. **Hours of Daylight** According to the *Old Farmer's Almanac*, in Anchorage, Alaska, the number of hours of sunlight on the summer solstice of 2007 was 19.42 and the number of hours of sunlight on the winter solstice was 5.48.
(a) Find a sinusoidal function of the form
$y = A \sin(\omega x - \phi) + B$ that models the data.
(b) Use the function found in part (a) to predict the number of hours of sunlight on April 1, the 91st day of the year.
(c) Draw a graph of the function found in part (a).
(d) Look up the number of hours of sunlight for April 1 in the *Old Farmer's Almanac*, and compare the actual hours of daylight to the results found in part (c).

38. **Hours of Daylight** According to the *Old Farmer's Almanac*, in Honolulu, Hawaii, the number of hours of sunlight on the summer solstice of 2007 was 13.43 and the number of hours of sunlight on the winter solstice was 10.85.
(a) Find a sinusoidal function of the form
$y = A \sin(\omega x - \phi) + B$ that models the data.
(b) Use the function found in part (a) to predict the number of hours of sunlight on April 1, the 91st day of the year.
(c) Draw a graph of the function found in part (a).
(d) Look up the number of hours of sunlight for April 1 in the *Old Farmer's Almanac*, and compare the actual hours of daylight to the results found in part (c).

Discussion and Writing

39. Explain how the amplitude and period of a sinusoidal graph are used to establish the scale on each coordinate axis.

40. Find an application in your major field that leads to a sinusoidal graph. Write a paper about your findings.

CHAPTER REVIEW

Things to Know

Definitions

Angle in standard position (p. 104)	Vertex is at the origin; initial side is along the positive x-axis.
1 Degree (1°) (p. 105)	$1° = \dfrac{1}{360}$ revolution
1 Radian (p. 107)	The measure of a central angle of a circle whose rays subtend an arc whose length is the radius of the circle
Acute angle (p. 117)	An angle θ whose measure is $0° < \theta < 90°$ or $0 < \theta < \dfrac{\pi}{2}$, θ in radians
Complementary angles (p. 123)	Two acute angles whose sum is $90°$ $\left(\dfrac{\pi}{2} \text{ radians}\right)$
Cofunction (p. 124)	The following pairs of functions are cofunctions of each other: sine and cosine; tangent and cotangent; secant and cosecant.
Trigonometric functions of a general angle (p. 141)	$P = (a, b)$ is the point on the terminal side of θ a distance r from the origin:

$$\sin\theta = \frac{b}{r} \qquad \cos\theta = \frac{a}{r} \qquad \tan\theta = \frac{b}{a}, \;\; a \neq 0$$

$$\csc\theta = \frac{r}{b}, \;\; b \neq 0 \qquad \sec\theta = \frac{r}{a}, \;\; a \neq 0 \qquad \cot\theta = \frac{a}{b}, \;\; b \neq 0$$

Reference angle of θ (p. 145)	The acute angle formed by the terminal side of θ and either the positive or negative x-axis
Periodic function (p. 157)	$f(\theta + p) = f(\theta)$, for all θ, $p > 0$, where the smallest such p is the fundamental period

Formulas

1 revolution $= 360°$ (p. 105)	
$\qquad\qquad\quad = 2\pi$ radians (p. 109)	
$s = r\theta$ (p. 108)	θ is measured in radians; s is the length of the arc subtended by the central angle θ of the circle of radius r.
$A = \dfrac{1}{2}r^2\theta$ (p. 111)	A is the area of the sector of a circle of radius r formed by a central angle of θ radians.
$v = r\omega$ (p. 112)	v is the linear speed along the circle of radius r; ω is the angular speed (measured in radians per unit time).

Table of Values

θ (Radians)	θ (Degrees)	$\sin\theta$	$\cos\theta$	$\tan\theta$	$\csc\theta$	$\sec\theta$	$\cot\theta$
0	0°	0	1	0	Not defined	1	Not defined
$\dfrac{\pi}{6}$	30°	$\dfrac{1}{2}$	$\dfrac{\sqrt{3}}{2}$	$\dfrac{\sqrt{3}}{3}$	2	$\dfrac{2\sqrt{3}}{3}$	$\sqrt{3}$
$\dfrac{\pi}{4}$	45°	$\dfrac{\sqrt{2}}{2}$	$\dfrac{\sqrt{2}}{2}$	1	$\sqrt{2}$	$\sqrt{2}$	1
$\dfrac{\pi}{3}$	60°	$\dfrac{\sqrt{3}}{2}$	$\dfrac{1}{2}$	$\sqrt{3}$	$\dfrac{2\sqrt{3}}{3}$	2	$\dfrac{\sqrt{3}}{3}$
$\dfrac{\pi}{2}$	90°	1	0	Not defined	1	Not defined	0
π	180°	0	-1	0	Not defined	-1	Not defined
$\dfrac{3\pi}{2}$	270°	-1	0	Not defined	-1	Not defined	0

Fundamental identities (p. 120)

$$\tan \theta = \frac{\sin \theta}{\cos \theta} \qquad \cot \theta = \frac{\cos \theta}{\sin \theta}$$

$$\csc \theta = \frac{1}{\sin \theta} \qquad \sec \theta = \frac{1}{\cos \theta} \qquad \cot \theta = \frac{1}{\tan \theta}$$

$$\sin^2 \theta + \cos^2 \theta = 1 \qquad \tan^2 \theta + 1 = \sec^2 \theta \qquad \cot^2 \theta + 1 = \csc^2 \theta$$

Properties of the trigonometric functions

$y = \sin x$ Domain: $-\infty < x < \infty$
(p. 162)   Range: $-1 \le y \le 1$
   Periodic: period $= 2\pi$ (360°)
   Odd function

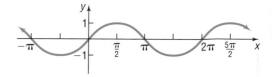

$y = \cos x$ Domain: $-\infty < x < \infty$
(p. 164)   Range: $-1 \le y \le 1$
   Periodic: period $= 2\pi$ (360°)
   Even function

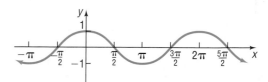

$y = \tan x$ Domain: $-\infty < x < \infty$, except odd integer multiples of $\dfrac{\pi}{2}$ (90°)
(pp. 178–179)   Range: $-\infty < y < \infty$
   Periodic: period $= \pi$ (180°)
   Odd function
   Vertical asymptotes at odd integer multiples of $\dfrac{\pi}{2}$

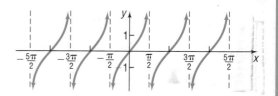

$y = \cot x$ Domain: $-\infty < x < \infty$, except integer multiples of π (180°)
(p. 180)   Range: $-\infty < y < \infty$
   Periodic: period $= \pi$ (180°)
   Odd function
   Vertical asymptotes at integer multiples of π

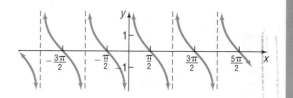

$y = \csc x$ Domain: $-\infty < x < \infty$, except integer multiples of π (180°)
(p. 181)   Range: $|y| \ge 1$
   Periodic: period $= 2\pi$ (360°)
   Odd function
   Vertical asymptotes at integer multiples of π

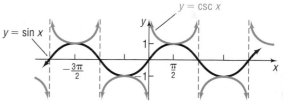

$y = \sec x$ Domain: $-\infty < x < \infty$, except odd integer multiples of $\dfrac{\pi}{2}$ (90°)
(p. 181)   Range: $|y| \ge 1$
   Periodic: period $= 2\pi$ (360°)
   Even function
   Vertical asymptotes at odd integer multiples of $\dfrac{\pi}{2}$

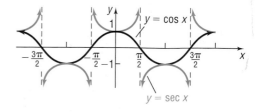

Sinusoidal graphs

$y = A \sin(\omega x) + B, \quad \omega > 0$ Period $= \dfrac{2\pi}{\omega}$ (pp. 167, 185)

$y = A \cos(\omega x) + B, \quad \omega > 0$ Amplitude $= |A|$ (pp. 167, 185)

$y = A \sin(\omega x - \phi) + B = A \sin\left[\omega\left(x - \dfrac{\phi}{\omega}\right)\right] + B$ Phase shift $= \dfrac{\phi}{\omega}$ (p. 185)

$y = A \cos(\omega x - \phi) + B = A \cos\left[\omega\left(x - \dfrac{\phi}{\omega}\right)\right] + B$

Objectives

Section		You should be able to:	Example(s)	Review Exercises
2.1	1	Convert between decimals and degrees, minutes, seconds forms for angles (p. 106)	2	86
	2	Find the arc length of a circle (p. 108)	3	87, 88
	3	Convert from degrees to radians and from radians to degrees (p. 108)	4–6	1–8
	4	Find the area of a sector of a circle (p. 111)	7	87
	5	Find the linear speed of an object traveling in circular motion (p. 112)	8	89–92
2.2	1	Find the values of trigonometric functions of acute angles (p. 117)	1	79
	2	Use the Fundamental Identities (p. 119)	2, 3	21–24
	3	Find the values of the remaining trigonometric functions, given the value of one of them (p. 121)	4, 5	31–32
	4	Use the Complementary Angle Theorem (p. 123)	6, 7	25–26
2.3	1	Find the exact values of the trigonometric functions of $\frac{\pi}{4} = 45°$ (p. 129)	1, 2	9, 11
	2	Find the exact values of the trigonometric functions of $\frac{\pi}{6} = 30°$ and $\frac{\pi}{3} = 60°$ (p. 130)	3, 4	9–12
	3	Use a calculator to approximate the values of the trigonometric functions of acute angles (p. 132)	5	80
	4	Model and solve applied problems involving right triangles (p. 132)	6–9	93–97
2.4	1	Find the exact values of the trigonometric functions for general angles (p. 140)	1, 2	81
	2	Use coterminal angles to find the exact value of a trigonometric function (p. 143)	3	19, 20
	3	Determine the signs of the trigonometric functions of an angle in a given quadrant (p. 144)	4	82
	4	Find the reference angle of a general angle (p. 145)	5	83
	5	Use a reference angle to find the exact value of a trigonometric function (p. 146)	6	13–16, 19
	6	Find the exact values of trigonometric functions of an angle, given information about the functions (p. 147)	7, 8	31–46
2.5	1	Find the exact values of the trigonometric functions using the unit circle (p. 151)	1	84
	2	Know the domain and range of the trigonometric functions (p. 155)	pp. 155–156	85
	3	Use the periodic properties to find the exact values of the trigonometric functions (p. 156)	3	19, 20
	4	Use even–odd properties to find the exact values of the trigonometric functions (p. 158)	4	13, 15, 16, 18–20, 27–30
2.6	1	Graph functions of the form $y = A\sin(\omega x)$ using transformations (p. 163)	1, 2	47
	2	Graph functions of the form $y = A\cos(\omega x)$ using transformations (p. 164)	3	48
	3	Determine the amplitude and period of sinusoidal functions (p. 166)	4	63–68
	4	Graph sinusoidal functions using key points (p. 167)	5–7	47, 48, 67, 68, 98
	5	Find an equation for a sinusoidal graph (p. 171)	8, 9	75–78
2.7	1	Graph functions of the form $y = A\tan(\omega x) + B$ and $y = A\cot(\omega x) + B$ (p. 179)	1, 2	53, 54, 56
	2	Graph functions of the form $y = A\csc(\omega x) + B$ and $y = A\sec(\omega x) + B$ (p. 181)	3	57
2.8	1	Graph sinusoidal functions of the form $y = A\sin(\omega x - \phi) + B$ (p. 185)	1, 2	49, 50, 59, 60, 99
	2	Build sinusoidal models from data (p. 189)	3	100, 101

Review Exercises

In Problems 1–4, convert each angle in degrees to radians. Express your answer as a multiple of π.

1. $135°$ **2.** $210°$ **3.** $18°$ **4.** $15°$

In Problems 5–8, convert each angle in radians to degrees.

5. $\dfrac{3\pi}{4}$ **6.** $\dfrac{2\pi}{3}$ **7.** $-\dfrac{5\pi}{2}$ **8.** $-\dfrac{3\pi}{2}$

In Problems 9–30, find the exact value of each expression. Do not use a calculator.

9. $\tan\dfrac{\pi}{4} - \sin\dfrac{\pi}{6}$

10. $\cos\dfrac{\pi}{3} + \sin\dfrac{\pi}{2}$

11. $3\sin 45° - 4\tan\dfrac{\pi}{6}$

12. $4\cos 60° + 3\tan\dfrac{\pi}{3}$

13. $6\cos\dfrac{3\pi}{4} + 2\tan\left(-\dfrac{\pi}{3}\right)$

14. $3\sin\dfrac{2\pi}{3} - 4\cos\dfrac{5\pi}{2}$

15. $\sec\left(-\dfrac{\pi}{3}\right) - \cot\left(-\dfrac{5\pi}{4}\right)$

16. $4\csc\dfrac{3\pi}{4} - \cot\left(-\dfrac{\pi}{4}\right)$

17. $\tan\pi + \sin\pi$

18. $\cos\dfrac{\pi}{2} - \csc\left(-\dfrac{\pi}{2}\right)$

19. $\cos 540° - \tan(-405°)$

20. $\sin 270° + \cos(-180°)$

21. $\sin^2 20° + \dfrac{1}{\sec^2 20°}$

22. $\dfrac{1}{\cos^2 40°} - \dfrac{1}{\cot^2 40°}$

23. $\sec 50° \cos 50°$

24. $\tan 10° \cot 10°$

25. $\dfrac{\sin 50°}{\cos 40°}$

26. $\dfrac{\tan 20°}{\cot 70°}$

27. $\dfrac{\sin(-40°)}{\cos 50°}$

28. $\tan(-20°)\cot 20°$

29. $\sin 400° \sec(-50°)$

30. $\cot 200° \cot(-70°)$

In Problems 31–46, find the exact value of each of the remaining trigonometric functions.

31. $\sin\theta = \dfrac{4}{5}, \quad \theta$ is acute

32. $\tan\theta = \dfrac{1}{4}, \quad \theta$ is acute

33. $\tan\theta = \dfrac{12}{5}, \quad \sin\theta < 0$

34. $\cot\theta = \dfrac{12}{5}, \quad \cos\theta < 0$

35. $\sec\theta = -\dfrac{5}{4}, \quad \tan\theta < 0$

36. $\csc\theta = -\dfrac{5}{3}, \quad \cot\theta < 0$

37. $\sin\theta = \dfrac{12}{13}, \quad \theta$ in quadrant II

38. $\cos\theta = -\dfrac{3}{5}, \quad \theta$ in quadrant III

39. $\sin\theta = -\dfrac{5}{13}, \quad \dfrac{3\pi}{2} < \theta < 2\pi$

40. $\cos\theta = \dfrac{12}{13}, \quad \dfrac{3\pi}{2} < \theta < 2\pi$

41. $\tan\theta = \dfrac{1}{3}, \quad 180° < \theta < 270°$

42. $\tan\theta = -\dfrac{2}{3}, \quad 90° < \theta < 180°$

43. $\sec\theta = 3, \quad \dfrac{3\pi}{2} < \theta < 2\pi$

44. $\csc\theta = -4, \quad \pi < \theta < \dfrac{3\pi}{2}$

45. $\cot\theta = -2, \quad \dfrac{\pi}{2} < \theta < \pi$

46. $\tan\theta = -2, \quad \dfrac{3\pi}{2} < \theta < 2\pi$

In Problems 47–62, graph each function. Each graph should contain at least two periods. Verify the graph using a graphing utility. Use the graph to determine the domain and the range of each function.

47. $y = 2\sin(4x)$

48. $y = -3\cos(2x)$

49. $y = -2\cos\left(x + \dfrac{\pi}{2}\right)$

50. $y = 3\sin(x - \pi)$

51. $y = \tan(x + \pi)$

52. $y = -\tan\left(x - \dfrac{\pi}{2}\right)$

53. $y = -2\tan(3x)$

54. $y = 4\tan(2x)$

55. $y = \cot\left(x + \dfrac{\pi}{4}\right)$

56. $y = -4\cot(2x)$

57. $y = 4\sec(2x)$

58. $y = \csc\left(x + \dfrac{\pi}{4}\right)$

59. $y = 4\sin(2x + 4) - 2$

60. $y = 3\cos(4x + 2) + 1$

61. $y = 4\tan\left(\dfrac{x}{2} + \dfrac{\pi}{4}\right)$

62. $y = 5\cot\left(\dfrac{x}{3} - \dfrac{\pi}{4}\right)$

In Problems 63–66, determine the amplitude and period of each function without graphing.

63. $y = 4\cos x$

64. $y = \sin(2x)$

65. $y = -8\sin\left(\dfrac{\pi}{2}x\right)$

66. $y = -2\cos(3\pi x)$

In Problems 67–74, find the amplitude, period, and phase shift of each function. Graph each function. Show at least two periods. Verify the graph using a graphing utility.

67. $y = 4\sin(3x)$

68. $y = 2\cos\left(\dfrac{1}{3}x\right)$

69. $y = 2\sin(2x - \pi)$

70. $y = -\cos\left(\dfrac{1}{2}x + \dfrac{\pi}{2}\right)$

71. $y = \dfrac{1}{2}\sin\left(\dfrac{3}{2}x - \pi\right)$

72. $y = \dfrac{3}{2}\cos(6x + 3\pi)$

73. $y = -\dfrac{2}{3}\cos(\pi x - 6)$

74. $y = -7\sin\left(\dfrac{\pi}{3}x - \dfrac{4}{3}\right)$

In Problems 75–78, find a function whose graph is given.

75.

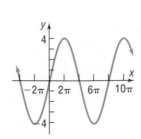

76.

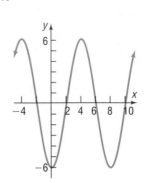

77.

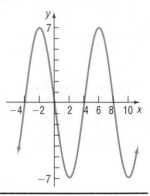

78.

79. Find the value of each of the six trigonometric functions of the angle θ in the illustration.

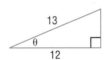

80. Use a calculator to approximate sec 10°. Round the answer to two decimal places.

81. Find the exact value of each of the six trigonometric functions of an angle θ if (3, −4) is a point on the terminal side of θ.

82. Name the quadrant θ lies in if cos θ > 0 and tan θ < 0.

83. Find the reference angle of $-\dfrac{4\pi}{5}$.

84. Find the exact value of sin *t*, cos *t*, and tan *t* if $P = \left(-\dfrac{3}{5}, \dfrac{4}{5}\right)$ is the point on the unit circle that corresponds to *t*.

85. What is the domain and the range of the secant function?

86. (a) Convert the angle 32°20′35″ to a decimal in degrees. Round the answer to two decimal places.
(b) Convert the angle 63.18° to D°M′S″ form. Express the answer to the nearest second.

87. Find the length of the arc subtended by a central angle of 30° on a circle of radius 2 feet. What is the area of the sector?

88. The minute hand of a clock is 8 inches long. How far does the tip of the minute hand move in 30 minutes? How far does it move in 20 minutes?

89. Angular Speed of a Race Car A race car is driven around a circular track at a constant speed of 180 miles per hour. If the diameter of the track is $\dfrac{1}{2}$ mile, what is the angular speed of the car? Express your answer in revolutions per hour (which is equivalent to laps per hour).

90. Merry-Go-Rounds A neighborhood carnival has a merry-go-round whose radius is 25 feet. If the time for one revolution is 30 seconds, how fast is the merry-go-round going? Give the linear speed and the angular speed.

91. Lighthouse Beacons The Montauk Point Lighthouse on Long Island has dual beams (two light sources opposite each other). Ships at sea observe a blinking light every 5 seconds. What rotation speed is required to do this?

92. Spin Balancing Tires The radius of each wheel of a car is 16 inches. At how many revolutions per minute should a spin balancer be set to balance the tires at a speed of 90 miles per hour? Is the setting different for a wheel of radius 14 inches? If so, what is this setting?

93. Measuring the Length of a Lake From a stationary hot-air balloon 500 feet above the ground, two sightings of a lake are made (see the figure). How long is the lake?

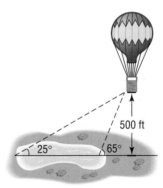

94. Finding the Speed of a Glider From a glider 200 feet above the ground, two sightings of a stationary object directly in front are taken 1 second apart (see the figure). What is the speed of the glider?

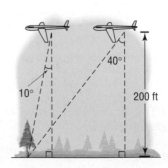

95. Finding the Width of a River Find the distance from A to C across the river illustrated in the figure.

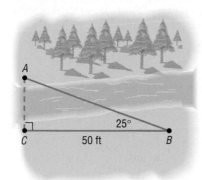

96. Finding the Height of a Building Find the height of the building shown in the figure.

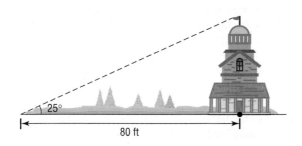

97. Finding the Distance to Shore The Sears Tower in Chicago is 1454 feet tall and is situated about 1 mile inland from the shore of Lake Michigan, as indicated in the figure. An observer in a pleasure boat on the lake directly in front of the Sears Tower looks at the top of the tower and measures the angle of elevation as 5°. How far offshore is the boat?

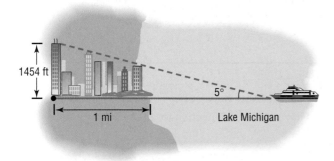

98. Alternating Voltage The electromotive force E, in volts, in a certain ac circuit obeys the equation

$$E = 120 \sin(120\pi t) \qquad t \geq 0$$

where t is measured in seconds.
(a) What is the maximum value of E?
(b) What is the period?
(c) Graph this function over two periods.

99. Alternating Current The current I, in amperes, flowing through an ac (alternating current) circuit at time t is

$$I = 220 \sin\left(30\pi t + \frac{\pi}{6}\right) \qquad t \geq 0$$

(a) What is the period?
(b) What is the amplitude?
(c) What is the phase shift?
(d) Graph this function over two periods.

100. Monthly Temperature The following data represent the average monthly temperatures for Phoenix, Arizona.

Month, m	Average Monthly Temperature, T
January, 1	51
February, 2	55
March, 3	63
April, 4	67
May, 5	77
June, 6	86
July, 7	90
August, 8	90
September, 9	84
October, 10	71
November, 11	59
December, 12	52

Source: U.S. National Oceanic and Atmospheric Administration

(a) Draw a scatter diagram of the data for one period.
(b) Find a sinusoidal function of the form
$y = A \sin(\omega x - \phi) + B$ that models the data.
(c) Draw the sinusoidal function found in part (b) on the scatter diagram.
(d) Use a graphing utility to find the sinusoidal function of best fit.
(e) Graph the sinusoidal function of best fit on the scatter diagram.

101. Hours of Daylight According to the *Old Farmer's Almanac,* in Las Vegas, Nevada, the number of hours of sunlight on the summer solstice is 14.63 and the number of hours of sunlight on the winter solstice is 9.70.
(a) Find a sinusoidal function of the form
$y = A \sin(\omega x - \phi) + B$ that models the data.
(b) Use the function found in part (a) to predict the number of hours of sunlight on April 1, the 91st day of the year.
(c) Draw a graph of the function found in part (a).
(d) Look up the number of hours of sunlight for April 1 in the *Old Farmer's Almanac* and compare the actual hours of daylight to the results found in part (c).

CHAPTER TEST

In Problems 1–3, convert each angle in degrees to radians. Express your answer as a multiple of π.

1. 260°   **2.** −400°   **3.** 13°

In Problems 4–6, convert each angle in radians to degrees.

4. $-\dfrac{\pi}{8}$ **5.** $\dfrac{9\pi}{2}$ **6.** $\dfrac{3\pi}{4}$

In Problems 7–12, find the exact value of each expression.

7. $\sin\dfrac{\pi}{6}$

8. $\cos\left(-\dfrac{5\pi}{4}\right) - \cos\dfrac{3\pi}{4}$

9. $\cos(-120°)$

10. $\tan 330°$

11. $\sin\dfrac{\pi}{2} - \tan\dfrac{19\pi}{4}$

12. $2\sin^2 60° - 3\cos 45°$

In Problems 13–16, use a calculator to evaluate each expression. Round your answer to three decimal places.

13. $\sin 17°$ **14.** $\cos\dfrac{2\pi}{5}$ **15.** $\sec 229°$ **16.** $\cot\dfrac{28\pi}{9}$

17. Fill in each table entry with the sign of each function.

	sin θ	cos θ	tan θ	sec θ	csc θ	cot θ
θ in QI						
θ in QII						
θ in QIII						
θ in QIV						

18. If $f(x) = \sin x$ and $f(a) = \dfrac{3}{5}$, find $f(-a)$.

In Problems 19–21 find the value of the remaining five trigonometric functions of θ.

19. $\sin\theta = \dfrac{5}{7}$, θ in quadrant II   **20.** $\cos\theta = \dfrac{2}{3}$, $\dfrac{3\pi}{2} < \theta < 2\pi$

21. $\tan\theta = -\dfrac{12}{5}$, $\dfrac{\pi}{2} < \theta < \pi$

In Problems 22–24, the point (x, y) is on the terminal side of angle θ in standard position. Find the exact value of the given trigonometric function.

22. $(2, 7)$, $\sin\theta$ **23.** $(-5, 11)$, $\cos\theta$ **24.** $(6, -3)$, $\tan\theta$

In Problems 25 and 26, graph the function.

25. $y = 2\sin\left(\dfrac{x}{3} - \dfrac{\pi}{6}\right)$ **26.** $y = \tan\left(-x + \dfrac{\pi}{4}\right) + 2$

27. Write an equation for a sinusoidal graph with the following properties:

$$A = -3 \qquad \text{period} = \dfrac{2\pi}{3} \qquad \text{phase shift} = -\dfrac{\pi}{4}$$

28. Logan has a garden in the shape of a sector of a circle; the outer rim of the garden is 25 feet long and the central angle of the sector is 50°. She wants to add a 3-foot-wide walk to the outer rim; how many square feet of paving blocks will she need to build the walk?

29. Hungarian Adrian Annus won the gold medal for the hammer throw at the 2004 Olympics in Athens with a winning distance of 83.19 meters.* The event consists of swinging a 16-pound weight attached to a wire 190 centimeters long in a circle and then releasing it. Assuming his release is at a 45° angle to the ground, the hammer will travel a distance of $\dfrac{v_0^2}{g}$ meters, where $g = 9.8$ meters/second² and v_0 is the linear speed of the hammer when released. At what rate (rpm) was he swinging the hammer upon release?

30. A ship is just offshore of New York City. A sighting is taken of the Statue of Liberty, which is about 305 feet tall. If the angle of elevation to the top of the statue is 20°, how far is the ship from the base of the statue?

31. To measure the height of a building, two sightings are taken a distance of 50 feet apart. If the first angle of elevation is 40° and the second is 32°, what is the height of the building?

*Annus was stripped of his medal after refusing to cooperate with postmedal drug testing.

CUMULATIVE REVIEW

1. Find the real solutions, if any, of the equation $2x^2 + x - 1 = 0$.

2. Find an equation for the line with slope −3 containing the point $(-2, 5)$.

3. Find an equation for a circle of radius 4 and center at the point $(0, -2)$.

4. Discuss the equation $2x - 3y = 12$. Graph it.

5. Discuss the equation $x^2 + y^2 - 2x + 4y - 4 = 0$. Graph it.

6. Use transformations to graph the function $y = (x - 3)^2 + 2$.

7. Sketch a graph of each of the following functions. Label at least three points on each graph.
   (a) $y = x^2$ (b) $y = x^3$
   (c) $y = \sin x$ (d) $y = \tan x$

8. Find the inverse function of $f(x) = 3x - 2$.

9. Find the exact value of $(\sin 14°)^2 + (\cos 14°)^2 - 3$.

10. Graph $y = 3\sin(2x)$.

11. Find the exact value of $\tan\dfrac{\pi}{4} - 3\cos\dfrac{\pi}{6} + \csc\dfrac{\pi}{6}$.

12. Find a sinusoidal function for the following graph.

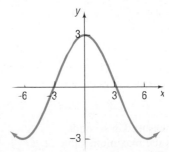

13. Find a linear function that contains the points $(-2, 3)$ and $(1, -6)$. What is the slope? What are the intercepts of the function? Graph the function. Be sure to label the intercepts.

CHAPTER PROJECTS

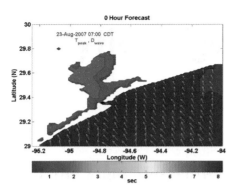

I. Tides The given table is a partial tide table for November 2006 for the Sabine Bank Lighthouse, a shoal located offshore from Texas where the Sabine River empties into the Gulf of Mexico.

1. On November 15, when was the tide high? This is called *high tide*. On November 19, when was the tide low? This is called *low tide*. Most days will have two high tides and two low tides.

2. Why do you think there is a negative height for the low tide on November 20? What is the height measured against?

3. On your graphing utility, draw a scatter diagram for the data in the table. Let t (time, in hours) be the independent variable, with $t = 0$ being 12:00 AM on November 14, $t = 24$ being 12:00 AM on November 15, and so on. Let h be the height in feet. Remember that there are 60 minutes in an hour. Also, make sure your graphing utility is in radian mode.

4. What shape does the data take? What is the period of the data? What is the amplitude? Is the amplitude constant? Explain.

5. Find a sine curve that models the data. Is there a vertical shift? Is there a phase shift?

6. Using your graphing utility, find the sinusoidal function of best fit. How does this model compare to the model from part 5?

7. Using the model found in part (5) and the sinusoidal equation of best fit found in part (6), predict the high tides and the low tides on November 21.

8. Looking at the times of day that the low tides occur, what do you think causes the low tides to vary so much each day? Explain. Does this seem to have the same type of effect on the high tides? Explain.

Nov	Low Tide Time	Ht (ft)	Low Tide Time	Ht (ft)	High Tide Time	Ht (ft)	High Tide Time	Ht (ft)	Sun/Moon Phase Sunrise/set	Moonrise/set
14	6:26a	2.0	4:38p	1.4	9:29a	2.2	11:14p	2.8	6:40a/5:20p	1:05a/2:02p
15	6:22a	1.6	5:34p	1.8	11:18a	2.4	11:15p	2.6	6:41a/5:20p	1:58a/2:27p
16	6:28a	1.2	6:25p	2.0	12:37p	2.6	11:16p	2.6	6:41a/5:19p	2:50a/2:52p
17	6:40a	0.8	7:12p	2.4	1:38p	2.8	11:16p	2.6	6:42a/5:19p	3:43a/3:19p
18	6:56a	0.4	7:57p	2.6	2:27p	3.0	11:14p	2.8	6:43a/5:19p	4:38a/3:47p
19	7:17a	0.0	8:38p	2.6	3:10p	3.2	11:05p	2.8	6:44a/5:18p	5:35a/4:20p
20	7:43a	−0.2			3:52p	3.4			6:45a/5:18p	6:34a/4:57p

[**Note:** a, AM; p, PM.]

Sources: National Oceanic and Atmospheric Administration (http://tidesandcurrents.noaa.gov) and U.S. Naval Observatory (http://aa.usno.navy.mil)

The following projects are available on the Instructor's Resource Center (IRC):

II. Project at Motorola *Digital Transmission over the Air* Learn how Motorola Corporation transmits digital sequences by modulating the phase of the carrier waves.

III. Identifying Mountain Peaks in Hawaii The visibility of a mountain is affected by its altitude, distance from the viewer, and the curvature of Earth's surface. Trigonometry can be used to determine whether a distant object can be seen.

IV. CBL Experiment Technology is used to model and study the effects of damping on sound waves.

3.1 The Inverse Sine, Cosine, and Tangent Functions

PREPARING FOR THIS SECTION *Before getting started, review the following:*

- Inverse Functions (Section 1.8, pp. 81–91)
- Values of the Trigonometric Functions (Section 2.3, pp. 129–131, and Section 2.4, pp. 140–143)
- Properties of the Sine, Cosine, and Tangent Functions (Section 2.5, pp. 150–159)
- Graphs of the Sine, Cosine, and Tangent Functions (Section 2.6, pp. 161–164, and Section 2.7, pp. 177–179)

Now Work the 'Are You Prepared?' problems on page 216.

OBJECTIVES 1 Find the Exact Value of an Inverse Sine Function (p. 207)
2 Find an Approximate Value of an Inverse Sine Function (p. 209)
3 Use Properties of Inverse Functions to Find Exact Values of Certain Composite Functions (p. 209)
4 Find the Inverse Function of a Trigonometric Function (p. 215)
5 Solve Equations Involving Inverse Trigonometric Functions (p. 216)

In Section 1.8 we discussed inverse functions, and we noted that if a function is one-to-one it will have an inverse function. We also observed that if a function is not one-to-one it may be possible to restrict its domain in some suitable manner so that the restricted function is one-to-one. For example, the function $y = x^2$ is not one-to-one; however, if we restrict the domain to $x \geq 0$, the function is one-to-one.

Other properties of a one-to-one function f and its inverse function f^{-1} that we discussed in Section 1.8 are summarized next.

1. $f^{-1}(f(x)) = x$ for every x in the domain of f and $f(f^{-1}(x)) = x$ for every x in the domain of f^{-1}.
2. Domain of f = range of f^{-1} and range of f = domain of f^{-1}.
3. The graph of f and the graph of f^{-1} are symmetric with respect to the line $y = x$.
4. If a function $y = f(x)$ has an inverse function, the implicit equation of the inverse function is $x = f(y)$. If we solve this equation for y, we obtain the explicit equation $y = f^{-1}(x)$.

The Inverse Sine Function

In Figure 1, we show the graph of $y = \sin x$. Because every horizontal line $y = b$, where b is between -1 and 1, intersects the graph of $y = \sin x$ infinitely many times, it follows from the horizontal-line test that the function $y = \sin x$ is not one-to-one.

Figure 1
$y = \sin x$, $-\infty < x < \infty$, $-1 \leq y \leq 1$

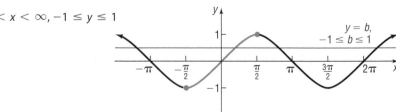

Figure 2
$y = \sin x$, $-\dfrac{\pi}{2} \leq x \leq \dfrac{\pi}{2}$, $-1 \leq y \leq 1$

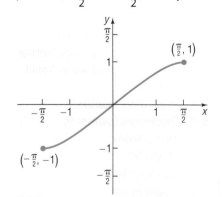

However, if we restrict the domain of $y = \sin x$ to the interval $\left[-\dfrac{\pi}{2}, \dfrac{\pi}{2}\right]$, the restricted function

$$y = \sin x \qquad -\dfrac{\pi}{2} \leq x \leq \dfrac{\pi}{2}$$

is one-to-one and so will have an inverse function.* See Figure 2.

---

*Although there are many other ways to restrict the domain and obtain a one-to-one function, mathematicians have agreed to use the interval $\left[-\dfrac{\pi}{2}, \dfrac{\pi}{2}\right]$ to define the inverse of $y = \sin x$.

CHAPTER PROJECTS

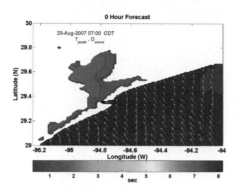

I. Tides The given table is a partial tide table for November 2006 for the Sabine Bank Lighthouse, a shoal located offshore from Texas where the Sabine River empties into the Gulf of Mexico.

1. On November 15, when was the tide high? This is called *high tide*. On November 19, when was the tide low? This is called *low tide*. Most days will have two high tides and two low tides.

2. Why do you think there is a negative height for the low tide on November 20? What is the height measured against?

3. On your graphing utility, draw a scatter diagram for the data in the table. Let t (time, in hours) be the independent variable, with $t = 0$ being 12:00 AM on November 14, $t = 24$ being 12:00 AM on November 15, and so on. Let h be the height in feet. Remember that there are 60 minutes in an hour. Also, make sure your graphing utility is in radian mode.

4. What shape does the data take? What is the period of the data? What is the amplitude? Is the amplitude constant? Explain.

5. Find a sine curve that models the data. Is there a vertical shift? Is there a phase shift?

6. Using your graphing utility, find the sinusoidal function of best fit. How does this model compare to the model from part 5?

7. Using the model found in part (5) and the sinusoidal equation of best fit found in part (6), predict the high tides and the low tides on November 21.

8. Looking at the times of day that the low tides occur, what do you think causes the low tides to vary so much each day? Explain. Does this seem to have the same type of effect on the high tides? Explain.

Nov	Low Tide Time	Ht (ft)	Low Tide Time	Ht (ft)	High Tide Time	Ht (ft)	High Tide Time	Ht (ft)	Sun/Moon Phase Sunrise/set	Moonrise/set
14	6:26a	2.0	4:38p	1.4	9:29a	2.2	11:14p	2.8	6:40a/5:20p	1:05a/2:02p
15	6:22a	1.6	5:34p	1.8	11:18a	2.4	11:15p	2.6	6:41a/5:20p	1:58a/2:27p
16	6:28a	1.2	6:25p	2.0	12:37p	2.6	11:16p	2.6	6:41a/5:19p	2:50a/2:52p
17	6:40a	0.8	7:12p	2.4	1:38p	2.8	11:16p	2.6	6:42a/5:19p	3:43a/3:19p
18	6:56a	0.4	7:57p	2.6	2:27p	3.0	11:14p	2.8	6:43a/5:19p	4:38a/3:47p
19	7:17a	0.0	8:38p	2.6	3:10p	3.2	11:05p	2.8	6:44a/5:18p	5:35a/4:20p
20	7:43a	−0.2			3:52p	3.4			6:45a/5:18p	6:34a/4:57p

[**Note:** a, AM; p, PM.]

Sources: National Oceanic and Atmospheric Administration (http://tidesandcurrents.noaa.gov) and U.S. Naval Observatory (http://aa.usno.navy.mil)

The following projects are available on the Instructor's Resource Center (IRC):

II. Project at Motorola *Digital Transmission over the Air* Learn how Motorola Corporation transmits digital sequences by modulating the phase of the carrier waves.

III. Identifying Mountain Peaks in Hawaii The visibility of a mountain is affected by its altitude, distance from the viewer, and the curvature of Earth's surface. Trigonometry can be used to determine whether a distant object can be seen.

IV. CBL Experiment Technology is used to model and study the effects of damping on sound waves.

Analytic Trigonometry

3

A Look Back

In Chapter 1, we defined inverse functions and developed their properties, particularly the relationship between the domain and range of a function and its inverse. We learned that the graph of a function and its inverse are symmetric with respect to the line $y = x$.

In Chapter 2, we defined the six trigonometric functions and looked at their properties.

A Look Ahead

In the first two sections of this chapter, we define the six inverse trigonometric functions and investigate their properties. In Sections 3.3 through 3.6 of this chapter, we continue the derivation of identities. These identities play an important role in calculus, the physical and life sciences, and economics, where they are used to simplify complicated expressions. The last two sections of this chapter discuss how to solve equations that contain trigonometric functions.

Outline

3.1 The Inverse Sine, Cosine, and Tangent Functions

PREPARING FOR THIS SECTION   *Before getting started, review the following:*

- Inverse Functions (Section 1.8, pp. 81–91)
- Values of the Trigonometric Functions (Section 2.3, pp. 129–131, and Section 2.4, pp. 140–143)
- Properties of the Sine, Cosine, and Tangent Functions (Section 2.5, pp. 150–159)
- Graphs of the Sine, Cosine, and Tangent Functions (Section 2.6, pp. 161–164, and Section 2.7, pp. 177–179)

Now Work the 'Are You Prepared?' problems on page 216.

OBJECTIVES   **1** Find the Exact Value of an Inverse Sine Function (p. 207)
   2 Find an Approximate Value of an Inverse Sine Function (p. 209)
   3 Use Properties of Inverse Functions to Find Exact Values of Certain Composite Functions (p. 209)
   4 Find the Inverse Function of a Trigonometric Function (p. 215)
   5 Solve Equations Involving Inverse Trigonometric Functions (p. 216)

In Section 1.8 we discussed inverse functions, and we noted that if a function is one-to-one it will have an inverse function. We also observed that if a function is not one-to-one it may be possible to restrict its domain in some suitable manner so that the restricted function is one-to-one. For example, the function $y = x^2$ is not one-to-one; however, if we restrict the domain to $x \geq 0$, the function is one-to-one.

Other properties of a one-to-one function f and its inverse function f^{-1} that we discussed in Section 1.8 are summarized next.

1. $f^{-1}(f(x)) = x$ for every x in the domain of f and $f(f^{-1}(x)) = x$ for every x in the domain of f^{-1}.
2. Domain of f = range of f^{-1} and range of f = domain of f^{-1}.
3. The graph of f and the graph of f^{-1} are symmetric with respect to the line $y = x$.
4. If a function $y = f(x)$ has an inverse function, the implicit equation of the inverse function is $x = f(y)$. If we solve this equation for y, we obtain the explicit equation $y = f^{-1}(x)$.

The Inverse Sine Function

In Figure 1, we show the graph of $y = \sin x$. Because every horizontal line $y = b$, where b is between -1 and 1, intersects the graph of $y = \sin x$ infinitely many times, it follows from the horizontal-line test that the function $y = \sin x$ is not one-to-one.

Figure 1
$y = \sin x, -\infty < x < \infty, -1 \leq y \leq 1$

Figure 2
$y = \sin x, -\dfrac{\pi}{2} \leq x \leq \dfrac{\pi}{2}, -1 \leq y \leq 1$

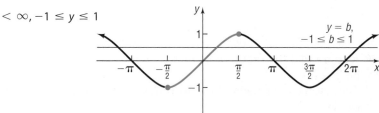

However, if we restrict the domain of $y = \sin x$ to the interval $\left[-\dfrac{\pi}{2}, \dfrac{\pi}{2}\right]$, the restricted function

$$y = \sin x \qquad -\frac{\pi}{2} \leq x \leq \frac{\pi}{2}$$

is one-to-one and so will have an inverse function.* See Figure 2.

*Although there are many other ways to restrict the domain and obtain a one-to-one function, mathematicians have agreed to use the interval $\left[-\dfrac{\pi}{2}, \dfrac{\pi}{2}\right]$ to define the inverse of $y = \sin x$.

An equation for the inverse of $y = f(x) = \sin x$ is obtained by interchanging x and y. The implicit form of the inverse function is $x = \sin y, -\dfrac{\pi}{2} \le y \le \dfrac{\pi}{2}$. The explicit form is called the **inverse sine** of x and is symbolized by $y = f^{-1}(x) = \sin^{-1} x$.

DEFINITION

> $$y = \sin^{-1} x \quad \text{means} \quad x = \sin y$$
> $$\text{where} \quad -1 \le x \le 1 \quad \text{and} \quad -\frac{\pi}{2} \le y \le \frac{\pi}{2} \qquad \textbf{(1)}$$

NOTE Remember, the domain of a function f equals the range of its inverse, f^{-1}, and the range of a function f equals the domain of its inverse, f^{-1}. Because the restricted domain of the sine function is $\left[-\dfrac{\pi}{2}, \dfrac{\pi}{2}\right]$, the range of the inverse sine function is $\left[-\dfrac{\pi}{2}, \dfrac{\pi}{2}\right]$; because the range of the sine function is $[-1, 1]$, the domain of the inverse sine function is $[-1, 1]$. ∎

Because $y = \sin^{-1} x$ means $x = \sin y$, we read $y = \sin^{-1} x$ as "y is the angle or real number whose sine equals x." Alternatively, we can say that "y is the inverse sine of x." Be careful about the notation used. The superscript -1 that appears in $y = \sin^{-1} x$ is not an exponent, but is reminiscent of the symbolism f^{-1} used to denote the inverse function of f. (To avoid this notation, some books use the notation $y = \text{Arcsin } x$ instead of $y = \sin^{-1} x$.)

The inverse of a function f receives as input an element from the range of f and returns as output an element in the domain of f. The restricted sine function, $y = f(x) = \sin x$, receives as input an angle or real number x in the interval $\left[-\dfrac{\pi}{2}, \dfrac{\pi}{2}\right]$ and outputs a real number in the interval $[-1, 1]$. Therefore, the inverse sine function $y = \sin^{-1} x$ receives as input a real number in the interval $[-1, 1]$ or $-1 \le x \le 1$, its domain, and outputs an angle or real number in the interval $\left[-\dfrac{\pi}{2}, \dfrac{\pi}{2}\right]$ or $-\dfrac{\pi}{2} \le y \le \dfrac{\pi}{2}$, its range.

The graph of the inverse sine function can be obtained by reflecting the restricted portion of the graph of $y = f(x) = \sin x$ about the line $y = x$, as shown in Figure 3(a). Figure 3(b) shows the graph using a graphing utility.

Figure 3

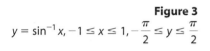

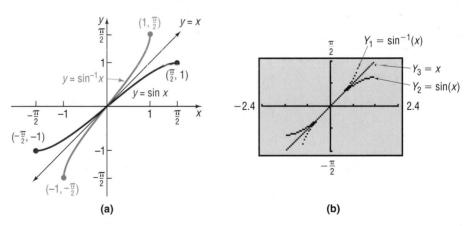

(a)   (b)

1 Find the Exact Value of an Inverse Sine Function

For some numbers x, it is possible to find the exact value of $y = \sin^{-1} x$.

EXAMPLE 1   **Finding the Exact Value of an Inverse Sine Function**

Find the exact value of: $\sin^{-1} 1$

Solution   Let $\theta = \sin^{-1} 1$. We seek the angle $\theta, -\dfrac{\pi}{2} \le \theta \le \dfrac{\pi}{2}$, whose sine equals 1.

$$\theta = \sin^{-1} 1 \qquad -\frac{\pi}{2} \le \theta \le \frac{\pi}{2}$$

$$\sin \theta = 1 \qquad -\frac{\pi}{2} \le \theta \le \frac{\pi}{2} \qquad \text{By definition of } y = \sin^{-1} x$$

Now look at Table 1 and Figure 4.

Table 1

θ	$-\dfrac{\pi}{2}$	$-\dfrac{\pi}{3}$	$-\dfrac{\pi}{4}$	$-\dfrac{\pi}{6}$	0	$\dfrac{\pi}{6}$	$\dfrac{\pi}{4}$	$\dfrac{\pi}{3}$	$\dfrac{\pi}{2}$
$\sin\theta$	-1	$-\dfrac{\sqrt{3}}{2}$	$-\dfrac{\sqrt{2}}{2}$	$-\dfrac{1}{2}$	0	$\dfrac{1}{2}$	$\dfrac{\sqrt{2}}{2}$	$\dfrac{\sqrt{3}}{2}$	1

Figure 4

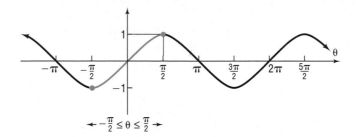

We see that the only angle θ within the interval $\left[-\dfrac{\pi}{2}, \dfrac{\pi}{2}\right]$ whose sine is 1 is $\dfrac{\pi}{2}$. (Note that $\sin\dfrac{5\pi}{2}$ also equals 1, but $\dfrac{5\pi}{2}$ lies outside the interval $\left[-\dfrac{\pi}{2}, \dfrac{\pi}{2}\right]$ and hence is not admissible.) So, since $\sin\dfrac{\pi}{2} = 1$ and $\dfrac{\pi}{2}$ is in $\left[-\dfrac{\pi}{2}, \dfrac{\pi}{2}\right]$, we conclude that

$$\sin^{-1} 1 = \dfrac{\pi}{2}$$

Now Work PROBLEM 13

EXAMPLE 2	**Finding the Exact Value of an Inverse Sine Function**

Find the exact value of: $\sin^{-1}\left(-\dfrac{1}{2}\right)$

Solution　Let $\theta = \sin^{-1}\left(-\dfrac{1}{2}\right)$. We seek the angle θ, $-\dfrac{\pi}{2} \le \theta \le \dfrac{\pi}{2}$, whose sine equals $-\dfrac{1}{2}$.

$$\theta = \sin^{-1}\left(-\dfrac{1}{2}\right) \qquad -\dfrac{\pi}{2} \le \theta \le \dfrac{\pi}{2}$$

$$\sin\theta = -\dfrac{1}{2} \qquad -\dfrac{\pi}{2} \le \theta \le \dfrac{\pi}{2}$$

(Refer to Table 1 and Figure 4, if necessary.) The only angle within the interval $\left[-\dfrac{\pi}{2}, \dfrac{\pi}{2}\right]$ whose sine is $-\dfrac{1}{2}$ is $-\dfrac{\pi}{6}$. So, since $\sin\left(-\dfrac{\pi}{6}\right) = -\dfrac{1}{2}$ and $-\dfrac{\pi}{6}$ is in the interval $\left[-\dfrac{\pi}{2}, \dfrac{\pi}{2}\right]$, we conclude that

$$\sin^{-1}\left(-\dfrac{1}{2}\right) = -\dfrac{\pi}{6}$$

Now Work PROBLEM 19

2 Find an Approximate Value of an Inverse Sine Function

For most numbers x, the value $y = \sin^{-1} x$ must be approximated.

EXAMPLE 3   **Finding an Approximate Value of an Inverse Sine Function**

Find an approximate value of:

(a) $\sin^{-1} \dfrac{1}{3}$ (b) $\sin^{-1}\left(-\dfrac{1}{4}\right)$

Express the answer in radians rounded to two decimal places.

Solution   Because we want the angle measured in radians, we first set the mode to radians.

(a) Figure 5(a) shows the solution using a TI-84 Plus graphing calculator.

(b) Figure 5(b) shows the solution using a TI-84 Plus graphing calculator.

Figure 5a

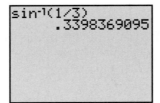
```
sin-1(1/3)
          .3398369095
```

Figure 5b

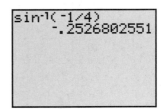
```
sin-1(-1/4)
          -.2526802551
```

We have $\sin^{-1} \dfrac{1}{3} = 0.34$, rounded to two decimal places.

We have $\sin^{-1}\left(-\dfrac{1}{4}\right) = -0.25$, rounded to two decimal places.

■

══ **Now Work** PROBLEM 25

3 Use Properties of Inverse Functions to Find Exact Values of Certain Composite Functions

When we discussed functions and their inverses in Section 1.8, we found that $f^{-1}(f(x)) = x$ for all x in the domain of f and $f(f^{-1}(x)) = x$ for all x in the domain of f^{-1}. In terms of the sine function and its inverse, these properties are of the form

$$f^{-1}(f(x)) = \sin^{-1}(\sin x) = x \qquad \text{where } -\dfrac{\pi}{2} \le x \le \dfrac{\pi}{2} \qquad \textbf{(2a)}$$

$$f(f^{-1}(x)) = \sin(\sin^{-1} x) = x \qquad \text{where } -1 \le x \le 1 \qquad \textbf{(2b)}$$

EXAMPLE 4   **Finding the Exact Value of Certain Composite Functions**

Find the exact value of each of the following composite functions:

(a) $\sin^{-1}\left(\sin \dfrac{\pi}{8}\right)$ (b) $\sin^{-1}\left(\sin \dfrac{5\pi}{8}\right)$

Solution   (a) The composite function $\sin^{-1}\left(\sin \dfrac{\pi}{8}\right)$ follows the form of equation (2a).

Because $\dfrac{\pi}{8}$ is in the interval $\left[-\dfrac{\pi}{2}, \dfrac{\pi}{2}\right]$, we can use (2a). Then

$$\sin^{-1}\left(\sin \dfrac{\pi}{8}\right) = \dfrac{\pi}{8}$$

Figure 6 shows the calculation using a graphing calculator.

Figure 6

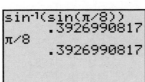
```
sin-1(sin(π/8))
          .3926990817
π/8
          .3926990817
```

(b) The composite function $\sin^{-1}\left(\sin\dfrac{5\pi}{8}\right)$ follows the form of equation (2a), but

$\dfrac{5\pi}{8}$ is not in the interval $\left[-\dfrac{\pi}{2}, \dfrac{\pi}{2}\right]$. To use (2a) we need to find an angle θ in

the interval $\left[-\dfrac{\pi}{2}, \dfrac{\pi}{2}\right]$ for which $\sin\theta = \sin\dfrac{5\pi}{8}$. Then, using (2a),

$\sin^{-1}\left(\sin\dfrac{5\pi}{8}\right) = \sin^{-1}(\sin\theta) = \theta$, and we are finished.

Figure 7

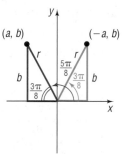

Look at Figure 7. The angle $\dfrac{5\pi}{8}$ is in quadrant II. The reference angle of $\dfrac{5\pi}{8}$

is $\dfrac{3\pi}{8}$ and $\sin\dfrac{5\pi}{8} = \dfrac{b}{r} = \sin\dfrac{3\pi}{8}$. Since $\dfrac{3\pi}{8}$ is in the interval $\left[-\dfrac{\pi}{2}, \dfrac{\pi}{2}\right]$, we have

$$\sin^{-1}\left(\sin\dfrac{5\pi}{8}\right) = \sin^{-1}\left(\sin\dfrac{3\pi}{8}\right) = \dfrac{3\pi}{8}$$

↑ Apply (2a).

We can verify the results using a graphing calculator. Figure 8(a) shows that $\sin^{-1}\left(\sin\dfrac{5\pi}{8}\right) \neq \dfrac{5\pi}{8}$. In Figure 8(b), we see that

$$\sin^{-1}\left(\sin\dfrac{5\pi}{8}\right) = \sin^{-1}\left(\sin\dfrac{3\pi}{8}\right) = \dfrac{3\pi}{8}$$

Figure 8

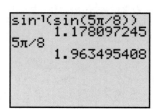

(a)

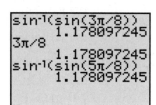

(b)

━━ **Now Work** PROBLEM 41

| EXAMPLE 5 | **Finding the Exact Value of Certain Composite Functions** |

Find the exact value, if any, of each composite function.

(a) $\sin(\sin^{-1}0.8)$ (b) $\sin(\sin^{-1}1.8)$

Solution

(a) The composite function $\sin(\sin^{-1}0.8)$ follows the form of equation (2b) and 0.8 is in the interval $[-1, 1]$. So we use (2b):

$$\sin(\sin^{-1}0.8) = 0.8$$

Figure 9

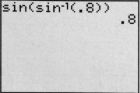

Figure 9 verifies the result.

(b) The composite function $\sin(\sin^{-1}1.8)$ follows the form of equation (2b), but 1.8 is not in the domain of the inverse sine function. This composite function is not defined. See Figure 10. Can you explain why the error occurs?

Figure 10

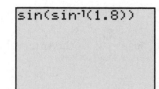 

━━ **Now Work** PROBLEM 45

The Inverse Cosine Function

In Figure 11 we show the graph of $y = \cos x$. Because every horizontal line $y = b$, where b is between -1 and 1, intersects the graph of $y = \cos x$ infinitely many times, it follows that the cosine function is not one-to-one.

Figure 11
$y = \cos x, -\infty < x < \infty, -1 \le y \le 1$.

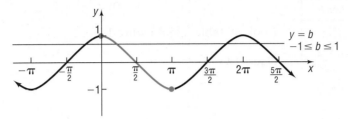

Figure 12
$y = \cos x, 0 \le x \le \pi, -1 \le y \le 1$

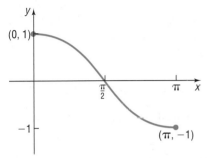

However, if we restrict the domain of $y = \cos x$ to the interval $[0, \pi]$, the restricted function

$$y = \cos x \qquad 0 \le x \le \pi$$

is one-to-one and hence will have an inverse function.* See Figure 12.

An equation for the inverse of $y = f(x) = \cos x$ is obtained by interchanging x and y. The implicit form of the inverse function is $x = \cos y, 0 \le y \le \pi$. The explicit form is called the **inverse cosine** of x and is symbolized by $y = f^{-1}(x) = \cos^{-1} x$ (or by $y = \text{Arccos } x$).

DEFINITION

> $$y = \cos^{-1} x \quad \text{means} \quad x = \cos y$$
> $$\text{where} \quad -1 \le x \le 1 \quad \text{and} \quad 0 \le y \le \pi \tag{3}$$

Here y is the angle whose cosine is x. Because the range of the cosine function, $y = \cos x$, is $-1 \le y \le 1$, the domain of the inverse function $y = \cos^{-1} x$ is $-1 \le x \le 1$. Because the restricted domain of the cosine function, $y = \cos x$, is $0 \le x \le \pi$, the range of the inverse function $y = \cos^{-1} x$ is $0 \le y \le \pi$.

The graph of $y = \cos^{-1} x$ can be obtained by reflecting the restricted portion of the graph of $y = \cos x$ about the line $y = x$, as shown in Figure 13.

Figure 13
$y = \cos^{-1}x, -1 \le x \le 1, 0 \le y \le \pi$

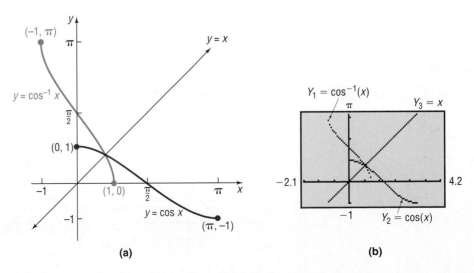

(a)   (b)

*This is the generally accepted restriction to define the inverse cosine function.

| EXAMPLE 6 | **Finding the Exact Value of an Inverse Cosine Function** |

Find the exact value of: $\cos^{-1} 0$

Solution   Let $\theta = \cos^{-1} 0$. We seek the angle $\theta, 0 \leq \theta \leq \pi$, whose cosine equals 0.

$$\theta = \cos^{-1} 0 \qquad 0 \leq \theta \leq \pi$$
$$\cos \theta = 0 \qquad 0 \leq \theta \leq \pi$$

Look at Table 2 and Figure 14.

Table 2

θ	$\cos \theta$
0	1
$\dfrac{\pi}{6}$	$\dfrac{\sqrt{3}}{2}$
$\dfrac{\pi}{4}$	$\dfrac{\sqrt{2}}{2}$
$\dfrac{\pi}{3}$	$\dfrac{1}{2}$
$\dfrac{\pi}{2}$	0
$\dfrac{2\pi}{3}$	$-\dfrac{1}{2}$
$\dfrac{3\pi}{4}$	$-\dfrac{\sqrt{2}}{2}$
$\dfrac{5\pi}{6}$	$-\dfrac{\sqrt{3}}{2}$
π	-1

Figure 14

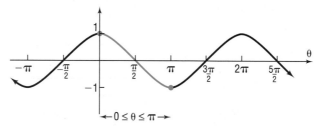

We see that the only angle θ within the interval $[0, \pi]$ whose cosine is 0 is $\dfrac{\pi}{2}$. [Note that $\cos \dfrac{3\pi}{2}$ and $\cos \left(-\dfrac{\pi}{2}\right)$ also equal 0, but they lie outside the interval $[0, \pi]$ and hence are not admissible.] Since $\cos \dfrac{\pi}{2} = 0$ and $\dfrac{\pi}{2}$ is in the interval $[0, \pi]$, we conclude that

$$\cos^{-1} 0 = \dfrac{\pi}{2}$$

| EXAMPLE 7 | **Finding the Exact Value of an Inverse Cosine Function** |

Find the exact value of: $\cos^{-1}\left(-\dfrac{\sqrt{2}}{2}\right)$

Solution   Let $\theta = \cos^{-1}\left(-\dfrac{\sqrt{2}}{2}\right)$. We seek the angle $\theta, 0 \leq \theta \leq \pi$, whose cosine equals $-\dfrac{\sqrt{2}}{2}$.

$$\theta = \cos^{-1}\left(-\dfrac{\sqrt{2}}{2}\right) \qquad 0 \leq \theta \leq \pi$$
$$\cos \theta = -\dfrac{\sqrt{2}}{2} \qquad 0 \leq \theta \leq \pi$$

Look at Table 2 and Figure 15.

Figure 15

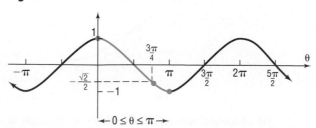

We see that the only angle θ within the interval $[0, \pi]$ whose cosine is $-\dfrac{\sqrt{2}}{2}$ is $\dfrac{3\pi}{4}$. So, since $\cos\dfrac{3\pi}{4} = -\dfrac{\sqrt{2}}{2}$ and $\dfrac{3\pi}{4}$ is in the interval $[0, \pi]$, we conclude that

$$\cos^{-1}\left(-\dfrac{\sqrt{2}}{2}\right) = \dfrac{3\pi}{4}$$

— **Now Work** PROBLEM 23

For the cosine function and its inverse, the following properties hold:

$$
\begin{array}{llr}
f^{-1}(f(x)) = \cos^{-1}(\cos x) = x & \text{where } 0 \le x \le \pi & \textbf{(4a)} \\
f(f^{-1}(x)) = \cos(\cos^{-1} x) = x & \text{where } -1 \le x \le 1 & \textbf{(4b)}
\end{array}
$$

| EXAMPLE 8 | **Using Properties of Inverse Functions to Find the Exact Value of Certain Composite Functions** |

Find the exact value of:

(a) $\cos^{-1}\left(\cos\dfrac{\pi}{12}\right)$ (b) $\cos[\cos^{-1}(-0.4)]$ (c) $\cos^{-1}\left[\cos\left(-\dfrac{2\pi}{3}\right)\right]$ (d) $\cos(\cos^{-1}\pi)$

Solution   (a) $\cos^{-1}\left(\cos\dfrac{\pi}{12}\right) = \dfrac{\pi}{12}$ $\dfrac{\pi}{12}$ is in the interval $[0, \pi]$; use Property (4a).

(b) $\cos[\cos^{-1}(-0.4)] = -0.4$ -0.4 is in the interval $[-1, 1]$; use Property (4b).

(c) The angle $-\dfrac{2\pi}{3}$ is not in the interval $[0, \pi]$, so we cannot use (4a). However, because the cosine function is even, $\cos\left(-\dfrac{2\pi}{3}\right) = \cos\dfrac{2\pi}{3}$. Since $\dfrac{2\pi}{3}$ is in the interval $[0, \pi]$, we have

$$\cos^{-1}\left[\cos\left(-\dfrac{2\pi}{3}\right)\right] = \cos^{-1}\left(\cos\dfrac{2\pi}{3}\right) = \dfrac{2\pi}{3}$$ $\dfrac{2\pi}{3}$ is in the interval $[0, \pi]$; apply (4a).

(d) Because π is not in the interval $[-1, 1]$, the domain of the inverse cosine function, $\cos^{-1}\pi$ is not defined. This means the composite function $\cos(\cos^{-1}\pi)$ is also not defined.

— **Now Work** PROBLEMS 37 AND 49

The Inverse Tangent Function

In Figure 16 on page 214 we show the graph of $y = \tan x$. Because every horizontal line intersects the graph infinitely many times, it follows that the tangent function is not one-to-one.

However, if we restrict the domain of $y = \tan x$ to the interval $\left(-\dfrac{\pi}{2}, \dfrac{\pi}{2}\right)$, the restricted function

$$y = \tan x \qquad -\dfrac{\pi}{2} < x < \dfrac{\pi}{2}$$

is one-to-one and hence has an inverse function.* See Figure 17.

*This is the generally accepted restriction.

Figure 16

$y = \tan x, -\infty < x < \infty, x$ not equal to odd multiples of $\dfrac{\pi}{2}, -\infty < y < \infty$

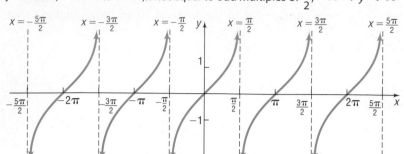

Figure 17

$y = \tan x, -\dfrac{\pi}{2} < x < \dfrac{\pi}{2}, -\infty < y < \infty$

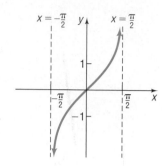

An equation for the inverse of $y = f(x) = \tan x$ is obtained by interchanging x and y. The implicit form of the inverse function is $x = \tan y, -\dfrac{\pi}{2} < y < \dfrac{\pi}{2}$. The explicit form is called the **inverse tangent** of x and is symbolized by $y = f^{-1}(x) = \tan^{-1} x$ (or by $y = \text{Arctan } x$).

DEFINITION

$$y = \tan^{-1} x \quad \text{means} \quad x = \tan y$$
$$\text{where} \quad -\infty < x < \infty \quad \text{and} \quad -\dfrac{\pi}{2} < y < \dfrac{\pi}{2} \qquad \textbf{(5)}$$

Here y is the angle whose tangent is x. The domain of the function $y = \tan^{-1} x$ is $-\infty < x < \infty$, and its range is $-\dfrac{\pi}{2} < y < \dfrac{\pi}{2}$. The graph of $y = \tan^{-1} x$ can be obtained by reflecting the restricted portion of the graph of $y = \tan x$ about the line $y = x$, as shown in Figure 18.

Figure 18

$y = \tan^{-1} x,$

$-\infty < x < \infty,$

$-\dfrac{\pi}{2} < y < \dfrac{\pi}{2}$

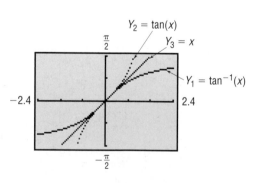

(a)

(b)

EXAMPLE 9   **Finding the Exact Value of an Inverse Tangent Function**

Find the exact value of:

(a) $\tan^{-1} 1$ (b) $\tan^{-1}\left(-\sqrt{3}\right)$

Solution

(a) Let $\theta = \tan^{-1} 1$. We seek the angle $\theta, -\dfrac{\pi}{2} < \theta < \dfrac{\pi}{2}$, whose tangent equals 1.

$$\theta = \tan^{-1} 1 \qquad -\dfrac{\pi}{2} < \theta < \dfrac{\pi}{2}$$

$$\tan \theta = 1 \qquad -\dfrac{\pi}{2} < \theta < \dfrac{\pi}{2}$$

Look at Table 3. The only angle θ within the interval $\left(-\dfrac{\pi}{2}, \dfrac{\pi}{2} \right)$ whose tangent is 1 is $\dfrac{\pi}{4}$. Since $\tan \dfrac{\pi}{4} = 1$ and $\dfrac{\pi}{4}$ is in the interval $\left(-\dfrac{\pi}{2}, \dfrac{\pi}{2} \right)$, we conclude that

$$\tan^{-1} 1 = \dfrac{\pi}{4}$$

(b) Let $\theta = \tan^{-1}\left(-\sqrt{3} \right)$. We seek the angle $\theta, -\dfrac{\pi}{2} < \theta < \dfrac{\pi}{2}$, whose tangent equals $-\sqrt{3}$.

$$\theta = \tan^{-1}\left(-\sqrt{3} \right) \qquad -\dfrac{\pi}{2} < \theta < \dfrac{\pi}{2}$$

$$\tan \theta = -\sqrt{3} \qquad -\dfrac{\pi}{2} < \theta < \dfrac{\pi}{2}$$

Look at Table 3. The only angle θ within the interval $\left(-\dfrac{\pi}{2}, \dfrac{\pi}{2} \right)$ whose tangent is $-\sqrt{3}$ is $-\dfrac{\pi}{3}$. Since $\tan\left(-\dfrac{\pi}{3} \right) = -\sqrt{3}$ and $-\dfrac{\pi}{3}$ is in the interval $\left(-\dfrac{\pi}{2}, \dfrac{\pi}{2} \right)$, we conclude that

$$\tan^{-1}\left(-\sqrt{3} \right) = -\dfrac{\pi}{3}$$

Table 3

θ	$\tan \theta$
$-\dfrac{\pi}{2}$	Undefined
$-\dfrac{\pi}{3}$	$-\sqrt{3}$
$-\dfrac{\pi}{4}$	-1
$-\dfrac{\pi}{6}$	$-\dfrac{\sqrt{3}}{3}$
0	0
$\dfrac{\pi}{6}$	$\dfrac{\sqrt{3}}{3}$
$\dfrac{\pi}{4}$	1
$\dfrac{\pi}{3}$	$\sqrt{3}$
$\dfrac{\pi}{2}$	Undefined

 Now Work PROBLEM 17

For the tangent function and its inverse, the following properties hold:

$$f^{-1}(f(x)) = \tan^{-1}(\tan x) = x \qquad \text{where } -\dfrac{\pi}{2} < x < \dfrac{\pi}{2}$$

$$f(f^{-1}(x)) = \tan(\tan^{-1} x) = x \qquad \text{where } -\infty < x < \infty$$

Now Work PROBLEM 43

4 Find the Inverse Function of a Trigonometric Function

EXAMPLE 10 | **Finding the Inverse Function of a Trigonometric Function**

Find the inverse function f^{-1} of $f(x) = 2 \sin x - 1$, $-\dfrac{\pi}{2} \le x \le \dfrac{\pi}{2}$. State the domain of f^{-1}.

Solution

We follow the steps on page 89 for finding the inverse function.

$$y = 2 \sin x - 1$$
$$x = 2 \sin y - 1 \qquad \text{Interchange } x \text{ and } y.$$
$$x + 1 = 2 \sin y \qquad \text{Proceed to solve for } y.$$
$$\sin y = \dfrac{x + 1}{2}$$
$$y = \sin^{-1} \dfrac{x + 1}{2} \qquad \text{Apply the definition (1).}$$

The inverse function is $f^{-1}(x) = \sin^{-1} \dfrac{x + 1}{2}$.

The graph of $f(x) = 2 \sin x - 1$ can be obtained by vertically stretching $y = \sin x$ by a factor of 2 and then shifting the resulting graph down 1 unit. So the range of $f(x) = 2 \sin x - 1$ is $[-3, 1]$. Therefore, the domain of f^{-1} is $[-3, 1]$.

We could also find the domain of f^{-1} by noting that the argument of the inverse sine function is $\dfrac{x+1}{2}$ and that it must lie in the interval $[-1, 1]$. That is,

$$-1 \le \frac{x+1}{2} \le 1$$

$$-2 \le x + 1 \le 2 \quad \text{Multiply each part by 2.}$$

$$-3 \le x \le 1 \quad \text{Add } -1 \text{ to each part.}$$

The domain of f^{-1} is $\{x \mid -3 \le x \le 1\}$.

━━━━━ **Now Work** PROBLEM 55

5 Solve Equations Involving Inverse Trigonometric Functions

Equations that contain inverse trigonometric functions are called **inverse trigonometric equations.**

EXAMPLE 11 | **Solving an Equation Involving an Inverse Trigonometric Function**

Solve the equation: $3 \sin^{-1} x = \pi$

Solution   To solve an equation involving a single inverse trigonometric function, first isolate the inverse trigonometric function.

$$3 \sin^{-1} x = \pi$$

$$\sin^{-1} x = \frac{\pi}{3} \quad \text{Divide both sides by 3.}$$

$$x = \sin \frac{\pi}{3} \quad y = \sin^{-1} x \text{ means } x = \sin y.$$

$$x = \frac{\sqrt{3}}{2}$$

The solution set is $\left\{ \dfrac{\sqrt{3}}{2} \right\}$.

━━━━━ **Now Work** PROBLEM 61

3.1 Assess Your Understanding

'Are You Prepared?' *Answers are given at the end of these exercises. If you get a wrong answer, read the pages listed in red.*

1. What is the domain and the range of $y = \sin x$? (pp. 155–156)

2. A suitable restriction on the domain of the function $f(x) = (x - 1)^2$ to make it one-to-one would be _____. (pp. 82–84, 90)

3. If the domain of a one-to-one function is $[3, \infty)$, the range of its inverse is _____. (pp. 84–91)

4. **True or False**   The graph of $y = \cos x$ is decreasing on the interval $[0, \pi]$. (p. 164)

5. $\tan \dfrac{\pi}{4} = $ _____; $\sin \dfrac{\pi}{3} = $ _____ (pp. 129–131)

6. $\sin\left(-\dfrac{\pi}{6}\right) = $ _____; $\cos \pi = $ _____ (pp. 142–147)

Concepts and Vocabulary

7. $y = \sin^{-1} x$ means _____,

   where $-1 \le x \le 1$ and $-\dfrac{\pi}{2} \le y \le \dfrac{\pi}{2}$.

8. The value of $\sin^{-1}\left[\sin \dfrac{\pi}{2}\right]$ is _____.

9. $\cos^{-1}\left[\cos\dfrac{\pi}{5}\right] = $ _____.

10. True or False   The domain of $y = \sin^{-1} x$ is $-\dfrac{\pi}{2} \le x \le \dfrac{\pi}{2}$.

11. True or False $\sin(\sin^{-1} 0) = 0$ and $\cos(\cos^{-1} 0) = 0$.

12. True or False $y = \tan^{-1} x$ means $x = \tan y$, where $-\infty < x < \infty$ and $-\dfrac{\pi}{2} < y < \dfrac{\pi}{2}$.

Skill Building

In Problems 13–24, find the exact value of each expression.

13. $\sin^{-1} 0$

14. $\cos^{-1} 1$

15. $\sin^{-1}(-1)$

16. $\cos^{-1}(-1)$

17. $\tan^{-1} 0$

18. $\tan^{-1}(-1)$

19. $\sin^{-1}\dfrac{\sqrt{2}}{2}$

20. $\tan^{-1}\dfrac{\sqrt{3}}{3}$

21. $\tan^{-1}\sqrt{3}$

22. $\sin^{-1}\left(-\dfrac{\sqrt{3}}{2}\right)$

23. $\cos^{-1}\left(-\dfrac{\sqrt{3}}{2}\right)$

24. $\sin^{-1}\left(-\dfrac{\sqrt{2}}{2}\right)$

In Problems 25–36, use a calculator to find the value of each expression rounded to two decimal places.

25. $\sin^{-1} 0.1$

26. $\cos^{-1} 0.6$

27. $\tan^{-1} 5$

28. $\tan^{-1} 0.2$

29. $\cos^{-1}\dfrac{7}{8}$

30. $\sin^{-1}\dfrac{1}{8}$

31. $\tan^{-1}(-0.4)$

32. $\tan^{-1}(-3)$

33. $\sin^{-1}(-0.12)$

34. $\cos^{-1}(-0.44)$

35. $\cos^{-1}\dfrac{\sqrt{2}}{3}$

36. $\sin^{-1}\dfrac{\sqrt{3}}{5}$

In Problems 37–44, find the exact value of each expression. Do not use a calculator.

37. $\cos^{-1}\left(\cos\dfrac{4\pi}{5}\right)$

38. $\sin^{-1}\left[\sin\left(-\dfrac{\pi}{10}\right)\right]$

39. $\tan^{-1}\left[\tan\left(-\dfrac{3\pi}{8}\right)\right]$

40. $\sin^{-1}\left[\sin\left(-\dfrac{3\pi}{7}\right)\right]$

41. $\sin^{-1}\left(\sin\dfrac{9\pi}{8}\right)$

42. $\cos^{-1}\left[\cos\left(-\dfrac{5\pi}{3}\right)\right]$

43. $\tan^{-1}\left(\tan\dfrac{4\pi}{5}\right)$

44. $\tan^{-1}\left[\tan\left(-\dfrac{2\pi}{3}\right)\right]$

In Problems 45–52, find the exact value, if any, of each composite function. If there is no value, say it is "not defined." Do not use a calculator.

45. $\sin\left(\sin^{-1}\dfrac{1}{4}\right)$

46. $\cos\left[\cos^{-1}\left(-\dfrac{2}{3}\right)\right]$

47. $\tan\left(\tan^{-1} 4\right)$

48. $\tan\left[\tan^{-1}(-2)\right]$

49. $\cos\left(\cos^{-1} 1.2\right)$

50. $\sin\left[\sin^{-1}(-2)\right]$

51. $\tan\left(\tan^{-1}\pi\right)$

52. $\sin\left[\sin^{-1}(-1.5)\right]$

In Problems 53–60, determine the appropriate restriction on the domain of f to make it one-to-one. Then, find the inverse function f^{-1} of each function f. State the domain of f^{-1}.

53. $f(x) = 5\sin x + 2$

54. $f(x) = 2\tan x - 3$

55. $f(x) = -2\cos(3x)$

56. $f(x) = 3\sin(2x)$

57. $f(x) = -\tan(x + 1) - 3$

58. $f(x) = \cos(x + 2) + 1$

59. $f(x) = 3\sin(2x + 1)$

60. $f(x) = 2\cos(3x + 2)$

In Problems 61–68, find the exact solution of each equation.

61. $4\sin^{-1}(x) = \pi$

62. $2\cos^{-1} x = \pi$

63. $3\cos^{-1}(2x) = 2\pi$

64. $-6\sin^{-1}(3x) = \pi$

65. $3\tan^{-1} x = \pi$

66. $-4\tan^{-1} x = \pi$

67. $4\cos^{-1} x - 2\pi = 2\cos^{-1} x$

68. $5\sin^{-1} x - 2\pi = 2\sin^{-1} x - 3\pi$

Applications and Extensions

In Problems 69–74, use the following discussion. The formula

$$D = 24\left[1 - \dfrac{\cos^{-1}(\tan i \tan \theta)}{\pi}\right]$$

can be used to approximate the number of hours of daylight D when the declination of the Sun is $i°$ at a location $\theta°$ north latitude for any date between the vernal equinox and autumnal equinox. The declination of the Sun is defined as the angle i between the equatorial plane and any ray of light from the Sun. The latitude of

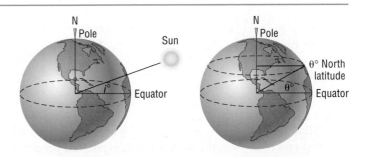

a location is the angle θ between the Equator and the location on the surface of Earth, with the vertex of the angle located at the center of Earth. See the figure. To use the formula, $\cos^{-1}(\tan i \tan \theta)$ *must be expressed in radians.*

69. Approximate the number of hours of daylight in Houston, Texas (29°45′ north latitude), for the following dates:
    (a) Summer solstice ($i = 23.5°$)
    (b) Vernal equinox ($i = 0°$)
    (c) July 4 ($i = 22°48′$)

70. Approximate the number of hours of daylight in New York, New York (40°45′ north latitude), for the following dates:
    (a) Summer solstice ($i = 23.5°$)
    (b) Vernal equinox ($i = 0°$)
    (c) July 4 ($i = 22°48′$)

71. Approximate the number of hours of daylight in Honolulu, Hawaii (21°18′ north latitude), for the following dates:
    (a) Summer solstice ($i = 23.5°$)
    (b) Vernal equinox ($i = 0°$)
    (c) July 4 ($i = 22°48′$)

72. Approximate the number of hours of daylight in Anchorage, Alaska (61°10′ north latitude), for the following dates:
    (a) Summer solstice ($i = 23.5°$)
    (b) Vernal equinox ($i = 0°$)
    (c) July 4 ($i = 22°48′$)

73. Approximate the number of hours of daylight at the Equator (0° north latitude) for the following dates:
    (a) Summer solstice ($i = 23.5°$)
    (b) Vernal equinox ($i = 0°$)
    (c) July 4 ($i = 22°48′$)
    (d) What do you conclude about the number of hours of daylight throughout the year for a location at the Equator?

74. Approximate the number of hours of daylight for any location that is 66°30′ north latitude for the following dates:
    (a) Summer solstice ($i = 23.5°$)
    (b) Vernal equinox ($i = 0°$)
    (c) July 4 ($i = 22°48′$)
    (d) The number of hours of daylight on the winter solstice may be found by computing the number of hours of daylight on the summer solstice and subtracting this result from 24 hours, due to the symmetry of the orbital path of Earth around the Sun. Compute the number of hours of daylight for this location on the winter solstice. What do you conclude about daylight for a location at 66°30′ north latitude?

75. **Being the First to See the Rising Sun** Cadillac Mountain, elevation 1530 feet, is located in Acadia National Park, Maine, and is the highest peak on the east coast of the United States. It is said that a person standing on the summit will be the first person in the United States to see the rays of the rising Sun. How much sooner would a person atop Cadillac Mountain see the first rays than a person standing below at sea level?

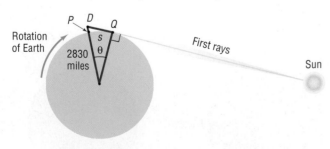

[**Hint:** Consult the figure. When the person at D sees the first rays of the Sun, the person at P does not. The person at

P sees the first rays of the Sun only after Earth has rotated so that P is at location Q. Compute the length of the arc subtended by the central angle θ. Then use the fact that, at the latitude of Cadillac Mountain, in 24 hours a length of $2\pi(2830) \approx 17,781.4$ miles is subtended, and find the time that it takes to subtend this length.]

76. **Movie Theater Screens** Suppose that a movie theater has a screen that is 28 feet tall. When you sit down, the bottom of the screen is 6 feet above your eye level. The angle formed by drawing a line from your eye to the bottom of the screen and from your eye to the top of the screen is called the **viewing angle**. In the figure, θ is the viewing angle. Suppose that you sit x feet from the screen. The viewing angle θ is given by the function

$$\theta(x) = \tan^{-1}\left(\frac{34}{x}\right) - \tan^{-1}\left(\frac{6}{x}\right)$$

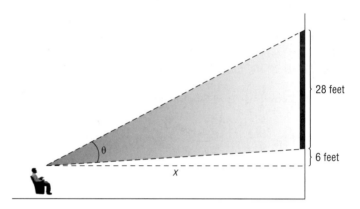

(a) What is your viewing angle if you sit 10 feet from the screen? 15 feet? 20 feet?
(b) If there is 5 feet between the screen and the first row of seats and there is 3 feet between each row, which row results in the largest viewing angle?
(c) Using a graphing utility, graph

$$\theta(x) = \tan^{-1}\left(\frac{34}{x}\right) - \tan^{-1}\left(\frac{6}{x}\right)$$

What value of x results in the largest viewing angle?

77. **Area under a Curve** The area under the graph of $y = \dfrac{1}{1 + x^2}$ and above the x-axis between $x = a$ and $x = b$ is given by

$$\tan^{-1} b - \tan^{-1} a$$

See the figure.

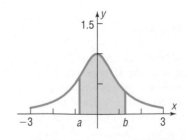

(a) Find the exact area under the graph of $y = \dfrac{1}{1 + x^2}$ and above the x-axis between $x = 0$ and $x = \sqrt{3}$.

(b) Find the exact area under the graph of $y = \dfrac{1}{1 + x^2}$ and above the x-axis between $x = -\dfrac{\sqrt{3}}{3}$ and $x = 1$.

78. Area under a Curve The area under the graph of

$$y = \dfrac{1}{\sqrt{1 - x^2}}$$ and above the x-axis between $x = a$ and $x = b$ is given by

$$\sin^{-1} b - \sin^{-1} a$$

See the figure.
(a) Find the exact area under the graph of $y = \dfrac{1}{\sqrt{1 - x^2}}$ and above the x-axis between $x = 0$ and $x = \dfrac{\sqrt{3}}{2}$.

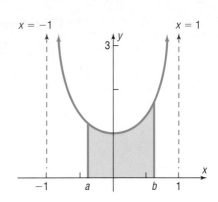

(b) Find the exact area under the graph of $y = \dfrac{1}{\sqrt{1 - x^2}}$ and above the x-axis between $x = -\dfrac{1}{2}$ and $x = \dfrac{1}{2}$.

Problems 79 and 80 require the following discussion:
The shortest distance between two points on Earth's surface can be determined from the latitude and longitude of the two locations. For example, if location 1 has (lat, lon) $= (\alpha_1, \beta_1)$ *and location 2 has* (lat, lon) $= (\alpha_2, \beta_2)$, *the shortest distance between the two locations is approximately*

$$d = \dfrac{2\pi}{360} r \cos^{-1}[(\cos \alpha_1 \cos \beta_1 \cos \alpha_2 \cos \beta_2) + (\cos \alpha_1 \sin \beta_1 \cos \alpha_2 \sin \beta_2) + (\sin \alpha_1 \sin \alpha_2)]$$

where r = radius of Earth ≈ 3960 miles and the inverse cosine function is expressed in radians. Also N latitude and E longitude are positive angles while S latitude and W longitude are negative angles.

Source: *www.infoplease.com*

City	Latitude	Longitude
Chicago, IL	41°50′N	87°37′W
Honolulu, HI	21°18′N	157°50′W
Melbourne, Australia	37°47′S	144°58′E

79. Shortest Distance from Chicago to Honolulu Find the shortest distance from Chicago, latitude 41°50′N, longitude 87°37′W to Honolulu, latitude 21°18′N, longitude 157°50′W. Round your answer to the nearest mile.

80. Shortest Distance from Honolulu to Melbourne, Australia Find the shortest distance from Honolulu to Melbourne, Australia, latitude 37°47′S, longitude 144°58′E. Round your answer to the nearest mile.

'Are You Prepared?' Answers

1. Domain: the set of all real numbers; range: $-1 \le y \le 1$

2. Two answers are possible: $x \le 1$ or $x \ge 1$

3. $[3, \infty)$

4. True

5. $1; \dfrac{\sqrt{3}}{2}$

6. $-\dfrac{1}{2}; -1$

3.2 The Inverse Trigonometric Functions (Continued)

PREPARING FOR THIS SECTION *Before getting started, review the following concepts:*

- Finding Exact Values Given the Value of a Trigonometric Function and the Quadrant of the Angle (Section 2.4, pp. 147–148)
- Graphs of the Secant, Cosecant, and Cotangent Functions (Section 2.7, pp. 180–183)

- Domain and Range of the Secant, Cosecant, and Cotangent Functions (Section 2.5, pp. 155–156)

Now Work the 'Are You Prepared?' problems on page 223.

OBJECTIVES
1 Find the Exact Value of Expressions Involving the Inverse Sine, Cosine, and Tangent Functions (p. 220)
2 Know the Definitions of the Inverse Secant, Cosecant, and Cotangent Functions (p. 221)
3 Use a Calculator to Evaluate $\sec^{-1} x$, $\csc^{-1} x$, and $\cot^{-1} x$ (p. 221)
4 Write a Trigonometric Expression as an Algebraic Expression (p. 222)

1 Find the Exact Value of Expressions Involving the Inverse Sine, Cosine, and Tangent Functions

EXAMPLE 1	Finding the Exact Value of Expressions Involving Inverse Trigonometric Functions

Find the exact value of: $\sin\left(\tan^{-1}\dfrac{1}{2}\right)$

Solution   Let $\theta = \tan^{-1}\dfrac{1}{2}$. Then $\tan\theta = \dfrac{1}{2}$, where $-\dfrac{\pi}{2} < \theta < \dfrac{\pi}{2}$. We seek $\sin\theta$. Because $\tan\theta > 0$, it follows that $0 < \theta < \dfrac{\pi}{2}$, so θ lies in quadrant I. Now, in Figure 19 we draw a triangle in quadrant I. Because $\tan\theta = \dfrac{b}{a} = \dfrac{1}{2}$, the side opposite θ is $b = 1$ and the side adjacent to θ is $a = 2$. The hypotenuse of this triangle is found using $a^2 + b^2 = c^2$ or $2^2 + 1^2 = c^2$, so $c = \sqrt{5}$. Then $\sin\theta = \dfrac{b}{c} = \dfrac{1}{\sqrt{5}}$, and

$$\sin\left(\tan^{-1}\dfrac{1}{2}\right) = \sin\theta = \dfrac{1}{\sqrt{5}} = \dfrac{\sqrt{5}}{5}$$

Figure 19

$\tan\theta = \dfrac{1}{2}$

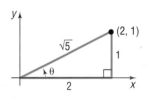

EXAMPLE 2	Finding the Exact Value of Expressions Involving Inverse Trigonometric Functions

Find the exact value of: $\cos\left[\sin^{-1}\left(-\dfrac{1}{3}\right)\right]$

Solution   Let $\theta = \sin^{-1}\left(-\dfrac{1}{3}\right)$. Then $\sin\theta = -\dfrac{1}{3}$ and $-\dfrac{\pi}{2} \le \theta \le \dfrac{\pi}{2}$. We seek $\cos\theta$. Because $\sin\theta < 0$, it follows that $-\dfrac{\pi}{2} \le \theta < 0$, so θ lies in quadrant IV. Figure 20 illustrates $\sin\theta = -\dfrac{1}{3}$ for θ in quadrant IV. Then

$$\cos\left[\sin^{-1}\left(-\dfrac{1}{3}\right)\right] = \cos\theta = \dfrac{2\sqrt{2}}{3}$$

Figure 20

$\sin\theta = -\dfrac{1}{3}$

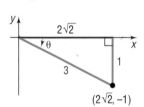

EXAMPLE 3	Finding the Exact Value of Expressions Involving Inverse Trigonometric Functions

Find the exact value of: $\tan\left[\cos^{-1}\left(-\dfrac{1}{3}\right)\right]$

Figure 21

$\cos\theta = -\dfrac{1}{3}$

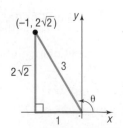

Solution   Let $\theta = \cos^{-1}\left(-\dfrac{1}{3}\right)$. Then $\cos\theta = -\dfrac{1}{3}$ and $0 \le \theta \le \pi$. We seek $\tan\theta$. Because $\cos\theta < 0$, it follows that $\dfrac{\pi}{2} < \theta \le \pi$, so θ lies in quadrant II. Figure 21 illustrates $\cos\theta = -\dfrac{1}{3}$ for θ in quadrant II. Then

$$\tan\left[\cos^{-1}\left(-\dfrac{1}{3}\right)\right] = \tan\theta = \dfrac{2\sqrt{2}}{-1} = -2\sqrt{2}$$

Now Work PROBLEMS **9** AND **27**

2 Know the Definitions of the Inverse Secant, Cosecant, and Cotangent Functions

The inverse secant, inverse cosecant, and inverse cotangent functions are defined as follows:

DEFINITION

$$y = \sec^{-1} x \quad \text{means} \quad x = \sec y \tag{1}$$

$$\text{where} \quad |x| \geq 1 \quad \text{and} \quad 0 \leq y \leq \pi, \quad y \neq \frac{\pi}{2}*$$

$$y = \csc^{-1} x \quad \text{means} \quad x = \csc y \tag{2}$$

$$\text{where} \quad |x| \geq 1 \quad \text{and} \quad -\frac{\pi}{2} \leq y \leq \frac{\pi}{2}, \quad y \neq 0^\dagger$$

$$y = \cot^{-1} x \quad \text{means} \quad x = \cot y \tag{3}$$

$$\text{where} \quad -\infty < x < \infty \quad \text{and} \quad 0 < y < \pi$$

You are encouraged to review the graphs of the cotangent, cosecant, and secant functions in Figures 104, 105, and 106 in Section 2.7 to help you to see the basis for these definitions.

EXAMPLE 4   Finding the Exact Value of an Inverse Cosecant Function

Find the exact value of: $\csc^{-1} 2$

Solution   Let $\theta = \csc^{-1} 2$. We seek the angle θ, $-\frac{\pi}{2} \leq \theta \leq \frac{\pi}{2}$, $\theta \neq 0$, whose cosecant equals 2 $\left(\text{or, equivalently, whose sine equals } \frac{1}{2}\right)$.

$$\theta = \csc^{-1} 2 \qquad -\frac{\pi}{2} \leq \theta \leq \frac{\pi}{2}, \quad \theta \neq 0$$

$$\csc \theta = 2 \qquad -\frac{\pi}{2} \leq \theta \leq \frac{\pi}{2}, \quad \theta \neq 0 \quad \sin \theta = \frac{1}{2}$$

The only angle θ in the interval $-\frac{\pi}{2} \leq \theta \leq \frac{\pi}{2}, \theta \neq 0$, whose cosecant is 2 $\left[\sin \theta = \frac{1}{2}\right]$ is $\frac{\pi}{6}$, so $\csc^{-1} 2 = \frac{\pi}{6}$.

■

Now Work PROBLEM 39

---

3 Use a Calculator to Evaluate $\sec^{-1} x$, $\csc^{-1} x$, and $\cot^{-1} x$

Most calculators do not have keys for evaluating the inverse cotangent, cosecant, and secant functions. The easiest way to evaluate them is to convert to an inverse trigonometric function whose range is the same as the one to be evaluated. In this regard, notice that $y = \cot^{-1} x$ and $y = \sec^{-1} x$, except where undefined, each have the same range as $y = \cos^{-1} x$; $y = \csc^{-1} x$, except where undefined, has the same range as $y = \sin^{-1} x$.

NOTE Remember, the range of $y = \sin^{-1} x$ is $\left[-\frac{\pi}{2}, \frac{\pi}{2}\right]$; the range of $y = \cos^{-1} x$ is $[0, \pi]$.   ■

---

*Most books use this definition. A few use the restriction $0 \leq y < \frac{\pi}{2}, \pi \leq y < \frac{3\pi}{2}$.

†Most books use this definition. A few use the restriction $-\pi < y \leq -\frac{\pi}{2}, 0 < y \leq \frac{\pi}{2}$.

| EXAMPLE 5 | **Approximating the Value of Inverse Trigonometric Functions** |

Use a calculator to approximate each expression in radians rounded to two decimal places.

(a) $\sec^{-1} 3$ (b) $\csc^{-1}(-4)$

(c) $\cot^{-1} \dfrac{1}{2}$ (d) $\cot^{-1}(-2)$

Solution   First, set your calculator to radian mode.

(a) Let $\theta = \sec^{-1} 3$. Then $\sec \theta = 3$ and $0 \le \theta \le \pi, \theta \ne \dfrac{\pi}{2}$. We seek $\cos \theta$ because $y = \cos^{-1}x$ has the same range as $y = \sec^{-1}x$, except where undefined. Since $\sec \theta = \dfrac{1}{\cos \theta} = 3$, we have $\cos \theta = \dfrac{1}{3}$. Then $\theta = \cos^{-1}\dfrac{1}{3}$ and

$$\sec^{-1} 3 = \theta = \cos^{-1}\frac{1}{3} \approx 1.23$$
$$\uparrow$$

Use a calculator.

(b) Let $\theta = \csc^{-1}(-4)$. Then $\csc \theta = -4, -\dfrac{\pi}{2} \le \theta \le \dfrac{\pi}{2}, \theta \ne 0$. We seek $\sin \theta$ because $y = \sin^{-1}x$ has the same range as $y = \csc^{-1}x$, except where undefined. Since $\csc \theta = \dfrac{1}{\sin \theta} = -4$, we have $\sin \theta = -\dfrac{1}{4}$. Then $\theta = \sin^{-1}\left(-\dfrac{1}{4}\right)$, and

$$\csc^{-1}(-4) = \theta = \sin^{-1}\left(-\frac{1}{4}\right) \approx -0.25$$

Figure 22

$\cot \theta = \dfrac{1}{2}, 0 < \theta < \pi$

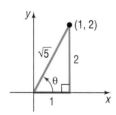

(c) We proceed as before. Let $\theta = \cot^{-1}\dfrac{1}{2}$. Then $\cot \theta = \dfrac{1}{2}, 0 < \theta < \pi$. From these facts we know that θ lies in quadrant I. We seek $\cos \theta$ because $y = \cos^{-1}x$ has the same range as $y = \cot^{-1}x$, except where undefined. To find $\cos \theta$, we use Figure 22. Then $\cos \theta = \dfrac{1}{\sqrt{5}}, 0 < \theta < \dfrac{\pi}{2}, \theta = \cos^{-1}\left(\dfrac{1}{\sqrt{5}}\right)$, and

$$\cot^{-1}\frac{1}{2} = \theta = \cos^{-1}\left(\frac{1}{\sqrt{5}}\right) \approx 1.11$$

(d) Let $\theta = \cot^{-1}(-2)$. Then $\cot \theta = -2, 0 < \theta < \pi$. From these facts we know that θ lies in quadrant II. We seek $\cos \theta$. To find it, we use Figure 23. Then $\cos \theta = -\dfrac{2}{\sqrt{5}}, \dfrac{\pi}{2} < \theta < \pi, \theta = \cos^{-1}\left(-\dfrac{2}{\sqrt{5}}\right)$, and

Figure 23
$\cot \theta = -2, 0 < \theta < \pi$

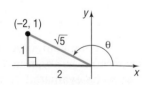

$$\cot^{-1}(-2) = \theta = \cos^{-1}\left(-\frac{2}{\sqrt{5}}\right) \approx 2.68$$

━ **Now Work** PROBLEM 45

4 Write a Trigonometric Expression as an Algebraic Expression

| EXAMPLE 6 | **Writing a Trigonometric Expression as an Algebraic Expression** |

Write $\sin(\tan^{-1}u)$ as an algebraic expression containing u.

Solution   Let $\theta = \tan^{-1}u$ so that $\tan\theta = u, -\dfrac{\pi}{2} < \theta < \dfrac{\pi}{2}, -\infty < u < \infty$. As a result, we know that $\sec\theta > 0$. Then

$$\sin(\tan^{-1}u) = \sin\theta = \sin\theta \cdot \frac{\cos\theta}{\cos\theta} = \tan\theta\cos\theta = \frac{\tan\theta}{\sec\theta} = \frac{\tan\theta}{\sqrt{1+\tan^2\theta}} = \frac{u}{\sqrt{1+u^2}}$$

$\uparrow$ Multiply by 1: $\dfrac{\cos\theta}{\cos\theta}$. $\uparrow$ $\dfrac{\sin\theta}{\cos\theta} = \tan\theta$ $\uparrow$ $\sec^2\theta = 1 + \tan^2\theta$

$\sec\theta > 0$

──────▶ **Now Work** PROBLEM 57

3.2 Assess Your Understanding

'Are You Prepared?' *Answers are given at the end of these exercises. If you get a wrong answer, read the pages listed in red.*

1. What is the domain and the range of $y = \sec x$? (pp. 155–156)

2. **True or False** The graph of $y = \sec x$ is increasing on the interval $\left[0, \dfrac{\pi}{2}\right)$ and on the interval $\left(\dfrac{\pi}{2}, \pi\right]$. (p. 181)

3. If $\cot\theta = -2$ and $0 < \theta < \pi$, then $\cos\theta =$ _____. (pp. 147–148)

Concepts and Vocabulary

4. $y = \sec^{-1}x$ means _____, where $|x|$ _____ and _____ $\leq y \leq$ _____, $y \neq \dfrac{\pi}{2}$.

5. $\cos(\tan^{-1}1) =$ ____.

6. **True or False** It is impossible to obtain exact values for the inverse secant function.

7. **True or False** $\csc^{-1} 0.5$ is not defined.

8. **True or False** The domain of the inverse cotangent function is the set of real numbers.

Skill Building

In Problems 9–36, find the exact value of each expression.

9. $\cos\left(\sin^{-1}\dfrac{\sqrt{2}}{2}\right)$

10. $\sin\left(\cos^{-1}\dfrac{1}{2}\right)$

11. $\tan\left[\cos^{-1}\left(-\dfrac{\sqrt{3}}{2}\right)\right]$

12. $\tan\left[\sin^{-1}\left(-\dfrac{1}{2}\right)\right]$

13. $\sec\left(\cos^{-1}\dfrac{1}{2}\right)$

14. $\cot\left[\sin^{-1}\left(-\dfrac{1}{2}\right)\right]$

15. $\csc(\tan^{-1}1)$

16. $\sec\left(\tan^{-1}\sqrt{3}\right)$

17. $\sin[\tan^{-1}(-1)]$

18. $\cos\left[\sin^{-1}\left(-\dfrac{\sqrt{3}}{2}\right)\right]$

19. $\sec\left[\sin^{-1}\left(-\dfrac{1}{2}\right)\right]$

20. $\csc\left[\cos^{-1}\left(-\dfrac{\sqrt{3}}{2}\right)\right]$

21. $\cos^{-1}\left(\cos\dfrac{5\pi}{4}\right)$

22. $\tan^{-1}\left(\tan\dfrac{2\pi}{3}\right)$

23. $\sin^{-1}\left[\sin\left(-\dfrac{7\pi}{6}\right)\right]$

24. $\cos^{-1}\left[\cos\left(-\dfrac{\pi}{3}\right)\right]$

25. $\tan\left(\sin^{-1}\dfrac{1}{3}\right)$

26. $\tan\left(\cos^{-1}\dfrac{1}{3}\right)$

27. $\sec\left(\tan^{-1}\dfrac{1}{2}\right)$

28. $\cos\left(\sin^{-1}\dfrac{\sqrt{2}}{3}\right)$

29. $\cot\left[\sin^{-1}\left(-\dfrac{\sqrt{2}}{3}\right)\right]$

30. $\csc[\tan^{-1}(-2)]$

31. $\sin[\tan^{-1}(-3)]$

32. $\cot\left[\cos^{-1}\left(-\dfrac{\sqrt{3}}{3}\right)\right]$

33. $\sec\left(\sin^{-1}\dfrac{2\sqrt{5}}{5}\right)$

34. $\csc\left(\tan^{-1}\dfrac{1}{2}\right)$

35. $\sin^{-1}\left(\cos\dfrac{3\pi}{4}\right)$

36. $\cos^{-1}\left(\sin\dfrac{7\pi}{6}\right)$

In Problems 37–44, find the exact value of each expression.

37. $\cot^{-1}\sqrt{3}$

38. $\cot^{-1}1$

39. $\csc^{-1}(-1)$

40. $\csc^{-1}\sqrt{2}$

41. $\sec^{-1}\dfrac{2\sqrt{3}}{3}$

42. $\sec^{-1}(-2)$

43. $\cot^{-1}\left(-\dfrac{\sqrt{3}}{3}\right)$

44. $\csc^{-1}\left(-\dfrac{2\sqrt{3}}{3}\right)$

In Problems 45–56, use a calculator to find the value of each expression rounded to two decimal places.

45. $\sec^{-1}4$

46. $\csc^{-1}5$

47. $\cot^{-1}2$

48. $\sec^{-1}(-3)$

49. $\csc^{-1}(-3)$

50. $\cot^{-1}\left(-\dfrac{1}{2}\right)$

51. $\cot^{-1}\left(-\sqrt{5}\right)$

52. $\cot^{-1}(-8.1)$

53. $\csc^{-1}\left(-\dfrac{3}{2}\right)$

54. $\sec^{-1}\left(-\dfrac{4}{3}\right)$

55. $\cot^{-1}\left(-\dfrac{3}{2}\right)$

56. $\cot^{-1}\left(-\sqrt{10}\right)$

In Problems 57–66, write each trigonometric expression as an algebraic expression in u.

57. $\cos(\tan^{-1}u)$

58. $\sin(\cos^{-1}u)$

59. $\tan(\sin^{-1}u)$

60. $\tan(\cos^{-1}u)$

61. $\sin(\sec^{-1}u)$

62. $\sin(\cot^{-1}u)$

63. $\cos(\csc^{-1}u)$

64. $\cos(\sec^{-1}u)$

65. $\tan(\cot^{-1}u)$

66. $\tan(\sec^{-1}u)$

Mixed Practice

In Problems 67–78, $f(x) = \sin x$, $g(x) = \cos x$, and $h(x) = \tan x$. Find the exact value of each composite function.

67. $g\left(f^{-1}\left(\dfrac{12}{13}\right)\right)$

68. $f\left(g^{-1}\left(\dfrac{5}{13}\right)\right)$

69. $g^{-1}\left(f\left(\dfrac{7\pi}{4}\right)\right)$

70. $f^{-1}\left(g\left(\dfrac{5\pi}{6}\right)\right)$

71. $h\left(f^{-1}\left(-\dfrac{3}{5}\right)\right)$

72. $h\left(g^{-1}\left(-\dfrac{4}{5}\right)\right)$

73. $g\left(h^{-1}\left(\dfrac{12}{5}\right)\right)$

74. $f\left(h^{-1}\left(\dfrac{5}{12}\right)\right)$

75. $g^{-1}\left(f\left(-\dfrac{4\pi}{3}\right)\right)$

76. $g^{-1}\left(f\left(-\dfrac{5\pi}{6}\right)\right)$

77. $h\left(g^{-1}\left(-\dfrac{1}{4}\right)\right)$

78. $h\left(f^{-1}\left(-\dfrac{2}{5}\right)\right)$

Applications and Extensions

*Problems 79 and 80 require the following discussion: When granular materials are allowed to fall freely, they form conical (cone-shaped) piles. The naturally occurring angle of slope, measured from the horizontal, at which the loose material comes to rest is called the **angle of repose** and varies for different materials. The angle of repose θ is related to the height h and base radius r of the conical pile by the equation $\theta = \cot^{-1}\dfrac{r}{h}$. See the illustration.*

79. Angle of Repose: Deicing Salt Because of potential transportation issues (for example, frozen waterways), the deicing salt used by highway departments in the Midwest must be ordered early and stored for future use. When deicing salt is stored in a pile 14 feet high, the diameter of the base of the pile is 45 feet.
(a) Find the angle of repose for deicing salt.
(b) What is the base diameter of a pile that is 17 feet high?
(c) What is the height of a pile that has a base diameter of approximately 122 feet?

Source: Salt Institute, The Salt Storage Handbook, 2006

80. Angle of Repose: Bunker Sand The steepness of sand bunkers on a golf course is affected by the angle of repose of the sand (a larger angle of repose allows for steeper bunkers). A freestanding pile of loose sand from a United States Golf Association (USGA) bunker had a height of 4 feet and a base diameter of approximately 6.68 feet.
(a) Find the angle of repose for USGA bunker sand.
(b) What is the height of such a pile if the diameter of the base is 8 feet?
(c) A 6-foot high pile of loose Tour Grade 50/50 sand has a base diameter of approximately 8.44 feet. Which type of sand (USGA or Tour Grade 50/50) would be better suited for steep bunkers?

Source: 2004 Annual Report, Purdue University Turfgrass Science Program

81. Artillery A projectile fired into the first quadrant from the origin of a coordinate system will pass through the point (x, y) at time t according to the relationship $\cot \theta = \dfrac{2x}{2y + gt^2}$ where θ = the angle of elevation of the launcher and g = the acceleration due to gravity = 32.2 feet/second². An artilleryman is firing at an enemy bunker located 2450 feet up the side of a hill that is 6175 feet away. He fires a round, and exactly 2.27 seconds later he scores a direct hit.
(a) What angle of elevation did he use?

(b) If the angle of elevation is also given by $\sec \theta = \dfrac{v_0 t}{x}$, where v_0 is the muzzle velocity of the weapon, find the muzzle velocity of the artillery piece he used.

Source: www.egwald.com/geometry/projectile3d.php

82. Using a graphing utility, graph $y = \cot^{-1}x$.

83. Using a graphing utility, graph $y = \sec^{-1}x$.

84. Using a graphing utility, graph $y = \csc^{-1}x$.

Discussion and Writing

85. Explain in your own words how you would use your calculator to find the value of $\cot^{-1}10$.

86. Consult three books on calculus and write down the definition in each of $y = \sec^{-1} x$ and $y = \csc^{-1} x$. Compare these with the definitions given in this book.

'Are You Prepared?' Answers

1. Domain: $\left\{x\,\middle|\,x \neq \text{odd integer multiples of } \dfrac{\pi}{2}\right\}$; range: $\{y \leq -1 \text{ or } y \geq 1\}$

2. True

3. $\dfrac{-2\sqrt{5}}{5}$

3.3 Trigonometric Identities

PREPARING FOR THIS SECTION   *Before getting started, review the following:*

- Fundamental Identities (Section 2.2, p. 120)
- Even–Odd Properties (Section 2.5, p. 158)

Now Work the 'Are You Prepared?' problems on page 229.

OBJECTIVES   **1** Use Algebra to Simplify Trigonometric Expressions (p. 226)
              2 Establish Identities (p. 227)

We saw in the previous chapter that the trigonometric functions lend themselves to a wide variety of identities. Before establishing some additional identities, let's review the definition of an *identity*.

DEFINITION   Two functions f and g are said to be **identically equal** if

$$f(x) = g(x)$$

for every value of x for which both functions are defined. Such an equation is referred to as an **identity.** An equation that is not an identity is called a **conditional equation.**

For example, the following are identities:

$$(x + 1)^2 = x^2 + 2x + 1 \qquad \sin^2 x + \cos^2 x = 1 \qquad \csc x = \frac{1}{\sin x}$$

The following are conditional equations:

$2x + 5 = 0$ True only if $x = -\dfrac{5}{2}$

$\sin x = 0$ True only if $x = k\pi$, k an integer

$\sin x = \cos x$ True only if $x = \dfrac{\pi}{4} + 2k\pi$ or $x = \dfrac{5\pi}{4} + 2k\pi$, k an integer

The following summarizes the trigonometric identities that we have established thus far.

Quotient Identities

$$\tan\theta = \frac{\sin\theta}{\cos\theta} \qquad \cot\theta = \frac{\cos\theta}{\sin\theta}$$

Reciprocal Identities

$$\csc\theta = \frac{1}{\sin\theta} \qquad \sec\theta = \frac{1}{\cos\theta} \qquad \cot\theta = \frac{1}{\tan\theta}$$

Pythagorean Identities

$$\sin^2\theta + \cos^2\theta = 1 \qquad \tan^2\theta + 1 = \sec^2\theta$$
$$\cot^2\theta + 1 = \csc^2\theta$$

Even–Odd Identities

$$\sin(-\theta) = -\sin\theta \qquad \cos(-\theta) = \cos\theta \qquad \tan(-\theta) = -\tan\theta$$
$$\csc(-\theta) = -\csc\theta \qquad \sec(-\theta) = \sec\theta \qquad \cot(-\theta) = -\cot\theta$$

This list of identities comprises what we shall refer to as the **basic trigonometric identities.** These identities should not merely be memorized, but should be *known* (just as you know your name rather than have it memorized). In fact, minor variations of a basic identity are often used. For example, we might want to use

$$\sin^2 \theta = 1 - \cos^2 \theta \quad \text{or} \quad \cos^2 \theta = 1 - \sin^2 \theta$$

instead of $\sin^2 \theta + \cos^2 \theta = 1$. For this reason, among others, you need to know these relationships and be very comfortable with variations of them.

1 Use Algebra to Simplify Trigonometric Expressions

The ability to use algebra to manipulate trigonometric expressions is a key skill that one must have to establish identities. Some of the techniques used in establishing identities are multiplying by a "well-chosen 1," writing a trigonometric expression over a common denominator, rewriting a trigonometric expression in terms of sine and cosine only, and factoring.

EXAMPLE 1 **Using Algebraic Techniques to Simplify Trigonometric Expressions**

(a) Simplify $\dfrac{\cot \theta}{\csc \theta}$ by rewriting each trigonometric function in terms of sine and cosine functions.

(b) Show that $\dfrac{\cos \theta}{1 + \sin \theta} = \dfrac{1 - \sin \theta}{\cos \theta}$ by multiplying the numerator and denominator by $1 - \sin \theta$.

(c) Simplify $\dfrac{1 + \sin u}{\sin u} + \dfrac{\cot u - \cos u}{\cos u}$ by rewriting the expression over a common denominator.

(d) Simplify $\dfrac{\sin^2 v - 1}{\tan v \sin v - \tan v}$ by factoring.

Solution (a) $\dfrac{\cot \theta}{\csc \theta} = \dfrac{\dfrac{\cos \theta}{\sin \theta}}{\dfrac{1}{\sin \theta}} = \dfrac{\cos \theta}{\sin \theta} \cdot \dfrac{\sin \theta}{1} = \cos \theta$

(b) $\dfrac{\cos \theta}{1 + \sin \theta} = \dfrac{\cos \theta}{1 + \sin \theta} \cdot \dfrac{1 - \sin \theta}{1 - \sin \theta} = \dfrac{\cos \theta (1 - \sin \theta)}{1 - \sin^2 \theta}$

↑ Multiply by a well-chosen 1: $\dfrac{1 - \sin \theta}{1 - \sin \theta}$.

$= \dfrac{\cos \theta (1 - \sin \theta)}{\cos^2 \theta} = \dfrac{1 - \sin \theta}{\cos \theta}$

(c) $\dfrac{1 + \sin u}{\sin u} + \dfrac{\cot u - \cos u}{\cos u} = \dfrac{1 + \sin u}{\sin u} \cdot \dfrac{\cos u}{\cos u} + \dfrac{\cot u - \cos u}{\cos u} \cdot \dfrac{\sin u}{\sin u}$

$= \dfrac{\cos u + \sin u \cos u + \cot u \sin u - \cos u \sin u}{\sin u \cos u} = \dfrac{\cos u + \dfrac{\cos u}{\sin u} \cdot \sin u}{\sin u \cos u}$

↑ $\cot u = \dfrac{\cos u}{\sin u}$

$= \dfrac{\cos u + \cos u}{\sin u \cos u} = \dfrac{2 \cos u}{\sin u \cos u} = \dfrac{2}{\sin u}$

(d) $\dfrac{\sin^2 v - 1}{\tan v \sin v - \tan v} = \dfrac{(\sin v + 1)(\sin v - 1)}{\tan v (\sin v - 1)} = \dfrac{\sin v + 1}{\tan v}$

🖊 **Now Work** PROBLEMS 9, 11, AND 13

2 Establish Identities

In the examples that follow, the directions will read "Establish the identity. . . . " As you will see, this is accomplished by starting with one side of the given equation (usually the one containing the more complicated expression) and, using appropriate basic identities and algebraic manipulations, arriving at the other side. The selection of appropriate basic identities to obtain the desired result is learned only through experience and lots of practice.

EXAMPLE 2 **Establishing an Identity**

Establish the identity: $\csc\theta \cdot \tan\theta = \sec\theta$

Solution We start with the left side, because it contains the more complicated expression, and apply a reciprocal identity and a quotient identity.

$$\csc\theta \cdot \tan\theta = \frac{1}{\sin\theta} \cdot \frac{\sin\theta}{\cos\theta} = \frac{1}{\cos\theta} = \sec\theta$$

Having arrived at the right side, the identity is established.

NOTE A graphing utility can be used to provide evidence of an identity. For example, if we graph $Y_1 = \csc x \cdot \tan x$ and $Y_2 = \sec x$, the graphs appear to be the same. This provides evidence that $Y_1 = Y_2$. However, it does not prove their equality. A graphing utility *cannot* be used to establish an identity—identities must be established algebraically. ∎

Now Work PROBLEM 19

EXAMPLE 3 **Establishing an Identity**

Establish the identity: $\sin^2(-\theta) + \cos^2(-\theta) = 1$

Solution We begin with the left side and, because the arguments are $-\theta$, apply Even–Odd Identities.

$$\begin{aligned}
\sin^2(-\theta) + \cos^2(-\theta) &= [\sin(-\theta)]^2 + [\cos(-\theta)]^2 \\
&= (-\sin\theta)^2 + (\cos\theta)^2 \quad \text{Even–Odd Identities} \\
&= (\sin\theta)^2 + (\cos\theta)^2 \\
&= 1 \quad \text{Pythagorean Identity}
\end{aligned}$$

EXAMPLE 4 **Establishing an Identity**

Establish the identity: $\dfrac{\sin^2(-\theta) - \cos^2(-\theta)}{\sin(-\theta) - \cos(-\theta)} = \cos\theta - \sin\theta$

Solution We begin with two observations: The left side contains the more complicated expression. Also, the left side contains expressions with the argument $-\theta$, whereas the right side contains expressions with the argument θ. We decide, therefore, to start with the left side and apply Even–Odd Identities.

$$\begin{aligned}
\frac{\sin^2(-\theta) - \cos^2(-\theta)}{\sin(-\theta) - \cos(-\theta)} &= \frac{[\sin(-\theta)]^2 - [\cos(-\theta)]^2}{\sin(-\theta) - \cos(-\theta)} \\
&= \frac{(-\sin\theta)^2 - (\cos\theta)^2}{-\sin\theta - \cos\theta} \quad \text{Even–Odd Identities} \\
&= \frac{(\sin\theta)^2 - (\cos\theta)^2}{-\sin\theta - \cos\theta} \quad \text{Simplify.} \\
&= \frac{(\sin\theta - \cos\theta)(\sin\theta + \cos\theta)}{-(\sin\theta + \cos\theta)} \quad \text{Factor.} \\
&= \frac{\sin\theta - \cos\theta}{-1} = -(\sin\theta - \cos\theta) = -\sin\theta + \cos\theta \\
&= \cos\theta - \sin\theta \quad \text{Cancel and simplify.}
\end{aligned}$$

EXAMPLE 5   **Establishing an Identity**

Establish the identity: $\dfrac{1 + \tan u}{1 + \cot u} = \tan u$

Solution $\dfrac{1 + \tan u}{1 + \cot u} = \dfrac{1 + \tan u}{1 + \dfrac{1}{\tan u}} = \dfrac{1 + \tan u}{\dfrac{\tan u + 1}{\tan u}}$

$= \dfrac{\tan u(1 + \tan u)}{\tan u + 1} = \tan u$

■

🖉 **Now Work** PROBLEMS 23 AND 27

When sums or differences of quotients appear, it is usually best to rewrite them as a single quotient, especially if the other side of the identity consists of only one term.

EXAMPLE 6   **Establishing an Identity**

Establish the identity: $\dfrac{\sin\theta}{1 + \cos\theta} + \dfrac{1 + \cos\theta}{\sin\theta} = 2\csc\theta$

Solution   The left side is more complicated, so we start with it and proceed to add.

$\dfrac{\sin\theta}{1 + \cos\theta} + \dfrac{1 + \cos\theta}{\sin\theta} = \dfrac{\sin^2\theta + (1 + \cos\theta)^2}{(1 + \cos\theta)(\sin\theta)}$ Add the quotients.

$= \dfrac{\sin^2\theta + 1 + 2\cos\theta + \cos^2\theta}{(1 + \cos\theta)(\sin\theta)}$ Remove parentheses in the numerator.

$= \dfrac{(\sin^2\theta + \cos^2\theta) + 1 + 2\cos\theta}{(1 + \cos\theta)(\sin\theta)}$ Regroup.

$= \dfrac{2 + 2\cos\theta}{(1 + \cos\theta)(\sin\theta)}$ Pythagorean Identity

$= \dfrac{2(1 + \cos\theta)}{(1 + \cos\theta)(\sin\theta)}$ Factor and cancel.

$= \dfrac{2}{\sin\theta}$

$= 2\csc\theta$ Reciprocal Identity

■

🖉 **Now Work** PROBLEM 49

Sometimes it helps to write one side in terms of sine and cosine functions only.

EXAMPLE 7   **Establishing an Identity**

Establish the identity: $\dfrac{\tan v + \cot v}{\sec v \csc v} = 1$

Solution

$$\frac{\tan v + \cot v}{\sec v \csc v} = \frac{\dfrac{\sin v}{\cos v} + \dfrac{\cos v}{\sin v}}{\dfrac{1}{\cos v} \cdot \dfrac{1}{\sin v}} = \frac{\dfrac{\sin^2 v + \cos^2 v}{\cos v \sin v}}{\dfrac{1}{\cos v \sin v}}$$

↑ Change to sines and cosines. ↑ Add the quotients in the numerator.

$$= \frac{1}{\cos v \sin v} \cdot \frac{\cos v \sin v}{1} = 1$$

↑ Divide the quotients; $\sin^2 v + \cos^2 v = 1$.

⬛

━━━ **Now Work** PROBLEM 69

Sometimes, multiplying the numerator and denominator by an appropriate factor will result in a simplification.

EXAMPLE 8   **Establishing an Identity**

Establish the identity: $\dfrac{1 - \sin \theta}{\cos \theta} = \dfrac{\cos \theta}{1 + \sin \theta}$

Solution

We start with the left side and multiply the numerator and the denominator by $1 + \sin \theta$. (Alternatively, we could multiply the numerator and denominator of the right side by $1 - \sin \theta$.)

$$\frac{1 - \sin \theta}{\cos \theta} = \frac{1 - \sin \theta}{\cos \theta} \cdot \frac{1 + \sin \theta}{1 + \sin \theta} \qquad \text{Multiply the numerator and denominator by } 1 + \sin \theta.$$

$$= \frac{1 - \sin^2 \theta}{\cos \theta (1 + \sin \theta)}$$

$$= \frac{\cos^2 \theta}{\cos \theta (1 + \sin \theta)} \qquad 1 - \sin^2 \theta = \cos^2 \theta$$

$$= \frac{\cos \theta}{1 + \sin \theta} \qquad \text{Cancel.}$$

━━━ **Now Work** PROBLEM 53

⬛

Although a lot of practice is the only real way to learn how to establish identities, the following guidelines should prove helpful.

WARNING Be careful not to handle identities to be established as if they were conditional equations. You *cannot* establish an identity by such methods as adding the same expression to each side and obtaining a true statement. This practice is not allowed, because the original statement is precisely the one that you are trying to establish. You do not know until it has been established that it is, in fact, true.   ■

Guidelines for Establishing Identities

1. It is almost always preferable to start with the side containing the more complicated expression.
2. Rewrite sums or differences of quotients as a single quotient.
3. Sometimes rewriting one side in terms of sine and cosine functions only will help.
4. Always keep your goal in mind. As you manipulate one side of the expression, you must keep in mind the form of the expression on the other side.

3.3 Assess Your Understanding

'Are You Prepared?' *Answers are given at the end of these exercises. If you get a wrong answer, read the pages listed in red.*

1. True or False $\sin^2 \theta = 1 - \cos^2 \theta$. (p. 120)

2. True or False $\sin(-\theta) + \cos(-\theta) = \cos \theta - \sin \theta$. (p. 158)

Concepts and Vocabulary

3. Suppose that f and g are two functions with the same domain. If $f(x) = g(x)$ for every x in the domain, the equation is called a(n) _____. Otherwise, it is called a(n) _____ equation.

4. $\tan^2 \theta - \sec^2 \theta =$ _____.

5. $\cos(-\theta) - \cos \theta =$ _____.

6. *True or False* $\sin(-\theta) + \sin \theta = 0$ for any value of θ.

7. *True or False* In establishing an identity, it is often easiest to just multiply both sides by a well-chosen nonzero expression involving the variable.

8. *True or False* $\tan \theta \cdot \cos \theta = \sin \theta$ for any $\theta \neq (2k + 1)\dfrac{\pi}{2}$.

Skill Building

In Problems 9–18, simplify each trigonometric expression by following the indicated direction.

9. Rewrite in terms of sine and cosine functions: $\tan \theta \cdot \csc \theta$

10. Rewrite in terms of sine and cosine functions: $\cot \theta \cdot \sec \theta$

11. Multiply $\dfrac{\cos \theta}{1 - \sin \theta}$ by $\dfrac{1 + \sin \theta}{1 + \sin \theta}$.

12. Multiply $\dfrac{\sin \theta}{1 + \cos \theta}$ by $\dfrac{1 - \cos \theta}{1 - \cos \theta}$.

13. Rewrite over a common denominator:
$$\frac{\sin \theta + \cos \theta}{\cos \theta} + \frac{\cos \theta - \sin \theta}{\sin \theta}$$

14. Rewrite over a common denominator:
$$\frac{1}{1 - \cos v} + \frac{1}{1 + \cos v}$$

15. Multiply and simplify: $\dfrac{(\sin \theta + \cos \theta)(\sin \theta + \cos \theta) - 1}{\sin \theta \cos \theta}$

16. Multiply and simplify: $\dfrac{(\tan \theta + 1)(\tan \theta + 1) - \sec^2 \theta}{\tan \theta}$

17. Factor and simplify: $\dfrac{3 \sin^2 \theta + 4 \sin \theta + 1}{\sin^2 \theta + 2 \sin \theta + 1}$

18. Factor and simplify: $\dfrac{\cos^2 \theta - 1}{\cos^2 \theta - \cos \theta}$

In Problems 19–98, establish each identity.

19. $\csc \theta \cdot \cos \theta = \cot \theta$

20. $\sec \theta \cdot \sin \theta = \tan \theta$

21. $1 + \tan^2(-\theta) = \sec^2 \theta$

22. $1 + \cot^2(-\theta) = \csc^2 \theta$

23. $\cos \theta(\tan \theta + \cot \theta) = \csc \theta$

24. $\sin \theta(\cot \theta + \tan \theta) = \sec \theta$

25. $\tan u \cot u - \cos^2 u = \sin^2 u$

26. $\sin u \csc u - \cos^2 u = \sin^2 u$

27. $(\sec \theta - 1)(\sec \theta + 1) = \tan^2 \theta$

28. $(\csc \theta - 1)(\csc \theta + 1) = \cot^2 \theta$

29. $(\sec \theta + \tan \theta)(\sec \theta - \tan \theta) = 1$

30. $(\csc \theta + \cot \theta)(\csc \theta - \cot \theta) = 1$

31. $\cos^2 \theta(1 + \tan^2 \theta) = 1$

32. $(1 - \cos^2 \theta)(1 + \cot^2 \theta) = 1$

33. $(\sin \theta + \cos \theta)^2 + (\sin \theta - \cos \theta)^2 = 2$

34. $\tan^2 \theta \cos^2 \theta + \cot^2 \theta \sin^2 \theta = 1$

35. $\sec^4 \theta - \sec^2 \theta = \tan^4 \theta + \tan^2 \theta$

36. $\csc^4 \theta - \csc^2 \theta = \cot^4 \theta + \cot^2 \theta$

37. $\sec u - \tan u = \dfrac{\cos u}{1 + \sin u}$

38. $\csc u - \cot u = \dfrac{\sin u}{1 + \cos u}$

39. $3 \sin^2 \theta + 4 \cos^2 \theta = 3 + \cos^2 \theta$

40. $9 \sec^2 \theta - 5 \tan^2 \theta = 5 + 4 \sec^2 \theta$

41. $1 - \dfrac{\cos^2 \theta}{1 + \sin \theta} = \sin \theta$

42. $1 - \dfrac{\sin^2 \theta}{1 - \cos \theta} = -\cos \theta$

43. $\dfrac{1 + \tan v}{1 - \tan v} = \dfrac{\cot v + 1}{\cot v - 1}$

44. $\dfrac{\csc v - 1}{\csc v + 1} = \dfrac{1 - \sin v}{1 + \sin v}$

45. $\dfrac{\sec \theta}{\csc \theta} + \dfrac{\sin \theta}{\cos \theta} = 2 \tan \theta$

46. $\dfrac{\csc \theta - 1}{\cot \theta} = \dfrac{\cot \theta}{\csc \theta + 1}$

47. $\dfrac{1 + \sin \theta}{1 - \sin \theta} = \dfrac{\csc \theta + 1}{\csc \theta - 1}$

48. $\dfrac{\cos \theta + 1}{\cos \theta - 1} = \dfrac{1 + \sec \theta}{1 - \sec \theta}$

49. $\dfrac{1 - \sin v}{\cos v} + \dfrac{\cos v}{1 - \sin v} = 2 \sec v$

50. $\dfrac{\cos v}{1 + \sin v} + \dfrac{1 + \sin v}{\cos v} = 2 \sec v$

51. $\dfrac{\sin \theta}{\sin \theta - \cos \theta} = \dfrac{1}{1 - \cot \theta}$

52. $1 - \dfrac{\sin^2 \theta}{1 + \cos \theta} = \cos \theta$

53. $\dfrac{1 - \sin \theta}{1 + \sin \theta} = (\sec \theta - \tan \theta)^2$

54. $\dfrac{1 - \cos \theta}{1 + \cos \theta} = (\csc \theta - \cot \theta)^2$

55. $\dfrac{\cos \theta}{1 - \tan \theta} + \dfrac{\sin \theta}{1 - \cot \theta} = \sin \theta + \cos \theta$

56. $\dfrac{\cot \theta}{1 - \tan \theta} + \dfrac{\tan \theta}{1 - \cot \theta} = 1 + \tan \theta + \cot \theta$

57. $\tan \theta + \dfrac{\cos \theta}{1 + \sin \theta} = \sec \theta$

58. $\dfrac{\sin \theta \cos \theta}{\cos^2 \theta - \sin^2 \theta} = \dfrac{\tan \theta}{1 - \tan^2 \theta}$

59. $\dfrac{\tan \theta + \sec \theta - 1}{\tan \theta - \sec \theta + 1} = \tan \theta + \sec \theta$

60. $\dfrac{\sin \theta - \cos \theta + 1}{\sin \theta + \cos \theta - 1} = \dfrac{\sin \theta + 1}{\cos \theta}$

61. $\dfrac{\tan \theta - \cot \theta}{\tan \theta + \cot \theta} = \sin^2 \theta - \cos^2 \theta$

62. $\dfrac{\sec \theta - \cos \theta}{\sec \theta + \cos \theta} = \dfrac{\sin^2 \theta}{1 + \cos^2 \theta}$

63. $\dfrac{\tan u - \cot u}{\tan u + \cot u} + 1 = 2 \sin^2 u$

64. $\dfrac{\tan u - \cot u}{\tan u + \cot u} + 2 \cos^2 u = 1$

65. $\dfrac{\sec \theta + \tan \theta}{\cot \theta + \cos \theta} = \tan \theta \sec \theta$

66. $\dfrac{\sec \theta}{1 + \sec \theta} = \dfrac{1 - \cos \theta}{\sin^2 \theta}$

67. $\dfrac{1 - \tan^2 \theta}{1 + \tan^2 \theta} + 1 = 2 \cos^2 \theta$

68. $\dfrac{1 - \cot^2 \theta}{1 + \cot^2 \theta} + 2 \cos^2 \theta = 1$

69. $\dfrac{\sec \theta - \csc \theta}{\sec \theta \csc \theta} = \sin \theta - \cos \theta$

70. $\dfrac{\sin^2 \theta - \tan \theta}{\cos^2 \theta - \cot \theta} = \tan^2 \theta$

71. $\sec \theta - \cos \theta = \sin \theta \tan \theta$

72. $\tan \theta + \cot \theta = \sec \theta \csc \theta$

73. $\dfrac{1}{1 - \sin \theta} + \dfrac{1}{1 + \sin \theta} = 2 \sec^2 \theta$

74. $\dfrac{1 + \sin \theta}{1 - \sin \theta} - \dfrac{1 - \sin \theta}{1 + \sin \theta} = 4 \tan \theta \sec \theta$

75. $\dfrac{\sec \theta}{1 - \sin \theta} = \dfrac{1 + \sin \theta}{\cos^3 \theta}$

76. $\dfrac{1 + \sin \theta}{1 - \sin \theta} = (\sec \theta + \tan \theta)^2$

77. $\dfrac{(\sec v - \tan v)^2 + 1}{\csc v (\sec v - \tan v)} = 2 \tan v$

78. $\dfrac{\sec^2 v - \tan^2 v + \tan v}{\sec v} = \sin v + \cos v$

79. $\dfrac{\sin \theta + \cos \theta}{\cos \theta} - \dfrac{\sin \theta - \cos \theta}{\sin \theta} = \sec \theta \csc \theta$

80. $\dfrac{\sin \theta + \cos \theta}{\sin \theta} - \dfrac{\cos \theta - \sin \theta}{\cos \theta} = \sec \theta \csc \theta$

81. $\dfrac{\sin^3 \theta + \cos^3 \theta}{\sin \theta + \cos \theta} = 1 - \sin \theta \cos \theta$

82. $\dfrac{\sin^3 \theta + \cos^3 \theta}{1 - 2 \cos^2 \theta} = \dfrac{\sec \theta - \sin \theta}{\tan \theta - 1}$

83. $\dfrac{\cos^2 \theta - \sin^2 \theta}{1 - \tan^2 \theta} = \cos^2 \theta$

84. $\dfrac{\cos \theta + \sin \theta - \sin^3 \theta}{\sin \theta} = \cot \theta + \cos^2 \theta$

85. $\dfrac{(2 \cos^2 \theta - 1)^2}{\cos^4 \theta - \sin^4 \theta} = 1 - 2 \sin^2 \theta$

86. $\dfrac{1 - 2 \cos^2 \theta}{\sin \theta \cos \theta} = \tan \theta - \cot \theta$

87. $\dfrac{1 + \sin \theta + \cos \theta}{1 + \sin \theta - \cos \theta} = \dfrac{1 + \cos \theta}{\sin \theta}$

88. $\dfrac{1 + \cos \theta + \sin \theta}{1 + \cos \theta - \sin \theta} = \sec \theta + \tan \theta$

89. $(a \sin \theta + b \cos \theta)^2 + (a \cos \theta - b \sin \theta)^2 = a^2 + b^2$

90. $(2a \sin \theta \cos \theta)^2 + a^2 (\cos^2 \theta - \sin^2 \theta)^2 = a^2$

91. $\dfrac{\tan \alpha + \tan \beta}{\cot \alpha + \cot \beta} = \tan \alpha \tan \beta$

92. $(\tan \alpha + \tan \beta)(1 - \cot \alpha \cot \beta) + (\cot \alpha + \cot \beta)(1 - \tan \alpha \tan \beta) = 0$

93. $(\sin \alpha + \cos \beta)^2 + (\cos \beta + \sin \alpha)(\cos \beta - \sin \alpha) = 2 \cos \beta (\sin \alpha + \cos \beta)$

94. $(\sin \alpha - \cos \beta)^2 + (\cos \beta + \sin \alpha)(\cos \beta - \sin \alpha) = -2 \cos \beta (\sin \alpha - \cos \beta)$

95. $\ln |\sec \theta| = -\ln |\cos \theta|$

96. $\ln |\tan \theta| = \ln |\sin \theta| - \ln |\cos \theta|$

97. $\ln |1 + \cos \theta| + \ln |1 - \cos \theta| = 2 \ln |\sin \theta|$

98. $\ln |\sec \theta + \tan \theta| + \ln |\sec \theta - \tan \theta| = 0$

In Problems 99–102, show that the functions f and g are identically equal.

99. $f(x) = \sin x \cdot \tan x, \quad g(x) = \sec x - \cos x$

100. $f(x) = \cos x \cdot \cot x, \quad g(x) = \csc x - \sin x$

101. $f(\theta) = \dfrac{1 - \sin \theta}{\cos \theta} - \dfrac{\cos \theta}{1 + \sin \theta}, \quad g(\theta) = 0$

102. $f(\theta) = \tan \theta + \sec \theta, \quad g(\theta) = \dfrac{\cos \theta}{1 - \sin \theta}$

Applications and Extensions

103. Searchlights A searchlight at the grand opening of a new car dealership casts a spot of light on a wall located 75 meters from the searchlight. The acceleration $\ddot{r}$ of the spot of light is found to be $\ddot{r} = 1200 \sec\theta(2\sec^2\theta - 1)$. Show that this is equivalent to $\ddot{r} = 1200\left(\dfrac{1 + \sin^2\theta}{\cos^3\theta}\right)$.

Source: Adapted from Hibbeler, *Engineering Mechanics: Dynamics,* 10th ed. © 2004.

104. Optical Measurement Optical methods of measurement often rely on the interference of two light waves. If two light waves, identical except for a phase lag, are mixed together, the resulting intensity, or irradiance, is given by

$$I_t = 4A^2\frac{(\csc\theta - 1)(\sec\theta + \tan\theta)}{\csc\theta\sec\theta}.$$ Show that this is equivalent to $I_t = (2A\cos\theta)^2$.

Source: Experimental Techniques, July/August 2002.

Discussion and Writing

105. Write a few paragraphs outlining your strategy for establishing identities.

106. Write down the three Pythagorean Identities.

107. Why do you think it is usually preferable to start with the side containing the more complicated expression when establishing an identity?

108. Make up an identity that is not a Fundamental Identity.

'Are You Prepared?' Answers

1. True   **2.** True

3.4 Sum and Difference Formulas

PREPARING FOR THIS SECTION *Before getting started, review the following:*

- Distance Formula (Section 1.1, pp. 4–6)
- Values of the Trigonometric Functions (Section 2.3, pp. 129–131, and Section 2.4, pp. 142–143)
- Finding Exact Values Given the Value of a Trigonometric Function and the Quadrant of the Angle (Section 2.4, pp. 147–148)

Now Work the 'Are You Prepared?' problems on page 239.

OBJECTIVES 1 Use Sum and Difference Formulas to Find Exact Values (p. 233)

2 Use Sum and Difference Formulas to Establish Identities (p. 234)

3 Use Sum and Difference Formulas Involving Inverse Trigonometric Functions (p. 238)

In this section, we continue our derivation of trigonometric identities by obtaining formulas that involve the sum or difference of two angles, such as $\cos(\alpha + \beta)$, $\cos(\alpha - \beta)$, or $\sin(\alpha + \beta)$. These formulas are referred to as the **sum and difference formulas.** We begin with the formulas for $\cos(\alpha + \beta)$ and $\cos(\alpha - \beta)$.

THEOREM

Sum and Difference Formulas for the Cosine Function

$$\cos(\alpha + \beta) = \cos\alpha\cos\beta - \sin\alpha\sin\beta \tag{1}$$

$$\cos(\alpha - \beta) = \cos\alpha\cos\beta + \sin\alpha\sin\beta \tag{2}$$

In Words

Formula (1) states that the cosine of the sum of two angles equals the cosine of the first angle times the cosine of the second angle minus the sine of the first angle times the sine of the second angle.

Proof We will prove formula (2) first. Although this formula is true for all numbers α and β, we shall assume in our proof that $0 < \beta < \alpha < 2\pi$. We begin with the unit circle and place the angles α and β in standard position, as shown in Figure 24(a). The point P_1 lies on the terminal side of β, so its coordinates are $(\cos\beta, \sin\beta)$; and the point P_2 lies on the terminal side of α, so its coordinates are $(\cos\alpha, \sin\alpha)$.

Now place the angle $\alpha - \beta$ in standard position, as shown in Figure 24(b). The point A has coordinates $(1, 0)$, and the point P_3 is on the terminal side of the angle $\alpha - \beta$, so its coordinates are $(\cos(\alpha - \beta), \sin(\alpha - \beta))$.

Figure 24

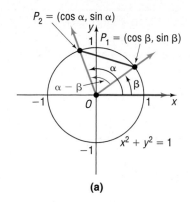

(a)

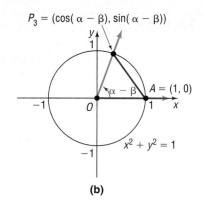

(b)

Looking at triangle OP_1P_2 in Figure 24(a) and triangle OAP_3 in Figure 24(b), we see that these triangles are congruent. (Do you see why? We have SAS: two sides and the included angle, $\alpha - \beta$, are equal.) As a result, the unknown side of each triangle must be equal; that is,

$$d(A, P_3) = d(P_1, P_2)$$

Using the distance formula, we find that

$$\sqrt{[\cos(\alpha - \beta) - 1]^2 + [\sin(\alpha - \beta) - 0]^2} = \sqrt{(\cos \alpha - \cos \beta)^2 + (\sin \alpha - \sin \beta)^2} \quad d(A, P_3) = d(P_1, P_2)$$

$$[\cos(\alpha - \beta) - 1]^2 + \sin^2(\alpha - \beta) = (\cos \alpha - \cos \beta)^2 + (\sin \alpha - \sin \beta)^2 \quad \text{Square both sides.}$$

$$\cos^2(\alpha - \beta) - 2\cos(\alpha - \beta) + 1 + \sin^2(\alpha - \beta) = \cos^2 \alpha - 2 \cos \alpha \cos \beta + \cos^2 \beta \quad \text{Multiply out the squared terms.}$$
$$+ \sin^2 \alpha - 2 \sin \alpha \sin \beta + \sin^2 \beta$$

$$2 - 2\cos(\alpha - \beta) = 2 - 2 \cos \alpha \cos \beta - 2 \sin \alpha \sin \beta \quad \begin{array}{l}\text{Apply a Pythagorean Identity} \\ \text{(3 times).}\end{array}$$

$$-2\cos(\alpha - \beta) = -2 \cos \alpha \cos \beta - 2 \sin \alpha \sin \beta \quad \text{Subtract 2 from each side.}$$

$$\cos(\alpha - \beta) = \cos \alpha \cos \beta + \sin \alpha \sin \beta \quad \text{Divide each side by } -2.$$

This is formula (2). ∎

The proof of formula (1) follows from formula (2) and the Even–Odd Identities. We use the fact that $\alpha + \beta = \alpha - (-\beta)$. Then

$$\cos(\alpha + \beta) = \cos[\alpha - (-\beta)]$$
$$= \cos \alpha \cos(-\beta) + \sin \alpha \sin(-\beta) \quad \text{Use formula (2).}$$
$$= \cos \alpha \cos \beta - \sin \alpha \sin \beta \quad \text{Even–Odd Identities}$$

1 Use Sum and Difference Formulas to Find Exact Values

One use of formulas (1) and (2) is to obtain the exact value of the cosine of an angle that can be expressed as the sum or difference of angles whose sine and cosine are known exactly.

EXAMPLE 1   **Using the Sum Formula to Find Exact Values**

Find the exact value of $\cos 75°$.

Solution   Since $75° = 45° + 30°$, we use formula (1) to obtain

$$\cos 75° = \cos(45° + 30°) = \cos 45° \cos 30° - \sin 45° \sin 30°$$
$$\uparrow$$
$$\text{Formula (1)}$$

$$= \frac{\sqrt{2}}{2} \cdot \frac{\sqrt{3}}{2} - \frac{\sqrt{2}}{2} \cdot \frac{1}{2} = \frac{1}{4}(\sqrt{6} - \sqrt{2})$$

| EXAMPLE 2 | Using the Difference Formula to Find Exact Values |

Find the exact value of $\cos\dfrac{\pi}{12}$.

Solution

$$\cos\frac{\pi}{12} = \cos\left(\frac{3\pi}{12} - \frac{2\pi}{12}\right) = \cos\left(\frac{\pi}{4} - \frac{\pi}{6}\right)$$

$$= \cos\frac{\pi}{4}\cos\frac{\pi}{6} + \sin\frac{\pi}{4}\sin\frac{\pi}{6} \qquad \text{Use formula (2).}$$

$$= \frac{\sqrt{2}}{2}\cdot\frac{\sqrt{3}}{2} + \frac{\sqrt{2}}{2}\cdot\frac{1}{2} = \frac{1}{4}\left(\sqrt{6} + \sqrt{2}\right)$$

■

━━━ **Now Work** PROBLEM 11

2 Use Sum and Difference Formulas to Establish Identities

Another use of formulas (1) and (2) is to establish other identities. Two important identities are given next.

$$\cos\left(\frac{\pi}{2} - \theta\right) = \sin\theta \qquad\qquad \textbf{(3a)}$$

$$\sin\left(\frac{\pi}{2} - \theta\right) = \cos\theta \qquad\qquad \textbf{(3b)}$$

Proof To prove formula (3a), we use the formula for $\cos(\alpha - \beta)$ with $\alpha = \dfrac{\pi}{2}$ and $\beta = \theta$.

$$\cos\left(\frac{\pi}{2} - \theta\right) = \cos\frac{\pi}{2}\cos\theta + \sin\frac{\pi}{2}\sin\theta$$

$$= 0\cdot\cos\theta + 1\cdot\sin\theta$$

$$= \sin\theta$$

To prove formula (3b), we make use of the identity (3a) just established.

$$\sin\left(\frac{\pi}{2} - \theta\right) = \cos\left[\frac{\pi}{2} - \left(\frac{\pi}{2} - \theta\right)\right] = \cos\theta$$
$$\underset{\text{Use (3a)}}{\uparrow}$$

■

Formulas (3a) and (3b) should look familiar. They are the basis for the theorem stated in Chapter 2: Cofunctions of complementary angles are equal. Also, since

$$\cos\left(\frac{\pi}{2} - \theta\right) = \cos\left[-\left(\theta - \frac{\pi}{2}\right)\right] = \cos\left(\theta - \frac{\pi}{2}\right)$$
$$\underset{\text{Even property of cosine}}{\uparrow}$$

and since

$$\cos\left(\frac{\pi}{2} - \theta\right) = \sin\theta$$
$$\underset{3(a)}{\uparrow}$$

it follows that $\cos\left(\theta - \dfrac{\pi}{2}\right) = \sin\theta$. The graphs of $y = \cos\left(\theta - \dfrac{\pi}{2}\right)$ and $y = \sin\theta$ are identical.

Having established the identities in formulas (3a) and (3b), we now can derive the sum and difference formulas for $\sin(\alpha + \beta)$ and $\sin(\alpha - \beta)$.

Proof
$$\sin(\alpha + \beta) = \cos\left[\frac{\pi}{2} - (\alpha + \beta)\right] \qquad \text{Formula (3a)}$$

$$= \cos\left[\left(\frac{\pi}{2} - \alpha\right) - \beta\right]$$

$$= \cos\left(\frac{\pi}{2} - \alpha\right)\cos\beta + \sin\left(\frac{\pi}{2} - \alpha\right)\sin\beta \qquad \text{Formula (2)}$$

$$= \sin\alpha\cos\beta + \cos\alpha\sin\beta \qquad \text{Formulas (3a) and (3b)}$$

$$\sin(\alpha - \beta) = \sin[\alpha + (-\beta)]$$

$$= \sin\alpha\cos(-\beta) + \cos\alpha\sin(-\beta) \qquad \text{Use the sum formula for the sine just obtained.}$$

$$= \sin\alpha\cos\beta + \cos\alpha(-\sin\beta) \qquad \text{Even–Odd Identities}$$

$$= \sin\alpha\cos\beta - \cos\alpha\sin\beta \qquad \blacksquare$$

In Words

Formula (4) states that the sine of the sum of two angles equals the sine of the first angle times the cosine of the second angle plus the cosine of the first angle times the sine of the second angle.

THEOREM

Sum and Difference Formulas for the Sine Function

$$\sin(\alpha + \beta) = \sin\alpha\cos\beta + \cos\alpha\sin\beta \qquad \textbf{(4)}$$
$$\sin(\alpha - \beta) = \sin\alpha\cos\beta - \cos\alpha\sin\beta \qquad \textbf{(5)}$$

EXAMPLE 3 **Using the Sum Formula to Find Exact Values**

Find the exact value of $\sin\dfrac{7\pi}{12}$.

Solution
$$\sin\frac{7\pi}{12} = \sin\left(\frac{3\pi}{12} + \frac{4\pi}{12}\right) = \sin\left(\frac{\pi}{4} + \frac{\pi}{3}\right)$$

$$= \sin\frac{\pi}{4}\cos\frac{\pi}{3} + \cos\frac{\pi}{4}\sin\frac{\pi}{3} \qquad \text{Formula (4)}$$

$$= \frac{\sqrt{2}}{2}\cdot\frac{1}{2} + \frac{\sqrt{2}}{2}\cdot\frac{\sqrt{3}}{2} = \frac{1}{4}\left(\sqrt{2} + \sqrt{6}\right)$$

✏️ **Now Work** PROBLEM 17

EXAMPLE 4 **Using the Difference Formula to Find Exact Values**

Find the exact value of $\sin 80° \cos 20° - \cos 80° \sin 20°$.

Solution The form of the expression $\sin 80° \cos 20° - \cos 80° \sin 20°$ is that of the right side of formula (5) for $\sin(\alpha - \beta)$ with $\alpha = 80°$ and $\beta = 20°$. That is,

$$\sin 80° \cos 20° - \cos 80° \sin 20° = \sin(80° - 20°) = \sin 60° = \frac{\sqrt{3}}{2}$$

✏️ **Now Work** PROBLEMS 23 AND 27

EXAMPLE 5 **Finding Exact Values**

If it is known that $\sin\alpha = \dfrac{4}{5}$, $\dfrac{\pi}{2} < \alpha < \pi$, and that $\sin\beta = -\dfrac{2}{\sqrt{5}} = -\dfrac{2\sqrt{5}}{5}$, $\pi < \beta < \dfrac{3\pi}{2}$, find the exact value of

(a) $\cos\alpha$ (b) $\cos\beta$ (c) $\cos(\alpha + \beta)$ (d) $\sin(\alpha + \beta)$

Solution

(a) Since $\sin \alpha = \dfrac{4}{5} = \dfrac{b}{r}$ and $\dfrac{\pi}{2} < \alpha < \pi$, we let $b = 4$ and $r = 5$ and place α in quadrant II. See Figure 25. Since $(a, 4)$ is in quadrant II, we have $a < 0$. The distance from $(a, 4)$ to $(0, 0)$ is 5, so

$$a^2 + 4^2 = 5^2$$
$$a^2 + 16 = 25$$
$$a^2 = 25 - 16 = 9$$
$$a = -3 \qquad a < 0$$

Figure 25

$\sin \alpha = \dfrac{4}{5}, \dfrac{\pi}{2} < \alpha < \pi$

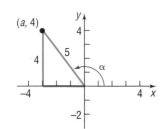

Then

$$\cos \alpha = \frac{a}{r} = -\frac{3}{5}$$

Alternatively, we can find $\cos \alpha$ using identities, as follows:

$$\cos \alpha \underset{\substack{\uparrow \\ \alpha \text{ in quadrant II,} \\ \cos \alpha < 0}}{=} -\sqrt{1 - \sin^2 \alpha} = -\sqrt{1 - \left(\frac{4}{5}\right)^2} = -\sqrt{\frac{9}{25}} = -\frac{3}{5}$$

(b) Since $\sin \beta = \dfrac{-2}{\sqrt{5}} = \dfrac{b}{r}$ and $\pi < \beta < \dfrac{3\pi}{2}$, we let $b = -2$ and $r = \sqrt{5}$ and place β in quadrant III. See Figure 26. Since $(a, -2)$ is in quadrant III, we have $a < 0$. The distance from $(a, -2)$ to $(0, 0)$ is $\sqrt{5}$, so

$$a^2 + (-2)^2 = \left(\sqrt{5}\right)^2$$
$$a^2 = 5 - 4 = 1$$
$$a = -1 \qquad a < 0$$

Figure 26

$\sin \beta = \dfrac{-2}{\sqrt{5}}, \pi < \beta < \dfrac{3\pi}{2}$

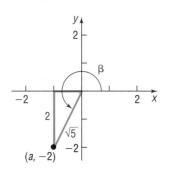

Then

$$\cos \beta = \frac{a}{r} = \frac{-1}{\sqrt{5}} = -\frac{\sqrt{5}}{5}$$

Alternatively, we can find $\cos \beta$ using identities, as follows:

$$\cos \beta = -\sqrt{1 - \sin^2 \beta} = -\sqrt{1 - \frac{4}{5}} = -\sqrt{\frac{1}{5}} = -\frac{\sqrt{5}}{5}$$

(c) Using the results found in parts (a) and (b) and formula (1), we have

$$\cos(\alpha + \beta) = \cos \alpha \cos \beta - \sin \alpha \sin \beta$$

$$= -\frac{3}{5}\left(-\frac{\sqrt{5}}{5}\right) - \frac{4}{5}\left(-\frac{2\sqrt{5}}{5}\right) = \frac{11\sqrt{5}}{25}$$

(d) $\sin(\alpha + \beta) = \sin \alpha \cos \beta + \cos \alpha \sin \beta$

$$= \frac{4}{5}\left(-\frac{\sqrt{5}}{5}\right) + \left(-\frac{3}{5}\right)\left(-\frac{2\sqrt{5}}{5}\right) = \frac{2\sqrt{5}}{25}$$

Now Work PROBLEMS 31(a), (b), AND (c)

EXAMPLE 6   **Establishing an Identity**

Establish the identity: $\dfrac{\cos(\alpha - \beta)}{\sin \alpha \sin \beta} = \cot \alpha \cot \beta + 1$

Solution

$$\frac{\cos(\alpha - \beta)}{\sin \alpha \sin \beta} = \frac{\cos \alpha \cos \beta + \sin \alpha \sin \beta}{\sin \alpha \sin \beta}$$

$$= \frac{\cos \alpha \cos \beta}{\sin \alpha \sin \beta} + \frac{\sin \alpha \sin \beta}{\sin \alpha \sin \beta}$$

$$= \frac{\cos \alpha}{\sin \alpha} \cdot \frac{\cos \beta}{\sin \beta} + 1$$

$$= \cot \alpha \cot \beta + 1$$

———— **Now Work** PROBLEMS 45 AND 57

We use the identity $\tan \theta = \dfrac{\sin \theta}{\cos \theta}$ and the sum formulas for $\sin(\alpha + \beta)$ and $\cos(\alpha + \beta)$ to derive a formula for $\tan(\alpha + \beta)$.

Proof $\quad \tan(\alpha + \beta) = \dfrac{\sin(\alpha + \beta)}{\cos(\alpha + \beta)} = \dfrac{\sin \alpha \cos \beta + \cos \alpha \sin \beta}{\cos \alpha \cos \beta - \sin \alpha \sin \beta}$

Now we divide the numerator and denominator by $\cos \alpha \cos \beta$.

$$\tan(\alpha + \beta) = \frac{\dfrac{\sin \alpha \cos \beta + \cos \alpha \sin \beta}{\cos \alpha \cos \beta}}{\dfrac{\cos \alpha \cos \beta - \sin \alpha \sin \beta}{\cos \alpha \cos \beta}} = \frac{\dfrac{\sin \alpha \cos \beta}{\cos \alpha \cos \beta} + \dfrac{\cos \alpha \sin \beta}{\cos \alpha \cos \beta}}{\dfrac{\cos \alpha \cos \beta}{\cos \alpha \cos \beta} - \dfrac{\sin \alpha \sin \beta}{\cos \alpha \cos \beta}}$$

$$= \frac{\dfrac{\sin \alpha}{\cos \alpha} + \dfrac{\sin \beta}{\cos \beta}}{1 - \dfrac{\sin \alpha}{\cos \alpha} \cdot \dfrac{\sin \beta}{\cos \beta}} = \frac{\tan \alpha + \tan \beta}{1 - \tan \alpha \tan \beta} \quad ■$$

Proof We use the sum formula for $\tan(\alpha + \beta)$ and Even–Odd Properties to get the difference formula.

$$\tan(\alpha - \beta) = \tan[\alpha + (-\beta)] = \frac{\tan \alpha + \tan(-\beta)}{1 - \tan \alpha \tan(-\beta)} = \frac{\tan \alpha - \tan \beta}{1 + \tan \alpha \tan \beta} \quad ■$$

We have proved the following results:

THEOREM

Sum and Difference Formulas for the Tangent Function

> $$\tan(\alpha + \beta) = \frac{\tan \alpha + \tan \beta}{1 - \tan \alpha \tan \beta} \tag{6}$$
>
> $$\tan(\alpha - \beta) = \frac{\tan \alpha - \tan \beta}{1 + \tan \alpha \tan \beta} \tag{7}$$

In Words
Formula (6) states that the tangent of the sum of two angles equals the tangent of the first angle plus the tangent of the second angle, all divided by 1 minus their product.

———— **Now Work** PROBLEM 31(d)

EXAMPLE 7

Establishing an Identity

Prove the identity: $\quad \tan(\theta + \pi) = \tan \theta$

Solution

$$\tan(\theta + \pi) = \frac{\tan \theta + \tan \pi}{1 - \tan \theta \tan \pi} = \frac{\tan \theta + 0}{1 - \tan \theta \cdot 0} = \tan \theta$$

The result obtained in Example 7 verifies that the tangent function is periodic with period π, a fact that we discussed earlier.

EXAMPLE 8 **Establishing an Identity**

Prove the identity: $\tan\left(\theta + \dfrac{\pi}{2}\right) = -\cot\theta$

Solution We cannot use formula (6), since $\tan\dfrac{\pi}{2}$ is not defined. Instead, we proceed as follows:

WARNING Be careful when using formulas (6) and (7). These formulas can be used only for angles α and β for which $\tan\alpha$ and $\tan\beta$ are defined, that is, all angles except odd integer multiples of $\dfrac{\pi}{2}$. ∎

$$\tan\left(\theta + \frac{\pi}{2}\right) = \frac{\sin\left(\theta + \dfrac{\pi}{2}\right)}{\cos\left(\theta + \dfrac{\pi}{2}\right)} = \frac{\sin\theta\cos\dfrac{\pi}{2} + \cos\theta\sin\dfrac{\pi}{2}}{\cos\theta\cos\dfrac{\pi}{2} - \sin\theta\sin\dfrac{\pi}{2}}$$

$$= \frac{(\sin\theta)(0) + (\cos\theta)(1)}{(\cos\theta)(0) - (\sin\theta)(1)} = \frac{\cos\theta}{-\sin\theta} = -\cot\theta$$

---

3 Use Sum and Difference Formulas Involving Inverse Trigonometric Functions

EXAMPLE 9 **Finding the Exact Value of an Expression Involving Inverse Trigonometric Functions**

Find the exact value of: $\sin\left(\cos^{-1}\dfrac{1}{2} + \sin^{-1}\dfrac{3}{5}\right)$

Solution We seek the sine of the sum of two angles, $\alpha = \cos^{-1}\dfrac{1}{2}$ and $\beta = \sin^{-1}\dfrac{3}{5}$. Then

$$\cos\alpha = \frac{1}{2} \quad 0 \le \alpha \le \pi \quad \text{and} \quad \sin\beta = \frac{3}{5} \quad -\frac{\pi}{2} \le \beta \le \frac{\pi}{2}$$

NOTE In Example 9, we could also find $\sin\alpha$ by using $\cos\alpha = \dfrac{1}{2} = \dfrac{a}{r}$, so $a = 1$ and $r = 2$. Then $b = \sqrt{3}$ and $\sin\alpha = \dfrac{b}{r} = \dfrac{\sqrt{3}}{2}$. We could find $\cos\beta$ in a similar fashion. ∎

We use Pythagorean Identities to obtain $\sin\alpha$ and $\cos\beta$. Since $\sin\alpha \ge 0$ and $\cos\beta \ge 0$ (do you know why?), we find

$$\sin\alpha = \sqrt{1 - \cos^2\alpha} = \sqrt{1 - \frac{1}{4}} = \sqrt{\frac{3}{4}} = \frac{\sqrt{3}}{2}$$

$$\cos\beta = \sqrt{1 - \sin^2\beta} = \sqrt{1 - \frac{9}{25}} = \sqrt{\frac{16}{25}} = \frac{4}{5}$$

As a result,

$$\sin\left(\cos^{-1}\frac{1}{2} + \sin^{-1}\frac{3}{5}\right) = \sin(\alpha + \beta) = \sin\alpha\cos\beta + \cos\alpha\sin\beta$$

$$= \frac{\sqrt{3}}{2}\cdot\frac{4}{5} + \frac{1}{2}\cdot\frac{3}{5} = \frac{4\sqrt{3} + 3}{10}$$

—— **Now Work** PROBLEM 73

EXAMPLE 10 **Writing a Trigonometric Expression as an Algebraic Expression**

Write $\sin(\sin^{-1}u + \cos^{-1}v)$ as an algebraic expression containing u and v (that is, without any trigonometric functions). Give the restrictions on u and v.

Solution First, for $\sin^{-1}u$, we have $-1 \le u \le 1$ and for $\cos^{-1}v$, we have $-1 \le v \le 1$. Now let $\alpha = \sin^{-1}u$ and $\beta = \cos^{-1}v$. Then

$$\sin\alpha = u \quad -\frac{\pi}{2} \le \alpha \le \frac{\pi}{2} \quad -1 \le u \le 1$$

$$\cos\beta = v \quad 0 \le \beta \le \pi \quad -1 \le v \le 1$$

Since $-\dfrac{\pi}{2} \le \alpha \le \dfrac{\pi}{2}$, we know that $\cos \alpha \ge 0$. As a result,

$$\cos \alpha = \sqrt{1 - \sin^2 \alpha} = \sqrt{1 - u^2}$$

Similarly, since $0 \le \beta \le \pi$, we know that $\sin \beta \ge 0$. Then

$$\sin \beta = \sqrt{1 - \cos^2 \beta} = \sqrt{1 - v^2}$$

As a result,

$$\sin(\sin^{-1} u + \cos^{-1} v) = \sin(\alpha + \beta) = \sin \alpha \cos \beta + \cos \alpha \sin \beta$$

$$= uv + \sqrt{1 - u^2}\,\sqrt{1 - v^2}$$

━ **Now Work** PROBLEM 83

SUMMARY   Sum and Difference Formulas

$$\cos(\alpha + \beta) = \cos \alpha \cos \beta - \sin \alpha \sin \beta \qquad \cos(\alpha - \beta) = \cos \alpha \cos \beta + \sin \alpha \sin \beta$$

$$\sin(\alpha + \beta) = \sin \alpha \cos \beta + \cos \alpha \sin \beta \qquad \sin(\alpha - \beta) = \sin \alpha \cos \beta - \cos \alpha \sin \beta$$

$$\tan(\alpha + \beta) = \dfrac{\tan \alpha + \tan \beta}{1 - \tan \alpha \tan \beta} \qquad\qquad \tan(\alpha - \beta) = \dfrac{\tan \alpha - \tan \beta}{1 + \tan \alpha \tan \beta}$$

3.4 Assess Your Understanding

'Are You Prepared?' *Answers are given at the end of these exercises. If you get a wrong answer, read the pages listed in red.*

1. The distance d from the point $(2, -3)$ to the point $(5, 1)$ is _____. (pp. 1 6)

2. If $\sin \theta = \dfrac{4}{5}$ and θ is in quadrant II, then $\cos \theta =$ _____. (pp. 147–148)

3. (a) $\sin \dfrac{\pi}{4} \cdot \cos \dfrac{\pi}{3} -$ _____. (pp. 129–131)

  (b) $\tan \dfrac{\pi}{4} - \sin \dfrac{\pi}{6} =$ _____ . (pp. 129–131)

Concepts and Vocabulary

4. $\cos(\alpha + \beta) = \cos \alpha \cos \beta$ ____ $\sin \alpha \sin \beta$

5. $\sin(\alpha - \beta) = \sin \alpha \cos \beta$ ____ $\cos \alpha \sin \beta$

6. *True or False* $\sin(\alpha + \beta) = \sin \alpha + \sin \beta + 2 \sin \alpha \sin \beta$

7. *True or False* $\tan 75° = \tan 30° + \tan 45°$

8. *True or False* $\cos\left(\dfrac{\pi}{2} - \theta\right) = \cos \theta$

Skill Building

In Problems 9–20, find the exact value of each expression.

9. $\sin \dfrac{5\pi}{12}$

10. $\sin \dfrac{\pi}{12}$

11. $\cos \dfrac{7\pi}{12}$

12. $\tan \dfrac{7\pi}{12}$

13. $\cos 165°$

14. $\sin 105°$

15. $\tan 15°$

16. $\tan 195°$

17. $\sin \dfrac{17\pi}{12}$

18. $\tan \dfrac{19\pi}{12}$

19. $\sec\left(-\dfrac{\pi}{12}\right)$

20. $\cot\left(-\dfrac{5\pi}{12}\right)$

In Problems 21–30, find the exact value of each expression.

21. $\sin 20° \cos 10° + \cos 20° \sin 10°$

22. $\sin 20° \cos 80° - \cos 20° \sin 80°$

23. $\cos 70° \cos 20° - \sin 70° \sin 20°$

24. $\cos 40° \cos 10° + \sin 40° \sin 10°$

25. $\dfrac{\tan 20° + \tan 25°}{1 - \tan 20° \tan 25°}$

26. $\dfrac{\tan 40° - \tan 10°}{1 + \tan 40° \tan 10°}$

27. $\sin \dfrac{\pi}{12} \cos \dfrac{7\pi}{12} - \cos \dfrac{\pi}{12} \sin \dfrac{7\pi}{12}$

28. $\cos \dfrac{5\pi}{12} \cos \dfrac{7\pi}{12} - \sin \dfrac{5\pi}{12} \sin \dfrac{7\pi}{12}$

29. $\cos \dfrac{\pi}{12} \cos \dfrac{5\pi}{12} + \sin \dfrac{5\pi}{12} \sin \dfrac{\pi}{12}$

30. $\sin \dfrac{\pi}{18} \cos \dfrac{5\pi}{18} + \cos \dfrac{\pi}{18} \sin \dfrac{5\pi}{18}$

In Problems 31–36, find the exact value of each of the following under the given conditions:

   (a) $\sin(\alpha + \beta)$ (b) $\cos(\alpha + \beta)$ (c) $\sin(\alpha - \beta)$ (d) $\tan(\alpha - \beta)$

31. $\sin \alpha = \dfrac{3}{5}, 0 < \alpha < \dfrac{\pi}{2}$; $\cos \beta = \dfrac{2\sqrt{5}}{5}, -\dfrac{\pi}{2} < \beta < 0$ **32.** $\cos \alpha = \dfrac{\sqrt{5}}{5}, 0 < \alpha < \dfrac{\pi}{2}$; $\sin \beta = -\dfrac{4}{5}, -\dfrac{\pi}{2} < \beta < 0$

33. $\tan \alpha = -\dfrac{4}{3}, \dfrac{\pi}{2} < \alpha < \pi$; $\cos \beta = \dfrac{1}{2}, 0 < \beta < \dfrac{\pi}{2}$ **34.** $\tan \alpha = \dfrac{5}{12}, \pi < \alpha < \dfrac{3\pi}{2}$; $\sin \beta = -\dfrac{1}{2}, \pi < \beta < \dfrac{3\pi}{2}$

35. $\sin \alpha = \dfrac{5}{13}, -\dfrac{3\pi}{2} < \alpha < -\pi$; $\tan \beta = -\sqrt{3}, \dfrac{\pi}{2} < \beta < \pi$ **36.** $\cos \alpha = \dfrac{1}{2}, -\dfrac{\pi}{2} < \alpha < 0$; $\sin \beta = \dfrac{1}{3}, 0 < \beta < \dfrac{\pi}{2}$

37. If $\sin \theta = \dfrac{1}{3}$, θ in quadrant II, find the exact value of: **38.** If $\cos \theta = \dfrac{1}{4}$, θ in quadrant IV, find the exact value of:

   (a) $\cos \theta$ (b) $\sin\left(\theta + \dfrac{\pi}{6}\right)$ (a) $\sin \theta$ (b) $\sin\left(\theta - \dfrac{\pi}{6}\right)$

   (c) $\cos\left(\theta - \dfrac{\pi}{3}\right)$ (d) $\tan\left(\theta + \dfrac{\pi}{4}\right)$ (c) $\cos\left(\theta + \dfrac{\pi}{3}\right)$ (d) $\tan\left(\theta - \dfrac{\pi}{4}\right)$

In Problems 39–44, use the figures to evaluate each function if $f(x) = \sin x$, $g(x) = \cos x$, and $h(x) = \tan x$.

39. $f(\alpha + \beta)$ **40.** $g(\alpha + \beta)$ **41.** $g(\alpha - \beta)$

42. $f(\alpha - \beta)$ **43.** $h(\alpha + \beta)$ **44.** $h(\alpha - \beta)$

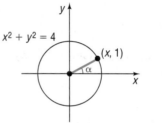 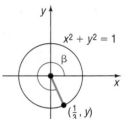

In Problems 45–70, establish each identity.

45. $\sin\left(\dfrac{\pi}{2} + \theta\right) = \cos \theta$ **46.** $\cos\left(\dfrac{\pi}{2} + \theta\right) = -\sin \theta$ **47.** $\sin(\pi - \theta) = \sin \theta$

48. $\cos(\pi - \theta) = -\cos \theta$ **49.** $\sin(\pi + \theta) = -\sin \theta$ **50.** $\cos(\pi + \theta) = -\cos \theta$

51. $\tan(\pi - \theta) = -\tan \theta$ **52.** $\tan(2\pi - \theta) = -\tan \theta$ **53.** $\sin\left(\dfrac{3\pi}{2} + \theta\right) = -\cos \theta$

54. $\cos\left(\dfrac{3\pi}{2} + \theta\right) = \sin \theta$ **55.** $\sin(\alpha + \beta) + \sin(\alpha - \beta) = 2 \sin \alpha \cos \beta$

56. $\cos(\alpha + \beta) + \cos(\alpha - \beta) = 2 \cos \alpha \cos \beta$ **57.** $\dfrac{\sin(\alpha + \beta)}{\sin \alpha \cos \beta} = 1 + \cot \alpha \tan \beta$

58. $\dfrac{\sin(\alpha + \beta)}{\cos \alpha \cos \beta} = \tan \alpha + \tan \beta$ **59.** $\dfrac{\cos(\alpha + \beta)}{\cos \alpha \cos \beta} = 1 - \tan \alpha \tan \beta$

60. $\dfrac{\cos(\alpha - \beta)}{\sin \alpha \cos \beta} = \cot \alpha + \tan \beta$ **61.** $\dfrac{\sin(\alpha + \beta)}{\sin(\alpha - \beta)} = \dfrac{\tan \alpha + \tan \beta}{\tan \alpha - \tan \beta}$

62. $\dfrac{\cos(\alpha + \beta)}{\cos(\alpha - \beta)} = \dfrac{1 - \tan \alpha \tan \beta}{1 + \tan \alpha \tan \beta}$ **63.** $\cot(\alpha + \beta) = \dfrac{\cot \alpha \cot \beta - 1}{\cot \beta + \cot \alpha}$

64. $\cot(\alpha - \beta) = \dfrac{\cot \alpha \cot \beta + 1}{\cot \beta - \cot \alpha}$ **65.** $\sec(\alpha + \beta) = \dfrac{\csc \alpha \csc \beta}{\cot \alpha \cot \beta - 1}$

66. $\sec(\alpha - \beta) = \dfrac{\sec \alpha \sec \beta}{1 + \tan \alpha \tan \beta}$ **67.** $\sin(\alpha - \beta) \sin(\alpha + \beta) = \sin^2 \alpha - \sin^2 \beta$

68. $\cos(\alpha - \beta) \cos(\alpha + \beta) = \cos^2 \alpha - \sin^2 \beta$ **69.** $\sin(\theta + k\pi) = (-1)^k \sin \theta$, k any integer

70. $\cos(\theta + k\pi) = (-1)^k \cos \theta$, k any integer

In Problems 71–82, find the exact value of each expression.

71. $\sin\left(\sin^{-1}\dfrac{1}{2} + \cos^{-1} 0\right)$ **72.** $\sin\left(\sin^{-1}\dfrac{\sqrt{3}}{2} + \cos^{-1} 1\right)$ **73.** $\sin\left[\sin^{-1}\dfrac{3}{5} - \cos^{-1}\left(-\dfrac{4}{5}\right)\right]$

74. $\sin\left[\sin^{-1}\left(-\dfrac{4}{5}\right) - \tan^{-1}\dfrac{3}{4}\right]$ **75.** $\cos\left(\tan^{-1}\dfrac{4}{3} + \cos^{-1}\dfrac{5}{13}\right)$ **76.** $\cos\left[\tan^{-1}\dfrac{5}{12} - \sin^{-1}\left(-\dfrac{3}{5}\right)\right]$

77. $\cos\left(\sin^{-1}\dfrac{5}{13} - \tan^{-1}\dfrac{3}{4}\right)$ **78.** $\cos\left(\tan^{-1}\dfrac{4}{3} + \cos^{-1}\dfrac{12}{13}\right)$ **79.** $\tan\left(\sin^{-1}\dfrac{3}{5} + \dfrac{\pi}{6}\right)$

80. $\tan\left(\dfrac{\pi}{4} - \cos^{-1}\dfrac{3}{5}\right)$ **81.** $\tan\left(\sin^{-1}\dfrac{4}{5} + \cos^{-1} 1\right)$ **82.** $\tan\left(\cos^{-1}\dfrac{4}{5} + \sin^{-1} 1\right)$

In Problems 83–88, write each trigonometric expression as an algebraic expression containing u and v. Give the restrictions required on u and v.

83. $\cos(\cos^{-1} u + \sin^{-1} v)$

84. $\sin(\sin^{-1} u - \cos^{-1} v)$

85. $\sin(\tan^{-1} u - \sin^{-1} v)$

86. $\cos(\tan^{-1} u + \tan^{-1} v)$

87. $\tan(\sin^{-1} u - \cos^{-1} v)$

88. $\sec(\tan^{-1} u + \cos^{-1} v)$

Applications and Extensions

89. Show that $\sin^{-1} v + \cos^{-1} v = \dfrac{\pi}{2}$.

90. Show that $\tan^{-1} v + \cot^{-1} v = \dfrac{\pi}{2}$.

91. Show that $\tan^{-1}\left(\dfrac{1}{v}\right) = \dfrac{\pi}{2} - \tan^{-1} v$, if $v > 0$.

92. Show that $\cot^{-1} e^{v} = \tan^{-1} e^{-v}$.

93. Show that $\sin(\sin^{-1} v + \cos^{-1} v) = 1$.

94. Show that $\cos(\sin^{-1} v + \cos^{-1} v) = 0$.

95. **Calculus**   Show that the difference quotient for $f(x) = \sin x$ is given by
$$\frac{f(x + h) - f(x)}{h} = \frac{\sin(x + h) - \sin x}{h}$$
$$= \cos x \cdot \frac{\sin h}{h} - \sin x \cdot \frac{1 - \cos h}{h}$$

96. **Calculus**   Show that the difference quotient for $f(x) = \cos x$ is given by
$$\frac{f(x + h) - f(x)}{h} = \frac{\cos(x + h) - \cos x}{h}$$
$$= -\sin x \cdot \frac{\sin h}{h} - \cos x \cdot \frac{1 - \cos h}{h}$$

97. **One, Two, Three**
    (a) Show that $\tan(\tan^{-1} 1 + \tan^{-1} 2 + \tan^{-1} 3) = 0$.
    (b) Conclude from part (a) that
    $$\tan^{-1} 1 + \tan^{-1} 2 + \tan^{-1} 3 = \pi$$
    Source: College Mathematics Journal, Vol. 37, No. 3, May 2006

98. **Electric Power**   In an alternating current (ac) circuit, the instantaneous power p at time t is given by
$$p(t) = V_m I_m \cos \phi \sin^2(\omega t) - V_m I_m \sin \phi \sin(\omega t) \cos(\omega t)$$
    Show that this is equivalent to
    $$p(t) = V_m I_m \sin(\omega t) \sin(\omega t - \phi)$$
    Source: HyperPhysics, hosted by Georgia State University

99. **Geometry: Angle Between Two Lines**   Let L_1 and L_2 denote two nonvertical intersecting lines, and let θ denote the acute angle between L_1 and L_2 (see the figure). Show that
$$\tan \theta = \frac{m_2 - m_1}{1 + m_1 m_2}$$
    where m_1 and m_2 are the slopes of L_1 and L_2, respectively. **[Hint:** Use the facts that $\tan \theta_1 = m_1$ and $\tan \theta_2 = m_2$.**]**

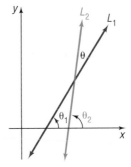

100. If $\alpha + \beta + \gamma = 180°$ and
     $$\cot \theta = \cot \alpha + \cot \beta + \cot \gamma, \quad 0 < \theta < 90°$$
     show that
     $$\sin^3 \theta = \sin(\alpha - \theta) \sin(\beta - \theta) \sin(\gamma - \theta)$$

101. If $\tan \alpha = x + 1$ and $\tan \beta = x - 1$, show that
     $$2 \cot(\alpha - \beta) = x^2$$

Discussion and Writing

102. Discuss the following derivation:
$$\tan\left(\theta + \frac{\pi}{2}\right) = \frac{\tan \theta + \tan \dfrac{\pi}{2}}{1 - \tan \theta \tan \dfrac{\pi}{2}} = \frac{\dfrac{\tan \theta}{\tan \dfrac{\pi}{2}} + 1}{\dfrac{1}{\tan \dfrac{\pi}{2}} - \tan \theta} = \frac{0 + 1}{0 - \tan \theta} = \frac{1}{-\tan \theta} = -\cot \theta$$

Can you justify each step?

103. Explain why formula (7) cannot be used to show that
$$\tan\left(\frac{\pi}{2} - \theta\right) = \cot \theta$$
Establish this identity by using formulas (3a) and (3b).

'Are You Prepared?' Answers

1. 5 2. $-\dfrac{3}{5}$ 3. (a) $\dfrac{\sqrt{2}}{4}$ (b) $\dfrac{1}{2}$

3.5 Double-angle and Half-angle Formulas

OBJECTIVES **1** Use Double-angle Formulas to Find Exact Values (p. 242)
2 Use Double-angle Formulas to Establish Identities (p. 243)
3 Use Half-angle Formulas to Find Exact Values (p. 245)

In this section we derive formulas for $\sin(2\theta)$, $\cos(2\theta)$, $\sin\left(\dfrac{1}{2}\theta\right)$, and $\cos\left(\dfrac{1}{2}\theta\right)$ in terms of $\sin\theta$ and $\cos\theta$. They are derived using the sum formulas.

In the sum formulas for $\sin(\alpha + \beta)$ and $\cos(\alpha + \beta)$, let $\alpha = \beta = \theta$. Then

$$\sin(\alpha + \beta) = \sin\alpha\cos\beta + \cos\alpha\sin\beta$$
$$\sin(\theta + \theta) = \sin\theta\cos\theta + \cos\theta\sin\theta$$
$$\sin(2\theta) = 2\sin\theta\cos\theta$$

and

$$\cos(\alpha + \beta) = \cos\alpha\cos\beta - \sin\alpha\sin\beta$$
$$\cos(\theta + \theta) = \cos\theta\cos\theta - \sin\theta\sin\theta$$
$$\cos(2\theta) = \cos^2\theta - \sin^2\theta$$

An application of the Pythagorean Identity $\sin^2\theta + \cos^2\theta = 1$ results in two other ways to express $\cos(2\theta)$.

$$\cos(2\theta) = \cos^2\theta - \sin^2\theta = (1 - \sin^2\theta) - \sin^2\theta = 1 - 2\sin^2\theta$$

and

$$\cos(2\theta) = \cos^2\theta - \sin^2\theta = \cos^2\theta - (1 - \cos^2\theta) = 2\cos^2\theta - 1$$

We have established the following **Double-angle Formulas:**

THEOREM **Double-angle Formulas**

$$\sin(2\theta) = 2\sin\theta\cos\theta \tag{1}$$
$$\cos(2\theta) = \cos^2\theta - \sin^2\theta \tag{2}$$
$$\cos(2\theta) = 1 - 2\sin^2\theta \tag{3}$$
$$\cos(2\theta) = 2\cos^2\theta - 1 \tag{4}$$

1 Use Double-angle Formulas to Find Exact Values

EXAMPLE 1 **Finding Exact Values Using the Double-angle Formula**

If $\sin\theta = \dfrac{3}{5}$, $\dfrac{\pi}{2} < \theta < \pi$, find the exact value of:

(a) $\sin(2\theta)$ (b) $\cos(2\theta)$

Solution (a) Because $\sin(2\theta) = 2\sin\theta\cos\theta$ and we already know that $\sin\theta = \dfrac{3}{5}$, we only need to find $\cos\theta$. Since $\sin\theta = \dfrac{3}{5} = \dfrac{b}{r}$, $\dfrac{\pi}{2} < \theta < \pi$, we let $b = 3$ and $r = 5$

Figure 27

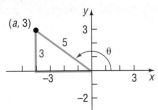

and place θ in quadrant II. See Figure 27. The point $(a, 3)$ is in quadrant II, so $a < 0$. The distance from $(a, 3)$ to $(0, 0)$ is 5, so

$$a^2 + 3^2 = 5^2 \qquad a^2 + b^2 = r^2$$
$$a^2 + 9 = 25$$
$$a^2 = 25 - 9 = 16$$
$$a = -4 \qquad a < 0$$

We find that $\cos \theta = \dfrac{a}{r} = -\dfrac{4}{5}$. Now we use formula (1) to obtain

$$\sin(2\theta) = 2 \sin \theta \cos \theta = 2\left(\frac{3}{5}\right)\left(-\frac{4}{5}\right) = -\frac{24}{25}$$

(b) Because we are given $\sin \theta = \dfrac{3}{5}$, it is easiest to use formula (3) to get $\cos(2\theta)$.

$$\cos(2\theta) = 1 - 2 \sin^2 \theta = 1 - 2\left(\frac{9}{25}\right) = 1 - \frac{18}{25} = \frac{7}{25}$$

WARNING In finding $\cos(2\theta)$ in Example 1(b), we chose to use a version of the Double-angle Formula, formula (3). Note that we are unable to use the Pythagorean Identity $\cos(2\theta) = \pm\sqrt{1 - \sin^2(2\theta)}$, with $\sin(2\theta) = -\dfrac{24}{25}$, because we have no way of knowing which sign to choose. ∎

──── **Now Work** PROBLEMS 7(a) AND (b)

2 Use Double-angle Formulas to Establish Identities

EXAMPLE 2	**Establishing Identities**

(a) Develop a formula for $\tan(2\theta)$ in terms of $\tan \theta$.
(b) Develop a formula for $\sin(3\theta)$ in terms of $\sin \theta$ and $\cos \theta$.

Solution (a) In the sum formula for $\tan(\alpha + \beta)$, let $\alpha = \beta = \theta$. Then

$$\tan(\alpha + \beta) = \frac{\tan \alpha + \tan \beta}{1 - \tan \alpha \tan \beta}$$

$$\tan(\theta + \theta) = \frac{\tan \theta + \tan \theta}{1 - \tan \theta \tan \theta}$$

$$\boxed{\tan(2\theta) = \frac{2 \tan \theta}{1 - \tan^2 \theta}} \qquad \textbf{(5)}$$

(b) To get a formula for $\sin(3\theta)$, we use the sum formula and write 3θ as $2\theta + \theta$.

$$\sin(3\theta) = \sin(2\theta + \theta) = \sin(2\theta) \cos \theta + \cos(2\theta) \sin \theta$$

Now use the Double-angle Formulas to get

$$\sin(3\theta) = (2 \sin \theta \cos \theta)(\cos \theta) + (\cos^2 \theta - \sin^2 \theta)(\sin \theta)$$
$$= 2 \sin \theta \cos^2 \theta + \sin \theta \cos^2 \theta - \sin^3 \theta$$
$$= 3 \sin \theta \cos^2 \theta - \sin^3 \theta$$

The formula obtained in Example 2(b) can also be written as

$$\sin(3\theta) = 3 \sin \theta \cos^2 \theta - \sin^3 \theta = 3 \sin \theta (1 - \sin^2 \theta) - \sin^3 \theta$$
$$= 3 \sin \theta - 4 \sin^3 \theta$$

That is, $\sin(3\theta)$ is a third-degree polynomial in the variable $\sin\theta$. In fact, $\sin(n\theta)$, n a positive odd integer, can always be written as a polynomial of degree n in the variable $\sin\theta$.*

▬▬▬▬ **Now Work** PROBLEM 65

By rearranging the Double-angle Formulas (3) and (4), we obtain other formulas that we will use later in this section.

We begin with formula (3) and proceed to solve for $\sin^2\theta$.

$$\cos(2\theta) = 1 - 2\sin^2\theta$$

$$2\sin^2\theta = 1 - \cos(2\theta)$$

$$\boxed{\sin^2\theta = \frac{1 - \cos(2\theta)}{2}} \qquad\qquad (6)$$

Similarly, using formula (4), we proceed to solve for $\cos^2\theta$.

$$\cos(2\theta) = 2\cos^2\theta - 1$$

$$2\cos^2\theta = 1 + \cos(2\theta)$$

$$\boxed{\cos^2\theta = \frac{1 + \cos(2\theta)}{2}} \qquad\qquad (7)$$

Formulas (6) and (7) can be used to develop a formula for $\tan^2\theta$.

$$\tan^2\theta = \frac{\sin^2\theta}{\cos^2\theta} = \frac{\dfrac{1 - \cos(2\theta)}{2}}{\dfrac{1 + \cos(2\theta)}{2}}$$

$$\boxed{\tan^2\theta = \frac{1 - \cos(2\theta)}{1 + \cos(2\theta)}} \qquad\qquad (8)$$

Formulas (6) through (8) do not have to be memorized since their derivations are so straightforward.

 Formulas (6) and (7) are important in calculus. The next example illustrates a problem that arises in calculus requiring the use of formula (7).

EXAMPLE 3 | Establishing an Identity

Write an equivalent expression for $\cos^4\theta$ that does not involve any powers of sine or cosine greater than 1.

Solution The idea here is to apply formula (7) twice.

$$\cos^4\theta = (\cos^2\theta)^2 = \left(\frac{1 + \cos(2\theta)}{2}\right)^2 \qquad\text{Formula (7)}$$

$$= \frac{1}{4}[1 + 2\cos(2\theta) + \cos^2(2\theta)]$$

*Because of the work done by P. L. Chebyshëv, these polynomials are sometimes called *Chebyshëv polynomials*.

$$= \frac{1}{4} + \frac{1}{2}\cos(2\theta) + \frac{1}{4}\cos^2(2\theta)$$

$$= \frac{1}{4} + \frac{1}{2}\cos(2\theta) + \frac{1}{4}\left\{\frac{1 + \cos[2(2\theta)]}{2}\right\} \quad \text{Formula (7)}$$

$$= \frac{1}{4} + \frac{1}{2}\cos(2\theta) + \frac{1}{8}[1 + \cos(4\theta)]$$

$$= \frac{3}{8} + \frac{1}{2}\cos(2\theta) + \frac{1}{8}\cos(4\theta)$$

Now Work PROBLEM 41

Identities, such as the Double-angle Formulas, can sometimes be used to rewrite expressions in a more suitable form. Let's look at an example.

EXAMPLE 4   **Projectile Motion**

Figure 28

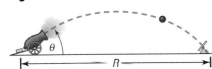

An object is propelled upward at an angle θ to the horizontal with an initial velocity of v_0 feet per second. See Figure 28. If air resistance is ignored, the **range** R, the horizontal distance that the object travels, is given by the function

$$R(\theta) = \frac{1}{16}v_0^2 \sin\theta \cos\theta$$

(a) Show that $R(\theta) = \frac{1}{32}v_0^2 \sin(2\theta)$.

(b) Find the angle θ for which R is a maximum.

Solution   (a) We rewrite the given expression for the range using the Double-angle Formula $\sin(2\theta) = 2\sin\theta\cos\theta$. Then

$$R(\theta) = \frac{1}{16}v_0^2 \sin\theta\cos\theta = \frac{1}{16}v_0^2 \frac{2\sin\theta\cos\theta}{2} = \frac{1}{32}v_0^2 \sin(2\theta)$$

(b) In this form, the largest value for the range R can be found. For a fixed initial speed v_0, the angle θ of inclination to the horizontal determines the value of R. Since the largest value of a sine function is 1, occurring when the argument 2θ is $90°$, it follows that for maximum R we must have

$$2\theta = 90°$$
$$\theta = 45°$$

An inclination to the horizontal of $45°$ results in the maximum range.

---

3 Use Half-angle Formulas to Find Exact Values

Another important use of formulas (6) through (8) is to prove the *Half-angle Formulas*. In formulas (6) through (8), let $\theta = \dfrac{\alpha}{2}$. Then

$$\sin^2\frac{\alpha}{2} = \frac{1 - \cos\alpha}{2} \qquad \cos^2\frac{\alpha}{2} = \frac{1 + \cos\alpha}{2} \qquad \tan^2\frac{\alpha}{2} = \frac{1 - \cos\alpha}{1 + \cos\alpha} \quad \textbf{(9)}$$

COMMENT The identities in box (9) will prove useful in integral calculus. ∎

If we solve for the trigonometric functions on the left sides of equations (9), we obtain the Half-angle Formulas.

THEOREM **Half-angle Formulas**

$$\sin\frac{\alpha}{2} = \pm\sqrt{\frac{1 - \cos\alpha}{2}} \qquad \text{(10a)}$$

$$\cos\frac{\alpha}{2} = \pm\sqrt{\frac{1 + \cos\alpha}{2}} \qquad \text{(10b)}$$

$$\tan\frac{\alpha}{2} = \pm\sqrt{\frac{1 - \cos\alpha}{1 + \cos\alpha}} \qquad \text{(10c)}$$

where the $+$ or $-$ sign is determined by the quadrant of the angle $\dfrac{\alpha}{2}$.

| EXAMPLE 5 | **Finding Exact Values Using Half-angle Formulas** |

Use a Half-angle Formula to find the exact value of:

(a) $\cos 15°$ (b) $\sin(-15°)$

Solution (a) Because $15° = \dfrac{30°}{2}$, we can use the Half-angle Formula for $\cos\dfrac{\alpha}{2}$ with $\alpha = 30°$.
Also, because $15°$ is in quadrant I, $\cos 15° > 0$, we choose the $+$ sign in using formula (10b):

$$\cos 15° = \cos\frac{30°}{2} = \sqrt{\frac{1 + \cos 30°}{2}}$$

$$= \sqrt{\frac{1 + \sqrt{3}/2}{2}} = \sqrt{\frac{2 + \sqrt{3}}{4}} = \frac{\sqrt{2 + \sqrt{3}}}{2}$$

(b) We use the fact that $\sin(-15°) = -\sin 15°$ and then apply formula (10a).

$$\sin(-15°) = -\sin\frac{30°}{2} = -\sqrt{\frac{1 - \cos 30°}{2}}$$

$$= -\sqrt{\frac{1 - \sqrt{3}/2}{2}} = -\sqrt{\frac{2 - \sqrt{3}}{4}} = -\frac{\sqrt{2 - \sqrt{3}}}{2}$$

It is interesting to compare the answer found in Example 5(a) with the answer to Example 2 of Section 3.4. There we calculated

$$\cos\frac{\pi}{12} = \cos 15° = \frac{1}{4}\left(\sqrt{6} + \sqrt{2}\right)$$

Based on this and the result of Example 5(a), we conclude that

$$\frac{1}{4}\left(\sqrt{6} + \sqrt{2}\right) \quad \text{and} \quad \frac{\sqrt{2 + \sqrt{3}}}{2}$$

are equal. (Since each expression is positive, you can verify this equality by squaring each expression.) Two very different looking, yet correct, answers can be obtained, depending on the approach taken to solve a problem.

Now Work PROBLEM 19

| **EXAMPLE 6** | **Finding Exact Values Using Half-angle Formulas** |

If $\cos \alpha = -\dfrac{3}{5}$, $\pi < \alpha < \dfrac{3\pi}{2}$, find the exact value of:

(a) $\sin \dfrac{\alpha}{2}$ (b) $\cos \dfrac{\alpha}{2}$ (c) $\tan \dfrac{\alpha}{2}$

Solution   First, we observe that if $\pi < \alpha < \dfrac{3\pi}{2}$ then $\dfrac{\pi}{2} < \dfrac{\alpha}{2} < \dfrac{3\pi}{4}$. As a result, $\dfrac{\alpha}{2}$ lies in quadrant II.

(a) Because $\dfrac{\alpha}{2}$ lies in quadrant II, $\sin \dfrac{\alpha}{2} > 0$, so we use the $+$ sign in formula (10a) to get

$$\sin \dfrac{\alpha}{2} = \sqrt{\dfrac{1 - \cos \alpha}{2}} = \sqrt{\dfrac{1 - \left(-\dfrac{3}{5}\right)}{2}}$$

$$= \sqrt{\dfrac{\dfrac{8}{5}}{2}} = \sqrt{\dfrac{4}{5}} = \dfrac{2}{\sqrt{5}} = \dfrac{2\sqrt{5}}{5}$$

(b) Because $\dfrac{\alpha}{2}$ lies in quadrant II, $\cos \dfrac{\alpha}{2} < 0$, so we use the $-$ sign in formula (10b) to get

$$\cos \dfrac{\alpha}{2} = -\sqrt{\dfrac{1 + \cos \alpha}{2}} = -\sqrt{\dfrac{1 + \left(-\dfrac{3}{5}\right)}{2}}$$

$$= -\sqrt{\dfrac{\dfrac{2}{5}}{2}} = -\dfrac{1}{\sqrt{5}} = -\dfrac{\sqrt{5}}{5}$$

(c) Because $\dfrac{\alpha}{2}$ lies in quadrant II, $\tan \dfrac{\alpha}{2} < 0$, so we use the $-$ sign in formula (10c) to get

$$\tan \dfrac{\alpha}{2} = -\sqrt{\dfrac{1 - \cos \alpha}{1 + \cos \alpha}} = -\sqrt{\dfrac{1 - \left(-\dfrac{3}{5}\right)}{1 + \left(-\dfrac{3}{5}\right)}} = -\sqrt{\dfrac{\dfrac{8}{5}}{\dfrac{2}{5}}} = -2$$

Another way to solve Example 6(c) is to use the solutions found in parts (a) and (b).

$$\tan \dfrac{\alpha}{2} = \dfrac{\sin \dfrac{\alpha}{2}}{\cos \dfrac{\alpha}{2}} = \dfrac{\dfrac{2\sqrt{5}}{5}}{-\dfrac{\sqrt{5}}{5}} = -2$$

━━━ **Now Work** PROBLEMS 7(c) AND (d)

There is a formula for $\tan \dfrac{\alpha}{2}$ that does not contain $+$ and $-$ signs, making it more useful than formula 10(c). To derive it, we use the formulas

$$1 - \cos \alpha = 2 \sin^2 \dfrac{\alpha}{2} \quad \text{Formula (9)}$$

and

$$\sin \alpha = \sin\left[2\left(\frac{\alpha}{2}\right)\right] = 2 \sin \frac{\alpha}{2} \cos \frac{\alpha}{2} \quad \text{Double-angle Formula}$$

Then

$$\frac{1 - \cos \alpha}{\sin \alpha} = \frac{2 \sin^2 \dfrac{\alpha}{2}}{2 \sin \dfrac{\alpha}{2} \cos \dfrac{\alpha}{2}} = \frac{\sin \dfrac{\alpha}{2}}{\cos \dfrac{\alpha}{2}} = \tan \frac{\alpha}{2}$$

Since it also can be shown that

$$\frac{1 - \cos \alpha}{\sin \alpha} = \frac{\sin \alpha}{1 + \cos \alpha}$$

we have the following two Half-angle Formulas:

Half-angle Formulas for $\tan \dfrac{\alpha}{2}$

$$\tan \frac{\alpha}{2} = \frac{1 - \cos \alpha}{\sin \alpha} = \frac{\sin \alpha}{1 + \cos \alpha} \qquad \textbf{(11)}$$

With this formula, the solution to Example 6(c) can be obtained as follows:

$$\cos \alpha = -\frac{3}{5} \quad \pi < \alpha < \frac{3\pi}{2}$$

$$\sin \alpha = -\sqrt{1 - \cos^2 \alpha} = -\sqrt{1 - \frac{9}{25}} = -\sqrt{\frac{16}{25}} = -\frac{4}{5}$$

Then, by equation (11),

$$\tan \frac{\alpha}{2} = \frac{1 - \cos \alpha}{\sin \alpha} = \frac{1 - \left(-\dfrac{3}{5}\right)}{-\dfrac{4}{5}} = \frac{\dfrac{8}{5}}{-\dfrac{4}{5}} = -2$$

3.5 Assess Your Understanding

Concepts and Vocabulary

1. $\cos(2\theta) = \cos^2 \theta - \underline{\qquad} = \underline{\qquad} - 1 = 1 - \underline{\qquad}.$

2. $\sin^2 \dfrac{\theta}{2} = \dfrac{\underline{\quad}}{2}.$

3. $\tan \dfrac{\theta}{2} = \dfrac{1 - \cos \theta}{\underline{\quad}}.$

4. *True or False* $\cos(2\theta)$ has three equivalent forms:
$\cos^2 \theta - \sin^2 \theta, \quad 1 - 2 \sin^2 \theta, \quad 2 \cos^2 \theta - 1$

5. *True or False* $\sin(2\theta)$ has two equivalent forms:
$2 \sin \theta \cos \theta \quad \text{and} \quad \sin^2 \theta - \cos^2 \theta$

6. *True or False* $\tan(2\theta) + \tan(2\theta) = \tan(4\theta)$

Skill Building

In Problems 7–18, use the information given about the angle $\theta, 0 \leq \theta < 2\pi$, to find the exact value of

(a) $\sin(2\theta)$ (b) $\cos(2\theta)$ (c) $\sin \dfrac{\theta}{2}$ (d) $\cos \dfrac{\theta}{2}$

7. $\sin \theta = \dfrac{3}{5} \quad 0 < \theta < \dfrac{\pi}{2}$

8. $\cos \theta = \dfrac{3}{5} \quad 0 < \theta < \dfrac{\pi}{2}$

9. $\tan \theta = \dfrac{4}{3} \quad \pi < \theta < \dfrac{3\pi}{2}$

10. $\tan \theta = \dfrac{1}{2} \quad \pi < \theta < \dfrac{3\pi}{2}$

11. $\cos \theta = -\dfrac{\sqrt{6}}{3} \quad \dfrac{\pi}{2} < \theta < \pi$

12. $\sin \theta = -\dfrac{\sqrt{3}}{3} \quad \dfrac{3\pi}{2} < \theta < 2\pi$

13. $\sec \theta = 3 \quad \sin \theta > 0$

14. $\csc \theta = -\sqrt{5} \quad \cos \theta < 0$

15. $\cot \theta = -2 \quad \sec \theta < 0$

16. $\sec \theta = 2 \quad \csc \theta < 0$

17. $\tan \theta = -3 \quad \sin \theta < 0$

18. $\cot \theta = 3 \quad \cos \theta < 0$

In Problems 19–28, use the Half-angle Formulas to find the exact value of each expression.

19. $\sin 22.5°$

20. $\cos 22.5°$

21. $\tan \dfrac{7\pi}{8}$

22. $\tan \dfrac{9\pi}{8}$

23. $\cos 165°$

24. $\sin 195°$

25. $\sec \dfrac{15\pi}{8}$

26. $\csc \dfrac{7\pi}{8}$

27. $\sin\left(-\dfrac{\pi}{8}\right)$

28. $\cos\left(-\dfrac{3\pi}{8}\right)$

In Problems 29–40, use the figures to evaluate each function given that $f(x) = \sin x$, $g(x) = \cos x$, and $h(x) = \tan x$.

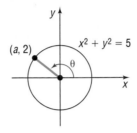

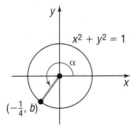

29. $f(2\theta)$

30. $g(2\theta)$

31. $g\left(\dfrac{\theta}{2}\right)$

32. $f\left(\dfrac{\theta}{2}\right)$

33. $h(2\theta)$

34. $h\left(\dfrac{\theta}{2}\right)$

35. $g(2\alpha)$

36. $f(2\alpha)$

37. $f\left(\dfrac{\alpha}{2}\right)$

38. $g\left(\dfrac{\alpha}{2}\right)$

39. $h\left(\dfrac{\alpha}{2}\right)$

40. $h(2\alpha)$

41. Show that $\sin^4 \theta = \dfrac{3}{8} - \dfrac{1}{2}\cos(2\theta) + \dfrac{1}{8}\cos(4\theta)$.

42. Show that $\sin(4\theta) = (\cos \theta)(4 \sin \theta - 8 \sin^3 \theta)$.

43. Develop a formula for $\cos(3\theta)$ as a third-degree polynomial in the variable $\cos \theta$.

44. Develop a formula for $\cos(4\theta)$ as a fourth-degree polynomial in the variable $\cos \theta$.

45. Find an expression for $\sin(5\theta)$ as a fifth-degree polynomial in the variable $\sin \theta$.

46. Find an expression for $\cos(5\theta)$ as a fifth-degree polynomial in the variable $\cos \theta$.

In Problems 47–68, establish each identity.

47. $\cos^4 \theta - \sin^4 \theta = \cos(2\theta)$

48. $\dfrac{\cot \theta - \tan \theta}{\cot \theta + \tan \theta} = \cos(2\theta)$

49. $\cot(2\theta) = \dfrac{\cot^2 \theta - 1}{2 \cot \theta}$

50. $\cot(2\theta) = \dfrac{1}{2}(\cot \theta - \tan \theta)$

51. $\sec(2\theta) = \dfrac{\sec^2 \theta}{2 - \sec^2 \theta}$

52. $\csc(2\theta) = \dfrac{1}{2}\sec \theta \csc \theta$

53. $\cos^2(2u) - \sin^2(2u) = \cos(4u)$

54. $(4 \sin u \cos u)(1 - 2 \sin^2 u) = \sin(4u)$

55. $\dfrac{\cos(2\theta)}{1 + \sin(2\theta)} = \dfrac{\cot \theta - 1}{\cot \theta + 1}$

56. $\sin^2 \theta \cos^2 \theta = \dfrac{1}{8}[1 - \cos(4\theta)]$

57. $\sec^2 \dfrac{\theta}{2} = \dfrac{2}{1 + \cos \theta}$

58. $\csc^2 \dfrac{\theta}{2} = \dfrac{2}{1 - \cos \theta}$

59. $\cot^2 \dfrac{v}{2} = \dfrac{\sec v + 1}{\sec v - 1}$

60. $\tan \dfrac{v}{2} = \csc v - \cot v$

61. $\cos \theta = \dfrac{1 - \tan^2 \dfrac{\theta}{2}}{1 + \tan^2 \dfrac{\theta}{2}}$

62. $1 - \dfrac{1}{2}\sin(2\theta) = \dfrac{\sin^3 \theta + \cos^3 \theta}{\sin \theta + \cos \theta}$

63. $\dfrac{\sin(3\theta)}{\sin \theta} - \dfrac{\cos(3\theta)}{\cos \theta} = 2$

64. $\dfrac{\cos \theta + \sin \theta}{\cos \theta - \sin \theta} - \dfrac{\cos \theta - \sin \theta}{\cos \theta + \sin \theta} = 2 \tan(2\theta)$

65. $\tan(3\theta) = \dfrac{3 \tan \theta - \tan^3 \theta}{1 - 3 \tan^2 \theta}$

66. $\tan \theta + \tan(\theta + 120°) + \tan(\theta + 240°) = 3 \tan(3\theta)$

67. $\ln|\sin \theta| = \dfrac{1}{2}(\ln|1 - \cos(2\theta)| - \ln 2)$

68. $\ln|\cos \theta| = \dfrac{1}{2}(\ln|1 + \cos(2\theta)| - \ln 2)$

Mixed Practice

In Problems 69–80, find the exact value of each expression.

69. $\sin\left(2\sin^{-1}\dfrac{1}{2}\right)$

70. $\sin\left[2\sin^{-1}\dfrac{\sqrt{3}}{2}\right]$

71. $\cos\left(2\sin^{-1}\dfrac{3}{5}\right)$

72. $\cos\left(2\cos^{-1}\dfrac{4}{5}\right)$

73. $\tan\left[2\cos^{-1}\left(-\dfrac{3}{5}\right)\right]$

74. $\tan\left(2\tan^{-1}\dfrac{3}{4}\right)$

75. $\sin\left(2\cos^{-1}\dfrac{4}{5}\right)$

76. $\cos\left[2\tan^{-1}\left(-\dfrac{4}{3}\right)\right]$

77. $\sin^2\left(\dfrac{1}{2}\cos^{-1}\dfrac{3}{5}\right)$

78. $\cos^2\left(\dfrac{1}{2}\sin^{-1}\dfrac{3}{5}\right)$

79. $\sec\left(2\tan^{-1}\dfrac{3}{4}\right)$

80. $\csc\left[2\sin^{-1}\left(-\dfrac{3}{5}\right)\right]$

Applications and Extensions

81. Laser Projection In a laser projection system, the **optical** or **scanning angle** θ is related to the throw distance D from the scanner to the screen and the projected image width W by the equation

$$D = \frac{\frac{1}{2}W}{\csc\theta - \cot\theta}$$

(a) Show that the projected image width is given by

$$W = 2D\tan\frac{\theta}{2}$$

(b) Find the optical angle if the throw distance is 15 feet and the projected image width is 6.5 feet.

Source: Pangolin Laser Systems, Inc.

82. Product of Inertia The **product of inertia** for an area about inclined axes is given by the formula

$$I_{uv} = I_x\sin\theta\cos\theta - I_y\sin\theta\cos\theta + I_{xy}(\cos^2\theta - \sin^2\theta)$$

Show that this is equivalent to

$$I_{uv} = \frac{I_x - I_y}{2}\sin(2\theta) + I_{xy}\cos(2\theta)$$

Source: Adapted from Hibbeler, *Engineering Mechanics: Statics,* 10th ed., Prentice Hall © 2004.

83. Projectile Motion An object is propelled upward at an angle θ, $45° < \theta < 90°$, to the horizontal with an initial velocity of v_0 feet per second from the base of a plane that makes an angle of $45°$ with the horizontal. See the illustration. If air resistance is ignored, the distance R that it travels up the inclined plane is given by the function

$$R(\theta) = \frac{v_0^2\sqrt{2}}{16}\cos\theta(\sin\theta - \cos\theta)$$

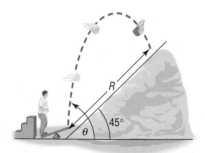

(a) Show that

$$R(\theta) = \frac{v_0^2\sqrt{2}}{32}[\sin(2\theta) - \cos(2\theta) - 1]$$

(b) Graph $R = R(\theta)$. (Use $v_0 = 32$ feet per second.)
(c) What value of θ makes R the largest? (Use $v_0 = 32$ feet per second.)

84. Sawtooth Curve An oscilloscope often displays a sawtooth curve. This curve can be approximated by sinusoidal curves of varying periods and amplitudes. A first approximation to the sawtooth curve is given by

$$y = \frac{1}{2}\sin(2\pi x) + \frac{1}{4}\sin(4\pi x)$$

Show that $y = \sin(2\pi x)\cos^2(\pi x)$.

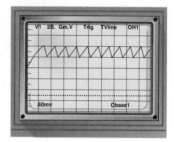

85. Area of an Isosceles Triangle Show that the area A of an isosceles triangle whose equal sides are of length s and θ is the angle between them is

$$A = \frac{1}{2}s^2\sin\theta$$

[Hint: See the illustration. The height h bisects the angle θ and is the perpendicular bisector of the base.**]**

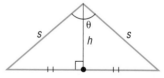

86. Geometry A rectangle is inscribed in a semicircle of radius 1. See the illustration.

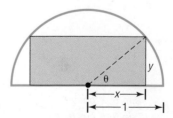

(a) Express the area A of the rectangle as a function of the angle θ shown in the illustration.
(b) Show that $A(\theta) = \sin(2\theta)$.
(c) Find the angle θ that results in the largest area A.
(d) Find the dimensions of this largest rectangle.

87. If $x = 2\tan\theta$, express $\sin(2\theta)$ as a function of x.

88. If $x = 2\tan\theta$, express $\cos(2\theta)$ as a function of x.

89. Find the value of the number C:
$$\frac{1}{2}\sin^2 x + C = -\frac{1}{4}\cos(2x)$$

90. Find the value of the number C:
$$\frac{1}{2}\cos^2 x + C = \frac{1}{4}\cos(2x)$$

91. If $z = \tan\dfrac{\alpha}{2}$, show that $\sin\alpha = \dfrac{2z}{1+z^2}$.

92. If $z = \tan\dfrac{\alpha}{2}$, show that $\cos\alpha = \dfrac{1-z^2}{1+z^2}$.

93. Graph $f(x) = \sin^2 x = \dfrac{1-\cos(2x)}{2}$ for $0 \le x \le 2\pi$ by using transformations.

94. Repeat Problem 93 for $g(x) = \cos^2 x$.

95. Use the fact that
$$\cos\frac{\pi}{12} = \frac{1}{4}(\sqrt{6}+\sqrt{2})$$
to find $\sin\dfrac{\pi}{24}$ and $\cos\dfrac{\pi}{24}$.

96. Show that
$$\cos\frac{\pi}{8} = \frac{\sqrt{2+\sqrt{2}}}{2}$$
and use it to find $\sin\dfrac{\pi}{16}$ and $\cos\dfrac{\pi}{16}$.

97. Show that
$$\sin^3\theta + \sin^3(\theta+120°) + \sin^3(\theta+240°) = -\frac{3}{4}\sin(3\theta)$$

98. If $\tan\theta = a\tan\dfrac{\theta}{3}$, express $\tan\dfrac{\theta}{3}$ in terms of a.

Discussion and Writing

99. Research Chebyshëv polynomials. Write a report on your findings.

3.6 Product-to-Sum and Sum-to-Product Formulas

> **OBJECTIVES** **1** Express Products as Sums (p. 251)
> **2** Express Sums as Products (p. 252)

1 Express Products as Sums

Sum and difference formulas can be used to derive formulas for writing the products of sines and/or cosines as sums or differences. These identities are usually called the **Product-to-Sum Formulas.**

THEOREM   **Product-to-Sum Formulas**

$$\sin\alpha\sin\beta = \frac{1}{2}[\cos(\alpha-\beta) - \cos(\alpha+\beta)] \qquad (1)$$

$$\cos\alpha\cos\beta = \frac{1}{2}[\cos(\alpha-\beta) + \cos(\alpha+\beta)] \qquad (2)$$

$$\sin\alpha\cos\beta = \frac{1}{2}[\sin(\alpha+\beta) + \sin(\alpha-\beta)] \qquad (3)$$

These formulas do not have to be memorized. Instead, you should remember how they are derived. Then, when you want to use them, either look them up or derive them, as needed.

To derive formulas (1) and (2), write down the sum and difference formulas for the cosine:

$$\cos(\alpha-\beta) = \cos\alpha\cos\beta + \sin\alpha\sin\beta \qquad (4)$$
$$\cos(\alpha+\beta) = \cos\alpha\cos\beta - \sin\alpha\sin\beta \qquad (5)$$

Subtract equation (5) from equation (4) to get

$$\cos(\alpha-\beta) - \cos(\alpha+\beta) = 2\sin\alpha\sin\beta$$

from which

$$\sin\alpha\sin\beta = \frac{1}{2}[\cos(\alpha-\beta) - \cos(\alpha+\beta)]$$

Now add equations (4) and (5) to get

$$\cos(\alpha - \beta) + \cos(\alpha + \beta) = 2 \cos \alpha \cos \beta$$

from which

$$\cos \alpha \cos \beta = \frac{1}{2}[\cos(\alpha - \beta) + \cos(\alpha + \beta)]$$

To derive Product-to-Sum Formula (3), use the sum and difference formulas for sine in a similar way. (You are asked to do this in Problem 43.)

| EXAMPLE 1 | **Expressing Products as Sums** |

Express each of the following products as a sum containing only sines or only cosines.

(a) $\sin(6\theta) \sin(4\theta)$ (b) $\cos(3\theta) \cos \theta$ (c) $\sin(3\theta) \cos(5\theta)$

Solution   (a) We use formula (1) to get

$$\sin(6\theta) \sin(4\theta) = \frac{1}{2}[\cos(6\theta - 4\theta) - \cos(6\theta + 4\theta)]$$

$$= \frac{1}{2}[\cos(2\theta) - \cos(10\theta)]$$

(b) We use formula (2) to get

$$\cos(3\theta) \cos \theta = \frac{1}{2}[\cos(3\theta - \theta) + \cos(3\theta + \theta)]$$

$$= \frac{1}{2}[\cos(2\theta) + \cos(4\theta)]$$

(c) We use formula (3) to get

$$\sin(3\theta) \cos(5\theta) = \frac{1}{2}[\sin(3\theta + 5\theta) + \sin(3\theta - 5\theta)]$$

$$= \frac{1}{2}[\sin(8\theta) + \sin(-2\theta)] = \frac{1}{2}[\sin(8\theta) - \sin(2\theta)]$$

 Now Work PROBLEM 1

2 Express Sums as Products

The **Sum-to-Product Formulas** are given next.

THEOREM   **Sum-to-Product Formulas**

$$\sin \alpha + \sin \beta = 2 \sin \frac{\alpha + \beta}{2} \cos \frac{\alpha - \beta}{2} \qquad (6)$$

$$\sin \alpha - \sin \beta = 2 \sin \frac{\alpha - \beta}{2} \cos \frac{\alpha + \beta}{2} \qquad (7)$$

$$\cos \alpha + \cos \beta = 2 \cos \frac{\alpha + \beta}{2} \cos \frac{\alpha - \beta}{2} \qquad (8)$$

$$\cos \alpha - \cos \beta = -2 \sin \frac{\alpha + \beta}{2} \sin \frac{\alpha - \beta}{2} \qquad (9)$$

We will derive formula (6) and leave the derivations of formulas (7) through (9) as exercises (see Problems 44 through 46).

Proof

$$2 \sin \frac{\alpha + \beta}{2} \cos \frac{\alpha - \beta}{2} \underset{\uparrow}{=} 2 \cdot \frac{1}{2} \left[\sin \left(\frac{\alpha + \beta}{2} + \frac{\alpha - \beta}{2} \right) + \sin \left(\frac{\alpha + \beta}{2} - \frac{\alpha - \beta}{2} \right) \right]$$

Product-to-Sum Formula (3)

$$= \sin \frac{2\alpha}{2} + \sin \frac{2\beta}{2} = \sin \alpha + \sin \beta \qquad \blacksquare$$

EXAMPLE 2	**Expressing Sums (or Differences) as a Product**

Express each sum or difference as a product of sines and/or cosines.

(a) $\sin(5\theta) - \sin(3\theta)$ (b) $\cos(3\theta) + \cos(2\theta)$

Solution (a) We use formula (7) to get

$$\sin(5\theta) - \sin(3\theta) = 2 \sin \frac{5\theta - 3\theta}{2} \cos \frac{5\theta + 3\theta}{2}$$

$$= 2 \sin \theta \cos(4\theta)$$

(b) $\cos(3\theta) + \cos(2\theta) = 2 \cos \dfrac{3\theta + 2\theta}{2} \cos \dfrac{3\theta - 2\theta}{2}$ Formula (8)

$$= 2 \cos \frac{5\theta}{2} \cos \frac{\theta}{2}$$

──── **Now Work** PROBLEM 11

3.6 Assess Your Understanding

Skill Building

In Problems 1–10, express each product as a sum containing only sines or only cosines.

1. $\sin(4\theta) \sin(2\theta)$ **2.** $\cos(4\theta) \cos(2\theta)$ **3.** $\sin(4\theta) \cos(2\theta)$ **4.** $\sin(3\theta) \sin(5\theta)$ **5.** $\cos(3\theta) \cos(5\theta)$

6. $\sin(4\theta) \cos(6\theta)$ **7.** $\sin \theta \sin(2\theta)$ **8.** $\cos(3\theta) \cos(4\theta)$ **9.** $\sin \dfrac{3\theta}{2} \cos \dfrac{\theta}{2}$ **10.** $\sin \dfrac{\theta}{2} \cos \dfrac{5\theta}{2}$

In Problems 11–18, express each sum or difference as a product of sines and/or cosines.

11. $\sin(4\theta) - \sin(2\theta)$ **12.** $\sin(4\theta) + \sin(2\theta)$ **13.** $\cos(2\theta) + \cos(4\theta)$ **14.** $\cos(5\theta) - \cos(3\theta)$

15. $\sin \theta + \sin(3\theta)$ **16.** $\cos \theta + \cos(3\theta)$ **17.** $\cos \dfrac{\theta}{2} - \cos \dfrac{3\theta}{2}$ **18.** $\sin \dfrac{\theta}{2} - \sin \dfrac{3\theta}{2}$

In Problems 19–36, establish each identity.

19. $\dfrac{\sin \theta + \sin(3\theta)}{2 \sin(2\theta)} = \cos \theta$ **20.** $\dfrac{\cos \theta + \cos(3\theta)}{2 \cos(2\theta)} = \cos \theta$ **21.** $\dfrac{\sin(4\theta) + \sin(2\theta)}{\cos(4\theta) + \cos(2\theta)} = \tan(3\theta)$

22. $\dfrac{\cos \theta - \cos(3\theta)}{\sin(3\theta) - \sin \theta} = \tan(2\theta)$ **23.** $\dfrac{\cos \theta - \cos(3\theta)}{\sin \theta + \sin(3\theta)} = \tan \theta$ **24.** $\dfrac{\cos \theta - \cos(5\theta)}{\sin \theta + \sin(5\theta)} = \tan(2\theta)$

25. $\sin \theta[\sin \theta + \sin(3\theta)] = \cos \theta[\cos \theta - \cos(3\theta)]$ **26.** $\sin \theta[\sin(3\theta) + \sin(5\theta)] = \cos \theta[\cos(3\theta) - \cos(5\theta)]$

27. $\dfrac{\sin(4\theta) + \sin(8\theta)}{\cos(4\theta) + \cos(8\theta)} = \tan(6\theta)$ **28.** $\dfrac{\sin(4\theta) - \sin(8\theta)}{\cos(4\theta) - \cos(8\theta)} = -\cot(6\theta)$

29. $\dfrac{\sin(4\theta) + \sin(8\theta)}{\sin(4\theta) - \sin(8\theta)} = -\dfrac{\tan(6\theta)}{\tan(2\theta)}$ **30.** $\dfrac{\cos(4\theta) - \cos(8\theta)}{\cos(4\theta) + \cos(8\theta)} = \tan(2\theta) \tan(6\theta)$

31. $\dfrac{\sin \alpha + \sin \beta}{\sin \alpha - \sin \beta} = \tan \dfrac{\alpha + \beta}{2} \cot \dfrac{\alpha - \beta}{2}$ **32.** $\dfrac{\cos \alpha + \cos \beta}{\cos \alpha - \cos \beta} = -\cot \dfrac{\alpha + \beta}{2} \cot \dfrac{\alpha - \beta}{2}$

33. $\dfrac{\sin \alpha + \sin \beta}{\cos \alpha + \cos \beta} = \tan \dfrac{\alpha + \beta}{2}$

34. $\dfrac{\sin \alpha - \sin \beta}{\cos \alpha - \cos \beta} = -\cot \dfrac{\alpha + \beta}{2}$

35. $1 + \cos(2\theta) + \cos(4\theta) + \cos(6\theta) = 4 \cos \theta \cos(2\theta) \cos(3\theta)$

36. $1 - \cos(2\theta) + \cos(4\theta) - \cos(6\theta) = 4 \sin \theta \cos(2\theta) \sin(3\theta)$

Applications and Extensions

37. **Touch-Tone Phones** On a Touch-Tone phone, each button produces a unique sound. The sound produced is the sum of two tones, given by

$$y = \sin(2\pi l t) \quad \text{and} \quad y = \sin(2\pi h t)$$

where l and h are the low and high frequencies (cycles per second) shown on the illustration. For example, if you touch 7, the low frequency is $l = 852$ cycles per second and the high frequency is $h = 1209$ cycles per second. The sound emitted by touching 7 is

$$y = \sin[2\pi(852)t] + \sin[2\pi(1209)t]$$

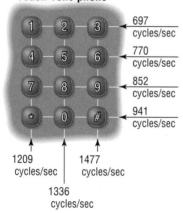

Touch-Tone phone

697 cycles/sec
770 cycles/sec
852 cycles/sec
941 cycles/sec

1209 cycles/sec   1477 cycles/sec

1336 cycles/sec

(a) Write this sound as a product of sines and/or cosines.
(b) Determine the maximum value of y.
(c) Graph the sound emitted by touching 7.

38. **Touch-Tone Phones**
(a) Write the sound emitted by touching the # key as a product of sines and/or cosines.
(b) Determine the maximum value of y.
(c) Graph the sound emitted by touching the # key.

39. **Moment of Inertia** The moment of inertia I of an object is a measure of how easy it is to rotate the object about some fixed point. In engineering mechanics, it is sometimes necessary to compute moments of inertia with respect

to a set of rotated axes. These moments are given by the equations

$$I_u = I_x \cos^2 \theta + I_y \sin^2 \theta - 2I_{xy} \sin \theta \cos \theta$$

$$I_v = I_x \sin^2 \theta + I_y \cos^2 \theta + 2I_{xy} \sin \theta \cos \theta$$

Use Product-to-Sum Formulas to show that

$$I_u = \frac{I_x + I_y}{2} + \frac{I_x - I_y}{2} \cos(2\theta) - I_{xy} \sin(2\theta)$$

and

$$I_v = \frac{I_x + I_y}{2} - \frac{I_x - I_y}{2} \cos(2\theta) + I_{xy} \sin(2\theta)$$

Source: Adapted from Hibbeler, *Engineering Mechanics: Statics,* 10th ed., Prentice Hall © 2004.

40. **Projectile Motion** The range R of a projectile propelled downward from the top of an inclined plane at an angle θ to the inclined plane is given by

$$R(\theta) = \frac{2v_0^2 \sin \theta \cos(\theta - \phi)}{g \cos^2 \phi}$$

where v_0 is the initial velocity of the projectile, ϕ is the angle the plane makes with respect to the horizontal, and g is acceleration due to gravity.

(a) Show that for fixed v_0 and ϕ the maximum range down the incline is given by $R_{\max} = \dfrac{v_0^2}{g(1 - \sin \phi)}$.

(b) Determine the maximum range if the projectile has an initial velocity of 50 meters/second, the angle of the plane is $\phi = 35°$, and $g = 9.8$ meters/second^2.

41. If $\alpha + \beta + \gamma = \pi$, show that

$$\sin(2\alpha) + \sin(2\beta) + \sin(2\gamma) = 4 \sin \alpha \sin \beta \sin \gamma$$

42. If $\alpha + \beta + \gamma = \pi$, show that

$$\tan \alpha + \tan \beta + \tan \gamma = \tan \alpha \tan \beta \tan \gamma$$

43. Derive formula (3).

44. Derive formula (7).

45. Derive formula (8).

46. Derive formula (9).

3.7 Trigonometric Equations (I)

PREPARING FOR THIS SECTION *Before getting started, review the following:*

- Solving Equations Algebraically (Appendix, Section A.4, pp. A25–A37)
- Values of the Trigonometric Functions (Section 2.3, pp. 129–132, and Section 2.4, pp. 142–147)

Now Work the 'Are You Prepared?' problems on page 258.

OBJECTIVE **1** Solve Equations Involving a Single Trigonometric Function (p. 255)

1 Solve Equations Involving a Single Trigonometric Function

The previous four sections of this chapter were devoted to trigonometric identities, that is, equations involving trigonometric functions that are satisfied by every value in the domain of the variable. In the remaining two sections, we discuss **trigonometric equations,** that is, equations involving trigonometric functions that are satisfied only by some values of the variable (or, possibly, are not satisfied by any values of the variable). The values that satisfy the equation are called **solutions** of the equation.

EXAMPLE 1   **Checking Whether a Given Number Is a Solution of a Trigonometric Equation**

Determine whether $\theta = \dfrac{\pi}{4}$ is a solution of the equation $2\sin\theta - 1 = 0$. Is $\theta = \dfrac{\pi}{6}$ a solution?

Solution   Replace θ by $\dfrac{\pi}{4}$ in the given equation. The result is

$$2\sin\frac{\pi}{4} - 1 = 2\cdot\frac{\sqrt{2}}{2} - 1 = \sqrt{2} - 1 \neq 0$$

We conclude that $\dfrac{\pi}{4}$ is not a solution.

Next replace θ by $\dfrac{\pi}{6}$ in the equation. The result is

$$2\sin\frac{\pi}{6} - 1 = 2\cdot\frac{1}{2} - 1 = 1 - 1 = 0$$

We conclude that $\dfrac{\pi}{6}$ is a solution of the given equation.

■

The equation given in Example 1 has other solutions besides $\theta = \dfrac{\pi}{6}$. For example, $\theta = \dfrac{5\pi}{6}$ is also a solution, as is $\theta = \dfrac{13\pi}{6}$. (You should check this for yourself.) In fact, the equation has an infinite number of solutions due to the periodicity of the sine function, as can be seen in Figure 29, which shows the graph of $Y_1 = 2\sin x - 1$. Each x-intercept represents a solution of the equation $2\sin x - 1 = 0$.

Figure 29

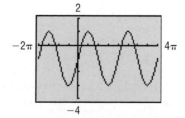

Unless the domain of the variable is restricted, we need to find *all* the solutions of a trigonometric equation. As the next example illustrates, finding all the solutions can be accomplished by first finding solutions over an interval whose length equals the period of the function and then adding multiples of that period to the solutions found. Let's look at some examples.

EXAMPLE 2   **Finding All the Solutions of a Trigonometric Equation**

Solve the equation: $\cos\theta = \dfrac{1}{2}$

Give a general formula for all the solutions. List eight of the solutions.

Solution

Figure 30

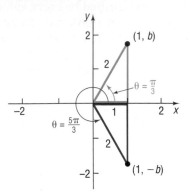

Figure 31

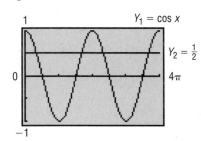

The period of the cosine function is 2π. In the interval $[0, 2\pi)$, there are two angles θ for which $\cos\theta = \dfrac{1}{2}$: $\theta = \dfrac{\pi}{3}$ and $\theta = \dfrac{5\pi}{3}$. See Figure 30. Because the cosine function has period 2π, all the solutions of $\cos\theta = \dfrac{1}{2}$ may be given by the general formula

$$\theta = \frac{\pi}{3} + 2k\pi \quad \text{or} \quad \theta = \frac{5\pi}{3} + 2k\pi \quad \text{\small k any integer}$$

Eight of the solutions are

$$\underbrace{-\frac{5\pi}{3}, \quad -\frac{\pi}{3}}_{k\,=\,-1}, \quad \underbrace{\frac{\pi}{3}, \quad \frac{5\pi}{3}}_{k\,=\,0}, \quad \underbrace{\frac{7\pi}{3}, \quad \frac{11\pi}{3}}_{k\,=\,1}, \quad \underbrace{\frac{13\pi}{3}, \quad \frac{17\pi}{3}}_{k\,=\,2}$$

✓ **Check:** We can verify the solutions by graphing $Y_1 = \cos x$ and $Y_2 = \dfrac{1}{2}$ to determine where the graphs intersect. (Be sure to graph in radian mode.) See Figure 31. The graph of Y_1 intersects the graph of Y_2 at $x = 1.05\left(\approx\dfrac{\pi}{3}\right)$, $5.24\left(\approx\dfrac{5\pi}{3}\right)$, $7.33\left(\approx\dfrac{7\pi}{3}\right)$, and $11.52\left(\approx\dfrac{11\pi}{3}\right)$, rounded to two decimal places.

━━ **Now Work** PROBLEM 31

In most of our work, we shall be interested only in finding solutions of trigonometric equations for $0 \le \theta < 2\pi$.

EXAMPLE 3 **Solving a Linear Trigonometric Equation**

Solve the equation: $2\sin\theta + \sqrt{3} = 0$, $0 \le \theta < 2\pi$

Solution We solve the equation for $\sin\theta$.

$$2\sin\theta + \sqrt{3} = 0$$
$$2\sin\theta = -\sqrt{3} \quad \text{\small Subtract } \sqrt{3} \text{ from both sides.}$$
$$\sin\theta = -\frac{\sqrt{3}}{2} \quad \text{\small Divide both sides by 2.}$$

In the interval $[0, 2\pi)$, there are two angles θ for which $\sin\theta = -\dfrac{\sqrt{3}}{2}$: $\theta = \dfrac{4\pi}{3}$ and $\theta = \dfrac{5\pi}{3}$. The solution set is $\left\{\dfrac{4\pi}{3}, \dfrac{5\pi}{3}\right\}$.

━━ **Now Work** PROBLEM 7

When the argument of the trigonometric function in an equation is a multiple of θ, the general formula must be used to solve the equation.

EXAMPLE 4 **Solving a Trigonometric Equation**

Solve the equation: $\sin(2\theta) = \dfrac{1}{2}$, $0 \le \theta < 2\pi$

Solution In the interval $[0, 2\pi)$, the sine function has a value $\dfrac{1}{2}$ at $\dfrac{\pi}{6}$ and $\dfrac{5\pi}{6}$. See Figure 32. Since the period of the sine function is 2π and the argument is 2θ in the equation $\sin(2\theta) = \dfrac{1}{2}$, we write the general formula that gives all the solutions.

Figure 32

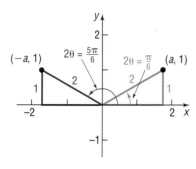

$$2\theta = \frac{\pi}{6} + 2k\pi \quad \text{or} \quad 2\theta = \frac{5\pi}{6} + 2k\pi \quad \text{\small{k any integer}}$$

$$\theta = \frac{\pi}{12} + k\pi \quad \text{or} \quad \theta = \frac{5\pi}{12} + k\pi \quad \text{\small{Divide by 2.}}$$

Then

$$\theta = \frac{\pi}{12} + (-1)\pi = \frac{-11\pi}{12} \quad \text{\small{$k = -1$}} \qquad \theta = \frac{5\pi}{12} + (-1)\pi = \frac{-7\pi}{12}$$

$$\theta = \frac{\pi}{12} + (0)\pi = \frac{\pi}{12} \quad \text{\small{$k = 0$}} \qquad \theta = \frac{5\pi}{12} + (0)\pi = \frac{5\pi}{12}$$

$$\theta = \frac{\pi}{12} + (1)\pi = \frac{13\pi}{12} \quad \text{\small{$k = 1$}} \qquad \theta = \frac{5\pi}{12} + (1)\pi = \frac{17\pi}{12}$$

$$\theta = \frac{\pi}{12} + (2)\pi = \frac{25\pi}{12} \quad \text{\small{$k = 2$}} \qquad \theta = \frac{5\pi}{12} + (2)\pi = \frac{29\pi}{12}$$

In the interval $[0, 2\pi)$, the solutions of $\sin(2\theta) = \dfrac{1}{2}$ are $\theta = \dfrac{\pi}{12}, \theta = \dfrac{5\pi}{12}, \theta = \dfrac{13\pi}{12}$, and $\theta = \dfrac{17\pi}{12}$. The solution set is $\left\{ \dfrac{\pi}{12}, \dfrac{5\pi}{12}, \dfrac{13\pi}{12}, \dfrac{17\pi}{12} \right\}$.

✓**Check:** Verify these solutions by graphing $Y_1 = \sin(2x)$ and $Y_2 = \dfrac{1}{2}$ for $0 \le x \le 2\pi$.

WARNING In solving a trigonometric equation for θ, $0 \le \theta < 2\pi$, in which the argument is not θ (as in Example 4), you must write down all the solutions first and then list those that are in the interval $[0, 2\pi)$. Otherwise, solutions may be lost. For example, in solving $\sin(2\theta) = \dfrac{1}{2}$, if you write the solutions $2\theta = \dfrac{\pi}{6}$ and $2\theta = \dfrac{5\pi}{6}$, you will find only $\theta = \dfrac{\pi}{12}$ and $\theta = \dfrac{5\pi}{12}$ and miss the other solutions. ■

━━━ **Now Work** PROBLEM 13

EXAMPLE 5 | **Solving a Trigonometric Equation**

Solve the equation: $\tan\left(\theta - \dfrac{\pi}{2} \right) = 1$, $\quad 0 \le \theta < 2\pi$

Solution The period of the tangent function is π. In the interval $[0, \pi)$, the tangent function has the value 1 when the argument is $\dfrac{\pi}{4}$. Because the argument is $\theta - \dfrac{\pi}{2}$ in the given equation, we write the general formula that gives all the solutions.

$$\theta - \frac{\pi}{2} = \frac{\pi}{4} + k\pi \quad \text{\small{k any integer}}$$

$$\theta = \frac{3\pi}{4} + k\pi$$

In the interval $[0, 2\pi)$, $\theta = \dfrac{3\pi}{4}$ and $\theta = \dfrac{3\pi}{4} + \pi = \dfrac{7\pi}{4}$ are the only solutions. The solution set is $\left\{ \dfrac{3\pi}{4}, \dfrac{7\pi}{4} \right\}$.

The next example illustrates how to solve trigonometric equations using a calculator.

EXAMPLE 6 | **Solving a Trigonometric Equation with a Calculator**

Use a calculator to solve the equation $\sin\theta = 0.3$, $0 \le \theta < 2\pi$. Express any solutions in radians, rounded to two decimal places.

Solution

Figure 33

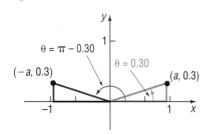

To solve $\sin\theta = 0.3$ on a calculator, first set the mode to radians. Then use the $\boxed{\sin^{-1}}$ key to obtain

$$\theta = \sin^{-1}(0.3) \approx 0.3046927$$

Rounded to two decimal places, $\theta = \sin^{-1}(0.3) = 0.30$ radian. Because of the definition of $y = \sin^{-1}x$, the angle θ that we obtain is the angle $-\dfrac{\pi}{2} \le \theta \le \dfrac{\pi}{2}$ for which $\sin\theta = 0.3$. Another angle for which $\sin\theta = 0.3$ is $\pi - 0.30$. See Figure 33. The angle $\pi - 0.30$ is the angle in quadrant II, where $\sin\theta = 0.3$. The solutions for $\sin\theta = 0.3$, $0 \le \theta < 2\pi$, are

$$\theta = 0.30 \text{ radian} \quad \text{and} \quad \theta = \pi - 0.30 \approx 2.84 \text{ radians}$$

The solution set is $\{0.30, 2.84\}$.

WARNING Example 6 illustrates that caution must be exercised when solving trigonometric equations on a calculator. Remember that the calculator supplies an angle only within the restrictions of the definition of the inverse trigonometric function. To find the remaining solutions, you must identify other quadrants, if any, in which a solution may be located. ∎

Now Work PROBLEM 41

3.7 Assess Your Understanding

'Are You Prepared?' *Answers are given at the end of these exercises. If you get a wrong answer, read the pages listed in red.*

1. Solve: $3x - 5 = -x + 1$ (pp. A27–A28)

2. $\sin\left(\dfrac{\pi}{4}\right) = $ _____; $\cos\left(\dfrac{8\pi}{3}\right) = $ _____ (pp. 129–130 and 146–147)

Concepts and Vocabulary

3. Two solutions of the equation $\sin\theta = \dfrac{1}{2}$ are _____ and _____.

4. All the solutions of the equation $\sin\theta = \dfrac{1}{2}$ are _____.

5. **True or False** Most trigonometric equations have unique solutions.

6. **True or False** The equation $\sin\theta = 2$ has a real solution that can be found using a calculator.

Skill Building

In Problems 7–30, solve each equation on the interval $0 \le \theta < 2\pi$.

7. $2\sin\theta + 3 = 2$

8. $1 - \cos\theta = \dfrac{1}{2}$

9. $4\cos^2\theta = 1$

10. $\tan^2\theta = \dfrac{1}{3}$

11. $2\sin^2\theta - 1 = 0$

12. $4\cos^2\theta - 3 = 0$

13. $\sin(3\theta) = -1$

14. $\tan\dfrac{\theta}{2} = \sqrt{3}$

15. $\cos(2\theta) = -\dfrac{1}{2}$

16. $\tan(2\theta) = -1$

17. $\sec\dfrac{3\theta}{2} = -2$

18. $\cot\dfrac{2\theta}{3} = -\sqrt{3}$

19. $2\sin\theta + 1 = 0$

20. $\cos\theta + 1 = 0$

21. $\tan\theta + 1 = 0$

22. $\sqrt{3}\cot\theta + 1 = 0$

23. $4\sec\theta + 6 = -2$

24. $5\csc\theta - 3 = 2$

25. $3\sqrt{2}\cos\theta + 2 = -1$

26. $4\sin\theta + 3\sqrt{3} = \sqrt{3}$

27. $\cos\left(2\theta - \dfrac{\pi}{2}\right) = -1$

28. $\sin\left(3\theta + \dfrac{\pi}{18}\right) = 1$

29. $\tan\left(\dfrac{\theta}{2} + \dfrac{\pi}{3}\right) = 1$

30. $\cos\left(\dfrac{\theta}{3} - \dfrac{\pi}{4}\right) = \dfrac{1}{2}$

In Problems 31–40, solve each equation. Give a general formula for all the solutions. List six solutions.

31. $\sin \theta = \dfrac{1}{2}$

32. $\tan \theta = 1$

33. $\tan \theta = -\dfrac{\sqrt{3}}{3}$

34. $\cos \theta = -\dfrac{\sqrt{3}}{2}$

35. $\cos \theta = 0$

36. $\sin \theta = \dfrac{\sqrt{2}}{2}$

37. $\cos(2\theta) = -\dfrac{1}{2}$

38. $\sin(2\theta) = -1$

39. $\sin \dfrac{\theta}{2} = -\dfrac{\sqrt{3}}{2}$

40. $\tan \dfrac{\theta}{2} = -1$

In Problems 41–52, use a calculator to solve each equation on the interval $0 \le \theta < 2\pi$. Round answers to two decimal places.

41. $\sin \theta = 0.4$

42. $\cos \theta = 0.6$

43. $\tan \theta = 5$

44. $\cot \theta = 2$

45. $\cos \theta = -0.9$

46. $\sin \theta = -0.2$

47. $\sec \theta = -4$

48. $\csc \theta = -3$

49. $5 \tan \theta + 9 = 0$

50. $4 \cot \theta = -5$

51. $3 \sin \theta - 2 = 0$

52. $4 \cos \theta + 3 = 0$

Mixed Practice

53. What are the zeros of $f(x) = 4 \sin^2 x - 3$ on the interval $[0, 2\pi]$?

54. What are the zeros of $f(x) = 2 \cos(3x) + 1$ on the interval $[0, \pi]$?

55. $f(x) = 3 \sin x$.
(a) Find the zeros of f on the interval $[-2\pi, 4\pi]$.
(b) Graph $f(x) = 3 \sin x$ on the interval $[-2\pi, 4\pi]$.
(c) Solve $f(x) = \dfrac{3}{2}$ on the interval $[-2\pi, 4\pi]$. What points are on the graph of f? Label these points on the graph drawn in part (b).
(d) Use the graph drawn in part (b) along with the results of part (c) to determine the values of x such that $f(x) > \dfrac{3}{2}$ on the interval $[-2\pi, 4\pi]$.

56. $f(x) = 2 \cos x$.
(a) Find the zeros of f on the interval $[-2\pi, 4\pi]$.
(b) Graph $f(x) = 2 \cos x$ on the interval $[-2\pi, 4\pi]$.
(c) Solve $f(x) = -\sqrt{3}$ on the interval $[-2\pi, 4\pi]$. What points are on the graph of f? Label these points on the graph drawn in part (b).
(d) Use the graph drawn in part (b) along with the results of part (c) to determine the values of x such that $f(x) < -\sqrt{3}$ on the interval $[-2\pi, 4\pi]$.

57. $f(x) = 4 \tan x$.
(a) Solve $f(x) = -4$.
(b) For what values of x is $f(x) < -4$ on the interval $\left(-\dfrac{\pi}{2}, \dfrac{\pi}{2}\right)$?

58. $f(x) = \cot x$.
(a) Solve $f(x) = -\sqrt{3}$.
(b) For what values of x is $f(x) > -\sqrt{3}$ on the interval $(0, \pi)$?

59. (a) Graph $f(x) = 3 \sin(2x) + 2$ and $g(x) = \dfrac{7}{2}$ on the same Cartesian plane for the interval $[0, \pi]$.
(b) Solve $f(x) = g(x)$ on the interval $[0, \pi]$ and label the points of intersection on the graph drawn in part (b).
(c) Solve $f(x) > g(x)$ on the interval $[0, \pi]$.
(d) Shade the region bounded by $f(x) = 3 \sin(2x) + 2$ and $g(x) = \dfrac{7}{2}$ between the two points found in part (b) on the graph drawn in part (a).

60. (a) Graph $f(x) = 2 \cos \dfrac{x}{2} + 3$ and $g(x) = 4$ on the same Cartesian plane for the interval $[0, 4\pi]$.
(b) Solve $f(x) = g(x)$ on the interval $[0, 4\pi]$ and label the points of intersection on the graph drawn in part (b).
(c) Solve $f(x) < g(x)$ on the interval $[0, 4\pi]$.
(d) Shade the region bounded by $f(x) = 2 \cos \dfrac{x}{2} + 3$ and $g(x) = 4$ between the two points found in part (b) on the graph drawn in part (a).

61. (a) Graph $f(x) = -4 \cos x$ and $g(x) = 2 \cos x + 3$ on the same Cartesian plane for the interval $[0, 2\pi]$.
(b) Solve $f(x) = g(x)$ on the interval $[0, 2\pi]$ and label the points of intersection on the graph drawn in part (b).
(c) Solve $f(x) > g(x)$ on the interval $[0, 2\pi]$.
(d) Shade the region bounded by $f(x) = -4 \cos x$ and $g(x) = 2 \cos x + 3$ between the two points found in part (b) on the graph drawn in part (a).

62. (a) Graph $f(x) = 2 \sin x$ and $g(x) = -2 \sin x + 2$ on the same Cartesian plane for the interval $[0, 2\pi]$.
(b) Solve $f(x) = g(x)$ on the interval $[0, 2\pi]$ and label the points of intersection on the graph drawn in part (b).
(c) Solve $f(x) > g(x)$ on the interval $[0, 2\pi]$.
(d) Shade the region bounded by $f(x) = 2 \sin x$ and $g(x) = -2 \sin x + 2$ between the two points found in part (b) on the graph drawn in part (a).

Applications and Extensions

63. Blood Pressure   Blood pressure is a way of measuring the amount of force exerted on the walls of blood vessels. It is measured using two numbers: systolic (as the heart beats) blood pressure and diastolic (as the heart rests) blood pressure. Blood pressures vary substantially from person to person, but a typical blood pressure is 120/80, which means the systolic blood pressure is 120 mm Hg and the diastolic blood pressure is 80 mmHg. Assuming that a person's heart beats 70 times per minute, the blood pressure of an individual P after t seconds can be modeled by the function

$$P(t) = 100 + 20 \sin\left(\dfrac{7\pi}{3}t\right)$$

(a) In the interval $[0, 1]$, determine the times at which the blood pressure is 100 mmHg.

(b) In the interval [0, 1], determine the times at which the blood pressure is 120 mmHg.

(c) In the interval [0, 1], determine the times at which the blood pressures is between 100 and 105 mmHg.

64. **The Ferris Wheel** In 1893, George Ferris engineered the Ferris Wheel. It was 250 feet in diameter. If the wheel makes 1 revolution every 40 seconds, then the function

$$h(t) = 125 \sin\left(0.157t - \frac{\pi}{2}\right) + 125$$

represents the height h, in feet, of a seat on the wheel as a function of time t, where t is measured in seconds. The ride begins when $t = 0$.

(a) During the first 40 seconds of the ride, at what time t is an individual on the Ferris Wheel exactly 125 feet above the ground?

(b) During the first 80 seconds of the ride, at what time t is an individual on the Ferris Wheel exactly 250 feet above the ground?

(c) During the first 40 seconds of the ride, over what interval of time t is an individual on the Ferris Wheel more than 125 feet above the ground?

65. **Holding Pattern** An airplane is asked to stay within a holding pattern near Chicago's O'Hare International Airport. The func-

tion $d(x) = 70 \sin(0.65x) + 150$ represents the distance d, in miles, that the airplane is from the airport at time x, in minutes.

(a) When the plane enters the holding pattern, $x = 0$, how far is it from O'Hare?

(b) During the first 20 minutes after the plane enters the holding pattern, at what time x is the plane exactly 100 miles from the airport?

(c) During the first 20 minutes after the plane enters the holding pattern, at what time x is the plane more than 100 miles from the airport?

(d) While the plane is in the holding pattern, will it ever be within 70 miles of the airport? Why?

66. **Projectile Motion** A golfer hits a golf ball with an initial velocity of 100 miles per hour. The range R of the ball as a function of the angle θ to the horizontal is given by $R(\theta) = 672 \sin(2\theta)$, where R is measured in feet.

(a) At what angle θ should the ball be hit if the golfer wants the ball to travel 450 feet (150 yards)?

(b) At what angle θ should the ball be hit if the golfer wants the ball to travel 540 feet (180 yards)?

(c) At what angle θ should the ball be hit if the golfer wants the ball to travel at least 480 feet (160 yards)?

(d) Can the golfer hit the ball 720 feet (240 yards)?

 The following discussion of Snell's Law of Refraction (named after Willebrord Snell, 1580–1626) is needed for Problems 67–73. Light, sound, and other waves travel at different speeds, depending on the media (air, water, wood, and so on) through which they pass. Suppose that light travels from a point A in one medium, where its speed is v_1, to a point B in another medium, where its speed is v_2. Refer to the figure, where the angle θ_1 is called the **angle of incidence** and the angle θ_2 is the **angle of refraction**. Snell's Law, which can be proved using calculus, states that*

$$\frac{\sin \theta_1}{\sin \theta_2} = \frac{v_1}{v_2}$$

*The ratio $\dfrac{v_1}{v_2}$ is called the **index of refraction**. Some values are given in the following table.*

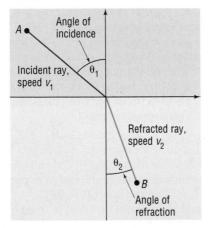

Some Indexes of Refraction	
Medium	**Index of Refraction**[†]
Water	1.33
Ethyl alcohol (20°C)	1.36
Carbon disulfide	1.63
Air (1 atm and 0°C)	1.00029
Diamond	2.42
Fused quartz	1.46
Glass, crown	1.52
Glass, dense flint	1.66
Sodium chloride	1.54

67. The index of refraction of light in passing from a vacuum into water is 1.33. If the angle of incidence is 40°, determine the angle of refraction.

68. The index of refraction of light in passing from a vacuum into dense flint glass is 1.66. If the angle of incidence is 50°, determine the angle of refraction.

69. Ptolemy, who lived in the city of Alexandria in Egypt during the second century AD, gave the measured values in the following table for the angle of incidence θ_1 and the angle of refraction θ_2 for a light beam passing from air into water. Do

these values agree with Snell's Law? If so, what index of refraction results? (These data are of interest as the oldest recorded physical measurements.)[†]

θ_1	θ_2	θ_1	θ_2
10°	8°	50°	35°0′
20°	15°30′	60°	40°30′
30°	22°30′	70°	45°30′
40°	29°0′	80°	50°0′

*Because this law was also deduced by René Descartes in France, it is also known as Descartes's Law.

[†]For light of wavelength 589 nanometers, measured with respect to a vacuum. The index with respect to air is negligibly different in most cases.

70. Bending Light The speed of yellow sodium light (wavelength, 589 nanometers) in a certain liquid is measured to be 1.92×10^8 meters per second. What is the index of refraction of this liquid, with respect to air, for sodium light?* [**Hint:** The speed of light in air is approximately 2.998×10^8 meters per second.]

71. Bending Light A beam of light with a wavelength of 589 nanometers traveling in air makes an angle of incidence of 40° on a slab of transparent material, and the refracted beam makes an angle of refraction of 26°. Find the index of refraction of the material.*

72. Bending Light A light ray with a wavelength of 589 nanometers (produced by a sodium lamp) traveling through air makes an angle of incidence of 30° on a smooth, flat slab of crown glass. Find the angle of refraction.*

73. A light beam passes through a thick slab of material whose index of refraction is n_2. Show that the emerging beam is parallel to the incident beam.*

74. Brewster's Law If the angle of incidence and the angle of refraction are complementary angles, the angle of incidence is referred to as the Brewster angle θ_B. The Brewster angle is related to the index of refractions of the two media, n_1 and n_2, by the equation $n_1 \sin \theta_B = n_2 \cos \theta_B$, where n_1 is the index of refraction of the incident medium and n_2 is the index of refraction of the refractive medium. Determine the Brewster angle for a light beam traveling through water (at 20°C) that makes an angle of incidence with a smooth, flat slab of crown glass.

*Adapted from Halliday and Resnick, *Fundamentals of Physics,* 7th ed., 2005, John Wiley & Sons.

Discussion and Writing

75. Explain in your own words how you would use your calculator to solve the equation $\cos x = -0.6, 0 \le x < 2\pi$. How would you modify your approach to solve the equation $\cot x = 5, 0 < x < 2\pi$?

'Are You Prepared?' Answers

1. $\left\{ \dfrac{3}{2} \right\}$ **2.** $\dfrac{\sqrt{2}}{2}; -\dfrac{1}{2}$

3.8 Trigonometric Equations (II)

PREPARING FOR THIS SECTION *Before getting started, review the following:*

- Solving Quadratic Equations by Factoring (Appendix, Section A.4, pp. A29–A30)
- The Quadratic Formula (Appendix, Section A.4, pp. A32–A35)

- Solving Equations Quadratic in Form (Appendix, Section A.4, pp. A35–A36)
- Using a Graphing Utility to Solve Equations (Appendix, Section A.5, pp. A39–A41)

Now Work the 'Are You Prepared?' problems on page 266.

OBJECTIVES **1** Solve Trigonometric Equations Quadratic in Form (p. 261)
2 Solve Trigonometric Equations Using Identities (p. 262)
3 Solve Trigonometric Equations Linear in Sine and Cosine (p. 264)
4 Solve Trigonometric Equations Using a Graphing Utility (p. 266)

1 Solve Trigonometric Equations Quadratic in Form

In this section we continue our study of trigonometric equations. Many trigonometric equations can be solved by applying techniques that we already know, such as applying the quadratic formula (if the equation is a second-degree polynomial) or factoring.

EXAMPLE 1 **Solving a Trigonometric Equation Quadratic in Form**

Solve the equation: $2 \sin^2 \theta - 3 \sin \theta + 1 = 0, \quad 0 \le \theta < 2\pi$

Solution This equation is a quadratic equation (in $\sin \theta$) that can be factored.

$$2 \sin^2 \theta - 3 \sin \theta + 1 = 0 \quad 2x^2 - 3x + 1 = 0, \quad x = \sin \theta$$
$$(2 \sin \theta - 1)(\sin \theta - 1) = 0 \quad (2x - 1)(x - 1) = 0$$

$$2 \sin \theta - 1 = 0 \quad \text{or} \quad \sin \theta - 1 = 0 \quad \text{Use the Zero-Product Property.}$$

$$\sin \theta = \frac{1}{2} \quad \text{or} \quad \sin \theta = 1$$

Solving each equation in the interval $[0, 2\pi)$, we obtain

$$\theta = \frac{\pi}{6} \qquad \theta = \frac{5\pi}{6} \qquad \theta = \frac{\pi}{2}$$

The solution set is $\left\{ \dfrac{\pi}{6}, \dfrac{5\pi}{6}, \dfrac{\pi}{2} \right\}$.

 **Now Work** PROBLEM 7

2 Solve Trigonometric Equations Using Identities

When a trigonometric equation contains more than one trigonometric function, identities sometimes can be used to obtain an equivalent equation that contains only one trigonometric function.

| EXAMPLE 2 | **Solving a Trigonometric Equation Using Identities** |

Solve the equation: $\quad 3 \cos \theta + 3 = 2 \sin^2 \theta, \quad 0 \le \theta < 2\pi$

Solution The equation in its present form contains a sine and a cosine. However, a form of the Pythagorean Identity can be used to transform the equation into an equivalent expression containing only cosines.

$$3 \cos \theta + 3 = 2 \sin^2 \theta$$
$$3 \cos \theta + 3 = 2(1 - \cos^2 \theta) \qquad \sin^2 \theta = 1 - \cos^2 \theta$$
$$3 \cos \theta + 3 = 2 - 2 \cos^2 \theta$$
$$2 \cos^2 \theta + 3 \cos \theta + 1 = 0 \qquad \text{Quadratic in } \cos \theta$$
$$(2 \cos \theta + 1)(\cos \theta + 1) = 0 \qquad \text{Factor.}$$
$$2 \cos \theta + 1 = 0 \quad \text{or} \quad \cos \theta + 1 = 0 \qquad \text{Use the Zero-Product Property.}$$
$$\cos \theta = -\frac{1}{2} \quad \text{or} \quad \cos \theta = -1$$

Solving each equation in the interval $[0, 2\pi)$, we obtain

$$\theta = \frac{2\pi}{3} \qquad \theta = \frac{4\pi}{3} \qquad \theta = \pi$$

The solution set is $\left\{ \dfrac{2\pi}{3}, \pi, \dfrac{4\pi}{3} \right\}$.

✓ Check: Graph $Y_1 = 3 \cos x + 3$ and $Y_2 = 2 \sin^2 x, 0 \le x \le 2\pi$, and approximate the points of intersection.

When a trigonometric equation contains trigonometric functions with different arguments, identities can sometimes be used to obtain an equivalent equation with the same argument.

| EXAMPLE 3 | **Solving a Trigonometric Equation Using Identities** |

Solve the equation: $\quad \cos(2\theta) + 3 = 5 \cos \theta, \quad 0 \le \theta < 2\pi$

Solution First, we observe that the given equation contains two cosine functions, but with different arguments, θ and 2θ. We use the Double-angle Formula $\cos(2\theta) = 2 \cos^2 \theta - 1$ to obtain an equivalent equation containing only $\cos \theta$.

$$\cos(2\theta) + 3 = 5 \cos \theta$$
$$(2 \cos^2 \theta - 1) + 3 = 5 \cos \theta \qquad \cos(2\theta) = 2 \cos^2 \theta - 1$$
$$2 \cos^2 \theta - 5 \cos \theta + 2 = 0 \qquad \text{Place in standard form.}$$

$$(\cos \theta - 2)(2 \cos \theta - 1) = 0 \quad \text{Factor.}$$

$$\cos \theta = 2 \quad \text{or} \quad \cos \theta = \frac{1}{2} \quad \text{Solve by using the Zero-Product property.}$$

For any angle θ, $-1 \leq \cos \theta \leq 1$; therefore, the equation $\cos \theta = 2$ has no solution. The solutions of $\cos \theta = \frac{1}{2}$, $0 \leq \theta < 2\pi$, are

$$\theta = \frac{\pi}{3} \qquad \theta = \frac{5\pi}{3}$$

The solution set is $\left\{ \dfrac{\pi}{3}, \dfrac{5\pi}{3} \right\}$.

✓ **Check:** Graph $Y_1 = \cos(2x) + 3$ and $Y_2 = 5 \cos x$, $0 \leq x \leq 2\pi$, and approximate the points of intersection.

━━━━━ **Now Work** PROBLEM 23

EXAMPLE 4	**Solving a Trigonometric Equation Using Identities**

Solve the equation: $\cos^2 \theta + \sin \theta = 2$, $0 \leq \theta < 2\pi$

Solution This equation involves two trigonometric functions, sine and cosine. We use a form of the Pythagorean Identity, $\sin^2 \theta + \cos^2 \theta = 1$, to rewrite the equation in terms of $\sin \theta$.

$$\cos^2 \theta + \sin \theta = 2$$
$$(1 - \sin^2 \theta) + \sin \theta = 2 \quad \cos^2 \theta = 1 - \sin^2 \theta$$
$$\sin^2 \theta - \sin \theta + 1 = 0$$

This is a quadratic equation in $\sin \theta$. The discriminant is $b^2 - 4ac = (-1)^2 - 4 \cdot 1 \cdot 1 = -3 < 0$. Therefore, the equation has no real solution. The solution set is the empty set, $\varnothing$.

✓ **Check:** Graph $Y_1 = \cos^2 x + \sin x$ and $Y_2 = 2$. See Figure 34. The two graphs never intersect, so the equation $Y_1 = Y_2$ has no real solution.

Figure 34

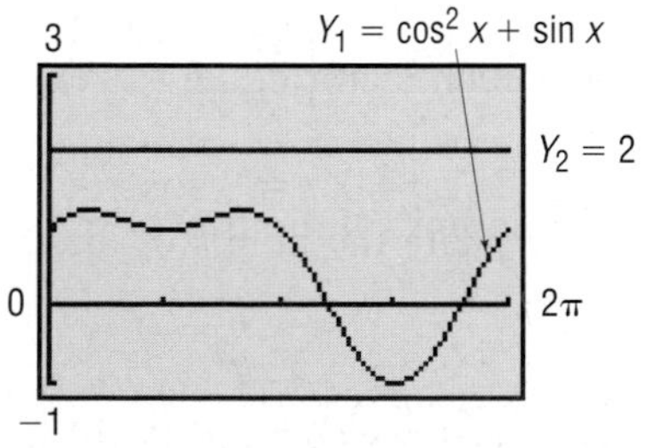

EXAMPLE 5	**Solving a Trigonometric Equation Using Identities**

Solve the equation: $\sin \theta \cos \theta = -\dfrac{1}{2}$, $0 \leq \theta < 2\pi$

Solution The left side of the given equation is in the form of the Double-angle Formula $2 \sin \theta \cos \theta = \sin(2\theta)$, except for a factor of 2. We multiply each side by 2.

$$\sin \theta \cos \theta = -\frac{1}{2}$$
$$2 \sin \theta \cos \theta = -1 \quad \text{Multiply each side by 2.}$$
$$\sin(2\theta) = -1 \quad \text{Double-angle Formula}$$

The argument here is 2θ. So we need to write all the solutions of this equation and then list those that are in the interval $[0, 2\pi)$. Because $\sin\left(\dfrac{3\pi}{2} + 2\pi k\right) = -1$ for any integer k, we have

$$2\theta = \frac{3\pi}{2} + 2k\pi \quad \text{\small\textit{k any integer}}$$

$$\theta = \frac{3\pi}{4} + k\pi$$

$$\theta = \frac{3\pi}{4} + (-1)\pi = -\frac{\pi}{4}, \quad \theta = \frac{3\pi}{4} + (0)\pi = \frac{3\pi}{4}, \quad \theta = \frac{3\pi}{4} + (1)\pi = \frac{7\pi}{4}, \quad \theta = \frac{3\pi}{4} + (2)\pi = \frac{11\pi}{4}$$

$$\uparrow \qquad\qquad\qquad\quad \uparrow \qquad\qquad\qquad \uparrow \qquad\qquad\qquad\quad \uparrow$$

$$\text{\small\textit{k = -1}} \qquad\qquad\quad \text{\small\textit{k = 0}} \qquad\qquad\quad \text{\small\textit{k = 1}} \qquad\qquad\quad \text{\small\textit{k = 2}}$$

The solutions in the interval $[0, 2\pi)$ are

$$\theta = \frac{3\pi}{4} \qquad \theta = \frac{7\pi}{4}$$

The solution set is $\left\{\dfrac{3\pi}{4}, \dfrac{7\pi}{4}\right\}$.

3 Solve Trigonometric Equations Linear in Sine and Cosine

Sometimes it is necessary to square both sides of an equation to obtain expressions that allow the use of identities. Remember, squaring both sides of an equation may introduce extraneous solutions. As a result, apparent solutions must be checked.

EXAMPLE 6 | **Solving a Trigonometric Equation Linear in Sine and Cosine**

Solve the equation: $\sin\theta + \cos\theta = 1, \quad 0 \le \theta < 2\pi$

Solution A Attempts to use available identities do not lead to equations that are easy to solve. (Try it yourself.) Given the form of this equation, we decide to square each side.

$$\sin\theta + \cos\theta = 1$$
$$(\sin\theta + \cos\theta)^2 = 1 \quad \text{\small Square each side.}$$
$$\sin^2\theta + 2\sin\theta\cos\theta + \cos^2\theta = 1 \quad \text{\small Remove parentheses.}$$
$$2\sin\theta\cos\theta = 0 \quad \text{\small $\sin^2\theta + \cos^2\theta = 1$}$$
$$\sin\theta\cos\theta = 0$$

Setting each factor equal to zero, we obtain

$$\sin\theta = 0 \quad \text{or} \quad \cos\theta = 0$$

The apparent solutions are

$$\theta = 0 \qquad \theta = \pi \qquad \theta = \frac{\pi}{2} \qquad \theta = \frac{3\pi}{2}$$

Because we squared both sides of the original equation, we must check these apparent solutions to see if any are extraneous.

$$\theta = 0: \quad \sin 0 + \cos 0 = 0 + 1 = 1 \qquad \text{\small A solution}$$
$$\theta = \pi: \quad \sin\pi + \cos\pi = 0 + (-1) = -1 \qquad \text{\small Not a solution}$$
$$\theta = \frac{\pi}{2}: \quad \sin\frac{\pi}{2} + \cos\frac{\pi}{2} = 1 + 0 = 1 \qquad \text{\small A solution}$$
$$\theta = \frac{3\pi}{2}: \quad \sin\frac{3\pi}{2} + \cos\frac{3\pi}{2} = -1 + 0 = -1 \qquad \text{\small Not a solution}$$

The values $\theta = \pi$ and $\theta = \dfrac{3\pi}{2}$ are extraneous. The solution set is $\left\{0, \dfrac{\pi}{2}\right\}$.

Solution B   We start with the equation

$$\sin \theta + \cos \theta = 1$$

and divide each side by $\sqrt{2}$. (The reason for this choice will become apparent shortly.) Then

$$\frac{1}{\sqrt{2}} \sin \theta + \frac{1}{\sqrt{2}} \cos \theta = \frac{1}{\sqrt{2}}$$

The left side now resembles the formula for the sine of the sum of two angles, one of which is θ. The other angle is unknown (call it ϕ.) Then

$$\sin(\theta + \phi) = \sin \theta \cos \phi + \cos \theta \sin \phi = \frac{1}{\sqrt{2}} = \frac{\sqrt{2}}{2} \qquad \textbf{(1)}$$

where

$$\cos \phi = \frac{1}{\sqrt{2}} = \frac{\sqrt{2}}{2}, \qquad \sin \phi = \frac{1}{\sqrt{2}} = \frac{\sqrt{2}}{2}, \qquad 0 \le \phi < 2\pi$$

The angle ϕ is therefore $\frac{\pi}{4}$. As a result, equation (1) becomes

$$\sin\left(\theta + \frac{\pi}{4}\right) = \frac{\sqrt{2}}{2}$$

Figure 35

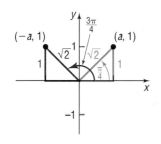

In the interval $[0, 2\pi)$, there are two angles whose sine is $\frac{\sqrt{2}}{2}$: $\frac{\pi}{4}$ and $\frac{3\pi}{4}$. See Figure 35. As a result,

$$\theta + \frac{\pi}{4} = \frac{\pi}{4} \quad \text{or} \quad \theta + \frac{\pi}{4} = \frac{3\pi}{4}$$

$$\theta = 0 \quad \text{or} \qquad \theta = \frac{\pi}{2}$$

The solution set is $\left\{0, \dfrac{\pi}{2}\right\}$.

This second method of solution can be used to solve any linear equation in the variables $\sin \theta$ and $\cos \theta$. Let's look at an example.

EXAMPLE 7   **Solving a Trigonometric Equation Linear in $\sin \theta$ and $\cos \theta$**

Solve:

$$a \sin \theta + b \cos \theta = c \qquad \textbf{(2)}$$

where a, b, and c are constants and either $a \ne 0$ or $b \ne 0$.

Solution   We divide each side of equation (2) by $\sqrt{a^2 + b^2}$. Then

$$\frac{a}{\sqrt{a^2 + b^2}} \sin \theta + \frac{b}{\sqrt{a^2 + b^2}} \cos \theta = \frac{c}{\sqrt{a^2 + b^2}} \qquad \textbf{(3)}$$

There is a unique angle ϕ, $0 \le \phi < 2\pi$, for which

Figure 36

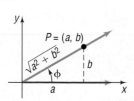

$$\cos \phi = \frac{a}{\sqrt{a^2 + b^2}} \quad \text{and} \quad \sin \phi = \frac{b}{\sqrt{a^2 + b^2}} \qquad \textbf{(4)}$$

Figure 36 shows the situation for $a > 0$ and $b > 0$. Equation (3) may be written as

$$\sin \theta \cos \phi + \cos \theta \sin \phi = \frac{c}{\sqrt{a^2 + b^2}}$$

or, equivalently,

$$\sin(\theta + \phi) = \frac{c}{\sqrt{a^2 + b^2}} \tag{5}$$

where ϕ satisfies the equations in (4).

If $|c| > \sqrt{a^2 + b^2}$, then $\sin(\theta + \phi) > 1$ or $\sin(\theta + \phi) < -1$, and equation (5) has no solution.

If $|c| \leq \sqrt{a^2 + b^2}$, then all the solutions of equation (5) are

$$\theta + \phi = \sin^{-1}\frac{c}{\sqrt{a^2 + b^2}} + 2k\pi \quad \text{or} \quad \theta + \phi = \pi - \sin^{-1}\frac{c}{\sqrt{a^2 + b^2}} + 2k\pi$$

Because the angle ϕ is determined by the equations in (4), these provide the solutions to equation (2).

Now Work PROBLEM 41

4 Solve Trigonometric Equations Using a Graphing Utility

The techniques introduced in this section apply only to certain types of trigonometric equations. Solutions for other types are usually studied in calculus, using numerical methods. In the next example, we show how a graphing utility may be used to obtain solutions.

EXAMPLE 8 | **Solving Trigonometric Equations Using a Graphing Utility**

Solve: $5 \sin x + x = 3$

Express the solution(s) rounded to two decimal places.

Solution This type of trigonometric equation cannot be solved by previous methods. A graphing utility, though, can be used here. The solution(s) of this equation is the same as the points of intersection of the graphs of $Y_1 = 5 \sin x + x$ and $Y_2 = 3$. See Figure 37.

There are three points of intersection; the x-coordinates are the solutions that we seek. Using INTERSECT, we find

$$x = 0.52 \qquad x = 3.18 \qquad x = 5.71$$

The solution set is $\{0.52, 3.18, 5.71\}$.

Figure 37

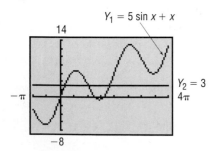

Now Work PROBLEM 47

3.8 Assess Your Understanding

'Are You Prepared?' *Answers are given at the end of these exercises. If you get a wrong answer, read the pages listed in red.*

1. Find the real solutions of $4x^2 - x - 5 = 0$. (pp. A29–A30)

2. Find the real solutions of $x^2 - x - 1 = 0$. (pp. A32–A35)

3. Find the real solutions of $(2x - 1)^2 - 3(2x - 1) - 4 = 0$. (pp. A35–A36)

4. Use a graphing utility to solve $5x^3 - 2 = x - x^2$. Round answers to two decimal places. (pp. A39–A41)

Sum and difference formulas (pp. 232, 235, and 237)

$$\cos(\alpha + \beta) = \cos\alpha\cos\beta - \sin\alpha\sin\beta \qquad \cos(\alpha - \beta) = \cos\alpha\cos\beta + \sin\alpha\sin\beta$$

$$\sin(\alpha + \beta) = \sin\alpha\cos\beta + \cos\alpha\sin\beta \qquad \sin(\alpha - \beta) = \sin\alpha\cos\beta - \cos\alpha\sin\beta$$

$$\tan(\alpha + \beta) = \frac{\tan\alpha + \tan\beta}{1 - \tan\alpha\tan\beta} \qquad \tan(\alpha - \beta) = \frac{\tan\alpha - \tan\beta}{1 + \tan\alpha\tan\beta}$$

Double-angle Formulas (pp. 242 and 243)

$$\sin(2\theta) = 2\sin\theta\cos\theta \qquad \cos(2\theta) = \cos^2\theta - \sin^2\theta \qquad \tan(2\theta) = \frac{2\tan\theta}{1 - \tan^2\theta}$$

$$\cos(2\theta) = 2\cos^2\theta - 1 \qquad \cos(2\theta) = 1 - 2\sin^2\theta$$

Half-angle Formulas (pp. 245, 246, and 248)

$$\sin^2\frac{\alpha}{2} = \frac{1 - \cos\alpha}{2} \qquad \cos^2\frac{\alpha}{2} = \frac{1 + \cos\alpha}{2} \qquad \tan^2\frac{\alpha}{2} = \frac{1 - \cos\alpha}{1 + \cos\alpha}$$

$$\sin\frac{\alpha}{2} = \pm\sqrt{\frac{1 - \cos\alpha}{2}} \qquad \cos\frac{\alpha}{2} = \pm\sqrt{\frac{1 + \cos\alpha}{2}} \qquad \tan\frac{\alpha}{2} = \pm\sqrt{\frac{1 - \cos\alpha}{1 + \cos\alpha}} = \frac{1 - \cos\alpha}{\sin\alpha} = \frac{\sin\alpha}{1 + \cos\alpha}$$

where the $+$ or $-$ is determined by the quadrant of $\dfrac{\alpha}{2}$.

Product-to-Sum Formulas (p. 251)

$$\sin\alpha\sin\beta = \frac{1}{2}[\cos(\alpha - \beta) - \cos(\alpha + \beta)]$$

$$\cos\alpha\cos\beta = \frac{1}{2}[\cos(\alpha - \beta) + \cos(\alpha + \beta)]$$

$$\sin\alpha\cos\beta = \frac{1}{2}[\sin(\alpha + \beta) + \sin(\alpha - \beta)]$$

Sum-to-Product Formulas (p. 252)

$$\sin\alpha + \sin\beta = 2\sin\frac{\alpha + \beta}{2}\cos\frac{\alpha - \beta}{2} \qquad \sin\alpha - \sin\beta = 2\sin\frac{\alpha - \beta}{2}\cos\frac{\alpha + \beta}{2}$$

$$\cos\alpha + \cos\beta = 2\cos\frac{\alpha + \beta}{2}\cos\frac{\alpha - \beta}{2} \qquad \cos\alpha - \cos\beta = -2\sin\frac{\alpha + \beta}{2}\sin\frac{\alpha - \beta}{2}$$

Objectives

Section		You should be able to ...	Example(s)	Review Exercises
3.1	1	Find the exact value of an inverse sine function (p. 207)	1, 2, 6, 7, 9	1–6
	2	Find an approximate value of an inverse sine function (p. 209)	3	121
	3	Use properties of inverse functions to find exact values of certain composite functions (p. 209)	4, 5, 8	9–20
	4	Find the inverse function of a trigonometric function (p. 215)	10	33–36
	5	Solve equations involving inverse trigonometric functions (p. 216)	11	133, 134
3.2	1	Find the exact value of expressions involving the inverse sine, cosine, and tangent functions (p. 220)	1–3	21–32
	2	Know the definitions of the inverse secant, cosecant, and cotangent functions (p. 221)	4	7, 8, 29, 30
	3	Use a calculator to evaluate $\sec^{-1}x$, $\csc^{-1}x$, and $\cot^{-1}x$ (p. 221)	5	125, 126
	4	Write a trigonometric expression as an algebraic expression (p. 222)	6	37–40
3.3	1	Use algebra to simplify trigonometric expressions (p. 226)	1	41–72
	2	Establish identities (p. 227)	2–8	41–58
3.4	1	Use sum and difference formulas to find exact values (p. 233)	1, 2	73–80, 81–90(a)–(d), 135
	2	Use sum and difference formulas to establish identities (p. 234)	3–8	59–62
	3	Use sum and difference formulas involving inverse trigonometric functions (p. 238)	9, 10	91–94

3.5	**1** Use double-angle formulas to find exact values (p. 242)	1	81–90(e), (f), 95, 96
	2 Use double-angle formulas to establish identities (p. 243)	2–4	65–67
	3 Use half-angle formulas to find exact values (p. 245)	5, 6	81–90(g), (h), 135
3.6	**1** Express products as sums (p. 251)	1	68
	2 Express sums as products (p. 252)	2	69–72
3.7	**1** Solve equations involving a single trigonometric function (p. 255)	1–6	97–106
3.8	**1** Solve trigonometric equations quadratic in form (p. 261)	1	113, 114
	2 Solve trigonometric equations using identities (p. 262)	2–5	97–112, 115–118
	3 Solve trigonometric equations linear in sine and cosine (p. 264)	6, 7	119, 120
	4 Solve trigonometric equations using a graphing utility (p. 266)	8	127–132

Review Exercises

In Problems 1–8, find the exact value of each expression. Do not use a calculator.

1. $\sin^{-1} 1$

2. $\cos^{-1} 0$

3. $\tan^{-1} 1$

4. $\sin^{-1}\left(-\dfrac{1}{2}\right)$

5. $\cos^{-1}\left(-\dfrac{\sqrt{3}}{2}\right)$

6. $\tan^{-1}\left(-\sqrt{3}\right)$

7. $\sec^{-1}\sqrt{2}$

8. $\cot^{-1}(-1)$

In Problems 9–32, find the exact value, if any, of each composite function. If there is no value, say it is "not defined." Do not use a calculator.

9. $\sin^{-1}\left(\sin\dfrac{3\pi}{8}\right)$

10. $\cos^{-1}\left(\cos\dfrac{3\pi}{4}\right)$

11. $\tan^{-1}\left(\tan\dfrac{2\pi}{3}\right)$

12. $\sin^{-1}\left[\sin\left(-\dfrac{\pi}{8}\right)\right]$

13. $\cos^{-1}\left(\cos\dfrac{15\pi}{7}\right)$

14. $\sin^{-1}\left[\sin\left(-\dfrac{8\pi}{9}\right)\right]$

15. $\sin(\sin^{-1} 0.9)$

16. $\cos(\cos^{-1} 0.6)$

17. $\cos[\cos^{-1}(-0.3)]$

18. $\tan[\tan^{-1} 5]$

19. $\cos[\cos^{-1}(-1.6)]$

20. $\sin(\sin^{-1} 1.6)$

21. $\sin^{-1}\left(\cos\dfrac{2\pi}{3}\right)$

22. $\cos^{-1}\left(\tan\dfrac{3\pi}{4}\right)$

23. $\tan^{-1}\left(\tan\dfrac{7\pi}{4}\right)$

24. $\cos^{-1}\left(\cos\dfrac{7\pi}{6}\right)$

25. $\tan\left[\sin^{-1}\left(-\dfrac{\sqrt{3}}{2}\right)\right]$

26. $\tan\left[\cos^{-1}\left(-\dfrac{1}{2}\right)\right]$

27. $\sec\left(\tan^{-1}\dfrac{\sqrt{3}}{3}\right)$

28. $\csc\left(\sin^{-1}\dfrac{\sqrt{3}}{2}\right)$

29. $\sin\left(\cot^{-1}\dfrac{3}{4}\right)$

30. $\cos\left(\csc^{-1}\dfrac{5}{3}\right)$

31. $\tan\left[\sin^{-1}\left(-\dfrac{4}{5}\right)\right]$

32. $\tan\left[\cos^{-1}\left(-\dfrac{3}{5}\right)\right]$

In Problems 33–36, find the inverse function f^{-1} of each function f. State the domain and the range of f^{-1} and f.

33. $f(x) = 2\sin(3x)$

34. $f(x) = \tan(2x + 3) - 1$

35. $f(x) = -\cos x + 3$

36. $f(x) = 2\sin(-x + 1)$

In Problems 37–40, write each trigonometric expression as an algebraic expression in u.

37. $\cos(\sin^{-1} u)$

38. $\cos(\csc^{-1} u)$

39. $\sin(\csc^{-1} u)$

40. $\tan(\csc^{-1} u)$

In Problems 41–72, establish each identity.

41. $\tan\theta\cot\theta - \sin^2\theta = \cos^2\theta$

42. $\sin\theta\csc\theta - \sin^2\theta = \cos^2\theta$

43. $\sin^2\theta(1 + \cot^2\theta) = 1$

44. $(1 - \sin^2\theta)(1 + \tan^2\theta) = 1$

45. $5\cos^2\theta + 3\sin^2\theta = 3 + 2\cos^2\theta$

46. $4\sin^2\theta + 2\cos^2\theta = 4 - 2\cos^2\theta$

47. $\dfrac{1 - \cos\theta}{\sin\theta} + \dfrac{\sin\theta}{1 - \cos\theta} = 2\csc\theta$

48. $\dfrac{\sin\theta}{1 + \cos\theta} + \dfrac{1 + \cos\theta}{\sin\theta} = 2\csc\theta$

49. $\dfrac{\cos\theta}{\cos\theta - \sin\theta} = \dfrac{1}{1 - \tan\theta}$

50. $1 - \dfrac{\sin^2\theta}{1 + \cos\theta} = \cos\theta$

51. $\dfrac{\csc\theta}{1 + \csc\theta} = \dfrac{1 - \sin\theta}{\cos^2\theta}$

52. $\dfrac{1 + \sec\theta}{\sec\theta} = \dfrac{\sin^2\theta}{1 - \cos\theta}$

53. $\csc\theta - \sin\theta = \cos\theta\cot\theta$

54. $\dfrac{\csc\theta}{1 - \cos\theta} = \dfrac{1 + \cos\theta}{\sin^3\theta}$

55. $\dfrac{1 - \sin\theta}{\sec\theta} = \dfrac{\cos^3\theta}{1 + \sin\theta}$

56. $\dfrac{1 - \cos\theta}{1 + \cos\theta} = (\csc\theta - \cot\theta)^2$

57. $\dfrac{1 - 2\sin^2\theta}{\sin\theta\cos\theta} = \cot\theta - \tan\theta$

58. $\dfrac{(2\sin^2\theta - 1)^2}{\sin^4\theta - \cos^4\theta} = 1 - 2\cos^2\theta$

59. $\dfrac{\cos(\alpha + \beta)}{\cos\alpha\sin\beta} = \cot\beta - \tan\alpha$

60. $\dfrac{\sin(\alpha - \beta)}{\sin\alpha\cos\beta} = 1 - \cot\alpha\tan\beta$

61. $\dfrac{\cos(\alpha - \beta)}{\cos\alpha\cos\beta} = 1 + \tan\alpha\tan\beta$

62. $\dfrac{\cos(\alpha + \beta)}{\sin\alpha\cos\beta} = \cot\alpha - \tan\beta$

63. $(1 + \cos\theta)\tan\dfrac{\theta}{2} = \sin\theta$

64. $\sin\theta\tan\dfrac{\theta}{2} = 1 - \cos\theta$

65. $2 \cot \theta \cot(2\theta) = \cot^2 \theta - 1$ **66.** $2 \sin(2\theta)(1 - 2 \sin^2 \theta) = \sin(4\theta)$ **67.** $1 - 8 \sin^2 \theta \cos^2 \theta = \cos(4\theta)$

68. $\dfrac{\sin(3\theta) \cos \theta - \sin \theta \cos(3\theta)}{\sin(2\theta)} = 1$ **69.** $\dfrac{\sin(2\theta) + \sin(4\theta)}{\cos(2\theta) + \cos(4\theta)} = \tan(3\theta)$ **70.** $\dfrac{\sin(2\theta) + \sin(4\theta)}{\sin(2\theta) - \sin(4\theta)} + \dfrac{\tan(3\theta)}{\tan \theta} = 0$

71. $\dfrac{\cos(2\theta) - \cos(4\theta)}{\cos(2\theta) + \cos(4\theta)} - \tan \theta \tan(3\theta) = 0$ **72.** $\cos(2\theta) - \cos(10\theta) = \tan(4\theta)[\sin(2\theta) + \sin(10\theta)]$

In Problems 73–80, find the exact value of each expression.

73. $\sin 165°$ **74.** $\tan 105°$ **75.** $\cos \dfrac{5\pi}{12}$ **76.** $\sin\left(-\dfrac{\pi}{12}\right)$

77. $\cos 80° \cos 20° + \sin 80° \sin 20°$ **78.** $\sin 70° \cos 40° - \cos 70° \sin 40°$

79. $\tan \dfrac{\pi}{8}$ **80.** $\sin \dfrac{5\pi}{8}$

In Problems 81–90, use the information given about the angles α and β to find the exact value of:

(a) $\sin(\alpha + \beta)$ (b) $\cos(\alpha + \beta)$ (c) $\sin(\alpha - \beta)$ (d) $\tan(\alpha + \beta)$

(e) $\sin(2\alpha)$ (f) $\cos(2\beta)$ (g) $\sin \dfrac{\beta}{2}$ (h) $\cos \dfrac{\alpha}{2}$

81. $\sin \alpha = \dfrac{4}{5}, 0 < \alpha < \dfrac{\pi}{2}; \sin \beta = \dfrac{5}{13}, \dfrac{\pi}{2} < \beta < \pi$ **82.** $\cos \alpha = \dfrac{4}{5}, 0 < \alpha < \dfrac{\pi}{2}; \cos \beta = \dfrac{5}{13}, -\dfrac{\pi}{2} < \beta < 0$

83. $\sin \alpha = -\dfrac{3}{5}, \pi < \alpha < \dfrac{3\pi}{2}; \cos \beta = \dfrac{12}{13}, \dfrac{3\pi}{2} < \beta < 2\pi$ **84.** $\sin \alpha = -\dfrac{4}{5}, -\dfrac{\pi}{2} < \alpha < 0; \cos \beta = -\dfrac{5}{13}, \dfrac{\pi}{2} < \beta < \pi$

85. $\tan \alpha = \dfrac{3}{4}, \pi < \alpha < \dfrac{3\pi}{2}; \tan \beta = \dfrac{12}{5}, 0 < \beta < \dfrac{\pi}{2}$ **86.** $\tan \alpha = -\dfrac{4}{3}, \dfrac{\pi}{2} < \alpha < \pi; \cot \beta = \dfrac{12}{5}, \pi < \beta < \dfrac{3\pi}{2}$

87. $\sec \alpha = 2, -\dfrac{\pi}{2} < \alpha < 0; \sec \beta = 3, \dfrac{3\pi}{2} < \beta < 2\pi$ **88.** $\csc \alpha = 2, \dfrac{\pi}{2} < \alpha < \pi; \sec \beta = -3, \dfrac{\pi}{2} < \beta < \pi$

89. $\sin \alpha = -\dfrac{2}{3}, \pi < \alpha < \dfrac{3\pi}{2}; \cos \beta = -\dfrac{2}{3}, \pi < \beta < \dfrac{3\pi}{2}$ **90.** $\tan \alpha = -2, \dfrac{\pi}{2} < \alpha < \pi; \cot \beta = -2, \dfrac{\pi}{2} < \beta < \pi$

In Problems 91–96, find the exact value of each expression.

91. $\cos\left(\sin^{-1} \dfrac{3}{5} - \cos^{-1} \dfrac{1}{2}\right)$ **92.** $\sin\left(\cos^{-1} \dfrac{5}{13} - \cos^{-1} \dfrac{4}{5}\right)$ **93.** $\tan\left[\sin^{-1}\left(-\dfrac{1}{2}\right) - \tan^{-1} \dfrac{3}{4}\right]$

94. $\cos\left[\tan^{-1}(-1) + \cos^{-1}\left(-\dfrac{4}{5}\right)\right]$ **95.** $\sin\left[2 \cos^{-1}\left(-\dfrac{3}{5}\right)\right]$ **96.** $\cos\left(2 \tan^{-1} \dfrac{4}{3}\right)$

In Problems 97–120, solve each equation on the interval $0 \le \theta < 2\pi$. Verify your results using a graphing utility.

97. $\cos \theta = \dfrac{1}{2}$ **98.** $\sin \theta = -\dfrac{\sqrt{3}}{2}$ **99.** $2 \cos \theta + \sqrt{2} = 0$

100. $\tan \theta + \sqrt{3} = 0$ **101.** $\sin(2\theta) + 1 = 0$ **102.** $\cos(2\theta) = 0$

103. $\tan(2\theta) = 0$ **104.** $\sin(3\theta) = 1$ **105.** $\sec^2 \theta = 4$

106. $\csc^2 \theta = 1$ **107.** $\sin \theta = \tan \theta$ **108.** $\cos \theta = \sec \theta$

109. $\sin \theta + \sin(2\theta) = 0$ **110.** $\cos(2\theta) = \sin \theta$ **111.** $\sin(2\theta) - \cos \theta - 2 \sin \theta + 1 = 0$

112. $\sin(2\theta) - \sin \theta - 2 \cos \theta + 1 = 0$ **113.** $2 \sin^2 \theta - 3 \sin \theta + 1 = 0$ **114.** $2 \cos^2 \theta + \cos \theta - 1 = 0$

115. $4 \sin^2 \theta = 1 + 4 \cos \theta$ **116.** $8 - 12 \sin^2 \theta = 4 \cos^2 \theta$ **117.** $\sin(2\theta) = \sqrt{2} \cos \theta$

118. $1 + \sqrt{3} \cos \theta + \cos(2\theta) = 0$ **119.** $\sin \theta - \cos \theta = 1$ **120.** $\sin \theta - \sqrt{3} \cos \theta = 2$

In Problems 121–126, use a calculator to find an approximate value for each expression, rounded to two decimal places.

121. $\sin^{-1} 0.7$ **122.** $\cos^{-1} \dfrac{4}{5}$ **123.** $\tan^{-1}(-2)$

124. $\cos^{-1}(-0.2)$ **125.** $\sec^{-1} 3$ **126.** $\cot^{-1}(-4)$

In Problems 127–132, use a graphing utility to solve each equation on the interval $0 \le x < 2\pi$. Approximate any solutions rounded to two decimal places.

127. $2x = 5 \cos x$ **128.** $2x = 5 \sin x$ **129.** $2 \sin x + 3 \cos x = 4x$

130. $3 \cos x + x = \sin x$ **131.** $\sin x = \ln x$ **132.** $\sin x = e^{-x}$

In Problems 133 and 134, find the exact solution of each equation.

133. $-3 \sin^{-1} x = \pi$ **134.** $2 \cos^{-1} x + \pi = 4 \cos^{-1} x$

135. Use a Half-angle Formula to find the exact value of $\sin 15°$. Then use a difference formula to find the exact value of $\sin 15°$. Show that the answers found are the same.

136. If you are given the value of $\cos \theta$ and want the exact value of $\cos(2\theta)$, what form of the Double-angle Formula for $\cos(2\theta)$ is most efficient to use?

CHAPTER TEST

In Problems 1–6, find the exact value of each expression. Express angles in radians.

1. $\sec^{-1}\left(\dfrac{2}{\sqrt{3}}\right)$

2. $\sin^{-1}\left(-\dfrac{\sqrt{2}}{2}\right)$

3. $\sin^{-1}\left(\sin\dfrac{11\pi}{5}\right)$

4. $\tan\left(\tan^{-1}\dfrac{7}{3}\right)$

5. $\cot\left(\csc^{-1}\sqrt{10}\right)$

6. $\sec\left(\cos^{-1}\left(-\dfrac{3}{4}\right)\right)$

In Problems 7–10, use a calculator to evaluate each expression. Express angles in radians rounded to two decimal places.

7. $\sin^{-1}0.382$

8. $\sec^{-1}1.4$

9. $\tan^{-1}3$

10. $\cot^{-1}5$

In Problems 11–16 establish each identity.

11. $\dfrac{\csc\theta + \cot\theta}{\sec\theta + \tan\theta} = \dfrac{\sec\theta - \tan\theta}{\csc\theta - \cot\theta}$

12. $\sin\theta\tan\theta + \cos\theta = \sec\theta$

13. $\tan\theta + \cot\theta = 2\csc(2\theta)$

14. $\dfrac{\sin(\alpha + \beta)}{\tan\alpha + \tan\beta} = \cos\alpha\cos\beta$

15. $\sin(3\theta) = 3\sin\theta - 4\sin^3\theta$

16. $\dfrac{\tan\theta - \cot\theta}{\tan\theta + \cot\theta} = 1 - 2\cos^2\theta$

In Problems 17–24 use sum, difference, product, or half-angle formulas to find the exact value of each expression.

17. $\cos 15°$

18. $\tan 75°$

19. $\sin\left(\dfrac{1}{2}\cos^{-1}\dfrac{3}{5}\right)$

20. $\tan\left(2\sin^{-1}\dfrac{6}{11}\right)$

21. $\cos\left(\sin^{-1}\dfrac{2}{3} + \tan^{-1}\dfrac{3}{2}\right)$

22. $\sin 75°\cos 15°$

23. $\sin 75° + \sin 15°$

24. $\cos 65°\cos 20° + \sin 65°\sin 20°$

In Problems 25–29, solve each equation on $0 \le \theta < 2\pi$.

25. $4\sin^2\theta - 3 = 0$

26. $-3\cos\left(\dfrac{\pi}{2} - \theta\right) = \tan\theta$

27. $\cos^2\theta + 2\sin\theta\cos\theta - \sin^2\theta = 0$

28. $\sin(\theta + 1) = \cos\theta$

29. $4\sin^2\theta + 7\sin\theta = 2$

30. Stage 16 of the 2004 Tour de France was a time trial from Bourg d'Oisans to L'Alpe d'Huez. The average grade (slope as a percent) for most of the 15-kilometer mountainous trek was 7.9%. What was the change in elevation from the beginning to the end of the route?

CUMULATIVE REVIEW

1. Find the real solutions, if any, of the equation $3x^2 + x - 1 = 0$.

2. Find an equation for the line containing the points $(-2, 5)$ and $(4, -1)$. What is the distance between these points? What is their midpoint?

3. Test the equation $3x + y^2 = 9$ for symmetry with respect to the x-axis, y-axis, and origin. List the intercepts.

4. Use transformations to graph the equation $y = |x - 3| + 2$.

5. Use transformations to graph the equation
$$y = \cos\left(x - \dfrac{\pi}{2}\right) - 1.$$

6. Sketch a graph of each of the following functions. Label at least three points on each graph. Name the inverse function of each and show its graph.
   (a) $y = x^3$
   (b) $y = \sin x, \quad -\dfrac{\pi}{2} \le x \le \dfrac{\pi}{2}$
   (c) $y = \cos x, \quad 0 \le x \le \pi$

7. If $\sin\theta = -\dfrac{1}{3}$ and $\pi < \theta < \dfrac{3\pi}{2}$, find the exact value of:
   (a) $\cos\theta$ (b) $\tan\theta$ (c) $\sin(2\theta)$
   (d) $\cos(2\theta)$ (e) $\sin\left(\dfrac{1}{2}\theta\right)$ (f) $\cos\left(\dfrac{1}{2}\theta\right)$

8. Find the exact value of $\cos(\tan^{-1}2)$.

9. If $\sin\alpha = \dfrac{1}{3}, \dfrac{\pi}{2} < \alpha < \pi$, and $\cos\beta = -\dfrac{1}{3}, \pi < \beta < \dfrac{3\pi}{2}$, find the exact value of:
   (a) $\cos\alpha$ (b) $\sin\beta$ (c) $\cos(2\alpha)$
   (d) $\cos(\alpha + \beta)$ (e) $\sin\dfrac{\beta}{2}$

CHAPTER PROJECTS

I. Waves A stretched string that is attached at both ends, pulled in a direction perpendicular to the string, and released has motion that is described as wave motion. If we assume no friction and a length such that there are no "echoes" (that is, the wave does not bounce back), the transverse motion (motion perpendicular to the string) can be modeled by the equation

$$y = y_m \sin(kx - \omega t)$$

where y_m is the amplitude measured in meters and k and ω are constants. The height of the wave depends on the distance x from one endpoint of the string and on the time t, so a typical wave has horizontal and vertical motion over time.

(a) What is the amplitude of the wave

$$y = 0.00421 \sin(68.3x - 2.68t)?$$

(b) The value of ω is the angular frequency measured in radians per second. What is the angular frequency of the wave given in part (a)?

(c) The frequency f is the number of vibrations per second (hertz) made by the wave as it passes a certain point. Its value is found using the formula $f = \dfrac{\omega}{2\pi}$. What is the frequency of the wave given in part (a)?

(d) The wavelength, λ, of a wave is the shortest distance at which the wave pattern repeats itself for a constant t. Thus, $\lambda = \dfrac{2\pi}{k}$. What is the wavelength of the wave given in part (a)?

(e) Graph the height of the string a distance $x = 1$ meter from an endpoint.

(f) If two waves travel simultaneously along the same stretched string, the vertical displacement of the string when both waves act is $y = y_1 + y_2$, where y_1 is the vertical displacement of the first wave and y_2 is the vertical displacement of the second wave. This result is called the Principle of Superposition and was analyzed by the French mathematician Jean Baptiste Fourier (1768–1830). When two waves travel along the same string, one wave will differ from the other wave by a phase constant ϕ. That is,

$$y_1 = y_m \sin(kx - \omega t)$$
$$y_2 = y_m \sin(kx - \omega t + \phi)$$

assuming that each wave has the same amplitude. Write $y_1 + y_2$ as a product using the Sum-to-Product Formulas.

(g) Suppose that two waves are moving in the same direction along a stretched string. The amplitude of each wave is 0.0045 meter, and the phase difference between them is 2.5 radians. The wavelength, λ, of each wave is 0.09 meter, and the frequency, f, is 2.3 hertz. Find y_1, y_2, and $y_1 + y_2$.

(h) Using a graphing utility, graph y_1, y_2, and $y_1 + y_2$ on the same viewing window.

(i) Redo parts (g) and (h) with the phase difference between the waves being 0.4 radian.

(j) What effect does the phase difference have on the amplitude of $y_1 + y_2$?

The following projects are available on the Instructor's Resource Center (IRC):

II. Project at Motorola *Sending Pictures Wirelessly* The electronic transmission of pictures is made practical by image compression, mathematical methods that greatly reduce the number of bits of data used to compose the picture.

III. Calculus of Differences Finding consecutive difference quotients is called finding finite differences and is used to analyze the graph of an unknown function.

Applications of Trigonometric Functions

4

From Lewis and Clark to Landsat

For $140, you can buy a handheld Global Positioning System receiver that will gauge your latitude and longitude to within a couple of meters. But in 1804, when Meriwether Lewis and William Clark ventured across the Louisiana Territory, a state of the art positioning system consisted of an octant, a pocket chronometer, and a surveyor's compass.

But somehow, Clark—the cartographer in the group—made do. When San Francisco map collector David Rumsey took his copy of Lewis and Clark's published map of their journey, scanned it into a computer, and matched landmarks such as river junctions against corresponding features on today's maps, he found that it took only a slight amount of digital stretching and twisting to make Clark's map conform to modern coordinates. In fact, Rumsey was able to combine Clark's depiction of his party's route to the Pacific with pages from government atlases from the 1870s and 1970s and photos from NASA Landsat satellites, creating a digital composite that documents not only a historic adventure, but also the history of mapmaking itself.

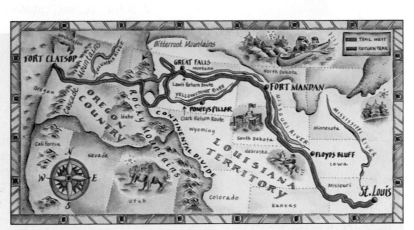

Source: Wade Roush, "From Lewis and Clark to Landsat: David Rumsey's Digital Maps Marry Past and Present," *Technology Review*, 108, no. 7 (July 2005): 26–27.

—See Chapter Project I—

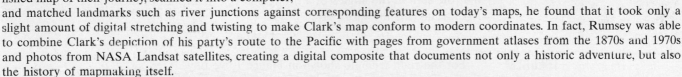

A Look Back

In Chapter 2 we defined the six trigonometric functions using right triangles and then extended this definition to include general angles. In particular, we learned to evaluate the trigonometric functions. We also learned how to graph sinusoidal functions. In Chapter 3, we defined the inverse trigonometric functions and solved equations involving the trigonometric functions.

A Look Ahead

In this chapter, we use the trigonometric functions to solve applied problems. The first four sections deal with applications involving right triangles and *oblique triangles*, triangles that do not have a right angle. To solve problems involving oblique triangles, we will develop the Law of Sines and the Law of Cosines. We will also develop formulas for finding the area of a triangle.

The final section deals with applications of sinusoidal functions involving simple harmonic motion and damped motion.

Outline

4.1 Applications Involving Right Triangles

PREPARING FOR THIS SECTION   *Before getting started, review the following:*

- Pythagorean Theorem (Appendix, Section A.2, pp. A14–A15)
- Trigonometric Equations (I) (Section 3.7, pp. 254–258)
- Complementary Angles (Section 2.2, pp. 123–125)

Now Work the 'Are You Prepared?' problems on page 279.

OBJECTIVES   **1** Solve Right Triangles (p. 276)
                     2 Solve Applied Problems (p. 277)

1 Solve Right Triangles

Figure 1

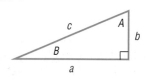

In the discussion that follows, we will always label a right triangle so that side a is opposite angle A, side b is opposite angle B, and side c is the hypotenuse, as shown in Figure 1. **To solve a right triangle** means to find the missing lengths of its sides and the measurements of its angles. We shall follow the practice of expressing the lengths of the sides rounded to two decimal places and expressing angles in degrees rounded to one decimal place. (Be sure that your calculator is in degree mode.)

To solve a right triangle, we need to know one of the acute angles A or B and a side, or else two sides. Then we make use of the Pythagorean Theorem and the fact that the sum of the angles of a triangle is $180°$. The sum of the angles A and B in a right triangle is therefore $90°$.

THEOREM

For the right triangle shown in Figure 1, we have

$$c^2 = a^2 + b^2 \qquad A + B = 90°$$

EXAMPLE 1   **Solving a Right Triangle**

Use Figure 2. If $b = 2$ and $A = 40°$, find a, c, and B.

Solution   Since $A = 40°$ and $A + B = 90°$, we find that $B = 50°$. To find the sides a and c, we use the facts that

Figure 2

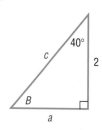

$$\tan 40° = \frac{a}{2} \quad \text{and} \quad \cos 40° = \frac{2}{c}$$

Now solve for a and c.

$$a = 2 \tan 40° \approx 1.68 \quad \text{and} \quad c = \frac{2}{\cos 40°} \approx 2.61$$

Now Work PROBLEM 9

EXAMPLE 2   **Solving a Right Triangle**

Figure 3

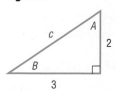

Use Figure 3. If $a = 3$ and $b = 2$, find c, A, and B.

Solution   Since $a = 3$ and $b = 2$, then, by the Pythagorean Theorem, we have

$$c^2 = a^2 + b^2 = 3^2 + 2^2 = 9 + 4 = 13$$

$$c = \sqrt{13} \approx 3.61$$

To find angle A, we use the fact that

$$\tan A = \frac{3}{2} \quad \text{so} \quad A = \tan^{-1}\frac{3}{2}$$

Set the mode on your calculator to degrees. Then, rounded to one decimal place, we find that $A = 56.3°$. Since $A + B = 90°$, we find that $B = 33.7°$.

◾━━━━━ **Now Work** PROBLEM 19

2 Solve Applied Problems

In Section 2.3, we used right triangle trigonometry to find the lengths of unknown sides of a right triangle given the measure of an angle and the length of a side. Now that we understand the concept of inverse trigonometric functions and know how to solve trigonometric equations, we can solve applied problems that require finding the measure of an angle given the lengths of two sides in a right triangle.

EXAMPLE 3	**Finding the Inclination of a Mountain Trail**

A straight trail leads from the Alpine Hotel, elevation 8000 feet, to a scenic overlook, elevation 11,100 feet. The length of the trail is 14,100 feet. What is the inclination (grade) of the trail? That is, what is the angle B in Figure 4?

Solution   As we can see in Figure 4, we know the length of the side opposite angle B and the length of the hypotenuse. The angle B obeys the equation

Figure 4

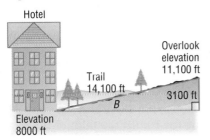

$$\sin B = \frac{3100}{14,100}$$

Using a calculator,

$$B = \sin^{-1}\frac{3100}{14,100} \approx 12.7°$$

The inclination (grade) of the trail is approximately $12.7°$.

◾━━━━━ **Now Work** PROBLEM 25

EXAMPLE 4	**The Gibb's Hill Lighthouse, Southampton, Bermuda**

In operation since 1846, the Gibb's Hill Lighthouse stands 117 feet high on a hill 245 feet high, so its beam of light is 362 feet above sea level. A brochure states that the light can be seen on the horizon about 26 miles distant. Verify the accuracy of this statement.

Solution   Figure 5 illustrates the situation. The central angle θ, positioned at the center of Earth, radius 3960 miles, obeys the equation

Figure 5

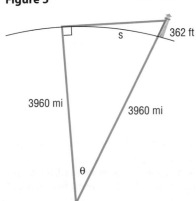

$$\cos \theta = \frac{3960}{3960 + \dfrac{362}{5280}} \approx 0.999982687 \quad \text{1 mile = 5280 feet}$$

Solving for θ, we find

$$\theta = \cos^{-1}0.999982687 \approx 0.33715° \approx 20.23'$$

The brochure does not indicate whether the distance is measured in nautical miles or statute miles. Let's calculate both distances.

The distance s in nautical miles (refer to Problem 114, p. 116) is the measure of the angle θ in minutes, so $s \approx 20.23$ nautical miles.

The distance s in statute miles is given by the formula $s = r\theta$, where θ is measured in radians. Then, since

$$\theta \approx 20.23' \approx 0.33715° \approx 0.00588 \text{ radian}$$

$$\uparrow \qquad\qquad\qquad \uparrow$$
$$1' = \frac{1°}{60} \qquad 1° = \frac{\pi}{180} \text{ radian}$$

we find that

$$s = r\theta \approx (3960)(0.00588) \approx 23.3 \text{ miles}$$

In either case, it would seem that the brochure overstated the distance somewhat.

Figure 6

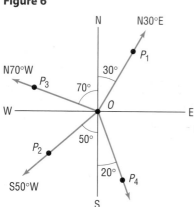

In navigation and surveying, the **direction** or **bearing** from a point O to a point P equals the acute angle θ between the ray OP and the vertical line through O, the north–south line.

Figure 6 illustrates some bearings. Notice that the bearing from O to P_1 is denoted by the symbolism N30°E, indicating that the bearing is 30° east of north. In writing the bearing from O to P, the direction north or south always appears first, followed by an acute angle, followed by east or west. In Figure 6, the bearing from O to P_2 is S50°W, and from O to P_3 it is N70°W.

EXAMPLE 5   Finding the Bearing of an Object

In Figure 6, what is the bearing from O to an object at P_4?

Solution   The acute angle between the ray OP_4 and the north–south line through O is given as 20°. The bearing from O to P_4 is S20°E.

EXAMPLE 6   Finding the Bearing of an Airplane

A Boeing 777 aircraft takes off from Nashville International Airport on runway 2 LEFT, which has a bearing of N20°E.* After flying for 1 mile, the pilot of the aircraft requests permission to turn 90° and head toward the northwest. The request is granted. After the plane goes 2 miles in this direction, what bearing should the control tower use to locate the aircraft?

Solution   Figure 7 illustrates the situation. After flying 1 mile from the airport O (the control tower), the aircraft is at P. After turning 90° toward the northwest and flying 2 miles, the aircraft is at the point Q. In triangle OPQ, the angle θ obeys the equation

Figure 7

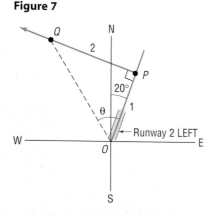

$$\tan\theta = \frac{2}{1} = 2 \quad \text{so} \quad \theta = \tan^{-1} 2 \approx 63.4°$$

The acute angle between north and the ray OQ is $63.4° - 20° = 43.4°$. The bearing of the aircraft from O to Q is N43.4°W.

➡ **Now Work** PROBLEM 31

---

*In air navigation, the term **azimuth** denotes the positive angle measured clockwise from the north (N) to a ray OP. In Figure 6, the azimuth from O to P_1 is 30°; the azimuth from O to P_2 is 230°; the azimuth from O to P_3 is 290°. In naming runways, the units digit is left off the azimuth. Runway 2 LEFT means the left runway with a direction of azimuth 20° (bearing N20°E). Runway 23 is the runway with azimuth 230° and bearing S50°W.

4.1 Assess Your Understanding

'Are You Prepared?' *Answers are given at the end of these exercises. If you get a wrong answer, read the pages listed in red.*

1. In a right triangle, if the length of the hypotenuse is 5 and the length of one of the other sides is 3, what is the length of the third side? (p. A14)

2. **True or False** $\sin 52° = \cos 48°$. (pp. 123–124)

3. If θ is an acute angle, solve the equation $\tan \theta = \frac{1}{2}$. Express your answer in degrees, rounded to one decimal place. (pp. 255–256)

4. If θ is an acute angle, solve the equation $\sin \theta = \frac{1}{2}$. (pp. 255–256)

Concepts and Vocabulary

5. **True or False**   In a right triangle, one of the angles is 90° and the sum of the other two angles is 90°.

6. In navigation or surveying, the _____ from a point O to a point P equals the acute angle θ between ray OP and the vertical line through O, the north–south line.

7. **True or False**   In a right triangle, if two sides are known, we can solve the triangle.

8. **True or False**   In a right triangle, if we know the two acute angles, we can solve the triangle.

Skill Building

In Problems 9–22, use the right triangle shown below. Then, using the given information, solve the triangle.

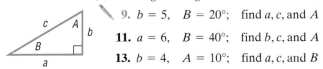

9. $b = 5$, $B = 20°$;   find a, c, and A

10. $b = 4$, $B = 10°$;   find a, c, and A

11. $a = 6$, $B = 40°$;   find b, c, and A

12. $a = 7$, $B = 50°$;   find b, c, and A

13. $b = 4$, $A = 10°$;   find a, c, and B

14. $b = 6$, $A = 20°$;   find a, c, and B

15. $a = 5$, $A = 25°$;   find b, c, and B

16. $a = 6$, $A = 40°$;   find b, c, and B

17. $c = 9$, $B = 20°$;   find b, a, and A

18. $c = 10$, $A = 40°$;   find b, a, and B

19. $a = 5$, $b = 3$;   find c, A, and B

20. $a = 2$, $b - 8$;   find c, A, and B

21. $a = 2$, $c = 5$;   find b, A, and B

22. $b = 4$, $c = 6$;   find a, A, and B

Applications and Extensions

23. **Geometry**   The hypotenuse of a right triangle is 5 inches. If one leg is 2 inches, find the degree measure of each angle.

24. **Geometry**   The hypotenuse of a right triangle is 3 feet. If one leg is 1 foot, find the degree measure of each angle.

25. **Finding the Angle of Elevation of the Sun**   At 10 AM on April 26, 2006, a building 300 feet high casts a shadow 50 feet long. What is the angle of elevation of the Sun?

26. **Directing a Laser Beam**   A laser beam is to be directed through a small hole in the center of a circle of radius 10 feet. The origin of the beam is 35 feet from the circle (see the figure). At what angle of elevation should the beam be aimed to ensure that it goes through the hole?

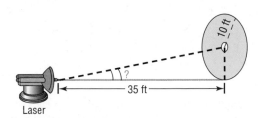

27. **Finding the Speed of a Truck**   A state trooper is hidden 30 feet from a highway. One second after a truck passes, the angle θ between the highway and the line of observation from the patrol car to the truck is measured. See the illustration.

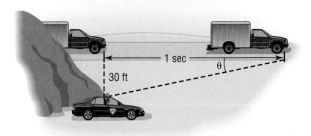

(a) If the angle measures 15°, how fast is the truck traveling? Express the answer in feet per second and in miles per hour.

(b) If the angle measures 20°, how fast is the truck traveling? Express the answer in feet per second and in miles per hour.

(c) If the speed limit is 55 miles per hour and a speeding ticket is issued for speeds of 5 miles per hour or more over the limit, for what angles should the trooper issue a ticket?

28. **Security**   A security camera in a neighborhood bank is mounted on a wall 9 feet above the floor. What angle of depression should be used if the camera is to be directed to a spot 6 feet above the floor and 12 feet from the wall?

29. Parallax One method of measuring the distance from Earth to a star is the parallax method. The idea behind computing this distance is to measure the angle formed between the Earth and the star at two different points in time. Typically, the measurements are taken so that the side opposite the angle is as large as possible. Therefore, the optimal approach is to measure the angle when Earth is on opposite sides of the Sun, as shown in the figure.

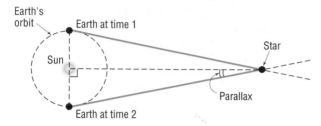

(a) Proxima Centauri is 4.22 light-years from Earth. If 1 light-year is about 5.9 trillion miles, how many miles is Proxima Centauri from Earth?

(b) The mean distance from Earth to the Sun is 93,000,000 miles. What is the parallax of Proxima Centauri?

30. Parallax See Problem 29. 61 Cygni, sometimes called Bessel's Star (after Friedrich Bessel, who measured the distance from Earth to the star in 1838), is a star in the constellation Cygnus.

(a) 61 Cygni is 11.14 light-years from Earth. If 1 light-year is about 5.9 trillion miles, how many miles is 61 Cygni from Earth?

(b) The mean distance from Earth to the Sun is 93,000,000 miles. What is the parallax of 61 Cygni?

31. Finding the Bearing of an Aircraft A DC-9 aircraft leaves Midway Airport from runway 4 RIGHT, whose bearing is N40°E. After flying for $\frac{1}{2}$ mile, the pilot requests permission to turn 90° and head toward the southeast. The permission is granted. After the airplane goes 1 mile in this direction, what bearing should the control tower use to locate the aircraft?

32. Finding the Bearing of a Ship A ship leaves the port of Miami with a bearing of S80°E and a speed of 15 knots. After 1 hour, the ship turns 90° toward the south. After 2 hours, maintaining the same speed, what is the bearing to the ship from the port?

33. Niagara Falls Incline Railway Situated between Portage Road and the Niagara Parkway directly across from the Canadian Horseshoe Falls, the Falls Incline Railway is a funicular that carries passengers up an embankment to Table Rock Observation Point. If the length of the track is 51.8 meters and the angle of inclination is 36°2′, determine the height of the embankment.

Source: www.niagaraparks.com

34. Sears Tower The Sears Tower in Chicago is the third tallest building in the world and is topped by a high antenna. A surveyor on the ground makes the following measurement:

1. The angle of elevation from her position to the top of the building is 34°.

2. The distance from her position to the top of the building is 2593 feet.

3. The distance from her position to the top of the antenna is 2743 feet.

(a) How far away from the base of the building is the surveyor located?

(b) How tall is the building?

(c) What is the angle of elevation from the surveyor to the top of the antenna?

(d) How tall is the antenna?

Source: www.infoplease.com/ce6/us/A0844218.html

35. Chicago Skyscrapers The angle of inclination from the base of the John Hancock Center to the top of the main structure of the Sears Tower is approximately 10.3°. If the main structure of the Sears Tower is 1451 feet tall, how far apart are the two skyscrapers? Assume the bases of the two buildings are at the same elevation.

Source: www.emporis.com

36. Estimating the Width of the Mississippi River A tourist at the top of the Gateway Arch (height, 630 feet) in St. Louis, Missouri, observes a boat moored on the Illinois side of the Mississippi River 2070 feet directly across from the Arch. She also observes a boat moored on the Missouri side directly across from the first boat (see diagram). Given that $B = \cot^{-1}\frac{67}{55}$, estimate the width of the Mississippi River at the St. Louis riverfront.

Source: U.S. Army Corps of Engineers

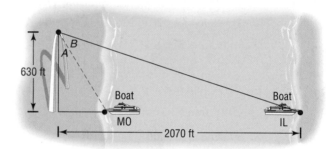

37. Finding the Pitch of a Roof A carpenter is preparing to put a roof on a garage that is 20 feet by 40 feet by 20 feet. A steel support beam 46 feet in length is positioned in the center of the garage. To support the roof, another beam will be attached to the top of the center beam (see the figure). At what angle of elevation is the new beam? In other words, what is the pitch of the roof?

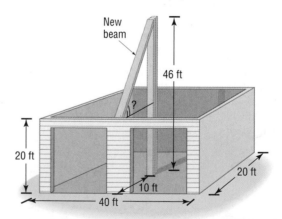

38. Shooting Free Throws in Basketball The eyes of a basketball player are 6 feet above the floor. The player is at the free-throw line, which is 15 feet from the center of the basket rim (see the figure). What is the angle of elevation from the player's eyes to the center of the rim?

[**Hint:** The rim is 10 feet above the floor.]

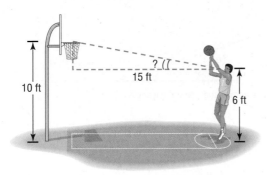

39. Geometry Find the value of the angle θ (see the figure) in degrees rounded to the nearest tenth of a degree.

40. Surveillance Satellites A surveillance satellite circles Earth at a height of h miles above the surface. Suppose that d is the distance, in miles, on the surface of Earth that can be observed from the satellite. See the illustration.
  (a) Find an equation that relates the central angle θ to the height h.
  (b) Find an equation that relates the observable distance d and θ.
  (c) Find an equation that relates d and h.
  (d) If d is to be 2500 miles, how high must the satellite orbit above Earth?
  (e) If the satellite orbits at a height of 300 miles, what distance d on the surface can be observed?

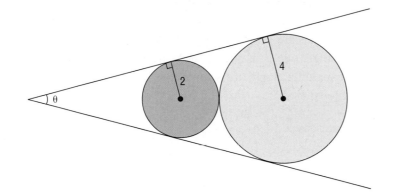

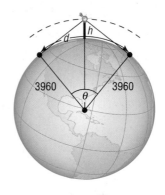

Discussion and Writing

41. The Gibb's Hill Lighthouse, Southampton, Bermuda In operation since 1846, the Gibb's Hill Lighthouse stands 117 feet high on a hill 245 feet high, so its beam of light is 362 feet above sea level. A brochure states that ships 40 miles away can see the light and planes flying at 10,000 feet can see it 120 miles away. Verify the accuracy of these statements. What assumption did the brochure make about the height of the ship?

'Are You Prepared?' Answers

1. 4   **2.** False   **3.** 26.6°   **4.** 30°

4.2 The Law of Sines

PREPARING FOR THIS SECTION *Before getting started, review the following:*

- Trigonometric Equations (I) (Section 3.7, pp. 254–258)
- Difference Formula for the Sine Function (Section 3.4, p. 235)
- Geometry Essentials (Appendix, Section A.2, pp. A14–A19)

Now Work the 'Are You Prepared?' problems on page 288.

OBJECTIVES **1** Solve SAA or ASA Triangles (p. 283)
  2 Solve SSA Triangles (p. 284)
  3 Solve Applied Problems (p. 286)

If none of the angles of a triangle is a right angle, the triangle is called **oblique.** An oblique triangle will have either three acute angles or two acute angles and one obtuse angle (an angle between 90° and 180°). See Figure 8.

Figure 8

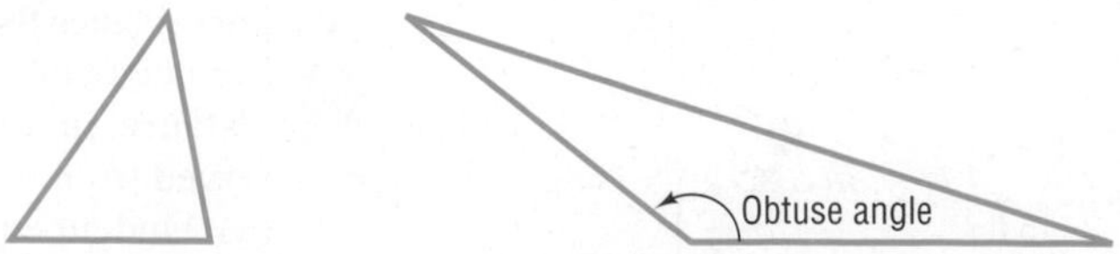

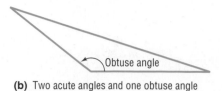

(a) All angles are acute (b) Two acute angles and one obtuse angle

Figure 9

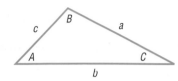

In the discussion that follows, we will always label an oblique triangle so that side a is opposite angle A, side b is opposite angle B, and side c is opposite angle C, as shown in Figure 9.

To **solve an oblique triangle** means to find the lengths of its sides and the measurements of its angles. To do this, we shall need to know the length of one side* along with (i) two angles; (ii) one angle and one other side; or (iii) the other two sides. There are four possibilities to consider:

Case 1: One side and two angles are known (ASA or SAA).

Case 2: Two sides and the angle opposite one of them are known (SSA).

Case 3: Two sides and the included angle are known (SAS).

Case 4: Three sides are known (SSS).

Figure 10 illustrates the four cases.

Figure 10

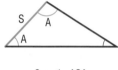 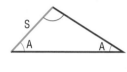 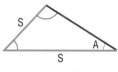 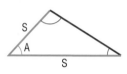 

Case 1: ASA   Case 1: SAA   Case 2: SSA   Case 3: SAS   Case 4: SSS

WARNING Oblique triangles cannot be solved using the methods of Section 4.1. Do you know why? ■

The **Law of Sines** is used to solve triangles for which Case 1 or 2 holds. Cases 3 and 4 are considered when we study the Law of Cosines in the next section.

THEOREM

Law of Sines

For a triangle with sides a, b, c and opposite angles A, B, C, respectively,

$$\frac{\sin A}{a} = \frac{\sin B}{b} = \frac{\sin C}{c} \qquad (1)$$

A proof of the Law of Sines is given at the end of this section. The Law of Sines actually consists of three equalities:

$$\frac{\sin A}{a} = \frac{\sin B}{b} \qquad \frac{\sin A}{a} = \frac{\sin C}{c} \qquad \frac{\sin B}{b} = \frac{\sin C}{c}$$

Formula (1) is a compact way to write these three equations.

In applying the Law of Sines to solve triangles, we use the fact that the sum of the angles of any triangle equals 180°; that is,

$$A + B + C = 180° \qquad (2)$$

*The reason we need to know the length of one side is that, if we only know the angles, this will result in a family of *similar triangles*.

✓ 1 Solve SAA or ASA Triangles

Our first two examples show how to solve a triangle when one side and two angles are known (Case 1: SAA or ASA).

| EXAMPLE 1 | **Using the Law of Sines to Solve an SAA Triangle** |

Solve the triangle: $A = 40°, B = 60°, a = 4$

Solution   Figure 11 shows the triangle that we want to solve. The third angle C is found using equation (2).

Figure 11

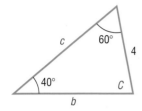

$$A + B + C = 180°$$
$$40° + 60° + C = 180°$$
$$C = 80°$$

Now we use the Law of Sines (twice) to find the unknown sides b and c.

$$\frac{\sin A}{a} = \frac{\sin B}{b} \qquad \frac{\sin A}{a} = \frac{\sin C}{c}$$

Because $a = 4$, $A = 40°$, $B = 60°$, and $C = 80°$, we have

$$\frac{\sin 40°}{4} = \frac{\sin 60°}{b} \qquad \frac{\sin 40°}{4} = \frac{\sin 80°}{c}$$

Solving for b and c, we find that

$$b = \frac{4 \sin 60°}{\sin 40°} \approx 5.39 \qquad c = \frac{4 \sin 80°}{\sin 40°} \approx 6.13$$

NOTE Although not a check, we can verify the reasonableness of our answer by determining if the longest side is opposite the largest angle and the shortest side is opposite the smallest angle. ∎

Notice in Example 1 that we found b and c by working with the given side a. This is better than finding b first and working with a rounded value of b to find c.

✏— **Now Work** PROBLEM 9

| EXAMPLE 2 | **Using the Law of Sines to Solve an ASA Triangle** |

Solve the triangle: $A = 35°, B = 15°, c = 5$

Solution   Figure 12 illustrates the triangle that we want to solve. Because we know two angles ($A = 35°$ and $B = 15°$), we find the third angle using equation (2).

Figure 12

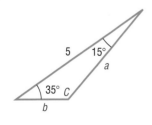

$$A + B + C = 180°$$
$$35° + 15° + C = 180°$$
$$C = 130°$$

Now we know the three angles and one side ($c = 5$) of the triangle. To find the remaining two sides a and b, we use the Law of Sines (twice).

$$\frac{\sin A}{a} = \frac{\sin C}{c} \qquad \frac{\sin B}{b} = \frac{\sin C}{c}$$

$$\frac{\sin 35°}{a} = \frac{\sin 130°}{5} \qquad \frac{\sin 15°}{b} = \frac{\sin 130°}{5}$$

$$a = \frac{5 \sin 35°}{\sin 130°} \approx 3.74 \qquad b = \frac{5 \sin 15°}{\sin 130°} \approx 1.69$$

✏— **Now Work** PROBLEM 23

2 Solve SSA Triangles

Figure 13

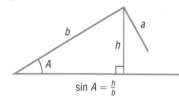

$\sin A = \frac{h}{b}$

Case 2 (SSA), which applies to triangles for which two sides and the angle opposite one of them are known, is referred to as the **ambiguous case,** because the known information may result in one triangle, two triangles, or no triangle at all. Suppose that we are given sides a and b and angle A, as illustrated in Figure 13. The key to determining the possible triangles, if any, that may be formed from the given information lies primarily with the relative size of side a, the height h, and the fact that $h = b \sin A$.

No Triangle If $a < h = b \sin A$, then side a is not sufficiently long to form a triangle. See Figure 14.

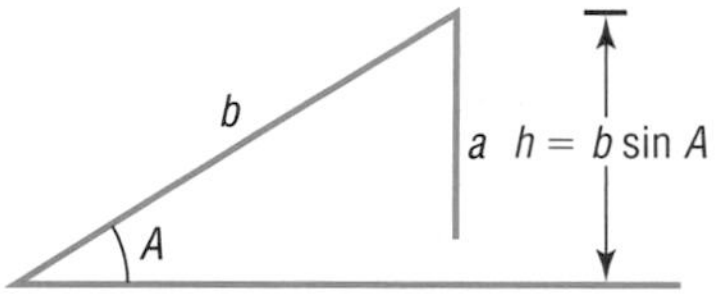

One Right Triangle If $a = h = b \sin A$, then side a is just long enough to form a right triangle. See Figure 15.

Figure 14
$a < h = b \sin A$

Figure 15
$a = h = b \sin A$

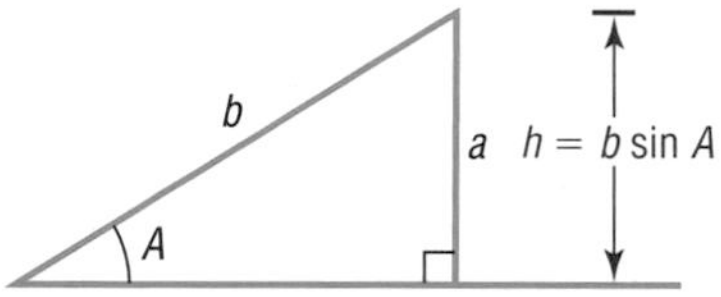

Two Triangles If $h = b \sin A < a$, and $a < b$, two distinct triangles can be formed from the given information. See Figure 16.

One Triangle If $a \geq b$, only one triangle can be formed. See Figure 17.

Figure 16
$b \sin A < a$ and $a < b$

Figure 17
$a \geq b$

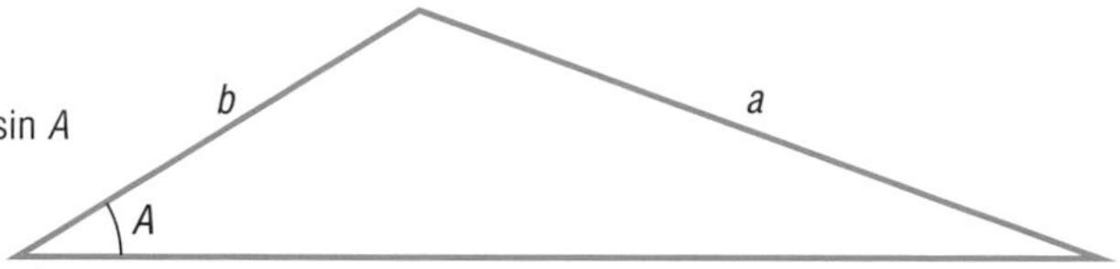

Fortunately, we do not have to rely on an illustration or complicated relationships to draw the correct conclusion in the ambiguous case. The Law of Sines will lead us to the correct determination. Let's see how.

EXAMPLE 3   Using the Law of Sines to Solve an SSA Triangle (One Solution)

Solve the triangle: $a = 3, b = 2, A = 40°$

Solution

Figure 18(a)

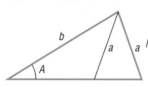

See Figure 18(a). Because we know an angle ($A = 40°$), the side opposite the known angle ($a = 3$), and the side opposite angle B ($b = 2$), we use the Law of Sines to find the angle B.

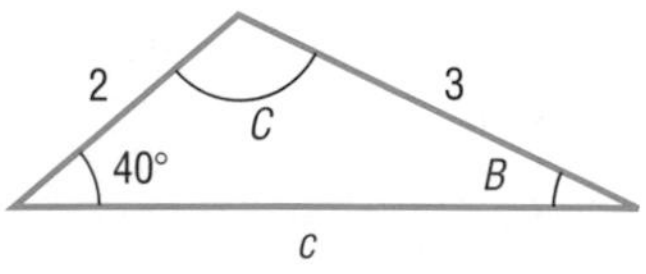

$$\frac{\sin A}{a} = \frac{\sin B}{b}$$

Then

$$\frac{\sin 40°}{3} = \frac{\sin B}{2}$$

$$\sin B = \frac{2 \sin 40°}{3} \approx 0.43$$

NOTE Here we computed B_1 by determining the value of $\sin^{-1}\left(\frac{2 \sin 40°}{3}\right)$. If you use the rounded value and evaluate $\sin^{-1}(0.43)$, you will obtain a slightly different result. ∎

There are two angles B, $0° < B < 180°$, for which $\sin B \approx 0.43$.

$$B_1 \approx 25.4° \quad \text{and} \quad B_2 \approx 180° - 25.4° = 154.6°$$

The second possibility, $B_2 \approx 154.6°$, is ruled out, because $A = 40°$, making $A + B_2 \approx 194.6° > 180°$. Now, using $B_1 \approx 25.4°$, we find that

$$C = 180° - A - B_1 \approx 180° - 40° - 25.4° = 114.6°$$

The third side c may now be determined using the Law of Sines.

$$\frac{\sin A}{a} = \frac{\sin C}{c}$$

$$\frac{\sin 40°}{3} = \frac{\sin 114.6°}{c}$$

$$c = \frac{3 \sin 114.6°}{\sin 40°} \approx 4.24$$

Figure 18(b) illustrates the solved triangle.

Figure 18(b)

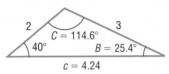

| EXAMPLE 4 | **Using the Law of Sines to Solve an SSA Triangle (Two Solutions)** |

Solve the triangle: $\quad a = 6, b = 8, A = 35°$

Solution — See Figure 19(a). Because $a = 6$, $b = 8$, and $A = 35°$ are known, we use the Law of Sines to find the angle B.

Figure 19(a)

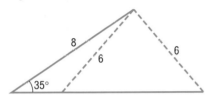

$$\frac{\sin A}{a} = \frac{\sin B}{b}$$

Then

$$\frac{\sin 35°}{6} = \frac{\sin B}{8}$$

$$\sin B = \frac{8 \sin 35°}{6} \approx 0.76$$

$$B_1 \approx 49.9° \quad \text{or} \quad B_2 \approx 180° - 49.9° = 130.1°$$

For both choices of B, we have $A + B < 180°$. There are two triangles, one containing the angle $B_1 \approx 49.9°$ and the other containing the angle $B_2 \approx 130.1°$. The third angle C is either

$$C_1 = 180° - A - B_1 \approx 95.1° \quad \text{or} \quad C_2 = 180° - A - B_2 \approx 14.9°$$

$$\uparrow \qquad\qquad\qquad\qquad\qquad \uparrow$$

$$A = 35° \qquad\qquad\qquad\qquad A = 35°$$
$$B_1 = 49.9° \qquad\qquad\qquad B_2 = 130.1°$$

The third side c obeys the Law of Sines, so we have

Figure 19(b)

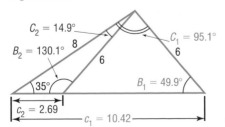

$$\frac{\sin A}{a} = \frac{\sin C_1}{c_1} \qquad\qquad \frac{\sin A}{a} = \frac{\sin C_2}{c_2}$$

$$\frac{\sin 35°}{6} = \frac{\sin 95.1°}{c_1} \qquad\qquad \frac{\sin 35°}{6} = \frac{\sin 14.9°}{c_2}$$

$$c_1 = \frac{6 \sin 95.1°}{\sin 35°} \approx 10.42 \qquad c_2 = \frac{6 \sin 14.9°}{\sin 35°} \approx 2.69$$

The two solved triangles are illustrated in Figure 19(b).

| EXAMPLE 5 | **Using the Law of Sines to Solve a SSA Triangle (No Solution)** |

Solve the triangle: $\quad a = 2, c = 1, C = 50°$

Solution   Because $a = 2$, $c = 1$, and $C = 50°$ are known, we use the Law of Sines to find the angle A.

$$\frac{\sin A}{a} = \frac{\sin C}{c}$$

$$\frac{\sin A}{2} = \frac{\sin 50°}{1}$$

$$\sin A = 2 \sin 50° \approx 1.53$$

Figure 20

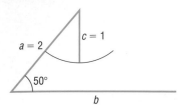

Since there is no angle A for which $\sin A > 1$, there can be no triangle with the given measurements. Figure 20 illustrates the measurements given. Notice that, no matter how we attempt to position side c, it will never touch side b to form a triangle.

═══ **Now Work** PROBLEMS 25 AND 31

3 Solve Applied Problems

The Law of Sines is particularly useful for solving certain applied problems.

EXAMPLE 6   **Finding the Height of a Mountain**

To measure the height of a mountain, a surveyor takes two sightings of the peak at a distance 900 meters apart on a direct line to the mountain.* See Figure 21(a). The first observation results in an angle of elevation of $47°$, and the second results in an angle of elevation of $35°$. If the transit is 2 meters high, what is the height h of the mountain?

Figure 21

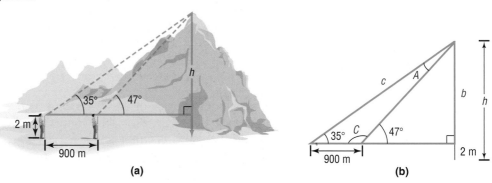

(a)   (b)

Solution   Figure 21(b) shows the triangles that replicate the illustration in Figure 21(a). Since $C + 47° = 180°$, we find that $C = 133°$. Also, since $A + C + 35° = 180°$, we find that $A = 180° - 35° - C = 145° - 133° = 12°$. We use the Law of Sines to find c.

$$\frac{\sin A}{a} = \frac{\sin C}{c} \qquad A = 12°, C = 133°, a = 900$$

$$c = \frac{900 \sin 133°}{\sin 12°} \approx 3165.86$$

Using the larger right triangle, we have

$$\sin 35° = \frac{b}{c} \qquad c = 3165.86$$

$$b = 3165.86 \sin 35° \approx 1815.86 \approx 1816 \text{ meters}$$

The height of the peak from ground level is approximately $1816 + 2 = 1818$ meters.

═══ **Now Work** PROBLEM 39

*For simplicity, we assume that these sightings are at the same level.

| EXAMPLE 7 | Rescue at Sea |

Coast Guard Station Zulu is located 120 miles due west of Station X-ray. A ship at sea sends an SOS call that is received by each station. The call to Station Zulu indicates that the bearing of the ship from Zulu is N40°E (40° east of north). The call to Station X-ray indicates that the bearing of the ship from X-ray is N30°W (30° west of north).

(a) How far is each station from the ship?

(b) If a helicopter capable of flying 200 miles per hour is dispatched from the nearest station to the ship, how long will it take to reach the ship?

Solution

(a) Figure 22 illustrates the situation. The angle C is found to be

$$C = 180° - 50° - 60° = 70°$$

The Law of Sines can now be used to find the two distances a and b that we seek.

Figure 22

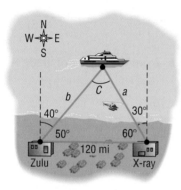

$$\frac{\sin 50°}{a} = \frac{\sin 70°}{120}$$

$$a = \frac{120 \sin 50°}{\sin 70°} \approx 97.82 \text{ miles}$$

$$\frac{\sin 60°}{b} = \frac{\sin 70°}{120}$$

$$b = \frac{120 \sin 60°}{\sin 70°} \approx 110.59 \text{ miles}$$

Station Zulu is about 111 miles from the ship, and Station X-ray is about 98 miles from the ship.

(b) The time t needed for the helicopter to reach the ship from Station X-ray is found by using the formula

$$(\text{Rate}, r)(\text{Time}, t) = \text{Distance}, a$$

Then

$$t = \frac{a}{r} = \frac{97.82}{200} \approx 0.49 \text{ hour} \approx 29 \text{ minutes}$$

It will take about 29 minutes for the helicopter to reach the ship.

Now Work PROBLEM 37

Proof of the Law of Sines To prove the Law of Sines, we construct an altitude of length h from one of the vertices of a triangle. Figure 23(a) shows h for a triangle with three acute angles, and Figure 23(b) shows h for a triangle with an obtuse angle. In each case, the altitude is drawn from the vertex at B. Using either illustration, we have

Figure 23

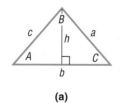

(a)

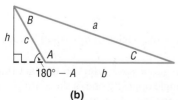

(b)

$$\sin C = \frac{h}{a}$$

from which

$$h = a \sin C \tag{3}$$

From Figure 23(a), it also follows that

$$\sin A = \frac{h}{c}$$

from which

$$h = c \sin A \tag{4}$$

From Figure 23(b), it follows that

$$\sin(180° - A) = \sin A = \frac{h}{c}$$

$\uparrow$

$\sin(180° - A) = \sin 180° \cos A - \cos 180° \sin A = \sin A$

which again gives

$$h = c \sin A$$

So, whether the triangle has three acute angles or has two acute angles and one obtuse angle, equations (3) and (4) hold. As a result, we may equate the expressions for h in equations (3) and (4) to get

$$a \sin C = c \sin A$$

from which

Figure 24

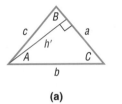

(a)

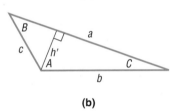

(b)

$$\frac{\sin A}{a} = \frac{\sin C}{c} \qquad \text{(5)}$$

In a similar manner, by constructing the altitude h' from the vertex of angle A as shown in Figure 24, we can show that

$$\sin B = \frac{h'}{c} \quad \text{and} \quad \sin C = \frac{h'}{b}$$

Equating the expressions for h', we find that

$$h' = c \sin B = b \sin C$$

from which

$$\frac{\sin B}{b} = \frac{\sin C}{c} \qquad \text{(6)}$$

When equations (5) and (6) are combined, we have equation (1), the Law of Sines. ∎

4.2 Assess Your Understanding

'Are You Prepared?'   *Answers are given at the end of these exercises. If you get a wrong answer, read the pages listed in red.*

1. The difference formula for the sine function is $\sin(A - B) = $ _____. (p. 235)

2. If θ is an acute angle, solve the equation $\cos\theta = \dfrac{\sqrt{3}}{2}$.
   (pp. 255–256)

3. The two triangles shown are similar. Find the missing length. (pp. A17–A19)

Concepts and Vocabulary

4. If none of the angles of a triangle is a right angle, the triangle is called _____.

5. For a triangle with sides a, b, c and opposite angles A, B, C, the Law of Sines states that _____.

6. **True or False**   An oblique triangle in which two sides and an angle are given always results in at least one triangle.

7. **True or False**   The sum of the angles of any triangle equals 180°.

8. **True or False**   The ambiguous case refers to the fact that, when two sides and the angle opposite one of them are given, sometimes the Law of Sines cannot be used.

Skill Building

In Problems 9–16, solve each triangle.

9.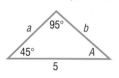

10.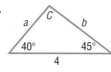

11.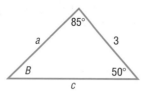

12.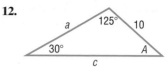

13.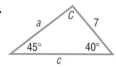

14.

15.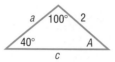

16.

In Problems 17–24, solve each triangle.

17. $A = 40°$, $B = 20°$, $a = 2$

18. $A = 50°$, $C = 20°$, $a = 3$

19. $B = 70°$, $C = 10°$, $b = 5$

20. $A = 70°$, $B = 60°$, $c = 4$

21. $A = 110°$, $C = 30°$, $c = 3$

22. $B = 10°$, $C = 100°$, $b = 2$

23. $A = 40°$, $B = 40°$, $c = 2$

24. $B = 20°$, $C = 70°$, $a = 1$

In Problems 25–36, two sides and an angle are given. Determine whether the given information results in one triangle, two triangles, or no triangle at all. Solve any triangle(s) that results.

25. $a = 3$, $b = 2$, $A = 50°$

26. $b = 4$, $c = 3$, $B = 40°$

27. $b = 5$, $c = 3$, $B = 100°$

28. $a = 2$, $c = 1$, $A = 120°$

29. $a = 4$, $b = 5$, $A = 60°$

30. $b = 2$, $c = 3$, $B = 40°$

31. $b = 4$, $c = 6$, $B = 20°$

32. $a = 3$, $b = 7$, $A = 70°$

33. $a = 2$, $c = 1$, $C = 100°$

34. $b = 4$, $c = 5$, $B = 95°$

35. $a = 2$, $c = 1$, $C - 25°$

36. $b = 4$, $c - 5$, $B = 40°$

Applications and Extensions

37. Rescue at Sea Coast Guard Station Able is located 150 miles due south of Station Baker. A ship at sea sends an SOS call that is received by each station. The call to Station Able indicates that the ship is located N55°E; the call to Station Baker indicates that the ship is located S60°E.
(a) How far is each station from the ship?
(b) If a helicopter capable of flying 200 miles per hour is dispatched from the station nearest the ship, how long will it take to reach the ship?

38. Distance to the Moon At exactly the same time, Tom and Alice measured the angle of elevation to the moon while standing exactly 300 km apart. The angle of elevation to the moon for Tom was 49.8974° and the angle of elevation to the moon for Alice was 49.9312°. See the figure. To the nearest 1000 km, how far was the moon from Earth when the measurement was obtained?

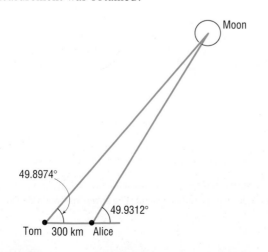

39. Finding the Length of a Ski Lift Consult the figure. To find the length of the span of a proposed ski lift from P to Q, a surveyor measures $\angle DPQ$ to be 25° and then walks off a distance of 1000 feet to R and measures $\angle PRQ$ to be 15°. What is the distance from P to Q?

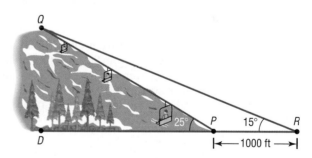

40. Finding the Height of a Mountain Use the illustration in Problem 39 to find the height QD of the mountain.

41. Finding the Height of an Airplane An aircraft is spotted by two observers who are 1000 feet apart. As the airplane passes over the line joining them, each observer takes a sighting of the angle of elevation to the plane, as indicated in the figure. How high is the airplane?

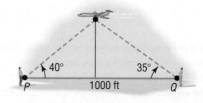

42. Finding the Height of the Bridge over the Royal Gorge The highest bridge in the world is the bridge over the Royal Gorge of the Arkansas River in Colorado. Sightings to the same point at water level directly under the bridge are taken from each side of the 880-foot-long bridge, as indicated in the figure. How high is the bridge?

Source: Guinness Book of World Records

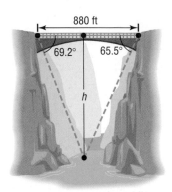

43. Landscaping Pat needs to determine the height of a tree before cutting it down to be sure that it will not fall on a nearby fence. The angle of elevation of the tree from one position on a flat path from the tree is 30°, and from a second position 40 feet farther along this path it is 20°. What is the height of the tree?

44. Construction A loading ramp 10 feet long that makes an angle of 18° with the horizontal is to be replaced by one that makes an angle of 12° with the horizontal. How long is the new ramp?

45. Commercial Navigation Adam must fly home to St. Louis from a business meeting in Oklahoma City. One flight option flies directly to St. Louis, a distance of about 461.1 miles. A second flight option flies first to Kansas City and then connects to St. Louis. The bearing from Oklahoma City to Kansas City is N29.6°E, and the bearing from Oklahoma City to St. Louis is N57.7°E. The bearing from St. Louis to Oklahoma City is S62.1°W, and the bearing from St. Louis to Kansas City is N79.4°W. How many more frequent flyer miles will Adam receive if he takes the connecting flight rather than the direct flight?

Source: www.landings.com

46. Time Lost due to a Navigation Error In attempting to fly from city *P* to city *Q*, an aircraft followed a course that was 10° in error, as indicated in the figure. After flying a distance of 50 miles, the pilot corrected the course by turning at point *R* and flying 70 miles farther. If the constant speed of the

aircraft was 250 miles per hour, how much time was lost due to the error?

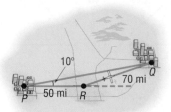

47. Finding the Lean of the Leaning Tower of Pisa The famous Leaning Tower of Pisa was originally 184.5 feet high.* At a distance of 123 feet from the base of the tower, the angle of elevation to the top of the tower is found to be 60°. Find ∠*RPQ* indicated in the figure. Also, find the perpendicular distance from *R* to *PQ*.

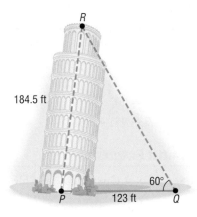

48. Crankshafts on Cars On a certain automobile, the crankshaft is 3 inches long and the connecting rod is 9 inches long (see the figure). At the time when ∠*OPQ* is 15°, how far is the piston (*P*) from the center (*O*) of the crankshaft?

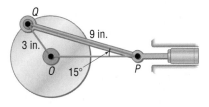

49. Constructing a Highway U.S. 41, a highway whose primary directions are north–south, is being constructed along the west coast of Florida. Near Naples, a bay obstructs the straight

---

*On February 27, 1964, the government of Italy requested aid in preventing the tower from toppling. A multinational task force of engineers, mathematicians, and historians was assigned and met on the Azores islands to discuss stabilization methods. After over two decades of work on the subject, the tower was closed to the public in January 1990. During the time that the tower was closed, the bells were removed to relieve some weight, and cables were cinched around the third level and anchored several hundred meters away. Apartments and houses in the path of the tower were vacated for safety concerns. After a decade of corrective reconstruction and stabilization efforts, the tower was reopened to the public on December 15, 2001. Many methods were proposed to stabilize the tower, including the addition of 800 metric tons of lead counterweights to the raised end of the base. The final solution to correcting the lean was to remove 38 cubic meters of soil from underneath the raised end. The tower has been declared stable for at least another 300 years.

Source: http://en.wikipedia.org/wiki/Leaning_Tower_of_Pisa, page last modified June 28, 2006

path of the road. Since the cost of a bridge is prohibitive, engineers decide to go around the bay. The illustration shows the path that they decide on and the measurements taken. What is the length of highway needed to go around the bay?

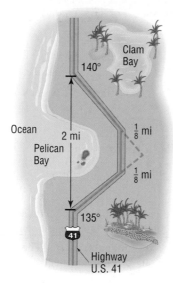

50. **Calculating Distances at Sea** The navigator of a ship at sea spots two lighthouses that she knows to be 3 miles apart along a straight seashore. She determines that the angles formed between two line-of-sight observations of the lighthouses and the line from the ship directly to shore are 15° and 35°. See the illustration.
    (a) How far is the ship from lighthouse *P*?
    (b) How far is the ship from lighthouse *Q*?
    (c) How far is the ship from shore?

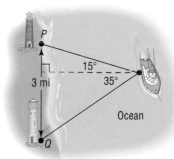

51. **Designing an Awning** An awning that covers a sliding glass door that is 88 inches tall forms an angle of 50° with the wall. The purpose of the awning is to prevent sunlight from entering the house when the angle of elevation of the Sun is more than 65°. See the figure. Find the length *L* of the awning.

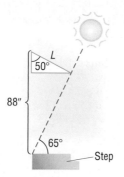

52. **Finding Distances** A forest ranger is walking on a path inclined at 5° to the horizontal directly toward a 100-foot-tall fire observation tower. The angle of elevation from the path to the top of the tower is 40°. How far is the ranger from the tower at this time?

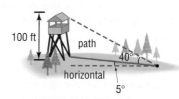

53. **Great Pyramid of Cheops** One of the original Seven Wonders of the World, the Great Pyramid of Cheops was built about 2580 BC. Its original height was 480 feet 11 inches, but owing to the loss of its topmost stones, it is now shorter. Find the current height of the Great Pyramid using the information given in the illustration.

    Source: Guinness Book of World Records

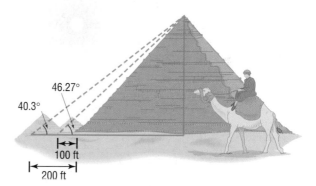

54. **Determining the Height of an Aircraft** Two sensors are spaced 700 feet apart along the approach to a small airport. When an aircraft is nearing the airport, the angle of elevation from the first sensor to the aircraft is 20°, and from the second sensor to the aircraft it is 15°. Determine how high the aircraft is at this time.

55. **Mercury** The distance from the Sun to Earth is approximately 149,600,000 kilometers (km). The distance from the Sun to Mercury is approximately 57,910,000 km. The **elongation angle** α is the angle formed between the line of sight from Earth to the Sun and the line of sight from Earth to Mercury. See the figure. Suppose that the elongation angle for Mercury is 15°. Use this information to find the possible distances between Earth and Mercury.

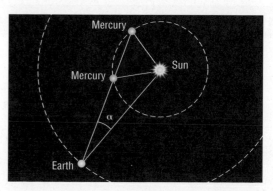

56. **Venus** The distance from the Sun to Earth is approximately 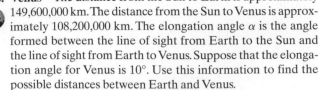 149,600,000 km. The distance from the Sun to Venus is approximately 108,200,000 km. The elongation angle α is the angle formed between the line of sight from Earth to the Sun and the line of sight from Earth to Venus. Suppose that the elongation angle for Venus is $10°$. Use this information to find the possible distances between Earth and Venus.

57. **The Original Ferris Wheel** George Washington Gale Ferris, Jr., designed the original Ferris wheel for the 1893 World's Columbian Exposition in Chicago, Illinois. The wheel had 36 equally spaced cars each the size of a school bus. The distance between adjacent cars was approximately 22 feet. Determine the diameter of the wheel to the nearest foot.

Source: Carnegie Library of Pittsburgh, www.clpgh.org

58. **Mollweide's Formula** For any triangle, Mollweide's Formula (named after Karl Mollweide, 1774–1825) states that

$$\frac{a + b}{c} = \frac{\cos\left[\frac{1}{2}(A - B)\right]}{\sin\left(\frac{1}{2}C\right)}$$

Derive it.

[**Hint:** Use the Law of Sines and then a Sum-to-Product Formula. Notice that this formula involves all six parts of a triangle. As a result, it is sometimes used to check the solution of a triangle.]

59. **Mollweide's Formula** Another form of Mollweide's Formula is

$$\frac{a - b}{c} = \frac{\sin\left[\frac{1}{2}(A - B)\right]}{\cos\left(\frac{1}{2}C\right)}$$

Derive it.

60. For any triangle, derive the formula

$$a = b \cos C + c \cos B$$

[**Hint:** Use the fact that $\sin A = \sin(180° - B - C)$.]

61. **Law of Tangents** For any triangle, derive the Law of Tangents.

$$\frac{a - b}{a + b} = \frac{\tan\left[\frac{1}{2}(A - B)\right]}{\tan\left[\frac{1}{2}(A + B)\right]}$$

[**Hint:** Use Mollweide's Formula.]

62. **Circumscribing a Triangle** Show that

$$\frac{\sin A}{a} = \frac{\sin B}{b} = \frac{\sin C}{c} = \frac{1}{2r}$$

where r is the radius of the circle circumscribing the triangle PQR whose sides are a, b, and c, as shown in the figure.

[**Hint:** Draw the diameter PP'. Then $B = \angle PQR = \angle PP'R$, and angle $\angle PRP' = 90°$.]

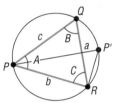

Discussion and Writing

63. Make up three problems involving oblique triangles. One should result in one triangle, the second in two triangles, and the third in no triangle.

64. What do you do first if you are asked to solve a triangle and are given one side and two angles?

65. What do you do first if you are asked to solve a triangle and are given two sides and the angle opposite one of them?

'Are You Prepared?' Answers

1. $\sin A \cos B - \cos A \sin B$ 2. $30°$ 3. $\dfrac{15}{2}$

4.3 The Law of Cosines

PREPARING FOR THIS SECTION *Before getting started, review the following:*

- Trigonometric Equations (I) (Section 3.7, pp. 254–258)
- Distance Formula (Section 1.1, p. 5)

Now Work the 'Are You Prepared?' problems on page 296.

OBJECTIVES **1** Solve SAS Triangles (p. 293)

2 Solve SSS Triangles (p. 294)

3 Solve Applied Problems (p. 295)

In the previous section, we used the Law of Sines to solve Case 1 (SAA or ASA) and Case 2 (SSA) of an oblique triangle. In this section, we derive the Law of Cosines and use it to solve the remaining cases, 3 and 4.

Case 3: Two sides and the included angle are known (SAS).
Case 4: Three sides are known (SSS).

THEOREM

Law of Cosines

For a triangle with sides a, b, c and opposite angles A, B, C, respectively,

$$c^2 = a^2 + b^2 - 2ab \cos C \qquad \textbf{(1)}$$
$$b^2 = a^2 + c^2 - 2ac \cos B \qquad \textbf{(2)}$$
$$a^2 = b^2 + c^2 - 2bc \cos A \qquad \textbf{(3)}$$

Figure 25

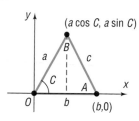

(a) Angle C is acute

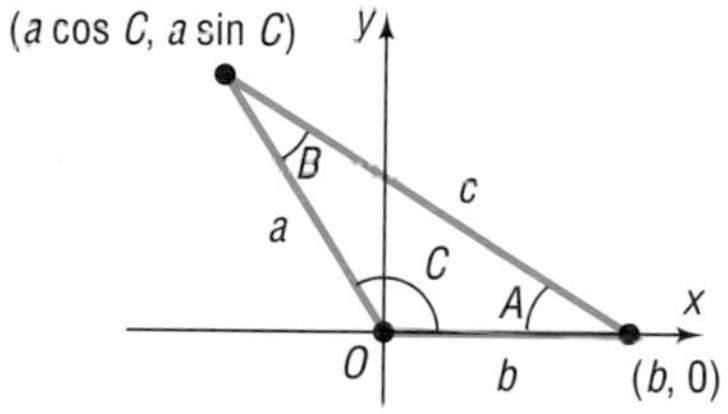

(b) Angle C is obtuse

Proof We will prove only formula (1) here. Formulas (2) and (3) may be proved using the same argument.

We begin by strategically placing a triangle on a rectangular coordinate system so that the vertex of angle C is at the origin and side b lies along the positive x-axis. Regardless of whether C is acute, as in Figure 25(a), or obtuse, as in Figure 25(b), the vertex of angle B has coordinates $(a \cos C, a \sin C)$. The vertex of angle A has coordinates $(b, 0)$.

We can now use the distance formula to compute c^2.

$$c^2 = (b - a \cos C)^2 + (0 - a \sin C)^2$$
$$= b^2 - 2ab \cos C + a^2 \cos^2 C + a^2 \sin^2 C$$
$$= b^2 - 2ab \cos C + a^2(\cos^2 C + \sin^2 C)$$
$$= a^2 + b^2 - 2ab \cos C \qquad ∎$$

Each of formulas (1), (2), and (3) may be stated in words as follows:

THEOREM

Law of Cosines

The square of one side of a triangle equals the sum of the squares of the other two sides minus twice their product times the cosine of their included angle.

Observe that if the triangle is a right triangle (so that, say, $C = 90°$), formula (1) becomes the familiar Pythagorean Theorem: $c^2 = a^2 + b^2$. The Pythagorean Theorem is a special case of the Law of Cosines!

1 Solve SAS Triangles

Let's see how to use the Law of Cosines to solve Case 3 (SAS), which applies to triangles for which two sides and the included angle are known.

EXAMPLE 1 Using the Law of Cosines to Solve an SAS Triangle

Solve the triangle: $a = 2$, $b = 3$, $C = 60°$

Solution See Figure 26. Because we know two sides, a and b, and the included angle, $C = 60°$, the Law of Cosines makes it easy to find the third side, c.

Figure 26

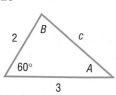

$$c^2 = a^2 + b^2 - 2ab \cos C$$
$$= 2^2 + 3^2 - 2 \cdot 2 \cdot 3 \cdot \cos 60° \qquad a = 2, b = 3, C = 60°$$
$$= 13 - \left(12 \cdot \frac{1}{2}\right) = 7$$
$$c = \sqrt{7}$$

Side c is of length $\sqrt{7}$. To find the angles A and B, we may use either the Law of Sines or the Law of Cosines. It is preferable to use the Law of Cosines, since it will lead to an equation with one solution. Using the Law of Sines would lead to an equation with two solutions that would need to be checked to determine which solution fits the given data. We choose to use formulas (2) and (3) of the Law of Cosines to find A and B.

For A:

$$a^2 = b^2 + c^2 - 2bc \cos A$$

$$2bc \cos A = b^2 + c^2 - a^2$$

$$\cos A = \frac{b^2 + c^2 - a^2}{2bc} = \frac{9 + 7 - 4}{2 \cdot 3\sqrt{7}} = \frac{12}{6\sqrt{7}} = \frac{2\sqrt{7}}{7}$$

$$A = \cos^{-1}\frac{2\sqrt{7}}{7} \approx 40.9°$$

For B:

$$b^2 = a^2 + c^2 - 2ac \cos B$$

$$\cos B = \frac{a^2 + c^2 - b^2}{2ac} = \frac{4 + 7 - 9}{4\sqrt{7}} = \frac{2}{4\sqrt{7}} = \frac{\sqrt{7}}{14}$$

$$B = \cos^{-1}\frac{\sqrt{7}}{14} \approx 79.1°$$

Notice that $A + B + C = 40.9° + 79.1° + 60° = 180°$, as required.

━━━━━ **Now Work** PROBLEM 9

2 Solve SSS Triangles

The next example illustrates how the Law of Cosines is used when three sides of a triangle are known, Case 4 (SSS).

EXAMPLE 2 | **Using the Law of Cosines to Solve an SSS Triangle**

Solve the triangle: $a = 4, b = 3, c = 6$

Solution See Figure 27. To find the angles A, B, and C, we proceed as we did in the latter part of the solution to Example 1.

For A:

$$\cos A = \frac{b^2 + c^2 - a^2}{2bc} = \frac{9 + 36 - 16}{2 \cdot 3 \cdot 6} = \frac{29}{36}$$

$$A = \cos^{-1}\frac{29}{36} \approx 36.3°$$

Figure 27

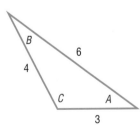

For B:

$$\cos B = \frac{a^2 + c^2 - b^2}{2ac} = \frac{16 + 36 - 9}{2 \cdot 4 \cdot 6} = \frac{43}{48}$$

$$B = \cos^{-1}\frac{43}{48} \approx 26.4°$$

Since we know A and B,

$$C = 180° - A - B \approx 180° - 36.3° - 26.4° = 117.3°$$

━━━━━ **Now Work** PROBLEM 15

3 Solve Applied Problems

EXAMPLE 3 | **Correcting a Navigational Error**

A motorized sailboat leaves Naples, Florida, bound for Key West, 150 miles away. Maintaining a constant speed of 15 miles per hour, but encountering heavy cross-winds and strong currents, the crew finds, after 4 hours, that the sailboat is off course by 20°.

(a) How far is the sailboat from Key West at this time?
(b) Through what angle should the sailboat turn to correct its course?
(c) How much time has been added to the trip because of this? (Assume that the speed remains at 15 miles per hour.)

Solution

See Figure 28. With a speed of 15 miles per hour, the sailboat has gone 60 miles af-ter 4 hours. We seek the distance x of the sailboat from Key West. We also seek the angle θ that the sailboat should turn through to correct its course.

(a) To find x, we use the Law of Cosines, since we know two sides and the included angle.

Figure 28

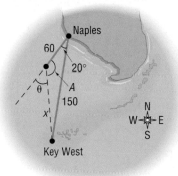

$$x^2 = 150^2 + 60^2 - 2(150)(60)\cos 20° \approx 9186.53$$

$$x \approx 95.8$$

The sailboat is about 96 miles from Key West.

(b) We now know three sides of the triangle, so we can use the Law of Cosines again to find the angle A opposite the side of length 150 miles.

$$150^2 = 96^2 + 60^2 - 2(96)(60)\cos A$$

$$9684 = -11{,}520 \cos A$$

$$\cos A \approx -0.8406$$

$$A \approx 147.2°$$

The sailboat should turn through an angle of

$$\theta = 180° - A \approx 180° - 147.2° = 32.8°$$

The sailboat should turn through an angle of about 33° to correct its course.

(c) The total length of the trip is now $60 + 96 = 156$ miles. The extra 6 miles will only require about 0.4 hour or 24 minutes more if the speed of 15 miles per hour is maintained.

━━━ **Now Work** PROBLEM 45

Historical Feature

The Law of Sines was known vaguely long before it was explicitly stated by Nasir Eddin (about AD 1250). Ptolemy (about AD 150) was aware of it in a form using a chord function instead of the sine function. But it was first clearly stated in Europe by Regiomontanus, writing in 1464.

The Law of Cosines appears first in Euclid's *Elements* (Book II), but in a well-disguised form in which squares built on the sides of triangles are added and a rectangle representing the cosine term is subtracted. It was thus known to all mathematicians because of their familiarity with Euclid's work. An early modern form of the Law of Cosines, that for find-ing the angle when the sides are known, was stated by François Viète (in 1593).

The Law of Tangents (see Problem 61 of Exercise 4.2) has become obsolete. In the past it was used in place of the Law of Cosines, because the Law of Cosines was very inconvenient for calculation with logarithms or slide rules. Mixing of addition and multiplication is now very easy on a calculator, however, and the Law of Tangents has been shelved along with the slide rule.

4.3 Assess Your Understanding

'Are You Prepared?' *Answers are given at the end of these exercises. If you get a wrong answer, read the pages listed in red.*

1. Write the formula for the distance d from $P_1 = (x_1, y_1)$ to $P_2 = (x_2, y_2)$. (p. 5)

2. If θ is an acute angle, solve the equation $\cos\theta = \dfrac{\sqrt{2}}{2}$. (pp. 255–256)

Concepts and Vocabulary

3. If three sides of a triangle are given, the Law of _____ is used to solve the triangle.

4. If one side and two angles of a triangle are given, the Law of _____ is used to solve the triangle.

5. If two sides and the included angle of a triangle are given, the Law of _____ is used to solve the triangle.

6. *True or False* Given only the three sides of a triangle, there is insufficient information to solve the triangle.

7. *True or False* Given two sides and the included angle, the first thing to do to solve the triangle is to use the Law of Sines.

8. *True or False* A special case of the Law of Cosines is the Pythagorean Theorem.

Skill Building

In Problems 9–16, solve each triangle.

9.
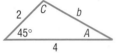

10.

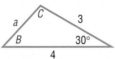

11.
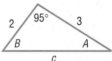

12.

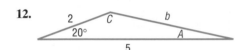

13.

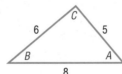

14.
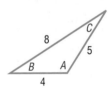

15.
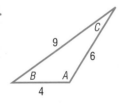

16.
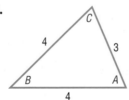

In Problems 17–32, solve each triangle.

17. $a = 3, \quad b = 4, \quad C = 40°$

18. $a = 2, \quad c = 1, \quad B = 10°$

19. $b = 1, \quad c = 3, \quad A = 80°$

20. $a = 6, \quad b = 4, \quad C = 60°$

21. $a = 3, \quad c = 2, \quad B = 110°$

22. $b = 4, \quad c = 1, \quad A = 120°$

23. $a = 2, \quad b = 2, \quad C = 50°$

24. $a = 3, \quad c = 2, \quad B = 90°$

25. $a = 12, \quad b = 13, \quad c = 5$

26. $a = 4, \quad b = 5, \quad c = 3$

27. $a = 2, \quad b = 2, \quad c = 2$

28. $a = 3, \quad b = 3, \quad c = 2$

29. $a = 5, \quad b = 8, \quad c = 9$

30. $a = 4, \quad b = 3, \quad c = 6$

31. $a = 10, \quad b = 8, \quad c = 5$

32. $a = 9, \quad b = 7, \quad c = 10$

Mixed Practice

In Problems 33–42, solve each triangle using either the Law of Sines or the Law of Cosines.

33. $B = 20°, \quad C = 75°, \quad b = 5$

34. $A = 50°, \quad B = 55°, \quad c = 9$

35. $a = 6, \quad b = 8, \quad c = 9$

36. $a = 14, \quad b = 7, \quad A = 85°$

37. $B = 35°, \quad C = 65°, \quad a = 15$

38. $a = 4, \quad c = 5, \quad B = 55°$

39. $A = 10°, \quad a = 3, \quad b = 10$

40. $A = 65°, \quad B = 72°, \quad b = 7$

41. $b = 5, \quad c = 12, \quad A = 60°$

42. $a = 10, \quad b = 10, \quad c = 15$

Applications and Extensions

43. Distance to the Green A golfer hits an errant tee shot that lands in the rough. A marker in the center of the fairway is 150 yards from the center of the green. While standing on the marker and facing the green, the golfer turns 110° toward his ball. He then paces off 35 yards to his ball. See the figure. How far is the ball from the center of the green?

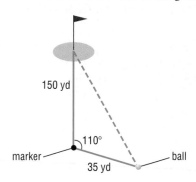

44. Navigation An airplane flies due north from Ft. Myers to Sarasota, a distance of 150 miles, and then turns through an angle of 50° and flies to Orlando, a distance of 100 miles. See the figure.
(a) How far is it directly from Ft. Myers to Orlando?
(b) What bearing should the pilot use to fly directly from Ft. Myers to Orlando?

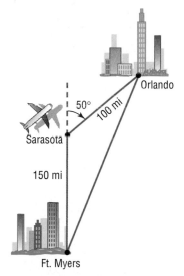

45. Avoiding a Tropical Storm A cruise ship maintains an average speed of 15 knots in going from San Juan, Puerto Rico, to Barbados, West Indies, a distance of 600 nautical miles. To avoid a tropical storm, the captain heads out of San Juan in a direction of 20° off a direct heading to Barbados. The captain maintains the 15-knot speed for 10 hours, after which time the path to Barbados becomes clear of storms.

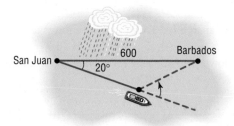

(a) Through what angle should the captain turn to head directly to Barbados?
(b) Once the turn is made, how long will it be before the ship reaches Barbados if the same 15-knot speed is maintained?

46. Revising a Flight Plan In attempting to fly from Chicago to Louisville, a distance of 330 miles, a pilot inadvertently took a course that was 10° in error, as indicated in the figure.
(a) If the aircraft maintains an average speed of 220 miles per hour and if the error in direction is discovered after 15 minutes, through what angle should the pilot turn to head toward Louisville?
(b) What new average speed should the pilot maintain so that the total time of the trip is 90 minutes?

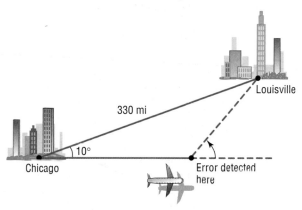

47. Major League Baseball Field A Major League baseball diamond is actually a square 90 feet on a side. The pitching rubber is located 60.5 feet from home plate on a line joining home plate and second base.
(a) How far is it from the pitching rubber to first base?
(b) How far is it from the pitching rubber to second base?
(c) If a pitcher faces home plate, through what angle does he need to turn to face first base?

48. Little League Baseball Field According to Little League baseball official regulations, the diamond is a square 60 feet on a side. The pitching rubber is located 46 feet from home plate on a line joining home plate and second base.
(a) How far is it from the pitching rubber to first base?
(b) How far is it from the pitching rubber to second base?
(c) If a pitcher faces home plate, through what angle does he need to turn to face first base?

49. Finding the Length of a Guy Wire The height of a radio tower is 500 feet, and the ground on one side of the tower slopes upward at an angle of 10° (see the figure).

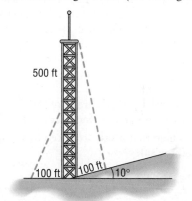

(a) How long should a guy wire be if it is to connect to the top of the tower and be secured at a point on the sloped side 100 feet from the base of the tower?

(b) How long should a second guy wire be if it is to connect to the middle of the tower and be secured at a point 100 feet from the base on the flat side?

50. Finding the Length of a Guy Wire A radio tower 500 feet high is located on the side of a hill with an inclination to the horizontal of 5°. See the figure. How long should two guy wires be if they are to connect to the top of the tower and be secured at two points 100 feet directly above and directly below the base of the tower?

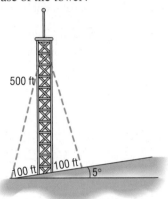

51. Wrigley Field, Home of the Chicago Cubs The distance from home plate to the fence in dead center in Wrigley Field is 400 feet (see the figure). How far is it from the fence in dead center to third base?

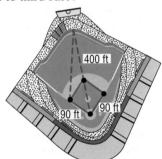

52. Little League Baseball The distance from home plate to the fence in dead center at the Oak Lawn Little League field is 280 feet. How far is it from the fence in dead center to third base?

[**Hint:** The distance between the bases in Little League is 60 feet.]

53. Building a Swing Set Clint is building a wooden swing set for his children. Each supporting end of the swing set is to be an A-frame constructed with two 10-foot-long 4 by 4s joined at a 45° angle. To prevent the swing set from tipping over, Clint wants to secure the base of each A-frame to concrete footings. How far apart should the footings for each A-frame be?

54. Rods and Pistons Rod OA rotates about the fixed point O so that point A travels on a circle of radius r. Connected to point A is another rod AB of length $L > 2r$, and point B is connected to a piston. See the figure. Show that the distance x between point O and point B is given by

$$x = r \cos \theta + \sqrt{r^2 \cos^2 \theta + L^2 - r^2}$$

where θ is the angle of rotation of rod OA.

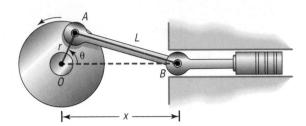

55. Geometry Show that the length d of a chord of a circle of radius r is given by the formula

$$d = 2r \sin \frac{\theta}{2}$$

where θ is the central angle formed by the radii to the ends of the chord. See the figure. Use this result to derive the fact that $\sin \theta < \theta$, where $\theta > 0$ is measured in radians.

56. For any triangle, show that

$$\cos \frac{C}{2} = \sqrt{\frac{s(s - c)}{ab}}$$

where $s = \frac{1}{2}(a + b + c)$.

[**Hint:** Use a Half-angle Formula and the Law of Cosines.]

57. For any triangle show that

$$\sin \frac{C}{2} = \sqrt{\frac{(s - a)(s - b)}{ab}}$$

where $s = \frac{1}{2}(a + b + c)$.

58. Use the Law of Cosines to prove the identity

$$\frac{\cos A}{a} + \frac{\cos B}{b} + \frac{\cos C}{c} = \frac{a^2 + b^2 + c^2}{2abc}$$

Discussion and Writing

59. What do you do first if you are asked to solve a triangle and are given two sides and the included angle?

60. What do you do first if you are asked to solve a triangle and are given three sides?

61. Make up an applied problem that requires using the Law of Cosines.

62. Write down your strategy for solving an oblique triangle.

'Are You Prepared?' Answers

1. $d = \sqrt{(x_2 - x_1)^2 + (y_2 - y_1)^2}$ **2.** $\theta = 45°$

4.4 Area of a Triangle

PREPARING FOR THIS SECTION *Before getting started, review the following:*

- Geometry Essentials (Appendix, Section A.2, pp. A14–A19)

 Now Work the 'Are You Prepared?' problems on page 301.

> **OBJECTIVES** **1** Find the Area of SAS Triangles (p. 299)
> **2** Find the Area of SSS Triangles (p. 300)

In this section, we will derive several formulas for calculating the area of a triangle. The most familiar of these is the following:

THEOREM

The area K of a triangle is

$$K = \frac{1}{2}bh \tag{1}$$

where b is the base and h is an altitude drawn to that base.

Proof The derivation of this formula is rather easy once a rectangle of base b and height h is constructed around the triangle. See Figures 29 and 30.

Triangles 1 and 2 in Figure 30 are equal in area, as are triangles 3 and 4. Consequently, the area of the triangle with base b and altitude h is exactly half the area of the rectangle, which is bh.

Figure 29

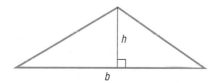

Figure 30

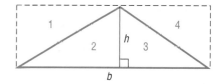

1 Find the Area of SAS Triangles

Figure 31

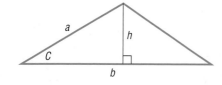

If the base b and altitude h to that base are known, then we can find the area of such a triangle using formula (1). Usually, though, the information required to use formula (1) is not given. Suppose, for example, that we know two sides a and b and the included angle C. See Figure 31. Then the altitude h can be found by noting that

$$\frac{h}{a} = \sin C$$

so that

$$h = a \sin C$$

Using this fact in formula (1) produces

$$K = \frac{1}{2}bh = \frac{1}{2}b(a \sin C) = \frac{1}{2}ab \sin C$$

We now have the formula

$$K = \frac{1}{2}ab \sin C \tag{2}$$

By dropping altitudes from the other two vertices of the triangle, we obtain the following corresponding formulas:

$$K = \frac{1}{2}bc \sin A \qquad\qquad (3)$$

$$K = \frac{1}{2}ac \sin B \qquad\qquad (4)$$

It is easiest to remember these formulas using the following wording:

THEOREM

The area K of a triangle equals one-half the product of two of its sides times the sine of their included angle.

| EXAMPLE 1 | Finding the Area of an SAS Triangle |

Find the area K of the triangle for which $a = 8$, $b = 6$, and $C = 30°$.

Solution   See Figure 32. We use formula (2) to get

Figure 32

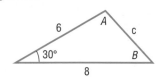

$$K = \frac{1}{2}ab \sin C = \frac{1}{2} \cdot 8 \cdot 6 \cdot \sin 30° = 12 \text{ square units}$$

Now Work PROBLEM 5

2 Find the Area of SSS Triangles

If the three sides of a triangle are known, another formula, called **Heron's Formula** (named after Heron of Alexandria), can be used to find the area of a triangle.

THEOREM

Heron's Formula

The area K of a triangle with sides a, b, and c is

$$K = \sqrt{s(s - a)(s - b)(s - c)} \qquad\qquad (5)$$

where $s = \frac{1}{2}(a + b + c)$.

A proof of Heron's Formula is given at the end of this section.

| EXAMPLE 2 | Finding the Area of an SSS Triangle |

Find the area of a triangle whose sides are 4, 5, and 7.

Solution   We let $a = 4$, $b = 5$, and $c = 7$. Then

$$s = \frac{1}{2}(a + b + c) = \frac{1}{2}(4 + 5 + 7) = 8$$

Heron's Formula then gives the area K as

$$K = \sqrt{s(s - a)(s - b)(s - c)} = \sqrt{8 \cdot 4 \cdot 3 \cdot 1} = \sqrt{96} = 4\sqrt{6} \text{ square units}$$

Now Work PROBLEM 11

Proof of Heron's Formula   The proof that we shall give uses the Law of Cosines and is quite different from the proof given by Heron.

From the Law of Cosines,

$$c^2 = a^2 + b^2 - 2ab \cos C$$

and the Half-angle Formula

$$\cos^2 \frac{C}{2} = \frac{1 + \cos C}{2}$$

we find that

$$\cos^2 \frac{C}{2} = \frac{1 + \cos C}{2} = \frac{1 + \dfrac{a^2 + b^2 - c^2}{2ab}}{2}$$

$$= \frac{a^2 + 2ab + b^2 - c^2}{4ab} = \frac{(a + b)^2 - c^2}{4ab}$$

$$= \underbrace{\frac{(a + b - c)(a + b + c)}{4ab}}_{\text{Factor}} = \frac{2(s - c) \cdot 2s}{4ab} = \frac{s(s - c)}{ab} \qquad \textbf{(6)}$$

$$a + b - c = a + b + c - 2c$$
$$= 2s - 2c = 2(s - c)$$

Similarly, using $\sin^2 \dfrac{C}{2} = \dfrac{1 - \cos C}{2}$, we find that

$$\sin^2 \frac{C}{2} = \frac{(s - a)(s - b)}{ab} \qquad \textbf{(7)}$$

Now we use formula (2) for the area.

$$K = \frac{1}{2} ab \sin C$$

$$= \frac{1}{2} ab \cdot 2 \sin \frac{C}{2} \cos \frac{C}{2} \qquad \sin C = \sin\left[2\left(\frac{C}{2}\right)\right] = 2 \sin \frac{C}{2} \cos \frac{C}{2}$$

$$= ab \sqrt{\frac{(s - a)(s - b)}{ab}} \sqrt{\frac{s(s - c)}{ab}} \qquad \text{Use equations (6) and (7).}$$

$$= \sqrt{s(s - a)(s - b)(s - c)} \qquad \blacksquare$$

Historical Feature

Heron's Formula (also known as *Hero's Formula*) is due to Heron of Alexandria (first century AD), who had, besides his mathematical talents, a good deal of engineering skills. In various temples his mechanical devices produced effects that seemed supernatural, and visitors presumably were thus influenced to generosity. Heron's book *Metrica*, on making such devices, has survived and was discovered in 1896 in the city of Constantinople.

Heron's Formulas for the area of a triangle caused some mild discomfort in Greek mathematics, because a product with two factors was an area, while one with three factors was a volume, but four factors seemed contradictory in Heron's time.

4.4 Assess Your Understanding

'Are You Prepared?' *Answer is given at the end of these exercises. If you get the wrong answer, read the page listed in red.*

1. The area of a triangle whose base is b and whose height is h is _____ (p. A15)

Concepts and Vocabulary

2. If three sides of a triangle are given, _____ Formula is used to find the area of the triangle.

3. *True or False* No formula exists for finding the area of a triangle when only three sides are given.

4. *True or False* Given two sides and the included angle, there is a formula that can be used to find the area of the triangle.

Skill Building

In Problems 5–12, find the area of each triangle. Round answers to two decimal places.

5.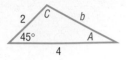

6.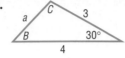

7.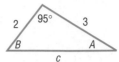

8.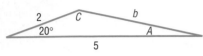

9.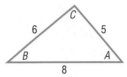

10.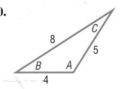

11.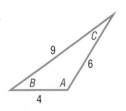

12.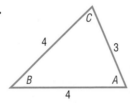

In Problems 13–24, find the area of each triangle. Round answers to two decimal places.

13. $a = 3$, $b = 4$, $C = 40°$

14. $a = 2$, $c = 1$, $B = 10°$

15. $b = 1$, $c = 3$, $A = 80°$

16. $a = 6$, $b = 4$, $C = 60°$

17. $a = 3$, $c = 2$, $B = 110°$

18. $b = 4$, $c = 1$, $A = 120°$

19. $a = 12$, $b = 13$, $c = 5$

20. $a = 4$, $b = 5$, $c = 3$

21. $a = 2$, $b = 2$, $c = 2$

22. $a = 3$, $b = 3$, $c = 2$

23. $a = 5$, $b = 8$, $c = 9$

24. $a = 4$, $b = 3$, $c = 6$

Applications and Extensions

25. Area of an ASA Triangle   If two angles and the included side are given, the third angle is easy to find. Use the Law of Sines to show that the area K of a triangle with side a and angles A, B, and C is

$$K = \frac{a^2 \sin B \sin C}{2 \sin A}$$

26. Area of a Triangle   Prove the two other forms of the formula given in Problem 25.

$$K = \frac{b^2 \sin A \sin C}{2 \sin B} \quad \text{and} \quad K = \frac{c^2 \sin A \sin B}{2 \sin C}$$

In Problems 27–32, use the results of Problem 25 or 26 to find the area of each triangle. Round answers to two decimal places.

27. $A = 40°$, $B = 20°$, $a = 2$

28. $A = 50°$, $C = 20°$, $a = 3$

29. $B = 70°$, $C = 10°$, $b = 5$

30. $A = 70°$, $B = 60°$, $c = 4$

31. $A = 110°$, $C = 30°$, $c = 3$

32. $B = 10°$, $C = 100°$, $b = 2$

---

33. Area of a Segment   Find the area of the segment (shaded in blue in the figure) of a circle whose radius is 8 feet, formed by a central angle of 70°.

[**Hint:** Subtract the area of the triangle from the area of the sector to obtain the area of the segment.]

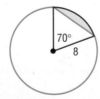

34. Area of a Segment   Find the area of the segment of a circle whose radius is 5 inches, formed by a central angle of 40°.

35. Cost of a Triangular Lot   The dimensions of a triangular lot are 100 feet by 50 feet by 75 feet. If the price of such land is $3 per square foot, how much does the lot cost?

36. Amount of Material to Make a Tent   A cone-shaped tent is made from a circular piece of canvas 24 feet in diameter by removing a sector with central angle 100° and connecting the ends. What is the surface area of the tent?

37. Dimensions of Home Plate   The dimensions of home plate at any major league baseball stadium are shown. Find the area of home plate.

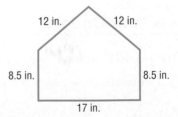

38. Computing Areas   Find the area of the shaded region enclosed in a semicircle of diameter 10 inches. The length of the chord PQ is 8 inches.

[**Hint:** Triangle PQR is a right triangle.]

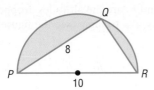

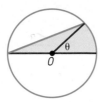

39. Geometry Consult the figure, which shows a circle of radius r with center at O. Find the area K of the shaded region as a function of the central angle θ.

40. Approximating the Area of a Lake To approximate the area of a lake, a surveyor walks around the perimeter of the lake, taking the measurements shown in the illustration. Using this technique, what is the approximate area of the lake?

[**Hint:** Use the Law of Cosines on the three triangles shown and then find the sum of their areas.]

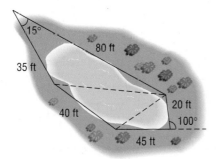

41. The Flatiron Building Completed in 1902 in New York City, the Flatiron Building is triangular shaped and bounded by 22nd Street, Broadway, and 5th Avenue. The building measures approximately 87 feet on the 22nd Street side, 190 feet on the Broadway side, and 173 feet on the 5th Avenue side. Approximate the ground area covered by the building.

Source: Sarah Bradford Landau and Carl W. Condit, *Rise of the New York Skyscraper: 1865–1913.* New Haven, CT: Yale University Press, 1996

42. Bermuda Triangle The Bermuda Triangle is roughly defined by Hamilton, Bermuda; San Juan, Puerto Rico; and Fort Lauderdale, Florida. The distances from Hamilton to Fort Lauderdale, Fort Lauderdale to San Juan, and San Juan to Hamilton are approximately 1028, 1046, and 965 miles respectively. Ignoring the curvature of Earth, approximate the area of the Bermuda Triangle.

Source: www.worldatlas.com

43. Geometry Refer to the figure. If $|OA| = 1$, show that:

(a) Area $\triangle OAC = \dfrac{1}{2} \sin \alpha \cos \alpha$

(b) Area $\triangle OCB = \dfrac{1}{2} |OB|^2 \sin \beta \cos \beta$

(c) Area $\triangle OAB = \dfrac{1}{2} |OB| \sin(\alpha + \beta)$

(d) $|OB| = \dfrac{\cos \alpha}{\cos \beta}$

(e) $\sin(\alpha + \beta) = \sin \alpha \cos \beta + \cos \alpha \sin \beta$

[**Hint:** Area $\triangle OAB$ = Area $\triangle OAC$ + Area $\triangle OCB$.]

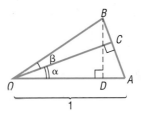

44. Geometry Refer to the figure, in which a unit circle is drawn. The line segment DB is tangent to the circle and θ is acute.

(a) Express the area of $\triangle OBC$ in terms of $\sin \theta$ and $\cos \theta$.

(b) Express the area of $\triangle OBD$ in terms of $\sin \theta$ and $\cos \theta$.

(c) The area of the sector $\overset{\frown}{OBC}$ of the circle is $\dfrac{1}{2}\theta$, where θ is measured in radians. Use the results of parts (a) and (b) and the fact that

$$\text{Area } \triangle OBC < \text{Area } \overset{\frown}{OBC} < \text{Area } \triangle OBD$$

to show that

$$1 < \frac{\theta}{\sin \theta} < \frac{1}{\cos \theta}$$

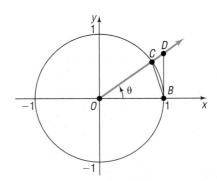

45. The Cow Problem* A cow is tethered to one corner of a square barn, 10 feet by 10 feet, with a rope 100 feet long. What is the maximum grazing area for the cow?

[**Hint:** See the illustration on page 304.]

*Suggested by Professor Teddy Koukounas of Suffolk Community College, who learned of it from an old farmer in Virginia. Solution provided by Professor Kathleen Miranda of SUNY at Old Westbury.

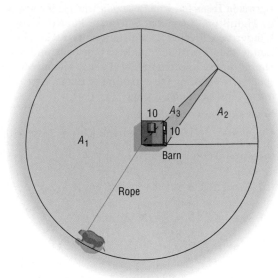

10 A_3 A_2

10

A_1

Barn

Rope

46. Another Cow Problem If the barn in Problem 45 is rectangular, 10 feet by 20 feet, what is the maximum grazing area for the cow?

47. Perfect Triangles A *perfect triangle* is one having natural number sides for which the area is numerically equal to the perimeter. Show that the triangles with the given side lengths are perfect.
(a) 9, 10, 17   (b) 6, 25, 29

Source: M.V. Bonsangue, G. E. Gannon, E. Buchman, and N. Gross, "In Search of Perfect Triangles," *Mathematics Teacher,* Vol. 92, No. 1, 1999: 56–61

48. If h_1, h_2, and h_3 are the altitudes dropped from $P, Q,$ and R, respectively, in a triangle (see the figure), show that

$$\frac{1}{h_1} + \frac{1}{h_2} + \frac{1}{h_3} = \frac{s}{K}$$

where K is the area of the triangle and $s = \frac{1}{2}(a + b + c)$.

[**Hint:** $h_1 = \frac{2K}{a}$.]

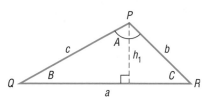

49. Show that a formula for the altitude h from a vertex to the opposite side a of a triangle is

$$h = \frac{a \sin B \sin C}{\sin A}$$

Inscribed Circle *For Problems 50–53, the lines that bisect each angle of a triangle meet in a single point O, and the perpendicular distance r from O to each side of the triangle is the same. The circle with center at O and radius r is called the* **inscribed circle** *of the triangle (see the figure).*

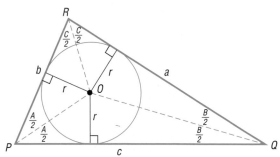

50. Apply the formula from Problem 49 to triangle OPQ to show that

$$r = \frac{c \sin \dfrac{A}{2} \sin \dfrac{B}{2}}{\cos \dfrac{C}{2}}$$

51. Use the result of Problem 50 and the result of Problems 56 and 57 in Section 4.3 to show that

$$\cot \frac{C}{2} = \frac{s - c}{r}$$

where $s = \frac{1}{2}(a + b + c)$.

52. Show that

$$\cot \frac{A}{2} + \cot \frac{B}{2} + \cot \frac{C}{2} = \frac{s}{r}$$

53. Show that the area K of triangle PQR is $K = rs$, where $s = \frac{1}{2}(a + b + c)$. Then show that

$$r = \sqrt{\frac{(s - a)(s - b)(s - c)}{s}}$$

Discussion and Writing

54. What do you do first if you are asked to find the area of a triangle and are given two sides and the included angle?

55. What do you do first if you are asked to find the area of a triangle and are given three sides?

'Are You Prepared?' Answer

1. $K = \frac{1}{2}bh$

4.5 Simple Harmonic Motion; Damped Motion; Combining Waves

PREPARING FOR THIS SECTION *Before getting started, review the following:*

- Sinusoidal Graphs (Section 2.6, pp. 165–172)

 Now Work the 'Are You Prepared?' problem on page 311.

OBJECTIVES
1. Build a Model for an Object in Simple Harmonic Motion (p. 305)
2. Analyze Simple Harmonic Motion (p. 307)
3. Analyze an Object in Damped Motion (p. 308)
4. Graph the Sum of Two Functions (p. 309)

1 Build a Model for an Object in Simple Harmonic Motion

Many physical phenomena can be described as simple harmonic motion. Radio and television waves, light waves, sound waves, and water waves exhibit motion that is simple harmonic.

The swinging of a pendulum, the vibrations of a tuning fork, and the bobbing of a weight attached to a coiled spring are examples of vibrational motion. In this type of motion, an object swings back and forth over the same path. In Figure 33, the point B is the **equilibrium (rest) position** of the vibrating object. The **amplitude** is the distance from the object's rest position to its point of greatest displacement (either point A or point C in Figure 33). The **period** is the time required to complete one vibration, that is, the time it takes to go from, say, point A through B to C and back to A.

> **Simple harmonic motion** is a special kind of vibrational motion in which the acceleration a of the object is directly proportional to the negative of its displacement d from its rest position. That is, $a = -kd$, $k > 0$.

For example, when the mass hanging from the spring in Figure 33 is pulled down from its rest position B to the point C, the force of the spring tries to restore the mass to its rest position. Assuming that there is no frictional force* to retard the motion, the amplitude will remain constant. The force increases in direct proportion to the distance that the mass is pulled from its rest position. Since the force increases directly, the acceleration of the mass of the object must do likewise, because (by Newton's Second Law of Motion) force is directly proportional to acceleration. As a result, the acceleration of the object varies directly with its displacement, and the motion is an example of simple harmonic motion.

Simple harmonic motion is related to circular motion. To see this relationship, consider a circle of radius a, with center at $(0, 0)$. See Figure 34. Suppose that an

Vibrating tuning fork

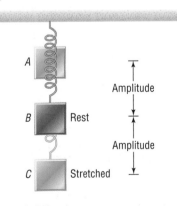

Figure 33

A

Amplitude

B Rest

Amplitude

C Stretched

Coiled spring

Figure 34

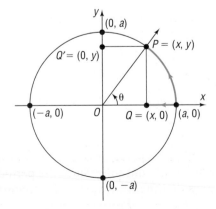

*If friction is present, the amplitude will decrease with time to 0. This type of motion is an example of **damped motion**, which is discussed later in this section.

object initially placed at $(a, 0)$ moves counterclockwise around the circle at a constant angular speed ω. Suppose further that after time t has elapsed the object is at the point $P = (x, y)$ on the circle. The angle θ, in radians, swept out by the ray $\overrightarrow{OP}$ in this time t is

$$\theta = \omega t$$

The coordinates of the point P at time t are

$$x = a \cos \theta = a \cos(\omega t)$$

$$y = a \sin \theta = a \sin(\omega t)$$

Corresponding to each position $P = (x, y)$ of the object moving about the circle, there is the point $Q = (x, 0)$, called the **projection of P on the x-axis.** As P moves around the circle at a constant rate, the point Q moves back and forth between the points $(a, 0)$ and $(-a, 0)$ along the x-axis with a motion that is simple harmonic. Similarly, for each point P there is a point $Q' = (0, y)$, called the **projection of P on the y-axis.** As P moves around the circle, the point Q' moves back and forth between the points $(0, a)$ and $(0, -a)$ on the y-axis with a motion that is simple harmonic. Simple harmonic motion can be described as the projection of constant circular motion on a coordinate axis.

To put it another way, again consider a mass hanging from a spring where the mass is pulled down from its rest position to the point C and then released. See Figure 35(a). The graph shown in Figure 35(b) describes the displacement d of the object from its rest position as a function of time t, assuming that no frictional force is present.

Figure 35

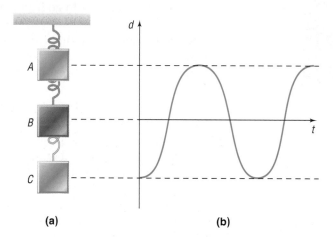

(a)   (b)

THEOREM   **Simple Harmonic Motion**

An object that moves on a coordinate axis so that the distance d from its rest position at time t is given by either

$$d = a \cos(\omega t) \quad \text{or} \quad d = a \sin(\omega t)$$

where a and $\omega > 0$ are constants, moves with simple harmonic motion. The motion has amplitude $|a|$ and period $\dfrac{2\pi}{\omega}$.

The **frequency** f of an object in simple harmonic motion is the number of oscillations per unit time. Since the period is the time required for one oscillation, it follows that the frequency is the reciprocal of the period; that is,

$$f = \frac{\omega}{2\pi} \qquad \omega > 0$$

| EXAMPLE 1 | Build a Model for an Object in Harmonic Motion |

Figure 36

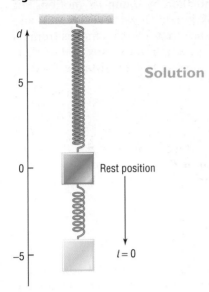

$d = 0$

Rest position

$t = 0$

NOTE In the solution to Example 1, we let $a = -5$, since the initial motion is down. If the initial direction were up, we would let $a = 5$. ■

Suppose that an object attached to a coiled spring is pulled down a distance of 5 inches from its rest position and then released. If the time for one oscillation is 3 seconds, develop a model that relates the displacement d of the object from its rest position after time t (in seconds). Assume no friction.

Solution   The motion of the object is simple harmonic. See Figure 36. When the object is released ($t = 0$), the displacement of the object from the rest position is -5 units (since the object was pulled down). Because $d = -5$ when $t = 0$, it is easier to use the cosine function*

$$d = a\cos(\omega t)$$

to describe the motion. Now the amplitude is $|-5| = 5$ and the period is 3, so

$$a = -5 \quad \text{and} \quad \frac{2\pi}{\omega} = \text{period} = 3, \quad \omega = \frac{2\pi}{3}$$

An equation that models the motion of the object is

$$d = -5\cos\left[\frac{2\pi}{3}t\right]$$

━━━ **Now Work** PROBLEM 5

2 Analyze Simple Harmonic Motion

| EXAMPLE 2 | Analyzing the Motion of an Object |

Suppose that the displacement d (in meters) of an object at time t (in seconds) satisfies the equation

$$d = 10\sin(5t)$$

(a) Describe the motion of the object.
(b) What is the maximum displacement from its resting position?
(c) What is the time required for one oscillation?
(d) What is the frequency?

Solution   We observe that the given equation is of the form

$$d = a\sin(\omega t) \quad d = 10 \sin(5t)$$

where $a = 10$ and $\omega = 5$.

(a) The motion is simple harmonic.
(b) The maximum displacement of the object from its resting position is the amplitude: $|a| = 10$ meters.
(c) The time required for one oscillation is the period:

$$\text{Period} = \frac{2\pi}{\omega} = \frac{2\pi}{5} \text{ seconds}$$

(d) The frequency is the reciprocal of the period. Thus,

$$\text{Frequency} = f = \frac{5}{2\pi} \text{ oscillations per second}$$

━━━ **Now Work** PROBLEM 13

*No phase shift is required if a cosine function is used.

3 Analyze an Object in Damped Motion

Most physical phenomena are affected by friction or other resistive forces. These forces remove energy from a moving system and thereby damp its motion. For example, when a mass hanging from a spring is pulled down a distance a and released, the friction in the spring causes the distance that the mass moves from its at-rest position to decrease over time. As a result, the amplitude of any real oscillating spring or swinging pendulum decreases with time due to air resistance, friction, and so forth. See Figure 37.

Figure 37

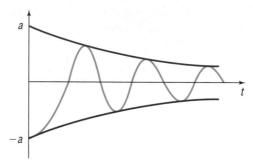

A model that describes this phenomenon maintains a sinusoidal component, but the amplitude of this component will decrease with time to account for the damping effect. In addition, the period of the oscillating component will be affected by the damping. The next result, from physics, describes damped motion.

THEOREM

Damped Motion

The displacement d of an oscillating object from its at-rest position at time t is given by

$$d(t) = ae^{-bt/(2m)} \cos\left(\sqrt{\omega^2 - \frac{b^2}{4m^2}}\, t\right)$$

where b is the **damping factor** or **damping coefficient** and m is the mass of the oscillating object. Here $|a|$ is the displacement at $t = 0$, and $\dfrac{2\pi}{\omega}$ is the period under simple harmonic motion (no damping).

Notice for $b = 0$ (zero damping) that we have the formula for simple harmonic motion with amplitude $|a|$ and period $\dfrac{2\pi}{\omega}$.

EXAMPLE 3 Analyzing the Motion of a Pendulum with Damped Motion

Figure 38

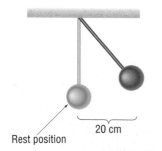

20 cm

Rest position

Suppose that a simple pendulum with a bob of mass 10 grams and a damping factor of 0.8 gram/second is pulled 20 centimeters from its at-rest position and released. See Figure 38. The period of the pendulum without the damping effect is 4 seconds.

(a) Develop a model that describes the position of the pendulum bob.

(b) Using a graphing utility, graph the function found in part (a).

(c) What is the displacement of the bob at the start of the second oscillation?

(d) What happens to the displacement of the bob as time increases without bound?

Solution

(a) We have $m = 10$, $a = 20$, and $b = 0.8$. Since the period of the pendulum under simple harmonic motion is 4 seconds, we have

$$4 = \frac{2\pi}{\omega}$$

$$\omega = \frac{2\pi}{4} = \frac{\pi}{2}$$

Figure 39

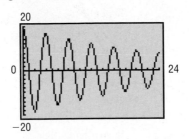

Substituting these values into the equation for damped motion, we obtain

$$d(t) = 20e^{-0.8t/2(10)} \cos\left(\sqrt{\left(\frac{\pi}{2}\right)^2 - \frac{0.8^2}{4(10)^2}}\, t\right)$$

$$= 20e^{-0.8t/20} \cos\left(\sqrt{\frac{\pi^2}{4} - \frac{0.64}{400}}\, t\right)$$

(b) See Figure 39 for the graph of $d = d(t)$.

(c) See Figure 40. At the start of the second oscillation the displacement is approximately 17.05 centimeters.

(d) As t increases without bound, $e^{-0.8t/20} \rightarrow 0$, so the displacement of the bob approaches zero. As a result, the pendulum will eventually come to rest.

Figure 40

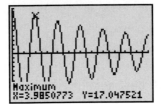

Maximum
X=3.9850773 Y=17.047521

━━━━ **Now Work** PROBLEM 21

4 Graph the Sum of Two Functions

Many physical and biological applications require the graph of the sum of two functions, such as

$$f(x) = x + \sin x \quad \text{or} \quad g(x) = \sin x + \cos(2x)$$

For example, if two tones are emitted, the sound produced is the sum of the waves produced by the two tones. See Problem 47 for an explanation of Touch-Tone phones.

To graph the sum of two (or more) functions, we can use the method of adding y-coordinates described next.

EXAMPLE 4

Graphing the Sum of Two Functions

Use the method of adding y-coordinates to graph $f(x) = x + \sin x$.

Solution

First, we graph the component functions,

$$y = f_1(x) = x \qquad y = f_2(x) = \sin x$$

in the same coordinate system. See Figure 41(a). Now, select several values of x, say, $x = 0$, $x = \dfrac{\pi}{2}$, $x = \pi$, $x = \dfrac{3\pi}{2}$, and $x = 2\pi$, at which we compute $f(x) = f_1(x) + f_2(x)$. Table 1 shows the computation. We plot these points and connect them to get the graph, as shown in Figure 41(b).

Table 1

x	0	$\dfrac{\pi}{2}$	π	$\dfrac{3\pi}{2}$	2π
$y = f_1(x) = x$	0	$\dfrac{\pi}{2}$	π	$\dfrac{3\pi}{2}$	2π
$y = f_2(x) = \sin x$	0	1	0	-1	0
$f(x) = x + \sin x$	0	$\dfrac{\pi}{2} + 1 \approx 2.57$	π	$\dfrac{3\pi}{2} - 1 \approx 3.71$	2π
Point on graph of f	$(0, 0)$	$\left(\dfrac{\pi}{2}, 2.57\right)$	(π, π)	$\left(\dfrac{3\pi}{2}, 3.71\right)$	$(2\pi, 2\pi)$

Figure 41

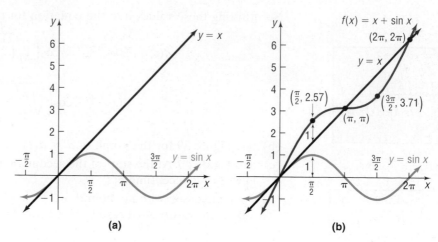

(a)　　　　　　　　　　　(b)

In Figure 41(b), notice that the graph of $f(x) = x + \sin x$ intersects the line $y = x$ whenever $\sin x = 0$. Also, notice that the graph of f is not periodic.

✓ **Check:** Graph $Y_1 = x$, $Y_2 = \sin x$, and $Y_3 = x + \sin x$ and compare the result with Figure 41(b). Use INTERSECT to verify that the graphs of Y_1 and Y_3 intersect when $\sin x = 0$.

■

The next example shows a periodic graph.

EXAMPLE 5 | **Graphing the Sum of Two Sinusoidal Functions**

Use the method of adding y-coordinates to graph

$$f(x) = \sin x + \cos(2x)$$

Solution　Table 2 shows the steps for computing several points on the graph of f. Figure 42 illustrates the graphs of the component functions, $y = f_1(x) = \sin x$ and $y = f_2(x) = \cos(2x)$, and the graph of $f(x) = \sin x + \cos(2x)$, which is shown in red.

Table 2

x	$-\dfrac{\pi}{2}$	0	$\dfrac{\pi}{2}$	π	$\dfrac{3\pi}{2}$	2π
$y = f_1(x) = \sin x$	-1	0	1	0	-1	0
$y = f_2(x) = \cos(2x)$	-1	1	-1	1	-1	1
$f(x) = \sin x + \cos(2x)$	-2	1	0	1	-2	1
Point on graph of f	$\left(-\dfrac{\pi}{2}, -2\right)$	$(0, 1)$	$\left(\dfrac{\pi}{2}, 0\right)$	$(\pi, 1)$	$\left(\dfrac{3\pi}{2}, -2\right)$	$(2\pi, 1)$

Figure 42

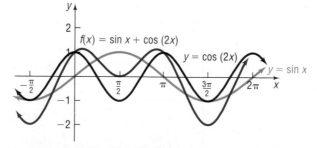

Notice that f is periodic, with period 2π.

✓ **Check:** Graph $Y_1 = \sin x$, $Y_2 = \cos(2x)$, and $Y_3 = \sin x + \cos(2x)$ and compare the result with Figure 42.

■

⟜━━━ **Now Work** PROBLEM 33

4.5 Assess Your Understanding

'Are You Prepared?' *Answer given at the end of these exercises. If you get a wrong answer, read the pages listed in red.*

1. The amplitude A and period T of $f(x) = 5 \sin(4x)$ are _____ and _____. (pp. 166–167)

Concepts and Vocabulary

2. The motion of an object obeys the equation $d = 4 \cos(6t)$. Such motion is described as _____ _____. The number 4 is called the _____.

3. When a mass hanging from a spring is pulled down and then released, the motion is called _____ _____ if there is no

frictional force to retard the motion, and the motion is called _____ if there is friction.

4. **True or False** If the distance d of an object from its rest position at time t is given by a sinusoidal graph, the motion of the object is simple harmonic motion.

Skill Building

In Problems 5–8, an object attached to a coiled spring is pulled down a distance a from its rest position and then released. Assuming that the motion is simple harmonic with period T, develop a model that relates the displacement d of the object from its rest position after t seconds. Also assume that the positive direction of the motion is up.

5. $a = 5$; $T = 2$ seconds

6. $a = 10$; $T - 3$ seconds

7. $a = 6$; $T = \pi$ seconds

8. $a = 4$; $T = \dfrac{\pi}{2}$ seconds

9. Rework Problem 5 under the same conditions except that, at time $t = 0$, the object is at its resting position and moving down.

10. Rework Problem 6 under the same conditions except that, at time $t = 0$, the object is at its resting position and moving down.

11. Rework Problem 7 under the same conditions except that, at time $t = 0$, the object is at its resting position and moving down.

12. Rework Problem 8 under the same conditions except that, at time $t = 0$, the object is at its resting position and moving down.

In Problems 13–20, the displacement d (in meters) of an object at time t (in seconds) is given.
 (a) Describe the motion of the object.
 (b) What is the maximum displacement from its resting position?
 (c) What is the time required for one oscillation?
 (d) What is the frequency?

13. $d = 5 \sin(3t)$

14. $d = 4 \sin(2t)$

15. $d = 6 \cos(\pi t)$

16. $d = 5 \cos\left(\dfrac{\pi}{2} t\right)$

17. $d = -3 \sin\left(\dfrac{1}{2} t\right)$

18. $d = -2 \cos(2t)$

19. $d = 6 + 2 \cos(2\pi t)$

20. $d = 4 + 3 \sin(\pi t)$

In Problems 21–26, an object of mass m (in grams) attached to a coiled spring with damping factor b (in grams per second) is pulled down a distance a (in centimeters) from its rest position and then released. Assume that the positive direction of the motion is up and the period is T (in seconds) under simple harmonic motion.
 (a) Develop a model that relates the distance d of the object from its rest position after t seconds.
 (b) Graph the equation found in part (a) for 5 oscillations using a graphing utility.

21. $m = 25$, $a = 10$, $b = 0.7$, $T = 5$

22. $m = 20$, $a = 15$, $b = 0.75$, $T = 6$

23. $m = 30$, $a = 18$, $b = 0.6$, $T = 4$

24. $m = 15$, $a = 16$, $b = 0.65$, $T = 5$

25. $m = 10$, $a = 5$, $b = 0.8$, $T = 3$

26. $m = 10$, $a = 5$, $b = 0.7$, $T = 3$

In Problems 27–32, the distance d (in meters) of the bob of a pendulum of mass m (in kilograms) from its rest position at time t (in seconds) is given. The bob is released from the left of its rest position, which represents a negative direction.
 (a) Describe the motion of the object. Be sure to give the mass and damping factor.
 (b) What is the initial displacement of the bob? That is, what is the displacement at t = 0?
 (c) Graph the motion using a graphing utility.
 (d) What is the displacement of the bob at the start of the second oscillation?
 (e) What happens to the displacement of the bob as time increases without bound?

27. $d = -20 e^{-0.7t/40} \cos\left(\sqrt{\left(\dfrac{2\pi}{5}\right)^2 - \dfrac{0.49}{1600}}\, t\right)$

28. $d = -20 e^{-0.8t/40} \cos\left(\sqrt{\left(\dfrac{2\pi}{5}\right)^2 - \dfrac{0.64}{1600}}\, t\right)$

29. $d = -30 e^{-0.6t/80} \cos\left(\sqrt{\left(\dfrac{2\pi}{7}\right)^2 - \dfrac{0.36}{6400}}\, t\right)$

30. $d = -30 e^{-0.5t/70} \cos\left(\sqrt{\left(\dfrac{\pi}{2}\right)^2 - \dfrac{0.25}{4900}}\, t\right)$

31. $d = -15 e^{-0.9t/30} \cos\left(\sqrt{\left(\dfrac{\pi}{3}\right)^2 - \dfrac{0.81}{900}}\, t\right)$

32. $d = -10 e^{-0.8t/50} \cos\left(\sqrt{\left(\dfrac{2\pi}{3}\right)^2 - \dfrac{0.64}{2500}}\, t\right)$

In Problems 33–40, use the method of adding y-coordinates to graph each function.

33. $f(x) = x + \cos x$

34. $f(x) = x + \cos(2x)$

35. $f(x) = x - \sin x$

36. $f(x) = x - \cos x$

37. $f(x) = \sin x + \cos x$

38. $f(x) = \sin(2x) + \cos x$

39. $g(x) = \sin x + \sin(2x)$

40. $g(x) = \cos(2x) + \cos x$

Applications and Extensions

41. Loudspeaker A loudspeaker diaphragm is oscillating in simple harmonic motion described by the equation $d = a \cos(\omega t)$ with a frequency of 520 hertz (cycles per second) and a maximum displacement of 0.80 millimeter. Find ω and then determine a model that describes the movement of the diaphragm.

42. Colossus Added to Six Flags St. Louis in 1986, the Colossus is a giant Ferris wheel. Its diameter is 165 feet, it rotates at a rate of about 1.6 revolutions per minute, and the bottom of the wheel is 15 feet above the ground. Determine a model that relates a rider's height above the ground at time t. Assume the passenger begins the ride at the bottom of the wheel.

Source: Six Flags Theme Parks, Inc.

43. Tuning Fork The end of a tuning fork moves in simple harmonic motion described by the equation $d = a \sin(\omega t)$. If a tuning fork for the note A above middle C on an even-tempered scale (A_4, the tone by which an orchestra tunes itself) has a frequency of 440 hertz (cycles per second), find ω. If the maximum displacement of the end of the tuning fork is 0.01 millimeter, determine a model that describes the movement of the tuning fork.

Source: David Lapp. Physics of Music and Musical Instruments. Medford, MA: Tufts University, 2003

44. Tuning Fork The end of a tuning fork moves in simple harmonic motion described by the equation $d = a \sin(\omega t)$. If a tuning fork for the note E above middle C on an even-tempered scale (E_4) has a frequency of approximately 329.63 hertz (cycles per second), find ω. If the maximum displacement of the end of the tuning fork is 0.025 millimeter, determine a model that describes the movement of the tuning fork.

Source: David Lapp. Physics of Music and Musical Instruments. Medford, MA: Tufts University, 2003

45. Charging a Capacitor If a charged capacitor is connected to a coil by closing a switch, energy is transferred to the coil and then back to the capacitor in an oscillatory motion. See the illustration. The voltage V (in volts) across the capacitor will gradually diminish to 0 with time t (in seconds).

(a) Graph the function relating V and t:

$$V(t) = e^{-t/3} \cos(\pi t), \quad 0 \le t \le 3$$

(b) At what times t will the graph of V touch the graph of $y = e^{-t/3}$? When does the graph of V touch the graph of $y = -e^{-t/3}$?

(c) When will the voltage V be between -0.4 and 0.4 volt?

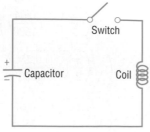

46. The Sawtooth Curve An oscilloscope often displays a *sawtooth curve*. This curve can be approximated by sinusoidal curves of varying periods and amplitudes.

(a) Use a graphing utility to graph the following function, which can be used to approximate the sawtooth curve.

$$f(x) = \frac{1}{2} \sin(2\pi x) + \frac{1}{4} \sin(4\pi x), \quad 0 \le x \le 4$$

(b) A better approximation to the sawtooth curve is given by

$$f(x) = \frac{1}{2} \sin(2\pi x) + \frac{1}{4} \sin(4\pi x) + \frac{1}{8} \sin(8\pi x)$$

Use a graphing utility to graph this function for $0 \le x \le 4$ and compare the result to the graph obtained in part (a).

(c) A third and even better approximation to the sawtooth curve is given by

$$f(x) = \frac{1}{2} \sin(2\pi x) + \frac{1}{4} \sin(4\pi x) + \frac{1}{8} \sin(8\pi x) + \frac{1}{16} \sin(16\pi x)$$

Use a graphing utility to graph this function for $0 \le x \le 4$ and compare the result to the graphs obtained in parts (a) and (b).

(d) What do you think the next approximation to the sawtooth curve is?

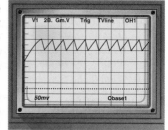

47. Touch-Tone Phones On a Touch-Tone phone, each button produces a unique sound. The sound produced is the sum of two tones, given by

$$y = \sin(2\pi l t) \quad \text{and} \quad y = \sin(2\pi h t)$$

where l and h are the low and high frequencies (cycles per second) shown in the illustration. For example, if you touch 7, the low frequency is $l = 852$ cycles per second and the high frequency is $h = 1209$ cycles per second. The sound emitted by touching 7 is

$$y = \sin[2\pi(852)t] + \sin[2\pi(1209)t]$$

Use a graphing utility to graph the sound emitted by touching 7.

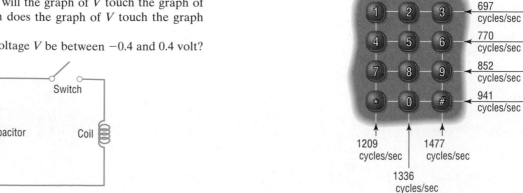

Touch-Tone phone

48. Use a graphing utility to graph the sound emitted by the * key on a Touch-Tone phone. See Problem 47.

49. CBL Experiment Pendulum motion is analyzed to estimate simple harmonic motion. A plot is generated with the position of the pendulum over time. The graph is used to find a sinusoidal curve of the form $y = A \cos[B(x - C)] + D$. Determine the amplitude, period, and frequency. (Activity 16, Real-World Math with the CBL System.)

50. CBL Experiment The sound from a tuning fork is collected over time. Determine the amplitude, frequency, and period of the graph. A model of the form $y = A \cos[B(x - C)]$ is fitted to the data. (Activity 23, Real-World Math with the CBL System.)

Discussion and Writing

51. Use a graphing utility to graph the function $f(x) = \dfrac{\sin x}{x}$, $x > 0$. Based on the graph, what do you conjecture about the value of $\dfrac{\sin x}{x}$ for x close to 0?

52. Use a graphing utility to graph $y = x \sin x$, $y = x^2 \sin x$, and $y = x^3 \sin x$ for $x > 0$. What patterns do you observe?

53. Use a graphing utility to graph $y = \dfrac{1}{x} \sin x$, $y = \dfrac{1}{x^2} \sin x$, and $y = \dfrac{1}{x^3} \sin x$ for $x > 0$. What patterns do you observe?

54. How would you explain to a friend what simple harmonic motion is? How would you explain damped motion?

'Are You Prepared?' Answer

1. $A = 5; T = \dfrac{\pi}{2}$

CHAPTER REVIEW

Things to Know

Formulas

Law of Sines (p. 282)	$\dfrac{\sin A}{a} = \dfrac{\sin B}{b} = \dfrac{\sin C}{c}$
Law of Cosines (p. 293)	$c^2 = a^2 + b^2 - 2ab \cos C$ $b^2 = a^2 + c^2 - 2ac \cos B$ $a^2 = b^2 + c^2 - 2bc \cos A$
Area of a triangle (pp. 299–300)	$K = \dfrac{1}{2}bh \quad K = \dfrac{1}{2}ab \sin C \quad K = \dfrac{1}{2}bc \sin A \quad K = \dfrac{1}{2}ac \sin B$ $K = \sqrt{s(s-a)(s-b)(s-c)} \quad$ where $\quad s = \dfrac{1}{2}(a + b + c)$

Objectives

Section		You should be able to . . .	Example(s)	Review Exercises
4.1	1	Solve right triangles (p. 276)	1, 2	1–4
	2	Solve applied problems (p. 277)	3–6	35, 36, 45–47
4.2	1	Solve SAA or ASA triangles (p. 283)	1, 2	5, 6, 22
	2	Solve SSA triangles (p. 284)	3–5	7–10, 12, 17, 18, 21
	3	Solve applied problems (p. 286)	6, 7	37–39
4.3	1	Solve SAS triangles (p. 293)	1	11, 15, 16, 23, 24
	2	Solve SSS triangles (p. 294)	2	13, 14, 19, 20
	3	Solve applied problems (p. 295)	3	40, 41
4.4	1	Find the area of SAS triangles (p. 299)	1	25–28, 43, 44
	2	Find the area of SSS triangles (p. 300)	2	29–32, 42
4.5	1	Build a model for an object in simple harmonic motion (p. 305)	1	48, 49
	2	Analyze simple harmonic motion (p. 307)	2	50–53
	3	Analyze an object in damped motion (p. 308)	3	54–57
	4	Graph the sum of two functions (p. 309)	4, 5	58, 59

Review Exercises

In Problems 1–4, solve each triangle.

1.

2.

3.

4.
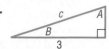

In Problems 5–24, find the remaining angle(s) and side(s) of each triangle, if it (they) exists. If no triangle exists, say "No triangle."

5. $A = 50°$, $B = 30°$, $a = 1$

6. $A = 10°$, $C = 40°$, $c = 2$

7. $A = 100°$, $a = 5$, $c = 2$

8. $a = 2$, $c = 5$, $A = 60°$

9. $a = 3$, $c = 1$, $C = 110°$

10. $a = 3$, $c = 1$, $C = 20°$

11. $a = 3$, $c = 1$, $B = 100°$

12. $a = 3$, $b = 5$, $B = 80°$

13. $a = 2$, $b = 3$, $c = 1$

14. $a = 10$, $b = 7$, $c = 8$

15. $a = 1$, $b = 3$, $C = 40°$

16. $a = 4$, $b = 1$, $C = 100°$

17. $a = 5$, $b = 3$, $A = 80°$

18. $a = 2$, $b = 3$, $A = 20°$

19. $a = 1$, $b = \frac{1}{2}$, $c = \frac{4}{3}$

20. $a = 3$, $b = 2$, $c = 2$

21. $a = 3$, $A = 10°$, $b = 4$

22. $a = 4$, $A = 20°$, $B = 100°$

23. $c = 5$, $b = 4$, $A = 70°$

24. $a = 1$, $b = 2$, $C = 60°$

In Problems 25–34, find the area of each triangle.

25. $a = 2$, $b = 3$, $C = 40°$

26. $b = 5$, $c = 5$, $A = 20°$

27. $b = 4$, $c = 10$, $A = 70°$

28. $a = 2$, $b = 1$, $C = 100°$

29. $a = 4$, $b = 3$, $c = 5$

30. $a = 10$, $b = 7$, $c = 8$

31. $a = 4$, $b = 2$, $c = 5$

32. $a = 3$, $b = 2$, $c = 2$

33. $A = 50°$, $B = 30°$, $a = 1$

34. $A = 10°$, $C = 40°$, $c = 3$

---

35. Finding the Grade of a Mountain Trail A straight trail with a uniform inclination leads from a hotel, elevation 5000 feet, to a lake in a valley, elevation 4100 feet. The length of the trail is 4100 feet. What is the inclination (grade) of the trail?

36. Geometry The hypotenuse of a right triangle is 12 feet. If one leg is 8 feet, find the degree measure of each angle.

37. Finding the Height of a Helicopter Two observers simultaneously measure the angle of elevation of a helicopter. One angle is measured as 25°, the other as 40° (see the figure). If the observers are 100 feet apart and the helicopter lies over the line joining them, how high is the helicopter?

38. Determining Distances at Sea Rebecca, the navigator of a ship at sea, spots two lighthouses that she knows to be 2 miles apart along a straight shoreline. She determines that the angles formed between two line-of-sight observations of the lighthouses and the line from the ship directly to shore are 12° and 30°. See the illustration.

(a) How far is the ship from lighthouse L_1?
(b) How far is the ship from lighthouse L_2?
(c) How far is the ship from shore?

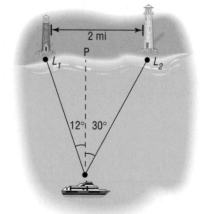

39. Constructing a Highway A highway whose primary directions are north–south is being constructed along the west coast of Florida. Near Naples, a bay obstructs the straight path of the road. Since the cost of a bridge is prohibitive, engineers decide to go around the bay. The illustration shows the path that they decide on and the measurements taken. What is the length of highway needed to go around the bay?

40. **Correcting a Navigational Error** A sailboat leaves St. Thomas bound for an island in the British West Indies, 200 miles away. Maintaining a constant speed of 18 miles per hour, but encountering heavy crosswinds and strong currents, the crew finds after 4 hours that the sailboat is off course by 15°.
    (a) How far is the sailboat from the island at this time?
    (b) Through what angle should the sailboat turn to correct its course?
    (c) How much time has been added to the trip because of this? (Assume that the speed remains at 18 miles per hour.)

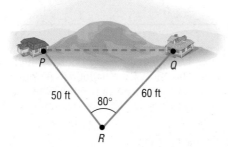

41. **Surveying** Two homes are located on opposite sides of a small hill. See the illustration. To measure the distance between them, a surveyor walks a distance of 50 feet from house P to point R, uses a transit to measure $\angle PRQ$, which is found to be 80°, and then walks to house Q, a distance of 60 feet. How far apart are the houses?

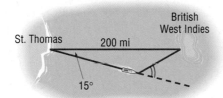

42. **Approximating the Area of a Lake** To approximate the area of a lake, Cindy walks around the perimeter of the lake, taking the measurements shown in the illustration. Using this technique, what is the approximate area of the lake?

[**Hint:** Use the Law of Cosines on the three triangles shown and then find the sum of their areas.]

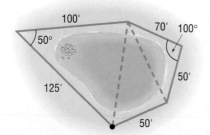

43. **Calculating the Cost of Land** The irregular parcel of land shown in the figure is being sold for $100 per square foot. What is the cost of this parcel?

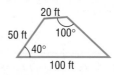

44. **Area of a Segment** Find the area of the segment of a circle whose radius is 6 inches formed by a central angle of 50°.

45. **Finding the Bearing of a Ship** The *Majesty* leaves the Port at Boston for Bermuda with a bearing of S80°E at an average speed of 10 knots. After 1 hour, the ship turns 90° toward the southwest. After 2 hours at an average speed of 20 knots, what is the bearing of the ship from Boston?

46. **Drive Wheels of an Engine** The drive wheel of an engine is 13 inches in diameter, and the pulley on the rotary pump is 5 inches in diameter. If the shafts of the drive wheel and the pulley are 2 feet apart, what length of belt is required to join them as shown in the figure?

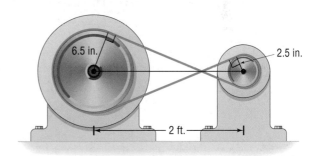

47. Rework Problem 46 if the belt is crossed, as shown in the figure.

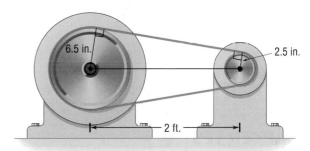

In Problems 48 and 49, an object attached to a coiled spring is pulled down a distance a from its rest position and then released. Assuming that the motion is simple harmonic with period T, develop a model that relates the displacement d of the object from its rest position after t seconds. Also assume that the positive direction of the motion is up.

48. $a = 3$; $T = 4$ seconds

49. $a = 5$; $T = 6$ seconds

In Problems 50–53, the distance d (in feet) that an object travels in time t (in seconds) is given.
(a) Describe the motion of the object.
(b) What is the maximum displacement from its rest position?
(c) What is the time required for one oscillation?
(d) What is the frequency?

50. $d = 6 \sin(2t)$ **51.** $d = 2 \cos(4t)$ **52.** $d = -2 \cos(\pi t)$ **53.** $d = -3 \sin\left[\dfrac{\pi}{2}t\right]$

In Problems 54 and 55, an object of mass m attached to a coiled spring with damping factor b is pulled down a distance a from its rest position and then released. Assume that the positive direction of the motion is up and the period is T under simple harmonic motion.
(a) Develop a model that relates the distance d of the object from its rest position after t seconds.
(b) Graph the equation found in part (a) for 5 oscillations.

54. $m = 40$ grams; $a = 15$ centimeters; $b = 0.75$ gram/second; $T = 5$ seconds

55. $m = 25$ grams; $a = 13$ centimeters; $b = 0.65$ gram/second; $T = 4$ seconds

In Problems 56 and 57, the distance d (in meters) of the bob of a pendulum of mass m (in kilograms) from its rest position at time t (in seconds) is given.
(a) Describe the motion of the object.
(b) What is the initial displacement of the bob? That is, what is the displacement at t = 0?
(c) Graph the motion using a graphing utility.
(d) What is the displacement of the bob at the start of the second oscillation?
(e) What happens to the displacement of the bob as time increases without bound?

56. $d = -15e^{-0.6t/40} \cos\left(\sqrt{\left(\dfrac{2\pi}{5}\right)^2 - \dfrac{0.36}{1600}}\, t\right)$ **57.** $d = -20e^{-0.5t/60} \cos\left(\sqrt{\left(\dfrac{2\pi}{3}\right)^2 - \dfrac{0.25}{3600}}\, t\right)$

In Problems 58 and 59, use the method of adding y-coordinates to graph each function.

58. $y = 2 \sin x + \cos(2x)$ **59.** $y = 2 \cos(2x) + \sin\dfrac{x}{2}$

CHAPTER TEST

1. A 12-foot ladder leans against a building. The top of the ladder leans against the wall 10.5 feet from the ground. What is the angle formed by the ground and the ladder?

2. A hot-air balloon is flying at a height of 600 feet and is directly above the Marshall Space Flight Center in Huntsville, Alabama. The pilot of the balloon looks down at the airport that is known to be 5 miles from the Marshall Space Flight Center. What is the angle of depression from the balloon to the airport?

In Problems 3–5, use the given information to determine the three remaining parts of each triangle.

3.
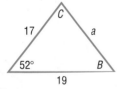

4.
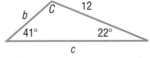

5.
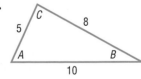

In Problems 6–8, solve each triangle.

6. $A = 55°$, $C = 20°$, $a = 4$ **7.** $a = 3$, $b = 7$, $A = 40°$ **8.** $a = 8$, $b = 4$, $C = 70°$

9. Find the area of the triangle described in Problem 8.

10. Find the area of the triangle described in Problem 5.

11. Find the area of the shaded region enclosed in a semicircle of diameter 8 centimeters. The length of the chord AB is 6 centimeters.

   [**Hint:** Triangle ABC is a right triangle.]

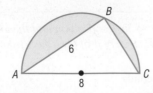

12. Find the area of the quadrilateral shown.

13. Madison wants to swim across Lake William from the fishing lodge (A) to the boat ramp (B), but she wants to know the distance first. Highway 20 goes right past the boat ramp, and County Road 3 goes to the lodge. The two roads

intersect at point (C), 4.2 miles from the ramp and 3.5 miles from the lodge. Madison uses a transit to measure the angle of intersection of the two roads to be 32°. How far will she need to swim?

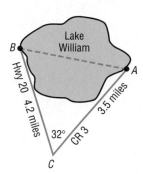

14. Given that $\triangle OAB$ is an isosceles triangle and the shaded sector is a semicircle, find the area of the entire region. Express your answer as a decimal rounded to two places.

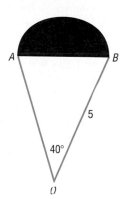

15. The area of the triangle shown is $54\sqrt{6}$ square units; find the lengths of the sides.

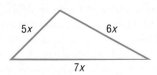

16. Logan is playing on her swing. One full swing (front to back to front) takes 6 seconds, and at the peak of her swing she is at an angle of 42° with the vertical. If her swing is 5 feet long and we ignore all resistive forces, build a model that relates her horizontal displacement (from the rest position) after time t.

CUMULATIVE REVIEW

1. Find the real solutions, if any, of the equation $3x^2 + 1 = 4x$.

2. Find an equation for the circle with center at the point $(-5, 1)$ and radius 3. Graph this circle.

3. Determine the domain of the function
$$f(x) = \sqrt{x^2 - 3x - 4}$$

4. Graph the function $y = 3\sin(\pi x)$.

5. Graph the function $y = -2\cos(2x - \pi)$.

6. If $\tan\theta = -2$ and $\dfrac{3\pi}{2} < \theta < 2\pi$, find the exact value of:
   - (a) $\sin\theta$
   - (b) $\cos\theta$
   - (c) $\sin(2\theta)$
   - (d) $\cos(2\theta)$
   - (e) $\sin\left(\dfrac{1}{2}\theta\right)$
   - (f) $\cos\left(\dfrac{1}{2}\theta\right)$

7. Graph each of the following functions on the interval $[0, 4]$:
   - (a) $y = \sin x$
   - (b) $y = 2x + \sin x$

8. Sketch the graph of each of the following functions:
   - (a) $y = x$
   - (b) $y = x^2$
   - (c) $y = \sqrt{x}$
   - (d) $y = x^3$
   - (e) $y = \sin x$
   - (f) $y = \cos x$
   - (g) $y = \tan x$

9. Solve the triangle:

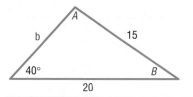

CHAPTER PROJECTS

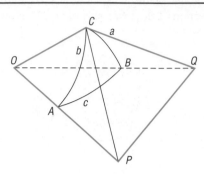

I. Spherical Trigonometry When the distance between two locations on the surface of Earth is small, we can compute the distance in statutory miles. Using this assumption, we can use the Law of Sines and the Law of Cosines to approximate distances and angles. However, if you look at a globe, you notice that Earth is a sphere, so, as the distance between two points on its surface increases, the linear distance is less accurate because of curvature. Under this circumstance, we need to take into account the curvature of Earth when using the Law of Sines and the Law of Cosines.

1. Draw a spherical triangle and label each vertex by A, B, and C. Then connect each vertex by a radius to the center O of the sphere. Now draw tangent lines to the sides a and b of the triangle that go through C. Extend the lines OA and OB to intersect the tangent lines at P and Q, respectively. See the figure. List the plane right triangles. Determine the measures of the central angles.

2. Apply the Law of Cosines to triangles OPQ and CPQ to find two expressions for the length of PQ.

3. Subtract the expressions in part (2) from each other. Solve for the term containing $\cos c$.

4. Use the Pythagorean Theorem to find another value for $OQ^2 - CQ^2$ and $OP^2 - CP^2$. Now solve for $\cos c$.

5. Replacing the ratios in part (4) by the cosines of the sides of the spherical triangle, you should now have the Law of Cosines for spherical triangles:

$$\cos c = \cos a \cos b + \sin a \sin b \cos C$$

Source: For the spherical Law of Cosines; see *Mathematics from the Birth of Numbers* by Jan Gullberg. W. W. Norton & Co., Publishers, 1996, pp. 491–494.

II. The Lewis and Clark Expedition Lewis and Clark followed several rivers in their trek from what is now Great Falls, Montana, to the Pacific coast. First, they went down the Missouri and Jefferson rivers from Great Falls to Lemhi, Idaho. Because the two cities are on different longitudes and different latitudes, we must account for the curvature of Earth when computing the distance that they traveled. Assume that the radius of Earth is 3960 miles.

1. Great Falls is at approximately 47.5°N and 111.3°W. Lemhi is at approximately 45.5°N and 113.5°W. (We will assume that the rivers flow straight from Great Falls to Lemhi on the surface of Earth.) This line is called a geodesic line. Apply the Law of Cosines for a spherical triangle to find the angle between Great Falls and Lemhi. (The central angles are found by using the differences in the latitudes and longitudes of the towns. See the diagram.) Then find the length of the arc joining the two towns. (Recall $s = r\theta$.)

2. From Lemhi, they went up the Bitteroot River and the Snake River to what is now Lewiston and Clarkston on the border of Idaho and Washington. Although this is not really a side to a triangle, we will make a side that goes from Lemhi to Lewiston and Clarkston. If Lewiston and Clarkston are at about 46.5°N 117.0°W, find the distance from Lemhi using the Law of Cosines for a spherical triangle and the arc length.

3. How far did the explorers travel just to get that far?

4. Draw a plane triangle connecting the three towns. If the distance from Lewiston to Great Falls is 282 miles and the angle at Great Falls is 42° and the angle at Lewiston is 48.5°, find the distance from Great Falls to Lemhi and from Lemhi to Lewiston. How do these distances compare with the ones computed in parts (1) and (2)?

Source: For Lewis and Clark Expedition: *American Journey: The Quest for Liberty to 1877, Texas Edition.* Prentice Hall, 1992, p. 345.

Source: For map coordinates: *National Geographic Atlas of the World,* published by National Geographic Society, 1981, pp. 74–75.

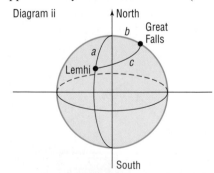

Diagram ii

The following projects are available at the Instructor's Resource Center (IRC):

III. Project at Motorola: ***How Can You Build or Analyze a Vibration Profile?*** Fourier functions are not only important to analyze vibrations, but they are also what a mathematician would call interesting. Complete the project to see why.

IV. Leaning Tower of Pisa Trigonometry is used to analyze the apparent height and tilt of the Leaning Tower of Pisa.

V. Locating Lost Treasure Clever treasure seekers who know the Law of Sines are able to efficiently find a buried treasure.

VI. Jacob's Field Angles of elevation and the Law of Sines are used to determine the height of the stadium wall and the distance from home plate to the top of the wall.

Polar Coordinates; Vectors

5

How Do Airplanes Fly?

Have you ever watched a big jetliner lumber into position on the runway for takeoff and wonder, "How does that thing ever get off the ground?" You know it's because of the wing that it stays up in the air, but how does it really work?

When air flows around a wing, it creates lift. The way it creates lift is based on the wing's movement through the air and the air pressure created around the wing. An airplane's wing, in varying degrees depending on the type and design of the airplane, is curved over the top of the wing and straighter underneath the wing. As air hits the wing, it is "split in two," with air moving both over and under the wing. Since the top of the wing has more curve than the bottom of the wing, the air moving over the top of the wing has farther to travel, and thus must move faster than the air moving underneath the wing. The air moving over the top of the wing now exerts less air pressure on the wing than the slower-moving air under the wing. Lift is created.

The difference in air pressure is the primary force creating lift on a wing, but one other force exerted on the wing also helps to produce lift. This is the force of deflection. Air moving along the underside of the wing is deflected downward. Remember the Newtonian principle: For every action, there is an equal and opposite reaction. The air that is deflected downward (action) helps to push the wing upward (reaction), producing more lift.

These two natural forces on the wing, pressure and deflection, produce lift. The faster the wing moves through the air, the greater the forces become, and the greater the lift.

Source: Thomas Schueneman. How do airplanes fly? http://meme.essortment.com/howdoairplane_rlmi.htm, accessed August 2006. © 2002 by Pagewise, Inc. Used with permission.

—See Chapter Project I—

A Look Back, A Look Ahead

This chapter is in two parts: Polar Coordinates, Sections 5.1–5.3, and Vectors, Sections 5.4–5.7. They are independent of each other and may be covered in any order.

Sections 5.1–5.3: In Chapter 1 we introduced rectangular coordinates x and y and discussed the graph of an equation in two variables involving x and y. In Sections 5.1 and 5.2, we introduce polar coordinates, an alternative to rectangular coordinates, and discuss graphing equations that involve polar coordinates. In Section 5.3 we discuss raising a complex number to a real power. As it turns out, polar coordinates are useful for the discussion.

Sections 5.4–5.7: We have seen in many chapters that we are often required to solve an equation to obtain a solution to applied problems. In the last four sections of this chapter, we develop the notion of a vector and show how it can be used to model applied problems in physics and engineering.

Outline

5.1 Polar Coordinates

PREPARING FOR THIS SECTION   *Before getting started, review the following:*

- Rectangular Coordinates (Section 1.1, pp. 2–4)
- Definitions of the Sine and Cosine Functions (Section 2.4, pp. 140–143)
- Inverse Tangent Function (Section 3.1, pp. 213–215)
- Completing the Square (Appendix, Section A.4, p. A31)

Now Work the 'Are You Prepared?' problems on page 327.

OBJECTIVES   **1** Plot Points Using Polar Coordinates (p. 320)
2 Convert from Polar Coordinates to Rectangular Coordinates (p. 322)
3 Convert from Rectangular Coordinates to Polar Coordinates (p. 324)
4 Transform Equations from Polar to Rectangular Form (p. 326)

Figure 1

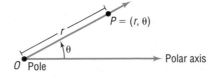

So far, we have always used a system of rectangular coordinates to plot points in the plane. Now we are ready to describe another system, called *polar coordinates*. As we shall soon see, in many instances polar coordinates offer certain advantages over rectangular coordinates.

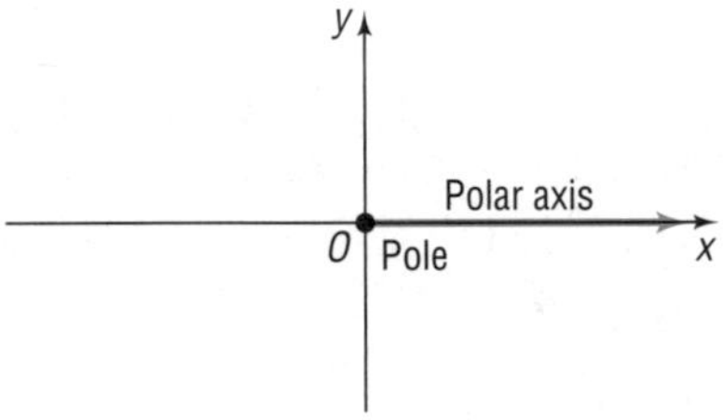

In a rectangular coordinate system, you will recall, a point in the plane is represented by an ordered pair of numbers (x, y), where x and y equal the signed distance of the point from the y-axis and x-axis, respectively. In a polar coordinate system, we select a point, called the **pole,** and then a ray with vertex at the pole, called the **polar axis.** See Figure 1. Comparing the rectangular and polar coordinate systems, we see that the origin in rectangular coordinates coincides with the pole in polar coordinates, and the positive x-axis in rectangular coordinates coincides with the polar axis in polar coordinates.

1 Plot Points Using Polar Coordinates

A point P in a polar coordinate system is represented by an ordered pair of numbers (r, θ). If $r > 0$, then r is the distance of the point from the pole; θ is an angle (in degrees or radians) formed by the polar axis and a ray from the pole through the point. We call the ordered pair (r, θ) the **polar coordinates** of the point. See Figure 2.

As an example, suppose that the polar coordinates of a point P are $\left(2, \dfrac{\pi}{4}\right)$. We locate P by first drawing an angle of $\dfrac{\pi}{4}$ radian, placing its vertex at the pole and its initial side along the polar axis. Then we go out a distance of 2 units along the terminal side of the angle to reach the point P. See Figure 3.

Figure 2

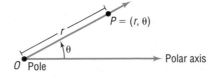

Figure 3

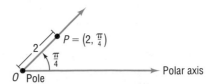

In using polar coordinates (r, θ), it is possible for r to be negative. When this happens, instead of the point being on the terminal side of θ, it is on the ray from the pole extending in the direction *opposite* the terminal side of θ at a distance $|r|$ units from the pole. See Figure 4 for an illustration.

For example, to plot the point $\left(-3, \dfrac{2\pi}{3}\right)$, we use the ray in the opposite direction of $\dfrac{2\pi}{3}$ and go out $|-3| = 3$ units along that ray. See Figure 5.

Figure 4

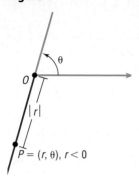

Figure 5

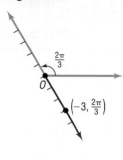

EXAMPLE 1

Plotting Points Using Polar Coordinates

Plot the points with the following polar coordinates:

(a) $\left(3, \dfrac{5\pi}{3}\right)$ (b) $\left(2, -\dfrac{\pi}{4}\right)$ (c) $(3, 0)$ (d) $\left(-2, \dfrac{\pi}{4}\right)$

Solution   Figure 6 shows the points.

Figure 6

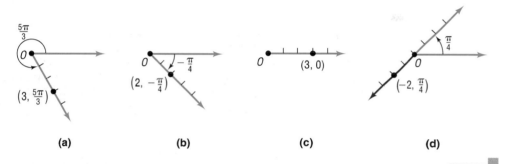

(a)   (b)   (c)   (d)

Now Work PROBLEMS **11, 19,** AND **27**

Recall that an angle measured counterclockwise is positive and an angle measured clockwise is negative. This convention has some interesting consequences relating to polar coordinates. Let's see what these consequences are.

EXAMPLE 2

Finding Several Polar Coordinates of a Single Point

Consider again the point P with polar coordinates $\left(2, \dfrac{\pi}{4}\right)$, as shown in Figure 7(a). Because $\dfrac{\pi}{4}$, $\dfrac{9\pi}{4}$, and $-\dfrac{7\pi}{4}$ all have the same terminal side, we also could have located this point P by using the polar coordinates $\left(2, \dfrac{9\pi}{4}\right)$ or $\left(2, -\dfrac{7\pi}{4}\right)$, as shown in Figures 7(b) and (c). The point $\left(2, \dfrac{\pi}{4}\right)$ can also be represented by the polar coordinates $\left(-2, \dfrac{5\pi}{4}\right)$. See Figure 7(d).

Figure 7

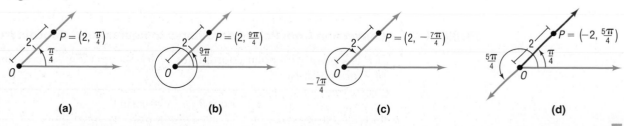

(a)   (b)   (c)   (d)

| EXAMPLE 3 | **Finding Other Polar Coordinates of a Given Point** |

Plot the point P with polar coordinates $\left(3, \dfrac{\pi}{6}\right)$, and find other polar coordinates (r, θ) of this same point for which:

(a) $r > 0, \quad 2\pi \leq \theta < 4\pi$ (b) $r < 0, \quad 0 \leq \theta < 2\pi$

(c) $r > 0, \quad -2\pi \leq \theta < 0$

Solution   The point $\left(3, \dfrac{\pi}{6}\right)$ is plotted in Figure 8.

Figure 8

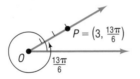

$P = \left(3, \frac{\pi}{6}\right)$

$\dfrac{\pi}{6}$

0

(a) We add 1 revolution (2π radians) to the angle $\dfrac{\pi}{6}$ to get

$$P = \left(3, \frac{\pi}{6} + 2\pi\right) = \left(3, \frac{13\pi}{6}\right). \text{ See Figure 9.}$$

(b) We add $\dfrac{1}{2}$ revolution (π radians) to the angle $\dfrac{\pi}{6}$ and replace 3 by -3 to get

$$P = \left(-3, \frac{\pi}{6} + \pi\right) = \left(-3, \frac{7\pi}{6}\right). \text{ See Figure 10.}$$

(c) We subtract 2π from the angle $\dfrac{\pi}{6}$ to get $P = \left(3, \dfrac{\pi}{6} - 2\pi\right) = \left(3, -\dfrac{11\pi}{6}\right)$. See Figure 11.

Figure 9   **Figure 10**   **Figure 11**

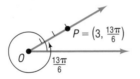
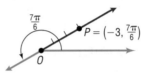
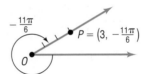

These examples show a major difference between rectangular coordinates and polar coordinates. In the former, each point has exactly one pair of rectangular coordinates; in the latter, a point can have infinitely many pairs of polar coordinates.

SUMMARY   A point with polar coordinates (r, θ) can also be represented by either of the following:

$$(r, \theta + 2\pi k) \quad \text{or} \quad (-r, \theta + \pi + 2\pi k) \qquad k \text{ any integer}$$

The polar coordinates of the pole are $(0, \theta)$, where θ can be any angle.

Now Work PROBLEM 31

2 Convert from Polar Coordinates to Rectangular Coordinates

Sometimes we need to convert coordinates or equations in rectangular form to polar form, and vice versa. To do this, we recall that the origin in rectangular coordinates is the pole in polar coordinates and that the positive x-axis in rectangular coordinates is the polar axis in polar coordinates.

THEOREM   **Conversion from Polar Coordinates to Rectangular Coordinates**

If P is a point with polar coordinates (r, θ), the rectangular coordinates (x, y) of P are given by

$$x = r \cos \theta \qquad y = r \sin \theta \tag{1}$$

Figure 12

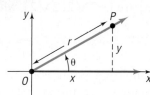

Proof Suppose that P has the polar coordinates (r, θ). We seek the rectangular coordinates (x, y) of P. Refer to Figure 12.

If $r = 0$, then, regardless of θ, the point P is the pole, for which the rectangular coordinates are $(0, 0)$. Formula (1) is valid for $r = 0$.

If $r > 0$, the point P is on the terminal side of θ, and $r = d(O, P) = \sqrt{x^2 + y^2}$. Since

$$\cos \theta = \frac{x}{r} \qquad \sin \theta = \frac{y}{r}$$

we have

$$x = r \cos \theta \qquad y = r \sin \theta$$

If $r < 0$, the point $P = (r, \theta)$ can be represented as $(-r, \pi + \theta)$, where $-r > 0$. Since

$$\cos(\pi + \theta) = -\cos \theta = \frac{x}{-r} \qquad \sin(\pi + \theta) = -\sin \theta = \frac{y}{-r}$$

we have

$$x = r \cos \theta \qquad y = r \sin \theta \qquad \blacksquare$$

EXAMPLE 4	**Converting from Polar Coordinates to Rectangular Coordinates**

Find the rectangular coordinates of the points with the following polar coordinates:

(a) $\left(6, \dfrac{\pi}{6} \right)$ (b) $\left(-4, -\dfrac{\pi}{4} \right)$

Solution We use formula (1): $x = r \cos \theta$ and $y = r \sin \theta$.

Figure 13

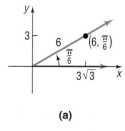

(a)

(b)

(a) Figure 13(a) shows $\left(6, \dfrac{\pi}{6} \right)$ plotted. Notice that $\left(6, \dfrac{\pi}{6} \right)$ lies in quadrant I of the rectangular coordinate system. So we expect both the x-coordinate and the y-coordinate to be positive. With $r = 6$ and $\theta = \dfrac{\pi}{6}$, we have

$$x = r \cos \theta = 6 \cos \frac{\pi}{6} = 6 \cdot \frac{\sqrt{3}}{2} = 3\sqrt{3}$$

$$y = r \sin \theta = 6 \sin \frac{\pi}{6} = 6 \cdot \frac{1}{2} = 3$$

The rectangular coordinates of the point $\left(6, \dfrac{\pi}{6} \right)$ are $\left(3\sqrt{3}, 3 \right)$, which lies in quadrant I, as expected.

(b) Figure 13(b) shows $\left(-4, -\dfrac{\pi}{4} \right)$ plotted. Notice that $\left(-4, -\dfrac{\pi}{4} \right)$ lies in quadrant II of the rectangular coordinate system. With $r = -4$ and $\theta = -\dfrac{\pi}{4}$, we have

$$x = r \cos \theta = -4 \cos\left(-\frac{\pi}{4} \right) = -4 \cdot \frac{\sqrt{2}}{2} = -2\sqrt{2}$$

$$y = r \sin \theta = -4 \sin\left(-\frac{\pi}{4} \right) = -4\left(-\frac{\sqrt{2}}{2} \right) = 2\sqrt{2}$$

The rectangular coordinates of the point $\left(-4, -\dfrac{\pi}{4}\right)$ are $\left(-2\sqrt{2}, 2\sqrt{2}\right)$, which lies in quadrant II, as expected.

Figure 14

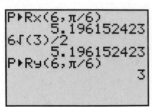

Most calculators have the capability of converting from polar coordinates to rectangular coordinates. Consult your owner's manual for the proper key strokes. Since in most cases this procedure is tedious, you will find that using formula (1) is faster.

Figure 14 verifies the result obtained in Example 4(a) using a TI-84 Plus. Note that the calculator is in radian mode.

━━━━ **Now Work** PROBLEMS 39 AND 51

3 Convert from Rectangular Coordinates to Polar Coordinates

Converting from rectangular coordinates (x, y) to polar coordinates (r, θ) is a little more complicated. Notice that we begin each example by plotting the given rectangular coordinates.

| EXAMPLE 5 | How to Convert from Rectangular Coordinates to Polar Coordinates with the Point on a Coordinate Axis |

Find polar coordinates of a point whose rectangular coordinates are $(0, 3)$.

Step-by-Step Solution

STEP 1 Plot the point (x, y) and note the quadrant the point lies in or the coordinate axis the point lies on.

We plot the point $(0, 3)$ in a rectangular coordinate system. See Figure 15. The point lies on the positive y-axis

Figure 15

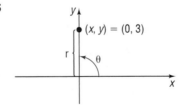

STEP 2 Determine the distance r from the origin to the point.

The point $(0, 3)$ lies on the y-axis a distance of 3 units from the origin (pole), so $r = 3$.

STEP 3 Determine θ.

A ray with vertex at the pole through $(0, 3)$ forms an angle $\theta = \dfrac{\pi}{2}$ with the polar axis.

Polar coordinates for this point can be given by $\left(3, \dfrac{\pi}{2}\right)$. Other possible representations include $\left(-3, -\dfrac{\pi}{2}\right)$ and $\left(3, \dfrac{5\pi}{2}\right)$.

Figure 16

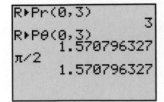

Most graphing calculators have the capability of converting from rectangular coordinates to polar coordinates. Consult your owner's manual for the proper keystrokes. Figure 16 verifies the results obtained in Example 5 using a TI-84 Plus. Note that the calculator is in radian mode.

Figure 17 shows polar coordinates of points that lie on either the x-axis or the y-axis. In each illustration, $a > 0$.

Figure 17

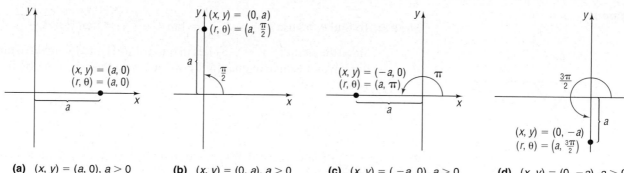

(a) $(x, y) = (a, 0), a > 0$ **(b)** $(x, y) = (0, a), a > 0$ **(c)** $(x, y) = (-a, 0), a > 0$ **(d)** $(x, y) = (0, -a), a > 0$

$\longrightarrow$ **Now Work** PROBLEM 55

| EXAMPLE 6 | **How to Convert from Rectangular Coordinates to Polar Coordinates with the Point in a Quadrant** |

Find the polar coordinates of a point whose rectangular coordinates are $(2, -2)$.

Step-by-Step Solution

STEP 1 Plot the point (x, y) and note the quadrant the point lies in or the coordinate axis the point lies on.

We plot the point $(2, -2)$ in a rectangular coordinate system. See Figure 18. The point lies in quadrant IV.

Figure 18

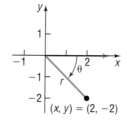

$(x, y) = (2, -2)$

STEP 2 Determine the distance r from the origin to the point using $r = \sqrt{x^2 + y^2}$.

$$r = \sqrt{x^2 + y^2} = \sqrt{(2)^2 + (-2)^2} = \sqrt{8} = 2\sqrt{2}$$

STEP 3 Determine θ.

We find θ by recalling that $\tan \theta = \dfrac{y}{x}$, so $\theta = \tan^{-1}\dfrac{y}{x}$, $-\dfrac{\pi}{2} < \theta < \dfrac{\pi}{2}$. Since $(2, -2)$ lies in quadrant IV, we know that $-\dfrac{\pi}{2} < \theta < 0$. As a result,

$$\theta = \tan^{-1}\frac{y}{x} = \tan^{-1}\left(\frac{-2}{2}\right) = \tan^{-1}(-1) = -\frac{\pi}{4}$$

A set of polar coordinates for the point $(2, -2)$ is $\left(2\sqrt{2}, -\dfrac{\pi}{4}\right)$. Other possible representations include $\left(2\sqrt{2}, \dfrac{7\pi}{4}\right)$ and $\left(-2\sqrt{2}, \dfrac{3\pi}{4}\right)$.

| EXAMPLE 7 | **Converting from Rectangular Coordinates to Polar Coordinates** |

Find polar coordinates of a point whose rectangular coordinates are $\left(-1, -\sqrt{3}\right)$.

Solution   **STEP 1:** See Figure 19. The point lies in quadrant III.

STEP 2: The distance r from the origin to the point $\left(-1, -\sqrt{3}\right)$ is

$$r = \sqrt{(-1)^2 + \left(-\sqrt{3}\right)^2} = \sqrt{4} = 2$$

Figure 19

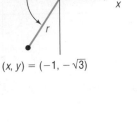

$(x, y) = (-1, -\sqrt{3})$

STEP 3: To find θ, we use $\theta = \tan^{-1}\dfrac{y}{x} = \tan^{-1}\dfrac{-\sqrt{3}}{-1} = \tan^{-1}\sqrt{3}, -\dfrac{\pi}{2} < \theta < \dfrac{\pi}{2}$.

Since the point $\left(-1, -\sqrt{3}\right)$ lies in quadrant III and the inverse tangent function gives an angle in quadrant I, we add π to the result to obtain an angle in quadrant III.

$$\theta = \pi + \tan^{-1}\left(\dfrac{-\sqrt{3}}{-1}\right) = \pi + \tan^{-1}\sqrt{3} = \pi + \dfrac{\pi}{3} = \dfrac{4\pi}{3}$$

A set of polar coordinates for this point is $\left(2, \dfrac{4\pi}{3}\right)$. Other possible representations include $\left(-2, \dfrac{\pi}{3}\right)$ and $\left(2, -\dfrac{2\pi}{3}\right)$.

Figure 20 shows how to find polar coordinates of a point that lies in a quadrant when its rectangular coordinates (x, y) are given.

Figure 20

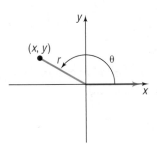
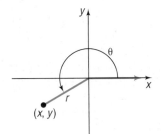
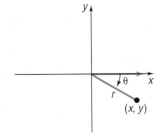

(a) $r = \sqrt{x^2 + y^2}$
$\theta = \tan^{-1}\dfrac{y}{x}$

(b) $r = \sqrt{x^2 + y^2}$
$\theta = \pi + \tan^{-1}\dfrac{y}{x}$

(c) $r = \sqrt{x^2 + y^2}$
$\theta = \pi + \tan^{-1}\dfrac{y}{x}$

(d) $r = \sqrt{x^2 + y^2}$
$\theta = \tan^{-1}\dfrac{y}{x}$

Based on the preceding discussion, we have the formulas

$$r^2 = x^2 + y^2 \qquad \tan\theta = \dfrac{y}{x} \qquad \text{if } x \neq 0 \qquad \textbf{(2)}$$

To use formula (2) effectively, follow these steps:

Steps for Converting from Rectangular to Polar Coordinates

STEP 1: Always plot the point (x, y) first, as we did in Examples 5 and 6. Note the quadrant the point lies in or the coordinate axis the point lies on.

STEP 2: If $x = 0$ or $y = 0$, use your illustration to find r. See Figure 17. If $x \neq 0$ and $y \neq 0$, then $r = \sqrt{x^2 + y^2}$.

STEP 3: Find θ. If $x = 0$ or $y = 0$, use your illustration to find θ. See Figure 17. If $x \neq 0$ and $y \neq 0$, note the quadrant in which the point lies.

Quadrant I or IV: $\theta = \tan^{-1}\dfrac{y}{x}$

Quadrant II or III: $\theta = \pi + \tan^{-1}\dfrac{y}{x}$

See Figure 20.

━━ **Now Work** PROBLEM 59

4 Transform Equations from Polar to Rectangular Form

Formulas (1) and (2) may also be used to transform equations from polar form to rectangular form, and vice versa. Two common techniques for transforming an equation from polar form to rectangular form are the following:

1. Multiplying both sides of the equation by r

2. Squaring both sides of the equation

| EXAMPLE 8 | **Transforming an Equation from Polar to Rectangular Form** |

Transform the equation $r = 6 \cos \theta$ from polar coordinates to rectangular coordinates, and identify the graph.

Solution   If we multiply each side by r, it will be easier to apply formulas (1) and (2).

$$r = 6 \cos \theta$$
$$r^2 = 6r \cos \theta \quad \text{Multiply each side by } r.$$
$$x^2 + y^2 = 6x \quad \quad r^2 = x^2 + y^2; x = r\cos\theta$$

This is the equation of a circle, so we proceed to complete the square to obtain the standard form of the equation.

$$x^2 + y^2 = 6x$$
$$(x^2 - 6x) + y^2 = 0 \quad \text{General form}$$
$$(x^2 - 6x + 9) + y^2 = 9 \quad \text{Complete the square in } x.$$
$$(x - 3)^2 + y^2 = 9 \quad \text{Factor.}$$

This is the standard form of the equation of a circle with center $(3, 0)$ and radius 3.

══════ **Now Work** PROBLEM 75

| EXAMPLE 9 | **Transforming an Equation from Rectangular to Polar Form** |

Transform the equation $4xy = 9$ from rectangular coordinates to polar coordinates.

Solution   We use formula (1): $x = r \cos \theta$ and $y = r \sin \theta$.

$$4xy = 9$$
$$4(r \cos \theta)(r \sin \theta) = 9 \quad x = r\cos\theta, y = r\sin\theta$$
$$4r^2 \cos \theta \sin \theta = 9$$

This is the polar form of the equation. It can be simplified as shown next:

$$2r^2(2 \sin \theta \cos \theta) = 9 \quad \text{Factor out } 2r^2.$$
$$2r^2 \sin(2\theta) = 9 \quad \text{Double-angle Formula}$$

══════ **Now Work** PROBLEM 69

5.1 Assess Your Understanding

'Are You Prepared?' *Answers are given at the end of these exercises. If you get a wrong answer, read the pages listed in red.*

1. Plot the point whose rectangular coordinates are $(3, -1)$. What quadrant does the point lie in? (pp. 2–3)

2. To complete the square of $x^2 + 6x$, add _____ . (p. A31)

3. If $P = (a, b)$ is a point on the terminal side of the angle θ at a distance r from the origin, then $\sin \theta = $ _____ . (pp. 140–143)

4. $\tan^{-1}(-1) = $ _____ . (pp. 213–215)

Concepts and Vocabulary

5. In polar coordinates, the origin is called the _____ and the positive x-axis is referred to as the _____ _____ .

6. Another representation in polar coordinates for the point $\left(2, \dfrac{\pi}{3}\right)$ is $\left(\underline{\quad}, \dfrac{4\pi}{3}\right)$.

7. The polar coordinates $\left(-2, \dfrac{\pi}{6}\right)$ are represented in rectangular coordinates by $(\underline{\quad}, \underline{\quad})$.

8. *True or False* The polar coordinates of a point are unique.

9. *True or False* The rectangular coordinates of a point are unique.

10. *True or False* In (r, θ), the number r can be negative.

Skill Building

In Problems 11–18, match each point in polar coordinates with either A, B, C, or D on the graph.

11. $\left(2, -\dfrac{11\pi}{6}\right)$ **12.** $\left(-2, -\dfrac{\pi}{6}\right)$ **13.** $\left(-2, \dfrac{\pi}{6}\right)$ **14.** $\left(2, \dfrac{7\pi}{6}\right)$

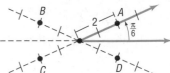

15. $\left(2, \dfrac{5\pi}{6}\right)$ **16.** $\left(-2, \dfrac{5\pi}{6}\right)$ **17.** $\left(-2, \dfrac{7\pi}{6}\right)$ **18.** $\left(2, \dfrac{11\pi}{6}\right)$

In Problems 19–30, plot each point given in polar coordinates.

19. $(3, 90°)$ **20.** $(4, 270°)$ **21.** $(-2, 0)$ **22.** $(-3, \pi)$

23. $\left(6, \dfrac{\pi}{6}\right)$ **24.** $\left(5, \dfrac{5\pi}{3}\right)$ **25.** $(-2, 135°)$ **26.** $(-3, 120°)$

27. $\left(-1, -\dfrac{\pi}{3}\right)$ **28.** $\left(-3, -\dfrac{3\pi}{4}\right)$ **29.** $(-2, -\pi)$ **30.** $\left(-3, -\dfrac{\pi}{2}\right)$

In Problems 31–38, plot each point given in polar coordinates, and find other polar coordinates (r, θ) of the point for which:

    (a) $r > 0, \quad -2\pi \le \theta < 0$ (b) $r < 0, \quad 0 \le \theta < 2\pi$ (c) $r > 0, \quad 2\pi \le \theta < 4\pi$

31. $\left(5, \dfrac{2\pi}{3}\right)$ **32.** $\left(4, \dfrac{3\pi}{4}\right)$ **33.** $(-2, 3\pi)$ **34.** $(-3, 4\pi)$

35. $\left(1, \dfrac{\pi}{2}\right)$ **36.** $(2, \pi)$ **37.** $\left(-3, -\dfrac{\pi}{4}\right)$ **38.** $\left(-2, -\dfrac{2\pi}{3}\right)$

In Problems 39–54, the polar coordinates of a point are given. Find the rectangular coordinates of each point.

39. $\left(3, \dfrac{\pi}{2}\right)$ **40.** $\left(4, \dfrac{3\pi}{2}\right)$ **41.** $(-2, 0)$ **42.** $(-3, \pi)$

43. $(6, 150°)$ **44.** $(5, 300°)$ **45.** $\left(-2, \dfrac{3\pi}{4}\right)$ **46.** $\left(-2, \dfrac{2\pi}{3}\right)$

47. $\left(-1, -\dfrac{\pi}{3}\right)$ **48.** $\left(-3, -\dfrac{3\pi}{4}\right)$ **49.** $(-2, -180°)$ **50.** $(-3, -90°)$

51. $(7.5, 110°)$ **52.** $(-3.1, 182°)$ **53.** $(6.3, 3.8)$ **54.** $(8.1, 5.2)$

In Problems 55–66, the rectangular coordinates of a point are given. Find polar coordinates for each point.

55. $(3, 0)$ **56.** $(0, 2)$ **57.** $(-1, 0)$ **58.** $(0, -2)$

59. $(1, -1)$ **60.** $(-3, 3)$ **61.** $\left(\sqrt{3}, 1\right)$ **62.** $\left(-2, -2\sqrt{3}\right)$

63. $(1.3, -2.1)$ **64.** $(-0.8, -2.1)$ **65.** $(8.3, 4.2)$ **66.** $(-2.3, 0.2)$

In Problems 67–74, the letters x and y represent rectangular coordinates. Write each equation using polar coordinates (r, θ).

67. $2x^2 + 2y^2 = 3$ **68.** $x^2 + y^2 = x$ **69.** $x^2 = 4y$ **70.** $y^2 = 2x$

71. $2xy = 1$ **72.** $4x^2y = 1$ **73.** $x = 4$ **74.** $y = -3$

In Problems 75–82, the letters r and θ represent polar coordinates. Write each equation using rectangular coordinates (x, y).

75. $r = \cos\theta$ **76.** $r = \sin\theta + 1$ **77.** $r^2 = \cos\theta$ **78.** $r = \sin\theta - \cos\theta$

79. $r = 2$ **80.** $r = 4$ **81.** $r = \dfrac{4}{1 - \cos\theta}$ **82.** $r = \dfrac{3}{3 - \cos\theta}$

Applications and Extensions

83. Chicago In Chicago, the road system is set up like a Cartesian plane, where streets are indicated by the number of blocks they are from Madison Street and State Street. For example, Wrigley Field in Chicago is located at 1060 West Addison, which is 10 blocks west of State Street and 36 blocks north of Madison Street. Treat the intersection of Madison Street and State Street as the origin of a coordinate system, with east being the positive x-axis.

(a) Write the location of Wrigley Field using rectangular coordinates.
(b) Write the location of Wrigley Field using polar coordinates. Use the east direction for the polar axis. Express θ in degrees.
(c) U.S. Cellular Field, home of the White Sox, is located at 35th and Princeton, which is 3 blocks west of State Street and 35 blocks south of Madison. Write the location of U.S. Cellular Field using rectangular coordinates.
(d) Write the location of U.S. Cellular Field using polar coordinates. Use the east direction for the polar axis. Express θ in degrees.

84. Show that the formula for the distance d between two points $P_1 = (r_1, \theta_1)$ and $P_2 = (r_2, \theta_2)$ is

$$d = \sqrt{r_1^2 + r_2^2 - 2r_1r_2\cos(\theta_2 - \theta_1)}$$

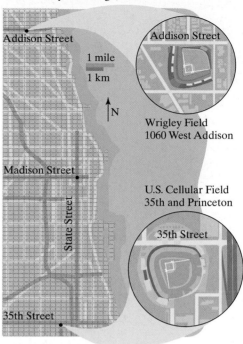

City of Chicago, Illinois

Discussion and Writing

85. In converting from polar coordinates to rectangular coordinates, what formulas will you use?

86. Explain how you proceed to convert from rectangular coordinates to polar coordinates.

87. Is the street system in your town based on a rectangular coordinate system, a polar coordinate system, or some other system? Explain.

'Are You Prepared?' Answers

1. 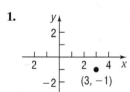 ; quadrant IV   2. 9   3. $\dfrac{b}{r}$ 4. $-\dfrac{\pi}{4}$

5.2 Polar Equations and Graphs

PREPARING FOR THIS SECTION *Before getting started, review the following:*

- Symmetry (Section 1.2, pp. 18–19)
- Circles (Section 1.2, pp. 22–26)
- Even–Odd Properties of Trigonometric Functions (Section 2.5, pp. 158–159)
- Difference Formulas for Sine and Cosine (Section 3.4, pp. 232 and 235)
- Value of the Sine and Cosine Functions at Certain Angles (Section 2.3, pp. 129–131; Section 2.4, pp. 140–147)

Now Work the 'Are You Prepared?' problems on page 344.

OBJECTIVES
1 Graph and Identify Polar Equations by Converting to Rectangular Equations (p. 330)
2 Graph Polar Equations Using a Graphing Utility (p. 331)
3 Test Polar Equations for Symmetry (p. 335)
4 Graph Polar Equations by Plotting Points (p. 336)

Just as a rectangular grid may be used to plot points given by rectangular coordinates, as in Figure 21(a), we can use a grid consisting of concentric circles (with centers at the pole) and rays (with vertices at the pole) to plot points given by polar coordinates, as shown in Figure 21(b). We shall use such **polar grids** to graph *polar equations*.

Figure 21

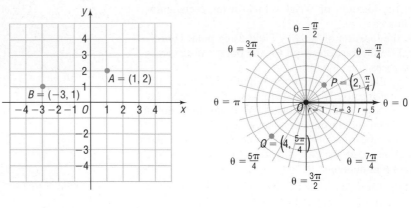

(a) Rectangular grid

(b) Polar grid

DEFINITION

An equation whose variables are polar coordinates is called a **polar equation.** The **graph of a polar equation** consists of all points whose polar coordinates satisfy the equation.

1 Graph and Identify Polar Equations by Converting to Rectangular Equations

One method that we can use to graph a polar equation is to convert the equation to rectangular coordinates. In the discussion that follows, (x, y) represent the rectangular coordinates of a point P, and (r, θ) represent polar coordinates of the point P.

EXAMPLE 1   Identifying and Graphing a Polar Equation (Circle)

Identify and graph the equation: $r = 3$

Solution   We convert the polar equation to a rectangular equation.

$$r = 3$$

$$r^2 = 9 \quad \text{Square both sides.}$$

$$x^2 + y^2 = 9 \quad r^2 = x^2 + y^2$$

The graph of $r = 3$ is a circle, with center at the pole and radius 3. See Figure 22.

Figure 22
$r = 3$ or $x^2 + y^2 = 9$

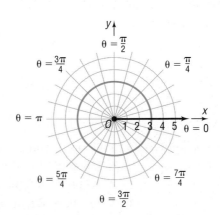

─ **Now Work** PROBLEM 1 3

EXAMPLE 2   Identifying and Graphing a Polar Equation (Line)

Identify and graph the equation: $\theta = \dfrac{\pi}{4}$

Solution   We convert the polar equation to a rectangular equation.

Figure 23

$\theta = \dfrac{\pi}{4}$ or $y = x$

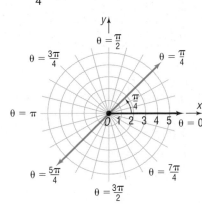

$$\theta = \frac{\pi}{4}$$

$$\tan \theta = \tan \frac{\pi}{4} \quad \text{Take the tangent of both sides.}$$

$$\frac{y}{x} = 1 \qquad \tan \theta = \frac{y}{x}; \ \tan \frac{\pi}{4} = 1$$

$$y = x$$

The graph of $\theta = \dfrac{\pi}{4}$ is a line passing through the pole making an angle of $\dfrac{\pi}{4}$ with the polar axis. See Figure 23.

Now Work PROBLEM 15

EXAMPLE 3   **Identifying and Graphing a Polar Equation (Horizontal Line)**

Identify and graph the equation: $r \sin \theta = 2$

Solution   Since $y = r \sin \theta$, we can write the equation as

$$y = 2$$

We conclude that the graph of $r \sin \theta = 2$ is a horizontal line 2 units above the pole. See Figure 24.

Figure 24
$r \sin \theta = 2$ or $y = 2$

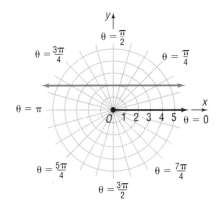

---

2 Graph Polar Equations Using a Graphing Utility

A second method we can use to graph a polar equation is to graph the equation using a graphing utility.

Most graphing utilities require the following steps to obtain the graph of an equation:

Graphing a Polar Equation Using a Graphing Utility

STEP 1: Solve the equation for r in terms of θ.

STEP 2: Select the viewing window in POLar mode. In addition to setting Xmin, Xmax, Xscl, and so forth, the viewing window in polar mode requires setting minimum and maximum values for θ and an increment setting for θ (θstep). Finally, a square screen and radian measure should be used.

STEP 3: Enter the expression involving θ that you found in Step 1. (Consult your manual for the correct way to enter the expression.)

STEP 4: Press graph.

| EXAMPLE 4 | Graphing a Polar Equation Using a Graphing Utility |

Use a graphing utility to graph the polar equation $r \sin \theta = 2$.

Solution   **STEP 1:** We solve the equation for r in terms of θ.

$$r \sin \theta = 2$$

$$r = \frac{2}{\sin \theta}$$

STEP 2: From the polar mode, select a square viewing window. We will use the one given next.

$$\theta\text{min} = 0 \qquad X\text{min} = -9 \qquad Y\text{min} = -6$$
$$\theta\text{max} = 2\pi \qquad X\text{max} = 9 \qquad Y\text{max} = 6$$
$$\theta\text{step} = \frac{\pi}{24} \qquad X\text{scl} = 1 \qquad Y\text{scl} = 1$$

θstep determines the number of points the graphing utility will plot. For example, if θstep is $\dfrac{\pi}{24}$, then the graphing utility will evaluate r at $\theta = 0$ (θmin), $\dfrac{\pi}{24}, \dfrac{2\pi}{24}, \dfrac{3\pi}{24}$, and so forth, up to 2π (θmax).

Figure 25

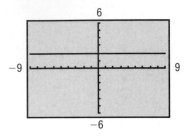

The smaller θstep is the more points the graphing utility will plot. The student is encouraged to experiment with different values for θmin, θmax, and θstep to see how the graph is affected.

STEP 3: Enter the expression $\dfrac{2}{\sin \theta}$ after the prompt $r =$.

STEP 4: Graph.

The graph is shown in Figure 25.

| EXAMPLE 5 | Identifying and Graphing a Polar Equation (Vertical Line) |

Identify and graph the equation: $r \cos \theta = -3$

Solution   Since $x = r \cos \theta$, we can write the equation as

$$x = -3$$

We conclude that the graph of $r \cos \theta = -3$ is a vertical line 3 units to the left of the pole. Figure 26(a) shows the graph drawn by hand. Figure 26(b) shows the graph using a graphing utility with θmin $= 0$, θmax $= 2\pi$, and θstep $= \dfrac{\pi}{24}$.

Figure 26
$r \cos \theta = -3$ or $x = -3$

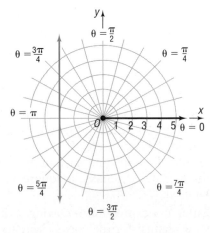

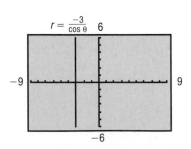

(a)   (b)

Based on Examples 3, 4, and 5, we are led to the following results. (The proofs are left as exercises. See Problems 81 and 82.)

THEOREM

Let a be a nonzero real number. Then the graph of the equation

$$r \sin \theta = a$$

is a horizontal line a units above the pole if $a > 0$ and $|a|$ units below the pole if $a < 0$.

The graph of the equation

$$r \cos \theta = a$$

is a vertical line a units to the right of the pole if $a > 0$ and $|a|$ units to the left of the pole if $a < 0$.

━━━━━ **Now Work** PROBLEM 19

EXAMPLE 6   Identifying and Graphing a Polar Equation (Circle)

Identify and graph the equation: $r = 4 \sin \theta$

Solution   To transform the equation to rectangular coordinates, we multiply each side by r.

$$r^2 = 4r \sin \theta$$

Now we use the facts that $r^2 = x^2 + y^2$ and $y = r \sin \theta$. Then

$$x^2 + y^2 = 4y$$

$$x^2 + (y^2 - 4y) = 0$$

$$x^2 + (y^2 - 4y + 4) = 4 \qquad \text{Complete the square in } y.$$

$$x^2 + (y - 2)^2 = 4 \qquad \text{Factor.}$$

This is the standard equation of a circle with center at $(0, 2)$ in rectangular coordinates and radius 2. Figure 27(a) shows the graph drawn by hand. Figure 27(b) shows the graph using a graphing utility with $\theta\text{min} = 0$, $\theta\text{max} = 2\pi$, and $\theta\text{step} = \dfrac{\pi}{24}$.

Figure 27
$r = 4 \sin \theta$ or $x^2 + (y - 2)^2 = 4$

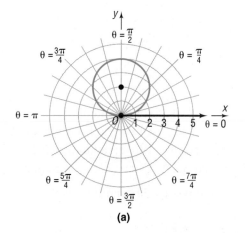

(a)

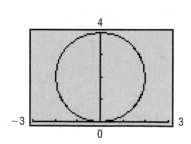

(b)

EXAMPLE 7	Identifying and Graphing a Polar Equation (Circle)

Identify and graph the equation: $r = -2 \cos \theta$

Solution   We proceed as in Example 6.

$$r^2 = -2r \cos \theta \qquad \text{Multiply both sides by } r.$$

$$x^2 + y^2 = -2x \qquad r^2 = x^2 + y^2; \quad x = r \cos \theta$$

$$x^2 + 2x + y^2 = 0$$

$$(x^2 + 2x + 1) + y^2 = 1 \qquad \text{Complete the square in } x.$$

$$(x + 1)^2 + y^2 = 1 \qquad \text{Factor.}$$

This is the standard equation of a circle with center at $(-1, 0)$ in rectangular coordinates and radius 1. Figure 28(a) shows the graph drawn by hand. Figure 28(b) shows the graph using a graphing utility with $\theta\text{min} = 0$, $\theta\text{max} = 2\pi$, and $\theta\text{step} = \dfrac{\pi}{24}$.

Figure 28
$r = -2 \cos \theta$ or $(x + 1)^2 + y^2 = 1$

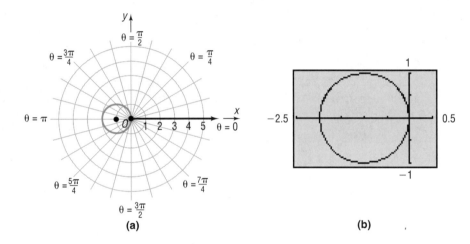

(a)   (b)

Exploration

Using a square screen, graph $r_1 = \sin \theta, r_2 = 2 \sin \theta$, and $r_3 = 3 \sin \theta$. Do you see the pattern? Clear the screen and graph $r_1 = -\sin \theta, r_2 = -2 \sin \theta$, and $r_3 = -3 \sin \theta$. Do you see the pattern? Clear the screen and graph $r_1 = \cos \theta, r_2 = 2 \cos \theta$, and $r_3 = 3 \cos \theta$. Do you see the pattern? Clear the screen and graph $r_1 = -\cos \theta, r_2 = -2 \cos \theta$, and $r_3 = -3 \cos \theta$. Do you see the pattern?

Based on Examples 6 and 7 and the preceding Exploration, we are led to the following results. (The proofs are left as exercises. See Problems 83–86.)

THEOREM   Let a be a positive real number. Then

Equation	Description
(a) $r = 2a \sin \theta$	Circle: radius a; center at $(0, a)$ in rectangular coordinates
(b) $r = -2a \sin \theta$	Circle: radius a; center at $(0, -a)$ in rectangular coordinates
(c) $r = 2a \cos \theta$	Circle: radius a; center at $(a, 0)$ in rectangular coordinates
(d) $r = -2a \cos \theta$	Circle: radius a; center at $(-a, 0)$ in rectangular coordinates

Each circle passes through the pole.

═══➤ **Now Work** PROBLEM 21

The method of converting a polar equation to an identifiable rectangular equation to obtain the graph is not always helpful, nor is it always necessary. Usually, we set up a table that lists several points on the graph. By checking for symmetry, it may be possible to reduce the number of points needed to draw the graph.

3 Test Polar Equations for Symmetry

In polar coordinates, the points (r, θ) and $(r, -\theta)$ are symmetric with respect to the polar axis (and to the x-axis). See Figure 29(a). The points (r, θ) and $(r, \pi - \theta)$ are symmetric with respect to the line $\theta = \dfrac{\pi}{2}$ (the y-axis). See Figure 29(b). The points (r, θ) and $(-r, \theta)$ are symmetric with respect to the pole (the origin). See Figure 29(c).

Figure 29

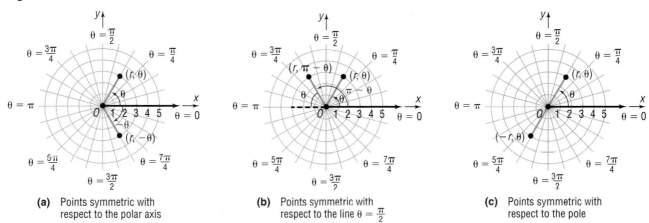

(a) Points symmetric with respect to the polar axis

(b) Points symmetric with respect to the line $\theta = \dfrac{\pi}{2}$

(c) Points symmetric with respect to the pole

The following tests are a consequence of these observations.

THEOREM

Tests for Symmetry

Symmetry with Respect to the Polar Axis (x-Axis)

In a polar equation, replace θ by $-\theta$. If an equivalent equation results, the graph is symmetric with respect to the polar axis.

Symmetry with Respect to the Line $\theta = \dfrac{\pi}{2}$ (y-Axis)

In a polar equation, replace θ by $\pi - \theta$. If an equivalent equation results, the graph is symmetric with respect to the line $\theta = \dfrac{\pi}{2}$.

Symmetry with Respect to the Pole (Origin)

In a polar equation, replace r by $-r$. If an equivalent equation results, the graph is symmetric with respect to the pole.

The three tests for symmetry given here are *sufficient* conditions for symmetry, but they are not *necessary* conditions. That is, an equation may fail these tests and still have a graph that is symmetric with respect to the polar axis, the line $\theta = \dfrac{\pi}{2}$, or the pole. For example, the graph of $r = \sin(2\theta)$ turns out to be symmetric with respect to the polar axis, the line $\theta = \dfrac{\pi}{2}$, and the pole, but all three tests given here fail. See also Problems 87–89.

4 Graph Polar Equations by Plotting Points

EXAMPLE 8	Graphing a Polar Equation (Cardioid)

Graph the equation: $r = 1 - \sin\theta$

Solution   We check for symmetry first.

Polar Axis: Replace θ by $-\theta$. The result is

$$r = 1 - \sin(-\theta) = 1 + \sin\theta \qquad \sin(-\theta) = -\sin\theta$$

The test fails, so the graph may or may not be symmetric with respect to the polar axis.

The Line $\theta = \dfrac{\pi}{2}$: Replace θ by $\pi - \theta$. The result is

$$r = 1 - \sin(\pi - \theta) = 1 - (\sin\pi\cos\theta - \cos\pi\sin\theta)$$
$$= 1 - [0\cdot\cos\theta - (-1)\sin\theta] = 1 - \sin\theta$$

The test is satisfied, so the graph is symmetric with respect to the line $\theta = \dfrac{\pi}{2}$.

The Pole: Replace r by $-r$. Then the result is $-r = 1 - \sin\theta$, so $r = -1 + \sin\theta$. The test fails, so the graph may or may not be symmetric with respect to the pole.

Next, we identify points on the graph by assigning values to the angle θ and calculating the corresponding values of r. Due to the symmetry with respect to the line $\theta = \dfrac{\pi}{2}$, we only need to assign values to θ from $-\dfrac{\pi}{2}$ to $\dfrac{\pi}{2}$, as given in Table 1.

Now we plot the points (r, θ) from Table 1 and trace out the graph, beginning at the point $\left(2, -\dfrac{\pi}{2}\right)$ and ending at the point $\left(0, \dfrac{\pi}{2}\right)$. Then we reflect this portion of the graph about the line $\theta = \dfrac{\pi}{2}$ (the y-axis) to obtain the complete graph. Figure 30(a) shows the graph drawn by hand. Figure 30(b) shows the graph using a graphing utility with $\theta\min = 0$, $\theta\max = 2\pi$, and $\theta\text{step} = \dfrac{\pi}{24}$.

Table 1

θ	$r = 1 - \sin\theta$
$-\dfrac{\pi}{2}$	$1 - (-1) = 2$
$-\dfrac{\pi}{3}$	$1 - \left(-\dfrac{\sqrt{3}}{2}\right) \approx 1.87$
$-\dfrac{\pi}{6}$	$1 - \left(-\dfrac{1}{2}\right) = \dfrac{3}{2}$
0	$1 - 0 = 1$
$\dfrac{\pi}{6}$	$1 - \dfrac{1}{2} = \dfrac{1}{2}$
$\dfrac{\pi}{3}$	$1 - \dfrac{\sqrt{3}}{2} \approx 0.13$
$\dfrac{\pi}{2}$	$1 - 1 = 0$

Figure 30

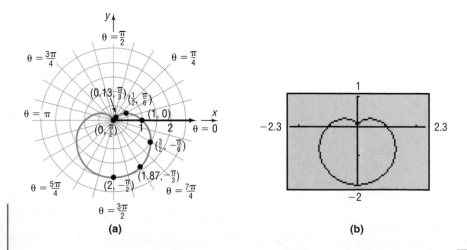

(a)   (b)

Exploration

Graph $r_1 = 1 + \sin\theta$. Clear the screen and graph $r_1 = 1 - \cos\theta$. Clear the screen and graph $r_1 = 1 + \cos\theta$. Do you see a pattern?

The curve in Figure 30 is an example of a *cardioid* (a heart-shaped curve).

DEFINITION

Cardioids are characterized by equations of the form

$$r = a(1 + \cos \theta) \qquad r = a(1 + \sin \theta)$$
$$r = a(1 - \cos \theta) \qquad r = a(1 - \sin \theta)$$

where $a > 0$. The graph of a cardioid passes through the pole.

Now Work PROBLEM 37

EXAMPLE 9

Graphing a Polar Equation (Limaçon without Inner Loop)

Graph the equation: $r = 3 + 2 \cos \theta$

Solution

We check for symmetry first.

Polar Axis: Replace θ by $-\theta$. The result is

$$r = 3 + 2 \cos(-\theta) = 3 + 2 \cos \theta \qquad \cos(-\theta) = \cos \theta$$

The test is satisfied, so the graph is symmetric with respect to the polar axis.

The Line $\theta = \dfrac{\pi}{2}$: Replace θ by $\pi - \theta$. The result is

$$r = 3 + 2 \cos(\pi - \theta) = 3 + 2(\cos \pi \cos \theta + \sin \pi \sin \theta)$$
$$= 3 - 2 \cos \theta$$

The test fails, so the graph may or may not be symmetric with respect to the line $\theta = \dfrac{\pi}{2}$.

The Pole: Replace r by $-r$. The test fails, so the graph may or may not be symmetric with respect to the pole.

Next, we identify points on the graph by assigning values to the angle θ and calculating the corresponding values of r. Due to the symmetry with respect to the polar axis, we only need to assign values to θ from 0 to π, as given in Table 2.

Now we plot the points (r, θ) from Table 2 and trace out the graph, beginning at the point $(5, 0)$ and ending at the point $(1, \pi)$. Then we reflect this portion of the graph about the polar axis (the x-axis) to obtain the complete graph. Figure 31(a) shows the graph drawn by hand. Figure 31(b) shows the graph using a graphing utility with θmin $= 0$, θmax $= 2\pi$, and θstep $= \dfrac{\pi}{24}$.

Table 2

θ	$r = 3 + 2 \cos \theta$
0	$3 + 2(1) = 5$
$\dfrac{\pi}{6}$	$3 + 2\left(\dfrac{\sqrt{3}}{2}\right) \approx 4.73$
$\dfrac{\pi}{3}$	$3 + 2\left(\dfrac{1}{2}\right) = 4$
$\dfrac{\pi}{2}$	$3 + 2(0) = 3$
$\dfrac{2\pi}{3}$	$3 + 2\left(-\dfrac{1}{2}\right) = 2$
$\dfrac{5\pi}{6}$	$3 + 2\left(-\dfrac{\sqrt{3}}{2}\right) \approx 1.27$
π	$3 + 2(-1) = 1$

Figure 31

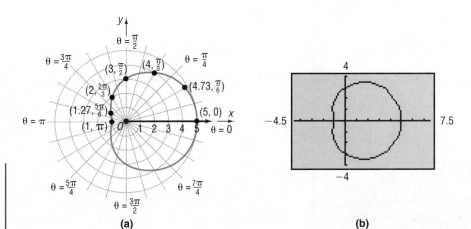

(a)

(b)

Exploration

Graph $r_1 = 3 - 2 \cos \theta$. Clear the screen and graph $r_1 = 3 + 2 \sin \theta$. Clear the screen and graph $r_1 = 3 - 2 \sin \theta$. Do you see a pattern?

The curve in Figure 31 is an example of a *limaçon* (the French word for *snail*) *without an inner loop*.

DEFINITION

Limaçons without an inner loop are characterized by equations of the form

$$r = a + b \cos \theta \qquad r = a + b \sin \theta$$
$$r = a - b \cos \theta \qquad r = a - b \sin \theta$$

where $a > 0, b > 0$, and $a > b$. The graph of a limaçon without an inner loop does not pass through the pole.

⬛▶ **Now Work** PROBLEM 43

EXAMPLE 10

Graphing a Polar Equation (Limaçon with Inner Loop)

Graph the equation: $r = 1 + 2 \cos \theta$

Solution

First, we check for symmetry.

Polar Axis: Replace θ by $-\theta$. The result is

$$r = 1 + 2 \cos(-\theta) = 1 + 2 \cos \theta$$

The test is satisfied, so the graph is symmetric with respect to the polar axis.

The Line $\theta = \dfrac{\pi}{2}$: Replace θ by $\pi - \theta$. The result is

$$r = 1 + 2 \cos(\pi - \theta) = 1 + 2(\cos \pi \cos \theta + \sin \pi \sin \theta)$$
$$= 1 - 2 \cos \theta$$

The test fails, so the graph may or may not be symmetric with respect to the line $\theta = \dfrac{\pi}{2}$.

The Pole: Replace r by $-r$. The test fails, so the graph may or may not be symmetric with respect to the pole.

Next, we identify points on the graph of $r = 1 + 2 \cos \theta$ by assigning values to the angle θ and calculating the corresponding values of r. Due to the symmetry with respect to the polar axis, we only need to assign values to θ from 0 to π, as given in Table 3.

Now we plot the points (r, θ) from Table 3, beginning at $(3, 0)$ and ending at $(-1, \pi)$. See Figure 32(a). Finally, we reflect this portion of the graph about the polar axis (the x-axis) to obtain the complete graph. See Figure 32(b). Figure 32(c) shows the graph using a graphing utility with θmin = 0, θmax = 2π, and θstep = $\dfrac{\pi}{24}$.

Table 3

θ	$r = 1 + 2 \cos \theta$
0	$1 + 2(1) = 3$
$\dfrac{\pi}{6}$	$1 + 2\left(\dfrac{\sqrt{3}}{2}\right) \approx 2.73$
$\dfrac{\pi}{3}$	$1 + 2\left(\dfrac{1}{2}\right) = 2$
$\dfrac{\pi}{2}$	$1 + 2(0) = 1$
$\dfrac{2\pi}{3}$	$1 + 2\left(-\dfrac{1}{2}\right) = 0$
$\dfrac{5\pi}{6}$	$1 + 2\left(-\dfrac{\sqrt{3}}{2}\right) \approx -0.73$
π	$1 + 2(-1) = -1$

Figure 32

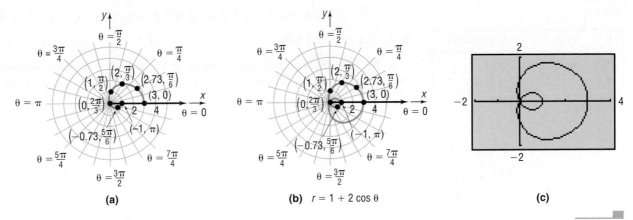

(a)

(b) $r = 1 + 2 \cos \theta$

(c)

Exploration

Graph $r_1 = 1 - 2 \cos \theta$. Clear the screen and graph $r_1 = 1 + 2 \sin \theta$. Clear the screen and graph $r_1 = 1 - 2 \sin \theta$. Do you see a pattern?

The curve in Figure 32(b) or 32(c) is an example of a *limaçon with an inner loop.*

DEFINITION

Limaçons with an inner loop are characterized by equations of the form

$$r = a + b \cos \theta \qquad r = a + b \sin \theta$$
$$r = a - b \cos \theta \qquad r = a - b \sin \theta$$

where $a > 0, b > 0$, and $a < b$. The graph of a limaçon with an inner loop will pass through the pole twice.

———— **Now Work** PROBLEM 45

EXAMPLE 11   Graphing a Polar Equation (Rose)

Graph the equation: $r = 2 \cos(2\theta)$

Solution   We check for symmetry.

Polar Axis: If we replace θ by $-\theta$, the result is

$$r = 2 \cos[2(-\theta)] = 2 \cos(2\theta)$$

The test is satisfied, so the graph is symmetric with respect to the polar axis.

The Line $\theta = \dfrac{\pi}{2}$: If we replace θ by $\pi - \theta$, we obtain

$$r = 2 \cos[2(\pi - \theta)] = 2 \cos(2\pi - 2\theta) = 2 \cos(2\theta)$$

The test is satisfied, so the graph is symmetric with respect to the line $\theta = \dfrac{\pi}{2}$.

The Pole: Since the graph is symmetric with respect to both the polar axis and the line $\theta = \dfrac{\pi}{2}$, it must be symmetric with respect to the pole.

Table 4

θ	$r = 2\cos(2\theta)$
0	$2(1) = 2$
$\dfrac{\pi}{6}$	$2\left(\dfrac{1}{2}\right) = 1$
$\dfrac{\pi}{4}$	$2(0) = 0$
$\dfrac{\pi}{3}$	$2\left(-\dfrac{1}{2}\right) = -1$
$\dfrac{\pi}{2}$	$2(-1) = -2$

Next, we construct Table 4. Due to the symmetry with respect to the polar axis, the line $\theta = \dfrac{\pi}{2}$, and the pole, we consider only values of θ from 0 to $\dfrac{\pi}{2}$.

We plot and connect these points in Figure 33(a). Finally, because of symmetry, we reflect this portion of the graph first about the polar axis (the x-axis) and then about the line $\theta = \dfrac{\pi}{2}$ (the y-axis) to obtain the complete graph. See Figure 33(b). Figure 33(c) shows the graph using a graphing utility with $\theta\text{min} = 0$, $\theta\text{max} = 2\pi$, and $\theta\text{step} = \dfrac{\pi}{24}$.

Figure 33

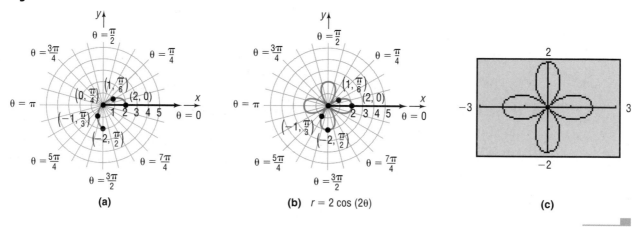

(a)   (b) $r = 2\cos(2\theta)$ (c)

Exploration

Graph $r_1 = 2\cos(4\theta)$; clear the screen and graph $r_1 = 2\cos(6\theta)$. How many petals did each of these graphs have?

Clear the screen and graph, in order, each on a clear screen, $r_1 = 2\cos(3\theta)$, $r_1 = 2\cos(5\theta)$, and $r_1 = 2\cos(7\theta)$. What do you notice about the number of petals?

The curve in Figure 33(b) or 33(c) is called a *rose* with four petals.

DEFINITION

Rose curves are characterized by equations of the form

$$r = a\cos(n\theta), \qquad r = a\sin(n\theta), \qquad a \neq 0$$

and have graphs that are rose shaped. If $n \neq 0$ is even, the rose has $2n$ petals; if $n \neq \pm 1$ is odd, the rose has n petals.

═══ **Now Work** PROBLEM 49

EXAMPLE 12   **Graphing a Polar Equation (Lemniscate)**

Graph the equation: $r^2 = 4\sin(2\theta)$

Table 5

θ	$r^2 = 4\sin(2\theta)$	r
0	$4(0) = 0$	0
$\dfrac{\pi}{6}$	$4\left(\dfrac{\sqrt{3}}{2}\right) = 2\sqrt{3}$	± 1.9
$\dfrac{\pi}{4}$	$4(1) = 4$	± 2
$\dfrac{\pi}{3}$	$4\left(\dfrac{\sqrt{3}}{2}\right) = 2\sqrt{3}$	± 1.9
$\dfrac{\pi}{2}$	$4(0) = 0$	0

Solution We leave it to you to verify that the graph is symmetric with respect to the pole. Because of the symmetry with respect to the pole, we only need to consider values of θ between $\theta = 0$ and $\theta = \pi$. Note that there are no points on the graph for $\dfrac{\pi}{2} < \theta < \pi$ (quadrant II), since $\sin(2\theta) < 0$ for such values. Table 5 lists points on the graph for values of $\theta = 0$ through $\theta = \dfrac{\pi}{2}$. The points from Table 5 where $r \geq 0$ are plotted in Figure 34(a). The remaining points on the graph may be obtained by using symmetry. Figure 34(b) shows the final graph drawn by hand. Figure 34(c) shows the graph using a graphing utility with $\theta\text{min} = 0$, $\theta\text{max} = 2\pi$, and $\theta\text{step} = \dfrac{\pi}{24}$.

Figure 34

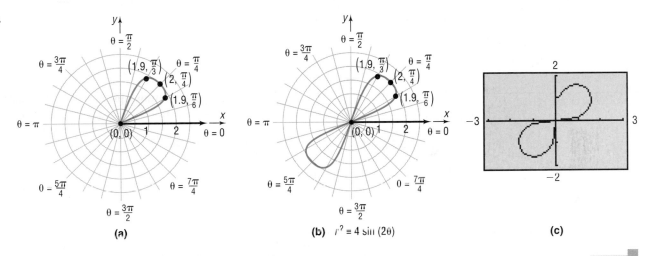

(a)   (b) $r^2 = 4\sin(2\theta)$ (c)

The curve in Figure 34(b) or 34(c) is an example of a *lemniscate* (from the Greek word *ribbon*).

DEFINITION **Lemniscates** are characterized by equations of the form

$$r^2 = a^2 \sin(2\theta) \qquad r^2 = a^2 \cos(2\theta)$$

where $a \neq 0$, and have graphs that are propeller shaped.

━━━ **Now Work** PROBLEM 53

EXAMPLE 13 **Graphing a Polar Equation (Spiral)**

Graph the equation: $r = e^{\theta/5}$

Solution The tests for symmetry with respect to the pole, the polar axis, and the line $\theta = \dfrac{\pi}{2}$ fail. Furthermore, there is no number θ for which $r = 0$, so the graph does not pass through the pole. We observe that r is positive for all θ, r increases as θ increases, $r \to 0$ as $\theta \to -\infty$, and $r \to \infty$ as $\theta \to \infty$. With the help of a calculator, we obtain the values in Table 6. See Figure 35(a) for the graph drawn by hand. Figure 35(b)

Table 6

θ	$r = e^{\theta/5}$
$-\dfrac{3\pi}{2}$	0.39
$-\pi$	0.53
$-\dfrac{\pi}{2}$	0.73
$-\dfrac{\pi}{4}$	0.85
0	1
$\dfrac{\pi}{4}$	1.17
$\dfrac{\pi}{2}$	1.37
π	1.87
$\dfrac{3\pi}{2}$	2.57
2π	3.51

shows the graph using a graphing utility with $\theta\min = -4\pi$, $\theta\max = 3\pi$, and $\theta\text{step} = \dfrac{\pi}{24}$.

Figure 35
$r = e^{\theta/5}$

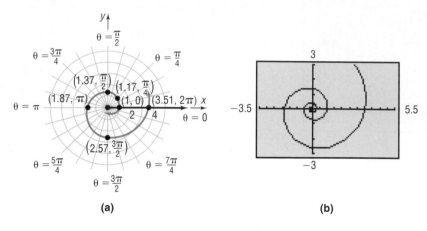

(a)   (b)

The curve in Figure 35 is called a **logarithmic spiral,** since its equation may be written as $\theta = 5 \ln r$ and it spirals infinitely both toward the pole and away from it.

Classification of Polar Equations

The equations of some lines and circles in polar coordinates and their corresponding equations in rectangular coordinates are given in Table 7. Also included are the names and graphs of a few of the more frequently encountered polar equations.

Table 7

Lines			
Description	Line passing through the pole making an angle α with the polar axis	Vertical line	Horizontal line
Rectangular equation	$y = (\tan \alpha)x$	$x = a$	$y = b$
Polar equation	$\theta = \alpha$	$r \cos \theta = a$	$r \sin \theta = b$
Typical graph			

Circles			
Description	Center at the pole, radius a	Passing through the pole, tangent to the line $\theta = \dfrac{\pi}{2}$, center on the polar axis, radius a	Passing through the pole, tangent to the polar axis, center on the line $\theta = \dfrac{\pi}{2}$, radius a
Rectangular equation	$x^2 + y^2 = a^2$, $a > 0$	$x^2 + y^2 = \pm 2ax$, $a > 0$	$x^2 + y^2 = \pm 2ay$, $a > 0$
Polar equation	$r = a$, $a > 0$	$r = \pm 2a \cos \theta$, $a > 0$	$r = \pm 2a \sin \theta$, $a > 0$
Typical graph			

Other Equations

Name	Cardioid	Limaçon without inner loop	Limaçon with inner loop
Polar equations	$r = a \pm a \cos \theta, \quad a > 0$ $r = a \pm a \sin \theta, \quad a > 0$	$r = a \pm b \cos \theta, \quad 0 < b < a$ $r = a \pm b \sin \theta, \quad 0 < b < a$	$r = a \pm b \cos \theta, \quad 0 < a < b$ $r = a \pm b \sin \theta, \quad 0 < a < b$
Typical graph			

Name	Lemniscate	Rose with three petals	Rose with four petals
Polar equations	$r^2 = a^2 \cos(2\theta), \quad a > 0$ $r^2 = a^2 \sin(2\theta), \quad a > 0$	$r = a \sin(3\theta), \quad a > 0$ $r = a \cos(3\theta), \quad a > 0$	$r = a \sin(2\theta), \quad a > 0$ $r = a \cos(2\theta), \quad a > 0$
Typical graph			

Sketching Quickly

If a polar equation involves only a sine (or cosine) function, you can quickly obtain a sketch of its graph by making use of Table 7, periodicity, and a short table.

EXAMPLE 14   **Sketching the Graph of a Polar Equation Quickly**

Graph the equation: $r = 2 + 2 \sin \theta$

Solution   We recognize the polar equation: Its graph is a cardioid. The period of $\sin \theta$ is 2π, so we form a table using $0 \leq \theta \leq 2\pi$, compute r, plot the points (r, θ), and sketch the graph of a cardioid as θ varies from 0 to 2π. See Table 8 and Figure 36.

Table 8

θ	$r = 2 + 2 \sin \theta$
0	$2 + 2(0) = 2$
$\dfrac{\pi}{2}$	$2 + 2(1) = 4$
π	$2 + 2(0) = 2$
$\dfrac{3\pi}{2}$	$2 + 2(-1) = 0$
2π	$2 + 2(0) = 2$

Figure 36

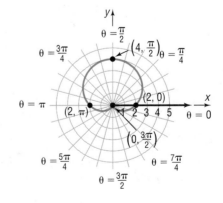

 Now Work PROBLEM **37**

Calculus Comment For those of you who are planning to study calculus, a comment about one important role of polar equations is in order.

In rectangular coordinates, the equation $x^2 + y^2 = 1$, whose graph is the unit circle, is not the graph of a function. In fact, it requires two functions to obtain the graph of the unit circle:

$$y_1 = \sqrt{1 - x^2} \quad \text{Upper semicircle} \qquad y_2 = -\sqrt{1 - x^2} \quad \text{Lower semicircle}$$

In polar coordinates, the equation $r = 1$, whose graph is also the unit circle, does define a function. For each choice of θ, there is only one corresponding value of r, that is, $r = 1$. Since many problems in calculus require the use of functions, the opportunity to express nonfunctions in rectangular coordinates as functions in polar coordinates becomes extremely useful.

Note also that the vertical-line test for functions is valid only for equations in rectangular coordinates.

Historical Feature

Jakob Bernoulli (1654–1705)

Polar coordinates seem to have been invented by Jakob Bernoulli (1654–1705) in about 1691, although, as with most such ideas, earlier traces of the notion exist. Early users of calculus remained committed to rectangular coordinates, and polar coordinates did not become widely used until the early 1800s. Even then, it was mostly geometers who used them for describing odd curves. Finally, about the mid-1800s, applied mathematicians realized the tremendous simplification that polar coordinates make possible in the description of objects with circular or cylindrical symmetry. From then on their use became widespread.

5.2 Assess Your Understanding

'Are You Prepared?   *Answers are given at the end of these exercises. If you get a wrong answer, read the pages listed in red.*

1. If the rectangular coordinates of a point are $(4, -6)$, the point symmetric to it with respect to the origin is _____. (pp. 18–19)

2. The difference formula for cosine is $\cos(A - B) =$ _____. (p. 232)

3. The standard equation of a circle with center at $(-2, 5)$ and radius 3 is _____. (pp. 22–23)

4. Is the sine function even, odd, or neither? (p. 158)

5. $\sin \dfrac{5\pi}{4} =$ _____. (pp. 146–147)

6. $\cos \dfrac{2\pi}{3} =$ _____. (pp. 146–147)

Concepts and Vocabulary

7. An equation whose variables are polar coordinates is called a(n) _____ _____.

8. Using polar coordinates (r, θ), the circle $x^2 + y^2 = 2x$ takes the form _____.

9. A polar equation is symmetric with respect to the pole if an equivalent equation results when r is replaced by _____.

10. *True or False*   The tests for symmetry in polar coordinates are necessary, but not sufficient.

11. *True or False*   The graph of a cardioid never passes through the pole.

12. *True or False*   All polar equations have a symmetric feature.

Skill Building

In Problems 13–28, transform each polar equation to an equation in rectangular coordinates. Then identify and graph the equation. Verify your graph using a graphing utility.

13. $r = 4$

14. $r = 2$

15. $\theta = \dfrac{\pi}{3}$

16. $\theta = -\dfrac{\pi}{4}$

17. $r \sin \theta = 4$

18. $r \cos \theta = 4$

19. $r \cos \theta = -2$

20. $r \sin \theta = -2$

21. $r = 2 \cos \theta$

22. $r = 2 \sin \theta$

23. $r = -4 \sin \theta$

24. $r = -4 \cos \theta$

25. $r \sec \theta = 4$

26. $r \csc \theta = 8$

27. $r \csc \theta = -2$

28. $r \sec \theta = -4$

In Problems 29–36, match each of the graphs (A) through (H) to one of the following polar equations.

29. $r = 2$

30. $\theta = \dfrac{\pi}{4}$

31. $r = 2 \cos \theta$

32. $r \cos \theta = 2$

33. $r = 1 + \cos \theta$

34. $r = 2 \sin \theta$

35. $\theta = \dfrac{3\pi}{4}$

36. $r \sin \theta = 2$

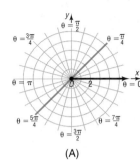

(A)

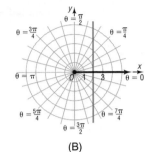

(B)

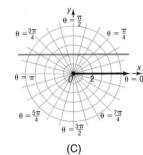

(C)

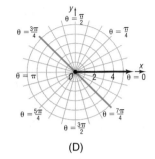

(D)

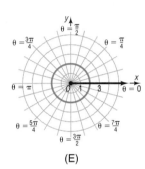

(E)

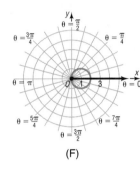

(F)

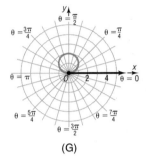

(G)

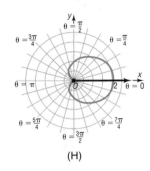

(H)

In Problems 37–60, identify and graph each polar equation. Verify your graph using a graphing utility.

37. $r = 2 + 2 \cos \theta$

38. $r = 1 + \sin \theta$

39. $r = 3 - 3 \sin \theta$

40. $r = 2 - 2 \cos \theta$

41. $r = 2 + \sin \theta$

42. $r = 2 - \cos \theta$

43. $r = 4 - 2 \cos \theta$

44. $r = 4 + 2 \sin \theta$

45. $r = 1 + 2 \sin \theta$

46. $r = 1 - 2 \sin \theta$

47. $r = 2 - 3 \cos \theta$

48. $r = 2 + 4 \cos \theta$

49. $r = 3 \cos(2\theta)$

50. $r = 2 \sin(3\theta)$

51. $r = 4 \sin(5\theta)$

52. $r = 3 \cos(4\theta)$

53. $r^2 = 9 \cos(2\theta)$

54. $r^2 = \sin(2\theta)$

55. $r = 2^\theta$

56. $r = 3^\theta$

57. $r = 1 - \cos \theta$

58. $r = 3 + \cos \theta$

59. $r = 1 - 3 \cos \theta$

60. $r = 4 \cos(3\theta)$

Mixed Practice

In Problems 61–66, graph each pair of polar equations on the same polar grid. Find the polar coordinates of the point(s) of intersection and label the point(s) on the graph.

61. $r = 8 \cos \theta; r = 2 \sec \theta$

62. $r = 8 \sin \theta; r = 4 \csc \theta$

63. $r = \sin \theta; r = 1 + \cos \theta$

64. $r = 3; r = 2 + 2 \cos \theta$

65. $r = 1 + \sin \theta; r = 1 + \cos \theta$

66. $r = 1 + \cos \theta; r = 3 \cos \theta$

Applications and Extensions

In Problems 67–70, the polar equation for each graph is either $r = a + b \cos \theta$ or $r = a + b \sin \theta, a > 0, b > 0$. Select the correct equation and find the values of a and b.

67.

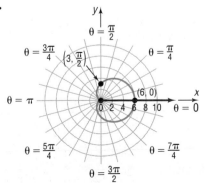

68.

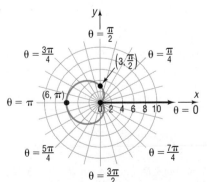

69.

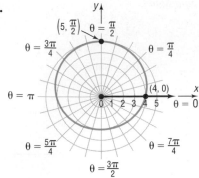

70.

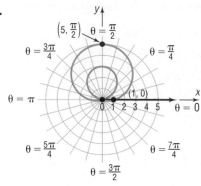

In Problems 71–80, graph each polar equation by hand. Verify your graph using a graphing utility.

71. $r = \dfrac{2}{1 - \cos \theta}$ *(parabola)*

72. $r = \dfrac{2}{1 - 2 \cos \theta}$ *(hyperbola)*

73. $r = \dfrac{1}{3 - 2 \cos \theta}$ *(ellipse)*

74. $r = \dfrac{1}{1 - \cos \theta}$ *(parabola)*

75. $r = \theta, \quad \theta \geq 0$ *(spiral of Archimedes)*

76. $r = \dfrac{3}{\theta}$ *(reciprocal spiral)*

77. $r = \csc \theta - 2, \quad 0 < \theta < \pi$ *(conchoid)*

78. $r = \sin \theta \tan \theta$ *(cissoid)*

79. $r = \tan \theta, \quad -\dfrac{\pi}{2} < \theta < \dfrac{\pi}{2}$ *(kappa curve)*

80. $r = \cos \dfrac{\theta}{2}$

81. Show that the graph of the equation $r \sin \theta = a$ is a horizontal line a units above the pole if $a > 0$ and $|a|$ units below the pole if $a < 0$.

82. Show that the graph of the equation $r \cos \theta = a$ is a vertical line a units to the right of the pole if $a > 0$ and $|a|$ units to the left of the pole if $a < 0$.

83. Show that the graph of the equation $r = 2a \sin \theta, a > 0$, is a circle of radius a with center at $(0, a)$ in rectangular coordinates.

84. Show that the graph of the equation $r = -2a \sin \theta, a > 0$, is a circle of radius a with center at $(0, -a)$ in rectangular coordinates.

85. Show that the graph of the equation $r = 2a \cos \theta, a > 0$, is a circle of radius a with center at $(a, 0)$ in rectangular coordinates.

86. Show that the graph of the equation $r = -2a \cos \theta, a > 0$, is a circle of radius a with center at $(-a, 0)$ in rectangular coordinates.

Discussion and Writing

87. Explain why the following test for symmetry is valid: Replace r by $-r$ and θ by $-\theta$ in a polar equation. If an equivalent equation results, the graph is symmetric with respect to the line $\theta = \dfrac{\pi}{2}$ (*y*-axis).

    (a) Show that the test on page 335 fails for $r^2 = \cos \theta$, yet this new test works.

    (b) Show that the test on page 335 works for $r^2 = \sin \theta$, yet this new test fails.

88. Develop a new test for symmetry with respect to the pole.

    (a) Find a polar equation for which this new test fails, yet the test on page 335 works.

    (b) Find a polar equation for which the test on page 335 fails, yet the new test works.

89. Write down two different tests for symmetry with respect to the polar axis. Find examples in which one test works and the other fails. Which test do you prefer to use? Justify your answer.

90. The tests for symmetry given on page 335 are sufficient, but not necessary. Explain what this means.

'Are You Prepared?' Answers

1. $(-4, 6)$ **2.** $\cos A \cos B + \sin A \sin B$ **3.** $(x + 2)^2 + (y - 5)^2 = 9$ **4.** Odd   **5.** $-\dfrac{\sqrt{2}}{2}$ **6.** $-\dfrac{1}{2}$

5.3 The Complex Plane; De Moivre's Theorem

PREPARING FOR THIS SECTION   *Before getting started, review the following:*

- Complex Numbers (Appendix, Section A.6, pp. A42–A46)
- Value of the Sine and Cosine Functions at Certain Angles (Section 2.3, pp. 129–131; Section 2.4, pp. 140–147)
- Sum and Difference Formulas for Sine and Cosine (Section 3.4, pp. 232 and 235)

 Now Work the 'Are You Prepared?' problems on page 353.

OBJECTIVES   1 Plot Points in the Complex Plane (p. 347)
   2 Convert a Complex Number from Rectangular Form to Polar Form (p. 348)
   3 Find Products and Quotients of Complex Numbers in Polar Form (p. 349)
   4 Use De Moivre's Theorem (p. 350)
   5 Find Complex Roots (p. 351)

1 Plot Points in the Complex Plane

Figure 37
Complex plane

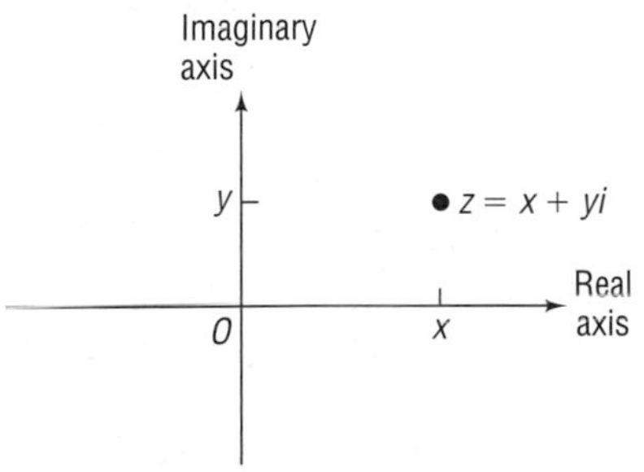

Imaginary axis

y • $z = x + yi$

Real axis

0 x

Complex numbers are discussed in the Appendix, Section A.6. In that discussion, we were not prepared to give a geometric interpretation of a complex number. Now we are ready. Although we could give several interpretations, the one that follows is the easiest to understand.

A complex number $z = x + yi$ can be interpreted geometrically as the point (x, y) in the xy-plane. Each point in the plane corresponds to a complex number and, conversely, each complex number corresponds to a point in the plane. We shall refer to the collection of such points as the **complex plane.** The x-axis will be referred to as the **real axis,** because any point that lies on the real axis is of the form $z = x + 0i = x$, a real number. The y-axis is called the **imaginary axis,** because any point that lies on it is of the form $z = 0 + yi = yi$, a pure imaginary number. See Figure 37.

EXAMPLE 1   Plotting a Point in the Complex Plane

Plot the point corresponding to $z = \sqrt{3} - i$ in the complex plane.

Solution   The point corresponding to $z = \sqrt{3} - i$ has the rectangular coordinates $(\sqrt{3}, -1)$. The point, located in quadrant IV, is plotted in Figure 38.

Figure 38

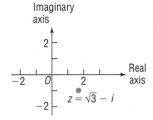

DEFINITION   Let $z = x + yi$ be a complex number. The **magnitude** or **modulus** of z, denoted by $|z|$, is defined as the distance from the origin to the point (x, y). That is,

$$|z| = \sqrt{x^2 + y^2} \qquad (1)$$

Figure 39

Imaginary axis
$|z| = \sqrt{x^2 + y^2}$
$z = x + yi$
y
Real axis
0
x

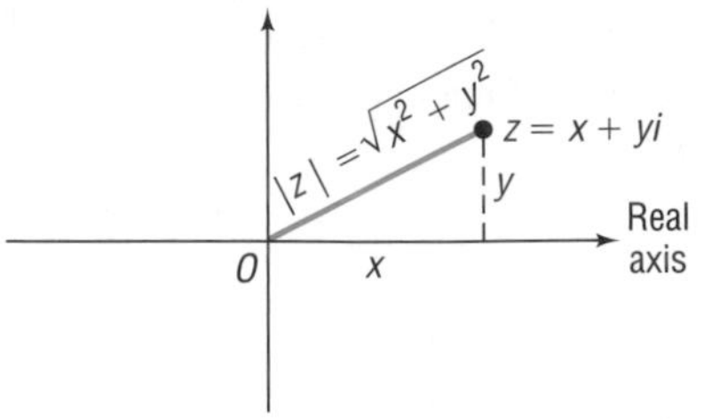

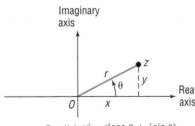

See Figure 39 for an illustration.

This definition for $|z|$ is consistent with the definition for the absolute value of a real number: If $z = x + yi$ is real, then $z = x + 0i$ and

$$|z| = \sqrt{x^2 + 0^2} = \sqrt{x^2} = |x|$$

For this reason, the magnitude of z is sometimes called the **absolute value of** z.

Recall that if $z = x + yi$ then its **conjugate,** denoted by $\bar{z}$, is $\bar{z} = x - yi$. Because $z\bar{z} = x^2 + y^2$, which is a nonnegative real number, it follows from equation (1) that the magnitude of z can be written as

$$|z| = \sqrt{z\bar{z}} \tag{2}$$

2 Convert a Complex Number from Rectangular Form to Polar Form

When a complex number is written in the standard form $z = x + yi$, we say that it is in **rectangular,** or **Cartesian, form,** because (x, y) are the rectangular coordinates of the corresponding point in the complex plane. Suppose that (r, θ) are the polar coordinates of this point. Then

$$x = r\cos\theta, \qquad y = r\sin\theta \tag{3}$$

DEFINITION

If $r \geq 0$ and $0 \leq \theta < 2\pi$, the complex number $z = x + yi$ may be written in **polar form** as

$$z = x + yi = (r\cos\theta) + (r\sin\theta)i = r(\cos\theta + i\sin\theta) \tag{4}$$

Figure 40

Imaginary axis
z
r
θ
y
Real axis
0
x

$z = x + yi = r(\cos\theta + i\sin\theta)$, $r \geq 0$, $0 \leq \theta < 2\pi$

See Figure 40.

If $z = r(\cos\theta + i\sin\theta)$ is the polar form of a complex number, the angle θ, $0 \leq \theta < 2\pi$, is called the **argument of** z.

Also, because $r \geq 0$, we have $r = \sqrt{x^2 + y^2}$. From equation (1), it follows that the magnitude of $z = r(\cos\theta + i\sin\theta)$ is

$$|z| = r$$

EXAMPLE 2 Writing a Complex Number in Polar Form

Write an expression for $z = \sqrt{3} - i$ in polar form.

Solution The point, located in quadrant IV, is plotted in Figure 38. Because $x = \sqrt{3}$ and $y = -1$, it follows that

$$r = \sqrt{x^2 + y^2} = \sqrt{\left(\sqrt{3}\right)^2 + (-1)^2} = \sqrt{4} = 2$$

So

$$\sin\theta = \frac{y}{r} = \frac{-1}{2}, \qquad \cos\theta = \frac{x}{r} = \frac{\sqrt{3}}{2}, \qquad 0 \leq \theta < 2\pi$$

Then $\theta = \dfrac{11\pi}{6}$ and $r = 2$, so the polar form of $z = \sqrt{3} - i$ is

$$z = r(\cos\theta + i\sin\theta) = 2\left(\cos\frac{11\pi}{6} + i\sin\frac{11\pi}{6}\right)$$

Now Work PROBLEM 11

| EXAMPLE 3 | **Plotting a Point in the Complex Plane and Converting from Polar to Rectangular Form** |

Plot the point corresponding to $z = 2(\cos 30° + i \sin 30°)$ in the complex plane, and write an expression for z in rectangular form.

Solution

Figure 41

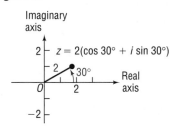

To plot the complex number $z = 2(\cos 30° + i \sin 30°)$, we plot the point whose polar coordinates are $(r, \theta) = (2, 30°)$, as shown in Figure 41. In rectangular form,

$$z = 2(\cos 30° + i \sin 30°) = 2\left(\frac{\sqrt{3}}{2} + \frac{1}{2}i\right) = \sqrt{3} + i$$

Now Work PROBLEM 23

3 Find Products and Quotients of Complex Numbers in Polar Form

The polar form of a complex number provides an alternative method for finding products and quotients of complex numbers.

THEOREM

Let $z_1 = r_1(\cos \theta_1 + i \sin \theta_1)$ and $z_2 = r_2(\cos \theta_2 + i \sin \theta_2)$ be two complex numbers. Then

$$z_1 z_2 = r_1 r_2 [\cos(\theta_1 + \theta_2) + i \sin(\theta_1 + \theta_2)] \qquad (5)$$

If $z_2 \neq 0$, then

$$\frac{z_1}{z_2} = \frac{r_1}{r_2} [\cos(\theta_1 - \theta_2) + i \sin(\theta_1 - \theta_2)] \qquad (6)$$

In Words

The magnitude of a complex number z is r and its argument is θ, so when

$$z = r(\cos \theta + i \sin \theta)$$

the magnitude of the product (quotient) of two complex numbers equals the product (quotient) of their magnitudes; the argument of the product (quotient) of two complex numbers is determined by the sum (difference) of their arguments.

Proof We will prove formula (5). The proof of formula (6) is left as an exercise (see Problem 66).

$$z_1 z_2 = [r_1(\cos \theta_1 + i \sin \theta_1)][r_2(\cos \theta_2 + i \sin \theta_2)]$$
$$= r_1 r_2 [(\cos \theta_1 + i \sin \theta_1)(\cos \theta_2 + i \sin \theta_2)]$$
$$= r_1 r_2 [(\cos \theta_1 \cos \theta_2 - \sin \theta_1 \sin \theta_2) + i(\sin \theta_1 \cos \theta_2 + \cos \theta_1 \sin \theta_2)]$$
$$= r_1 r_2 [\cos(\theta_1 + \theta_2) + i \sin(\theta_1 + \theta_2)] \qquad ■$$

Let's look at an example of how this theorem can be used.

| EXAMPLE 4 | **Finding Products and Quotients of Complex Numbers in Polar Form** |

If $z = 3(\cos 20° + i \sin 20°)$ and $w = 5(\cos 100° + i \sin 100°)$, find the following (leave your answers in polar form):

(a) zw (b) $\dfrac{z}{w}$

Solution

(a) $zw = [3(\cos 20° + i \sin 20°)][5(\cos 100° + i \sin 100°)]$
$$= (3 \cdot 5)[\cos(20° + 100°) + i \sin(20° + 100°)]$$
$$= 15(\cos 120° + i \sin 120°) \qquad \text{Apply equation (5).}$$

(b) $\dfrac{z}{w} = \dfrac{3(\cos 20° + i \sin 20°)}{5(\cos 100° + i \sin 100°)}$

$= \dfrac{3}{5}[\cos(20° - 100°) + i \sin(20° - 100°)]$ Apply equation (6).

$= \dfrac{3}{5}[\cos(-80°) + i \sin(-80°)]$

$= \dfrac{3}{5}(\cos 280° + i \sin 280°)$ The argument must lie between $0°$ and $360°$.

━━━━━ **Now Work** PROBLEM 33

4 Use De Moivre's Theorem

De Moivre's Theorem, stated by Abraham De Moivre (1667–1754) in 1730, but already known to many people by 1710, is important for the following reason: The fundamental processes of algebra are the four operations of addition, subtraction, multiplication, and division, together with powers and the extraction of roots. De Moivre's Theorem allows these latter fundamental algebraic operations to be applied to complex numbers.

De Moivre's Theorem, in its most basic form, is a formula for raising a complex number z to the power n, where $n \geq 1$ is a positive integer. Let's see if we can conjecture the form of the result.

Let $z = r(\cos \theta + i \sin \theta)$ be a complex number. Then, based on equation (5), we have

$n = 2$: $z^2 = r^2[\cos(2\theta) + i \sin(2\theta)]$ Equation (5)

$n = 3$: $z^3 = z^2 \cdot z$

$= \{r^2[\cos(2\theta) + i \sin(2\theta)]\}[r(\cos \theta + i \sin \theta)]$

$= r^3[\cos(3\theta) + i \sin(3\theta)]$ Equation (5)

$n = 4$: $z^4 = z^3 \cdot z$

$= \{r^3[\cos(3\theta) + i \sin(3\theta)]\}[r(\cos \theta + i \sin \theta)]$

$= r^4[\cos(4\theta) + i \sin(4\theta)]$ Equation (5)

Do you see the pattern?

THEOREM | **De Moivre's Theorem**

If $z = r(\cos \theta + i \sin \theta)$ is a complex number, then

$$z^n = r^n[\cos(n\theta) + i \sin(n\theta)] \tag{7}$$

where $n \geq 1$ is a positive integer.

The proof of De Moivre's Theorem requires mathematical induction (which is not discussed in this text), so it is omitted here.

Let's look at some examples.

EXAMPLE 5 | **Using De Moivre's Theorem**

Write $[2(\cos 20° + i \sin 20°)]^3$ in the standard form $a + bi$.

Solution $[2(\cos 20° + i \sin 20°)]^3 = 2^3[\cos(3 \cdot 20°) + i \sin(3 \cdot 20°)]$ Apply De Moivre's Theorem.

$= 8(\cos 60° + i \sin 60°)$

$= 8\left(\dfrac{1}{2} + \dfrac{\sqrt{3}}{2}i\right) = 4 + 4\sqrt{3}i$

━━━━━ **Now Work** PROBLEM 41

EXAMPLE 6	Using De Moivre's Theorem

Write $(1 + i)^5$ in the standard form $a + bi$.

Algebraic Solution

To apply De Moivre's Theorem, we must first write the complex number in polar form. Since the magnitude of $1 + i$ is $\sqrt{1^2 + 1^2} = \sqrt{2}$, we begin by writing

$$1 + i = \sqrt{2}\left(\frac{1}{\sqrt{2}} + \frac{1}{\sqrt{2}}i\right) = \sqrt{2}\left(\cos\frac{\pi}{4} + i\sin\frac{\pi}{4}\right)$$

Now

$$(1 + i)^5 = \left[\sqrt{2}\left(\cos\frac{\pi}{4} + i\sin\frac{\pi}{4}\right)\right]^5$$

$$= (\sqrt{2})^5\left[\cos\left(5\cdot\frac{\pi}{4}\right) + i\sin\left(5\cdot\frac{\pi}{4}\right)\right]$$

$$= 4\sqrt{2}\left(\cos\frac{5\pi}{4} + i\sin\frac{5\pi}{4}\right)$$

$$= 4\sqrt{2}\left[-\frac{1}{\sqrt{2}} + \left(-\frac{1}{\sqrt{2}}\right)i\right] = -4 - 4i$$

Graphing Solution

Using a TI-84 Plus graphing calculator, we obtain the solution shown in Figure 42.

Figure 42

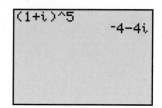

5 Find Complex Roots

Let w be a given complex number, and let $n \geq 2$ denote a positive integer. Any complex number z that satisfies the equation

$$z^n = w$$

is called a **complex nth root** of w. In keeping with previous usage, if $n = 2$, the solutions of the equation $z^2 = w$ are called **complex square roots** of w, and if $n = 3$, the solutions of the equation $z^3 = w$ are called **complex cube roots** of w.

THEOREM

Finding Complex Roots

Let $w = r(\cos\theta_0 + i\sin\theta_0)$ be a complex number, and let $n \geq 2$ be an integer. If $w \neq 0$, there are n distinct complex roots of w, given by the formula

$$z_k = \sqrt[n]{r}\left[\cos\left(\frac{\theta_0}{n} + \frac{2k\pi}{n}\right) + i\sin\left(\frac{\theta_0}{n} + \frac{2k\pi}{n}\right)\right] \qquad (8)$$

where $k = 0, 1, 2, \ldots, n - 1$.

Proof (Outline)   We will not prove this result in its entirety. Instead, we shall show only that each z_k in equation (8) satisfies the equation $z_k^n = w$, proving that each z_k is a complex nth root of w.

$$z_k^n = \left\{\sqrt[n]{r}\left[\cos\left(\frac{\theta_0}{n} + \frac{2k\pi}{n}\right) + i\sin\left(\frac{\theta_0}{n} + \frac{2k\pi}{n}\right)\right]\right\}^n$$

$$= (\sqrt[n]{r})^n\left\{\cos\left[n\left(\frac{\theta_0}{n} + \frac{2k\pi}{n}\right)\right] + i\sin\left[n\left(\frac{\theta_0}{n} + \frac{2k\pi}{n}\right)\right]\right\} \quad \text{Apply De Moivre's Theorem.}$$

$$= r[\cos(\theta_0 + 2k\pi) + i\sin(\theta_0 + 2k\pi)] \qquad\qquad \text{Simplify.}$$

$$= r(\cos\theta_0 + i\sin\theta_0) = w \qquad\qquad\qquad \text{The Periodic Property}$$

So each z_k, $k = 0, 1, \ldots, n - 1$, is a complex nth root of w. To complete the proof, we would need to show that each z_k, $k = 0, 1, \ldots, n - 1$, is, in fact, distinct and that there are no complex nth roots of w other than those given by equation (8). ■

EXAMPLE 7 **Finding Complex Cube Roots**

Find the complex cube roots of $-1 + \sqrt{3}i$. Leave your answers in polar form, with the argument in degrees.

Solution First, we express $-1 + \sqrt{3}i$ in polar form using degrees.

$$-1 + \sqrt{3}i = 2\left(-\frac{1}{2} + \frac{\sqrt{3}}{2}i\right) = 2(\cos 120° + i \sin 120°)$$

The three complex cube roots of $-1 + \sqrt{3}i = 2(\cos 120° + i \sin 120°)$ are

$$z_k = \sqrt[3]{2}\left[\cos\left(\frac{120°}{3} + \frac{360°k}{3}\right) + i \sin\left(\frac{120°}{3} + \frac{360°k}{3}\right)\right]$$

$$= \sqrt[3]{2}[\cos(40° + 120°k) + i \sin(40° + 120°k)] \qquad k = 0, 1, 2$$

So

WARNING Most graphing utilities will only provide the answer z_0 to the calculation $(-1 + \sqrt{3}i) \wedge \left(\dfrac{1}{3}\right)$. The paragraph following Example 7 explains how to obtain z_1 and z_2 from z_0. ■

$$z_0 = \sqrt[3]{2}[\cos(40° + 120°\cdot 0) + i \sin(40° + 120°\cdot 0)] = \sqrt[3]{2}(\cos 40° + i \sin 40°)$$

$$z_1 = \sqrt[3]{2}[\cos(40° + 120°\cdot 1) + i \sin(40° + 120°\cdot 1)] = \sqrt[3]{2}(\cos 160° + i \sin 160°)$$

$$z_2 = \sqrt[3]{2}[\cos(40° + 120°\cdot 2) + i \sin(40° + 120°\cdot 2)] = \sqrt[3]{2}(\cos 280° + i \sin 280°)$$

Notice that each of the three complex roots of $-1 + \sqrt{3}i$ has the same magnitude, $\sqrt[3]{2}$. This means that the points corresponding to each cube root lie the same distance from the origin; that is, the three points lie on a circle with center at the origin and radius $\sqrt[3]{2}$. Furthermore, the arguments of these cube roots are 40°, 160°, and 280°, the difference of consecutive pairs being $120° = \dfrac{360°}{3}$. This means that the three points are equally spaced on the circle, as shown in Figure 43. These results are not coincidental. In fact, you are asked to show that these results hold for complex nth roots in Problems 63 through 65.

Figure 43

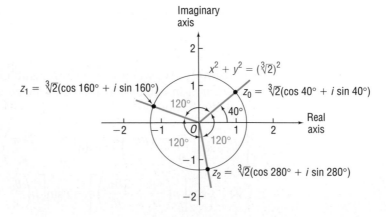

$z_1 = \sqrt[3]{2}(\cos 160° + i \sin 160°)$

$x^2 + y^2 = (\sqrt[3]{2})^2$

$z_0 = \sqrt[3]{2}(\cos 40° + i \sin 40°)$

$z_2 = \sqrt[3]{2}(\cos 280° + i \sin 280°)$

Historical Feature

John Wallis

The Babylonians, Greeks, and Arabs considered square roots of negative quantities to be impossible and equations with complex solutions to be unsolvable. The first hint that there was some connection between real solutions of equations and complex numbers came when Girolamo Cardano (1501–1576) and Tartaglia (1499–1557) found *real* roots of cubic equations by taking cube roots of *complex* quantities. For centuries thereafter, mathematicians worked with complex numbers without much belief in their actual existence. In 1673, John Wallis appears to have been the first to suggest the graphical representation of complex numbers, a truly significant idea that was not pursued further until about 1800. Several people, including Karl Friedrich Gauss (1777–1855), then rediscovered the idea, and graphical representation helped to establish complex numbers as equal members of the number family. In practical applications, complex numbers have found their greatest uses in the study of alternating current, where they are a commonplace tool, and in the field of subatomic physics.

Historical Problems

1. The quadratic formula will work perfectly well if the coefficients are complex numbers. Solve the following. [**Hint:** The answers are "nice."]

   (a) $z^2 - (2 + 5i)z - 3 + 5i = 0$
   (b) $z^2 - (1 + i)z - 2 - i = 0$

5.3 Assess Your Understanding

'Are You Prepared?' *Answers are given at the end of these exercises. If you get a wrong answer, read the pages listed in red.*

1. The conjugate of $-4 - 3i$ is _____. (p. A44)
2. The sum formula for the sine function is $\sin(A + B) =$ _____. (p. 235)
3. The sum formula for the cosine function is $\cos(A + B) =$ _____. (p. 232)
4. $\sin 120° =$ _____; $\cos 240° =$ _____. (pp. 146–147)

Concepts and Vocabulary

5. When a complex number z is written in the polar form $z = r(\cos\theta + i\sin\theta)$, the nonnegative number r is the _____ or _____ of z, and the angle $\theta, 0 \leq \theta < 2\pi$, is the _____ of z.

6. _____ Theorem can be used to raise a complex number to a power.

7. Every nonzero complex number will have exactly _____ cube roots.

8. **True or False** De Moivre's Theorem is useful for raising a complex number to a positive integer power.

9. **True or False** Using De Moivre's Theorem, the square of a complex number will have two answers.

10. **True or False** The polar form of a complex number is unique.

Skill Building

In Problems 11–22, plot each complex number in the complex plane and write it in polar form. Express the argument in degrees.

11. $1 + i$
12. $-1 + i$
13. $\sqrt{3} - i$
14. $1 - \sqrt{3}i$

15. $-3i$
16. -2
17. $4 - 4i$
18. $9\sqrt{3} + 9i$

19. $3 - 4i$
20. $2 + \sqrt{3}i$
21. $-2 + 3i$
22. $\sqrt{5} - i$

In Problems 23–32, write each complex number in rectangular form.

23. $2(\cos 120° + i\sin 120°)$
24. $3(\cos 210° + i\sin 210°)$
25. $4\left(\cos\dfrac{7\pi}{4} + i\sin\dfrac{7\pi}{4}\right)$

26. $2\left(\cos\dfrac{5\pi}{6} + i\sin\dfrac{5\pi}{6}\right)$
27. $3\left(\cos\dfrac{3\pi}{2} + i\sin\dfrac{3\pi}{2}\right)$
28. $4\left(\cos\dfrac{\pi}{2} + i\sin\dfrac{\pi}{2}\right)$

29. $0.2(\cos 100° + i\sin 100°)$
30. $0.4(\cos 200° + i\sin 200°)$

31. $2\left(\cos\dfrac{\pi}{18} + i\sin\dfrac{\pi}{18}\right)$
32. $3\left(\cos\dfrac{\pi}{10} + i\sin\dfrac{\pi}{10}\right)$

In Problems 33–40, find zw and $\dfrac{z}{w}$. Leave your answers in polar form.

33. $z = 2(\cos 40° + i\sin 40°)$
    $w = 4(\cos 20° + i\sin 20°)$

34. $z = \cos 120° + i\sin 120°$
    $w = \cos 100° + i\sin 100°$

35. $z = 3(\cos 130° + i\sin 130°)$
    $w = 4(\cos 270° + i\sin 270°)$

36. $z = 2(\cos 80° + i \sin 80°)$
$w = 6(\cos 200° + i \sin 200°)$

37. $z = 2\left(\cos \dfrac{\pi}{8} + i \sin \dfrac{\pi}{8}\right)$
$w = 2\left(\cos \dfrac{\pi}{10} + i \sin \dfrac{\pi}{10}\right)$

38. $z = 4\left(\cos \dfrac{3\pi}{8} + i \sin \dfrac{3\pi}{8}\right)$
$w = 2\left(\cos \dfrac{9\pi}{16} + i \sin \dfrac{9\pi}{16}\right)$

39. $z = 2 + 2i$
$w = \sqrt{3} - i$

40. $z = 1 - i$
$w = 1 - \sqrt{3}i$

In Problems 41–52, write each expression in the standard form $a + bi$.

41. $[4(\cos 40° + i \sin 40°)]^3$

42. $[3(\cos 80° + i \sin 80°)]^3$

43. $\left[2\left(\cos \dfrac{\pi}{10} + i \sin \dfrac{\pi}{10}\right)\right]^5$

44. $\left[\sqrt{2}\left(\cos \dfrac{5\pi}{16} + i \sin \dfrac{5\pi}{16}\right)\right]^4$

45. $[\sqrt{3}(\cos 10° + i \sin 10°)]^6$

46. $\left[\dfrac{1}{2}(\cos 72° + i \sin 72°)\right]^5$

47. $\left[\sqrt{5}\left(\cos \dfrac{3\pi}{16} + i \sin \dfrac{3\pi}{16}\right)\right]^4$

48. $\left[\sqrt{3}\left(\cos \dfrac{5\pi}{18} + i \sin \dfrac{5\pi}{18}\right)\right]^6$

49. $(1 - i)^5$

50. $\left(\sqrt{3} - i\right)^6$

51. $\left(\sqrt{2} - i\right)^6$

52. $\left(1 - \sqrt{5}i\right)^8$

In Problems 53–60, find all the complex roots. Leave your answers in polar form with the argument in degrees.

53. The complex cube roots of $1 + i$

54. The complex fourth roots of $\sqrt{3} - i$

55. The complex fourth roots of $4 - 4\sqrt{3}i$

56. The complex cube roots of $-8 - 8i$

57. The complex fourth roots of $-16i$

58. The complex cube roots of -8

59. The complex fifth roots of i

60. The complex fifth roots of $-i$

Applications and Extensions

61. Find the four complex fourth roots of unity (1) and plot them.

62. Find the six complex sixth roots of unity (1) and plot them.

63. Show that each complex nth root of a nonzero complex number w has the same magnitude.

64. Use the result of Problem 63 to draw the conclusion that each complex nth root lies on a circle with center at the origin. What is the radius of this circle?

65. Refer to Problem 64. Show that the complex nth roots of a nonzero complex number w are equally spaced on the circle.

66. Prove formula (6).

67. Mandelbrot Sets
(a) Consider the expression $a_n = (a_{n-1})^2 + z$, where z is some complex number (called the **seed**) and $a_0 = z$. Compute $a_1 (=a_0^2 + z)$, $a_2 (=a_1^2 + z)$, $a_3 (=a_2^2 + z)$, a_4, a_5, and a_6 for the following seeds: $z_1 = 0.1 - 0.4i$, $z_2 = 0.5 + 0.8i$, $z_3 = -0.9 + 0.7i$, $z_4 = -1.1 + 0.1i$, $z_5 = 0 - 1.3i$, and $z_6 = 1 + 1i$.
(b) The dark portion of the graph represents the set of all values $z = x + yi$ that are in the Mandelbrot set. Determine which complex numbers in part (a) are in this set by plotting them on the graph. Do the complex

numbers that are not in the Mandelbrot set have any common characteristics regarding the values of a_6 found in part (a)?
(c) Compute $|z| = \sqrt{x^2 + y^2}$ for each of the complex numbers in part (a). Now compute $|a_6|$ for each of the complex numbers in part (a). For which complex numbers is $|a_6| \le |z|$ and $|z| \le 2$? Conclude that the criterion for a complex number to be in the Mandelbrot set is that $|a_n| \le |z|$ and $|z| \le 2$.

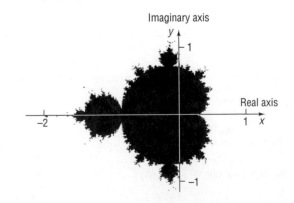

'Are You Prepared?' Answers

1. $-4 + 3i$ **2.** $\sin A \cos B + \cos A \sin B$ **3.** $\cos A \cos B - \sin A \sin B$ **4.** $\dfrac{\sqrt{3}}{2}; -\dfrac{1}{2}$

5.4 Vectors

Figure 44

In simple terms, a **vector** (derived from the Latin *vehere,* meaning "to carry") is a quantity that has both magnitude and direction. It is customary to represent a vector by using an arrow. The length of the arrow represents the **magnitude** of the vector, and the arrowhead indicates the **direction** of the vector.

Many quantities in physics can be represented by vectors. For example, the velocity of an aircraft can be represented by an arrow that points in the direction of movement; the length of the arrow represents speed. If the aircraft speeds up, we lengthen the arrow; if the aircraft changes direction, we introduce an arrow in the new direction. See Figure 44. Based on this representation, it is not surprising that vectors and *directed line segments* are somehow related.

Geometric Vectors

If P and Q are two distinct points in the xy-plane, there is exactly one line containing both P and Q [Figure 45(a)]. The points on that part of the line that joins P to Q, including P and Q, form what is called the **line segment** $\overline{PQ}$ [Figure 45(b)]. If we order the points so that they proceed from P to Q, we have a **directed line segment** from P to Q, or a **geometric vector,** which we denote by $\overrightarrow{PQ}$. In a directed line segment $\overrightarrow{PQ}$, we call P the **initial point** and Q the **terminal point,** as indicated in Figure 45(c).

Figure 45

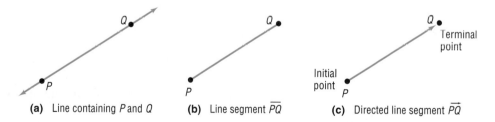

(a) Line containing P and Q **(b)** Line segment $\overline{PQ}$ **(c)** Directed line segment $\overrightarrow{PQ}$

The magnitude of the directed line segment $\overrightarrow{PQ}$ is the distance from the point P to the point Q; that is, it is the length of the line segment. The direction of $\overrightarrow{PQ}$ is from P to Q. If a vector **v*** has the same magnitude and the same direction as the directed line segment $\overrightarrow{PQ}$, we write

$$\mathbf{v} = \overrightarrow{PQ}$$

The vector **v** whose magnitude is 0 is called the **zero vector, 0.** The zero vector is assigned no direction.

Two vectors **v** and **w** are **equal,** written

$$\mathbf{v} = \mathbf{w}$$

if they have the same magnitude and the same direction.

For example, the three vectors shown in Figure 46 have the same magnitude and the same direction, so they are equal, even though they have different initial points

Figure 46

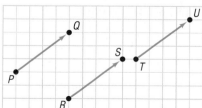

*Boldface letters will be used to denote vectors, to distinguish them from numbers. For handwritten work, an arrow is placed over the letter to signify a vector.

and different terminal points. As a result, we find it useful to think of a vector simply as an arrow, keeping in mind that two arrows (vectors) are equal if they have the same direction and the same magnitude (length).

Adding Vectors Geometrically

Figure 47

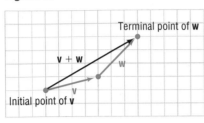

The **sum v + w** of two vectors is defined as follows: We position the vectors **v** and **w** so that the terminal point of **v** coincides with the initial point of **w**, as shown in Figure 47. The vector **v + w** is then the unique vector whose initial point coincides with the initial point of **v** and whose terminal point coincides with the terminal point of **w**.

Vector addition is **commutative.** That is, if **v** and **w** are any two vectors, then

$$\mathbf{v} + \mathbf{w} = \mathbf{w} + \mathbf{v}$$

Figure 48

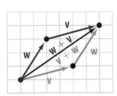

Figure 48 illustrates this fact. (Observe that the commutative property is another way of saying that opposite sides of a parallelogram are equal and parallel.)

Vector addition is also **associative.** That is, if **u**, **v**, and **w** are vectors, then

$$\mathbf{u} + (\mathbf{v} + \mathbf{w}) = (\mathbf{u} + \mathbf{v}) + \mathbf{w}$$

Figure 49
$(\mathbf{u} + \mathbf{v}) + \mathbf{w} = \mathbf{u} + (\mathbf{v} + \mathbf{w})$

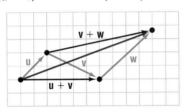

Figure 49 illustrates the associative property for vectors.

The zero vector **0** has the property that

$$\mathbf{v} + \mathbf{0} = \mathbf{0} + \mathbf{v} = \mathbf{v}$$

for any vector **v**.

Figure 50

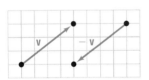

If **v** is a vector, then −**v** is the vector having the same magnitude as **v**, but whose direction is opposite to **v**, as shown in Figure 50.
Furthermore,

$$\mathbf{v} + (-\mathbf{v}) = \mathbf{0}$$

Figure 51

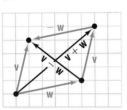

If **v** and **w** are two vectors, we define the **difference v − w** as

$$\mathbf{v} - \mathbf{w} = \mathbf{v} + (-\mathbf{w})$$

Figure 51 illustrates the relationships among **v**, **w**, **v + w**, and **v − w**.

Multiplying Vectors by Numbers Geometrically

When dealing with vectors, we refer to real numbers as **scalars.** Scalars are quantities that have only magnitude. Examples from physics of scalar quantities are temperature, speed, and time. We now define how to multiply a vector by a scalar.

DEFINITION If α is a scalar and **v** is a vector, the **scalar multiple** $\alpha\mathbf{v}$ is defined as follows:

1. If $\alpha > 0$, $\alpha\mathbf{v}$ is the vector whose magnitude is α times the magnitude of **v** and whose direction is the same as **v**.
2. If $\alpha < 0$, $\alpha\mathbf{v}$ is the vector whose magnitude is $|\alpha|$ times the magnitude of **v** and whose direction is opposite that of **v**.
3. If $\alpha = 0$ or if **v** = **0**, then $\alpha\mathbf{v} = \mathbf{0}$.

Figure 52

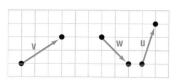

See Figure 52 for some illustrations.

For example, if **a** is the acceleration of an object of mass m due to a force **F** being exerted on it, then, by Newton's second law of motion, $\mathbf{F} = m\mathbf{a}$. Here, $m\mathbf{a}$ is the product of the scalar m and the vector **a**.

Scalar multiples have the following properties:

$$0\mathbf{v} = \mathbf{0} \qquad 1\mathbf{v} = \mathbf{v} \qquad -1\mathbf{v} = -\mathbf{v}$$

$$(\alpha + \beta)\mathbf{v} = \alpha\mathbf{v} + \beta\mathbf{v} \qquad \alpha(\mathbf{v} + \mathbf{w}) = \alpha\mathbf{v} + \alpha\mathbf{w}$$

$$\alpha(\beta\mathbf{v}) = (\alpha\beta)\mathbf{v}$$

1 Graph Vectors

> **EXAMPLE 1** Graphing Vectors
>
> Use the vectors illustrated in Figure 53 to graph each of the following vectors:
>
> (a) $\mathbf{v} - \mathbf{w}$ (b) $2\mathbf{v} + 3\mathbf{w}$ (c) $2\mathbf{v} - \mathbf{w} + \mathbf{u}$
>
> **Solution** Figure 54 illustrates each graph.

Figure 53

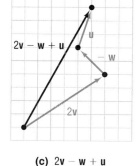

Figure 54

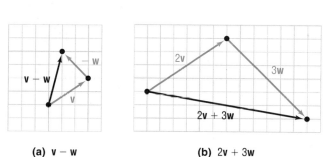

(a) $\mathbf{v} - \mathbf{w}$ (b) $2\mathbf{v} + 3\mathbf{w}$ (c) $2\mathbf{v} - \mathbf{w} + \mathbf{u}$

──── **Now Work** PROBLEMS 7 AND 9

Magnitudes of Vectors

If **v** is a vector, we use the symbol $\|\mathbf{v}\|$ to represent the **magnitude** of **v**. Since $\|\mathbf{v}\|$ equals the length of a directed line segment, it follows that $\|\mathbf{v}\|$ has the following properties:

> **THEOREM** **Properties of $\|\mathbf{v}\|$**
>
> If **v** is a vector and if α is a scalar, then
>
> (a) $\|\mathbf{v}\| \geq 0$ (b) $\|\mathbf{v}\| = 0$ if and only if $\mathbf{v} = \mathbf{0}$
>
> (c) $\|-\mathbf{v}\| = \|\mathbf{v}\|$ (d) $\|\alpha\mathbf{v}\| = |\alpha|\|\mathbf{v}\|$

Property (a) is a consequence of the fact that distance is a nonnegative number. Property (b) follows, because the length of the directed line segment $\overrightarrow{PQ}$ is positive unless P and Q are the same point, in which case the length is 0. Property (c) follows because the length of the line segment $\overline{PQ}$ equals the length of the line segment $\overline{QP}$. Property (d) is a direct consequence of the definition of a scalar multiple.

DEFINITION A vector **u** for which $\|\mathbf{u}\| = 1$ is called a **unit vector**.

2 Find a Position Vector

To compute the magnitude and direction of a vector, we need an algebraic way of representing vectors.

DEFINITION

An **algebraic vector v** is represented as

$$\mathbf{v} = \langle a, b \rangle$$

where a and b are real numbers (scalars) called the **components** of the vector **v**.

Figure 55

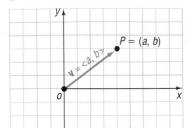

We use a rectangular coordinate system to represent algebraic vectors in the plane. If $\mathbf{v} = \langle a, b \rangle$ is an algebraic vector whose initial point is at the origin, then **v** is called a **position vector.** See Figure 55. Notice that the terminal point of the position vector $\mathbf{v} = \langle a, b \rangle$ is $P = (a, b)$.

The next result states that any vector whose initial point is not at the origin is equal to a unique position vector.

THEOREM

Suppose that **v** is a vector with initial point $P_1 = (x_1, y_1)$, not necessarily the origin, and terminal point $P_2 = (x_2, y_2)$. If $\mathbf{v} = \overrightarrow{P_1 P_2}$, then **v** is equal to the position vector

$$\mathbf{v} = \langle x_2 - x_1, y_2 - y_1 \rangle \qquad \textbf{(1)}$$

To see why this is true, look at Figure 56.

Figure 56
$\mathbf{v} = \langle a, b \rangle = \langle x_2 - x_1, y_2 - y_1 \rangle$

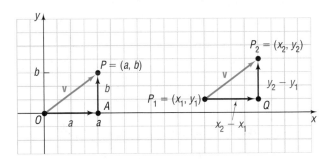

Triangle OPA and triangle P_1P_2Q are congruent. [Do you see why? The line segments have the same magnitude, so $d(O, P) = d(P_1, P_2)$; and they have the same direction, so $\angle POA = \angle P_2P_1Q$. Since the triangles are right triangles, we have angle–side–angle.] It follows that corresponding sides are equal. As a result, $x_2 - x_1 = a$ and $y_2 - y_1 = b$, so **v** may be written as

$$\mathbf{v} = \langle a, b \rangle = \langle x_2 - x_1, y_2 - y_1 \rangle$$

Because of this result, we can replace any algebraic vector by a unique position vector, and vice versa. This flexibility is one of the main reasons for the wide use of vectors.

EXAMPLE 2 Finding a Position Vector

Find the position vector of the vector $\mathbf{v} = \overrightarrow{P_1 P_2}$ if $P_1 = (-1, 2)$ and $P_2 = (4, 6)$.

Solution By equation (1), the position vector equal to **v** is

$$\mathbf{v} = \langle 4 - (-1), 6 - 2 \rangle = \langle 5, 4 \rangle$$

See Figure 57.

Figure 57

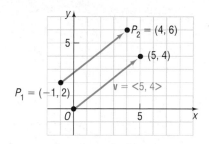

Two position vectors **v** and **w** are equal if and only if the terminal point of **v** is the same as the terminal point of **w**. This leads to the following result:

THEOREM

Equality of Vectors

Two vectors **v** and **w** are equal if and only if their corresponding components are equal. That is,

$$\text{If } \mathbf{v} = \langle a_1, b_1 \rangle \quad \text{and} \quad \mathbf{w} = \langle a_2, b_2 \rangle$$
$$\text{then} \quad \mathbf{v} = \mathbf{w} \quad \text{if and only if} \quad a_1 = a_2 \quad \text{and} \quad b_1 = b_2.$$

Figure 58

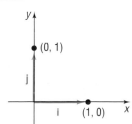

We now present an alternative representation of a vector in the plane that is common in the physical sciences. Let **i** denote the unit vector whose direction is along the positive x-axis; let **j** denote the unit vector whose direction is along the positive y-axis. Then $\mathbf{i} = \langle 1, 0 \rangle$ and $\mathbf{j} = \langle 0, 1 \rangle$, as shown in Figure 58. Any vector $\mathbf{v} = \langle a, b \rangle$ can be written using the unit vectors **i** and **j** as follows:

$$\mathbf{v} = \langle a, b \rangle = a\langle 1, 0 \rangle + b\langle 0, 1 \rangle = a\mathbf{i} + b\mathbf{j}$$

We call a and b the **horizontal** and **vertical components** of **v**, respectively. For example, if $\mathbf{v} = \langle 5, 4 \rangle = 5\mathbf{i} + 4\mathbf{j}$, then 5 is the horizontal component and 4 is the vertical component.

═══ **Now Work** PROBLEM 27

3 Add and Subtract Vectors Algebraically

We define addition, subtraction, scalar multiple, and magnitude of algebraic vectors in terms of their components.

DEFINITION

Let $\mathbf{v} = a_1\mathbf{i} + b_1\mathbf{j} = \langle a_1, b_1 \rangle$ and $\mathbf{w} = a_2\mathbf{i} + b_2\mathbf{j} = \langle a_2, b_2 \rangle$ be two vectors, and let α be a scalar. Then

$$\mathbf{v} + \mathbf{w} = (a_1 + a_2)\mathbf{i} + (b_1 + b_2)\mathbf{j} = \langle a_1 + a_2, b_1 + b_2 \rangle \qquad (2)$$

$$\mathbf{v} - \mathbf{w} = (a_1 - a_2)\mathbf{i} + (b_1 - b_2)\mathbf{j} = \langle a_1 - a_2, b_1 - b_2 \rangle \qquad (3)$$

$$\alpha\mathbf{v} = (\alpha a_1)\mathbf{i} + (\alpha b_1)\mathbf{j} = \langle \alpha a_1, \alpha b_1 \rangle \qquad (4)$$

$$\|\mathbf{v}\| = \sqrt{a_1^2 + b_1^2} \qquad (5)$$

In Words
To add two vectors, add corresponding components. To subtract two vectors, subtract corresponding components.

These definitions are compatible with the geometric definitions given earlier in this section. See Figure 59 on the next page.

Figure 59

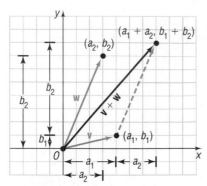

(a) Illustration of property (2)

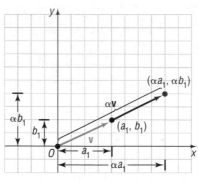

(b) Illustration of property (4), $\alpha > 0$

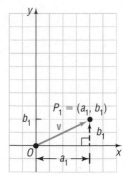

(c) Illustration of property (5):
$\| \mathbf{v} \|$ = Distance from O to P_1
$\| \mathbf{v} \| = \sqrt{a_1^2 + b_1^2}$

EXAMPLE 3	**Adding and Subtracting Vectors**

If $\mathbf{v} = 2\mathbf{i} + 3\mathbf{j} = \langle 2, 3 \rangle$ and $\mathbf{w} = 3\mathbf{i} - 4\mathbf{j} = \langle 3, -4 \rangle$, find:

(a) $\mathbf{v} + \mathbf{w}$ (b) $\mathbf{v} - \mathbf{w}$

Solution (a) $\mathbf{v} + \mathbf{w} = (2\mathbf{i} + 3\mathbf{j}) + (3\mathbf{i} - 4\mathbf{j}) = (2 + 3)\mathbf{i} + (3 - 4)\mathbf{j} = 5\mathbf{i} - \mathbf{j}$

or

$\mathbf{v} + \mathbf{w} = \langle 2, 3 \rangle + \langle 3, -4 \rangle = \langle 2 + 3, 3 + (-4) \rangle = \langle 5, -1 \rangle$

(b) $\mathbf{v} - \mathbf{w} = (2\mathbf{i} + 3\mathbf{j}) - (3\mathbf{i} - 4\mathbf{j}) = (2 - 3)\mathbf{i} + [3 - (-4)]\mathbf{j} = -\mathbf{i} + 7\mathbf{j}$

or

$\mathbf{v} - \mathbf{w} = \langle 2, 3 \rangle - \langle 3, -4 \rangle = \langle 2 - 3, 3 - (-4) \rangle = \langle -1, 7 \rangle$

4 Find a Scalar Multiple and the Magnitude of a Vector

EXAMPLE 4	**Finding Scalar Multiples and Magnitudes of Vectors**

If $\mathbf{v} = 2\mathbf{i} + 3\mathbf{j} = \langle 2, 3 \rangle$ and $\mathbf{w} = 3\mathbf{i} - 4\mathbf{j} = \langle 3, -4 \rangle$, find:

(a) $3\mathbf{v}$ (b) $2\mathbf{v} - 3\mathbf{w}$ (c) $\| \mathbf{v} \|$

Solution (a) $3\mathbf{v} = 3(2\mathbf{i} + 3\mathbf{j}) = 6\mathbf{i} + 9\mathbf{j}$

or

$3\mathbf{v} = 3\langle 2, 3 \rangle = \langle 6, 9 \rangle$

(b) $2\mathbf{v} - 3\mathbf{w} = 2(2\mathbf{i} + 3\mathbf{j}) - 3(3\mathbf{i} - 4\mathbf{j}) = 4\mathbf{i} + 6\mathbf{j} - 9\mathbf{i} + 12\mathbf{j}$

$= -5\mathbf{i} + 18\mathbf{j}$

or

$2\mathbf{v} - 3\mathbf{w} = 2\langle 2, 3 \rangle - 3\langle 3, -4 \rangle = \langle 4, 6 \rangle - \langle 9, -12 \rangle$

$= \langle 4 - 9, 6 - (-12) \rangle = \langle -5, 18 \rangle$

(c) $\| \mathbf{v} \| = \| 2\mathbf{i} + 3\mathbf{j} \| = \sqrt{2^2 + 3^2} = \sqrt{13}$

Now Work PROBLEMS **33** AND **39**

For the remainder of the section, we will express a vector $\mathbf{v}$ in the form $a\mathbf{i} + b\mathbf{j}$.

5 Find a Unit Vector

Recall that a unit vector $\mathbf{u}$ is a vector for which $\| \mathbf{u} \| = 1$. In many applications, it is useful to be able to find a unit vector $\mathbf{u}$ that has the same direction as a given vector $\mathbf{v}$.

THEOREM   **Unit Vector in the Direction of v**

For any nonzero vector $\mathbf{v}$, the vector

$$\mathbf{u} = \frac{\mathbf{v}}{\|\mathbf{v}\|}$$

is a unit vector that has the same direction as $\mathbf{v}$.

Proof Let $\mathbf{v} = a\mathbf{i} + b\mathbf{j}$. Then $\|\mathbf{v}\| = \sqrt{a^2 + b^2}$ and

$$\mathbf{u} = \frac{\mathbf{v}}{\|\mathbf{v}\|} = \frac{a\mathbf{i} + b\mathbf{j}}{\sqrt{a^2 + b^2}} = \frac{a}{\sqrt{a^2 + b^2}}\mathbf{i} + \frac{b}{\sqrt{a^2 + b^2}}\mathbf{j}$$

The vector $\mathbf{u}$ is in the same direction as $\mathbf{v}$, since $\|\mathbf{v}\| > 0$. Furthermore,

$$\|\mathbf{u}\| = \sqrt{\frac{a^2}{a^2 + b^2} + \frac{b^2}{a^2 + b^2}} = \sqrt{\frac{a^2 + b^2}{a^2 + b^2}} = 1$$

That is, $\mathbf{u}$ is a unit vector in the direction of $\mathbf{v}$. ∎

As a consequence of this theorem, if $\mathbf{u}$ is a unit vector in the same direction as a vector $\mathbf{v}$, then $\mathbf{v}$ may be expressed as

$$\mathbf{v} = \|\mathbf{v}\|\mathbf{u} \qquad\qquad (6)$$

This way of expressing a vector is useful in many applications.

EXAMPLE 5   Finding a Unit Vector

Find a unit vector in the same direction as $\mathbf{v} = 4\mathbf{i} - 3\mathbf{j}$.

Solution   We find $\|\mathbf{v}\|$ first.

$$\|\mathbf{v}\| = \|4\mathbf{i} - 3\mathbf{j}\| = \sqrt{16 + 9} = 5$$

Now we multiply $\mathbf{v}$ by the scalar $\dfrac{1}{\|\mathbf{v}\|} = \dfrac{1}{5}$. A unit vector in the same direction as $\mathbf{v}$ is

$$\frac{\mathbf{v}}{\|\mathbf{v}\|} = \frac{4\mathbf{i} - 3\mathbf{j}}{5} = \frac{4}{5}\mathbf{i} - \frac{3}{5}\mathbf{j}$$

✓ **Check:** This vector is, in fact, a unit vector because

$$\left(\frac{4}{5}\right)^2 + \left(-\frac{3}{5}\right)^2 = \frac{16}{25} + \frac{9}{25} = \frac{25}{25} = 1$$

━ **Now Work** PROBLEM 49

6 Find a Vector from Its Direction and Magnitude

If a vector represents the speed and direction of an object, it is called a **velocity vector.** If a vector represents the direction and amount of a force acting on an object, it is called a **force vector.** In many applications, a vector is described in terms of its magnitude and direction, rather than in terms of its components. For example, a ball thrown with an initial speed of 25 miles per hour at an angle 30° to the horizontal is a velocity vector.

Suppose that we are given the magnitude $\|\mathbf{v}\|$ of a nonzero vector $\mathbf{v}$ and the angle α, $0° \leq \alpha < 360°$, between $\mathbf{v}$ and $\mathbf{i}$. To express $\mathbf{v}$ in terms of $\|\mathbf{v}\|$ and α, we first find the unit vector $\mathbf{u}$ having the same direction as $\mathbf{v}$.

$$\mathbf{u} = \frac{\mathbf{v}}{\|\mathbf{v}\|} \quad \text{or} \quad \mathbf{v} = \|\mathbf{v}\|\mathbf{u} \qquad\qquad (7)$$

Figure 60

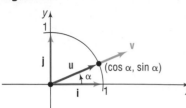

Look at Figure 60. The coordinates of the terminal point of **u** are $(\cos \alpha, \sin \alpha)$. Then $\mathbf{u} = \cos \alpha \mathbf{i} + \sin \alpha \mathbf{j}$ and, from (7),

$$\mathbf{v} = \|\mathbf{v}\|(\cos \alpha \mathbf{i} + \sin \alpha \mathbf{j}) \qquad (8)$$

where α is the angle between **v** and **i**.

EXAMPLE 6 **Writing a Vector When Its Magnitude and Direction Are Given**

A ball is thrown with an initial speed of 25 miles per hour in a direction that makes an angle of 30° with the positive x-axis. Express the velocity vector **v** in terms of **i** and **j**. What is the initial speed in the horizontal direction? What is the initial speed in the vertical direction?

Solution The magnitude of **v** is $\|\mathbf{v}\| = 25$ miles per hour, and the angle between the direction of **v** and **i**, the positive x-axis, is $\alpha = 30°$. By equation (8),

Figure 61

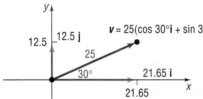

$$\mathbf{v} = \|\mathbf{v}\|(\cos \alpha \mathbf{i} + \sin \alpha \mathbf{j}) = 25(\cos 30°\mathbf{i} + \sin 30°\mathbf{j}) = 25\left(\frac{\sqrt{3}}{2}\mathbf{i} + \frac{1}{2}\mathbf{j}\right) = \frac{25\sqrt{3}}{2}\mathbf{i} + \frac{25}{2}\mathbf{j}$$

The initial speed of the ball in the horizontal direction is the horizontal component of **v**, $\frac{25\sqrt{3}}{2} \approx 21.65$ miles per hour. The initial speed in the vertical direction is the vertical component of **v**, $\frac{25}{2} = 12.5$ miles per hour. See Figure 61.

Now Work PROBLEM 63

Figure 62

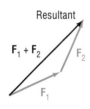

7 Analyze Objects in Static Equilibrium

Because forces can be represented by vectors, two forces "combine" the way that vectors "add." If $\mathbf{F}_1$ and $\mathbf{F}_2$ are two forces simultaneously acting on an object, the vector sum $\mathbf{F}_1 + \mathbf{F}_2$ is the **resultant force**. The resultant force produces the same effect on the object as that obtained when the two forces $\mathbf{F}_1$ and $\mathbf{F}_2$ act on the object. See Figure 62. An application of this concept is *static equilibrium*. An object is said to be in **static equilibrium** if (1) the object is at rest and (2) the sum of all forces acting on the object is zero, that is, if the resultant force is 0.

EXAMPLE 7 **An Object in Static Equilibrium**

Figure 63

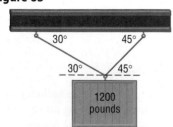

A box of supplies that weighs 1200 pounds is suspended by two cables attached to the ceiling, as shown in Figure 63. What are the tensions in the two cables?

Solution We draw a force diagram using the vectors shown in Figure 64. The tensions in the cables are the magnitudes $\|\mathbf{F}_1\|$ and $\|\mathbf{F}_2\|$ of the force vectors $\mathbf{F}_1$ and $\mathbf{F}_2$. The magnitude of the force vector $\mathbf{F}_3$ equals 1200 pounds, the weight of the box. Now write each force vector in terms of the unit vectors **i** and **j**. For $\mathbf{F}_1$ and $\mathbf{F}_2$, we use equation (8). Remember that α is the angle between the vector and the positive x-axis.

Figure 64

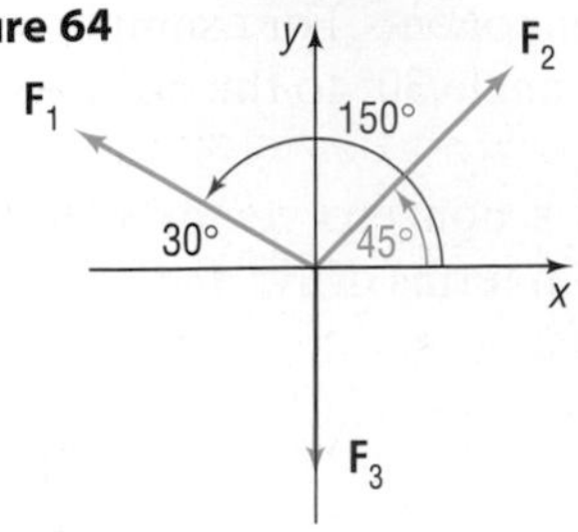

$$\mathbf{F}_1 = \|\mathbf{F}_1\|(\cos 150°\mathbf{i} + \sin 150°\mathbf{j}) = \|\mathbf{F}_1\|\left(-\frac{\sqrt{3}}{2}\mathbf{i} + \frac{1}{2}\mathbf{j}\right) = -\frac{\sqrt{3}}{2}\|\mathbf{F}_1\|\mathbf{i} + \frac{1}{2}\|\mathbf{F}_1\|\mathbf{j}$$

$$\mathbf{F}_2 = \|\mathbf{F}_2\|(\cos 45°\mathbf{i} + \sin 45°\mathbf{j}) = \|\mathbf{F}_2\|\left(\frac{\sqrt{2}}{2}\mathbf{i} + \frac{\sqrt{2}}{2}\mathbf{j}\right) = \frac{\sqrt{2}}{2}\|\mathbf{F}_2\|\mathbf{i} + \frac{\sqrt{2}}{2}\|\mathbf{F}_2\|\mathbf{j}$$

$$\mathbf{F}_3 = -1200\mathbf{j}$$

For static equilibrium, the sum of the force vectors must equal zero.

$$\mathbf{F}_1 + \mathbf{F}_2 + \mathbf{F}_3 = -\frac{\sqrt{3}}{2}\|\mathbf{F}_1\|\mathbf{i} + \frac{1}{2}\|\mathbf{F}_1\|\mathbf{j} + \frac{\sqrt{2}}{2}\|\mathbf{F}_2\|\mathbf{i} + \frac{\sqrt{2}}{2}\|\mathbf{F}_2\|\mathbf{j} - 1200\mathbf{j} = \mathbf{0}$$

The $\mathbf{i}$ component and $\mathbf{j}$ component will each equal zero. This results in the two equations

$$-\frac{\sqrt{3}}{2}\|\mathbf{F}_1\| + \frac{\sqrt{2}}{2}\|\mathbf{F}_2\| = 0 \tag{9}$$

$$\frac{1}{2}\|\mathbf{F}_1\| + \frac{\sqrt{2}}{2}\|\mathbf{F}_2\| - 1200 = 0 \tag{10}$$

We solve equation (9) for $\|\mathbf{F}_2\|$ and obtain

$$\|\mathbf{F}_2\| = \frac{\sqrt{3}}{\sqrt{2}}\|\mathbf{F}_1\| \tag{11}$$

Substituting into equation (10) and solving for $\|\mathbf{F}_1\|$, we obtain

$$\frac{1}{2}\|\mathbf{F}_1\| + \frac{\sqrt{2}}{2}\left(\frac{\sqrt{3}}{\sqrt{2}}\|\mathbf{F}_1\|\right) - 1200 = 0$$

$$\frac{1}{2}\|\mathbf{F}_1\| + \frac{\sqrt{3}}{2}\|\mathbf{F}_1\| - 1200 = 0$$

$$\frac{1 + \sqrt{3}}{2}\|\mathbf{F}_1\| = 1200$$

$$\|\mathbf{F}_1\| = \frac{2400}{1 + \sqrt{3}} \approx 878.5 \text{ pounds}$$

Substituting this value into equation (11) yields $\|\mathbf{F}_2\|$.

$$\|\mathbf{F}_2\| = \frac{\sqrt{3}}{\sqrt{2}}\|\mathbf{F}_1\| = \frac{\sqrt{3}}{\sqrt{2}} \cdot \frac{2400}{1 + \sqrt{3}} \approx 1075.9 \text{ pounds}$$

The left cable has tension of approximately 878.5 pounds and the right cable has tension of approximately 1075.9 pounds.

━━━━━▶ **Now Work** PROBLEM **67**

Historical Feature

Josiah Gibbs (1839–1903)

The history of vectors is surprisingly complicated for such a natural concept. In the xy-plane, complex numbers do a good job of imitating vectors. About 1840, mathematicians became interested in finding a system that would do for three dimensions what the complex numbers do for two dimensions. Hermann Grassmann (1809–1877), in Germany, and William Rowan Hamilton (1805–1865), in Ireland, both attempted to find solutions.

Hamilton's system was the *quaternions*, which are best thought of as a real number plus a vector, and do for four dimensions what complex numbers do for two dimensions. In this system the order of multiplication matters; that is, **ab** ≠ **ba**. Also, two products of vectors emerged, the scalar (or dot) product and the vector (or cross) product.

Grassmann's abstract style, although easily read today, was almost impenetrable during the previous century, and only a few of his ideas were appreciated. Among those few were the same scalar and vector products that Hamilton had found.

About 1880, the American physicist Josiah Willard Gibbs (1839–1903) worked out an algebra involving only the simplest concepts: the vectors and the two products. He then added some calculus, and the resulting system was simple, flexible, and well adapted to expressing a large number of physical laws. This system remains in use essentially unchanged. Hamilton's and Grassmann's more extensive systems each gave birth to much interesting mathematics, but little of this mathematics is seen at elementary levels.

5.4 Assess Your Understanding

Concepts and Vocabulary

1. A vector whose magnitude is 1 is called a(n) _____ vector.

2. The product of a vector by a number is called a(n) _____ multiple.

3. If $\mathbf{v} = a\mathbf{i} + b\mathbf{j}$, then a is called the _____ component of $\mathbf{v}$ and b is the _____ component of $\mathbf{v}$.

4. *True or False* Vectors are quantities that have magnitude and direction.

5. *True or False* Force is a physical example of a vector.

6. *True or False* Mass is a physical example of a vector.

Skill Building

In Problems 7–14, use the vectors in the figure at the right to graph each of the following vectors.

7. $\mathbf{v} + \mathbf{w}$

8. $\mathbf{u} + \mathbf{v}$

9. $3\mathbf{v}$

10. $4\mathbf{w}$

11. $\mathbf{v} - \mathbf{w}$

12. $\mathbf{u} - \mathbf{v}$

13. $3\mathbf{v} + \mathbf{u} - 2\mathbf{w}$

14. $2\mathbf{u} - 3\mathbf{v} + \mathbf{w}$

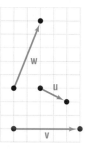

In Problems 15–22, use the figure at the right. Determine whether the given statement is true or false.

15. $\mathbf{A} + \mathbf{B} = \mathbf{F}$

16. $\mathbf{K} + \mathbf{G} = \mathbf{F}$

17. $\mathbf{C} = \mathbf{D} - \mathbf{E} + \mathbf{F}$

18. $\mathbf{G} + \mathbf{H} + \mathbf{E} = \mathbf{D}$

19. $\mathbf{E} + \mathbf{D} = \mathbf{G} + \mathbf{H}$

20. $\mathbf{H} - \mathbf{C} = \mathbf{G} - \mathbf{F}$

21. $\mathbf{A} + \mathbf{B} + \mathbf{K} + \mathbf{G} = \mathbf{0}$

22. $\mathbf{A} + \mathbf{B} + \mathbf{C} + \mathbf{H} + \mathbf{G} = \mathbf{0}$

23. If $\|\mathbf{v}\| = 4$, what is $\|3\mathbf{v}\|$?

24. If $\|\mathbf{v}\| = 2$, what is $\|-4\mathbf{v}\|$?

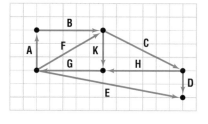

In Problems 25–32, the vector $\mathbf{v}$ has initial point P and terminal point Q. Write $\mathbf{v}$ in the form $a\mathbf{i} + b\mathbf{j}$; that is, find its position vector.

25. $P = (0, 0);\quad Q = (3, 4)$

26. $P = (0, 0);\quad Q = (-3, -5)$

27. $P = (3, 2);\quad Q = (5, 6)$

28. $P = (-3, 2);\quad Q = (6, 5)$

29. $P = (-2, -1);\quad Q = (6, -2)$

30. $P = (-1, 4);\quad Q = (6, 2)$

31. $P = (1, 0);\quad Q = (0, 1)$

32. $P = (1, 1);\quad Q = (2, 2)$

In Problems 33–38, find $\|\mathbf{v}\|$.

33. $\mathbf{v} = 3\mathbf{i} - 4\mathbf{j}$

34. $\mathbf{v} = -5\mathbf{i} + 12\mathbf{j}$

35. $\mathbf{v} = \mathbf{i} - \mathbf{j}$

36. $\mathbf{v} = -\mathbf{i} - \mathbf{j}$

37. $\mathbf{v} = -2\mathbf{i} + 3\mathbf{j}$

38. $\mathbf{v} = 6\mathbf{i} + 2\mathbf{j}$

In Problems 39–44, find each quantity if $\mathbf{v} = 3\mathbf{i} - 5\mathbf{j}$ and $\mathbf{w} = -2\mathbf{i} + 3\mathbf{j}$.

39. $2\mathbf{v} + 3\mathbf{w}$

40. $3\mathbf{v} - 2\mathbf{w}$

41. $\|\mathbf{v} - \mathbf{w}\|$

42. $\|\mathbf{v} + \mathbf{w}\|$

43. $\|\mathbf{v}\| - \|\mathbf{w}\|$

44. $\|\mathbf{v}\| + \|\mathbf{w}\|$

In Problems 45–50, find the unit vector in the same direction as $\mathbf{v}$.

45. $\mathbf{v} = 5\mathbf{i}$

46. $\mathbf{v} = -3\mathbf{j}$

47. $\mathbf{v} = 3\mathbf{i} - 4\mathbf{j}$

48. $\mathbf{v} = -5\mathbf{i} + 12\mathbf{j}$

49. $\mathbf{v} = \mathbf{i} - \mathbf{j}$

50. $\mathbf{v} = 2\mathbf{i} - \mathbf{j}$

51. Find a vector $\mathbf{v}$ whose magnitude is 4 and whose component in the $\mathbf{i}$ direction is twice the component in the $\mathbf{j}$ direction.

52. Find a vector $\mathbf{v}$ whose magnitude is 3 and whose component in the $\mathbf{i}$ direction is equal to the component in the $\mathbf{j}$ direction.

53. If $\mathbf{v} = 2\mathbf{i} - \mathbf{j}$ and $\mathbf{w} = x\mathbf{i} + 3\mathbf{j}$, find all numbers x for which $\|\mathbf{v} + \mathbf{w}\| = 5$.

54. If $P = (-3, 1)$ and $Q = (x, 4)$, find all numbers x such that the vector represented by $\overrightarrow{PQ}$ has length 5.

*In Problems 55–60, write the vector **v** in the form a**i** + b**j**, given its magnitude ‖**v**‖ and the angle α it makes with the positive x-axis.*

55. ‖**v**‖ = 5, α = 60°

56. ‖**v**‖ = 8, α = 45°

57. ‖**v**‖ = 14, α = 120°

58. ‖**v**‖ = 3, α = 240°

59. ‖**v**‖ = 25, α = 330°

60. ‖**v**‖ = 15, α = 315°

Applications and Extensions

61. Computer Graphics The field of computer graphics utilizes vectors to compute translations of points. For example, if the point $(-3, 2)$ is to be translated by $\mathbf{v} = \langle 5, 2 \rangle$, then the new location will be $\mathbf{u}' = \mathbf{u} + \mathbf{v} = \langle -3, 2 \rangle + \langle 5, 2 \rangle = \langle 2, 4 \rangle$. As illustrated in the figure, the point $(-3, 2)$ is translated to $(2, 4)$ by **v**.

Source: Phil Dadd. Vectors and Matrices: A Primer. www.gamedev.net/reference/articles/article1832.asp

(a) Determine the new coordinates of $(3, -1)$ if it is translated by $\mathbf{v} = \langle -4, 5 \rangle$.

(b) Illustrate this translation graphically.

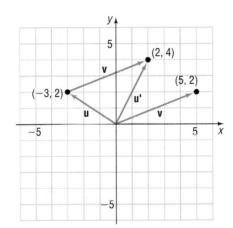

62. Computer Graphics Refer to Problem 61. The points $(-3, 0), (-1, -2), (3, 1)$ and $(1, 3)$ are the vertices of a parallelogram $ABCD$.

(a) Find the new vertices of a parallelogram $A'B'C'D'$ if it is translated by $\mathbf{v} - \langle 3, -2 \rangle$.

(b) Find the new vertices of a parallelogram $A'B'C'D'$ if it is translated by $-\dfrac{1}{2}\mathbf{v}$.

63. Force Vectors A child pulls a wagon with a force of 40 pounds. The handle of the wagon makes an angle of 30° with the ground. Express the force vector **F** in terms of **i** and **j**.

64. Force Vectors A man pushes a wheelbarrow up an incline of 20° with a force of 100 pounds. Express the force vector **F** in terms of **i** and **j**.

65. Resultant Force Two forces of magnitude 40 newtons (N) and 60 N act on an object at angles of 30° and −45° with the positive x-axis, as shown in the figure. Find the direction and magnitude of the resultant force; that is, find $\mathbf{F}_1 + \mathbf{F}_2$.

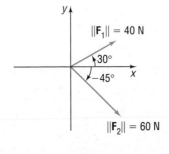

66. Resultant Force Two forces of magnitude 30 newtons (N) and 70 N act on an object at angles of 45° and 120° with the positive x-axis, as shown in the figure. Find the direction and magnitude of the resultant force; that is, find $\mathbf{F}_1 + \mathbf{F}_2$.

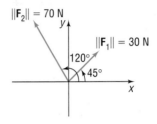

67. Static Equilibrium A weight of 1000 pounds is suspended from two cables as shown in the figure. What are the tensions in the two cables?

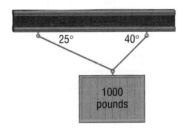

68. Static Equilibrium A weight of 800 pounds is suspended from two cables, as shown in the figure. What are the tensions in the two cables?

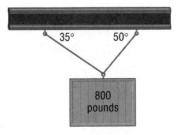

69. Static Equilibrium A tightrope walker located at a certain point deflects the rope as indicated in the figure. If the weight of the tightrope walker is 150 pounds, how much tension is in each part of the rope?

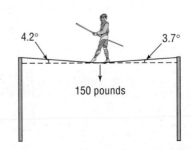

70. Static Equilibrium Repeat Problem 69 if the left angle is 3.8°, the right angle is 2.6°, and the weight of the tightrope walker is 135 pounds.

71. Truck Pull At a county fair truck pull, two pickup trucks are attached to the back end of a monster truck as illustrated in the figure. One of the pickups pulls with a force of 2000 pounds and the other pulls with a force of 3000 pounds with an angle of 45° between them. With how much force must the monster truck pull in order to remain unmoved?
[**Hint:** Find the resultant force of the two trucks.]

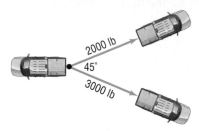

72. Removing a Stump A farmer wishes to remove a stump from a field by pulling it out with his tractor. Having removed many stumps before, he estimates that he will need 6 tons (12,000 pounds) of force to remove the stump. However, his tractor is only capable of pulling with a force of 7000 pounds, so he asks his neighbor to help. His neighbor's tractor can pull with a force of 5500 pounds. They attach the two tractors to the stump with a 40° angle between the forces as shown in the figure.

(a) Assuming the farmer's estimate of a needed 6-ton force is correct, will the farmer be successful in removing the stump? Explain.
(b) Had the farmer arranged the tractors with a 25° angle between the forces, would he have been successful in removing the stump? Explain.

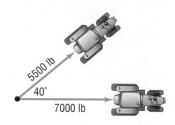

73. Static Equilibrium Show on the following graph the force needed for the object at P to be in static equilibrium.

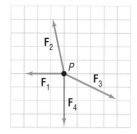

Discussion and Writing

74. Explain in your own words what a vector is. Give an example of a vector.

75. Write a brief paragraph comparing the algebra of complex numbers and the algebra of vectors.

5.5 The Dot Product

PREPARING FOR THIS SECTION *Before getting started, review the following:*

- Law of Cosines (Section 4.3, p. 293)

Now Work the 'Are You Prepared?' problem on page 372.

OBJECTIVES
1 Find the Dot Product of Two Vectors (p. 366)
2 Find the Angle between Two Vectors (p. 367)
3 Determine Whether Two Vectors Are Parallel (p. 369)
4 Determine Whether Two Vectors Are Orthogonal (p. 369)
5 Decompose a Vector into Two Orthogonal Vectors (p. 370)
6 Compute Work (p. 371)

1 Find the Dot Product of Two Vectors

The definition for a product of two vectors is somewhat unexpected. However, such a product has meaning in many geometric and physical applications.

DEFINITION If $\mathbf{v} = a_1\mathbf{i} + b_1\mathbf{j}$ and $\mathbf{w} = a_2\mathbf{i} + b_2\mathbf{j}$ are two vectors, the **dot product** $\mathbf{v} \cdot \mathbf{w}$ is defined as

$$\mathbf{v} \cdot \mathbf{w} = a_1a_2 + b_1b_2 \qquad (1)$$

EXAMPLE 1	**Finding Dot Products**

If $\mathbf{v} = 2\mathbf{i} - 3\mathbf{j}$ and $\mathbf{w} = 5\mathbf{i} + 3\mathbf{j}$, find:

(a) $\mathbf{v} \cdot \mathbf{w}$ (b) $\mathbf{w} \cdot \mathbf{v}$ (c) $\mathbf{v} \cdot \mathbf{v}$

(d) $\mathbf{w} \cdot \mathbf{w}$ (e) $\|\mathbf{v}\|$ (f) $\|\mathbf{w}\|$

Solution

(a) $\mathbf{v} \cdot \mathbf{w} = 2(5) + (-3)3 = 1$ (b) $\mathbf{w} \cdot \mathbf{v} = 5(2) + 3(-3) = 1$

(c) $\mathbf{v} \cdot \mathbf{v} = 2(2) + (-3)(-3) = 13$ (d) $\mathbf{w} \cdot \mathbf{w} = 5(5) + 3(3) = 34$

(e) $\|\mathbf{v}\| = \sqrt{2^2 + (-3)^2} = \sqrt{13}$ (f) $\|\mathbf{w}\| = \sqrt{5^2 + 3^2} = \sqrt{34}$

COMMENT A scalar multiple $\alpha\mathbf{v}$ is a vector. A dot product $\mathbf{u} \cdot \mathbf{v}$ is a scalar (real number). ■

Since the dot product $\mathbf{v} \cdot \mathbf{w}$ of two vectors $\mathbf{v}$ and $\mathbf{w}$ is a real number (scalar), we sometimes refer to it as the **scalar product.**

The results obtained in Example 1 suggest some general properties.

THEOREM

Properties of the Dot Product

If $\mathbf{u}$, $\mathbf{v}$, and $\mathbf{w}$ are vectors, then

Commutative Property

$$\mathbf{u} \cdot \mathbf{v} = \mathbf{v} \cdot \mathbf{u} \tag{2}$$

Distributive Property

$$\mathbf{u} \cdot (\mathbf{v} + \mathbf{w}) = \mathbf{u} \cdot \mathbf{v} + \mathbf{u} \cdot \mathbf{w} \tag{3}$$

$$\mathbf{v} \cdot \mathbf{v} = \|\mathbf{v}\|^2 \tag{4}$$

$$\mathbf{0} \cdot \mathbf{v} = 0 \tag{5}$$

Proof We will prove properties (2) and (4) here and leave properties (3) and (5) as exercises (see Problems 40 and 41).

To prove property (2), we let $\mathbf{u} = a_1\mathbf{i} + b_1\mathbf{j}$ and $\mathbf{v} = a_2\mathbf{i} + b_2\mathbf{j}$. Then

$$\mathbf{u} \cdot \mathbf{v} = a_1a_2 + b_1b_2 = a_2a_1 + b_2b_1 = \mathbf{v} \cdot \mathbf{u}$$

To prove property (4), we let $\mathbf{v} = a\mathbf{i} + b\mathbf{j}$. Then

$$\mathbf{v} \cdot \mathbf{v} = a^2 + b^2 = \|\mathbf{v}\|^2 \qquad ■$$

2 Find the Angle between Two Vectors

One use of the dot product is to calculate the angle between two vectors. We proceed as follows.

Let $\mathbf{u}$ and $\mathbf{v}$ be two vectors with the same initial point A. Then the vectors $\mathbf{u}$, $\mathbf{v}$, and $\mathbf{u} - \mathbf{v}$ form a triangle. The angle θ at vertex A of the triangle is the angle between the vectors $\mathbf{u}$ and $\mathbf{v}$. See Figure 65. We wish to find a formula for calculating the angle θ.

The sides of the triangle have lengths $\|\mathbf{v}\|$, $\|\mathbf{u}\|$, and $\|\mathbf{u} - \mathbf{v}\|$, and θ is the included angle between the sides of length $\|\mathbf{v}\|$ and $\|\mathbf{u}\|$. The Law of Cosines (Section 4.3) can be used to find the cosine of the included angle.

$$\|\mathbf{u} - \mathbf{v}\|^2 = \|\mathbf{u}\|^2 + \|\mathbf{v}\|^2 - 2\|\mathbf{u}\|\|\mathbf{v}\| \cos\theta$$

Figure 65

Now we use property (4) to rewrite this equation in terms of dot products.

$$(\mathbf{u} - \mathbf{v}) \cdot (\mathbf{u} - \mathbf{v}) = \mathbf{u} \cdot \mathbf{u} + \mathbf{v} \cdot \mathbf{v} - 2\|\mathbf{u}\|\|\mathbf{v}\| \cos\theta \tag{6}$$

Then we apply the distributive property (3) twice on the left side of (6) to obtain

$$(\mathbf{u} - \mathbf{v}) \cdot (\mathbf{u} - \mathbf{v}) = \mathbf{u} \cdot (\mathbf{u} - \mathbf{v}) - \mathbf{v} \cdot (\mathbf{u} - \mathbf{v})$$
$$= \mathbf{u} \cdot \mathbf{u} - \mathbf{u} \cdot \mathbf{v} - \mathbf{v} \cdot \mathbf{u} + \mathbf{v} \cdot \mathbf{v}$$
$$= \mathbf{u} \cdot \mathbf{u} + \mathbf{v} \cdot \mathbf{v} - 2\,\mathbf{u} \cdot \mathbf{v} \qquad \text{(7)}$$
$$\uparrow$$
Property (2)

Combining equations (6) and (7), we have

$$\mathbf{u} \cdot \mathbf{u} + \mathbf{v} \cdot \mathbf{v} - 2\,\mathbf{u} \cdot \mathbf{v} = \mathbf{u} \cdot \mathbf{u} + \mathbf{v} \cdot \mathbf{v} - 2\,\|\mathbf{u}\|\|\mathbf{v}\| \cos\theta$$
$$\mathbf{u} \cdot \mathbf{v} = \|\mathbf{u}\|\|\mathbf{v}\| \cos\theta$$

We have proved the following result:

THEOREM

Angle between Vectors

If $\mathbf{u}$ and $\mathbf{v}$ are two nonzero vectors, the angle $\theta, 0 \le \theta \le \pi$, between $\mathbf{u}$ and $\mathbf{v}$ is determined by the formula

$$\cos\theta = \frac{\mathbf{u} \cdot \mathbf{v}}{\|\mathbf{u}\|\|\mathbf{v}\|} \qquad \text{(8)}$$

EXAMPLE 2 | **Finding the Angle θ between Two Vectors**

Find the angle θ between $\mathbf{u} = 4\mathbf{i} - 3\mathbf{j}$ and $\mathbf{v} = 2\mathbf{i} + 5\mathbf{j}$.

Solution | We compute the quantities $\mathbf{u} \cdot \mathbf{v}$, $\|\mathbf{u}\|$, and $\|\mathbf{v}\|$.

Figure 66

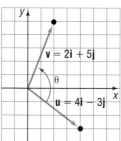

$$\mathbf{u} \cdot \mathbf{v} = 4(2) + (-3)(5) = -7$$
$$\|\mathbf{u}\| = \sqrt{4^2 + (-3)^2} = 5$$
$$\|\mathbf{v}\| = \sqrt{2^2 + 5^2} = \sqrt{29}$$

By formula (8), if θ is the angle between $\mathbf{u}$ and $\mathbf{v}$, then

$$\cos\theta = \frac{\mathbf{u} \cdot \mathbf{v}}{\|\mathbf{u}\|\|\mathbf{v}\|} = \frac{-7}{5\sqrt{29}} \approx -0.26$$

We find that $\theta \approx 105°$. See Figure 66.

Now Work PROBLEMS 7(a) AND (b)

EXAMPLE 3 | **Finding the Actual Speed and Direction of an Aircraft**

A Boeing 737 aircraft maintains a constant airspeed of 500 miles per hour in the direction due south. The velocity of the jet stream is 80 miles per hour in a northeasterly direction. Find the actual speed and direction of the aircraft relative to the ground.

Solution | We set up a coordinate system in which north (N) is along the positive y-axis. See Figure 67. Let

$$\mathbf{v}_a = \text{velocity of aircraft relative to the air} = -500\mathbf{j}$$
$$\mathbf{v}_w = \text{velocity of jet stream}$$
$$\mathbf{v}_g = \text{velocity of aircraft relative to ground}$$

The velocity of the jet stream $\mathbf{v}_w$ has magnitude 80 and direction NE (northeast), so the angle between $\mathbf{v}_w$ and $\mathbf{i}$ is 45°. We express $\mathbf{v}_w$ in terms of $\mathbf{i}$ and $\mathbf{j}$ as

$$\mathbf{v}_w = 80(\cos 45°\mathbf{i} + \sin 45°\mathbf{j}) = 80\left(\frac{\sqrt{2}}{2}\mathbf{i} + \frac{\sqrt{2}}{2}\mathbf{j}\right) = 40\sqrt{2}(\mathbf{i} + \mathbf{j})$$

Figure 67

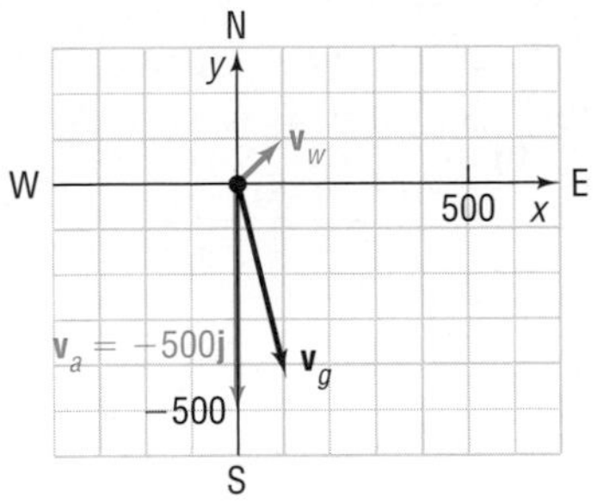

The velocity of the aircraft relative to the ground is

$$\mathbf{v}_g = \mathbf{v}_a + \mathbf{v}_w = -500\mathbf{j} + 40\sqrt{2}\,(\mathbf{i} + \mathbf{j}) = 40\sqrt{2}\,\mathbf{i} + \left(40\sqrt{2} - 500\right)\mathbf{j}$$

The actual speed of the aircraft is

$$\|\mathbf{v}_g\| = \sqrt{\left(40\sqrt{2}\right)^2 + \left(40\sqrt{2} - 500\right)^2} \approx 447 \text{ miles per hour}$$

The angle θ between $\mathbf{v}_g$ and the vector $\mathbf{v}_a = -500\mathbf{j}$ (the velocity of the aircraft relative to the air) is determined by the equation

$$\cos\theta = \frac{\mathbf{v}_g \cdot \mathbf{v}_a}{\|\mathbf{v}_g\|\|\mathbf{v}_a\|} \approx \frac{\left(40\sqrt{2} - 500\right)(-500)}{(447)(500)} \approx 0.9920$$

$$\theta \approx 7.3°$$

The direction of the aircraft relative to the ground is approximately S7.3°E (about 7.3° east of south).

────── **Now Work** PROBLEM 29

3 Determine Whether Two Vectors Are Parallel

Two vectors $\mathbf{v}$ and $\mathbf{w}$ are said to be **parallel** if there is a nonzero scalar α so that $\mathbf{v} = \alpha\mathbf{w}$. In this case, the angle θ between $\mathbf{v}$ and $\mathbf{w}$ is 0 or π.

EXAMPLE 4   Determining Whether Vectors Are Parallel

The vectors $\mathbf{v} = 3\mathbf{i} - \mathbf{j}$ and $\mathbf{w} = 6\mathbf{i} - 2\mathbf{j}$ are parallel, since $\mathbf{v} = \dfrac{1}{2}\mathbf{w}$. Furthermore, since

$$\cos\theta = \frac{\mathbf{v} \cdot \mathbf{w}}{\|\mathbf{v}\|\|\mathbf{w}\|} = \frac{18 + 2}{\sqrt{10}\,\sqrt{40}} = \frac{20}{\sqrt{400}} = 1$$

the angle θ between $\mathbf{v}$ and $\mathbf{w}$ is 0.

4 Determine Whether Two Vectors Are Orthogonal

Figure 68
v is orthogonal to **w**.

If the angle θ between two nonzero vectors $\mathbf{v}$ and $\mathbf{w}$ is $\dfrac{\pi}{2}$, the vectors $\mathbf{v}$ and $\mathbf{w}$ are called **orthogonal**.* See Figure 68.

Since $\cos\dfrac{\pi}{2} = 0$, it follows from formula (8) that if $\mathbf{v}$ and $\mathbf{w}$ are orthogonal then $\mathbf{v} \cdot \mathbf{w} = 0$.

On the other hand, if $\mathbf{v} \cdot \mathbf{w} = 0$, then either $\mathbf{v} = 0$ or $\mathbf{w} = 0$ or $\cos\theta = 0$. In the latter case, $\theta = \dfrac{\pi}{2}$, and $\mathbf{v}$ and $\mathbf{w}$ are orthogonal. If $\mathbf{v}$ or $\mathbf{w}$ is the zero vector, then, since the zero vector has no specific direction, we adopt the convention that the zero vector is orthogonal to every vector.

THEOREM

Two vectors $\mathbf{v}$ and $\mathbf{w}$ are orthogonal if and only if

$$\mathbf{v} \cdot \mathbf{w} = 0$$

Orthogonal, perpendicular, and *normal* are all terms that mean "meet at a right angle." It is customary to refer to two vectors as being *orthogonal,* two lines as being *perpendicular,* and a line and a plane or a vector and a plane as being *normal.*

| EXAMPLE 5 | Determining Whether Two Vectors Are Orthogonal |

The vectors

$$\mathbf{v} = 2\mathbf{i} - \mathbf{j} \quad \text{and} \quad \mathbf{w} = 3\mathbf{i} + 6\mathbf{j}$$

are orthogonal, since

$$\mathbf{v} \cdot \mathbf{w} = 6 - 6 = 0$$

See Figure 69.

Figure 69

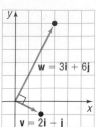

Now Work PROBLEM 7(c)

5 Decompose a Vector into Two Orthogonal Vectors

Figure 70

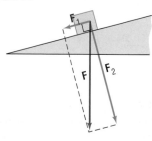

In many physical applications, it is necessary to find "how much" of a vector is applied in a given direction. Look at Figure 70. The force **F** due to gravity is pulling straight down (toward the center of Earth) on the block. To study the effect of gravity on the block, it is necessary to determine how much of **F** is actually pushing the block down the incline ($\mathbf{F}_1$) and how much is pressing the block against the incline ($\mathbf{F}_2$), at a right angle to the incline. Knowing the **decomposition** of **F** often will allow us to determine when friction (the force holding the block in place on the incline) is overcome and the block will slide down the incline.

Figure 71

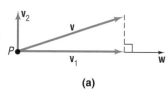

(a)

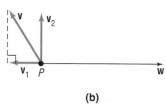

(b)

Suppose that **v** and **w** are two nonzero vectors with the same initial point P. We seek to decompose **v** into two vectors: $\mathbf{v}_1$, which is parallel to **w**, and $\mathbf{v}_2$, which is orthogonal to **w**. See Figure 71(a) and (b). The vector $\mathbf{v}_1$ is called the **vector projection of v onto w.**

The vector $\mathbf{v}_1$ is obtained as follows: From the terminal point of **v**, drop a perpendicular to the line containing **w**. The vector $\mathbf{v}_1$ is the vector from P to the foot of this perpendicular. The vector $\mathbf{v}_2$ is given by $\mathbf{v}_2 = \mathbf{v} - \mathbf{v}_1$. Note that $\mathbf{v} = \mathbf{v}_1 + \mathbf{v}_2$, the vector $\mathbf{v}_1$ is parallel to **w**, and the vector $\mathbf{v}_2$ is orthogonal to **w**. This is the decomposition of **v** that we wanted.

Now we seek a formula for $\mathbf{v}_1$ that is based on a knowledge of the vectors **v** and **w**. Since $\mathbf{v} = \mathbf{v}_1 + \mathbf{v}_2$, we have

$$\mathbf{v} \cdot \mathbf{w} = (\mathbf{v}_1 + \mathbf{v}_2) \cdot \mathbf{w} = \mathbf{v}_1 \cdot \mathbf{w} + \mathbf{v}_2 \cdot \mathbf{w} \qquad (9)$$

Since $\mathbf{v}_2$ is orthogonal to **w**, we have $\mathbf{v}_2 \cdot \mathbf{w} = 0$. Since $\mathbf{v}_1$ is parallel to **w**, we have $\mathbf{v}_1 = \alpha\mathbf{w}$ for some scalar α. Equation (9) can be written as

$$\mathbf{v} \cdot \mathbf{w} = \alpha\mathbf{w} \cdot \mathbf{w} = \alpha\|\mathbf{w}\|^2 \qquad \mathbf{v}_1 = \alpha\mathbf{w}; \mathbf{v}_2 \cdot \mathbf{w} = 0$$

$$\alpha = \frac{\mathbf{v} \cdot \mathbf{w}}{\|\mathbf{w}\|^2}$$

Then

$$\mathbf{v}_1 = \alpha\mathbf{w} = \frac{\mathbf{v} \cdot \mathbf{w}}{\|\mathbf{w}\|^2}\mathbf{w}$$

THEOREM

If **v** and **w** are two nonzero vectors, the vector projection of **v** onto **w** is

$$\mathbf{v}_1 = \frac{\mathbf{v} \cdot \mathbf{w}}{\|\mathbf{w}\|^2}\mathbf{w} \qquad (10)$$

The decomposition of **v** into $\mathbf{v}_1$ and $\mathbf{v}_2$, where $\mathbf{v}_1$ is parallel to **w** and $\mathbf{v}_2$ is orthogonal to **w**, is

$$\mathbf{v}_1 = \frac{\mathbf{v} \cdot \mathbf{w}}{\|\mathbf{w}\|^2}\mathbf{w} \qquad \mathbf{v}_2 = \mathbf{v} - \mathbf{v}_1 \qquad (11)$$

| EXAMPLE 6 | Decomposing a Vector into Two Orthogonal Vectors |

Find the vector projection of $\mathbf{v} = \mathbf{i} + 3\mathbf{j}$ onto $\mathbf{w} = \mathbf{i} + \mathbf{j}$. Decompose $\mathbf{v}$ into two vectors, $\mathbf{v}_1$ and $\mathbf{v}_2$, where $\mathbf{v}_1$ is parallel to $\mathbf{w}$ and $\mathbf{v}_2$ is orthogonal to $\mathbf{w}$.

Solution   We use formulas (10) and (11).

$$\mathbf{v}_1 = \frac{\mathbf{v} \cdot \mathbf{w}}{\|\mathbf{w}\|^2} \mathbf{w} = \frac{1 + 3}{\left(\sqrt{2}\right)^2} \mathbf{w} = 2\mathbf{w} = 2(\mathbf{i} + \mathbf{j})$$

$$\mathbf{v}_2 = \mathbf{v} - \mathbf{v}_1 = (\mathbf{i} + 3\mathbf{j}) - 2(\mathbf{i} + \mathbf{j}) = -\mathbf{i} + \mathbf{j}$$

Figure 72

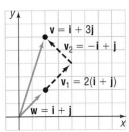

See Figure 72.

━━━━ **Now Work** PROBLEM 19

6 Compute Work

Figure 73

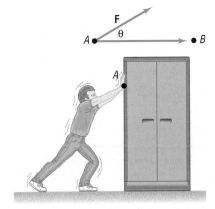

In elementary physics, the **work** W done by a constant force $\mathbf{F}$ in moving an object from a point A to a point B is defined as

$$W = (\text{magnitude of force})(\text{distance}) = \|\mathbf{F}\|\|\overrightarrow{AB}\|$$

Work is commonly measured in foot-pounds or in newton-meters (joules).

In this definition, it is assumed that the force $\mathbf{F}$ is applied along the line of motion. If the constant force $\mathbf{F}$ is not along the line of motion, but, instead, is at an angle θ to the direction of the motion, as illustrated in Figure 73, the **work** W **done by F** in moving an object from A to B is defined as

$$W = \mathbf{F} \cdot \overrightarrow{AB} \qquad (12)$$

This definition is compatible with the force times distance definition, since

$$W = (\text{amount of force in the direction of } \overrightarrow{AB})(\text{distance})$$

$$= \|\text{projection of } \mathbf{F} \text{ on } AB\|\|\overrightarrow{AB}\| = \frac{\mathbf{F} \cdot \overrightarrow{AB}}{\|\overrightarrow{AB}\|^2}\|\overrightarrow{AB}\|\|\overrightarrow{AB}\| = \mathbf{F} \cdot \overrightarrow{AB}$$

$\uparrow$
Use formula (10).

| EXAMPLE 7 | Computing Work |

Figure 74(a) shows a girl pulling a wagon with a force of 50 pounds. How much work is done in moving the wagon 100 feet if the handle makes an angle of $30°$ with the ground?

Figure 74

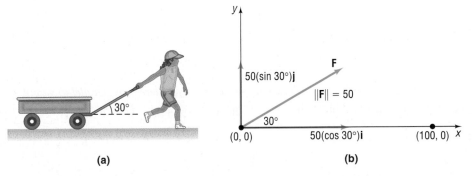

(a)   (b)

Solution   We position the vectors in a coordinate system in such a way that the wagon is moved from $(0, 0)$ to $(100, 0)$. The motion is from $A = (0, 0)$ to $B = (100, 0)$, so $\overrightarrow{AB} = 100\mathbf{i}$. The force vector $\mathbf{F}$, as shown in Figure 74(b), is

$$\mathbf{F} = 50(\cos 30°\mathbf{i} + \sin 30°\mathbf{j}) = 50\left(\frac{\sqrt{3}}{2}\mathbf{i} + \frac{1}{2}\mathbf{j}\right) = 25\left(\sqrt{3}\mathbf{i} + \mathbf{j}\right)$$

By formula (12), the work done is

$$W = \mathbf{F} \cdot \overrightarrow{AB} = 25\left(\sqrt{3}\mathbf{i} + \mathbf{j}\right) \cdot 100\mathbf{i} = 2500\sqrt{3} \text{ foot-pounds}$$

➤ **Now Work** PROBLEM 25

Historical Feature

We stated in an earlier Historical Feature that complex numbers were used as vectors in the plane before the general notion of a vector was clarified. Suppose that we make the correspondence

Vector ↔ Complex number

$a\mathbf{i} + b\mathbf{j} \leftrightarrow a + bi$

$c\mathbf{i} + d\mathbf{j} \leftrightarrow c + di$

Show that

$$(a\mathbf{i} + b\mathbf{j}) \cdot (c\mathbf{i} + d\mathbf{j}) = \text{real part } [\overline{(a + bi)}(c + di)]$$

This is how the dot product was found originally. The imaginary part is also interesting. It is a determinant and represents the area of the parallelogram whose edges are the vectors. This is close to some of Hermann Grassmann's ideas and is also connected with the scalar triple product of three-dimensional vectors.

5.5 Assess Your Understanding

'Are You Prepared?' *The answer is given at the end of these exercises. If you get the wrong answer, read the page listed in red.*

1. In a triangle with sides a, b, c and angles A, B, C, the Law of Cosines states that _____. (p. 293)

Concepts and Vocabulary

2. If $\mathbf{v} \cdot \mathbf{w} = 0$, then the two vectors $\mathbf{v}$ and $\mathbf{w}$ are _____.

3. If $\mathbf{v} = 3\mathbf{w}$, then the two vectors $\mathbf{v}$ and $\mathbf{w}$ are _____.

4. *True or False* If $\mathbf{v}$ and $\mathbf{w}$ are parallel vectors, then $\mathbf{v} \cdot \mathbf{w} = 0$.

5. *True or False* Given two nonzero vectors $\mathbf{v}$ and $\mathbf{w}$, it is always possible to decompose $\mathbf{v}$ into two vectors, one parallel to $\mathbf{w}$ and the other perpendicular to $\mathbf{w}$.

6. *True or False* Work is a physical example of a vector.

Skill Building

In Problems 7–16, (a) find the dot product $\mathbf{v} \cdot \mathbf{w}$; (b) find the angle between $\mathbf{v}$ and $\mathbf{w}$; (c) state whether the vectors are parallel, orthogonal, or neither.

7. $\mathbf{v} = \mathbf{i} - \mathbf{j}, \quad \mathbf{w} = \mathbf{i} + \mathbf{j}$

8. $\mathbf{v} = \mathbf{i} + \mathbf{j}, \quad \mathbf{w} = -\mathbf{i} + \mathbf{j}$

9. $\mathbf{v} = 2\mathbf{i} + \mathbf{j}, \quad \mathbf{w} = \mathbf{i} - 2\mathbf{j}$

10. $\mathbf{v} = 2\mathbf{i} + 2\mathbf{j}, \quad \mathbf{w} = \mathbf{i} + 2\mathbf{j}$

11. $\mathbf{v} = \sqrt{3}\mathbf{i} - \mathbf{j}, \quad \mathbf{w} = \mathbf{i} + \mathbf{j}$

12. $\mathbf{v} = \mathbf{i} + \sqrt{3}\mathbf{j}, \quad \mathbf{w} = \mathbf{i} - \mathbf{j}$

13. $\mathbf{v} = 3\mathbf{i} + 4\mathbf{j}, \quad \mathbf{w} = -6\mathbf{i} - 8\mathbf{j}$

14. $\mathbf{v} = 3\mathbf{i} - 4\mathbf{j}, \quad \mathbf{w} = 9\mathbf{i} - 12\mathbf{j}$

15. $\mathbf{v} = 4\mathbf{i}, \quad \mathbf{w} = \mathbf{j}$

16. $\mathbf{v} = \mathbf{i}, \quad \mathbf{w} = -3\mathbf{j}$

17. Find a so that the vectors $\mathbf{v} = \mathbf{i} - a\mathbf{j}$ and $\mathbf{w} = 2\mathbf{i} + 3\mathbf{j}$ are orthogonal.

18. Find b so that the vectors $\mathbf{v} = \mathbf{i} + \mathbf{j}$ and $\mathbf{w} = \mathbf{i} + b\mathbf{j}$ are orthogonal.

In Problems 19–24, decompose $\mathbf{v}$ into two vectors $\mathbf{v}_1$ and $\mathbf{v}_2$, where $\mathbf{v}_1$ is parallel to $\mathbf{w}$ and $\mathbf{v}_2$ is orthogonal to $\mathbf{w}$.

19. $\mathbf{v} = 2\mathbf{i} - 3\mathbf{j}, \quad \mathbf{w} = \mathbf{i} - \mathbf{j}$

20. $\mathbf{v} = -3\mathbf{i} + 2\mathbf{j}, \quad \mathbf{w} = 2\mathbf{i} + \mathbf{j}$

21. $\mathbf{v} = \mathbf{i} - \mathbf{j}, \quad \mathbf{w} = -\mathbf{i} - 2\mathbf{j}$

22. $\mathbf{v} = 2\mathbf{i} - \mathbf{j}, \quad \mathbf{w} = \mathbf{i} - 2\mathbf{j}$

23. $\mathbf{v} = 3\mathbf{i} + \mathbf{j}, \quad \mathbf{w} = -2\mathbf{i} - \mathbf{j}$

24. $\mathbf{v} = \mathbf{i} - 3\mathbf{j}, \quad \mathbf{w} = 4\mathbf{i} - \mathbf{j}$

Applications and Extensions

25. **Computing Work** Find the work done by a force of 3 pounds acting in the direction 60° to the horizontal in moving an object 6 feet from $(0, 0)$ to $(6, 0)$.

26. **Computing Work** A wagon is pulled horizontally by exerting a force of 20 pounds on the handle at an angle of 30° with

the horizontal. How much work is done in moving the wagon 100 feet?

27. **Solar Energy** The amount of energy collected by a solar panel depends on the intensity of the sun's rays and the area of the panel. Let the vector $\mathbf{I}$ represent the intensity, in watts

per square centimeter, having the direction of the sun's rays. Let the vector **A** represent the area, in square centimeters, whose direction is the orientation of a solar panel. See the figure. The total number of watts collected by the panel is given by $W = |\mathbf{I} \cdot \mathbf{A}|$.

Suppose $\mathbf{I} = \langle -0.02, -0.01 \rangle$ and $\mathbf{A} = \langle 300, 400 \rangle$.

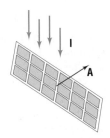

(a) Find $\|\mathbf{I}\|$ and $\|\mathbf{A}\|$ and interpret the meaning of each.
(b) Compute W and interpret its meaning.
(c) If the solar panel is to collect the maximum number of watts, what must be true about **I** and **A**?

28. **Rainfall Measurement** Let the vector **R** represent the amount of rainfall, in inches, whose direction is the inclination of the rain to a rain gauge. Let the vector **A** represent the area, in square inches, whose direction is the orientation of the opening of the rain gauge. See the figure. The volume of rain collected in the gauge, in cubic inches, is given by $V = |\mathbf{R} \cdot \mathbf{A}|$, even when the rain falls in a slanted direction or the gauge is not perfectly vertical.

Suppose $\mathbf{R} = \langle 0.75, -1.75 \rangle$ and $\mathbf{A} = \langle 0.3, 1 \rangle$.

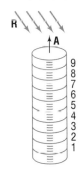

(a) Find $\|\mathbf{R}\|$ and $\|\mathbf{A}\|$ and interpret the meaning of each.
(b) Compute V and interpret its meaning
(c) If the gauge is to collect the maximum volume of rain, what must be true about **R** and **A**?

29. **Finding the Actual Speed and Direction of an Aircraft** A Boeing 747 jumbo jet maintains an airspeed of 550 miles per hour in a southwesterly direction. The velocity of the jet stream is a constant 80 miles per hour from the west. Find the actual speed and direction of the aircraft.

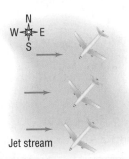

Jet stream

30. **Finding the Correct Compass Heading** The pilot of an aircraft wishes to head directly east, but is faced with a wind speed of 40 miles per hour from the northwest. If the pilot maintains an airspeed of 250 miles per hour, what compass heading should be maintained to head directly east? What is the actual speed of the aircraft?

31. **Correct Direction for Crossing a River** A river has a constant current of 3 kilometers per hour. At what angle to a boat dock should a motorboat, capable of maintaining a constant speed of 20 kilometers per hour, be headed in order to reach a point directly opposite the dock? If the river is $\frac{1}{2}$ kilometer wide, how long will it take to cross?

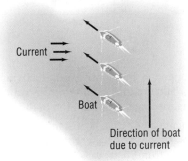

32. **Crossing a River** A small motorboat in still water maintains a speed of 20 miles per hour. In heading directly across a river (that is, perpendicular to the current) whose current is 3 miles per hour, find a vector representing the speed and direction of the motorboat. What is the true speed of the motorboat? What is its direction?

33. **Braking Load** A Toyota Sienna with a gross weight of 5300 pounds is parked on a street with a slope of 8°. See the figure. Find the force required to keep the Sienna from rolling down the hill. What is the force perpendicular to the hill?

Weight = 5300 pounds

34. **Braking Load** A Pontiac Bonneville with a gross weight of 4500 pounds is parked on a street with a slope of 10°. Find the force required to keep the Bonneville from rolling down the hill. What is the force perpendicular to the hill?

35. **Ground Speed and Direction of an Airplane** An airplane has an airspeed of 500 kilometers per hour (km/hr) bearing N45°E. The wind velocity is 60 km/hr in the direction N30°W. Find the resultant vector representing the path of the plane relative to the ground. What is the ground speed of the plane? What is its direction?

36. **Ground Speed and Direction of an Airplane** An airplane has an airspeed of 600 km/hr bearing S30°E. The wind velocity is 40 km/hr in the direction S45°E. Find the resultant vector representing the path of the plane relative to the ground. What is the ground speed of the plane? What is its direction?

37. **Ramp Angle** Billy and Timmy are using a ramp to load furniture into a truck. While rolling a 250-pound piano up the ramp, they discover that the truck is too full of other

furniture for the piano to fit. Timmy holds the piano in place on the ramp while Billy repositions other items to make room for it in the truck. If the angle of inclination of the ramp is 20°, how many pounds of force must Timmy exert to hold the piano in position?

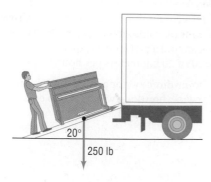

38. Incline Angle A bulldozer exerts 1000 pounds of force to prevent a 5000-pound boulder from rolling down a hill. Determine the angle of inclination of the hill.

39. Find the acute angle that a constant unit force vector makes with the positive x-axis if the work done by the force in moving a particle from $(0, 0)$ to $(4, 0)$ equals 2.

40. Prove the distributive property:

$$\mathbf{u} \cdot (\mathbf{v} + \mathbf{w}) = \mathbf{u} \cdot \mathbf{v} + \mathbf{u} \cdot \mathbf{w}$$

41. Prove property (5), $\mathbf{0} \cdot \mathbf{v} = 0$.

42. If $\mathbf{v}$ is a unit vector and the angle between $\mathbf{v}$ and $\mathbf{i}$ is α, show that $\mathbf{v} = \cos \alpha \mathbf{i} + \sin \alpha \mathbf{j}$.

43. Suppose that $\mathbf{v}$ and $\mathbf{w}$ are unit vectors. If the angle between $\mathbf{v}$ and $\mathbf{i}$ is α and that between $\mathbf{w}$ and $\mathbf{i}$ is β, use the idea of the dot product $\mathbf{v} \cdot \mathbf{w}$ to prove that

$$\cos(\alpha - \beta) = \cos \alpha \cos \beta + \sin \alpha \sin \beta$$

44. Show that the projection of $\mathbf{v}$ onto $\mathbf{i}$ is $(\mathbf{v} \cdot \mathbf{i})\mathbf{i}$. Then show that we can always write a vector $\mathbf{v}$ as

$$\mathbf{v} = (\mathbf{v} \cdot \mathbf{i})\mathbf{i} + (\mathbf{v} \cdot \mathbf{j})\mathbf{j}$$

45. (a) If $\mathbf{u}$ and $\mathbf{v}$ have the same magnitude, show that $\mathbf{u} + \mathbf{v}$ and $\mathbf{u} - \mathbf{v}$ are orthogonal.
(b) Use this to prove that an angle inscribed in a semicircle is a right angle (see the figure).

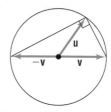

46. Let $\mathbf{v}$ and $\mathbf{w}$ denote two nonzero vectors. Show that the vector $\mathbf{v} - \alpha\mathbf{w}$ is orthogonal to $\mathbf{w}$ if $\alpha = \dfrac{\mathbf{v} \cdot \mathbf{w}}{\|\mathbf{w}\|^2}$.

47. Let $\mathbf{v}$ and $\mathbf{w}$ denote two nonzero vectors. Show that the vectors $\|\mathbf{w}\|\mathbf{v} + \|\mathbf{v}\|\mathbf{w}$ and $\|\mathbf{w}\|\mathbf{v} - \|\mathbf{v}\|\mathbf{w}$ are orthogonal.

48. In the definition of work given in this section, what is the work done if $\mathbf{F}$ is orthogonal to $\overrightarrow{AB}$?

49. Prove the **polarization identity**,

$$\|\mathbf{u} + \mathbf{v}\|^2 - \|\mathbf{u} - \mathbf{v}\|^2 = 4(\mathbf{u} \cdot \mathbf{v})$$

Discussion and Writing

50. Create an application different from any found in the text that requires a dot product.

'Are You Prepared?' Answer

1. $c^2 = a^2 + b^2 - 2ab \cos C$

5.6 Vectors in Space

PREPARING FOR THIS SECTION *Before getting started, review the following:*

- Distance Formula (Section 1.1, p. 5)

Now Work the 'Are You Prepared?' problem on page 382.

OBJECTIVES
1 Find the Distance between Two Points in Space (p. 376)
2 Find Position Vectors in Space (p. 376)
3 Perform Operations on Vectors (p. 377)
4 Find the Dot Product (p. 378)
5 Find the Angle between Two Vectors (p. 379)
6 Find the Direction Angles of a Vector (p. 379)

Rectangular Coordinates in Space

Figure 75

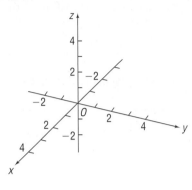

In the plane, each point is associated with an ordered pair of real numbers. In space, each point is associated with an ordered triple of real numbers. Through a fixed point, called the **origin** O, draw three mutually perpendicular lines, the x-axis, the y-axis, and the z-axis. On each of these axes, select an appropriate scale and the positive direction. See Figure 75.

The direction chosen for the positive z-axis in Figure 75 makes the system *right-handed*. This conforms to the *right-hand rule*, which states that, if the index finger of the right hand points in the direction of the positive x-axis and the middle finger points in the direction of the positive y-axis, then the thumb will point in the direction of the positive z-axis. See Figure 76.

We associate with each point P an ordered triple (x, y, z) of real numbers, the **coordinates of P.** For example, the point $(2, 3, 4)$ is located by starting at the origin and moving 2 units along the positive x-axis, 3 units in the direction of the positive y axis, and 4 units in the direction of the positive z-axis. See Figure 77.

Figure 76

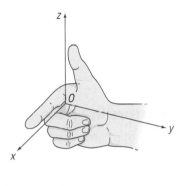

Figure 77

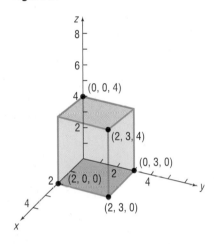

Figure 77 also shows the location of the points $(2, 0, 0)$, $(0, 3, 0)$, $(0, 0, 4)$, and $(2, 3, 0)$. Points of the form $(x, 0, 0)$ lie on the x-axis, while points of the form $(0, y, 0)$ and $(0, 0, z)$ lie on the y-axis and z-axis, respectively. Points of the form $(x, y, 0)$ lie in a plane, called the **xy-plane.** Its equation is $z = 0$. Similarly, points of the form $(x, 0, z)$ lie in the **xz-plane** (equation $y = 0$) and points of the form $(0, y, z)$ lie in the **yz-plane** (equation $x = 0$). See Figure 78(a). By extension of these ideas, all points obeying the equation $z = 3$ will lie in a plane parallel to and 3 units above the xy-plane. The equation $y = 4$ represents a plane parallel to the xz-plane and 4 units to the right of the plane $y = 0$. See Figure 78(b).

Figure 78

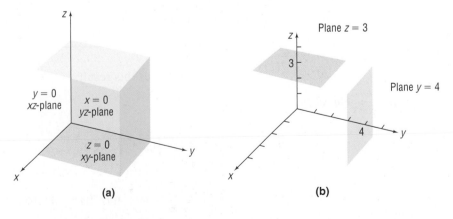

(a)   (b)

Now Work PROBLEM 9

1 Find the Distance between Two Points in Space

The formula for the distance between two points in space is an extension of the Distance Formula for points in the plane given in Chapter 1.

THEOREM

Distance Formula in Space

If $P_1 = (x_1, y_1, z_1)$ and $P_2 = (x_2, y_2, z_2)$ are two points in space, the distance d from P_1 to P_2 is

$$d = \sqrt{(x_2 - x_1)^2 + (y_2 - y_1)^2 + (z_2 - z_1)^2} \qquad (1)$$

The proof, which we omit, utilizes a double application of the Pythagorean Theorem.

EXAMPLE 1 | Using the Distance Formula

Find the distance from $P_1 = (-1, 3, 2)$ to $P_2 = (4, -2, 5)$.

Solution

$$d = \sqrt{[4 - (-1)]^2 + [-2 - 3]^2 + [5 - 2]^2} = \sqrt{25 + 25 + 9} = \sqrt{59}$$

— **Now Work** PROBLEM 15

2 Find Position Vectors in Space

To represent vectors in space, we introduce the unit vectors $\mathbf{i}, \mathbf{j}$, and $\mathbf{k}$ whose directions are along the positive x-axis, positive y-axis, and positive z-axis, respectively. If $\mathbf{v}$ is a vector with initial point at the origin O and terminal point at $P = (a, b, c)$, we can represent $\mathbf{v}$ in terms of the vectors $\mathbf{i}, \mathbf{j}$, and $\mathbf{k}$ as

$$\mathbf{v} = a\mathbf{i} + b\mathbf{j} + c\mathbf{k}$$

Figure 79

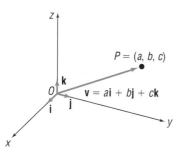

See Figure 79.

The scalars a, b, and c are called the **components** of the vector $\mathbf{v} = a\mathbf{i} + b\mathbf{j} + c\mathbf{k}$, with a being the component in the direction $\mathbf{i}$, b the component in the direction $\mathbf{j}$, and c the component in the direction $\mathbf{k}$.

A vector whose initial point is at the origin is called a **position vector.** The next result states that any vector whose initial point is not at the origin is equal to a unique position vector.

THEOREM

Suppose that $\mathbf{v}$ is a vector with initial point $P_1 = (x_1, y_1, z_1)$, not necessarily the origin, and terminal point $P_2 = (x_2, y_2, z_2)$. If $\mathbf{v} = \overrightarrow{P_1P_2}$, then $\mathbf{v}$ is equal to the position vector

$$\mathbf{v} = (x_2 - x_1)\mathbf{i} + (y_2 - y_1)\mathbf{j} + (z_2 - z_1)\mathbf{k} \qquad (2)$$

Figure 80 illustrates this result.

Figure 80

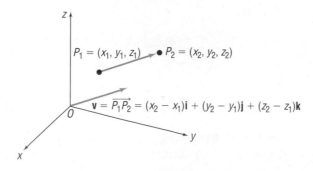

EXAMPLE 2	Finding a Position Vector

Find the position vector of the vector $\mathbf{v} = \overrightarrow{P_1P_2}$ if $P_1 = (-1, 2, 3)$ and $P_2 = (4, 6, 2)$.

Solution By equation (2), the position vector equal to $\mathbf{v}$ is

$$\mathbf{v} = [4 - (-1)]\mathbf{i} + (6 - 2)\mathbf{j} + (2 - 3)\mathbf{k} = 5\mathbf{i} + 4\mathbf{j} - \mathbf{k}$$

===== **Now Work** PROBLEM 29

3 Perform Operations on Vectors

We can define equality, addition, subtraction, scalar product, and magnitude in terms of the components of a vector.

DEFINITION Let $\mathbf{v} = a_1\mathbf{i} + b_1\mathbf{j} + c_1\mathbf{k}$ and $\mathbf{w} = a_2\mathbf{i} + b_2\mathbf{j} + c_2\mathbf{k}$ be two vectors, and let α be a scalar. Then

$$\mathbf{v} = \mathbf{w} \quad \text{if and only if} \quad a_1 = a_2, b_1 = b_2, \text{ and } c_1 = c_2$$
$$\mathbf{v} + \mathbf{w} = (a_1 + a_2)\mathbf{i} + (b_1 + b_2)\mathbf{j} + (c_1 + c_2)\mathbf{k}$$
$$\mathbf{v} - \mathbf{w} = (a_1 - a_2)\mathbf{i} + (b_1 - b_2)\mathbf{j} + (c_1 - c_2)\mathbf{k}$$
$$\alpha\mathbf{v} = (\alpha a_1)\mathbf{i} + (\alpha b_1)\mathbf{j} + (\alpha c_1)\mathbf{k}$$
$$\|\mathbf{v}\| = \sqrt{a_1^2 + b_1^2 + c_1^2}$$

These definitions are compatible with the geometric definitions given earlier in Section 5.4.

EXAMPLE 3	Adding and Subtracting Vectors

If $\mathbf{v} = 2\mathbf{i} + 3\mathbf{j} - 2\mathbf{k}$ and $\mathbf{w} = 3\mathbf{i} - 4\mathbf{j} + 5\mathbf{k}$, find:

(a) $\mathbf{v} + \mathbf{w}$ (b) $\mathbf{v} - \mathbf{w}$

Solution (a) $\mathbf{v} + \mathbf{w} = (2\mathbf{i} + 3\mathbf{j} - 2\mathbf{k}) + (3\mathbf{i} - 4\mathbf{j} + 5\mathbf{k})$
$$= (2 + 3)\mathbf{i} + (3 - 4)\mathbf{j} + (-2 + 5)\mathbf{k}$$
$$= 5\mathbf{i} - \mathbf{j} + 3\mathbf{k}$$

(b) $\mathbf{v} - \mathbf{w} = (2\mathbf{i} + 3\mathbf{j} - 2\mathbf{k}) - (3\mathbf{i} - 4\mathbf{j} + 5\mathbf{k})$
$$= (2 - 3)\mathbf{i} + [3 - (-4)]\mathbf{j} + [-2 - 5]\mathbf{k}$$
$$= -\mathbf{i} + 7\mathbf{j} - 7\mathbf{k}$$

EXAMPLE 4	Finding Scalar Products and Magnitudes

If $\mathbf{v} = 2\mathbf{i} + 3\mathbf{j} - 2\mathbf{k}$ and $\mathbf{w} = 3\mathbf{i} - 4\mathbf{j} + 5\mathbf{k}$, find:

(a) $3\mathbf{v}$ (b) $2\mathbf{v} - 3\mathbf{w}$ (c) $\|\mathbf{v}\|$

Solution (a) $3\mathbf{v} = 3(2\mathbf{i} + 3\mathbf{j} - 2\mathbf{k}) = 6\mathbf{i} + 9\mathbf{j} - 6\mathbf{k}$

(b) $2\mathbf{v} - 3\mathbf{w} = 2(2\mathbf{i} + 3\mathbf{j} - 2\mathbf{k}) - 3(3\mathbf{i} - 4\mathbf{j} + 5\mathbf{k})$
$$= 4\mathbf{i} + 6\mathbf{j} - 4\mathbf{k} - 9\mathbf{i} + 12\mathbf{j} - 15\mathbf{k} = -5\mathbf{i} + 18\mathbf{j} - 19\mathbf{k}$$

(c) $\|\mathbf{v}\| = \|2\mathbf{i} + 3\mathbf{j} - 2\mathbf{k}\| = \sqrt{2^2 + 3^2 + (-2)^2} = \sqrt{17}$

===== **Now Work** PROBLEMS 33 AND 39

Recall that a unit vector **u** is one for which $\|\mathbf{u}\| = 1$. In many applications, it is useful to be able to find a unit vector **u** that has the same direction as a given vector **v**.

THEOREM **Unit Vector in the Direction of v**

For any nonzero vector **v**, the vector

$$\mathbf{u} = \frac{\mathbf{v}}{\|\mathbf{v}\|}$$

is a unit vector that has the same direction as **v**.

As a consequence of this theorem, if **u** is a unit vector in the same direction as a vector **v**, then **v** may be expressed as

$$\mathbf{v} = \|\mathbf{v}\|\mathbf{u}$$

EXAMPLE 5 Finding a Unit Vector

Find the unit vector in the same direction as $\mathbf{v} = 2\mathbf{i} - 3\mathbf{j} - 6\mathbf{k}$.

Solution We find $\|\mathbf{v}\|$ first.

$$\|\mathbf{v}\| = \|2\mathbf{i} - 3\mathbf{j} - 6\mathbf{k}\| = \sqrt{4 + 9 + 36} = \sqrt{49} = 7$$

Now we multiply **v** by the scalar $\dfrac{1}{\|\mathbf{v}\|} = \dfrac{1}{7}$. The result is the unit vector

$$\mathbf{u} = \frac{\mathbf{v}}{\|\mathbf{v}\|} = \frac{2\mathbf{i} - 3\mathbf{j} - 6\mathbf{k}}{7} = \frac{2}{7}\mathbf{i} - \frac{3}{7}\mathbf{j} - \frac{6}{7}\mathbf{k}$$

 Now Work PROBLEM 47

4 Find the Dot Product

The definition of *dot product* is an extension of the definition given for vectors in the plane.

DEFINITION If $\mathbf{v} = a_1\mathbf{i} + b_1\mathbf{j} + c_1\mathbf{k}$ and $\mathbf{w} = a_2\mathbf{i} + b_2\mathbf{j} + c_2\mathbf{k}$ are two vectors, the **dot product** $\mathbf{v} \cdot \mathbf{w}$ is defined as

$$\mathbf{v} \cdot \mathbf{w} = a_1 a_2 + b_1 b_2 + c_1 c_2 \tag{3}$$

EXAMPLE 6 Finding Dot Products

If $\mathbf{v} = 2\mathbf{i} - 3\mathbf{j} + 6\mathbf{k}$ and $\mathbf{w} = 5\mathbf{i} + 3\mathbf{j} - \mathbf{k}$, find:

(a) $\mathbf{v} \cdot \mathbf{w}$ (b) $\mathbf{w} \cdot \mathbf{v}$ (c) $\mathbf{v} \cdot \mathbf{v}$

(d) $\mathbf{w} \cdot \mathbf{w}$ (e) $\|\mathbf{v}\|$ (f) $\|\mathbf{w}\|$

Solution (a) $\mathbf{v} \cdot \mathbf{w} = 2(5) + (-3)3 + 6(-1) = -5$

(b) $\mathbf{w} \cdot \mathbf{v} = 5(2) + 3(-3) + (-1)(6) = -5$

(c) $\mathbf{v} \cdot \mathbf{v} = 2(2) + (-3)(-3) + 6(6) = 49$

(d) $\mathbf{w} \cdot \mathbf{w} = 5(5) + 3(3) + (-1)(-1) = 35$

(e) $\|\mathbf{v}\| = \sqrt{2^2 + (-3)^2 + 6^2} = \sqrt{49} = 7$

(f) $\|\mathbf{w}\| = \sqrt{5^2 + 3^2 + (-1)^2} = \sqrt{35}$

The dot product in space has the same properties as the dot product in the plane.

THEOREM

Properties of the Dot Product

If **u**, **v**, and **w** are vectors, then

Commutative Property

$$\mathbf{u} \cdot \mathbf{v} = \mathbf{v} \cdot \mathbf{u}$$

Distributive Property

$$\mathbf{u} \cdot (\mathbf{v} + \mathbf{w}) = \mathbf{u} \cdot \mathbf{v} + \mathbf{u} \cdot \mathbf{w}$$

$$\mathbf{v} \cdot \mathbf{v} = \|\mathbf{v}\|^2$$
$$\mathbf{0} \cdot \mathbf{v} = 0$$

5 Find the Angle between Two Vectors

The angle θ between two vectors in space follows the same formula as for two vectors in the plane.

THEOREM

Angle between Vectors

If **u** and **v** are two nonzero vectors, the angle $\theta, 0 \le \theta \le \pi$, between **u** and **v** is determined by the formula

$$\cos \theta = \frac{\mathbf{u} \cdot \mathbf{v}}{\|\mathbf{u}\| \, \|\mathbf{v}\|} \tag{4}$$

EXAMPLE 7	Finding the Angle θ between Two Vectors

Find the angle θ between $\mathbf{u} = 2\mathbf{i} - 3\mathbf{j} + 6\mathbf{k}$ and $\mathbf{v} = 2\mathbf{i} + 5\mathbf{j} - \mathbf{k}$.

Solution   We compute the quantities $\mathbf{u} \cdot \mathbf{v}$, $\|\mathbf{u}\|$, and $\|\mathbf{v}\|$.

$$\mathbf{u} \cdot \mathbf{v} = 2(2) + (-3)(5) + 6(-1) = -17$$
$$\|\mathbf{u}\| = \sqrt{2^2 + (-3)^2 + 6^2} = \sqrt{49} = 7$$
$$\|\mathbf{v}\| = \sqrt{2^2 + 5^2 + (-1)^2} = \sqrt{30}$$

By formula (4), if θ is the angle between **u** and **v**, then

$$\cos \theta = \frac{\mathbf{u} \cdot \mathbf{v}}{\|\mathbf{u}\| \, \|\mathbf{v}\|} = \frac{-17}{7\sqrt{30}} \approx -0.443$$

We find that $\theta \approx 116.3°$.

⟶ **Now Work**   PROBLEM 51

6 Find the Direction Angles of a Vector

A nonzero vector **v** in space can be described by specifying its magnitude and its three **direction angles** α, β, and γ. These direction angles are defined as

α = the angle between **v** and **i**, the positive x-axis, $0 \le \alpha \le \pi$

β = the angle between **v** and **j**, the positive y-axis, $0 \le \beta \le \pi$

γ = the angle between **v** and **k**, the positive z-axis, $0 \le \gamma \le \pi$

See Figure 81.

Figure 81

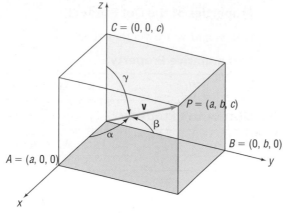

$$0 \le \alpha \le \pi, 0 \le \beta \le \pi, 0 \le \gamma \le \pi$$

Our first goal is to find expressions for α, β, and γ in terms of the components of a vector. Let $\mathbf{v} = a\mathbf{i} + b\mathbf{j} + c\mathbf{k}$ denote a nonzero vector. The angle α between $\mathbf{v}$ and $\mathbf{i}$, the positive x-axis, obeys

$$\cos \alpha = \frac{\mathbf{v} \cdot \mathbf{i}}{\|\mathbf{v}\| \|\mathbf{i}\|} = \frac{a}{\|\mathbf{v}\|}$$

Similarly,

$$\cos \beta = \frac{b}{\|\mathbf{v}\|} \qquad \cos \gamma = \frac{c}{\|\mathbf{v}\|}$$

Since $\|\mathbf{v}\| = \sqrt{a^2 + b^2 + c^2}$, we have the following result:

THEOREM

Direction Angles

If $\mathbf{v} = a\mathbf{i} + b\mathbf{j} + c\mathbf{k}$ is a nonzero vector in space, the direction angles α, β, and γ obey

$$\cos \alpha = \frac{a}{\sqrt{a^2 + b^2 + c^2}} = \frac{a}{\|\mathbf{v}\|} \qquad \cos \beta = \frac{b}{\sqrt{a^2 + b^2 + c^2}} = \frac{b}{\|\mathbf{v}\|}$$

$$\cos \gamma = \frac{c}{\sqrt{a^2 + b^2 + c^2}} = \frac{c}{\|\mathbf{v}\|} \qquad \qquad \textbf{(5)}$$

The numbers $\cos \alpha$, $\cos \beta$, and $\cos \gamma$ are called the **direction cosines** of the vector $\mathbf{v}$.

EXAMPLE 8 Finding the Direction Angles of a Vector

Find the direction angles of $\mathbf{v} = -3\mathbf{i} + 2\mathbf{j} - 6\mathbf{k}$.

Solution

$$\|\mathbf{v}\| = \sqrt{(-3)^2 + 2^2 + (-6)^2} = \sqrt{49} = 7$$

Using the formulas in equation (5), we have

$$\cos \alpha = \frac{-3}{7} \qquad \cos \beta = \frac{2}{7} \qquad \cos \gamma = \frac{-6}{7}$$

$$\alpha \approx 115.4° \qquad \beta \approx 73.4° \qquad \gamma \approx 149.0°$$

THEOREM   **Property of the Direction Cosines**

If α, β, and γ are the direction angles of a nonzero vector $\mathbf{v}$ in space, then

$$\cos^2\alpha + \cos^2\beta + \cos^2\gamma = 1 \qquad \textbf{(6)}$$

The proof is a direct consequence of the equations in (5).

Based on equation (6), when two direction cosines are known, the third is determined up to its sign. Knowing two direction cosines is not sufficient to uniquely determine the direction of a vector in space.

EXAMPLE 9   Finding the Direction Angle of a Vector

The vector $\mathbf{v}$ makes an angle of $\alpha = \dfrac{\pi}{3}$ with the positive x-axis, an angle of $\beta = \dfrac{\pi}{3}$ with the positive y-axis, and an acute angle γ with the positive z-axis. Find γ.

Solution   By equation (6), we have

$$\cos^2\left(\frac{\pi}{3}\right) + \cos^2\left(\frac{\pi}{3}\right) + \cos^2\gamma = 1 \qquad 0 < \gamma < \frac{\pi}{2}$$

$$\left(\frac{1}{2}\right)^2 + \left(\frac{1}{2}\right)^2 + \cos^2\gamma = 1$$

$$\cos^2\gamma = \frac{1}{2}$$

$$\cos\gamma = \frac{\sqrt{2}}{2} \quad \text{or} \quad \cos\gamma = -\frac{\sqrt{2}}{2}$$

$$\gamma = \frac{\pi}{4} \quad \text{or} \quad \gamma = \frac{3\pi}{4}$$

Since we are requiring that γ be acute, $\gamma = \dfrac{\pi}{4}$. ∎

The direction cosines of a vector give information about only the direction of the vector; they provide no information about its magnitude. For example, *any* vector parallel to the *xy*-plane and making an angle of $\dfrac{\pi}{4}$ radian with the positive *x*-axis and *y*-axis has direction cosines

$$\cos\alpha = \frac{\sqrt{2}}{2} \qquad \cos\beta = \frac{\sqrt{2}}{2} \qquad \cos\gamma = 0$$

However, if the direction angles *and* the magnitude of a vector are known, the vector is uniquely determined.

EXAMPLE 10   Writing a Vector in Terms of Its Magnitude and Direction Cosines

Show that any nonzero vector $\mathbf{v}$ in space can be written in terms of its magnitude and direction cosines as

$$\mathbf{v} = \|\mathbf{v}\|[(\cos\alpha)\mathbf{i} + (\cos\beta)\mathbf{j} + (\cos\gamma)\mathbf{k}] \qquad \textbf{(7)}$$

Solution   Let $\mathbf{v} = a\mathbf{i} + b\mathbf{j} + c\mathbf{k}$. From the equations in (5), we see that

$$a = \|\mathbf{v}\|\cos\alpha \qquad b = \|\mathbf{v}\|\cos\beta \qquad c = \|\mathbf{v}\|\cos\gamma$$

Substituting, we find that

$$\mathbf{v} = a\mathbf{i} + b\mathbf{j} + c\mathbf{k} = \|\mathbf{v}\|(\cos \alpha)\mathbf{i} + \|\mathbf{v}\|(\cos \beta)\mathbf{j} + \|\mathbf{v}\|(\cos \gamma)\mathbf{k}$$
$$= \|\mathbf{v}\|[(\cos \alpha)\mathbf{i} + (\cos \beta)\mathbf{j} + (\cos \gamma)\mathbf{k}]$$

━━━━━▶ **Now Work** PROBLEM 59

Example 10 shows that the direction cosines of a vector **v** are also the components of the unit vector in the direction of **v**.

5.6 Assess Your Understanding

'Are You Prepared?' *Answer is given at the end of these exercises. If you get the wrong answer, read the page listed in red.*

1. The distance d from $P_1 = (x_1, y_1)$ to $P_2 = (x_2, y_2)$ is $d =$ _____ (p. 5)

Concepts and Vocabulary

2. In space, points of the form $(x, y, 0)$ lie in a plane called the _____.

3. If $\mathbf{v} = a\mathbf{i} + b\mathbf{j} + c\mathbf{k}$ is a vector in space, the scalars a, b, c are called the _____ of **v**.

4. The sum of the squares of the direction cosines of a vector in space add up to _____.

5. *True or False* In space, the dot product of two vectors is a positive number.

6. *True or False* A vector in space may be described by specifying its magnitude and its direction angles.

Skill Building

In Problems 7–14, describe the set of points (x, y, z) defined by the equation.

7. $y = 0$　　　　**8.** $x = 0$　　　　**9.** $z = 2$　　　　**10.** $y = 3$

11. $x = -4$　　　**12.** $z = -3$　　　**13.** $x = 1$ and $y = 2$　　**14.** $x = 3$ and $z = 1$

In Problems 15–20, find the distance from P_1 to P_2.

15. $P_1 = (0, 0, 0)$ and $P_2 = (4, 1, 2)$　　　　　**16.** $P_1 = (0, 0, 0)$ and $P_2 = (1, -2, 3)$

17. $P_1 = (-1, 2, -3)$ and $P_2 = (0, -2, 1)$　　　**18.** $P_1 = (-2, 2, 3)$ and $P_2 = (4, 0, -3)$

19. $P_1 = (4, -2, -2)$ and $P_2 = (3, 2, 1)$　　　　**20.** $P_1 = (2, -3, -3)$ and $P_2 = (4, 1, -1)$

In Problems 21–26, opposite vertices of a rectangular box whose edges are parallel to the coordinate axes are given. List the coordinates of the other six vertices of the box.

21. $(0, 0, 0);$　$(2, 1, 3)$　　**22.** $(0, 0, 0);$　$(4, 2, 2)$　　**23.** $(1, 2, 3);$　$(3, 4, 5)$

24. $(5, 6, 1);$　$(3, 8, 2)$　　**25.** $(-1, 0, 2);$　$(4, 2, 5)$　　**26.** $(-2, -3, 0);$　$(-6, 7, 1)$

*In Problems 27–32, the vector **v** has initial point P and terminal point Q. Write **v** in the form $a\mathbf{i} + b\mathbf{j} + c\mathbf{k}$; that is, find its position vector.*

27. $P = (0, 0, 0);$　$Q = (3, 4, -1)$　　　　**28.** $P = (0, 0, 0);$　$Q = (-3, -5, 4)$

29. $P = (3, 2, -1);$　$Q = (5, 6, 0)$　　　　**30.** $P = (-3, 2, 0);$　$Q = (6, 5, -1)$

31. $P = (-2, -1, 4);$　$Q = (6, -2, 4)$　　　**32.** $P = (-1, 4, -2);$　$Q = (6, 2, 2)$

In Problems 33–38, find $\|\mathbf{v}\|$.

33. $\mathbf{v} = 3\mathbf{i} - 6\mathbf{j} - 2\mathbf{k}$　　　**34.** $\mathbf{v} = -6\mathbf{i} + 12\mathbf{j} + 4\mathbf{k}$　　　**35.** $\mathbf{v} = \mathbf{i} - \mathbf{j} + \mathbf{k}$

36. $\mathbf{v} = -\mathbf{i} - \mathbf{j} + \mathbf{k}$　　　**37.** $\mathbf{v} = -2\mathbf{i} + 3\mathbf{j} - 3\mathbf{k}$　　　**38.** $\mathbf{v} = 6\mathbf{i} + 2\mathbf{j} - 2\mathbf{k}$

In Problems 39–44, find each quantity if $\mathbf{v} = 3\mathbf{i} - 5\mathbf{j} + 2\mathbf{k}$ and $\mathbf{w} = -2\mathbf{i} + 3\mathbf{j} - 2\mathbf{k}$.

39. $2\mathbf{v} + 3\mathbf{w}$　　　**40.** $3\mathbf{v} - 2\mathbf{w}$　　　**41.** $\|\mathbf{v} - \mathbf{w}\|$

42. $\|\mathbf{v} + \mathbf{w}\|$　　　**43.** $\|\mathbf{v}\| - \|\mathbf{w}\|$　　　**44.** $\|\mathbf{v}\| + \|\mathbf{w}\|$

*In Problems 45–50, find the unit vector in the same direction as **v**.*

45. $\mathbf{v} = 5\mathbf{i}$　　　**46.** $\mathbf{v} = -3\mathbf{j}$　　　**47.** $\mathbf{v} = 3\mathbf{i} - 6\mathbf{j} - 2\mathbf{k}$

48. $\mathbf{v} = -6\mathbf{i} + 12\mathbf{j} + 4\mathbf{k}$　　**49.** $\mathbf{v} = \mathbf{i} + \mathbf{j} + \mathbf{k}$　　**50.** $\mathbf{v} = 2\mathbf{i} - \mathbf{j} + \mathbf{k}$

*In Problems 51–58, find the dot product $\mathbf{v} \cdot \mathbf{w}$ and the angle between **v** and **w**.*

51. $\mathbf{v} = \mathbf{i} - \mathbf{j},$　$\mathbf{w} = \mathbf{i} + \mathbf{j} + \mathbf{k}$　　　　**52.** $\mathbf{v} = \mathbf{i} + \mathbf{j},$　$\mathbf{w} = -\mathbf{i} + \mathbf{j} - \mathbf{k}$

53. $\mathbf{v} = 2\mathbf{i} + \mathbf{j} - 3\mathbf{k},$　$\mathbf{w} = \mathbf{i} + 2\mathbf{j} + 2\mathbf{k}$　　**54.** $\mathbf{v} = 2\mathbf{i} + 2\mathbf{j} - \mathbf{k},$　$\mathbf{w} = \mathbf{i} + 2\mathbf{j} + 3\mathbf{k}$

55. $\mathbf{v} = 3\mathbf{i} - \mathbf{j} + 2\mathbf{k}, \quad \mathbf{w} = \mathbf{i} + \mathbf{j} - \mathbf{k}$

56. $\mathbf{v} = \mathbf{i} + 3\mathbf{j} + 2\mathbf{k}, \quad \mathbf{w} = \mathbf{i} - \mathbf{j} + \mathbf{k}$

57. $\mathbf{v} = 3\mathbf{i} + 4\mathbf{j} + \mathbf{k}, \quad \mathbf{w} = 6\mathbf{i} + 8\mathbf{j} + 2\mathbf{k}$

58. $\mathbf{v} = 3\mathbf{i} - 4\mathbf{j} + \mathbf{k}, \quad \mathbf{w} = 6\mathbf{i} - 8\mathbf{j} + 2\mathbf{k}$

In Problems 59–66, find the direction angles of each vector. Write each vector in the form of equation (7).

59. $\mathbf{v} = 3\mathbf{i} - 6\mathbf{j} - 2\mathbf{k}$

60. $\mathbf{v} = -6\mathbf{i} + 12\mathbf{j} + 4\mathbf{k}$

61. $\mathbf{v} = \mathbf{i} + \mathbf{j} + \mathbf{k}$

62. $\mathbf{v} = \mathbf{i} - \mathbf{j} - \mathbf{k}$

63. $\mathbf{v} = \mathbf{i} + \mathbf{j}$

64. $\mathbf{v} = \mathbf{j} + \mathbf{k}$

65. $\mathbf{v} = 3\mathbf{i} - 5\mathbf{j} + 2\mathbf{k}$

66. $\mathbf{v} = 2\mathbf{i} + 3\mathbf{j} - 4\mathbf{k}$

Applications and Extensions

67. Robotic Arm Consider the double-jointed robotic arm shown in the figure. Let the lower arm be modeled by $\mathbf{a} = \langle 2, 3, 4 \rangle$, the middle arm be modeled by $\mathbf{b} = \langle 1, -1, 3 \rangle$, and the upper arm by $\mathbf{c} = \langle 4, -1, -2 \rangle$, where units are in feet.

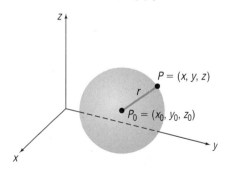

(a) Find a vector $\mathbf{d}$ that represents the position of the hand.

(b) Determine the distance of the hand from the origin.

68. The Sphere In space, the collection of all points that are the same distance from some fixed point is called a **sphere**. See the illustration. The constant distance is called the **radius,** and the fixed point is the **center** of the sphere. Show that the equation of a sphere with center at (x_0, y_0, z_0) and radius r is

$$(x - x_0)^2 + (y - y_0)^2 + (z - z_0)^2 = r^2$$

[**Hint:** Use the Distance Formula (1).]

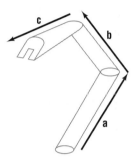

In Problems 69 and 70, find the equation of a sphere with radius r and center P_0.

69. $r = 1; \quad P_0 = (3, 1, 1)$

70. $r = 2; \quad P_0 = (1, 2, 2)$

In Problems 71–76, find the radius and center of each sphere.
[**Hint:** Complete the square in each variable.]

71. $x^2 + y^2 + z^2 + 2x - 2y = 2$

72. $x^2 + y^2 + z^2 + 2x - 2z = -1$

73. $x^2 + y^2 + z^2 - 4x + 4y + 2z = 0$

74. $x^2 + y^2 + z^2 - 4x = 0$

75. $2x^2 + 2y^2 + 2z^2 - 8x + 4z = -1$

76. $3x^2 + 3y^2 + 3z^2 + 6x - 6y = 3$

*The **work** W done by a constant force $\mathbf{F}$ in moving an object from a point A in space to a point B in space is defined as $W = \mathbf{F} \cdot \overrightarrow{AB}$. Use this definition in Problems 77–79.*

77. Work Find the work done by a force of 3 newtons acting in the direction $2\mathbf{i} + \mathbf{j} + 2\mathbf{k}$ in moving an object 2 meters from $(0, 0, 0)$ to $(0, 2, 0)$.

78. Work Find the work done by a force of 1 newton acting in the direction $2\mathbf{i} + 2\mathbf{j} + \mathbf{k}$ in moving an object 3 meters from $(0, 0, 0)$ to $(1, 2, 2)$.

79. Work Find the work done in moving an object along a vector $\mathbf{u} = 3\mathbf{i} + 2\mathbf{j} - 5\mathbf{k}$ if the applied force is $\mathbf{F} = 2\mathbf{i} - \mathbf{j} - \mathbf{k}$. Use meters for distance and newtons for force.

'Are You Prepared?' Answer

1. $d = \sqrt{(x_2 - x_1)^2 + (y_2 - y_1)^2}$

5.7 The Cross Product

OBJECTIVES **1** Find the Cross Product of Two Vectors (p. 384)

2 Know Algebraic Properties of the Cross Product (p. 385)

3 Know Geometric Properties of the Cross Product (p. 386)

4 Find a Vector Orthogonal to Two Given Vectors (p. 386)

5 Find the Area of a Parallelogram (p. 387)

1 Find the Cross Product of Two Vectors

For vectors in space, and only for vectors in space, a second product of two vectors is defined, called the *cross product*. The cross product of two vectors in space is, in fact, also a vector that has applications in both geometry and physics.

DEFINITION

If $\mathbf{v} = a_1\mathbf{i} + b_1\mathbf{j} + c_1\mathbf{k}$ and $\mathbf{w} = a_2\mathbf{i} + b_2\mathbf{j} + c_2\mathbf{k}$ are two vectors in space, the **cross product** $\mathbf{v} \times \mathbf{w}$ is defined as the vector

$$\mathbf{v} \times \mathbf{w} = (b_1c_2 - b_2c_1)\mathbf{i} - (a_1c_2 - a_2c_1)\mathbf{j} + (a_1b_2 - a_2b_1)\mathbf{k} \qquad \textbf{(1)}$$

Notice that the cross product $\mathbf{v} \times \mathbf{w}$ of two vectors is a vector. Because of this, it is sometimes referred to as the **vector product.**

EXAMPLE 1

Finding Cross Products Using Equation (1)

If $\mathbf{v} = 2\mathbf{i} + 3\mathbf{j} + 5\mathbf{k}$ and $\mathbf{w} = \mathbf{i} + 2\mathbf{j} + 3\mathbf{k}$, an application of equation (1) gives

$$\begin{aligned}
\mathbf{v} \times \mathbf{w} &= (3 \cdot 3 - 2 \cdot 5)\mathbf{i} - (2 \cdot 3 - 1 \cdot 5)\mathbf{j} + (2 \cdot 2 - 1 \cdot 3)\mathbf{k} \\
&= (9 - 10)\mathbf{i} - (6 - 5)\mathbf{j} + (4 - 3)\mathbf{k} \\
&= -\mathbf{i} - \mathbf{j} + \mathbf{k}
\end{aligned}$$

Determinants may be used as an aid in computing cross products. A **2 by 2 determinant,** symbolized by

$$\begin{vmatrix} a_1 & b_1 \\ a_2 & b_2 \end{vmatrix}$$

has the value $a_1b_2 - a_2b_1$; that is,

$$\begin{vmatrix} a_1 & b_1 \\ a_2 & b_2 \end{vmatrix} = a_1b_2 - a_2b_1$$

A **3 by 3 determinant** has the value

$$\begin{vmatrix} A & B & C \\ a_1 & b_1 & c_1 \\ a_2 & b_2 & c_2 \end{vmatrix} = \begin{vmatrix} b_1 & c_1 \\ b_2 & c_2 \end{vmatrix} A - \begin{vmatrix} a_1 & c_1 \\ a_2 & c_2 \end{vmatrix} B + \begin{vmatrix} a_1 & b_1 \\ a_2 & b_2 \end{vmatrix} C$$

EXAMPLE 2

Evaluating Determinants

(a) $\begin{vmatrix} 2 & 3 \\ 1 & 2 \end{vmatrix} = 2 \cdot 2 - 1 \cdot 3 = 4 - 3 = 1$

(b) $\begin{vmatrix} A & B & C \\ 2 & 3 & 5 \\ 1 & 2 & 3 \end{vmatrix} = \begin{vmatrix} 3 & 5 \\ 2 & 3 \end{vmatrix} A - \begin{vmatrix} 2 & 5 \\ 1 & 3 \end{vmatrix} B + \begin{vmatrix} 2 & 3 \\ 1 & 2 \end{vmatrix} C$

$$\begin{aligned}
&= (9 - 10)A - (6 - 5)B + (4 - 3)C \\
&= -A - B + C
\end{aligned}$$

Now Work PROBLEM 7

The cross product of the vectors $\mathbf{v} = a_1\mathbf{i} + b_1\mathbf{j} + c_1\mathbf{k}$ and $\mathbf{w} = a_2\mathbf{i} + b_2\mathbf{j} + c_2\mathbf{k}$, that is,

$$\mathbf{v} \times \mathbf{w} = (b_1c_2 - b_2c_1)\mathbf{i} - (a_1c_2 - a_2c_1)\mathbf{j} + (a_1b_2 - a_2b_1)\mathbf{k}$$

may be written symbolically using determinants as

$$\mathbf{v} \times \mathbf{w} = \begin{vmatrix} \mathbf{i} & \mathbf{j} & \mathbf{k} \\ a_1 & b_1 & c_1 \\ a_2 & b_2 & c_2 \end{vmatrix} = \begin{vmatrix} b_1 & c_1 \\ b_2 & c_2 \end{vmatrix}\mathbf{i} - \begin{vmatrix} a_1 & c_1 \\ a_2 & c_2 \end{vmatrix}\mathbf{j} + \begin{vmatrix} a_1 & b_1 \\ a_2 & b_2 \end{vmatrix}\mathbf{k}$$

| EXAMPLE 3 | Using Determinants to Find Cross Products |

If $\mathbf{v} = 2\mathbf{i} + 3\mathbf{j} + 5\mathbf{k}$ and $\mathbf{w} = \mathbf{i} + 2\mathbf{j} + 3\mathbf{k}$, find:

(a) $\mathbf{v} \times \mathbf{w}$　　　　(b) $\mathbf{w} \times \mathbf{v}$　　　　(c) $\mathbf{v} \times \mathbf{v}$　　　　(d) $\mathbf{w} \times \mathbf{w}$

Solution　(a) $\mathbf{v} \times \mathbf{w} = \begin{vmatrix} \mathbf{i} & \mathbf{j} & \mathbf{k} \\ 2 & 3 & 5 \\ 1 & 2 & 3 \end{vmatrix} = \begin{vmatrix} 3 & 5 \\ 2 & 3 \end{vmatrix}\mathbf{i} - \begin{vmatrix} 2 & 5 \\ 1 & 3 \end{vmatrix}\mathbf{j} + \begin{vmatrix} 2 & 3 \\ 1 & 2 \end{vmatrix}\mathbf{k} = -\mathbf{i} - \mathbf{j} + \mathbf{k}$

(b) $\mathbf{w} \times \mathbf{v} = \begin{vmatrix} \mathbf{i} & \mathbf{j} & \mathbf{k} \\ 1 & 2 & 3 \\ 2 & 3 & 5 \end{vmatrix} = \begin{vmatrix} 2 & 3 \\ 3 & 5 \end{vmatrix}\mathbf{i} - \begin{vmatrix} 1 & 3 \\ 2 & 5 \end{vmatrix}\mathbf{j} + \begin{vmatrix} 1 & 2 \\ 2 & 3 \end{vmatrix}\mathbf{k} = \mathbf{i} + \mathbf{j} - \mathbf{k}$

(c) $\mathbf{v} \times \mathbf{v} = \begin{vmatrix} \mathbf{i} & \mathbf{j} & \mathbf{k} \\ 2 & 3 & 5 \\ 2 & 3 & 5 \end{vmatrix}$

$= \begin{vmatrix} 3 & 5 \\ 3 & 5 \end{vmatrix}\mathbf{i} - \begin{vmatrix} 2 & 5 \\ 2 & 5 \end{vmatrix}\mathbf{j} + \begin{vmatrix} 2 & 3 \\ 2 & 3 \end{vmatrix}\mathbf{k} = 0\mathbf{i} - 0\mathbf{j} + 0\mathbf{k} = \mathbf{0}$

(d) $\mathbf{w} \times \mathbf{w} = \begin{vmatrix} \mathbf{i} & \mathbf{j} & \mathbf{k} \\ 1 & 2 & 3 \\ 1 & 2 & 3 \end{vmatrix}$

$= \begin{vmatrix} 2 & 3 \\ 2 & 3 \end{vmatrix}\mathbf{i} - \begin{vmatrix} 1 & 3 \\ 1 & 3 \end{vmatrix}\mathbf{j} + \begin{vmatrix} 1 & 2 \\ 1 & 2 \end{vmatrix}\mathbf{k} = 0\mathbf{i} - 0\mathbf{j} + 0\mathbf{k} = \mathbf{0}$

◾

═══ **Now Work** PROBLEM 15

2 Know Algebraic Properties of the Cross Product

Notice in Examples 3(a) and (b) that $\mathbf{v} \times \mathbf{w}$ and $\mathbf{w} \times \mathbf{v}$ are negatives of one another. From Examples 3(c) and (d), we might conjecture that the cross product of a vector with itself is the zero vector. These and other algebraic properties of the cross product are given next.

THEOREM　**Algebraic Properties of the Cross Product**

If $\mathbf{u}$, $\mathbf{v}$, and $\mathbf{w}$ are vectors in space and if α is a scalar, then

$$\mathbf{u} \times \mathbf{u} = \mathbf{0} \tag{2}$$
$$\mathbf{u} \times \mathbf{v} = -(\mathbf{v} \times \mathbf{u}) \tag{3}$$
$$\alpha(\mathbf{u} \times \mathbf{v}) = (\alpha\mathbf{u}) \times \mathbf{v} = \mathbf{u} \times (\alpha\mathbf{v}) \tag{4}$$
$$\mathbf{u} \times (\mathbf{v} + \mathbf{w}) = (\mathbf{u} \times \mathbf{v}) + (\mathbf{u} \times \mathbf{w}) \tag{5}$$

Proof　We will prove properties (2) and (4) here and leave properties (3) and (5) as exercises (see Problems 60 and 61).

To prove property (2), we let $\mathbf{u} = a_1\mathbf{i} + b_1\mathbf{j} + c_1\mathbf{k}$. Then

$$\mathbf{u} \times \mathbf{u} = \begin{vmatrix} \mathbf{i} & \mathbf{j} & \mathbf{k} \\ a_1 & b_1 & c_1 \\ a_1 & b_1 & c_1 \end{vmatrix} = \begin{vmatrix} b_1 & c_1 \\ b_1 & c_1 \end{vmatrix}\mathbf{i} - \begin{vmatrix} a_1 & c_1 \\ a_1 & c_1 \end{vmatrix}\mathbf{j} + \begin{vmatrix} a_1 & b_1 \\ a_1 & b_1 \end{vmatrix}\mathbf{k}$$

$$= 0\mathbf{i} - 0\mathbf{j} + 0\mathbf{k} = \mathbf{0}$$

To prove property (4), we let $\mathbf{u} = a_1\mathbf{i} + b_1\mathbf{j} + c_1\mathbf{k}$ and $\mathbf{v} = a_2\mathbf{i} + b_2\mathbf{j} + c_2\mathbf{k}$. Then

$$\alpha(\mathbf{u} \times \mathbf{v}) = \alpha[(b_1c_2 - b_2c_1)\mathbf{i} - (a_1c_2 - a_2c_1)\mathbf{j} + (a_1b_2 - a_2b_1)\mathbf{k}]$$
$$\uparrow$$
Apply (1).
$$= \alpha(b_1c_2 - b_2c_1)\mathbf{i} - \alpha(a_1c_2 - a_2c_1)\mathbf{j} + \alpha(a_1b_2 - a_2b_1)\mathbf{k} \qquad \textbf{(6)}$$

Since $\alpha\mathbf{u} = \alpha a_1\mathbf{i} + \alpha b_1\mathbf{j} + \alpha c_1\mathbf{k}$, we have

$$(\alpha\mathbf{u}) \times \mathbf{v} = (\alpha b_1c_2 - b_2\alpha c_1)\mathbf{i} - (\alpha a_1c_2 - a_2\alpha c_1)\mathbf{j} + (\alpha a_1b_2 - a_2\alpha b_1)\mathbf{k}$$
$$= \alpha(b_1c_2 - b_2c_1)\mathbf{i} - \alpha(a_1c_2 - a_2c_1)\mathbf{j} + \alpha(a_1b_2 - a_2b_1)\mathbf{k} \qquad \textbf{(7)}$$

Based on equations (6) and (7), the first part of property (4) follows. The second part can be proved in like fashion. ∎

Now Work PROBLEM 17

3 Know Geometric Properties of the Cross Product

The cross product has several interesting geometric properties.

THEOREM

Geometric Properties of the Cross Product

Let $\mathbf{u}$ and $\mathbf{v}$ be vectors in space.

$\mathbf{u} \times \mathbf{v}$ is orthogonal to both $\mathbf{u}$ and $\mathbf{v}$. $\qquad$ **(8)**

$\|\mathbf{u} \times \mathbf{v}\| = \|\mathbf{u}\|\,\|\mathbf{v}\| \sin\theta,$ $\qquad$ **(9)**
where θ is the angle between $\mathbf{u}$ and $\mathbf{v}$.

$\|\mathbf{u} \times \mathbf{v}\|$ is the area of the parallelogram $\qquad$ **(10)**
having $\mathbf{u} \neq \mathbf{0}$ and $\mathbf{v} \neq \mathbf{0}$ as adjacent sides.

$\mathbf{u} \times \mathbf{v} = \mathbf{0}$ if and only if $\mathbf{u}$ and $\mathbf{v}$ are parallel. $\qquad$ **(11)**

Proof of Property (8) Let $\mathbf{u} = a_1\mathbf{i} + b_1\mathbf{j} + c_1\mathbf{k}$ and $\mathbf{v} = a_2\mathbf{i} + b_2\mathbf{j} + c_2\mathbf{k}$. Then

$$\mathbf{u} \times \mathbf{v} = (b_1c_2 - b_2c_1)\mathbf{i} - (a_1c_2 - a_2c_1)\mathbf{j} + (a_1b_2 - a_2b_1)\mathbf{k}$$

Now we compute the dot product $\mathbf{u} \cdot (\mathbf{u} \times \mathbf{v})$.

$$\mathbf{u} \cdot (\mathbf{u} \times \mathbf{v}) = (a_1\mathbf{i} + b_1\mathbf{j} + c_1\mathbf{k}) \cdot [(b_1c_2 - b_2c_1)\mathbf{i} - (a_1c_2 - a_2c_1)\mathbf{j} + (a_1b_2 - a_2b_1)\mathbf{k}]$$
$$= a_1(b_1c_2 - b_2c_1) - b_1(a_1c_2 - a_2c_1) + c_1(a_1b_2 - a_2b_1) = 0$$

Since two vectors are orthogonal if their dot product is zero, it follows that $\mathbf{u}$ and $\mathbf{u} \times \mathbf{v}$ are orthogonal. Similarly, $\mathbf{v} \cdot (\mathbf{u} \times \mathbf{v}) = 0$, so $\mathbf{v}$ and $\mathbf{u} \times \mathbf{v}$ are orthogonal. ∎

Figure 82

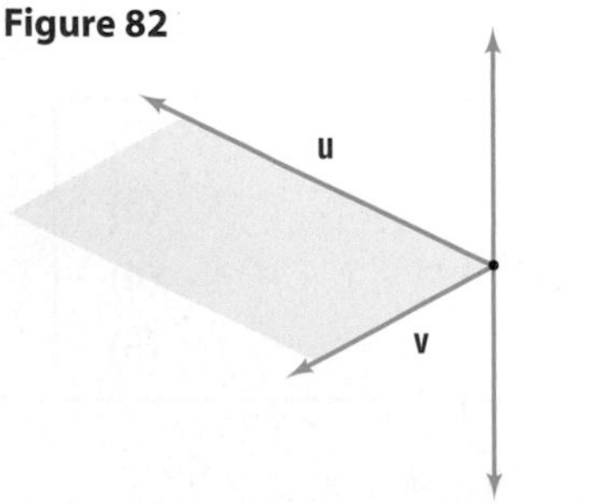

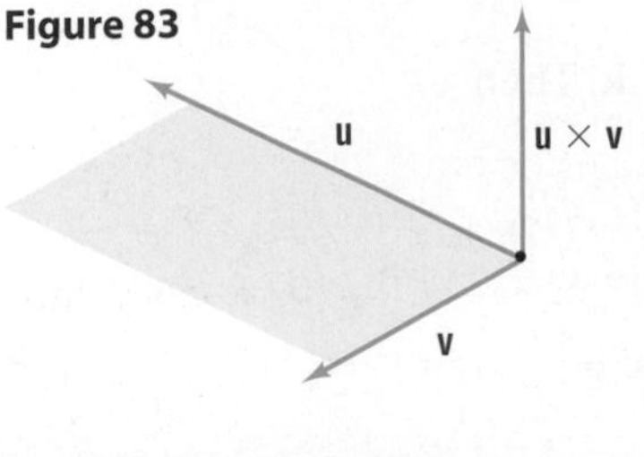

4 Find a Vector Orthogonal to Two Given Vectors

As long as the vectors $\mathbf{u}$ and $\mathbf{v}$ are not parallel, they will form a plane in space. See Figure 82. Based on property (8), the vector $\mathbf{u} \times \mathbf{v}$ is normal to this plane. As Figure 82 illustrates, there are essentially (without regard to magnitude) two vectors normal to the plane containing $\mathbf{u}$ and $\mathbf{v}$. It can be shown that the vector $\mathbf{u} \times \mathbf{v}$ is the one determined by the thumb of the right hand when the other fingers of the right hand are cupped so that they point in a direction from $\mathbf{u}$ to $\mathbf{v}$. See Figure 83.*

Figure 83

* This is a consequence of using a right-handed coordinate system.

| EXAMPLE 4 | Finding a Vector Orthogonal to Two Given Vectors |

Find a vector that is orthogonal to $\mathbf{u} = 3\mathbf{i} - 2\mathbf{j} + \mathbf{k}$ and $\mathbf{v} = -\mathbf{i} + 3\mathbf{j} - \mathbf{k}$.

Solution Based on property (8), such a vector is $\mathbf{u} \times \mathbf{v}$.

$$\mathbf{u} \times \mathbf{v} = \begin{vmatrix} \mathbf{i} & \mathbf{j} & \mathbf{k} \\ 3 & -2 & 1 \\ -1 & 3 & -1 \end{vmatrix} = (2 - 3)\mathbf{i} - [-3 - (-1)]\mathbf{j} + (9 - 2)\mathbf{k} = -\mathbf{i} + 2\mathbf{j} + 7\mathbf{k}$$

The vector $-\mathbf{i} + 2\mathbf{j} + 7\mathbf{k}$ is orthogonal to both $\mathbf{u}$ and $\mathbf{v}$.

✓**Check:** Two vectors are orthogonal if their dot product is zero.

$$\mathbf{u} \cdot (-\mathbf{i} + 2\mathbf{j} + 7\mathbf{k}) = (3\mathbf{i} - 2\mathbf{j} + \mathbf{k}) \cdot (-\mathbf{i} + 2\mathbf{j} + 7\mathbf{k}) = -3 - 4 + 7 = 0$$
$$\mathbf{v} \cdot (-\mathbf{i} + 2\mathbf{j} + 7\mathbf{k}) = (-\mathbf{i} + 3\mathbf{j} - \mathbf{k}) \cdot (-\mathbf{i} + 2\mathbf{j} + 7\mathbf{k}) = 1 + 6 - 7 = 0$$

══ **Now Work** PROBLEM 41

The proof of property (9) is left as an exercise. See Problem 62.

Figure 84

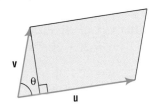

Proof of Property (10) Suppose that $\mathbf{u}$ and $\mathbf{v}$ are adjacent sides of a parallelogram. See Figure 84. Then the lengths of these sides are $\|\mathbf{u}\|$ and $\|\mathbf{v}\|$. If θ is the angle between $\mathbf{u}$ and $\mathbf{v}$, then the height of the parallelogram is $\|\mathbf{v}\| \sin \theta$ and its area is

$$\text{Area of parallelogram} = \text{Base} \times \text{Height} = \|\mathbf{u}\|[\|\mathbf{v}\| \sin \theta] = \|\mathbf{u} \times \mathbf{v}\|$$
$$\uparrow$$
$$\text{Property (9)} \qquad \blacksquare$$

5 Find the Area of a Parallelogram

| EXAMPLE 5 | Finding the Area of a Parallelogram |

Find the area of the parallelogram whose vertices are $P_1 = (0, 0, 0)$, $P_2 = (3, -2, 1)$, $P_3 = (-1, 3, -1)$, and $P_4 = (2, 1, 0)$.

Solution Two adjacent sides of this parallelogram are

$$\mathbf{u} = \overrightarrow{P_1 P_2} = 3\mathbf{i} - 2\mathbf{j} + \mathbf{k} \quad \text{and} \quad \mathbf{v} = \overrightarrow{P_1 P_3} = -\mathbf{i} + 3\mathbf{j} - \mathbf{k}$$

WARNING Not all pairs of vertices give rise to a side. For example, $\overrightarrow{P_1 P_4}$ is a diagonal of the parallelogram since $\overrightarrow{P_1 P_3} + \overrightarrow{P_3 P_4} = \overrightarrow{P_1 P_4}$. Also, $\overrightarrow{P_1 P_3}$ and $\overrightarrow{P_2 P_4}$ are not adjacent sides; they are parallel sides. ∎

Since $\mathbf{u} \times \mathbf{v} = -\mathbf{i} + 2\mathbf{j} + 7\mathbf{k}$ (Example 4), the area of the parallelogram is

$$\text{Area of parallelogram} = \|\mathbf{u} \times \mathbf{v}\| = \sqrt{1 + 4 + 49} = \sqrt{54} = 3\sqrt{6} \text{ square units}$$

══ **Now Work** PROBLEM 49

Proof of Property (11) The proof requires two parts. If $\mathbf{u}$ and $\mathbf{v}$ are parallel, then there is a scalar α such that $\mathbf{u} = \alpha \mathbf{v}$. Then

$$\mathbf{u} \times \mathbf{v} = (\alpha \mathbf{v}) \times \mathbf{v} = \alpha(\mathbf{v} \times \mathbf{v}) = \mathbf{0}$$
$$\qquad\qquad \uparrow \qquad\qquad \uparrow$$
$$\text{Property (4)} \qquad \text{Property (2)}$$

If $\mathbf{u} \times \mathbf{v} = \mathbf{0}$, then, by property (9), we have

$$\|\mathbf{u} \times \mathbf{v}\| = \|\mathbf{u}\| \, \|\mathbf{v}\| \sin \theta = 0$$

Since $\mathbf{u} \neq \mathbf{0}$ and $\mathbf{v} \neq \mathbf{0}$, we must have $\sin \theta = 0$, so $\theta = 0$ or $\theta = \pi$. In either case, since θ is the angle between $\mathbf{u}$ and $\mathbf{v}$, then $\mathbf{u}$ and $\mathbf{v}$ are parallel. ∎

5.7 Assess Your Understanding

Concepts and Vocabulary

1. *True or False* If **u** and **v** are parallel vectors, then **u** × **v** = **0**.

2. *True or False* For any vector **v**, **v** × **v** = **0**.

3. *True or False* If **u** and **v** are vectors, then **u** × **v** + **v** × **u** = **0**.

4. *True or False* **u** × **v** is a vector that is parallel to both **u** and **v**.

5. *True or False* $\|\mathbf{u} \times \mathbf{v}\| = \|\mathbf{u}\| \|\mathbf{v}\| \cos \theta$, where θ is the angle between **u** and **v**.

6. *True or False* The area of the parallelogram having **u** and **v** as adjacent sides is the magnitude of the cross product of **u** and **v**.

Skill Building

In Problems 7–14, find the value of each determinant.

7. $\begin{vmatrix} 3 & 4 \\ 1 & 2 \end{vmatrix}$

8. $\begin{vmatrix} -2 & 5 \\ 2 & -3 \end{vmatrix}$

9. $\begin{vmatrix} 6 & 5 \\ -2 & -1 \end{vmatrix}$

10. $\begin{vmatrix} -4 & 0 \\ 5 & 3 \end{vmatrix}$

11. $\begin{vmatrix} A & B & C \\ 2 & 1 & 4 \\ 1 & 3 & 1 \end{vmatrix}$

12. $\begin{vmatrix} A & B & C \\ 0 & 2 & 4 \\ 3 & 1 & 3 \end{vmatrix}$

13. $\begin{vmatrix} A & B & C \\ -1 & 3 & 5 \\ 5 & 0 & -2 \end{vmatrix}$

14. $\begin{vmatrix} A & B & C \\ 1 & -2 & -3 \\ 0 & 2 & -2 \end{vmatrix}$

In Problems 15–22, find (a) **v** × **w**, *(b)* **w** × **v**, *(c)* **w** × **w**, *and (d)* **v** × **v**.

15. **v** = 2**i** − 3**j** + **k**
    w = 3**i** − 2**j** − **k**

16. **v** = −**i** + 3**j** + 2**k**
    w = 3**i** − 2**j** − **k**

17. **v** = **i** + **j**
    w = 2**i** + **j** + **k**

18. **v** = **i** − 4**j** + 2**k**
    w = 3**i** + 2**j** + **k**

19. **v** = 2**i** − **j** + 2**k**
    w = **j** − **k**

20. **v** = 3**i** + **j** + 3**k**
    w = **i** − **k**

21. **v** = **i** − **j** − **k**
    w = 4**i** − 3**k**

22. **v** = 2**i** − 3**j**
    w = 3**j** − 2**k**

In Problems 23–44, use the given vectors **u**, **v**, *and* **w** *to find each expression.*

$$\mathbf{u} = 2\mathbf{i} - 3\mathbf{j} + \mathbf{k}, \qquad \mathbf{v} = -3\mathbf{i} + 3\mathbf{j} + 2\mathbf{k}, \qquad \mathbf{w} = \mathbf{i} + \mathbf{j} + 3\mathbf{k}$$

23. **u** × **v**

24. **v** × **w**

25. **v** × **u**

26. **w** × **v**

27. **v** × **v**

28. **w** × **w**

29. (3**u**) × **v**

30. **v** × (4**w**)

31. **u** × (2**v**)

32. (−3**v**) × **w**

33. **u** · (**u** × **v**)

34. **v** · (**v** × **w**)

35. **u** · (**v** × **w**)

36. (**u** × **v**) · **w**

37. **v** · (**u** × **w**)

38. (**v** × **u**) · **w**

39. **u** × (**v** × **v**)

40. (**w** × **w**) × **v**

41. Find a vector orthogonal to both **u** and **v**.

42. Find a vector orthogonal to both **u** and **w**.

43. Find a vector orthogonal to both **u** and **i** + **j**.

44. Find a vector orthogonal to both **u** and **j** + **k**.

In Problems 45–48, find the area of the parallelogram with one corner at P_1 and adjacent sides $\overrightarrow{P_1P_2}$ and $\overrightarrow{P_1P_3}$.

45. $P_1 = (0, 0, 0)$, $P_2 = (1, 2, 3)$, $P_3 = (-2, 3, 0)$

46. $P_1 = (0, 0, 0)$, $P_2 = (2, 3, 1)$, $P_3 = (-2, 4, 1)$

47. $P_1 = (1, 2, 0)$, $P_2 = (-2, 3, 4)$, $P_3 = (0, -2, 3)$

48. $P_1 = (-2, 0, 2)$, $P_2 = (2, 1, -1)$, $P_3 = (2, -1, 2)$

In Problems 49–52, find the area of the parallelogram with vertices P_1, P_2, P_3, and P_4.

49. $P_1 = (1, 1, 2)$, $P_2 = (1, 2, 3)$, $P_3 = (-2, 3, 0)$,
    $P_4 = (-2, 4, 1)$

50. $P_1 = (2, 1, 1)$, $P_2 = (2, 3, 1)$, $P_3 = (-2, 4, 1)$,
    $P_4 = (-2, 6, 1)$

51. $P_1 = (1, 2, -1)$, $P_2 = (4, 2, -3)$, $P_3 = (6, -5, 2)$,
    $P_4 = (9, -5, 0)$

52. $P_1 = (-1, 1, 1)$, $P_2 = (-1, 2, 2)$, $P_3 = (-3, 4, -5)$,
    $P_4 = (-3, 5, -4)$

Applications and Extensions

53. Find a unit vector normal to the plane containing **v** = **i** + 3**j** − 2**k** and **w** = −2**i** + **j** + 3**k**.

54. Find a unit vector normal to the plane containing **v** = 2**i** + 3**j** − **k** and **w** = −2**i** − 4**j** − 3**k**.

55. Volume of a Parallelepiped A **parallelepiped** is a prism whose faces are all parallelograms. Let $\mathbf{A}$, $\mathbf{B}$, and $\mathbf{C}$ be the vectors that define the parallelepiped shown in the figure. The volume V of the parallelepiped is given by the formula $V = |(\mathbf{A} \times \mathbf{B}) \cdot \mathbf{C}|$.

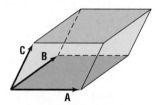

Find the volume of a parallelepiped if the defining vectors are $\mathbf{A} = 3\mathbf{i} - 2\mathbf{j} + 4\mathbf{k}$, $\mathbf{B} = 2\mathbf{i} + \mathbf{j} - 2\mathbf{k}$, and $\mathbf{C} = 3\mathbf{i} - 6\mathbf{j} - 2\mathbf{k}$.

56. Volume of a Parallelepiped Refer to Problems 55. Find the volume of a parallelepiped whose defining vectors are $\mathbf{A} = \langle 1, 0, 6 \rangle$, $\mathbf{B} = \langle 2, 3, -8 \rangle$, and $\mathbf{C} = \langle 8, -5, 6 \rangle$.

57. Prove for vectors $\mathbf{u}$ and $\mathbf{v}$ that
$$\|\mathbf{u} \times \mathbf{v}\|^2 = \|\mathbf{u}\|^2 \|\mathbf{v}\|^2 - (\mathbf{u} \cdot \mathbf{v})^2$$
[**Hint:** Proceed as in the proof of property (4), computing first the left side and then the right side.]

58. Show that if $\mathbf{u}$ and $\mathbf{v}$ are orthogonal then
$$\|\mathbf{u} \times \mathbf{v}\| = \|\mathbf{u}\| \, \|\mathbf{v}\|$$

59. Show that if $\mathbf{u}$ and $\mathbf{v}$ are orthogonal unit vectors then $\mathbf{u} \times \mathbf{v}$ is also a unit vector.

60. Prove property (3).

61. Prove property (5).

62. Prove property (9).

[**Hint:** Use the result of Problem 57 and the fact that if θ is the angle between $\mathbf{u}$ and $\mathbf{v}$ then $\mathbf{u} \cdot \mathbf{v} = \|\mathbf{u}\| \, \|\mathbf{v}\| \cos \theta$.]

Discussion and Writing

63. If $\mathbf{u} \cdot \mathbf{v} = 0$ and $\mathbf{u} \times \mathbf{v} = \mathbf{0}$, what, if anything, can you conclude about $\mathbf{u}$ and $\mathbf{v}$?

CHAPTER REVIEW

Things to Know

Polar Coordinates (pp. 320–326)

Relationship between polar coordinates (r, θ) and rectangular coordinates (x, y) (pp. 322 and 326)	$x = r \cos \theta$, $y = r \sin \theta$ $r^2 = x^2 + y^2$, $\tan \theta = \dfrac{y}{x}$, $x \neq 0$
Polar form of a complex number (p. 348)	If $z = x + yi$, then $z = r(\cos \theta + i \sin \theta)$, where $r = |z| = \sqrt{x^2 + y^2}$, $\sin \theta = \dfrac{y}{r}$, $\cos \theta = \dfrac{x}{r}$, $0 \le \theta < 2\pi$.
De Moivre's Theorem (p. 350)	If $z = r(\cos \theta + i \sin \theta)$, then $z^n = r^n[\cos(n\theta) + i \sin(n\theta)]$, where $n \ge 1$ is a positive integer.
nth root of a complex number $w = r(\cos \theta_0 + i \sin \theta_0)$ (p. 351)	$z_k = \sqrt[n]{r}\left[\cos\left(\dfrac{\theta_0}{n} + \dfrac{2k\pi}{n}\right) + i \sin\left(\dfrac{\theta_0}{n} + \dfrac{2k\pi}{n}\right)\right]$, $k = 0, \ldots, n - 1$, where $n \ge 2$ is an integer.

Vector (pp. 355–362, 375–382)

	Quantity having magnitude and direction; equivalent to a directed line segment $\overrightarrow{PQ}$
Position vector (pp. 358 and 376)	Vector whose initial point is at the origin
Unit vector (pp. 361 and 378)	Vector whose magnitude is 1
Dot product (pp. 366 and 378)	If $\mathbf{v} = a_1\mathbf{i} + b_1\mathbf{j}$ and $\mathbf{w} = a_2\mathbf{i} + b_2\mathbf{j}$, then $\mathbf{v} \cdot \mathbf{w} = a_1 a_2 + b_1 b_2$. If $\mathbf{v} = a_1\mathbf{i} + b_1\mathbf{j} + c_1\mathbf{k}$ and $\mathbf{w} = a_2\mathbf{i} + b_2\mathbf{j} + c_2\mathbf{k}$, then $\mathbf{v} \cdot \mathbf{w} = a_1 a_2 + b_1 b_2 + c_1 c_2$.
Angle θ between two nonzero vectors $\mathbf{u}$ and $\mathbf{v}$ (pp. 368 and 379)	$\cos \theta = \dfrac{\mathbf{u} \cdot \mathbf{v}}{\|\mathbf{u}\| \, \|\mathbf{v}\|}$
Direction angles of a vector in space (pp. 380–381)	If $\mathbf{v} = a\mathbf{i} + b\mathbf{j} + c\mathbf{k}$, then $\mathbf{v} = \|\mathbf{v}\|[(\cos \alpha)\mathbf{i} + (\cos \beta)\mathbf{j} + (\cos \gamma)\mathbf{k}]$, where $\cos \alpha = \dfrac{a}{\|\mathbf{v}\|}$, $\cos \beta = \dfrac{b}{\|\mathbf{v}\|}$, $\cos \gamma = \dfrac{c}{\|\mathbf{v}\|}$.
Cross product (p. 384)	If $\mathbf{v} = a_1\mathbf{i} + b_1\mathbf{j} + c_1\mathbf{k}$ and $\mathbf{w} = a_2\mathbf{i} + b_2\mathbf{j} + c_2\mathbf{k}$, then $\mathbf{v} \times \mathbf{w} = [b_1 c_2 - b_2 c_1]\mathbf{i} - [a_1 c_2 - a_2 c_1]\mathbf{j} + [a_1 b_2 - a_2 b_1]\mathbf{k}$.
Area of a parallelogram (pp. 386 and 387)	$\|\mathbf{u} \times \mathbf{v}\| = \|\mathbf{u}\| \, \|\mathbf{v}\| \sin \theta$, where θ is the angle between $\mathbf{u}$ and $\mathbf{v}$.

Objectives

Section		You should be able to . . .	Example(s)	Review Exercises
5.1	1	Plot points using polar coordinates (p. 320)	1–3	1–6
	2	Convert from polar coordinates to rectangular coordinates (p. 322)	4	1–6
	3	Convert from rectangular coordinates to polar coordinates (p. 324)	5–7	7–12
	4	Transform equations from polar to rectangular form (p. 326)	8, 9	13(a)–18(a)
5.2	1	Graph and identify polar equations by converting to rectangular equations (p. 330)	1–7	13(b)–18(b)
	2	Graph polar equations using a graphing utility (p. 331)	4–7	19–24
	3	Test polar equations for symmetry (p. 335)	8–11	19–24
	4	Graph polar equations by plotting points (p. 336)	8–13	19–24
5.3	1	Plot points in the complex plane (p. 347)	1	25–28
	2	Convert a complex number from rectangular form to polar form (p. 348)	2, 3	29–34
	3	Find products and quotients of complex numbers in polar form (p. 349)	4	35–40
	4	Use De Moivre's Theorem (p. 350)	5, 6	41–48
	5	Find complex roots (p. 351)	7	49–50
5.4	1	Graph vectors (p. 357)	1	51–54
	2	Find a position vector (p. 358)	2	55–58
	3	Add and subtract vectors algebraically (p. 359)	3	59, 60
	4	Find a scalar multiple and the magnitude of a vector (p. 360)	4	55–58, 61–66
	5	Find a unit vector (p. 360)	5	67, 68
	6	Find a vector from its direction and magnitude (p. 361)	6	69, 70
	7	Analyze objects in static equilibrium (p. 362)	7	111
5.5	1	Find the dot product of two vectors (p. 366)	1	85–88
	2	Find the angle between two vectors (p. 367)	2, 3	85–88, 109, 110, 112
	3	Determine whether two vectors are parallel (p. 369)	4	93–98
	4	Determine whether two vectors are orthogonal (p. 369)	5	93–98
	5	Decompose a vector into two orthogonal vectors (p. 370)	6	99–102
	6	Compute work (p. 371)	7	113
5.6	1	Find the distance between two points in space (p. 376)	1	71, 72
	2	Find position vectors in space (p. 376)	2	73, 74
	3	Perform operations on vectors (p. 377)	3–5	75–80
	4	Find the dot product (p. 378)	6	89–92
	5	Find the angle between two vectors (p. 379)	7	89–92
	6	Find the direction angles of a vector (p. 379)	8–10	103, 104
5.7	1	Find the cross product of two vectors (p. 384)	1–3	81, 82
	2	Know algebraic properties of the cross product (p. 385)	pp. 385–386	107, 108
	3	Know geometric properties of the cross product (p. 386)	p. 386	105, 106
	4	Find a vector orthogonal to two given vectors (p. 386)	4	84
	5	Find the area of a parallelogram (p. 387)	5	105, 106

Review Exercises

In Problems 1–6, plot each point given in polar coordinates, and find its rectangular coordinates.

1. $\left(3, \dfrac{\pi}{6}\right)$ **2.** $\left(4, \dfrac{2\pi}{3}\right)$ **3.** $\left(-2, \dfrac{4\pi}{3}\right)$ **4.** $\left(-1, \dfrac{5\pi}{4}\right)$ **5.** $\left(-3, -\dfrac{\pi}{2}\right)$ **6.** $\left(-4, -\dfrac{\pi}{4}\right)$

In Problems 7–12, the rectangular coordinates of a point are given. Find two pairs of polar coordinates (r, θ) for each point, one with $r > 0$ and the other with $r < 0$. Express θ in radians.

7. $(-3, 3)$ **8.** $(1, -1)$ **9.** $(0, -2)$ **10.** $(2, 0)$ **11.** $(3, 4)$ **12.** $(-5, 12)$

In Problems 13–18, the variables r and θ represent polar coordinates. (a) Write each polar equation as an equation in rectangular coordinates (x, y). (b) Identify the equation and graph it.

13. $r = 2 \sin \theta$

14. $3r = \sin \theta$

15. $r = 5$

16. $\theta = \dfrac{\pi}{4}$

17. $r \cos \theta + 3r \sin \theta = 6$

18. $r^2 + 4r \sin \theta - 8r \cos \theta = 5$

In Problems 19–24, by hand sketch the graph of each polar equation. Be sure to test for symmetry. Verify your graph using a graphing utility.

19. $r = 4 \cos \theta$

20. $r = 3 \sin \theta$

21. $r = 3 - 3 \sin \theta$

22. $r = 2 + \cos \theta$

23. $r = 4 - \cos \theta$

24. $r = 1 - 2 \sin \theta$

In Problems 25–28, write each complex number in polar form. Express each argument in degrees.

25. $-1 - i$

26. $-\sqrt{3} + i$

27. $4 - 3i$

28. $3 - 2i$

In Problems 29–34, write each complex number in the standard form $a + bi$ and plot each in the complex plane.

29. $2(\cos 150° + i \sin 150°)$

30. $3(\cos 60° + i \sin 60°)$

31. $3\left(\cos \dfrac{2\pi}{3} + i \sin \dfrac{2\pi}{3} \right)$

32. $4\left(\cos \dfrac{3\pi}{4} + i \sin \dfrac{3\pi}{4} \right)$

33. $0.1(\cos 350° + i \sin 350°)$

34. $0.5(\cos 160° + i \sin 160°)$

In Problems 35–40, find zw and $\dfrac{z}{w}$. Leave your answers in polar form.

35. $z = \cos 80° + i \sin 80°$
$w = \cos 50° + i \sin 50°$

36. $z = \cos 205° + i \sin 205°$
$w = \cos 85° + i \sin 85°$

37. $z = 3\left(\cos \dfrac{9\pi}{5} + i \sin \dfrac{9\pi}{5} \right)$
$w = 2\left(\cos \dfrac{\pi}{5} + i \sin \dfrac{\pi}{5} \right)$

38. $z = 2\left(\cos \dfrac{5\pi}{3} + i \sin \dfrac{5\pi}{3} \right)$
$w = 3\left(\cos \dfrac{\pi}{3} + i \sin \dfrac{\pi}{3} \right)$

39. $z = 5(\cos 10° + i \sin 10°)$
$w = \cos 355° + i \sin 355°$

40. $z = 4(\cos 50° + i \sin 50°)$
$w = \cos 340° + i \sin 340°$

In Problems 41–48, write each expression in the standard form $a + bi$.

41. $[3(\cos 20° + i \sin 20°)]^3$

42. $[2(\cos 50° + i \sin 50°)]^3$

43. $\left[\sqrt{2}\left(\cos \dfrac{5\pi}{8} + i \sin \dfrac{5\pi}{8} \right) \right]^4$

44. $\left[2\left(\cos \dfrac{5\pi}{16} + i \sin \dfrac{5\pi}{16} \right) \right]^4$

45. $\left(1 - \sqrt{3}i \right)^6$

46. $(2 - 2i)^8$

47. $(3 + 4i)^4$

48. $(1 - 2i)^4$

49. Find all the complex cube roots of 27.

50. Find all the complex fourth roots of -16.

In Problems 51–54, use the figure to graph each of the following:

51. $\mathbf{u} + \mathbf{v}$

52. $\mathbf{v} + \mathbf{w}$

53. $2\mathbf{u} + 3\mathbf{v}$

54. $5\mathbf{v} - 2\mathbf{w}$

In Problems 55–58, the vector $\mathbf{v}$ is represented by the directed line segment $\overrightarrow{PQ}$. Write $\mathbf{v}$ in the form $a\mathbf{i} + b\mathbf{j}$ and find $\|\mathbf{v}\|$.

55. $P = (1, -2);\quad Q = (3, -6)$

56. $P = (-3, 1);\quad Q = (4, -2)$

57. $P = (0, -2);\quad Q = (-1, 1)$

58. $P = (3, -4);\quad Q = (-2, 0)$

In Problems 59–68, use the vectors $\mathbf{v} = -2\mathbf{i} + \mathbf{j}$ and $\mathbf{w} = 4\mathbf{i} - 3\mathbf{j}$ to find:

59. $\mathbf{v} + \mathbf{w}$

60. $\mathbf{v} - \mathbf{w}$

61. $4\mathbf{v} - 3\mathbf{w}$

62. $-\mathbf{v} + 2\mathbf{w}$

63. $\|\mathbf{v}\|$

64. $\|\mathbf{v} + \mathbf{w}\|$

65. $\|\mathbf{v}\| + \|\mathbf{w}\|$

66. $\|2\mathbf{v}\| - 3\|\mathbf{w}\|$

67. Find a unit vector in the same direction as $\mathbf{v}$.

68. Find a unit vector in the opposite direction of $\mathbf{w}$.

69. Find the vector $\mathbf{v}$ in the xy-plane with magnitude 3 if the angle between $\mathbf{v}$ and $\mathbf{i}$ is 60°.

70. Find the vector $\mathbf{v}$ in the xy-plane with magnitude 5 if the angle between $\mathbf{v}$ and $\mathbf{i}$ is 150°.

71. Find the distance from $P_1 = (1, 3, -2)$ to $P_2 = (4, -2, 1)$.

72. Find the distance from $P_1 = (0, -4, 3)$ to $P_2 = (6, -5, -1)$.

73. A vector **v** has initial point $P = (1, 3, -2)$ and terminal point $Q = (4, -2, 1)$. Write **v** in the form $\mathbf{v} = a\mathbf{i} + b\mathbf{j} + c\mathbf{k}$.

74. A vector **v** has initial point $P = (0, -4, 3)$ and terminal point $Q = (6, -5, -1)$. Write **v** in the form $\mathbf{v} = a\mathbf{i} + b\mathbf{j} + c\mathbf{k}$.

In Problems 75–84, use the vectors $\mathbf{v} = 3\mathbf{i} + \mathbf{j} - 2\mathbf{k}$ *and* $\mathbf{w} = -3\mathbf{i} + 2\mathbf{j} - \mathbf{k}$ *to find each expression.*

75. $4\mathbf{v} - 3\mathbf{w}$ **76.** $-\mathbf{v} + 2\mathbf{w}$ **77.** $\|\mathbf{v} - \mathbf{w}\|$ **78.** $\|\mathbf{v} + \mathbf{w}\|$

79. $\|\mathbf{v}\| - \|\mathbf{w}\|$ **80.** $\|\mathbf{v}\| + \|\mathbf{w}\|$ **81.** $\mathbf{v} \times \mathbf{w}$ **82.** $\mathbf{v} \cdot (\mathbf{v} \times \mathbf{w})$

83. Find a unit vector in the same direction as **v** and then in the opposite direction of **v**.

84. Find a unit vector orthogonal to both **v** and **w**.

In Problems 85–92, find the dot product $\mathbf{v} \cdot \mathbf{w}$ *and the angle between* **v** *and* **w**.

85. $\mathbf{v} = -2\mathbf{i} + \mathbf{j}, \quad \mathbf{w} = 4\mathbf{i} - 3\mathbf{j}$

86. $\mathbf{v} = 3\mathbf{i} - \mathbf{j}, \quad \mathbf{w} = \mathbf{i} + \mathbf{j}$

87. $\mathbf{v} = \mathbf{i} - 3\mathbf{j}, \quad \mathbf{w} = -\mathbf{i} + \mathbf{j}$

88. $\mathbf{v} = \mathbf{i} + 4\mathbf{j}, \quad \mathbf{w} = 3\mathbf{i} - 2\mathbf{j}$

89. $\mathbf{v} = \mathbf{i} + \mathbf{j} + \mathbf{k}, \quad \mathbf{w} = \mathbf{i} - \mathbf{j} + \mathbf{k}$

90. $\mathbf{v} = \mathbf{i} - \mathbf{j} + \mathbf{k}, \quad \mathbf{w} = 2\mathbf{i} + \mathbf{j} + \mathbf{k}$

91. $\mathbf{v} = 4\mathbf{i} - \mathbf{j} + 2\mathbf{k}, \quad \mathbf{w} = \mathbf{i} - 2\mathbf{j} - 3\mathbf{k}$

92. $\mathbf{v} = -\mathbf{i} - 2\mathbf{j} + 3\mathbf{k}, \quad \mathbf{w} = 5\mathbf{i} + \mathbf{j} + \mathbf{k}$

In Problems 93–98, determine whether **v** *and* **w** *are parallel, orthogonal, or neither.*

93. $\mathbf{v} = 2\mathbf{i} + 3\mathbf{j}; \quad \mathbf{w} = -4\mathbf{i} - 6\mathbf{j}$

94. $\mathbf{v} = -2\mathbf{i} - \mathbf{j}; \quad \mathbf{w} = 2\mathbf{i} + \mathbf{j}$

95. $\mathbf{v} = 3\mathbf{i} - 4\mathbf{j}; \quad \mathbf{w} = -3\mathbf{i} + 4\mathbf{j}$

96. $\mathbf{v} = -2\mathbf{i} + 2\mathbf{j}; \quad \mathbf{w} = -3\mathbf{i} + 2\mathbf{j}$

97. $\mathbf{v} = 3\mathbf{i} - 2\mathbf{j}; \quad \mathbf{w} = 4\mathbf{i} + 6\mathbf{j}$

98. $\mathbf{v} = -4\mathbf{i} + 2\mathbf{j}; \quad \mathbf{w} = 2\mathbf{i} + 4\mathbf{j}$

In Problems 99–102, decompose **v** *into two vectors, one parallel to* **w** *and the other orthogonal to* **w**.

99. $\mathbf{v} = 2\mathbf{i} + \mathbf{j}; \quad \mathbf{w} = -4\mathbf{i} + 3\mathbf{j}$

100. $\mathbf{v} = -3\mathbf{i} + 2\mathbf{j}; \quad \mathbf{w} = -2\mathbf{i} + \mathbf{j}$

101. $\mathbf{v} = 2\mathbf{i} + 3\mathbf{j}; \quad \mathbf{w} = 3\mathbf{i} + \mathbf{j}$

102. $\mathbf{v} = -\mathbf{i} + 2\mathbf{j}; \quad \mathbf{w} = 3\mathbf{i} - \mathbf{j}$

---

103. Find the direction angles of the vector $\mathbf{v} = 3\mathbf{i} - 4\mathbf{j} + 2\mathbf{k}$.

104. Find the direction angles of the vector $\mathbf{v} = \mathbf{i} - \mathbf{j} + 2\mathbf{k}$.

105. Find the area of the parallelogram with vertices $P_1 = (1, 1, 1)$, $P_2 = (2, 3, 4)$, $P_3 = (6, 5, 2)$, and $P_4 = (7, 7, 5)$.

106. Find the area of the parallelogram with vertices $P_1 = (2, -1, 1)$, $P_2 = (5, 1, 4)$, $P_3 = (0, 1, 1)$, and $P_4 = (3, 3, 4)$.

107. If $\mathbf{u} \times \mathbf{v} = 2\mathbf{i} - 3\mathbf{j} + \mathbf{k}$, what is $\mathbf{v} \times \mathbf{u}$?

108. Suppose that $\mathbf{u} = 3\mathbf{v}$. What is $\mathbf{u} \times \mathbf{v}$?

109. Actual Speed and Direction of a Swimmer A swimmer can maintain a constant speed of 5 miles per hour. If the swimmer heads directly across a river that has a current moving at the rate of 2 miles per hour, what is the actual speed of the swimmer? (See the figure.) If the river is 1 mile wide, how far downstream will the swimmer end up from the point directly across the river from the starting point?

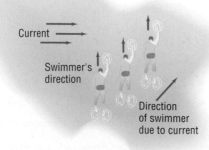

110. Actual Speed and Direction of an Airplane An airplane has an airspeed of 500 kilometers per hour in a northerly direction. The wind velocity is 60 kilometers per hour in a southeasterly direction. Find the actual speed and direction of the plane relative to the ground.

111. Static Equilibrium A weight of 2000 pounds is suspended from two cables, as shown in the figure. What are the tensions in each cable?

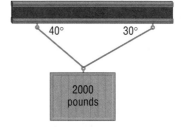

112. Actual Speed and Distance of a Motorboat A small motorboat is moving at a true speed of 11 miles per hour in a southerly direction. The current is known to be from the northeast at 3 miles per hour. What is the speed of the motorboat relative to the water? In what direction does the compass indicate that the boat is headed?

113. Computing Work Find the work done by a force of 5 pounds acting in the direction $60°$ to the horizontal in moving an object 20 feet from $(0, 0)$ to $(20, 0)$.

CHAPTER TEST

In Problems 1–3, plot each point given in polar coordinates.

1. $\left(2, \dfrac{3\pi}{4}\right)$

2. $\left(3, -\dfrac{\pi}{6}\right)$

3. $\left(-4, \dfrac{\pi}{3}\right)$

4. Convert $\left(2, 2\sqrt{3}\right)$ from rectangular coordinates to polar coordinates (r, θ), where $r > 0$ and $0 \le \theta < 2\pi$.

In Problems 5–7, convert the polar equation to a rectangular equation. Graph the equation.

5. $r = 7$

6. $\tan \theta = 3$

7. $r \sin^2 \theta + 8 \sin \theta = r$

In Problems 8 and 9, test the polar equation for symmetry with respect to the pole, the polar axis, and the line $\theta = \dfrac{\pi}{2}$.

8. $r^2 \cos \theta = 5$

9. $r = 5 \sin \theta \cos^2 \theta$

In Problems 10–12, perform the given operation, where $z = 2(\cos 85° + i \sin 85°)$ and $w = 3(\cos 22° + i \sin 22°)$. Write your answer in polar form.

10. $z \cdot w$

11. $\dfrac{w}{z}$

12. w^5

13. Find all the cube roots of $-8 + 8\sqrt{3}i$. Write all answers in the form $a + bi$ and then plot them in rectangular coordinates.

In Problems 14–18, $P_1 = \left(3\sqrt{2}, 7\sqrt{2}\right)$ and $P_2 = \left(8\sqrt{2}, 2\sqrt{2}\right)$.

14. Find the position vector $\mathbf{v}$ equal to $\overrightarrow{P_1 P_2}$.

15. Find $\|\mathbf{v}\|$.

16. Find the unit vector in the direction of $\mathbf{v}$.

17. Find the angle between $\mathbf{v}$ and $\mathbf{i}$.

18. Decompose $\mathbf{v}$ into its vertical and horizontal components.

In Problems 19–22, $\mathbf{v}_1 = \langle 4, 6 \rangle$, $\mathbf{v}_2 = \langle -3, -6 \rangle$, $\mathbf{v}_3 = \langle -8, 4 \rangle$, and $\mathbf{v}_4 = \langle 10, 15 \rangle$.

19. Find the vector $\mathbf{v}_1 + 2\mathbf{v}_2 - \mathbf{v}_3$.

20. Which two vectors are parallel?

21. Which two vectors are orthogonal?

22. Find the angle between vectors $\mathbf{v}_1$ and $\mathbf{v}_2$.

In Problems 23–25, use the vectors $\mathbf{u} = 2\mathbf{i} - 3\mathbf{j} + \mathbf{k}$ and $\mathbf{v} = -\mathbf{i} + 3\mathbf{j} + 2\mathbf{k}$.

23. Find $\mathbf{u} \times \mathbf{v}$.

24. Find the direction angles for $\mathbf{u}$.

25. Find the area of the parallelogram that has $\mathbf{u}$ and $\mathbf{v}$ as adjacent sides.

26. A 1200-pound chandelier is to be suspended over a large ballroom; the chandelier will be hung on a cable whose ends will be attached to the ceiling, 16 feet apart. The chandelier will be free hanging so that the ends of the cable will make equal angles with the ceiling. If the top of the chandelier is to be 16 feet from the ceiling, what is the minimum tension the cable must be able to endure?

CUMULATIVE REVIEW

1. Find an equation for the line containing the origin that makes an angle of $30°$ with the positive x-axis.

2. Find an equation for the circle with center at the point $(0, 1)$ and radius 3. Graph this circle.

3. Test the equation $x^2 + y^3 = 2x^4$ for symmetry with respect to the x-axis, the y-axis, and the origin.

4. Graph the function $y = |\sin x|$.

5. Graph the function $y = \sin|x|$.

6. Find the exact value of $\sin^{-1}\left(-\dfrac{1}{2}\right)$.

7. Graph the equations $x = 3$ and $y = 4$ on the same set of rectangular coordinates.

8. Graph the equations $r = 2$ and $\theta = \dfrac{\pi}{3}$ on the same set of polar coordinates.

9. What is the amplitude and period of $y = -4 \cos(\pi x)$?

CHAPTER PROJECTS

Source: www.aeromuseum.org/eduHowtoFly.html

I. Modeling Aircraft Motion   Four aerodynamic forces act on an airplane in flight: lift, weight, thrust, and drag. While an aircraft is in flight, these four forces continuously battle each other. Weight opposes lift and drag opposes thrust. See the figure. In balanced flight at constant speed, both the lift and weight are equal and the thrust and drag are equal.

1. What will happen to the aircraft if the lift is held constant while the weight is decreased (say from burning off fuel)?

2. What will happen to the aircraft if the lift is decreased while the weight is held constant?

3. What will happen to the aircraft if the thrust is increased while the drag is held constant?

4. What will happen to the aircraft if the drag is increased while the thrust is held constant?

In 1903 the Wright brothers made the first controlled powered flight. The weight of their plane was approximately 700 pounds (lb). Newton's Second Law of motion states that force = mass × acceleration ($F = ma$). If the mass is measured in kilograms (kg) and acceleration in meters per second squared (m/sec²), then the force will be measured in newtons (N). [**Note:** $1\,N = 1\,kg \cdot m/sec^2$.]

5. If 1 kg = 2.205 lb, convert the weight of the Wright brother's plane to kilograms.

6. If acceleration due to gravity is $a = 9.80$ m/sec², determine the force due to weight on the Wright brother's plane.

7. What must be true about the lift force of the Wright brother's plane in order for it to get off the ground?

8. The weight of a fully loaded Cessna 170B is 2200 lb. What lift force is required to get this plane off the ground?

9. The maximum gross weight of a Boeing 747 is 255,000 lb. What lift force is required to get this jet off the ground?

The following projects are available at the Instructor's Resource Center (IRC):

II. Project at Motorola   *Signal Fades Due to Interference* Complex trigonometric functions are used to assure that a cellphone has optimal reception as the user travels up and down an elevator.

III. Compound Interest   The effect of continuously compounded interest is analyzed using polar coordinates.

IV. Complex Equations   Analysis of complex equations illustrates the connections between complex and real equations. At times, using complex equations is more efficient for proving mathematical theorems.

Analytic Geometry

Pluto's Unusual Orbit

Pluto is about 39 times as far from the Sun as Earth is. Its average distance from the Sun is about 3,647,240,000 miles (5,869,660,000 kilometers). Pluto travels around the Sun in an elliptical (oval-shaped) orbit. At some point in its orbit, it comes closer to the Sun than Neptune, the outermost planet. It stays inside Neptune's orbit for about 20 Earth-years. This event occurs every 248 Earth-years, which is about the same number of Earth-years it takes Pluto to travel once around the Sun. Pluto entered Neptune's orbit on January 23, 1979, and remained there until February 11, 1999.

Source: Spinrad, Hyron. "Pluto." World Book Online Reference Center. 2004. World Book, Inc. http://www.worldbookonline.com/wb/Article?id=ar435500. Used with permission of World Book, Inc.

—See Chapter Project I—

A Look Back

In Chapter 1, we introduced rectangular coordinates and showed how geometry problems can be solved algebraically. In Section 1.2, we defined a circle geometrically and then used the distance formula and rectangular coordinates to obtain an equation for a circle.

A Look Ahead

In this chapter we give geometric definitions for the conics and use the distance formula and rectangular coordinates to obtain their equations. We will learn that a circle is a conic.

Historically, Apollonius (200 BC) was among the first to study *conics* and discover some of their interesting properties. Today, conics are still studied because of their many uses. *Paraboloids of revolution* (parabolas rotated about their axes of symmetry) are used as signal collectors (the satellite dishes used with radar and dish TV, for example), as solar energy collectors, and as reflectors (telescopes, light projection, and so on). The planets circle the Sun in approximately *elliptical* orbits. Elliptical surfaces can be used to reflect signals such as light and sound from one place to another. A third conic, the *hyperbola,* can be used to determine the location of lightning strikes.

The Greeks used the methods of Euclidean geometry to study conics. However, we shall use the more powerful methods of analytic geometry, bringing to bear both algebra and geometry, for our study of conics.

This chapter concludes with a section on equations of conics in polar coordinates and a section on plane curves and parametric equations.

Outline

6.1 Conics

OBJECTIVE 1 Know the Names of the Conics (p. 396)

1 Know the Names of the Conics

The word *conic* derives from the word *cone*, which is a geometric figure that can be constructed in the following way: Let *a* and *g* be two distinct lines that intersect at a point *V*. Keep the line *a* fixed. Now rotate the line *g* about *a* while maintaining the same angle between *a* and *g*. The collection of points swept out (generated) by the line *g* is called a **(right circular) cone.** See Figure 1. The fixed line *a* is called the **axis** of the cone; the point *V* is its **vertex;** the lines that pass through *V* and make the same angle with *a* as *g* are **generators** of the cone. Each generator is a line that lies entirely on the cone. The cone consists of two parts, called **nappes,** that intersect at the vertex.

Figure 1

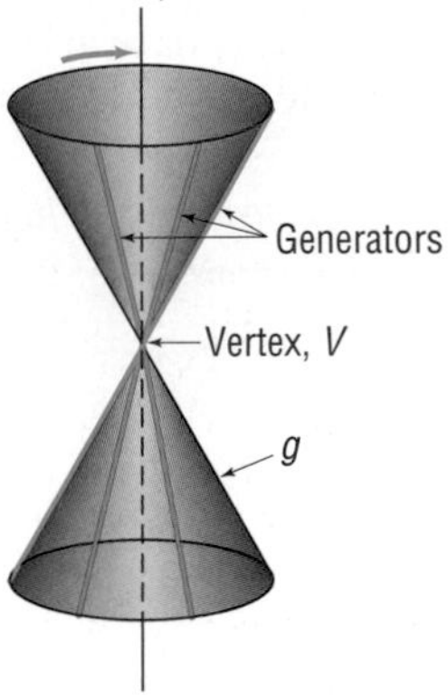

Axis, *a*

Generators

Vertex, *V*

g

Conics, an abbreviation for **conic sections,** are curves that result from the intersection of a right circular cone and a plane. The conics we shall study arise when the plane does not contain the vertex, as shown in Figure 2. These conics are **circles** when the plane is perpendicular to the axis of the cone and intersects each generator; **ellipses** when the plane is tilted slightly so that it intersects each generator, but intersects only one nappe of the cone; **parabolas** when the plane is tilted farther so that it is parallel to one (and only one) generator and intersects only one nappe of the cone; and **hyperbolas** when the plane intersects both nappes.

If the plane does contain the vertex, the intersection of the plane and the cone is a point, a line, or a pair of intersecting lines. These are usually called **degenerate conics.**

Figure 2

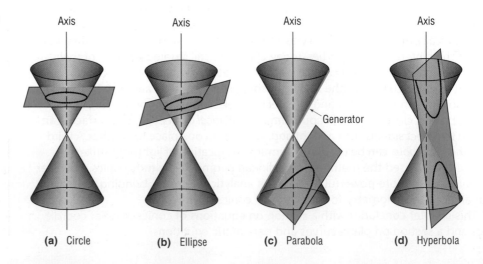

(a) Circle   (b) Ellipse   (c) Parabola   (d) Hyperbola

6.2 The Parabola

PREPARING FOR THIS SECTION *Before getting started, review the following:*

- Distance Formula (Section 1.1, p. 5)
- Symmetry (Section 1.2, pp. 18–19)
- Square Root Method (Appendix, Section A.4, pp. A30–A31)
- Completing the Square (Appendix, Section A.4, p. A31)
- Graphing Techniques: Transformations (Section 1.7, pp. 67–76)

Now Work the 'Are You Prepared?' problems on page 404.

OBJECTIVES 1 Analyze Parabolas with Vertex at the Origin (p. 397)
2 Analyze Parabolas with Vertex at (h, k) (p. 401)
3 Solve Applied Problems Involving Parabolas (p. 402)

The graph of a quadratic function is a parabola. In this section, we begin with a geometric definition of a parabola and use it to obtain an equation.

DEFINITION

A **parabola** is the collection of all points P in the plane that are the same distance from a fixed point F as they are from a fixed line D. The point F is called the **focus** of the parabola, and the line D is its **directrix.** As a result, a parabola is the set of points P for which

$$d(F, P) = d(P, D) \qquad (1)$$

Figure 3

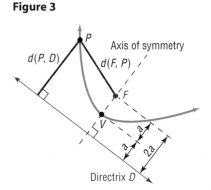

Figure 3 shows a parabola (in blue). The line through the focus F and perpendicular to the directrix D is called the **axis of symmetry** of the parabola. The point of intersection of the parabola with its axis of symmetry is called the **vertex** V.

Because the vertex V lies on the parabola, it must satisfy equation (1): $d(F, V) = d(V, D)$. The vertex is midway between the focus and the directrix. We shall let a equal the distance $d(F, V)$ from F to V. Now we are ready to derive an equation for a parabola. To do this, we use a rectangular system of coordinates positioned so that the vertex V, focus F, and directrix D of the parabola are conveniently located.

1 Analyze Parabolas with Vertex at the Origin

If we choose to locate the vertex V at the origin $(0, 0)$, we can conveniently position the focus F on either the x-axis or the y-axis. First, we consider the case where the focus F is on the positive x-axis, as shown in Figure 4. Because the distance from F to V is a, the coordinates of F will be $(a, 0)$ with $a > 0$. Similarly, because the distance from V to the directrix D is also a and, because D must be perpendicular to the x-axis (since the x-axis is the axis of symmetry), the equation of the directrix D must be $x = -a$.

Now, if $P = (x, y)$ is any point on the parabola, P must obey equation (1):

$$d(F, P) = d(P, D)$$

Figure 4

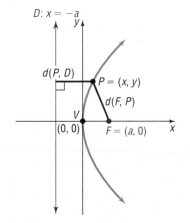

So we have

$$\sqrt{(x - a)^2 + (y - 0)^2} = |x + a| \qquad \text{Use the Distance Formula.}$$

$$(x - a)^2 + y^2 = (x + a)^2 \qquad \text{Square both sides.}$$

$$x^2 - 2ax + a^2 + y^2 = x^2 + 2ax + a^2 \qquad \text{Remove parentheses.}$$

$$y^2 = 4ax \qquad \text{Simplify.}$$

THEOREM | **Equation of a Parabola: Vertex at (0, 0), Focus at (a, 0), a > 0**

The equation of a parabola with vertex at $(0, 0)$, focus at $(a, 0)$, and directrix $x = -a, a > 0$, is

$$y^2 = 4ax \tag{2}$$

EXAMPLE 1 | **Finding the Equation of a Parabola and Graphing It**

Find an equation of the parabola with vertex at $(0, 0)$ and focus at $(3, 0)$. Graph the equation.

Solution

The distance from the vertex $(0, 0)$ to the focus $(3, 0)$ is $a = 3$. Based on equation (2), the equation of this parabola is

$$y^2 = 4ax$$
$$y^2 = 12x \quad a = 3$$

To graph this parabola, it is helpful to plot the points on the graph directly above or below the focus. To locate these two points, we let $x = 3$. Then

$$y^2 = 12x = 12(3) = 36$$
$$y = \pm 6 \qquad \text{Solve for } y.$$

The points on the parabola directly above or below the focus are $(3, 6)$ and $(3, -6)$. These points help in graphing the parabola because they determine the "opening." See Figure 5.

Figure 5

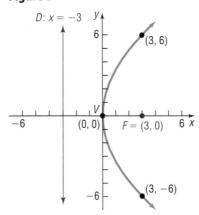

In general, the points on a parabola $y^2 = 4ax$ that lie above and below the focus $(a, 0)$ are each at a distance $2a$ from the focus. This follows from the fact that if $x = a$ then $y^2 = 4ax = 4a^2$, so $y = \pm 2a$. The line segment joining these two points is called the **latus rectum;** its length is $4a$.

Now Work PROBLEM 19

EXAMPLE 2 | **Graphing a Parabola Using a Graphing Utility**

Graph the parabola $y^2 = 12x$.

Solution

To graph the parabola $y^2 = 12x$, we need to graph the two functions $Y_1 = \sqrt{12x}$ and $Y_2 = -\sqrt{12x}$ on a square screen. Figure 6 shows the graph of $y^2 = 12x$. Notice that the graph fails the vertical line test, so $y^2 = 12x$ is not a function.

Figure 6

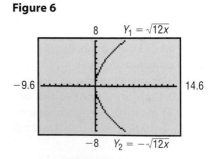

By reversing the steps we used to obtain equation (2), it follows that the graph of an equation of the form of equation (2), $y^2 = 4ax$, is a parabola; its vertex is at $(0, 0)$, its focus is at $(a, 0)$, its directrix is the line $x = -a$, and its axis of symmetry is the x-axis.

For the remainder of this section, the direction **"Analyze the equation"** will mean to find the vertex, focus, and directrix of the parabola and graph it.

EXAMPLE 3 | **Analyzing the Equation of a Parabola**

Analyze the equation: $y^2 = 8x$

Solution

The equation $y^2 = 8x$ is of the form $y^2 = 4ax$, where $4a = 8$, so $a = 2$. Consequently, the graph of the equation is a parabola with vertex at $(0, 0)$ and focus on the positive x-axis at $(2, 0)$. The directrix is the vertical line $x = -2$. The two points defining the latus rectum are obtained by letting $x = 2$. Then $y^2 = 16$, so $y = \pm 4$. See Figure 7(a) for the graph drawn by hand. Figure 7(b) shows the graph obtained using a graphing utility.

Figure 7

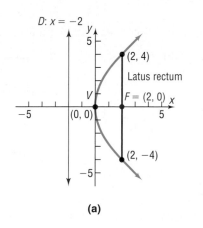

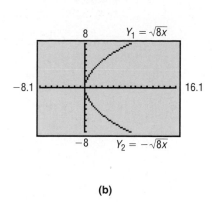

(a)   (b)

Recall that we arrived at equation (2) after placing the focus on the positive x-axis. If the focus is placed on the negative x-axis, positive y-axis, or negative y-axis, a different form of the equation for the parabola results. The four forms of the equation of a parabola with vertex at $(0, 0)$ and focus on a coordinate axis a distance a from $(0, 0)$ are given in Table 1, and their graphs are given in Figure 8. Notice that each graph is symmetric with respect to its axis of symmetry.

Table 1

EQUATIONS OF A PARABOLA VERTEX AT (0,0); FOCUS ON AN AXIS; $a > 0$				
Vertex	**Focus**	**Directrix**	**Equation**	**Description**
$(0,0)$	$(a, 0)$	$x = -a$	$y^2 = 4ax$	Parabola, axis of symmetry is the x-axis, opens right
$(0,0)$	$(-a, 0)$	$x = a$	$y^2 = -4ax$	Parabola, axis of symmetry is the x-axis, opens left
$(0,0)$	$(0, a)$	$y = -a$	$x^2 = 4ay$	Parabola, axis of symmetry is the y-axis, opens up
$(0,0)$	$(0, -a)$	$y = a$	$x^2 = -4ay$	Parabola, axis of symmetry is the y-axis, opens down

Figure 8

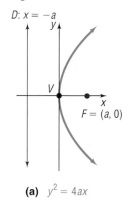

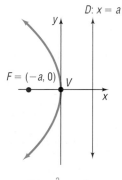

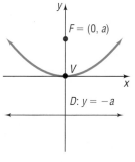

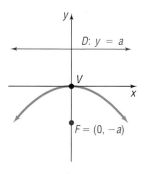

(a) $y^2 = 4ax$ (b) $y^2 = -4ax$ (c) $x^2 = 4ay$ (d) $x^2 = -4ay$

EXAMPLE 4	**Analyzing the Equation of a Parabola**

Analyze the equation: $x^2 = -12y$

Solution

The equation $x^2 = -12y$ is of the form $x^2 = -4ay$, with $a = 3$. Consequently, the graph of the equation is a parabola with vertex at $(0, 0)$, focus at $(0, -3)$ and directrix the line $y = 3$. The parabola opens down, and its axis of symmetry is the y-axis. To obtain the points defining the latus rectum, let $y = -3$. Then $x^2 = 36$, so $x = \pm 6$. See Figure 9(a) for the graph drawn by hand. Figure 9(b) shows the graph using a graphing utility.

Figure 9

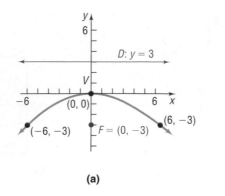

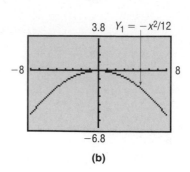

(a)

(b)

Now Work PROBLEM 39

EXAMPLE 5 | Finding the Equation of a Parabola

Find the equation of the parabola with focus at $(0, 4)$ and directrix the line $y = -4$. Graph the equation.

Figure 10

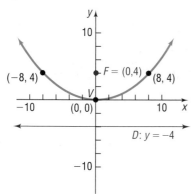

Solution

A parabola whose focus is at $(0, 4)$ and whose directrix is the horizontal line $y = -4$ will have its vertex at $(0, 0)$. (Do you see why? The vertex is midway between the focus and the directrix.) Since the focus is on the positive y-axis at $(0, 4)$, the equation of this parabola is of the form $x^2 = 4ay$, with $a = 4$; that is,

$$x^2 = 4ay = 4(4)y = 16y$$
$$\uparrow$$
$$a = 4$$

Letting $y = 4$, we find $x^2 = 64$, so $x = \pm 8$. The points $(8, 4)$ and $(-8, 4)$ determine the latus rectum. Figure 10 shows the graph of $x^2 = 16y$.

✓Check: Verify the graph drawn in Figure 10 by graphing $Y_1 = \dfrac{x^2}{16}$ using a graphing utility.

EXAMPLE 6 | Finding the Equation of a Parabola

Find the equation of a parabola with vertex at $(0, 0)$ if its axis of symmetry is the x-axis and its graph contains the point $\left(-\dfrac{1}{2}, 2\right)$. Find its focus and directrix, and graph the equation.

Solution

The vertex is at the origin, the axis of symmetry is the x-axis, and the graph contains a point in the second quadrant, so the parabola opens to the left. We see from Table 1 that the form of the equation is

$$y^2 = -4ax$$

Because the point $\left(-\dfrac{1}{2}, 2\right)$ is on the parabola, the coordinates $x = -\dfrac{1}{2}, y = 2$ must satisfy the equation $y^2 = -4ax$. Substituting $x = -\dfrac{1}{2}$ and $y = 2$ into the equation, we find that

$$2^2 = -4a\left(-\dfrac{1}{2}\right) \qquad y^2 = -4ax; x = -\dfrac{1}{2}, y = 2$$

$$a = 2$$

The equation of the parabola is

$$y^2 = -4(2)x = -8x$$

Figure 11

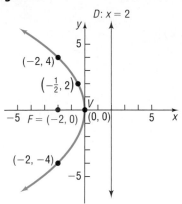

The focus is at $(-2, 0)$ and the directrix is the line $x = 2$. Letting $x = -2$, we find $y^2 = 16$, so $y = \pm 4$. The points $(-2, 4)$ and $(-2, -4)$ define the latus rectum. See Figure 11.

Now Work PROBLEM 27

2 Analyze Parabolas with Vertex at (*h, k*)

If a parabola with vertex at the origin and axis of symmetry along a coordinate axis is shifted horizontally h units and then vertically k units, the result is a parabola with vertex at (h, k) and axis of symmetry parallel to a coordinate axis. The equations of such parabolas have the same forms as those in Table 1, but with x replaced by $x - h$ (the horizontal shift) and y replaced by $y - k$ (the vertical shift). Table 2 gives the forms of the equations of such parabolas. Figures 12(a)–(d) illustrate the graphs for $h > 0, k > 0$.

Table 2

| PARABOLAS WITH VERTEX AT (h, k); AXIS OF SYMMETRY PARALLEL TO A COORDINATE AXIS; $a > 0$ | | | | |
Vertex	Focus	Directrix	Equation	Description
(h, k)	$(h + a, k)$	$x = h - a$	$(y - k)^2 = 4a(x - h)$	Parabola, axis of symmetry parallel to x-axis, opens right
(h, k)	$(h - a, k)$	$x = h + a$	$(y - k)^2 = -4a(x - h)$	Parabola, axis of symmetry parallel to x-axis, opens left
(h, k)	$(h, k + a)$	$y = k - a$	$(x - h)^2 = 4a(y - k)$	Parabola, axis of symmetry parallel to y-axis, opens up
(h, k)	$(h, k - a)$	$y = k + a$	$(x - h)^2 = -4a(y - k)$	Parabola, axis of symmetry parallel to y-axis, opens down

Figure 12

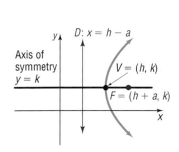

(a) $(y - k)^2 = 4a(x - h)$

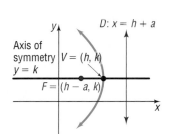

(b) $(y - k)^2 = -4a(x - h)$

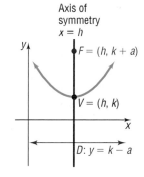

(c) $(x - h)^2 = 4a(y - k)$

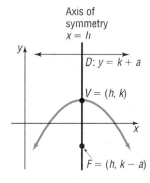

(d) $(x - h)^2 = -4a(y - k)$

EXAMPLE 7	Finding the Equation of a Parabola, Vertex Not at Origin

Find an equation of the parabola with vertex at $(-2, 3)$ and focus at $(0, 3)$. Graph the equation.

Solution The vertex $(-2, 3)$ and focus $(0, 3)$ both lie on the horizontal line $y = 3$ (the axis of symmetry). The distance a from the vertex $(-2, 3)$ to the focus $(0, 3)$ is $a = 2$. Also, because the focus lies to the right of the vertex, we know that the parabola opens to the right. Consequently, the form of the equation is

$$(y - k)^2 = 4a(x - h)$$

Figure 13

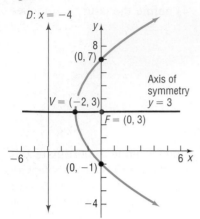

where $(h, k) = (-2, 3)$ and $a = 2$. Therefore, the equation is

$$(y - 3)^2 = 4 \cdot 2[x - (-2)]$$

$$(y - 3)^2 = 8(x + 2)$$

If $x = 0$, then $(y - 3)^2 = 16$. Then $y - 3 = \pm 4$, so $y = -1$ or $y = 7$. The points $(0, -1)$ and $(0, 7)$ define the latus rectum; the line $x = -4$ is the directrix. See Figure 13.

➡ **Now Work** PROBLEM 29

EXAMPLE 8 | Using a Graphing Utility to Graph a Parabola, Vertex Not at Origin

Using a graphing utility, graph the equation $(y - 3)^2 = 8(x + 2)$.

Solution

Figure 14

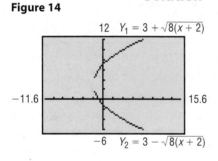

First, we must solve the equation for y.

$$(y - 3)^2 = 8(x + 2)$$

$$y - 3 = \pm\sqrt{8(x + 2)} \qquad \text{Use the Square Root Method.}$$

$$y = 3 \pm \sqrt{8(x + 2)} \qquad \text{Add 3 to both sides.}$$

Figure 14 shows the graphs of the equations $Y_1 = 3 + \sqrt{8(x + 2)}$ and $Y_2 = 3 - \sqrt{8(x + 2)}$.

Polynomial equations define parabolas whenever they involve two variables that are quadratic in one variable and linear in the other. To analyze this type of equation, we first complete the square of the variable that is quadratic.

EXAMPLE 9 | Analyzing the Equation of a Parabola

Analyze the equation: $x^2 + 4x - 4y = 0$

Solution

Figure 15

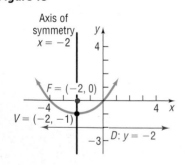

To analyze the equation $x^2 + 4x - 4y = 0$, we complete the square involving the variable x.

$$x^2 + 4x - 4y = 0$$

$$x^2 + 4x = 4y \qquad \text{Isolate the terms involving x on the left side.}$$

$$x^2 + 4x + 4 = 4y + 4 \qquad \text{Complete the square on the left side.}$$

$$(x + 2)^2 = 4(y + 1) \qquad \text{Factor.}$$

This equation is of the form $(x - h)^2 = 4a(y - k)$, with $h = -2, k = -1$, and $a = 1$. The graph is a parabola with vertex at $(h, k) = (-2, -1)$ that opens up. The focus is $a = 1$ unit above the vertex at $(-2, 0)$, and the directrix is the line $y = -2$. See Figure 15.

➡ **Now Work** PROBLEM 47

3 Solve Applied Problems Involving Parabolas

Parabolas find their way into many applications. For example, suspension bridges have cables in the shape of a parabola. Another property of parabolas that is used in applications is their reflecting property.

Suppose that a mirror is shaped like a **paraboloid of revolution,** a surface formed by rotating a parabola about its axis of symmetry. If a light (or any other emitting source) is placed at the focus of the parabola, all the rays emanating from the light will reflect off the mirror in lines parallel to the axis of symmetry. This principle is used in the design of searchlights, flashlights, certain automobile headlights, and other such devices. See Figure 16.

Conversely, suppose that rays of light (or other signals) emanate from a distant source so that they are essentially parallel. When these rays strike the surface of a parabolic mirror whose axis of symmetry is parallel to these rays, they are reflected to a single point at the focus. This principle is used in the design of some solar energy devices, satellite dishes, and the mirrors used in some types of telescopes. See Figure 17.

Figure 16
Searchlight

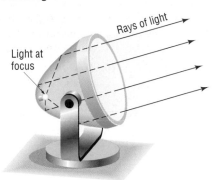

Figure 17
Telescope

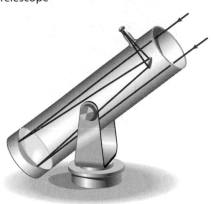

EXAMPLE 10 | **Satellite Dish**

A satellite dish is shaped like a paraboloid of revolution. The signals that cmanate from a satellite strike the surface of the dish and are reflected to a single point, where the receiver is located. If the dish is 8 feet across at its opening and 3 feet deep at its center, at what position should the receiver be placed? That is, where is the focus?

Solution

Figure 18(a) shows the satellite dish. We draw the parabola used to form the dish on a rectangular coordinate system so that the vertex of the parabola is at the origin and its focus is on the positive y-axis. See Figure 18(b).

Figure 18

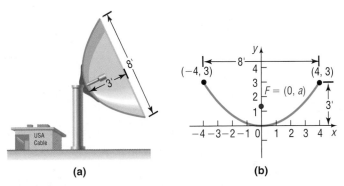

(a)   (b)

The form of the equation of the parabola is

$$x^2 = 4ay$$

and its focus is at $(0, a)$. Since $(4, 3)$ is a point on the graph, we have

$$4^2 = 4a(3) \qquad x = 4, y = 3$$

$$a = \frac{4}{3} \qquad \text{Solve for } a.$$

The receiver should be located $1\frac{1}{3}$ feet from the base of the dish, along its axis of symmetry.

Now Work PROBLEM 63

6.2 Assess Your Understanding

1. The formula for the distance d from $P_1 = (x_1, y_1)$ to $P_2 = (x_2, y_2)$ is $d =$ _____. (p. 5)

2. To complete the square of $x^2 - 4x$, add _____. (p. A31)

3. Use the Square Root Method to find the real solutions of $(x + 4)^2 = 9$. (pp. A30–A31)

4. The point that is symmetric with respect to the x-axis to the point $(-2, 5)$ is _____. (pp. 18–19)

5. To graph $y = (x - 3)^2 + 1$, shift the graph of $y = x^2$ to the right _____ units and then _____ 1 unit. (pp. 67–70)

Concepts and Vocabulary

6. A(n) _____ is the collection of all points in the plane such that the distance from each point to a fixed point equals its distance to a fixed line.

7. The surface formed by rotating a parabola about its axis of symmetry is called a _____ _____ _____.

8. *True or False* The vertex of a parabola is a point on the parabola that also is on its axis of symmetry.

9. *True or False* If a light is placed at the focus of a parabola, all the rays reflected off the parabola will be parallel to the axis of symmetry.

10. *True or False* The graph of a quadratic function is a parabola.

Skill Building

In Problems 11–18, the graph of a parabola is given. Match each graph to its equation.

(A) $y^2 = 4x$
(B) $x^2 = 4y$

(C) $y^2 = -4x$
(D) $x^2 = -4y$

(E) $(y - 1)^2 = 4(x - 1)$
(F) $(x + 1)^2 = 4(y + 1)$

(G) $(y - 1)^2 = -4(x - 1)$
(H) $(x + 1)^2 = -4(y + 1)$

11.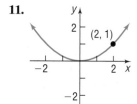

12.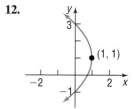

13.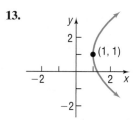

14.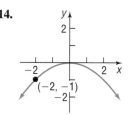

15.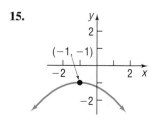

16.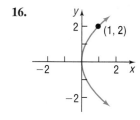

17.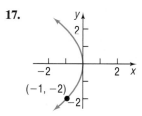

18.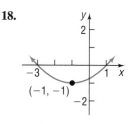

In Problems 19–36, find the equation of the parabola described. Find the two points that define the latus rectum, and graph the equation by hand.

19. Focus at $(4, 0)$; vertex at $(0, 0)$

20. Focus at $(0, 2)$; vertex at $(0, 0)$

21. Focus at $(0, -3)$; vertex at $(0, 0)$

22. Focus at $(-4, 0)$; vertex at $(0, 0)$

23. Focus at $(-2, 0)$; directrix the line $x = 2$

24. Focus at $(0, -1)$; directrix the line $y = 1$

25. Directrix the line $y = -\dfrac{1}{2}$; vertex at $(0, 0)$

26. Directrix the line $x = -\dfrac{1}{2}$; vertex at $(0, 0)$

27. Vertex at $(0, 0)$; axis of symmetry the y-axis; containing the point $(2, 3)$

28. Vertex at $(0, 0)$; axis of symmetry the x-axis; containing the point $(2, 3)$

29. Vertex at $(2, -3)$; focus at $(2, -5)$

30. Vertex at $(4, -2)$; focus at $(6, -2)$

31. Vertex at $(-1, -2)$; focus at $(0, -2)$

32. Vertex at $(3, 0)$; focus at $(3, -2)$

33. Focus at $(-3, 4)$; directrix the line $y = 2$

34. Focus at $(2, 4)$; directrix the line $x = -4$

35. Focus at $(-3, -2)$; directrix the line $x = 1$

36. Focus at $(-4, 4)$; directrix the line $y = -2$

In Problems 37–54, find the vertex, focus, and directrix of each parabola. Graph the equation by hand. Verify your graph using a graphing utility.

37. $x^2 = 4y$

38. $y^2 = 8x$

39. $y^2 = -16x$

40. $x^2 = -4y$

41. $(y - 2)^2 = 8(x + 1)$

42. $(x + 4)^2 = 16(y + 2)$

43. $(x - 3)^2 = -(y + 1)$

44. $(y + 1)^2 = -4(x - 2)$

45. $(y + 3)^2 = 8(x - 2)$

46. $(x - 2)^2 = 4(y - 3)$

47. $y^2 - 4y + 4x + 4 = 0$

48. $x^2 + 6x - 4y + 1 = 0$

49. $x^2 + 8x = 4y - 8$

50. $y^2 - 2y = 8x - 1$

51. $y^2 + 2y - x = 0$

52. $x^2 - 4x = 2y$

53. $x^2 - 4x = y + 4$

54. $y^2 + 12y = -x + 1$

In Problems 55–62, write an equation for each parabola.

55.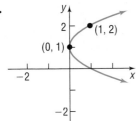

56.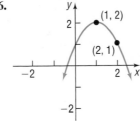

57.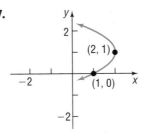

58.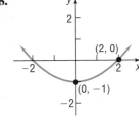

59.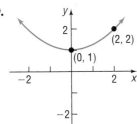

60.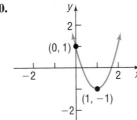

61.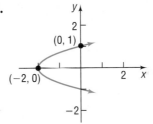

62.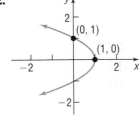

Applications and Extensions

63. Satellite Dish A satellite dish is shaped like a paraboloid of revolution. The signals that emanate from a satellite strike the surface of the dish and are reflected to a single point, where the receiver is located. If the dish is 10 feet across at its opening and 4 feet deep at its center, at what position should the receiver be placed?

64. Constructing a TV Dish A cable TV receiving dish is in the shape of a paraboloid of revolution. Find the location of the receiver, which is placed at the focus, if the dish is 6 feet across at its opening and 2 feet deep.

65. Constructing a Flashlight The reflector of a flashlight is in the shape of a paraboloid of revolution. Its diameter is 4 inches and its depth is 1 inch. How far from the vertex should the light bulb be placed so that the rays will be reflected parallel to the axis?

66. Constructing a Headlight A sealed-beam headlight is in the shape of a paraboloid of revolution. The bulb, which is placed at the focus, is 1 inch from the vertex. If the depth is to be 2 inches, what is the diameter of the headlight at its opening?

67. Suspension Bridge The cables of a suspension bridge are in the shape of a parabola, as shown in the figure. The towers supporting the cable are 600 feet apart and 80 feet high. If

the cables touch the road surface midway between the towers, what is the height of the cable from the road at a point 150 feet from the center of the bridge?

68. Suspension Bridge The cables of a suspension bridge are in the shape of a parabola. The towers supporting the cable are 400 feet apart and 100 feet high. If the cables are at a height of 10 feet midway between the towers, what is the height of the cable at a point 50 feet from the center of the bridge?

69. Searchlight A searchlight is shaped like a paraboloid of revolution. If the light source is located 2 feet from the base along the axis of symmetry and the opening is 5 feet across, how deep should the searchlight be?

70. Searchlight A searchlight is shaped like a paraboloid of revolution. If the light source is located 2 feet from the base along the axis of symmetry and the depth of the searchlight is 4 feet, what should the width of the opening be?

71. Solar Heat A mirror is shaped like a paraboloid of revolution and will be used to concentrate the rays of the sun at its focus, creating a heat source. See the figure. If the mirror is 20 feet across at its opening and is 6 feet deep, where will the heat source be concentrated?

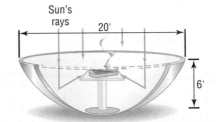

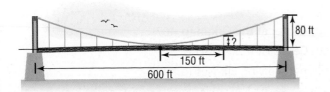

72. Reflecting Telescope A reflecting telescope contains a mirror shaped like a paraboloid of revolution. If the mirror is 4 inches across at its opening and is 3 inches deep, where will the collected light be concentrated?

73. Parabolic Arch Bridge A bridge is built in the shape of a parabolic arch. The bridge has a span of 120 feet and a maximum height of 25 feet. See the illustration. Choose a suitable rectangular coordinate system and find the height of the arch at distances of 10, 30, and 50 feet from the center.

74. Parabolic Arch Bridge A bridge is to be built in the shape of a parabolic arch and is to have a span of 100 feet. The height of the arch a distance of 40 feet from the center is to be 10 feet. Find the height of the arch at its center.

75. Gateway Arch The Gateway Arch in St. Louis is often mistaken to be parabolic is shape. In fact, it is a *catenary*, which has a more complicated formula than a parabola. The Arch is 625 feet high and 598 feet wide at its base.
(a) Find the equation of a parabola with the same dimensions. Let *x* equal the horizontal distance from the center of the arc.
(b) The table gives the height of the Arch at various widths; find the corresponding heights for the parabola found in part (a).

Width (ft)	Height (ft)
567	100
478	312.5
308	525

(c) Do the data support the notion that the Arch is in the shape of a parabola?

Source: Wikipedia, the free encyclopedia

76. Show that an equation of the form

$$Ax^2 + Ey = 0 \qquad A \neq 0, E \neq 0$$

is the equation of a parabola with vertex at $(0, 0)$ and axis of symmetry the *y*-axis. Find its focus and directrix.

77. Show that an equation of the form

$$Cy^2 + Dx = 0 \qquad C \neq 0, D \neq 0$$

is the equation of a parabola with vertex at $(0, 0)$ and axis of symmetry the *x*-axis. Find its focus and directrix.

78. Show that the graph of an equation of the form

$$Ax^2 + Dx + Ey + F = 0 \qquad A \neq 0$$

(a) Is a parabola if $E \neq 0$.
(b) Is a vertical line if $E = 0$ and $D^2 - 4AF = 0$.
(c) Is two vertical lines if $E = 0$ and $D^2 - 4AF > 0$.
(d) Contains no points if $E = 0$ and $D^2 - 4AF < 0$.

79. Show that the graph of an equation of the form

$$Cy^2 + Dx + Ey + F = 0 \qquad C \neq 0$$

(a) Is a parabola if $D \neq 0$.
(b) Is a horizontal line if $D = 0$ and $E^2 - 4CF = 0$.
(c) Is two horizontal lines if $D = 0$ and $E^2 - 4CF > 0$.
(d) Contains no points if $D = 0$ and $E^2 - 4CF < 0$.

'Are You Prepared?' Answers

1. $d = \sqrt{(x_2 - x_1)^2 + (y_2 - y_1)^2}$ **2.** 4 **3.** $x + 4 = \pm 3; \{-7, -1\}$ **4.** $(-2, -5)$ **5.** 3; up

6.3 The Ellipse

PREPARING FOR THIS SECTION *Before getting started, review the following:*

- Distance Formula (Section 1.1, p. 5)
- Completing the Square (Appendix, Section A.4, p. A31)
- Intercepts (Section 1.2, pp. 17–18)
- Symmetry (Section 1.2, pp. 18–19)
- Circles (Section 1.2, pp. 22–26)
- Graphing Techniques: Transformations (Section 1.7, pp. 67–76)

Now Work the 'Are You Prepared?' problems on page 415.

OBJECTIVES **1** Analyze Ellipses with Center at the Origin (p. 407)
2 Analyze Ellipses with Center at (h, k) (p. 411)
3 Solve Applied Problems Involving Ellipses (p. 414)

DEFINITION

An **ellipse** is the set of all points P in the plane the sum of whose distances from two fixed points, called the **foci,** is a constant.

The definition contains within it a physical means for drawing an ellipse. Find a piece of string (the length of this string is the constant referred to in the definition). Then take two thumbtacks (the foci) and stick them into a piece of cardboard so that the distance between them is less than the length of the string. Now attach the ends of the string to the thumbtacks and, using the point of a pencil, pull the string taut. See Figure 19. Keeping the string taut, rotate the pencil around the two thumbtacks. The pencil traces out an ellipse, as shown in Figure 19.

In Figure 19, the foci are labeled F_1 and F_2. The line containing the foci is called the **major axis.** The midpoint of the line segment joining the foci is the **center** of the ellipse. The line through the center and perpendicular to the major axis is the **minor axis.**

The two points of intersection of the ellipse and the major axis are the **vertices,** V_1 and V_2, of the ellipse. The distance from one vertex to the other is the **length of the major axis.** The ellipse is symmetric with respect to its major axis, with respect to its minor axis, and with respect to its center.

Figure 19

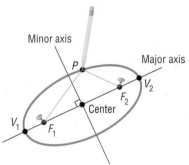

Figure 20

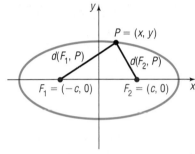

1 Analyze Ellipses with Center at the Origin

With these ideas in mind, we are now ready to find the equation of an ellipse in a rectangular coordinate system. First, we place the center of the ellipse at the origin. Second, we position the ellipse so that its major axis coincides with a coordinate axis. Suppose that the major axis coincides with the x-axis, as shown in Figure 20. If c is the distance from the center to a focus, one focus will be at $F_1 = (-c, 0)$ and the other at $F_2 = (c, 0)$. As we shall see, it is convenient to let $2a$ denote the constant distance referred to in the definition. Then, if $P = (x, y)$ is any point on the ellipse, we have

$$d(F_1, P) + d(F_2, P) = 2a$$

Sum of the distances from P to the foci equals a constant, $2a$.

$$\sqrt{(x + c)^2 + (y - 0)^2} + \sqrt{(x - c)^2 + (y - 0)^2} = 2a$$

Use the Distance Formula.

$$\sqrt{(x + c)^2 + y^2} = 2a - \sqrt{(x - c)^2 + y^2}$$

Isolate one radical.

$$(x + c)^2 + y^2 = 4a^2 - 4a\sqrt{(x - c)^2 + y^2} + (x - c)^2 + y^2$$

Square both sides.

$$x^2 + 2cx + c^2 + y^2 = 4a^2 - 4a\sqrt{(x - c)^2 + y^2} + x^2 - 2cx + c^2 + y^2$$

Remove parentheses.

$$4cx - 4a^2 = -4a\sqrt{(x - c)^2 + y^2}$$

Simplify; isolate the radical.

$$cx - a^2 = -a\sqrt{(x - c)^2 + y^2}$$

Divide each side by 4.

$$(cx - a^2)^2 = a^2[(x - c)^2 + y^2]$$

Square both sides again.

$$c^2x^2 - 2a^2cx + a^4 = a^2(x^2 - 2cx + c^2 + y^2)$$

Remove parentheses.

$$(c^2 - a^2)x^2 - a^2y^2 = a^2c^2 - a^4$$

Rearrange the terms.

$$(a^2 - c^2)x^2 + a^2y^2 = a^2(a^2 - c^2)$$

Multiply each side by -1;   **(1)**
factor a^2 on the right side.

To obtain points on the ellipse off the x-axis, it must be that $a > c$. To see why, look again at Figure 20. Then

$$d(F_1, P) + d(F_2, P) > d(F_1, F_2)$$

The sum of the lengths of two sides of a triangle is greater than the length of the third side.

$$2a > 2c$$

$d(F_1, P) + d(F_2, P) = 2a$, $d(F_1, F_2) = 2c$

$$a > c$$

Since $a > c$, we also have $a^2 > c^2$, so $a^2 - c^2 > 0$. Let $b^2 = a^2 - c^2, b > 0$. Then $a > b$ and equation (1) can be written as

$$b^2 x^2 + a^2 y^2 = a^2 b^2$$

$$\frac{x^2}{a^2} + \frac{y^2}{b^2} = 1 \qquad \text{Divide each side by } a^2 b^2.$$

As you can verify, this equation is symmetric with respect to the x-axis, y-axis, and origin.

Because the major axis is the x-axis, we find the vertices of this ellipse by letting $y = 0$. The vertices satisfy the equation $\frac{x^2}{a^2} = 1$, the solutions of which are $x = \pm a$. Consequently, the vertices of this ellipse are $V_1 = (-a, 0)$ and $V_2 = (a, 0)$. The y-intercepts of the ellipse, found by letting $x = 0$, have coordinates $(0, -b)$ and $(0, b)$. These four intercepts, $(a, 0)$, $(-a, 0)$, $(0, b)$, and $(0, -b)$, are used to graph the ellipse.

Figure 21

THEOREM

Equation of an Ellipse: Center at (0, 0); Major Axis along the x-Axis

An equation of the ellipse with center at $(0, 0)$, foci at $(-c, 0)$ and $(c, 0)$, and vertices at $(-a, 0)$ and $(a, 0)$ is

$$\frac{x^2}{a^2} + \frac{y^2}{b^2} = 1 \qquad \text{where } a > b > 0 \text{ and } b^2 = a^2 - c^2 \qquad \textbf{(2)}$$

The major axis is the x-axis. See Figure 21.

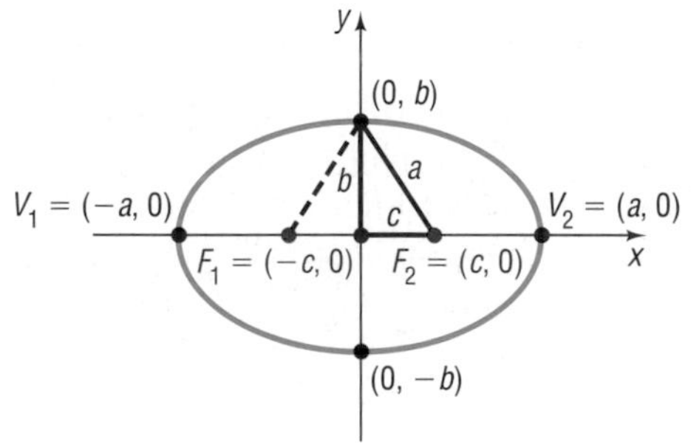

Notice in Figure 21 the right triangle formed with the points $(0, 0)$, $(c, 0)$, and $(0, b)$. Because $b^2 = a^2 - c^2$ (or $b^2 + c^2 = a^2$), the distance from the focus at $(c, 0)$ to the point $(0, b)$ is a.

This can be seen another way. Look at the two right triangles in Figure 21. They are congruent. Do you see why? Because the sum of the distances from the foci to a point on the ellipse is $2a$, it follows that the distance from $(c, 0)$ to $(0, b)$ is a.

EXAMPLE 1 | **Finding an Equation of an Ellipse**

Find an equation of the ellipse with center at the origin, one focus at $(3, 0)$, and a vertex at $(-4, 0)$. Graph the equation.

Figure 22

$$\frac{x^2}{16} + \frac{y^2}{7} = 1$$

Solution

The ellipse has its center at the origin and, since the given focus and vertex lie on the x-axis, the major axis is the x-axis. The distance from the center, $(0, 0)$, to one of the foci, $(3, 0)$, is $c = 3$. The distance from the center, $(0, 0)$, to one of the vertices, $(-4, 0)$, is $a = 4$. Because the center is the origin, the other focus is $(-3, 0)$ and the other vertex is $(4, 0)$. With $a = 4$ and $c = 3$, from equation (2), it follows that

$$b^2 = a^2 - c^2 = 16 - 9 = 7$$

so an equation of the ellipse is

$$\frac{x^2}{16} + \frac{y^2}{7} = 1$$

The y-intercepts are $(0, \pm b) = (0, \pm\sqrt{7})$. Figure 22 shows the graph.

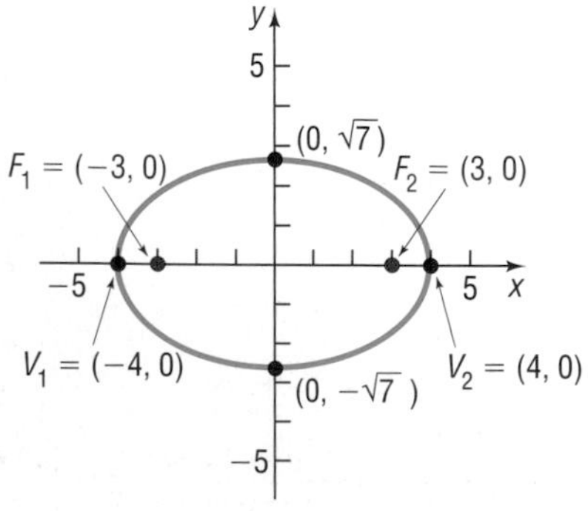

Notice in Figure 22 how we used the intercepts of the equation to graph the ellipse. Following this practice will make it easier for you to obtain an accurate graph of an ellipse when graphing by hand. The intercepts also tell you how to set the initial viewing window when using a graphing utility.

Now Work PROBLEM **27**

EXAMPLE 2   Graphing an Ellipse Using a Graphing Utility

Use a graphing utility to graph the ellipse $\dfrac{x^2}{16} + \dfrac{y^2}{7} = 1$.

Solution   First, we must solve $\dfrac{x^2}{16} + \dfrac{y^2}{7} = 1$ for y.

Figure 23

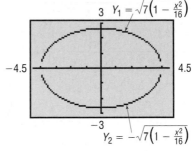

$$\frac{y^2}{7} = 1 - \frac{x^2}{16} \qquad \text{Subtract } \frac{x^2}{16} \text{ from each side.}$$

$$y^2 = 7\left(1 - \frac{x^2}{16}\right) \qquad \text{Multiply both sides by 7.}$$

$$y = \pm\sqrt{7\left(1 - \frac{x^2}{16}\right)} \qquad \text{Apply the Square Root Method.}$$

Figure 23* shows the graphs of $Y_1 = \sqrt{7\left(1 - \dfrac{x^2}{16}\right)}$ and $Y_2 = -\sqrt{7\left(1 - \dfrac{x^2}{16}\right)}$.

Notice in Figure 23 that we used a square screen. As with circles and parabolas, this is done to avoid a distorted view of the graph.

An equation of the form of equation (2), with $a > b$, is the equation of an ellipse with center at the origin, foci on the x-axis at $(-c, 0)$ and $(c, 0)$, where $c^2 = a^2 - b^2$, and major axis along the x-axis.

For the remainder of this section, the direction **"Analyze the equation"** will mean to find the center, major axis, foci, and vertices of the ellipse and graph it.

EXAMPLE 3   Analyzing the Equation of an Ellipse

Analyze the equation: $\dfrac{x^2}{25} + \dfrac{y^2}{9} = 1$

Solution   The given equation is of the form of equation (2), with $a^2 = 25$ and $b^2 = 9$. The equation is that of an ellipse with center $(0, 0)$ and major axis along the x-axis. The vertices are at $(\pm a, 0) = (\pm 5, 0)$. Because $b^2 = a^2 - c^2$, we find that

$$c^2 = a^2 - b^2 = 25 - 9 = 16$$

The foci are at $(\pm c, 0) = (\pm 4, 0)$. The y-intercepts are $(0, \pm b) = (0, \pm 3)$. Figure 24(a) shows the graph drawn by hand. Figure 24(b) shows the graph obtained using a graphing utility.

Figure 24

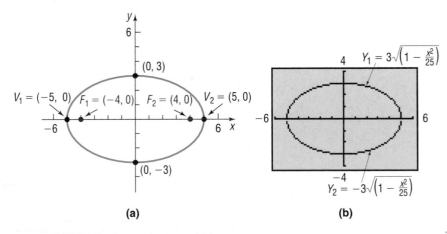

(a)   (b)

Now Work PROBLEM 17

*The initial viewing window selected was $X\text{min} = -4$, $X\text{max} = 4$, $Y\text{min} = -3$, $Y\text{max} = 3$. Then we used the ZOOM-SQUARE option to obtain the window shown.

If the major axis of an ellipse with center at $(0, 0)$ lies on the y-axis, then the foci are at $(0, -c)$ and $(0, c)$. Using the same steps as before, the definition of an ellipse leads to the following result:

THEOREM

Equation of an Ellipse; Center at (0, 0); Major Axis along the y-Axis

An equation of the ellipse with center at $(0, 0)$, foci at $(0, -c)$ and $(0, c)$, and vertices at $(0, -a)$ and $(0, a)$ is

$$\frac{x^2}{b^2} + \frac{y^2}{a^2} = 1 \qquad \text{where } a > b > 0 \text{ and } b^2 = a^2 - c^2 \qquad \textbf{(3)}$$

The major axis is the y-axis.

Figure 25

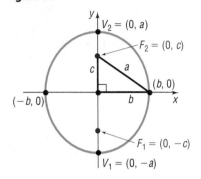

Figure 25 illustrates the graph of such an ellipse. Again, notice the right triangle with the points at $(0, 0)$, $(b, 0)$, and $(0, c)$.

Look closely at equations (2) and (3). Although they may look alike, there is a difference! In equation (2), the larger number, a^2, is in the denominator of the x^2-term, so the major axis of the ellipse is along the x-axis. In equation (3), the larger number, a^2, is in the denominator of the y^2-term, so the major axis is along the y-axis.

EXAMPLE 4

Analyzing the Equation of an Ellipse

Analyze the equation: $9x^2 + y^2 = 9$

Solution

To put the equation in proper form, we divide each side by 9.

$$x^2 + \frac{y^2}{9} = 1$$

The larger number, 9, is in the denominator of the y^2 term so, based on equation (3), this is the equation of an ellipse with center at the origin and major axis along the y-axis. Also, we conclude that $a^2 = 9$, $b^2 = 1$, and $c^2 = a^2 - b^2 = 9 - 1 = 8$. The vertices are at $(0, \pm a) = (0, \pm 3)$, and the foci are at $(0, \pm c) = (0, \pm 2\sqrt{2})$. The x-intercepts are at $(\pm b, 0) = (\pm 1, 0)$. Figure 26(a) shows the graph drawn by hand. Figure 26(b) shows the graph obtained using a graphing utility.

Figure 26

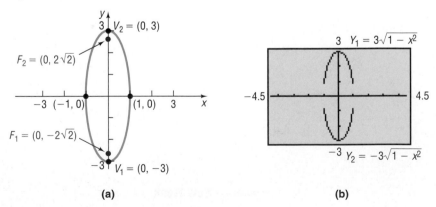

(a)   (b)

Now Work PROBLEM 21

EXAMPLE 5   Finding an Equation of an Ellipse

Find an equation of the ellipse having one focus at $(0, 2)$ and vertices at $(0, -3)$ and $(0, 3)$. Graph the equation by hand.

Solution

Figure 27

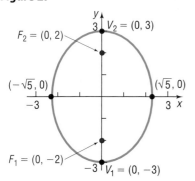

Because the vertices are at $(0, -3)$ and $(0, 3)$, the center of this ellipse is at their midpoint, the origin. The vertices and foci lie on the y-axis, so the major axis lies on the y-axis. The distance from the center, $(0, 0)$, to one of the foci, $(0, 2)$, is $c = 2$. The distance from the center, $(0, 0)$, to one of the vertices, $(0, 3)$, is $a = 3$. So $b^2 = a^2 - c^2 = 9 - 4 = 5$. The form of the equation of this ellipse is given by equation (3).

$$\frac{x^2}{b^2} + \frac{y^2}{a^2} = 1$$

$$\frac{x^2}{5} + \frac{y^2}{9} = 1$$

Figure 27 shows the graph.

━━━ **Now Work** PROBLEM 29

The circle may be considered a special kind of ellipse. To see why, let $a = b$ in equation (2) or (3). Then

$$\frac{x^2}{a^2} + \frac{y^2}{a^2} = 1$$

$$x^2 + y^2 = a^2$$

This is the equation of a circle with center at the origin and radius a. The value of c is

$$c^2 = a^2 - b^2 = 0$$

We conclude that the closer the two foci of an ellipse are to the center the more the ellipse will look like a circle.

2 Analyze Ellipses with Center at (h, k)

If an ellipse with center at the origin and major axis coinciding with a coordinate axis is shifted horizontally h units and then vertically k units, the result is an ellipse with center at (h, k) and major axis parallel to a coordinate axis. The equations of such ellipses have the same forms as those given in equations (2) and (3), except that x is replaced by $x - h$ (the horizontal shift) and y is replaced by $y - k$ (the vertical shift). Table 3 gives the forms of the equations of such ellipses and Figure 28 on the next page shows their graphs.

Table 3

ELLIPSES WITH CENTER AT (h, k) AND MAJOR AXIS PARALLEL TO A COORDINATE AXIS				
Center	Major Axis	Foci	Vertices	Equation
(h, k)	Parallel to x-axis	$(h + c, k)$	$(h + a, k)$	$\dfrac{(x - h)^2}{a^2} + \dfrac{(y - k)^2}{b^2} = 1$,
		$(h - c, k)$	$(h - a, k)$	$a > b$ and $b^2 = a^2 - c^2$
(h, k)	Parallel to y-axis	$(h, k + c)$	$(h, k + a)$	$\dfrac{(x - h)^2}{b^2} + \dfrac{(y - k)^2}{a^2} = 1$,
		$(h, k - c)$	$(h, k - a)$	$a > b$ and $b^2 = a^2 - c^2$

3 Solve Applied Problems Involving Ellipses

Ellipses are found in many applications in science and engineering. For example, the orbits of the planets around the Sun are elliptical, with the Sun's position at a focus. See Figure 32.

Figure 32

Stone and concrete bridges are often shaped as semielliptical arches. Elliptical gears are used in machinery when a variable rate of motion is required.

Ellipses also have an interesting reflection property. If a source of light (or sound) is placed at one focus, the waves transmitted by the source will reflect off the ellipse and concentrate at the other focus. This is the principle behind *whispering galleries*, which are rooms designed with elliptical ceilings. A person standing at one focus of the ellipse can whisper and be heard by a person standing at the other focus, because all the sound waves that reach the ceiling are reflected to the other person.

| EXAMPLE 9 | A Whispering Gallery |

The whispering gallery in the Museum of Science and Industry in Chicago is 47.3 feet long. The distance from the center of the room to the foci is 20.3 feet. Find an equation that describes the shape of the room. How high is the room at its center?

Source: Chicago Museum of Science and Industry Web site; www.msichicago.org

Solution We set up a rectangular coordinate system so that the center of the ellipse is at the origin and the major axis is along the *x*-axis. The equation of the ellipse is

$$\frac{x^2}{a^2} + \frac{y^2}{b^2} = 1$$

Since the length of the room is 47.3 feet, the distance from the center of the room to each vertex (the end of the room) will be $\dfrac{47.3}{2} = 23.65$ feet; so $a = 23.65$ feet. The distance from the center of the room to each focus is $c = 20.3$ feet. See Figure 33.

Since $b^2 = a^2 - c^2$, we find $b^2 = 23.65^2 - 20.3^2 = 147.2325$. An equation that describes the shape of the room is given by

$$\frac{x^2}{23.65^2} + \frac{y^2}{147.2325} = 1$$

The height of the room at its center is $b = \sqrt{147.2325} \approx 12.1$ feet.

Figure 33

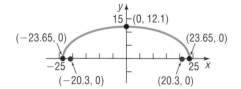

6.3 Assess Your Understanding

'Are You Prepared?' *Answers are given at the end of these exercises. If you get a wrong answer, read the pages listed in red.*

1. The distance d from $P_1 = (2, -5)$ to $P_2 = (4, -2)$ is $d =$ _____. (pp. 4–6)

2. To complete the square of $x^2 - 3x$, add _____. (p. A31)

3. Find the intercepts of the equation $y^2 = 16 - 4x^2$. (pp. 17–18)

4. The point that is symmetric with respect to the y-axis to the point $(-2, 5)$ is _____. (pp. 18–19)

5. To graph $y = (x + 1)^2 - 4$, shift the graph of $y = x^2$ to the (left/right) _____ unit(s) and then (up/down) _____ unit(s). (pp. 67–70)

6. The standard equation of a circle with center at $(2, -3)$ and radius 1 is _____. (pp. 22–23)

Concepts and Vocabulary

7. A(n) _____ is the collection of all points in the plane the sum of whose distances from two fixed points is a constant.

8. For an ellipse, the foci lie on a line called the _____ axis.

9. For the ellipse $\dfrac{x^2}{4} + \dfrac{y^2}{25} = 1$, the vertices are the points _____ and _____.

10. *True or False* The foci, vertices, and center of an ellipse lie on a line called the axis of symmetry.

11. *True or False* If the center of an ellipse is at the origin and the foci lie on the y-axis, the ellipse is symmetric with respect to the x-axis, the y-axis, and the origin.

12. *True or False* A circle is a certain type of ellipse.

Skill Building

In Problems 13–16, the graph of an ellipse is given. Match each graph to its equation.

(A) $\dfrac{x^2}{4} + y^2 = 1$ (B) $x^2 + \dfrac{y^2}{4} = 1$ (C) $\dfrac{x^2}{16} + \dfrac{y^2}{4} = 1$ (D) $\dfrac{x^2}{4} + \dfrac{y^2}{16} = 1$

13.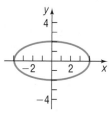

14.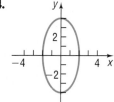

15.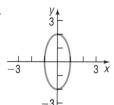

16.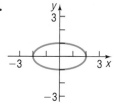

In Problems 17–26, find the vertices and foci of each ellipse. Graph each equation by hand. Verify your graph using a graphing utility.

17. $\dfrac{x^2}{25} + \dfrac{y^2}{4} = 1$

18. $\dfrac{x^2}{9} + \dfrac{y^2}{4} = 1$

19. $\dfrac{x^2}{9} + \dfrac{y^2}{25} = 1$

20. $x^2 + \dfrac{y^2}{16} = 1$

21. $4x^2 + y^2 = 16$

22. $x^2 + 9y^2 = 18$

23. $4y^2 + x^2 = 8$

24. $4y^2 + 9x^2 = 36$

25. $x^2 + y^2 = 16$

26. $x^2 + y^2 = 4$

In Problems 27–38, find an equation for each ellipse. Graph the equation by hand.

27. Center at $(0, 0)$; focus at $(3, 0)$; vertex at $(5, 0)$

28. Center at $(0, 0)$; focus at $(-1, 0)$; vertex at $(3, 0)$

29. Center at $(0, 0)$; focus at $(0, -4)$; vertex at $(0, 5)$

30. Center at $(0, 0)$; focus at $(0, 1)$; vertex at $(0, -2)$

31. Foci at $(\pm 2, 0)$; length of the major axis is 6

32. Foci at $(0, \pm 2)$; length of the major axis is 8

33. Focus at $(-4, 0)$; vertices at $(\pm 5, 0)$

34. Focus at $(0, -4)$; vertices at $(0, \pm 8)$

35. Foci at $(0, \pm 3)$; x-intercepts are ± 2

36. Vertices at $(\pm 4, 0)$; y-intercepts are ± 1

37. Center at $(0, 0)$; vertex at $(0, 4)$; $b = 1$

38. Vertices at $(\pm 5, 0)$; $c = 2$

In Problems 39–42, write an equation for each ellipse.

39.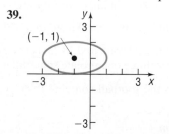

40.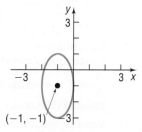

41.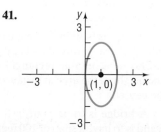

42.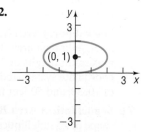

In Problems 43–54, analyze each equation; that is, find the center, foci, and vertices of each ellipse. Graph each equation by hand. Verify your graph using a graphing utility.

43. $\dfrac{(x-3)^2}{4} + \dfrac{(y+1)^2}{9} = 1$

44. $\dfrac{(x+4)^2}{9} + \dfrac{(y+2)^2}{4} = 1$

45. $(x+5)^2 + 4(y-4)^2 = 16$

46. $9(x-3)^2 + (y+2)^2 = 18$

47. $x^2 + 4x + 4y^2 - 8y + 4 = 0$

48. $x^2 + 3y^2 - 12y + 9 = 0$

49. $2x^2 + 3y^2 - 8x + 6y + 5 = 0$

50. $4x^2 + 3y^2 + 8x - 6y = 5$

51. $9x^2 + 4y^2 - 18x + 16y - 11 = 0$

52. $x^2 + 9y^2 + 6x - 18y + 9 = 0$

53. $4x^2 + y^2 + 4y = 0$

54. $9x^2 + y^2 - 18x = 0$

In Problems 55–64, find an equation for each ellipse. Graph the equation by hand.

55. Center at $(2, -2)$; vertex at $(7, -2)$; focus at $(4, -2)$

56. Center at $(-3, 1)$; vertex at $(-3, 3)$; focus at $(-3, 0)$

57. Vertices at $(4, 3)$ and $(4, 9)$; focus at $(4, 8)$

58. Foci at $(1, 2)$ and $(-3, 2)$; vertex at $(-4, 2)$

59. Foci at $(5, 1)$ and $(-1, 1)$; length of the major axis is 8

60. Vertices at $(2, 5)$ and $(2, -1)$; $c = 2$

61. Center at $(1, 2)$; focus at $(4, 2)$; contains the point $(1, 3)$

62. Center at $(1, 2)$; focus at $(1, 4)$; contains the point $(2, 2)$

63. Center at $(1, 2)$; vertex at $(4, 2)$; contains the point $(1, 3)$

64. Center at $(1, 2)$; vertex at $(1, 4)$; contains the point $(2, 2)$

In Problems 65–68, graph each function. Be sure to label all the intercepts.
[Hint: *Notice that each function is half an ellipse.*]

65. $f(x) = \sqrt{16 - 4x^2}$

66. $f(x) = \sqrt{9 - 9x^2}$

67. $f(x) = -\sqrt{64 - 16x^2}$

68. $f(x) = -\sqrt{4 - 4x^2}$

Applications and Extensions

69. Semielliptical Arch Bridge An arch in the shape of the upper half of an ellipse is used to support a bridge that is to span a river 20 meters wide. The center of the arch is 6 meters above the center of the river. See the figure. Write an equation for the ellipse in which the *x*-axis coincides with the water level and the *y*-axis passes through the center of the arch.

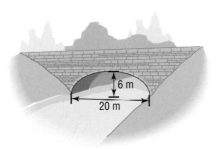

70. Semielliptical Arch Bridge The arch of a bridge is a semi-ellipse with a horizontal major axis. The span is 30 feet, and the top of the arch is 10 feet above the major axis. The roadway is horizontal and is 2 feet above the top of the arch. Find the vertical distance from the roadway to the arch at 5-foot intervals along the roadway.

71. Whispering Gallery A hall 100 feet in length is to be designed as a whispering gallery. If the foci are located 25 feet from the center, how high will the ceiling be at the center?

72. Whispering Gallery Jim, standing at one focus of a whispering gallery, is 6 feet from the nearest wall. His friend is standing at the other focus, 100 feet away. What is the length of this whispering gallery? How high is its elliptical ceiling at the center?

73. Semielliptical Arch Bridge A bridge is built in the shape of a semielliptical arch. The bridge has a span of 120 feet and a maximum height of 25 feet. Choose a suitable rectangular coordinate system and find the height of the arch at distances of 10, 30, and 50 feet from the center.

74. Semielliptical Arch Bridge A bridge is to be built in the shape of a semielliptical arch and is to have a span of 100 feet.

The height of the arch, at a distance of 40 feet from the center, is to be 10 feet. Find the height of the arch at its center.

75. Racetrack Design Consult the figure. A racetrack is in the shape of an ellipse, 100 feet long and 50 feet wide. What is the width 10 feet from a vertex?

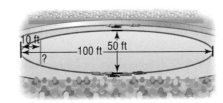

76. Semielliptical Arch Bridge An arch for a bridge over a highway is in the form of half an ellipse. The top of the arch is 20 feet above the ground level (the major axis). The highway has four lanes, each 12 feet wide; a center safety strip 8 feet wide; and two side strips, each 4 feet wide. What should the span of the bridge be (the length of its major axis) if the height 28 feet from the center is to be 13 feet?

77. Installing a Vent Pipe A homeowner is putting in a fireplace that has a 4-inch-radius vent pipe. He needs to cut an elliptical hole in his roof to accommodate the pipe. If the pitch of his roof is $\dfrac{5}{4}$ (a rise of 5, run of 4), what are the dimensions of the hole?

Source: www.pen.k12.va.us

78. Volume of a Football A football is in the shape of a **prolate spheroid,** which is simply a solid obtained by rotating an ellipse $\left(\dfrac{x^2}{a^2} + \dfrac{y^2}{b^2} = 1\right)$ about its major axis. An inflated NFL football averages 11.125 inches in length and 28.25 inches in center circumference. If the volume of a prolate spheroid is $\dfrac{4}{3}\pi ab^2$, how much air does the football contain? (Neglect material thickness.)

Source: www.answerbag.com

In Problems 79–82, use the fact that the orbit of a planet about the Sun is an ellipse, with the Sun at one focus. The **aphelion** of a planet is its greatest distance from the Sun, and the **perihelion** is its shortest distance. The **mean distance** of a planet from the Sun is the length of the semimajor axis of the elliptical orbit. See the illustration.

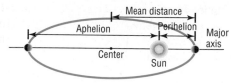

79. Earth The mean distance of Earth from the Sun is 93 million miles. If the aphelion of Earth is 94.5 million miles, what is the perihelion? Write an equation for the orbit of Earth around the Sun.

80. Mars The mean distance of Mars from the Sun is 142 million miles. If the perihelion of Mars is 128.5 million miles, what is the aphelion? Write an equation for the orbit of Mars about the Sun.

81. Jupiter The aphelion of Jupiter is 507 million miles. If the distance from the center of its elliptical orbit to the Sun is 23.2 million miles, what is the perihelion? What is the mean distance? Write an equation for the orbit of Jupiter around the Sun.

82. Pluto The perihelion of Pluto is 4551 million miles, and the distance from the center of its elliptical orbit to the Sun is 897.5 million miles. Find the aphelion of Pluto. What is the mean distance of Pluto from the Sun? Write an equation for the orbit of Pluto about the Sun.

83. Show that an equation of the form
$$Ax^2 + Cy^2 + F = 0 \qquad A \neq 0, C \neq 0, F \neq 0$$
where A and C are of the same sign and F is of opposite sign,
(a) Is the equation of an ellipse with center at $(0, 0)$ if $A \neq C$.
(b) Is the equation of a circle with center $(0, 0)$ if $A = C$.

84. Show that the graph of an equation of the form
$$Ax^2 + Cy^2 + Dx + Ey + F = 0 \qquad A \neq 0, C \neq 0$$
where A and C are of the same sign,
(a) Is an ellipse if $\dfrac{D^2}{4A} + \dfrac{E^2}{4C} - F$ is the same sign as A.
(b) Is a point if $\dfrac{D^2}{4A} + \dfrac{E^2}{4C} - F = 0$.
(c) Contains no points if $\dfrac{D^2}{4A} + \dfrac{E^2}{4C} - F$ is of opposite sign to A.

Discussion and Writing

85. The **eccentricity** e of an ellipse is defined as the number $\dfrac{c}{a}$, where a and c are the numbers given in equation (2). Because $a > c$, it follows that $e < 1$. Write a brief paragraph about the general shape of each of the following ellipses. Be sure to justify your conclusions.
(a) Eccentricity close to 0   (b) Eccentricity = 0.5   (c) Eccentricity close to 1

'Are You Prepared?' Answers

1. $\sqrt{13}$ **2.** $\dfrac{9}{4}$ **3.** $(-2, 0), (2, 0), (0, -4), (0, 4)$ **4.** $(2, 5)$ **5.** left; 1; down: 4   **6.** $(x - 2)^2 + (y + 3)^2 = 1$

6.4 The Hyperbola

PREPARING FOR THIS SECTION *Before getting started, review the following:*

- Distance Formula (Section 1.1, p. 5)
- Completing the Square (Appendix, Section A.4, p. A31)
- Intercepts (Section 1.2, pp. 17–18)
- Symmetry (Section 1.2, pp. 18–19)
- Asymptotes (Section 1.7, pp. 76–77)
- Graphing Techniques: Transformations (Section 1.7, pp. 67–76)
- Square Root Method (Appendix, Section A.4, pp. A30–A31)

Now Work the 'Are You Prepared?' problems on page 428.

OBJECTIVES 1 Analyze Hyperbolas with Center at the Origin (p. 418)
2 Find the Asymptotes of a Hyperbola (p. 423)
3 Analyze Hyperbolas with Center at (h, k) (p. 424)
4 Solve Applied Problems Involving Hyperbolas (p. 426)

DEFINITION A **hyperbola** is the set of all points P in the plane, the difference of whose distances from two fixed points, called the **foci**, is a constant.

Figure 34

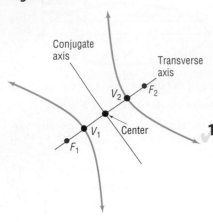

Figure 34 illustrates a hyperbola with foci F_1 and F_2. The line containing the foci is called the **transverse axis.** The midpoint of the line segment joining the foci is the **center** of the hyperbola. The line through the center and perpendicular to the transverse axis is the **conjugate axis.** The hyperbola consists of two separate curves, called **branches,** that are symmetric with respect to the transverse axis, conjugate axis, and center. The two points of intersection of the hyperbola and the transverse axis are the **vertices,** V_1 and V_2, of the hyperbola.

1 Analyze Hyperbolas with Center at the Origin

With these ideas in mind, we are now ready to find the equation of a hyperbola in the rectangular coordinate system. First, we place the center at the origin. Next, we position the hyperbola so that its transverse axis coincides with a coordinate axis. Suppose that the transverse axis coincides with the x-axis, as shown in Figure 35.

If c is the distance from the center to a focus, one focus will be at $F_1 = (-c, 0)$ and the other at $F_2 = (c, 0)$. Now we let the constant difference of the distances from any point $P = (x, y)$ on the hyperbola to the foci F_1 and F_2 be denoted by $\pm 2a$. (If P is on the right branch, the $+$ sign is used; if P is on the left branch, the $-$ sign is used.) The coordinates of P must satisfy the equation

Figure 35
$d(F_1, P) - d(F_2, P) = \pm 2a$

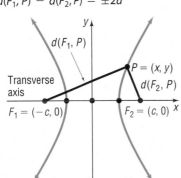

$$d(F_1, P) - d(F_2, P) = \pm 2a \qquad \text{Difference of the distances from } P \text{ to the foci equals } \pm 2a.$$

$$\sqrt{(x+c)^2 + y^2} - \sqrt{(x-c)^2 + y^2} = \pm 2a \qquad \text{Use the Distance Formula.}$$

$$\sqrt{(x+c)^2 + y^2} = \pm 2a + \sqrt{(x-c)^2 + y^2} \qquad \text{Isolate one radical.}$$

$$(x+c)^2 + y^2 = 4a^2 \pm 4a\sqrt{(x-c)^2 + y^2} \qquad \text{Square both sides.}$$
$$+ (x-c)^2 + y^2$$

Next we remove the parentheses by expanding the binomials.

$$x^2 + 2cx + c^2 + y^2 = 4a^2 \pm 4a\sqrt{(x-c)^2 + y^2} + x^2 - 2cx + c^2 + y^2$$

$$4cx - 4a^2 = \pm 4a\sqrt{(x-c)^2 + y^2} \qquad \text{Simplify; isolate the radical.}$$

$$cx - a^2 = \pm a\sqrt{(x-c)^2 + y^2} \qquad \text{Divide each side by 4.}$$

$$(cx - a^2)^2 = a^2[(x-c)^2 + y^2] \qquad \text{Square both sides.}$$

$$c^2x^2 - 2ca^2x + a^4 = a^2(x^2 - 2cx + c^2 + y^2) \qquad \text{Simplify.}$$

$$c^2x^2 + a^4 = a^2x^2 + a^2c^2 + a^2y^2 \qquad \text{Distribute and simplify.}$$

$$(c^2 - a^2)x^2 - a^2y^2 = a^2c^2 - a^4 \qquad \text{Rearrange terms.}$$

$$(c^2 - a^2)x^2 - a^2y^2 = a^2(c^2 - a^2) \qquad \text{Factor } a^2 \text{ on the right side.} \quad \textbf{(1)}$$

To obtain points on the hyperbola off the x-axis, it must be that $a < c$. To see why, look again at Figure 35.

$$d(F_1, P) < d(F_2, P) + d(F_1, F_2) \quad \text{Use triangle } F_1PF_2.$$

$$d(F_1, P) - d(F_2, P) < d(F_1, F_2)$$

$$2a < 2c \qquad \qquad \qquad P \text{ is on the right branch, so}$$
$$d(F_1, P) - d(F_2, P) = 2a;$$
$$d(F_1, F_2) = 2c.$$

$$a < c$$

Since $a < c$, we also have $a^2 < c^2$, so $c^2 - a^2 > 0$. Let $b^2 = c^2 - a^2, b > 0$. Then equation (1) can be written as

$$b^2 x^2 - a^2 y^2 = a^2 b^2$$

$$\frac{x^2}{a^2} - \frac{y^2}{b^2} = 1 \qquad \text{Divide each side by } a^2 b^2.$$

To find the vertices of the hyperbola defined by this equation, let $y = 0$. The vertices satisfy the equation $\dfrac{x^2}{a^2} = 1$, the solutions of which are $x = \pm a$. Consequently, the vertices of the hyperbola are $V_1 = (-a, 0)$ and $V_2 = (a, 0)$. Notice that the distance from the center $(0, 0)$ to either vertex is a.

THEOREM

Equation of a Hyperbola: Center at (0, 0); Transverse Axis along the x-Axis

An equation of the hyperbola with center at $(0, 0)$, foci at $(-c, 0)$ and $(c, 0)$, and vertices at $(-a, 0)$ and $(a, 0)$ is

$$\frac{x^2}{a^2} - \frac{y^2}{b^2} = 1 \qquad \text{where } b^2 = c^2 - a^2 \qquad \qquad \textbf{(2)}$$

The transverse axis is the x-axis.

Figure 36

$\dfrac{x^2}{a^2} - \dfrac{y^2}{b^2} = 1, \quad b^2 = c^2 - a^2$

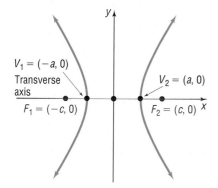

See Figure 36. As you can verify, the hyperbola defined by equation (2) is symmetric with respect to the x-axis, y-axis, and origin. To find the y-intercepts, if any, let $x = 0$ in equation (2). This results in the equation $\dfrac{y^2}{b^2} = -1$, which has no real solution. We conclude that the hyperbola defined by equation (2) has no y-intercepts. In fact, since $\dfrac{x^2}{a^2} - 1 = \dfrac{y^2}{b^2} \geq 0$, it follows that $\dfrac{x^2}{a^2} \geq 1$. There are no points on the graph for $-a < x < a$.

EXAMPLE 1

Finding and Graphing an Equation of a Hyperbola

Find an equation of the hyperbola with center at the origin, one focus at $(3, 0)$, and one vertex at $(-2, 0)$. Graph the equation.

Solution

The hyperbola has its center at the origin. Because the focus and vertex lie on the x-axis, the transverse axis coincides with the x-axis. One focus is at $(c, 0) = (3, 0)$, so $c = 3$. One vertex is at $(-a, 0) = (-2, 0)$, so $a = 2$. From equation (2), it follows that $b^2 = c^2 - a^2 = 9 - 4 = 5$, so an equation of the hyperbola is

$$\frac{x^2}{4} - \frac{y^2}{5} = 1$$

To graph a hyperbola, it is helpful to locate and plot other points on the graph. For example, to find the points above and below the foci, we let $x = \pm 3$. Then

$$\frac{x^2}{4} - \frac{y^2}{5} = 1$$

$$\frac{(\pm 3)^2}{4} - \frac{y^2}{5} = 1 \qquad x = \pm 3$$

$$\frac{9}{4} - \frac{y^2}{5} = 1$$

Figure 37

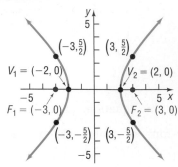

$$\frac{y^2}{5} = \frac{5}{4}$$

$$y^2 = \frac{25}{4}$$

$$y = \pm\frac{5}{2}$$

The points above and below the foci are $\left(\pm 3, \dfrac{5}{2}\right)$ and $\left(\pm 3, -\dfrac{5}{2}\right)$. These points determine the "opening" of the hyperbola. See Figure 37.

━━━ **Now Work** PROBLEM 17

| EXAMPLE 2 | Using a Graphing Utility to Graph a Hyperbola |

Using a graphing utility, graph the hyperbola: $\dfrac{x^2}{4} - \dfrac{y^2}{5} = 1$

Solution

Figure 38

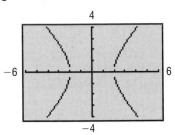

To graph the hyperbola $\dfrac{x^2}{4} - \dfrac{y^2}{5} = 1$, we need to graph the two functions $Y_1 = \sqrt{5}\sqrt{\dfrac{x^2}{4} - 1}$ and $Y_2 = -\sqrt{5}\sqrt{\dfrac{x^2}{4} - 1}$. As with graphing circles, parabolas, and ellipses on a graphing utility, we use a square screen setting so that the graph is not distorted. Figure 38 shows the graph of the hyperbola.

━━━

An equation of the form of equation (2) is the equation of a hyperbola with center at the origin, foci on the x-axis at $(-c, 0)$ and $(c, 0)$, where $c^2 = a^2 + b^2$, and transverse axis along the x-axis.

For the next two examples of this section, the direction **"Analyze the equation"** will mean to find the center, transverse axis, vertices, and foci of the hyperbola and graph it.

| EXAMPLE 3 | Analyzing the Equation of a Hyperbola |

Analyze the equation: $\dfrac{x^2}{16} - \dfrac{y^2}{4} = 1$

Solution

The given equation is of the form of equation (2), with $a^2 = 16$ and $b^2 = 4$. The graph of the equation is a hyperbola with center at $(0, 0)$ and transverse axis along the x-axis. Also, we know that $c^2 = a^2 + b^2 = 16 + 4 = 20$. The vertices are at $(\pm a, 0) = (\pm 4, 0)$, and the foci are at $(\pm c, 0) = (\pm 2\sqrt{5}, 0)$.

To locate the points on the graph above and below the foci, we let $x = \pm 2\sqrt{5}$. Then

$$\frac{x^2}{16} - \frac{y^2}{4} = 1$$

$$\frac{\left(\pm 2\sqrt{5}\right)^2}{16} - \frac{y^2}{4} = 1 \qquad x = \pm 2\sqrt{5}$$

$$\frac{20}{16} - \frac{y^2}{4} = 1$$

$$\frac{5}{4} - \frac{y^2}{4} = 1$$

$$\frac{y^2}{4} = \frac{1}{4}$$

$$y = \pm 1$$

The points above and below the foci are $(\pm 2\sqrt{5}, 1)$ and $(\pm 2\sqrt{5}, -1)$. See Figure 39(a) for the graph drawn by hand. Figure 39(b) shows the graph obtained using a graphing utility, where $Y_1 = \sqrt{4\left(\dfrac{x^2}{16} - 1\right)}$ and $Y_2 = -\sqrt{4\left(\dfrac{x^2}{16} - 1\right)}$.

Figure 39

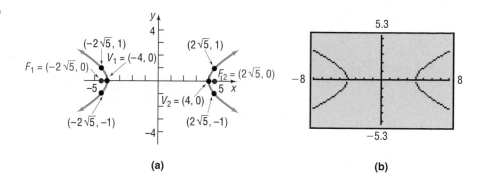

(a)   (b)

The next result gives the form of the equation of a hyperbola with center at the origin and transverse axis along the y-axis.

THEOREM

Equation of a Hyperbola; Center at (0, 0); Transverse Axis along the y-Axis

An equation of the hyperbola with center at $(0, 0)$, foci at $(0, -c)$ and $(0, c)$, and vertices at $(0, -a)$ and $(0, a)$ is

$$\frac{y^2}{a^2} - \frac{x^2}{b^2} = 1 \qquad \text{where } b^2 = c^2 - a^2 \qquad \textbf{(3)}$$

The transverse axis is the y-axis.

Figure 40

$\dfrac{y^2}{a^2} - \dfrac{x^2}{b^2} = 1, b^2 = c^2 - a^2$

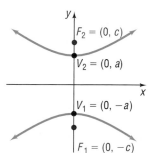

Figure 40 shows the graph of a typical hyperbola defined by equation (3).

An equation of the form of equation (2), $\dfrac{x^2}{a^2} - \dfrac{y^2}{b^2} = 1$, is the equation of a hyperbola with center at the origin, foci on the x-axis at $(-c, 0)$ and $(c, 0)$, where $c^2 = a^2 + b^2$, and transverse axis along the x-axis.

An equation of the form of equation (3), $\dfrac{y^2}{a^2} - \dfrac{x^2}{b^2} = 1$, is the equation of a hyperbola with center at the origin, foci on the y-axis at $(0, -c)$ and $(0, c)$, where $c^2 = a^2 + b^2$, and transverse axis along the y-axis.

Notice the difference in the forms of equations (2) and (3). When the y^2-term is subtracted from the x^2-term, the transverse axis is along the x-axis. When the x^2-term is subtracted from the y^2-term, the transverse axis is along the y-axis.

EXAMPLE 4

Analyzing the Equation of a Hyperbola

Analyze the equation: $y^2 - 4x^2 = 4$

Solution

To put the equation in proper form, we divide each side by 4:

$$\frac{y^2}{4} - x^2 = 1$$

Since the x^2-term is subtracted from the y^2-term, the equation is that of a hyperbola with center at the origin and transverse axis along the y-axis. Also, comparing the above equation to equation (3), we find $a^2 = 4, b^2 = 1$, and $c^2 = a^2 + b^2 = 5$. The vertices are at $(0, \pm a) = (0, \pm 2)$, and the foci are at $(0, \pm c) = (0, \pm\sqrt{5})$.

To locate other points on the graph, we let $x = \pm 2$. Then

$$y^2 - 4x^2 = 4$$

$$y^2 - 4(\pm 2)^2 = 4 \qquad x = \pm 2$$

$$y^2 - 16 = 4$$

$$y^2 = 20$$

$$y = \pm 2\sqrt{5}$$

Four other points on the graph are $(\pm 2, 2\sqrt{5})$ and $(\pm 2, -2\sqrt{5})$. See Figure 41(a) for the graph drawn by hand. Figure 41(b) shows the graph obtained using a graphing utility.

Figure 41

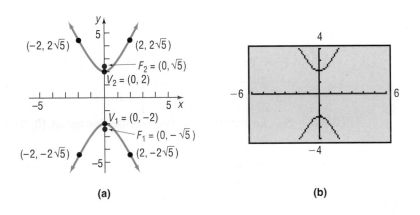

(a)

(b)

EXAMPLE 5	**Finding an Equation of a Hyperbola**

Find an equation of the hyperbola having one vertex at $(0, 2)$ and foci at $(0, -3)$ and $(0, 3)$. Graph the equation.

Solution

Since the foci are at $(0, -3)$ and $(0, 3)$, the center of the hyperbola, which is at their midpoint, is the origin. Because the foci lie on the y-axis, the transverse axis is along the y-axis. The given information also reveals that $c = 3, a = 2$, and $b^2 = c^2 - a^2 = 9 - 4 = 5$. The form of the equation of the hyperbola is given by equation (3):

$$\frac{y^2}{a^2} - \frac{x^2}{b^2} = 1$$

$$\frac{y^2}{4} - \frac{x^2}{5} = 1$$

Figure 42

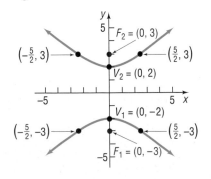

Let $y = \pm 3$ to obtain points on the graph on either side of the foci. See Figure 42.

- **Now Work** PROBLEM 19

Look at the equations of the hyperbolas in Examples 3 and 5. For the hyperbola in Example 3, $a^2 = 16$ and $b^2 = 4$, so $a > b$; for the hyperbola in Example 5, $a^2 = 4$ and $b^2 = 5$, so $a < b$. We conclude that, for hyperbolas, there are no requirements involving the relative sizes of a and b. Contrast this situation to the case of an ellipse, in which the relative sizes of a and b dictate which axis is the major axis. Hyperbolas have another feature to distinguish them from ellipses and parabolas: Hyperbolas have asymptotes.

2 Find the Asymptotes of a Hyperbola

Recall from Section 1.7 that a horizontal or oblique asymptote of a graph is a line with the property that the distance from the line to points on the graph approaches 0 as $x \to -\infty$ or as $x \to \infty$. The asymptotes provide information about the end behavior of the graph of a hyperbola.

THEOREM

Asymptotes of a Hyperbola

The hyperbola $\dfrac{x^2}{a^2} - \dfrac{y^2}{b^2} = 1$ has the two oblique asymptotes

$$y = \frac{b}{a}x \quad \text{and} \quad y = -\frac{b}{a}x \tag{4}$$

Proof   We begin by solving for y in the equation of the hyperbola.

$$\frac{x^2}{a^2} - \frac{y^2}{b^2} = 1$$

$$\frac{y^2}{b^2} = \frac{x^2}{a^2} - 1$$

$$y^2 = b^2\left(\frac{x^2}{a^2} - 1\right)$$

Since $x \neq 0$, we can rearrange the right side in the form

$$y^2 = \frac{b^2 x^2}{a^2}\left(1 - \frac{a^2}{x^2}\right)$$

$$y = \pm\frac{bx}{a}\sqrt{1 - \frac{a^2}{x^2}}$$

Now, as $x \to -\infty$ or as $x \to \infty$, the term $\dfrac{a^2}{x^2}$ approaches 0, so the expression under the radical approaches 1. So, as $x \to -\infty$ or as $x \to \infty$, the value of y approaches $\pm\dfrac{bx}{a}$; that is, the graph of the hyperbola approaches the lines

$$y = -\frac{b}{a}x \quad \text{and} \quad y = \frac{b}{a}x$$

These lines are oblique asymptotes of the hyperbola.   ■

The asymptotes of a hyperbola are not part of the hyperbola, but they do serve as a guide for graphing a hyperbola. For example, suppose that we want to graph the equation

$$\frac{x^2}{a^2} - \frac{y^2}{b^2} = 1$$

We begin by plotting the vertices $(-a, 0)$ and $(a, 0)$. Then we plot the points $(0, -b)$ and $(0, b)$ and use these four points to construct a rectangle, as shown in Figure 43. The diagonals of this rectangle have slopes $\dfrac{b}{a}$ and $-\dfrac{b}{a}$, and their extensions are the asymptotes $y = \dfrac{b}{a}x$ and $y = -\dfrac{b}{a}x$ of the hyperbola. If we graph the asymptotes, we can use them to establish the "opening" of the hyperbola and avoid plotting other points.

Figure 43

$\dfrac{x^2}{a^2} - \dfrac{y^2}{b^2} = 1$

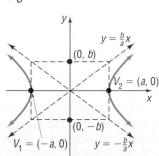

THEOREM   **Asymptotes of a Hyperbola**

The hyperbola $\dfrac{y^2}{a^2} - \dfrac{x^2}{b^2} = 1$ has the two oblique asymptotes:

$$y = \frac{a}{b}x \quad \text{and} \quad y = -\frac{a}{b}x \qquad \text{(5)}$$

You are asked to prove this result in Problem 82.

For the remainder of this section, the direction **"Analyze the equation"** will mean to find the center, transverse axis, vertices, foci, and asymptotes of the hyperbola and graph it.

EXAMPLE 6   Analyzing the Equation of a Hyperbola

Analyze the equation: $9x^2 - 4y^2 = 36$

Solution   Divide each side of the equation by 36 to put the equation in proper form.

$$\frac{x^2}{4} - \frac{y^2}{9} = 1$$

We now proceed to analyze the equation. The center of the hyperbola is the origin. Since the x^2-term is first in the equation, we know that the transverse axis is along the x-axis and the vertices and foci will lie on the x-axis. Using equation (2), we find $a^2 = 4, b^2 = 9$, and $c^2 = a^2 + b^2 = 13$. The vertices are $a = 2$ units left and right of the center at $(\pm a, 0) = (\pm 2, 0)$, the foci are $c = \sqrt{13}$ units left and right of the center at $(\pm c, 0) = (\pm\sqrt{13}, 0)$, and the asymptotes have the equations

$$y = \frac{b}{a}x = \frac{3}{2}x \quad \text{and} \quad y = -\frac{b}{a}x = -\frac{3}{2}x$$

To graph the hyperbola by hand, form the rectangle containing the points $(\pm a, 0)$ and $(0, \pm b)$, that is, $(-2, 0)$, $(2, 0)$, $(0, -3)$, and $(0, 3)$. The extensions of the diagonals of this rectangle are the asymptotes. See Figure 44(a) for the graph drawn by hand. Figure 44(b) shows the graph obtained using a graphing utility.

Figure 44

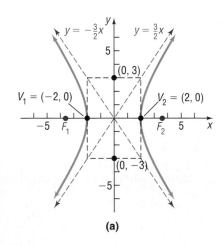

(a)

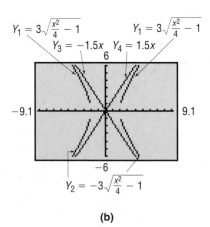

(b)

Seeing the Concept

Refer to Figure 44(b). Create a TABLE using Y_1 and Y_4 with $x = 10, 100, 1000$, and 10,000. Compare the values of Y_1 and Y_4. Repeat for Y_1 and Y_3, Y_2 and Y_3, and Y_2 and Y_4.

— **Now Work** PROBLEM 29

---

3 Analyze Hyperbolas with Center at (h, k)

If a hyperbola with center at the origin and transverse axis coinciding with a coordinate axis is shifted horizontally h units and then vertically k units, the result is a hyperbola with center at (h, k) and transverse axis parallel to a coordinate axis. The equations of such hyperbolas have the same forms as those given in equations (2) and (3), except that x is replaced by $x - h$ (the horizontal shift) and y is replaced by $y - k$ (the vertical shift). Table 4 gives the forms of the equations of such hyperbolas. See Figure 45 for typical graphs.

Table 4

HYPERBOLAS WITH CENTER AT (h, k) AND TRANSVERSE AXIS PARALLEL TO A COORDINATE AXIS					
Center	Transverse Axis	Foci	Vertices	Equation	Asymptotes
(h, k)	Parallel to the x-axis	$(h \pm c, k)$	$(h \pm a, k)$	$\dfrac{(x-h)^2}{a^2} - \dfrac{(y-k)^2}{b^2} = 1, \quad b^2 = c^2 - a^2$	$y - k = \pm\dfrac{b}{a}(x - h)$
(h, k)	Parallel to the y-axis	$(h, k \pm c)$	$(h, k \pm a)$	$\dfrac{(y-k)^2}{a^2} - \dfrac{(x-h)^2}{b^2} = 1, \quad b^2 = c^2 - a^2$	$y - k = \pm\dfrac{a}{b}(x - h)$

Figure 45

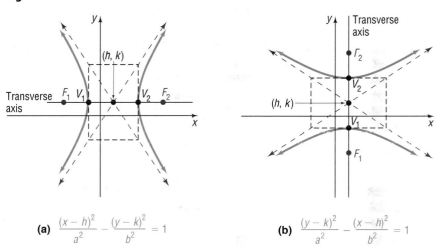

(a) $\dfrac{(x-h)^2}{a^2} - \dfrac{(y-k)^2}{b^2} = 1$

(b) $\dfrac{(y-k)^2}{a^2} - \dfrac{(x-h)^2}{b^2} = 1$

EXAMPLE 7 Finding an Equation of a Hyperbola, Center Not at the Origin

Find an equation for the hyperbola with center at $(1, -2)$, one focus at $(4, -2)$, and one vertex at $(3, -2)$. Graph the equation by hand.

Figure 46

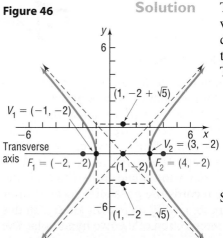

Solution The center is at $(h, k) = (1, -2)$, so $h = 1$ and $k = -2$. Since the center, focus, and vertex all lie on the line $y = -2$, the transverse axis is parallel to the x-axis. The distance from the center $(1, -2)$ to the focus $(4, -2)$ is $c = 3$; the distance from the center $(1, -2)$ to the vertex $(3, -2)$ is $a = 2$. Thus, $b^2 = c^2 - a^2 = 9 - 4 = 5$. The equation is

$$\frac{(x-h)^2}{a^2} - \frac{(y-k)^2}{b^2} = 1$$

$$\frac{(x-1)^2}{4} - \frac{(y+2)^2}{5} = 1$$

See Figure 46.

Now Work PROBLEM 39

| EXAMPLE 8 | Analyzing the Equation of a Hyperbola |

Analyze the equation: $-x^2 + 4y^2 - 2x - 16y + 11 = 0$

Solution   We complete the squares in x and in y.

$$-x^2 + 4y^2 - 2x - 16y + 11 = 0$$

$$-(x^2 + 2x) + 4(y^2 - 4y) = -11 \qquad \text{Group terms.}$$

$$-(x^2 + 2x + 1) + 4(y^2 - 4y + 4) = -11 - 1 + 16 \qquad \text{Complete each square.}$$

$$-(x + 1)^2 + 4(y - 2)^2 = 4$$

$$(y - 2)^2 - \frac{(x + 1)^2}{4} = 1 \qquad \text{Divide each side by 4.}$$

This is the equation of a hyperbola with center at $(-1, 2)$ and transverse axis parallel to the y-axis. Also, $a^2 = 1$ and $b^2 = 4$, so $c^2 = a^2 + b^2 = 5$. Since the transverse axis is parallel to the y-axis, the vertices and foci are located a and c units above and below the center, respectively. The vertices are at $(h, k \pm a) = (-1, 2 \pm 1)$, or $(-1, 1)$ and $(-1, 3)$. The foci are at $(h, k \pm c) = \left(-1, 2 \pm \sqrt{5}\right)$. The asymptotes are $y - 2 = \frac{1}{2}(x + 1)$ and $y - 2 = -\frac{1}{2}(x + 1)$. Figure 47(a) shows the graph drawn by hand. Figure 47(b) shows the graph obtained using a graphing utility.

Figure 47

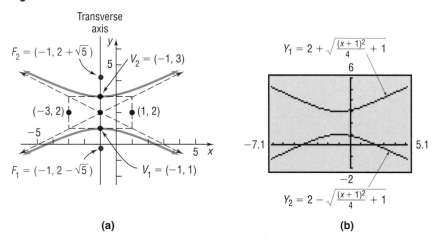

(a)   (b)

— **Now Work**   PROBLEM 53

4 Solve Applied Problems Involving Hyperbolas

Figure 48

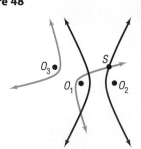

Look at Figure 48. Suppose that three microphones are located at points O_1, O_2, and O_3 (the foci of the two hyperbolas). In addition, suppose that a gun is fired at S and the microphone at O_1 records the gun shot 1 second after the microphone at O_2. Because sound travels at about 1100 feet per second, we conclude that the microphone at O_1 is 1100 feet farther from the gunshot than O_2. We can model this situation by saying that S lies on a branch of a hyperbola with foci at O_1 and O_2. (Do you see why? The difference of the distances from S to O_1 and from S to O_2 is the constant 1100.) If the third microphone at O_3 records the gunshot 2 seconds after O_1, then S will lie on a branch of a second hyperbola with foci at O_1 and O_3. In this case, the constant difference will be 2200. The intersection of the two hyperbolas will identify the location of S.

EXAMPLE 9	Lightning Strikes

Suppose that two people standing 1 mile apart both see a flash of lightning. After a period of time, the person standing at point A hears the thunder. One second later, the person standing at point B hears the thunder. If the person at B is due west of the person at A and the lightning strike is known to occur due north of the person standing at point A, where did the lightning strike?

Solution   See Figure 49 in which the ordered pair (x, y) represents the location of the lightning strike. We know that sound travels at 1100 feet per second, so the person at point A is 1100 feet closer to the lightning strike than the person at point B. Since the difference of the distance from (x, y) to A and the distance from (x, y) to B is the constant 1100, the point (x, y) lies on a hyperbola whose foci are at A and B.

Figure 49

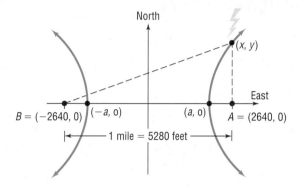

An equation of the hyperbola is

$$\frac{x^2}{a^2} - \frac{y^2}{b^2} = 1$$

where $2a = 1100$, so $a = 550$.

Because the distance between the two people is 1 mile (5280 feet) and each person is at a focus of the hyperbola, we have

$$2c = 5280$$

$$c = \frac{5280}{2} = 2640$$

Since $b^2 = c^2 - a^2 = 2640^2 - 550^2 = 6{,}667{,}100$, the equation of the hyperbola that describes the location of the lightning strike is

$$\frac{x^2}{550^2} - \frac{y^2}{6{,}667{,}100} = 1$$

Refer again to Figure 49. Since the lightning strikes due north of the individual at the point $A = (2640, 0)$, we let $x = 2640$ and solve the resulting equation.

$$\frac{2640^2}{550^2} - \frac{y^2}{6{,}667{,}100} = 1$$

$$-\frac{y^2}{6{,}667{,}100} = -22.04 \qquad \text{Subtract } \frac{2640^2}{550^2} \text{ from both sides.}$$

$$y^2 = 146{,}942{,}884 \qquad \text{Multiply both sides by } -6{,}667{,}100.$$

$$y = 12{,}122 \qquad \text{$y > 0$ since the lightning strike occured in quadrant I.}$$

✓Check: The difference between the distance from $(2640, 12{,}122)$ to the person at the point $B = (-2640, 0)$ and the distance from $(2640, 12{,}122)$ to the person at the point $A = (2640, 0)$ should be 1100. Using the distance formula, we find the difference in the distances is

$$\sqrt{[2640 - (-2640)]^2 + (12{,}122 - 0)^2} - \sqrt{(2640 - 2640)^2 + (12{,}122 - 0)^2} = 1100$$

as required.

The lightning strike is 12,122 feet north of the person standing at point A.

Now Work PROBLEM 73

6.4 Assess Your Understanding

'Are You Prepared?' *Answers are given at the end of these exercises. If you get a wrong answer, read the pages listed in red.*

1. The distance d from $P_1 = (3, -4)$ to $P_2 = (-2, 1)$ is $d =$ _____. (pp. 4–6)

2. To complete the square of $x^2 + 5x$, add _____. (p. A31)

3. Find the intercepts of the equation $y^2 = 9 + 4x^2$. (pp. 17–18)

4. **True or False** The equation $y^2 = 9 + x^2$ is symmetric with respect to the x-axis, the y-axis, and the origin. (pp. 18–19)

5. To graph $y = (x - 5)^3 - 4$, shift the graph of $y = x^3$ to the (left/right) _____ unit(s) and then (up/down) _____ unit(s). (pp. 67–70)

6. Find the vertical asymptotes, if any, and the horizontal or oblique asymptotes, if any, of $y = \dfrac{x^2 - 9}{x^2 - 4}$. (pp. 76–77)

Concepts and Vocabulary

7. A(n) _____ is the collection of points in the plane the difference of whose distances from two fixed points is a constant.

8. For a hyperbola, the foci lie on a line called the _____ _____.

9. The asymptotes of the hyperbola $\dfrac{x^2}{4} - \dfrac{y^2}{9} = 1$ are _____ and _____.

10. **True or False** The foci of a hyperbola lie on a line called the axis of symmetry.

11. **True or False** Hyperbolas always have asymptotes.

12. **True or False** A hyperbola will never intersect its transverse axis.

Skill Building

In Problems 13–16, the graph of a hyperbola is given. Match each graph to its equation.

(A) $\dfrac{x^2}{4} - y^2 = 1$ (B) $x^2 - \dfrac{y^2}{4} = 1$ (C) $\dfrac{y^2}{4} - x^2 = 1$ (D) $y^2 - \dfrac{x^2}{4} = 1$

13.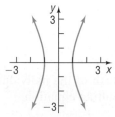

14.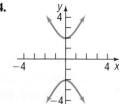

15.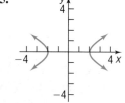

16.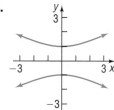

In Problems 17–26, find an equation for the hyperbola described. Graph the equation by hand.

17. Center at $(0, 0)$; focus at $(3, 0)$; vertex at $(1, 0)$

18. Center at $(0, 0)$; focus at $(0, 5)$; vertex at $(0, 3)$

19. Center at $(0, 0)$; focus at $(0, -6)$; vertex at $(0, 4)$

20. Center at $(0, 0)$; focus at $(-3, 0)$; vertex at $(2, 0)$

21. Foci at $(-5, 0)$ and $(5, 0)$; vertex at $(3, 0)$

22. Focus at $(0, 6)$; vertices at $(0, -2)$ and $(0, 2)$

23. Vertices at $(0, -6)$ and $(0, 6)$; asymptote the line $y = 2x$

24. Vertices at $(-4, 0)$ and $(4, 0)$; asymptote the line $y = 2x$

25. Foci at $(-4, 0)$ and $(4, 0)$; asymptote the line $y = -x$

26. Foci at $(0, -2)$ and $(0, 2)$; asymptote the line $y = -x$

In Problems 27–34, find the center, transverse axis, vertices, foci, and asymptotes. Graph each equation by hand. Verify your graph using a graphing utility.

27. $\dfrac{x^2}{25} - \dfrac{y^2}{9} = 1$

28. $\dfrac{y^2}{16} - \dfrac{x^2}{4} = 1$

29. $4x^2 - y^2 = 16$

30. $4y^2 - x^2 = 16$

31. $y^2 - 9x^2 = 9$

32. $x^2 - y^2 = 4$

33. $y^2 - x^2 = 25$

34. $2x^2 - y^2 = 4$

In Problems 35–38, write an equation for each hyperbola.

35.
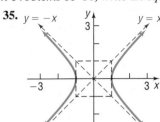

36.
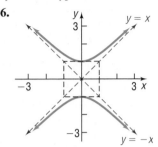

37.
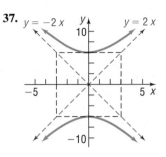

38.
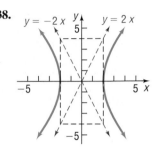

In Problems 39–46, find an equation for the hyperbola described. Graph each equation by hand.

39. Center at $(4, -1)$;   focus at $(7, -1)$;   vertex at $(6, -1)$

40. Center at $(-3, 1)$;   focus at $(-3, 6)$;   vertex at $(-3, 4)$

41. Center at $(-3, -4)$;   focus at $(-3, -8)$;   vertex at $(-3, -2)$

42. Center at $(1, 4)$;   focus at $(-2, 4)$;   vertex at $(0, 4)$

43. Foci at $(3, 7)$ and $(7, 7)$;   vertex at $(6, 7)$

44. Focus at $(-4, 0)$ vertices at $(-4, 4)$ and $(-4, 2)$

45. Vertices at $(-1, -1)$ and $(3, -1)$;   asymptote the line $y + 1 = \dfrac{3}{2}(x - 1)$

46. Vertices at $(1, -3)$ and $(1, 1)$;   asymptote the line $y + 1 = \dfrac{3}{2}(x - 1)$

In Problems 47–60, find the center, transverse axis, vertices, foci, and asymptotes. Graph each equation by hand. Verify your graph using a graphing utility.

47. $\dfrac{(x - 2)^2}{4} - \dfrac{(y + 3)^2}{9} = 1$

48. $\dfrac{(y + 3)^2}{4} - \dfrac{(x - 2)^2}{9} = 1$

49. $(y - 2)^2 - 4(x + 2)^2 = 4$

50. $(x + 4)^2 - 9(y - 3)^2 = 9$

51. $(x + 1)^2 - (y + 2)^2 = 4$

52. $(y - 3)^2 - (x + 2)^2 = 4$

53. $x^2 - y^2 - 2x - 2y - 1 = 0$

54. $y^2 - x^2 - 4y + 4x - 1 = 0$

55. $y^2 - 4x^2 - 4y - 8x - 4 = 0$

56. $2x^2 - y^2 + 4x + 4y - 4 = 0$

57. $4x^2 - y^2 - 24x - 4y + 16 = 0$

58. $2y^2 - x^2 + 2x + 8y + 3 = 0$

59. $y^2 - 4x^2 - 16x - 2y - 19 = 0$

60. $x^2 - 3y^2 + 8x - 6y + 4 = 0$

In Problems 61–64, graph each function. Be sure to label any intercepts.
[**Hint:** Notice that each function is half a hyperbola.]

61. $f(x) = \sqrt{16 + 4x^2}$

62. $f(x) = -\sqrt{9 + 9x^2}$

63. $f(x) = -\sqrt{-25 + x^2}$

64. $f(x) = \sqrt{-1 + x^2}$

Mixed Practice

In Problems 65–72, analyze each conic section.

65. $\dfrac{(x - 3)^2}{4} - \dfrac{y^2}{25} = 1$

66. $\dfrac{(y + 2)^2}{16} - \dfrac{(x - 2)^2}{4} = 1$

67. $x^2 = 16(y - 3)$

68. $y^2 = -12(x + 1)$

69. $25x^2 + 9y^2 - 250x + 400 = 0$

70. $x^2 + 36y^2 - 2x + 288y + 541 = 0$

71. $x^2 - 6x - 8y - 31 = 0$

72. $9x^2 - y^2 - 18x - 8y - 88 = 0$

Applications and Extensions

73. Fireworks Display   Suppose that two people standing 2 miles apart both see the burst from a fireworks display. After a period of time, the first person standing at point A hears the burst. One second later, the second person standing at point B hears the burst. If the person at point B is due west of the person at point A and if the display is known to occur due north of the person at point A, where did the fireworks display occur?

74. Lightning Strikes Suppose that two people standing 1 mile apart both see a flash of lightning. After a period of time, the first person standing at point A hears the thunder. Two seconds later, the second person standing at point B hears the thunder. If the person at point B is due west of the person at point A and if the lightning strike is known to occur due north of the person standing at point A, where did the lightning strike?

75. Nuclear Power Plant Some nuclear power plants utilize "natural draft" cooling towers in the shape of a **hyperboloid,** a solid obtained by rotating a hyperbola about its conjugate axis. Suppose such a cooling tower has a base diameter of 400 feet and the diameter at its narrowest point, 360 feet above the ground, is 200 feet. If the diameter at the top of the tower is 300 feet, how tall is the tower?

Source: Bay Area Air Quality Management District

76. An Explosion Two recording devices are set 2400 feet apart, with the device at point A to the west of the device at point B. At a point between the devices, 300 feet from point B, a small amount of explosive is detonated. The recording devices record the time until the sound reaches each. How far directly north of point B should a second explosion be done so that the measured time difference recorded by the devices is the same as that for the first detonation?

77. Rutherford's Experiment In May 1911, Ernest Rutherford published a paper in *Philosophical Magazine*. In this article, he described the motion of alpha particles as they are shot at a piece of gold foil 0.00004 cm thick. Before conducting this experiment, Rutherford expected that the alpha particles would shoot through the foil just as a bullet would shoot through snow. Instead, a small fraction of the alpha particles bounced off the foil. This led to the conclusion that the nucleus of an atom is dense, while the remainder of the atom is sparse. Only the density of the nucleus could cause the alpha particles to deviate from their path. The figure shows a diagram from Rutherford's paper that indicates that the deflected alpha particles follow the path of one branch of a hyperbola.

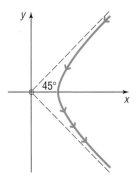

(a) Find an equation of the asymptotes under this scenario.
(b) If the vertex of the path of the alpha particles is 10 cm from the center of the hyperbola, find a model that describes the path of the particle.

78. Hyperbolic Mirrors Hyperbolas have interesting reflective properties that make them useful for lenses and mirrors. For example, if a ray of light strikes a convex hyperbolic mirror on a line that would (theoretically) pass through its rear focus, it is reflected through the front focus. This property, and that of the parabola, were used to develop the *Cassegrain* telescope in 1672. The focus of the parabolic mirror and the rear focus of the hyperbolic mirror are the same point. The rays are collected by the parabolic mirror, reflected toward the (common) focus, and thus are reflected by the hyperbolic mirror through the opening to its front focus, where the eyepiece is located. If the equation of the hyperbola is $\dfrac{y^2}{9} - \dfrac{x^2}{16} = 1$ and the focal length (distance from the vertex to the focus) of the parabola is 6, find the equation of the parabola.

Source: www.enchantedlearning.com

79. The **eccentricity** e of a hyperbola is defined as the number $\dfrac{c}{a}$, where a and c are the numbers given in equation (2). Because $c > a$, it follows that $e > 1$. Describe the general shape of a hyperbola whose eccentricity is close to 1. What is the shape if e is very large?

80. A hyperbola for which $a = b$ is called an **equilateral hyperbola.** Find the eccentricity e of an equilateral hyperbola.

[**Note:** The eccentricity of a hyperbola is defined in Problem 79.]

81. Two hyperbolas that have the same set of asymptotes are called **conjugate.** Show that the hyperbolas

$$\frac{x^2}{4} - y^2 = 1 \quad \text{and} \quad y^2 - \frac{x^2}{4} = 1$$

are conjugate. Graph each hyperbola on the same set of coordinate axes.

82. Prove that the hyperbola

$$\frac{y^2}{a^2} - \frac{x^2}{b^2} = 1$$

has the two oblique asymptotes

$$y = \frac{a}{b}x \quad \text{and} \quad y = -\frac{a}{b}x$$

83. Show that the graph of an equation of the form

$$Ax^2 + Cy^2 + F = 0 \qquad A \neq 0, C \neq 0, F \neq 0$$

where A and C are of opposite sign, is a hyperbola with center at $(0, 0)$.

84. Show that the graph of an equation of the form

$$Ax^2 + Cy^2 + Dx + Ey + F = 0 \qquad A \neq 0, C \neq 0$$

where A and C are of opposite sign,

(a) is a hyperbola if $\dfrac{D^2}{4A} + \dfrac{E^2}{4C} - F \neq 0$

(b) is two intersecting lines if $\dfrac{D^2}{4A} + \dfrac{E^2}{4C} - F = 0$

'Are You Prepared?' Answers

1. $5\sqrt{2}$ **2.** $\dfrac{25}{4}$ **3.** $(0, -3), (0, 3)$ **4.** True   **5.** right; 5; down; 4   **6.** Vertical: $x = -2, x = 2$; horizontal: $y = 1$

6.5 Rotation of Axes; General Form of a Conic

PREPARING FOR THIS SECTION *Before getting started, review the following:*

- Sum Formulas for Sine and Cosine (Section 3.4, pp. 232 and 235)
- Half-angle Formulas for Sine and Cosine (Section 3.5, p. 246)
- Double-angle Formulas for Sine and Cosine (Section 3.5, p. 242)

Now Work the 'Are You Prepared?' problems on page 437.

> **OBJECTIVES**
> 1 Identify a Conic (p. 431)
> 2 Use a Rotation of Axes to Transform Equations (p. 432)
> 3 Analyze an Equation Using a Rotation of Axes (p. 434)
> 4 Identify Conics without a Rotation of Axes (p. 436)

In this section, we show that the graph of a general second-degree polynomial containing two variables x and y, that is, an equation of the form

$$Ax^2 + Bxy + Cy^2 + Dx + Ey + F = 0 \tag{1}$$

where A, B, and C are not simultaneously 0, is a conic. We shall not concern ourselves here with the degenerate cases of equation (1), such as $x^2 + y^2 = 0$, whose graph is a single point $(0, 0)$; or $x^2 + 3y^2 + 3 = 0$, whose graph contains no points; or $x^2 - 4y^2 = 0$, whose graph is two lines, $x - 2y = 0$ and $x + 2y = 0$.

We begin with the case where $B = 0$. In this case, the term containing xy is not present, so equation (1) has the form

$$Ax^2 + Cy^2 + Dx + Ey + F = 0$$

where either $A \neq 0$ or $C \neq 0$.

1 Identify a Conic

We have already discussed the procedure for identifying the graph of this kind of equation; we complete the squares of the quadratic expressions in x or y, or both. Once this has been done, the conic can be identified by comparing it to one of the forms studied in Sections 6.2 through 6.4.

In fact, though, we can identify the conic directly from the equation without completing the squares.

THEOREM

Identifying Conics without Completing the Squares

Excluding degenerate cases, the equation

$$Ax^2 + Cy^2 + Dx + Ey + F = 0 \tag{2}$$

where A and C cannot both equal zero:

(a) Defines a parabola if $AC = 0$.
(b) Defines an ellipse (or a circle) if $AC > 0$.
(c) Defines a hyperbola if $AC < 0$.

Proof

(a) If $AC = 0$, then either $A = 0$ or $C = 0$, but not both, so the form of equation (2) is either

$$Ax^2 + Dx + Ey + F = 0, \qquad A \neq 0$$

or

$$Cy^2 + Dx + Ey + F = 0, \qquad C \neq 0$$

Using the results of Problems 78 and 79 in Exercise 6.2, it follows that, except for the degenerate cases, the equation is a parabola.

(b) If $AC > 0$, then A and C are of the same sign. Using the results of Problem 84 in Exercise 6.3, except for the degenerate cases, the equation is an ellipse if $A \neq C$ or a circle if $A = C$.

(c) If $AC < 0$, then A and C are of opposite sign. Using the results of Problem 84 in Exercise 6.4, except for the degenerate cases, the equation is a hyperbola. ∎

We will not be concerned with the degenerate cases of equation (2). However, in practice, you should be alert to the possibility of degeneracy.

| EXAMPLE 1 | Identifying a Conic without Completing the Squares |

Identify each equation without completing the squares.

(a) $3x^2 + 6y^2 + 6x - 12y = 0$ (b) $2x^2 - 3y^2 + 6y + 4 = 0$

(c) $y^2 - 2x + 4 = 0$

Solution

(a) We compare the given equation to equation (2) and conclude that $A = 3$ and $C = 6$. Since $AC = 18 > 0$, the equation is an ellipse.

(b) Here $A = 2$ and $C = -3$, so $AC = -6 < 0$. The equation is a hyperbola.

(c) Here $A = 0$ and $C = 1$, so $AC = 0$. The equation is a parabola.

∎

=➤ **Now Work** PROBLEM 11

Although we can now identify the type of conic represented by any equation of the form of equation (2) without completing the squares, we will still need to complete the squares if we desire additional information about the conic, such as its graph.

Now we turn our attention to equations of the form of equation (1), where $B \neq 0$. To discuss this case, we first need to investigate a new procedure: *rotation of axes*.

2 Use a Rotation of Axes to Transform Equations

In a **rotation of axes,** the origin remains fixed while the x-axis and y-axis are rotated through an angle θ to a new position; the new positions of the x-axis and the y-axis are denoted by x' and y', respectively, as shown in Figure 50(a).

Now look at Figure 50(b). There the point P has the coordinates (x, y) relative to the xy-plane, while the same point P has coordinates (x', y') relative to the $x'y'$-plane. We seek relationships that will enable us to express x and y in terms of x', y', and θ.

As Figure 50(b) shows, r denotes the distance from the origin O to the point P, and α denotes the angle between the positive x'-axis and the ray from O through P. Then, using the definitions of sine and cosine, we have

Figure 50

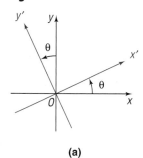

(a)

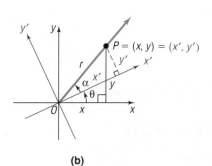

(b)

$$x' = r \cos \alpha \qquad y' = r \sin \alpha \tag{3}$$
$$x = r \cos(\theta + \alpha) \qquad y = r \sin(\theta + \alpha) \tag{4}$$

Now

$$
\begin{aligned}
x &= r \cos(\theta + \alpha) \\
&= r(\cos \theta \cos \alpha - \sin \theta \sin \alpha) &&\text{Apply the Sum Formula for cosine.}\\
&= (r \cos \alpha)(\cos \theta) - (r \sin \alpha)(\sin \theta) \\
&= x' \cos \theta - y' \sin \theta &&\text{By equation (3)}
\end{aligned}
$$

Similarly,

$$
\begin{aligned}
y &= r \sin(\theta + \alpha) \\
&= r(\sin \theta \cos \alpha + \cos \theta \sin \alpha) &&\text{Apply the Sum Formula for sine.}\\
&= x' \sin \theta + y' \cos \theta &&\text{By equation (3)}
\end{aligned}
$$

THEOREM

Rotation Formulas

If the x- and y-axes are rotated through an angle θ, the coordinates (x, y) of a point P relative to the xy-plane and the coordinates (x', y') of the same point relative to the new x'- and y'-axes are related by the formulas

$$x = x' \cos \theta - y' \sin \theta \qquad y = x' \sin \theta + y' \cos \theta \qquad (5)$$

EXAMPLE 2 | **Rotating Axes**

Express the equation $xy = 1$ in terms of new $x'y'$-coordinates by rotating the axes through a 45° angle. Discuss the new equation.

Solution Let $\theta = 45°$ in equation (5). Then

$$x = x' \cos 45° - y' \sin 45° = x' \frac{\sqrt{2}}{2} - y' \frac{\sqrt{2}}{2} = \frac{\sqrt{2}}{2}(x' - y')$$

$$y = x' \sin 45° + y' \cos 45° = x' \frac{\sqrt{2}}{2} + y' \frac{\sqrt{2}}{2} = \frac{\sqrt{2}}{2}(x' + y')$$

Figure 51

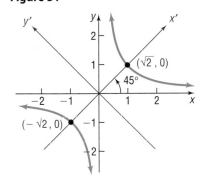

Substituting these expressions for x and y in $xy = 1$ gives

$$\left[\frac{\sqrt{2}}{2}(x' - y') \right]\left[\frac{\sqrt{2}}{2}(x' + y') \right] = 1$$

$$\frac{1}{2}(x'^2 - y'^2) = 1$$

$$\frac{x'^2}{2} - \frac{y'^2}{2} = 1$$

This is the equation of a hyperbola with center at $(0, 0)$ and transverse axis along the x'-axis. The vertices are at $(\pm\sqrt{2}, 0)$ on the x'-axis; the asymptotes are $y' = x'$ and $y' = -x'$ (which correspond to the original x- and y-axes). See Figure 51 for the graph.

As Example 2 illustrates, a rotation of axes through an appropriate angle can transform a second degree equation in x and y containing an xy-term into one in x' and y' in which no $x'y'$-term appears. In fact, we will show that a rotation of axes through an appropriate angle will transform any equation of the form of equation (1) into an equation in x' and y' without an $x'y'$-term.

To find the formula for choosing an appropriate angle θ through which to rotate the axes, we begin with equation (1),

$$Ax^2 + Bxy + Cy^2 + Dx + Ey + F = 0 \qquad B \neq 0$$

Next we rotate through an angle θ using rotation formulas (5).

$$A(x' \cos \theta - y' \sin \theta)^2 + B(x' \cos \theta - y' \sin \theta)(x' \sin \theta + y' \cos \theta)$$
$$+ C(x' \sin \theta + y' \cos \theta)^2 + D(x' \cos \theta - y' \sin \theta)$$
$$+ E(x' \sin \theta + y' \cos \theta) + F = 0$$

By expanding and collecting like terms, we obtain

$$(A \cos^2 \theta + B \sin \theta \cos \theta + C \sin^2 \theta)x'^2 + [B(\cos^2 \theta - \sin^2 \theta) + 2(C - A)(\sin \theta \cos \theta)]x'y'$$
$$+ (A \sin^2 \theta - B \sin \theta \cos \theta + C \cos^2 \theta)y'^2$$
$$+ (D \cos \theta + E \sin \theta)x'$$
$$+ (-D \sin \theta + E \cos \theta)y' + F = 0 \qquad (6)$$

In equation (6), the coefficient of $x'y'$ is

$$B(\cos^2 \theta - \sin^2 \theta) + 2(C - A)(\sin \theta \cos \theta)$$

Since we want to eliminate the $x'y'$-term, we select an angle θ so that this coefficient is 0.

$$B(\cos^2 \theta - \sin^2 \theta) + 2(C - A)(\sin \theta \cos \theta) = 0$$
$$B \cos(2\theta) + (C - A) \sin(2\theta) = 0 \qquad \text{Double-angle Formulas}$$
$$B \cos(2\theta) = (A - C) \sin(2\theta)$$
$$\cot(2\theta) = \frac{A - C}{B} \qquad B \neq 0$$

THEOREM

To transform the equation

$$Ax^2 + Bxy + Cy^2 + Dx + Ey + F = 0 \qquad B \neq 0$$

into an equation in x' and y' without an $x'y'$-term, rotate the axes through an angle θ that satisfies the equation

$$\cot(2\theta) = \frac{A - C}{B} \qquad \qquad \textbf{(7)}$$

Equation (7) has an infinite number of solutions for θ. We shall adopt the convention of choosing the acute angle θ that satisfies (7). Then we have the following two possibilities:

If $\cot(2\theta) \geq 0$, then $0° < 2\theta \leq 90°$, so $0° < \theta \leq 45°$.

If $\cot(2\theta) < 0$, then $90° < 2\theta < 180°$, so $45° < \theta < 90°$.

Each of these results in a counterclockwise rotation of the axes through an acute angle θ.*

WARNING Be careful if you use a calculator to solve equation (7).

1. If $\cot(2\theta) = 0$, then $2\theta = 90°$ and $\theta = 45°$.
2. If $\cot(2\theta) \neq 0$, first find $\cos(2\theta)$. Then use the inverse cosine function key(s) to obtain $2\theta, 0° < 2\theta < 180°$. Finally, divide by 2 to obtain the correct acute angle θ. ∎

3 Analyze an Equation Using a Rotation of Axes

For the remainder of this section, the direction **"Analyze the equation"** will mean to transform the given equation so that it contains no xy-term and to graph the equation.

EXAMPLE 3 | **Analyzing an Equation Using a Rotation of Axes**

Analyze the equation: $x^2 + \sqrt{3}xy + 2y^2 - 10 = 0$

Solution

Since an xy-term is present, we must rotate the axes. Using $A = 1, B = \sqrt{3}$, and $C = 2$ in equation (7), the appropriate acute angle θ through which to rotate the axes satisfies the equation

$$\cot(2\theta) = \frac{A - C}{B} = \frac{-1}{\sqrt{3}} = -\frac{\sqrt{3}}{3} \qquad 0° < 2\theta < 180°$$

---

* Any rotation through an angle θ that satisfies $\cot(2\theta) = \dfrac{A - C}{B}$ will eliminate the $x'y'$-term. However, the final form of the transformed equation may be different (but equivalent), depending on the angle chosen.

Since $\cot(2\theta) = -\dfrac{\sqrt{3}}{3}$, we find $2\theta = 120°$, so $\theta = 60°$. Using $\theta = 60°$ in rotation formulas (5), we find

$$x = x'\cos 60° - y'\sin 60° = \frac{1}{2}x' - \frac{\sqrt{3}}{2}y' = \frac{1}{2}\left(x' - \sqrt{3}y'\right)$$

$$y = x'\sin 60° + y'\cos 60° = \frac{\sqrt{3}}{2}x' + \frac{1}{2}y' = \frac{1}{2}\left(\sqrt{3}x' + y'\right)$$

Substituting these values into the original equation and simplifying, we have

$$x^2 + \sqrt{3}xy + 2y^2 - 10 = 0$$

$$\frac{1}{4}\left(x' - \sqrt{3}y'\right)^2 + \sqrt{3}\left[\frac{1}{2}\left(x' - \sqrt{3}y'\right)\right]\left[\frac{1}{2}\left(\sqrt{3}x' + y'\right)\right] + 2\left[\frac{1}{4}\left(\sqrt{3}x' + y'\right)^2\right] = 10$$

Multiply both sides by 4 and expand to obtain

$$x'^2 - 2\sqrt{3}x'y' + 3y'^2 + \sqrt{3}\left(\sqrt{3}x'^2 - 2x'y' - \sqrt{3}y'^2\right) + 2\left(3x'^2 + 2\sqrt{3}x'y' + y'^2\right) = 40$$

$$10x'^2 + 2y'^2 = 40$$

$$\frac{x'^2}{4} + \frac{y'^2}{20} = 1$$

This is the equation of an ellipse with center at $(0,0)$ and major axis along the y'-axis. The vertices are at $(0, \pm 2\sqrt{5})$ on the y'-axis. See Figure 52 for the graph.

Figure 52

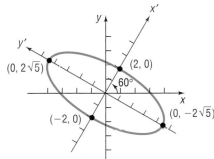

To graph the equation $x^2 + \sqrt{3}xy + 2y^2 - 10 = 0$ using a graphing utility, we need to solve the equation for y. Rearranging the terms, we observe that the equation is quadratic in the variable y: $2y^2 + \sqrt{3}xy + (x^2 - 10) = 0$. We can solve the equation for y using the quadratic formula with $a = 2, b = \sqrt{3}x$, and $c = x^2 - 10$.

$$Y_1 = \frac{-\sqrt{3}x + \sqrt{\left(\sqrt{3}x\right)^2 - 4(2)(x^2 - 10)}}{2(2)} = \frac{-\sqrt{3}x + \sqrt{-5x^2 + 80}}{4}$$

and

$$Y_2 = \frac{-\sqrt{3}x - \sqrt{\left(\sqrt{3}x\right)^2 - 4(2)(x^2 - 10)}}{2(2)} = \frac{-\sqrt{3}x - \sqrt{5x^2 + 80}}{4}$$

Figure 53 shows the graph of Y_1 and Y_2.

Figure 53

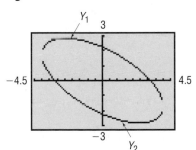

⬤▬▬▬ **Now Work** PROBLEM 31

In Example 3, the acute angle θ through which to rotate the axes was easy to find because of the numbers that we used in the given equation. In general, the equation $\cot(2\theta) = \dfrac{A - C}{B}$ will not have such a "nice" solution. As the next example shows, we can still find the appropriate rotation formulas without using a calculator approximation by applying Half-angle Formulas.

EXAMPLE 4 | **Analyzing an Equation Using a Rotation of Axes**

Analyze the equation: $4x^2 - 4xy + y^2 + 5\sqrt{5}x + 5 = 0$

Solution   Letting $A = 4, B = -4$, and $C = 1$ in equation (7), the appropriate angle θ through which to rotate the axes satisfies

$$\cot(2\theta) = \frac{A - C}{B} = \frac{3}{-4} = -\frac{3}{4}$$

To use rotation formulas (5), we need to know the values of $\sin\theta$ and $\cos\theta$. Since we seek an acute angle θ, we know that $\sin\theta > 0$ and $\cos\theta > 0$. We use the Half-angle Formulas in the form

$$\sin\theta = \sqrt{\frac{1 - \cos(2\theta)}{2}} \qquad \cos\theta = \sqrt{\frac{1 + \cos(2\theta)}{2}}$$

Now we need to find the value of $\cos(2\theta)$. Since $\cot(2\theta) = -\dfrac{3}{4}$, then $90° < 2\theta < 180°$ (Do you know why?), so $\cos(2\theta) = -\dfrac{3}{5}$. Then

$$\sin\theta = \sqrt{\frac{1 - \cos(2\theta)}{2}} = \sqrt{\frac{1 - \left(-\dfrac{3}{5}\right)}{2}} = \sqrt{\frac{4}{5}} = \frac{2}{\sqrt{5}} = \frac{2\sqrt{5}}{5}$$

$$\cos\theta = \sqrt{\frac{1 + \cos(2\theta)}{2}} = \sqrt{\frac{1 + \left(-\dfrac{3}{5}\right)}{2}} = \sqrt{\frac{1}{5}} = \frac{1}{\sqrt{5}} = \frac{\sqrt{5}}{5}$$

With these values, the rotation formulas (5) are

$$x = \frac{\sqrt{5}}{5}x' - \frac{2\sqrt{5}}{5}y' = \frac{\sqrt{5}}{5}(x' - 2y')$$

$$y = \frac{2\sqrt{5}}{5}x' + \frac{\sqrt{5}}{5}y' = \frac{\sqrt{5}}{5}(2x' + y')$$

Substituting these values in the original equation and simplifying, we obtain

$$4x^2 - 4xy + y^2 + 5\sqrt{5}x + 5 = 0$$

$$4\left[\frac{\sqrt{5}}{5}(x' - 2y')\right]^2 - 4\left[\frac{\sqrt{5}}{5}(x' - 2y')\right]\left[\frac{\sqrt{5}}{5}(2x' + y')\right]$$

$$+ \left[\frac{\sqrt{5}}{5}(2x' + y')\right]^2 + 5\sqrt{5}\left[\frac{\sqrt{5}}{5}(x' - 2y')\right] = -5$$

Multiply both sides by 5 and expand to obtain

$$4(x'^2 - 4x'y' + 4y'^2) - 4(2x'^2 - 3x'y' - 2y'^2)$$

$$+ 4x'^2 + 4x'y' + y'^2 + 25(x' - 2y') = -25$$

$$25y'^2 - 50y' + 25x' = -25 \qquad \textit{Combine like terms.}$$

$$y'^2 - 2y' + x' = -1 \qquad \textit{Divide by 25.}$$

$$y'^2 - 2y' + 1 = -x' \qquad \textit{Complete the square in y'.}$$

$$(y' - 1)^2 = -x'$$

Figure 54

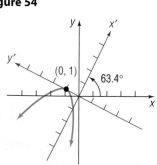

This is the equation of a parabola with vertex at $(0, 1)$ in the $x'y'$-plane. The axis of symmetry is parallel to the x'-axis. Using a calculator to solve $\sin\theta = \dfrac{2\sqrt{5}}{5}$, we find that $\theta \approx 63.4°$. See Figure 54 for the graph.

Now Work PROBLEM 37

4 Identify Conics without a Rotation of Axes

Suppose that we are required only to identify (rather than analyze) an equation of the form

$$Ax^2 + Bxy + Cy^2 + Dx + Ey + F = 0 \qquad B \neq 0 \qquad \textbf{(8)}$$

If we apply the rotation formulas (5) to this equation, we obtain an equation of the form

$$A'x'^2 + B'x'y' + C'y'^2 + D'x' + E'y' + F' = 0 \qquad \textbf{(9)}$$

where A', B', C', D', E', and F' can be expressed in terms of A, B, C, D, E, F and the angle θ of rotation (see Problem 53). It can be shown that the value of $B^2 - 4AC$ in equation (8) and the value of $B'^2 - 4A'C'$ in equation (9) are equal no matter what angle θ of rotation is chosen (see Problem 55). In particular, if the angle θ of rotation satisfies equation (7), then $B' = 0$ in equation (9), and $B^2 - 4AC = -4A'C'$. Since equation (9) then has the form of equation (2),

$$A'x'^2 + C'y'^2 + D'x' + E'y' + F' = 0$$

we can identify it without completing the squares, as we did in the beginning of this section. In fact, now we can identify the conic described by any equation of the form of equation (8) without a rotation of axes.

THEOREM	**Identifying Conics without a Rotation of Axes**

Except for degenerate cases, the equation

$$\boxed{Ax^2 + Bxy + Cy^2 + Dx + Ey + F = 0}$$

(a) Defines a parabola if $B^2 - 4AC = 0$.
(b) Defines an ellipse (or a circle) if $B^2 - 4AC < 0$.
(c) Defines a hyperbola if $B^2 - 4AC > 0$.

You are asked to prove this theorem in Problem 56.

EXAMPLE 5	**Identifying a Conic without a Rotation of Axes**

Identify the equation: $8x^2 - 12xy + 17y^2 - 4\sqrt{5}x - 2\sqrt{5}y - 15 = 0$

Solution   Here $A = 8$, $B = -12$, and $C = 17$, so $B^2 - 4AC = -400$. Since $B^2 - 4AC < 0$, the equation defines an ellipse.

Now Work PROBLEM 43

6.5 Assess Your Understanding

'Are You Prepared?' *Answers are given at the end of these exercises. If you get a wrong answer, read the pages listed in red.*

1. The sum formula for the sine function is $\sin(A + B) = $ _____. (p. 235)

2. The Double-angle Formula for the sine function is $\sin(2\theta) = $ _____. (p. 242)

3. If θ is acute, the Half-angle Formula for the sine function is $\sin\dfrac{\theta}{2} = $ _____. (p. 246)

4. If θ is acute, the Half-angle Formula for the cosine function is $\cos\dfrac{\theta}{2} = $ _____. (p. 246)

Concepts and Vocabulary

5. To transform the equation

    $$Ax^2 + Bxy + Cy^2 + Dx + Ey + F = 0, \qquad B \neq 0$$

    into one in x' and y' without an $x'y'$-term, rotate the axes through an acute angle θ that satisfies the equation _____.

6. Identify the conic: $x^2 - 2y^2 - x - y - 18 = 0$ _____.

7. Identify the conic: $x^2 + 2xy + 3y^2 - 2x + 4y + 10 = 0$ _____.

8. **True or False**   The equation $ax^2 + 6y^2 - 12y = 0$ defines an ellipse if $a > 0$.

9. **True or False**   The equation $3x^2 + bxy + 12y^2 = 10$ defines a parabola if $b = -12$.

10. **True or False**   To eliminate the xy-term from the equation $x^2 - 2xy + y^2 - 2x + 3y + 5 = 0$, rotate the axes through an angle θ, where $\cot \theta = B^2 - 4AC$.

Skill Building

In Problems 11–20, identify each equation without completing the squares.

11. $x^2 + 4x + y + 3 = 0$

12. $2y^2 - 3y + 3x = 0$

13. $6x^2 + 3y^2 - 12x + 6y = 0$

14. $2x^2 + y^2 - 8x + 4y + 2 = 0$

15. $3x^2 - 2y^2 + 6x + 4 = 0$

16. $4x^2 - 3y^2 - 8x + 6y + 1 = 0$

17. $2y^2 - x^2 - y + x = 0$

18. $y^2 - 8x^2 - 2x - y = 0$

19. $x^2 + y^2 - 8x + 4y = 0$

20. $2x^2 + 2y^2 - 8x + 8y = 0$

In Problems 21–30, determine the appropriate rotation formulas to use so that the new equation contains no xy-term.

21. $x^2 + 4xy + y^2 - 3 = 0$

22. $x^2 - 4xy + y^2 - 3 = 0$

23. $5x^2 + 6xy + 5y^2 - 8 = 0$

24. $3x^2 - 10xy + 3y^2 - 32 = 0$

25. $13x^2 - 6\sqrt{3}xy + 7y^2 - 16 = 0$

26. $11x^2 + 10\sqrt{3}xy + y^2 - 4 = 0$

27. $4x^2 - 4xy + y^2 - 8\sqrt{5}x - 16\sqrt{5}y = 0$

28. $x^2 + 4xy + 4y^2 + 5\sqrt{5}y + 5 = 0$

29. $25x^2 - 36xy + 40y^2 - 12\sqrt{13}x - 8\sqrt{13}y = 0$

30. $34x^2 - 24xy + 41y^2 - 25 = 0$

In Problems 31–42, rotate the axes so that the new equation contains no xy-term. Analyze and graph the new equation. Refer to Problems 21–30 for Problems 31–40. Verify your graph using a graphing utility.

31. $x^2 + 4xy + y^2 - 3 = 0$

32. $x^2 - 4xy + y^2 - 3 = 0$

33. $5x^2 + 6xy + 5y^2 - 8 = 0$

34. $3x^2 - 10xy + 3y^2 - 32 = 0$

35. $13x^2 - 6\sqrt{3}xy + 7y^2 - 16 = 0$

36. $11x^2 + 10\sqrt{3}xy + y^2 - 4 = 0$

37. $4x^2 - 4xy + y^2 - 8\sqrt{5}x - 16\sqrt{5}y = 0$

38. $x^2 + 4xy + 4y^2 + 5\sqrt{5}y + 5 = 0$

39. $25x^2 - 36xy + 40y^2 - 12\sqrt{13}x - 8\sqrt{13}y = 0$

40. $34x^2 - 24xy + 41y^2 - 25 = 0$

41. $16x^2 + 24xy + 9y^2 - 130x + 90y = 0$

42. $16x^2 + 24xy + 9y^2 - 60x + 80y = 0$

In Problems 43–52, identify each equation without applying a rotation of axes.

43. $x^2 + 3xy - 2y^2 + 3x + 2y + 5 = 0$

44. $2x^2 - 3xy + 4y^2 + 2x + 3y - 5 = 0$

45. $x^2 - 7xy + 3y^2 - y - 10 = 0$

46. $2x^2 - 3xy + 2y^2 - 4x - 2 = 0$

47. $9x^2 + 12xy + 4y^2 - x - y = 0$

48. $10x^2 + 12xy + 4y^2 - x - y + 10 = 0$

49. $10x^2 - 12xy + 4y^2 - x - y - 10 = 0$

50. $4x^2 + 12xy + 9y^2 - x - y = 0$

51. $3x^2 - 2xy + y^2 + 4x + 2y - 1 = 0$

52. $3x^2 + 2xy + y^2 + 4x - 2y + 10 = 0$

Applications and Extensions

In Problems 53–56, apply the rotation formulas (5) to

$$Ax^2 + Bxy + Cy^2 + Dx + Ey + F = 0$$

to obtain the equation

$$A'x'^2 + B'x'y' + C'y'^2 + D'x' + E'y' + F' = 0$$

53. Express A', B', C', D', E', and F' in terms of A, B, C, D, E, F, and the angle θ of rotation.

    [**Hint:** Refer to equation (6).]

54. Show that $A + C = A' + C'$, and thus show that $A + C$ is **invariant;** that is, its value does not change under a rotation of axes.

55. Refer to Problem 54. Show that $B^2 - 4AC$ is invariant.

56. Prove that, except for degenerate cases, the equation

$$Ax^2 + Bxy + Cy^2 + Dx + Ey + F = 0$$

    (a) Defines a parabola if $B^2 - 4AC = 0$.
    (b) Defines an ellipse (or a circle) if $B^2 - 4AC < 0$.
    (c) Defines a hyperbola if $B^2 - 4AC > 0$.

57. Use rotation formulas (5) to show that distance is invariant under a rotation of axes. That is, show that the distance from $P_1 = (x_1, y_1)$ to $P_2 = (x_2, y_2)$ in the xy-plane equals the distance from $P_1 = (x_1', y_1')$ to $P_2 = (x_2', y_2')$ in the $x'y'$-plane.

58. Show that the graph of the equation $x^{1/2} + y^{1/2} = a^{1/2}$ is part of the graph of a parabola.

Discussion and Writing

59. Formulate a strategy for discussing and graphing an equation of the form

$$Ax^2 + Cy^2 + Dx + Ey + F = 0$$

60. How does your strategy change if the equation is of the following form?

$$Ax^2 + Bxy + Cy^2 + Dx + Ey + F = 0$$

1. $\sin A \cos B + \cos A \sin B$ **2.** $2 \sin \theta \cos \theta$ **3.** $\sqrt{\dfrac{1 - \cos \theta}{2}}$ **4.** $\sqrt{\dfrac{1 + \cos \theta}{2}}$

6.6 Polar Equations of Conics

PREPARING FOR THIS SECTION *Before getting started, review the following:*

- Polar Coordinates (Section 5.1, pp. 320–327)

✎ **Now Work** the 'Are You Prepared?' problems on page 443.

    OBJECTIVES   **1** Analyze and Graph Polar Equations of Conics (p. 439)
                         2 Convert the Polar Equation of a Conic to a Rectangular Equation (p. 443)

1 Analyze and Graph Polar Equations of Conics

In Sections 6.2 through 6.4, we gave separate definitions for the parabola, ellipse, and hyperbola based on geometric properties and the distance formula. In this section, we present an alternative definition that simultaneously defines all these conics. As we shall see, this approach is well suited to polar coordinate representation. (Refer to Section 5.1.)

DEFINITION   Let D denote a fixed line called the **directrix;** let F denote a fixed point called the **focus,** which is not on D; and let e be a fixed positive number called the **eccentricity.** A **conic** is the set of points P in the plane such that the ratio of the distance from F to P to the distance from D to P equals e. That is, a conic is the collection of points P for which

$$\frac{d(F, P)}{d(D, P)} = e \tag{1}$$

If $e = 1$, the conic is a **parabola.**
If $e < 1$, the conic is an **ellipse.**
If $e > 1$, the conic is a **hyperbola.**

Observe that if $e = 1$ the definition of a parabola in equation (1) is exactly the same as the definition used earlier in Section 6.2.

In the case of an ellipse, the **major axis** is a line through the focus perpendicular to the directrix. In the case of a hyperbola, the **transverse axis** is a line through the focus perpendicular to the directrix. For both an ellipse and a hyperbola, the eccentricity e satisfies

$$e = \frac{c}{a} \tag{2}$$

where c is the distance from the center to the focus and a is the distance from the center to a vertex.

Just as we did earlier using rectangular coordinates, we derive equations for the conics in polar coordinates by choosing a convenient position for the focus F and the directrix D. The focus F is positioned at the pole, and the directrix D is either parallel or perpendicular to the polar axis.

Suppose that we start with the directrix D perpendicular to the polar axis at a distance p units to the left of the pole (the focus F). See Figure 55.

Figure 55

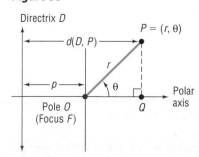

If $P = (r, \theta)$ is any point on the conic, then, by equation (1),

$$\frac{d(F, P)}{d(D, P)} = e \quad \text{or} \quad d(F, P) = e \cdot d(D, P) \qquad \textbf{(3)}$$

Now we use the point Q obtained by dropping the perpendicular from P to the polar axis to calculate $d(D, P)$.

$$d(D, P) = p + d(O, Q) = p + r \cos \theta$$

Using this expression and the fact that $d(F, P) = d(O, P) = r$ in equation (3), we get

$$d(F, P) = e \cdot d(D, P)$$
$$r = e(p + r \cos \theta)$$
$$r = ep + er \cos \theta$$
$$r - er \cos \theta = ep$$
$$r(1 - e \cos \theta) = ep$$
$$r = \frac{ep}{1 - e \cos \theta}$$

THEOREM

Polar Equation of a Conic; Focus at the Pole; Directrix Perpendicular to the Polar Axis a Distance p to the Left of the Pole

The polar equation of a conic with focus at the pole and directrix perpendicular to the polar axis at a distance p to the left of the pole is

$$r = \frac{ep}{1 - e \cos \theta} \qquad \textbf{(4)}$$

where e is the eccentricity of the conic.

EXAMPLE 1   **Analyzing and Graphing the Polar Equation of a Conic**

Analyze and graph the equation: $r = \dfrac{4}{2 - \cos \theta}$

Solution   The given equation is not quite in the form of equation (4), since the first term in the denominator is 2 instead of 1. We divide the numerator and denominator by 2 to obtain

$$r = \frac{2}{1 - \frac{1}{2}\cos \theta} \qquad \qquad r = \frac{ep}{1 - e \cos \theta}$$

This equation is in the form of equation (4), with

$$e = \frac{1}{2} \quad \text{and} \quad ep = 2$$

Then

$$\frac{1}{2}p = 2, \quad \text{so} \quad p = 4$$

We conclude that the conic is an ellipse, since $e = \dfrac{1}{2} < 1$. One focus is at the pole, and the directrix is perpendicular to the polar axis, a distance of $p = 4$ units to the left of the pole. It follows that the major axis is along the polar axis. To find the vertices, we let $\theta = 0$ and $\theta = \pi$. The vertices of the ellipse are $(4, 0)$ and $\left(\dfrac{4}{3}, \pi\right)$. The midpoint of the vertices, $\left(\dfrac{4}{3}, 0\right)$ in polar coordinates, is the center of the ellipse.

Figure 56

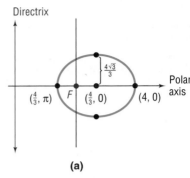

(a)

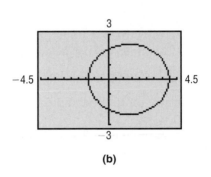

(b)

[Do you see why? The vertices $(4, 0)$ and $\left(\dfrac{4}{3}, \pi\right)$ in polar coordinates are $(4, 0)$ and $\left(-\dfrac{4}{3}, 0\right)$ in rectangular coordinates. The midpoint in rectangular coordinates is $\left(\dfrac{4}{3}, 0\right)$, which is also $\left(\dfrac{4}{3}, 0\right)$ in polar coordinates.] Then a = distance from the center to a vertex = $\dfrac{8}{3}$. Using $a = \dfrac{8}{3}$ and $e = \dfrac{1}{2}$ in equation (2), $e = \dfrac{c}{a}$, we find $c = ae = \dfrac{4}{3}$. Finally, using $a = \dfrac{8}{3}$ and $c = \dfrac{4}{3}$ in $b^2 = a^2 - c^2$, we have

$$b^2 = a^2 - c^2 = \frac{64}{9} - \frac{16}{9} = \frac{48}{9}$$

$$b = \frac{4\sqrt{3}}{3}$$

Figure 56(a) shows the graph drawn by hand.

Figure 56(b) shows the graph of the equation obtained using a graphing utility in POLar mode with θmin $= 0$, θmax $= 2\pi$, and θstep $= \dfrac{\pi}{24}$.

Exploration

Graph $r_1 = \dfrac{4}{2 + \cos \theta}$ and compare the result with Figure 56. What do you conclude? Clear the screen and graph $r_1 = \dfrac{4}{2 - \sin \theta}$ and then $r_1 = \dfrac{4}{2 + \sin \theta}$. Compare each of these graphs with Figure 56. What do you conclude?

Now Work PROBLEM 11

Equation (4) was obtained under the assumption that the directrix was perpendicular to the polar axis at a distance p units to the left of the pole. A similar derivation (see Problem 43), in which the directrix is perpendicular to the polar axis at a distance p units to the right of the pole, results in the equation

$$r = \frac{ep}{1 + e \cos \theta}$$

In Problems 44 and 45, you are asked to derive the polar equations of conics with focus at the pole and directrix parallel to the polar axis. Table 5 summarizes the polar equations of conics.

Table 5

POLAR EQUATIONS OF CONICS (FOCUS AT THE POLE, ECCENTRICITY e)	
Equation	**Description**
(a) $r = \dfrac{ep}{1 - e \cos \theta}$	Directrix is perpendicular to the polar axis at a distance p units to the left of the pole.
(b) $r = \dfrac{ep}{1 + e \cos \theta}$	Directrix is perpendicular to the polar axis at a distance p units to the right of the pole.
(c) $r = \dfrac{ep}{1 + e \sin \theta}$	Directrix is parallel to the polar axis at a distance p units above the pole.
(d) $r = \dfrac{ep}{1 - e \sin \theta}$	Directrix is parallel to the polar axis at a distance p units below the pole.

Eccentricity

If $e = 1$, the conic is a parabola; the axis of symmetry is perpendicular to the directrix.

If $e < 1$, the conic is an ellipse; the major axis is perpendicular to the directrix.

If $e > 1$, the conic is a hyperbola; the transverse axis is perpendicular to the directrix.

| EXAMPLE 2 | Analyzing and Graphing the Polar Equation of a Conic |

Analyze and graph the equation: $r = \dfrac{6}{3 + 3 \sin \theta}$

Solution

Figure 57

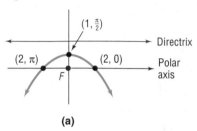

(a)

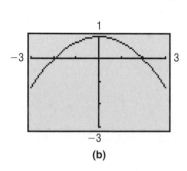

(b)

To place the equation in proper form, we divide the numerator and denominator by 3 to get

$$r = \frac{2}{1 + \sin \theta}$$

Referring to Table 5, we conclude that this equation is in the form of equation (c) with

$$e = 1 \quad \text{and} \quad ep = 2$$

$$p = 2 \quad e = 1$$

The conic is a parabola with focus at the pole. The directrix is parallel to the polar axis at a distance 2 units above the pole; the axis of symmetry is perpendicular to the polar axis. The vertex of the parabola is at $\left(1, \dfrac{\pi}{2}\right)$. (Do you see why?) See Figure 57(a) for the graph drawn by hand. Notice that we plotted two additional points, $(2, 0)$ and $(2, \pi)$, to assist in graphing.

Figure 57(b) shows the graph of the equation using a graphing utility in POLar mode with $\theta\min = 0$, $\theta\max = 2\pi$, and $\theta\text{step} = \dfrac{\pi}{24}$.

Now Work PROBLEM 13

| EXAMPLE 3 | Analyzing and Graphing the Polar Equation of a Conic |

Analyze and graph the equation: $r = \dfrac{3}{1 + 3 \cos \theta}$

Solution

This equation is in the form of equation (b) in Table 5. We conclude that

$$e = 3 \quad \text{and} \quad ep = 3$$

$$p = 1 \quad e = 3$$

This is the equation of a hyperbola with a focus at the pole. The directrix is perpendicular to the polar axis, 1 unit to the right of the pole. The transverse axis is along the polar axis. To find the vertices, we let $\theta = 0$ and $\theta = \pi$. The vertices are $\left(\dfrac{3}{4}, 0\right)$ and $\left(-\dfrac{3}{2}, \pi\right)$. The center, which is at the midpoint of $\left(\dfrac{3}{4}, 0\right)$ and $\left(-\dfrac{3}{2}, \pi\right)$, is $\left(\dfrac{9}{8}, 0\right)$. Then $c = $ distance from the center to a focus $= \dfrac{9}{8}$. Since $e = 3$, it follows from equation (2), $e = \dfrac{c}{a}$, that $a = \dfrac{3}{8}$. Finally, using $a = \dfrac{3}{8}$ and $c = \dfrac{9}{8}$ in $b^2 = c^2 - a^2$, we find

$$b^2 = c^2 - a^2 = \frac{81}{64} - \frac{9}{64} = \frac{72}{64} = \frac{9}{8}$$

$$b = \frac{3}{2\sqrt{2}} = \frac{3\sqrt{2}}{4}$$

Figure 58(a) shows the graph drawn by hand. Notice that we plotted two additional points, $\left(3, \dfrac{\pi}{2}\right)$ and $\left(3, \dfrac{3\pi}{2}\right)$, on the left branch and used symmetry to obtain the right branch. The asymptotes of this hyperbola were found in the usual way by constructing the rectangle shown.

Figures 58(b) and (c) show the graph of the equation using a graphing utility in POLar mode with θmin $= 0$, θmax $= 2\pi$, and θstep $= \dfrac{\pi}{24}$, using both dot mode and connected mode. Notice the extraneous asymptotes in connected mode.

Figure 58

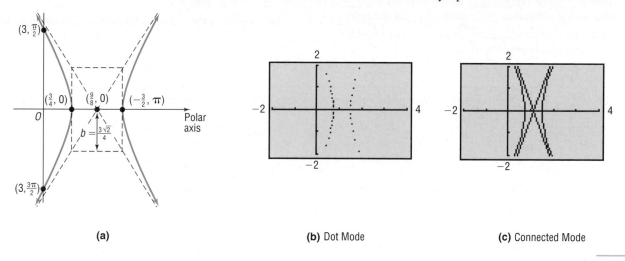

(a)

(b) Dot Mode

(c) Connected Mode

$\longrightarrow$ **Now Work** PROBLEM 17

2 Convert the Polar Equation of a Conic to a Rectangular Equation

EXAMPLE 4   **Converting a Polar Equation to a Rectangular Equation**

Convert the polar equation

$$r = \frac{1}{3 - 3\cos\theta}$$

to a rectangular equation.

Solution   The strategy here is first to rearrange the equation and square each side before using the transformation equations.

$$r = \frac{1}{3 - 3\cos\theta}$$

$$3r - 3r\cos\theta = 1$$

$$3r = 1 + 3r\cos\theta \qquad \text{Rearrange the equation.}$$

$$9r^2 = (1 + 3r\cos\theta)^2 \qquad \text{Square each side.}$$

$$9(x^2 + y^2) = (1 + 3x)^2 \qquad x^2 + y^2 = r^2; x = r\cos\theta$$

$$9x^2 + 9y^2 = 9x^2 + 6x + 1$$

$$9y^2 = 6x + 1$$

This is the equation of a parabola in rectangular coordinates.

$\longrightarrow$ **Now Work** PROBLEM 25

6.6 Assess Your Understanding

'Are You Prepared?' *Answers are given at the end of these exercises. If you get a wrong answer, read the pages listed in red.*

1. If (x, y) are the rectangular coordinates of a point P and (r, θ) are its polar coordinates, then $x = $ _____ and $y = $ _____. (pp. 322–324)

2. Transform the equation $r = 6\cos\theta$ from polar coordinates to rectangular coordinates. (pp. 326–327)

Concepts and Vocabulary

3. The polar equation $r = \dfrac{8}{4 - 2\sin\theta}$ is a conic whose eccentricity is _____. It is a(n) _____ whose directrix is _____ to the polar axis at a distance _____ units _____ of the pole.

4. The eccentricity e of a parabola is _____, of an ellipse it is _____, and of a hyperbola it is _____.

5. **True or False**   If (r, θ) are polar coordinates, the equation $r = \dfrac{2}{2 + 3\sin\theta}$ defines a hyperbola.

6. **True or False**   The eccentricity of any parabola is 1.

Skill Building

In Problems 7–12, identify the conic that each polar equation represents. Also, give the position of the directrix.

7. $r = \dfrac{1}{1 + \cos\theta}$

8. $r = \dfrac{3}{1 - \sin\theta}$

9. $r = \dfrac{4}{2 - 3\sin\theta}$

10. $r = \dfrac{2}{1 + 2\cos\theta}$

11. $r = \dfrac{3}{4 - 2\cos\theta}$

12. $r = \dfrac{6}{8 + 2\sin\theta}$

In Problems 13–24, analyze each equation and graph it by hand. Verify your graph using a graphing utility.

13. $r = \dfrac{1}{1 + \cos\theta}$

14. $r = \dfrac{3}{1 - \sin\theta}$

15. $r = \dfrac{8}{4 + 3\sin\theta}$

16. $r = \dfrac{10}{5 + 4\cos\theta}$

17. $r = \dfrac{9}{3 - 6\cos\theta}$

18. $r = \dfrac{12}{4 + 8\sin\theta}$

19. $r = \dfrac{8}{2 - \sin\theta}$

20. $r = \dfrac{8}{2 + 4\cos\theta}$

21. $r(3 - 2\sin\theta) = 6$

22. $r(2 - \cos\theta) = 2$

23. $r = \dfrac{6\sec\theta}{2\sec\theta - 1}$

24. $r = \dfrac{3\csc\theta}{\csc\theta - 1}$

In Problems 25–36, convert each polar equation to a rectangular equation.

25. $r = \dfrac{1}{1 + \cos\theta}$

26. $r = \dfrac{3}{1 - \sin\theta}$

27. $r = \dfrac{8}{4 + 3\sin\theta}$

28. $r = \dfrac{10}{5 + 4\cos\theta}$

29. $r = \dfrac{9}{3 - 6\cos\theta}$

30. $r = \dfrac{12}{4 + 8\sin\theta}$

31. $r = \dfrac{8}{2 - \sin\theta}$

32. $r = \dfrac{8}{2 + 4\cos\theta}$

33. $r(3 - 2\sin\theta) = 6$

34. $r(2 - \cos\theta) = 2$

35. $r = \dfrac{6\sec\theta}{2\sec\theta - 1}$

36. $r = \dfrac{3\csc\theta}{\csc\theta - 1}$

In Problems 37–42, find a polar equation for each conic. For each, a focus is at the pole.

37. $e = 1$; directrix is parallel to the polar axis 1 unit above the pole.

38. $e = 1$; directrix is parallel to the polar axis 2 units below the pole.

39. $e = \dfrac{4}{5}$; directrix is perpendicular to the polar axis 3 units to the left of the pole.

40. $e = \dfrac{2}{3}$; directrix is parallel to the polar axis 3 units above the pole.

41. $e = 6$; directrix is parallel to the polar axis 2 units below the pole.

42. $e = 5$; directrix is perpendicular to the polar axis 5 units to the right of the pole.

Applications and Extensions

43. Derive equation (b) in Table 5:
$$r = \frac{ep}{1 + e\cos\theta}$$

44. Derive equation (c) in Table 5:
$$r = \frac{ep}{1 + e\sin\theta}$$

45. Derive equation (d) in Table 5:
$$r = \frac{ep}{1 - e\sin\theta}$$

46. **Orbit of Mercury**   The planet Mercury travels around the Sun in an elliptical orbit given approximately by
$$r = \frac{(3.442)10^7}{1 - 0.206\cos\theta}$$

where r is measured in miles and the Sun is at the pole. Find the distance from Mercury to the Sun at *aphelion* (greatest distance from the Sun) and at *perihelion* (shortest distance from the Sun). See the figure. Use the aphelion and perihelion to graph the orbit of Mercury using a graphing utility.

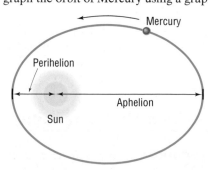

'Are You Prepared?' Answers

1. $r\cos\theta$; $r\sin\theta$

2. $x^2 + y^2 = 6x$ or $(x - 3)^2 + y^2 = 9$

6.7 Plane Curves and Parametric Equations

PREPARING FOR THIS SECTION   *Before getting started, review the following:*

- Amplitude and Period of Sinusoidal Graphs (Section 2.6, pp. 165–172)

Now Work the 'Are You Prepared?' problem on page 455.

OBJECTIVES   **1** Graph Parametric Equations by Hand (p. 445)
  2 Graph Parametric Equations Using a Graphing Utility (p. 446)
  3 Find a Rectangular Equation for a Curve Defined Parametrically (p. 447)
  4 Use Time as a Parameter in Parametric Equations (p. 449)
  5 Find Parametric Equations for Curves Defined by Rectangular Equations (p. 452)

Equations of the form $y = f(x)$, where f is a function, have graphs that are intersected no more than once by any vertical line. The graphs of many of the conics and certain other, more complicated, graphs do not have this characteristic. Yet each graph, like the graph of a function, is a collection of points (x, y) in the xy-plane; that is, each is a *plane curve*. In this section, we discuss another way of representing such graphs.

Let $x = f(t)$ and $y = g(t)$, where f and g are two functions whose common domain is some interval I. The collection of points defined by

$$(x, y) = (f(t), g(t))$$

is called a **plane curve.** The equations

$$x = f(t) \qquad y = g(t)$$

where t is in I, are called **parametric equations** of the curve. The variable t is called a **parameter.**

1 Graph Parametric Equations by Hand

Parametric equations are particularly useful in describing movement along a curve. Suppose that a curve is defined by the parametric equations

Figure 59

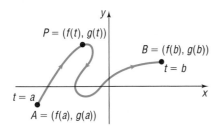

$$x = f(t), \qquad y = g(t), \qquad a \le t \le b$$

where f and g are each defined over the interval $a \le t \le b$. For a given value of t, we can find the value of $x = f(t)$ and $y = g(t)$, obtaining a point (x, y) on the curve. In fact, as t varies over the interval from $t = a$ to $t = b$, successive values of t give rise to a directed movement along the curve; that is, the curve is traced out in a certain direction by the corresponding succession of points (x, y). See Figure 59. The arrows show the direction, or **orientation,** along the curve as t varies from a to b.

EXAMPLE 1   Graphing a Curve Defined by Parametric Equations

Graph the curve defined by the parametric equations

$$x = 3t^2, \qquad y = 2t, \qquad -2 \le t \le 2$$

Solution   For each number t, $-2 \le t \le 2$, there corresponds a number x and a number y. For example, when $t = -2$, then $x = 3(-2)^2 = 12$ and $y = 2(-2) = -4$. When $t = 0$, then $x = 0$ and $y = 0$. We can set up a table listing various choices of the parameter t and the corresponding values for x and y, as shown in Table 6. Plotting these points and connecting them with a smooth curve leads to Figure 60. The arrows in Figure 60 are used to indicate the orientation.

Table 6

t	x	y	(x, y)
−2	12	−4	(12, −4)
−1	3	−2	(3, −2)
0	0	0	(0, 0)
1	3	2	(3, 2)
2	12	4	(12, 4)

Figure 60

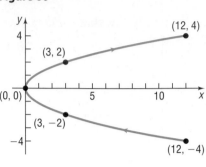

2 Graph Parametric Equations Using a Graphing Utility

Most graphing utilities have the capability of graphing parametric equations. The following steps are usually required to obtain the graph of parametric equations. Check your owner's manual to see how yours works.

Graphing Parametric Equations Using a Graphing Utility

STEP 1: Set the mode to PARametric, Enter $x(t)$ and $y(t)$.
STEP 2: Select the viewing window. In addition to setting Xmin, Xmax, Xscl, and so on, the viewing window in parametric mode requires setting minimum and maximum values for the parameter t and an increment setting for $t(T$step$)$.
STEP 3: Graph.

EXAMPLE 2

Graphing a Curve Defined by Parametric Equations Using a Graphing Utility

Graph the curve defined by the parametric equations

$$x = 3t^2, \qquad y = 2t, \qquad -2 \le t \le 2 \qquad \textbf{(1)}$$

Solution

STEP 1: Enter the equations $x(t) = 3t^2$, $y(t) = 2t$ with the graphing utility in PARametric mode.

STEP 2: Select the viewing window. The interval I is $-2 \le t \le 2$, so we select the following square viewing window:

$$T\text{min} = -2 \qquad X\text{min} = 0 \qquad Y\text{min} = -5$$
$$T\text{max} = 2 \qquad X\text{max} = 15 \qquad Y\text{max} = 5$$
$$T\text{step} = 0.1 \qquad X\text{scl} = 1 \qquad Y\text{scl} = 1$$

We choose Tmin $= -2$ and Tmax $= 2$ because $-2 \le t \le 2$. Finally, the choice for Tstep will determine the number of points the graphing utility will plot. For example, with Tstep at 0.1, the graphing utility will evaluate x and y at $t = -2, -1.9, -1.8$, and so on. The smaller the Tstep, the more points the graphing utility will plot. The reader is encouraged to experiment with different values of Tstep to see how the graph is affected.

STEP 3: Graph. Notice the direction the graph is drawn in. This direction shows the orientation of the curve. See Figure 61.

Figure 61

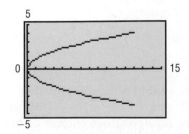

Exploration

Graph the following parametric equations using a graphing utility with $X\text{min} = 0, X\text{max} = 15$, $Y\text{min} = -5, Y\text{max} = 5$, and $T\text{step} = 0.1$:

1. $x = \dfrac{3t^2}{4}, y = t, -4 \le t \le 4$

2. $x = 3t^2 + 12t + 12, y = 2t + 4, -4 \le t \le 0$

3. $x = 3t^{\frac{2}{3}}, y = 2\sqrt[3]{t}, -8 \le t \le 8$

Compare these graphs to the graph in Figure 61. Conclude that parametric equations defining a curve are not unique; that is, different parametric equations can represent the same graph.

3 Find a Rectangular Equation for a Curve Defined Parametrically

The curve given in Examples 1 and 2 should be familiar. To identify it accurately, we find the corresponding rectangular equation by eliminating the parameter t from the parametric equations given in Example 1:

$$x = 3t^2, \qquad y = 2t, \qquad -2 \le t \le 2$$

Noting that we can easily solve for t in $y = 2t$, obtaining $t = \dfrac{y}{2}$, we substitute this expression in the other equation.

$$x = 3t^2 = 3\left(\frac{y}{2}\right)^2 = \frac{3y^2}{4}$$

$$\uparrow$$
$$t = \frac{y}{2}$$

This equation, $x = \dfrac{3y^2}{4}$, is the equation of a parabola with vertex at $(0, 0)$ and axis of symmetry along the x-axis.

Exploration

In FUNCtion mode, graph $x = \dfrac{3y^2}{4}\left(Y_1 = \sqrt{\dfrac{4x}{3}} \text{ and } Y_2 = -\sqrt{\dfrac{4x}{3}}\right)$ with $X\text{min} = 0, X\text{max} = 15$, $Y\text{min} - -5, Y\text{max} = 5$. Compare this graph with Figure 61. Why do the graphs differ?

Note that the parameterized curve defined by equation (1) and shown in Figure 60 (or 61) is only a part of the parabola $x = \dfrac{3y^2}{4}$. The graph of the rectangular equation obtained by eliminating the parameter will, in general, contain more points than the original parameterized curve. Care must therefore be taken when a parameterized curve is sketched by hand after eliminating the parameter. Even so, the process of eliminating the parameter t of a parameterized curve to identify it accurately is sometimes a better approach than plotting points. However, the elimination process sometimes requires a little ingenuity.

EXAMPLE 3

Finding the Rectangular Equation of a Curve Defined Parametrically

Find the rectangular equation of the curve whose parametric equations are

$$x = a \cos t, \qquad y - a \sin t, \qquad -\infty < t < \infty$$

where $a > 0$ is a constant. By hand, graph this curve, indicating its orientation.

Solution

The presence of sines and cosines in the parametric equations suggests that we use a Pythagorean Identity. In fact, since

$$\cos t = \frac{x}{a} \qquad \sin t = \frac{y}{a}$$

we find that

$$\cos^2 t + \sin^2 t = 1$$

$$\left(\frac{x}{a}\right)^2 + \left(\frac{y}{a}\right)^2 = 1$$

$$x^2 + y^2 = a^2$$

Figure 62

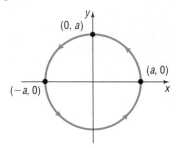

The curve is a circle with center at $(0, 0)$ and radius a. As the parameter t increases, say from $t = 0$ [the point $(a, 0)$] to $t = \dfrac{\pi}{2}$ [the point $(0, a)$] to $t = \pi$ [the point $(-a, 0)$], we see that the corresponding points are traced in a counterclockwise direction around the circle. The orientation is as indicated in Figure 62.

■

Now Work PROBLEMS **7** AND **19**

Let's analyze the curve in Example 3 further. The domain of each parametric equation is $-\infty < t < \infty$. Thus, the graph in Figure 62 is actually being repeated each time that t increases by 2π.

If we wanted the curve to consist of exactly 1 revolution in the counterclockwise direction, we could write

$$x = a \cos t, \qquad y = a \sin t, \qquad 0 \le t \le 2\pi$$

This curve starts at $t = 0$ [the point $(a, 0)$] and, proceeding counterclockwise around the circle, ends at $t = 2\pi$ [also the point $(a, 0)$].

If we wanted the curve to consist of exactly three revolutions in the counterclockwise direction, we could write

$$x = a \cos t, \qquad y = a \sin t, \qquad -2\pi \le t \le 4\pi$$

or

$$x = a \cos t, \qquad y = a \sin t, \qquad 0 \le t \le 6\pi$$

or

$$x = a \cos t, \qquad y = a \sin t, \qquad 2\pi \le t \le 8\pi$$

EXAMPLE 4 **Describing Parametric Equations**

Find rectangular equations for the following curves defined by parametric equations. Graph each curve.

(a) $x = a \cos t, \quad y = a \sin t, \quad 0 \le t \le \pi, \quad a > 0$

(b) $x = -a \sin t, \quad y = -a \cos t, \quad 0 \le t \le \pi, \quad a > 0$

Solution

(a) We eliminate the parameter t using a Pythagorean Identity.

$$\cos^2 t + \sin^2 t = 1$$

$$\left(\frac{x}{a}\right)^2 + \left(\frac{y}{a}\right)^2 = 1$$

$$x^2 + y^2 = a^2$$

The curve defined by these parametric equations is a circle, with radius a and center at $(0, 0)$. The circle begins at the point $(a, 0)$, $t = 0$; passes through the point $(0, a)$, $t = \dfrac{\pi}{2}$; and ends at the point $(-a, 0)$, $t = \pi$.

Figure 63

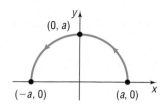

The parametric equations define an upper semicircle of radius a with a counterclockwise orientation. See Figure 63. The rectangular equation is

$$y = \sqrt{a^2 - x^2} \qquad -a \le x \le a$$

(b) We eliminate the parameter t using a Pythagorean Identity.

$$\sin^2 t + \cos^2 t = 1$$

$$\left(\frac{x}{-a}\right)^2 + \left(\frac{y}{-a}\right)^2 = 1$$

$$x^2 + y^2 = a^2$$

Figure 64

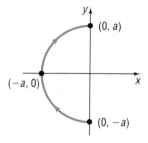

The curve defined by these parametric equations is a circle, with radius a and center at $(0, 0)$. The circle begins at the point $(0, -a)$, $t = 0$; passes through the point $(-a, 0)$, $t = \dfrac{\pi}{2}$; and ends at the point $(0, a)$, $t = \pi$. The parametric equations define a left semicircle of radius a with a clockwise orientation. See Figure 64. The rectangular equation is

$$x = -\sqrt{a^2 - y^2} \qquad -a \le y \le a$$

Example 4 illustrates the versatility of parametric equations for replacing complicated rectangular equations, while providing additional information about orientation. These characteristics make parametric equations very useful in applications, such as projectile motion.

4 Use Time as a Parameter in Parametric Equations

If we think of the parameter t as time, then the parametric equations $x = f(t)$ and $y = g(t)$ of a curve C specify how the x- and y-coordinates of a moving point vary with time.

For example, we can use parametric equations to model the motion of an object, sometimes referred to as **curvilinear motion.** Using parametric equations, we can specify not only where the object travels, that is, its location (x, y), but also when it gets there, that is, the time t.

When an object is propelled upward at an inclination θ to the horizontal with initial speed v_0, the resulting motion is called **projectile motion.** See Figure 65(a).

In calculus it is shown that the parametric equations of the path of a projectile fired at an inclination θ to the horizontal, with an initial speed v_0, from a height h above the horizontal are

$$x = (v_0 \cos \theta)t \qquad y = -\frac{1}{2}gt^2 + (v_0 \sin \theta)t + h \qquad \text{(2)}$$

where t is the time and g is the constant acceleration due to gravity (approximately 32 ft/sec/sec or 9.8 m/sec/sec). See Figure 65(b).

Figure 65

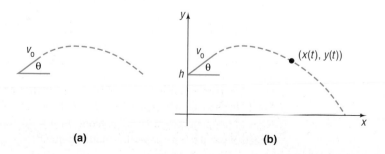

(a)   (b)

| EXAMPLE 5 | Projectile Motion |

Figure 66

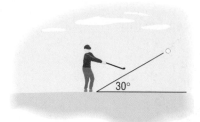

Suppose that Jim hit a golf ball with an initial velocity of 150 feet per second at an angle of 30° to the horizontal. See Figure 66.

(a) Find parametric equations that describe the position of the ball as a function of time.
(b) How long is the golf ball in the air?
(c) When is the ball at its maximum height? Determine the maximum height of the ball.
(d) Determine the distance that the ball traveled.
(e) Using a graphing utility, simulate the motion of the golf ball by simultaneously graphing the equations found in part (a).

Solution
(a) We have $v_0 = 150$ ft/sec, $\theta = 30°$, $h = 0$ (the ball is on the ground), and $g = 32$ (since the units are in feet and seconds). Substituting these values into equations (2), we find that

$$x = (v_0 \cos \theta)t = (150 \cos 30°)t = 75\sqrt{3}\,t$$

$$y = -\frac{1}{2}gt^2 + (v_0 \sin \theta)t + h = -\frac{1}{2}(32)t^2 + (150 \sin 30°)t + 0$$

$$= -16t^2 + 75t$$

(b) To determine the length of time that the ball is in the air, we solve the equation $y = 0$.

$$-16t^2 + 75t = 0$$

$$t(-16t + 75) = 0$$

$$t = 0 \text{ sec} \quad \text{or} \quad t = \frac{75}{16} = 4.6875 \text{ sec}$$

The ball will strike the ground after 4.6875 seconds.

(c) Notice that the height y of the ball is a quadratic function of t, so the maximum height of the ball can be found by determining the vertex of $y = -16t^2 + 75t$. The value of t at the vertex is

$$t = \frac{-b}{2a} = \frac{-75}{-32} = 2.34375 \text{ sec}$$

The ball is at its maximum height after 2.34375 seconds. The maximum height of the ball is found by evaluating the function y at $t = 2.34375$ seconds.

$$\text{Maximum height} = -16(2.34375)^2 + (75)2.34375 \approx 87.89 \text{ feet}$$

Figure 67

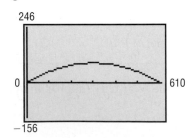

(d) Since the ball is in the air for 4.6875 seconds, the horizontal distance that the ball travels is

$$x = \left(75\sqrt{3}\right)4.6875 \approx 608.92 \text{ feet}$$

(e) We enter the equations from part (a) into a graphing utility with $T\text{min} = 0$, $T\text{max} = 4.7$, and $T\text{step} = 0.1$. We use ZOOM-SQUARE to avoid any distortion to the angle of elevation. See Figure 67.

Exploration

Simulate the motion of a ball thrown straight up with an initial speed of 100 feet per second from a height of 5 feet above the ground. Use PARametric mode with $T\text{min} = 0$, $T\text{max} = 6.5$, $T\text{step} = 0.1$, $X\text{min} = 0$, $X\text{max} = 5$, $Y\text{min} = 0$, and $Y\text{max} = 180$. What happens to the speed with which the graph is drawn as the ball goes up and then comes back down? How do you interpret this physically? Repeat the experiment using other values for $T\text{step}$. How does this affect the experiment?

[**Hint:** In the projectile motion equations, let $\theta = 90°$, $v_0 = 100$, $h = 5$, and $g = 32$. We use $x = 3$ instead of $x = 0$ to see the vertical motion better.]

Result   See Figure 68. In Figure 68(a) the ball is going up. In Figure 68(b) the ball is near its highest point. Finally, in Figure 68(c) the ball is coming back down.

Figure 68

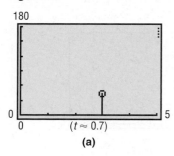

(a)

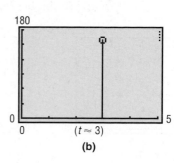

(b)

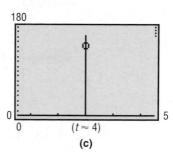

(c)

Notice that, as the ball goes up, its speed decreases, until at the highest point it is zero. Then the speed increases as the ball comes back down.

Now Work   P R O B L E M   **49**

A graphing utility can be used to simulate other kinds of motion as well.

EXAMPLE 6   **Simulating Motion**

Tanya, who is a long distance runner, runs at an average speed of 8 miles per hour. Two hours after Tanya leaves your house, you leave in your Honda and follow the same route. If your average speed is 40 miles per hour, how long will it be before you catch up to Tanya? See Figure 69. Use a simulation of the two motions to verify the answer.

Figure 69

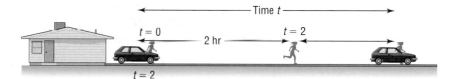

Solution   We begin with two sets of parametric equations: one to describe Tanya's motion, the other to describe the motion of the Honda. We choose time $t = 0$ to be when Tanya leaves the house. If we choose $y_1 = 2$ as Tanya's path, then we can use $y_2 = 4$ as the parallel path of the Honda. The horizontal distances traversed in time t (Distance = Rate × Time) are

$$\text{Tanya:} \quad x_1 = 8t \qquad \text{Honda:} \quad x_2 = 40(t - 2)$$

The Honda catches up to Tanya when $x_1 = x_2$.

$$8t = 40(t - 2)$$
$$8t = 40t - 80$$
$$-32t = -80$$
$$t = \frac{-80}{-32} = 2.5$$

The Honda catches up to Tanya 2.5 hours after Tanya leaves the house.
In PARametric mode with Tstep = 0.01, we simultaneously graph

$$\text{Tanya:} \quad x_1 = 8t \qquad \text{Honda:} \quad x_2 = 40(t - 2)$$
$$y_1 = 2 \qquad\qquad\qquad y_2 = 4$$

for $0 \leq t \leq 3$.

Figure 70 shows the relative position of Tanya and the Honda for $t = 0, t = 2, t = 2.25, t = 2.5,$ and $t = 2.75$.

Figure 70

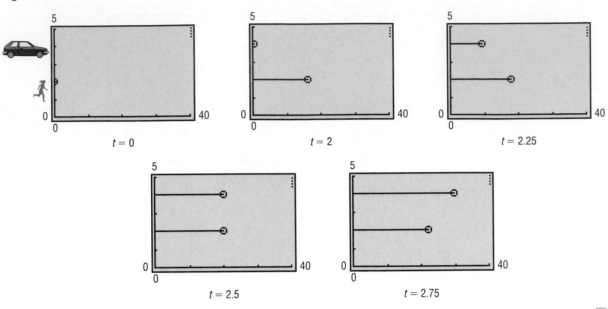

$t = 0$

$t = 2$

$t = 2.25$

$t = 2.5$

$t = 2.75$

5 Find Parametric Equations for Curves Defined by Rectangular Equations

We now take up the question of how to find parametric equations of a given curve. If a curve is defined by the equation $y = f(x)$, where f is a function, one way of finding parametric equations is to let $x = t$. Then $y = f(t)$ and

$$x = t, \quad y = f(t), \qquad t \text{ in the domain of } f$$

are parametric equations of the curve.

EXAMPLE 7 **Finding Parametric Equations for a Curve Defined by a Rectangular Equation**

Find two different parametric equations for the equation $y = x^2 - 4$.

Solution For the first parametric equation, let $x = t$. Then the parametric equations are

$$x = t, \quad y = t^2 - 4, \qquad -\infty < t < \infty$$

A second parametric equation is found by letting $x = t^3$. Then the parametric equations become

$$x = t^3, \quad y = t^6 - 4, \qquad -\infty < t < \infty$$

Care must be taken when using the second approach in Example 7, since the substitution for x must be a function that allows x to take on all the values stipulated by the domain of f. For example, letting $x = t^2$ so that $y = t^4 - 4$ does not result in equivalent parametric equations for $y = x^2 - 4$, since only points for which $x \geq 0$ are obtained; yet the domain of $y = x^2 - 4$ is $\{x \mid x \text{ is any real number}\}$.

Now Work PROBLEM 33

| EXAMPLE 8 | Finding Parametric Equations for an Object in Motion |

Find parametric equations for the ellipse

$$x^2 + \frac{y^2}{9} = 1$$

where the parameter t is time (in seconds) and

(a) The motion around the ellipse is clockwise, begins at the point $(0, 3)$, and requires 1 second for a complete revolution.
(b) The motion around the ellipse is counterclockwise, begins at the point $(1, 0)$, and requires 2 seconds for a complete revolution.

Figure 71

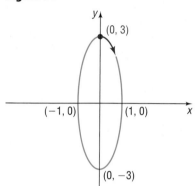

Solution   (a) See Figure 71. Since the motion begins at the point $(0, 3)$, we want $x = 0$ and $y = 3$ when $t = 0$. Furthermore, since the given equation is an ellipse, we begin by letting

$$x = \sin(\omega t) \qquad \frac{y}{3} = \cos(\omega t)$$

for some constant ω. These parametric equations satisfy the equation of the ellipse. Furthermore, with this choice, when $t = 0$, we have $x = 0$ and $y = 3$.

For the motion to be clockwise, the motion will have to begin with the value of x increasing and y decreasing as t increases. This requires that $\omega > 0$. [Do you know why? If $\omega > 0$, then $x = \sin(\omega t)$ is increasing when $t > 0$ is near zero and $y = 3\cos(\omega t)$ is decreasing when $t > 0$ is near zero.] See the red part of the graph in Figure 71.

Finally, since 1 revolution requires 1 second, the period $\dfrac{2\pi}{\omega} = 1$, so $\omega = 2\pi$.

Parametric equations that satisfy the conditions stipulated are

$$x = \sin(2\pi t), \qquad y = 3\cos(2\pi t), \qquad 0 \le t \le 1 \qquad \textbf{(3)}$$

Figure 72

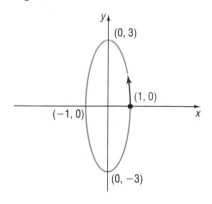

(b) See Figure 72. Since the motion begins at the point $(1, 0)$, we want $x = 1$ and $y = 0$ when $t = 0$. The given equation is an ellipse, so we begin by letting

$$x = \cos(\omega t) \qquad \frac{y}{3} = \sin(\omega t)$$

for some constant ω. These parametric equations satisfy the equation of the ellipse. Furthermore, with this choice, when $t = 0$, we have $x = 1$ and $y = 0$.

For the motion to be counterclockwise, the motion will have to begin with the value of x decreasing and y increasing as t increases. This requires that $\omega > 0$. [Do you know why?] Finally, since 1 revolution requires 2 seconds, the period is $\dfrac{2\pi}{\omega} = 2$, so $\omega = \pi$. The parametric equations that satisfy the conditions stipulated are

$$x = \cos(\pi t), \qquad y = 3\sin(\pi t), \qquad 0 \le t \le 2 \qquad \textbf{(4)}$$

Either equation (3) or (4) can serve as a parametric equation for the ellipse $x^2 + \dfrac{y^2}{9} = 1$ given in Example 8. The direction of the motion, the beginning point, and the time for 1 revolution serve to help us arrive at a particular parametric representation.

Now Work PROBLEM 39

The Cycloid

Suppose that a circle of radius a rolls along a horizontal line without slipping. As the circle rolls along the line, a point P on the circle will trace out a curve called a **cycloid** (see Figure 73). We now seek parametric equations* for a cycloid.

Figure 73

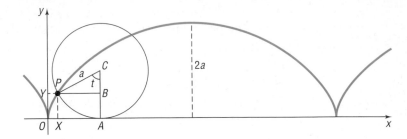

We begin with a circle of radius a and take the fixed line on which the circle rolls as the x-axis. Let the origin be one of the points at which the point P comes in contact with the x-axis. Figure 73 illustrates the position of this point P after the circle has rolled somewhat. The angle t (in radians) measures the angle through which the circle has rolled.

Since we require no slippage, it follows that

$$\text{Arc } AP = d(O, A)$$

The length of the arc AP is given by $s = r\theta$, where $r = a$ and $\theta = t$ radians. Then

$$at = d(O, A) \qquad s = r\theta, \text{ where } r = a \text{ and } \theta = t$$

The x-coordinate of the point P is

$$d(O, X) = d(O, A) - d(X, A) = at - a \sin t = a(t - \sin t)$$

The y-coordinate of the point P is equal to

$$d(O, Y) = d(A, C) - d(B, C) = a - a \cos t = a(1 - \cos t)$$

The parametric equations of the cycloid are

$$x = a(t - \sin t) \qquad y = a(1 - \cos t) \tag{5}$$

Exploration

Graph $x = t - \sin t, y = 1 - \cos t,$ $0 \le t \le 3\pi$, using your graphing utility with $T\text{step} = \dfrac{\pi}{36}$ and a square screen. Compare your results with Figure 73.

Applications to Mechanics

If a is negative in equation (5), we obtain an inverted cycloid, as shown in Figure 74(a). The inverted cycloid occurs as a result of some remarkable applications in the field of mechanics. We shall mention two of them: the *brachistochrone* and the *tautochrone*.†

Figure 74

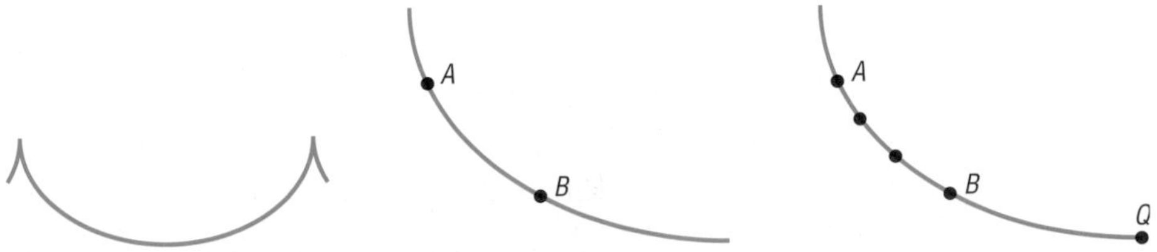

(a) Inverted cycloid   (b) Curve of quickest descent   (c) All reach Q at the same time

*Any attempt to derive the rectangular equation of a cycloid would soon demonstrate how complicated the task is.

†In Greek, *brachistochrone* means "the shortest time" and *tautochrone* means "equal time."

The **brachistochrone** is the curve of quickest descent. If a particle is constrained to follow some path from one point A to a lower point B (not on the same vertical line) and is acted on only by gravity, the time needed to make the descent is least if the path is an inverted cycloid. See Figure 74(b). This remarkable discovery, which is attributed to many famous mathematicians (including Johann Bernoulli and Blaise Pascal), was a significant step in creating the branch of mathematics known as the *calculus of variations*.

Figure 75

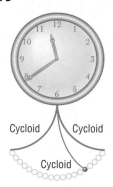

To define the **tautochrone**, let Q be the lowest point on an inverted cycloid. If several particles placed at various positions on an inverted cycloid simultaneously begin to slide down the cycloid, they will reach the point Q at the same time, as indicated in Figure 74(c). The tautochrone property of the cycloid was used by Christiaan Huygens (1629–1695), the Dutch mathematician, physicist, and astronomer, to construct a pendulum clock with a bob that swings along a cycloid (see Figure 75). In Huygen's clock, the bob was made to swing along a cycloid by suspending the bob on a thin wire constrained by two plates shaped like cycloids. In a clock of this design, the period of the pendulum is independent of its amplitude.

6.7 Assess Your Understanding

'Are You Prepared?' *Answers are given at the end of this exercise. If you get a wrong answer, read the page listed in red.*

1. The function $f(x) = 3\sin(4x)$ has amplitude _____ and period _____. (p. 167)

Concepts and Vocabulary

2. Let $x = f(t)$ and $y = g(t)$, where f and g are two functions whose common domain is some interval I. The collection of points defined by $(x, y) = (f(t), g(t))$ is called a(n) _____ _____. The variable t is called a(n) _____.

3. The parametric equations $x = 2\sin t$, $y = 3\cos t$ define a(n) _____.

4. If a circle rolls along a horizontal line without slippage, a point P on the circle will trace out a curve called a(n) _____.

5. *True or False* Parametric equations defining a curve are unique.

6. *True or False* Curves defined using parametric equations have an orientation.

Skill Building

In Problems 7–26, graph the curve whose parametric equations are given and show its orientation. Find the rectangular equation of each curve. Verify your graph using a graphing utility.

7. $x = 3t + 2$, $y = t + 1$; $0 \le t \le 4$

8. $x = t - 3$, $y = 2t + 4$; $0 \le t \le 2$

9. $x = t + 2$, $y = \sqrt{t}$; $t \ge 0$

10. $x = \sqrt{2t}$, $y = 4t$; $t \ge 0$

11. $x = t^2 + 4$, $y = t^2 - 4$; $-\infty < t < \infty$

12. $x = \sqrt{t} + 4$, $y = \sqrt{t} - 4$; $t \ge 0$

13. $x = 3t^2$, $y = t + 1$; $-\infty < t < \infty$

14. $x = 2t - 4$, $y = 4t^2$; $-\infty < t < \infty$

15. $x = 2e^t$, $y = 1 + e^t$; $t \ge 0$

16. $x = e^t$, $y = e^{-t}$; $t \ge 0$

17. $x = \sqrt{t}$, $y = t^{3/2}$; $t \ge 0$

18. $x = t^{3/2} + 1$, $y = \sqrt{t}$; $t \ge 0$

19. $x = 2\cos t$, $y = 3\sin t$; $0 \le t \le 2\pi$

20. $x = 2\cos t$, $y = 3\sin t$; $0 \le t \le \pi$

21. $x = 2\cos t$, $y = 3\sin t$; $-\pi \le t \le 0$

22. $x = 2\cos t$, $y = \sin t$; $0 \le t \le \dfrac{\pi}{2}$

23. $x = \sec t$, $y = \tan t$; $0 \le t \le \dfrac{\pi}{4}$

24. $x = \csc t$, $y = \cot t$; $\dfrac{\pi}{4} \le t \le \dfrac{\pi}{2}$

25. $x = \sin^2 t$, $y = \cos^2 t$; $0 \le t \le 2\pi$

26. $x = t^2$, $y = \ln t$; $t > 0$

In Problems 27–34, find two different parametric equations for each rectangular equation.

27. $y = 4x - 1$

28. $y = -8x + 3$

29. $y = x^2 + 1$

30. $y = -2x^2 + 1$

31. $y = x^3$

32. $y = x^4 + 1$

33. $x = y^{3/2}$

34. $x = \sqrt{y}$

In Problems 35–38, find parametric equations that define the curve shown.

35.
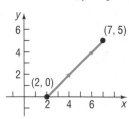

36.
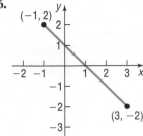

37.
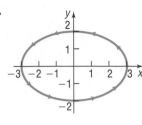

38.
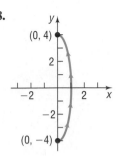

In Problems 39–42, find parametric equations for an object that moves along the ellipse $\dfrac{x^2}{4} + \dfrac{y^2}{9} = 1$ with the motion described.

39. The motion begins at $(2, 0)$, is clockwise, and requires 2 seconds for a complete revolution.

40. The motion begins at $(0, 3)$, is counterclockwise, and requires 1 second for a complete revolution.

41. The motion begins at $(0, 3)$, is clockwise, and requires 1 second for a complete revolution.

42. The motion begins at $(2, 0)$, is counterclockwise, and requires 3 seconds for a complete revolution.

In Problems 43 and 44, the parametric equations of four curves are given. Graph each of them, indicating the orientation.

43. C_1: $x = t$, $y = t^2$; $-4 \le t \le 4$

C_2: $x = \cos t$, $y = 1 - \sin^2 t$; $0 \le t \le \pi$

C_3: $x = e^t$, $y = e^{2t}$; $0 \le t \le \ln 4$

C_4: $x = \sqrt{t}$, $y = t$; $0 \le t \le 16$

44. C_1: $x = t$, $y = \sqrt{1 - t^2}$; $-1 \le t \le 1$

C_2: $x = \sin t$, $y = \cos t$; $0 \le t \le 2\pi$

C_3: $x = \cos t$, $y = \sin t$; $0 \le t \le 2\pi$

C_4: $x = \sqrt{1 - t^2}$, $y = t$; $-1 \le t \le 1$

In Problems 45–48, use a graphing utility to graph the curve defined by the given parametric equations.

45. $x = t \sin t$, $y = t \cos t$, $t > 0$

46. $x = \sin t + \cos t$, $y = \sin t - \cos t$

47. $x = 4 \sin t - 2 \sin(2t)$
$y = 4 \cos t - 2 \cos(2t)$

48. $x = 4 \sin t + 2 \sin(2t)$
$y = 4 \cos t + 2 \cos(2t)$

Applications and Extensions

49. Projectile Motion Bob throws a ball straight up with an initial speed of 50 feet per second from a height of 6 feet.
(a) Find parametric equations that model the motion of the ball as a function of time.
(b) How long is the ball in the air?
(c) When is the ball at its maximum height? Determine the maximum height of the ball.
(d) Simulate the motion of the ball by graphing the equations found in part (a).

50. Projectile Motion Alice throws a ball straight up with an initial speed of 40 feet per second from a height of 5 feet.
(a) Find parametric equations that model the motion of the ball as a function of time.
(b) How long is the ball in the air?
(c) When is the ball at its maximum height? Determine the maximum height of the ball.
(d) Simulate the motion of the ball by graphing the equations found in part (a).

51. Catching a Train Bill's train leaves at 8:06 AM and accelerates at the rate of 2 meters per second per second. Bill, who can run 5 meters per second, arrives at the train station 5 seconds after the train has left and runs for the train.
(a) Find parametric equations that model the motions of the train and Bill as a function of time.

[**Hint:** The position s at time t of an object having acceleration a is $s = \dfrac{1}{2}at^2$.]

(b) Determine algebraically whether Bill will catch the train. If so, when?
(c) Simulate the motion of the train and Bill by simultaneously graphing the equations found in part (a).

52. Catching a Bus Jodi's bus leaves at 5:30 PM and accelerates at the rate of 3 meters per second per second. Jodi, who can run 5 meters per second, arrives at the bus station 2 seconds after the bus has left and runs for the bus.
(a) Find parametric equations that model the motions of the bus and Jodi as a function of time.

[**Hint:** The position s at time t of an object having acceleration a is $s = \dfrac{1}{2}at^2$.]

(b) Determine algebraically whether Jodi will catch the bus. If so, when?
(c) Simulate the motion of the bus and Jodi by simultaneously graphing the equations found in part (a).

53. Projectile Motion Ichiro throws a baseball with an initial speed of 145 feet per second at an angle of 20° to the horizontal. The ball leaves Ichiro's hand at a height of 5 feet.
(a) Find parametric equations that model the position of the ball as a function of time.
(b) How long is the ball in the air?
(c) Determine the horizontal distance that the ball traveled.
(d) When is the ball at its maximum height? Determine the maximum height of the ball.
(e) Using a graphing utility, simultaneously graph the equations found in part (a).

54. Projectile Motion Barry Bonds hit a baseball with an initial speed of 125 feet per second at an angle of 40° to the horizontal. The ball was hit at a height of 3 feet off the ground.
(a) Find parametric equations that model the position of the ball as a function of time.
(b) How long is the ball in the air?
(c) Determine the horizontal distance that the ball traveled.
(d) When is the ball at its maximum height? Determine the maximum height of the ball.
(e) Using a graphing utility, simultaneously graph the equations found in part (a).

55. Projectile Motion Suppose that Adam hits a golf ball off a cliff 300 meters high with an initial speed of 40 meters per second at an angle of 45° to the horizontal.
(a) Find parametric equations that model the position of the ball as a function of time.
(b) How long is the ball in the air?
(c) Determine the horizontal distance that the ball traveled.
(d) When is the ball at its maximum height? Determine the maximum height of the ball.
(e) Using a graphing utility, simultaneously graph the equations found in part (a).

56. Projectile Motion Suppose that Karla hits a golf ball off a cliff 300 meters high with an initial speed of 40 meters per second at an angle of 45° to the horizontal on the Moon (gravity on the Moon is one-sixth of that on Earth).
(a) Find parametric equations that model the position of the ball as a function of time.
(b) How long is the ball in the air?
(c) Determine the horizontal distance that the ball traveled.
(d) When is the ball at its maximum height? Determine the maximum height of the ball.
(e) Using a graphing utility, simultaneously graph the equations found in part (a).

57. Uniform Motion A Toyota Camry (traveling east at 40 mph) and a Chevy Impala (traveling north at 30 mph) are heading toward the same intersection. The Camry is 5 miles from the intersection when the Impala is 4 miles from the intersection. See the figure.

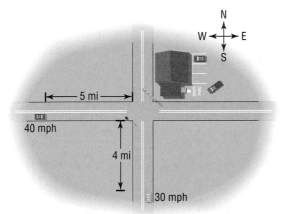

(a) Find parametric equations that model the motion of the Camry and Impala.
(b) Find a formula for the distance between the cars as a function of time.
(c) Graph the function in part (b) using a graphing utility.
(d) What is the minimum distance between the cars? When are the cars closest?
(e) Simulate the motion of the cars by simultaneously graphing the equations found in part (a).

58. Uniform Motion A Cessna (heading south at 120 mph) and a Boeing 747 (heading west at 600 mph) are flying toward the same point at the same altitude. The Cessna is 100 miles from the point where the flight patterns intersect, and the 747 is 550 miles from this intersection point. See the figure.

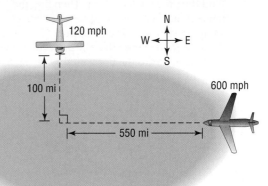

(a) Find parametric equations that model the motion of the Cessna and the 747.
(b) Find a formula for the distance between the planes as a function of time.
(c) Graph the function in part (b) using a graphing utility.
(d) What is the minimum distance between the planes? When are the planes closest?
(e) Simulate the motion of the planes by simultaneously graphing the equations found in part (a).

59. The Green Monster The left field wall at Fenway Park is 310 feet from home plate; the wall itself (affectionately named the Green Monster) is 37 feet high. A batted ball must clear the wall to be a home run. Suppose a ball leaves the bat 3 feet off the ground, at an angle of 45°. Use $g = 32$ feet per second^2 as the acceleration due to gravity and ignore any air resistance.
(a) Find parametric equations that model the position of the ball as a function of time.
(b) What is the maximum height of the ball if it leaves the bat with a speed of 90 miles per hour? Give your answer in feet.
(c) How far is the ball from home plate at its maximum height? Give your answer in feet.
(d) If the ball is hit straight down the left field wall, will it clear the Green Monster? If it does, by how much does it clear the wall?

Source: The Boston Red Sox

60. Projectile Motion The position of a projectile fired with an initial velocity v_0 feet per second and at an angle θ to the horizontal at the end of t seconds is given by the parametric equations

$$x = (v_0 \cos \theta)t \qquad y = (v_0 \sin \theta)t - 16t^2$$

See the illustration.

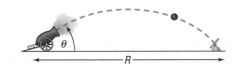

(a) Obtain the rectangular equation of the trajectory and identify the curve.

(b) Show that the projectile hits the ground ($y = 0$) when

$$t = \frac{1}{16}v_0 \sin \theta.$$

(c) How far has the projectile traveled (horizontally) when it strikes the ground? In other words, find the range R.

(d) Find the time t when $x = y$. Then find the horizontal distance x and the vertical distance y traveled by the projectile in this time. Then compute $\sqrt{x^2 + y^2}$. This is the distance R, the range, that the projectile travels up a plane inclined at $45°$ to the horizontal ($x = y$). See the following illustration. (See also Problem 83 in Section 3.5.)

61. Show that the parametric equations for a line passing through the points (x_1, y_1) and (x_2, y_2) are

$$x = (x_2 - x_1)t + x_1$$

$$y = (y_2 - y_1)t + y_1, \quad -\infty < t < \infty$$

What is the orientation of this line?

62. Hypocycloid   The hypocycloid is a curve defined by the parametric equations

$$x(t) = \cos^3 t, \quad y(t) = \sin^3 t, \quad 0 \le t \le 2\pi$$

(a) Graph the hypocycloid using a graphing utility.
(b) Find rectangular equations of the hypocycloid.

Discussion and Writing

63. In Problem 62, we graphed the hypocycloid. Now graph the rectangular equations of the hypocycloid. Did you obtain a complete graph? If not, experiment until you do.

64. Look up the curves called *hypocycloid* and *epicycloid*. Write a report on what you find. Be sure to draw comparisons with the cycloid.

'Are You Prepared?' Answers

1. $3; \dfrac{\pi}{2}$

CHAPTER REVIEW

Things to Know

Equations

Parabola (pp. 397–402)	See Tables 1 and 2 (pp. 399 and 401).
Ellipse (pp. 406–413)	See Table 3 (p. 411).
Hyperbola (pp. 417–426)	See Table 4 (p. 425).

General equation of a conic (p. 437) $Ax^2 + Bxy + Cy^2 + Dx + Ey + F = 0$ Parabola if $B^2 - 4AC = 0$
Ellipse (or circle) if $B^2 - 4AC < 0$
Hyperbola if $B^2 - 4AC > 0$

Polar equations of a conic with focus at the pole (pp. 439–443)   See Table 5 (p. 441).

Parametric equations of a curve (p. 445) $x = f(t), y = g(t), t$ is the parameter

Definitions

Parabola (p. 397)   Set of points P in the plane for which $d(F, P) = d(P, D)$, where F is the focus and D is the directrix

Ellipse (p. 407)   Set of points P in the plane, the sum of whose distances from two fixed points (the foci) is a constant

Hyperbola (p. 417)   Set of points P in the plane, the difference of whose distances from two fixed points (the foci) is a constant

Conic in polar coordinates (p. 439) $\dfrac{d(F, P)}{d(P, D)} = e$ Parabola if $e = 1$
Ellipse if $e < 1$
Hyperbola if $e > 1$

Formulas

Rotation formulas (p. 433) $x = x' \cos \theta - y' \sin \theta$
$y = x' \sin \theta + y' \cos \theta$

Angle θ of rotation that eliminates the $x'y'$-term (p. 434) $\cot(2\theta) = \dfrac{A - C}{B}, \quad 0° < \theta < 90°$

Objectives

Section		You should be able to . . .	Example(s)	Review Exercises
6.1	1	Know the names of the conics (p. 396)		1–32
6.2	1	Analyze parabolas with vertex at the origin (p. 397)	1–6	1, 2, 21, 24
	2	Analyze parabolas with vertex at (h, k) (p. 401)	7–9	7, 11, 12, 17, 18, 27, 30
	3	Solve applied problems involving parabolas (p. 402)	10	77, 78
6.3	1	Analyze ellipses with center at the origin (p. 407)	1–5	5, 6, 10, 22, 25
	2	Analyze ellipses with center at (h, k) (p. 411)	6–8	14–16, 19, 28, 31
	3	Solve applied problems involving ellipses (p. 414)	9	79, 80
6.4	1	Analyze hyperbolas with center at the origin (p. 418)	1–5	3, 4, 8, 9, 23, 26
	2	Find the asymptotes of a hyperbola (p. 423)	6	3, 4, 8, 9
	3	Analyze hyperbolas with center at (h, k) (p. 424)	7, 8	13, 20, 29, 32–36
	4	Solve applied problems involving hyperbolas (p. 426)	9	81
6.5	1	Identify a conic (p. 431)	1	37–40
	2	Use a rotation of axes to transform equations (p. 432)	2	47–52
	3	Analyze an equation using a rotation of axes (p. 434)	3, 4	47–52
	4	Identify conics without a rotation of axes (p. 436)	5	41–46
6.6	1	Analyze and graph polar equations of conics (p. 439)	1–3	53–58
	2	Convert the polar equation of a conic to a rectangular equation (p. 443)	4	59–62
6.7	1	Graph parametric equations by hand (p. 445)	1	63–68
	2	Graph parametric equations using a graphing utility (p. 446)	2	63–68
	3	Find a rectangular equation for a curve defined parametrically (p. 447)	3, 4	63–68
	4	Use time as a parameter in parametric equations (p. 449)	5, 6	82, 83
	5	Find parametric equations for curves defined by rectangular equations (p. 452)	7, 8	69–72

Review Exercises

In Problems 1–20, identify each equation. If it is a parabola, give its vertex, focus, and directrix; if it is an ellipse, give its center, vertices, and foci; if it is a hyperbola, give its center, vertices, foci, and asymptotes.

1. $y^2 = -16x$ **2.** $16x^2 = y$ **3.** $\dfrac{x^2}{25} - y^2 = 1$ **4.** $\dfrac{y^2}{25} - x^2 = 1$

5. $\dfrac{y^2}{25} + \dfrac{x^2}{16} = 1$ **6.** $\dfrac{x^2}{9} + \dfrac{y^2}{16} = 1$ **7.** $x^2 + 4y = 4$ **8.** $3y^2 - x^2 = 9$

9. $4x^2 - y^2 = 8$ **10.** $9x^2 + 4y^2 = 36$ **11.** $x^2 - 4x = 2y$ **12.** $2y^2 - 4y = x - 2$

13. $y^2 - 4y - 4x^2 + 8x = 4$ **14.** $4x^2 + y^2 + 8x - 4y + 4 = 0$ **15.** $4x^2 + 9y^2 - 16x - 18y = 11$

16. $4x^2 + 9y^2 - 16x + 18y = 11$ **17.** $4x^2 - 16x + 16y + 32 = 0$ **18.** $4y^2 + 3x - 16y + 19 = 0$

19. $9x^2 + 4y^2 - 18x + 8y = 23$ **20.** $x^2 - y^2 - 2x - 2y = 1$

In Problems 21–36, find an equation of the conic described. Graph the equation by hand.

21. Parabola; focus at $(-2, 0)$; directrix the line $x = 2$

22. Ellipse; center at $(0, 0)$; focus at $(0, 3)$; vertex at $(0, 5)$

23. Hyperbola; center at $(0,0)$; focus at $(0,4)$; vertex at $(0, -2)$

24. Parabola; vertex at $(0, 0)$; directrix the line $y = -3$

25. Ellipse; foci at $(-3, 0)$ and $(3, 0)$; vertex at $(4, 0)$

26. Hyperbola; vertices at $(-2, 0)$ and $(2, 0)$; focus at $(4, 0)$

27. Parabola; vertex at $(2, -3)$; focus at $(2, -4)$

28. Ellipse; center at $(-1, 2)$; focus at $(0, 2)$; vertex at $(2, 2)$

29. Hyperbola; center at $(-2, -3)$; focus at $(-4, -3)$; vertex at $(-3, -3)$

30. Parabola; focus at $(3, 6)$; directrix the line $y = 8$

31. Ellipse; foci at $(-4, 2)$ and $(-4, 8)$; vertex at $(-4, 10)$

32. Hyperbola; vertices at $(-3, 3)$ and $(5, 3)$; focus at $(7, 3)$

33. Center at $(-1, 2)$; $a = 3$; $c = 4$; transverse axis parallel to the x-axis

34. Center at $(4, -2)$; $a = 1$; $c = 4$; transverse axis parallel to the y-axis

35. Vertices at $(0, 1)$ and $(6, 1)$; asymptote the line $3y + 2x = 9$

36. Vertices at $(4, 0)$ and $(4, 4)$; asymptote the line $y + 2x = 10$

In Problems 37–46, identify each conic without completing the squares and without applying a rotation of axes.

37. $y^2 + 4x + 3y - 8 = 0$

38. $2x^2 - y + 8x = 0$

39. $x^2 + 2y^2 + 4x - 8y + 2 = 0$

40. $x^2 - 8y^2 - x - 2y = 0$

41. $9x^2 - 12xy + 4y^2 + 8x + 12y = 0$

42. $4x^2 + 4xy + y^2 - 8\sqrt{5}x + 16\sqrt{5}y = 0$

43. $4x^2 + 10xy + 4y^2 - 9 = 0$

44. $4x^2 - 10xy + 4y^2 - 9 = 0$

45. $x^2 - 2xy + 3y^2 + 2x + 4y - 1 = 0$

46. $4x^2 + 12xy - 10y^2 + x + y - 10 = 0$

In Problems 47–52, rotate the axes so that the new equation contains no xy-term. Analyze and graph the new equation.

47. $2x^2 + 5xy + 2y^2 - \dfrac{9}{2} = 0$

48. $2x^2 - 5xy + 2y^2 - \dfrac{9}{2} = 0$

49. $6x^2 + 4xy + 9y^2 - 20 = 0$

50. $x^2 + 4xy + 4y^2 + 16\sqrt{5}x - 8\sqrt{5}y = 0$

51. $4x^2 - 12xy + 9y^2 + 12x + 8y = 0$

52. $9x^2 - 24xy + 16y^2 + 80x + 60y = 0$

In Problems 53–58, identify the conic that each polar equation represents and graph it by hand. Verify your graph using a graphing utility.

53. $r = \dfrac{4}{1 - \cos\theta}$

54. $r = \dfrac{6}{1 + \sin\theta}$

55. $r = \dfrac{6}{2 - \sin\theta}$

56. $r = \dfrac{2}{3 + 2\cos\theta}$

57. $r = \dfrac{8}{4 + 8\cos\theta}$

58. $r = \dfrac{10}{5 + 20\sin\theta}$

In Problems 59–62, convert each polar equation to a rectangular equation.

59. $r = \dfrac{4}{1 - \cos\theta}$

60. $r = \dfrac{6}{2 - \sin\theta}$

61. $r = \dfrac{8}{4 + 8\cos\theta}$

62. $r = \dfrac{2}{3 + 2\cos\theta}$

In Problems 63–68, by hand, graph the curve whose parametric equations are given and show its orientation. Find the rectangular equation of each curve. Verify your graph using a graphing utility.

63. $x = 4t - 2$, $y = 1 - t$; $-\infty < t < \infty$

64. $x = 2t^2 + 6$, $y = 5 - t$; $-\infty < t < \infty$

65. $x = 3\sin t$, $y = 4\cos t + 2$; $0 \le t \le 2\pi$

66. $x = \ln t$, $y = t^3$; $t > 0$

67. $x = \sec^2 t$, $y = \tan^2 t$; $0 \le t \le \dfrac{\pi}{4}$

68. $x = t^{\frac{3}{2}}$, $y = 2t + 4$; $t \ge 0$

In Problems 69 and 70, find two different parametric equations for each rectangular equation.

69. $y = -2x + 4$

70. $y = 2x^2 - 8$

In Problems 71 and 72, find parametric equations for an object that moves along the ellipse $\dfrac{x^2}{16} + \dfrac{y^2}{9} = 1$ with the motion described.

71. The motion begins at $(4, 0)$, is counterclockwise, and requires 4 seconds for a complete revolution.

72. The motion begins at $(0, 3)$, is clockwise, and requires 5 seconds for a complete revolution.

---

73. Find an equation of the hyperbola whose foci are the vertices of the ellipse $4x^2 + 9y^2 = 36$ and whose vertices are the foci of this ellipse.

74. Find an equation of the ellipse whose foci are the vertices of the hyperbola $x^2 - 4y^2 = 16$ and whose vertices are the foci of this hyperbola.

75. Describe the collection of points in a plane so that the distance from each point to the point $(3, 0)$ is three-fourths of its distance from the line $x = \dfrac{16}{3}$.

76. Describe the collection of points in a plane so that the distance from each point to the point $(5, 0)$ is five-fourths of its distance from the line $x = \dfrac{16}{5}$.

77. Searchlight A searchlight is shaped like a paraboloid of revolution. If a light source is located 1 foot from the vertex

along the axis of symmetry and the opening is 2 feet across, how deep should the mirror be in order to reflect the light rays parallel to the axis of symmetry?

78. Parabolic Arch Bridge A bridge is built in the shape of a parabolic arch. The bridge has a span of 60 feet and a maximum height of 20 feet. Find the height of the arch at distances of 5, 10, and 20 feet from the center.

79. Semielliptical Arch Bridge A bridge is built in the shape of a semielliptical arch. The bridge has a span of 60 feet and a maximum height of 20 feet. Find the height of the arch at distances of 5, 10, and 20 feet from the center.

80. Whispering Gallery The figure shows the specifications for an elliptical ceiling in a hall designed to be a whispering gallery. Where are the foci located in the hall?

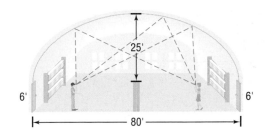

81. Calibrating Instruments In a test of their recording devices, a team of seismologists positioned two of the devices 2000 feet apart, with the device at point A to the west of the device at point B. At a point between the devices and 200 feet from point B, a small amount of explosive was detonated and a note made of the time at which the sound reached each device. A second explosion is to be carried out at a point directly north of point B. How far north should the site of the second explosion be chosen so that the measured time difference recorded by the devices for the second detonation is the same as that recorded for the first detonation?

82. Uniform Motion Mary's train leaves at 7:15 AM and accelerates at the rate of 3 meters per second per second. Mary, who can run 6 meters per second, arrives at the train station 2 seconds after the train has left.

(a) Find parametric equations that model the motion of the train and Mary as a function of time.

   [**Hint:** The position s at time t of an object having acceleration a is $s = \dfrac{1}{2}at^2$.]

(b) Determine algebraically whether Mary will catch the train. If so, when?

(c) Simulate the motions of the train and Mary by simultaneously graphing the equations found in part (a).

83. Projectile Motion Drew Brees throws a football with an initial speed of 80 feet per second at an angle of 35° to the horizontal. The ball leaves Brees's hand at a height of 6 feet.

(a) Find parametric equations that model the position of the ball as a function of time.

(b) How long is the ball in the air?

(c) When is the ball at its maximum height? Determine the maximum height of the ball.

(d) Determine the horizontal distance that the ball travels.

(e) Using a graphing utility, simultaneously graph the equations found in part (a).

84. Formulate a strategy for discussing and graphing an equation of the form

$$Ax^2 + Bxy + Cy^2 + Dx + Ey + F = 0$$

CHAPTER TEST

In Problems 1–3, identify each equation. If it is a parabola, give its vertex, focus, and directrix; if an ellipse, give its center, vertices, and foci; if a hyperbola, give its center, vertices, foci, and asymptotes.

1. $\dfrac{(x+1)^2}{4} - \dfrac{y^2}{9} = 1$

2. $8y = (x-1)^2 - 4$

3. $2x^2 + 3y^2 + 4x - 6y = 13$

In Problems 4–6, find an equation of the conic described; graph the equation by hand.

4. Parabola: focus $(-1, 4.5)$, vertex $(-1, 3)$

5. Ellipse: center $(0, 0)$, vertex $(0, -4)$, focus $(0, 3)$

6. Hyperbola: center $(2, 2)$, vertex $(2, 4)$, contains the point $\left(2 + \sqrt{10}, 5\right)$

In Problems 7–9, identify each conic without completing the square or rotating axes.

7. $2x^2 + 5xy + 3y^2 + 3x - 7 = 0$

8. $3x^2 - xy + 2y^2 + 3y + 1 = 0$

9. $x^2 - 6xy + 9y^2 + 2x - 3y - 2 = 0$

10. Given the equation $41x^2 - 24xy + 34y^2 - 25 = 0$, rotate the axes so that there is no xy-term. Analyze and graph the new equation.

11. Identify the conic represented by the polar equation $r = \dfrac{3}{1 - 2\cos\theta}$. Find the rectangular equation.

12. Graph the curve whose parametric equations are given and show its orientation. Find the rectangular equation for the curve.

$$x = 3t - 2, \quad y = 1 - \sqrt{t}, \quad 0 \le t \le 9$$

13. A parabolic reflector (paraboloid of revolution) is used by TV crews at football games to pick up the referee's announcements, quarterback signals, and so on. A microphone is placed at the focus of the parabola. If a certain reflector is 4 feet wide and 1.5 feet deep, where should the microphone be placed?

CUMULATIVE REVIEW

1. Find an equation for each of the following graphs:

(a) Line: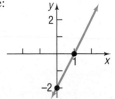

(b) Circle: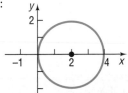

(c) Ellipse: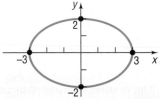

(d) Parabola:

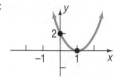

(e) Hyperbola:

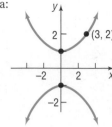

2. Find all the solutions of the equation $\sin(2\theta) = 0.5$.

3. Find a polar equation for the line containing the origin that makes an angle of 30° with the positive x-axis.

4. Find a polar equation for the circle with center at the point $(0, 4)$ and radius 4. Graph this circle.

5. What is the domain of the function $f(x) = \dfrac{3}{\sin x + \cos x}$?

6. Solve the equation $\cot(2\theta) = 1$, where $0° < \theta < 90°$.

7. Find the rectangular equation of the curve

$$x = 5 \tan t, \quad y = 5 \sec^2 t, \quad -\frac{\pi}{2} < t < \frac{\pi}{2}$$

CHAPTER PROJECTS

I. **The Orbits of Neptune and Pluto** The orbit of a planet about the Sun is an ellipse, with the Sun at one focus. The **aphelion** of a planet is its greatest distance from the Sun and the **perihelion** is its shortest distance. The **mean distance** of a planet from the Sun is the length of the semimajor axis of the elliptical orbit. See the illustration.

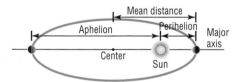

1. The aphelion of Neptune is 4532.2×10^6 kilometers (km) and its perihelion is 4458.0×10^6 km. Find a model for the orbit of Neptune around the Sun.

2. The aphelion of Pluto* is 7381.2×10^6 km and its perihelion is 4445.8×10^6 km. Find a model for the orbit of Pluto around the Sun.

3. Graph the orbits of Pluto and Neptune on a graphing utility. Knowing that the orbits of the planets intersect, what is wrong with the graphs you obtained?

4. The graphs of the orbits drawn in part 3 have the same center, but their foci lie in different locations. To see an accurate representation, the location of the Sun (a focus) needs to be the same for both graphs. This can be accomplished by shifting Pluto's orbit to the left. The shift amount is equal to Pluto's distance from the center (in the graph in part 3) to the Sun minus Neptune's distance from the center to the Sun. Find the new model representing the orbit of Pluto.

5. Graph the equation for the orbit of Pluto found in part 4 along with the equation of the orbit of Neptune. Do you see that Pluto's orbit is sometimes inside Neptune's?

6. Find the point(s) of intersection of the two orbits.

7. Do you think two planets will ever collide?

The following projects can be found on the Instructor's Resource Center (IRC):

II. **Project at Motorola** *Distorted Deployable Space Reflector Antennas* An engineer designs an antenna that will deploy in space to collect sunlight.

III. **Constructing a Bridge over the East River** The size of ships using a river and fluctuations in water height due to tides or flooding must be considered when designing a bridge that will cross a major waterway.

IV. **Systems of Parametric Equations** Choosing an approach to use when solving a system of equations depends on the form of the system and on the domains of the equations.

*Pluto's status was reduced to a dwarf planet in September 2006.

Exponential and Logarithmic Functions

7

The McDonald's Scalding Coffee Case

April 3, 1996—There is a lot of hype about the McDonald's scalding coffee case. No one is in favor of frivolous cases or outlandish results; however, it is important to understand some points that were not reported in most of the stories about the case. McDonald's coffee was not only hot, it was scalding, capable of almost instantaneous destruction of skin, flesh, and muscle.

Plaintiff's expert, a scholar in thermodynamics applied to human skin burns, testified that liquids at 180 degrees will cause a full thickness burn to human skin in two to seven seconds. Other testimony showed that, as the temperature decreases toward 155 degrees, the extent of the burn relative to that temperature decreases exponentially. Thus, if (the) spill had involved coffee at 155 degrees, the liquid would have cooled and given her time to avoid a serious burn.

Source: ATLA fact sheet © 1995, 1996 Consumer Attorneys of CA. Used with permission of the Association of Trial Lawyers of America.

—See Chapter Project I—

A Look Back

Polynomial and rational functions belong to the class of **algebraic functions,** that is, functions that can be expressed in terms of sums, differences, products, quotients, powers, or roots of polynomials. Functions that are not algebraic are termed **transcendental** (they transcend, or go beyond, algebraic functions). Trigonometric functions are part of this class of functions.

A Look Ahead

In this chapter, we study two more transcendental functions: the exponential function and the logarithmic function. These functions occur frequently in a wide variety of applications, such as biology, chemistry, economics, and psychology.

Outline

7.1 Exponential Functions

PREPARING FOR THIS SECTION   *Before getting started, review the following:*

- Exponents (Appendix, Section A.1, pp. A7–A9, and Section A.8, pp. A64–A65)
- Graphing Techniques: Transformations (Section 1.7, pp. 67–76)
- Average Rate of Change (Section 1.5, pp. 55–56)

- Solving Linear and Quadratic Equations (Appendix, Section A.4, pp. A27–A28 and A29–A35)
- Horizontal Asymptotes (Section 1.7, pp. 76–77)

Now Work the 'Are You Prepared?' problems on page 474.

OBJECTIVES   1   Evaluate Exponential Functions (p. 464)
2   Graph Exponential Functions (p. 468)
3   Define the Number *e* (p. 471)
4   Solve Exponential Equations (p. 473)

1 Evaluate Exponential Functions

In the Appendix, Section A.8, we give a definition for raising a real number *a* to a rational power. Based on that discussion, we gave meaning to expressions of the form

$$a^r$$

where the base *a* is a positive real number and the exponent *r* is a rational number. But what is the meaning of a^x, where the base *a* is a positive real number and the exponent *x* is an irrational number? Although a rigorous definition requires methods discussed in calculus, the basis for the definition is easy to follow: Select a rational number *r* that is formed by truncating (removing) all but a finite number of digits from the irrational number *x*. Then it is reasonable to expect that

$$a^x \approx a^r$$

For example, take the irrational number $\pi = 3.14159\ldots$. Then an approximation to a^π is

$$a^\pi \approx a^{3.14}$$

where the digits after the hundredths position have been removed from the value for π. A better approximation would be

$$a^\pi \approx a^{3.14159}$$

where the digits after the hundred-thousandths position have been removed. Continuing in this way, we can obtain approximations to a^π to any desired degree of accuracy.

Most calculators have an $\boxed{x^y}$ key or a caret key $\boxed{\wedge}$ for working with exponents. To evaluate expressions of the form a^x, enter the base *a*, then press the $\boxed{x^y}$ key (or the $\boxed{\wedge}$ key), enter the exponent *x*, and press $\boxed{=}$ (or $\boxed{\text{ENTER}}$).

EXAMPLE 1   Using a Calculator to Evaluate Powers of 2

Using a calculator, evaluate:

(a) $2^{1.4}$ (b) $2^{1.41}$ (c) $2^{1.414}$

(d) $2^{1.4142}$ (e) $2^{\sqrt{2}}$

Figure 1

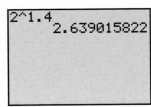

Solution Figure 1 shows the solution to part (a) using a TI-84 Plus graphing calculator.

(a) $2^{1.4} \approx 2.639015822$

(b) $2^{1.41} \approx 2.657371628$

(c) $2^{1.414} \approx 2.66474965$

(d) $2^{1.4142} \approx 2.665119089$

(e) $2^{\sqrt{2}} \approx 2.665144143$

────── **Now Work** PROBLEM 11

It can be shown that the familiar laws for rational exponents hold for real exponents.

THEOREM

Laws of Exponents

If $s, t, a,$ and b are real numbers with $a > 0$ and $b > 0$, then

$$a^s \cdot a^t = a^{s+t} \qquad (a^s)^t = a^{st} \qquad (ab)^s = a^s \cdot b^s$$

$$1^s = 1 \qquad a^{-s} = \frac{1}{a^s} = \left(\frac{1}{a}\right)^s \qquad a^0 = 1 \qquad \textbf{(1)}$$

Introduction to Exponential Growth

Let's examine a function f that has the following two properties:

1. The value of f doubles with every 1-unit increase in the independent variable x.

2. The value of f at $x = 0$ is 5, so $f(0) = 5$.

Table 1 shows values of the function f for $x = 0, 1, 2, 3,$ and 4.

We seek an equation $y = f(x)$ that describes this function f. The key is the fact that the value of f doubles for every 1-unit increase in x.

Table 1

x	$f(x)$
0	5
1	10
2	20
3	40
4	80

$f(0) = 5$

$f(1) = 2f(0) = 2 \cdot 5 = 5 \cdot 2^1$ *We double the value of f at 0 to get the value at 1.*

$f(2) = 2f(1) = 2(5 \cdot 2) = 5 \cdot 2^2$ *We double the value of f at 1 to get the value at 2.*

$f(3) = 2f(2) = 2(5 \cdot 2^2) = 5 \cdot 2^3$

$f(4) = 2f(3) = 2(5 \cdot 2^3) = 5 \cdot 2^4$

The pattern leads us to

$$f(x) = 2f(x - 1) = 2(5 \cdot 2^{x-1}) = 5 \cdot 2^x$$

We are now ready for the following definition.

DEFINITION

An **exponential function** is a function of the form

$$f(x) = Ca^x$$

where a is a positive real number ($a > 0$) and $a \neq 1$, and $C \neq 0$ is a real number. The domain of f is the set of all real numbers. The base a is the **growth factor,** and because $f(0) = Ca^0 = C$, we call C the **initial value.**

In the definition of an exponential function, we exclude the base $a = 1$ because this function is simply the constant function $f(x) = C \cdot 1^x = C$. We also need to exclude bases that are negative; otherwise, we would have to exclude many values

WARNING It is important to distinguish a power function, $g(x) = ax^n$, $n \geq 2$, an integer, from an exponential function, $f(x) = C \cdot a^x$, $a \neq 1$, $a > 0$. In a power function, the base is a variable and the exponent is a constant. In an exponential function, the base is a constant and the exponent is a variable. ∎

of x from the domain, such as $x = \dfrac{1}{2}$ and $x = \dfrac{3}{4}$. [Recall that $(-2)^{1/2} = \sqrt{-2}$, $(-3)^{3/4} = \sqrt[4]{(-3)^3} = \sqrt[4]{-27}$, and so on, are not defined in the set of real numbers.]

Some examples of exponential functions are

$$f(x) = 2^x \qquad F(x) = \left(\frac{1}{3}\right)^x \qquad G(x) = 2 \cdot 3^x$$

Notice for each function that the base is a constant and the exponent is a variable.

In the function $f(x) = 5 \cdot 2^x$, notice that the ratio of consecutive outputs is constant for 1-unit increases in the input. This ratio equals the constant 2, the base of the exponential function. In other words,

$$\frac{f(1)}{f(0)} = \frac{5 \cdot 2^1}{5} = 2 \qquad \frac{f(2)}{f(1)} = \frac{5 \cdot 2^2}{5 \cdot 2^1} = 2 \qquad \frac{f(3)}{f(2)} = \frac{5 \cdot 2^3}{5 \cdot 2^2} = 2 \quad \text{and so on}$$

This leads to the following result.

THEOREM

For an exponential function $f(x) = C \cdot a^x$, $a > 0$, $a \neq 1$, if x is any real number, then

$$\frac{f(x+1)}{f(x)} = a \quad \text{or} \quad f(x+1) = af(x)$$

In Words
For 1-unit changes in the input x of an exponential function $f(x) = C \cdot a^x$, the ratio of consecutive outputs is the constant a.

Proof

$$\frac{f(x+1)}{f(x)} = \frac{Ca^{x+1}}{Ca^x} = a^{x+1-x} = a^1 = a \qquad ∎$$

EXAMPLE 2   **Identifying Linear or Exponential Functions**

Determine whether the given function is linear, exponential, or neither. For those that are linear, find a linear function that models the data. For those that are exponential, find an exponential function that models the data.

(a)

x	y
−1	5
0	2
1	−1
2	−4
3	−7

(b)

x	y
−1	2
0	4
1	7
2	11
3	15

(c)

x	y
−1	32
0	16
1	8
2	4
3	2

Solution

For each function we compute the average rate of change of y with respect to x. If the average rate of change is constant, then the function is linear. Also, we compute the ratio of consecutive outputs. If the ratio is constant, then the function is exponential.

(a) See Table 2(a). The average rate of change for every 1-unit increase in x is –3. Therefore, the function is a linear function. In a linear function the average rate of change is the slope m, so $m = -3$. The y-intercept b is the value of the function at $x = 0$, so $b = 2$. The linear function that models the data is $f(x) = mx + b = -3x + 2$.

Table 2

x	y	Average Rate of Change	Ratio of Consecutive Outputs
−1	5	$\dfrac{\Delta y}{\Delta x} = \dfrac{2-5}{0-(-1)} = -3$	$\dfrac{2}{5}$
0	2	−3	$-\dfrac{1}{2}$
1	−1	−3	4
2	−4	−3	$\dfrac{7}{4}$
3	−7		

(a)

x	y	Average Rate of Change	Ratio of Consecutive Outputs
−1	2	$\dfrac{\Delta y}{\Delta x} = \dfrac{4-2}{0-(-1)} = 2$	2
0	4	3	$\dfrac{7}{4}$
1	7	4	$\dfrac{11}{7}$
2	11	5	$\dfrac{16}{11}$
3	16		

(b)

x	y	Average Rate of Change	Ratio of Consecutive Outputs
−1	32	$\dfrac{\Delta y}{\Delta x} = \dfrac{16-32}{0-(-1)} = -16$	$\dfrac{16}{32} = \dfrac{1}{2}$
0	16	−8	$\dfrac{8}{16} = \dfrac{1}{2}$
1	8	−4	$\dfrac{4}{8} = \dfrac{1}{2}$
2	4	−2	$\dfrac{2}{4} = \dfrac{1}{2}$
3	2		

(c)

(b) See Table 2(b). In this function the average rate of change from −1 to 0 is 2, and the average rate of change from 0 to 1 is 3. Because the average rate of change is not constant, the function is not a linear function. In this function, the ratio of consecutive outputs from −1 to 0 is 2, and the ratio of consecutive outputs from 0 to 1 is $\dfrac{7}{4}$. Because the ratio of consecutive outputs is not a constant, the function is not an exponential function.

(c) See Table 2(c). In this function the average rate of change from −1 to 0 is −16, and the average rate of change from 0 to 1 is −8. Because the average rate of change is not constant, the function is not a linear function. The ratio of consecutive outputs for a 1-unit increase in the inputs is a constant, $\dfrac{1}{2}$. Because the ratio of consecutive outputs is constant, the function is an exponential function with growth factor $a = \dfrac{1}{2}$. The initial value of the exponential function is $C = 16$. Therefore, the exponential function that models the data is $g(x) = Ca^x = 16 \cdot \left(\dfrac{1}{2}\right)^x$.

Now Work PROBLEM 21

2 Graph Exponential Functions

If we know how to graph an exponential function of the form $f(x) = a^x$, then a compression, stretch, and perhaps reflection about the x-axis will enable us to obtain the graph of $f(x) = C \cdot a^x$.

First, we graph the exponential function $f(x) = 2^x$.

| EXAMPLE 3 | Graphing an Exponential Function |

Graph the exponential function: $f(x) = 2^x$

Solution

Table 3

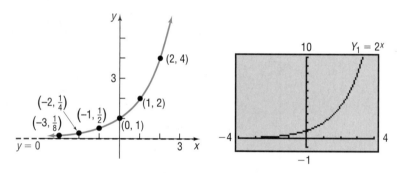

The domain of $f(x) = 2^x$ is the set of all real numbers. We begin by locating some points on the graph of $f(x) = 2^x$, as listed in Table 3.

Since $2^x > 0$ for all x, the range of f is $(0, \infty)$. From this, we conclude that the graph has no x-intercepts, and, in fact, the graph will lie above the x-axis for all x. As Table 3 indicates, the y-intercept is 1. Table 3 also indicates that as $x \to -\infty$ the value of $f(x) = 2^x$ gets closer and closer to 0. We conclude that the x-axis ($y = 0$) is a horizontal asymptote to the graph as $x \to -\infty$. This gives us the end behavior for x large and negative.

To determine the end behavior for x large and positive, look again at Table 3. As $x \to \infty$, $f(x) = 2^x$ grows very quickly, causing the graph of $f(x) = 2^x$ to rise very rapidly. It is apparent that f is an increasing function and hence is one-to-one.

Figure 2 shows the graph of $f(x) = 2^x$. Notice that all the conclusions given earlier are confirmed by the graph.

Figure 2

As we shall see, graphs that look like the one in Figure 2 occur very frequently in a variety of situations. For example, look at the graph in Figure 3, which illustrates the number of cellular telephone subscribers at the end of each year from 1985 to 2006. We might conclude from this graph that the number of cellular telephone subscribers is growing *exponentially*.

Figure 3

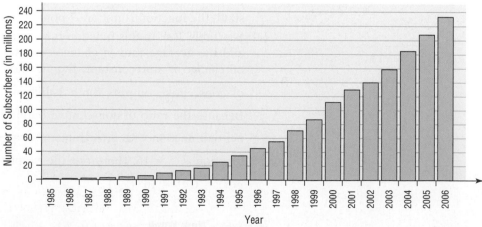

Number of Cellular Phone Subscribers at Year End

We shall have more to say about situations that lead to exponential growth later in this chapter. For now, we continue to seek properties of the exponential functions.

The graph of $f(x) = 2^x$ in Figure 2 is typical of all exponential functions of the form $f(x) = a^x$ with $a > 1$. Such functions are increasing functions and hence are one-to-one. Their graphs lie above the x-axis, pass through the point $(0, 1)$, and thereafter rise rapidly as $x \to \infty$. As $x \to -\infty$, the x-axis ($y = 0$) is a horizontal asymptote. There are no vertical asymptotes. Finally, the graphs are smooth and continuous, with no corners or gaps.

Figures 4 and 5 illustrate the graphs of two more exponential functions whose bases are larger than 1. Notice the larger the base, the steeper the graph is when $x > 0$ and when $x < 0$, the larger the base, the closer the graph of the equation is to the x-axis.

Figure 4

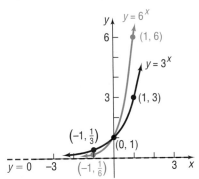

Figure 5

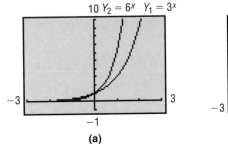
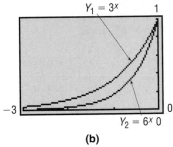

(a)   (b)

The following display summarizes the information that we have about $f(x) = a^x, a > 1$.

Properties of the Exponential Function $f(x) = a^x, a > 1$

1. The domain is the set of all real numbers; the range is the set of positive real numbers.

2. There are no x-intercepts; the y-intercept is 1.

3. The x-axis ($y = 0$) is a horizontal asymptote as $x \to -\infty$.

4. $f(x) = a^x, a > 1$, is an increasing function and is one-to-one.

5. The graph of f contains the points $(0, 1)$, $(1, a)$, and $\left(-1, \dfrac{1}{a}\right)$.

6. The graph of f is smooth and continuous, with no corners or gaps. See Figure 6.

Figure 6
$f(x) = a^x, a > 1$

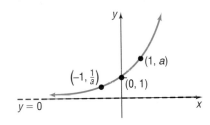

Now we consider $f(x) = a^x$ when $0 < a < 1$.

EXAMPLE 4 | Graphing an Exponential Function

Graph the exponential function: $f(x) = \left(\dfrac{1}{2}\right)^x$

Solution

The domain of $f(x) = \left(\dfrac{1}{2}\right)^x$ consists of all real numbers. As before, we locate some points on the graph, as listed in Table 4. Since $\left(\dfrac{1}{2}\right)^x > 0$ for all x, the range of f is the interval $(0, \infty)$. The graph lies above the x-axis and so has no x-intercepts. The y-intercept is 1. As $x \to -\infty$, $f(x) = \left(\dfrac{1}{2}\right)^x$ grows very quickly. As $x \to \infty$, the values of $f(x)$ approach 0. The x-axis ($y = 0$) is a horizontal asymptote as $x \to \infty$. It is apparent that f is a decreasing function and hence is one-to-one. Figure 7 illustrates the graph.

Table 4

X	Y1
-10	1024
-3	8
-1	2
0	1
1	.5
3	.125
10	9.8E-4

Y1 ◻ (1/2)^X

Figure 7

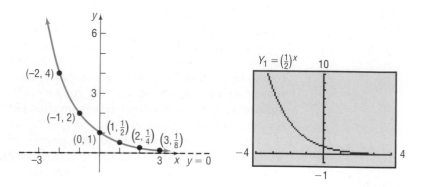

Seeing the Concept
Using a graphing utility, simultaneously graph:

(a) $Y_1 = 3^x$, $Y_2 = \left(\dfrac{1}{3}\right)^x$

(b) $Y_1 = 6^x$, $Y_2 = \left(\dfrac{1}{6}\right)^x$

Conclude that the graph of $Y_2 = \left(\dfrac{1}{a}\right)^x$, for $a > 0$, is the reflection about the y-axis of the graph of $Y_1 = a^x$.

We could have obtained the graph of $y = \left(\dfrac{1}{2}\right)^x$ from the graph of $y = 2^x$ using transformations. The graph of $y = \left(\dfrac{1}{2}\right)^x = 2^{-x}$ is a reflection about the y-axis of the graph of $y = 2^x$. Compare Figures 2 and 7.

The graph of $f(x) = \left(\dfrac{1}{2}\right)^x$ in Figure 7 is typical of all exponential functions of the form $f(x) = a^x$ with $0 < a < 1$. Such functions are decreasing and one-to-one. Their graphs lie above the x-axis and pass through the point $(0, 1)$. The graphs rise rapidly as $x \to -\infty$. As $x \to \infty$, the x-axis $(y = 0)$ is a horizontal asymptote. There are no vertical asymptotes. Finally, the graphs are smooth and continuous, with no corners or gaps.

Figures 8 and 9 illustrate the graphs of two more exponential functions whose bases are between 0 and 1. Notice the smaller the base, the steeper the graph when $x < 0$ and when $x > 0$, the smaller the base, the closer the graph of the equation is to the x-axis.

The following display summarizes the information that we have about the function $f(x) = a^x$, $0 < a < 1$.

Figure 8

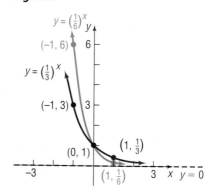

Figure 9

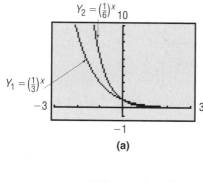

(a)

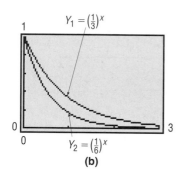

(b)

Figure 10
$f(x) = a^x, 0 < a < 1$

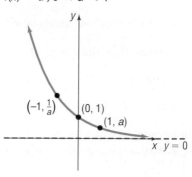

Properties of the Exponential Function $f(x) = a^x$, $0 < a < 1$

1. The domain is the set of all real numbers; the range is the set of positive real numbers.

2. There are no x-intercepts; the y-intercept is 1.

3. The x-axis $(y = 0)$ is a horizontal asymptote as $x \to \infty$.

4. $f(x) = a^x$, $0 < a < 1$, is a decreasing function and is one-to-one.

5. The graph of f contains the points $(0, 1)$, $(1, a)$, and $\left(-1, \dfrac{1}{a}\right)$.

6. The graph of f is smooth and continuous, with no corners or gaps. See Figure 10.

| EXAMPLE 5 | **Graphing Exponential Functions Using Transformations** |

Graph $f(x) = 2^{-x} - 3$ and determine the domain, range, and horizontal asymptote of f.

Solution We begin with the graph of $y = 2^x$. Figure 11 shows the stages.

Figure 11

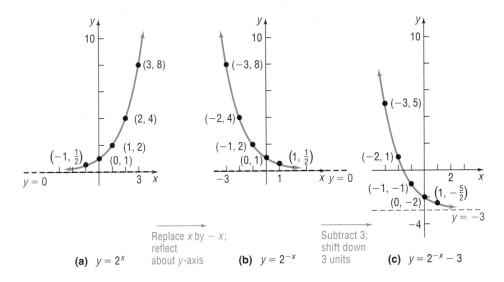

(a) $y = 2^x$ Replace x by $-x$; reflect about y-axis

(b) $y = 2^{-x}$ Subtract 3; shift down 3 units

(c) $y = 2^{-x} - 3$

As Figure 11(c) illustrates, the domain of $f(x) = 2^{-x} - 3$ is the interval $(-\infty, \infty)$, and the range is the interval $(-3, \infty)$. The horizontal asymptote of f is the line $y = -3$.

✓ **Check:** Graph $Y_1 = 2^{-x} - 3$ to verify the graph obtained in Figure 11(c).

■

━ **Now Work** PROBLEM 37

3 Define the Number *e*

As we shall see shortly, many problems that occur in nature require the use of an exponential function whose base is a certain irrational number, symbolized by the letter e.

Let's look at one way of arriving at this important number e.

DEFINITION

The **number *e*** is defined as the number that the expression

$$\left(1 + \frac{1}{n}\right)^n \tag{2}$$

 approaches as $n \to \infty$. In calculus, this is expressed using limit notation as

$$e = \lim_{n \to \infty} \left(1 + \frac{1}{n}\right)^n$$

Table 5 illustrates what happens to the defining expression (2) as n takes on increasingly large values. The last number in the right column in the table is correct to nine decimal places and is the same as the entry given for e on your calculator (if expressed correctly to nine decimal places).

Table 5

n	$\dfrac{1}{n}$	$1 + \dfrac{1}{n}$	$\left(1 + \dfrac{1}{n}\right)^n$
1	1	2	2
2	0.5	1.5	2.25
5	0.2	1.2	2.48832
10	0.1	1.1	2.59374246
100	0.01	1.01	2.704813829
1,000	0.001	1.001	2.716923932
10,000	0.0001	1.0001	2.718145927
100,000	0.00001	1.00001	2.718268237
1,000,000	0.000001	1.000001	2.718280469
1,000,000,000	10^{-9}	$1 + 10^{-9}$	2.718281827

Table 6

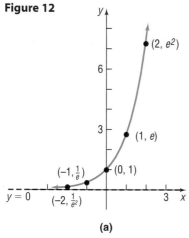

Figure 12

The exponential function $f(x) = e^x$, whose base is the number e, occurs with such frequency in applications that it is usually referred to as *the* exponential function. Graphing calculators have the key $\boxed{e^x}$ or $\boxed{\exp(x)}$, which may be used to evaluate the exponential function for a given value of x. Use your calculator to find e^x for $x = -2$, $x = -1$, $x = 0$, $x = 1$, and $x = 2$, as we have done to create Table 6. The graph of the exponential function $f(x) = e^x$ is given in Figures 12(a) and 12(b). Since $2 < e < 3$, the graph of $y = e^x$ is increasing and lies between the graphs of $y = 2^x$ and $y = 3^x$. [See Figure 12(c).]

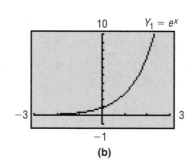

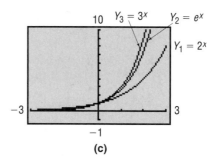

(a)   (b)   (c)

EXAMPLE 6   **Graphing Exponential Functions Using Transformations**

Graph $f(x) = -e^{x-3}$ and determine the domain, range, and horizontal asymptote of f.

Solution   We begin with the graph of $y = e^x$. Figure 13 shows the stages.

Figure 13

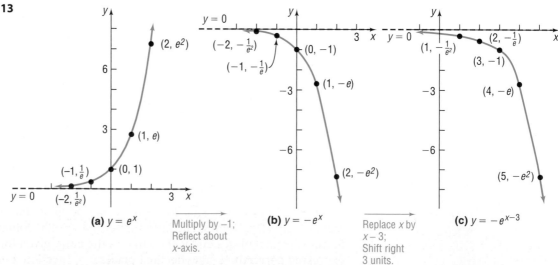

(a) $y = e^x$ Multiply by -1; Reflect about x-axis.   (b) $y = -e^x$ Replace x by $x - 3$; Shift right 3 units.   (c) $y = -e^{x-3}$

As Figure 13(c) illustrates, the domain of $f(x) = -e^{x-3}$ is the interval $(-\infty, \infty)$, and the range is the interval $(-\infty, 0)$. The horizontal asymptote is the line $y = 0$.

✓ Check: Graph $Y_1 = -e^{x-3}$ to verify the graph obtained in Figure 13(c).

━━━━➤ **Now Work** PROBLEM 49

4 Solve Exponential Equations

Equations that involve terms of the form a^x, $a > 0$, $a \neq 1$, are referred to as **exponential equations.** Such equations can sometimes be solved by appropriately applying the Laws of Exponents and property (3):

> In Words
> When two exponential expressions with the same base are equal, then their exponents are equal.

$$\text{If} \quad a^u = a^v, \quad \text{then} \quad u = v \qquad \textbf{(3)}$$

Property (3) is a consequence of the fact that exponential functions are one-to-one. To use property (3), each side of the equality must be written with the same base.

| EXAMPLE 7 | Solving an Exponential Equation |

Solve: $3^{x+1} = 81$

Algebraic Solution

$$3^{x+1} = 81$$
$$3^{x+1} = 3^4 \qquad 81 = 3^4$$

Now we have the same base, 3, on each side of the equation, so we can set the exponents equal to each other to obtain

$$x + 1 = 4$$
$$x = 3$$

The solution set is {3}.

Graphing Solution

Graph $Y_1 = 3^{x+1}$ and $Y_2 = 81$. Use INTERSECT to determine the point of intersection. See Figure 14.

Figure 14

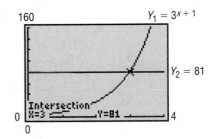

The graphs intersect at $(3, 81)$, so the solution set is {3}.

━━━━➤ **Now Work** PROBLEM 59

| EXAMPLE 8 | Solving an Exponential Equation |

Solve: $e^{-x^2} = (e^x)^2 \cdot \dfrac{1}{e^3}$

Solution We use the Laws of Exponents first to get the base e on the right side.

$$(e^x)^2 \cdot \frac{1}{e^3} = e^{2x} \cdot e^{-3} = e^{2x-3}$$

As a result,

$$e^{-x^2} = e^{2x-3}$$
$$-x^2 = 2x - 3 \qquad \text{Apply Property (3).}$$
$$x^2 + 2x - 3 = 0 \qquad \text{Place the quadratic equation in standard form.}$$
$$(x + 3)(x - 1) = 0 \qquad \text{Factor.}$$
$$x = -3 \quad \text{or} \quad x = 1 \qquad \text{Use the Zero-Product Property.}$$

The solution set is $\{-3, 1\}$.

✓ Check: Verify these solutions using a graphing utility.

Many applications involve the exponential function. Let's look at one.

| EXAMPLE 9 | **Exponential Probability** |

Between 9:00 PM and 10:00 PM cars arrive at Burger King's drive-thru at the rate of 12 cars per hour (0.2 car per minute). The following formula from statistics can be used to determine the probability that a car will arrive within t minutes of 9:00 PM.

$$F(t) = 1 - e^{-0.2t}$$

(a) Determine the probability that a car will arrive within 5 minutes of 9 PM (that is, before 9:05 PM).

(b) Determine the probability that a car will arrive within 30 minutes of 9 PM (before 9:30 PM).

(c) Graph F using your graphing utility.

(d) What value does F approach as t becomes unbounded in the positive direction?

Solution

(a) The probability that a car will arrive within 5 minutes is found by evaluating $F(t)$ at $t = 5$.

$$F(5) = 1 - e^{-0.2(5)} \approx 0.63212$$

Figure 15

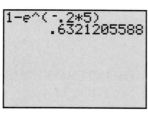

We evaluate this expression in Figure 15. We conclude that there is a 63% probability that a car will arrive within 5 minutes.

(b) The probability that a car will arrive within 30 minutes is found by evaluating $F(t)$ at $t = 30$.

$$F(30) = 1 - e^{-0.2(30)} \approx 0.9975$$

Use a calculator.

There is a 99.75% probability that a car will arrive within 30 minutes.

Figure 16

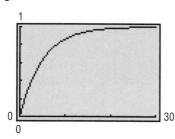

(c) See Figure 16 for the graph of F.

(d) As time passes, the probability that a car will arrive increases. The value that F approaches can be found by letting $t \to \infty$. Since $e^{-0.2t} = \dfrac{1}{e^{0.2t}}$, it follows that $e^{-0.2t} \to 0$ as $t \to \infty$. We conclude that F approaches 1 as t gets large.

━ **Now Work** PROBLEM 101

SUMMARY   Properties of the Exponential Function

$f(x) = a^x, \quad a > 1$	Domain: the interval $(-\infty, \infty)$; Range: the interval $(0, \infty)$ x-intercepts: none; y-intercept: 1 Horizontal asymptote: x-axis $(y = 0)$ as $x \to -\infty$ Increasing; one-to-one; smooth; continuous See Figure 6 for a typical graph.
$f(x) = a^x, \quad 0 < a < 1$	Domain: the interval $(-\infty, \infty)$; Range: the interval $(0, \infty)$ x-intercepts: none; y-intercept: 1 Horizontal asymptote: x-axis $(y = 0)$ as $x \to \infty$ Decreasing; one-to-one; smooth; continuous See Figure 10 for a typical graph.

If $a^u = a^v$, then $u = v$.

7.1 Assess Your Understanding

'Are You Prepared?' *Answers are given at the end of these exercises. If you get a wrong answer, read the pages listed in red.*

1. (a) $4^3 =$ _____   (b) $8^{2/3} =$ _____   (c) $3^{-2} =$ _____   **2.** Solve: (a) $5x - 2 = 3$ (pp. A27–A28)

(d) $(x^2)^3 =$ _____   (e) $a^3 \cdot a^5 =$ _____ (pp. A7–A9 and p. A64)   (b) $2x^2 + 3x = 14$ (pp. A29–A30)

3. **True or False**   To graph $y = (x - 2)^3$, shift the graph of $y = x^3$ to the left 2 units. (pp. 67–70)

4. Find the average rate of change of $f(x) = 3x - 5$ from $x = 0$ to $x = 4$. (pp. 55–56)

5. **True or False**   The function $f(x) = \dfrac{2x}{x - 3}$ has $y = 2$ as a horizontal asymptote. (pp. 76–77)

Concepts and Vocabulary

6. The graph of every exponential function $f(x) = a^x$, $a > 0, a \neq 1$, passes through three points: _____, _____, and _____.

7. If the graph of the exponential function $f(x) = a^x$, $a > 0, a \neq 1$, is decreasing, then a must be less than _____.

8. If $3^x = 3^4$, then $x =$ _____.

9. **True or False**   The graphs of $y = 3^x$ and $y = \left(\dfrac{1}{3}\right)^x$ are identical.

10. **True or False**   The range of the exponential function $f(x) = a^x, a > 0, a \neq 1$, is the set of all real numbers.

Skill Building

In Problems 11–20, approximate each number using a calculator. Express your answer rounded to three decimal places.

11. (a) $3^{2.2}$ (b) $3^{2.23}$ (c) $3^{2.236}$ (d) $3^{\sqrt{5}}$

12. (a) $5^{1.7}$ (b) $5^{1.73}$ (c) $5^{1.732}$ (d) $5^{\sqrt{3}}$

13. (a) $2^{3.14}$ (b) $2^{3.141}$ (c) $2^{3.1415}$ (d) 2^{π}

14. (a) $2^{2.7}$ (b) $2^{2.71}$ (c) $2^{2.718}$ (d) 2^e

15. (a) $3.1^{2.7}$ (b) $3.14^{2.71}$ (c) $3.141^{2.718}$ (d) π^e

16. (a) $2.7^{3.1}$ (b) $2.71^{3.14}$ (c) $2.718^{3.141}$ (d) e^{π}

17. $e^{1.2}$

18. $e^{-1.3}$

19. $e^{-0.85}$

20. $e^{2.1}$

In Problems 21–28, determine whether the given function is linear, exponential, or neither. For those that are linear functions, find a linear function that models the data; for those that are exponential, find an exponential function that models the data.

21.

x	f(x)
−1	3
0	6
1	12
2	18
3	30

22.

x	g(x)
−1	2
0	5
1	8
2	11
3	14

23.

x	H(x)
−1	$\frac{1}{4}$
0	1
1	4
2	16
3	64

24.

x	F(x)
−1	$\frac{2}{3}$
0	1
1	$\frac{3}{2}$
2	$\frac{9}{4}$
3	$\frac{27}{8}$

25.

x	f(x)
−1	$\frac{3}{2}$
0	3
1	6
2	12
3	24

26.

x	g(x)
−1	6
0	1
1	0
2	3
3	10

27.

x	H(x)
−1	2
0	4
1	6
2	8
3	10

28.

x	F(x)
−1	$\frac{1}{2}$
0	$\frac{1}{4}$
1	$\frac{1}{8}$
2	$\frac{1}{16}$
3	$\frac{1}{32}$

In Problems 29–36, the graph of an exponential function is given. Match each graph to one of the following functions.

(a) $y = 3^x$
(b) $y = 3^{-x}$
(c) $y = -3^x$
(d) $y = -3^{-x}$
(e) $y = 3^x - 1$
(f) $y = 3^{x-1}$
(g) $y = 3^{1-x}$
(h) $y = 1 - 3^x$

29.

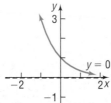

30.

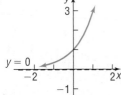

31.

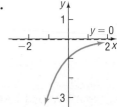

32.

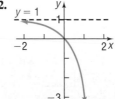

33.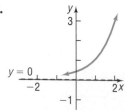

34.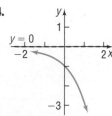

35.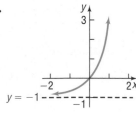

36.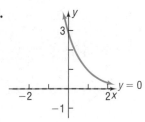

In Problems 37–48, use transformations to graph each function. Determine the domain, range, and horizontal asymptote of each function.

37. $f(x) = 2^x + 1$

38. $f(x) = 3^x - 2$

39. $f(x) = 3^{x-1}$

40. $f(x) = 2^{x+2}$

41. $f(x) = 3 \cdot \left(\dfrac{1}{2}\right)^x$

42. $f(x) = 4 \cdot \left(\dfrac{1}{3}\right)^x$

43. $f(x) = 3^{-x} - 2$

44. $f(x) = -3^x + 1$

45. $f(x) = 2 + 4^{x-1}$

46. $f(x) = 1 - 2^{x+3}$

47. $f(x) = 2 + 3^{x/2}$

48. $f(x) = 1 - 2^{-x/3}$

In Problems 49–56, begin with the graph of $y = e^x$ (Figure 30(a)) and use transformations to graph each function. Determine the domain, range, and horizontal asymptote of each function.

49. $f(x) = e^{-x}$

50. $f(x) = -e^x$

51. $f(x) = e^{x+2}$

52. $f(x) = e^x - 1$

53. $f(x) = 5 - e^{-x}$

54. $f(x) = 9 - 3e^{-x}$

55. $f(x) = 2 - e^{-x/2}$

56. $f(x) = 7 - 3e^{2x}$

In Problems 57–76, solve each equation. Verify your results using a graphing utility.

57. $7^x = 7^3$

58. $5^x = 5^{-6}$

59. $2^{-x} = 16$

60. $3^{-x} = 81$

61. $\left(\dfrac{1}{5}\right)^x = \dfrac{1}{25}$

62. $\left(\dfrac{1}{4}\right)^x = \dfrac{1}{64}$

63. $2^{2x-1} = 4$

64. $5^{x+3} = \dfrac{1}{5}$

65. $3^{x^3} = 9^x$

66. $4^{x^2} = 2^x$

67. $8^{-x+14} = 16^x$

68. $9^{-x+15} = 27^x$

69. $3^{x^2-7} = 27^{2x}$

70. $5^{x^2+8} = 125^{2x}$

71. $4^x \cdot 2^{x^2} = 16^2$

72. $9^{2x} \cdot 27^{x^2} = 3^{-1}$

73. $e^x = e^{3x+8}$

74. $e^{3x} = e^{2-x}$

75. $e^{x^2} = e^{3x} \cdot \dfrac{1}{e^2}$

76. $(e^4)^x \cdot e^{x^2} = e^{12}$

77. If $4^x = 7$, what does 4^{-2x} equal?

78. If $2^x = 3$, what does 4^{-x} equal?

79. If $3^{-x} = 2$, what does 3^{2x} equal?

80. If $5^{-x} = 3$, what does 5^{3x} equal?

In Problems 81–84, determine the exponential function whose graph is given.

81.

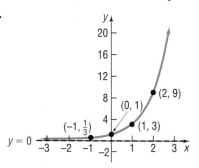

82.

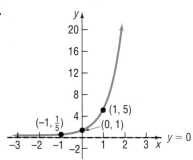

83.

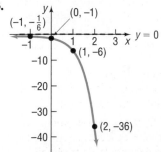

84.

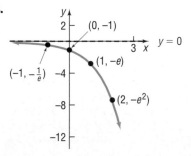

Mixed Practice

85. Suppose that $f(x) = 2^x$.
    (a) What is $f(4)$? What point is on the graph of f?
    (b) If $f(x) = \dfrac{1}{16}$, what is x? What point is on the graph of f?

86. Suppose that $f(x) = 3^x$.
    (a) What is $f(4)$? What point is on the graph of f?
    (b) If $f(x) = \dfrac{1}{9}$, what is x? What point is on the graph of f?

87. Suppose that $g(x) = 4^x + 2$.
    (a) What is $g(-1)$? What point is on the graph of g?
    (b) If $g(x) = 66$, what is x? What point is on the graph of g?

88. Suppose that $g(x) = 5^x - 3$.
    (a) What is $g(-1)$? What point is on the graph of g?
    (b) If $g(x) = 122$, what is x? What point is on the graph of g?

89. Suppose that $H(x) = \left(\dfrac{1}{2}\right)^x - 4$.
    (a) What is $H(-6)$? What point is on the graph of H?
    (b) If $H(x) = 12$, what is x? What point is on the graph of H?
    (c) Find the zero of H.

90. Suppose that $F(x) = \left(\dfrac{1}{3}\right)^x - 3$.
    (a) What is $F(-5)$? What point is on the graph of F?
    (b) If $F(x) = 24$, what is x? What point is on the graph of F?
    (c) Find the zero of F.

In Problems 91–94, graph each function. Based on the graph, state the domain and the range and find any intercepts.

91. $f(x) = \begin{cases} e^{-x} & \text{if } x < 0 \\ e^x & \text{if } x \geq 0 \end{cases}$

92. $f(x) = \begin{cases} e^x & \text{if } x < 0 \\ e^{-x} & \text{if } x \geq 0 \end{cases}$

93. $f(x) = \begin{cases} -e^x & \text{if } x < 0 \\ -e^{-x} & \text{if } x \geq 0 \end{cases}$

94. $f(x) = \begin{cases} -e^{-x} & \text{if } x < 0 \\ -e^x & \text{if } x \geq 0 \end{cases}$

Applications and Extensions

95. **Optics** If a single pane of glass obliterates 3% of the light passing through it, the percent p of light that passes through n successive panes is given approximately by the function
$$p(n) = 100(0.97)^n$$
    (a) What percent of light will pass through 10 panes?
    (b) What percent of light will pass through 25 panes?

96. **Atmospheric Pressure** The atmospheric pressure p on a balloon or plane decreases with increasing height. This pressure, measured in millimeters of mercury, is related to the height h (in kilometers) above sea level by the function
$$p(h) = 760e^{-0.145h}$$
    (a) Find the atmospheric pressure at a height of 2 kilometers (over a mile).
    (b) What is it at a height of 10 kilometers (over 30,000 feet)?

97. **Depreciation** The price p, in dollars, of a Honda Civic DX Sedan that is x years old is modeled by
$$p(x) = 16{,}630(0.90)^x$$
    (a) How much does a 3-year-old Civic DX Sedan cost?
    (b) How much does a 9-year-old Civic DX Sedan cost?

98. **Healing of Wounds** The normal healing of wounds can be modeled by an exponential function. If A_0 represents the original area of the wound and if A equals the area of the wound, then the function
$$A(n) = A_0 e^{-0.35n}$$
    describes the area of a wound after n days following an injury when no infection is present to retard the healing. Suppose that a wound initially had an area of 100 square millimeters.
    (a) If healing is taking place, how large will the area of the wound be after 3 days?
    (b) How large will it be after 10 days?

99. **Drug Medication** The function
$$D(h) = 5e^{-0.4h}$$
    can be used to find the number of milligrams D of a certain drug that is in a patient's bloodstream h hours after the drug has been administered. How many milligrams will be present after 1 hour? After 6 hours?

100. **Spreading of Rumors** A model for the number N of people in a college community who have heard a certain rumor is
$$N = P(1 - e^{-0.15d})$$
     where P is the total population of the community and d is the number of days that have elapsed since the rumor began. In a community of 1000 students, how many students will have heard the rumor after 3 days?

101. **Exponential Probability** Between 12:00 PM and 1:00 PM, cars arrive at Citibank's drive-thru at the rate of 6 cars per hour (0.1 car per minute). The following formula from probability can be used to determine the probability that a car will arrive within t minutes of 12:00 PM:
$$F(t) = 1 - e^{-0.1t}$$
     (a) Determine the probability that a car will arrive within 10 minutes of 12:00 PM (that is, before 12:10 PM).
     (b) Determine the probability that a car will arrive within 40 minutes of 12:00 PM (before 12:40 PM).
     (c) What value does F approach as t becomes unbounded in the positive direction?
     (d) Graph F using a graphing utility.
     (e) Using INTERSECT, determine how many minutes are needed for the probability to reach 50%.

102. **Exponential Probability** Between 5:00 PM and 6:00 PM, cars arrive at Jiffy Lube at the rate of 9 cars per hour (0.15 car per minute). The following formula from probability can be

used to determine the probability that a car will arrive within t minutes of 5:00 PM:

$$F(t) = 1 - e^{-0.15t}$$

(a) Determine the probability that a car will arrive within 15 minutes of 5:00 PM (that is, before 5:15 PM).
(b) Determine the probability that a car will arrive within 30 minutes of 5:00 PM (before 5:30 PM).
(c) What value does F approach as t becomes unbounded in the positive direction?
(d) Graph F using a graphing utility.
(e) Using INTERSECT, determine how many minutes are needed for the probability to reach 60%.

103. Poisson Probability Between 5:00 PM and 6:00 PM, cars arrive at McDonald's drive-thru at the rate of 20 cars per hour. The following formula from probability can be used to determine the probability that x cars will arrive between 5:00 PM and 6:00 PM.

$$P(x) = \frac{20^x e^{-20}}{x!}$$

where

$$x! = x \cdot (x-1) \cdot (x-2) \cdot \cdots \cdot 3 \cdot 2 \cdot 1$$

(a) Determine the probability that $x = 15$ cars will arrive between 5:00 PM and 6:00 PM.
(b) Determine the probability that $x = 20$ cars will arrive between 5:00 PM and 6:00 PM.

104. Poisson Probability People enter a line for the *Demon Roller Coaster* at the rate of 4 per minute. The following formula from probability can be used to determine the probability that x people will arrive within the next minute.

$$P(x) = \frac{4^x e^{-4}}{x!}$$

where

$$x! = x \cdot (x-1) \cdot (x-2) \cdot \cdots \cdot 3 \cdot 2 \cdot 1$$

(a) Determine the probability that $x = 5$ people will arrive within the next minute.
(b) Determine the probability that $x = 8$ people will arrive within the next minute.

105. Relative Humidity The relative humidity is the ratio (expressed as a percent) of the amount of water vapor in the air to the maximum amount that it can hold at a specific temperature. The relative humidity, R, is found using the following formula:

$$R = 10^{\left(\frac{4221}{T+459.4} - \frac{4221}{D+459.4} + 2\right)}$$

where T is the air temperature (in °F) and D is the dew point temperature (in °F).

(a) Determine the relative humidity if the air temperature is 50° Fahrenheit and the dew point temperature is 41° Fahrenheit.
(b) Determine the relative humidity if the air temperature is 68° Fahrenheit and the dew point temperature is 59° Fahrenheit.
(c) What is the relative humidity if the air temperature and the dew point temperature are the same?

106. Learning Curve Suppose that a student has 500 vocabulary words to learn. If the student learns 15 words after 5 minutes, the function

$$L(t) = 500(1 - e^{-0.0061t})$$

approximates the number of words L that the student will learn after t minutes.

(a) How many words will the student learn after 30 minutes?
(b) How many words will the student learn after 60 minutes?

107. Current in a RL Circuit The equation governing the amount of current I (in amperes) after time t (in seconds) in a single RL circuit consisting of a resistance R (in ohms), an inductance L (in henrys), and an electromotive force E (in volts) is

$$I = \frac{E}{R}[1 - e^{-(R/L)t}]$$

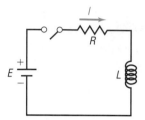

(a) If $E = 120$ volts, $R = 10$ ohms, and $L = 5$ henrys, how much current I_1 is flowing after 0.3 second? After 0.5 second? After 1 second?
(b) What is the maximum current?
(c) Graph the function $I = I_1(t)$, measuring I along the y-axis and t along the x-axis.
(d) If $E = 120$ volts, $R = 5$ ohms, and $L = 10$ henrys, how much current I_2 is flowing after 0.3 second? After 0.5 second? After 1 second?
(e) What is the maximum current?
(f) Graph the function $I = I_2(t)$ on the same coordinate axes as $I_1(t)$.

108. Current in a RC Circuit The equation governing the amount of current I (in amperes) after time t (in microseconds) in a single RC circuit consisting of a resistance R (in ohms), a capacitance C (in microfarads), and an electromotive force E (in volts) is

$$I = \frac{E}{R}e^{-t/(RC)}$$

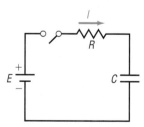

(a) If $E = 120$ volts, $R = 2000$ ohms, and $C = 1.0$ microfarad, how much current I_1 is flowing initially $(t = 0)$? After 1000 microseconds? After 3000 microseconds?
(b) What is the maximum current?
(c) Graph the function $I = I_1(t)$, measuring I along the y-axis and t along the x-axis.
(d) If $E = 120$ volts, $R = 1000$ ohms, and $C = 2.0$ microfarads, how much current I_2 is flowing initially? After 1000 microseconds? After 3000 microseconds?
(e) What is the maximum current?
(f) Graph the function $I = I_2(t)$ on the same coordinate axes as $I_1(t)$.

109. Another Formula for e Use a calculator to compute the values of

$$2 + \frac{1}{2!} + \frac{1}{3!} + \cdots + \frac{1}{n!}$$

for $n = 4, 6, 8,$ and 10. Compare each result with e.
[**Hint:** $1! = 1, 2! = 2 \cdot 1, 3! = 3 \cdot 2 \cdot 1,$
$n! = n(n - 1) \cdots (3)(2)(1).$]

110. Another Formula for e Use a calculator to compute the various values of the expression. Compare the values to e.

$$2 + \cfrac{1}{1 + \cfrac{1}{2 + \cfrac{2}{3 + \cfrac{3}{4 + \cfrac{4}{\text{etc.}}}}}}$$

111. Difference Quotient If $f(x) = a^x$, show that

$$\frac{f(x + h) - f(x)}{h} = a^x \cdot \frac{a^h - 1}{h} \quad h \neq 0$$

112. If $f(x) = a^x$, show that $f(A + B) = f(A) \cdot f(B)$.

113. If $f(x) = a^x$, show that $f(-x) = \dfrac{1}{f(x)}$.

114. If $f(x) = a^x$, show that $f(\alpha x) = [f(x)]^\alpha$.

Problems 115 and 116 provide definitions for two other transcendental functions.

115. The **hyperbolic sine function,** designated by $\sinh x$, is defined as

$$\sinh x = \frac{1}{2}(e^x - e^{-x})$$

(a) Show that $f(x) = \sinh x$ is an odd function.
(b) Graph $f(x) = \sinh x$ using a graphing utility.

116. The **hyperbolic cosine function,** designated by $\cosh x$, is defined as

$$\cosh x = \frac{1}{2}(e^x + e^{-x})$$

(a) Show that $f(x) = \cosh x$ is an even function.
(b) Graph $f(x) = \cosh x$ using a graphing utility.
(c) Refer to Problem 115. Show that, for every x,

$$(\cosh x)^2 - (\sinh x)^2 = 1$$

117. Historical Problem Pierre de Fermat (1601–1665) conjectured that the function

$$f(x) = 2^{(2^x)} + 1$$

for $x = 1, 2, 3, \ldots$, would always have a value equal to a prime number. But Leonhard Euler (1707–1783) showed that this formula fails for $x = 5$. Use a calculator to determine the prime numbers produced by f for $x = 1, 2, 3, 4$. Then show that $f(5) = 641 \times 6,700,417$, which is not prime.

Discussion and Writing

118. The bacteria in a 4-liter container double every minute. After 60 minutes the container is full. How long did it take to fill half the container?

119. Explain in your own words what the number e is. Provide at least two applications that use this number.

120. Do you think that there is a power function that increases more rapidly than an exponential function whose base is greater than 1? Explain.

121. As the base a of an exponential function $f(x) = a^x, a > 1$, increases, what happens to the behavior of its graph for $x > 0$? What happens to the behavior of its graph for $x < 0$?

122. The graphs of $y = a^{-x}$ and $y = \left(\dfrac{1}{a}\right)^x$ are identical. Why?

'Are You Prepared?' Answers

1. (a) 64 (b) 4 (c) $\dfrac{1}{9}$ (d) x^6 (e) a^8 **2.** (a) $\{1\}$ (b) $\left\{-\dfrac{7}{2}, 2\right\}$ **3.** False   **4.** 3   **5.** True

7.2 Logarithmic Functions

PREPARING FOR THIS SECTION *Before getting started, review the following:*

• Solving Inequalities (Appendix, Section A.7, pp. A51–A55)

Now Work the 'Are You Prepared?' problems on page 488.

OBJECTIVES **1** Change Exponential Statements to Logarithmic Statements and Logarithmic Statements to Exponential Statements (p. 480)
2 Evaluate Logarithmic Expressions (p. 480)
3 Determine the Domain of a Logarithmic Function (p. 481)
4 Graph Logarithmic Functions (p. 482)
5 Solve Logarithmic Equations (p. 486)

Recall that a one-to-one function $y = f(x)$ has an inverse function that is defined (implicitly) by the equation $x = f(y)$. In particular, the exponential function $y = f(x) = a^x$, $a > 0$, $a \neq 1$, is one-to-one and hence has an inverse function that is defined implicitly by the equation

$$x = a^y, \qquad a > 0, \qquad a \neq 1$$

This inverse function is so important that it is given a name, the *logarithmic function*.

DEFINITION

The **logarithmic function to the base a,** where $a > 0$ and $a \neq 1$, is denoted by $y = \log_a x$ (read as "y is the logarithm to the base a of x") and is defined by

$$y = \log_a x \quad \text{if and only if} \quad x = a^y$$

The domain of the logarithmic function $y = \log_a x$ is $x > 0$.

As this definition illustrates, a *logarithm* is a name for a certain exponent.

| EXAMPLE 1 | Relating Logarithms to Exponents |

(a) If $y = \log_3 x$, then $x = 3^y$. For example, $4 = \log_3 81$ is equivalent to $81 = 3^4$.

(b) If $y = \log_5 x$, then $x = 5^y$. For example, $-1 = \log_5\left(\dfrac{1}{5}\right)$ is equivalent to $\dfrac{1}{5} = 5^{-1}$.

✓1 Change Exponential Statements to Logarithmic Statements and Logarithmic Statements to Exponential Statements

We can use the definition of a logarithm to convert from exponential form to logarithmic form, and vice versa, as the following two examples illustrate.

| EXAMPLE 2 | Changing Exponential Statements to Logarithmic Statements |

Change each exponential statement to an equivalent statement involving a logarithm.
(a) $1.2^3 = m$ \qquad (b) $e^b = 9$ \qquad (c) $a^4 = 24$

Solution
We use the fact that $y = \log_a x$ and $x = a^y$, $a > 0$, $a \neq 1$, are equivalent.
(a) If $1.2^3 = m$, then $3 = \log_{1.2} m$. \qquad (b) If $e^b = 9$, then $b = \log_e 9$.
(c) If $a^4 = 24$, then $4 = \log_a 24$.

➤ **Now Work** PROBLEM 9

| EXAMPLE 3 | Changing Logarithmic Statements to Exponential Statements |

Change each logarithmic statement to an equivalent statement involving an exponent.
(a) $\log_a 4 = 5$ \qquad (b) $\log_e b = -3$ \qquad (c) $\log_3 5 = c$

Solution
(a) If $\log_a 4 = 5$, then $a^5 = 4$. \qquad (b) If $\log_e b = -3$, then $e^{-3} = b$.
(c) If $\log_3 5 = c$, then $3^c = 5$.

➤ **Now Work** PROBLEM 17

---

2 Evaluate Logarithmic Expressions

To find the exact value of a logarithm, we write the logarithm in exponential notation and use the fact that if $a^u = a^v$ then $u = v$.

| EXAMPLE 4 | Finding the Exact Value of a Logarithmic Expression |

Find the exact value of:

(a) $\log_2 16$

(b) $\log_3 \dfrac{1}{27}$

Solution

(a) $y = \log_2 16$

$2^y = 16$ *Change to exponential form.*

$2^y = 2^4$ *$16 = 2^4$*

$y = 4$ *Equate exponents.*

Therefore, $\log_2 16 = 4$.

(b) $y = \log_3 \dfrac{1}{27}$

$3^y = \dfrac{1}{27}$ *Change to exponential form.*

$3^y = 3^{-3}$ *$\dfrac{1}{27} = \dfrac{1}{3^3} = 3^{-3}$*

$y = -3$ *Equate exponents.*

Therefore, $\log_3 \dfrac{1}{27} = -3$.

> ○ **In Words**
> ○ The expression $\log_2 16$ means,
> ○ "give me the exponent y such
> ○ that $2^y = 16$."

━━━━ **Now Work** PROBLEM 25

3 Determine the Domain of a Logarithmic Function

The logarithmic function $y = \log_a x$ has been defined as the inverse of the exponential function $y = a^x$. That is, if $f(x) = a^x$, then $f^{-1}(x) = \log_a x$. Based on the discussion given in Section 1.8 on inverse functions, for a function f and its inverse f^{-1}, we have

$$\text{Domain of } f^{-1} = \text{Range of } f \quad \text{and} \quad \text{Range of } f^{-1} = \text{Domain of } f$$

Consequently, it follows that

> Domain of the logarithmic function = Range of the exponential function = $(0, \infty)$
>
> Range of the logarithmic function = Domain of the exponential function = $(-\infty, \infty)$

In the next box, we summarize some properties of the logarithmic function:

> $y = \log_a x$ (defining equation: $x = a^y$)
>
> Domain: $0 < x < \infty$ Range: $-\infty < y < \infty$

The domain of a logarithmic function consists of the *positive* real numbers, so the argument of a logarithmic function must be greater than zero.

| EXAMPLE 5 | Finding the Domain of a Logarithmic Function |

Find the domain of each logarithmic function.

(a) $F(x) = \log_2(x + 3)$ (b) $g(x) = \log_5\left(\dfrac{1 + x}{1 - x}\right)$ (c) $h(x) = \log_{1/2}|x|$

Solution

(a) The domain of F consists of all x for which $x + 3 > 0$, that is, $x > -3$. Using interval notation, the domain of f is $(-3, \infty)$.

(b) The domain of g is restricted to

$$\frac{1 + x}{1 - x} > 0$$

Solving this inequality, we find that the domain of g consists of all x between -1 and 1, that is, $-1 < x < 1$ or, using interval notation, $(-1, 1)$.

(c) Since $|x| > 0$, provided that $x \neq 0$, the domain of h consists of all real numbers except zero or, using interval notation, $(-\infty, 0) \cup (0, \infty)$.

━━━━ **Now Work** PROBLEMS 39 AND 45

4 Graph Logarithmic Functions

Since exponential functions and logarithmic functions are inverses of each other, the graph of the logarithmic function $y = \log_a x$ is the reflection about the line $y = x$ of the graph of the exponential function $y = a^x$, as shown in Figure 17.

Figure 17

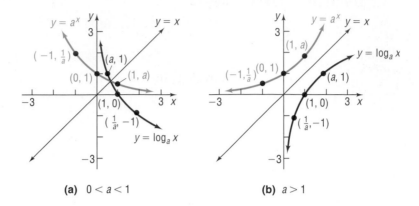

(a) $0 < a < 1$ (b) $a > 1$

For example, to graph $y = \log_2 x$, graph $y = 2^x$ and reflect it about the line $y = x$. See Figure 18. To graph $y = \log_{1/3} x$, graph $y = \left(\dfrac{1}{3}\right)^x$ and reflect it about the line $y = x$. See Figure 19.

Figure 18

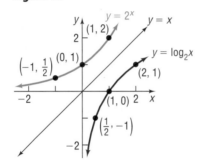

Figure 19

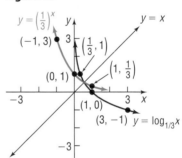

━━━━━ **Now Work** PROBLEM 59

The graphs of $y = \log_a x$ in Figures 17(a) and (b) lead to the following properties.

Properties of the Logarithmic Function $f(x) = \log_a x$

1. The domain is the set of positive real numbers; the range is the set of all real numbers.
2. The x-intercept of the graph is 1. There is no y-intercept.
3. The y-axis ($x = 0$) is a vertical asymptote of the graph.
4. A logarithmic function is decreasing if $0 < a < 1$ and increasing if $a > 1$.
5. The graph of f contains the points $(1, 0)$, $(a, 1)$, and $\left(\dfrac{1}{a}, -1\right)$.
6. The graph is smooth and continuous, with no corners or gaps.

If the base of a logarithmic function is the number e, then we have the **natural logarithm function.** This function occurs so frequently in applications that it is given a special symbol, **ln** (from the Latin, *logarithmus naturalis*). That is,

In Words
$y = \log_e x$ is written $y = \ln x$

$$y = \ln x \quad \text{if and only if} \quad x = e^y \tag{1}$$

Since $y = \ln x$ and the exponential function $y = e^x$ are inverse functions, we can obtain the graph of $y = \ln x$ by reflecting the graph of $y = e^x$ about the line $y = x$. See Figure 20.

Table 7 displays other points on the graph of $f(x) = \ln x$. Notice for $x \le 0$ that we obtain an error message. Do you recall why?

Figure 20

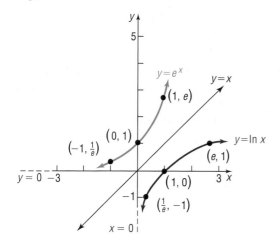

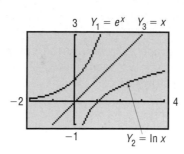

Table 7

X	Y2
-1	ERROR
0.5	-.6931
1	0
2	.69315
2.7183	1
3	1.0986

Y2◻ln(X)

EXAMPLE 6 | **Graphing a Logarithmic Function and Its Inverse**

(a) Find the domain of the logarithmic function $f(x) = -\ln(x - 2)$.
(b) Graph f.
(c) From the graph, determine the range and vertical asymptote of f.
(d) Find f^{-1}, the inverse of f.
(e) Use f^{-1} to confirm the range of f found in part (c). From the domain of f, find the range of f^{-1}.
(f) Graph f^{-1}.

Solution

(a) The domain of f consists of all x for which $x - 2 > 0$ or, equivalently, $x > 2$. The domain of f is $\{x \mid x > 2\}$ or $(2, \infty)$.

(b) To obtain the graph of $y = -\ln(x - 2)$, we begin with the graph of $y = \ln x$ and use transformations. See Figure 21.

Figure 21

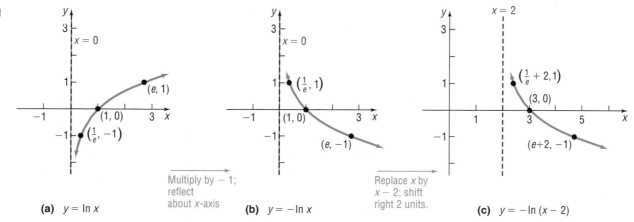

(c) The range of $f(x) = -\ln(x - 2)$ is the set of all real numbers. The vertical asymptote is $x = 2$. [Do you see why? The original asymptote ($x = 0$) is shifted to the right 2 units.]

(d) We begin with $y = -\ln(x - 2)$. The inverse function is defined (implicitly) by the equation

$$x = -\ln(y - 2)$$

We proceed to solve for y.

$$-x = \ln(y - 2) \quad \text{Isolate the logarithm.}$$
$$e^{-x} = y - 2 \quad \text{Change to an exponential expression.}$$
$$y = e^{-x} + 2 \quad \text{Solve for } y.$$

The inverse of f is $f^{-1}(x) = e^{-x} + 2$.

(e) The range of f is the domain of f^{-1}, which is the set of all real numbers, confirming what we found from the graph of f. The range of f^{-1} is the domain of f, which is $(2, \infty)$

(f) To graph f^{-1}, we use the graph of f in Figure 21(c) and reflect it about the line $y = x$. See Figure 22. We could also graph $f^{-1}(x) = e^{-x} + 2$ using transformations.

Figure 22

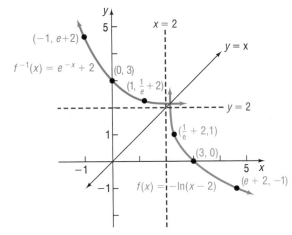

Figure 23

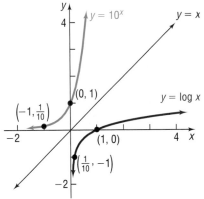

Now Work PROBLEM **71**

If the base of a logarithmic function is the number 10, then we have the **common logarithm function.** If the base a of the logarithmic function is not indicated, it is understood to be 10. That is,

$$y = \log x \quad \text{if and only if} \quad x = 10^y$$

Since $y = \log x$ and the exponential function $y = 10^x$ are inverse functions, we can obtain the graph of $y = \log x$ by reflecting the graph of $y = 10^x$ about the line $y = x$. See Figure 23.

EXAMPLE 7 | **Graphing a Logarithmic Function and Its Inverse**

(a) Find the domain of the logarithmic function $f(x) = 3\log(x - 1)$.
(b) Graph f.
(c) From the graph, determine the range and vertical asymptote of f.
(d) Find f^{-1}, the inverse of f.
(e) Use f^{-1} to confirm the range of f found in part (c). From the domain of f, find the range of f^{-1}.
(f) Graph f^{-1}.

Solution (a) The domain of f consists of all x for which $x - 1 > 0$ or, equivalently, $x > 1$. The domain of f is $\{x \mid x > 1\}$ or $(1, \infty)$.

(b) To obtain the graph of $y = 3\log(x - 1)$, we begin with the graph of $y = \log x$ and use transformations. See Figure 24 on page 485.

Figure 24

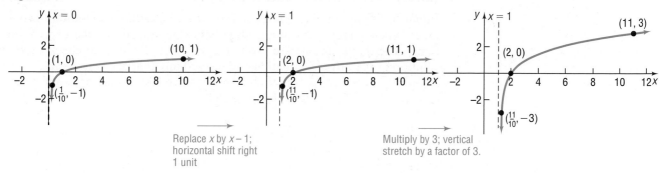

(a) $y = \log x$ Replace x by $x - 1$; horizontal shift right 1 unit   **(b)** $y = \log (x - 1)$ Multiply by 3; vertical stretch by a factor of 3.   **(c)** $y = 3 \log (x - 1)$

(c) The range of $f(x) = 3 \log(x - 1)$ is the set of all real numbers. The vertical asymptote is $x = 1$.

(d) We begin with $y = 3 \log(x - 1)$. The inverse function is defined (implicitly) by the equation

$$x = 3 \log(y \quad 1)$$

We proceed to solve for y.

$$\frac{x}{3} = \log (y - 1) \qquad \text{Isolate the logarithm.}$$
$$10^{x/3} = y - 1 \qquad \text{Change to an exponential expression.}$$
$$y = 10^{x/3} + 1 \qquad \text{Solve for y.}$$

The inverse of f is $f^{-1}(x) = 10^{x/3} + 1$.

(e) The range of f is the domain of f^{-1}, which is the set of all real numbers, confirming what we found from the graph of f. The range of f^{-1} is the domain of f, which is $(1, \infty)$.

(f) To graph f^{-1}, we use the graph of f in Figure 24(c) and reflect it about the line $y = x$. See Figure 25. We could also graph $f^{-1}(x) = 10^{x/3} + 1$ using transformations.

Figure 25

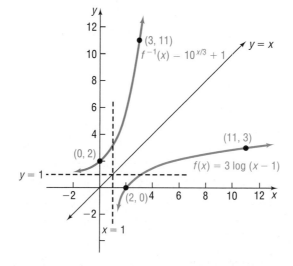

5 Solve Logarithmic Equations

Equations that contain logarithms are called **logarithmic equations.** Care must be taken when solving logarithmic equations algebraically. In the expression $\log_a M$, remember that a and M are positive and $a \neq 1$. Be sure to check each apparent solution in the original equation and discard any that are extraneous.

Some logarithmic equations can be solved by changing from a logarithmic expression to an exponential expression.

| EXAMPLE 8 | Solving a Logarithmic Equation |

Solve: (a) $\log_3(4x - 7) = 2$ (b) $\log_x 64 = 2$

Solution   (a) We can obtain an exact solution by changing the logarithmic equation to exponential form.

$$\log_3(4x - 7) = 2$$
$$4x - 7 = 3^2 \quad \text{Change to an exponential equation.}$$
$$4x - 7 = 9$$
$$4x = 16$$
$$x = 4$$

✓ Check: $\log_3(4x - 7) = \log_3(4 \cdot 4 - 7) = \log_3 9 = 2 \quad 3^2 = 9$

The solution set is {4}.

(b) We can obtain an exact solution by changing the logarithmic equation to exponential form.

$$\log_x 64 = 2$$
$$x^2 = 64 \qquad \text{Change to an exponential equation.}$$
$$x = \pm\sqrt{64} = \pm 8 \quad \text{Square Root Method}$$

The base of a logarithm is always positive. As a result, we discard -8. We check the solution 8.

✓ Check: $\log_8 64 = 2 \quad 8^2 = 64$

The solution set is {8}.

| EXAMPLE 9 | Using Logarithms to Solve Exponential Equations |

Solve: $e^{2x} = 5$

Solution   We can obtain an exact solution by changing the exponential equation to logarithmic form.

$$e^{2x} = 5$$
$$\ln 5 = 2x \qquad \begin{array}{l}\text{Change to a logarithmic equation using the}\\\text{fact that if } e^y = x \text{ then } y = \ln x.\end{array}$$

$$x = \frac{\ln 5}{2} \qquad \text{Exact solution}$$

$$\approx 0.805 \qquad \text{Approximate solution}$$

The solution set is $\left\{\dfrac{\ln 5}{2}\right\}$.

✏️ **Now Work** PROBLEMS 87 AND 99

| EXAMPLE 10 | Alcohol and Driving |

The blood alcohol concentration (BAC) is the amount of alcohol in a person's bloodstream. A BAC of 0.04% means that a person has 4 parts alcohol per 10,000 parts blood in the body. Relative risk is defined as the likelihood of one event occurring divided by the likelihood of a second event occurring. For example, if an individual with a BAC of 0.02% is 1.4 times as likely to have a car accident as an individual that has not been drinking, the relative risk of an accident with a BAC of 0.02% is 1.4. Recent medical research suggests that the relative risk R of having an accident while driving a car can be modeled by the equation

$$R = e^{kx}$$

where x is the percent of concentration of alcohol in the bloodstream and k is a constant.

(a) Research indicates that the relative risk of a person having an accident with a BAC of 0.02% is 1.4. Find the constant k in the equation.

(b) Using this value of k, what is the relative risk if the concentration is 0.17%?

(c) Using this same value of k, what BAC corresponds to a relative risk of 100?

(d) If the law asserts that anyone with a relative risk of 5 or more should not have driving privileges, at what concentration of alcohol in the bloodstream should a driver be arrested and charged with a DUI (driving under the influence)?

Solution

(a) For a concentration of alcohol in the blood of 0.02% and a relative risk of 1.4, we let $x = 0.02$ and $R = 1.4$ in the equation and solve for k.

$$R = e^{kx}$$
$$1.4 = e^{k(0.02)} \qquad R = 1.4; x = 0.02$$
$$0.02k = \ln 1.4 \qquad \text{Change to a logarithmic expression.}$$
$$k = \frac{\ln 1.4}{0.02} \approx 16.82 \quad \text{Solve for } k.$$

(b) For a concentration of 0.17%, we have $x = 0.17$. Using $k = 16.82$ in the equation, we find the relative risk R to be

$$R = e^{kx} = e^{(16.82)(0.17)} \approx 17.5$$

For a concentration of alcohol in the blood of 0.17%, the relative risk of an accident is about 17.5. That is, a person with a BAC of 0.17% is 17.5 times as likely to have a car accident as a person with no alcohol in the bloodstream.

(c) For a relative risk of 100, we have $R = 100$. Using $k = 16.82$ in the equation $R = e^{kx}$, we find the concentration x of alcohol in the blood obeys

$$100 = e^{16.82x} \qquad R = e^{kx}, R = 100; k = 16.82$$
$$16.82x = \ln 100 \qquad \text{Change to a logarithmic expression.}$$
$$x = \frac{\ln 100}{16.82} \approx 0.27 \quad \text{Solve for } x.$$

NOTE A BAC of 0.30% results in a loss of consciousness in most people. ∎

For a concentration of alcohol in the blood of 0.27%, the relative risk of an accident is 100.

(d) For a relative risk of 5, we have $R = 5$. Using $k = 16.82$ in the equation $R = e^{kx}$, we find the concentration x of alcohol in the bloodstream obeys

$$5 = e^{16.82x}$$
$$16.82x = \ln 5$$
$$x = \frac{\ln 5}{16.82} \approx 0.096$$

NOTE Most states use 0.08% or 0.10% as the blood alcohol content at which a DUI citation is given. ∎

A driver with a BAC of 0.096% or more should be arrested and charged with DUI.

SUMMARY   Properties of the Logarithmic Function

$f(x) = \log_a x$, $\quad a > 1$ ($y = \log_a x$ means $x = a^y$)	Domain: the interval $(0, \infty)$; Range: the interval $(-\infty, \infty)$ x-intercept: 1; y-intercept: none; vertical asymptote: $x = 0$ (y-axis); increasing; one-to-one See Figure 26(a) for a typical graph.
$f(x) = \log_a x$, $\quad 0 < a < 1$ ($y = \log_a x$ means $x = a^y$)	Domain: the interval $(0, \infty)$; Range: the interval $(-\infty, \infty)$ x-intercept: 1; y-intercept: none; vertical asymptote: $x = 0$ (y-axis); decreasing; one-to-one See Figure 26(b) for a typical graph.

Figure 26

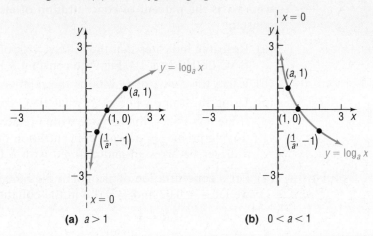

(a) $a > 1$ **(b)** $0 < a < 1$

7.2 Assess Your Understanding

'Are You Prepared?'   *Answers are given at the end of these exercises. If you get a wrong answer, read the pages listed in red.*

1. Solve the inequality: $3x - 7 \leq 8 - 2x$ (pp. A54–A55)

2. Solve the inequality: $x^2 - x - 6 > 0$ (pp. A54–A55)

3. Solve the inequality: $\dfrac{x - 1}{x + 4} > 0$ (pp. A54–A55)

Concepts and Vocabulary

4. The domain of the logarithmic function $f(x) = \log_a x$ is _____.

5. The graph of every logarithmic function $f(x) = \log_a x$, $a > 0, a \neq 1$, passes through three points: _____, _____, and _____.

6. If the graph of a logarithmic function $f(x) = \log_a x$, $a > 0, a \neq 1$, is increasing, then its base must be larger than _____.

7. *True or False*   If $y = \log_a x$, then $y = a^x$.

8. *True or False*   The graph of $f(x) = \log_a x, a > 0, a \neq 1$, has an x-intercept equal to 1 and no y-intercept.

Skill Building

In Problems 9–16, change each exponential statement to an equivalent statement involving a logarithm.

9. $9 = 3^2$

10. $16 = 4^2$

11. $a^2 = 1.6$

12. $a^3 = 2.1$

13. $2^x = 7.2$

14. $3^x = 4.6$

15. $e^x = 8$

16. $e^{2.2} = M$

In Problems 17–24, change each logarithmic statement to an equivalent statement involving an exponent.

17. $\log_2 8 = 3$

18. $\log_3\left(\dfrac{1}{9}\right) = -2$

19. $\log_a 3 = 6$

20. $\log_b 4 = 2$

21. $\log_3 2 = x$

22. $\log_2 6 = x$

23. $\ln 4 = x$

24. $\ln x = 4$

In Problems 25–36, find the exact value of each logarithm without using a calculator.

25. $\log_2 1$ 　　　　 **26.** $\log_8 8$ 　　　　 **27.** $\log_5 25$ 　　　　 **28.** $\log_3\left(\dfrac{1}{9}\right)$

29. $\log_{1/2} 16$ 　　　 **30.** $\log_{1/3} 9$ 　　　 **31.** $\log_{10}\sqrt{10}$ 　　　 **32.** $\log_5 \sqrt[3]{25}$

33. $\log_{\sqrt{2}} 4$ 　　　 **34.** $\log_{\sqrt{3}} 9$ 　　　 **35.** $\ln\sqrt{e}$ 　　　　 **36.** $\ln e^3$

In Problems 37–48, find the domain of each function.

37. $f(x) = \ln(x - 3)$ 　　　 **38.** $g(x) = \ln(x - 1)$ 　　　 **39.** $F(x) = \log_2 x^2$

40. $H(x) = \log_5 x^3$ 　　　 **41.** $f(x) = 3 - 2\log_4\left[\dfrac{x}{2} - 5\right]$ 　　　 **42.** $g(x) = 8 + 5\ln(2x + 3)$

43. $f(x) = \ln\left(\dfrac{1}{x + 1}\right)$ 　　 **44.** $g(x) = \ln\left(\dfrac{1}{x - 5}\right)$ 　　 **45.** $g(x) = \log_5\left(\dfrac{x + 1}{x}\right)$

46. $h(x) = \log_3\left(\dfrac{x}{x - 1}\right)$ 　　 **47.** $f(x) = \sqrt{\ln x}$ 　　 **48.** $g(x) = \dfrac{1}{\ln x}$

In Problems 49–56, use a calculator to evaluate each expression. Round your answer to three decimal places.

49. $\ln\dfrac{5}{3}$ 　　 **50.** $\dfrac{\ln 5}{3}$ 　　 **51.** $\dfrac{\ln\dfrac{10}{3}}{0.04}$ 　　 **52.** $\dfrac{\ln\dfrac{2}{3}}{-0.1}$

53. $\dfrac{\ln 4 + \ln 2}{\log 4 + \log 2}$ 　　 **54.** $\dfrac{\log 15 + \log 20}{\ln 15 + \ln 20}$ 　　 **55.** $\dfrac{2\ln 5 + \log 50}{\log 4 - \ln 2}$ 　　 **56.** $\dfrac{3\log 80 - \ln 5}{\log 5 + \ln 20}$

57. Find a so that the graph of $f(x) = \log_a x$ contains the point $(2, 2)$.

58. Find a so that the graph of $f(x) = \log_a x$ contains the point $\left(\dfrac{1}{2}, -4\right)$.

In Problems 59–62, graph each function and its inverse on the same Cartesian plane.

59. $f(x) = 3^x; f^{-1}(x) = \log_3 x$ 　　　　 **60.** $f(x) = 4^x; f^{-1}(x) = \log_4 x$

61. $f(x) = \left(\dfrac{1}{2}\right)^x; f^{-1}(x) = \log_{\frac{1}{2}} x$ 　　　 **62.** $f(x) = \left(\dfrac{1}{3}\right)^x; f^{-1}(x) = \log_{\frac{1}{3}} x$

In Problems 63–70, the graph of a logarithmic function is given. Match each graph to one of the following functions:

(a) $y = \log_3 x$ 　　　 (b) $y = \log_3(-x)$ 　　　 (c) $y = -\log_3 x$ 　　　 (d) $y = -\log_3(-x)$

(e) $y = \log_3 x - 1$ 　 (f) $y = \log_3(x - 1)$ 　 (g) $y = \log_3(1 - x)$ 　 (h) $y = 1 - \log_3 x$

63. 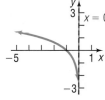　　**64.** 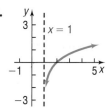　　**65.** 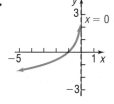　　**66.**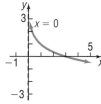

67. 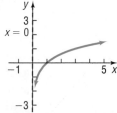　　**68.** 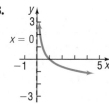　　**69.** 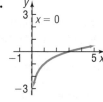　　**70.**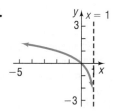

In Problems 71–86, use the given function f to:
(a) *Find the domain of f.*
(b) *Graph f.*
(c) *From the graph, determine the range and any asymptotes of f.*

(d) Find f^{-1}, the inverse of f.
(e) Use f^{-1} to confirm the range of f found in part (c). From the domain of f, find the range of f^{-1}.
(f) Graph f^{-1}.

71. $f(x) = \ln(x + 4)$ **72.** $f(x) = \ln(x - 3)$ **73.** $f(x) = 2 + \ln x$ **74.** $f(x) = -\ln(-x)$

75. $f(x) = \ln(2x) - 3$ **76.** $f(x) = -2\ln(x + 1)$ **77.** $f(x) = \log(x - 4) + 2$ **78.** $f(x) = \frac{1}{2}\log x - 5$

79. $f(x) = \frac{1}{2}\log(2x)$ **80.** $f(x) = \log(-2x)$ **81.** $f(x) = 3 + \log_3(x + 2)$ **82.** $f(x) = 2 - \log_3(x + 1)$

83. $f(x) = e^{x+2} - 3$ **84.** $f(x) = 3e^x + 2$ **85.** $f(x) = 2^{x/3} + 4$ **86.** $f(x) = -3^{x+1}$

In Problems 87–110, solve each equation.

87. $\log_3 x = 2$ **88.** $\log_5 x = 3$ **89.** $\log_2(2x + 1) = 3$ **90.** $\log_3(3x - 2) = 2$

91. $\log_x 4 = 2$ **92.** $\log_x\left(\frac{1}{8}\right) = 3$ **93.** $\ln e^x = 5$ **94.** $\ln e^{-2x} = 8$

95. $\log_4 64 = x$ **96.** $\log_5 625 = x$ **97.** $\log_3 243 = 2x + 1$ **98.** $\log_6 36 = 5x + 3$

99. $e^{3x} = 10$ **100** $e^{-2x} = \frac{1}{3}$ **101.** $e^{2x+5} = 8$ **102.** $e^{-2x+1} = 13$

103. $\log_3(x^2 + 1) = 2$ **104.** $\log_5(x^2 + x + 4) = 2$ **105.** $\log_2 8^x = -3$ **106.** $\log_3 3^x = -1$

107. $5e^{0.2x} = 7$ **108.** $8 \cdot 10^{2x-7} = 3$ **109.** $2 \cdot 10^{2-x} = 5$ **110.** $4e^{x+1} = 5$

Mixed Practice

111. Suppose that $G(x) = \log_3(2x + 1) - 2$.
  (a) What is the domain of G?
  (b) What is $G(40)$? What point is on the graph of G?
  (c) If $G(x) = 3$, what is x? What point is on the graph of G?
  (d) What is the zero of G?

112. Suppose that $F(x) = \log_2(x + 1) - 3$.
  (a) What is the domain of F?
  (b) What is $F(7)$? What point is on the graph of F?
  (c) If $F(x) = -1$, what is x? What point is on the graph of F?
  (d) What is the zero of F?

In Problems 113–116, graph each function. Based on the graph, state the domain and the range and find any intercepts.

113. $f(x) = \begin{cases} \ln(-x) & \text{if } x < 0 \\ \ln x & \text{if } x > 0 \end{cases}$

114. $f(x) = \begin{cases} \ln(-x) & \text{if } x \leq -1 \\ -\ln(-x) & \text{if } -1 < x < 0 \end{cases}$

115. $f(x) = \begin{cases} -\ln x & \text{if } 0 < x < 1 \\ \ln x & \text{if } x \geq 1 \end{cases}$

116. $f(x) = \begin{cases} \ln x & \text{if } 0 < x < 1 \\ -\ln x & \text{if } x \geq 1 \end{cases}$

Applications and Extensions

117. Chemistry The pH of a chemical solution is given by the formula

$$pH = -\log_{10}[H^+]$$

where $[H^+]$ is the concentration of hydrogen ions in moles per liter. Values of pH range from 0 (acidic) to 14 (alkaline).
(a) What is the pH of a solution for which $[H^+]$ is 0.1?
(b) What is the pH of a solution for which $[H^+]$ is 0.01?
(c) What is the pH of a solution for which $[H^+]$ is 0.001?
(d) What happens to pH as the hydrogen ion concentration decreases?
(e) Determine the hydrogen ion concentration of an orange (pH = 3.5).
(f) Determine the hydrogen ion concentration of human blood (pH = 7.4).

118. Diversity Index **Shannon's diversity index** is a measure of the diversity of a population. The diversity index is given by the formula

$$H = -(p_1 \log p_1 + p_2 \log p_2 + \cdots + p_n \log p_n)$$

where p_1 is the proportion of the population that is species 1, p_2 is the proportion of the population that is species 2, and so on.
(a) According to the U.S. Census Bureau, the distribution of race in the United States in 2000 was as follows:

Race	Proportion
American Indian or Native Alaskan	0.014
Asian	0.041
Black or African American	0.128
Hispanic	0.124
Native Hawaiian or Pacific Islander	0.003
White	0.690

Source: U.S. Census Bureau

Compute the diversity index of the United States in 2000.

(b) The largest value of the diversity index is given by $H_{\max} = \log(S)$, where S is the number of categories of race. Compute $H_{\max}$.

(c) The **evenness ratio** is given by $E_H = \dfrac{H}{H_{\max}}$, where $0 \le E_H \le 1$. If $E_H = 1$, there is complete evenness. Compute the evenness ratio for the United States.

(d) Obtain the distribution of race for the United States in 1990 from the Census Bureau. Compute Shannon's diversity index. Is the United States becoming more diverse? Why?

119. **Atmospheric Pressure** The atmospheric pressure p on a balloon or an aircraft decreases with increasing height. This pressure, measured in millimeters of mercury, is related to the height h (in kilometers) above sea level by the function

$$p(h) = 760e^{-0.145h}$$

(a) Find the height of an aircraft if the atmospheric pressure is 320 millimeters of mercury.

(b) Find the height of a mountain if the atmospheric pressure is 667 millimeters of mercury.

120. **Healing of Wounds** The normal healing of wounds can be modeled by an exponential function. If A_0 represents the original area of the wound and if A equals the area of the wound, then the function

$$A(n) = A_0e^{-0.35n}$$

describes the area of a wound after n days following an injury when no infection is present to retard the healing. Suppose that a wound initially had an area of 100 square millimeters.

(a) If healing is taking place, after how many days will the wound be one-half its original size?

(b) How long before the wound is 10% of its original size?

121. **Exponential Probability** Between 12:00 PM and 1:00 PM, cars arrive at Citibank's drive-thru at the rate of 6 cars per hour (0.1 car per minute). The following formula from statistics can be used to determine the probability that a car will arrive within t minutes of 12:00 PM.

$$F(t) = 1 - e^{-0.1t}$$

(a) Determine how many minutes are needed for the probability to reach 50%.

(b) Determine how many minutes are needed for the probability to reach 80%.

(c) Is it possible for the probability to equal 100%? Explain.

122. **Exponential Probability** Between 5:00 PM and 6:00 PM, cars arrive at Jiffy Lube at the rate of 9 cars per hour (0.15 car per minute). The following formula from statistics can be used to determine the probability that a car will arrive within t minutes of 5:00 PM.

$$F(t) = 1 - e^{-0.15t}$$

(a) Determine how many minutes are needed for the probability to reach 50%.

(b) Determine how many minutes are needed for the probability to reach 80%.

123. **Drug Medication** The formula

$$D = 5e^{-0.4h}$$

can be used to find the number of milligrams D of a certain drug that is in a patient's bloodstream h hours after the drug was administered. When the number of milligrams reaches 2, the drug is to be administered again. What is the time between injections?

124. **Spreading of Rumors** A model for the number N of people in a college community who have heard a certain rumor is

$$N = P(1 - e^{-0.15d})$$

where P is the total population of the community and d is the number of days that have elapsed since the rumor began. In a community of 1000 students, how many days will elapse before 450 students have heard the rumor?

125. **Current in a *RL* Circuit** The equation governing the amount of current I (in amperes) after time t (in seconds) in a simple *RL* circuit consisting of a resistance R (in ohms), an inductance L (in henrys), and an electromotive force E (in volts) is

$$I = \frac{E}{R}\left[1 - e^{-(R/L)t}\right]$$

If $E = 12$ volts, $R = 10$ ohms, and $L = 5$ henrys, how long does it take to obtain a current of 0.5 ampere? Of 1.0 ampere? Graph the equation.

126. **Learning Curve** Psychologists sometimes use the function

$$L(t) = A(1 - e^{-kt})$$

to measure the amount L learned at time t. The number A represents the amount to be learned, and the number k measures the rate of learning. Suppose that a student has an amount A of 200 vocabulary words to learn. A psychologist determines that the student learned 20 vocabulary words after 5 minutes.

(a) Determine the rate of learning k.

(b) Approximately how many words will the student have learned after 10 minutes?

(c) After 15 minutes?

(d) How long does it take for the student to learn 180 words?

Loudness of Sound *Problems 127–130 use the following discussion: The **loudness** $L(x)$, measured in decibels, of a sound of intensity x, measured in watts per square meter, is defined as $L(x) = 10 \log \dfrac{x}{I_0}$, where $I_0 = 10^{-12}$ watt per square meter is the least intense sound that a human ear can detect. Determine the loudness, in decibels, of each of the following sounds.*

127. Normal conversation: intensity of $x = 10^{-7}$ watt per square meter

128. Amplified rock music: intensity of 10^{-1} watt per square meter

129. Heavy city traffic: intensity of $x = 10^{-3}$ watt per square meter

130. Diesel truck traveling 40 miles per hour 50 feet away: intensity 10 times that of a passenger car traveling 50 miles per hour 50 feet away whose loudness is 70 decibels

The Richter Scale *Problems 131 and 132 use the following discussion: The **Richter scale** is one way of converting seismographic readings into numbers that provide an easy reference for measuring the magnitude M of an earthquake. All earthquakes are compared to a **zero-level earthquake** whose seismographic reading measures 0.001 millimeter at a distance of 100 kilometers from the epicenter. An earthquake whose seismographic reading measures x millimeters has **magnitude** M(x), given by*

$$M(x) = \log\left(\frac{x}{x_0}\right)$$

where $x_0 = 10^{-3}$ is the reading of a zero-level earthquake the same distance from its epicenter. In Problems 131 and 132, determine the magnitude of each earthquake.

131. **Magnitude of an Earthquake** Mexico City in 1985: seismographic reading of 125,892 millimeters 100 kilometers from the center

132. **Magnitude of an Earthquake** San Francisco in 1906: seismographic reading of 50,119 millimeters 100 kilometers from the center

133. **Alcohol and Driving** The concentration of alcohol in a person's bloodstream is measurable. Suppose that the relative risk R of having an accident while driving a car can be modeled by the equation

$$R = e^{kx}$$

where x is the percent of concentration of alcohol in the bloodstream and k is a constant.

(a) Suppose that a concentration of alcohol in the bloodstream of 0.03 percent results in a relative risk of an accident of 1.4. Find the constant k in the equation.

(b) Using this value of k, what is the relative risk if the concentration is 0.17 percent?

(c) Using the same value of k, what concentration of alcohol corresponds to a relative risk of 100?

(d) If the law asserts that anyone with a relative risk of having an accident of 5 or more should not have driving privileges, at what concentration of alcohol in the bloodstream should a driver be arrested and charged with a DUI?

(e) Compare this situation with that of Example 10. If you were a lawmaker, which situation would you support? Give your reasons.

Discussion and Writing

134. Is there any function of the form $y = x^\alpha, 0 < \alpha < 1$, that increases more slowly than a logarithmic function whose base is greater than 1? Explain.

135. In the definition of the logarithmic function, the base a is not allowed to equal 1. Why?

136. **Critical Thinking** In buying a new car, one consideration might be how well the price of the car holds up over time. Different makes of cars have different depreciation rates. One way to compute a depreciation rate for a car is given here. Suppose that the current prices of a certain automobile are as shown in the table.

| | Age in Years | | | | |
New	1	2	3	4	5
$38,000	$36,600	$32,400	$28,750	$25,400	$21,200

Use the formula New = Old(e^{Rt}) to find R, the annual depreciation rate, for a specific time t. When might be the best time to trade in the car? Consult the NADA ("blue") book and compare two like models that you are interested in. Which has the better depreciation rate?

'Are You Prepared?' Answers

1. $x \leq 3$ 2. $x < -2$ or $x > 3$ 3. $x < -4$ or $x > 1$

7.3 Properties of Logarithms

OBJECTIVES 1 Work with the Properties of Logarithms (p. 492)

2 Write a Logarithmic Expression as a Sum or Difference of Logarithms (p. 494)

3 Write a Logarithmic Expression as a Single Logarithm (p. 495)

4 Evaluate Logarithms Whose Base Is Neither 10 Nor e (p. 496)

5 Graph Logarithmic Functions Whose Base Is Neither 10 Nor e (p. 498)

1 Work with the Properties of Logarithms

Logarithms have some very useful properties that can be derived directly from its definition and the laws of exponents.

EXAMPLE 1   Establishing Properties of Logarithms

(a) Show that $\log_a 1 = 0$.   (b) Show that $\log_a a = 1$.

Solution   (a) This fact was established when we graphed $y = \log_a x$ (see Figure 17). To show the result algebraically, let $y = \log_a 1$. Then

$$y = \log_a 1$$
$$a^y = 1 \qquad \text{Change to an exponential expression.}$$
$$a^y = a^0 \qquad a^0 = 1 \text{ since } a > 0, a \neq 1$$
$$y = 0 \qquad \text{Solve for } y.$$
$$\log_a 1 = 0 \qquad y = \log_a 1$$

(b) Let $y = \log_a a$. Then

$$y = \log_a a$$
$$a^y = a \qquad \text{Change to an exponential expression.}$$
$$a^y = a^1 \qquad a = a^1$$
$$y = 1 \qquad \text{Solve for } y.$$
$$\log_a a = 1 \qquad y = \log_a a$$

To summarize:

$$\log_a 1 = 0 \qquad \log_a a = 1$$

THEOREM

Properties of Logarithms

In the properties given next, M and a are positive real numbers, $a \neq 1$, and r is any real number.

The number $\log_a M$ is the exponent to which a must be raised to obtain M. That is,

$$a^{\log_a M} = M \tag{1}$$

The logarithm to the base a of a raised to a power equals that power. That is,

$$\log_a a^r = r \tag{2}$$

The proof uses the fact that $y = a^x$ and $y = \log_a x$ are inverses.

Proof of Property (1)   For inverse functions,

$$f(f^{-1}(x)) = x \quad \text{for all } x \text{ in the domain of } f^{-1}$$

Using $f(x) = a^x$ and $f^{-1}(x) = \log_a x$, we find

$$f(f^{-1}(x)) = a^{\log_a x} = x \quad \text{for } x > 0$$

Now let $x = M$ to obtain $a^{\log_a M} = M$, where $M > 0$. ∎

Proof of Property (2)   For inverse functions,

$$f^{-1}(f(x)) = x \quad \text{for all } x \text{ in the domain of } f$$

Using $f(x) = a^x$ and $f^{-1}(x) = \log_a x$, we find

$$f^{-1}(f(x)) = \log_a a^x = x \quad \text{for all real numbers } x$$

Now let $x = r$ to obtain $\log_a a^r = r$, where r is any real number. ∎

EXAMPLE 2	Using Properties (1) and (2)

(a) $2^{\log_2 \pi} = \pi$ (b) $\log_{0.2} 0.2^{-\sqrt{2}} = -\sqrt{2}$ (c) $\ln e^{kt} = kt$

◼

— Now Work PROBLEM 9

Other useful properties of logarithms are given next.

THEOREM

Properties of Logarithms

In the following properties, M, N, and a are positive real numbers, $a \neq 1$, and r is any real number.

The Log of a Product Equals the Sum of the Logs

$$\log_a(MN) = \log_a M + \log_a N \tag{3}$$

The Log of a Quotient Equals the Difference of the Logs

$$\log_a\left(\frac{M}{N}\right) = \log_a M - \log_a N \tag{4}$$

The Log of a Power Equals the Product of the Power and the Log

$$\log_a M^r = r \log_a M \tag{5}$$

We shall derive properties (3) and (5) and leave the derivation of property (4) as an exercise (see Problem 103).

Proof of Property (3) Let $A = \log_a M$ and let $B = \log_a N$. These expressions are equivalent to the exponential expressions

$$a^A = M \quad \text{and} \quad a^B = N$$

Now

$$
\begin{aligned}
\log_a(MN) = \log_a(a^A a^B) &= \log_a a^{A+B} && \text{Law of Exponents} \\
&= A + B && \text{Property (2) of logarithms} \\
&= \log_a M + \log_a N
\end{aligned}
$$

◼

Proof of Property (5) Let $A = \log_a M$. This expression is equivalent to

$$a^A = M$$

Now

$$
\begin{aligned}
\log_a M^r = \log_a(a^A)^r &= \log_a a^{rA} && \text{Law of Exponents} \\
&= rA && \text{Property (2) of logarithms} \\
&= r \log_a M
\end{aligned}
$$

◼

— Now Work PROBLEM 13

2 Write a Logarithmic Expression as a Sum or Difference of Logarithms

Logarithms can be used to transform products into sums, quotients into differences, and powers into factors. Such transformations prove useful in certain types of calculus problems.

EXAMPLE 3 **Writing a Logarithmic Expression as a Sum of Logarithms**

Write $\log_a\left(x\sqrt{x^2 + 1}\right)$, $x > 0$, as a sum of logarithms. Express all powers as factors.

Solution
$$\log_a\left(x\sqrt{x^2 + 1}\right) = \log_a x + \log_a \sqrt{x^2 + 1} \quad {\scriptstyle \log_a(M \cdot N) = \log_a M + \log_a N}$$
$$= \log_a x + \log_a(x^2 + 1)^{1/2}$$
$$= \log_a x + \frac{1}{2}\log_a(x^2 + 1) \quad {\scriptstyle \log_a M^r = r\log_a M}$$

EXAMPLE 4 **Writing a Logarithmic Expression as a Difference of Logarithms**

Write

$$\ln \frac{x^2}{(x - 1)^3} \qquad x > 1$$

as a difference of logarithms. Express all powers as factors.

Solution
$$\ln \frac{x^2}{(x - 1)^3} = \ln x^2 - \ln(x - 1)^3 = 2 \ln x - 3 \ln(x - 1)$$

$$\underset{\uparrow}{\scriptstyle \log_a\left(\frac{M}{N}\right) = \log_a M - \log_a N} \qquad \underset{\uparrow}{\scriptstyle \log_a M^r = r\log_a M}$$

EXAMPLE 5 **Writing a Logarithmic Expression as a Sum and Difference of Logarithms**

Write

$$\log_a \frac{\sqrt{x^2 + 1}}{x^3(x + 1)^4} \qquad x > 0$$

as a sum and difference of logarithms. Express all powers as factors.

Solution
$$\log_a \frac{\sqrt{x^2 + 1}}{x^3(x + 1)^4} = \log_a \sqrt{x^2 + 1} - \log_a[x^3(x + 1)^4] \qquad {\scriptstyle Property\ (4)}$$

$$= \log_a \sqrt{x^2 + 1} - [\log_a x^3 + \log_a(x + 1)^4] \qquad {\scriptstyle Property\ (3)}$$

$$= \log_a(x^2 + 1)^{1/2} - \log_a x^3 - \log_a(x + 1)^4 \qquad {\scriptstyle Property\ (5)}$$

$$= \frac{1}{2}\log_a(x^2 + 1) - 3 \log_a x - 4 \log_a(x + 1) \qquad {\scriptstyle Property\ (5)}$$

WARNING In using properties (3) through (5), be careful about the values that the variable may assume. For example, the domain of the variable for $\log_a x$ is $x > 0$ and for $\log_a(x - 1)$ it is $x > 1$. If we add these functions, the domain is $x > 1$. That is, the equality

$$\log_a x + \log_a(x - 1) = \log_a[x(x - 1)]$$

is true only for $x > 1$. ∎

Now Work PROBLEM 45

3 Write a Logarithmic Expression as a Single Logarithm

Another use of properties (3) through (5) is to write sums and/or differences of logarithms with the same base as a single logarithm. This skill will be needed to solve certain logarithmic equations discussed in the next section.

EXAMPLE 6 **Writing Expressions as a Single Logarithm**

Write each of the following as a single logarithm.

(a) $\log_a 7 + 4 \log_a 3$

(b) $\frac{2}{3}\ln 8 - \ln(3^4 - 8)$

(c) $\log_a x + \log_a 9 + \log_a(x^2 + 1) - \log_a 5$

Solution (a) $\log_a 7 + 4\log_a 3 = \log_a 7 + \log_a 3^4$ $r\log_a M = \log_a M^r$

$$= \log_a 7 + \log_a 81$$

$$= \log_a(7 \cdot 81)$$ $\log_a M + \log_a N = \log_a(M \cdot N)$

$$= \log_a 567$$

(b) $\dfrac{2}{3}\ln 8 - \ln(3^4 - 8) = \ln 8^{2/3} - \ln(81 - 8)$ $r\log_a M = \log_a M^r$

$$= \ln 4 - \ln 73$$ $8^{2/3} = (\sqrt[3]{8})^2 = 2^2 = 4$

$$= \ln\left(\frac{4}{73}\right)$$ $\log_a M - \log_a N = \log_a\left(\dfrac{M}{N}\right)$

(c) $\log_a x + \log_a 9 + \log_a(x^2 + 1) - \log_a 5 = \log_a(9x) + \log_a(x^2 + 1) - \log_a 5$

$$= \log_a[9x(x^2 + 1)] - \log_a 5$$

$$= \log_a\left[\frac{9x(x^2 + 1)}{5}\right]$$

WARNING A common error made by some students is to express the logarithm of a sum as the sum of logarithms.

$$\log_a(M + N) \quad \textit{is not equal to} \quad \log_a M + \log_a N$$

Correct statement $\log_a(MN) = \log_a M + \log_a N$ Property (3)

Another common error is to express the difference of logarithms as the quotient of logarithms.

$$\log_a M - \log_a N \quad \textit{is not equal to} \quad \frac{\log_a M}{\log_a N}$$

Correct statement $\log_a M - \log_a N = \log_a\left(\dfrac{M}{N}\right)$ Property (4)

A third common error is to express a logarithm raised to a power as the product of the power times the logarithm.

$$(\log_a M)^r \quad \textit{is not equal to} \quad r\log_a M$$

Correct statement $\log_a M^r = r\log_a M$ Property (5) ∎

Now Work PROBLEM 51

Two other properties of logarithms that we need to know are consequences of the fact that the logarithmic function $y = \log_a x$ is a one-to-one function.

THEOREM

Properties of Logarithms

In the following properties, M, N, and a are positive real numbers, $a \neq 1$.

> If $M = N$, then $\log_a M = \log_a N$.   **(6)**
>
> If $\log_a M = \log_a N$, then $M = N$.   **(7)**

When property (6) is used, we start with the equation $M = N$ and say "take the logarithm of both sides" to obtain $\log_a M = \log_a N$.

Properties (6) and (7) are useful for solving *exponential and logarithmic equations*, a topic discussed in the next section.

4 Evaluate Logarithms Whose Base Is Neither 10 Nor *e*

Logarithms to the base 10, common logarithms, were used to facilitate arithmetic computations before the widespread use of calculators. (See the Historical Feature at the end of this section.) Natural logarithms, that is, logarithms whose base is the number *e*, remain very important because they arise frequently in the study of natural phenomena.

Common logarithms are usually abbreviated by writing **log,** with the base understood to be 10, just as natural logarithms are abbreviated by **ln,** with the base understood to be e.

Most calculators have both $\boxed{\log}$ and $\boxed{\ln}$ keys to calculate the common logarithm and natural logarithm of a number. Let's look at an example to see how to approximate logarithms having a base other than 10 or e.

EXAMPLE 7 **Approximating a Logarithm Whose Base Is Neither 10 Nor e**

Approximate $\log_2 7$. Round the answer to four decimal places.

Solution Let $y = \log_2 7$. Then $2^y = 7$, so

$$2^y = 7$$

$$\ln 2^y = \ln 7 \qquad \text{Property (6)}$$

$$y \ln 2 = \ln 7 \qquad \text{Property (5)}$$

$$y = \frac{\ln 7}{\ln 2} \qquad \text{Exact value}$$

$$y \approx 2.8074 \qquad \text{Approximate value rounded to four decimal places}$$

Example 7 shows how to approximate a logarithm whose base is 2 by changing to logarithms involving the base e. In general, we use the **Change-of-Base Formula.**

THEOREM **Change-of-Base Formula**

If $a \neq 1$, $b \neq 1$, and M are positive real numbers, then

$$\log_a M = \frac{\log_b M}{\log_b a} \tag{8}$$

Proof We derive this formula as follows: Let $y = \log_a M$. Then

$$a^y = M$$

$$\log_b a^y = \log_b M \qquad \text{Property (6)}$$

$$y \log_b a = \log_b M \qquad \text{Property (5)}$$

$$y = \frac{\log_b M}{\log_b a} \qquad \text{Solve for y.}$$

$$\log_a M = \frac{\log_b M}{\log_b a} \qquad y = \log_a M \qquad \blacksquare$$

Since calculators have keys only for $\boxed{\log}$ and $\boxed{\ln}$, in practice, the Change-of-Base Formula uses either $b = 10$ or $b = e$. That is,

$$\log_a M = \frac{\log M}{\log a} \quad \text{and} \quad \log_a M = \frac{\ln M}{\ln a} \tag{9}$$

EXAMPLE 8 **Using the Change-of-Base Formula**

Approximate: (a) $\log_5 89$

(b) $\log_{\sqrt{2}} \sqrt{5}$

Round answers to four decimal places.

Solution   (a) $\log_5 89 = \dfrac{\log 89}{\log 5} \approx \dfrac{1.949390007}{0.6989700043} \approx 2.7889$

or

$\log_5 89 = \dfrac{\ln 89}{\ln 5} \approx \dfrac{4.48863637}{1.609437912} \approx 2.7889$

(b) $\log_{\sqrt{2}} \sqrt{5} = \dfrac{\log \sqrt{5}}{\log \sqrt{2}} = \dfrac{\frac{1}{2}\log 5}{\frac{1}{2}\log 2} = \dfrac{\log 5}{\log 2} \approx 2.3219$

or

$\log_{\sqrt{2}} \sqrt{5} = \dfrac{\ln \sqrt{5}}{\ln \sqrt{2}} = \dfrac{\frac{1}{2}\ln 5}{\frac{1}{2}\ln 2} = \dfrac{\ln 5}{\ln 2} \approx 2.3219$

━━━━ Now Work PROBLEMS 17 AND 65

5 Graph Logarithmic Functions Whose Base Is Neither 10 Nor *e*

We also use the Change-of-Base Formula to graph logarithmic functions whose base is neither 10 nor *e*.

EXAMPLE 9   **Graphing a Logarithmic Function Whose Base Is Neither 10 Nor *e***

Use a graphing utility to graph $y = \log_2 x$.

Solution   Since graphing utilities only have logarithms with the base 10 or the base *e*, we need to use the Change-of-Base Formula to express $y = \log_2 x$ in terms of logarithms with base 10 or base *e*. We can graph either $y = \dfrac{\ln x}{\ln 2}$ or $y = \dfrac{\log x}{\log 2}$ to obtain the graph of $y = \log_2 x$. See Figure 27.

Figure 27

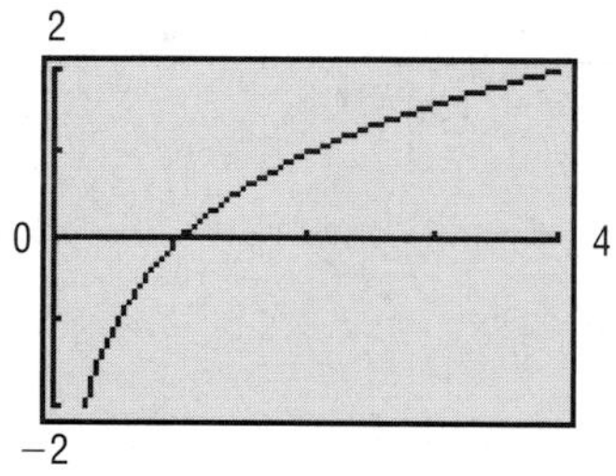

✓Check: Verify that $y = \dfrac{\ln x}{\ln 2}$ and $y = \dfrac{\log x}{\log 2}$ result in the same graph by graphing each on the same screen.

━━━━ Now Work PROBLEM 73

SUMMARY   Properties of Logarithms

In the list that follows, a, b, M, N, and r are real numbers. Also, $a > 0$, $a \neq 1$, $b > 0$, $b \neq 1$, $M > 0$, and $N > 0$.

Definition	$y = \log_a x$ means $x = a^y$
Properties of logarithms	$\log_a 1 = 0;\quad \log_a a = 1$
	$a^{\log_a M} = M;\quad \log_a a^r = r$
	$\log_a(MN) = \log_a M + \log_a N$
	$\log_a\left(\dfrac{M}{N}\right) = \log_a M - \log_a N$
Change-of-Base Formula	$\log_a M = \dfrac{\log_b M}{\log_b a}$

$\log_a M^r = r \log_a M$

If $M = N$, then $\log_a M = \log_a N$.

If $\log_a M = \log_a N$, then $M = N$.

Historical Feature

John Napier
(1550–1617)

Logarithms were invented about 1590 by John Napier (1550–1617) and Joost Bürgi (1552–1632), working independently. Napier, whose work had the greater influence, was a Scottish lord, a secretive man whose neighbors were inclined to believe him to be in league with the devil. His approach to logarithms was very different from ours; it was based on the relationship between arithmetic and geometric sequences, discussed in a later chapter, and not on the inverse function relationship of logarithms to exponential functions (described in

Section 7.2). Napier's tables, published in 1614, listed what would now be called *natural logarithms* of sines and were rather difficult to use. A London professor, Henry Briggs, became interested in the tables and visited Napier. In their conversations, they developed the idea of common logarithms, which were published in 1617. Their importance for calculation was immediately recognized, and by 1650 they were being printed as far away as China. They remained an important calculation tool until the advent of the inexpensive handheld calculator about 1972, which has decreased their calculational, but not their theoretical, importance.

A side effect of the invention of logarithms was the popularization of the decimal system of notation for real numbers.

7.3 Assess Your Understanding

Concepts and Vocabulary

1. The logarithm of a product equals the _____ of the logarithms.

2. If $\log_8 M = \dfrac{\log_5 7}{\log_5 8}$, then $M -$ _____.

3. $\log_a M^r =$ _____.

4. True or False $\ln(x + 3) - \ln(2x) = \dfrac{\ln(x + 3)}{\ln(2x)}$

5. True or False $\log_2(3x^4) = 4 \log_2(3x)$

6. True or False $\log_2 16 = \dfrac{\ln 16}{\ln 2}$

Skill Building

In Problems 7–22, use properties of logarithms to find the exact value of each expression. Do not use a calculator.

7. $\log_3 3^{71}$

8. $\log_2 2^{-13}$

9. $\ln e^{-4}$

10. $\ln e^{\sqrt{2}}$

11. $2^{\log_2 7}$

12. $e^{\ln 8}$

13. $\log_8 2 + \log_8 4$

14. $\log_6 9 + \log_6 4$

15. $\log_6 18 - \log_6 3$

16. $\log_8 16 - \log_8 2$

17. $\log_2 6 \cdot \log_6 4$

18. $\log_3 8 \cdot \log_8 9$

19. $3^{\log_3 5 - \log_3 4}$

20. $5^{\log_5 6 + \log_5 7}$

21. $e^{\log_{e^2} 16}$

22. $e^{\log_{e^2} 9}$

In Problems 23–30, suppose that $\ln 2 = a$ and $\ln 3 = b$. Use properties of logarithms to write each logarithm in terms of a and b.

23. $\ln 6$

24. $\ln \dfrac{2}{3}$

25. $\ln 1.5$

26. $\ln 0.5$

27. $\ln 8$

28. $\ln 27$

29. $\ln \sqrt[5]{6}$

30. $\ln \sqrt[4]{\dfrac{2}{3}}$

In Problems 31–50, write each expression as a sum and/or difference of logarithms. Express powers as factors.

31. $\log_5(25x)$

32. $\log_3 \dfrac{x}{9}$

33. $\log_2 z^3$

34. $\log_7(x^5)$

35. $\ln(ex)$

36. $\ln \dfrac{e}{x}$

37. $\ln(xe^x)$

38. $\ln \dfrac{x}{e^x}$

39. $\log_a(u^2 v^3)$ $u > 0, v > 0$

40. $\log_2 \left(\dfrac{a}{b^2}\right)$ $a > 0, b > 0$

41. $\ln\left(x^2 \sqrt{1 - x}\right)$ $0 < x < 1$

42. $\ln\left(x\sqrt{1 + x^2}\right)$ $x > 0$

43. $\log_2 \left(\dfrac{x^3}{x - 3}\right)$ $x > 3$

44. $\log_5 \left(\dfrac{\sqrt[3]{x^2 + 1}}{x^2 - 1}\right)$ $x > 1$

45. $\log\left[\dfrac{x(x + 2)}{(x + 3)^2}\right]$ $x > 0$

46. $\log\left[\dfrac{x^3\sqrt{x + 1}}{(x - 2)^2}\right]$ $x > 2$

47. $\ln\left[\dfrac{x^2 - x - 2}{(x + 4)^2}\right]^{1/3}$ $x > 2$

48. $\ln\left[\dfrac{(x - 4)^2}{x^2 - 1}\right]^{2/3}$ $x > 4$

49. $\ln \dfrac{5x\sqrt{1 + 3x}}{(x - 4)^3}$ $x > 4$

50. $\ln\left[\dfrac{5x^2\sqrt[3]{1 - x}}{4(x + 1)^2}\right]$ $0 < x < 1$

In Problems 51–64, write each expression as a single logarithm.

51. $3 \log_5 u + 4 \log_5 v$

52. $2 \log_3 u - \log_3 v$

53. $\log_3 \sqrt{x} - \log_3 x^3$

54. $\log_2 \left(\dfrac{1}{x}\right) + \log_2 \left(\dfrac{1}{x^2}\right)$

55. $\log_4(x^2 - 1) - 5 \log_4(x + 1)$

56. $\log(x^2 + 3x + 2) - 2 \log(x + 1)$

57. $\ln\left(\dfrac{x}{x-1}\right) + \ln\left(\dfrac{x+1}{x}\right) - \ln(x^2 - 1)$ **58.** $\log\left(\dfrac{x^2 + 2x - 3}{x^2 - 4}\right) - \log\left(\dfrac{x^2 + 7x + 6}{x + 2}\right)$ **59.** $8\log_2\sqrt{3x - 2} - \log_2\left(\dfrac{4}{x}\right) + \log_2 4$

60. $21\log_3\sqrt[3]{x} + \log_3(9x^2) - \log_3 9$ **61.** $2\log_a(5x^3) - \dfrac{1}{2}\log_a(2x + 3)$ **62.** $\dfrac{1}{3}\log(x^3 + 1) + \dfrac{1}{2}\log(x^2 + 1)$

63. $2\log_2(x + 1) - \log_2(x + 3) - \log_2(x - 1)$ **64.** $3\log_5(3x + 1) - 2\log_5(2x - 1) - \log_5 x$

In Problems 65–72, use the Change-of-Base Formula and a calculator to evaluate each logarithm. Round your answer to three decimal places.

65. $\log_3 21$ **66.** $\log_5 18$ **67.** $\log_{1/3} 71$ **68.** $\log_{1/2} 15$

69. $\log_{\sqrt{2}} 7$ **70.** $\log_{\sqrt{5}} 8$ **71.** $\log_\pi e$ **72.** $\log_\pi \sqrt{2}$

In Problems 73–78, graph each function using a graphing utility and the Change-of-Base Formula.

73. $y = \log_4 x$ **74.** $y = \log_5 x$ **75.** $y = \log_2(x + 2)$

76. $y = \log_4(x - 3)$ **77.** $y = \log_{x-1}(x + 1)$ **78.** $y = \log_{x+2}(x - 2)$

Mixed Practice

79. If $f(x) = \ln x$, $g(x) = e^x$, and $h(x) = x^2$, find:
(a) $(f \circ g)(x)$. What is the domain of $f \circ g$?
(b) $(g \circ f)(x)$. What is the domain of $g \circ f$?
(c) $(f \circ g)(5)$
(d) $(f \circ h)(x)$. What is the domain of $f \circ h$?
(e) $(f \circ h)(e)$

80. If $f(x) = \log_2 x$, $g(x) = 2^x$, and $h(x) = 4x$, find:
(a) $(f \circ g)(x)$. What is the domain of $f \circ g$?
(b) $(g \circ f)(x)$. What is the domain of $g \circ f$?
(c) $(f \circ g)(3)$
(d) $(f \circ h)(x)$. What is the domain of $f \circ h$?
(e) $(f \circ h)(8)$

Applications and Extensions

In Problems 81–90, express y as a function of x. The constant C is a positive number.

81. $\ln y = \ln x + \ln C$

82. $\ln y = \ln(x + C)$

83. $\ln y = \ln x + \ln(x + 1) + \ln C$

84. $\ln y = 2\ln x - \ln(x + 1) + \ln C$

85. $\ln y = 3x + \ln C$

86. $\ln y = -2x + \ln C$

87. $\ln(y - 3) = -4x + \ln C$

88. $\ln(y + 4) = 5x + \ln C$

89. $3\ln y = \dfrac{1}{2}\ln(2x + 1) - \dfrac{1}{3}\ln(x + 4) + \ln C$

90. $2\ln y = -\dfrac{1}{2}\ln x + \dfrac{1}{3}\ln(x^2 + 1) + \ln C$

91. Find the value of $\log_2 3 \cdot \log_3 4 \cdot \log_4 5 \cdot \log_5 6 \cdot \log_6 7 \cdot \log_7 8$.

92. Find the value of $\log_2 4 \cdot \log_4 6 \cdot \log_6 8$.

93. Find the value of $\log_2 3 \cdot \log_3 4 \cdot \cdots \cdot \log_n(n + 1) \cdot \log_{n+1} 2$.

94. Find the value of $\log_2 2 \cdot \log_2 4 \cdot \cdots \cdot \log_2 2^n$.

95. Show that $\log_a\left(x + \sqrt{x^2 - 1}\right) + \log_a\left(x - \sqrt{x^2 - 1}\right) = 0$.

96. Show that $\log_a\left(\sqrt{x} + \sqrt{x - 1}\right) + \log_a\left(\sqrt{x} - \sqrt{x - 1}\right) = 0$.

97. Show that $\ln(1 + e^{2x}) = 2x + \ln(1 + e^{-2x})$.

98. Difference Quotient If $f(x) = \log_a x$, show that $\dfrac{f(x + h) - f(x)}{h} = \log_a\left(1 + \dfrac{h}{x}\right)^{1/h}$, $h \neq 0$.

99. If $f(x) = \log_a x$, show that $-f(x) = \log_{1/a} x$.

100. If $f(x) = \log_a x$, show that $f(AB) = f(A) + f(B)$.

101. If $f(x) = \log_a x$, show that $f\left(\dfrac{1}{x}\right) = -f(x)$.

102. If $f(x) = \log_a x$, show that $f(x^\alpha) = \alpha f(x)$.

103. Show that $\log_a\left(\dfrac{M}{N}\right) = \log_a M - \log_a N$, where a, M, and N are positive real numbers and $a \neq 1$.

104. Show that $\log_a\left(\dfrac{1}{N}\right) = -\log_a N$, where a and N are positive real numbers and $a \neq 1$.

Discussion and Writing

105. Graph $Y_1 = \log(x^2)$ and $Y_2 = 2\log(x)$ using a graphing utility. Are they equivalent? What might account for any differences in the two functions?

106. Write an example that illustrates why $(\log_a x)^r \neq r\log_a x$.

107. Write an example that illustrates why $\log_2(x + y) \neq \log_2 x + \log_2 y$.

108. Does $3^{\log_3(-5)} = -5$? Why or why not?

7.4 Logarithmic and Exponential Equations

PREPARING FOR THIS SECTION *Before getting started, review the following:*

- Solving Equations Using a Graphing Utility (Appendix, Section A.5, pp. A39–A41)
- Solving Quadratic Equations (Appendix, Section A.4, pp. A29–A35)
- Equations Quadratic in Form (Appendix, Section A.4, pp. A35–A36)

Now Work the 'Are You Prepared?' problems on page 505.

> **OBJECTIVES** 1 Solve Logarithmic Equations (p. 501)
> 2 Solve Exponential Equations (p. 503)
> 3 Solve Logarithmic and Exponential Equations Using a Graphing Utility (p. 505)

1 Solve Logarithmic Equations

In Section 7.2 we solved logarithmic equations by changing a logarithmic equation to an exponential equation. That is, we used the definition of a logarithm:

$$y = \log_a x \quad \text{is equivalent to} \quad x = a^y \qquad a > 0, a \neq 1$$

For example, to solve the equation $\log_2(1 - 2x) = 3$, we use the equivalent exponential equation $1 - 2x = 2^3$ and solve for x.

$$\log_2(1 - 2x) = 3$$
$$1 - 2x = 2^3 \qquad \text{Change to an exponential expression.}$$
$$-2x = 7 \qquad \text{Simplify.}$$
$$x = -\frac{7}{2} \qquad \text{Divide both sides by } -2.$$

You should check this solution for yourself.

For most logarithmic equations, some manipulation of the equation (usually using properties of logarithms) is required to obtain a solution. Also, to avoid extraneous solutions with logarithmic equations, we determine the domain of the variable first.

Our practice will be to solve equations, whenever possible, by finding exact solutions using algebraic methods and exact or approximate solutions using a graphing utility. When algebraic methods cannot be used, approximate solutions will be obtained using a graphing utility. The reader is encouraged to pay particular attention to the form of equations for which exact solutions are possible.

We begin with an example of a logarithmic equation that requires using the fact that a logarithmic function is a one-to-one function.

> If $\log_a M = \log_a N$, then $M = N$ $\qquad M, N,$ and a are positive and $a \neq 1$

EXAMPLE 1	Solving a Logarithmic Equation

Solve: $2 \log_5 x = \log_5 9$

Algebraic Solution

We note that the domain of the variable in this equation is $x > 0$. Because each logarithm is to the same base, 5, we can obtain an exact solution as follows:

$$2 \log_5 x = \log_5 9$$

$$\log_5 x^2 = \log_5 9 \quad \log_a M^r = r \log_a M$$

$$x^2 = 9 \quad \text{If } \log_a M = \log_a N, \text{ then } M = N.$$

$$x = 3 \quad \text{or} \quad \cancel{x = -3} \quad \text{Recall that the domain of the variable is } x > 0.$$
$$\text{Therefore, } -3 \text{ is extraneous and we discard it.}$$

✓ Check:
$$2 \log_5 3 \overset{?}{=} \log_5 9$$
$$\log_5 3^2 \overset{?}{=} \log_5 9 \quad r \log_a M = \log_a M^r$$
$$\log_5 9 = \log_5 9$$

The solution set is {3}.

Graphing Solution

To solve the equation using a graphing utility, graph

$$Y_1 = 2 \log_5 x = \frac{2 \log x}{\log 5} \quad \text{and} \quad Y_2 = \log_5 9 = \frac{\log 9}{\log 5},$$

and determine the point of intersection. See Figure 28. The point of intersection is $(3, 1.3652124)$; so $x = 3$ is the only solution. The solution set is {3}.

Figure 28

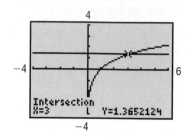

Now Work PROBLEM 13

| EXAMPLE 2 | Solving a Logarithmic Equation |

Solve: $\log_5(x + 6) + \log_5(x + 2) = 1$

Algebraic Solution

The domain of the variable requires that $x + 6 > 0$ and $x + 2 > 0$, so $x > -6$ and $x > -2$. This means any solution must satisfy $x > -2$. To obtain an exact solution, we express the left side as a single logarithm. Then we will change the equation to exponential form.

$$\log_5(x + 6) + \log_5(x + 2) = 1$$

$$\log_5[(x + 6)(x + 2)] = 1 \quad \substack{\log_a M + \log_a N = \\ \log_a(MN)}$$

$$(x + 6)(x + 2) = 5^1 = 5 \quad \substack{\text{Change to an} \\ \text{exponential equation.}}$$

$$x^2 + 8x + 12 = 5 \quad \text{Simplify.}$$

$$x^2 + 8x + 7 = 0 \quad \substack{\text{Place the quadratic} \\ \text{equation in standard} \\ \text{form.}}$$

$$(x + 7)(x + 1) = 0 \quad \text{Factor.}$$

$$x = -7 \quad \text{or} \quad x = -1 \quad \text{Zero-Product Property}$$

Only $x = -1$ satisfies the restriction that $x > -2$, so $x = -7$ is extraneous. The solution set is {−1}, which you should check.

Graphing Solution

Graph $Y_1 = \log_5(x + 6) + \log_5(x + 2) = \dfrac{\log(x + 6)}{\log 5} + \dfrac{\log(x + 2)}{\log 5}$ and $Y_2 = 1$ and determine the point(s) of intersection. See Figure 29. The point of intersection is $(-1, 1)$, so $x = -1$ is the only solution. The solution set is {−1}.

Figure 29

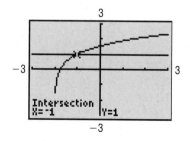

WARNING A negative solution is not automatically extraneous. You must determine whether the potential solution causes the argument of any logarithmic expression in the equation to be negative. ∎

Now Work PROBLEM 21

| EXAMPLE 3 | Solving a Logarithmic Equation |

Solve: $\ln x + \ln(x - 4) = \ln(x + 6)$

Algebraic Solution

The domain of the variable requires that $x > 0$, $x - 4 > 0$, and $x + 6 > 0$. As a result, the domain of the variable is $x > 4$. We begin the solution using the log of a product property.

$$\ln x + \ln(x - 4) = \ln(x + 6)$$

$$\ln[x(x - 4)] = \ln(x + 6) \qquad \text{In } M + \text{In } N = \text{In}(MN)$$

$$x(x - 4) = x + 6 \qquad \text{If In } M = \text{In } N, \text{ then } M = N.$$

$$x^2 - 4x = x + 6 \qquad \text{Simplify.}$$

$$x^2 - 5x - 6 = 0 \qquad \text{Place the quadratic equation in standard form.}$$

$$(x - 6)(x + 1) = 0 \qquad \text{Factor.}$$

$$x = 6 \quad \text{or} \quad x = -1 \qquad \text{Zero-Product Property}$$

Since the domain of the variable is $x > 4$, we discard -1 as extraneous. The solution set is $\{6\}$, which you should check.

Graphing Solution

Graph $Y_1 = \ln x + \ln(x - 4)$ and $Y_2 = \ln(x + 6)$ and determine the point(s) of intersection. See Figure 30. The x-coordinate of the point of intersection is 6, so the solution set is $\{6\}$.

Figure 30

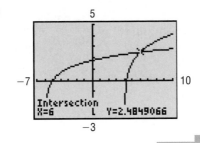

Intersection
X=6 Y=2.4849066

→ **Now Work** PROBLEM 31

2 Solve Exponential Equations

In Sections 7.1 and 7.2, we solved exponential equations algebraically by expressing each side of the equation using the same base. That is, we used the one-to-one property of the exponential function:

$$\text{If } a^u = a^v, \quad \text{then} \quad u = v \quad a > 0, a \neq 1$$

For example, to solve the exponential equation $4^{2x+1} = 16$, we notice that $16 = 4^2$ and apply the property above to obtain $2x + 1 = 2$, from which we find $x = \dfrac{1}{2}$.

For most exponential equations, we cannot express each side of the equation using the same base. In such cases, algebraic techniques can sometimes be used to obtain exact solutions. When algebraic techniques cannot be used, we use a graphing utility to obtain approximate solutions. You should pay particular attention to the form of equations for which exact solutions are obtained.

In the next two examples we solve exponential equations by changing the exponential expression to a logarithmic expression.

| EXAMPLE 4 | Solving an Exponential Equation |

Solve: $2^x = 5$

Algebraic Solution

Since 5 cannot be written as an integer power of 2, we write the exponential equation as the equivalent logarithmic equation.

$$2^x = 5$$

$$x = \log_2 5 = \frac{\ln 5}{\ln 2}$$
$$\uparrow$$
Change-of-Base Formula (9), Section 7.3

Graphing Solution

Graph $Y_1 = 2^x$ and $Y_2 = 5$ and determine the x-coordinate of the point of intersection. See Figure 31.

Figure 31

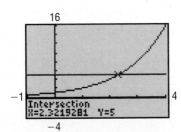

Intersection
X=2.3219281 Y=5

Alternatively, we can solve the equation $2^x = 5$ by taking the natural logarithm (or common logarithm) of each side. Taking the natural logarithm,

$$2^x = 5$$

$$\ln 2^x = \ln 5 \qquad \text{If } M = N, \text{ then } \ln M = \ln N.$$

$$x \ln 2 = \ln 5 \qquad \ln M^r = r \ln M$$

$$x = \frac{\ln 5}{\ln 2} \qquad \text{Exact solution}$$

$$\approx 2.322 \qquad \text{Approximate solution}$$

The solution set is $\left\{\dfrac{\ln 5}{\ln 2}\right\}$.

The approximate solution, rounded to three decimal places, is 2.322.

━━━━━━ **Now Work**   PROBLEM 35

| **EXAMPLE 5** | **Solving an Exponential Equation** |

Solve: $8 \cdot 3^x = 5$

Algebraic Solution

We wish to isolate the exponential expression and then rewrite the statement as an equivalent logarithm.

$$8 \cdot 3^x = 5$$

$$3^x = \frac{5}{8} \qquad \text{Solve for } 3^x.$$

$$x = \log_3\left(\frac{5}{8}\right) = \frac{\ln\left(\frac{5}{8}\right)}{\ln 3} \qquad \text{Exact solution}$$

$$\approx -0.428 \qquad \text{Approximate solution}$$

The solution set is $\left\{\log_3\left(\dfrac{5}{8}\right)\right\}$.

Graphing Solution

Graph $Y_1 = 8 \cdot 3^x$ and $Y_2 = 5$ and determine the x-coordinate of the point of intersection. See Figure 32.

Figure 32

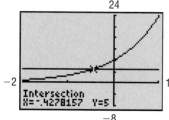

The approximate solution, rounded to three decimal places, is −0.428.

| **EXAMPLE 6** | **Solving an Exponential Equation** |

Solve: $5^{x-2} = 3^{3x+2}$

Algebraic Solution

Because the bases are different, we first apply Property (6), Section 7.3 (take the natural logarithm of each side), and then use appropriate properties of logarithms. The result is a linear equation in x that we can solve.

$$5^{x-2} = 3^{3x+2}$$

$$\ln 5^{x-2} = \ln 3^{3x+2} \qquad \text{If } M = N, \ln M = \ln N.$$

$$(x - 2)\ln 5 = (3x + 2)\ln 3 \qquad \ln M^r = r \ln M$$

$$(\ln 5)x - 2\ln 5 = (3\ln 3)x + 2\ln 3 \qquad \begin{array}{l}\text{Distribute. The equation is now} \\ \text{linear in } x.\end{array}$$

$$(\ln 5)x - (3\ln 3)x = 2\ln 3 + 2\ln 5 \qquad \text{Place terms involving } x \text{ on the left.}$$

$$(\ln 5 - 3\ln 3)x = 2(\ln 3 + \ln 5) \qquad \text{Factor.}$$

$$x = \frac{2(\ln 3 + \ln 5)}{\ln 5 - 3\ln 3} \qquad \text{Exact solution}$$

$$\approx -3.212 \qquad \text{Approximate solution}$$

Graphing Solution

Graph $Y_1 = 5^{x-2}$ and $Y_2 = 3^{3x+2}$ and determine the x-coordinate of the point of intersection. See Figure 33.

Figure 33

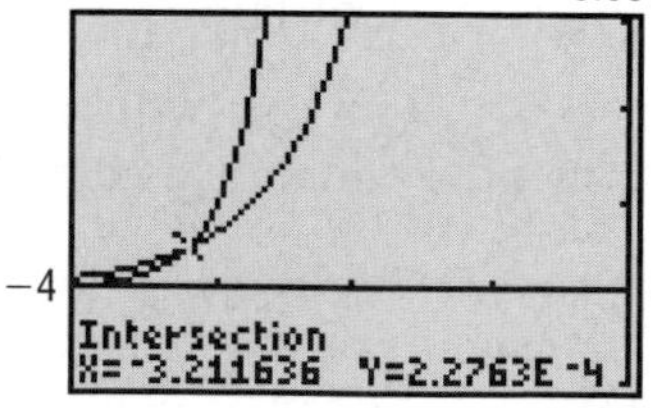

The approximate solution, rounded to three decimal places, is −3.212.

━━━━━━ **Now Work**   PROBLEM 41

The next example deals with an exponential equation that is quadratic in form.

| EXAMPLE 7 | Solving an Exponential Equation That Is Quadratic in Form |

Solve: $4^x - 2^x - 12 = 0$

Algebraic Solution

We note that $4^x = (2^2)^x = 2^{2x} = (2^x)^2$, so the equation is actually quadratic in form, and we can rewrite it as

$$(2^x)^2 - 2^x - 12 = 0 \quad \text{Let } u = 2^x; \text{ then } u^2 - u - 12 = 0.$$

Now we can factor as usual.

$$(2^x - 4)(2^x + 3) = 0 \quad (u - 4)(u + 3) = 0$$
$$2^x - 4 = 0 \quad \text{or} \quad 2^x + 3 = 0 \quad u - 4 = 0 \quad \text{or} \quad u + 3 = 0$$
$$2^x = 4 \qquad\qquad 2^x = -3 \quad u = 2^x = 4 \qquad u = 2^x = -3$$

The equation on the left has the solution $x = 2$, since $2^x = 4 = 2^2$; the equation on the right has no solution, since $2^x > 0$ for all x. The only solution is 2. The solution set is $\{2\}$.

Graphing Solution

Graph $Y_1 = 4^x - 2^x - 12$ and determine the x-intercept. See Figure 34. The x-intercept is 2, so the solution set is $\{2\}$.

Figure 34

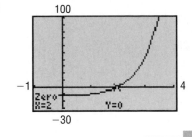

━━━━ **Now Work** PROBLEM 53

3 Solve Logarithmic and Exponential Equations Using a Graphing Utility

 The algebraic techniques introduced in this section to obtain exact solutions apply only to certain types of logarithmic and exponential equations. Solutions for other types are usually studied in calculus, using numerical methods. For such types, we can use a graphing utility to approximate the solution.

| EXAMPLE 8 | Solving Equations Using a Graphing Utility |

Solve: $x + e^x = 2$

Express the solution(s) rounded to two decimal places.

Solution   The solution is found by graphing $Y_1 = x + e^x$ and $Y_2 = 2$. Since Y_1 is an increasing function (do you know why?), there is only one point of intersection for Y_1 and Y_2. Figure 35 shows the graphs of Y_1 and Y_2. Using the INTERSECT command, the solution is 0.44 rounded to two decimal places.

Figure 35

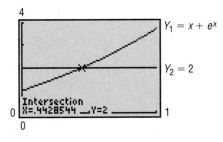

━━━━ **Now Work** PROBLEM 63

7.4 Assess Your Understanding

'Are You Prepared?'   *Answers are given at the end of these exercises. If you get a wrong answer, read the pages listed in red.*

1. Solve $x^2 - 7x - 30 = 0$. (pp. A29–A30)

2. Solve $(x + 3)^2 - 4(x + 3) + 3 = 0$. (pp. A35–A36)

3. Approximate the solution(s) to $x^3 = x^2 - 5$ using a graphing utility. (pp. A40–A41)

4. Approximate the solution(s) to $x^3 - 2x + 2 = 0$ using a graphing utility. (pp. A40–A41)

Skill Building

In Problems 5–32, solve each logarithmic equation. Express irrational solutions in exact form and as a decimal rounded to 3 decimal places. Verify your results using a graphing utility.

5. $\log_4 x = 2$

6. $\log(x + 6) = 1$

7. $\log_2(5x) = 4$

8. $\log_3(3x - 1) = 2$

9. $\log_4(x + 2) = \log_4 8$

10. $\log_5(2x + 3) = \log_5 3$

11. $\dfrac{1}{2}\log_3 x = 2\log_3 2$

12. $-2\log_4 x = \log_4 9$

13. $3\log_2 x = -\log_2 27$

14. $2\log_5 x = 3\log_5 4$

15. $3\log_2(x - 1) + \log_2 4 = 5$

16. $2\log_3(x + 4) - \log_3 9 = 2$

17. $\log x + \log(x + 15) = 2$

18. $\log x + \log(x - 21) = 2$

19. $\log(2x + 1) = 1 + \log(x - 2)$

20. $\log(2x) - \log(x - 3) = 1$

21. $\log_2(x + 7) + \log_2(x + 8) = 1$

22. $\log_6(x + 4) + \log_6(x + 3) = 1$

23. $\log_8(x + 6) = 1 - \log_8(x + 4)$

24. $\log_5(x + 3) = 1 - \log_5(x - 1)$

25. $\ln x + \ln(x + 2) = 4$

26. $\ln(x + 1) - \ln x = 2$

27. $\log_3(x + 1) + \log_3(x + 4) = 2$

28. $\log_2(x + 1) + \log_2(x + 7) = 3$

29. $\log_{1/3}(x^2 + x) - \log_{1/3}(x^2 - x) = -1$

30. $\log_4(x^2 - 9) - \log_4(x + 3) = 3$

31. $\log_a(x - 1) - \log_a(x + 6) = \log_a(x - 2) - \log_a(x + 3)$

32. $\log_a x + \log_a(x - 2) = \log_a(x + 4)$

In Problems 33–60, solve each exponential equation. Express irrational solutions in exact form and as a decimal rounded to 3 decimal places. Verify your results using a graphing utility.

33. $2^{x-5} = 8$

34. $5^{-x} = 25$

35. $2^x = 10$

36. $3^x = 14$

37. $8^{-x} = 1.2$

38. $2^{-x} = 1.5$

39. $5(2^{3x}) = 8$

40. $0.3(4^{0.2x}) = 0.2$

41. $3^{1-2x} = 4^x$

42. $2^{x+1} = 5^{1-2x}$

43. $\left(\dfrac{3}{5}\right)^x = 7^{1-x}$

44. $\left(\dfrac{4}{3}\right)^{1-x} = 5^x$

45. $1.2^x = (0.5)^{-x}$

46. $0.3^{1+x} = 1.7^{2x-1}$

47. $\pi^{1-x} = e^x$

48. $e^{x+3} = \pi^x$

49. $2^{2x} + 2^x - 12 = 0$

50. $3^{2x} + 3^x - 2 = 0$

51. $3^{2x} + 3^{x+1} - 4 = 0$

52. $2^{2x} + 2^{x+2} - 12 = 0$

53. $16^x + 4^{x+1} - 3 = 0$

54. $9^x - 3^{x+1} + 1 = 0$

55. $25^x - 8 \cdot 5^x = -16$

56. $36^x - 6 \cdot 6^x = -9$

57. $3 \cdot 4^x + 4 \cdot 2^x + 8 = 0$

58. $2 \cdot 49^x + 11 \cdot 7^x + 5 = 0$

59. $4^x - 10 \cdot 4^{-x} = 3$

60. $3^x - 14 \cdot 3^{-x} = 5$

In Problems 61–74, use a graphing utility to solve each equation. Express your answer rounded to two decimal places.

61. $\log_5(x + 1) - \log_4(x - 2) = 1$

62. $\log_2(x - 1) - \log_6(x + 2) = 2$

63. $e^x = -x$

64. $e^{2x} = x + 2$

65. $e^x = x^2$

66. $e^x = x^3$

67. $\ln x = -x$

68. $\ln(2x) = -x + 2$

69. $\ln x = x^3 - 1$

70. $\ln x = -x^2$

71. $e^x + \ln x = 4$

72. $e^x - \ln x = 4$

73. $e^{-x} = \ln x$

74. $e^{-x} = -\ln x$

In Problems 75–86, solve each equation. Express irrational solutions in exact form and as a decimal rounded to 3 decimal places. Verify your results using a graphing utility.

75. $\log_2(x + 1) - \log_4 x = 1$
[**Hint:** Change $\log_4 x$ to base 2.]

76. $\log_2(3x + 2) - \log_4 x = 3$

77. $\log_{16} x + \log_4 x + \log_2 x = 7$

78. $\log_9 x + 3\log_3 x = 14$

79. $\left(\sqrt[3]{2}\right)^{2-x} = 2^{x^2}$

80. $\log_2 x^{\log_2 x} = 4$

81. $\dfrac{e^x + e^{-x}}{2} = 1$
[**Hint:** Multiply each side by e^x.]

82. $\dfrac{e^x + e^{-x}}{2} = 3$

83. $\dfrac{e^x - e^{-x}}{2} = 2$

84. $\dfrac{e^x - e^{-x}}{2} = -2$

85. $\log_5 x + \log_3 x = 1$
[**Hint:** Use the Change-of-Base Formula and factor out $\ln x$.]

86. $\log_2 x + \log_6 x = 3$

Mixed Practice

87. $f(x) = \log_2(x + 3)$ and $g(x) = \log_2(3x + 1)$.
  (a) Solve $f(x) = 3$. What point is on the graph of f?
  (b) Solve $g(x) = 4$. What point is on the graph of g?
  (c) Solve $f(x) = g(x)$. Do the graphs of f and g intersect? If so, where?
  (d) Solve $(f + g)(x) = 7$.
  (e) Solve $(f - g)(x) = 2$.

88. $f(x) = \log_3(x + 5)$ and $g(x) = \log_3(x - 1)$.
  (a) Solve $f(x) = 2$. What point is on the graph of f?
  (b) Solve $g(x) = 3$. What point is on the graph of g?
  (c) Solve $f(x) = g(x)$. Do the graphs of f and g intersect? If so, where?
  (d) Solve $(f + g)(x) = 3$.
  (e) Solve $(f - g)(x) = 2$.

89. (a) If $f(x) = 3^{x+1}$ and $g(x) = 2^{x+2}$, graph f and g on the same Cartesian plane.
  (b) Find the point(s) of intersection of the graphs of f and g by solving $f(x) = g(x)$. Round answers to three decimal places. Label this point on the graph drawn in part (a).
  (c) Based on the graph, solve $f(x) > g(x)$.

90. (a) If $f(x) = 5^{x-1}$ and $g(x) = 2^{x+1}$, graph f and g on the same Cartesian plane.
  (b) Find the point(s) of intersection of the graphs of f and g by solving $f(x) = g(x)$. Label this point on the graph drawn in part (a).
  (c) Based on the graph, solve $f(x) > g(x)$.

91. (a) Graph $f(x) = 3^x$ and $g(x) = 10$ on the same Cartesian plane.
  (b) Shade the region bounded by the y-axis, $f(x) = 3^x$, and $g(x) = 10$ on the graph drawn in part (a).
  (c) Solve $f(x) = g(x)$ and label the point of intersection on the graph drawn in part (a).

92. (a) Graph $f(x) = 2^x$ and $g(x) = 12$ on the same Cartesian plane.
  (b) Shade the region bounded by the y-axis, $f(x) = 2^x$, and $g(x) = 12$ on the graph drawn in part (a).
  (c) Solve $f(x) = g(x)$ and label the point of intersection on the graph drawn in part (a).

93. (a) Graph $f(x) = 2^{x+1}$ and $g(x) = 2^{-x+2}$ on the same Cartesian plane.
  (b) Shade the region bounded by the y-axis, $f(x) = 2^{x+1}$, and $g(x) = 2^{-x+2}$ on the graph drawn in part (a).
  (c) Solve $f(x) = g(x)$ and label the point of intersection on the graph drawn in part (a).

94. (a) Graph $f(x) = 3^{-x+1}$ and $g(x) = 3^{x-2}$ on the same Cartesian plane.
  (b) Shade the region bounded by the y-axis, $f(x) = 3^{-x+1}$, and $g(x) = 3^{x-2}$ on the graph drawn in part (a).
  (c) Solve $f(x) = g(x)$ and label the point of intersection on the graph drawn in part (a).

95. (a) Graph $f(x) = 2^x - 4$.
  (b) Find the zero of f.
  (c) Based on the graph, solve $f(x) < 0$.

96. (a) Graph $g(x) = 3^x - 9$.
  (b) Find the zero of g.
  (c) Based on the graph, solve $g(x) > 0$.

Applications and Extensions

97. A Population Model The resident population of the United States in 2006 was 298 million people and was growing at a rate of 0.9% per year. Assuming that this growth rate continues, the model $P(t) = 298(1.009)^{t-2006}$ represents the population P (in millions of people) in year t.
  (a) According to this model, when will the population of the United States be 310 million people?
  (b) According to this model, when will the population of the United States be 360 million people?
  Source: Statistical Abstract of the United States, 125th ed., 2006.

98. A Population Model The population of the world in 2006 was 6.53 billion people and was growing at a rate of 1.14% per year. Assuming that this growth rate continues, the model $P(t) = 6.53(1.0114)^{t-2006}$ represents the population P (in billions of people) in year t.
  (a) According to this model, when will the population of the world be 9.25 billion people?
  (b) According to this model, when will the population of the world be 11.75 billion people?
  Source: U.S. Census Bureau.

99. Depreciation The value V of a Chevy Cobalt that is t years old can be modeled by $V(t) = 14{,}512(0.82)^t$.
  (a) According to the model, when will the car be worth $9000?
  (b) According to the model, when will the car be worth $4000?
  (c) According to the model, when will the car be worth $2000?
  Source: Kelley Blue Book

100. Depreciation   The value V of a Dodge Stratus that is t years old can be modeled by $V(t) = 19{,}282(0.84)^t$.
   (a) According to the model, when will the car be worth $15,000?
   (b) According to the model, when will the car be worth $8000?
   (c) According to the model, when will the car be worth $2000?
   Source: Kelley Blue Book

Discussion and Writing

101. Fill in reasons for each step in the following two solutions.

   Solve: $\log_3(x - 1)^2 = 2$

Solution A	**Solution B**
$\log_3(x - 1)^2 = 2$	$\log_3(x - 1)^2 = 2$
$(x - 1)^2 = 3^2 = 9$ _____	$2\log_3(x - 1) = 2$ _____
$(x - 1) = \pm 3$ _____	$\log_3(x - 1) = 1$ _____
$x - 1 = -3$ or $x - 1 = 3$ _____	$x - 1 = 3^1 = 3$ _____
$x = -2$ or $x = 4$ _____	$x = 4$ _____

   Both solutions given in Solution A check. Explain what caused the solution $x = -2$ to be lost in Solution B.

'Are You Prepared?' Answers

   1. $\{-3, 10\}$ **2.** $\{-2, 0\}$ **3.** $\{-1.43\}$ **4.** $\{-1.77\}$

7.5 Financial Models

   OBJECTIVES   **1** Determine the Future Value of a Lump Sum of Money (p. 508)
                     2 Calculate Effective Rates of Return (p. 512)
                     3 Determine the Present Value of a Lump Sum of Money (p. 513)
                     4 Determine the Rate of Interest or Time Required to Double a Lump Sum of Money (p. 514)

1 Determine the Future Value of a Lump Sum of Money

Interest is money paid for the use of money. The total amount borrowed (whether by an individual from a bank in the form of a loan or by a bank from an individual in the form of a savings account) is called the **principal.** The **rate of interest,** expressed as a percent, is the amount charged for the use of the principal for a given period of time, usually on a yearly (that is, per annum) basis.

THEOREM   **Simple Interest Formula**

   If a principal of P dollars is borrowed for a period of t years at a per annum interest rate r, expressed as a decimal, the interest I charged is

$$I = Prt \tag{1}$$

   Interest charged according to formula (1) is called **simple interest.**

In working with problems involving interest, we define the term **payment period** as follows:

Annually:	Once per year	**Monthly:**	12 times per year
Semiannually:	Twice per year	**Daily:**	365 times per year*
Quarterly:	Four times per year		

When the interest due at the end of a payment period is added to the principal so that the interest computed at the end of the next payment period is based on this new principal amount (old principal + interest), the interest is said to have been **compounded. Compound interest** is interest paid on principal and previously earned interest.

EXAMPLE 1	Computing Compound Interest

A credit union pays interest of 8% per annum compounded quarterly on a certain savings plan. If $1000 is deposited in such a plan and the interest is left to accumulate, how much is in the account after 1 year?

Solution   We use the simple interest formula, $I = Prt$. The principal P is $1000 and the rate of interest is 8% = 0.08. After the first quarter of a year, the time t is $\frac{1}{4}$ year, so the interest earned is

$$I = Prt = (\$1000)(0.08)\left(\frac{1}{4}\right) = \$20$$

The new principal is $P + I = \$1000 + \$20 = \$1020$. At the end of the second quarter, the interest on this principal is

$$I = (\$1020)(0.08)\left(\frac{1}{4}\right) = \$20.40$$

At the end of the third quarter, the interest on the new principal of $1020 + $20.40 = $1040.40 is

$$I = (\$1040.40)(0.08)\left(\frac{1}{4}\right) = \$20.81$$

Finally, after the fourth quarter, the interest is

$$I = (\$1061.21)(0.08)\left(\frac{1}{4}\right) = \$21.22$$

After 1 year the account contains $1061.21 + $21.22 = $1082.43. ∎

The pattern of the calculations performed in Example 1 leads to a general formula for compound interest. To fix our ideas, let P represent the principal to be invested at a per annum interest rate r that is compounded n times per year, so the time of each compounding period is $\frac{1}{n}$ years. (For computing purposes, r is expressed as a decimal.) The interest earned after each compounding period is given by formula (1).

$$\text{Interest} = \text{principal} \times \text{rate} \times \text{time} = P \cdot r \cdot \frac{1}{n} = P \cdot \left(\frac{r}{n}\right)$$

The amount A after one compounding period is

$$A = P + P \cdot \left(\frac{r}{n}\right) = P \cdot \left(1 + \frac{r}{n}\right)$$

*Most banks use a 360-day "year." Why do you think they do?

After two compounding periods, the amount A, based on the new principal $P \cdot \left(1 + \dfrac{r}{n}\right)$, is

$$A = \underbrace{P \cdot \left(1 + \frac{r}{n}\right)}_{\substack{\text{New} \\ \text{principal}}} + \underbrace{P \cdot \left(1 + \frac{r}{n}\right)\left(\frac{r}{n}\right)}_{\substack{\text{Interest on} \\ \text{new principal}}} = \underset{\substack{\uparrow \\ \text{Factor out } P \cdot (1 + \frac{r}{n})}}{P \cdot \left(1 + \frac{r}{n}\right)\left(1 + \frac{r}{n}\right)} = P \cdot \left(1 + \frac{r}{n}\right)^2$$

After three compounding periods, the amount A is

$$A = P \cdot \left(1 + \frac{r}{n}\right)^2 + P \cdot \left(1 + \frac{r}{n}\right)^2\left(\frac{r}{n}\right) = P \cdot \left(1 + \frac{r}{n}\right)^2 \cdot \left(1 + \frac{r}{n}\right) = P \cdot \left(1 + \frac{r}{n}\right)^3$$

Continuing this way, after n compounding periods (1 year), the amount A is

$$A = P \cdot \left(1 + \frac{r}{n}\right)^n$$

Because t years will contain $n \cdot t$ compounding periods, after t years we have

$$A = P \cdot \left(1 + \frac{r}{n}\right)^{nt}$$

THEOREM

Exploration
To see the effects of compounding interest monthly on an initial deposit of \$1, graph $Y_1 = \left(1 + \dfrac{r}{12}\right)^{12x}$ with $r = 0.06$ and $r = 0.12$ for $0 \leq x \leq 30$. What is the future value of \$1 in 30 years when the interest rate per annum is $r = 0.06$ (6%)? What is the future value of \$1 in 30 years when the interest rate per annum is $r = 0.12$ (12%)? Does doubling the interest rate double the future value?

Compound Interest Formula

The amount A after t years due to a principal P invested at an annual interest rate r compounded n times per year is

$$A = P \cdot \left(1 + \frac{r}{n}\right)^{nt} \tag{2}$$

For example, to rework Example 1, we would use $P = \$1000$, $r = 0.08$, $n = 4$ (quarterly compounding), and $t = 1$ year to obtain

$$A = P \cdot \left(1 + \frac{r}{n}\right)^{nt} = 1000\left(1 + \frac{0.08}{4}\right)^{4 \cdot 1} = \$1082.43$$

In equation (2), the amount A is typically referred to as the **future value** of the account, while P is called the **present value.**

— Now Work PROBLEM 3

EXAMPLE 2 **Comparing Investments Using Different Compounding Periods**

Investing \$1000 at an annual rate of 10% compounded annually, semiannually, quarterly, monthly, and daily will yield the following amounts after 1 year:

Annual compounding ($n = 1$):

$$A = P \cdot (1 + r)$$
$$= (\$1000)(1 + 0.10) = \$1100.00$$

Semiannual compounding ($n = 2$):

$$A = P \cdot \left(1 + \frac{r}{2}\right)^2$$
$$= (\$1000)(1 + 0.05)^2 = \$1102.50$$

Quarterly compounding ($n = 4$):

$$A = P \cdot \left(1 + \frac{r}{4}\right)^4$$
$$= (\$1000)(1 + 0.025)^4 = \$1103.81$$

Monthly compounding ($n = 12$):

$$A = P \cdot \left(1 + \frac{r}{12}\right)^{12}$$
$$= (\$1000)\left(1 + \frac{0.10}{12}\right)^{12} = \$1104.71$$

Daily compounding ($n = 365$):
$$A = P \cdot \left(1 + \frac{r}{365}\right)^{365}$$
$$= (\$1000)\left(1 + \frac{0.10}{365}\right)^{365} = \$1105.16$$

From Example 2, we can see that the effect of compounding more frequently is that the amount after 1 year is higher: \$1000 compounded 4 times a year at 10% results in \$1103.81, \$1000 compounded 12 times a year at 10% results in \$1104.71, and \$1000 compounded 365 times a year at 10% results in \$1105.16. This leads to the following question: What would happen to the amount after 1 year if the number of times that the interest is compounded were increased without bound?

Let's find the answer. Suppose that P is the principal, r is the per annum interest rate, and n is the number of times that the interest is compounded each year. The amount after 1 year is

$$A = P \cdot \left(1 + \frac{r}{n}\right)^n$$

Rewrite this expression as follows:

$$A = P \cdot \left(1 + \frac{r}{n}\right)^n = P \cdot \left(1 + \frac{1}{\frac{n}{r}}\right)^n = P \cdot \left[\left(1 + \frac{1}{\frac{n}{r}}\right)^{n/r}\right]^r = P \cdot \left[\left(1 + \frac{1}{h}\right)^h\right]^r \qquad \textbf{(3)}$$

$$\underset{h = \frac{n}{r}}{\uparrow}$$

Now suppose that the number n of times that the interest is compounded per year gets larger and larger; that is, suppose that $n \to \infty$. Then $h = \frac{n}{r} \to \infty$, and the expression in brackets in Equation (3) equals e. That is, $A \to Pe^r$.

Table 8 compares $\left(1 + \frac{r}{n}\right)^n$, for large values of n, to e^r for $r = 0.05$, $r = 0.10$, $r = 0.15$, and $r = 1$. The larger that n gets, the closer $\left(1 + \frac{r}{n}\right)^n$ gets to e^r.

No matter how frequent the compounding, the amount after 1 year has the definite ceiling Pe^r.

Table 8

	$\left(1 + \frac{r}{n}\right)^n$			
	$n = 100$	$n = 1000$	$n = 10{,}000$	e^r
$r = 0.05$	1.0512580	1.0512698	1.051271	1.0512711
$r = 0.10$	1.1051157	1.1051654	1.1051704	1.1051709
$r = 0.15$	1.1617037	1.1618212	1.1618329	1.1618342
$r = 1$	2.7048138	2.7169239	2.7181459	2.7182818

When interest is compounded so that the amount after 1 year is Pe^r, we say the interest is **compounded continuously.**

THEOREM

Continuous Compounding

The amount A after t years due to a principal P invested at an annual interest rate r compounded continuously is

$$A = Pe^{rt} \qquad \textbf{(4)}$$

EXAMPLE 3	Using Continuous Compounding

The amount A that results from investing a principal P of $1000 at an annual rate r of 10% compounded continuously for a time t of 1 year is

$$A = \$1000e^{0.10} = (\$1000)(1.10517) = \$1105.17$$

■

Now Work PROBLEM 11

2 Calculate Effective Rates of Return

Suppose you have $1000 and a bank offers to pay you 3% annual interest on a savings account compounded monthly. What annual interest rate would you need to earn if the interest was compounded annually (once per year)? To answer this question, we first determine the value of the $1000 in the account that earns 3% compounded monthly.

$$A = \$1000\left(1 + \frac{0.03}{12}\right)^{12} \qquad \text{Use } A = P\left(1 + \frac{r}{n}\right)^{n} \text{ with } P = \$1000, r = 0.03, n = 12.$$

$$= \$1030.42$$

So the interest earned is $30.42. Using $I = Prt$ with $t = 1$, $I = \$30.42$, and $P = \$1000$, we find the annual simple interest rate is $0.03042 = 3.042\%$. This interest rate is known as the *effective rate of interest*.

The **effective rate of interest** is the equivalent annual simple interest rate that would yield the same amount as compounding n times per year, or continuously, after 1 year. We can determine the effective rate of interest using the following.

THEOREM	**Effective Rate of Interest**

The effective rate of interest r_e of an investment earning an annual interest rate r is given by

$$\text{Compounding } n \text{ times per year:} \quad r_e = \left(1 + \frac{r}{n}\right)^{n} - 1$$

$$\text{Continuous compounding:} \quad r_e = e^{r} - 1$$

⌐

EXAMPLE 4	Computing the Effective Rate of Interest—Which Is the Best Deal?

Suppose you want to open a money market account. You visit three banks to determine their money market rates. Bank A offers you 6% annual interest compounded daily and Bank B offers you 6.02% compounded quarterly. Bank C offers 5.98% compounded continuously. Determine which bank is offering the best deal.

Solution The bank that offers the best deal is the one with the highest effective interest rate.

Bank A	**Bank B**	**Bank C**
$r_e = \left(1 + \dfrac{0.06}{365}\right)^{365} - 1$	$r_e = \left(1 + \dfrac{0.0602}{4}\right)^{4} - 1$	$r_e = e^{0.0598} - 1$
$\approx 1.06183 - 1$	$\approx 1.06157 - 1$	$\approx 1.06162 - 1$
$= 0.06183$	$= 0.06157$	$= 0.06162$
$= 6.183\%$	$= 6.157\%$	$= 6.162\%$

Since the effective rate of interest is highest for Bank A, Bank A is offering the best deal.

■

Now Work PROBLEM 23

3 Determine the Present Value of a Lump Sum of Money

When people engaged in finance speak of the "time value of money," they are usually referring to the *present value* of money. The **present value** of A dollars to be received at a future date is the principal that you would need to invest now so that it will grow to A dollars in the specified time period. The present value of money to be received at a future date is always less than the amount to be received, since the amount to be received will equal the present value (money invested now) *plus* the interest accrued over the time period.

We use the compound interest formula (2) to get a formula for present value. If P is the present value of A dollars to be received after t years at a per annum interest rate r compounded n times per year, then, by formula (2),

$$A = P \cdot \left(1 + \frac{r}{n}\right)^{nt}$$

To solve for P, we divide both sides by $\left(1 + \frac{r}{n}\right)^{nt}$. The result is

$$\frac{A}{\left(1 + \frac{r}{n}\right)^{nt}} = P \quad \text{or} \quad P = A \cdot \left(1 + \frac{r}{n}\right)^{-nt}$$

THEOREM

Present Value Formulas

The present value P of A dollars to be received after t years, assuming a per annum interest rate r compounded n times per year, is

$$P = A \cdot \left(1 + \frac{r}{n}\right)^{-nt} \tag{5}$$

If the interest is compounded continuously,

$$P = Ae^{-rt} \tag{6}$$

To prove (6), solve formula (4) for P.

EXAMPLE 5 | **Computing the Value of a Zero-coupon Bond**

A zero-coupon (noninterest-bearing) bond can be redeemed in 10 years for $1000. How much should you be willing to pay for it now if you want a return of
(a) 8% compounded monthly?
(b) 7% compounded continuously?

Solution (a) We are seeking the present value of $1000. We use formula (5) with $A = \$1000$, $n = 12, r = 0.08$, and $t = 10$.

$$P = A \cdot \left(1 + \frac{r}{n}\right)^{-nt} = \$1000\left(1 + \frac{0.08}{12}\right)^{-12(10)} = \$450.52$$

For a return of 8% compounded monthly, you should pay $450.52 for the bond.
(b) Here we use formula (6) with $A = \$1000, r = 0.07$, and $t = 10$.

$$P = Ae^{-rt} = \$1000e^{-(0.07)(10)} = \$496.59$$

For a return of 7% compounded continuously, you should pay $496.59 for the bond.

Now Work PROBLEM 13

4 Determine the Rate of Interest or Time Required to Double a Lump Sum of Money

| EXAMPLE 6 | Rate of Interest Required to Double an Investment |

What annual rate of interest compounded annually should you seek if you want to double your investment in 5 years?

Solution   If P is the principal and we want P to double, the amount A will be $2P$. We use the compound interest formula with $n = 1$ and $t = 5$ to find r.

$$A = P \cdot \left(1 + \frac{r}{n}\right)^{nt}$$

$$2P = P \cdot (1 + r)^5 \qquad\qquad A = 2P, n = 1, t = 5$$

$$2 = (1 + r)^5 \qquad\qquad \text{Divide both sides by } P.$$

$$1 + r = \sqrt[5]{2} \qquad\qquad \text{Take the fifth root of each side.}$$

$$r = \sqrt[5]{2} - 1 \approx 1.148698 - 1 = 0.148698$$

The annual rate of interest needed to double the principal in 5 years is 14.87%.

 Now Work   PROBLEM 31

| EXAMPLE 7 | Time Required to Double or Triple an Investment |

(a) How long will it take for an investment to double in value if it earns 5% compounded continuously?

(b) How long will it take to triple at this rate?

Solution   (a) If P is the initial investment and we want P to double, the amount A will be $2P$. We use formula (4) for continuously compounded interest with $r = 0.05$. Then

$$A = Pe^{rt}$$

$$2P = Pe^{0.05t} \qquad\qquad A = 2P, r = 0.05$$

$$2 = e^{0.05t} \qquad\qquad \text{Divide both sides by } P$$

$$0.05t = \ln 2 \qquad\qquad \text{Rewrite as a logarithm.}$$

$$t = \frac{\ln 2}{0.05} \approx 13.86 \qquad \text{Solve for } t.$$

It will take about 14 years to double the investment.

(b) To triple the investment, we set $A = 3P$ in formula (4).

$$A = Pe^{rt}$$

$$3P = Pe^{0.05t} \qquad\qquad A = 3P, r = 0.05$$

$$3 = e^{0.05t} \qquad\qquad \text{Divide both sides by } P$$

$$0.05t = \ln 3 \qquad\qquad \text{Rewrite as a logarithm.}$$

$$t = \frac{\ln 3}{0.05} \approx 21.97 \qquad \text{Solve for } t.$$

It will take about 22 years to triple the investment.

 Now Work   PROBLEM 35

7.5 Assess Your Understanding

Skill Building

1. What is the interest due if $500 is borrowed for 6 months at a simple interest rate of 6% per annum?

2. If you borrow $5000 and, after 9 months, pay off the loan in the amount of $5500, what per annum rate of interest was charged?

In Problems 3–12, find the amount that results from each investment.

3. $100 invested at 4% compounded quarterly after a period of 2 years

4. $50 invested at 6% compounded monthly after a period of 3 years

5. $500 invested at 8% compounded quarterly after a period of $2\frac{1}{2}$ years

6. $300 invested at 12% compounded monthly after a period of $1\frac{1}{2}$ years

7. $600 invested at 5% compounded daily after a period of 3 years

8. $700 invested at 6% compounded daily after a period of 2 years

9. $10 invested at 11% compounded continuously after a period of 2 years

10. $40 invested at 7% compounded continuously after a period of 3 years

11. $100 invested at 10% compounded continuously after a period of $2\frac{1}{4}$ years

12. $100 invested at 12% compounded continuously after a period of $3\frac{3}{4}$ years

In Problems 13–22, find the principal needed now to get each amount; that is, find the present value.

13. To get $100 after 2 years at 6% compounded monthly

14. To get $75 after 3 years at 8% compounded quarterly

15. To get $1000 after $2\frac{1}{2}$ years at 6% compounded daily

16. To get $800 after $3\frac{1}{2}$ years at 7% compounded monthly

17. To get $600 after 2 years at 4% compounded quarterly

18. To get $300 after 4 years at 3% compounded daily

19. To get $80 after $3\frac{1}{4}$ years at 9% compounded continuously

20. To get $800 after $2\frac{1}{2}$ years at 8% compounded continuously

21. To get $400 after 1 year at 10% compounded continuously

22. To get $1000 after 1 year at 12% compounded continuously

In Problems 23–26, find the effective rate of interest.

23. For 5% compounded quarterly

24. For 6% compounded monthly

25. For 5% compounded continuously

26. For 6% compounded continuously

In Problems 27–30, determine the rate that represents the better deal.

27. 6% compounded quarterly or $6\frac{1}{4}$% compounded annually

28. 9% compounded quarterly of $9\frac{1}{4}$% compounded annually

29. 9% compounded monthly or 8.8% compounded daily

30. 8% compounded semiannually or 7.9% compounded daily

31. What rate of interest compounded annually is required to double an investment in 3 years?

32. What rate of interest compounded annually is required to double an investment in 6 years?

33. What rate of interest compounded annually is required to triple an investment in 5 years?

34. What rate of interest compounded annually is required to triple an investment in 10 years?

35. (a) How long does it take for an investment to double in value if it is invested at 8% compounded monthly?
  (b) How long does it take if the interest is compounded continuously?

36. (a) How long does it take for an investment to triple in value if it is invested at 6% compounded monthly?
  (b) How long does it take if the interest is compounded continuously?

37. What rate of interest compounded quarterly will yield an effective interest rate of 7%?

38. What rate of interest compounded continuously will yield an effective interest rate of 6%?

Applications and Extensions

39. Time Required to Reach a Goal If Tanisha has $100 to invest at 8% per annum compounded monthly, how long will it be before she has $150? If the compounding is continuous, how long will it be?

40. Time Required to Reach a Goal If Angela has $100 to invest at 10% per annum compounded monthly, how long will it be before she has $175? If the compounding is continuous, how long will it be?

41. **Time Required to Reach a Goal** How many years will it take for an initial investment of $10,000 to grow to $25,000? Assume a rate of interest of 6% compounded continuously.

42. **Time Required to Reach a Goal** How many years will it take for an initial investment of $25,000 to grow to $80,000? Assume a rate of interest of 7% compounded continuously.

43. **Price Appreciation of Homes** What will a $90,000 condominium cost 5 years from now if the price appreciation for condos over that period averages 3% compounded annually?

44. **Credit Card Interest** A department store charges 1.25% per month on the unpaid balance for customers with charge accounts (interest is compounded monthly). A customer charges $200 and does not pay her bill for 6 months. What is the bill at that time?

45. **Saving for a Car** Jerome will be buying a used car for $15,000 in 3 years. How much money should he ask his parents for now so that, if he invests it at 5% compounded continuously, he will have enough to buy the car?

46. **Paying off a Loan** John requires $3000 in 6 months to pay off a loan that has no prepayment privileges. If he has the $3000 now, how much of it should he save in an account paying 3% compounded monthly so that in 6 months he will have exactly $3000?

47. **Return on a Stock** George contemplates the purchase of 100 shares of a stock selling for $15 per share. The stock pays no dividends. The history of the stock indicates that it should grow at an annual rate of 15% per year. How much will the 100 shares of stock be worth in 5 years?

48. **Return on an Investment** A business purchased for $650,000 in 2005 is sold in 2008 for $850,000. What is the annual rate of return for this investment?

49. **Comparing Savings Plans** Jim places $1000 in a bank account that pays 5.6% compounded continuously. After 1 year, will he have enough money to buy a computer system that costs $1060? If another bank will pay Jim 5.9% compounded monthly, is this a better deal?

50. **Savings Plans** On January 1, Kim places $1000 in a certificate of deposit that pays 6.8% compounded continuously and matures in 3 months. Then Kim places the $1000 and the interest in a passbook account that pays 5.25% compounded monthly. How much does Kim have in the passbook account on May 1?

51. **Comparing IRA Investments** Will invests $2000 in his IRA in a bond trust that pays 9% interest compounded semiannually. His friend Henry invests $2000 in his IRA in a certificate of deposit that pays $8\frac{1}{2}$% compounded continuously. Who has more money after 20 years, Will or Henry?

52. **Comparing Two Alternatives** Suppose that April has access to an investment that will pay 10% interest compounded continuously. Which is better: to be given $1000 now so that she can take advantage of this investment opportunity or to be given $1325 after 3 years?

53. **College Costs** The average annual cost of college at 4-year private colleges was $29,026 in 2005. This was a 5.5% increase from the previous year.

    Source: Trends in College Pricing 2005, College Board

    (a) If the cost of college increases by 5.5% each year, what will be the average cost of college at a 4-year private college in 2015?

    (b) College savings plans, such as a 529 plan, allow individuals to put money aside now to help pay for college later. If one such plan offers a rate of 4% compounded continuously, how much should be put in a college savings plan in 2005 to pay for 1 year of the cost of college at a 4-year private college in 2015?

54. **Analyzing Interest Rates on a Mortgage** Colleen and Bill have just purchased a house for $650,000, with the seller holding a second mortgage of $100,000. They promise to pay the seller $100,000 plus all accrued interest 5 years from now. The seller offers them three interest options on the second mortgage:

    (a) Simple interest at 12% per annum

    (b) $11\frac{1}{2}$% interest compounded monthly

    (c) $11\frac{1}{4}$% interest compounded continuously

    Which option is best; that is, which results in the least interest on the loan?

55. **Federal Deficit** At the end of fiscal year 2005, the federal budget deficit was $319 billion. At that time, 20-year Series EE bonds had a fixed rate of 3.2% compounded semiannually. If the federal government financed this deficit through EE bonds, how much would it have to pay back in 2025?

    Source: U.S. Treasury Department

56. **Federal Deficit** On February 6, 2006, President Bush proposed the fiscal year 2007 federal budget. The proposal projected a fiscal year 2006 deficit of $423 billion and a fiscal year 2007 deficit of $354 billion. Assuming the deficit decreases at the same rate each year, when will the deficit be cut to $100 billion?

    Source: Office of Management and Budget

Inflation *Problems 57–62 require the following discussion.* **Inflation** *is a term used to describe the erosion of the purchasing power of money. For example, suppose the annual inflation rate is 3%. Then $1000 worth of purchasing power now will have only $970 worth of purchasing power in 1 year because 3% of the original $1000 (0.03 × 1000 = 30) has been eroded due to inflation. In general, if the rate of inflation averages r% over n years, the amount A that $P will purchase after n years is*

$$A = P \cdot (1 - r)^n$$

where r is expressed as a decimal.

57. **Inflation** If the inflation rate averages 3%, how much will $1000 purchase in 2 years?

58. **Inflation** If the inflation rate averages 2%, how much will $1000 purchase in 3 years?

59. **Inflation** If the amount that $1000 will purchase is only $950 after 2 years, what was the average inflation rate?

60. **Inflation** If the amount that $1000 will purchase is only $930 after 2 years, what was the average inflation rate?

61. **Inflation** If the average inflation rate is 2%, how long is it until purchasing power is cut in half?

62. **Inflation** If the average inflation rate is 4%, how long is it until purchasing power is cut in half?

*Problems 63–66 involve zero-coupon bonds. A **zero-coupon bond** is a bond that is sold now at a discount and will pay its face value at the time when it matures; no interest payments are made.*

63. Zero-Coupon Bonds A zero-coupon bond can be redeemed in 20 years for $10,000. How much should you be willing to pay for it now if you want a return of:

(a) 10% compounded monthly?

(b) 10% compounded continuously?

64. Zero-Coupon Bonds A child's grandparents are considering buying a $40,000 face-value, zero-coupon bond at birth so that she will have enough money for her college education 17 years later. If they want a rate of return of 8% compounded annually, what should they pay for the bond?

65. Zero-Coupon Bonds How much should a $10,000 face-value, zero-coupon bond, maturing in 10 years, be sold for now if its rate of return is to be 8% compounded annually?

66. Zero-Coupon Bonds If Pat pays $12,485.52 for a $25,000 face-value, zero-coupon bond that matures in 8 years, what is his annual rate of return?

67. Time to Double or Triple an Investment The formula

$$t = \frac{\ln m}{n \ln\left(1 + \dfrac{r}{n}\right)}$$

can be used to find the number of years t required to multiply an investment m times when r is the per annum interest rate compounded n times a year.

(a) How many years will it take to double the value of an IRA that compounds annually at the rate of 12%?

(b) How many years will it take to triple the value of a savings account that compounds quarterly at an annual rate of 6%?

(c) Give a derivation of this formula.

68. Time to Reach an Investment Goal The formula

$$t = \frac{\ln A - \ln P}{r}$$

can be used to find the number of years t required for an investment P to grow to a value A when compounded continuously at an annual rate r.

(a) How long will it take to increase an initial investment of $1000 to $8000 at an annual rate of 10%?

(b) What annual rate is required to increase the value of a $2000 IRA to $30,000 in 35 years?

(c) Give a derivation of this formula.

*Problems 69–72, require the following discussion. The **Consumer Price Index (CPI)** indicates the relative change in price over time for a fixed basket of goods and services. It is a cost of living index that helps measure the effect of inflation on the cost of goods and services. The CPI uses the base period 1982–1984 for comparison (the CPI for this period is 100). The CPI for January 2006 was 198.3. This means that $100 in the period 1982–1984 had the same purchasing power as $198.30 in January 2006. In general, if the rate of inflation averages r% over n years, then the CPI index after n years is*

$$\text{CPI} = \text{CPI}_0\left(1 + \frac{r}{100}\right)^n$$

where CPI_0 is the CPI index at the beginning of the n-year period. Source: U.S. Bureau of Labor Statistics

69. Consumer Price Index

(a) The CPI was 152.4 for 1995 and 195.3 for 2005. Assuming that annual inflation remained constant for this time period, determine the average annual inflation rate.

(b) Using the inflation rate from part (a), in what year will the CPI reach 300?

70. Consumer Price Index If the current CPI is 234.2 and the average annual inflation rate is 2.8%, what will be the CPI in 5 years?

71. Consumer Price Index If the average annual inflation rate is 3.1%, how long will it take for the CPI index to double? (A doubling of the CPI index means purchasing power is cut in half).

72. Consumer Price Index The base period for the CPI changed in 1998. Under the previous weight and item structure, the CPI for 1995 was 456.5. If the average annual inflation rate was 5.57%, what year was used as the base period for the CPI?

Discussion and Writing

73. Explain in your own words what the term *compound interest* means. What does *continuous compounding* mean?

74. Explain in your own words the meaning of *present value*.

75. Critical Thinking You have just contracted to buy a house and will seek financing in the amount of $100,000. You go to several banks. Bank 1 will lend you $100,000 at the rate of 8.75% amortized over 30 years with a loan origination fee of 1.75%. Bank 2 will lend you $100,000 at the rate of 8.375% amortized over 15 years with a loan origination fee of 1.5%. Bank 3 will lend you $100,000 at the rate of 9.125% amortized over 30 years with no loan origination fee. Bank 4 will lend you $100,000 at the rate of 8.625% amortized over 15 years with no loan origination fee. Which loan would you take? Why? Be sure to have sound reasons for your choice.

Use the information in the table to assist you. If the amount of the monthly payment does not matter to you, which loan would you take? Again, have sound reasons for your choice. Compare your final decision with others in the class. Discuss.

	Monthly Payment	Loan Origination Fee
Bank 1	$786.70	$1,750.00
Bank 2	$977.42	$1,500.00
Bank 3	$813.63	$0.00
Bank 4	$990.68	$0.00

7.6 Exponential Growth and Decay Models; Newton's Law; Logistic Growth and Decay Models

> **OBJECTIVES** 1 Find Equations of Populations That Obey the Law of Uninhibited Growth (p. 518)
> 2 Find Equations of Populations That Obey the Law of Decay (p. 520)
> 3 Use Newton's Law of Cooling (p. 522)
> 4 Use Logistic Models (p. 523)

1 Find Equations of Populations That Obey the Law of Uninhibited Growth

Many natural phenomena have been found to follow the law that an amount A varies with time t according to the function

$$A(t) = A_0 e^{kt} \qquad \qquad \textbf{(1)}$$

Here A_0 is the original amount ($t = 0$) and $k \neq 0$ is a constant.

If $k > 0$, then equation (1) states that the amount A is increasing over time; if $k < 0$, the amount A is decreasing over time. In either case, when an amount A varies over time according to equation (1), it is said to follow the **exponential law** or the **law of uninhibited growth** ($k > 0$) **or decay** ($k < 0$). See Figure 36.

Figure 36

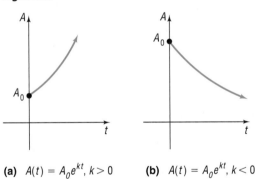

(a) $A(t) = A_0 e^{kt},\ k > 0$ **(b)** $A(t) = A_0 e^{kt},\ k < 0$

For example, we saw in Section 7.5 that continuously compounded interest follows the law of uninhibited growth. In this section we shall look at three additional phenomena that follow the exponential law.

Cell division is the growth process of many living organisms, such as amoebas, plants, and human skin cells. Based on an ideal situation in which no cells die and no by-products are produced, the number of cells present at a given time follows the law of uninhibited growth. Actually, however, after enough time has passed, growth at an exponential rate will cease due to the influence of factors such as lack of living space and dwindling food supply. The law of uninhibited growth accurately models only the early stages of the cell division process.

The cell division process begins with a culture containing N_0 cells. Each cell in the culture grows for a certain period of time and then divides into two identical cells. We assume that the time needed for each cell to divide in two is constant and does not change as the number of cells increases. These new cells then grow, and eventually each divides in two, and so on.

Uninhibited Growth of Cells

A model that gives the number N of cells in a culture after a time t has passed (in the early stages of growth) is

$$N(t) = N_0 e^{kt} \qquad k > 0 \tag{2}$$

where N_0 is the initial number of cells and k is a positive constant that represents the growth rate of the cells.

In using formula (2) to model the growth of cells, we are using a function that yields positive real numbers, even though we are counting the number of cells, which must be an integer. This is a common practice in many applications.

EXAMPLE 1	**Bacterial Growth**

A colony of bacteria grows according to the law of uninhibited growth according to the function $N(t) = 100e^{0.045t}$, where N is measured in grams and t is measured in days.

(a) Determine the initial amount of bacteria.
(b) What is the growth rate of the bacteria?
(c) Graph the function using a graphing utility.
(d) What is the population after 5 days?
(e) How long will it take for the population to reach 140 grams?
(f) What is the doubling time for the population?

Solution (a) The initial amount of bacteria, N_0, is obtained when $t = 0$, so

$$N_0 = N(0) = 100e^{0.045(0)} = 100 \text{ grams}$$

(b) Compare $N(t) = 100e^{0.045t}$ to $N(t) = 100e^{kt}$. The value of k, 0.045, indicates a growth rate of 4.5%.

(c) Figure 37 shows the graph of $N(t) = 100e^{0.045t}$.

(d) The population after 5 days is $N(5) = 100e^{0.045(5)} \approx 125.2$ grams.

(e) To find how long it takes for the population to reach 140 grams, we solve the equation $N(t) = 140$.

$$100e^{0.045t} = 140$$

$$e^{0.045t} = 1.4 \qquad \text{Divide both sides of the equation by 100.}$$

$$0.045t = \ln 1.4 \qquad \text{Rewrite as a logarithm.}$$

$$t = \frac{\ln 1.4}{0.045} \qquad \text{Divide both sides of the equation by 0.045.}$$

$$\approx 7.5 \text{ days}$$

Figure 37

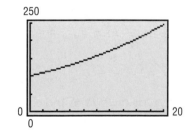

(f) The population doubles when $N(t) = 200$ grams, so we find the doubling time by solving the equation $200 = 100e^{0.045t}$ for t.

$$200 = 100e^{0.045t}$$

$$2 = e^{0.045t} \qquad \text{Divide both sides of the equation by 100.}$$

$$\ln 2 = 0.045t \qquad \text{Rewrite as a logarithm.}$$

$$t = \frac{\ln 2}{0.045} \qquad \text{Divide both sides of the equation by 0.045.}$$

$$\approx 15.4 \text{ days}$$

The population doubles approximately every 15.4 days.

Now Work PROBLEM 1

| EXAMPLE 2 | **Bacterial Growth** |

A colony of bacteria grows according to the law of uninhibited growth.

(a) If N is the number of cells and t is the time in hours, express N as a function of t.

(b) If the number of bacteria doubles in 3 hours, find the function that gives the number of cells in the culture.

(c) How long will it take for the size of the colony to triple?

(d) How long will it take for the population to double a second time (that is, increase four times)?

Solution
(a) Using formula (2), the number N of cells at a time t is

$$N(t) = N_0 e^{kt}$$

where N_0 is the initial number of bacteria present and k is a positive number.

(b) We seek the number k. The number of cells doubles in 3 hours, so we have

$$N(3) = 2N_0$$

But $N(3) = N_0 e^{k(3)}$, so

$$N_0 e^{k(3)} = 2N_0$$
$$e^{3k} = 2 \qquad \text{Divide both sides by } N_0.$$
$$3k = \ln 2 \qquad \text{Write the exponential equation as a logarithm.}$$
$$k = \frac{1}{3}\ln 2$$

The function that models this growth process is therefore

$$N(t) = N_0 e^{\left(\frac{1}{3}\ln 2\right)t}$$

(c) The time t needed for the size of the colony to triple requires that $N = 3N_0$. We substitute $3N_0$ for N to get

$$3N_0 = N_0 e^{\left(\frac{1}{3}\ln 2\right)t}$$
$$3 = e^{\left(\frac{1}{3}\ln 2\right)t} \qquad \text{Divide both sides by } N_0.$$
$$\left(\frac{1}{3}\ln 2\right)t = \ln 3 \qquad \text{Write the exponential equation as a logarithm.}$$
$$t = \frac{3\ln 3}{\ln 2} \approx 4.755 \text{ hours}$$

It will take about 4.755 hours or 4 hours, 45 minutes for the size of the colony to triple.

(d) If a population doubles in 3 hours, it will double a second time in 3 more hours, for a total time of 6 hours.

2 Find Equations of Populations That Obey the Law of Decay

Radioactive materials follow the law of uninhibited decay.

Uninhibited Radioactive Decay

The amount A of a radioactive material present at time t is given by

$$A(t) = A_0 e^{kt} \qquad k < 0 \tag{3}$$

where A_0 is the original amount of radioactive material and k is a negative number that represents the rate of decay.

All radioactive substances have a specific **half-life,** which is the time required for half of the radioactive substance to decay. In **carbon dating,** we use the fact that all living organisms contain two kinds of carbon, carbon 12 (a stable carbon) and carbon 14 (a radioactive carbon with a half-life of 5600 years). While an organism is living, the ratio of carbon 12 to carbon 14 is constant. But when an organism dies, the original amount of carbon 12 present remains unchanged, whereas the amount of carbon 14 begins to decrease. This change in the amount of carbon 14 present relative to the amount of carbon 12 present makes it possible to calculate when the organism died.

EXAMPLE 3	**Estimating the Age of Ancient Tools**

Traces of burned wood along with ancient stone tools in an archeological dig in Chile were found to contain approximately 1.67% of the original amount of carbon 14.

(a) If the half-life of carbon 14 is 5600 years, approximately when was the tree cut and burned?

(b) Using a graphing utility, graph the relation between the percentage of carbon 14 remaining and time.

(c) Determine the time that elapses until half of the carbon 14 remains. This answer should equal the half-life of carbon 14.

(d) Use a graphing utility to verify the answer found in part (a).

Solution (a) Using formula (3), the amount A of carbon 14 present at time t is

$$A(t) = A_0 e^{kt}$$

where A_0 is the original amount of carbon 14 present and k is a negative number. We first seek the number k. To find it, we use the fact that after 5600 years half of the original amount of carbon 14 remains, so $A(5600) = \dfrac{1}{2} A_0$. Then

$$\frac{1}{2} A_0 = A_0 e^{k(5600)}$$

$$\frac{1}{2} = e^{5600k} \qquad \text{Divide both sides of the equation by } A_0.$$

$$5600k = \ln \frac{1}{2} \qquad \text{Rewrite as a logarithm.}$$

$$k = \frac{1}{5600} \ln \frac{1}{2} \approx -0.000124$$

Formula (3) therefore becomes

$$A(t) = A_0 e^{\frac{\ln \frac{1}{2}}{5600} t}$$

If the amount A of carbon 14 now present is 1.67% of the original amount, it follows that

$$0.0167 A_0 = A_0 e^{\frac{\ln \frac{1}{2}}{5600} t}$$

$$0.0167 = e^{\frac{\ln \frac{1}{2}}{5600} t} \qquad \text{Divide both sides of the equation by } A_0.$$

$$\frac{\ln \frac{1}{2}}{5600} t = \ln 0.0167 \qquad \text{Rewrite as a logarithm.}$$

$$t = \frac{5600}{\ln \frac{1}{2}} \ln 0.0167 \approx 33{,}062 \text{ years}$$

The tree was cut and burned about 33,062 years ago. Some archeologists use this conclusion to argue that humans lived in the Americas 33,000 years ago, much earlier than is generally accepted.

(b) Figure 38 shows the graph of $y = e^{\frac{\ln\frac{1}{2}}{5600}x}$, where y is the fraction of carbon 14 present and x is the time.

(c) By graphing $Y_1 = 0.5$ and $Y_2 = e^{\frac{\ln\frac{1}{2}}{5600}x}$, where x is time, and using INTERSECT, we find that it takes 5600 years until half the carbon 14 remains. The half-life of carbon 14 is 5600 years.

(d) By graphing $Y_1 = 0.0167$ and $Y_2 = e^{\frac{\ln\frac{1}{2}}{5600}x}$, where x is time, and using INTERSECT, we find that it takes 33,062 years until 1.67% of the carbon 14 remains.

Figure 38

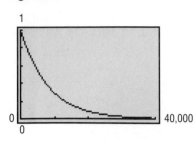

■— **Now Work** PROBLEM 3

3 Use Newton's Law of Cooling

Newton's Law of Cooling* states that the temperature of a heated object decreases exponentially over time toward the temperature of the surrounding medium.

Newton's Law of Cooling

The temperature u of a heated object at a given time t can be modeled by the following function:

$$u(t) = T + (u_0 - T)e^{kt} \qquad k < 0 \qquad \textbf{(4)}$$

where T is the constant temperature of the surrounding medium, u_0 is the initial temperature of the heated object, and k is a negative constant.

EXAMPLE 4   **Using Newton's Law of Cooling**

An object is heated to 100°C (degrees Celsius) and is then allowed to cool in a room whose air temperature is 30°C.

(a) If the temperature of the object is 80°C after 5 minutes, when will its temperature be 50°C?

(b) Using a graphing utility, graph the relation found between the temperature and time.

(c) Using a graphing utility, verify the results from part (a).

(d) Using a graphing utility, determine the elapsed time before the object is 35°C.

(e) What do you notice about the temperature as time passes?

Solution   (a) Using formula (4) with $T = 30$ and $u_0 = 100$, the temperature (in degrees Celsius) of the object at time t (in minutes) is

$$u(t) = 30 + (100 - 30)e^{kt} = 30 + 70e^{kt}$$

where k is a negative constant. To find k, we use the fact that $u = 80$ when $t = 5$. Then

$$80 = 30 + 70e^{k(5)}$$
$$50 = 70e^{5k}$$
$$e^{5k} = \frac{50}{70}$$
$$5k = \ln\frac{5}{7}$$
$$k = \frac{1}{5}\ln\frac{5}{7} \approx -0.0673$$

*Named after Sir Isaac Newton (1642–1727), one of the cofounders of calculus.

Formula (4) therefore becomes

$$u(t) = 30 + 70e^{\frac{\ln\frac{5}{7}}{5}t}$$

We want to find t when $u = 50°C$, so

$$50 = 30 + 70e^{\frac{\ln\frac{5}{7}}{5}t}$$

$$20 = 70e^{\frac{\ln\frac{5}{7}}{5}t}$$

$$e^{\frac{\ln\frac{5}{7}}{5}t} = \frac{20}{70}$$

$$\frac{\ln\frac{5}{7}}{5}t = \ln\frac{2}{7}$$

$$t = \frac{5}{\ln\frac{5}{7}}\ln\frac{2}{7} \approx 18.6 \text{ minutes}$$

The temperature of the object will be 50°C after about 18.6 minutes or 18 minutes, 37 seconds.

(b) Figure 39 shows the graph of $y = 30 + 70e^{\frac{\ln\frac{5}{7}}{5}x}$, where y is the temperature and x is the time.

(c) By graphing $Y_1 = 50$ and $Y_2 = 30 + 70e^{\frac{\ln\frac{5}{7}}{5}x}$, where x is time, and using INTERSECT, we find that it takes $x = 18.6$ minutes (18 minutes, 37 seconds) for the temperature to cool to 50°C.

(d) By graphing $Y_1 = 35$ and $Y_2 = 30 + 70e^{\frac{\ln\frac{5}{7}}{5}x}$, where x is time, and using INTERSECT, we find that it takes $x = 39.22$ minutes (39 minutes, 13 seconds) for the temperature to cool to 35°C.

(e) As x increases, the value of $e^{\frac{\ln\frac{5}{7}}{5}x}$ approaches zero, so the value of y, the temperature of the object, approaches 30°C, the air temperature of the room.

Figure 39

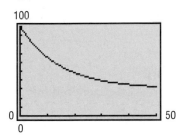

— **Now Work** PROBLEM 13

4 Use Logistic Models

The exponential growth model $A(t) = A_0e^{kt}$, $k > 0$, assumes uninhibited growth, meaning that the value of the function grows without limit. Recall that we stated that cell division could be modeled using this function, assuming that no cells die and no by-products are produced. However, cell division eventually is limited by factors such as living space and food supply. The **logistic model,** given next, can describe situations where the growth or decay of the dependent variable is limited.

Logistic Model

In a logistic model, the population P after time t is given by the function

$$P(t) = \frac{c}{1 + ae^{-bt}} \tag{5}$$

where a, b, and c are constants with $c > 0$. The model is a growth model if $b > 0$; the model is a decay model if $b < 0$.

The number c is called the **carrying capacity** (for growth models) because the value $P(t)$ approaches c as t approaches infinity; that is, $\lim\limits_{t \to \infty} P(t) = c$. The number $|b|$ is the growth rate for $b > 0$ and the decay rate for $b < 0$. Figure 40(a) shows the graph of a typical logistic growth function, and Figure 40(b) shows the graph of a typical logistic decay function.

Figure 40

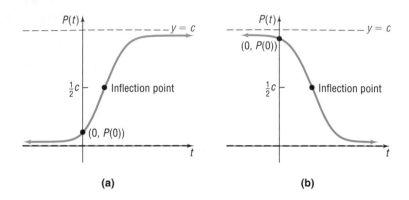

(a)   (b)

Based on the figures, we have the following properties of logistic growth functions.

Properties of the Logistic Model

1. The domain is the set of all real numbers. The range is the interval $(0, c)$, where c is the carrying capacity.
2. There are no x-intercepts; the y-intercept is $P(0)$.
3. There are two horizontal asymptotes: $y = 0$ and $y = c$.
4. $P(t)$ is an increasing function if $b > 0$ and a decreasing function if $b < 0$.
5. There is an **inflection point** where $P(t)$ equals $\dfrac{1}{2}$ of the carrying capacity. The inflection point is the point on the graph where the graph changes from being curved upward to curved downward for growth functions and the point where the graph changes from being curved downward to curved upward for decay functions.
6. The graph is smooth and continuous, with no corners or gaps.

EXAMPLE 5 | **Fruit Fly Population**

Fruit flies are placed in a half-pint milk bottle with a banana (for food) and yeast plants (for food and to provide a stimulus to lay eggs). Suppose that the fruit fly population after t days is given by

$$P(t) = \frac{230}{1 + 56.5e^{-0.37t}}$$

(a) State the carrying capacity and the growth rate.
(b) Determine the initial population.
(c) What is the population after 5 days?
(d) How long does it take for the population to reach 180?
(e) Use a graphing utility to determine how long it takes for the population to reach one-half of the carrying capacity.

Solution (a) As $t \to \infty$, $e^{-0.37t} \to 0$ and $P(t) \to \dfrac{230}{1}$. The carrying capacity of the half-pint bottle is 230 fruit flies. The growth rate is $|b| = |0.37| = 37\%$ per day.

(b) To find the initial number of fruit flies in the half-pint bottle, we evaluate $P(0)$.

$$P(0) = \frac{230}{1 + 56.5e^{-0.37(0)}}$$

$$= \frac{230}{1 + 56.5}$$

$$= 4$$

So, initially, there were 4 fruit flies in the half-pint bottle.

(c) To find the number of fruit flies in the half-pint bottle after 5 days, we evaluate $P(5)$.

$$P(5) = \frac{230}{1 + 56.5e^{-0.37(5)}} \approx 23 \text{ fruit flies}$$

After 5 days, there are approximately 23 fruit flies in the bottle.

(d) To determine when the population of fruit flies will be 180, we solve the equation $P(t) = 180$.

$$\frac{230}{1 + 56.5e^{-0.37t}} = 180$$

$$230 = 180(1 + 56.5e^{-0.37t})$$

$$1.2778 = 1 + 56.5e^{-0.37t} \qquad \text{Divide both sides by 180.}$$

$$0.2778 = 56.5e^{-0.37t} \qquad \text{Subtract 1 from both sides.}$$

$$0.0049 = e^{-0.37t} \qquad \text{Divide both sides by 56.5.}$$

$$\ln(0.0049) = -0.37t \qquad \text{Rewrite as a logarithmic expression.}$$

$$t \approx 14.4 \text{ days} \qquad \text{Divide both sides by } -0.37.$$

It will take approximately 14.4 days (14 days, 10 hours) for the population to reach 180 fruit flies.

(e) One-half of the carrying capacity is 115 fruit flies. We solve $P(t) = 115$ by graphing $Y_1 = \dfrac{230}{1 + 56.5e^{-0.37x}}$ and $Y_2 = 115$ and using INTERSECT. See Figure 41. The population will reach one-half of the carrying capacity in about 10.9 days (10 days, 22 hours).

Figure 41

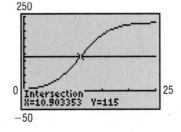

Look back at Figure 41. Notice the point where the graph reaches 115 fruit flies (one-half of the carrying capacity): the graph changes from being curved upward to being curved downward. Using the language of calculus, we say the graph changes from increasing at an increasing rate to increasing at a decreasing rate. For any logistic growth function, when the population reaches one-half the carrying capacity, the population growth starts to slow down.

Exploration

On the same viewing rectangle, graph

$$Y_1 = \frac{500}{1 + 24e^{-0.03x}} \text{ and } Y_2 = \frac{500}{1 + 24e^{-0.08x}}$$

What effect does the growth rate $|b|$ have on the logistic growth function?

Now Work PROBLEM 23

| EXAMPLE 6 | **Wood Products** |

The EFISCEN wood product model classifies wood products according to their life-span. There are four classifications: short (1 year), medium short (4 years), medium long (16 years), and long (50 years). Based on data obtained from the European Forest Institute, the percentage of remaining wood products after t years for wood products with long life-spans (such as those used in the building industry) is given by

$$P(t) = \frac{100.3952}{1 + 0.0316e^{0.0581t}}$$

(a) What is the decay rate?

(b) What is the percentage of remaining wood products after 10 years?

(c) How long does it take for the percentage of remaining wood products to reach 50%?

(d) Explain why the numerator given in the model is reasonable.

Solution

(a) The decay rate is $|b| = |-0.0581| = 5.81\%$.

(b) Evaluate $P(10)$.

$$P(10) = \frac{100.3952}{1 + 0.0316e^{0.0581(10)}} \approx 95.0$$

So 95% of long-life-span wood products remains after 10 years.

(c) Solve the equation $P(t) = 50$.

$$\frac{100.3952}{1 + 0.0316e^{0.0581t}} = 50$$

$$100.3952 = 50(1 + 0.0316e^{0.0581t})$$

$2.0079 = 1 + 0.0316e^{0.0581t}$ Divide both sides by 50.

$1.0079 = 0.0316e^{0.0581t}$ Subtract 1 from both sides.

$31.8957 = e^{0.0581t}$ Divide both sides by 0.0316.

$\ln(31.8957) = 0.0581t$ Rewrite as a logarithmic expression.

$t \approx 59.6$ years   Divide both sides by 0.0581.

It will take approximately 59.6 years for the percentage of long-life-span wood products remaining to reach 50%.

(d) The numerator of 100.3952 is reasonable because the maximum percentage of wood products remaining that is possible is 100%.

 **Now Work** PROBLEM 27

7.6 Assess Your Understanding

Applications and Extensions

1. **Growth of an Insect Population** The size P of a certain insect population at time t (in days) obeys the model $P(t) = 500e^{0.02t}$.
   (a) Determine the number of insects at $t = 0$ days.
   (b) What is the growth rate of the insect population?
   (c) Graph the function using a graphing utility.
   (d) What is the population after 10 days?
   (e) When will the insect population reach 800?
   (f) When will the insect population double?

2. **Growth of Bacteria** The number N of bacteria present in a culture at time t (in hours) obeys the model $N(t) = 1000e^{0.01t}$.
   (a) Determine the number of bacteria at $t = 0$ hours.
   (b) What is the growth rate of the bacteria?

   (c) Graph the function using a graphing utility.
   (d) What is the population after 4 hours?
   (e) When will the number of bacteria reach 1700?
   (f) When will the number of bacteria double?

3. **Radioactive Decay** Strontium 90 is a radioactive material that decays according to the function $A(t) = A_0 e^{-0.0244t}$, where A_0 is the initial amount present and A is the amount present at time t (in years). Assume that a scientist has a sample of 500 grams of strontium 90.
   (a) What is the decay rate of strontium 90?
   (b) Graph the function using a graphing utility.
   (c) How much strontium 90 is left after 10 years?
   (d) When will 400 grams of strontium 90 be left?
   (e) What is the half-life of strontium 90?

4. Radioactive Decay Iodine 131 is a radioactive material that decays according to the function $A(t) = A_0 e^{-0.087t}$, where A_0 is the initial amount present and A is the amount present at time t (in days). Assume that a scientist has a sample of 100 grams of iodine 131.
(a) What is the decay rate of iodine 131?
(b) Graph the function using a graphing utility.
(c) How much iodine 131 is left after 9 days?
(d) When will 70 grams of iodine 131 be left?
(e) What is the half-life of iodine 131?

5. Growth of a Colony of Mosquitoes The population of a colony of mosquitoes obeys the law of uninhibited growth.
(a) If N is the population of the colony and t is the time in days, express N as a function of t.
(b) If there are 1000 mosquitoes initially and there are 1800 after 1 day, what is the size of the colony after 3 days?
(c) How long is it until there are 10,000 mosquitoes?

6. Bacterial Growth A culture of bacteria obeys the law of uninhibited growth.
(a) If N is the number of bacteria in the culture and t is the time in hours, express N as a function of t.
(b) If 500 bacteria are present initially and there are 800 after 1 hour, how many will be present in the culture after 5 hours?
(c) How long is it until there are 20,000 bacteria?

7. Population Growth The population of a southern city follows the exponential law.
(a) If N is the population of the city and t is the time in years, express N as a function of t.
(b) If the population doubled in size over an 18-month period and the current population is 10,000, what will the population be 2 years from now?

8. Population Decline The population of a midwestern city follows the exponential law.
(a) If N is the population of the city and t is the time in years, express N as a function of t.
(b) If the population decreased from 900,000 to 800,000 from 2005 to 2007, what will the population be in 2009?

9. Radioactive Decay The half-life of radium is 1690 years. If 10 grams are present now, how much will be present in 50 years?

10. Radioactive Decay The half-life of radioactive potassium is 1.3 billion years. If 10 grams are present now, how much will be present in 100 years? In 1000 years?

11. Estimating the Age of a Tree A piece of charcoal is found to contain 30% of the carbon 14 that it originally had.
(a) When did the tree from which the charcoal came die? Use 5600 years as the half-life of carbon 14.
(b) Using a graphing utility, graph the relation between the percentage of carbon 14 remaining and time.
(c) Using INTERSECT, determine the time that elapses until half of the carbon 14 remains.
(d) Verify the answer found in part (a).

12. Estimating the Age of a Fossil A fossilized leaf contains 70% of its normal amount of carbon 14.
(a) How old is the fossil?
(b) Using a graphing utility, graph the relation between the percentage of carbon 14 remaining and time.
(c) Using INTERSECT, determine the time that elapses until half of the carbon 14 remains.
(d) Verify the answer found in part (a).

13. Cooling Time of a Pizza A pizza baked at 450°F is removed from the oven at 5:00 PM into a room that is a constant 70°F. After 5 minutes, the pizza is at 300°F.
(a) At what time can you begin eating the pizza if you want its temperature to be 135°F?
(b) Using a graphing utility, graph the relation between temperature and time.
(c) Using INTERSECT, determine the time that needs to elapse before the pizza is 160°F.
(d) TRACE the function for large values of time. What do you notice about y, the temperature?

14. Newton's Law of Cooling A thermometer reading 72°F is placed in a refrigerator where the temperature is a constant 38°F.
(a) If the thermometer reads 60°F after 2 minutes, what will it read after 7 minutes?
(b) How long will it take before the thermometer reads 39°F?
(c) Using a graphing utility, graph the relation between temperature and time.
(d) Using INTERSECT, determine the time needed to elapse before the thermometer reads 45°F.
(e) TRACE the function for large values of time. What do you notice about y, the temperature?

15. Newton's Law of Heating A thermometer reading 8°C is brought into a room with a constant temperature of 35°C.
(a) If the thermometer reads 15°C after 3 minutes, what will it read after being in the room for 5 minutes? For 10 minutes?
(b) Graph the relation between temperature and time. TRACE to verify that your answers are correct.
[**Hint:** You need to construct a formula similar to equation (4).]

16. Warming Time of a Beer Stein A beer stein has a temperature of 28°F. It is placed in a room with a constant temperature of 70°F. After 10 minutes, the temperature of the stein has risen to 35°F. What will the temperature of the stein be after 30 minutes? How long will it take the stein to reach a temperature of 45°F? (See the hint given for Problem 15.)

17. Decomposition of Chlorine in a Pool Under certain water conditions, the free chlorine (hypochlorous acid, HOCl) in a swimming pool decomposes according to the law of uninhibited decay. After shocking his pool, Ben tested the water and found the amount of free chlorine to be 2.5 parts per million (ppm). Twenty-four hours later, Ben tested the water again and found the amount of free chlorine to be 2.2 ppm. What will be the reading after 3 days (that is, 72 hours)? When the chlorine level reaches 1.0 ppm, Ben must shock the pool again. How long can Ben go before he must shock the pool again?

18. Decomposition of Dinitrogen Pentoxide At 45°C, dinitrogen pentoxide (N_2O_5) decomposes into nitrous dioxide (NO_2) and oxygen (O_2) according to the law of uninhibited decay. An initial amount of 0.25 mole of dinitrogen pentoxide decomposes to 0.15 mole in 17 minutes. How much dinitrogen pentoxide will remain after 30 minutes? How long will it take until 0.01 mole of dinitrogen pentoxide remains?

19. Decomposition of Sucrose Reacting with water in an acidic solution at 35°C, sucrose ($C_{12}H_{22}O_{11}$) decomposes into glucose ($C_6H_{12}O_6$) and fructose ($C_6H_{12}O_6$)* according to the law of uninhibited decay. An initial amount of 0.40 mole of sucrose decomposes to 0.36 mole in 30 minutes. How much sucrose will remain after 2 hours? How long will it take until 0.10 mole of sucrose remains?

20. Decomposition of Salt in Water Salt ($NaCl$) decomposes in water into sodium (Na^+) and chloride (Cl^-) ions according to the law of uninhibited decay. If the initial amount of salt is 25 kilograms and, after 10 hours, 15 kilograms of salt is left, how much salt is left after 1 day? How long does it take until $\frac{1}{2}$ kilogram of salt is left?

21. Radioactivity from Chernobyl After the release of radioactive material into the atmosphere from a nuclear power plant at Chernobyl (Ukraine) in 1986, the hay in Austria was contaminated by iodine 131 (half-life 8 days). If it is safe to feed the hay to cows when 10% of the iodine 131 remains, how long did the farmers need to wait to use this hay?

22. Pig Roasts The hotel Bora-Bora is having a pig roast. At noon, the chef put the pig in a large earthen oven. The pig's original temperature was 75°F. At 2:00 PM the chef checked the pig's temperature and was upset because it had reached only 100°F. If the oven's temperature remains a constant 325°F, at what time may the hotel serve its guests, assuming that pork is done when it reaches 175°F?

23. Population of a Bacteria Culture The logistic growth model

$$P(t) = \frac{1000}{1 + 32.33e^{-0.439t}}$$

represents the population (in grams) of a bacterium after t hours.
(a) Determine the carrying capacity of the environment.
(b) What is the growth rate of the bacteria?
(c) Determine the initial population size.
(d) Use a graphing utility to graph $P = P(t)$.
(e) What is the population after 9 hours?
(f) When will the population be 700 grams?
(g) How long does it take for the population to reach one-half of the carrying capacity?

24. Population of an Endangered Species Often environmentalists capture an endangered species and transport the species to a controlled environment where the species can produce offspring and regenerate its population. Suppose

*Author's Note: Surprisingly, the chemical formulas for glucose and fructose are the same. This is not a typo.

that six American bald eagles are captured, transported to Montana, and set free. Based on experience, the environmentalists expect the population to grow according to the model

$$P(t) = \frac{500}{1 + 82.33e^{-0.162t}}$$

where t is measured in years.

(a) Determine the carrying capacity of the environment.
(b) What is the growth rate of the bald eagle?
(c) Use a graphing utility to graph $P = P(t)$.
(d) What is the population after 3 years?
(e) When will the population be 300 eagles?
(f) How long does it take for the population to reach one-half of the carrying capacity?

25. The *Challenger* Disaster After the *Challenger* disaster in 1986, a study was made of the 23 launches that preceded the fatal flight. A mathematical model was developed involving the relationship between the Fahrenheit temperature x around the O-rings and the number y of eroded or leaky primary O-rings. The model stated that

$$y = \frac{6}{1 + e^{-(5.085 - 0.1156x)}}$$

where the number 6 indicates the 6 primary O-rings on the spacecraft.
(a) What is the predicted number of eroded or leaky primary O-rings at a temperature of 100°F?
(b) What is the predicted number of eroded or leaky primary O-rings at a temperature of 60°F?
(c) What is the predicted number of eroded or leaky primary O-rings at a temperature of 30°F?
(d) Graph the equation. At what temperature is the predicted number of eroded or leaky O-rings 1? 3? 5?

Source: Linda Tappin, "Analyzing Data Relating to the Challenger Disaster," Mathematics Teacher, Vol. 87, No. 6, September 1994, pp. 423–426.

26. Word Users According to a survey by Olsten Staffing Services, the percentage of companies reporting usage of Microsoft Word t years since 1984 is given by

$$P(t) = \frac{99.744}{1 + 3.014e^{-0.799t}}$$

(a) What is the growth rate in the percentage of Microsoft Word users?
(b) Use a graphing utility to graph $P = P(t)$.
(c) What was the percentage of Microsoft Word users in 1990?
(d) During what year did the percentage of Microsoft Word users reach 90%?
(e) Explain why the numerator given in the model is reasonable. What does it imply?

27. Home Computers The logistic model

$$P(t) = \frac{95.4993}{1 + 0.0405e^{0.1968t}}$$

represents the percentage of households that do not own a personal computer t years since 1984.
(a) Evaluate and interpret $P(0)$.
(b) Use a graphing utility to graph $P = P(t)$.
(c) What percentage of households did not own a personal computer in 1995?
(d) In what year will the percentage of households that do not own a personal computer reach 10%?

Source: U.S. Department of Commerce

28. Farmers The logistic model

$$W(t) = \frac{14,656.248}{1 + 0.059e^{0.057t}}$$

represents the number of farm workers in the United States t years after 1910.
(a) Evaluate and interpret $W(0)$.
(b) Use a graphing utility to graph $W = W(t)$.

(c) How many farm workers were there in the United States in 1990?
(d) When did the number of farm workers in the United States reach 10,000?
(e) According to this model, what happens to the number of farm workers in the United States as t approaches ∞? Based on this result, do you think that it is reasonable to use this model to predict the number of farm workers in the United States in 2060? Why?

Source: U.S. Department of Agriculture

29. Birthdays The logistic model

$$P(n) = \frac{113.3198}{1 + 0.115e^{0.0912n}}$$

models the probability that, in a room of n people, no two people share the same birthday.
(a) Use a graphing utility to graph $P = P(n)$.
(b) In a room of $n = 15$ people, what is the probability that no two share the same birthday?
(c) How many people must be in a room before the probability that no two people share the same birthday falls below 10%?
(d) What happens to the probability as n increases? Explain what this result means.

30. Bank Failures The logistic model

$$F(t) = \frac{130.118}{1 + 0.00022e^{3.042t}}$$

models the number of bank failures in the United States t years since 1991.
(a) Evaluate and interpret $F(0)$.
(b) Use a graphing utility to graph $F = F(t)$.
(c) How many bank failures were there in 1994 ($t = 3$)?
(d) In what year did the number of bank failures reach 10?
(e) What does the model imply will happen to the number of bank failures as time passes?

Source: Federal Deposit Insurance Corporation

7.7 Building Exponential, Logarithmic, and Logistic Models from Data

PREPARING FOR THIS SECTION *Before getting started, review the following:*

• Building Linear Models from Data (Appendix, Section A.10, pp. A84–A88)

OBJECTIVES 1 Build an Exponential Model from Data (p. 530)
2 Build a Logarithmic Model from Data (p. 531)
3 Build a Logistic Model from Data (p. 532)

In the Appendix, Section A.10, we discussed how to find the linear function of best fit ($y = ax + b$).

In this section we will discuss how to use a graphing utility to find equations of best fit that describe the relation between two variables when the relation is thought to be exponential ($y = ab^x$), logarithmic ($y = a + b \ln x$), or logistic $\left(y = \dfrac{c}{1 + ae^{-bx}} \right)$. As before, we draw a scatter diagram of the data to help to determine the appropriate model to use.

Figure 42 shows scatter diagrams that will typically be observed for the three models. Below each scatter diagram are any restrictions on the values of the parameters.

Figure 42

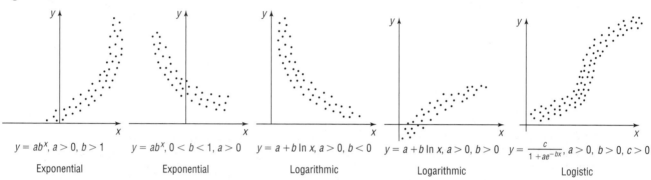

$y = ab^x, a > 0, b > 1$
Exponential

$y = ab^x, 0 < b < 1, a > 0$
Exponential

$y = a + b \ln x, a > 0, b < 0$
Logarithmic

$y = a + b \ln x, a > 0, b > 0$
Logarithmic

$y = \dfrac{c}{1 + ae^{-bx}}, a > 0, b > 0, c > 0$
Logistic

Most graphing utilities have REGression options that fit data to a specific type of curve. Once the data have been entered and a scatter diagram obtained, the type of curve that you want to fit to the data is selected. Then that REGression option is used to obtain the curve of *best fit* of the type selected.

The correlation coefficient r will appear only if the model can be written as a linear expression. As it turns out, r will appear for the linear, power, exponential, and logarithmic models, since these models can be written as a linear expression. Remember, the closer $|r|$ is to 1, the better the fit.

Let's look at some examples.

1 Build an Exponential Model from Data

We saw in Section 7.5 that the future value of money behaves exponentially, and we saw in Section 7.6 that growth and decay models also behave exponentially. The next example shows how data can lead to an exponential model.

EXAMPLE 1   Fitting an Exponential Function to Data

Kathleen is interested in finding a function that explains the growth of cell phone usage in the United States. She gathers data on the number (in millions) of U.S. cell phone subscribers from 1985 through 2005. The data are shown in Table 9.

(a) Using a graphing utility, draw a scatter diagram with year as the independent variable.

(b) Using a graphing utility, build an exponential model from the data.

(c) Express the function found in part (b) in the form $A = A_0 e^{kt}$.

(d) Graph the exponential function found in part (b) or (c) on the scatter diagram.

(e) Using the solution to part (b) or (c), predict the number of U.S. cell phone subscribers in 2009.

(f) Interpret the value of k found in part (c).

Solution

Table 9

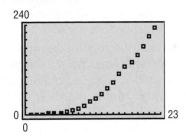

Year, x	Number of Subscribers (in millions), y
1985 (x = 1)	0.34
1986 (x = 2)	0.68
1987 (x = 3)	1.23
1988 (x = 4)	2.07
1989 (x = 5)	3.51
1990 (x = 6)	5.28
1991 (x = 7)	7.56
1992 (x = 8)	11.03
1993 (x = 9)	16.01
1994 (x = 10)	24.13
1995 (x = 11)	33.76
1996 (x = 12)	44.04
1997 (x = 13)	55.31
1998 (x = 14)	69.21
1999 (x = 15)	86.05
2000 (x = 16)	109.48
2001 (x = 17)	128.37
2002 (x = 18)	140.77
2003 (x = 19)	158.72
2004 (x = 20)	182.14
2005 (x = 21)	207.90
2006 (x = 22)	233.00

Source: ©2006 CTIA–The Wireless Association®. All Rights Reserved.

Figure 45

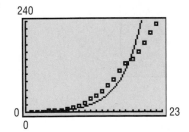

(a) Enter the data into the graphing utility, letting 1 represent 1985, 2 represent 1986, and so on. We obtain the scatter diagram shown in Figure 43.

(b) A graphing utility fits the data in Figure 43 to an exponential function of the form $y = ab^x$ using the EXPonential REGression option. From Figure 44 we find that $y = ab^x = 0.72673(1.3470)^x$. Notice that $|r|$ is close to 1, indicating a good fit.

Figure 43

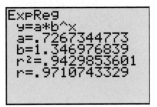

Figure 44

(c) To express $y = ab^x$ in the form $A = A_0e^{kt}$, where $x = t$ and $y = A$, we proceed as follows:

$$ab^x = A_0e^{kt} \quad x = t$$

When $x = t = 0$, we find that $a = A_0$. This leads to

$$a = A_0 \qquad b^x = e^{kt}$$
$$b^x = (e^k)^t$$
$$b = e^k \qquad x = t$$

Since $y = ab^x = 0.72673(1.3470)^x$, we find that $a = 0.72673$ and $b = 1.3470$.

$$a = A_0 = 0.72673 \quad \text{and} \quad b = e^k = 1.3470$$

We want to find k, so we rewrite $e^k = 1.3470$ as a logarithm and obtain

$$k = \ln(1.3470) \approx 0.2979$$

As a result, $A = A_0e^{kt} = 0.72673e^{0.2979t}$.

(d) See Figure 45 for the graph of the exponential function of best fit.

(e) Let $t = 25$ (end of 2009) in the function found in part (c). The predicted number (in millions) of cell phone subscribers in the United States in 2009 is

$$A_0e^{kt} = 0.72673e^{0.2979(25)} \approx 1247$$

This prediction (1247 millon) far exceeds what the U.S. population will be in 2009 (currently the U.S. population is about 300 million). See the answer in part (f).

(f) The value of $k = 0.2979$ represents the growth rate of the number of cell phone subscribers in the United States. Over the period 1985 through 2006, the number of cell phone subscribers has grown at an annual rate of 29.79% compounded continuously. This growth rate is not sustainable as we learned in part (e). In Problem 10 you are asked to build a better model from this data.

◢ **Now Work** PROBLEM 1

2 Build a Logarithmic Model from Data

Many relations between variables do not follow an exponential model; instead, the independent variable is related to the dependent variable using a logarithmic model.

EXAMPLE 2	Fitting a Logarithmic Function to Data

Jodi, a meteorologist, is interested in finding a function that explains the relation between the height of a weather balloon (in kilometers) and the atmospheric pressure (measured in millimeters of mercury) on the balloon. She collects the data shown in Table 10.

(a) Using a graphing utility, draw a scatter diagram of the data with atmospheric pressure as the independent variable.

(b) It is known that the relation between atmospheric pressure and height follows a logarithmic model. Using a graphing utility, build a logarithmic model from the data.

(c) Draw the logarithmic function found in part (b) on the scatter diagram.

(d) Use the function found in part (b) to predict the height of the weather balloon if the atmospheric pressure is 560 millimeters of mercury.

Table 10

Atmospheric Pressure, p	Height, h
760	0
740	0.184
725	0.328
700	0.565
650	1.079
630	1.291
600	1.634
580	1.862
550	2.235

Solution

(a) After entering the data into the graphing utility, we obtain the scatter diagram shown in Figure 46.

Figure 46

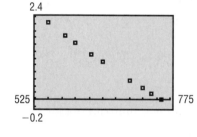

(b) A graphing utility fits the data in Figure 46 to a logarithmic function of the form $y = a + b \ln x$ by using the LOGarithm REGression option. See Figure 47. The logarithmic model from the data is

$$h(p) = 45.7863 - 6.9025 \ln p$$

where h is the height of the weather balloon and p is the atmospheric pressure. Notice that $|r|$ is close to 1, indicating a good fit.

(c) Figure 48 shows the graph of $h(p) = 45.7863 - 6.9025 \ln p$ on the scatter diagram.

Figure 47

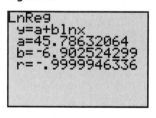

Figure 48

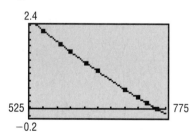

(d) Using the function found in part (b), Jodi predicts the height of the weather balloon when the atmospheric pressure is 560 to be

$$h(560) = 45.7863 - 6.9025 \ln 560$$
$$\approx 2.108 \text{ kilometers}$$

Now Work PROBLEM 5

3 Build a Logistic Model from Data

Logistic growth models can be used to model situations for which the value of the dependent variable is limited. Many real-world situations conform to this scenario. For example, the population of the human race is limited by the availability of natural resources such as food and shelter. When the value of the dependent variable is limited, a logistic growth model is often appropriate.

| EXAMPLE 3 | Fitting a Logistic Function to Data |

The data in Table 11 represent the amount of yeast biomass in a culture after t hours.

Table 11

Time (in hours)	Yeast Biomass	Time (in hours)	Yeast Biomass
0	9.6	10	513.3
1	18.3	11	559.7
2	29.0	12	594.8
3	47.2	13	629.4
4	71.1	14	640.8
5	119.1	15	651.1
6	174.6	16	655.9
7	257.3	17	659.6
8	350.7	18	661.8
9	441.0		

Source: Tor Carlson (Über Geschwindigkeit und Grösse der Hefevermehrung in Würze, Biochemische Zeitschrift, Bd. 57, pp. 313–334, 1913)

(a) Using a graphing utility, draw a scatter diagram of the data with time as the independent variable.

(b) Using a graphing utility, build a logistic model from the data.

(c) Using a graphing utility, graph the function found in part (b) on the scatter diagram.

(d) What is the predicted carrying capacity of the culture?

(e) Use the function found in part (b) to predict the population of the culture at $t = 19$ hours.

Solution

(a) See Figure 49 for a scatter diagram of the data.

(b) A graphing utility fits a logistic growth model of the form $y = \dfrac{c}{1 + ae^{-bx}}$ by using the LOGISTIC regression option. See Figure 50. The logistic model from the data is

$$y = \frac{663.0}{1 + 71.6e^{-0.5470x}}$$

where y is the amount of yeast biomass in the culture and x is the time.

(c) See Figure 51 for the graph of the logistic model.

Figure 49

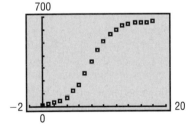

Figure 50

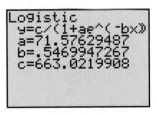

Figure 51

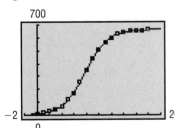

(d) Based on the logistic growth model found in part (b), the carrying capacity of the culture is 663.

(e) Using the logistic growth model found in part (b), the predicted amount of yeast biomass at $t = 19$ hours is

$$y = \frac{663.0}{1 + 71.6e^{-0.5470(19)}} \approx 661.5$$

Now Work PROBLEM 7

7.7 Assess Your Understanding

Applications and Extensions

1. Biology A strain of E-coli Beu 397-recA441 is placed into a nutrient broth at 30° Celsius and allowed to grow. The following data are collected. Theory states that the number of bacteria in the petri dish will initially grow according to the law of uninhibited growth. The population is measured using an optical device in which the amount of light that passes through the petri dish is measured.

Time (hours), x	Population, y
0	0.09
2.5	0.18
3.5	0.26
4.5	0.35
6	0.50

Source: Dr. Polly Lavery, Joliet Junior College

(a) Draw a scatter diagram treating time as the independent variable.
(b) Using a graphing utility, build an exponential model from the data.
(c) Express the function found in part (b) in the form $N(t) = N_0 e^{kt}$.
(d) Graph the exponential function found in part (b) or (c) on the scatter diagram.
(e) Use the exponential function from part (b) or (c) to predict the population at $x = 7$ hours.
(f) Use the exponential function from part (b) or (c) to predict when the population will reach 0.75.

2. Biology A strain of E-coli SC18del-recA718 is placed into a nutrient broth at 30° Celsius and allowed to grow. The following data are collected. Theory states that the number of bacteria in the petri dish will initially grow according to the law of uninhibited growth. The population is measured using an optical device in which the amount of light that passes through the petri dish is measured.

Time (hours), x	Population, y
2.5	0.175
3.5	0.38
4.5	0.63
4.75	0.76
5.25	1.20

Source: Dr. Polly Lavery, Joliet Junior College

(a) Draw a scatter diagram treating time as the independent variable.
(b) Using a graphing utility, build an exponential model from the data.
(c) Express the function found in part (b) in the form $N(t) = N_0 e^{kt}$.
(d) Graph the exponential function found in part (b) or (c) on the scatter diagram.

(e) Use the exponential function from part (b) or (c) to predict the population at $x = 6$ hours.
(f) Use the exponential function from part (b) or (c) to predict when the population will reach 2.1.

3. Chemistry A chemist has a 100-gram sample of a radioactive material. He records the amount of radioactive material every week for 6 weeks and obtains the following data:

Week	Weight (in Grams)
0	100.0
1	88.3
2	75.9
3	69.4
4	59.1
5	51.8
6	45.5

(a) Using a graphing utility, draw a scatter diagram with week as the independent variable.
(b) Using a graphing utility, build an exponential model from the data.
(c) Express the function found in part (b) in the form $A(t) = A_0 e^{kt}$.
(d) Graph the exponential function found in part (b) or (c) on the scatter diagram.
(e) From the result found in part (b), determine the half-life of the radioactive material.
(f) How much radioactive material will be left after 50 weeks?
(g) When will there be 20 grams of radioactive material?

4. Cigarette Exports The following data represent the number of cigarettes (in billions) exported from the United States by year.

Year	Cigarette Exports (in billions of pieces)
1995	231.1
1998	201.3
1999	151.4
2000	147.9
2001	133.9
2002	127.4
2003	121.5
2004	118.7

Source: Statistical Abstract of the United States, 2006

(a) Let $t =$ the number of years since 1995. Using a graphing utility, draw a scatter diagram of the data using t as the independent variable and number of cigarettes as the dependent variable.

(b) Using a graphing utility, build an exponential model from the data.

(c) Express the function found in part (b) in the form $A(t) = A_0 e^{kt}$.

(d) Graph the exponential function found in part (b) or (c) on the scatter diagram.

(e) Use the exponential function from part (b) or (c) to predict the number of cigarettes that will be exported from the United States in 2010.

(f) Use the exponential function from part (b) or (c) to predict when the number of cigarettes exported from the United States will decrease to 50 billion.

5. **Economics and Marketing** The following data represent the price and quantity demanded in 2007 for Dell personal computers.

Price ($/Computer)	Quantity Demanded
2300	152
2000	159
1700	164
1500	171
1300	176
1200	180
1000	189

(a) Using a graphing utility, draw a scatter diagram of the data with price as the dependent variable.

(b) Using a graphing utility, build a logarithmic model from the data.

(c) Using a graphing utility, draw the logarithmic function found in part (b) on the scatter diagram.

(d) Use the function found in part (b) to predict the number of Dell personal computers that will be demanded if the price is $1650.

6. **Economics and Marketing** The following data represent the price and quantity supplied in 2007 for Dell personal computers.

Price ($/Computer)	Quantity Supplied
2300	180
2000	173
1700	160
1500	150
1300	137
1200	130
1000	113

(a) Using a graphing utility, draw a scatter diagram of the data with price as the dependent variable.

(b) Using a graphing utility, build a logarithmic model from the data.

(c) Using a graphing utility, draw the logarithmic function found in part (b) on the scatter diagram.

(d) Use the function found in part (b) to predict the number of Dell personal computers that will be supplied if the price is $1650.

7. **Population Model** The following data represent the population of the United States. An ecologist is interested in building a model that describes the population of the United States.

Year	Population
1900	76,212,168
1910	92,228,496
1920	106,021,537
1930	123,202,624
1940	132,164,569
1950	151,325,798
1960	179,323,175
1970	203,302,031
1980	226,542,203
1990	248,709,873
2000	281,421,906

Source: U.S. Census Bureau

(a) Using a graphing utility, draw a scatter diagram of the data using years since 1900 as the independent variable and population as the dependent variable.

(b) Using a graphing utility, build a logistic model from the data.

(c) Using a graphing utility, draw the function found in part (b) on the scatter diagram.

(d) Based on the function found in part (b), what is the carrying capacity of the United States?

(e) Use the function found in part (b) to predict the population of the United States in 2004.

(f) When will the United States population be 300,000,000?

(g) Compare actual U.S. Census figures to the predictions found in parts (e) and (f). Discuss any differences.

8. **Population Model** The following data represent the world population. An ecologist is interested in building a model that describes the world population.

Year	Population (in Billions)
1993	5.531
1994	5.611
1995	5.691
1996	5.769
1997	5.847
1998	5.925
1999	6.003
2000	6.080
2001	6.157

Source: U.S. Census Bureau

(a) Using a graphing utility, draw a scatter diagram of the data using years since 1992 as the independent variable and population as the dependent variable.

(b) Using a graphing utility, build a logistic model from the data.

(c) Using a graphing utility, draw the function found in part (b) on the scatter diagram.

(d) Based on the function found in part (b), what is the carrying capacity of the world?

(e) Use the function found in part (b) to predict the population of the world in 2004.

(f) When will world population be 7 billion?

(g) Compare actual U.S. Census figures to the prediction found in part (e). Discuss any differences.

9. **Cable Subscribers** The following data represent the number of basic cable TV subscribers in the United States. A market

Year	Subscribers (1,000)
1975 ($t = 5$)	9,800
1980 ($t = 10$)	17,500
1985 ($t = 15$)	35,440
1990 ($t = 20$)	50,520
1992 ($t = 22$)	54,300
1994 ($t = 24$)	58,373
1996 ($t = 26$)	62,300
1998 ($t = 28$)	64,650
2000 ($t = 30$)	66,250
2002 ($t = 32$)	66,472
2004 ($t = 34$)	65,853

Source: Statistical Abstract of the United States, 2006

researcher believes that external factors, such as satellite TV, have affected the growth of cable subscribers. She is interested in building a model that can be used to describe the number of cable TV subscribers in the United States.

(a) Using a graphing utility, draw a scatter diagram of the data using the number of years after 1970, t, as the independent variable and number of subscribers as the dependent variable.

(b) Using a graphing utility, build a logistic model from the data.

(c) Using a graphing utility, draw the function found in part (b) on the scatter diagram.

(d) Based on the model found in part (b), what is the carrying capacity of the cable TV market in the United States?

(e) Use the model found in part (b) to predict the number of cable TV subscribers in the United States in 2015.

10. **Cell Phone Users** Refer to the data in Table 9 (p. 531).

(a) Using a graphing utility, build a logistic model from the data.

(b) Graph the logistic function found in part (b) on a scatter diagram of the data.

(c) What is the predicted carrying capacity of U.S. cell phone subscribers?

(d) Use the model found in part (b) to predict the number of U.S. cell phone subscribers at the end of 2009.

(e) Compare the answer to part (d) above with the answer to Example 1, part (e). How do you explain the different predictions?

Mixed Practice

11. **Age versus Total Cholesterol** The following data represent the age and average total cholesterol for adult males at various ages.

Age	Total Cholesterol
27	189
40	205
50	215
60	210
70	210
80	194

(a) Using a graphing utility, draw a scatter diagram of the data using age, x, as the independent variable and total cholesterol, y, as the dependent variable.

(b) Based on the scatter diagram drawn in part (a), decide on a model (linear, quadratic, cubic, exponential, logarithmic, or logistic) that you think best describes the relation between age and total cholesterol. Be sure to justify your choice of model.

(c) Using a graphing utility, find the model of best fit.

(d) Using a graphing utility, draw the model of best fit on the scatter diagram drawn in part (a).

(e) Use your model to predict the total cholesterol of a 35-year-old male.

12. **Income versus Crime Rate** The following data represent crime rate against individuals (crimes per 1000 population) and their income in the United States in 2003.

Income	Crime Rate
$5000	51.1
11,250	31.9
20,000	27.0
30,000	25.8
42,500	22.0
62,500	23.3
85,000	18.5

Source: Statistical Abstract of the United States, 2006

(a) Using a graphing utility, draw a scatter diagram of the data using income, x, as the independent variable and crime rate, y, as the dependent variable.

(b) Based on the scatter diagram drawn in part (a), decide on a model (linear, quadratic, cubic, exponential, logarithmic, or logistic) that you think best describes the relation between income and crime rate. Be sure to justify your choice of model.

(c) Using a graphing utility, find the model of best fit.

(d) Using a graphing utility, draw the model of best fit on the scatter diagram drawn in part (a).

(e) Use your model to predict the crime rate of an individual whose income is $55,000.

13. Depreciation of a Chevrolet Impala The following data represent the asking price and age of a Chevrolet Impala SS.

Age	Asking Price
1	$27,417
1	26,750
2	22,995
2	23,195
3	17,999
4	16,995
4	16,490

Source: cars.com

(a) Using a graphing utility, draw a scatter diagram of the data using age, x, as the independent variable and asking price, y, as the dependent variable.
(b) Based on the scatter diagram drawn in part (a), decide on a model (linear, quadratic, cubic, exponential, logarithmic, or logistic) that you think best describes the relation between age and asking price. Be sure to justify your choice of model.
(c) Using a graphing utility, find the model of best fit.
(d) Using a graphing utility, draw the model of best fit on the scatter diagram drawn in part (a).
(e) Use your model to predict the asking price of a Chevrolet Impala SS that is 5 years old.

CHAPTER REVIEW

Things to Know

Properties of the exponential function (pp. 469–470)

$f(x) = Ca^x, \quad a > 1$

Domain: the interval $(-\infty, \infty)$

Range: the interval $(0, \infty)$

x-intercepts: none; y-intercept: 1

Horizontal asymptote: x-axis $(y = 0)$ as $x \to -\infty$

Increasing; one-to-one; smooth; continuous

See Figure 6 for a typical graph.

$f(x) = Ca^x, \quad 0 < a < 1$

Domain: the interval $(-\infty, \infty)$

Range: the interval $(0, \infty)$

x-intercepts: none; y-intercept: 1

Horizontal asymptote: x-axis $(y = 0)$ as $x \to \infty$

Decreasing; one-to-one; smooth; continuous

See Figure 10 for a typical graph.

Number e (p. 471)

Value approached by the expression $\left(1 + \dfrac{1}{n}\right)^n$ as $n \to \infty$; that is, $\lim\limits_{n \to \infty} \left(1 + \dfrac{1}{n}\right)^n = e$

Property of exponents (p. 473)

If $a^u = a^v$, then $u = v$.

Properties of the logarithmic function (p. 488)

$f(x) = \log_a x, \quad a > 1$
$(y = \log_a x \text{ means } x = a^y)$

Domain: the interval $(0, \infty)$

Range: the interval $(-\infty, \infty)$

x-intercept: 1; y-intercept; none

Vertical asymptote: $x = 0$ (y-axis)

Increasing; one-to-one; smooth; continuous

See Figure 26(a) for a typical graph.

$f(x) = \log_a x, \quad 0 < a < 1$
$(y = \log_a x \text{ means } x = a^y)$

Domain: the interval $(0, \infty)$

Range: the interval $(-\infty, \infty)$

x-intercept: 1; y-intercept; none

Vertical asymptote: $x = 0$ (y-axis)

Decreasing; one-to-one; smooth; continuous

See Figure 26(b) for a typical graph.

Natural logarithm (p. 482)

$y = \ln x$ means $x = e^y$.

Properties of logarithms (pp. 492–494, 496)

$\log_a 1 = 0 \qquad \log_a a = 1 \qquad a^{\log_a M} = M \qquad \log_a a^r = r$

$\log_a(MN) = \log_a M + \log_a N \qquad \log_a\left(\dfrac{M}{N}\right) = \log_a M - \log_a N$

$$\log_a M^r = r \log_a M$$

If $M = N$, then $\log_a M = \log_a N$.

If $\log_a M = \log_a N$, then $M = N$.

Formulas

Change-of-Base Formula (p. 497) $\qquad \log_a M = \dfrac{\log_b M}{\log_b a}$

Compound Interest Formula (p. 510) $\qquad A = P \cdot \left(1 + \dfrac{r}{n}\right)^{nt}$

Continuous compounding (p. 511) $\qquad A = Pe^{rt}$

Effective rate of interest (p. 512) $\qquad$ Compounding n times per year: $\quad r_e = \left(1 + \dfrac{r}{n}\right)^n - 1$

$\qquad\qquad\qquad\qquad\qquad\qquad\qquad\qquad$ Continuous Compounding $r_e = e^r - 1$

Present Value Formulas (p. 514) $\qquad P = A \cdot \left(1 + \dfrac{r}{n}\right)^{-nt} \quad$ or $\quad P = Ae^{-rt}$

Growth and decay (p. 518) $\qquad A(t) = A_0 e^{kt}$

Newton's Law of Cooling (p. 522) $\qquad u(t) = T + (u_0 - T)e^{kt} \quad k < 0$

Logistic model (p. 523) $\qquad P(t) = \dfrac{c}{1 + ae^{-bt}}$

Objectives

Section		You should be able to ...	Example(s)	Review Exercises
7.1	1	Evaluate exponential functions (p. 464)	1	1(a), (c), 2(a), (c), 65(a)
	2	Graph exponential functions (p. 468)	3–6	33–38
	3	Define the number e (p. 471)	pg. 471	37, 38
	4	Solve exponential equations (p. 473)	7, 8	41–44, 49, 50, 52–54, 65(b)
7.2	1	Change exponential statements to logarithmic statements and logarithmic statements to exponential statements (p. 480)	2, 3	3–6
	2	Evaluate logarithmic expressions (p. 480)	4	1(b), (d), 1(b), (d), 11, 12, 61(b), 62(b), 63, 64, 66(a), 67
	3	Determine the domain of a logarithmic function (p. 481)	5	7–10, 39(a), 40(a)
	4	Graph logarithmic functions (p. 482)	6, 7	39, 40, 61(a), 62(a)
	5	Solve logarithmic equations (p. 486)	8, 9	45, 46, 51, 61(c), 62(c), 66(b)
7.3	1	Work with the properties of logarithms (p. 492)	1, 2	13–16
	2	Write a logarithmic expression as a sum or difference of logarithms (p. 494)	3–5	17–22
	3	Write a logarithmic expression as a single logarithm (p. 495)	6	23–28
	4	Evaluate logarithms whose base is neither 10 nor e (p. 496)	7, 8	29, 30
	5	Graph logarithmic functions whose base is neither 10 nor e (p. 498)	9	31, 32
7.4	1	Solve logarithmic equations (p. 501)	1–3	45, 55, 56
	2	Solve exponential equations (p. 503)	4–7	41–44, 47, 48, 52–54, 57–60
	3	Solve logarithmic and exponential equations using a graphing utility (p. 505)	8	47–60
7.5	1	Determine the future value of a lump sum of money (p. 508)	1–3	68, 70, 75
	2	Calculate effective rates of return (p. 512)	4	68
	3	Determine the present value of a lump sum of money (p. 513)	5	69
	4	Determine the rate of interest or time required to double a lump sum of money (p. 514)	6, 7	68

Review Exercises

In Problems 1 and 2, $f(x) = 3^x$ *and* $g(x) = \log_3 x$.

1. Evaluate: (a) $f(4)$ (b) $g(9)$ (c) $f(-2)$ (d) $g\left(\dfrac{1}{27}\right)$

2. Evaluate: (a) $f(1)$ (b) $g(81)$ (c) $f(-4)$ (d) $g\left(\dfrac{1}{243}\right)$

In Problems 3 and 4, convert each exponential statement to an equivalent statement involving a logarithm. In Problems 5 and 6, convert each logarithmic statement to an equivalent statement involving an exponent.

3. $5^2 = z$ **4.** $a^5 = m$ **5.** $\log_5 u = 13$ **6.** $\log_a 4 = 3$

In Problems 7–10, find the domain of each logarithmic function.

7. $f(x) = \log(3x - 2)$ **8.** $F(x) = \log_5(2x + 1)$ **9.** $H(x) = \log_2(x^2 - 3x + 2)$ **10.** $F(x) = \ln(x^2 - 9)$

In Problems 11–16, evaluate each expression. Do not use a calculator.

11. $\log_2\left(\dfrac{1}{8}\right)$ **12.** $\log_3 81$ **13.** $\ln e^{\sqrt{2}}$

14. $e^{\ln 0.1}$ **15.** $2^{\log_2 0.4}$ **16.** $\log_2 2^{\sqrt{3}}$

In Problems 17–22, write each expression as the sum and/or difference of logarithms. Express powers as factors.

17. $\log_3\left(\dfrac{uv^2}{w}\right)$, $u > 0, v > 0, w > 0$ **18.** $\log_2\left(a^2\sqrt{b}\right)^4$, $a > 0, b > 0$ **19.** $\log\left(x^2\sqrt{x^3 + 1}\right)$, $x > 0$

20. $\log_5\left(\dfrac{x^2 + 2x + 1}{x^2}\right)$, $x > 0$ **21.** $\ln\left(\dfrac{x\sqrt[3]{x^2 + 1}}{x - 3}\right)$, $x > 3$ **22.** $\ln\left(\dfrac{2x + 3}{x^2 - 3x + 2}\right)^2$, $x > 2$

In Problems 23–28, write each expression as a single logarithm.

23. $3\log_4 x^2 + \dfrac{1}{2}\log_4 \sqrt{x}$ **24.** $-2\log_3\left(\dfrac{1}{x}\right) + \dfrac{1}{3}\log_3 \sqrt{x}$

25. $\ln\left(\dfrac{x - 1}{x}\right) + \ln\left(\dfrac{x}{x + 1}\right) - \ln(x^2 - 1)$ **26.** $\log(x^2 - 9) - \log(x^2 + 7x + 12)$

27. $2\log 2 + 3\log x - \dfrac{1}{2}[\log(x + 3) + \log(x - 2)]$ **28.** $\dfrac{1}{2}\ln(x^2 + 1) - 4\ln\dfrac{1}{2} - \dfrac{1}{2}[\ln(x - 4) + \ln x]$

In Problems 29 and 30, use the Change-of-Base Formula and a calculator to evaluate each logarithm. Round your answer to three decimal places.

29. $\log_4 19$ **30.** $\log_2 21$

In Problems 31 and 32, graph each function using a graphing utility and the Change-of-Base Formula.

31. $y = \log_3 x$ **32.** $y = \log_7 x$

In Problems 33–40, use the given function f to:
    (a) Find the domain of f.
    (b) Graph f.
    (c) From the graph, determine the range and any asymptotes of f.
    (d) Find f^{-1}, the inverse of f.

(e) Use f^{-1} to find the range of f.
(f) Graph f^{-1}.

33. $f(x) = 2^{x-3}$
34. $f(x) = -2^x + 3$
35. $f(x) = \frac{1}{2}(3^{-x})$
36. $f(x) = 1 + 3^{-x}$

37. $f(x) = 1 - e^{-x}$
38. $f(x) = 3e^{x-2}$
39. $f(x) = \frac{1}{2}\ln(x + 3)$
40. $f(x) = 3 + \ln(2x)$

In Problems 41–60, solve each equation. Express irrational solutions in exact form and as a decimal rounded to 3 decimal places.

41. $4^{1-2x} = 2$
42. $8^{6+3x} = 4$
43. $3^{x^2+x} = \sqrt{3}$
44. $4^{x-x^2} = \frac{1}{2}$

45. $\log_x 64 = -3$
46. $\log_{\sqrt{2}} x = -6$
47. $5^x = 3^{x+2}$
48. $5^{x+2} = 7^{x-2}$

49. $9^{2x} = 27^{3x-4}$
50. $25^{2x} = 5^{x^2-12}$
51. $\log_3 \sqrt{x-2} = 2$
52. $2^{x+1} \cdot 8^{-x} = 4$

53. $8 = 4^{x^2} \cdot 2^{5x}$
54. $2^x \cdot 5 = 10^x$
55. $\log_6(x + 3) + \log_6(x + 4) = 1$

56. $\log(7x - 12) = 2 \log x$
57. $e^{1-x} = 5$
58. $e^{1-2x} = 4$

59. $9^x + 4 \cdot 3^x - 3 = 0$
60. $4^x - 14 \cdot 4^{-x} = 5$

61. Suppose that $f(x) = \log_2(x - 2) + 1$.
(a) Graph f.
(b) What is $f(6)$? What point is on the graph of f?
(c) Solve $f(x) = 4$. What point is on the graph of f?
(d) Based on the graph drawn in part (a), solve $f(x) > 0$.
(e) Find $f^{-1}(x)$. Graph f^{-1} on the same Cartesian plane as f.

62. Suppose that $f(x) = \log_3(x + 1) - 4$.
(a) Graph f.
(b) What is $f(8)$? What point is on the graph of f?
(c) Solve $f(x) = -3$. What point is on the graph of f?
(d) Based on the graph drawn in part (a), solve $f(x) < 0$.
(e) Find $f^{-1}(x)$. Graph f^{-1} on the same Cartesian plane as f.

In Problems 63 and 64, use the following result: If x is the atmospheric pressure (measured in millimeters of mercury), then the formula for the altitude $h(x)$ (measured in meters above sea level) is

$$h(x) = (30T + 8000) \log\left(\frac{P_0}{x}\right)$$

where T is the temperature (in degrees Celsius) and P_0 is the atmospheric pressure at sea level, which is approximately 760 millimeters of mercury.

63. Finding the Altitude of an Airplane At what height is a Piper Cub whose instruments record an outside temperature of 0°C and a barometric pressure of 300 millimeters of mercury?

64. Finding the Height of a Mountain How high is a mountain if instruments placed on its peak record a temperature of 5°C and a barometric pressure of 500 millimeters of mercury?

65. Amplifying Sound An amplifier's power output P (in watts) is related to its decibel voltage gain d by the formula

$$P = 25e^{0.1d}$$

(a) Find the power output for a decibel voltage gain of 4 decibels.
(b) For a power output of 50 watts, what is the decibel voltage gain?

66. Limiting Magnitude of a Telescope A telescope is limited in its usefulness by the brightness of the star that it is aimed at and by the diameter of its lens. One measure of a star's brightness is its *magnitude;* the dimmer the star, the larger its magnitude. A formula for the limiting magnitude L of a telescope, that is, the magnitude of the dimmest star that it can be used to view, is given by

$$L = 9 + 5.1 \log d$$

where d is the diameter (in inches) of the lens.
(a) What is the limiting magnitude of a 3.5-inch telescope?
(b) What diameter is required to view a star of magnitude 14?

67. Salvage Value The number of years n for a piece of machinery to depreciate to a known salvage value can be found using the formula

$$n = \frac{\log s - \log i}{\log(1 - d)}$$

where s is the salvage value of the machinery, i is its initial value, and d is the annual rate of depreciation.
(a) How many years will it take for a piece of machinery to decline in value from $90,000 to $10,000 if the annual rate of depreciation is 0.20 (20%)?
(b) How many years will it take for a piece of machinery to lose half of its value if the annual rate of depreciation is 15%?

68. Funding a College Education A child's grandparents purchase a $10,000 bond fund that matures in 18 years to be used for her college education. The bond fund pays 4% interest compounded semiannually. How much will the bond fund be worth at maturity? What is the effective rate of interest? How long will it take the bond to double in value under these terms?

69. Funding a College Education A child's grandparents wish to purchase a bond that matures in 18 years to be used for her college education. The bond pays 4% interest compounded semiannually. How much should they pay so that the bond will be worth $85,000 at maturity?

70. Funding an IRA First Colonial Bankshares Corporation advertised the following IRA investment plans.

Target IRA Plans

For each $5000 Maturity Value Desired	
Deposit:	**At a Term of:**
$620.17	20 Years
$1045.02	15 Years
$1760.92	10 Years
$2967.26	5 Years

(a) Assuming continuous compounding, what annual rate of interest did they offer?

(b) First Colonial Bankshares claims that $4000 invested today will have a value of over $32,000 in 20 years. Use the answer found in part (a) to find the actual value of $4000 in 20 years. Assume continuous compounding.

71. Estimating the Date That a Prehistoric Man Died The bones of a prehistoric man found in the desert of New Mexico contain approximately 5% of the original amount of carbon 14. If the half-life of carbon 14 is 5600 years, approximately how long ago did the man die?

72. Temperature of a Skillet A skillet is removed from an oven whose temperature is 450°F and placed in a room whose temperature is 70°F. After 5 minutes, the temperature of the skillet is 400°F. How long will it be until its temperature is 150°F?

73. World Population The annual growth rate of the world's population in 2005 was $k = 1.15\% = 0.0115$. The population of the world in 2005 was 6,451,058,790. Letting $t = 0$ represent 2005, use the uninhibited growth model to predict the world's population in the year 2015.

Source: U.S. Census Bureau

74. Radioactive Decay The half-life of radioactive cobalt is 5.27 years. If 100 grams of radioactive cobalt is present now, how much will be present in 20 years? In 40 years?

75. Federal Deficit In fiscal year 2005, the federal deficit was $319 billion. At that time, 10-year treasury notes were paying 4.25% interest per annum. If the federal government financed this deficit through 10-year notes, how much would it have to pay back in 2015?

Source: U.S. Treasury Department

76. Logistic Growth The logistic growth model

$$P(t) = \frac{0.8}{1 + 1.67e^{-0.16t}}$$

represents the proportion of new cars with a global positioning system (GPS). Let $t = 0$ represent 2006, $t = 1$ represent 2007, and so on.

(a) What proportion of new cars in 2006 had a GPS?

(b) Determine the maximum proportion of new cars that have a GPS.

(c) Using a graphing utility, graph $P = P(t)$.

(d) When will 75% of new cars have a GPS?

77. CBL Experiment The following data were collected by placing a temperature probe in a portable heater, removing the probe, and then recording temperature over time.

According to Newton's Law of Cooling, these data should follow an exponential model.

Time (sec.)	Temperature (°F)
0	165.07
1	164.77
2	163.99
3	163.22
4	162.82
5	161.96
6	161.20
7	160.45
8	159.35
9	158.61
10	157.89
11	156.83
12	156.11
13	155.08
14	154.40
15	153.72

(a) Using a graphing utility, draw a scatter diagram for the data.

(b) Using a graphing utility, build an exponential model from the data.

(c) Graph the exponential function found in part (b) on the scatter diagram.

(d) Predict how long it will take for the probe to reach a temperature of 110°F.

78. Wind Chill Factor The following data represent the wind speed (mph) and wind chill factor at an air temperature of 15°F.

Wind Speed (mph)	Wind Chill Factor (°F)
5	7
10	3
15	0
20	−2
25	−4
30	−5
35	−7

Source: U.S. National Weather Service

(a) Using a graphing utility, draw a scatter diagram with wind speed as the independent variable.

(b) Using a graphing utility, build a logarithmic model from the data.

(c) Using a graphing utility, draw the logarithmic function found in part (b) on the scatter diagram.

(d) Use the function found in part (b) to predict the wind chill factor if the air temperature is 15°F and the wind speed is 23 mph.

79. Spreading of a Disease Jack and Diane live in a small town of 50 people. Unfortunately, both Jack and Diane have a cold. Those who come in contact with someone who has this cold will themselves catch the cold. The following data represent the number of people in the small town who have caught the cold after t days.

Days, t	Number of People with Cold, C
0	2
1	4
2	8
3	14
4	22
5	30
6	37
7	42
8	44

(a) Using a graphing utility, draw a scatter diagram of the data. Comment on the type of relation that appears to exist between the days and number of people with a cold.

(b) Using a graphing utility, build a logistic model from the data.

(c) Graph the function found in part (b) on the scatter diagram.

(d) According to the function found in part (b), what is the maximum number of people who will catch the cold? In reality, what is the maximum number of people who could catch the cold?

(e) Sometime between the second and third day, 10 people in the town had a cold. According to the model found in part (b), when did 10 people have a cold?

(f) How long will it take for 46 people to catch the cold?

CHAPTER TEST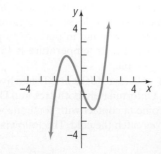

In Problems 1–3, solve each equation algebraically.

1. $3^x = 243$ **2.** $\log_b 16 = 2$ **3.** $\log_5 x = 4$

In Problems 4–7, use a calculator to evaluate each expression. Round your answer to three decimal places.

4. $e^3 + 2$ **5.** $\log 20$

6. $\log_3 21$ **7.** $\ln 133$

In Problems 8 and 9, use the given function f to:
(a) Find the domain of f.
(b) Graph f.
(c) From the graph, determine the range and any asymptotes of f.
(d) Find f^{-1}, the inverse of f.
(e) Use f^{-1} to find the range of f.
(f) Graph f^{-1}.

8. $f(x) = 4^{x+1} - 2$ **9.** $f(x) = 1 - \log_5(x - 2)$

In Problems 10–15, solve each equation.

10. $5^{x+2} = 125$ **11.** $\log(x + 9) = 2$

12. $8 - 2e^{-x} = 4$ **13.** $\log(x^2 + 3) = \log(x + 6)$

14. $7^{x+3} = e^x$ **15.** $\log_2(x - 4) + \log_2(x + 4) = 3$

16. Write $\log_2\left(\dfrac{4x^3}{x^2 - 3x - 18}\right)$ as the sum and/or difference of logarithms. Express powers as factors.

17. A 50-mg sample of a radioactive substance decays to 34 mg after 30 days. How long will it take for there to be 2 mg remaining?

18. (a) If $1000 is invested at 5% compounded monthly, how much is there after 8 months?

(b) If you want to have $1000 in 9 months, how much do you need to place in a savings account now that pays 5% compounded quarterly?

(c) How long does it take to double your money if you can invest it at 6% compounded annually?

19. The decibel level, D, of sound is given by the equation $D = 10\log\left(\dfrac{I}{I_0}\right)$, where I is the intensity of the sound and $I_0 = 10^{-12}$ watt per square meter.

(a) If the shout of a single person measures 80 decibels, how loud will the sound be if two people shout at the same time? That is, how loud would the sound be if the intensity doubled?

(b) The pain threshold for sound is 125 decibels. If the Athens Olympic Stadium 2004 (Olympiako Stadio Athinas 'Spyros Louis') can seat 74,400 people, how many people in the crowd need to shout at the same time for the resulting sound level to meet or exceed the pain threshold? (Ignore any possible sound dampening.)

CUMULATIVE REVIEW

1. Is the following graph the graph of a function? If it is, is the function one-to-one?

2. For the function $f(x) = 2x^2 - 3x + 1$, find the following:

(a) $f(3)$ (b) $f(-x)$ (c) $f(x + h)$

3. Determine which of the following points are on the graph of $x^2 + y^2 = 1$.

(a) $\left(\dfrac{1}{2}, \dfrac{1}{2}\right)$ (b) $\left(\dfrac{1}{2}, \dfrac{\sqrt{3}}{2}\right)$

4. Solve the equation $3(x - 2) = 4(x + 5)$.

5. Graph the line $2x - 4y = 16$.

6. Graph $f(x) = 3(x + 1)^3 - 2$ using transformations.

7. For the function $g(x) = 3^x + 2$:
  (a) Graph g using transformations. State the domain, range, and horizontal asymptote of g.
  (b) Determine the inverse of g. State the domain, range, and vertical asymptote of g^{-1}.
  (c) On the same graph as g, graph g^{-1}.

8. Solve the equation $4^{x-3} = 8^{2x}$.

9. Solve the equation: $\log_3(x + 1) + \log_3(2x - 3) = \log_9 9$

10. Suppose that $f(x) = \log_3(x + 2)$. Solve:
  (a) $f(x) = 0$ (b) $f(x) > 0$ (c) $f(x) = 3$

CHAPTER PROJECTS

I. Hot Coffee   A fast-food restaurant wants a special container to hold coffee. The restaurant wishes the container to quickly cool the coffee from 200° to 130°F and keep the liquid between 110° and 130°F as long as possible. The restaurant has three containers to select from.

1. The CentiKeeper Company has a container that reduces the temperature of a liquid from 200° to 100°F in 30 minutes by maintaining a constant temperature of 70°F.

2. The TempControl Company has a container that reduces the temperature of a liquid from 200° to 110°F in 25 minutes by maintaining a constant temperature of 60°F.

3. The Hot'n'Cold Company has a container that reduces the temperature of a liquid from 200° to 120°F in 20 minutes by maintaining a constant temperature of 65°F.

  You need to recommend which container the restaurant should purchase.

  (a) Use Newton's Law of Cooling to find a function relating the temperature of the liquid over time for each container.
  (b) How long does it take each container to lower the coffee temperature from 200° to 130°F?
  (c) How long will the coffee temperature remain between 110° and 130°F? This temperature is considered the optimal drinking temperature.
  (d) Graph each function using a graphing utility.
  (e) Which company would you recommend to the restaurant? Why?
  (f) How might the cost of the container affect your decision?

The following projects are available on the Instructor's Resource Center (IRC):

II. Project at Motorola   ***Thermal Fatigue of Solder Connections***   Product reliability is a major concern of a manufacturer. Here a logarithmic transformation is used to simplify the analysis of a cell phone's ability to withstand temperature change.

III. Depreciation of a New Car   Resale value is a factor to consider when purchasing a car, and exponential functions provide a way to compare the depreciation rates of different makes and models.

Review

Outline

A.1 Algebra Essentials

PREPARING FOR THIS BOOK *Before getting started, read "To the Student" at the beginning of this book on page xi.*

OBJECTIVES
1 Work with Sets (p. A1)
2 Graph Inequalities (p. A4)
3 Find Distance on the Real Number Line (p. A5)
4 Evaluate Algebraic Expressions (p. A6)
5 Determine the Domain of a Variable (p. A7)
6 Use the Laws of Exponents (p. A7)
7 Evaluate Square Roots (p. A9)
8 Use a Calculator to Evaluate Exponents (p. A10)

1 Work with Sets

A **set** is a well-defined collection of distinct objects. The objects of a set are called its **elements.** By **well-defined,** we mean that there is a rule that enables us to determine whether a given object is an element of the set. If a set has no elements, it is called the **empty set,** or **null set,** and is denoted by the symbol $\varnothing$.

For example, the set of *digits* consists of the collection of numbers 0, 1, 2, 3, 4, 5, 6, 7, 8, and 9. If we use the symbol D to denote the set of digits, then we can write

$$D = \{0, 1, 2, 3, 4, 5, 6, 7, 8, 9\}$$

In this notation, the braces $\{\ \}$ are used to enclose the objects, or **elements,** in the set. This method of denoting a set is called the **roster method.** A second way to denote a set is to use **set-builder notation,** where the set D of digits is written as

$$D = \{\quad x \quad | \quad x \text{ is a digit}\}$$

Read as "D is the set of all x such that x is a digit."

EXAMPLE 1	Using Set-builder Notation and the Roster Method

(a) $E = \{x | x \text{ is an even digit}\} = \{0, 2, 4, 6, 8\}$
(b) $O = \{x | x \text{ is an odd digit}\} = \{1, 3, 5, 7, 9\}$

Because the elements of a set are distinct, we never repeat elements. For example, we would never write $\{1, 2, 3, 2\}$; the correct listing is $\{1, 2, 3\}$. Because a set is a collection, the order in which the elements are listed is immaterial. $\{1, 2, 3\}$, $\{1, 3, 2\}$, $\{2, 1, 3\}$, and so on, all represent the same set.

If every element of a set A is also an element of a set B, then we say that A is a **subset** of B and write $A \subseteq B$. If two sets A and B have the same elements, then we say that A **equals** B and write $A = B$.

For example, $\{1, 2, 3\} \subseteq \{1, 2, 3, 4, 5\}$ and $\{1, 2, 3\} = \{2, 3, 1\}$.

DEFINITION If A and B are sets, the **intersection** of A with B, denoted $A \cap B$, is the set consisting of elements that belong to both A and B. The **union** of A with B, denoted $A \cup B$, is the set consisting of elements that belong to either A or B, or both.

EXAMPLE 2	Finding the Intersection and Union of Sets

Let $A = \{1, 3, 5, 8\}$, $B = \{3, 5, 7\}$, and $C = \{2, 4, 6, 8\}$. Find:

(a) $A \cap B$ (b) $A \cup B$ (c) $B \cap (A \cup C)$

Solution (a) $A \cap B = \{1, 3, 5, 8\} \cap \{3, 5, 7\} = \{3, 5\}$
(b) $A \cup B = \{1, 3, 5, 8\} \cup \{3, 5, 7\} = \{1, 3, 5, 7, 8\}$
(c) $B \cap (A \cup C) = \{3, 5, 7\} \cap (\{1, 3, 5, 8\} \cup \{2, 4, 6, 8\})$
$\qquad\qquad\qquad = \{3, 5, 7\} \cap \{1, 2, 3, 4, 5, 6, 8\} = \{3, 5\}$

Now Work PROBLEM 13

Usually, in working with sets, we designate a **universal set** U, the set consisting of all the elements that we wish to consider. Once a universal set has been designated, we can consider elements of the universal set not found in a given set.

DEFINITION If A is a set, the **complement** of A, denoted $\overline{A}$, is the set consisting of all the elements in the universal set that are not in A.*

EXAMPLE 3	Finding the Complement of a Set

If the universal set is $U = \{1, 2, 3, 4, 5, 6, 7, 8, 9\}$ and if $A = \{1, 3, 5, 7, 9\}$, then $\overline{A} = \{2, 4, 6, 8\}$.

It follows from the definition of complement that $A \cup \overline{A} = U$ and $A \cap \overline{A} = \varnothing$. Do you see why?

Now Work PROBLEM 17

It is often helpful to draw pictures of sets. Such pictures, called **Venn diagrams,** represent sets as circles enclosed in a rectangle, which represents the universal set. Such diagrams often help us to visualize various relationships among sets. See Figure 1.

Figure 1

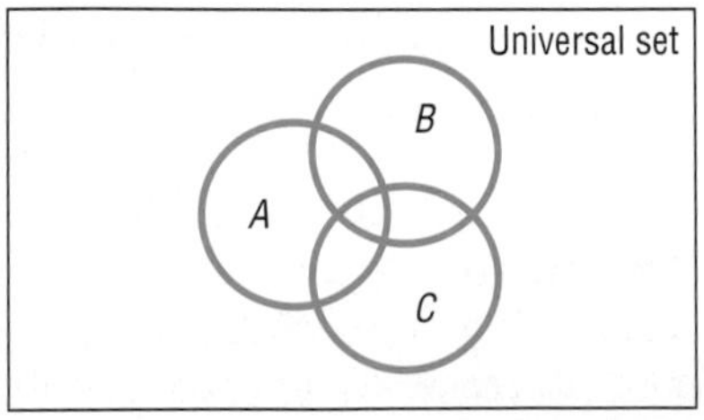

*Some books use the notation A' or A^c for the complement of A.

If we know that $A \subseteq B$, we might use the Venn diagram in Figure 2(a). If we know that A and B have no elements in common, that is, if $A \cap B = \varnothing$, we might use the Venn diagram in Figure 2(b). The sets A and B in Figure 2(b) are said to be **disjoint.**

Figure 2

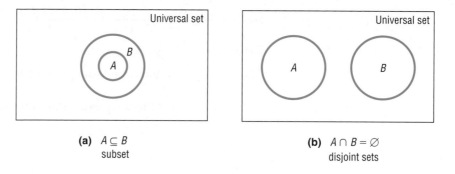

(a) $A \subseteq B$
subset

(b) $A \cap B = \varnothing$
disjoint sets

Figures 3(a), 3(b), and 3(c) use Venn diagrams to illustrate the definitions of intersection, union, and complement, respectively.

Figure 3

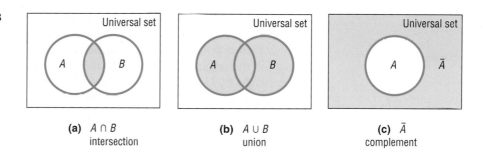

(a) $A \cap B$
intersection

(b) $A \cup B$
union

(c) $\bar{A}$
complement

Real Numbers

Real numbers are represented by symbols such as

$$25, \quad 0, \quad -3, \quad \frac{1}{2}, \quad -\frac{5}{4}, \quad 0.125, \quad \sqrt{2}, \quad \pi, \quad \sqrt[3]{-2}, \quad 0.666\ldots$$

The set of **counting numbers,** or **natural numbers,** contains the numbers in the set $\{1, 2, 3, 4, \ldots\}$. (The three dots, called an **ellipsis,** indicate that the pattern continues indefinitely.) The set of **integers** contains the numbers in the set $\{\ldots, -3, -2, -1, 0, 1, 2, 3, \ldots\}$. A **rational number** is a number that can be expressed as a *quotient* $\dfrac{a}{b}$ of two integers, where the integer b cannot be 0. Examples of rational numbers are $\dfrac{3}{4}, \dfrac{5}{2}, \dfrac{0}{4}$, and $-\dfrac{2}{3}$. Since $\dfrac{a}{1} = a$ for any integer a, every integer is also a rational number. Real numbers that are not rational are called **irrational.** Examples of irrational numbers are $\sqrt{2}$ and π (the Greek letter pi), which equals the constant ratio of the circumference to the diameter of a circle. See Figure 4.

Real numbers can be represented as **decimals.** Rational real numbers have decimal representations that either **terminate** or are nonterminating with **repeating** blocks of digits. For example, $\dfrac{3}{4} = 0.75$, which terminates; and $\dfrac{2}{3} = 0.666\ldots$, in which the digit 6 repeats indefinitely. Irrational real numbers have decimal representations that neither repeat nor terminate. For example, $\sqrt{2} = 1.414213\ldots$ and $\pi = 3.14159\ldots$. In practice, the decimal representation of an irrational number is given as an approximation. We use the symbol $\approx$ (read as "approximately equal to") to write $\sqrt{2} \approx 1.4142$ and $\pi \approx 3.1416$.

Figure 4 $\quad \pi = \dfrac{c}{d}$

Two properties of real numbers that we shall use often are given next. Suppose that a, b, and c are real numbers.

Distributive Property

$$a \cdot (b + c) = ab + ac$$

Zero-Product Property

If $ab = 0$, then either $a = 0$ or $b = 0$ or both equal 0.

The Distributive Property can be used to remove parentheses:

$$2(x + 3) = 2x + 2 \cdot 3 = 2x + 6$$

The Zero-Product Property will be used to solve equations (Section A.4). For example, if $2x = 0$, then $2 = 0$ or $x = 0$. Since $2 \neq 0$, it follows that $x = 0$.

The Real Number Line

The real numbers can be represented by points on a line called the **real number line.** There is a one-to-one correspondence between real numbers and points on a line. That is, every real number corresponds to a point on the line, and each point on the line has a unique real number associated with it.

Pick a point on the line somewhere in the center, and label it O. This point, called the **origin,** corresponds to the real number 0. See Figure 5. The point 1 unit to the right of O corresponds to the number 1. The distance between 0 and 1 determines the **scale** of the number line. For example, the point associated with the number 2 is twice as far from O as 1. Notice that an arrowhead on the right end of the line indicates the direction in which the numbers increase. Points to the left of the origin correspond to the real numbers -1, -2, and so on. Figure 5 also shows the points associated with the rational numbers $-\frac{1}{2}$ and $\frac{1}{2}$ and with the irrational numbers $\sqrt{2}$ and π.

Figure 5
Real number line

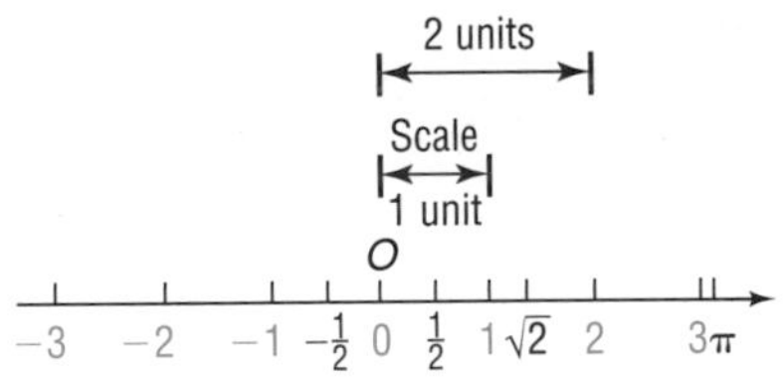

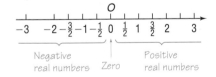

DEFINITION

The real number associated with a point P is called the **coordinate** of P, and the line whose points have been assigned coordinates is called the **real number line.**

The real number line consists of three classes of real numbers, as shown in Figure 6.

Figure 6

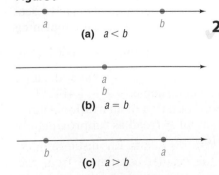

1. The **negative real numbers** are the coordinates of points to the left of the origin O.

2. The real number **zero** is the coordinate of the origin O.

3. The **positive real numbers** are the coordinates of points to the right of the origin O.

Now Work PROBLEM **21**

Figure 7

(a) $a < b$

(b) $a = b$

(c) $a > b$

2 Graph Inequalities

An important property of the real number line follows from the fact that, given two numbers (points) a and b, either a is to the left of b, or a is at the same location as b, or a is to the right of b. See Figure 7.

If a is to the left of b, we say that "a is less than b" and write $a < b$. If a is to the right of b, we say that "a is greater than b" and write $a > b$. If a is at the same location as b, then $a = b$. If a is either less than or equal to b, we write $a \leq b$. Similarly, $a \geq b$ means that a is either greater than or equal to b. Collectively, the symbols $<$, $>$, $\leq$, and $\geq$ are called **inequality symbols.**

Note that $a < b$ and $b > a$ mean the same thing. It does not matter whether we write $2 < 3$ or $3 > 2$.

Furthermore, if $a < b$ or if $b > a$, then the difference $b - a$ is positive. Do you see why?

An **inequality** is a statement in which two expressions are related by an inequality symbol. The expressions are referred to as the **sides** of the inequality. Statements of the form $a < b$ or $b > a$ are called **strict inequalities,** whereas statements of the form $a \leq b$ or $b \geq a$ are called **nonstrict inequalities.**

Based on the discussion so far, we conclude that

> $a > 0$ is equivalent to a is positive
>
> $a < 0$ is equivalent to a is negative

We sometimes read $a > 0$ by saying that "a is positive." If $a \geq 0$, then either $a > 0$ or $a = 0$, and we may read this as "a is nonnegative."

Now Work PROBLEMS 25 AND 35

EXAMPLE 4

Graphing Inequalities

(a) On the real number line, graph all numbers x for which $x > 4$.
(b) On the real number line, graph all numbers x for which $x \leq 5$.

Solution

(a) See Figure 8. Notice that we use a left parenthesis to indicate that the number 4 is *not* part of the graph.
(b) See Figure 9. Notice that we use a right bracket to indicate that the number 5 *is* part of the graph.

Figure 8

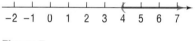

Figure 9

Now Work PROBLEM 41

3 Find Distance on the Real Number Line

Figure 10

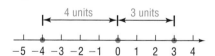

The *absolute value* of a number a is the distance from 0 to a on the number line. For example, -4 is 4 units from 0, and 3 is 3 units from 0. See Figure 10. Thus, the absolute value of -4 is 4, and the absolute value of 3 is 3.

A more formal definition of absolute value is given next.

DEFINITION

The **absolute value** of a real number a, denoted by the symbol $|a|$, is defined by the rules

$$|a| = a \quad \text{if } a \geq 0 \qquad \text{and} \qquad |a| = -a \quad \text{if } a < 0$$

For example, since $-4 < 0$, the second rule must be used to get $|-4| = -(-4) = 4$.

EXAMPLE 5

Computing Absolute Value

(a) $|8| = 8$ (b) $|0| = 0$ (c) $|-15| = -(-15) = 15$

Look again at Figure 10. The distance from -4 to 3 is 7 units. This distance is the difference $3 - (-4)$, obtained by subtracting the smaller coordinate from the larger. However, since $|3 - (-4)| = |7| = 7$ and $|-4 - 3| = |-7| = 7$, we can use absolute value to calculate the distance between two points without being concerned about which is smaller.

DEFINITION

If P and Q are two points on a real number line with coordinates a and b, respectively, the **distance between P and Q,** denoted by $d(P, Q)$, is

$$d(P, Q) = |b - a|$$

Since $|b - a| = |a - b|$, it follows that $d(P, Q) = d(Q, P)$.

EXAMPLE 6 | **Finding Distance on a Number Line**

Let P, Q, and R be points on a real number line with coordinates -5, 7, and -3, respectively. Find the distance

(a) between P and Q (b) between Q and R

Solution   See Figure 11.

Figure 11

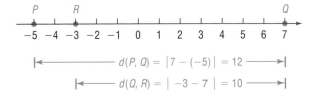

(a) $d(P, Q) = |7 - (-5)| = |12| = 12$
(b) $d(Q, R) = |-3 - 7| = |-10| = 10$

Now Work PROBLEM 47

4 Evaluate Algebraic Expressions

In algebra we use letters such as x, y, a, b, and c to represent numbers. If the letter used is to represent *any* number from a given set of numbers, it is called a **variable.** A **constant** is either a fixed number, such as 5 or $\sqrt{3}$, or a letter that represents a fixed (possibly unspecified) number.

Constants and variables are combined using the operations of addition, subtraction, multiplication, and division to form *algebraic expressions*. Examples of algebraic expressions include

$$x + 3 \qquad \frac{3}{1 - t} \qquad 7x - 2y$$

To evaluate an algebraic expression, substitute for each variable its numerical value.

EXAMPLE 7 | **Evaluating an Algebraic Expression**

Evaluate each expression if $x = 3$ and $y = -1$.

(a) $x + 3y$ (b) $5xy$ (c) $\dfrac{3y}{2 - 2x}$ (d) $|-4x + y|$

Solution   (a) Substitute 3 for x and -1 for y in the expression $x + 3y$.

$$x + 3y = 3 + 3(-1) = 3 + (-3) = 0$$
$$\underset{\underset{x = 3, y = -1}{\uparrow}}{}$$

(b) If $x = 3$ and $y = -1$, then

$$5xy = 5(3)(-1) = -15$$

(c) If $x = 3$ and $y = -1$, then

$$\frac{3y}{2 - 2x} = \frac{3(-1)}{2 - 2(3)} = \frac{-3}{2 - 6} = \frac{-3}{-4} = \frac{3}{4}$$

(d) If $x = 3$ and $y = -1$, then

$$\left|-4x + y\right| = \left|-4(3) + (-1)\right| = \left|-12 + (-1)\right| = \left|-13\right| = 13$$

➤ **Now Work** PROBLEMS 49 AND 57

5 Determine the Domain of a Variable

In working with expressions or formulas involving variables, the variables may be allowed to take on values from only a certain set of numbers. For example, in the formula for the area A of a circle of radius r, $A = \pi r^2$, the variable r is restricted to the positive real numbers since a radius cannot be negative or zero. In the expression $\dfrac{1}{x}$, the variable x cannot take on the value 0, since division by 0 is not defined.

DEFINITION　　The set of values that a variable may assume is called the **domain of the variable.**

EXAMPLE 8　　**Finding the Domain of a Variable**

The domain of the variable x in the expression

$$\frac{5}{x - 2}$$

is $\{x | x \neq 2\}$, since, if $x = 2$, the denominator becomes 0, which is not defined.

EXAMPLE 9　　**Circumference of a Circle**

In the formula for the circumference C of a circle of radius r,

$$C = 2\pi r$$

the domain of the variable r, representing the radius of the circle, is the set of positive real numbers. The domain of the variable C, representing the circumference of the circle, is also the set of positive real numbers.

In describing the domain of a variable, we may use either set notation or words, whichever is more convenient.

➤ **Now Work** PROBLEM 67

6 Use the Laws of Exponents

Integer exponents provide a shorthand notation for representing repeated multiplications of a real number. For example,

$$2^3 = 2 \cdot 2 \cdot 2 = 8 \qquad 3^4 = 3 \cdot 3 \cdot 3 \cdot 3 = 81$$

DEFINITION

If a is a real number and n is a positive integer, then the symbol a^n represents the product of n factors of a. That is,

$$a^n = \underbrace{a \cdot a \cdot \ldots \cdot a}_{n \text{ factors}} \qquad (1)$$

Here it is understood that $a^1 = a$.

Then $a^2 = a \cdot a$, $a^3 = a \cdot a \cdot a$, and so on. In the expression a^n, a is called the **base** and n is called the **exponent,** or **power.** We read a^n as "a raised to the power n" or as "a to the nth power." We usually read a^2 as "a squared" and a^3 as "a cubed."

In working with exponents, the operation of *raising to a power* is performed before any other operation. As examples,

COMMENT Be careful with negatives and exponents.

$$-2^4 = -1 \cdot 2^4 = -16$$

whereas

$$(-2)^4 = (-2)(-2)(-2)(-2) = 16 \quad \blacksquare$$

$$4 \cdot 3^2 = 4 \cdot 9 = 36 \qquad 2^2 + 3^2 = 4 + 9 = 13$$
$$-2^4 = -16 \qquad 5 \cdot 3^2 + 2 \cdot 4 = 5 \cdot 9 + 2 \cdot 4 = 45 + 8 = 53$$

Parentheses are used to indicate operations to be performed first. For example,

$$(-2)^4 = (-2)(-2)(-2)(-2) = 16 \qquad (2 + 3)^2 = 5^2 = 25$$

DEFINITION

If $a \neq 0$, we define

$$a^0 = 1 \quad \text{if } a \neq 0$$

DEFINITION

If $a \neq 0$ and if n is a positive integer, then we define

$$a^{-n} = \frac{1}{a^n} \quad \text{if } a \neq 0$$

Whenever you encounter a negative exponent, think "reciprocal."

EXAMPLE 10 Evaluating Expressions Containing Negative Exponents

(a) $2^{-3} = \dfrac{1}{2^3} = \dfrac{1}{8}$ (b) $-x^{-4} = \dfrac{-1}{x^4}$ (c) $\left(\dfrac{1}{5}\right)^{-2} = \dfrac{1}{\left(\dfrac{1}{5}\right)^2} = \dfrac{1}{\dfrac{1}{25}} = 25$

■

Now Work PROBLEMS 85 AND 105

The following properties, called the **Laws of Exponents,** can be proved using the preceding definitions. In the list, a and b are real numbers, and m and n are integers.

THEOREM **Laws of Exponents**

$$a^m a^n = a^{m+n} \qquad (a^m)^n = a^{mn} \qquad (ab)^n = a^n b^n$$
$$\frac{a^m}{a^n} = a^{m-n} = \frac{1}{a^{n-m}} \quad \text{if } a \neq 0 \qquad \left(\frac{a}{b}\right)^n = \frac{a^n}{b^n} \quad \text{if } b \neq 0$$

EXAMPLE 11 **Using the Laws of Exponents**

Write each expression so that all exponents are positive.

(a) $\dfrac{x^5 y^{-2}}{x^3 y}$ $x \neq 0,\quad y \neq 0$ (b) $\left(\dfrac{x^{-3}}{3y^{-1}}\right)^{-2}$ $x \neq 0,\quad y \neq 0$

Solution

(a) $\dfrac{x^5 y^{-2}}{x^3 y} = \dfrac{x^5}{x^3} \cdot \dfrac{y^{-2}}{y} = x^{5-3} \cdot y^{-2-1} = x^2 y^{-3} = x^2 \cdot \dfrac{1}{y^3} = \dfrac{x^2}{y^3}$

(b) $\left(\dfrac{x^{-3}}{3y^{-1}}\right)^{-2} = \dfrac{(x^{-3})^{-2}}{(3y^{-1})^{-2}} = \dfrac{x^6}{3^{-2}(y^{-1})^{-2}} = \dfrac{x^6}{\dfrac{1}{9}y^2} = \dfrac{9x^6}{y^2}$

Now Work   PROBLEMS 87 AND 97

7 Evaluate Square Roots

A real number is squared when it is raised to the power 2. The inverse of squaring is finding a **square root.** For example, since $6^2 = 36$ and $(-6)^2 = 36$, the numbers 6 and -6 are square roots of 36.

The symbol $\sqrt{}$, called a **radical sign,** is used to denote the **principal,** or nonnegative, square root. For example, $\sqrt{36} = 6$. We read $\sqrt{36}$ as "give me the nonnegative number whose square is 36."

In Words

The symbol $\sqrt{a}$ means "give me the non negative number whose square is a."

DEFINITION

If a is a nonnegative real number, the nonnegative number b, for which $b^2 = a$, is the **principal square root** of a and is denoted by $b = \sqrt{a}$.

The following comments are noteworthy:

1. Negative numbers do not have square roots (in the real number system), because the square of any real number is *nonnegative.* For example, $\sqrt{-4}$ is not a real number, because there is no real number whose square is -4.
2. The principal square root of 0 is 0, since $0^2 = 0$. That is, $\sqrt{0} = 0$.
3. The principal square root of a positive number is positive.
4. If $c \geq 0$, then $(\sqrt{c})^2 = c$. For example, $(\sqrt{2})^2 = 2$ and $(\sqrt{3})^2 = 3$.

EXAMPLE 12 **Evaluating Square Roots**

(a) $\sqrt{64} = 8$ (b) $\sqrt{\dfrac{1}{16}} = \dfrac{1}{4}$ (c) $\left(\sqrt{1.4}\right)^2 = 1.4$

Examples 12(a) and (b) are examples of square roots of perfect squares, since

$64 = 8^2$ and $\dfrac{1}{16} = \left(\dfrac{1}{4}\right)^2$.

Consider the expression $\sqrt{a^2}$. Since $a^2 \geq 0$, the principal square root of a^2 is defined whether $a > 0$ or $a < 0$. However, since the principal square root is nonnegative, we need an absolute value to ensure the nonnegative result. That is,

$$\sqrt{a^2} = |a| \qquad a \text{ any real number} \qquad (2)$$

EXAMPLE 13   Simplifying Expressions Using Equation (2)

(a) $\sqrt{(2.3)^2} = |2.3| = 2.3$ (b) $\sqrt{(-2.3)^2} = |-2.3| = 2.3$ (c) $\sqrt{x^2} = |x|$

 Now Work   PROBLEM 93

Calculators

Calculators are finite machines. As a result, they are incapable of displaying decimals that contain a large number of digits. For example, some calculators are capable of displaying only eight digits. When a number requires more than eight digits, the calculator either truncates or rounds. To see how your calculator handles decimals, divide 2 by 3. How many digits do you see? Is the last digit a 6 or a 7? If it is a 6, your calculator truncates; if it is a 7, your calculator rounds.

There are different kinds of calculators. An **arithmetic** calculator can only add, subtract, multiply, and divide numbers; therefore, this type is not adequate for this course. **Scientific** calculators have all the capabilities of arithmetic calculators and also contain **function keys** labeled ln, log, sin, cos, tan, x^y, inv, and so on. **Graphing** calculators have all the capabilities of scientific calculators and contain a screen on which graphs can be displayed.

8 Use a Calculator to Evaluate Exponents

Your calculator has a caret key, $\boxed{\wedge}$, which is used for computations involving exponents.

EXAMPLE 14   Exponents on a Graphing Calculator

Evaluate: $(2.3)^5$

Solution   Figure 12 shows the result using a TI-84 Plus graphing calculator.

Figure 12

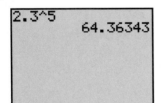

 Now Work   PROBLEM 123

A.1 Assess Your Understanding

Concepts and Vocabulary

1. A(n) _____ is a letter used in algebra to represent any number from a given set of numbers.

2. On the real number line, the real number zero is the coordinate of the _____.

3. An inequality of the form $a > b$ is called a(n) _____ inequality.

4. In the expression 2^4, the number 2 is called the _____ and 4 is called the _____.

5. *True or False* The product of two negative real numbers is always greater than zero.

6. *True or False* The distance between two distinct points on the real number line is always greater than zero.

7. *True or False* The absolute value of a real number is always greater than zero.

8. *True or False* To multiply two expressions having the same base, retain the base and multiply the exponents.

Skill Building

In Problems 9–20, use U = universal set = $\{0, 1, 2, 3, 4, 5, 6, 7, 8, 9\}$, $A = \{1, 3, 4, 5, 9\}$, $B = \{2, 4, 6, 7, 8\}$, and $C = \{1, 3, 4, 6\}$ to find each set.

9. $A \cup B$ **10.** $A \cup C$ **11.** $A \cap B$ **12.** $A \cap C$

13. $(A \cup B) \cap C$ **14.** $(A \cap B) \cup C$ **15.** $\overline{A}$ **16.** $\overline{C}$

17. $\overline{A \cap B}$ **18.** $\overline{B \cup C}$ **19.** $\overline{A} \cup \overline{B}$ **20.** $\overline{B} \cap \overline{C}$

21. On the real number line, label the points with coordinates 0, 1, -1, $\dfrac{5}{2}$, -2.5, $\dfrac{3}{4}$, and 0.25.

22. On the real number line, label the points with coordinates 0, -2, 2, -1.5, $\dfrac{3}{2}$, $\dfrac{1}{3}$, and $\dfrac{2}{3}$.

In Problems 23–32, replace the question mark by $<$, $>$, or $=$, whichever is correct.

23. $\dfrac{1}{2} \,?\, 0$ **24.** $5 \,?\, 6$ **25.** $-1 \,?\, -2$ **26.** $-3 \,?\, -\dfrac{5}{2}$ **27.** $\pi \,?\, 3.14$

28. $\sqrt{2} \,?\, 1.41$ **29.** $\dfrac{1}{2} \,?\, 0.5$ **30.** $\dfrac{1}{3} \,?\, 0.33$ **31.** $\dfrac{2}{3} \,?\, 0.67$ **32.** $\dfrac{1}{4} \,?\, 0.25$

In Problems 33–38, write each statement as an inequality.

33. x is positive   **34.** z is negative   **35.** x is less than 2

36. y is greater than -5 **37.** x is less than or equal to 1   **38.** x is greater than or equal to 2

In Problems 39–42, graph the numbers x on the real number line.

39. $x \geq -2$ **40.** $x < 4$ **41.** $x > -1$ **42.** $x \leq 7$

In Problems 43–48, use the given real number line to compute each distance.

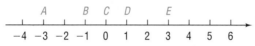

43. $d(C, D)$ **44.** $d(C, A)$ **45.** $d(D, E)$ **46.** $d(C, F)$ **47.** $d(A, E)$ **48.** $d(D, B)$

In Problems 49–56, evaluate each expression if $x = -2$ and $y = 3$.

49. $x + 2y$ **50.** $3x + y$ **51.** $5xy + 2$ **52.** $-2x + xy$

53. $\dfrac{2x}{x - y}$ **54.** $\dfrac{x + y}{x - y}$ **55.** $\dfrac{3x + 2y}{2 + y}$ **56.** $\dfrac{2x - 3}{y}$

In Problems 57–66, find the value of each expression if $x = 3$ and $y = -2$.

57. $|x + y|$ **58.** $|x - y|$ **59.** $|x| + |y|$ **60.** $|x| - |y|$ **61.** $\dfrac{|x|}{x}$

62. $\dfrac{|y|}{y}$ **63.** $|4x - 5y|$ **64.** $|3x + 2y|$ **65.** $||4x| - |5y||$ **66.** $3|x| + 2|y|$

In Problems 67–74, determine which of the value(s) (a) through (d), if any, must be excluded from the domain of the variable in each expression:

  (a) $x = 3$ (b) $x = 1$ (c) $x = 0$ (d) $x = -1$

67. $\dfrac{x^2 - 1}{x}$ **68.** $\dfrac{x^2 + 1}{x}$ **69.** $\dfrac{x}{x^2 - 9}$ **70.** $\dfrac{x}{x^2 + 9}$

71. $\dfrac{x^2}{x^2 + 1}$ **72.** $\dfrac{x^3}{x^2 - 1}$ **73.** $\dfrac{x^2 + 5x - 10}{x^3 - x}$ **74.** $\dfrac{-9x^2 - x + 1}{x^3 + x}$

In Problems 75–78, determine the domain of the variable x in each expression.

75. $\dfrac{4}{x-5}$

76. $\dfrac{-6}{x+4}$

77. $\dfrac{x}{x+4}$

78. $\dfrac{x-2}{x-6}$

In Problems 79–82, use the formula $C = \dfrac{5}{9}(F-32)$ for converting degrees Fahrenheit into degrees Celsius to find the Celsius measure of each Fahrenheit temperature.

79. $F = 32°$

80. $F = 212°$

81. $F = 77°$

82. $F = -4°$

In Problems 83–94, simplify each expression.

83. $(-4)^2$

84. -4^2

85. 4^{-2}

86. -4^{-2}

87. $3^{-6} \cdot 3^4$

88. $4^{-2} \cdot 4^3$

89. $(3^{-2})^{-1}$

90. $(2^{-1})^{-3}$

91. $\sqrt{25}$

92. $\sqrt{36}$

93. $\sqrt{(-4)^2}$

94. $\sqrt{(-3)^2}$

In Problems 95–104, simplify each expression. Express the answer so that all exponents are positive. Whenever an exponent is 0 or negative, we assume that the base is not 0.

95. $(8x^3)^2$

96. $(-4x^2)^{-1}$

97. $(x^2 y^{-1})^2$

98. $(x^{-1}y)^3$

99. $\dfrac{x^2 y^3}{xy^4}$

100. $\dfrac{x^{-2}y}{xy^2}$

101. $\dfrac{(-2)^3 x^4 (yz)^2}{3^2 xy^3 z}$

102. $\dfrac{4x^{-2}(yz)^{-1}}{2^3 x^4 y}$

103. $\left(\dfrac{3x^{-1}}{4y^{-1}}\right)^{-2}$

104. $\left(\dfrac{5x^{-2}}{6y^{-2}}\right)^{-3}$

In Problems 105–116, find the value of each expression if $x = 2$ and $y = -1$.

105. $2xy^{-1}$

106. $-3x^{-1}y$

107. $x^2 + y^2$

108. $x^2 y^2$

109. $(xy)^2$

110. $(x+y)^2$

111. $\sqrt{x^2}$

112. $\left(\sqrt{x}\right)^2$

113. $\sqrt{x^2 + y^2}$

114. $\sqrt{x^2} + \sqrt{y^2}$

115. x^y

116. y^x

117. Find the value of the expression $2x^3 - 3x^2 + 5x - 4$ if $x = 2$. What is the value if $x = 1$?

118. Find the value of the expression $4x^3 + 3x^2 - x + 2$ if $x = 1$. What is the value if $x = 2$?

119. What is the value of $\dfrac{(666)^4}{(222)^4}$?

120. What is the value of $(0.1)^3 (20)^3$?

In Problems 121–128, use a calculator to evaluate each expression. Round your answer to three decimal places.

121. $(8.2)^6$

122. $(3.7)^5$

123. $(6.1)^{-3}$

124. $(2.2)^{-5}$

125. $(-2.8)^6$

126. $-(2.8)^6$

127. $(-8.11)^{-4}$

128. $-(8.11)^{-4}$

Applications and Extensions

In Problems 129–138, express each statement as an equation involving the indicated variables.

129. Area of a Rectangle The area A of a rectangle is the product of its length l and its width w.

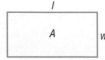

131. Circumference of a Circle The circumference C of a circle is the product of π and its diameter d.

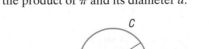

130. Perimeter of a Rectangle The perimeter P of a rectangle is twice the sum of its length l and its width w.

132. Area of a Triangle The area A of a triangle is one-half the product of its base b and its height h.

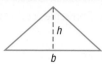

133. Area of an Equilateral Triangle The area A of an equilateral triangle is $\dfrac{\sqrt{3}}{4}$ times the square of the length x of one side.

134. Perimeter of an Equilateral Triangle The perimeter P of an equilateral triangle is 3 times the length x of one side.

135. Volume of a Sphere The volume V of a sphere is $\dfrac{4}{3}$ times π times the cube of the radius r.

136. Surface Area of a Sphere The surface area S of a sphere is 4 times π times the square of the radius r.

137. Volume of a Cube The volume V of a cube is the cube of the length x of a side.

138. Surface Area of a Cube The surface area S of a cube is 6 times the square of the length x of a side.

139. Manufacturing Cost The weekly production cost C of manufacturing x watches is given by the formula $C = 4000 + 2x$, where the variable C is in dollars.
(a) What is the cost of producing 1000 watches?
(b) What is the cost of producing 2000 watches?

140. Balancing a Checkbook At the beginning of the month, Mike had a balance of $210 in his checking account. During the next month, he deposited $80, wrote a check for $120, made another deposit of $25, and wrote two checks: one for $60 and the other for $32. He was also assessed a monthly service charge of $5. What was his balance at the end of the month?

In Problems 141 and 142, write an inequality using an absolute value to describe each statement.

141. x is at least 6 units from 4.

142. x is more than 5 units from 2.

143. U.S. Voltage In the United States, normal household voltage is 110 volts. It is acceptable for the actual voltage x to differ from normal by at most 5 volts. A formula that describes this is

$$|x - 110| \le 5$$

(a) Show that a voltage of 108 volts is acceptable.
(b) Show that a voltage of 104 volts is not acceptable.

144. Foreign Voltage In some countries, normal household voltage is 220 volts. It is acceptable for the actual voltage x to differ from normal by at most 8 volts. A formula that describes this is

$$|x - 220| \le 8$$

(a) Show that a voltage of 214 volts is acceptable.
(b) Show that a voltage of 209 volts is not acceptable.

145. Making Precision Ball Bearings The FireBall Company manufactures ball bearings for precision equipment. One of its products is a ball bearing with a stated radius of 3 centimeters (cm). Only ball bearings with a radius within 0.01 cm of this stated radius are acceptable. If x is the radius of a ball bearing, a formula describing this situation is

$$|x - 3| \le 0.01$$

(a) Is a ball bearing of radius $x = 2.999$ acceptable?
(b) Is a ball bearing of radius $x = 2.89$ acceptable?

146. Body Temperature Normal human body temperature is 98.6°F. A temperature x that differs from normal by at least 1.5°F is considered unhealthy. A formula that describes this is

$$|x - 98.6| \ge 1.5$$

(a) Show that a temperature of 97°F is unhealthy.
(b) Show that a temperature of 100°F is not unhealthy.

147. Does $\dfrac{1}{3}$ equal 0.333? If not, which is larger? By how much?

148. Does $\dfrac{2}{3}$ equal 0.666? If not, which is larger? By how much?

Discussion and Writing

149. Is there a positive real number "closest" to 0?

150. Number Game I'm thinking of a number! It lies between 1 and 10; its square is rational and lies between 1 and 10. The number is larger than π. Correct to two decimal places (that is, truncated to two decimal places) name the number. Now think of your own number, describe it, and challenge a fellow student to name it.

151. Write a brief paragraph that illustrates the similarities and differences between "less than" ($<$) and "less than or equal to" ($\le$).

152. Give a reason why the statement $5 < 8$ is true.

A.2 Geometry Essentials

OBJECTIVES 1 Use the Pythagorean Theorem and Its Converse (p. A14)
2 Know Geometry Formulas (p. A15)
3 Understand Congruent Triangles and Similar Triangles (p. A16)

In this section we review some topics studied in geometry that we shall need for our study of trigonometry.

Figure 13

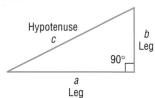

1 Use the Pythagorean Theorem and Its Converse

The *Pythagorean Theorem* is a statement about *right triangles*. A **right triangle** is one that contains a **right angle,** that is, an angle of 90°. The side of the triangle opposite the 90° angle is called the **hypotenuse;** the remaining two sides are called **legs.** In Figure 13 we have used c to represent the length of the hypotenuse and a and b to represent the lengths of the legs. Notice the use of the symbol $\ulcorner$ to show the 90° angle. We now state the Pythagorean Theorem.

PYTHAGOREAN THEOREM

In a right triangle, the square of the length of the hypotenuse is equal to the sum of the squares of the lengths of the legs. That is, in the right triangle shown in Figure 13,

$$c^2 = a^2 + b^2 \tag{1}$$

A proof of the Pythagorean Theorem is given at the end of this section.

EXAMPLE 1 Finding the Hypotenuse of a Right Triangle

In a right triangle, one leg has length 4 and the other has length 3. What is the length of the hypotenuse?

Solution Since the triangle is a right triangle, we use the Pythagorean Theorem with $a = 4$ and $b = 3$ to find the length c of the hypotenuse. From equation (1), we have

$$c^2 = a^2 + b^2$$
$$c^2 = 4^2 + 3^2 = 16 + 9 = 25$$
$$c = \sqrt{25} = 5$$

Now Work PROBLEM 13

The converse of the Pythagorean Theorem is also true.

CONVERSE OF THE PYTHAGOREAN THEOREM

In a triangle, if the square of the length of one side equals the sum of the squares of the lengths of the other two sides, the triangle is a right triangle. The 90° angle is opposite the longest side.

A proof is given at the end of this section.

EXAMPLE 2 Verifying That a Triangle Is a Right Triangle

Show that a triangle whose sides are of lengths 5, 12, and 13 is a right triangle. Identify the hypotenuse.

Solution We square the lengths of the sides.

$$5^2 = 25, \qquad 12^2 = 144, \qquad 13^2 = 169$$

Figure 14

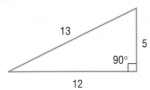

Notice that the sum of the first two squares (25 and 144) equals the third square (169). Hence, the triangle is a right triangle. The longest side, 13, is the hypotenuse. See Figure 14.

---- **Now Work** PROBLEM 21

| EXAMPLE 3 | **Applying the Pythagorean Theorem** |

Excluding antenna, the tallest inhabited building in the world is Taipei 101 in Taipei, Taiwan. See Figure 15. If the indoor observation deck is 1437 feet above ground level, how far can a person standing on the observation deck see (with the aid of a telescope)? Use 3960 miles for the radius of Earth.

Figure 15

Source: Council on Tall Buildings and Urban Habitat, 2006.

Solution From the center of Earth, draw two radii: one through Taipei 101 and the other to the farthest point a person can see from the observation deck. See Figure 16. Apply the Pythagorean Theorem to the right triangle.

Since 1 mile = 5280 feet, then 1437 feet = $\dfrac{1437}{5280}$ mile. So we have

$$d^2 + (3960)^2 = \left(3960 + \frac{1437}{5280}\right)^2$$

$$d^2 = \left(3960 + \frac{1437}{5280}\right)^2 - (3960)^2 \approx 2155.57$$

$$d \approx 46.43$$

A person can see about 46 miles from the observation tower.

Figure 16

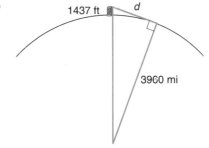

1437 ft   *d*

3960 mi

---- **Now Work** PROBLEM 53

2 Know Geometry Formulas

Certain formulas from geometry are useful in solving trigonometry problems. We list some of these next.

For a rectangle of length *l* and width *w*,

w

l

| Area = *lw*   Perimeter = 2*l* + 2*w* |

For a triangle with base *b* and altitude *h*,

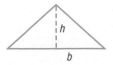

h

b

| Area = $\dfrac{1}{2}bh$ |

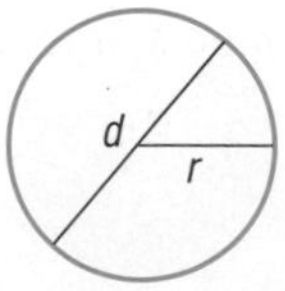

For a circle of radius r (diameter $d = 2r$),

Area $= \pi r^2$	Circumference $= 2\pi r = \pi d$

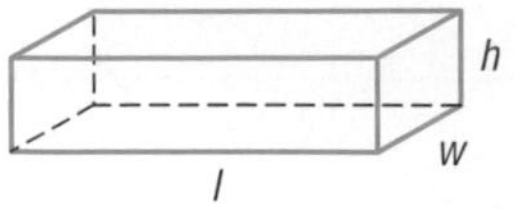

For a closed rectangular box of length l, width w, and height h,

Volume $= lwh$	Surface area $= 2lh + 2wh + 2lw$

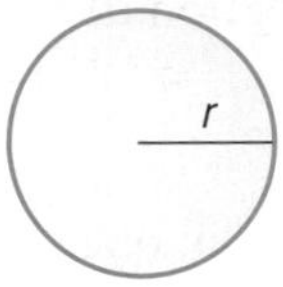

For a sphere of radius r,

Volume $= \dfrac{4}{3}\pi r^3$	Surface area $= 4\pi r^2$

For a right circular cylinder of height h and radius r,

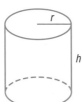

Volume $= \pi r^2 h$	Surface area $= 2\pi r^2 + 2\pi rh$

 Now Work PROBLEM 29

EXAMPLE 4 | **Using Geometry Formulas**

A Christmas tree ornament is in the shape of a semicircle on top of a triangle. How many square centimeters (cm) of copper is required to make the ornament if the height of the triangle is 6 cm and the base is 4 cm?

Solution

See Figure 17. The amount of copper required equals the shaded area. This area is the sum of the area of the triangle and the semicircle. The triangle has height $h = 6$ and base $b = 4$. The semicircle has diameter $d = 4$, so its radius is $r = 2$.

Figure 17

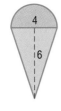

$$\text{Area} = \text{Area of triangle} + \text{Area of semicircle}$$
$$= \frac{1}{2}bh + \frac{1}{2}\pi r^2 = \frac{1}{2}(4)(6) + \frac{1}{2}\pi \cdot 2^2 \qquad b=4; h=6; r=2$$
$$= 12 + 2\pi \approx 18.28 \text{ cm}^2$$

About 18.28 cm² of copper is required.

 Now Work PROBLEM 47

3 Understand Congruent Triangles and Similar Triangles

Throughout the text we will make reference to triangles. We begin with a discussion of *congruent* triangles. According to dictionary.com, the word **congruent** means coinciding exactly when superimposed. For example, two angles are congruent if they have the same measure and two line segments are congruent if they have the same length.

> **In Words**
> Two triangles are congruent if they have the same size and shape.

DEFINITION

Two triangles are **congruent** if each of the corresponding angles is the same measure and each of the corresponding sides is the same length.

In Figure 18, corresponding angles are equal and the lengths of the corresponding sides are equal: $a = d$, $b = e$, and $c = f$. We conclude that these triangles are congruent.

Figure 18
Congruent triangles

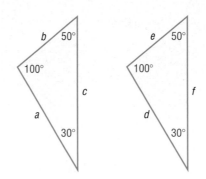

It is not necessary to verify that all three angles and all three sides are the same measure to determine whether two triangles are congruent.

Determining Congruent Triangles

1. **Angle–Side–Angle Case** Two triangles are congruent if two of the angles are equal and the lengths of the corresponding sides between the two angles are equal.

   For example, in Figure 19(a), the two triangles are congruent because two angles and the included side are equal.

2. **Side–Side–Side Case** Two triangles are congruent if the lengths of the corresponding sides of the triangles are equal.

   For example, in Figure 19(b), the two triangles are congruent because the three corresponding sides are all equal.

3. **Side–Angle–Side Case** Two triangles are congruent if the lengths of two corresponding sides are equal and the angles between the two sides are the same.

   For example, in Figure 19(c), the two triangles are congruent because two sides and the included angle are equal.

Figure 19

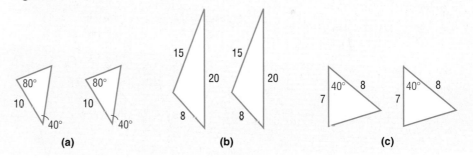

We contrast congruent triangles with *similar* triangles.

DEFINITION   Two triangles are **similar** if the corresponding angles are equal and the lengths of the corresponding sides are proportional.

For example, the triangles in Figure 20 on page A18 are similar because the corresponding angles are equal. In addition, the lengths of the corresponding sides are proportional because each side in the triangle on the right is twice as long as each corresponding side in the triangle on the left. That is, the ratio of the corresponding sides is a constant: $\dfrac{d}{a} = \dfrac{e}{b} = \dfrac{f}{c} = 2$.

Figure 20

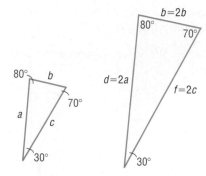

It is not necessary to verify that all three angles are equal and all three sides are proportional to determine whether two triangles are similar.

Determining Similar Triangles

1. **Angle–Angle Case**   Two triangles are similar if two of the corresponding angles are equal.

   For example, in Figure 21(a), the two triangles are similar because two angles are equal.

2. **Side–Side–Side Case**   Two triangles are similar if the lengths of all three sides of each triangle are proportional.

   For example, in Figure 21(b), the two triangles are similar because

   $$\frac{10}{30} = \frac{5}{15} = \frac{6}{18} = \frac{1}{3}$$

3. **Side–Angle–Side Case**   Two triangles are similar if two corresponding sides are proportional and the angles between the two sides are equal.

   For example, in Figure 21(c), the two triangles are similar because $\frac{4}{6} = \frac{12}{18} = \frac{2}{3}$ and the angle between the sides are equal.

Figure 21

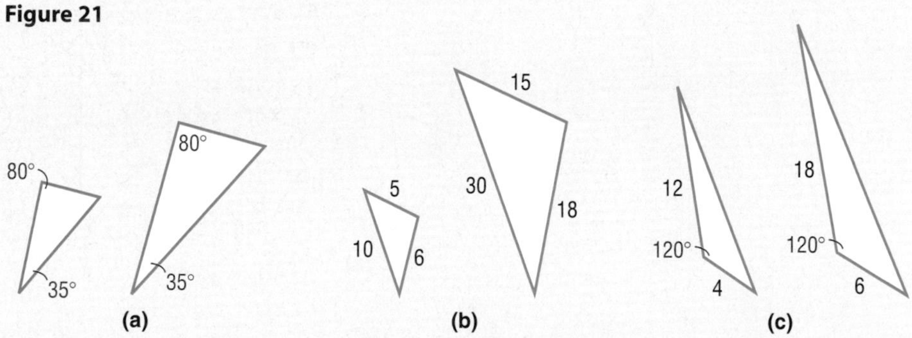

| (a) | (b) | (c) |

EXAMPLE 5 | **Using Similar Triangles**

Given that the triangles in Figure 22 are similar, find the missing length x and the angles A, B, and C.

Figure 22

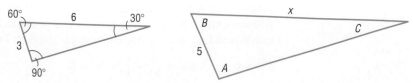

Solution   Because the triangles are similar, corresponding angles are equal. So $A = 90°$, $B = 60°$, and $C = 30°$. Also, the corresponding sides are proportional. That is, $\frac{3}{5} = \frac{6}{x}$. We solve this equation for x.

$$\frac{3}{5} = \frac{6}{x}$$

$$5x \cdot \frac{3}{5} = 5x \cdot \frac{6}{x} \qquad \text{Multiply both sides by 5x.}$$

$$3x = 30 \qquad \text{Simplify.}$$

$$x = 10 \qquad \text{Divide both sides by 3.}$$

The missing length is 10 units.

<hr>

— **Now Work** PROBLEM 41

Proof of the Pythagorean Theorem   We begin with a square, each side of length $a + b$. In this square, we can form four right triangles, each having legs equal in length to a and b. See Figure 23. All these triangles are congruent (two sides and their included angle are equal). As a result, the hypotenuse of each is the same, say c, and the pink shading in Figure 23 indicates a square with an area equal to c^2.

Figure 23

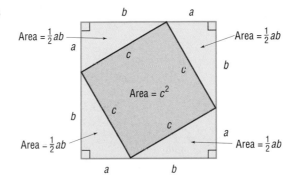

The area of the original square with sides $a + b$ equals the sum of the areas of the four triangles (each of area $\frac{1}{2}ab$) plus the area of the square with side c. That is,

$$(a + b)^2 = \frac{1}{2}ab + \frac{1}{2}ab + \frac{1}{2}ab + \frac{1}{2}ab + c^2$$

$$a^2 + 2ab + b^2 = 2ab + c^2$$

$$a^2 + b^2 = c^2$$

The proof is complete. ■

Proof of the Converse of the Pythagorean Theorem   We begin with two triangles: one a right triangle with legs a and b and the other a triangle with sides a, b, and c for which $c^2 = a^2 + b^2$. See Figure 24. By the Pythagorean Theorem, the length x of the third side of the first triangle is

$$x^2 = a^2 + b^2$$

But $c^2 = a^2 + b^2$. Hence,

$$x^2 = c^2$$

$$x = c$$

The two triangles have sides with the same length and are therefore congruent. This means corresponding angles are equal, so the angle opposite side c of the second triangle equals 90°.

The proof is complete. ■

Figure 24

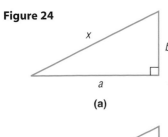

(a)

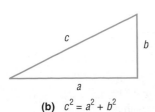

(b) $c^2 = a^2 + b^2$

A.2 Assess Your Understanding

Concepts and Vocabulary

1. A(n) _____ triangle is one that contains an angle of 90 degrees. The longest side is called the _____.

2. For a triangle with base b and altitude h, a formula for the area A is _____.

3. The formula for the circumference C of a circle of radius r is _____.

4. Two triangles are _____ if corresponding angles are equal and the lengths of the corresponding sides are proportional.

5. **True or False** In a right triangle, the square of the length of the longest side equals the sum of the squares of the lengths of the other two sides.

6. **True or False** The triangle with sides of length 6, 8, and 10 is a right triangle.

7. **True or False** The volume of a sphere of radius r is $\frac{4}{3}\pi r^2$.

8. **True or False** The triangles shown are congruent.

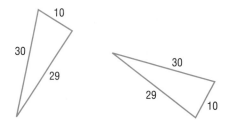

9. **True or False** The triangles shown are similar.

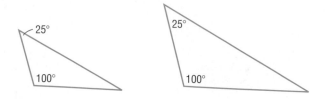

10. **True or False** The triangles shown are similar.

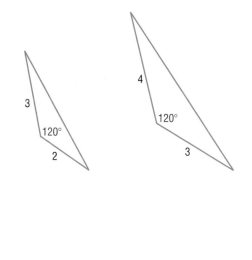

Skill Building

In Problems 11–16, the lengths of the legs of a right triangle are given. Find the length of the hypotenuse.

11. $a = 5,\quad b = 12$

12. $a = 6,\quad b = 8$

13. $a = 10,\quad b = 24$

14. $a = 4,\quad b = 3$

15. $a = 7,\quad b = 24$

16. $a = 14,\quad b = 48$

In Problems 17–24, the lengths of the sides of a triangle are given. Determine which are right triangles. For those that are, identify the hypotenuse.

17. 3, 4, 5

18. 6, 8, 10

19. 4, 5, 6

20. 2, 2, 3

21. 7, 24, 25

22. 10, 24, 26

23. 6, 4, 3

24. 5, 4, 7

25. Find the area A of a rectangle with length 4 inches and width 2 inches.

26. Find the area A of a rectangle with length 9 centimeters and width 4 centimeters.

27. Find the area A of a triangle with height 4 inches and base 2 inches.

28. Find the area A of a triangle with height 9 centimeters and base 4 centimeters.

29. Find the area A and circumference C of a circle of radius 5 meters.

30. Find the area A and circumference C of a circle of radius 2 feet.

31. Find the volume V and surface area S of a rectangular box with length 8 feet, width 4 feet, and height 7 feet.

32. Find the volume V and surface area S of a rectangular box with length 9 inches, width 4 inches, and height 8 inches.

33. Find the volume V and surface area S of a sphere of radius 4 centimeters.

34. Find the volume V and surface area S of a sphere of radius 3 feet.

35. Find the volume V and surface area S of a right circular cylinder with radius 9 inches and height 8 inches.

36. Find the volume V and surface area S of a right circular cylinder with radius 8 inches and height 9 inches.

In Problems 37–40, find the area of the shaded region.

37.

38.

39.

40.

In Problems 41–44, each pair of triangles is similar. Find the missing length x and the missing angles A, B, and C.

41.

42.

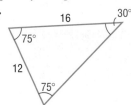

43.

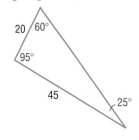

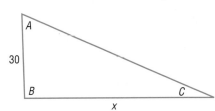

44.

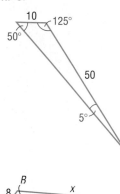

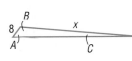

Applications and Extensions

45. How many feet does a wheel with a diameter of 16 inches travel after four revolutions?

46. How many revolutions will a circular disk with a diameter of 4 feet have completed after it has rolled 20 feet?

47. In the figure shown, *ABCD* is a square, with each side of length 6 feet. The width of the border (shaded portion) between the outer square *EFGH* and *ABCD* is 2 feet. Find the area of the border.

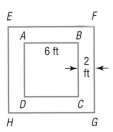

48. Refer to the figure. Square *ABCD* has an area of 100 square feet; square *BEFG* has an area of 16 square feet. What is the area of the triangle *CGF*?

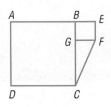

49. Architecture   A **Norman window** consists of a rectangle surmounted by a semicircle. Find the area of the Norman window shown in the illustration. How much wood frame is needed to enclose the window?

50. Construction   A circular swimming pool, 20 feet in diameter, is enclosed by a wooden deck that is 3 feet wide. What is the area of the deck? How much fence is required to enclose the deck?

51. How Tall Is the Great Pyramid? The ancient Greek philosopher Thales of Miletus is reported on one occasion to have visited Egypt and calculated the height of the Great Pyramid of Cheops by means of shadow reckoning. Thales knew that each side of the base of the pyramid was 252 paces and that his own height was 2 paces. He measured the length of the pyramid's shadow to be 114 paces and determined the length of his shadow to be 3 paces. See the illustration. Using similar triangles, determine the height of the Great Pyramid in terms of the number of paces.

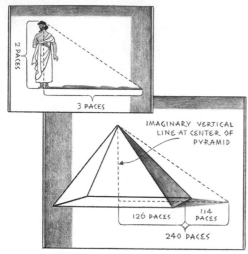

Source: www.anselm.edu/homepage/dbanach/ thales.htm. This site references another source: Selections, from Julia E. Diggins, *String, Straightedge, and Shadow,* Viking Press, New York, 1965, Illustrations by Corydon Bell.

52. The Bermuda Triangle Karen is doing research on the Bermuda Triangle which she defines roughly by Hamilton, Bermuda; San Juan, Puerto Rico; and Fort Lauderdale, Florida. On her atlas Karen measures the straight-line distances from Hamilton to Fort Lauderdale, Fort Lauderdale to San Juan, and San Juan to Hamilton to be approximately 57 millimeters (mm), 58 mm, and 53.5 mm respectively. If the actual distance from Fort Lauderdale to San Juan is 1046 miles, approximate the actual distances from San Juan to Hamilton and from Hamilton to Fort Lauderdale.

Source: www.worldatlas.com

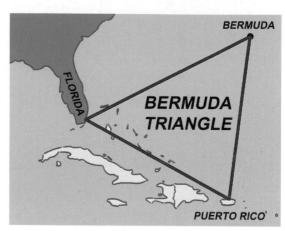

Source: www.en.wikipedia.org/wiki/Bermuda_Triangle.

In Problems 53–55, use the facts that the radius of Earth is 3960 miles and 1 mile = 5280 feet.

53. How Far Can You See? The conning tower of the U.S.S. *Silversides,* a World War II submarine now permanently stationed in Muskegon, Michigan, is approximately 20 feet above sea level. How far can you see from the conning tower?

54. How Far Can You See? A person who is 6 feet tall is standing on the beach in Fort Lauderdale, Florida, and looks out onto the Atlantic Ocean. Suddenly, a ship appears on the horizon. How far is the ship from shore?

55. How Far Can You See? The deck of a destroyer is 100 feet above sea level. How far can a person see from the deck? How far can a person see from the bridge, which is 150 feet above sea level?

56. Suppose that m and n are positive integers with $m > n$. If $a = m^2 - n^2$, $b = 2mn$, and $c = m^2 + n^2$, show that a, b, and c are the lengths of the sides of a right triangle. (This formula can be used to find the sides of a right triangle that are integers, such as 3, 4, 5; 5, 12, 13; and so on. Such triplets of integers are called **Pythagorean triples.**)

Discussion and Writing

57. You have 1000 feet of flexible pool siding and wish to construct a swimming pool. Experiment with rectangular-shaped pools with perimeters of 1000 feet. How do their areas vary? What is the shape of the rectangle with the largest area? Now compute the area enclosed by a circular pool with a perimeter (circumference) of 1000 feet. What would be your choice of shape for the pool? If rectangular, what is your preference for dimensions? Justify your choice. If your only consideration is to have a pool that encloses the most area, what shape should you use?

58. The Gibb's Hill Lighthouse, Southampton, Bermuda, in operation since 1846, stands 117 feet high on a hill 245 feet high, so its beam of light is 362 feet above sea level. A brochure states that the light itself can be seen on the horizon about 26 miles distant. Verify the correctness of this information. The brochure further states that ships 40 miles away can see the light and planes flying at 10,000 feet can see it 120 miles away. Verify the accuracy of these statements. What assumption did the brochure make about the height of the ship?

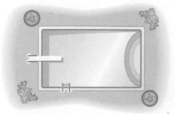

A.3 Factoring Polynomials

> **OBJECTIVE 1** Factor Polynomials (p. A23)

1 Factor Polynomials

Consider the following product:

$$(2x + 3)(x - 4) = 2x^2 - 5x - 12$$

The two polynomials on the left side are called **factors** of the polynomial on the right side. Expressing a given polynomial as a product of other polynomials, that is, finding the factors of a polynomial, is called **factoring.**

We shall restrict our discussion here to factoring polynomials in one variable into products of polynomials in one variable, where all coefficients are integers. We call this **factoring over the integers.**

Any polynomial can be written as the product of 1 times itself or as -1 times its additive inverse. If a polynomial cannot be written as the product of two other polynomials (excluding 1 and -1), then the polynomial is said to be **prime.** When a polynomial has been written as a product consisting only of prime factors, it is said to be **factored completely.** Examples of prime polynomials (over the integers) are

COMMENT Over the real numbers, $3x + 4$ factors into $3(x + \frac{4}{3})$. It is the noninteger $\frac{4}{3}$ that causes $3x + 4$ to be prime over the integers. ∎

$$2 \quad 3 \quad 5 \quad x \quad x + 1 \quad x - 1 \quad 3x + 4 \quad x^2 + 4$$

The first factor to look for in a factoring problem is a common monomial factor present in each term of the polynomial. If one is present, use the Distributive Property to factor it out.

EXAMPLE 1   Identifying Common Monomial Factors

Polynomial	Common Monomial Factor	Remaining Factor	Factored Form
$2x + 4$	2	$x + 2$	$2x + 4 = 2(x + 2)$
$3x - 6$	3	$x - 2$	$3x - 6 = 3(x - 2)$
$2x^2 - 4x + 8$	2	$x^2 - 2x + 4$	$2x^2 - 4x + 8 = 2(x^2 - 2x + 4)$
$8x - 12$	4	$2x - 3$	$8x - 12 = 4(2x - 3)$
$x^2 + x$	x	$x + 1$	$x^2 + x = x(x + 1)$
$x^3 - 3x^2$	x^2	$x - 3$	$x^3 - 3x^2 = x^2(x - 3)$
$6x^2 + 9x$	$3x$	$2x + 3$	$6x^2 + 9x = 3x(2x + 3)$

∎

Notice that, once all common monomial factors have been removed from a polynomial, the remaining factor is either a prime polynomial of degree 1 or a polynomial of degree 2 or higher. (Do you see why?)

EXAMPLE 2   Factoring Polynomials

Factor completely each polynomial.

(a) $x^4 - 16$ (b) $x^3 - 1$

(c) $9x^2 - 6x + 1$ (d) $x^2 + 4x - 12$

(e) $3x^2 + 10x - 8$ (f) $x^3 - 4x^2 + 2x - 8$

Solution (a) $x^4 - 16 = (x^2 - 4)(x^2 + 4) = (x - 2)(x + 2)(x^2 + 4)$

        ↑   ↑

    Difference of squares   Difference of squares

(b) $x^3 - 1 = (x - 1)(x^2 + x + 1)$

        ↑

    Difference of cubes

(c) $9x^2 - 6x + 1 = (3x - 1)^2$

          ↑

    Perfect square

(d) $x^2 + 4x - 12 = (x + 6)(x - 2)$

       ↑

The product of 6 and -2 is -12, and the sum of 6 and -2 is 4.

$12x - 2x = 10x$

(e) $3x^2 + 10x - 8 = (3x - 2)(x + 4)$

    $3x^2$ -8

NOTE The technique used in part (f) is called **factoring by grouping.** ■

(f) $x^3 - 4x^2 + 2x - 8 = (x^3 - 4x^2) + (2x - 8)$

                    ↑

                 Group terms

$= x^2(x - 4) + 2(x - 4) = (x^2 + 2)(x - 4)$

 ↑   ↑

Distributive property   Distributive property

Now Work PROBLEMS 13, 29, AND 57

A.3 Assess Your Understanding

Concepts and Vocabulary

1. If factored completely, $3x^3 - 12x =$ _____.
2. If a polynomial cannot be written as the product of two other polynomials (excluding 1 and -1), then the polynomial is said to be _____.

3. **True or False** The polynomial $x^2 + 4$ is prime.
4. **True or False** $3x^3 - 2x^2 - 6x + 4 = (3x - 2)(x^2 + 2)$.

Skill Building

In Problems 5–52, factor completely each polynomial. If the polynomial cannot be factored, say it is prime.

5. $x^2 - 36$
6. $x^2 - 9$
7. $2 - 8x^2$
8. $3 - 27x^2$

9. $x^2 + 11x + 10$
10. $x^2 + 5x + 4$
11. $x^2 - 10x + 21$
12. $x^2 - 6x + 8$

13. $4x^2 - 8x + 32$
14. $3x^2 - 12x + 15$
15. $x^2 + 4x + 16$
16. $x^2 + 12x + 36$

17. $15 + 2x - x^2$
18. $14 + 6x - x^2$
19. $3x^2 - 12x - 36$
20. $x^3 + 8x^2 - 20x$

21. $y^4 + 11y^3 + 30y^2$
22. $3y^3 - 18y^2 - 48y$
23. $4x^2 + 12x + 9$
24. $9x^2 - 12x + 4$

25. $6x^2 + 8x + 2$
26. $8x^2 + 6x - 2$
27. $x^4 - 81$
28. $x^4 - 1$

29. $x^6 - 2x^3 + 1$
30. $x^6 + 2x^3 + 1$
31. $x^7 - x^5$
32. $x^8 - x^5$

33. $16x^2 + 24x + 9$
34. $9x^2 - 24x + 16$
35. $5 + 16x - 16x^2$
36. $5 + 11x - 16x^2$

37. $4y^2 - 16y + 15$
38. $9y^2 + 9y - 4$
39. $1 - 8x^2 - 9x^4$
40. $4 - 14x^2 - 8x^4$

41. $x(x + 3) - 6(x + 3)$
42. $5(3x - 7) + x(3x - 7)$
43. $(x + 2)^2 - 5(x + 2)$
44. $(x - 1)^2 - 2(x - 1)$

45. $(3x - 2)^3 - 27$

46. $(5x + 1)^3 - 1$

47. $3(x^2 + 10x + 25) - 4(x + 5)$

48. $7(x^2 - 6x + 9) + 5(x - 3)$

49. $x^3 + 2x^2 - x - 2$

50. $x^3 - 3x^2 - x + 3$

51. $x^4 - x^3 + x - 1$

52. $x^4 + x^3 + x + 1$

Applications and Extensions

In Problems 53–64, expressions that occur in calculus are given. Factor completely each expression.

53. $2(3x + 4)^2 + (2x + 3) \cdot 2(3x + 4) \cdot 3$

54. $5(2x + 1)^2 + (5x - 6) \cdot 2(2x + 1) \cdot 2$

55. $2x(2x + 5) + x^2 \cdot 2$

56. $3x^2(8x - 3) + x^3 \cdot 8$

57. $2(x + 3)(x - 2)^3 + (x + 3)^2 \cdot 3(x - 2)^2$

58. $4(x + 5)^3(x - 1)^2 + (x + 5)^4 \cdot 2(x - 1)$

59. $(4x - 3)^2 + x \cdot 2(4x - 3) \cdot 4$

60. $3x^2(3x + 4)^2 + x^3 \cdot 2(3x + 4) \cdot 3$

61. $2(3x - 5) \cdot 3(2x + 1)^3 + (3x - 5)^2 \cdot 3(2x + 1)^2 \cdot 2$

62. $3(4x + 5)^2 \cdot 4(5x + 1)^2 + (4x + 5)^3 \cdot 2(5x + 1) \cdot 5$

63. Show that $x^2 + 4$ is prime.

64. Show that $x^2 + x + 1$ is prime.

Discussion and Writing

65. Make up a polynomial that factors into a perfect square.

66. Explain to a fellow student what you look for first when presented with a factoring problem. What do you do next?

A.4 Solving Equations Algebraically

PREPARING FOR THIS SECTION   *Before getting started, review the following:*

- Factoring Polynomials (Appendix, Section A.3, pp. A23–A24)
- Zero-Product Property (Appendix, Section A.1, p. A4)
- Square Roots (Appendix, Section A.1, pp. A9–A10)
- Absolute Value (Appendix, Section A.1, p. A5)

Now Work the 'Are You Prepared?' problems on page A37.

OBJECTIVES   **1** Solve Linear Equations (p. A27)

2 Solve Rational Equations (p. A28)

3 Solve Quadratic Equations by Factoring (p. A29)

4 Solve Quadratic Equations Using the Square Root Method (p. A30)

5 Solve Quadratic Equations by Completing the Square (p. A31)

6 Solve Quadratic Equations Using the Quadratic Formula (p. A32)

7 Solve Equations Quadratic in Form (p. A35)

8 Solve Absolute Value Equations (p. A36)

9 Solve Equations by Factoring (p. A36)

An **equation in one variable** is a statement in which two expressions, at least one containing the variable, are equal. The expressions are called the **sides** of the equation. Since an equation is a statement, it may be true or false, depending on the value of the variable. Unless otherwise restricted, the admissible values of the variable are those in the domain of the variable. Those admissible values of the variable, if any, that result in a true statement are called **solutions**, or **roots**, of the equation. To **solve an equation** means to find all the solutions of the equation.

For example, the following are all equations in one variable, x:

$$x + 5 = 9 \qquad x^2 + 5x = 2x - 2 \qquad \frac{x^2 - 4}{x + 1} = 0 \qquad \sqrt{x^2 + 9} = 5$$

The first of these statements, $x + 5 = 9$, is true when $x = 4$ and false for any other choice of x. Thus, 4 is a solution of the equation $x + 5 = 9$. We also say that 4 **satisfies** the equation $x + 5 = 9$, because, when we substitute 4 for x, a true statement results.

Sometimes an equation will have more than one solution. For example, the equation

$$\frac{x^2 - 4}{x + 1} = 0$$

has $x = -2$ and $x = 2$ as solutions.

Usually, we will write the solution of an equation in set notation. This set is called the **solution set** of the equation. For example, the solution set of the equation $x^2 - 9 = 0$ is $\{-3, 3\}$.

Some equations have no real solution. For example, $x^2 + 9 = 5$ has no real solution, because there is no real number whose square when added to 9 equals 5.

An equation that is satisfied for every choice of the variable for which both sides are defined is called an **identity**. For example, the equation

$$3x + 5 = x + 3 + 2x + 2$$

is an identity, because this statement is true for any real number x.

Solving Equations Algebraically

One method for solving equations algebraically requires that a series of *equivalent equations* be developed from the original equation until an obvious solution results.

For example, all the following equations are equivalent, because each has only the solution $x = 5$:

$$2x + 3 = 13$$
$$2x = 10$$
$$x = 5$$

The question, though, is "How do I obtain an equivalent equation?" In general, there are five ways to do so.

Procedures That Result in Equivalent Equations

1. Interchange the two sides of the equation:

   Replace $3 = x$ by $x = 3$

2. Simplify the sides of the equation by combining like terms, eliminating parentheses, and so on:

   Replace $x + 2 + 6 = 2x + 3(x + 1)$

   by $x + 8 = 5x + 3$

3. Add or subtract the same expression on both sides of the equation:

   Replace $3x - 5 = 4$

   by $(3x - 5) + 5 = 4 + 5$

4. Multiply or divide both sides of the equation by the same nonzero expression:

   Replace $\dfrac{3x}{x - 1} = \dfrac{6}{x - 1}$ $x \neq 1$

   by $\dfrac{3x}{x - 1} \cdot (x - 1) = \dfrac{6}{x - 1} \cdot (x - 1)$

5. If one side of the equation is 0 and the other side can be factored, then we may use the Zero-Product Property and set each factor equal to 0:

   Replace $x(x - 3) = 0$

   by $x = 0$ or $x - 3 = 0$

WARNING Squaring both sides of an equation does not necessarily lead to an equivalent equation. ∎

Whenever it is possible to solve an equation in your head, do so. For example:

The solution of $2x = 8$ is $x = 4$.

The solution of $3x - 15 = 0$ is $x = 5$.

— Now Work PROBLEM 13

We now introduce specific types of equations that can be solved algebraically to obtain exact solutions. We start with **linear equations.**

1 Solve Linear Equations

Linear equations are equations such as

$$3x + 12 = 0 \qquad \frac{3}{4}x - \frac{1}{5} = 0 \qquad 0.62x - 0.3 = 0$$

A **linear equation in one variable** is equivalent to an equation of the form

$$ax + b = 0$$

where a and b are real numbers and $a \neq 0$.

Sometimes a linear equation is called a **first-degree equation,** because the left side is a polynomial in x of degree 1.

EXAMPLE 1	Solving a Linear Equation

Solve the equation: $3(x - 2) = 5(x - 1)$

Solution

$$
\begin{aligned}
3(x - 2) &= 5(x - 1) \\
3x - 6 &= 5x - 5 &&\text{Use the Distributive Property.} \\
3x - 6 - 5x &= 5x - 5 - 5x &&\text{Subtract 5x from each side.} \\
-2x - 6 &= -5 &&\text{Simplify.} \\
-2x - 6 + 6 &= -5 + 6 &&\text{Add 6 to each side.} \\
-2x &= 1 &&\text{Simplify.} \\
\frac{-2x}{-2} &= \frac{1}{-2} &&\text{Divide each side by } -2. \\
x &= -\frac{1}{2} &&\text{Simplify.}
\end{aligned}
$$

✓**Check** Let $x = -\dfrac{1}{2}$ in the expression in x on the left side of the equation and simplify. Let $x = -\dfrac{1}{2}$ in the expression in x on the right side of the equation and simplify. If the two expressions are equal, the solution checks.

$$3(x - 2) = 3\left(-\frac{1}{2} - 2\right) = 3\left(-\frac{5}{2}\right) = -\frac{15}{2}$$

$$5(x - 1) = 5\left(-\frac{1}{2} - 1\right) = 5\left(-\frac{3}{2}\right) = -\frac{15}{2}$$

Since the two expressions are equal, the solution $x = -\dfrac{1}{2}$ checks. The solution set is $\left\{-\dfrac{1}{2}\right\}$.

— Now Work PROBLEM 23

The next example illustrates the solution of an equation that does not appear to be linear, but leads to a linear equation upon simplification.

| **EXAMPLE 2** | **Solving an Equation That Leads to a Linear Equation** |

Solve the equation: $(2x - 1)(x - 1) = (x - 5)(2x - 5)$

Solution

$$(2x - 1)(x - 1) = (x - 5)(2x - 5)$$

$$2x^2 - 3x + 1 = 2x^2 - 15x + 25 \qquad \text{Multiply and combine like terms.}$$

$$2x^2 - 3x + 1 - 2x^2 = 2x^2 - 15x + 25 - 2x^2 \qquad \text{Subtract } 2x^2 \text{ from each side.}$$

$$-3x + 1 = -15x + 25 \qquad \text{Simplify.}$$

$$-3x + 1 - 1 = -15x + 25 - 1 \qquad \text{Subtract 1 from each side.}$$

$$-3x = -15x + 24 \qquad \text{Simplify.}$$

$$-3x + 15x = -15x + 24 + 15x \qquad \text{Add 15x to each side.}$$

$$12x = 24 \qquad \text{Simplify.}$$

$$\frac{12x}{12} = \frac{24}{12} \qquad \text{Divide each side by 12.}$$

$$x = 2 \qquad \text{Simplify.}$$

✓**Check** $\quad (2x - 1)(x - 1) = (2 \cdot 2 - 1)(2 - 1) = (3)(1) = 3$

$\qquad\qquad (x - 5)(2x - 5) = (2 - 5)(2 \cdot 2 - 5) = (-3)(-1) = 3$

Since the two expressions are equal, the solution checks. The solution set is $\{2\}$.

∎

✏️ **Now Work** PROBLEM 33

2 Solve Rational Equations

We now introduce another type of equation, the *rational equation*. A **rational equation** is an equation that contains a rational expression. Examples of rational equations are

$$\frac{3}{x + 1} = \frac{2}{x - 1} + 7 \quad \text{and} \quad \frac{x - 5}{x - 4} = \frac{3}{x + 2}$$

To solve a rational equation, multiply both sides of the equation by the least common multiple of the denominators of the rational expressions that make up the rational equation.

| **EXAMPLE 3** | **Solving a Rational Equation** |

Solve the equation: $\dfrac{3}{x - 2} = \dfrac{1}{x - 1} + \dfrac{7}{(x - 1)(x - 2)}$

Solution

First, we note that the domain of the variable is $\{x \,|\, x \neq 1, x \neq 2\}$. We clear the equation of rational expressions by multiplying both sides by the least common multiple of the denominators of the three rational expressions, $(x - 1)(x - 2)$.

$$\frac{3}{x - 2} = \frac{1}{x - 1} + \frac{7}{(x - 1)(x - 2)}$$

$$(x - 1)(x - 2)\frac{3}{x - 2} = (x - 1)(x - 2)\left[\frac{1}{x - 1} + \frac{7}{(x - 1)(x - 2)}\right] \qquad \begin{array}{l}\text{Multiply both sides by} \\ (x - 1)(x - 2). \text{ Cancel on the left.}\end{array}$$

$$3x - 3 = \cancel{(x-1)}(x-2)\frac{1}{\cancel{x-1}} + \cancel{(x-1)}\cancel{(x-2)}\frac{7}{\cancel{(x-1)}\cancel{(x-2)}}$$

Use the Distributive Property on each side; cancel on the right.

$$3x - 3 = (x - 2) + 7$$

$$3x - 3 = x + 5$$

Combine like terms.

$$2x = 8$$

Add 3 to each side; subtract x from each side.

$$x = 4$$

Divide each side by 2.

✓Check $\dfrac{3}{x - 2} = \dfrac{3}{4 - 2} = \dfrac{3}{2}$

$$\frac{1}{x - 1} + \frac{7}{(x - 1)(x - 2)} = \frac{1}{4 - 1} + \frac{7}{(4 - 1)(4 - 2)} = \frac{1}{3} + \frac{7}{3 \cdot 2} = \frac{2}{6} + \frac{7}{6} = \frac{9}{6} = \frac{3}{2}$$

Since the two expressions are equal, the solution $x = 4$ checks. The solution set is $\{4\}$.

Now Work PROBLEM 45

Quadratic Equations

Quadratic equations are equations such as

$$2x^2 + x + 8 = 0 \qquad 3x^2 - 5x = 0 \qquad x^2 - 9 = 0$$

A general definition is given next.

A **quadratic equation** is an equation equivalent to one of the form

$$ax^2 + bx + c = 0 \qquad \qquad \text{(1)}$$

where a, b, and c are real numbers and $a \neq 0$.

A quadratic equation written in the form $ax^2 + bx + c = 0$ is said to be in **standard form.**

Sometimes, a quadratic equation is called a **second-degree equation,** because the left side is a polynomial of degree 2. We shall discuss three algebraic ways of solving quadratic equations: by factoring, by completing the square, and by using the quadratic formula.

3 Solve Quadratic Equations by Factoring

When a quadratic equation is written in standard form, $ax^2 + bx + c = 0$, it may be possible to factor the expression on the left side as the product of two first-degree polynomials. Then, by setting each factor equal to 0 and solving the resulting linear equations, we obtain the *exact* solutions of the quadratic equation. This approach leads us to a basic premise in mathematics. Whenever a problem is encountered, use techniques that reduce the problem to one you already know how to solve. In this instance, we are reducing quadratic equations to linear equations using the technique of factoring.

Let's look at an example.

EXAMPLE 4 | Solving a Quadratic Equation by Factoring

Solve the equation: $x^2 = 12 - x$

Solution   We put the equation in standard form by adding $x - 12$ to each side:

$$x^2 = 12 - x$$
$$x^2 + x - 12 = 0$$

The left side of the equation may now be factored as

$$(x + 4)(x - 3) = 0$$

Then, by the Zero-Product Property, we have

$$x + 4 = 0 \quad \text{or} \quad x - 3 = 0$$
$$x = -4 \qquad\qquad x = 3$$

We leave the check to you. The solution set is $\{-4, 3\}$.

─────── **Now Work** PROBLEM 67

When the left side factors into two linear equations with the same solution, the quadratic equation is said to have a **repeated solution.** We also call this solution a **root of multiplicity 2,** or a **double root.**

| EXAMPLE 5 | Solving a Quadratic Equation by Factoring |

Solve the equation: $x^2 - 6x + 9 = 0$

Solution   This equation is already in standard form, and the left side can be factored:

$$x^2 - 6x + 9 = 0$$
$$(x - 3)(x - 3) = 0$$

so

$$x = 3 \quad \text{or} \quad x = 3$$

This equation has only the repeated solution 3. The solution set is $\{3\}$.

4 Solve Quadratic Equations Using the Square Root Method

Suppose that we wish to solve the quadratic equation

$$x^2 = p \tag{2}$$

where $p \geq 0$ is a nonnegative number. We proceed as in the earlier examples:

$$x^2 - p = 0 \qquad \text{Put in standard form.}$$
$$(x - \sqrt{p})(x + \sqrt{p}) = 0 \qquad \text{Factor (over the real numbers).}$$
$$x = \sqrt{p} \quad \text{or} \quad x = -\sqrt{p} \qquad \text{Solve.}$$

We have the following result:

$$\text{If } x^2 = p \text{ and } p \geq 0, \text{ then } x = \sqrt{p} \text{ or } x = -\sqrt{p}. \tag{3}$$

When statement (3) is used, it is called the **Square Root Method.** In statement (3), note that if $p > 0$ the equation $x^2 = p$ has two solutions, $x = \sqrt{p}$ and $x = -\sqrt{p}$. We usually abbreviate these solutions as $x = \pm\sqrt{p}$, read as "x equals plus or minus the square root of p." For example, the two solutions of the equation

$$x^2 = 4$$

are

$$x = \pm\sqrt{4}$$

and, since $\sqrt{4} = 2$, we have

$$x = \pm 2$$

The solution set is $\{-2, 2\}$.

| EXAMPLE 6 | **Solving Quadratic Equations by Using the Square Root Method** |

Solve each equation.

(a) $x^2 = 5$ (b) $(x - 2)^2 = 16$

Solution (a) $x^2 = 5$

$$x = \pm\sqrt{5} \qquad \text{Use the Square Root Method.}$$

$$x = \sqrt{5} \quad \text{or} \quad x = -\sqrt{5}$$

The solution set is $\left\{-\sqrt{5}, \sqrt{5}\right\}$.

(b) $(x - 2)^2 = 16$

$$x - 2 = \pm\sqrt{16} \qquad \text{Use the Square Root Method.}$$

$$x - 2 = \sqrt{16} \quad \text{or} \quad x - 2 = -\sqrt{16}$$

$$x - 2 = 4 \qquad\qquad x - 2 = -4$$

$$x = 6 \qquad\qquad\quad x = -2$$

The solution set is $\{-2, 6\}$.

 Now Work PROBLEM 95

5 Solve Quadratic Equations by Completing the Square

We now introduce the method of **completing the square.** The idea behind this method is to "adjust" the left side of a quadratic equation, $ax^2 + bx + c = 0$, so that it becomes a perfect square, that is, the square of a first-degree polynomial. For example, $x^2 + 6x + 9$ and $x^2 - 4x + 4$ are perfect squares because

$$x^2 + 6x + 9 = (x + 3)^2 \quad \text{and} \quad x^2 - 4x + 4 = (x - 2)^2$$

How do we "adjust" the left side? We do it by adding the appropriate number to create a perfect square. For example, to make $x^2 + 6x$ a perfect square, we add 9.

Let's look at several examples of completing the square when the coefficient of x^2 is 1.

Start	Add	Result
$x^2 + 4x$	4	$x^2 + 4x + 4 = (x + 2)^2$
$x^2 + 12x$	36	$x^2 + 12x + 36 = (x + 6)^2$
$x^2 - 6x$	9	$x^2 - 6x + 9 = (x - 3)^2$
$x^2 + x$	$\dfrac{1}{4}$	$x^2 + x + \dfrac{1}{4} = \left(x + \dfrac{1}{2}\right)^2$

Do you see the pattern? Provided that the coefficient of x^2 is 1, we complete the square by adding the square of one-half of the coefficient of x.

Start	Add	Result
$x^2 + mx$	$\left(\dfrac{m}{2}\right)^2$	$x^2 + mx + \left(\dfrac{m}{2}\right)^2 = \left(x + \dfrac{m}{2}\right)^2$

 Now Work PROBLEM 99

The next example illustrates how the procedure of completing the square can be used to solve a quadratic equation.

| EXAMPLE 7 | Solving a Quadratic Equation by Completing the Square |

Solve by completing the square: $x^2 + 5x + 4 = 0$

Solution We always begin this procedure by rearranging the equation so that the constant is on the right side.

$$x^2 + 5x + 4 = 0$$
$$x^2 + 5x = -4$$

Since the coefficient of x^2 is 1, we can complete the square on the left side by adding $\left(\frac{1}{2} \cdot 5\right)^2 = \frac{25}{4}$. Of course, in an equation, whatever we add to the left side also must be added to the right side. We add $\frac{25}{4}$ to *both* sides.

$$x^2 + 5x + \frac{25}{4} = -4 + \frac{25}{4} \qquad \text{Add } \frac{25}{4} \text{ to both sides.}$$

$$\left(x + \frac{5}{2}\right)^2 = \frac{9}{4} \qquad \text{Factor; simplify.}$$

$$x + \frac{5}{2} = \pm\sqrt{\frac{9}{4}} \qquad \text{Use the Square Root Method.}$$

$$x + \frac{5}{2} = \pm\frac{3}{2}$$

$$x = -\frac{5}{2} \pm \frac{3}{2}$$

$$x = -\frac{5}{2} + \frac{3}{2} = -1 \quad \text{or} \quad x = -\frac{5}{2} - \frac{3}{2} = -4$$

The solution set is $\{-4, -1\}$.

The solution of the equation in Example 7 can also be obtained by factoring. Rework Example 7 using factoring.

Now Work PROBLEM 105

6 Solve Quadratic Equations Using the Quadratic Formula

We can use the method of completing the square to obtain a general formula for solving the quadratic equation

$$ax^2 + bx + c = 0 \qquad a > 0$$

NOTE There is no loss in generality to assume that $a > 0$, since if $a < 0$ we can multiply both sides by -1 to obtain an equivalent equation with a positive leading coefficient. ∎

As in Example 7, we begin by rearranging the terms as

$$ax^2 + bx = -c$$

Since $a > 0$, we can divide both sides by a to get

$$x^2 + \frac{b}{a}x = -\frac{c}{a}$$

Now the coefficient of x^2 is 1. To complete the square on the left side, add the square of $\frac{1}{2}$ of the coefficient of x; that is, add

$$\left(\frac{1}{2} \cdot \frac{b}{a}\right)^2 = \frac{b^2}{4a^2}$$

to each side. Then

$$x^2 + \frac{b}{a}x + \frac{b^2}{4a^2} = \frac{b^2}{4a^2} - \frac{c}{a}$$

$$\left(x + \frac{b}{2a}\right)^2 = \frac{b^2 - 4ac}{4a^2} \qquad \frac{b^2}{4a^2} - \frac{c}{a} = \frac{b^2}{4a^2} - \frac{4ac}{4a^2} = \frac{b^2 - 4ac}{4a^2} \qquad \textbf{(4)}$$

Provided that $b^2 - 4ac \geq 0$, we now can use the Square Root Method to get

$$x + \frac{b}{2a} = \pm\sqrt{\frac{b^2 - 4ac}{4a^2}}$$

$$x + \frac{b}{2a} = \frac{\pm\sqrt{b^2 - 4ac}}{2a} \qquad \text{\textit{The square root of a quotient equals the quotient of the square roots. Also,} } \sqrt{4a^2} = 2a \text{ \textit{since} } a > 0.$$

$$x = -\frac{b}{2a} \pm \frac{\sqrt{b^2 - 4ac}}{2a} \qquad \text{\textit{Add} } -\frac{b}{2a} \text{ \textit{to both sides.}}$$

$$= \frac{-b \pm \sqrt{b^2 - 4ac}}{2a} \qquad \text{\textit{Combine the quotients on the right.}}$$

What if $b^2 - 4ac$ is negative? Then equation (4) states that the left expression (a real number squared) equals the right expression (a negative number). Since this occurrence is impossible for real numbers, we conclude that if $b^2 - 4ac < 0$ the quadratic equation has no *real* solution.[*]

We now state the *quadratic formula*.

THEOREM

Quadratic Formula

Consider the quadratic equation

$$ax^2 + bx + c = 0 \qquad a \neq 0$$

If $b^2 - 4ac < 0$, this equation has no real solution.
If $b^2 - 4ac \geq 0$, the real solution(s) of this equation is (are) given by the **quadratic formula.**

$$x = \frac{-b \pm \sqrt{b^2 - 4ac}}{2a}$$

The quantity $b^2 - 4ac$ is called the **discriminant** of the quadratic equation, because its value tells us whether the equation has real solutions. In fact, it also tells us how many solutions to expect.

Discriminant of a Quadratic Equation

For a quadratic equation $ax^2 + bx + c = 0$:

1. If $b^2 - 4ac > 0$, there are two unequal real solutions.
2. If $b^2 - 4ac = 0$, there is a repeated real solution, a root of multiplicity 2.
3. If $b^2 - 4ac < 0$, there is no real solution.

[*]We consider quadratic equations where $b^2 - 4ac$ is negative in Section A.6.

When asked to find the real solutions, if any, of a quadratic equation, always evaluate the discriminant first to see how many real solutions there are.

EXAMPLE 8	**Solving a Quadratic Equation by Using the Quadratic Formula**

Find the real solutions, if any, of the equation $3x^2 - 5x + 1 = 0$.

Solution The equation is in standard form, so we compare it to $ax^2 + bx + c = 0$ to find a, b, and c.

$$3x^2 - 5x + 1 = 0$$
$$ax^2 + bx + c = 0 \qquad a = 3, b = -5, c = 1$$

With $a = 3$, $b = -5$, and $c = 1$, we evaluate the discriminant $b^2 - 4ac$.

$$b^2 - 4ac = (-5)^2 - 4(3)(1) = 25 - 12 = 13$$

Since $b^2 - 4ac > 0$, there are two real solutions.
We use the quadratic formula with $a = 3$, $b = -5$, $c = 1$, and $b^2 - 4ac = 13$.

$$x = \frac{-b \pm \sqrt{b^2 - 4ac}}{2a} = \frac{-(-5) \pm \sqrt{13}}{2(3)} = \frac{5 \pm \sqrt{13}}{6}$$

The solution set is $\left\{ \dfrac{5 - \sqrt{13}}{6}, \dfrac{5 + \sqrt{13}}{6} \right\}$.

━━━━━➤ **Now Work** PROBLEM 111

EXAMPLE 9	**Solving a Quadratic Equation by Using the Quadratic Formula**

Find the real solutions, if any, of the equation

$$3x^2 + 2 = 4x$$

Solution The equation, as given, is not in standard form.

$$3x^2 + 2 = 4x$$
$$3x^2 - 4x + 2 = 0 \qquad \text{Subtract 4x from both sides to put the equation in standard form.}$$
$$ax^2 + bx + c = 0 \qquad \text{Compare to standard form.}$$

With $a = 3$, $b = -4$, and $c = 2$, we find that

$$b^2 - 4ac = (-4)^2 - 4(3)(2)$$
$$= 16 - 24$$
$$= -8$$

Since $b^2 - 4ac < 0$, the equation has no real solution.

━━━━━➤ **Now Work** PROBLEM 117

SUMMARY   Procedure for Solving a Quadratic Equation Algebraically

To solve a quadratic equation, first put it in standard form:

$$ax^2 + bx + c = 0$$

Then:

STEP 1:   Identify a, b, and c.

STEP 2:   Evaluate the discriminant, $b^2 - 4ac$.

STEP 3:   (a) If the discriminant is negative, the equation has no real solution.

(b) If the discriminant is nonnegative, determine whether the left side can be factored. If you can easily spot factors, use the factoring method to solve the equation. Otherwise, use the quadratic formula or the method of completing the square.

7 Solve Equations Quadratic in Form

The equation $x^4 + x^2 - 12 = 0$ is not quadratic in x, but it is quadratic in x^2. That is, if we let $u = x^2$, we get $u^2 + u - 12 = 0$, a quadratic equation. This equation can be solved for u and, in turn, by using $u = x^2$, we can find the solutions x of the original equation.

In general, if an appropriate substitution u transforms an equation into one of the form

$$au^2 + bu + c = 0 \qquad a \neq 0$$

then the original equation is called an **equation of the quadratic type** or an **equation quadratic in form.**

The difficulty of solving such an equation lies in the determination that the equation is, in fact, quadratic in form. After you are told an equation is quadratic in form, it is easy enough to see it, but some practice is needed to enable you to recognize them on your own.

EXAMPLE 10   Solving Equations That Are Quadratic in Form

Find the real solutions of the equation: $(x + 2)^2 + 11(x + 2) - 12 = 0$

Solution   For this equation, let $u = x + 2$. Then $u^2 = (x + 2)^2$, and the original equation,

$$(x + 2)^2 + 11(x + 2) - 12 = 0$$

becomes

$$u^2 + 11u - 12 = 0 \quad \text{Let } u = x + 2.$$

$$(u + 12)(u - 1) = 0 \quad \text{Factor.}$$

$$u = -12 \quad \text{or} \quad u = 1 \quad \text{Solve.}$$

But we want to solve for x. Because $u = x + 2$, we have

$$x + 2 = -12 \quad \text{or} \quad x + 2 = 1$$

$$x = -14 \qquad\qquad x = -1$$

✓Check $x = -14$: $(-14 + 2)^2 + 11(-14 + 2) - 12$

$$= (-12)^2 + 11(-12) - 12 = 144 - 132 - 12 = 0$$

$x = -1$: $(-1 + 2)^2 + 11(-1 + 2) - 12 = 1 + 11 - 12 = 0$

The original equation has the solution set $\{-14, -1\}$.

The idea should now be clear. If an equation contains an expression and that same expression squared, make a substitution for the expression. You may get a quadratic equation.

Now Work PROBLEM 85

8 Solve Absolute Value Equations

Recall that, on the real number line, the absolute value of a equals the distance from the origin to the point whose coordinate is a. For example, there are two points whose distance from the origin is 5 units, -5 and 5. Thus the equation $|x| = 5$ will have the solution set $\{-5, 5\}$. This leads to the following result:

Equations Involving Absolute Value

If a is a positive real number and if u is any algebraic expression, then

$$|u| = a \quad \text{is equivalent to} \quad u = a \text{ or } u = -a \qquad (5)$$

EXAMPLE 11 | **Solving an Equation Involving Absolute Value**

Solve the equation $|x + 4| = 13$.

Solution   This follows the form of equation (5), where $u = x + 4$. There are two possibilities.

$$x + 4 = 13 \quad \text{or} \quad x + 4 = -13$$
$$x = 9 \qquad\qquad x = -17$$

The solution set is $\{-17, 9\}$.

Now Work PROBLEM 49

---

9 Solve Equations by Factoring

We have already solved certain quadratic equations using factoring. Let's look at examples of other kinds of equations that can be solved by factoring.

EXAMPLE 12 | **Solving Equations by Factoring**

Solve the equation: $x^3 - x^2 - 4x + 4 = 0$

Solution   Do you recall the method of factoring by grouping? (If not, review Example 2(f) on p. A24.) We group the terms of $x^3 - x^2 - 4x + 4 = 0$ as follows:

$$(x^3 - x^2) - (4x - 4) = 0$$

Factor out x^2 from the first grouping and 4 from the second.

$$x^2(x - 1) - 4(x - 1) = 0$$

This reveals the common factor $(x - 1)$, so we have

$$(x^2 - 4)(x - 1) = 0$$
$$(x - 2)(x + 2)(x - 1) = 0 \qquad \text{Factor again.}$$
$$x - 2 = 0 \quad \text{or} \quad x + 2 = 0 \quad \text{or} \quad x - 1 = 0 \quad \text{Set each factor equal to 0.}$$
$$x = 2 \qquad\qquad x = -2 \qquad\qquad x = 1 \quad \text{Solve.}$$

The solution set is $\{-2, 1, 2\}$.

✓Check

$x = -2$: $(-2)^3 - (-2)^2 - 4(-2) + 4 = -8 - 4 + 8 + 4 = 0$ -2 is a solution.

$x = 1$: $1^3 - 1^2 - 4(1) + 4 = 1 - 1 - 4 + 4 = 0$ 1 is a solution.

$x = 2$: $2^3 - 2^2 - 4(2) + 4 = 8 - 4 - 8 + 4 = 0$ 2 is a solution.

■

━━━━━ **Now Work** PROBLEM 89

A.4 Assess Your Understanding

'Are You Prepared?' *Answers are given at the end of these exercises. If you get a wrong answer, read the pages listed in red.*

1. $|-10| = $ _____. (p. A5)

2. Factor $2x^2 - x - 3$. (pp. A23–A24)

3. The solution set of the equation $(x - 3)(3x + 5) = 0$ is _____. (p. A4)

4. *True or False:* $\sqrt{x^2} = |x|$. (pp. A9–A10)

Concepts and Vocabulary

5. Two equations that have the same solution set are called _____.

6. An equation that is satisfied for every choice of the variable for which both sides are defined is called a(n) _____.

7. *True or False:* The solution of the equation $3x - 8 = 0$ is $\dfrac{3}{8}$.

8. *True or False:* Some equations have no solution.

9. To complete the square of the expression $x^2 + 5x$, you would _____ the number _____.

10. The quantity $b^2 - 4ac$ is called the _____ of a quadratic equation. If it is _____, the equation has no real solution.

11. *True or False:* Quadratic equations always have two real solutions.

12. *True or False:* If the discriminant of a quadratic equation is positive, then the equation has two solutions that are negatives of one another.

Skill Building

In Problems 13–92, solve each equation.

13. $3x = 21$

14. $3x = -24$

15. $5x + 15 = 0$

16. $3x + 18 = 0$

17. $2x - 3 = 5$

18. $3x + 4 = -8$

19. $\dfrac{1}{3}x = \dfrac{5}{12}$

20. $\dfrac{2}{3}x = \dfrac{9}{2}$

21. $6 - x = 2x + 9$

22. $3 - 2x = 2 - x$

23. $2(3 + 2x) = 3(x - 4)$

24. $3(2 - x) = 2x - 1$

25. $8x - (2x + 1) = 3x - 10$

26. $5 - (2x - 1) = 10$

27. $\dfrac{1}{2}x - 4 = \dfrac{3}{4}x$

28. $1 - \dfrac{1}{2}x = 5$

29. $0.9t = 0.4 + 0.1t$

30. $0.9t = 1 + t$

31. $\dfrac{2}{y} + \dfrac{4}{y} = 3$

32. $\dfrac{4}{y} - 5 = \dfrac{5}{2y}$

33. $(x + 7)(x - 1) = (x + 1)^2$

34. $(x + 2)(x - 3) = (x - 3)^2$

35. $z(z^2 + 1) = 3 + z^3$

36. $w(4 - w^2) = 8 - w^3$

37. $x^2 = 9x$

38. $x^3 = x^2$

39. $t^3 - 9t^2 = 0$

40. $4z^3 - 8z^2 = 0$

41. $\dfrac{3}{2x - 3} = \dfrac{2}{x + 5}$

42. $\dfrac{-2}{x + 4} = \dfrac{-3}{x + 1}$

43. $(x + 2)(3x) = (x + 2)(6)$

44. $(x - 5)(2x) = (x - 5)(4)$

45. $\dfrac{2}{x - 2} = \dfrac{3}{x + 5} + \dfrac{10}{(x + 5)(x - 2)}$

46. $\dfrac{1}{2x+3} + \dfrac{1}{x-1} = \dfrac{1}{(2x+3)(x-1)}$ **47.** $|2x| = 6$ **48.** $|3x| = 12$

49. $|2x+3| = 5$ **50.** $|3x-1| = 2$ **51.** $|1-4t| = 5$

52. $|1-2z| = 3$ **53.** $|-2x| = 8$ **54.** $|-x| = 1$

55. $|-2|x = 4$ **56.** $|3|x = 9$ **57.** $|x-2| = -\dfrac{1}{2}$

58. $|2-x| = -1$ **59.** $|x^2-4| = 0$ **60.** $|x^2-9| = 0$

61. $|x^2-2x| = 3$ **62.** $|x^2+x| = 12$ **63.** $|x^2+x-1| = 1$

64. $|x^2+3x-2| = 2$ **65.** $x^2 = 4x$ **66.** $x^2 = -8x$

67. $z^2 + 4z - 12 = 0$ **68.** $v^2 + 7v + 12 = 0$ **69.** $2x^2 - 5x - 3 = 0$

70. $3x^2 + 5x + 2 = 0$ **71.** $x(x-7) + 12 = 0$ **72.** $x(x+1) = 12$

73. $4x^2 + 9 = 12x$ **74.** $25x^2 + 16 = 40x$ **75.** $6x - 5 = \dfrac{6}{x}$

76. $x + \dfrac{12}{x} = 7$ **77.** $\dfrac{4(x-2)}{x-3} + \dfrac{3}{x} = \dfrac{-3}{x(x-3)}$ **78.** $\dfrac{5}{x+4} = 4 + \dfrac{3}{x-2}$

79. $\dfrac{x}{x^2-1} - \dfrac{x+3}{x^2-x} = \dfrac{-3}{x^2+x}$ **80.** $\dfrac{x+1}{x^2+2x} - \dfrac{x+4}{x^2+x} = \dfrac{-3}{x^2+3x+2}$ **81.** $x^4 - 5x^2 + 4 = 0$

82. $x^4 - 10x^2 + 25 = 0$ **83.** $(x+2)^2 + 7(x+2) + 12 = 0$ **84.** $(2x+5)^2 - (2x+5) - 6 = 0$

85. $2(s+1)^2 - 5(s+1) = 3$ **86.** $3(1-y)^2 + 5(1-y) + 2 = 0$ **87.** $x^3 + x^2 - 20x = 0$

88. $x^3 + 6x^2 - 7x = 0$ **89.** $x^3 + x^2 - x - 1 = 0$ **90.** $x^3 + 4x^2 - x - 4 = 0$

91. $2x^3 + 4 = x^2 + 8x$ **92.** $3x^3 + 4x^2 = 27x + 36$

In Problems 93–98, solve each equation by the Square Root Method.

93. $x^2 = 25$ **94.** $x^2 = 36$ **95.** $(x-1)^2 = 4$

96. $(x+2)^2 = 1$ **97.** $(2x+3)^2 = 9$ **98.** $(3x-2)^2 = 4$

In Problems 99–104, what number should be added to complete the square of each expression?

99. $x^2 + 8x$ **100.** $x^2 - 4x$ **101.** $x^2 + \dfrac{1}{2}x$

102. $x^2 - \dfrac{1}{3}x$ **103.** $x^2 - \dfrac{2}{3}x$ **104.** $x^2 - \dfrac{2}{5}x$

In Problems 105–110, solve each equation by completing the square.

105. $x^2 + 4x = 21$ **106.** $x^2 - 6x = 13$ **107.** $x^2 - \dfrac{1}{2}x - \dfrac{3}{16} = 0$

108. $x^2 + \dfrac{2}{3}x - \dfrac{1}{3} = 0$ **109.** $3x^2 + x - \dfrac{1}{2} = 0$ **110.** $2x^2 - 3x - 1 = 0$

In Problems 111–122, find the real solutions, if any, of each equation. Use the quadratic formula.

111. $x^2 - 4x + 2 = 0$ **112.** $x^2 + 4x + 2 = 0$ **113.** $x^2 - 5x - 1 = 0$

114. $x^2 + 5x + 3 = 0$ **115.** $2x^2 - 5x + 3 = 0$ **116.** $2x^2 + 5x + 3 = 0$

117. $4y^2 - y + 2 = 0$ **118.** $4t^2 + t + 1 = 0$ **119.** $4x^2 = 1 - 2x$

120. $2x^2 = 1 - 2x$ **121.** $x^2 + \sqrt{3}x - 3 = 0$ **122.** $x^2 + \sqrt{2}x - 2 = 0$

In Problems 123–128, use the discriminant to determine whether each quadratic equation has two unequal real solutions, a repeated real solution, or no real solution without solving the equation.

123. $x^2 - 5x + 7 = 0$ **124.** $x^2 + 5x + 7 = 0$ **125.** $9x^2 - 30x + 25 = 0$

126. $25x^2 - 20x + 4 = 0$ **127.** $3x^2 + 5x - 8 = 0$ **128.** $2x^2 - 3x - 4 = 0$

Problems 129–134 list some formulas that occur in applications. Solve each formula for the indicated variable.

129. Electricity $\dfrac{1}{R} = \dfrac{1}{R_1} + \dfrac{1}{R_2}$ for R

130. Finance $A = P(1 + rt)$ for r

131. Mechanics $F = \dfrac{mv^2}{R}$ for R

132. Chemistry $PV = nRT$ for T

133. Mathematics $S = \dfrac{a}{1 - r}$ for r

134. Mechanics $v = -gt + v_0$ for t

---

135. Show that the sum of the roots of a quadratic equation is $-\dfrac{b}{a}$.

136. Show that the product of the roots of a quadratic equation is $\dfrac{c}{a}$.

137. Find k so that the equation $kx^2 + x + k = 0$ has a repeated real solution.

138. Find k so that the equation $x^2 - kx + 4 = 0$ has a repeated real solution.

139. Show that the real solutions of the equation $ax^2 + bx + c = 0$ are the negatives of the real solutions of the equation $ax^2 - bx + c = 0$. Assume that $b^2 - 4ac \geq 0$.

140. Show that the real solutions of the equation $ax^2 + bx + c = 0$ are the reciprocals of the real solutions of the equation $cx^2 + bx + a = 0$. Assume that $b^2 - 4ac \geq 0$.

Discussion and Writing

141. Which of the following pairs of equations are equivalent? Explain.
(a) $x^2 = 9;\quad x = 3$
(b) $x = \sqrt{9};\quad x = 3$
(c) $(x - 1)(x - 2) = (x - 1)^2;\quad x - 2 = x - 1$

142. The equation

$$\frac{5}{x + 3} + 3 = \frac{8 + x}{x + 3}$$

has no solution, yet when we go through the process of solving it we obtain $x = -3$. Write a brief paragraph to explain what causes this to happen.

143. Make up an equation that has no solution and give it to a fellow student to solve. Ask the fellow student to write a critique of your equation.

144. Describe three ways you might solve a quadratic equation. State your preferred method; explain why you chose it.

145. Explain the benefits of evaluating the discriminant of a quadratic equation before attempting to solve it.

146. Make up three quadratic equations: one having two distinct solutions, one having no real solution, and one having exactly one real solution.

147. The word *quadratic* seems to imply four (*quad*), yet a quadratic equation is an equation that involves a polynomial of degree 2. Investigate the origin of the term *quadratic* as it is used in the expression *quadratic equation*. Write a brief essay on your findings.

148. The equation $|x| = -2$ has no real solution. Why?

'Are You Prepared?' Answers

1. 10   **2.** $(2x - 3)(x + 1)$ **3.** $\left\{-\dfrac{5}{3}, 3\right\}$ **4.** True

---

A.5 Solving Equations Using a Graphing Utility

PREPARING FOR THIS SECTION *Before getting started, review the following:*

- Solving Equations Algebraically (Appendix, Section A.4, pp. A25–A37)

Now Work the 'Are You Prepared?' problems on page A41.

OBJECTIVE 1 Solve Equations Using a Graphing Utility (p. A40)

In this text, we present two methods for solving equations: algebraic and graphical. We shall see as we proceed through this book that some equations can be solved using algebraic techniques that result in *exact* solutions. For other equations, however, there are no algebraic techniques that lead to an exact solution. For such equations, a graphing utility can often be used to investigate possible solutions.

One goal of this text is to determine when equations can be solved algebraically. If an algebraic method for solving an equation exists, we shall use it to obtain an exact solution. A graphing utility can then be used to support the algebraic result. However, if an equation must be solved for which no algebraic techniques are available, a graphing utility will be used to obtain approximate solutions.

1 Solve Equations Using a Graphing Utility

When a graphing utility is used to solve an equation, usually *approximate* solutions are obtained. Unless otherwise stated, we shall follow the practice of giving approximate solutions as decimals *rounded to two decimal places*.

The ZERO (or ROOT) feature of a graphing utility can be used to find the solutions of an equation when one side of the equation is 0. In using this feature to solve equations, we make use of the fact that when the graph of an equation crosses or touches the x-axis then $y = 0$. For this reason, any value of x for which $y = 0$ will be a solution to the equation. That is, solving an equation for x when one side of the equation is 0 is equivalent to finding the x-intercept of the corresponding graph of the equation.

| EXAMPLE 1 | Using ZERO (or ROOT) to Approximate Solutions of an Equation |

Find the solution(s) of the equation $x^3 - x + 1 = 0$. Round answers to two decimal places.

Solution The solutions of the equation $x^3 - x + 1 = 0$ are the same as the x-intercepts of the graph of $Y_1 = x^3 - x + 1$. Figure 25 shows the graph.

From the graph there appears to be one x-intercept (solution to the equation) between -2 and -1. Using the ZERO (or ROOT) feature of our graphing utility, we determine that the x-intercept, and thus the solution to the equation, is $x = -1.32$ rounded to two decimal places. See Figure 26.

Figure 25

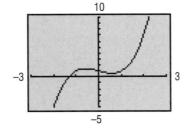

Figure 26

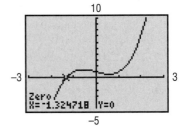

- **Now Work** PROBLEM 5

A second method for solving equations using a graphing utility involves the INTERSECT feature of the graphing utility. This feature is used most effectively when one side of the equation is not 0.

| EXAMPLE 2 | Using INTERSECT to Approximate Solutions of an Equation |

Find the solution(s) to the equation $4x^4 - 3 = 2x + 1$. Round answers to two decimal places.

Solution We begin by graphing each side of the equation as follows: graph $Y_1 = 4x^4 - 3$ and $Y_2 = 2x + 1$. See Figure 27.

At the point of intersection of the graphs, the value of the y-coordinate is the same. Thus, the x-coordinate of the point of intersection represents the solution to the equation. Do you see why? The INTERSECT feature on a graphing utility determines the point of intersection of the graphs. Using this feature, we find that the graphs intersect at $(-0.87, -0.73)$ and $(1.12, 3.23)$ rounded to two decimal places. See Figure 28(a) and (b). The solutions of the equation are $x = -0.87$ and $x = 1.12$ rounded to two decimal places.

Figure 27

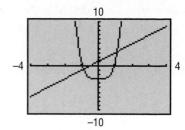

Figure 28

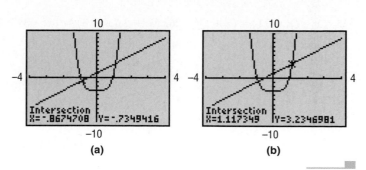

(a) (b)

Now Work PROBLEM 7

SUMMARY

Steps for Approximating Solutions of Equations Using ZERO (or ROOT)

Step 1: Write the equation in the form {expression in x} $= 0$.
Step 2: Graph $Y_1 = $ {expression in x}.
Step 3: Use ZERO (or ROOT) to determine each x-intercept of the graph.

Steps for Approximating Solutions of Equations Using INTERSECT

Step 1: Graph $Y_1 = $ {expression in x on the left side of the equation}
$Y_2 = $ {expression in x on the right side of the equation}
Step 2: Use INTERSECT to determine the x-coordinate of each point of intersection.

A.5 Assess Your Understanding

'Are You Prepared?' *Answers are given at the end of these exercises. If you get a wrong answer, read the pages listed in* red.

1. Solve the equation $2x^2 + 5x + 2 = 0$. (pp. A29–A30)

2. Solve the equation $2x + 3 = 4(x - 1) + 1$. (pp. A27–A28)

Concepts and Vocabulary

3. To solve an equation of the form {expression in x} $= 0$ using a graphing utility, we graph $Y_1 = $ {expression in x} and use _____ to determine each x-intercept of the graph.

4. *True or False* In using a graphing utility to solve an equation, exact solutions are always obtained.

Skill Building

In Problems 5–16, use a graphing utility to approximate the real solutions, if any, of each equation rounded to two decimal places. All solutions lie between −10 and 10.

5. $x^3 - 4x + 2 = 0$

6. $x^3 - 8x + 1 = 0$

7. $-2x^4 + 5 = 3x - 2$

8. $-x^4 + 1 = 2x^2 - 3$

9. $x^4 - 2x^3 + 3x - 1 = 0$

10. $3x^4 - x^3 + 4x^2 - 5 = 0$

11. $-x^3 - \frac{5}{3}x^2 + \frac{7}{2}x + 2 = 0$

12. $-x^4 + 3x^3 + \frac{7}{3}x^2 - \frac{15}{2}x + 2 = 0$

13. $-\frac{2}{3}x^4 - 2x^3 + \frac{5}{2}x = -\frac{2}{3}x^2 + \frac{1}{2}$

14. $\frac{1}{4}x^3 - 5x = \frac{1}{5}x^2 - 4$

15. $x^4 - 5x^2 + 2x + 11 = 0$

16. $-3x^4 + 8x^2 - 2x - 9 = 0$

In Problems 17–36, solve each equation algebraically. Verify your solution using a graphing utility.

17. $2(3 + 2x) = 3(x - 4)$

18. $3(2 - x) = 2x - 1$

19. $8x - (2x + 1) = 3x - 13$

20. $5 - (2x - 1) = 10 - x$

21. $\frac{x + 1}{3} + \frac{x + 2}{7} = 5$

22. $\frac{2x + 1}{3} + 16 = 3x$

23. $\frac{5}{y} + \frac{4}{y} = 3$

24. $\frac{4}{y} - 5 = \frac{18}{2y}$

25. $(x + 7)(x - 1) = (x + 1)^2$ **26.** $(x + 2)(x - 3) = (x - 3)^2$ **27.** $x^2 - 3x - 28 = 0$ **28.** $x^2 - 7x - 18 = 0$

29. $3x^2 = 4x + 4$ **30.** $5x^2 = 13x + 6$ **31.** $x^3 + x^2 - 4x - 4 = 0$ **32.** $x^3 + 2x^2 - 9x - 18 = 0$

33. $\sqrt{x + 1} = 4$ **34.** $\sqrt{x - 2} = 3$ **35.** $\dfrac{2}{x + 2} + \dfrac{3}{x - 1} = \dfrac{-8}{5}$ **36.** $\dfrac{1}{x + 1} - \dfrac{5}{x - 4} = \dfrac{21}{4}$

'Are You Prepared?' Answers

1. $\left\{ -2, -\dfrac{1}{2} \right\}$ **2.** $\{3\}$

A.6 Complex Numbers; Quadratic Equations in the Complex Number System

PREPARING FOR THIS SECTION *Before getting started, review the following:*

- Classification of Numbers (Appendix, Section A.1, p. A3)

Now Work the 'Are You Prepared?' problems on page A49.

OBJECTIVES 1 Add, Subtract, Multiply, and Divide Complex Numbers (p. A43)

2 Solve Quadratic Equations in the Complex Number System (p. A47)

Complex Numbers

One property of a real number is that its square is nonnegative (greater than or equal to 0). For example, there is no real number x for which

$$x^2 = -1$$

To remedy this situation, we introduce a number called the **imaginary unit,** which we denote by i, whose square is -1; that is,

$$i^2 = -1$$

Notice that each time a problem arises that can't be solved using an existing number system, we expanded the number system. If our universe were to consist only of integers, there would be no number x for which $2x = 1$. This unfortunate circumstance was remedied by introducing numbers such as $\dfrac{1}{2}$ and $\dfrac{2}{3}$, the *rational numbers*. If our universe were to consist only of rational numbers, there would be no x whose square equals 2. That is, there would be no number x for which $x^2 = 2$. To remedy this, we introduced numbers such as $\sqrt{2}$ and $\sqrt[3]{5}$, the *irrational numbers*. The *real numbers*, you will recall, consist of the rational numbers and the irrational numbers. Now, if our universe were to consist only of real numbers, then there would be no number x whose square is -1. To remedy this, we introduce a number i, whose square is -1.

In the progression outlined, each time that we encountered a situation that was unsuitable, we introduced a new number system to remedy this situation. And each new number system contained the earlier number system as a subset. The number system that results from introducing the number i is called the **complex number system.**

Complex numbers are numbers of the form $a + bi$, where a and b are real numbers. The real number a is called the **real part** of the number $a + bi$; the real number b is called the **imaginary part** of $a + bi$.

For example, the complex number $-5 + 6i$ has the real part -5 and the imaginary part 6.

When a complex number is written in the form $a + bi$, where a and b are real numbers, we say it is in **standard form.** However, if the imaginary part of a complex number is negative, such as in the complex number $3 + (-2)i$, we agree to write it instead in the form $3 - 2i$.

Also, the complex number $a + 0i$ is usually written simply as a. This serves to remind us that the real numbers are a subset of the complex numbers. The complex number $0 + bi$ is usually written as bi. Sometimes the complex number bi is called a **pure imaginary number.**

1 Add, Subtract, Multiply, and Divide Complex Numbers

Equality, addition, subtraction, and multiplication of complex numbers are defined so as to preserve the familiar rules of algebra for real numbers. Thus, two complex numbers are equal if and only if their real parts are equal and their imaginary parts are equal. That is,

Equality of Complex Numbers

$$a + bi = c + di \quad \text{if and only if} \quad a = c \text{ and } b = d \qquad \textbf{(1)}$$

Two complex numbers are added by forming the complex number whose real part is the sum of the real parts and whose imaginary part is the sum of the imaginary parts. That is,

Sum of Complex Numbers

$$(a + bi) + (c + di) = (a + c) + (b + d)i \qquad \textbf{(2)}$$

To subtract two complex numbers, we use this rule:

Difference of Complex Numbers

$$(a + bi) - (c + di) = (a - c) + (b - d)i \qquad \textbf{(3)}$$

EXAMPLE 1   Adding and Subtracting Complex Numbers

Figure 29

```
(3+5i)+(-2+3i)
              1+8i
(6+4i)-(3+6i)
              3-2i
```

(a) $(3 + 5i) + (-2 + 3i) = [3 + (-2)] + (5 + 3)i = 1 + 8i$

(b) $(6 + 4i) - (3 + 6i) = (6 - 3) + (4 - 6)i = 3 + (-2)i = 3 - 2i$

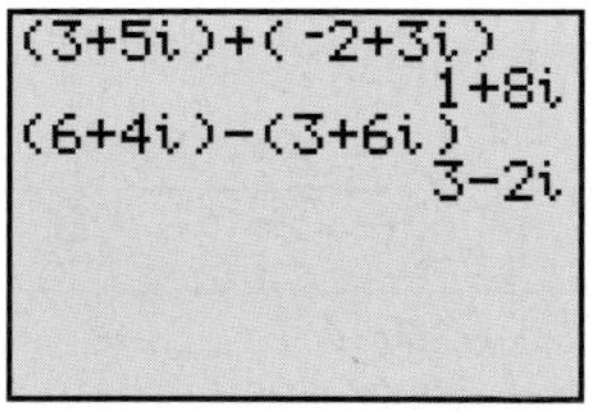

Some graphing calculators have the capability of handling complex numbers. For example, Figure 29 shows the results of Example 1 using a TI-84 Plus graphing calculator.

Now Work PROBLEM 13

Products of complex numbers are calculated as illustrated in Example 2.

EXAMPLE 2   Multiplying Complex Numbers

$$
\begin{aligned}
(5 + 3i) \cdot (2 + 7i) &= 5 \cdot (2 + 7i) + 3i(2 + 7i) &&\text{Distributive Property}\\
&= 10 + 35i + 6i + 21i^2 &&\text{Distributive Property}\\
&= 10 + 41i + 21(-1) &&i^2 = -1\\
&= -11 + 41i
\end{aligned}
$$

Based on the procedure of Example 2, we define the **product** of two complex numbers by the following formula:

Product of Complex Numbers

$$(a + bi) \cdot (c + di) = (ac - bd) + (ad + bc)i \qquad (4)$$

Figure 30

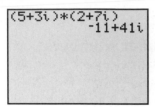

Do not bother to memorize formula (4). Instead, whenever it is necessary to multiply two complex numbers, follow the usual rules for multiplying two binomials, as in Example 2, remembering that $i^2 = -1$. For example,

$$(2i)(2i) = 4i^2 = -4$$
$$(2 + i)(1 - i) = 2 - 2i + i - i^2 = 3 - i$$

Graphing calculators may also be used to multiply complex numbers. Figure 30 shows the result obtained in Example 2 using a TI-84 Plus graphing calculator.

Now Work PROBLEM 19

Algebraic properties for addition and multiplication, such as the Commutative, Associative, and Distributive Properties, hold for complex numbers. However, the property that every nonzero complex number has a multiplicative inverse, or reciprocal, requires a closer look.

Conjugates

If $z = a + bi$ is a complex number, then its **conjugate,** denoted by $\bar{z}$, is defined as

$$\bar{z} = \overline{a + bi} = a - bi$$

For example, $\overline{2 + 3i} = 2 - 3i$ and $\overline{-6 - 2i} = -6 + 2i$.

EXAMPLE 3 | Multiplying a Complex Number by Its Conjugate

Find the product of the complex number $z = 3 + 4i$ and its conjugate $\bar{z}$.

Solution | Since $\bar{z} = 3 - 4i$, we have

$$z\bar{z} = (3 + 4i)(3 - 4i) = 9 - 12i + 12i - 16i^2 = 9 + 16 = 25$$

The result obtained in Example 3 has an important generalization.

THEOREM | The product of a complex number and its conjugate is a nonnegative real number. That is, if $z = a + bi$, then

$$z\bar{z} = a^2 + b^2 \qquad (5)$$

Proof If $z = a + bi$, then

$$z\bar{z} = (a + bi)(a - bi) = a^2 - abi + abi - (bi)^2 = a^2 - b^2i^2 = a^2 + b^2 \quad \blacksquare$$

To express the reciprocal of a nonzero complex number z in standard form, multiply the numerator and denominator of $\dfrac{1}{z}$ by its conjugate $\bar{z}$. That is, if $z = a + bi$ is a nonzero complex number, then

$$\frac{1}{a + bi} = \frac{1}{z} = \frac{1}{z} \cdot \frac{\bar{z}}{\bar{z}} = \frac{\bar{z}}{z\bar{z}} \underset{\uparrow}{=} \frac{a - bi}{a^2 + b^2} = \frac{a}{a^2 + b^2} - \frac{b}{a^2 + b^2}i$$

Use (5).

| EXAMPLE 4 | **Writing the Reciprocal of a Complex Number in Standard Form** |

Write $\dfrac{1}{3+4i}$ in standard form $a + bi$; that is, find the reciprocal of $3 + 4i$.

Solution Multiply the numerator and denominator of $\dfrac{1}{3+4i}$ by the conjugate of $3 + 4i$, the complex number $3 - 4i$. The result is

Figure 31

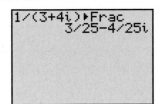

$$\frac{1}{3+4i} = \frac{1}{3+4i} \cdot \frac{3-4i}{3-4i} = \frac{3-4i}{9+16} = \frac{3}{25} - \frac{4}{25}i$$

A graphing calculator can be used to verify the result of Example 4. See Figure 31. To express the quotient of two complex numbers in standard form, we multiply the numerator and denominator of the quotient by the conjugate of the denominator.

| EXAMPLE 5 | **Writing the Quotient of a Complex Number in Standard Form** |

Write each of the following in standard form.

(a) $\dfrac{1+4i}{5-12i}$

(b) $\dfrac{2-3i}{4-3i}$

Solution (a) $\dfrac{1+4i}{5-12i} = \dfrac{1+4i}{5-12i} \cdot \dfrac{5+12i}{5+12i} = \dfrac{5+12i+20i+48i^2}{25+144}$

$$= \frac{-43+32i}{169} = \frac{-43}{169} + \frac{32}{169}i$$

(b) $\dfrac{2-3i}{4-3i} = \dfrac{2-3i}{4-3i} \cdot \dfrac{4+3i}{4+3i} = \dfrac{8+6i-12i-9i^2}{16+9}$

$$= \frac{17-6i}{25} = \frac{17}{25} - \frac{6}{25}i$$

Now Work PROBLEM 27

| EXAMPLE 6 | **Writing Other Expressions in Standard Form** |

If $z = 2 - 3i$ and $w = 5 + 2i$, write each of the following expressions in standard form.

(a) $\dfrac{z}{w}$

(b) $\overline{z+w}$

(c) $z + \overline{z}$

Solution (a) $\dfrac{z}{w} = \dfrac{z \cdot \overline{w}}{w \cdot \overline{w}} = \dfrac{(2-3i)(5-2i)}{(5+2i)(5-2i)} = \dfrac{10-4i-15i+6i^2}{25+4}$

$$= \frac{4-19i}{29} = \frac{4}{29} - \frac{19}{29}i$$

(b) $\overline{z+w} = \overline{(2-3i)+(5+2i)} = \overline{7-i} = 7+i$

(c) $z + \overline{z} = (2-3i) + (2+3i) = 4$

The conjugate of a complex number has certain general properties that we shall find useful later.

For a real number $a = a + 0i$, the conjugate is $\bar{a} = \overline{a + 0i} = a - 0i = a$. That is,

THEOREM   The conjugate of a real number is the real number itself.

Other properties that are direct consequences of the definition of the conjugate are given next. In each statement, z and w represent complex numbers.

THEOREM   The conjugate of the conjugate of a complex number is the complex number itself.

$$(\bar{\bar{z}}) = z \qquad (6)$$

The conjugate of the sum of two complex numbers equals the sum of their conjugates.

$$\overline{z + w} = \bar{z} + \bar{w} \qquad (7)$$

The conjugate of the product of two complex numbers equals the product of their conjugates.

$$\overline{z \cdot w} = \bar{z} \cdot \bar{w} \qquad (8)$$

We leave the proofs of equations (6), (7), and (8) as exercises. See Problems 88–90.

Powers of i

The powers of i follow a pattern that is useful to know.

$$i^1 = i \qquad\qquad i^5 = i^4 \cdot i = 1 \cdot i = i$$
$$i^2 = -1 \qquad\qquad i^6 = i^4 \cdot i^2 = -1$$
$$i^3 = i^2 \cdot i = -i \qquad\qquad i^7 = i^4 \cdot i^3 = -i$$
$$i^4 = i^2 \cdot i^2 = (-1)(-1) = 1 \qquad i^8 = i^4 \cdot i^4 = 1$$

And so on. The powers of i repeat with every fourth power.

EXAMPLE 7   Evaluating Powers of i

(a) $i^{27} = i^{24} \cdot i^3 = (i^4)^6 \cdot i^3 = 1^6 \cdot i^3 = -i$

(b) $i^{101} = i^{100} \cdot i^1 = (i^4)^{25} \cdot i = 1^{25} \cdot i = i$

▪

EXAMPLE 8   Writing the Power of a Complex Number in Standard Form

Write $(2 + i)^3$ in standard form.

Solution   We use the special product formula for $(x + a)^3$.

$$(x + a)^3 = x^3 + 3ax^2 + 3a^2x + a^3$$

Using this special product formula,

$$(2 + i)^3 = 2^3 + 3 \cdot i \cdot 2^2 + 3 \cdot i^2 \cdot 2 + i^3$$
$$= 8 + 12i + 6(-1) + (-i)$$
$$= 2 + 11i$$

NOTE If you did not remember the special product formula for $(x + a)^3$, you could find $(2 + i)^3$ by simplifying $(2 + i)^2(2 + i)$. ∎

▪

Now Work PROBLEMS 33 AND 41

2 Solve Quadratic Equations in the Complex Number System

Quadratic equations with a negative discriminant have no real number solution. However, if we extend our number system to allow complex numbers, quadratic equations will always have a solution. Since the solution to a quadratic equation involves the square root of the discriminant, we begin with a discussion of square roots of negative numbers.

DEFINITION

If N is a positive real number, we define the **principal square root of** $-N$, denoted by $\sqrt{-N}$, as

$$\sqrt{-N} = \sqrt{N}i$$

where i is the imaginary unit and $i^2 = -1$.

EXAMPLE 9 — Evaluating the Square Root of a Negative Number

(a) $\sqrt{-1} = \sqrt{1}i = i$ (b) $\sqrt{-16} = \sqrt{16}i = 4i$

(c) $\sqrt{-8} = \sqrt{8}i = 2\sqrt{2}i$

EXAMPLE 10 — Using the Square Root Method in the Complex Number System

Solve each equation in the complex number system.

(a) $x^2 = 4$ (b) $x^2 = -9$

Solution

(a) $x^2 = 4$

$$x = \pm\sqrt{4} = \pm 2$$

The equation has the solution set $\{-2, 2\}$.

(b) $x^2 = -9$

$$x = \pm\sqrt{-9} = \pm\sqrt{9}i = \pm 3i$$

The equation has the solution set $\{-3i, 3i\}$.

WARNING When working with square roots of negative numbers, do not set the square root of a product equal to the product of the square roots (which can be done with positive numbers). To see why, look at this calculation: We know that $\sqrt{100} = 10$. However, it is also true that $100 = (-25)(-4)$, so

$$10 = \sqrt{100}$$
$$= \sqrt{(-25)(-4)}$$
$$\neq \sqrt{-25}\,\sqrt{-4}$$

because $\sqrt{-25} \cdot \sqrt{-4}$

$$= \left(\sqrt{25}i\right)\left(\sqrt{4}i\right)$$
$$= (5i)(2i)$$
$$= 10i^2 = -10 \qquad ■$$

Now Work PROBLEM 53

Because we have defined the square root of a negative number, we can now restate the quadratic formula without restriction.

THEOREM

In the complex number system, the solutions of the quadratic equation $ax^2 + bx + c = 0$, where a, b, and c are real numbers and $a \neq 0$, are given by the formula

$$x = \frac{-b \pm \sqrt{b^2 - 4ac}}{2a} \qquad \text{(9)}$$

| EXAMPLE 11 | Solving a Quadratic Equation in the Complex Number System |

Solve the equation $x^2 - 4x + 8 = 0$ in the complex number system.

Solution Here $a = 1, b = -4, c = 8$, and $b^2 - 4ac = 16 - 4(1)(8) = -16$. Using equation (9), we find that

$$x = \frac{-(-4) \pm \sqrt{-16}}{2(1)} = \frac{4 \pm \sqrt{16}i}{2} = \frac{4 \pm 4i}{2} = 2 \pm 2i$$

The equation has the solution set $\{2 - 2i, 2 + 2i\}$.

✓ Check:

$$2 + 2i: \quad (2 + 2i)^2 - 4(2 + 2i) + 8 = 4 + 8i + 4i^2 - 8 - 8i + 8$$
$$= 4 - 4 = 0$$
$$2 - 2i: \quad (2 - 2i)^2 - 4(2 - 2i) + 8 = 4 - 8i + 4i^2 - 8 + 8i + 8$$
$$= 4 - 4 = 0$$

Figure 32

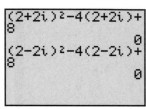

Figure 32 shows the check of the solution using a TI-84 Plus graphing calculator. Graph $Y_1 = x^2 - 4x + 8$. How many x-intercepts are there?

Now Work PROBLEM 59

The discriminant $b^2 - 4ac$ of a quadratic equation still serves as a way to determine the character of the solutions.

Character of the Solutions of a Quadratic Equation

In the complex number system, consider a quadratic equation $ax^2 + bx + c = 0$ with real coefficients.

1. If $b^2 - 4ac > 0$, the equation has two unequal real solutions.
2. If $b^2 - 4ac = 0$, the equation has a repeated real solution, a double root.
3. If $b^2 - 4ac < 0$, the equation has two complex solutions that are not real. The solutions are conjugates of each other.

The third conclusion is a consequence of the fact that if $b^2 - 4ac = -N < 0$ then, by the quadratic formula, the solutions are

$$x = \frac{-b + \sqrt{b^2 - 4ac}}{2a} = \frac{-b + \sqrt{-N}}{2a} = \frac{-b + \sqrt{N}i}{2a} = \frac{-b}{2a} + \frac{\sqrt{N}}{2a}i$$

and

$$x = \frac{-b - \sqrt{b^2 - 4ac}}{2a} = \frac{-b - \sqrt{-N}}{2a} = \frac{-b - \sqrt{N}i}{2a} = \frac{-b}{2a} - \frac{\sqrt{N}}{2a}i$$

which are conjugates of each other.

| EXAMPLE 12 | Determining the Character of the Solutions of a Quadratic Equation |

Without solving, determine the character of the solutions of each equation.

(a) $3x^2 + 4x + 5 = 0$ (b) $2x^2 + 4x + 1 = 0$

(c) $9x^2 - 6x + 1 = 0$

Solution (a) Here $a = 3, b = 4$, and $c = 5$, so $b^2 - 4ac = 4^2 - 4(3)(5) = -44$. The solutions are two complex numbers that are not real and are conjugates of each other.

(b) Here $a = 2, b = 4$, and $c = 1$, so $b^2 - 4ac = 4^2 - 4(2)(1) = 8$. The solutions are two unequal real numbers.

(c) Here $a = 9, b = -6$, and $c = 1$, so $b^2 - 4ac = (-6)^2 - 4(9)(1) = 0$. The solution is a repeated real number, that is, a double root.

Now Work PROBLEM 73

A.6 Assess Your Understanding

'Are You Prepared?' *Answers are given at the end of these exercises. If you get a wrong answer, read the pages listed in red.*

1. Name the integers and the rational numbers in the set $\left\{-3, 0, \sqrt{2}, \dfrac{6}{5}, \pi\right\}$. (p. A3)

2. *True or False* Rational numbers and irrational numbers are in the set of real numbers. (p. A3)

3. *True or False* All integers are rational numbers. (p. A3)

Concepts and Vocabulary

4. In the complex number $5 + 2i$, the number 5 is called the _____ part; the number 2 is called the _____ part; the number i is called the _____ _____.

5. The equation $x^2 = -4$ has the solution set _____.

6. *True or False* The conjugate of $2 + 5i$ is $-2 - 5i$.

7. *True or False* All real numbers are complex numbers.

8. *True or False* If $2 - 3i$ is a solution of a quadratic equation with real coefficients, then $-2 + 3i$ is also a solution.

Skill Building

In Problems 9–46, write each expression in the standard form $a + bi$. Verify your results using a graphing utility.

9. $(2 - 3i) + (6 + 8i)$

10. $(4 + 5i) + (-8 + 2i)$

11. $(-3 + 2i) - (4 - 4i)$

12. $(3 - 4i) - (-3 - 4i)$

13. $(2 - 5i) - (8 + 6i)$

14. $(-8 + 4i) - (2 - 2i)$

15. $3(2 - 6i)$

16. $-4(2 + 8i)$

17. $2i(2 - 3i)$

18. $3i(-3 + 4i)$

19. $(3 - 4i)(2 + i)$

20. $(5 + 3i)(2 - i)$

21. $(-6 + i)(-6 - i)$

22. $(-3 + i)(3 + i)$

23. $\dfrac{10}{3 - 4i}$

24. $\dfrac{13}{5 - 12i}$

25. $\dfrac{2 + i}{i}$

26. $\dfrac{2 - i}{-2i}$

27. $\dfrac{6 - i}{1 + i}$

28. $\dfrac{2 + 3i}{1 - i}$

29. $\left(\dfrac{1}{2} + \dfrac{\sqrt{3}}{2}i\right)^2$

30. $\left(\dfrac{\sqrt{3}}{2} - \dfrac{1}{2}i\right)^2$

31. $(1 + i)^2$

32. $(1 - i)^2$

33. i^{23}

34. i^{14}

35. i^{-15}

36. i^{-23}

37. $i^6 - 5$

38. $4 + i^3$

39. $6i^3 - 4i^5$

40. $4i^3 - 2i^2 + 1$

41. $(1 + i)^3$

42. $(3i)^4 + 1$

43. $i^7(1 + i^2)$

44. $2i^4(1 + i^2)$

45. $i^6 + i^4 + i^2 + 1$

46. $i^7 + i^5 + i^3 + i$

In Problems 47–52, perform the indicated operations and express your answer in the form a + bi.

47. $\sqrt{-4}$

48. $\sqrt{-9}$

49. $\sqrt{-25}$

50. $\sqrt{-64}$

51. $\sqrt{(3 + 4i)(4i - 3)}$

52. $\sqrt{(4 + 3i)(3i - 4)}$

In Problems 53–72, solve each equation in the complex number system. Check your results using a graphing utility.

53. $x^2 + 4 = 0$

54. $x^2 - 4 = 0$

55. $x^2 - 16 = 0$

56. $x^2 + 25 = 0$

57. $x^2 - 6x + 13 = 0$

58. $x^2 + 4x + 8 = 0$

59. $x^2 - 6x + 10 = 0$

60. $x^2 - 2x + 5 = 0$

61. $8x^2 - 4x + 1 = 0$

62. $10x^2 + 6x + 1 = 0$

63. $5x^2 + 2x + 1 = 0$

64. $13x^2 + 6x + 1 = 0$

65. $x^2 + x + 1 = 0$

66. $x^2 - x + 1 = 0$

67. $x^3 - 8 = 0$

68. $x^3 + 27 = 0$

69. $x^4 - 16 = 0$

70. $x^4 - 1 = 0$

71. $x^4 + 13x^2 + 36 = 0$

72. $x^4 + 3x^2 - 4 = 0$

In Problems 73–78, without solving, determine the character of the solutions of each equation. Verify your answer using a graphing utility.

73. $3x^2 - 3x + 4 = 0$

74. $2x^2 - 4x + 1 = 0$

75. $2x^2 + 3x - 4 = 0$

76. $x^2 + 2x + 6 = 0$

77. $9x^2 - 12x + 4 = 0$

78. $4x^2 + 12x + 9 = 0$

Applications and Extensions

79. $2 + 3i$ is a solution of a quadratic equation with real coefficients. Find the other solution.

80. $4 - i$ is a solution of a quadratic equation with real coefficients. Find the other solution.

In Problems 81–84, $z = 3 - 4i$ and $w = 8 + 3i$. Write each expression in the standard form a + bi.

81. $z + \bar{z}$

82. $w - \bar{w}$

83. $z\bar{z}$

84. $\overline{z - w}$

85. Electrical Circuits The impedance Z, in ohms, of a circuit element is defined as the ratio of the phasor voltage V, in volts, across the element to the phasor current I, in amperes, through the elements. That is, $Z = \dfrac{V}{I}$. If the voltage across a circuit element is $18 + i$ volts and the current through the element is $3 - 4i$ amperes, determine the impedance.

86. Parallel Circuits In an ac circuit with two parallel pathways, the total impedance Z, in ohms, satisfies the formula $\dfrac{1}{Z} = \dfrac{1}{Z_1} + \dfrac{1}{Z_2}$, where Z_1 is the impedance of the first pathway and Z_2 is the impedance of the second pathway. Determine the total impedance if the impedances of the two pathways are $Z_1 = 2 + i$ ohms and $Z_2 = 4 - 3i$ ohms.

87. Use $z = a + bi$ to show that $z + \bar{z} = 2a$ and that $z - \bar{z} = 2bi$.

88. Use $z = a + bi$ to show that $(\bar{\bar{z}}) = z$.

89. Use $z = a + bi$ and $w = c + di$ to show that $\overline{z + w} = \bar{z} + \bar{w}$.

90. Use $z = a + bi$ and $w = c + di$ to show that $\overline{z \cdot w} = \bar{z} \cdot \bar{w}$.

Discussion and Writing

91. Explain to a friend how you would add two complex numbers and how you would multiply two complex numbers. Explain any differences in the two explanations.

92. Write a brief paragraph that compares the method used to rationalize denominators and the method used to write the quotient of two complex numbers in standard form.

93. Use an Internet search engine to investigate the origins of complex numbers. Write a paragraph describing what you find and present it to the class.

94. Explain how the method of multiplying two complex numbers is related to multiplying two binomials.

95. What Went Wrong? A student multiplied $\sqrt{-9}$ and $\sqrt{-9}$ as follows:

$$\sqrt{-9} \cdot \sqrt{-9} = \sqrt{(-9)(-9)}$$
$$= \sqrt{81}$$
$$= 9$$

The instructor marked the problem incorrect. Why?

'Are You Prepared?' Answers

1. Integers: $\{-3, 0\}$; rational numbers: $\left\{-3, 0, \dfrac{6}{5}\right\}$

2. True

3. True

A.7 Interval Notation; Solving Inequalities

PREPARING FOR THIS SECTION *Before getting started, review the following:*

- Algebra Essentials (Appendix, Section A.1, pp. A4–A6)

Now Work the 'Are You Prepared?' problems on page A58.

OBJECTIVES **1** Use Interval Notation (p. A51)
2 Use Properties of Inequalities (p. A52)
3 Solve Inequalities (p. A54)
4 Solve Combined Inequalities (p. A55)
5 Solve Inequalities Involving Absolute Value (p. A56)

Suppose that a and b are two real numbers and $a < b$. We shall use the notation $a < x < b$ to mean that x is a number *between* a and b. The expression $a < x < b$ is equivalent to the two inequalities $a < x$ and $x < b$. Similarly, the expression $a \leq x \leq b$ is equivalent to the two inequalities $a \leq x$ and $x \leq b$. The remaining two possibilities, $a \leq x < b$ and $a < x \leq b$, are defined similarly.

Although it is acceptable to write $3 \geq x \geq 2$, it is preferable to reverse the inequality symbols and write instead $2 \leq x \leq 3$ so that, as you read from left to right, the values go from smaller to larger.

A statement such as $2 \leq x \leq 1$ is false because there is no number x for which $2 \leq x$ and $x \leq 1$. Finally, we never mix inequality symbols, as in $2 \leq x \geq 3$.

1 Use Interval Notation

Let a and b represent two real numbers with $a < b$.

DEFINITION

A **closed interval,** denoted by **[a, b],** consists of all real numbers x for which $a \leq x \leq b$.

An **open interval,** denoted by **(a, b),** consists of all real numbers x for which $a < x < b$.

The **half-open,** or **half-closed, intervals** are **(a, b],** consisting of all real numbers x for which $a < x \leq b$, and **[a, b),** consisting of all real numbers x for which $a \leq x < b$.

In each of these definitions, a is called the **left endpoint** and b the **right endpoint** of the interval.

The symbol ∞ (read as "infinity") is not a real number, but a notational device used to indicate unboundedness in the positive direction. The symbol $-\infty$ (read as "negative infinity") also is not a real number, but a notational device used to indicate unboundedness in the negative direction. Using the symbols ∞ and $-\infty$, we can define five other kinds of intervals:

[a, ∞)	Consists of all real numbers x for which $x \geq a$
(a, ∞)	Consists of all real numbers x for which $x > a$
(−∞, a]	Consists of all real numbers x for which $x \leq a$
(−∞, a)	Consists of all real numbers x for which $x < a$
(−∞, ∞)	Consists of all real numbers x

Note that ∞ and $-\infty$ are never included as endpoints, since neither is a real number.

Table 1 summarizes interval notation, corresponding inequality notation, and their graphs.

Table 1

Interval	Inequality	Graph
The open interval (a, b)	$a < x < b$	
The closed interval $[a, b]$	$a \le x \le b$	
The half-open interval $[a, b)$	$a \le x < b$	
The half-open interval $(a, b]$	$a < x \le b$	
The interval $[a, \infty)$	$x \ge a$	
The interval (a, ∞)	$x > a$	
The interval $(-\infty, a]$	$x \le a$	
The interval $(-\infty, a)$	$x < a$	
The interval $(-\infty, \infty)$	All real numbers	

EXAMPLE 1 **Writing Inequalities Using Interval Notation**

Write each inequality using interval notation.

(a) $1 \le x \le 3$ (b) $-4 < x < 0$ (c) $x > 5$ (d) $x \le 1$

Solution
(a) $1 \le x \le 3$ describes all numbers x between 1 and 3, inclusive. In interval notation, we write $[1, 3]$.
(b) In interval notation, $-4 < x < 0$ is written $(-4, 0)$.
(c) $x > 5$ consists of all numbers x greater than 5. In interval notation, we write $(5, \infty)$.
(d) In interval notation, $x \le 1$ is written $(-\infty, 1]$.

EXAMPLE 2 **Writing Intervals Using Inequality Notation**

Write each interval as an inequality involving x.

(a) $[1, 4)$ (b) $(2, \infty)$ (c) $[2, 3]$ (d) $(-\infty, -3]$

Solution
(a) $[1, 4)$ consists of all numbers x for which $1 \le x < 4$.
(b) $(2, \infty)$ consists of all numbers x for which $x > 2$.
(c) $[2, 3]$ consists of all numbers x for which $2 \le x \le 3$.
(d) $(-\infty, -3]$ consists of all numbers x for which $x \le -3$.

Now Work PROBLEMS 11, 23, AND 31

2 Use Properties of Inequalities

The product of two positive real numbers is positive, the product of two negative real numbers is positive, and the product of 0 and 0 is 0. For any real number a, the value of a^2 is 0 or positive; that is, a^2 is nonnegative. This is called the **nonnegative property.**

In Words
The square of a real number is never negative.

Nonnegative Property

For any real number a,

$$a^2 \ge 0 \tag{1}$$

If we add the same number to both sides of an inequality, we obtain an equivalent inequality. For example, since $3 < 5$, then $3 + 4 < 5 + 4$ or $7 < 9$. This is called the **addition property** of inequalities.

Addition Property of Inequalities

For real numbers a, b, and c,

$$\text{If } a < b, \text{ then } a + c < b + c. \tag{2a}$$

$$\text{If } a > b, \text{ then } a + c > b + c. \tag{2b}$$

The addition property states that the sense, or direction, of an inequality remains unchanged if the same number is added to each side.

Now let's see what happens if we multiply each side of an inequality by a nonzero number. We begin with $3 < 7$ and multiply each side by 2. The numbers 6 and 14 that result obey the inequality $6 < 14$.

Now start with $9 > 2$ and multiply each side by -4. The numbers -36 and -8 that result obey the inequality $-36 < -8$.

Note that the effect of multiplying both sides of $9 > 2$ by the negative number -4 is that the direction of the inequality symbol is reversed. We are led to the following general **multiplication properties** for inequalities:

Multiplication Properties for Inequalities

For real numbers a, b, and c,

$$\text{If } a < b \text{ and if } c > 0, \text{ then } ac < bc.$$
$$\text{If } a < b \text{ and if } c < 0, \text{ then } ac > bc. \tag{3a}$$

$$\text{If } a > b \text{ and if } c > 0, \text{ then } ac > bc.$$
$$\text{If } a > b \text{ and if } c < 0, \text{ then } ac < bc. \tag{3b}$$

In Words

Multiplying by a negative number reverses the inequality.

The multiplication properties state that the sense, or direction, of an inequality *remains the same* if each side is multiplied by a *positive* real number, whereas the direction is *reversed* if each side is multiplied by a *negative* real number.

| EXAMPLE 3 | Multiplication Property of Inequalities |

(a) If $2x < 6$, then $\dfrac{1}{2}(2x) < \dfrac{1}{2}(6)$ or $x < 3$.

(b) If $\dfrac{x}{-3} > 12$, then $-3\left(\dfrac{x}{-3}\right) < -3(12)$ or $x < -36$.

(c) If $-4x < -8$, then $\dfrac{-4x}{-4} > \dfrac{-8}{-4}$ or $x > 2$.

(d) If $-x > 8$, then $(-1)(-x) < (-1)(8)$ or $x < -8$.

Now Work PROBLEM 45

The **reciprocal property** states that the reciprocal of a positive real number is positive and that the reciprocal of a negative real number is negative.

Reciprocal Property for Inequalities

$$\text{If } a > 0, \text{ then } \frac{1}{a} > 0 \qquad \text{If } \frac{1}{a} > 0, \text{ then } a > 0 \qquad \textbf{(4a)}$$

$$\text{If } a < 0, \text{ then } \frac{1}{a} < 0 \qquad \text{If } \frac{1}{a} < 0, \text{ then } a < 0 \qquad \textbf{(4b)}$$

3 Solve Inequalities

An **inequality in one variable** is a statement involving two expressions, at least one containing the variable, separated by one of the inequality symbols $<$, $\leq$, $>$, or $\geq$. To **solve an inequality** means to find all values of the variable for which the statement is true. These values are called **solutions** of the inequality.

For example, the following are all inequalities involving one variable x:

$$x + 5 < 8 \qquad 2x - 3 \geq 4 \qquad x^2 - 1 \leq 3 \qquad \frac{x+1}{x-2} > 0$$

As with equations, one method for solving an inequality is to replace it by a series of equivalent inequalities until an inequality with an obvious solution, such as $x < 3$, is obtained. We obtain equivalent inequalities by applying some of the same properties as those used to find equivalent equations. The addition property and the multiplication properties form the basis for the following procedures.

Procedures That Leave the Inequality Symbol Unchanged

1. Simplify both sides of the inequality by combining like terms and eliminating parentheses:

$$\begin{aligned} \text{Replace} \quad & (x + 2) + 6 > 2x + 5(x + 1) \\ \text{by} \quad & x + 8 > 7x + 5 \end{aligned}$$

2. Add or subtract the same expression on both sides of the inequality:

$$\begin{aligned} \text{Replace} \quad & 3x - 5 < 4 \\ \text{by} \quad & (3x - 5) + 5 < 4 + 5 \end{aligned}$$

3. Multiply or divide both sides of the inequality by the same positive expression:

$$\text{Replace} \quad 4x > 16 \quad \text{by} \quad \frac{4x}{4} > \frac{16}{4}$$

Procedures That Reverse the Sense or Direction of the Inequality Symbol

1. Interchange the two sides of the inequality:

$$\text{Replace} \quad 3 < x \quad \text{by} \quad x > 3$$

2. Multiply or divide both sides of the inequality by the same *negative* expression:

$$\text{Replace} \quad -2x > 6 \quad \text{by} \quad \frac{-2x}{-2} < \frac{6}{-2}$$

As the examples that follow illustrate, we solve inequalities using many of the same steps that we would use to solve equations. In writing the solution of an inequality, we may use either set notation or interval notation, whichever is more convenient.

EXAMPLE 4 | **Solving an Inequality**

Solve the inequality: $4x + 7 \geq 2x - 3$
Graph the solution set.

Solution

$$4x + 7 \geq 2x - 3$$

$4x + 7 - 7 \geq 2x - 3 - 7$ Subtract 7 from both sides.

$4x \geq 2x - 10$ Simplify.

$4x - 2x \geq 2x - 10 - 2x$ Subtract 2x from both sides.

$2x \geq -10$ Simplify.

$\dfrac{2x}{2} \geq \dfrac{-10}{2}$ Divide both sides by 2. (The direction of the inequality symbol is unchanged.)

$x \geq -5$ Simplify.

Figure 33

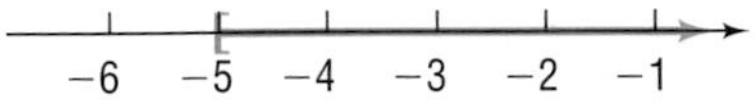

The solution set is $\{x \mid x \geq -5\}$ or, using interval notation, all numbers in the interval $[-5, \infty)$. See Figure 33 for the graph.

Now Work PROBLEM 57

4 Solve Combined Inequalities

EXAMPLE 5 | **Solving Combined Inequalities**

Solve the inequality: $-5 < 3x - 2 < 1$
Graph the solution set.

Solution

Recall that the inequality

$$-5 < 3x - 2 < 1$$

is equivalent to the two inequalities

$$-5 < 3x - 2 \quad \text{and} \quad 3x - 2 < 1$$

We will solve each of these inequalities separately.

$-5 < 3x - 2$		$3x - 2 < 1$
$-5 + 2 < 3x - 2 + 2$ Add 2 to both sides.	$3x - 2 + 2 < 1 + 2$
$-3 < 3x$ Simplify.	$3x < 3$
$\dfrac{-3}{3} < \dfrac{3x}{3}$ Divide both sides by 3.	$\dfrac{3x}{3} < \dfrac{3}{3}$
$-1 < x$ Simplify.	$x < 1$

The solution set of the original pair of inequalities consists of all x for which

$$-1 < x \quad \text{and} \quad x < 1$$

Figure 34

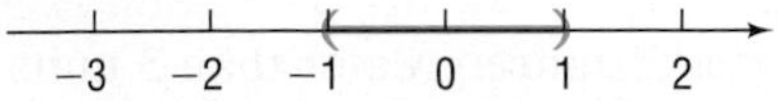

This may be written more compactly as $\{x \mid -1 < x < 1\}$. In interval notation, the solution is $(-1, 1)$. See Figure 34 for the graph.

We observe in the preceding process that the two inequalities we solved required exactly the same steps. A shortcut to solving the original inequality algebraically is to deal with the two inequalities at the same time, as follows:

$$-5 < \quad 3x - 2 \quad < 1$$
$$-5 + 2 < 3x - 2 + 2 < 1 + 2 \quad \text{Add 2 to each part.}$$
$$-3 < \quad 3x \quad < 3 \quad \text{Simplify.}$$
$$\frac{-3}{3} < \quad \frac{3x}{3} \quad < \frac{3}{3} \quad \text{Divide each part by 3.}$$
$$-1 < \quad x \quad < 1 \quad \text{Simplify.}$$

━━━━▶ **Now Work** PROBLEM 73

| EXAMPLE 6 | **Using the Reciprocal Property to Solve an Inequality** |

Solve the inequality: $(4x - 1)^{-1} > 0$
Graph the solution set.

Solution Since $(4x - 1)^{-1} = \dfrac{1}{4x - 1}$ and since the Reciprocal Property states that when $\dfrac{1}{a} > 0$ then $a > 0$, we have

$$(4x - 1)^{-1} > 0$$
$$\frac{1}{4x - 1} > 0$$
$$4x - 1 > 0 \quad \text{Reciprocal Property}$$
$$4x > 1$$
$$x > \frac{1}{4}$$

Figure 35

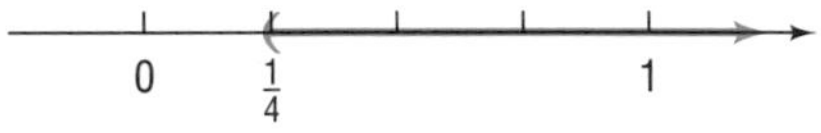

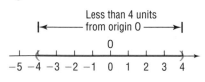

The solution set is $\left\{ x \middle| x > \dfrac{1}{4} \right\}$, that is, all x in the interval $\left(\dfrac{1}{4}, \infty \right)$. Figure 35 illustrates the graph.

━━━━▶ **Now Work** PROBLEM 83

---

5 Solve Inequalities Involving Absolute Value

| EXAMPLE 7 | **Solving an Inequality Involving Absolute Value** |

Solve the inequality $|x| < 4$, and graph the solution set.

Solution

Figure 36

Less than 4 units
from origin 0

0

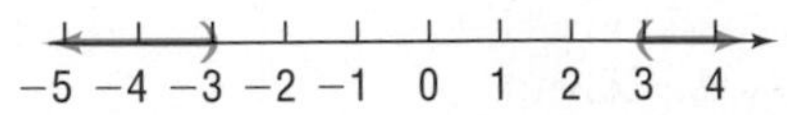

-5 -4 -3 -2 -1 0 1 2 3 4

We are looking for all points whose coordinate x is a distance less than 4 units from the origin. See Figure 36 for an illustration. Because any x between -4 and 4 satisfies the condition $|x| < 4$, the solution set consists of all numbers x for which $-4 < x < 4$, that is, all x in the interval $(-4, 4)$.

| EXAMPLE 8 | **Solving an Inequality Involving Absolute Value** |

Solve the inequality $|x| > 3$, and graph the solution set.

Figure 37

-5 -4 -3 -2 -1 0 1 2 3 4

Solution We are looking for all points whose coordinate x is a distance greater than 3 units from the origin. Figure 37 illustrates the situation. We conclude that any x less than -3 or greater than 3 satisfies the condition $|x| > 3$. Consequently, the solution set consists of all numbers x for which $x < -3$ or $x > 3$, that is, all x in $(-\infty, -3) \cup (3, \infty)$.*

We are led to the following results:

THEOREM

If a is any positive number, then

$\|u\| < a$ is equivalent to $-a < u < a$	**(5)**
$\|u\| \leq a$ is equivalent to $-a \leq u \leq a$	**(6)**
$\|u\| > a$ is equivalent to $u < -a$ or $u > a$	**(7)**
$\|u\| \geq a$ is equivalent to $u \leq -a$ or $u \geq a$	**(8)**

EXAMPLE 9 **Solving an Inequality Involving Absolute Value**

Solve the inequality $|2x + 4| \leq 3$, and graph the solution set.

Solution

$$|2x + 4| \leq 3$$ This follows the form of statement (6); the expression $u = 2x + 4$ is inside the absolute value bars.

$$-3 \leq 2x + 4 \leq 3$$ Apply statement (6).

$$-3 - 4 \leq 2x + 4 - 4 \leq 3 - 4$$ Subtract 4 from each part.

$$-7 \leq 2x \leq -1$$ Simplify.

$$\frac{-7}{2} \leq \frac{2x}{2} \leq \frac{-1}{2}$$ Divide each part by 2.

$$-\frac{7}{2} \leq x \leq -\frac{1}{2}$$ Simplify.

Figure 38

The solution set is $\left\{ x \mid -\dfrac{7}{2} \leq x \leq -\dfrac{1}{2} \right\}$, that is, all x in the interval $\left[-\dfrac{7}{2}, -\dfrac{1}{2} \right]$. See Figure 38 for a graph of the solution set.

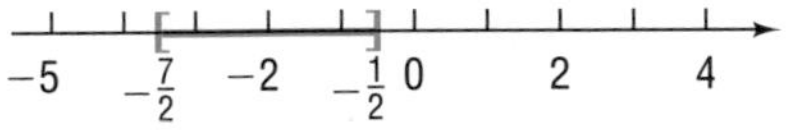

Now Work PROBLEM 89

EXAMPLE 10 **Solving an Inequality Involving Absolute Value**

Solve the inequality $|2x - 5| > 3$, and graph the solution set.

Solution

$$|2x - 5| > 3$$ This follows the form of statement (7); the expression $u = 2x - 5$ is inside the absolute value bars.

$$2x - 5 < -3 \quad \text{or} \quad 2x - 5 > 3$$ Apply statement (7).

$$2x - 5 + 5 < -3 + 5 \quad \text{or} \quad 2x - 5 + 5 > 3 + 5$$ Add 5 to each part.

$$2x < 2 \quad \text{or} \quad 2x > 8$$ Simplify.

$$\frac{2x}{2} < \frac{2}{2} \quad \text{or} \quad \frac{2x}{2} > \frac{8}{2}$$ Divide each part by 2.

$$x < 1 \quad \text{or} \quad x > 4$$ Simplify.

Figure 39

The solution set is $\{x \mid x < 1 \text{ or } x > 4\}$, that is, all x in $(-\infty, 1) \cup (4, \infty)$. See Figure 39 for a graph of the solution set.

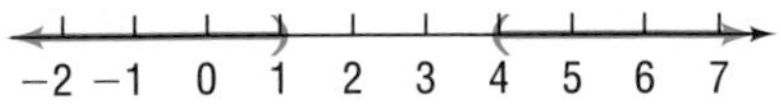

WARNING A common error to be avoided is to attempt to write the solution $x < 1$ or $x > 4$ as the combined inequality $1 > x > 4$, which is incorrect, since there are no numbers x for which $x < 1$ *and* $x > 4$. Another common error is to "mix" the symbols and write $1 < x > 4$, which makes no sense. ∎

Now Work PROBLEM 95

---

*The symbol $\cup$ stands for the union of two sets. Refer to page A2 if necessary.

A.7 Assess Your Understanding

1. Graph the inequality: $x \geq -2$. (pp. A4–A5)

2. *True or False* $-5 > -3$ (pp. A4–A5)

3. $|-2| =$ _____. (p. A5)

4. *True or False* $|x| \geq 0$ for any real number x. (p. A5)

Concepts and Vocabulary

5. If each side of an inequality is multiplied by a(n) _____ number, then the sense of the inequality symbol is reversed.

6. A(n) _____ _____, denoted $[a, b]$, consists of all real numbers x for which $a \leq x \leq b$.

7. The solution set of the equation $|x| = 5$ is {_____}.

8. The solution set of the inequality $|x| < 5$ is $\{x|$ _____ $\}$

9. *True or False* The equation $|x| = -2$ has no solution.

10. *True or False* The inequality $|x| \geq -2$ has the set of real numbers as solution set.

Skill Building

In Problems 11–16, express the graph shown in blue using interval notation. Also express each as an inequality involving x.

11.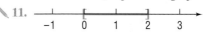

12.

13.

14.
-2   -1   0   1   2

15.
-1   0   1   2   3

16.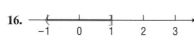

In Problems 17–22, an inequality is given. Write the inequality obtained by:
(a) Adding 3 to each side of the given inequality.
(b) Subtracting 5 from each side of the given inequality.
(c) Multiplying each side of the given inequality by 3.
(d) Multiplying each side of the given inequality by −2.

17. $3 < 5$
18. $2 > 1$
19. $4 > -3$
20. $-3 > -5$
21. $2x + 1 < 2$
22. $1 - 2x > 5$

In Problems 23–30, write each inequality using interval notation, and illustrate each inequality using the real number line.

23. $0 \leq x \leq 4$
24. $-1 < x < 5$
25. $4 \leq x < 6$
26. $-2 < x < 0$

27. $x \geq 4$
28. $x \leq 5$
29. $x < -4$
30. $x > 1$

In Problems 31–38, write each interval as an inequality involving x, and illustrate each inequality using the real number line.

31. $[2, 5]$
32. $(1, 2)$
33. $(-3, -2)$
34. $[0, 1)$

35. $[4, \infty)$
36. $(-\infty, 2]$
37. $(-\infty, -3)$
38. $(-8, \infty)$

In Problems 39–52, fill in the blank with the correct inequality symbol.

39. If $x < 5$, then $x - 5$ _____ 0.

40. If $x < -4$, then $x + 4$ _____ 0.

41. If $x > -4$, then $x + 4$ _____ 0.

42. If $x > 6$, then $x - 6$ _____ 0.

43. If $x \geq -4$, then $3x$ _____ -12.

44. If $x \leq 3$, then $2x$ _____ 6.

45. If $x > 6$, then $-2x$ _____ -12.

46. If $x > -2$, then $-4x$ _____ 8.

47. If $x \geq 5$, then $-4x$ _____ -20.

48. If $x \leq -4$, then $-3x$ _____ 12.

49. If $2x > 6$, then x _____ 3.

50. If $3x \leq 12$, then x _____ 4.

51. If $-\dfrac{1}{2}x \leq 3$, then x _____ -6.

52. If $-\dfrac{1}{4}x > 1$, then x _____ -4.

In Problems 53–100, solve each inequality. Express your answer using set notation or interval notation. Graph the solution set.

53. $x + 1 < 5$
54. $x - 6 < 1$
55. $1 - 2x \leq 3$

56. $2 - 3x \leq 5$
57. $3x - 7 > 2$
58. $2x + 5 > 1$

59. $3x - 1 \geq 3 + x$

60. $2x - 2 \geq 3 + x$

61. $-2(x + 3) < 8$

62. $-3(1 - x) < 12$

63. $4 - 3(1 - x) \leq 3$

64. $8 - 4(2 - x) \leq -2x$

65. $\frac{1}{2}(x - 4) > x + 8$

66. $3x + 4 > \frac{1}{3}(x - 2)$

67. $\frac{x}{2} \geq 1 - \frac{x}{4}$

68. $\frac{x}{3} \geq 2 + \frac{x}{6}$

69. $0 \leq 2x - 6 \leq 4$

70. $4 \leq 2x + 2 \leq 10$

71. $-5 \leq 4 - 3x \leq 2$

72. $-3 \leq 3 - 2x \leq 9$

73. $-3 < \frac{2x - 1}{4} < 0$

74. $0 < \frac{3x + 2}{2} < 4$

75. $1 < 1 - \frac{1}{2}x < 4$

76. $0 < 1 - \frac{1}{3}x < 1$

77. $(x + 2)(x - 3) > (x - 1)(x + 1)$

78. $(x - 1)(x + 1) > (x - 3)(x + 4)$

79. $x(4x + 3) \leq (2x + 1)^2$

80. $x(9x - 5) \leq (3x - 1)^2$

81. $\frac{1}{2} \leq \frac{x + 1}{3} < \frac{3}{4}$

82. $\frac{1}{3} < \frac{x + 1}{2} \leq \frac{2}{3}$

83. $(4x + 2)^{-1} < 0$

84. $(2x - 1)^{-1} > 0$

85. $0 < \frac{2}{x} < \frac{3}{5}$

86. $0 < \frac{4}{x} < \frac{2}{3}$

87. $0 < (2x - 4)^{-1} < \frac{1}{2}$

88. $0 < (3x + 6)^{-1} < \frac{1}{3}$

89. $|2x| < 8$

90. $|3x| < 12$

91. $|3x| > 12$

92. $|2x| > 6$

93. $|2x - 1| \leq 1$

94. $|2x + 5| \leq 7$

95. $|1 - 2x| > 3$

96. $|2 - 3x| > 1$

97. $|-4x| + |-5| \leq 9$

98. $|-x| - |4| \leq 2$

99. $|-2x| \geq |-4|$

100. $|-x - 2| \geq 1$

Applications and Extensions

101. Express the fact that x differs from 2 by less than $\frac{1}{2}$ as an inequality involving an absolute value. Solve for x.

102. Express the fact that x differs from -1 by less than 1 as an inequality involving an absolute value. Solve for x.

103. Express the fact that x differs from -3 by more than 2 as an inequality involving an absolute value. Solve for x.

104. Express the fact that x differs from 2 by more than 3 as an inequality involving an absolute value. Solve for x.

105. What is the domain of the variable in the expression $\sqrt{3x + 6}$?

106. What is the domain of the variable in the expression $\sqrt{8 + 2x}$?

107. A young adult may be defined as someone older than 21, but less than 30 years of age. Express this statement using inequalities.

108. Middle-aged may be defined as being 40 or more and less than 60. Express this statement using inequalities.

109. Life Expectancy The Social Security Administration determined that an average 30-year-old male in 2005 could expect to live at least 49.66 more years and an average 30-year-old female in 2005 could expect to live at least 53.58 more years.
 (a) To what age can an average 30-year-old male expect to live? Express your answer as an inequality.

 (b) To what age can an average 30-year-old female expect to live? Express your answer as an inequality.
 (c) Who can expect to live longer, a male or a female? By how many years?

Source: Actuarial Study No. 120, August 2005

110. General Chemistry For a certain ideal gas, the volume V (in cubic centimeters) equals 20 times the temperature T (in degrees Celsius). If the temperature varies from 80° to 120° C inclusive, what is the corresponding range of the volume of the gas?

111. Real Estate A real estate agent agrees to sell an apartment complex according to the following commission schedule: $45,000 plus 25% of the selling price in excess of $900,000. Assuming that the complex will sell at some price between $900,000 and $1,100,000 inclusive, over what range does the agent's commission vary? How does the commission vary as a percent of selling price?

112. Sales Commission A used car salesperson is paid a commission of $25 plus 40% of the selling price in excess of owner's cost. The owner claims that used cars typically sell for at least owner's cost plus $200 and at most owner's cost plus $3000. For each sale made, over what range can the salesperson expect the commission to vary?

113. Federal Tax Withholding The percentage method of withholding for federal income tax (2006) states that a single person whose weekly wages, after subtracting withholding allowances, are over $620, but not over $1409, shall have $78.30 plus 25% of the excess over $620 withheld. Over what range does the amount withheld vary if the weekly wages vary from $700 to $900 inclusive?

Source: Employer's Tax Guide. Department of the Treasury, Internal Revenue Service, Publication 2006.

114. Exercising Sue wants to lose weight. For healthy weight loss, the American College of Sports Medicine (ACSM) recommends 200 to 300 minutes of exercise per week. For the first six days of the week, Sue exercised 40, 45, 0, 50, 25, and 35 minutes. How long should Sue exercise on the seventh day in order to stay within the ACSM guidelines?

115. Electricity Rates Commonwealth Edison Company's charge for electricity in May 2006 is 8.275¢ per kilowatt-hour. In addition, each monthly bill contains a customer charge of $7.58. If last year's bills ranged from a low of $63.47 to a high of $214.53, over what range did usage vary (in kilowatt-hours)?

Source: Commonwealth Edison Co., Chicago, Illinois, 2006.

116. Water Bills The Village of Oak Lawn charges homeowners $28.84 per quarter-year plus $2.28 per 1000 gallons for water usage in excess of 12,000 gallons. In 2006 one homeowner's quarterly bill ranged from a high of $74.44 to a low of $42.52. Over what range did water usage vary?

Source: Village of Oak Lawn, Illinois, April 2006.

117. Markup of a New Car The markup over dealer's cost of a new car ranges from 12% to 18%. If the sticker price is $18,000, over what range will the dealer's cost vary?

118. IQ Tests A standard intelligence test has an average score of 100. According to statistical theory, of the people who take the test, the 2.5% with the highest scores will have scores of more than 1.96σ above the average, where σ (sigma, a number called the **standard deviation**) depends on the nature of the test. If $\sigma = 12$ for this test and there is (in principle) no upper limit to the score possible on the test, write the interval of possible test scores of the people in the top 2.5%.

119. Computing Grades In your Economics 101 class, you have scores of 68, 82, 87, and 89 on the first four of five tests. To get

a grade of B, the average of the first five test scores must be greater than or equal to 80 and less than 90.
(a) Solve an inequality to find the range of the score that you need on the last test to get a B.
(b) What score do you need if the fifth test counts double?

What do I need to get a B?

120. "Light" Foods For food products to be labeled "light," the U.S. Food and Drug Administration requires that the altered product must either contain one-third or fewer calories than the regular product or it must contain one-half or less fat than the regular product. If a serving of Miracle Whip® Light contains 20 calories and 1.5 grams of fat, then what must be true about either the number of calories or the grams of fat in a serving of regular Miracle Whip®?

121. Arithmetic Mean If $a < b$, show that $a < \dfrac{a+b}{2} < b$. The number $\dfrac{a+b}{2}$ is called the **arithmetic mean** of a and b.

122. Refer to Problem 121. Show that the arithmetic mean of a and b is equidistant from a and b.

123. Geometric Mean If $0 < a < b$, show that $a < 0\sqrt{ab} < b$. The number $\sqrt{ab}$ is called the **geometric mean** of a and b.

124. Refer to Problems 121 and 123. Show that the geometric mean of a and b is less than the arithmetic mean of a and b.

125. Harmonic Mean For $0 < a < b$, let h be defined by

$$\frac{1}{h} = \frac{1}{2}\left(\frac{1}{a} + \frac{1}{b}\right)$$

Show that $a < h < b$. The number h is called the **harmonic mean** of a and b.

126. Refer to Problems 121, 123, and 125. Show that the harmonic mean of a and b equals the geometric mean squared divided by the arithmetic mean.

127. Another Reciprocal Property Prove that if $0 < a < b$, then

$$0 < \frac{1}{b} < \frac{1}{a}.$$

Discussion and Writing

128. Make up an inequality that has no solution. Make up one that has exactly one solution.

129. The inequality $x^2 + 1 < -5$ has no real solution. Explain why.

130. Do you prefer to use inequality notation or interval notation to express the solution to an inequality? Give your reasons. Are there particular circumstances when you prefer one to the other? Cite examples.

131. How would you explain to a fellow student the underlying reason for the multiplication properties for inequalities (page A53), that is, the sense or direction of an inequality remains the same if each side is multiplied by a positive real number, whereas the direction is reversed if each side is multiplied by a negative real number.

'Are You Prepared?' Answers

1. [number line with points at −4, −2, 0]   2. False   3. 2   4. True

A.8 *nth* Roots; Rational Exponents

PREPARING FOR THIS SECTION *Before getting started, review the following:*

- Exponents, Square Roots (Appendix, Section A.1, pp. A7–A10)

Now Work the 'Are You Prepared?' problems on page A66.

OBJECTIVES **1** Work with *nth* Roots (p. A61)
2 Simplify Radicals (p. A62)
3 Rationalize Denominators (p. A63)
4 Simplify Expressions with Rational Exponents (p. A64)

1 Work with *nth* Roots

DEFINITION

The **principal *nth* root of a real number *a*, *n* $\geq$ 2** an integer, symbolized by $\sqrt[n]{a}$, is defined as follows:

$$\sqrt[n]{a} = b \quad \text{means} \quad a = b^n$$

where $a \geq 0$ and $b \geq 0$ if *n* is even and *a*, *b* are any real numbers if *n* is odd.

In Words
The symbol $\sqrt[n]{a}$ means "give me the number, when raised to the nth power, that results in *a*."

Notice that if *a* is negative and *n* is even then $\sqrt[n]{a}$ is not defined. When it is defined, the principal *nth* root of a number is unique.

The symbol $\sqrt[n]{a}$ for the principal *nth* root of *a* is called a **radical;** the integer *n* is called the **index,** and *a* is called the **radicand.** If the index of a radical is 2, we call $\sqrt[2]{a}$ the **square root** of *a* and omit the index 2 by writing $\sqrt{a}$. If the index is 3, we call $\sqrt[3]{a}$ the **cube root** of *a*.

EXAMPLE 1 **Simplifying Principal *nth* Roots**

(a) $\sqrt[3]{8} = \sqrt[3]{2^3} = 2$ The number whose cube is 8 is 2. (b) $\sqrt[3]{-64} = \sqrt[3]{(-4)^3} = -4$

(c) $\sqrt[4]{\dfrac{1}{16}} = \sqrt[4]{\left(\dfrac{1}{2}\right)^4} = \dfrac{1}{2}$ (d) $\sqrt[6]{(-2)^6} = |-2| = 2$

These are examples of **perfect roots,** since each simplifies to a rational number. Notice the absolute value in Example 1(d). If *n* is even, the principal *nth* root must be nonnegative.

In general, if *n* $\geq$ 2 is an integer and *a* is a real number, we have

$$\sqrt[n]{a^n} = a \qquad \text{if } n \geq 3 \text{ is odd} \tag{1a}$$
$$\sqrt[n]{a^n} = |a| \qquad \text{if } n \geq 2 \text{ is even} \tag{1b}$$

Now Work PROBLEM 7

Radicals provide a way of representing many irrational real numbers. For example, there is no rational number whose square is 2. Using radicals, we can say that $\sqrt{2}$ is the positive number whose square is 2.

| EXAMPLE 2 | Using a Calculator to Approximate Roots |

Use a calculator to approximate $\sqrt[5]{16}$.

Solution Figure 40 shows the result using a TI-84 Plus graphing calculator.

Figure 40

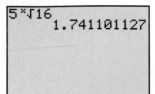

Now Work PROBLEM 101

2 Simplify Radicals

Let $n \geq 2$ and $m \geq 2$ denote positive integers, and let a and b represent real numbers. Assuming that all radicals are defined, we have the following properties:

Properties of Radicals

$$\sqrt[n]{ab} = \sqrt[n]{a}\,\sqrt[n]{b} \tag{2a}$$

$$\sqrt[n]{\dfrac{a}{b}} = \dfrac{\sqrt[n]{a}}{\sqrt[n]{b}} \tag{2b}$$

$$\sqrt[n]{a^m} = (\sqrt[n]{a})^m \tag{2c}$$

When used in reference to radicals, the direction to "simplify" will mean to remove from the radicals any perfect roots that occur as factors. Let's look at some examples of how the preceding rules are applied to simplify radicals.

| EXAMPLE 3 | Simplifying Radicals |

(a) $\sqrt{32} = \underset{\substack{\uparrow \\ \text{Factor out 16,} \\ \text{a perfect square.}}}{\sqrt{16 \cdot 2}} = \underset{\substack{\uparrow \\ (2a)}}{\sqrt{16} \cdot \sqrt{2}} = 4\sqrt{2}$

(b) $\sqrt[3]{16} = \underset{\substack{\uparrow \\ \text{Factor out 8,} \\ \text{a perfect cube.}}}{\sqrt[3]{8 \cdot 2}} = \underset{\substack{\uparrow \\ (2a)}}{\sqrt[3]{8} \cdot \sqrt[3]{2}} = \sqrt[3]{2^3} \cdot \sqrt[3]{2} = 2\sqrt[3]{2}$

(c) $\sqrt[3]{-16x^4} = \underset{\substack{\uparrow \\ \text{Factor perfect} \\ \text{cubes inside radical.}}}{\sqrt[3]{-8 \cdot 2 \cdot x^3 \cdot x}} = \underset{\substack{\uparrow \\ \text{Group perfect} \\ \text{cubes.}}}{\sqrt[3]{(-8x^3)(2x)}}$

$$= \sqrt[3]{(-2x)^3 \cdot 2x} = \underset{\substack{\uparrow \\ (2a)}}{\sqrt[3]{(-2x)^3} \cdot \sqrt[3]{2x}} = -2x\sqrt[3]{2x}$$

(d) $\sqrt[4]{\dfrac{16x^5}{81}} = \sqrt[4]{\dfrac{2^4 x^4 x}{3^4}} = \sqrt[4]{\left(\dfrac{2x}{3}\right)^4 \cdot x} = \sqrt[4]{\left(\dfrac{2x}{3}\right)^4} \cdot \sqrt[4]{x} = \left|\dfrac{2x}{3}\right| \sqrt[4]{x}$

Now Work PROBLEMS 11 AND 17

Two or more radicals can be combined, provided that they have the same index and the same radicand. Such radicals are called **like radicals.**

| EXAMPLE 4 | Combining Like Radicals |

(a) $-8\sqrt{12} + \sqrt{3} = -8\sqrt{4 \cdot 3} + \sqrt{3}$

$$= -8 \cdot \sqrt{4}\sqrt{3} + \sqrt{3}$$

$$= -16\sqrt{3} + \sqrt{3} = -15\sqrt{3}$$

(b) $\sqrt[3]{8x^4} + \sqrt[3]{-x} + 4\sqrt[3]{27x} = \sqrt[3]{2^3 x^3 x} + \sqrt[3]{-1 \cdot x} + 4\sqrt[3]{3^3 x}$

$$= \sqrt[3]{(2x)^3} \cdot \sqrt[3]{x} + \sqrt[3]{-1} \cdot \sqrt[3]{x} + 4\sqrt[3]{3^3} \cdot \sqrt[3]{x}$$

$$= 2x\sqrt[3]{x} - 1 \cdot \sqrt[3]{x} + 12\sqrt[3]{x}$$

$$= (2x + 11)\sqrt[3]{x}$$

➤ **Now Work** PROBLEM 33

3 Rationalize Denominators

When radicals occur in quotients, it is customary to rewrite the quotient so that the new denominator contains no radicals. This process is referred to as **rationalizing the denominator.**

The idea is to multiply by an appropriate expression so that the new denominator contains no radicals. For example:

If a Denominator Contains the Factor	Multiply by	To Obtain a Denominator Free of Radicals
$\sqrt{3}$	$\sqrt{3}$	$(\sqrt{3})^2 = 3$
$\sqrt{3} + 1$	$\sqrt{3} - 1$	$(\sqrt{3})^2 - 1^2 = 3 - 1 = 2$
$\sqrt{2} - 3$	$\sqrt{2} + 3$	$(\sqrt{2})^2 - 3^2 = 2 - 9 = -7$
$\sqrt{5} - \sqrt{3}$	$\sqrt{5} + \sqrt{3}$	$(\sqrt{5})^2 - (\sqrt{3})^2 = 5 - 3 = 2$
$\sqrt[3]{4}$	$\sqrt[3]{2}$	$\sqrt[3]{4} \cdot \sqrt[3]{2} = \sqrt[3]{8} = 2$

In rationalizing the denominator of a quotient, be sure to multiply both the numerator and the denominator by the expression.

EXAMPLE 5

Rationalizing Denominators

Rationalize the denominator of each expression:

(a) $\dfrac{1}{\sqrt{3}}$ 　　(b) $\dfrac{5}{4\sqrt{2}}$ 　　(c) $\dfrac{\sqrt{2}}{\sqrt{3} - 3\sqrt{2}}$

Solution

(a) The denominator contains the factor $\sqrt{3}$, so we multiply the numerator and denominator by $\sqrt{3}$ to obtain

$$\frac{1}{\sqrt{3}} = \frac{1}{\sqrt{3}} \cdot \frac{\sqrt{3}}{\sqrt{3}} = \frac{\sqrt{3}}{(\sqrt{3})^2} = \frac{\sqrt{3}}{3}$$

(b) The denominator contains the factor $\sqrt{2}$, so we multiply the numerator and denominator by $\sqrt{2}$ to obtain

$$\frac{5}{4\sqrt{2}} = \frac{5}{4\sqrt{2}} \cdot \frac{\sqrt{2}}{\sqrt{2}} = \frac{5\sqrt{2}}{4(\sqrt{2})^2} = \frac{5\sqrt{2}}{4 \cdot 2} = \frac{5\sqrt{2}}{8}$$

(c) The denominator contains the factor $\sqrt{3} - 3\sqrt{2}$, so we multiply the numerator and denominator by $\sqrt{3} + 3\sqrt{2}$ to obtain

$$\frac{\sqrt{2}}{\sqrt{3} - 3\sqrt{2}} = \frac{\sqrt{2}}{\sqrt{3} - 3\sqrt{2}} \cdot \frac{\sqrt{3} + 3\sqrt{2}}{\sqrt{3} + 3\sqrt{2}} = \frac{\sqrt{2}(\sqrt{3} + 3\sqrt{2})}{(\sqrt{3})^2 - (3\sqrt{2})^2}$$

$$= \frac{\sqrt{2}\sqrt{3} + 3(\sqrt{2})^2}{3 - 18} = \frac{\sqrt{6} + 6}{-15} = -\frac{6 + \sqrt{6}}{15}$$

➤ **Now Work** PROBLEM 47

4 Simplify Expressions with Rational Exponents

Radicals are used to define rational exponents.

DEFINITION

If a is a real number and $n \geq 2$ is an integer, then

$$a^{1/n} = \sqrt[n]{a} \qquad (3)$$

provided that $\sqrt[n]{a}$ exists.

Note that if n is even and $a < 0$ then $\sqrt[n]{a}$ and $a^{1/n}$ do not exist.

EXAMPLE 6 | **Writing Expressions Containing Fractional Exponents as Radicals**

(a) $4^{1/2} = \sqrt{4} = 2$ (b) $8^{1/2} = \sqrt{8} = 2\sqrt{2}$

(c) $(-27)^{1/3} = \sqrt[3]{-27} = -3$ (d) $16^{1/3} = \sqrt[3]{16} = 2\sqrt[3]{2}$

DEFINITION

If a is a real number and m and n are integers containing no common factors, with $n \geq 2$, then

$$a^{m/n} = \sqrt[n]{a^m} = (\sqrt[n]{a})^m \qquad (4)$$

provided that $\sqrt[n]{a}$ exists.

We have two comments about equation (4):

1. The exponent $\dfrac{m}{n}$ must be in lowest terms and n must be positive.

2. In simplifying the rational expression $a^{m/n}$, either $\sqrt[n]{a^m}$ or $(\sqrt[n]{a})^m$ may be used, the choice depending on which is easier to simplify. Generally, taking the root first, as in $(\sqrt[n]{a})^m$, is easier.

EXAMPLE 7 | **Using Equation (4)**

(a) $4^{3/2} = (\sqrt{4})^3 = 2^3 = 8$ (b) $(-8)^{4/3} = (\sqrt[3]{-8})^4 = (-2)^4 = 16$

(c) $(32)^{-2/5} = (\sqrt[5]{32})^{-2} = 2^{-2} = \dfrac{1}{4}$ (d) $25^{6/4} = 25^{3/2} = (\sqrt{25})^3 = 5^3 = 125$

Now Work PROBLEM 55

It can be shown that the Laws of Exponents hold for rational exponents. The next example illustrates using the Law of Exponents to simplify.

EXAMPLE 8 | **Simplifying Expressions Containing Rational Exponents**

Simplify each expression. Express your answer so that only positive exponents occur. Assume that the variables are positive.

(a) $(x^{2/3}y)(x^{-2}y)^{1/2}$ (b) $\left(\dfrac{2x^{1/3}}{y^{2/3}}\right)^{-3}$ (c) $\left(\dfrac{9x^2y^{1/3}}{x^{1/3}y}\right)^{1/2}$

Solution

(a)
$$(x^{2/3}y)(x^{-2}y)^{1/2} = (x^{2/3}y)[(x^{-2})^{1/2}y^{1/2}] \qquad (ab)^n = a^n b^n$$
$$= x^{2/3}yx^{-1}y^{1/2} \qquad (a^m)^n = a^{mn}$$
$$= (x^{2/3} \cdot x^{-1})(y \cdot y^{1/2})$$
$$= x^{-1/3}y^{3/2} \qquad a^m \cdot a^n = a^{m+n}$$
$$= \frac{y^{3/2}}{x^{1/3}}$$

(b) $\left(\dfrac{2x^{1/3}}{y^{2/3}}\right)^{-3} = \left(\dfrac{y^{2/3}}{2x^{1/3}}\right)^3 = \dfrac{(y^{2/3})^3}{(2x^{1/3})^3} = \dfrac{y^2}{2^3(x^{1/3})^3} = \dfrac{y^2}{8x}$

(c) $\left(\dfrac{9x^2 y^{1/3}}{x^{1/3}y}\right)^{1/2} = \left(\dfrac{9x^{2-(1/3)}}{y^{1-(1/3)}}\right)^{1/2} = \left(\dfrac{9x^{5/3}}{y^{2/3}}\right)^{1/2} = \dfrac{9^{1/2}(x^{5/3})^{1/2}}{(y^{2/3})^{1/2}} = \dfrac{3x^{5/6}}{y^{1/3}}$

Now Work PROBLEM 71

 The next two examples illustrate some algebra that you will need to know for certain calculus problems.

EXAMPLE 9   **Writing an Expression as a Single Quotient**

Write the following expression as a single quotient in which only positive exponents appear.

$$(x^2 + 1)^{1/2} + x \cdot \frac{1}{2}(x^2 + 1)^{-1/2} \cdot 2x$$

Solution

$$(x^2 + 1)^{1/2} + x \cdot \frac{1}{2}(x^2 + 1)^{-1/2} \cdot 2x = (x^2 + 1)^{1/2} + \frac{x^2}{(x^2 + 1)^{1/2}}$$
$$= \frac{(x^2 + 1)^{1/2}(x^2 + 1)^{1/2} + x^2}{(x^2 + 1)^{1/2}}$$
$$= \frac{(x^2 + 1) + x^2}{(x^2 + 1)^{1/2}}$$
$$= \frac{2x^2 + 1}{(x^2 + 1)^{1/2}}$$

Now Work PROBLEM 77

EXAMPLE 10   **Factoring an Expression Containing Rational Exponents**

Factor: $\dfrac{4}{3}x^{1/3}(2x + 1) + 2x^{4/3}$

Solution

We begin by writing $2x^{4/3}$ as a fraction with 3 as denominator.

$$\frac{4}{3}x^{1/3}(2x + 1) + 2x^{4/3} = \frac{4x^{1/3}(2x + 1)}{3} + \frac{6x^{4/3}}{3} = \frac{4x^{1/3}(2x + 1) + 6x^{4/3}}{3}$$

$\uparrow$
Add the two fractions

$$= \frac{2x^{1/3}[2(2x + 1) + 3x]}{3} = \frac{2x^{1/3}(7x + 2)}{3}$$

$\uparrow$ $\uparrow$
2 and $x^{1/3}$ are common factors   Simplify

Now Work PROBLEM 89

Historical Feature

The radical sign, $\sqrt{}$, was first used in print by Christoff Rudolff in 1525. It is thought to be the manuscript form of the letter r (for the Latin word *radix* = root), although this is not quite conclusively confirmed. It took a long time for $\sqrt{}$ to become the standard symbol for a square root and much longer to standardize $\sqrt[3]{}$, $\sqrt[4]{}$, $\sqrt[5]{}$, and so on. The indexes of the root were placed in every conceivable position, with

$$\sqrt[3]{8}, \quad \sqrt[3]{8}, \quad \text{and} \quad \underset{3}{\sqrt{8}}$$

all being variants for $\sqrt[3]{8}$. The notation $\sqrt{}\sqrt{16}$ was popular for $\sqrt[4]{16}$. By the 1700s, the index had settled where we now put it.

The bar on top of the present radical symbol, as follows,

$$\sqrt{a^2 + 2ab + b^2}$$

is the last survivor of the **vinculum**, a bar placed atop an expression to indicate what we would now indicate with parentheses. For example,

$$\overline{ab + c} = a(b + c)$$

A.8 Assess Your Understanding

'Are You Prepared?' *Answers are given at the end of these exercises. If you get a wrong answer, read the pages in red.*

1. $(-3)^2 = $ _____; $-3^2 = $ _____ (pp. A7–A8)

2. $\sqrt{16} = $ _____; $\sqrt{(-4)^2} = $ _____ (pp. A9–A10)

Concepts and Vocabulary

3. In the symbol $\sqrt[n]{a}$, the integer n is called the _____.

4. *True or False* $\sqrt[5]{-32} = -2$

5. We call $\sqrt[3]{a}$ the _____ _____ of a.

6. *True or False* $\sqrt[4]{(-3)^4} = -3$

Skill Building

In Problems 7–42, simplify each expression. Assume that all variables are positive when they appear.

7. $\sqrt[3]{27}$

8. $\sqrt[4]{16}$

9. $\sqrt[3]{-8}$

10. $\sqrt[3]{-1}$

11. $\sqrt{8}$

12. $\sqrt[3]{54}$

13. $\sqrt[3]{-8x^4}$

14. $\sqrt[4]{48x^5}$

15. $\sqrt[4]{x^{12}y^8}$

16. $\sqrt[5]{x^{10}y^5}$

17. $\sqrt[4]{\dfrac{x^9y^7}{xy^3}}$

18. $\sqrt[3]{\dfrac{3xy^2}{81x^4y^2}}$

19. $\sqrt{36x}$

20. $\sqrt{9x^5}$

21. $\sqrt{3x^2}\sqrt{12x}$

22. $\sqrt{5x}\sqrt{20x^3}$

23. $\left(\sqrt{5}\sqrt[3]{9}\right)^2$

24. $\left(\sqrt[3]{3}\sqrt{10}\right)^4$

25. $\left(3\sqrt{6}\right)\left(2\sqrt{2}\right)$

26. $\left(5\sqrt{8}\right)\left(-3\sqrt{3}\right)$

27. $3\sqrt{2} + 4\sqrt{2}$

28. $6\sqrt{5} - 4\sqrt{5}$

29. $-\sqrt{18} + 2\sqrt{8}$

30. $2\sqrt{12} - 3\sqrt{27}$

31. $\left(\sqrt{3} + 3\right)\left(\sqrt{3} - 1\right)$

32. $\left(\sqrt{5} - 2\right)\left(\sqrt{5} + 3\right)$

33. $5\sqrt[3]{2} - 2\sqrt[3]{54}$

34. $9\sqrt[3]{24} - \sqrt[3]{81}$

35. $\left(\sqrt{x} - 1\right)^2$

36. $\left(\sqrt{x} + \sqrt{5}\right)^2$

37. $\sqrt[3]{16x^4} - \sqrt[3]{2x}$

38. $\sqrt[4]{32x} + \sqrt[4]{2x^5}$

39. $\sqrt{8x^3} - 3\sqrt{50x}$

40. $3x\sqrt{9y} + 4\sqrt{25y}$

41. $\sqrt[3]{16x^4y} - 3x\sqrt[3]{2xy} + 5\sqrt[3]{-2xy^4}$

42. $8xy - \sqrt{25x^2y^2} + \sqrt[3]{8x^3y^3}$

In Problems 43–54, rationalize the denominator of each expression. Assume that all variables are positive when they appear.

43. $\dfrac{1}{\sqrt{2}}$

44. $\dfrac{2}{\sqrt{3}}$

45. $\dfrac{-\sqrt{3}}{\sqrt{5}}$

46. $\dfrac{-\sqrt{3}}{\sqrt{8}}$

47. $\dfrac{\sqrt{3}}{5 - \sqrt{2}}$

48. $\dfrac{\sqrt{2}}{\sqrt{7} + 2}$

49. $\dfrac{2 - \sqrt{5}}{2 + 3\sqrt{5}}$

50. $\dfrac{\sqrt{3} - 1}{2\sqrt{3} + 3}$

51. $\dfrac{5}{\sqrt[3]{2}}$ **52.** $\dfrac{-2}{\sqrt[3]{9}}$ **53.** $\dfrac{\sqrt{x+h}-\sqrt{x}}{\sqrt{x+h}+\sqrt{x}}$ **54.** $\dfrac{\sqrt{x+h}+\sqrt{x-h}}{\sqrt{x+h}-\sqrt{x-h}}$

In Problems 55–66, simplify each expression.

55. $8^{2/3}$ **56.** $4^{3/2}$ **57.** $(-27)^{1/3}$ **58.** $16^{3/4}$ **59.** $16^{3/2}$ **60.** $25^{3/2}$

61. $9^{-3/2}$ **62.** $16^{-3/2}$ **63.** $\left(\dfrac{9}{8}\right)^{3/2}$ **64.** $\left(\dfrac{27}{8}\right)^{2/3}$ **65.** $\left(\dfrac{8}{9}\right)^{-3/2}$ **66.** $\left(\dfrac{8}{27}\right)^{-2/3}$

In Problems 67–74, simplify each expression. Express your answer so that only positive exponents occur. Assume that the variables are positive.

67. $x^{3/4}x^{1/3}x^{-1/2}$ **68.** $x^{2/3}x^{1/2}x^{-1/4}$ **69.** $(x^3y^6)^{1/3}$ **70.** $(x^4y^8)^{3/4}$

71. $\dfrac{(x^2y)^{1/3}(xy^2)^{2/3}}{x^{2/3}y^{2/3}}$ **72.** $\dfrac{(xy)^{1/4}(x^2y^2)^{1/2}}{(x^2y)^{3/4}}$ **73.** $\dfrac{(16x^2y^{-1/3})^{3/4}}{(xy^2)^{1/4}}$ **74.** $\dfrac{(4x^{-1}y^{1/3})^{3/2}}{(xy)^{3/2}}$

Applications and Extensions

In Problems 75–88, expressions that occur in calculus are given. Write each expression as a single quotient in which only positive exponents and/or radicals appear.

75. $\dfrac{x}{(1+x)^{1/2}}+2(1+x)^{1/2}\quad x>-1$

76. $\dfrac{1+x}{2x^{1/2}}+x^{1/2}\quad x>0$

77. $2x(x^2+1)^{1/2}+x^2\cdot\dfrac{1}{2}(x^2+1)^{-1/2}\cdot 2x$

78. $(x+1)^{1/3}+x\cdot\dfrac{1}{3}(x+1)^{-2/3}\quad x\neq-1$

79. $\sqrt{4x+3}\cdot\dfrac{1}{2\sqrt{x-5}}+\sqrt{x-5}\cdot\dfrac{1}{5\sqrt{4x+3}}\quad x>5$

80. $\dfrac{\sqrt[3]{8x+1}}{3\sqrt[3]{(x-2)^2}}+\dfrac{\sqrt[3]{x-2}}{24\sqrt[3]{(8x+1)^2}}\quad x\neq2,\ x\neq-\dfrac{1}{8}$

81. $\dfrac{\sqrt{1+x}-x\cdot\dfrac{1}{2\sqrt{1+x}}}{1+x}\quad x>-1$

82. $\dfrac{\sqrt{x^2+1}-x\cdot\dfrac{2x}{2\sqrt{x^2+1}}}{x^2+1}$

83. $\dfrac{(x+4)^{1/2}-2x(x+4)^{-1/2}}{x+4}\quad x>-4$

84. $\dfrac{(9-x^2)^{1/2}+x^2(9-x^2)^{-1/2}}{9-x^2}\quad -3<x<3$

85. $\dfrac{\dfrac{x^2}{(x^2-1)^{1/2}}-(x^2-1)^{1/2}}{x^2}\quad x<-1\ \text{ or }\ x>1$

86. $\dfrac{(x^2+4)^{1/2}-x^2(x^2+4)^{-1/2}}{x^2+4}$

87. $\dfrac{\dfrac{1+x^2}{2\sqrt{x}}-2x\sqrt{x}}{(1+x^2)^2}\quad x>0$

88. $\dfrac{2x(1-x^2)^{1/3}+\dfrac{2}{3}x^3(1-x^2)^{-2/3}}{(1-x^2)^{2/3}}\quad x\neq-1,x\neq1$

In Problems 89–98, expressions that occur in calculus are given. Factor each expression. Express your answer so that only positive exponents occur.

89. $(x+1)^{3/2}+x\cdot\dfrac{3}{2}(x+1)^{1/2}\quad x\geq-1$

90. $(x^2+4)^{4/3}+x\cdot\dfrac{4}{3}(x^2+4)^{1/3}\cdot 2x$

91. $6x^{1/2}(x^2 + x) - 8x^{3/2} - 8x^{1/2}$ $x \geq 0$

92. $6x^{1/2}(2x + 3) + x^{3/2} \cdot 8$ $x \geq 0$

93. $3(x^2 + 4)^{4/3} + x \cdot 4(x^2 + 4)^{1/3} \cdot 2x$

94. $2x(3x + 4)^{4/3} + x^2 \cdot 4(3x + 4)^{1/3}$

95. $4(3x + 5)^{1/3}(2x + 3)^{3/2} + 3(3x + 5)^{4/3}(2x + 3)^{1/2}$ $x \geq -\dfrac{3}{2}$

96. $6(6x + 1)^{1/3}(4x - 3)^{3/2} + 6(6x + 1)^{4/3}(4x - 3)^{1/2}$ $x \geq \dfrac{3}{4}$

97. $3x^{-1/2} + \dfrac{3}{2}x^{1/2}$ $x > 0$

98. $8x^{1/3} - 4x^{-2/3}$ $x \neq 0$

In Problems 99–106, use a calculator to approximate each radical. Round your answer to two decimal places.

99. $\sqrt{2}$ **100.** $\sqrt{7}$ **101.** $\sqrt[3]{4}$ **102.** $\sqrt[3]{-5}$

103. $\dfrac{2 + \sqrt{3}}{3 - \sqrt{5}}$ **104.** $\dfrac{\sqrt{5} - 2}{\sqrt{2} + 4}$ **105.** $\dfrac{3\sqrt[3]{5} - \sqrt{2}}{\sqrt{3}}$ **106.** $\dfrac{2\sqrt{3} - \sqrt[3]{4}}{\sqrt{2}}$

Applications and Extensions

107. Calculating the Amount of Gasoline in a Tank   A Shell station stores its gasoline in underground tanks that are right circular cylinders lying on their sides. See the illustration. The volume V of gasoline in the tank (in gallons) is given by the formula

$$V = 40h^2\sqrt{\dfrac{96}{h} - 0.608}$$

where h is the height of the gasoline (in inches) as measured on a depth stick.
(a) If $h = 12$ inches, how many gallons of gasoline are in the tank?
(b) If $h = 1$ inch, how many gallons of gasoline are in the tank?

108. Inclined Planes   The final velocity v of an object in feet per second (ft/sec) after it slides down a frictionless inclined plane of height h feet is

$$v = \sqrt{64h + v_0^2}$$

where v_0 is the initial velocity (in ft/sec) of the object.
(a) What is the final velocity v of an object that slides down a frictionless inclined plane of height 4 feet? Assume that the initial velocity is 0.
(b) What is the final velocity v of an object that slides down a frictionless inclined plane of height 16 feet? Assume that the initial velocity is 0.
(c) What is the final velocity v of an object that slides down a frictionless inclined plane of height 2 feet with an initial velocity of 4 ft/sec?

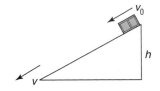

Problems 109–112 require the following information. Express your answer both as a square root and as a decimal.

Period of a Pendulum   *The period T, in seconds, of a pendulum of length l, in feet, may be approximated using the formula*

$$T = 2\pi\sqrt{\dfrac{l}{32}}$$

109. Find the period T of a pendulum whose length is 64 feet.
110. Find the period T of a pendulum whose length is 16 feet.
111. Find the period T of a pendulum whose length is 8 inches.
112. Find the period T of a pendulum whose length is 4 inches.

Discussion and Writing

113. Give an example to show that $\sqrt{a^2}$ is not equal to a. Use it to explain why $\sqrt{a^2} = |a|$.

'Are You Prepared?' Answers

1. $9; -9$ **2.** $4; 4$

A.9 Lines

OBJECTIVES
1 Calculate and Interpret the Slope of a Line (p. A69)
2 Graph Lines Given a Point and the Slope (p. A71)
3 Find the Equation of a Vertical Line (p. A72)
4 Use the Point–Slope Form of a Line; Identify Horizontal Lines (p. A73)
5 Find the Equation of a Line Given Two Points (p. A74)
6 Write the Equation of a Line in Slope–Intercept Form (p. A74)
7 Identify the Slope and *y*-Intercept of a Line from Its Equation (p. A75)
8 Graph Lines Written in General Form Using Intercepts (p. A76)
9 Find Equations of Parallel Lines (p. A77)
10 Find Equations of Perpendicular Lines (p. A78)

In this section we study a certain type of equation that contains two variables, called a *linear equation,* and its graph, a *line.*

Figure 41

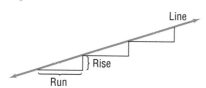

1 Calculate and Interpret the Slope of a Line

Consider the staircase illustrated in Figure 41. Each step contains exactly the same horizontal **run** and the same vertical **rise.** The ratio of the rise to the run, called the *slope,* is a numerical measure of the steepness of the staircase. For example, if the run is increased and the rise remains the same, the staircase becomes less steep. If the run is kept the same, but the rise is increased, the staircase becomes more steep. This important characteristic of a line is best defined using rectangular coordinates.

DEFINITION

Let $P = (x_1, y_1)$ and $Q = (x_2, y_2)$ be two distinct points. If $x_1 \neq x_2$, the **slope m** of the nonvertical line L containing P and Q is defined by the formula

$$m = \frac{y_2 - y_1}{x_2 - x_1} \qquad x_1 \neq x_2 \qquad \textbf{(1)}$$

If $x_1 = x_2$, L is a **vertical line** and the slope m of L is **undefined** (since this results in division by 0).

Figure 42(a) provides an illustration of the slope of a nonvertical line; Figure 42(b) illustrates a vertical line.

Figure 42

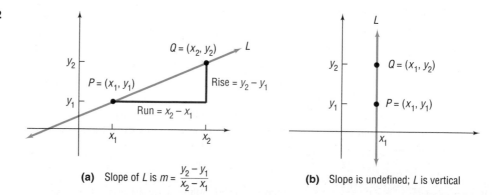

(a) Slope of *L* is $m = \frac{y_2 - y_1}{x_2 - x_1}$

(b) Slope is undefined; *L* is vertical

As Figure 42(a) illustrates, the slope m of a nonvertical line may be viewed as

$$m = \frac{y_2 - y_1}{x_2 - x_1} = \frac{\text{Rise}}{\text{Run}} \quad \text{or} \quad m = \frac{y_2 - y_1}{x_2 - x_1} = \frac{\text{Change in } y}{\text{Change in } x} = \frac{\Delta y}{\Delta x}$$

The slope m of a nonvertical line L measures the amount that y changes as x changes from x_1 to x_2. We call this the **average rate of change** of y with respect to x.

Two comments about computing the slope of a nonvertical line may prove helpful:

1. Any two distinct points on the line can be used to compute the slope of the line. (See Figure 43 for justification.)

> **In Words**
> The symbol Δ is the Greek letter delta. In mathematics, we read Δ as "change in," so $\dfrac{\Delta y}{\Delta x}$ is read, "change in y divided by change in x."

Figure 43
Triangles ABC and PQR are similar (equal angles). Hence, ratios of corresponding sides are proportional so that

Slope using P and $Q = \dfrac{y_2 - y_1}{x_2 - x_1}$

$= \dfrac{d(B, C)}{d(A, C)} = $ Slope using A and B

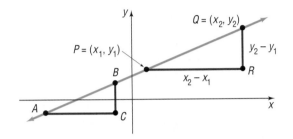

2. The slope of a line may be computed from $P = (x_1, y_1)$ to $Q = (x_2, y_2)$ or from Q to P because

$$\frac{y_2 - y_1}{x_2 - x_1} = \frac{y_1 - y_2}{x_1 - x_2}$$

EXAMPLE 1 | **Finding and Interpreting the Slope of a Line Containing Two Points**

The slope m of the line containing the points $(1, 2)$ and $(5, -3)$ may be computed as

$$m = \frac{-3 - 2}{5 - 1} = \frac{-5}{4} = -\frac{5}{4} \quad \text{or as} \quad m = \frac{2 - (-3)}{1 - 5} = \frac{5}{-4} = -\frac{5}{4}$$

For every 4-unit change in x, y will change by -5 units. That is, if x increases by 4 units, then y will decrease by 5 units. The average rate of change of y with respect to x is $-\dfrac{5}{4}$.

Figure 44

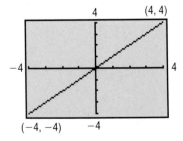

Now Work PROBLEMS 11 AND 17

Square Screens

To get an undistorted view of slope, the same scale must be used on each axis. However, most graphing utilities have a rectangular screen. Because of this, using the same interval for both x and y will result in a distorted view. For example, Figure 44 shows the graph of the line $y = x$ connecting the points $(-4, -4)$ and $(4, 4)$. We expect the line to bisect the first and third quadrants, but it doesn't. We need to adjust the selections for Xmin, Xmax, Ymin, and Ymax so that a **square screen** results. On most graphing utilities, this is accomplished by setting the ratio of x to y at $3:2$.*

Figure 45

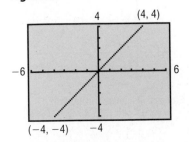

Figure 45 shows the graph of the line $y = x$ on a square screen using a TI-84 Plus. Notice that the line now bisects the first and third quadrants. Compare this illustration to Figure 44.

---

*Most graphing utilities have a feature that automatically squares the viewing window. Consult your owner's manual for the appropriate keystrokes.

To get a better idea of the meaning of the slope m of a line, consider the following:

Exploration

Figure 46

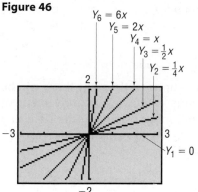

On the same square screen, graph the following equations:

$Y_1 = 0$ *Slope of line is 0.*

$Y_2 = \dfrac{1}{4}x$ *Slope of line is $\dfrac{1}{4}$.*

$Y_3 = \dfrac{1}{2}x$ *Slope of line is $\dfrac{1}{2}$.*

$Y_4 = x$ *Slope of line is 1.*

$Y_5 = 2x$ *Slope of line is 2.*

$Y_6 = 6x$ *Slope of line is 6.*

See Figure 46.

Exploration

Figure 47

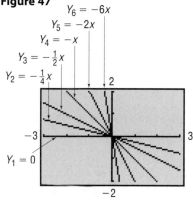

On the same square screen, graph the following equations:

$Y_1 = 0$ *Slope of line is 0.*

$Y_2 = -\dfrac{1}{4}x$ *Slope of line is $-\dfrac{1}{4}$.*

$Y_3 = -\dfrac{1}{2}x$ *Slope of line is $-\dfrac{1}{2}$.*

$Y_4 = -x$ *Slope of line is -1.*

$Y_5 = -2x$ *Slope of line is -2.*

$Y_6 = -6x$ *Slope of line is -6.*

See Figure 47.

Figures 46 and 47 illustrate the following facts:

1. When the slope of a line is positive, the line slants upward from left to right.
2. When the slope of a line is negative, the line slants downward from left to right.
3. When the slope is 0, the line is horizontal.

Figures 46 and 47 also illustrate that the closer the line is to the vertical position, the greater the magnitude of the slope. So, a line with slope 6 will be steeper than a line whose slope is 3.

2 Graph Lines Given a Point and the Slope

The next example illustrates how the slope of a line can be used to graph the line.

EXAMPLE 2   Graphing a Line Given a Point and a Slope

Draw a graph of the line that contains the point $(3, 2)$ and has a slope of:

(a) $\dfrac{3}{4}$ (b) $-\dfrac{4}{5}$

Figure 48

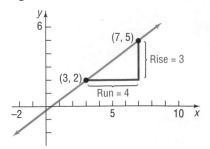

Solution (a) Slope $= \dfrac{\text{Rise}}{\text{Run}}$. The fact that the slope is $\dfrac{3}{4}$ means that for every horizontal movement (run) of 4 units to the right there will be a vertical movement (rise) of 3 units. If we start at the given point $(3, 2)$ and move 4 units to the right and 3 units up, we reach the point $(7, 5)$. By drawing the line through this point and the point $(3, 2)$, we have the graph. See Figure 48.

(b) The fact that the slope is

$$-\frac{4}{5} = \frac{-4}{5} = \frac{\text{Rise}}{\text{Run}}$$

means that for every horizontal movement of 5 units to the right there will be a corresponding vertical movement of -4 units (a downward movement of 4 units). If we start at the given point $(3, 2)$ and move 5 units to the right and then 4 units down, we arrive at the point $(8, -2)$. By drawing the line through these points, we have the graph. See Figure 49.

Alternatively, we can set

$$-\frac{4}{5} = \frac{4}{-5} = \frac{\text{Rise}}{\text{Run}}$$

so that for every horizontal movement of -5 units (a movement to the left of 5 units) there will be a corresponding vertical movement of 4 units (upward). This approach brings us to the point $(-2, 6)$, which is also on the graph shown in Figure 49.

Figure 49

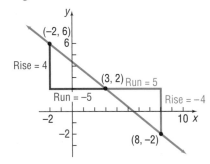

➤ **Now Work** PROBLEM 23

3 Find the Equation of a Vertical Line

Now that we have discussed the slope of a line, we are ready to derive equations of lines. As we shall see, there are several forms of the equation of a line. Let's start with an example.

EXAMPLE 3 **Graphing a Line**

Graph the equation: $x = 3$

Solution To graph $x = 3$ by hand, we are looking for all points (x, y) in the plane for which $x = 3$. No matter what y-coordinate is used, the corresponding x-coordinate always equals 3. Consequently, the graph of the equation $x = 3$ is a vertical line with x-intercept 3 and undefined slope. See Figure 50(a).

To use a graphing utility, we need to express the equation in the form $y = \{\text{expression in } x\}$. But $x = 3$ cannot be put into this form, so an alternative method must be used. Consult your manual to determine the key strokes required to draw vertical lines. Figure 50(b) shows the graph that you should obtain.

Figure 50

$x = 3$

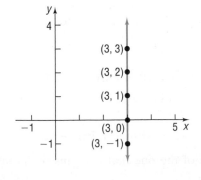

(a)

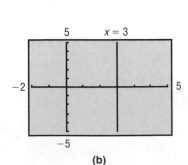

(b)

As suggested by Example 3, we have the following result:

THEOREM   **Equation of a Vertical Line**

A vertical line is given by an equation of the form

$$x = a$$

where a is the x-intercept.

Figure 51

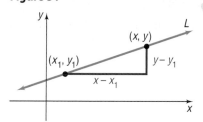

4 Use the Point–Slope Form of a Line; Identify Horizontal Lines

Now let L be a nonvertical line with slope m and containing the point (x_1, y_1). See Figure 51. For any other point (x, y) on L, we have

$$m = \frac{y - y_1}{x - x_1} \quad \text{or} \quad y - y_1 = m(x - x_1)$$

THEOREM   **Point–Slope Form of an Equation of a Line**

An equation of a nonvertical line of slope m that contains the point (x_1, y_1) is

$$y - y_1 = m(x - x_1) \tag{2}$$

EXAMPLE 4   Using the Point–Slope Form of a Line

Figure 52

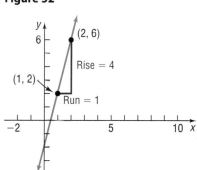

An equation of the line with slope 4 and containing the point $(1, 2)$ can be found by using the point–slope form with $m = 4$, $x_1 = 1$, and $y_1 = 2$.

$$
\begin{aligned}
y - y_1 &= m(x - x_1) \quad &\text{Point-slope form} \\
y - 2 &= 4(x - 1) \quad &m = 4,\, x_1 = 1,\, y_1 = 2 \\
y &= 4x - 2
\end{aligned}
$$

See Figure 52 for the graph.

EXAMPLE 5   Finding the Equation of a Horizontal Line

Find an equation of the horizontal line containing the point $(3, 2)$.

Solution   Because all the y-values are the same on a horizontal line, the slope of a horizontal line is 0. To get an equation, we use the point–slope form with $m = 0$, $x_1 = 3$, and $y_1 = 2$.

Figure 53
$y = 2$

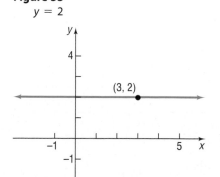

$$
\begin{aligned}
y - y_1 &= m(x - x_1) \quad &\text{Point-slope form} \\
y - 2 &= 0 \cdot (x - 3) \quad &m = 0,\, x_1 = 3,\, \text{and } y_1 = 2 \\
y - 2 &= 0 \\
y &= 2
\end{aligned}
$$

See Figure 53 for the graph.

As suggested by Example 5, we have the following result:

THEOREM **Equation of a Horizontal Line**

A horizontal line is given by an equation of the form

$$y = b$$

where b is the y-intercept.

5 Find the Equation of a Line Given Two Points

We use the slope formula and the point–slope form of a line to find an equation given two points.

EXAMPLE 6 **Finding an Equation of a Line Given Two Points**

Find an equation of the line containing the points $(2, 3)$ and $(-4, 5)$. Graph the line.

Solution We first compute the slope of the line with $(x_1, y_1) = (2, 3)$ and $(x_2, y_2) = (-4, 5)$.

$$m = \frac{5 - 3}{-4 - 2} = \frac{2}{-6} = -\frac{1}{3} \qquad m = \frac{y_2 - y_1}{x_2 - x_1}$$

Figure 54

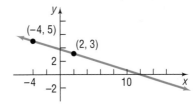

We use the point $(x_1, y_1) = (2, 3)$ and the slope $m = -\frac{1}{3}$ to get the point–slope form of the equation of the line.

$$y - 3 = -\frac{1}{3}(x - 2) \qquad y - y_1 = m(x - x_1)$$

See Figure 54 for the graph.

In the solution to Example 6, we could have used the other point, $(-4, 5)$, instead of the point $(2, 3)$. The equation that results, although it looks different, is equivalent to the equation that we obtained in the example. (Try it for yourself.)

Now Work PROBLEM 37

6 Write the Equation of a Line in Slope–Intercept Form

Another useful equation of a line is obtained when the slope m and y-intercept b are known. In this event, we know both the slope m of the line and a point $(0, b)$ on the line; thus, we may use the point–slope form, equation (2), to obtain the following equation:

$$y - b = m(x - 0) \quad \text{or} \quad y = mx + b$$

THEOREM **Slope–Intercept Form of an Equation of a Line**

An equation of a line with slope m and y-intercept b is

$$y = mx + b \tag{3}$$

Now Work PROBLEM 45

(EXPRESS ANSWER IN SLOPE–INTERCEPT FORM)

Seeing the Concept

Figure 55
$y = mx + 2$

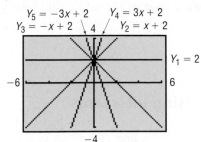

To see the role that the slope m plays, graph the following lines on the same square screen.

$$Y_1 = 2 \qquad\qquad m = 0$$
$$Y_2 = x + 2 \qquad\quad m = 1$$
$$Y_3 = -x + 2 \qquad m = -1$$
$$Y_4 = 3x + 2 \qquad m = 3$$
$$Y_5 = -3x + 2 \qquad m = -3$$

See Figure 55. What do you conclude about the lines $y = mx + 2$?

Seeing the Concept

Figure 56
$y = 2x + b$

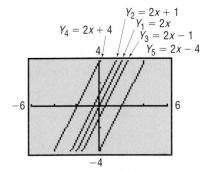

To see the role of the y-intercept b, graph the following lines on the same square screen.

$$Y_1 = 2x \qquad\qquad b = 0$$
$$Y_2 = 2x + 1 \qquad b = 1$$
$$Y_3 = 2x - 1 \qquad b = -1$$
$$Y_4 = 2x + 4 \qquad b = 4$$
$$Y_5 = 2x - 4 \qquad b = -4$$

See Figure 56. What do you conclude about the lines $y = 2x + b$?

7 Identify the Slope and *y*-Intercept of a Line from Its Equation

When the equation of a line is written in slope–intercept form, it is easy to find the slope m and y-intercept b of the line. For example, suppose that the equation of a line is

$$y = -2x + 3$$

Compare it to $y = mx + b$.

$$y = -2x + 3$$
$$\qquad\quad \uparrow \qquad \uparrow$$
$$y = \quad mx + b$$

The slope of this line is -2 and its y-intercept is 3.

━━ **Now Work** PROBLEM 71

EXAMPLE 7	**Finding the Slope and y-Intercept**

Find the slope m and y-intercept b of the equation $2x + 4y = 8$. Graph the equation.

Solution

To obtain the slope and y-intercept, we transform the equation into its slope–intercept form by solving for y.

$$2x + 4y = 8$$
$$4y = -2x + 8$$
$$y = -\frac{1}{2}x + 2 \quad y = mx + b$$

Figure 57

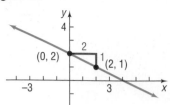

The coefficient of x, $-\dfrac{1}{2}$, is the slope, and the y-intercept is 2. We can graph the line using the fact that the y-intercept is 2 and the slope is $-\dfrac{1}{2}$. Starting at the point $(0, 2)$, go to the right 2 units and then down 1 unit to the point $(2, 1)$. See Figure 57.

Now Work PROBLEM 77

8 Graph Lines Written in General Form Using Intercepts

The form of the equation of the line in Example 7, $2x + 4y = 8$, is called the *general form*.

DEFINITION

The equation of a line is in **general form** when it is written as

$$Ax + By = C \qquad\qquad \textbf{(4)}$$

where A, B, and C are real numbers and A and B are not both 0.

When we want to graph a linear equation that is written in general form, we can solve the equation for y and write the equation in slope–intercept form as we did in Example 7. Another approach to graphing the equation would be to find its intercepts. Remember, the intercepts of the graph of an equation are the points where the graph crosses or touches a coordinate axis.

EXAMPLE 8 | **Graphing a Linear Equation Using Its Intercepts**

Graph the linear equation $2x + 4y = 8$ by finding its intercepts.

Solution

To obtain the x-intercept, let $y = 0$ in the equation and solve for x.

$$2x + 4y = 8$$
$$2x + 4(0) = 8 \quad \text{Let } y = 0.$$
$$2x = 8$$
$$x = 4 \quad \text{Divide both sides by 2.}$$

The x-intercept is 4 and the point $(4, 0)$ is on the graph of the equation.
To obtain the y-intercept, let $x = 0$ in the equation and solve for y.

$$2x + 4y = 8$$
$$2(0) + 4y = 8 \quad \text{Let } x = 0.$$
$$4y = 8$$
$$y = 2 \quad \text{Divide both sides by 4.}$$

Figure 58

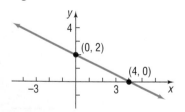

The y-intercept is 2 and the point $(0, 2)$ is on the graph of the equation.
We plot the points $(4, 0)$ and $(0, 2)$ in a Cartesian plane and draw a line through the points. See Figure 58.

Now Work PROBLEM 91

Every line has an equation that is equivalent to an equation written in general form. For example, a vertical line whose equation is

$$x = a$$

can be written in the general form

$$1 \cdot x + 0 \cdot y = a \quad A = 1, B = 0, C = a$$

A horizontal line whose equation is

$$y = b$$

can be written in the general form

$$0 \cdot x + 1 \cdot y = b \qquad A = 0, B = 1, C = b$$

Lines that are neither vertical nor horizontal have general equations of the form

$$Ax + By = C \qquad A \neq 0 \text{ and } B \neq 0$$

Because the equation of every line can be written in general form, any equation equivalent to equation (4) is called a **linear equation.**

9 Find Equations of Parallel Lines

Figure 59
The distinct lines are parallel if and only if their slopes are equal.

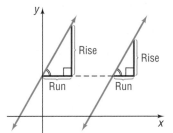

When two lines (in the plane) do not intersect (that is, they have no points in common), they are said to be **parallel.** Look at Figure 59. There we have drawn two lines and have constructed two right triangles by drawing sides parallel to the coordinate axes. These lines are parallel if and only if the right triangles are similar. (Do you see why? Two angles are equal.) And the triangles are similar if and only if the ratios of corresponding sides are equal.

This suggests the following result:

THEOREM

Criterion for Parallel Lines

Two nonvertical lines are parallel if and only if their slopes are equal and they have different y-intercepts.

The use of the words "if and only if" in the preceding theorem means that two statements are being made, one the converse of the other.

If two nonvertical lines are parallel, then their slopes are equal and they have different y-intercepts.

If two nonvertical lines have equal slopes and they have different y-intercepts, then they are parallel.

EXAMPLE 9

Showing That Two Lines Are Parallel

Show that the lines given by the following equations are parallel:

$$L_1: \ 2x + 3y = 6 \qquad L_2: \ 4x + 6y = 0$$

Solution

We need to determine whether these lines have equal slopes and different y-intercepts, so we write each equation in slope–intercept form:

$$
\begin{array}{ll}
L_1: \ 2x + 3y = 6 & L_2: \ 4x + 6y = 0 \\
\quad\ \ 3y = -2x + 6 & \quad\ \ 6y = -4x \\
\quad\ \ \ y = -\dfrac{2}{3}x + 2 & \quad\ \ \ y = -\dfrac{2}{3}x \\
\end{array}
$$

$$\text{Slope} = -\frac{2}{3}; \ y\text{-intercept} = 2 \qquad \text{Slope} = -\frac{2}{3}; \ y\text{-intercept} = 0$$

Because these lines have the same slope, $-\dfrac{2}{3}$, but different y-intercepts, the lines are parallel. See Figure 60.

Figure 60
Parallel lines

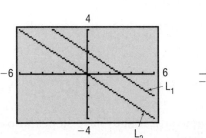

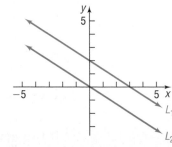

| EXAMPLE 10 | **Finding a Line That Is Parallel to a Given Line** |

Find an equation for the line that contains the point $(2, -3)$ and is parallel to the line $2x + y = 6$. By hand, graph the two lines.

Solution

Since the two lines are to be parallel, the slope of the line that we seek equals the slope of the line $2x + y = 6$. We begin by writing the equation of the line $2x + y = 6$ in slope–intercept form.

$$2x + y = 6$$
$$y = -2x + 6$$

The slope is -2. Since the line that we seek contains the point $(2, -3)$, we use the point–slope form to obtain

$$\begin{aligned} y - y_1 &= m(x - x_1) && \text{Point–slope form} \\ y - (-3) &= -2(x - 2) && m = -2, x_1 = 2, y_1 = -3 \\ y + 3 &= -2x + 4 && \text{Simplify.} \\ y &= -2x + 1 && \text{Slope–intercept form} \\ 2x + y &= 1 && \text{General form} \end{aligned}$$

This line is parallel to the line $2x + y = 6$ and contains the point $(2, -3)$. See Figure 61.

Figure 61

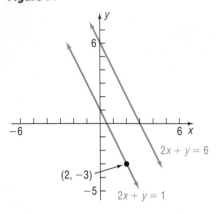

$2x + y = 6$

$(2, -3)$

$2x + y = 1$

➤ **Now Work** PROBLEM 59

10 Find Equations of Perpendicular Lines

When two lines intersect at a right angle (90°), they are said to be **perpendicular.** See Figure 62.

Figure 62
Perpendicular lines

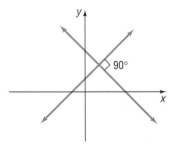

90°

The following result gives a condition, in terms of their slopes, for two lines to be perpendicular.

THEOREM

Criterion for Perpendicular Lines

Two nonvertical lines are perpendicular if and only if the product of their slopes is -1.

Here we shall prove the "only if" part of the statement:

If two nonvertical lines are perpendicular, then the product of their slopes is -1.

In Problem 128, you are asked to prove the "if" part of the theorem; that is:

If two nonvertical lines have slopes whose product is -1, then the lines are perpendicular.

Figure 63

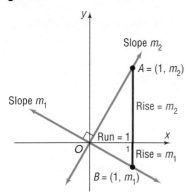

Proof   Let m_1 and m_2 denote the slopes of the two lines. There is no loss in generality (that is, neither the angle nor the slopes are affected) if we situate the lines so that they meet at the origin. See Figure 63. The point $A = (1, m_2)$ is on the line having slope m_2, and the point $B = (1, m_1)$ is on the line having slope m_1. (Do you see why this must be true?)

Suppose that the lines are perpendicular. Then triangle OAB is a right triangle. As a result of the Pythagorean Theorem, it follows that

$$[d(O, A)]^2 + [d(O, B)]^2 = [d(A, B)]^2 \qquad (5)$$

By the distance formula, we can write the squares of these distances as

$$[d(O, A)]^2 = (1 - 0)^2 + (m_2 - 0)^2 = 1 + m_2^2$$
$$[d(O, B)]^2 = (1 - 0)^2 + (m_1 - 0)^2 = 1 + m_1^2$$
$$[d(A, B)]^2 = (1 - 1)^2 + (m_2 - m_1)^2 = m_2^2 - 2m_1m_2 + m_1^2$$

Using these facts in equation (5), we get

$$(1 + m_2^2) + (1 + m_1^2) = m_2^2 - 2m_1m_2 + m_1^2$$

which, upon simplification, can be written as

$$m_1m_2 = -1$$

If the lines are perpendicular, the product of their slopes is -1. ∎

You may find it easier to remember the condition for two nonvertical lines to be perpendicular by observing that the equality $m_1m_2 = -1$ means that m_1 and m_2 are negative reciprocals of each other; that is, $m_1 = -\dfrac{1}{m_2}$ and $m_2 = -\dfrac{1}{m_1}$.

EXAMPLE 11   **Finding the Slope of a Line Perpendicular to Another Line**

If a line has slope $\dfrac{3}{2}$, any line having slope $-\dfrac{2}{3}$ is perpendicular to it.

EXAMPLE 12   **Finding the Equation of a Line Perpendicular to a Given Line**

Find an equation of the line that contains the point $(1, -2)$ and is perpendicular to the line $x + 3y = 6$. Graph the two lines.

Solution   We first write the equation of the given line in slope–intercept form to find its slope.

$$x + 3y = 6$$
$$3y = -x + 6 \qquad \text{Proceed to solve for } y.$$
$$y = -\frac{1}{3}x + 2 \qquad \text{Place in the form } y = mx + b.$$

The given line has slope $-\dfrac{1}{3}$. Any line perpendicular to this line will have slope 3.

Because we require the point $(1, -2)$ to be on this line with slope 3, we use the point–slope form of the equation of a line.

$$y - y_1 = m(x - x_1) \qquad \text{Point–slope form}$$
$$y - (-2) = 3(x - 1) \qquad m = 3, x_1 = 1, y_1 = -2$$

To obtain other forms of the equation, we proceed as follows:

$$y + 2 = 3(x - 1)$$
$$y + 2 = 3x - 3 \qquad \text{Simplify}$$
$$y = 3x - 5 \qquad \text{Slope–intercept form}$$
$$3x - y = 5 \qquad \text{General form}$$

Figure 64 on the next page shows the graphs.

Figure 64

WARNING Be sure to use a square screen when you graph perpendicular lines. Otherwise, the angle between the two lines will appear distorted. ∎

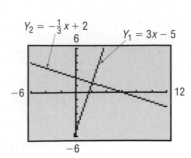

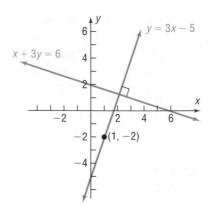

Now Work PROBLEM 65

A.9 Assess Your Understanding

Concepts and Vocabulary

1. The slope of a vertical line is _____; the slope of a horizontal line is _____.

2. For the line $2x + 3y = 6$, the x-intercept is _____ and the y-intercept is _____.

3. A horizontal line is given by an equation of the form _____, where b is the _____.

4. *True or False* Vertical lines have an undefined slope.

5. *True or False* The slope of the line $2y = 3x + 5$ is 3.

6. *True or False* The point $(1, 2)$ is on the line $2x + y = 4$.

7. Two nonvertical lines have slopes m_1 and m_2, respectively. The lines are parallel if _____ and the _____ are unequal; the lines are perpendicular if _____.

8. The lines $y = 2x + 3$ and $y = ax + 5$ are parallel if $a =$ _____.

9. The lines $y = 2x - 1$ and $y = ax + 2$ are perpendicular if $a =$ _____.

10. *True or False* Perpendicular lines have slopes that are reciprocals of one another.

Skill Building

In Problems 11–14, (a) find the slope of the line and (b) interpret the slope.

11.
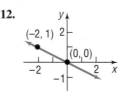

12.

13.
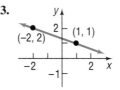

14.
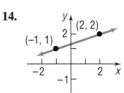

In Problems 15–22, plot each pair of points and determine the slope of the line containing them. Graph the line.

15. $(2, 3); (4, 0)$

16. $(4, 2); (3, 4)$

17. $(-2, 3); (2, 1)$

18. $(-1, 1); (2, 3)$

19. $(-3, -1); (2, -1)$

20. $(4, 2); (-5, 2)$

21. $(-1, 2); (-1, -2)$

22. $(2, 0); (2, 2)$

In Problems 23–30, graph the line containing the point P and having slope m.

23. $P = (1, 2); m = 3$

24. $P = (2, 1); m = 4$

25. $P = (2, 4); m = -\dfrac{3}{4}$

26. $P = (1, 3); m = -\dfrac{2}{5}$

27. $P = (-1, 3); m = 0$

28. $P = (2, -4); m = 0$

29. $P = (0, 3)$; slope undefined

30. $P = (-2, 0)$; slope undefined

In Problems 31–36, the slope and a point on a line are given. Use this information to locate three additional points on the line. Answers may vary.

[**Hint:** *It is not necessary to find the equation of the line. See Example 2.*]

31. Slope 4; point $(1, 2)$

32. Slope 2; point $(-2, 3)$

33. Slope $-\dfrac{3}{2}$; point $(2, -4)$

34. Slope $\dfrac{4}{3}$; point $(-3, 2)$

35. Slope -2; point $(-2, -3)$

36. Slope -1; point $(4, 1)$

In Problems 37–44, find an equation of the line L.

37.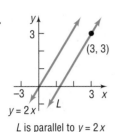

38.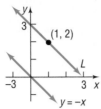

39.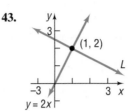

40.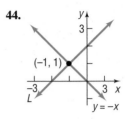

41.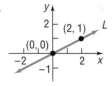

L is parallel to y = 2x

42.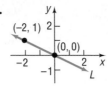

L is parallel to y = −x

43.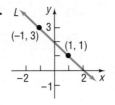

L is perpendicular to y = 2x

44.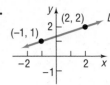

L is perpendicular to y = −x

In Problems 45–70, find an equation for the line with the given properties. Express your answer using either the general form or the slope–intercept form of the equation of a line, whichever you prefer.

45. Slope $= 3$; containing the point $(-2, 3)$

46. Slope $= 2$; containing the point $(4, -3)$

47. Slope $= -\dfrac{2}{3}$; containing the point $(1, -1)$

48. Slope $= \dfrac{1}{2}$; containing the point $(3, 1)$

49. Containing the points $(1, 3)$ and $(-1, 2)$

50. Containing the points $(-3, 4)$ and $(2, 5)$

51. Slope $= -3$; y-intercept $= 3$

52. Slope $= -2$; y-intercept $= -2$

53. x-intercept $= 2$; y-intercept $= -1$

54. x-intercept $= -4$; y-intercept $= 4$

55. Slope undefined; containing the point $(2, 4)$

56. Slope undefined; containing the point $(3, 8)$

57. Horizontal; containing the point $(-3, 2)$

58. Vertical; containing the point $(4, -5)$

59. Parallel to the line $y = 2x$; containing the point $(-1, 2)$

60. Parallel to the line $y = -3x$; containing the point $(-1, 2)$

61. Parallel to the line $2x - y = -2$; containing the point $(0, 0)$

62. Parallel to the line $x - 2y = -5$; containing the point $(0, 0)$

63. Parallel to the line $x = 5$; containing the point $(4, 2)$

64. Parallel to the line $y = 5$; containing the point $(4, 2)$

65. Perpendicular to the line $y = \dfrac{1}{2}x + 4$; containing the point $(1, -2)$

66. Perpendicular to the line $y = 2x - 3$; containing the point $(1, -2)$

67. Perpendicular to the line $2x + y = 2$; containing the point $(-3, 0)$

68. Perpendicular to the line $x - 2y = -5$; containing the point $(0, 4)$

69. Perpendicular to the line $x = 8$; containing the point $(3, 4)$

70. Perpendicular to the line $y = 8$; containing the point $(3, 4)$

In Problems 71–90, find the slope and y-intercept of each line. Graph the line by hand. Check your graph using a graphing utility.

71. $y = 2x + 3$

72. $y = -3x + 4$

73. $\dfrac{1}{2}y = x - 1$

74. $\dfrac{1}{3}x + y = 2$

75. $y = \dfrac{1}{2}x + 2$

76. $y = 2x + \dfrac{1}{2}$

77. $x + 2y = 4$

78. $-x + 3y = 6$

79. $2x - 3y = 6$

80. $3x + 2y = 6$

81. $x + y = 1$

82. $x - y = 2$

83. $x = -4$

84. $y = -1$

85. $y = 5$

86. $x = 2$

87. $y - x = 0$

88. $x + y = 0$

89. $2y - 3x = 0$

90. $3x + 2y = 0$

In Problems 91–100, (a) find the intercepts of the graph of each equation and (b) graph the equation.

 91. $2x + 3y = 6$

92. $3x - 2y = 6$

93. $-4x + 5y = 40$

94. $6x - 4y = 24$

95. $7x + 2y = 21$

96. $5x + 3y = 18$

97. $\dfrac{1}{2}x + \dfrac{1}{3}y = 1$

98. $x - \dfrac{2}{3}y = 4$

99. $0.2x - 0.5y = 1$

100. $-0.3x + 0.4y = 1.2$

101. Find an equation of the *x*-axis.

102. Find an equation of the *y*-axis.

In Problems 103–106, the equations of two lines are given. Determine if the lines are parallel, perpendicular, or neither.

103. $y = 2x - 3$
$y = 2x + 4$

104. $y = \dfrac{1}{2}x - 3$
$y = -2x + 4$

105. $y = 4x + 5$
$y = -4x + 2$

106. $y = -2x + 3$
$y = -\dfrac{1}{2}x + 2$

In Problems 107–110, match each graph with the correct equation:

(a) $y = x$ (b) $y = 2x$ (c) $y = \dfrac{x}{2}$ (d) $y = 4x$

107.

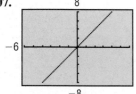

108.

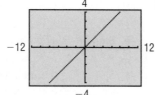

109.

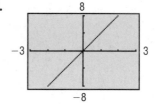

110.

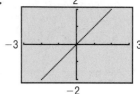

Applications and Extensions

111. Geometry Use slopes to show that the triangle whose vertices are $(-2, 5)$, $(1, 3)$, and $(-1, 0)$ is a right triangle.

112. Geometry Use slopes to show that the quadrilateral whose vertices are $(1, -1)$, $(4, 1)$, $(2, 2)$, and $(5, 4)$ is a parallelogram.

113. Geometry Use slopes to show that the quadrilateral whose vertices are $(-1, 0)$, $(2, 3)$, $(1, -2)$, and $(4, 1)$ is a rectangle.

114. Geometry Use slopes and the distance formula to show that the quadrilateral whose vertices are $(0, 0)$, $(1, 3)$, $(4, 2)$, and $(3, -1)$ is a square.

*115. Truck Rentals A truck rental company rents a moving truck for one day by charging $29 plus $0.20 per mile. Write a linear model that relates the cost *C*, in dollars, of renting the truck to the number *x* of miles driven. What is the cost of renting the truck if the truck is driven 110 miles? 230 miles?

116. Cost Equation The **fixed costs** of operating a business are the costs incurred regardless of the level of production. Fixed costs include rent, fixed salaries, and costs of leasing machinery. The **variable costs** of operating a business are the costs that change with the level of output. Variable costs include raw materials, hourly wages, and electricity. Suppose that a manufacturer of jeans has fixed daily costs of $500 and variable costs of $8 for each pair of jeans manufactured. Write a linear model that relates the daily cost *C*, in dollars, of manufacturing the jeans to the number *x* of jeans manufactured. What is the cost of manufacturing 400 pairs of jeans? 740 pairs?

117. Cost of Sunday Home Delivery The cost to the *Chicago Tribune* for Sunday home delivery is approximately $0.53 per newspaper with fixed costs of $1,070,000. Write a linear model that relates the cost *C* and the number *x* of copies delivered.

Source: Chicago Tribune, 2002.

118. Wages of a Car Salesperson Dan receives $375 per week for selling new and used cars at a car dealership in Oak Lawn, Illinois. In addition, he receives 5% of the profit on any sales that he generates. Write a linear model that relates Dan's weekly salary *S* when he has sales that generate a profit of *x* dollars.

119. Electricity Rates in Illinois Commonwealth Edison Company supplies electricity to residential customers for a monthly customer charge of $7.58 plus 8.275 cents per kilowatt-hour for up to 400 kilowatt-hours.

(a) Write a linear model that relates the monthly charge *C*, in dollars, to the number *x* of kilowatt-hours used in a month, $0 \leq x \leq 400$.

*This is a Model It! icon. It indicates that problems in purple require building a mathematical model.

(b) Graph this equation.

(c) What is the monthly charge for using 100 kilowatt-hours?

(d) What is the monthly charge for using 300 kilowatt-hours?

(e) Interpret the slope of the line.

Source: Commonwealth Edison Company, April 2006.

120. **Electricity Rates in Florida** Florida Power & Light Company supplies electricity to residential customers for a monthly customer charge of $5.17 plus 10.07 cents per kilowatt-hour for up to 1000 kilowatt-hours.

(a) Write a linear model that relates the monthly charge C, in dollars, to the number x of kilowatt-hours used in a month, $0 \le x \le 1000$.

(b) Graph this equation.

(c) What is the monthly charge for using 200 kilowatt-hours?

(d) What is the monthly charge for using 500 kilowatt-hours?

(e) Interpret the slope of the line.

Source: Florida Power & Light Company, April 2006.

121. **Measuring Temperature** The relationship between Celsius (°C) and Fahrenheit (°F) degrees for measuring temperature is linear. Find a linear equation relating °C and °F if 0°C corresponds to 32°F and 100°C corresponds to 212°F. Use the equation to find the Celsius measure of 70°F.

122. **Measuring Temperature** The Kelvin (K) scale for measuring temperature is obtained by adding 273 to the Celsius temperature.

(a) Write a linear equation relating K and °C.

(b) Write a linear equation relating K and °F (see Problem 121).

123. **Access Ramp** A wooden access ramp is being built to reach a platform that sits 30 inches above the floor. The proposed ramp drops 2 inches for every 25-inch run.

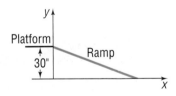

(a) Write a linear equation that relates the height y above the floor to the horizontal distance x from the platform.

(b) Find and interpret the x-intercept of the graph of your equation.

(c) Design requirements stipulate that the maximum run be 30 feet and that the maximum slope be a drop of 1 inch for each 12 inches of run. Will this ramp meet the requirements? Explain.

(d) What slopes could be used to obtain the 30-inch rise and still meet design requirements?

Source: www.adaptiveaccess.com/wood_ramps.php

124. **Cigarette Use** A study by the Partnership for a Drug-Free America indicated that, in 1998, 42% of teens in grades 7 through 12 had recently used cigarettes. A similar study in 2005 indicated that 22% of such teens had recently used cigarettes.

(a) Write a linear model that relates the percent of teens that recently used cigarettes y to the number of years after 1998, x.

(b) Find the intercepts of the graph of your equation.

(c) Do the intercepts have any meaningful interpretation?

(d) Use your equation to predict the percent for the year 2019. Is this result reasonable?

Source: www.drugfree.org

125. **Product Promotion** A cereal company finds that the number of people who will buy one of its products in the first month that it is introduced is linearly related to the amount of money it spends on advertising. If it spends $40,000 on advertising, then 100,000 boxes of cereal will be sold, and if it spends $60,000, then 200,000 boxes will be sold.

(a) Write a linear model that relates the amount A spent on advertising to the number x of boxes the company aims to sell.

(b) How much advertising is needed to sell 300,000 boxes of cereal?

(c) Interpret the slope.

126. Show that the line containing the points (a, b) and (b, a), $a \ne b$, is perpendicular to the line $y = x$. Also show that the midpoint of (a, b) and (b, a) lies on the line $y = x$.

127. The equation $2x - y = C$ defines a **family of lines,** one line for each value of C. On one set of coordinate axes, graph the members of the family when $C = -4$, $C = 0$, and $C = 2$. Can you draw a conclusion from the graph about each member of the family?

128. Prove that if two nonvertical lines have slopes whose product is -1 then the lines are perpendicular. [**Hint:** Refer to Figure 63 and use the converse of the Pythagorean Theorem.]

Discussion and Writing

129. Which of the following equations might have the graph shown? (More than one answer is possible.)

(a) $2x + 3y = 6$

(b) $-2x + 3y = 6$

(c) $3x - 4y = -12$

(d) $x - y = 1$

(e) $x - y = -1$

(f) $y = 3x - 5$

(g) $y = 2x + 3$

(h) $y = -3x + 3$

130. Which of the following equations might have the graph shown? (More than one answer is possible.)

(a) $2x + 3y = 6$

(b) $2x - 3y = 6$

(c) $3x + 4y = 12$

(d) $x - y = 1$

(e) $x - y = -1$

(f) $y = -2x - 1$

(g) $y = -\dfrac{1}{2}x + 10$

(h) $y = x + 4$

131. The figure shows the graph of two parallel lines. Which of the following pairs of equations might have such a graph?

(a) $x - 2y = 3$
    $x + 2y = 7$
(b) $x + y = 2$
    $x + y = -1$
(c) $x - y = -2$
    $x - y = 1$
(d) $x - y = -2$
    $2x - 2y = -4$
(e) $x + 2y = 2$
    $x + 2y = -1$

132. The figure shows the graph of two perpendicular lines. Which of the following pairs of equations might have such a graph?

(a) $y - 2x = 2$
    $y + 2x = -1$
(b) $y - 2x = 0$
    $2y + x = 0$
(c) $2y - x = 2$
    $2y + x = -2$
(d) $y - 2x = 2$
    $x + 2y = -1$
(e) $2x + y = -2$
    $2y + x = -2$

133. *m* **Is for Slope** The accepted symbol used to denote the slope of a line is the letter *m*. Investigate the origin of this symbolism. Begin by consulting a French dictionary and looking up the French word *monter*. Write a brief essay on your findings.

134. **Grade of a Road** The term *grade* is used to describe the inclination of a road. How does this term relate to the notion of slope of a line? Is a 4% grade very steep? Investigate the grades of some mountainous roads and determine their slopes. Write a brief essay on your findings.

135. **Carpentry** Carpenters use the term *pitch* to describe the steepness of staircases and roofs. How does pitch relate to slope? Investigate typical pitches used for stairs and for roofs. Write a brief essay on your findings.

136. Can the equation of every line be written in slope–intercept form? Why?

137. Does every line have exactly one *x*-intercept and one *y*-intercept? Are there any lines that have no intercepts?

138. What can you say about two lines that have equal slopes and equal *y*-intercepts?

139. What can you say about two lines with the same *x*-intercept and the same *y*-intercept? Assume that the *x*-intercept is not 0.

140. If two distinct lines have the same slope, but different *x*-intercepts, can they have the same *y*-intercept?

141. If two distinct lines have the same *y*-intercept, but different slopes, can they have the same *x*-intercept?

142. **What Went Wrong?** A student is asked to find the slope of the line joining $(-3, 2)$ and $(1, -4)$. He states that the slope is $\dfrac{3}{2}$. Is he correct? If not, what went wrong?

A.10 Building Linear Models from Data

OBJECTIVES 1 Draw and Interpret Scatter Diagrams (p. A84)

2 Distinguish between Linear and Nonlinear Relations (p. A85)

3 Use a Graphing Utility to Find the Line of Best Fit (p. A87)

1 Draw and Interpret Scatter Diagrams

Linear models can be constructed by fitting a linear function to data. The first step in finding this relation is to plot the ordered pairs using rectangular coordinates. The resulting graph is called a **scatter diagram.**

EXAMPLE 1   Drawing and Interpreting a Scatter Diagram

In baseball, the on-base percentage for a team represents the percentage of time that the players safely reach base. The data given in Table 2 represent the number of runs scored *y* and the on-base percentage *x* for teams in the National League during the 2006 baseball season.

Table 2

Team	On-Base Percentage, x	Runs Scored, y	(x, y)
Atlanta	33.7	849	(33.7, 849)
St. Louis	33.7	781	(33.7, 781)
Colorado	34.1	813	(34.1, 813)
Houston	33.2	735	(33.2, 735)
Philadelphia	34.7	865	(34.7, 865)
San Francisco	32.4	746	(32.4, 746)
Pittsburgh	32.7	691	(32.7, 691)
Florida	33.1	758	(33.1, 758)
Chicago Cubs	31.9	716	(31.9, 716)
Arizona	33.1	773	(33.1, 773)
Milwaukee	32.7	730	(32.7, 730)
Washington	33.8	746	(33.8, 746)
Cincinnati	33.6	749	(33.6, 749)
San Diego	33.2	731	(33.2, 731)
NY Mets	33.4	834	(33.4, 834)
Los Angeles	34.8	820	(34.8, 820)

Source: espn.com

(a) Draw a scatter diagram of the data, treating on-base percentage as the independent variable.

(b) Use a graphing utility to draw a scatter diagram.

(c) Describe what happens to runs scored as the on-base percentage increases.

Solution

(a) To draw a scatter diagram, we plot the ordered pairs listed in Table 2, with the on-base percentage as the *x*-coordinate and the runs scored as the *y*-coordinate. See Figure 65(a). Notice that the points in the scatter diagram are not connected.

(b) Figure 65(b) shows a scatter diagram using a TI-84 Plus graphing calculator.

(c) We see from the scatter diagrams that, as the on-base percentage increases, the trend is that the number of runs scored also increases.

Figure 65

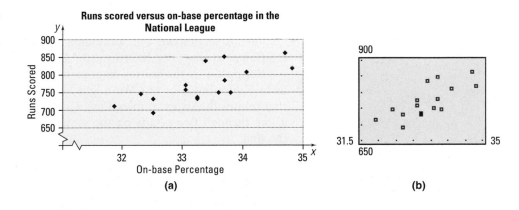

(a)

(b)

Now Work   PROBLEM 9(a)

2 Distinguish between Linear and Nonlinear Relations

You should notice that the points in Figure 65 do not follow a perfect linear relation. However, the data do exhibit a linear pattern. There are numerous explanations as to why the data are not perfectly linear, but one easy explanation is the fact that other variables besides on-base percentage play a role in determining runs scored, such as number of home runs hit.

Scatter diagrams are used to help us to see the type of relation that exists between two variables. In this text, we will discuss a variety of different relations that may exist between two variables. For now, we concentrate on distinguishing between linear and nonlinear relations. See Figure 66.

Figure 66

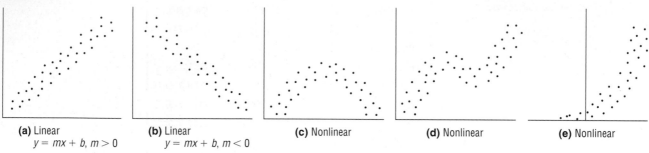

(a) Linear
$y = mx + b, m > 0$

(b) Linear
$y = mx + b, m < 0$

(c) Nonlinear

(d) Nonlinear

(e) Nonlinear

EXAMPLE 2   **Distinguishing between Linear and Nonlinear Relations**

Determine whether each relation between the two variables in Figure 67 is linear or nonlinear.

Figure 67

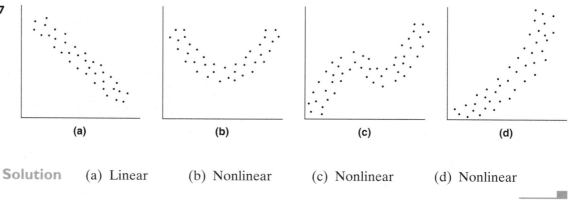

(a)   (b)   (c)   (d)

Solution   (a) Linear   (b) Nonlinear   (c) Nonlinear   (d) Nonlinear

 Now Work PROBLEM 3

In this section we will study data whose scatter diagrams imply that a linear relation exists between the two variables.

Suppose that the scatter diagram of a set of data appears to be linearly related as in Figure 66(a) or (b). We might wish to find an equation of a line (a model) that relates the two variables. One way to obtain a model for such data is to draw a line through two points on the scatter diagram and determine the equation of the line.

EXAMPLE 3   **Finding a Model for Linearly Related Data**

Use the data in Table 2 from Example 1 to:

(a) Select two points and find an equation of the line containing the points.

(b) Graph the line on the scatter diagram obtained in Example 1(b).

Solution   (a) Select two points, say (32.7, 730) and (34.7, 865). The slope of the line joining the points (32.7, 730) and (34.7, 865) is

$$m = \frac{865 - 730}{34.7 - 32.7} = \frac{135}{2} = 67.5$$

Figure 68

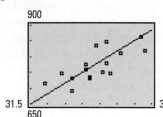

The equation of the line with slope 67.5 and passing through (32.7, 730) is found using the point–slope form with $m = 67.5$, $x_1 = 32.7$, and $y_1 = 730$.

$$y - y_1 = m(x - x_1) \qquad \text{Point–slope form of a line}$$
$$y - 730 = 67.5(x - 32.7) \qquad x_1 = 32.7, \ y_1 = 730, \ m = 67.5$$
$$y - 730 = 67.5x - 2207.25$$
$$y = 67.5x - 1477.25 \qquad \text{The Model}$$

(b) Figure 68 shows the scatter diagram with the graph of the line found in part (a).

Select two other points and complete the solution. Graph the line on the scatter diagram obtained in Figure 68.

⟍ **Now Work** PROBLEMS 9(b) AND (c)

3 Use a Graphing Utility to Find the Line of Best Fit

The model obtained in Example 3 depends on the selection of points, which will vary from person to person. So the model that we found might be different from the model you found. Although the model we found in Example 3 appears to fit the data well, there may be a model that "fits it better." Do you think your model fits the data better? Is there a *line of best fit*? As it turns out, there is a method for finding a model that best fits linearly related data (called the **line of best fit**).*

EXAMPLE 4	**Finding a Model for Linearly Related Data**

Use the data in Table 2 from Example 1.

(a) Use a graphing utility to find the line of best fit that models the relation between on-base percentage and runs scored.

(b) Graph the line of best fit on the scatter diagram obtained in Example 1(b).

(c) Interpret the slope.

(d) Use the line of best fit to predict the number of runs a team will score if their on-base percentage is 34.1.

Solution

Figure 69

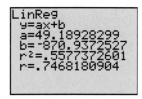

(a) Graphing utilities contain built-in programs that find the line of best fit for a collection of points in a scatter diagram. Upon executing the LINear REGression program, we obtain the results shown in Figure 69. The output that the utility provides shows us the equation $y = ax + b$, where a is the slope of the line and b is the y-intercept. The line of best fit that relates on-base percentage to runs scored may be expressed as the line

$$y = 49.19x - 870.94 \qquad \text{The Model}$$

(b) Figure 70 shows the graph of the line of best fit, along with the scatter diagram.

(c) The slope of the line of best fit is 49.19, which means that, for every 1 percent increase in the on-base percentage, runs scored increase 49.19, on average.

(d) Letting $x = 34.1$ in the equation of the line of best fit, we obtain $y = 49.19(34.1) - 870.94 \approx 806$ runs.

Figure 70

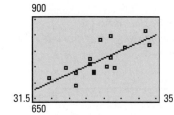

⟍ **Now Work** PROBLEMS 9(d) AND (e)

Does the line of best fit appear to be a good fit? In other words, does the line appear to accurately describe the relation between on-base percentage and runs scored?

---

* We shall not discuss the underlying mathematics of lines of best fit in this book.

And just how "good" is this line of best fit? Look again at Figure 69. The last line of output is $r = 0.747$. This number, called the **correlation coefficient,** r, $-1 \le r \le 1$, is a measure of the strength of the linear relation that exists between two variables. The closer that $|r|$ is to 1, the more perfect the linear relationship is. If r is close to 0, there is little or no linear relationship between the variables. A negative value of r, $r < 0$, indicates that as x increases y decreases; a positive value of r, $r > 0$, indicates that as x increases y does also. The data given in Table 2, having a correlation coefficient of 0.747, are indicative of a relatively strong linear relationship with positive slope.

A.10 Assess Your Understanding

Concepts and Vocabulary

1. A _____ _____ is used to help us to see the type of relation, if any, that may exist between two variables.

2. *True or False* The correlation coefficient is a measure of the strength of a linear relation between two variables and must lie between -1 and 1, inclusive.

Skill Building

In Problems 3–8, examine the scatter diagram and determine whether the type of relation is linear or nonlinear.

3.

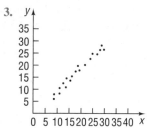

4.

5.

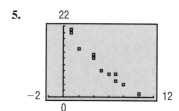

6.

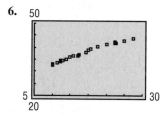

7.

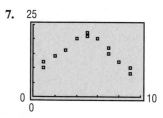

8.

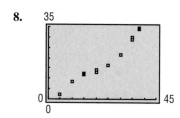 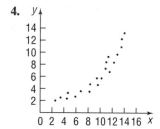

In Problems 9–14,

(a) Draw a scatter diagram.
(b) Select two points from the scatter diagram and find the equation of the line containing the points selected.
(c) Graph the line found in part (b) on the scatter diagram.

(d) Use a graphing utility to find the line of best fit.
(e) Use a graphing utility to draw the scatter diagram and graph the line of best fit on it.

9.

x	3	4	5	6	7	8	9
y	4	6	7	10	12	14	16

10.

x	3	5	7	9	11	13
y	0	2	3	6	9	11

11.

x	-2	-1	0	1	2
y	-4		0	1	4	5

12.

x	-2	-1	0	1	2
y	7	6	3	2	0

13.

x	-20	-17	-15	-14	-10
y	100	120	118	130	140

14.

x	-30	-27	-25	-20	-14
y	10	12	13	13	18

Applications and Extensions

15. Candy The following data represent the weight (in grams) of various candy bars and the corresponding number of calories.

Candy Bar	Weight, x	Calories, y
Hershey's Milk Chocolate®	44.28	230
Nestle's Crunch®	44.84	230
Butterfinger®	61.30	270
Baby Ruth®	66.45	280
Almond Joy®	47.33	220
Twix® (with Caramel)	58.00	280
Snickers®	61.12	280
Heath®	39.52	210

Source: Megan Pocius, Student at Joliet Junior College

(a) Draw a scatter diagram of the data treating weight as the independent variable.
(b) What type of relation appears to exist between the weight of a candy bar and the number of calories?
(c) Select two points and find a linear model that contains the points.
(d) Graph the line on the scatter diagram drawn in part (a).
(e) Use the linear model to predict the number of calories in a candy bar that weighs 62.3 grams.
(f) Interpret the slope of the line found in part (c).

16. Raisins The following data represent the weight (in grams) of a box of raisins and the number of raisins in the box.

Weight (in grams), w	Number of Raisins, N
42.3	87
42.7	91
42.8	93
42.4	87
42.6	89
42.4	90
42.3	82
42.5	86
42.7	86
42.5	86

Source: Jennifer Maxwell, Student at Joliet Junior College

(a) Does the relation defined by the set of ordered pairs (w, N) represent a function?
(b) Draw a scatter diagram of the data treating weight as the independent variable.
(c) Select two points and find a linear model that contains the points.
(d) Graph the line on the scatter diagram drawn in part (b).
(e) Express the relationship found in part (c) using function notation.
(f) Use the linear model to predict the number of raisins in a box that weighs 42.5 grams.
(g) Interpret the slope of the line found in part (c).

17. Height versus Head Circumference A pediatrician wanted to find a linear model that relates a child's height, H, to head circumference, C. She randomly selects nine children from her practice, measures their height and head circumference, and obtains the data shown. Let H represent the independent variable and C the dependent variable.

(a) Use a graphing utility to draw a scatter diagram.
(b) Use a graphing utility to find the line of best fit that models the relation between height and head circumference. Express the solution using function notation.
(c) Interpret the slope.
(d) Predict the head circumference of a child that is 26 inches tall.
(e) What is the height of a child whose head circumference is 17.4 inches?

Height, H (inches)	Head Circumference, C (inches)
25.25	16.4
25.75	16.9
25	16.9
27.75	17.6
26.5	17.3
27	17.5
26.75	17.3
26.75	17.5
27.5	17.5

Source: Denise Slucki, Student at Joliet Junior College

18. Gestation Period versus Life Expectancy A researcher would like to develop a linear model relating the gestation period of an animal, G, and its life expectancy, L. She collects the following data. Let G represent the independent variable and L the dependent variable.

Animal	Gestation (or incubation) Period, G (days)	Life Expectancy, L (years)
Cat	63	11
Chicken	22	7.5
Dog	63	11
Duck	28	10
Goat	151	12
Lion	108	10
Parakeet	18	8
Pig	115	10
Rabbit	31	7
Squirrel	44	9

Source: Time Almanac 2007

(a) Use a graphing utility to draw a scatter diagram.
(b) Use a graphing utility to find the line of best fit that models the relation between gestation period and life expectancy. Express the solution using function notation.
(c) Interpret the slope.
(d) Predict the life expectancy of an animal whose gestation period is 89 days.

Mixed Practice

19. Demand for Jeans The marketing manager at Levi-Strauss wishes to find a function that relates the demand D for men's jeans and p, the price of the jeans. The following data were obtained based on a price history of the jeans.

Price ($/Pair), p	Demand (Pairs of Jeans Sold per Day), D
20	60
22	57
23	56
23	53
27	52
29	49
30	44

(a) Does the relation defined by the set of ordered pairs (p, D) represent a function?
(b) Draw a scatter diagram of the data.
(c) Using a graphing utility, find the line of best fit that models the relation between price and quantity demanded.
(d) Interpret the slope.
(e) Express the relationship found in part (c) using function notation.
(f) What is the domain of the function?
(g) How many jeans will be demanded if the price is $28 a pair?

20. Advertising and Sales Revenue A marketing firm wishes to find a function that relates the sales S of a product and A, the amount spent on advertising the product. The data are obtained from past experience. Advertising and sales are measured in thousands of dollars.

Advertising Expenditures, A	Sales, S
20	335
22	339
22.5	338
24	343
24	341
27	350
28.3	351

(a) Does the relation defined by the set of ordered pairs (A, S) represent a function?
(b) Draw a scatter diagram of the data.
(c) Using a graphing utility, find the line of best fit that models the relation between advertising expenditures and sales.
(d) Interpret the slope.
(e) Express the relationship found in part (c) using function notation.
(f) What is the domain of the function?
(g) Predict sales if advertising expenditures are $25,000.

Discussion and Writing

21. Maternal Age versus Down Syndrome A biologist would like to know how the age of the mother affects the incidence rate of Down syndrome. The data to the right represent the age of the mother and the incidence rate of Down syndrome per 1000 pregnancies.

Draw a scatter diagram treating age of the mother as the independent variable. Would it make sense to find the line of best fit for these data? Why or why not?

22. Find the line of best fit for the ordered pairs $(1, 5)$ and $(3, 8)$. What is the correlation coefficient for these data? Why is this result reasonable?

23. What does a correlation coefficient of 0 imply?

Age of Mother, x	Incidence of Down Syndrome, y
33	2.4
34	3.1
35	4
36	5
37	6.7
38	8.3
39	10
40	13.3
41	16.7
42	22.2
43	28.6
44	33.3
45	50

Source: Hook, E.B., *Journal of the American Medical Association*, 249, 2034-2038, 1983.

Answers

CHAPTER 1 Graphs and Functions

1.1 Assess Your Understanding *(page 13)*

5. abscissa; ordinate **6.** quadrants **7.** midpoint **8.** F **9.** F **10.** T

11. (a) Quadrant II **(d)** Quadrant I
(b) Positive x-axis **(e)** Negative y-axis
(c) Quadrant III **(f)** Quadrant IV

13. The points will be on a vertical line that is 2 units to the right of the y-axis.

15. $(-1, 4)$; Quadrant II
17. $(3, 1)$; Quadrant I
19. $X\min = -11, X\max = 5, X\text{scl} = 1,$ $Y\min = -3, Y\max = 6, Y\text{scl} = 1$
21. $X\min = -30, X\max = 50, X\text{scl} = 10,$ $Y\min = -90, Y\max = 50, Y\text{scl} = 10$
23. $X\min = -10, X\max = 110, X\text{scl} = 10,$ $Y\min = -10, Y\max = 160, Y\text{scl} = 10$
25. $X\min = -6, X\max = 6, X\text{scl} = 2,$ $Y\min = -4, Y\max = 4, Y\text{scl} = 2$
27. $X\min = -6, X\max = 6, X\text{scl} = 2,$ $Y\min = -1, Y\max = 3, Y\text{scl} = 1$

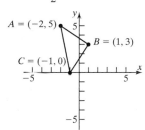
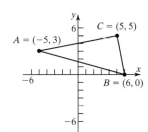

29. $X\min = 3, X\max = 9, X\text{scl} = 1, Y\min = 2, Y\max = 10, Y\text{scl} = 2$ **31.** $\sqrt{5}$ **33.** $\sqrt{10}$ **35.** $2\sqrt{17}$ **37.** 20 **39.** $\sqrt{53}$ **41.** $\sqrt{5.49} \approx 2.34$

43. $\sqrt{a^2 + b^2}$ **45.** $4\sqrt{10}$ **47.** $2\sqrt{65}$

49. $d(A, B) = \sqrt{13}$
$d(B, C) = \sqrt{13}$
$d(A, C) = \sqrt{26}$
$(\sqrt{13})^2 + (\sqrt{13})^2 = (\sqrt{26})^2$
Area $= \dfrac{13}{2}$ square units

51. $d(A, B) = \sqrt{130}$
$d(B, C) = \sqrt{26}$
$d(A, C) = 2\sqrt{26}$
$(\sqrt{26})^2 + (2\sqrt{26})^2 = (\sqrt{130})^2$
Area $= 26$ square units

53. $d(A, B) = 4$
$d(A, C) = 5$
$d(B, C) = \sqrt{41}$
$4^2 + 5^2 = (\sqrt{41})^2$
Area $= 10$ square units

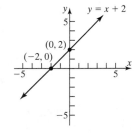
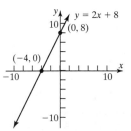
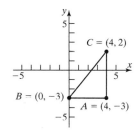

55. $(4, 0)$ **57.** $(3, 3)$ **59.** $(5, -1)$ **61.** $(0.55, 1.2)$ **63.** $\left(\dfrac{a}{2}, \dfrac{b}{2}\right)$ **65.** $(0, 0)$ is on the graph. **67.** $(0, 3)$ is on the graph.

69. $(0, 2)$ and $(\sqrt{2}, \sqrt{2})$ are on the graph. **71.** $(-1, 0), (1, 0)$ **73.** $\left(-\dfrac{\pi}{2}, 0\right), (0, 1), \left(\dfrac{\pi}{2}, 0\right)$ **75.** $(0, 2), (1, 0), (0, -2)$

77. $(-4, 0), (-1, 0), (0, -3), (4, 0)$ **79.**

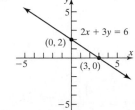

81.

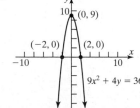

83.

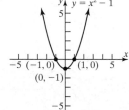

85.

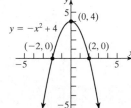

87.

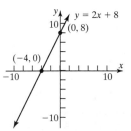

89.

AN1

91. *x*-intercept: 6.5
y-intercept: -13

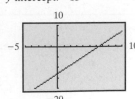

93. *x*-intercepts: $-2.74, 2.74$
y-intercept: -15

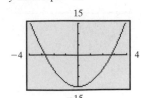

95. *x*-intercept: 14.33
y-intercept: -21.5

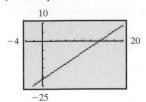

97. *x*-intercepts: $-2.72, 2.72$
y-intercept: 12.33

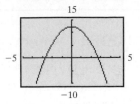

99. $\sqrt{17}; 2\sqrt{5}; \sqrt{29}$ **101.** $d(P_1, P_2) = 6; d(P_2, P_3) = 4; d(P_1, P_3) = 2\sqrt{13};$ right triangle   **103.** $d(P_1, P_2) = 2\sqrt{17}; d(P_2, P_3) = \sqrt{34};$
$d(P_1, P_3) = \sqrt{34};$ isosceles right triangle   **105.** $90\sqrt{2} \approx 127.28$ ft   **107. (a)** $(90, 0), (90, 90), (0, 90)$ **(b)** $5\sqrt{2161} \approx 232.43$ ft
(c) $30\sqrt{149} \approx 366.20$ ft   **109.** $d = 50t$ **111. (a)** $(2.65, 1.6)$ **(b)** ≈ 1.285 units   **113.** \$17,630.50   **115.** $B = (5, -2)$

1.2 Assess Your Understanding *(page 26)*

3. $y = 0$ **4.** *y*-axis   **5.** 4   **6.** $(-3, 4)$ **7.** T   **8.** F   **9.** F

11. $(-2, 0), (0, 2)$

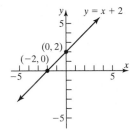

13. $(-4, 0), (0, 8)$

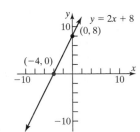

15. $(-1, 0), (1, 0), (0, -1)$

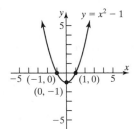

17. $(-2, 0), (2, 0), (0, 4)$

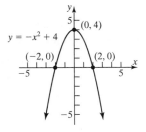

19. $(3, 0), (0, 2)$

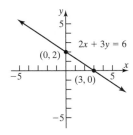

21. $(-2, 0), (2, 0), (0, 9)$

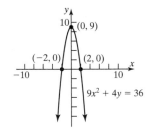

23.

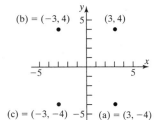

25.

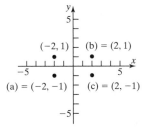

27.

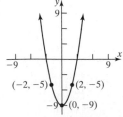

29.

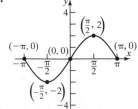

31.

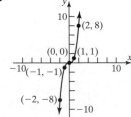

33. Symmetric with respect to the *x*-axis, the *y*-axis, and the origin   **35.** Symmetric with respect to the *y*-axis
37. Symmetric with respect to the *x*-axis   **39.** Symmetric with respect to the origin

41.

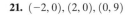

43.

45.

47.

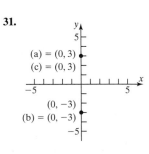

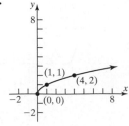

49. $(-4, 0)$, $(0, -2)$, $(0, 2)$; symmetric with respect to the x-axis   **51.** $(0, 0)$; symmetric with respect to the origin   **53.** $(0, -9)$, $(3, 0)$, $(-3, 0)$; symmetric with respect to the y-axis   **55.** $(-2, 0)$, $(2, 0)$, $(0, -3)$, $(0, 3)$; symmetric with respect to the x-axis, y-axis, and origin
57. $(0, -9)$, $(3, 0)$, $(-3, 0)$, $(-1, 0)$; no symmetry   **59.** $(0, -4)$, $(4, 0)$, $(-4, 0)$; symmetric with respect to the y-axis
61. $(0, 0)$; symmetric with respect to the origin   **63.** $(0, 0)$; symmetric with respect to the origin

65. Center $(2, 1)$; radius $= 2$; $(x - 2)^2 + (y - 1)^2 = 4$ **67.** Center $\left(\frac{5}{2}, 2\right)$; radius $= \frac{3}{2}$; $\left(x - \frac{5}{2}\right)^2 + (y - 2)^2 = \frac{9}{4}$

69. $x^2 + y^2 = 4$;
$x^2 + y^2 - 4 = 0$

71. $x^2 + (y - 2)^2 = 4$;
$x^2 + y^2 - 4y = 0$

73. $(x - 4)^2 + (y + 3)^2 = 25$;
$x^2 + y^2 - 8x + 6y = 0$

75. $(x + 2)^2 + (y - 1)^2 = 16$;
$x^2 + y^2 + 4x - 2y - 11 = 0$

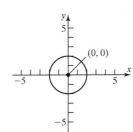
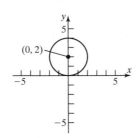
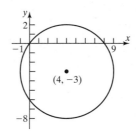
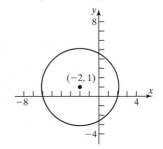

77. $\left(x - \frac{1}{2}\right)^2 + y^2 = \frac{1}{4}$;
$x^2 + y^2 - x = 0$

79. (a) $(h, k) = (0, 0)$; $r = 2$
(b)

81. (a) $(h, k) = (3, 0)$; $r = 2$
(b)

83. (a) $(h, k) = (1, 2)$, $r = 3$
(b)

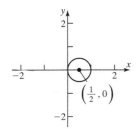
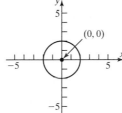
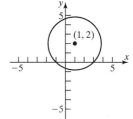

(c) $(\pm 2, 0)$; $(0, \pm 2)$

(c) $(1, 0)$; $(5, 0)$

(c) $\left(1 \pm \sqrt{5}, 0\right)$; $\left(0, 2 \pm 2\sqrt{2}\right)$

85. (a) $(h, k) = (-2, 2)$; $r = 3$
(b)

87. (a) $(h, k) = \left(\frac{1}{2}, -1\right)$; $r = \frac{1}{2}$
(b)

89. (a) $(h, k) = (3, -2)$; $r = 5$
(b)

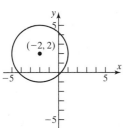
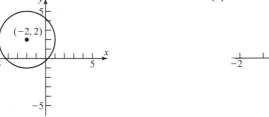
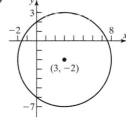

(c) $\left(-2 \pm \sqrt{5}, 0\right)$; $\left(0, 2 \pm \sqrt{5}\right)$

(c) $(0, -1)$

(c) $\left(3 \pm \sqrt{21}, 0\right)$; $(0, -6)$, $(0, 2)$

91. (a) $(h, k) = (-2, 0)$; $r = 2$
(b)

93. $x^2 + y^2 = 13$
95. $(x - 2)^2 + (y - 3)^2 = 9$
97. $(x + 1)^2 + (y - 3)^2 = 5$
99. $(x + 1)^2 + (y - 3)^2 = 1$

101. (a) $(0, -5)$, $(-\sqrt{5}, 0)$, $(\sqrt{5}, 0)$
(b) symmetric with respect to the y-axis
(c)

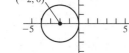
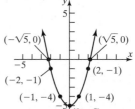

(c) $(0, 0)$, $(-4, 0)$

103. (a) $(-9, 0), (0, -3), (0, 3)$
(b) symmetric with respect to the x-axis
(c)

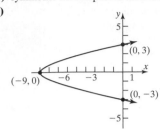

105. (a) $(0, 3), (0, -3), (-3, 0), (3, 0)$
(b) symmetric with respect to the x-axis,
y-axis, and origin
(c)

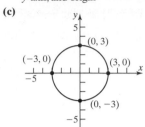

107. (a) $(0, 0), (-2, 0), (2, 0)$
(b) symmetric with respect to the origin
(c)

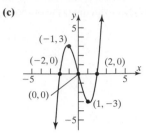

109. $(-1, -2)$ **111.** 4 **113. (a)** $(0, 0), (2, 0), (0, 1), (0, -1)$ **(b)** x-axis symmetry **115.** 18 units^2 **117.** $x^2 + (y - 139)^2 = 15{,}625$

119. $x^2 + y^2 + 2x + 4y - 4168.16 = 0$ **121.** $\sqrt{2}x + 4y = 9\sqrt{2}$ **123.** $(1, 0)$ **125.** $y = 2$ **129.** (b), (c), (e), (g)

1.3 Assess Your Understanding (page 39)

5. independent; dependent **6.** range **7.** $[0, 5]$ **8.** image **9.** $(g - f)(x)$ **10.** F **11.** T **12.** T **13.** F **14.** F

15. Function; Domain: {Elvis, Colleen, Kaleigh, Marissa}, Range: {January 8, March 15, September 17} **17.** Not a function **19.** Not a function

21. Function; Domain: $\{1, 2, 3, 4\}$; Range: $\{3\}$ **23.** Not a function **25.** Function; Domain: $\{-2, -1, 0, 1\}$, Range: $\{0, 1, 4\}$ **27.** Function

29. Function **31.** Not a function **33.** Not a function **35.** Function **37.** Not a function **39. (a)** -4 **(b)** 1 **(c)** -3 **(d)** $3x^2 - 2x - 4$

(e) $-3x^2 - 2x + 4$ **(f)** $3x^2 + 8x + 1$ **(g)** $12x^2 + 4x - 4$ **(h)** $3x^2 + 6xh + 3h^2 + 2x + 2h - 4$ **41. (a)** 0 **(b)** $\dfrac{1}{2}$ **(c)** $-\dfrac{1}{2}$ **(d)** $\dfrac{-x}{x^2 + 1}$

(e) $\dfrac{-x}{x^2 + 1}$ **(f)** $\dfrac{x + 1}{x^2 + 2x + 2}$ **(g)** $\dfrac{2x}{4x^2 + 1}$ **(h)** $\dfrac{x + h}{x^2 + 2xh + h^2 + 1}$ **43. (a)** 4 **(b)** 5 **(c)** 5 **(d)** $|x| + 4$ **(e)** $-|x| - 4$ **(f)** $|x + 1| + 4$

(g) $2|x| + 4$ **(h)** $|x + h| + 4$ **45. (a)** $-\dfrac{1}{5}$ **(b)** $-\dfrac{3}{2}$ **(c)** $\dfrac{1}{8}$ **(d)** $\dfrac{2x - 1}{3x + 5}$ **(e)** $\dfrac{-2x - 1}{3x - 5}$ **(f)** $\dfrac{2x + 3}{3x - 2}$ **(g)** $\dfrac{4x + 1}{6x - 5}$ **(h)** $\dfrac{2x + 2h + 1}{3x + 3h - 5}$

47. All real numbers **49.** All real numbers **51.** $\{x \mid x \neq -4, x \neq 4\}$ **53.** $\{x \mid x \neq 0\}$ **55.** $\{x \mid x \geq 4\}$ **57.** $\{x \mid x > 9\}$ **59.** $\{x \mid x > 1\}$

61. $A = -\dfrac{7}{2}$ **63.** $A = -4$ **65.** $A = 8$; undefined at $x = 3$ **67.** $A(x) = \dfrac{1}{2}x^2$ **69.** $G(x) = 10x$ **71. (a)** P is the dependent variable; a is the

independent variable. **(b)** $P(20) = 197.34$; In 2005, there are 197.34 million people 20 years of age or older. **(c)** $P(0) = 290.580$; In 2005, there

are 290.580 million people. **73. (a)** 15.1 m, 14.071 m, 12.944 m, 11.719 m **(b)** 1.01 s, 1.43 s, 1.75 s **(c)** 2.02 s **75. (a)** \$222 **(b)** \$225 **(c)** \$220

(d) \$230 **79.** $f(x) = \sqrt{3x + 5}$

1.4 Assess Your Understanding (page 46)

3. vertical **4.** 5; -3 **5.** $(-2, 7)$ **6.** F **7.** F **8.** T **9. (a)** $f(0) = 3; f(-6) = -3$ **(b)** $f(6) = 0; f(11) = 1$ **(c)** Positive **(d)** Negative

(e) $-3, 6,$ and 10 **(f)** $-3 < x < 6; 10 < x \leq 11$ **(g)** $\{x \mid -6 \leq x \leq 11\}$ **(h)** $\{y \mid -3 \leq y \leq 4\}$ **(i)** $-3, 6, 10$ **(j)** 3 **(k)** 3 times **(l)** Once

(m) $0, 4$ **(n)** $-5, 8$ **11.** Not a function **13.** Function **(a)** Domain: $\{x \mid -\pi \leq x \leq \pi\}$; Range: $\{y \mid -1 \leq y \leq 1\}$ **(b)** $\left(-\dfrac{\pi}{2}, 0\right), \left(\dfrac{\pi}{2}, 0\right), (0, 1)$

(c) y-axis **15.** Not a function **17.** Function **(a)** Domain: $\{x \mid x > 0\}$; Range: all real numbers **(b)** $(1, 0)$ **(c)** None

19. Function **(a)** Domain: all real numbers; Range: $\{y \mid y \leq 2\}$ **(b)** $(-3, 0), (3, 0), (0, 2)$ **(c)** y-axis **21.** Function **(a)** Domain: all real numbers;

Range: $\{y \mid y \geq -3\}$ **(b)** $(1, 0), (3, 0), (0, 9)$ **(c)** None **23. (a)** Yes **(b)** $f(-2) = 9; (-2, 9)$ **(c)** $0, \dfrac{1}{2}; (0, -1), \left(\dfrac{1}{2}, -1\right)$ **(d)** All real numbers

(e) $-\dfrac{1}{2}, 1$ **(f)** -1 **25. (a)** No **(b)** $f(4) = -3; (4, -3)$ **(c)** $14; (14, 2)$ **(d)** $\{x \mid x \neq 6\}$ **(e)** -2 **(f)** $-\dfrac{1}{3}$ **27. (a)** Yes **(b)** $f(2) = \dfrac{8}{17}; \left(2, \dfrac{8}{17}\right)$

(c) $-1, 1; (-1, 1), (1, 1)$ **(d)** All real numbers **(e)** 0 **(f)** 0

29. (a) Approximately 10.4 ft high **(b)** Approximately 9.9 ft high **31. (a)**
(c)

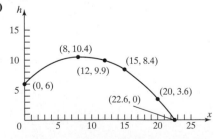

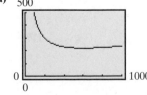

(b)

X	Y1
0	ERROR
50	825
100	470
150	355
200	300
250	269
300	250

X=0

(c) 600 mi/hr

(d) The ball will not go through the hoop; $h(15) \approx 8.4$ ft.
If $v = 30$ ft/s, $h(15) = 10$ ft.

33. (a) 3   **(b)** −2   **(c)** −1   **(d)** 1   **(e)** 2   **(f)** $-\dfrac{1}{3}$ **35. (a)** $80; it costs $80 if you use 0 minutes.   **(b)** $80; it costs $80 if you use 1000 minutes.

(c) $210; it costs $210 if you use 2000 minutes.   **(d)** $\{m|0 \le m \le 14{,}400\}$. There are at most 14,400 anytime minutes in a month.

37. The *x*-intercepts can number anywhere from 0 to infinitely many. There is at most one *y*-intercept.   **39. (a)** III   **(b)** IV   **(c)** I   **(d)** V   **(e)** II

41.

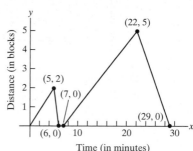

43. (a) 2 hr elapsed during which Kevin was between 0 and 3 mi from home.   **(b)** 0.5 hr elapsed during which Kevin was 3 mi from home.
   (c) 0.3 hr elapsed during which Kevin was between 0 and 3 mi from home.   **(d)** 0.2 hr elapsed during which Kevin was 0 mi from home.
   (e) 0.9 hr elapsed during which Kevin was between 0 and 2.8 mi from home.   **(f)** 0.3 hr elapsed during which Kevin was 2.8 mi from home.
   (g) 1.1 hr elapsed during which Kevin was between 0 and 2.8 mi from home.   **(h)** 3 mi   **(i)** 2 times
45. No points whose *x*-coordinate is 5 or whose *y*-coordinate is 0 can be on the graph.

1.5 Assess Your Understanding *(page 56)*

6. increasing   **7.** even; odd   **8.** T   **9.** T   **10.** F   **11.** Yes   **13.** No   **15.** $(-8, -2)$; $(0, 2)$; $(5, \infty)$ **17.** Yes; 10   **19.** $-2, 2$; 6, 10

21. (a) $(-2, 0), (0, 3), (2, 0)$ **(b)** Domain: $\{x|-4 \le x \le 4\}$ or $[-4, 4]$; Range: $\{y|0 \le y \le 3\}$ or $[0, 3]$
   (c) Increasing on $(-2, 0)$ and $(2, 4)$; Decreasing on $(-4, -2)$ and $(0, 2)$ **(d)** Even
23. (a) $(0, 1)$ **(b)** Domain: all real numbers; Range: $\{y|y > 0\}$ or $(0, \infty)$ **(c)** Increasing on $(-\infty, \infty)$ **(d)** Neither
25. (a) $(-\pi, 0), (0, 0), (\pi, 0)$ **(b)** Domain: $\{x|-\pi \le x \le \pi\}$ or $[-\pi, \pi]$; Range: $\{y|-1 \le y \le 1\}$ or $[-1, 1]$
   (c) Increasing on $\left(-\dfrac{\pi}{2}, \dfrac{\pi}{2}\right)$; Decreasing on $\left(-\pi, -\dfrac{\pi}{2}\right)$ and $\left(\dfrac{\pi}{2}, \pi\right)$ **(d)** Odd
27. (a) $\left(0, \dfrac{1}{2}\right), \left(\dfrac{1}{3}, 0\right), \left(\dfrac{5}{2}, 0\right)$ **(b)** Domain: $\{x|-3 \le x \le 3\}$ or $[-3, 3]$; Range: $\{y|-1 \le y \le 2\}$ or $[-1, 2]$
   (c) Increasing on $(2, 3)$; Decreasing on $(-1, 1)$; Constant on $(-3, -1)$ and $(1, 2)$ **(d)** Neither
29. (a) 0; 3   **(b)** $-2, 2$; 0, 0   **31. (a)** $\dfrac{\pi}{2}$; 1   **(b)** $-\dfrac{\pi}{2}$; −1   **33.** Odd   **35.** Even   **37.** Odd   **39.** Neither   **41.** Even   **43.** Odd

45.

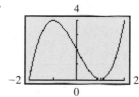

Increasing: $(-2, -1), (1, 2)$
Decreasing: $(-1, 1)$
Local maximum: 4 at $x = -1$
Local minimum: 0 at $x = 1$

47.

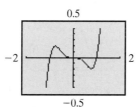

Increasing: $(-2, -0.77), (0.77, 2)$
Decreasing: $(-0.77, 0.77)$
Local maximum: 0.19 at $x = -0.77$
Local minimum: −0.19 at $x = 0.77$

49.

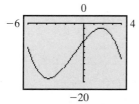

Increasing: $(-3.77, 1.77)$
Decreasing: $(-6, -3.77), (1.77, 4)$
Local maximum: −1.91 at $x = 1.77$
Local minimum: −18.89 at $x = -3.77$

51.

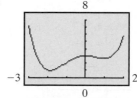

Increasing: $(-1.87, 0), (0.97, 2)$
Decreasing: $(-3, -1.87), (0, 0.97)$
Local maximum: 3 at $x = 0$
Local minimum: 0.95 at $x = -1.87$
Local minimum: 2.65 at $x = 0.97$

53. (a) −4   **(b)** −8   **(c)** −10   **55. (a)** 17   **(b)** −1   **(c)** 11

57. (a) Odd

(b)

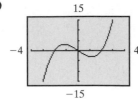

(c) Local minimum: -3.08 at $x = 1.15$
(d) Local maximum: 3.08 at $x = -1.15$

59. (a) Even

(b)

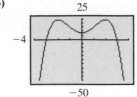

(c) Local maximum: 24 at $x = 2$
(d) Local maximum: 24 at $x = -2$
(e) 47.4 sq. units

61. (a)

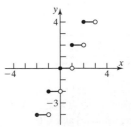

(b) 10 riding lawn mowers/hr
(c) $239/mower

63. (a) On average, the population is increasing at a rate of 0.036 g/hr from 0 to 2.5 hr.
   (b) On average, from 4.5 to 6 hr, the population is increasing at a rate of 0.1 g/hr.
   (c) The average rate of change is increasing over time.

65. (a) 1 **(b)** 0.5 **(c)** 0.1 **(d)** 0.01 **(e)** 0.001 **(f)** It is getting closer to 0. **69.** At most one

71. Yes; the function $f(x) = 0$ is both even and odd.

1.6 Assess Your Understanding *(page 65)*

4. $(-\infty, 0)$ **5.** piecewise-defined **6.** T **7.** F **8.** F **9.** C **11.** E **13.** B **15.** F

17.

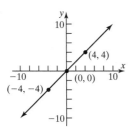

19.

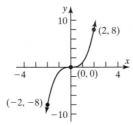

21.

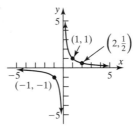

23.

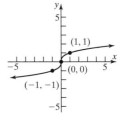

25. (a) 4 **(b)** 2 **(c)** 5 **27. (a)** -4 **(b)** -2 **(c)** 0 **(d)** 25

29. (a) All real numbers
   (b) $(0, 1)$
   (c)

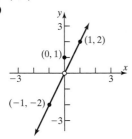

   (d) $\{y \mid y \neq 0\}$; $(-\infty, 0) \cup (0, \infty)$
   (e) Discontinuous at $x = 0$

31. (a) All real numbers
   (b) $(0, 3)$
   (c)

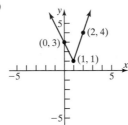

   (d) $\{y \mid y \geq 1\}$; $[1, \infty)$
   (e) Continuous

33. (a) $\{x \mid x \geq -2\}$; $[-2, \infty)$
   (b) $(0, 3), (2, 0)$
   (c)

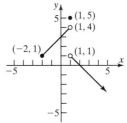

   (d) $\{y \mid y < 4, y = 5\}$; $(-\infty, 4) \cup \{5\}$
   (e) Discontinuous at $x = 1$

35. (a) All real numbers
   (b) $(-1, 0), (0, 0)$
   (c)

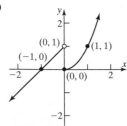

   (d) All real numbers
   (e) Discontinuous at $x = 0$

37. (a) $\{x \mid x \geq -2, x \neq 0\}$; $[-2, 0) \cup (0, \infty)$
   (b) No intercepts
   (c)

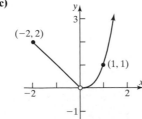

   (d) $\{y \mid y > 0\}$; $(0, \infty)$
   (e) Discontinuous at $x = 0$

39. (a) All real numbers
   (b) $(x, 0)$ for $0 \leq x < 1$
   (c)

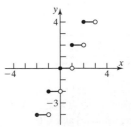

   (d) Set of even integers
   (e) Discontinuous at $\{x \mid x$ is an integer$\}$

41. $f(x) = \begin{cases} -x & \text{if } -1 \le x \le 0 \\ \frac{1}{2}x & \text{if } 0 < x \le 2 \end{cases}$ (Other answers are possible.)   **43.** $f(x) = \begin{cases} -x & \text{if } x \le 0 \\ -x + 2 & \text{if } 0 < x \le 2 \end{cases}$ (Other answers are possible.)

45. (a) 2   **(b)** 3   **(c)** −4   **47. (a)** $39.99   **(b)** $46.74   **(c)** $40.44

49. (a) $71.33   **(b)** $516.04   **(c)** $C(x) = \begin{cases} 1.23755x + 9.45 & \text{if } 0 \le x \le 50 \\ 0.98825x + 21.915 & \text{if } x > 50 \end{cases}$

(d)

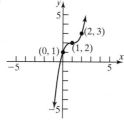

51. Each graph is that of $y = x^2$, but shifted vertically. If $y = x^2 + k$, $k > 0$, the shift is up k units; if $y = x^2 - k$, $k > 0$, the shift is down k units.
53. Each graph is that of $y = |x|$, but either compressed or stretched. If $y = k|x|$ and $k > 1$, the graph is stretched vertically; if $y = k|x|$, $0 < k < 1$, the graph is compressed vertically.
55. The graph of $y = f(-x)$ is the reflection about the y-axis of the graph of $y = f(x)$.
57. They are all U-shaped and open upward. All three go through the points $(-1, 1)$, $(0, 0)$ and $(1, 1)$. As the exponent increases, the steepness of the curve increases (except near $x = 0$).
59. y is a function. Domain: all real numbers; $(-\infty, \infty)$; range: $\{y \mid y = 0 \text{ or } y = 1\}$; y–intercept: $(0, 1)$; x–intercepts: $\{(x, 0) \, x \text{ is an irrational number}\}$. The function is even.

1.7 Assess Your Understanding (page 77)

1. horizontal; right   **2.** y **3.** $-5; -2; 2$ **4.** T   **5.** F   **6.** T   **7.** B **9.** H **11.** I **13.** L **15.** F **17.** G **19.** $y = (x \quad 4)^3$ **21.** $y = x^3 + 4$
23. $y = -x^3$ **25.** $y = 4x^3$ **27.** (1) $y = \sqrt{x} + 2$; (2) $y = -(\sqrt{x} + 2)$; (3) $y = -(\sqrt{-x} + 2)$ **29.** (1) $y = -\sqrt{x}$; (2) $y = -\sqrt{x} + 2$;
(3) $y = -\sqrt{x + 3} + 2$ **31.** (c)   **33.** (c)   **35. (a)** −7 and 1   **(b)** −3 and 5   **(c)** −5 and 3   **(d)** −3 and 5   **37 (a)** $(-3, 3)$ **(b)** $(4, 10)$
(c) Decreasing on $(-1, 5)$ **(d)** Decreasing on $(-5, 1)$

39.

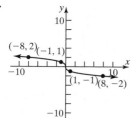

Domain: $(-\infty, \infty)$; Range: $[-1, \infty)$

41.

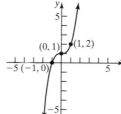

Domain: $(-\infty, \infty)$; Range: $(-\infty, \infty)$

43.

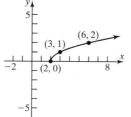

Domain: $[2, \infty)$; Range: $[0, \infty)$

45.

Domain: $(-\infty, \infty)$; Range: $(-\infty, \infty)$

47.

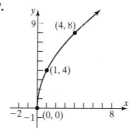

Domain: $[0, \infty)$; Range: $[0, \infty)$

49.

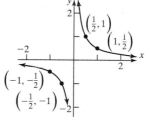

Domain: $(-\infty, 0) \, h \, (0, \infty)$;
Range: $(-\infty, 0) \, h \, (0, \infty)$

51.

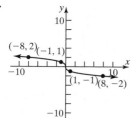

Domain: $(-\infty, \infty)$; Range: $(-\infty, \infty)$

53.

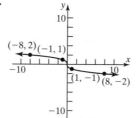

Domain: $(-\infty, \infty)$; Range: $(-\infty, \infty)$

55.

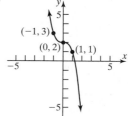

Domain: $(-\infty, \infty)$; Range: $(-\infty, \infty)$

57.

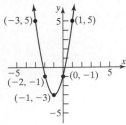

Domain: $(-\infty, \infty)$; Range: $[-3, \infty)$

59.

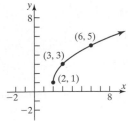

Domain: $[2, \infty)$; Range: $[1, \infty)$

61.

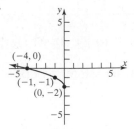

Domain: $(-\infty, 0]$; Range: $[-2, \infty)$

63.

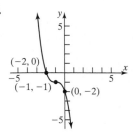

Domain: $(-\infty, \infty)$; Range: $(-\infty, \infty)$

65.

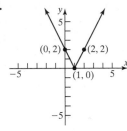

Domain: $(-\infty, \infty)$; Range: $[0, \infty)$

67.

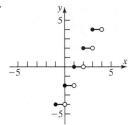

Domain: $(-\infty, \infty)$;

Range: $\{y | y \text{ is an even integer}\}$

69. (a) $F(x) = f(x) + 3$

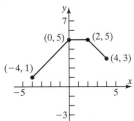

(b) $G(x) = f(x + 2)$

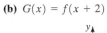
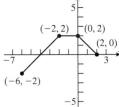

(c) $P(x) = -f(x)$

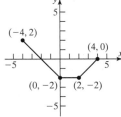

(d) $H(x) = f(x + 1) - 2$

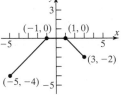

(e) $Q(x) = \dfrac{1}{2} f(x)$

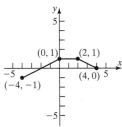

(f) $g(x) = f(-x)$

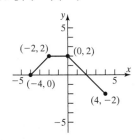

(g) $h(x) = f(2x)$

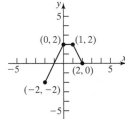
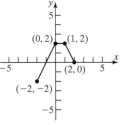

71. (a) $F(x) = f(x) + 3$

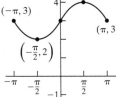

(b) $G(x) = f(x + 2)$

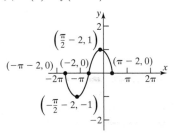

(c) $P(x) = -f(x)$

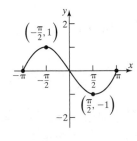

(d) $H(x) = f(x + 1) - 2$

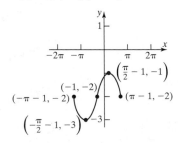

(e) $Q(x) = \dfrac{1}{2} f(x)$

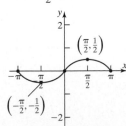

(f) $g(x) = f(-x)$

(g) $h(x) = f(2x)$

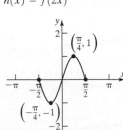

73. $x = 2; y = 1$
75. $x = -2, x = 2; y = 1$

77. (a)

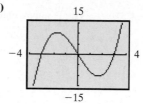

(b) $-3, 0, 3$

(c) Local maximum: 10.39 at $x = -1.73$;
Local minimum: -10.39 at $x = 1.73$

(d) Increasing: $(-4, -1.73)$, $(1.73, 4)$;
Decreasing: $(-1.73, 1.73)$

(e) Intercepts: $-5, -2, 1$;
Local maximum: 10.39 at $x = -3.73$;
Local minimum: -10.39 at $x = -0.27$;
Increasing: $(-6, -3.73)$, $(-0.27, 2)$;
Decreasing: $(-3.73, -0.27)$

(f) Intercepts: $-3, 0, 3$;
Local maximum: 20.78 at $x = -1.73$;
Local minimum: -20.78 at $x = 1.73$;
Increasing: $(-4, -1.73)$, $(1.73, 4)$;
Decreasing: $(-1.73, 1.73)$

(g) Intercepts: $-3, 0, 3$;
Local maximum: 10.39 at $x = 1.73$;
Local minimum: -10.39 at $x = -1.73$;
Increasing: $(-1.73, 1.73)$
Decreasing: $(-4, -1.73)$, $(1.73, 4)$;

79.

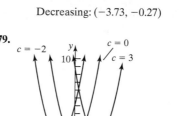

81. (a) $72°F$; $65°F$
(b) The temperature decreases by $2°$ to $70°F$ during the day and $63°F$ overnight

(c) The time at which the temperature adjusts between the daytime and overnight settings is moved to 1 hr sooner. It begins warming up at 5:00 AM instead of 6:00 AM, and it begins cooling down at 8:00 PM in stead of 9:00 PM.

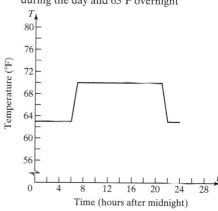

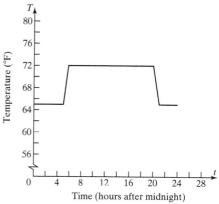

83.

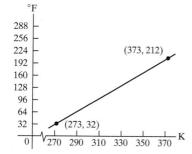

85. (a)

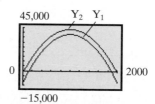

(b) 10% tax
(c) Y_1 is the graph of $p(x)$ shifted down vertically 10,000 units.
Y_2 is the graph of $p(x)$ vertically compressed by a factor of 0.9.
(d) 10% tax

87. (a)

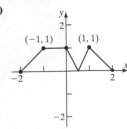

(b)

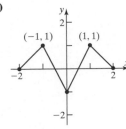

89. (a) $(-4, 2)$ **(b)** $(1, -12)$ **(c)** $(-4, 5)$ **91.** $\dfrac{16}{3}$ square units

1.8 Assess Your Understanding *(page 91)*

4. one-to-one **5.** $y = x$ **6.** $[4, \infty)$ **7.** F **8.** T **9.** one-to-one **11.** not one-to-one **13.** not one-to-one **15.** one-to-one **17.** one-to-one
19. not one-to-one **21.** one-to-one

23.

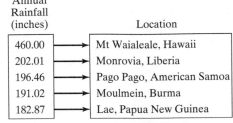

Domain: $\{460.00, 202.01, 196.46, 191.02, 182.87\}$

Range: $\{\text{Mt Waialeale, Monrovia, Pago Pago, Moulmein, Lae}\}$

25. Monthly Cost of
Life Insurance   Age

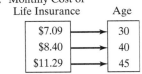

Domain: $\{\$7.09, \$8.40, \$11.29\}$

Range: $\{30, 40, 45\}$

27. $\{(5, -3), (9, -2), (2, -1), (11, 0), (-5, 1)\}$
Domain: $\{5, 9, 2, 11, -5\}$
Range: $\{-3, -2, -1, 0, 1\}$

29. $\{(1, -2), (2, -3), (0, -10), (9, 1), (4, 2)\}$
Domain: $\{1, 2, 0, 9, 4\}$
Range: $\{-2, -3, -10, 1, 2\}$

31. $f(g(x)) = f\left(\dfrac{1}{3}(x - 4)\right) = 3\left[\dfrac{1}{3}(x - 4)\right] + 4 = (x - 4) + 4 = x$

$g(f(x)) = g(3x + 4) = \dfrac{1}{3}[(3x + 4) - 4] = \dfrac{1}{3}(3x) = x$

33. $f(g(x)) = f\left(\dfrac{x}{4} + 2\right) = 4\left[\dfrac{x}{4} + 2\right] - 8 = (x + 8) - 8 = x$

$g(f(x)) = g(4x - 8) = \dfrac{4x - 8}{4} + 2 = (x - 2) + 2 = x$

35. $f(g(x)) = f\left(\sqrt[3]{x + 8}\right) = \left(\sqrt[3]{x + 8}\right)^3 - 8 = (x + 8) - 8 = x$
$g(f(x)) = g(x^3 - 8) = \sqrt[3]{(x^3 - 8) + 8} = \sqrt[3]{x^3} = x$

37. $f(g(x)) = f\left(\dfrac{1}{x}\right) = \dfrac{1}{\left(\dfrac{1}{x}\right)} = x$

$g(f(x)) = g\left(\dfrac{1}{x}\right) = \dfrac{1}{\left(\dfrac{1}{x}\right)} = x$

39. $f(g(x)) = f\left(\dfrac{4x - 3}{2 - x}\right) = \dfrac{2\left(\dfrac{4x - 3}{2 - x}\right) + 3}{\dfrac{4x - 3}{2 - x} + 4} = \dfrac{2(4x - 3) + 3(2 - x)}{4x - 3 + 4(2 - x)} = \dfrac{5x}{5} = x, x \neq 2$

$g(f(x)) = g\left(\dfrac{2x + 3}{x + 4}\right) = \dfrac{4\left(\dfrac{2x + 3}{x + 4}\right) - 3}{2 - \dfrac{2x + 3}{x + 4}} = \dfrac{4(2x + 3) - 3(x + 4)}{2(x + 4) - (2x + 3)} = \dfrac{5x}{5} = x, x \neq -4$

41.

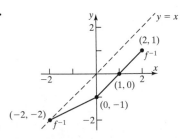

43.

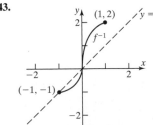

45.

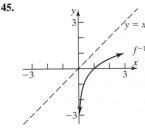

47. $f^{-1}(x) = \dfrac{1}{3}x$

$f(f^{-1}(x)) = f\left(\dfrac{1}{3}x\right) = 3\left(\dfrac{1}{3}x\right) = x$

$f^{-1}(f(x)) = f^{-1}(3x) = \dfrac{1}{3}(3x) = x$

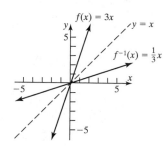

49. $f^{-1}(x) = \dfrac{x}{4} - \dfrac{1}{2}$

$f(f^{-1}(x)) = f\left(\dfrac{x}{4} - \dfrac{1}{2}\right) = 4\left(\dfrac{x}{4} - \dfrac{1}{2}\right) + 2$

$\qquad = (x - 2) + 2 = x$

$f^{-1}(f(x)) = f^{-1}(4x + 2) = \dfrac{4x + 2}{4} - \dfrac{1}{2}$

$\qquad = \left(x + \dfrac{1}{2}\right) - \dfrac{1}{2} = x$

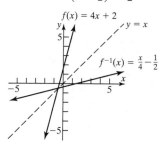

51. $f^{-1}(x) = \sqrt[3]{x + 1}$

$f(f^{-1}(x)) = f\left(\sqrt[3]{x + 1}\right)$

$\qquad = \left(\sqrt[3]{x + 1}\right)^3 - 1 = x$

$f^{-1}(f(x)) = f^{-1}(x^3 - 1)$

$\qquad = \sqrt[3]{(x^3 - 1) + 1} = x$

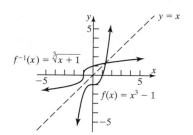

53. $f^{-1}(x) = \sqrt{x - 4}, x \ge 4$

$f(f^{-1}(x)) = f\left(\sqrt{x - 4}\right) = \left(\sqrt{x - 4}\right)^2 + 4 = x$

$f^{-1}(f(x)) = f^{-1}(x^2 + 4) = \sqrt{(x^2 + 4) - 4} = \sqrt{x^2} = x, x \ge 0$

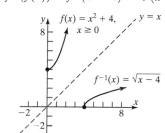

55. $f^{-1}(x) = \dfrac{4}{x}$

$f(f^{-1}(x)) = f\left(\dfrac{4}{x}\right) = \dfrac{4}{\left(\dfrac{4}{x}\right)} = x$

$f^{-1}(f(x)) = f^{-1}\left(\dfrac{4}{x}\right) = \dfrac{4}{\left(\dfrac{4}{x}\right)} = x$

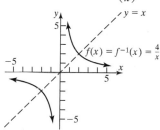

57. $f^{-1}(x) = \dfrac{2x + 1}{x}$

$f(f^{-1}(x)) = f\left(\dfrac{2x + 1}{x}\right) = \dfrac{1}{\dfrac{2x + 1}{x} - 2} = \dfrac{x}{(2x + 1) - 2x} = x$

$f^{-1}(f(x)) = f^{-1}\left(\dfrac{1}{x - 2}\right) = \dfrac{2\left(\dfrac{1}{x - 2}\right) + 1}{\dfrac{1}{x - 2}} = \dfrac{2 + (x - 2)}{1} = x$

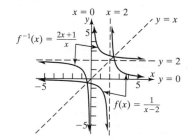

59. $f^{-1}(x) = \dfrac{2 - 3x}{x}$

$f(f^{-1}(x)) = f\left(\dfrac{2 - 3x}{x}\right) = \dfrac{2}{3 + \dfrac{2 - 3x}{x}} = \dfrac{2x}{3x + 2 - 3x} = \dfrac{2x}{2} = x$

$f^{-1}(f(x)) = f^{-1}\left(\dfrac{2}{3 + x}\right) = \dfrac{2 - 3\left(\dfrac{2}{3 + x}\right)}{\dfrac{2}{3 + x}} = \dfrac{2(3 + x) - 3 \cdot 2}{2} = \dfrac{2x}{2} = x$

Domain f = range of f^{-1} = all real numbers except -3

Range f = domain f^{-1} = all real numbers except 0

61. $f^{-1}(x) = \dfrac{-2x}{x-3}$

$$f(f^{-1}(x)) = f\left(\dfrac{-2x}{x-3}\right) = \dfrac{3\left(\dfrac{-2x}{x-3}\right)}{\dfrac{-2x}{x-3}+2} = \dfrac{3(-2x)}{-2x+2(x-3)} = \dfrac{-6x}{-6} = x$$

$$f^{-1}(f(x)) = f^{-1}\left(\dfrac{3x}{x+2}\right) = \dfrac{-2\left(\dfrac{3x}{x+2}\right)}{\dfrac{3x}{x+2}-3} = \dfrac{-2(3x)}{3x-3(x+2)} = \dfrac{-6x}{-6} = x$$

Domain f = range of f^{-1} = all real numbers except -2

Range f = domain f^{-1} = all real numbers except 3

63. $f^{-1}(x) = \dfrac{x}{3x-2}$

$$f(f^{-1}(x)) = f\left(\dfrac{x}{3x-2}\right) = \dfrac{2\left(\dfrac{x}{3x-2}\right)}{3\left(\dfrac{x}{3x-2}\right)-1} = \dfrac{2x}{3x-(3x-2)} = \dfrac{2x}{2} = x$$

$$f^{-1}(f(x)) = f^{-1}\left(\dfrac{2x}{3x-1}\right) = \dfrac{\dfrac{2x}{3x-1}}{3\left(\dfrac{2x}{3x-1}\right)-2} = \dfrac{2x}{6x-2(3x-1)} = \dfrac{2x}{2} = x$$

Domain f = range of f^{-1} = all real numbers except $\dfrac{1}{3}$

Range f = domain f^{-1} = all real numbers except $\dfrac{2}{3}$

65. $f^{-1}(x) = \dfrac{3x+4}{2x-3}$

$$f(f^{-1}(x)) = f\left(\dfrac{3x+4}{2x-3}\right) = \dfrac{3\left(\dfrac{3x+4}{2x-3}\right)+4}{2\left(\dfrac{3x+4}{2x-3}\right)-3}$$

$$= \dfrac{3(3x+4)+4(2x-3)}{2(3x+4)-3(2x-3)}$$

$$= \dfrac{17x}{17} = x$$

$$f^{-1}(f(x)) = f^{-1}\left(\dfrac{3x+4}{2x-3}\right) = \dfrac{3\left(\dfrac{3x+4}{2x-3}\right)+4}{2\left(\dfrac{3x+4}{2x-3}\right)-3}$$

$$= \dfrac{3(3x+4)+4(2x-3)}{2(3x+4)-3(2x-3)} = \dfrac{17x}{17} = x$$

Domain f = range of f^{-1} = all real numbers except $\dfrac{3}{2}$

Range f = domain f^{-1} = all real numbers except $\dfrac{3}{2}$

67. $f^{-1}(x) = \dfrac{-2x+3}{x-2}$

$$f(f^{-1}(x)) = f\left(\dfrac{-2x+3}{x-2}\right) = \dfrac{2\left(\dfrac{-2x+3}{x-2}\right)+3}{\dfrac{-2x+3}{x-2}+2}$$

$$= \dfrac{2(-2x+3)+3(x-2)}{-2x+3+2(x-2)} = \dfrac{-x}{-1} = x$$

$$f^{-1}(f(x)) = f^{-1}\left(\dfrac{2x+3}{x+2}\right) = \dfrac{-2\left(\dfrac{2x+3}{x+2}\right)+3}{\dfrac{2x+3}{x+2}-2}$$

$$= \dfrac{-2(2x+3)+3(x+2)}{2x+3-2(x+2)} = \dfrac{-x}{-1} = x$$

Domain f = range of f^{-1} = all real numbers except -2

Range f = domain f^{-1} = all real numbers except 2

69. $f^{-1}(x) = \dfrac{2}{\sqrt{1-2x}}$

$$f(f^{-1}(x)) = f\left(\dfrac{2}{\sqrt{1-2x}}\right) = \dfrac{\dfrac{4}{1-2x}-4}{2\cdot\dfrac{4}{1-2x}} = \dfrac{4-4(1-2x)}{2\cdot 4}$$

$$= \dfrac{8x}{8} = x$$

$$f^{-1}(f(x)) = f^{-1}\left(\dfrac{x^2-4}{2x^2}\right) = \dfrac{2}{\sqrt{1-2\left(\dfrac{x^2-4}{2x^2}\right)}} = \dfrac{2}{\sqrt{\dfrac{4}{x^2}}} = \sqrt{x^2}$$

$$= x, \text{ since } x > 0$$

Domain f = range of f^{-1} = $(0, \infty)$

Range f = domain f^{-1} = $\left(-\infty, \dfrac{1}{2}\right)$

71. (a) 0   (b) 2   (c) 0   (d) 1

73. 7   **75.** Domain of f^{-1}: $[-2, \infty)$; range of f^{-1}: $[5, \infty)$

77. Domain of g^{-1}: $[0, \infty)$; range of g^{-1}: all real numbers

79. Increasing on the interval $(f(0), f(5))$

81. $f^{-1}(x) = \dfrac{1}{m}(x-b), m \neq 0$

83. Quadrant I

85. Possible answer: $f(x) = |x|, x \geq 0$, is one-to-one; $f^{-1}(x) = x, x \geq 0$

87. (a) $r(d) = \dfrac{d+90.39}{6.97}$

(b) $r(d(r)) = \dfrac{6.97r - 90.39 + 90.39}{6.97} = \dfrac{6.97r}{6.97} = r$

$$d(r(d)) = 6.97\left(\dfrac{d+90.39}{6.97}\right) - 90.39 = d + 90.39 - 90.39 = d$$

(c) 56 miles per hour

89. (a) 77.6 kg **(b)** $h(W) = \dfrac{W - 50}{2.3} + 60 = \dfrac{W + 88}{2.3}$ **(c)** $h(W(h)) = \dfrac{50 + 2.3(h - 60) + 88}{2.3} = \dfrac{2.3h}{2.3} = h$ **(d)** 73 inches

$$W(h(W)) = 50 + 2.3\left(\dfrac{W + 88}{2.3} - 60\right)$$

$$= 50 + W + 88 - 138 = W$$

91. (a) $\{g | 31,850 \le g \le 77,100\}$ **(b)** $\{T | 4386.25 \le T \le 15,698.75\}$ **(c)** $g(T) = \dfrac{T - 4386.25}{0.25} + 31,850$

Domain: $\{T | 4386.25 \le T \le 15,698.75\}$
Range: $\{g | 31,850 \le g \le 77,100\}$

93. (a) t represents time, so $t \ge 0$. **(b)** $t(H) = \sqrt{\dfrac{H - 100}{-4.9}} = \sqrt{\dfrac{100 - H}{4.9}}$ **(c)** 2.02 seconds

95. $f^{-1}(x) = \dfrac{-dx + b}{cx - a}; f = f^{-1}$ if $a = -d$ **99.** No

Review Exercises (page 97)

1. (a) $2\sqrt{5}$ **(b)** $(2, 1)$ **3. (a)** 5 **(b)** $\left(-\dfrac{1}{2}, 1\right)$ **5. (a)** 12 **(b)** $(4, 2)$ **7.** $(-4, 0), (0, 2), (0, 0), (0, -2), (2, 0)$

9.

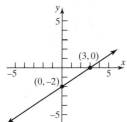

11.

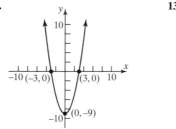

13.

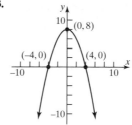

15. x-axis
17. x-axis, y-axis, origin
19. y-axis
21. no symmetry

23.

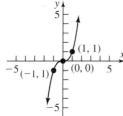

25. $(x + 2)^2 + (y - 3)^2 = 16$
27. $(x + 1)^2 + (y + 2)^2 = 1$

29. Center $(1, -2)$; radius $= 3$

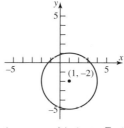

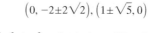

$\left(0, -2 \pm 2\sqrt{2}\right), \left(1 \pm \sqrt{5}, 0\right)$

31. Center $(1, -2)$; radius $= \sqrt{5}$

$(0, 0), (0, -4), (2, 0)$

33. $d(A, B) = \sqrt{13}; d(B, C) = \sqrt{13}$ **35.** Center $(1, -2)$; radius $= 4\sqrt{2}; x^2 + y^2 - 2x + 4y - 27 = 0$

37. Function; domain $\{-1, 2, 4\}$, range $\{0, 3\}$ **39. (a)** 2 **(b)** -2 **(c)** $-\dfrac{3x}{x^2 - 1}$ **(d)** $-\dfrac{3x}{x^2 - 1}$ **(e)** $\dfrac{3(x - 2)}{x^2 - 4x + 3}$ **(f)** $\dfrac{6x}{4x^2 - 1}$

41. (a) 0 **(b)** 0 **(c)** $\sqrt{x^2 - 4}$ **(d)** $-\sqrt{x^2 - 4}$ **(e)** $\sqrt{x^2 - 4x}$ **(f)** $2\sqrt{x^2 - 1}$

43. (a) 0 **(b)** 0 **(c)** $\dfrac{x^2 - 4}{x^2}$ **(d)** $-\dfrac{x^2 - 4}{x^2}$ **(e)** $\dfrac{x(x - 4)}{(x - 2)^2}$ **(f)** $\dfrac{x^2 - 1}{x^2}$

45. $\{x | x \ne -3, x \ne 3\}$ **47.** $\{x | x \le 2\}$ **49.** $\{x | x > 0\}$ **51.** $\{x | x \ne -3, x \ne 1\}$

53. (a) Domain: $\{x | -4 \le x \le 3\}$; Range: $\{y | -3 \le y \le 3\}$ **(b)** $(0, 0)$ **(c)** -1 **(d)** -4 **(e)** $\{x | 0 < x \le 3\}$

(f)

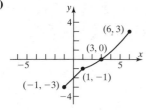

(g)

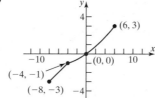

(h)

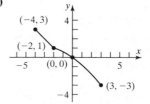

55. (a) Domain: $\{x | -4 \le x \le 4\}$ or $[-4, 4]$ **(b)** Increasing on $(-4, -1)$ and $(3, 4)$ **(c)** Local maximum is 1 and occurs at $x = -1$.

Range: $\{y | -3 \le y \le 1\}$ or $[-3, 1]$ Decreasing on $(-1, 3)$ Local minimum is -3 and occurs at $x = 3$.

(d) No symmetry **(e)** Neither **(f)** x-intercepts: $-2, 0, 4$; y-intercept: 0

57. Odd **59.** Even **61.** Neither **63.** Odd

65.

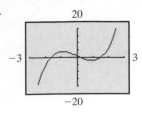

Local maximum: 4.04 at $x = -0.91$
Local minimum: -2.04 at $x = 0.91$

Increasing: $(-3, -0.91)$; $(0.91, 3)$
Decreasing: $(-0.91, 0.91)$

67.

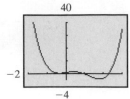

Local maximum: 1.53 at $x = 0.41$
Local minimum: 0.54 at $x = -0.34$
and -3.56 at $x = 1.80$
Increasing: $(-0.34, 0.41)$; $(1.80, 3)$
Decreasing: $(-2, -0.34)$; $(0.41, 1.80)$

69. (a) 23 **(b)** 7 **(c)** 47

71. -5

73. -17

75. No

77. Yes

79.

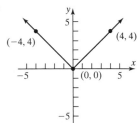

81.

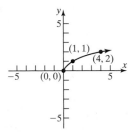

83.

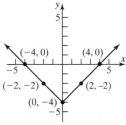

Intercepts: $(-4, 0), (4, 0), (0, -4)$
Domain: all real numbers
Range: $\{y | y \geq -4\}$ or $[-4, \infty)$

85.

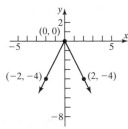

Intercept: $(0, 0)$
Domain: all real numbers
Range: $\{y | y \leq 0\}$ or $(-\infty, 0]$

87.

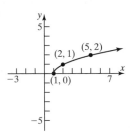

Intercept: $(1, 0)$

Domain: $\{x | x \geq 1\}$ or $[1, \infty)$

Range: $\{y | y \geq 0\}$ or $[0, \infty)$

89.

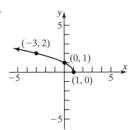

Intercepts: $(0, 1), (1, 0)$

Domain: $\{x | x \leq 1\}$ or $(-\infty, 1]$

Range: $\{y | y \geq 0\}$ or $[0, \infty)$

91.

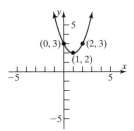

Intercept: $(0, 3)$

Domain: all real numbers

Range: $\{y | y \geq 2\}$ or $[2, \infty)$

93.

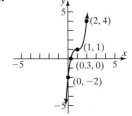

Intercepts: $(0, -2)$,

$\left(1 - \dfrac{\sqrt[3]{9}}{3}, 0\right)$ or about $(0.3, 0)$

Domain: all real numbers
Range: all real numbers

95. (a) $\{x | x > -2\}$ or $(-2, \infty)$

(b) $(0, 0)$

(c)

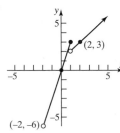

(d) $\{y | y > -6\}$ or $(-6, \infty)$

(e) Discontinuous at $x = 1$

97. (a) $\{x | x \geq -4\}$ or $[-4, \infty)$

(b) $(0, 1)$

(c)

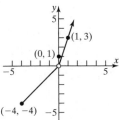

(d) $\{y | -4 \leq y < 0 \text{ or } y > 0\}$
or $[-4, 0) \cup (0, \infty)$

(e) Discontinuous at $x = 0$

99. $A = 11$

101. (a) one-to-one

(b) $\{(2, 1), (5, 3), (8, 5), (10, 6)\}$

103.

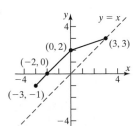

105. $f^{-1}(x) = \dfrac{2x + 3}{5x - 2}$

$$f(f^{-1}(x)) = \dfrac{2\left(\dfrac{2x + 3}{5x - 2}\right) + 3}{5\left(\dfrac{2x + 3}{5x - 2}\right) - 2} = x$$

$$f^{-1}(f(x)) = \dfrac{2\left(\dfrac{2x + 3}{5x - 2}\right) + 3}{5\left(\dfrac{2x + 3}{5x - 2}\right) - 2} = x$$

Domain of f = range of f^{-1}

= all real numbers except $\dfrac{2}{5}$

Range of f = domain of f^{-1}

= all real numbers except $\dfrac{2}{5}$

107. $f^{-1}(x) = \dfrac{x + 1}{x}$

$$f(f^{-1}(x)) = \dfrac{1}{\dfrac{x + 1}{x} - 1} = x$$

$$f^{-1}(f(x)) = \dfrac{\dfrac{1}{x - 1} + 1}{\dfrac{1}{x - 1}} = x$$

Domain of f = range of f^{-1}

= all real numbers except 1
Range of f = domain of f^{-1}

= all real numbers except 0

109. $f^{-1}(x) = \dfrac{27}{x^3}$

$$f(f^{-1}(x)) = \dfrac{3}{\left(\dfrac{27}{x^3}\right)^{1/3}} = x$$

$$f^{-1}(f(x)) = \dfrac{27}{\left(\dfrac{3}{x^{1/3}}\right)^3} = x$$

Domain of f = range of f^{-1}

= all real numbers except 0
Range of f = domain of f^{-1}

= all real numbers except 0

Chapter Test (page 101)

1. $d = 2\sqrt{13}$ **2.** $(2, 1)$

3.

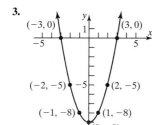

4.

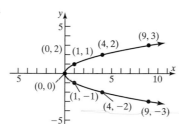

5. Intercepts: $(-3, 0), (3, 0), (0, 9)$;
symmetric with respect to the y-axis

6. $x^2 + y^2 - 8x + 6y = 0$

7. Center: $(-2, 1)$; radius: 3

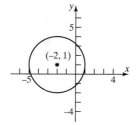

8. (a) Function; domain: $\{2, 4, 6, 8\}$; range: $\{5, 6, 7, 8\}$ **(b)** Not a function **(c)** Not a function

  (d) Function; domain; all real numbers; range: $\{y | y \geq 2\}$

9. Domain: $\left\{x | x \leq \dfrac{4}{5}\right\}$; $f(-1) = 3$ **10.** Domain: $\{x | x \neq -2\}$; $g(-1) = 1$ **11.** Domain: $\{x | x \neq -9, x \neq 4\}$; $h(-1) = \dfrac{1}{8}$

12. (a) Domain: $\{x | -5 \leq x \leq 5\}$; range: $\{y | -3 \leq y \leq 3\}$ **(b)** $(0, 2), (-2, 0)$, and $(2, 0)$ **(c)** $f(1) = 3$

  (d) $x = -5$ and $x = 3$ **(e)** $\{x | -5 \leq x < -2 \text{ or } 2 < x \leq 5\}$ or $[-5, -2) \cup (2, 5]$

13. Local maxima: $f(-0.85) \approx -0.86$; $f(2.35) \approx 15.55$; Local minima: $f(0) = -2$; the function is increasing on the intervals $(-5, -0.85)$
  and $(0, 2.35)$ and decreasing on the intervals $(-0.85, 0)$ and $(2.35, 5)$.

14. (a)

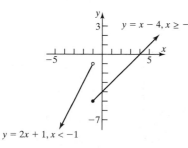

$y = x - 4, x \geq -1$

$y = 2x + 1, x < -1$

15. 19 **16. (a)**

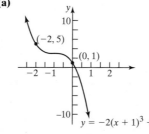

$(-2, 5)$

$(0, 1)$

$y = -2(x + 1)^3 + 3$

(b)

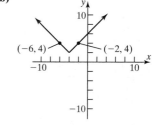

$(-6, 4)$ $(-2, 4)$

(b) $(0, -4), (4, 0)$ **(c)** $g(-5) = -9$ **(d)** $g(2) = -2$

17. $f^{-1}(x) = \dfrac{2 + 5x}{3x}$; domain of $f = \left\{x \middle| x \neq \dfrac{5}{3}\right\}$, range of $f = \{y | y \neq 0\}$; domain of $f^{-1} = \{x | x \neq 0\}$; range of $f^{-1} = \left\{y \middle| y \neq \dfrac{5}{3}\right\}$

18. The point $(-5, 3)$ must be on the graph of f^{-1}.

CHAPTER 2 Trigonometric Functions

2.1 Assess Your Understanding *(page 113)*

3. standard position **4.** $r\theta; \frac{1}{2}r^2\theta$ **5.** $\frac{s}{t}; \frac{\theta}{t}$ **6.** F **7.** T **8.** T **9.** T **10.** F

11. 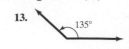 **13.**  **15.** 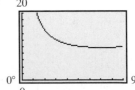 **17.** 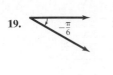 **19.** 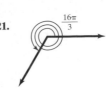 **21.**

23. $40.17°$ **25.** $1.03°$ **27.** $9.15°$ **29.** $40°19'12''$ **31.** $18°15'18''$ **33.** $19°59'24''$ **35.** $\frac{\pi}{6}$ **37.** $\frac{4\pi}{3}$ **39.** $-\frac{\pi}{3}$ **41.** π **43.** $-\frac{3\pi}{4}$ **45.** $-\frac{\pi}{2}$ **47.** $60°$

49. $-225°$ **51.** $90°$ **53.** $15°$ **55.** $-90°$ **57.** $-30°$ **59.** 0.30 **61.** -0.70 **63.** 2.18 **65.** $179.91°$ **67.** $114.59°$ **69.** $362.11°$ **71.** 5 m

73. 6 ft **75.** 0.6 radian **77.** $\frac{\pi}{3} \approx 1.047$ in. **79.** 25 m^2 **81.** $2\sqrt{3} \approx 3.464$ ft **83.** 0.24 radian **85.** $\frac{\pi}{3} \approx 1.047$ in.^{2} **87.** $s = 2.094$ ft; $A = 2.094$ ft^2

89. $s = 14.661$ yd; $A = 87.965$ yd^2 **91.** $3\pi \approx 9.42$ in; $5\pi \approx 15.71$ in. **93.** $2\pi \approx 6.28$ m^2 **95.** $\frac{675\pi}{2} \approx 1060.29$ ft^2 **97.** $\omega = \frac{1}{60}$ radian/s; $v = \frac{1}{12}$ cm/s

99. Approximately 452.5 rpm **101.** Approximately 359 mi **103.** Approximately 898 mi/h **105.** Approximately 2292 mi/h **107.** $\frac{3}{4}$ rpm

109. Approximately 2.86 mi/h **111.** Approximately 31.47 rpm **113.** Approximately 1037 mi/h **115.** Radius ≈ 3979 mi; circumference $\approx 25{,}000$ mi

117. $v_1 = r_1\omega_1$, $v_2 = r_2\omega_2$, and $v_1 = v_2$, so $r_1\omega_1 = r_2\omega_2 \Rightarrow \dfrac{r_1}{r_2} = \dfrac{\omega_2}{\omega_1}$.

2.2 Assess Your Understanding *(page 125)*

3. complementary **4.** cosine **5.** $62°$ **6.** 1 **7.** T **8.** F **9.** T **10.** F

11. $\sin\theta = \frac{5}{13}; \cos\theta = \frac{12}{13}; \tan\theta = \frac{5}{12}; \csc\theta = \frac{13}{5}; \sec\theta = \frac{13}{12}; \cot\theta = \frac{12}{5}$ **13.** $\sin\theta = \frac{2\sqrt{13}}{13}; \cos\theta = \frac{3\sqrt{13}}{13}; \tan\theta = \frac{2}{3}; \csc\theta = \frac{\sqrt{13}}{2}; \sec\theta = \frac{\sqrt{13}}{3};$

$\cot\theta = \frac{3}{2}$ **15.** $\sin\theta = \frac{\sqrt{3}}{2}; \cos\theta = \frac{1}{2}; \tan\theta = \sqrt{3}; \csc\theta = \frac{2\sqrt{3}}{3}; \sec\theta = 2; \cot\theta = \frac{\sqrt{3}}{3}$ **17.** $\sin\theta = \frac{\sqrt{6}}{3}; \cos\theta = \frac{\sqrt{3}}{3}; \tan\theta = \sqrt{2};$

$\csc\theta = \frac{\sqrt{6}}{2}; \sec\theta = \sqrt{3}; \cot\theta = \frac{\sqrt{2}}{2}$ **19.** $\sin\theta = \frac{\sqrt{5}}{5}; \cos\theta = \frac{2\sqrt{5}}{5}; \tan\theta = \frac{1}{2}; \csc\theta = \sqrt{5}; \sec\theta = \frac{\sqrt{5}}{2}; \cot\theta = 2$ **21.** $\tan\theta = \frac{\sqrt{3}}{3}; \csc\theta = 2;$

$\sec\theta = \frac{2\sqrt{3}}{3}; \cot\theta = \sqrt{3}$ **23.** $\tan\theta = \frac{2\sqrt{5}}{5}; \csc\theta = \frac{3}{2}; \sec\theta = \frac{3\sqrt{5}}{5}; \cot\theta = \frac{\sqrt{5}}{2}$ **25.** $\cos\theta = \frac{\sqrt{2}}{2}; \tan\theta = 1; \csc\theta = \sqrt{2}; \sec\theta = \sqrt{2}; \cot\theta = 1$

27. $\sin\theta = \frac{2\sqrt{2}}{3}; \tan\theta = 2\sqrt{2}; \csc\theta = \frac{3\sqrt{2}}{4}; \sec\theta = 3; \cot\theta = \frac{\sqrt{2}}{4}$ **29.** $\sin\theta = \frac{\sqrt{5}}{5}; \cos\theta = \frac{2\sqrt{5}}{5}; \csc\theta = \sqrt{5}; \sec\theta = \frac{\sqrt{5}}{2}; \cot\theta = 2$

31. $\sin\theta = \frac{2\sqrt{2}}{3}; \cos\theta = \frac{1}{3}; \tan\theta = 2\sqrt{2}; \csc\theta = \frac{3\sqrt{2}}{4}; \cot\theta = \frac{\sqrt{2}}{4}$ **33.** $\sin\theta = \frac{\sqrt{6}}{3}; \cos\theta = \frac{\sqrt{3}}{3}; \csc\theta = \frac{\sqrt{6}}{2}; \sec\theta = \sqrt{3}; \cot\theta = \frac{\sqrt{2}}{2}$

35. $\sin\theta = \frac{1}{2}; \cos\theta = \frac{\sqrt{3}}{2}; \tan\theta = \frac{\sqrt{3}}{3}; \sec\theta = \frac{2\sqrt{3}}{3}; \cot\theta = \sqrt{3}$ **37.** 1 **39.** 1 **41.** 0 **43.** 0 **45.** 1 **47.** 0 **49.** 0 **51.** 1 **53.** 1

55. (a) $\frac{1}{2}$ **(b)** $\frac{3}{4}$ **(c)** 2 **(d)** 2 **57. (a)** 17 **(b)** $\frac{1}{4}$ **(c)** 4 **(d)** $\frac{17}{16}$ **59. (a)** $\frac{1}{4}$ **(b)** 15 **(c)** 4 **(d)** $\frac{16}{15}$ **61. (a)** 0.78 **(b)** 0.79 **(c)** 1.27

(d) 1.28 **(e)** 1.61 **(f)** 0.78 **(g)** 0.62 **(h)** 1.27 **63.** 0.6 **65.** $20°$

67. (a) 10 min **(d)** Approximately 15.8 min **(f)**

(b) 20 min **(e)** Approximately 10.4 min

(c) $T(\theta) = 5\left(1 - \dfrac{1}{3\tan\theta} + \dfrac{1}{\sin\theta}\right)$

69. (a) $Z = 200\sqrt{13} \approx 721.1$ ohms **(b)** $\tan\phi = \frac{2}{3}; \sin\phi = \frac{2\sqrt{13}}{13}; \cos\phi = \frac{3\sqrt{13}}{13}; \cot\phi = \frac{3}{2}; \csc\phi = \frac{\sqrt{13}}{2}; \sec\phi = \frac{\sqrt{13}}{3}$

71. (a) $|OA| = |OC| = 1$; angle OAC = angle OCA; angle OAC + angle OAC + $180° - \theta = 180°$; angle $OAC = \dfrac{\theta}{2}$

(b) $\sin\theta = \dfrac{|CD|}{|OC|} = |CD|; \cos\theta = \dfrac{|OD|}{|OC|} = |OD|$ **(c)** $\tan\dfrac{\theta}{2} = \dfrac{|CD|}{|AD|} = \dfrac{\sin\theta}{1 + |OD|} = \dfrac{\sin\theta}{1 + \cos\theta}$

73. $h = x\tan\theta$ and $h = (1-x)\tan n\theta$; thus, $x\tan\theta = (1-x)\tan n\theta$, so $x = \dfrac{\tan n\theta}{\tan\theta + \tan n\theta}$.

75. (a) Area $\triangle OAC = \frac{1}{2}|AC||OC| = \frac{1}{2} \cdot \frac{|AC|}{1} \cdot \frac{|OC|}{1} = \frac{1}{2}\sin\alpha\cos\alpha$

(b) Area $\triangle OCB = \frac{1}{2}|BC||OC| = \frac{1}{2}|OB|^2 \frac{|BC|}{|OB|} \cdot \frac{|OC|}{|OB|} = \frac{1}{2}|OB|^2\sin\beta\cos\beta$

(c) Area $\triangle OAB = \frac{1}{2}|BD||OA| = \frac{1}{2}|OB|\frac{|BD|}{|OB|} = \frac{1}{2}|OB|\sin(\alpha+\beta)$

(d) $\dfrac{\cos\alpha}{\cos\beta} = \dfrac{\frac{|OC|}{1}}{\frac{|OC|}{|OB|}} = |OB|$

(e) $\quad$ Area$\triangle OAB$ = area $\triangle OAC$ + area $\triangle OCB$

$\frac{1}{2}|OB|\sin(\alpha+\beta) = \frac{1}{2}\sin\alpha\cos\alpha + \frac{1}{2}|OB|^2\sin\beta\cos\beta$

$\sin(\alpha+\beta) = \dfrac{\sin\alpha\cos\alpha + |OB|^2\sin\beta\cos\beta}{|OB|}$

$\sin(\alpha+\beta) = \dfrac{\sin\alpha(|OB|\cos\beta) + |OB|^2\sin\beta\left(\dfrac{\cos\alpha}{|OB|}\right)}{|OB|}$

$\sin(\alpha+\beta) = \sin\alpha\cos\beta + \cos\alpha\sin\beta$

77. $\sin\alpha = \tan\alpha\cos\alpha = \cos\beta\cos\alpha = \cos\beta\tan\beta = \sin\beta$;

$\sin^2\alpha + \cos^2\alpha = 1$

$\sin^2\alpha + \tan^2\beta = 1$

$\sin^2\alpha + \dfrac{\sin^2\beta}{\cos^2\beta} = 1$

$\sin^2\alpha + \dfrac{\sin^2\alpha}{1 - \sin^2\alpha} = 1$

$\sin^2\alpha - \sin^4\alpha + \sin^2\alpha = 1 - \sin^2\alpha$

$\sin^4\alpha - 3\sin^2\alpha + 1 = 0$

$\sin^2\alpha = \dfrac{3 \pm \sqrt{5}}{2}$

$\sin^2\alpha = \dfrac{3 - \sqrt{5}}{2}$

$\sin\alpha = \sqrt{\dfrac{3 - \sqrt{5}}{2}}$

79. Since $a^2 + b^2 = c^2, a > 0, b > 0$, then $0 < a^2 < c^2$ or $0 < a < c$. Thus $0 < \dfrac{a}{c} < 1$ and $0 < \cos\theta < 1$ or $\dfrac{\cos\theta}{\cos\theta} < \dfrac{1}{\cos\theta}$; therefore, $\sec\theta > 1$.

2.3 Assess Your Understanding *(page 136)*

1. $\dfrac{3}{2}$ **2.** 0.91 **3.** T **4.** F **5.** $\sin 45° = \dfrac{\sqrt{2}}{2}$; $\cos 45° = \dfrac{\sqrt{2}}{2}$; $\tan 45° = 1$; $\csc 45° = \sqrt{2}$; $\sec 45° = \sqrt{2}$; $\cot 45° = 1$ **7.** $\dfrac{\sqrt{3}}{2}$ **9.** $\dfrac{1}{2}$ **11.** $\dfrac{3}{4}$

13. $\sqrt{3}$ **15.** $\dfrac{\sqrt{3}}{4}$ **17.** $\sqrt{2}$ **19.** 2 **21.** $\sqrt{2} + \dfrac{4\sqrt{3}}{3}$ **23.** $-\dfrac{8}{3}$ **25.** $\dfrac{1}{2}$ **27.** 0 **29.** 0.47 **31.** 0.38 **33.** 1.33 **35.** 0.31 **37.** 3.73 **39.** 1.04

41. 0.84 **43.** 0.02 **45.** 0.31 **47.** $\dfrac{1 + \sqrt{3}}{2}$ **49.** $\dfrac{1}{2}$ **51.** $\dfrac{\sqrt{3}}{2}$ **53.** $\dfrac{\sqrt{2}}{4}$ **55. (a)** $\dfrac{\sqrt{2}}{2}$; $\left(\dfrac{\pi}{4}, \dfrac{\sqrt{2}}{2}\right)$ **(b)** $\left(\dfrac{\sqrt{2}}{2}, \dfrac{\pi}{4}\right)$ **(c)** $\left(\dfrac{\pi}{4}, -2\right)$

57. $R \approx 310.56$ ft; $H \approx 77.64$ ft **59.** $R \approx 19,541.95$ m; $H \approx 2278.14$ m **61. (a)** 1.20 s **(b)** 1.12 s **(c)** 1.20 s

63. (a) $T(\theta) = 1 + \dfrac{2}{3\sin\theta} - \dfrac{1}{4\tan\theta}$ **(b)** 1.9 h; 0.57 h **(c)** 1.69 h; 0.75 h **(d)** 1.63 h; 0.86 h **(e)** 1.67 h **(f)** 2.75 h

(g)

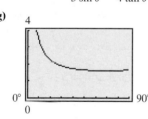

67.98°; 1.62 h; 0.9 h

65. 4.59 in.; 6.55 in. **67. (a)** 5.52 in. or 11.83 in. **69.** 70.02 ft **71.** 985.91 ft **73.** $50\dfrac{\sqrt{3}}{3} \approx 28.87$ m

75. 1978.09 ft **77.** 60.27 ft **79.** $190\sqrt{2} \approx 268.70$ ft **81.** 554.52 ft **83.** 3.83 mi

85. The white ball should hit the top cushion 4.125 ft from the upper-left corner.

87.

θ	0.5	0.4	0.2	0.1	0.01	0.001	0.0001	0.00001
$f(\theta) = \dfrac{\sin\theta}{\theta}$	0.9589	0.9735	0.9933	0.9983	1.0000	1.0000	1.0000	1.0000

$\dfrac{\sin\theta}{\theta}$ approaches 1 as $\theta \to 0$.

89. 1 **91.** $\dfrac{\sqrt{2}}{2}$

2.4 Assess Your Understanding *(page 148)*

1. tangent; cotangent **2.** coterminal **3.** 60° **4.** F **5.** T **6.** T **7.** 60° **8.** I, IV **9.** $\dfrac{\pi}{2}, \dfrac{3\pi}{2}$ **10.** $\dfrac{\pi}{3}$ **11.** $\sin\theta = \dfrac{4}{5}$; $\cos\theta = -\dfrac{3}{5}$;

$\tan\theta = -\dfrac{4}{3}$; $\csc\theta = \dfrac{5}{4}$; $\sec\theta = -\dfrac{5}{3}$; $\cot\theta = -\dfrac{3}{4}$ **13.** $\sin\theta = -\dfrac{3\sqrt{13}}{13}$; $\cos\theta = \dfrac{2\sqrt{13}}{13}$; $\tan\theta = -\dfrac{3}{2}$; $\csc\theta = -\dfrac{\sqrt{13}}{3}$; $\sec\theta = \dfrac{\sqrt{13}}{2}$; $\cot\theta = -\dfrac{2}{3}$

15. $\sin\theta = -\dfrac{\sqrt{2}}{2}$; $\cos\theta = -\dfrac{\sqrt{2}}{2}$; $\tan\theta = 1$; $\csc\theta = -\sqrt{2}$; $\sec\theta = -\sqrt{2}$; $\cot\theta = 1$

17. $\sin\theta = \dfrac{1}{2}$; $\cos\theta = \dfrac{\sqrt{3}}{2}$; $\tan\theta = \dfrac{\sqrt{3}}{3}$; $\csc\theta = 2$; $\sec\theta = \dfrac{2\sqrt{3}}{3}$; $\cot\theta = \sqrt{3}$ **19.** $\sin\theta = -\dfrac{\sqrt{2}}{2}$; $\cos\theta = \dfrac{\sqrt{2}}{2}$; $\tan\theta = -1$; $\csc\theta = -\sqrt{2}$;

$\sec\theta = \sqrt{2}$; $\cot\theta = -1$ **21.** $\dfrac{\sqrt{2}}{2}$ **23.** 1 **25.** 1 **27.** $\sqrt{3}$ **29.** $\dfrac{\sqrt{2}}{2}$ **31.** 0 **33.** II **35.** IV **37.** IV **39.** III **41.** 30° **43.** 60° **45.** 60° **47.** $\dfrac{\pi}{4}$

49. $\dfrac{\pi}{3}$ **51.** 45° **53.** $\dfrac{\pi}{3}$ **55.** 80° **57.** $\dfrac{\pi}{4}$ **59.** $\dfrac{1}{2}$ **61.** $\dfrac{1}{2}$ **63.** $\dfrac{\sqrt{2}}{2}$ **65.** -2 **67.** $-\sqrt{3}$ **69.** $\dfrac{\sqrt{2}}{2}$ **71.** $-\dfrac{\sqrt{2}}{2}$ **73.** $-\dfrac{\sqrt{3}}{2}$ **75.** $-\sqrt{3}$ **77.** 0 **79.** 0

81. -1 **83.** $\cos\theta = -\dfrac{5}{13}; \tan\theta = -\dfrac{12}{5}; \csc\theta = \dfrac{13}{12}; \sec\theta = -\dfrac{13}{5}; \cot\theta = -\dfrac{5}{12}$ **85.** $\sin\theta = -\dfrac{3}{5}; \tan\theta = \dfrac{3}{4}; \csc\theta = -\dfrac{5}{3}; \sec\theta = -\dfrac{5}{4}; \cot\theta = \dfrac{4}{3}$

87. $\cos\theta = -\dfrac{12}{13}; \tan\theta = -\dfrac{5}{12}; \csc\theta = \dfrac{13}{5}; \sec\theta = -\dfrac{13}{12}; \cot\theta = -\dfrac{12}{5}$ **89.** $\sin\theta = -\dfrac{2\sqrt{2}}{3}; \tan\theta = 2\sqrt{2}; \csc\theta = -\dfrac{3\sqrt{2}}{4}; \sec\theta = -3; \cot\theta = \dfrac{\sqrt{2}}{4}$

91. $\cos\theta = -\dfrac{\sqrt{5}}{3}; \tan\theta = -\dfrac{2\sqrt{5}}{5}; \csc\theta = \dfrac{3}{2}; \sec\theta = -\dfrac{3\sqrt{5}}{5}; \cot\theta = -\dfrac{\sqrt{5}}{2}$ **93.** $\sin\theta = -\dfrac{\sqrt{3}}{2}; \cos\theta = \dfrac{1}{2}; \tan\theta = -\sqrt{3}; \csc\theta = -\dfrac{2\sqrt{3}}{3};$

$\cot\theta = -\dfrac{\sqrt{3}}{3}$ **95.** $\sin\theta = -\dfrac{3}{5}; \cos\theta = -\dfrac{4}{5}; \csc\theta = -\dfrac{5}{3}; \sec\theta = -\dfrac{5}{4}; \cot\theta = \dfrac{4}{3}$ **97.** $\sin\theta = \dfrac{\sqrt{10}}{10}; \cos\theta = -\dfrac{3\sqrt{10}}{10};$

$\csc\theta = \sqrt{10}; \sec\theta = -\dfrac{\sqrt{10}}{3}; \cot\theta = -3$ **99.** $\sin\theta = -\dfrac{1}{2}; \cos\theta = -\dfrac{\sqrt{3}}{2}; \tan\theta = \dfrac{\sqrt{3}}{3}; \sec\theta = -\dfrac{2\sqrt{3}}{3}; \cot\theta = \sqrt{3}$ **101.** 0

103. (a) $-\dfrac{\sqrt{2}}{2}; \left(315°, -\dfrac{\sqrt{2}}{2}\right)$ (b) $\sqrt{2}; (315°, \sqrt{2})$ (c) $-1; (315°, -1)$ **105.** (a) $-\dfrac{\sqrt{3}}{2}; \left(\dfrac{7\pi}{6}, -\dfrac{\sqrt{3}}{2}\right)$ (b) $-2; \left(\dfrac{7\pi}{6}, -2\right)$ (c) $1; (-315°, 1)$

107. -0.2 **109.** 3 **111.** -5 **113.** 0 **115.** (a) Approximately 16.6 ft   (b)   (c) $67.5°$

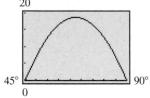

2.5 Assess Your Understanding (page 159)

4. $2\pi; \pi$ **5.** All real numbers except odd multiples of $\dfrac{\pi}{2}$ **6.** All real numbers from -1 to 1 inclusive   **7.** $-0.2; 0.2$ **8.** T

9. $\sin t = -\dfrac{1}{2}; \cos t = \dfrac{\sqrt{3}}{2}; \tan t = -\dfrac{\sqrt{3}}{3}; \csc t = -2; \sec t = \dfrac{2\sqrt{3}}{3}; \cot t = -\sqrt{3}$ **11.** $\sin t = -\dfrac{\sqrt{2}}{2}; \cos t = -\dfrac{\sqrt{2}}{2}; \tan t = 1;$

$\csc t = -\sqrt{2}; \sec t = -\sqrt{2}; \cot t = 1$ **13.** $\sin t = \dfrac{2}{3}; \cos t = \dfrac{\sqrt{5}}{3}; \tan t = \dfrac{2\sqrt{5}}{5}; \csc t = \dfrac{3}{2}; \sec t = \dfrac{3\sqrt{5}}{5}; \cot t = \dfrac{\sqrt{5}}{2}$

15. $\sin\theta = -\dfrac{4}{5}; \cos\theta = \dfrac{3}{5}; \tan\theta = -\dfrac{4}{3}; \csc\theta = -\dfrac{5}{4}; \sec\theta = \dfrac{5}{3}; \cot\theta = -\dfrac{3}{4}$

17. $\sin\theta = \dfrac{3\sqrt{13}}{13}; \cos\theta = -\dfrac{2\sqrt{13}}{13}; \tan\theta = -\dfrac{3}{2}; \csc\theta = \dfrac{\sqrt{13}}{3}; \sec\theta = -\dfrac{\sqrt{13}}{2}; \cot\theta = -\dfrac{2}{3}$

19. $\sin\theta = -\dfrac{\sqrt{2}}{2}; \cos\theta = -\dfrac{\sqrt{2}}{2}; \tan\theta = 1; \csc\theta = -\sqrt{2}; \sec\theta = -\sqrt{2}; \cot\theta = 1$ **21.** $\dfrac{\sqrt{2}}{2}$ **23.** 1 **25.** 1 **27.** $\sqrt{3}$ **29.** $\dfrac{\sqrt{2}}{2}$ **31.** 0 **33.** $\sqrt{2}$

35. $\dfrac{\sqrt{3}}{3}$ **37.** $-\dfrac{\sqrt{3}}{2}$ **39.** $-\dfrac{\sqrt{3}}{3}$ **41.** 2 **43.** -1 **45.** -1 **47.** $\dfrac{\sqrt{2}}{2}$ **49.** 0 **51.** $-\sqrt{2}$ **53.** $\dfrac{2\sqrt{3}}{3}$ **55.** -1 **57.** -2 **59.** $\dfrac{2-\sqrt{2}}{2}$ **61.** $(-\infty, \infty)$

63. At odd multiples of $\dfrac{\pi}{2}$ **65.** At odd multiples of $\dfrac{\pi}{2}$ **67.** $[-1, 1]$ **69.** $(-\infty, \infty)$ **71.** $(-\infty, -1] \, h \, [1, \infty)$ **73.** Odd; yes; origin

75. Odd; yes; origin   **77.** Even; yes; y-axis   **79.** 0.9 **81.** 9 **83.** (a) $-\dfrac{1}{3}$ (b) 1 **85.** (a) -2 (b) 6 **87.** (a) -4 (b) -12

89. (a) Using graph: $\sin 1 \approx 0.8, \cos 1 \approx 0.5, \tan 1 \approx 1.6, \csc 1 \approx 1.3, \sec 1 \approx 2, \cot 1 \approx 0.6;$
   Using calculator: $\sin 1 \approx 0.8, \cos 1 \approx 0.5, \tan 1 \approx 1.6, \csc 1 \approx 1.2, \sec 1 \approx 1.9, \cot 1 \approx 0.6$
   (b) Using graph: $\sin 5.1 \approx -0.9, \cos 5.1 \approx 0.4, \tan 5.1 \approx -2.3, \csc 5.1 \approx -1.1, \sec 5.1 \approx 2.5, \cot 5.1 \approx -0.4;$
   Using calculator: $\sin 5.1 \approx -0.9, \cos 5.1 \approx 0.4, \tan 5.1 \approx -2.4, \csc 5.1 \approx -1.1, \sec 5.1 \approx 2.6, \cot 5.1 \approx -0.4$

91. Let $P = (x, y)$ be the point on the unit circle that corresponds to t. Consider the equation $\tan t = \dfrac{y}{x} = a$. Then $y = ax$.

But $x^2 + y^2 = 1$, so $x^2 + a^2 x^2 = 1$. Thus $x = \pm\dfrac{1}{\sqrt{1 + a^2}}$ and $y = \pm\dfrac{a}{\sqrt{1 + a^2}}$; that is, for any real number a, there is a point $P = (x, y)$ on the

unit circle for which $\tan t = a$. In other words, $-\infty < \tan t < \infty$, and the range of the tangent function is the set of all real numbers.

93. Suppose that there is a number $p, 0 < p < 2\pi$, for which $\sin(\theta + p) = \sin\theta$ for all θ. If $\theta = 0$, then $\sin(0 + p) = \sin p = \sin 0 = 0$;

so $p = \pi$. If $\theta = \dfrac{\pi}{2}$, then $\sin\left(\dfrac{\pi}{2} + p\right) = \sin\left(\dfrac{\pi}{2}\right)$. But $p = \pi$. Thus $\sin\left(\dfrac{3\pi}{2}\right) = -1 = \sin\left(\dfrac{\pi}{2}\right) = 1$. This is impossible. Therefore, the smallest

positive number p for which $\sin(\theta + p) = \sin\theta$ for all θ is $p = 2\pi$.

95. $\sec\theta = \dfrac{1}{\cos\theta}$; since $\cos\theta$ has period 2π, so does $\sec\theta$.

97. If $P = (a, b)$ is the point on the unit circle corresponding to θ, then $Q = (-a, -b)$ is the point on the unit circle corresponding to $\theta + \pi$.

Thus $\tan(\theta + p) = \dfrac{(-b)}{(-a)} = \dfrac{b}{a} = \tan\theta$. If there exists a number $p, 0 < p < \pi$, for which $\tan(\theta + p) = \tan\theta$ for all θ, then, if $\theta = 0$,

$\tan(p) = \tan 0 = 0$. But this means that p is a multiple of π. Since no multiple of π exists in the interval $(0, \pi)$, this is a impossible. Therefore,

the fundamental period of $f(\theta) = \tan\theta$ is π.

99. $m = \dfrac{\sin\theta - 0}{\cos\theta - 0} = \dfrac{\sin\theta}{\cos\theta} = \tan\theta$

2.6 Assess Your Understanding *(page 172)*

3. $1; \frac{\pi}{2} + 2\pi k, k$ any integer **4.** $3; \pi$ **5.** $3; \frac{\pi}{3}$ **6.** T **7.** F **8.** T **9.** 0 **11.** $-\frac{\pi}{2} < x < \frac{\pi}{2}$ **13.** 1 **15.** $0, \pi, 2\pi$

17. $\sin x = 1$ for $x = -\frac{3\pi}{2}, \frac{\pi}{2}$; $\sin x = -1$ for $x = -\frac{\pi}{2}, \frac{3\pi}{2}$ **19.** Amplitude $= 2$; period $= 2\pi$ **21.** Amplitude $= 4$; period $= \pi$

23. Amplitude $= 6$; period $= 2$ **25.** Amplitude $= \frac{1}{2}$; period $= \frac{4\pi}{3}$ **27.** Amplitude $= \frac{5}{3}$; period $= 3$ **29.** F **31.** A **33.** H

35. C **37.** J **39.** A **41.** B

43.

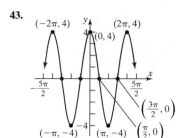

Domain: $(-\infty, \infty)$
Range: $[-4, 4]$

45.

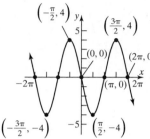

Domain: $(-\infty, \infty)$
Range: $[-4, 4]$

47.

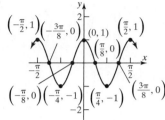

Domain: $(-\infty, \infty)$
Range: $[-1, 1]$

49.

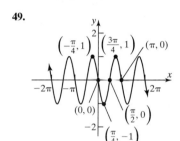

Domain: $(-\infty, \infty)$

Range: $[-1, 1]$

51.

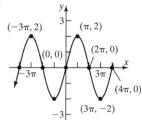

Domain: $(-\infty, \infty)$

Range: $[-2, 2]$

53.

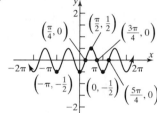

Domain: $(-\infty, \infty)$

Range: $\left[-\frac{1}{2}, \frac{1}{2}\right]$

55.

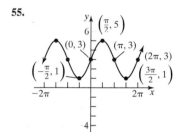

Domain: $(-\infty, \infty)$

Range: $[1, 5]$

57.

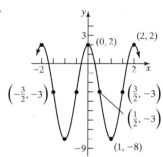

Domain: $(-\infty, \infty)$

Range: $[-8, 2]$

59.

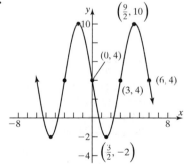

Domain: $(-\infty, \infty)$

Range: $[-2, 10]$

61.

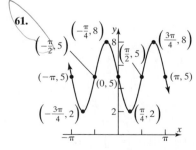

Domain: $(-\infty, \infty)$

Range: $[2, 8]$

63.

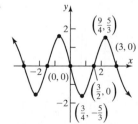

Domain: $(-\infty, \infty)$

Range: $\left[-\frac{5}{3}, \frac{5}{3}\right]$

65.

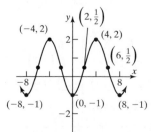

Domain: $(-\infty, \infty)$

Range: $[-1, 2]$

67. $y = \pm 3 \sin(2x)$ **69.** $y = \pm 3 \sin(\pi x)$ **71.** $y = 5 \cos\left(\dfrac{\pi}{4}x\right)$ **73.** $y = -3 \cos\left(\dfrac{1}{2}x\right)$ **75.** $y = \dfrac{3}{4}\sin(2\pi x)$ **77.** $y = -\sin\left(\dfrac{3}{2}x\right)$

79. $y = -\cos\left(\dfrac{4\pi}{3}x\right) + 1$ **81.** $y = 3 \sin\left(\dfrac{\pi}{2}x\right)$ **83.** $y = -4 \cos(3x)$ **85.** $\dfrac{2}{\pi}$ **87.** $\dfrac{\sqrt{2}}{\pi}$

89. $(f \circ g)(x) = \sin(4x)$ **91.** $(f \circ g)(x) = -2 \cos x$ **93.** Period $= \dfrac{1}{30}$ s; amplitude $= 220$ amp

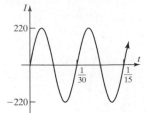

$(g \circ f)(x) = 4 \sin x$ $(g \circ f)(x) = \cos(-2x)$

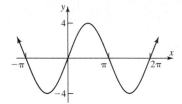

95. (a) Amplitude $= 220$ V; period $= \dfrac{1}{60}$ s   **97. (a)** $P(t) = \dfrac{[V_0 \sin(2\pi f t)]^2}{R} = \dfrac{V_0^2}{R}\sin^2(2\pi f t)$

(b), (e)

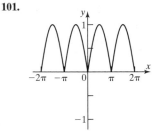

(b) Since the graph of P has amplitude $\dfrac{V_0^2}{2R}$ and period $\dfrac{1}{2f}$ and is of the form

$$y = A \cos(\omega t) + B, \text{ then } A = -\dfrac{V_0^2}{2R} \text{ and } B = \dfrac{V_0^2}{2R}. \text{ Since } \dfrac{1}{2f} = \dfrac{2\pi}{\omega}, \text{ then } \omega = 4\pi f.$$

$$\text{Therefore, } P(t) = -\dfrac{V_0^2}{2R}\cos(4\pi f t) + \dfrac{V_0^2}{2R} = \dfrac{V_0^2}{2R}[1 - \cos(4\pi f t)].$$

(c) $I(t) = 22 \sin(120\pi t)$

(d) Amplitude $= 22$ amp; period $= \dfrac{1}{60}$ s

99. (a) Physical potential: $\omega = \dfrac{2\pi}{23}$; emotional potential: $\omega = \dfrac{\pi}{14}$; intellectual potential: $\omega = \dfrac{2\pi}{33}$ **101.**

(b)

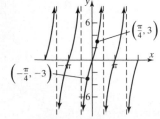

(c) No

(d) Physical potential peaks at 15 days after 20th birthday. Emotional potential is 50% at 17 days, with a maximum at 10 days and a minimum at 24 days. Intellectual potential starts fairly high, drops to a minimum at 13 days, and rises to a maximum at 29 days.

2.7 Assess Your Understanding *(page 183)*

3. origin; odd multiples of $\dfrac{\pi}{2}$ **4.** y-axis; odd multiples of $\dfrac{\pi}{2}$ **5.** $y = \cos x$ **6.** T   **7.** 0   **9.** 1

11. $\sec x = 1$ for $x = -2\pi, 0, 2\pi$; $\sec x = -1$ for $x = -\pi, \pi$ **13.** $-\dfrac{3\pi}{2}, -\dfrac{\pi}{2}, \dfrac{\pi}{2}, \dfrac{3\pi}{2}$ **15.** $-\dfrac{3\pi}{2}, -\dfrac{\pi}{2}, \dfrac{\pi}{2}, \dfrac{3\pi}{2}$

17.   **19.**   **21.**

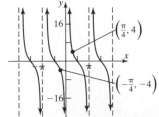   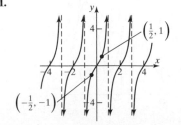

Domain: $\left\{x \mid x \neq \dfrac{k\pi}{2}, k \text{ is an odd integer}\right\}$ Domain: $\{x \mid x \neq k\pi, k \text{ is an integer}\}$ Domain: $\{x \mid x \text{ does not equal an odd integer}\}$

Range: $(-\infty, \infty)$ Range: $(-\infty, \infty)$ Range: $(-\infty, \infty)$

23.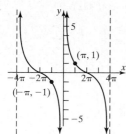

Domain: $\{x | x \neq 4k\pi, k \text{ is an integer}\}$

Range: $(-\infty, \infty)$

25.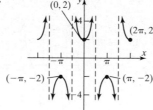

Domain: $\left\{x | x \neq \dfrac{k\pi}{2}, k \text{ is an odd integer}\right\}$

Range: $\{y | y \leq -2 \text{ or } y \geq 2\}$

27.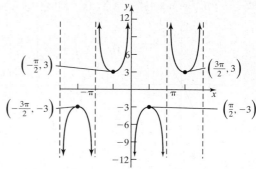

Domain: $\{x | x \neq k\pi, k \text{ is an integer}\}$

Range: $\{y | y \leq -3 \text{ or } y \geq 3\}$

29.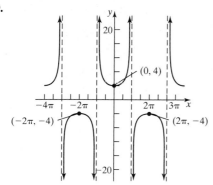

Domain: $\{x | x \neq k\pi, k \text{ is an odd integer}\}$

Range: $\{y | y \leq -4 \text{ or } y \geq 4\}$

31.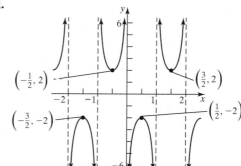

Domain: $\{x | x \text{ does not equal an integer}\}$

Range: $\{y | y \leq -2 \text{ or } y \geq 2\}$

33.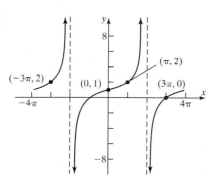

Domain: $\{x | x \neq 2\pi k, k \text{ is an odd integer}\}$

Range: $(-\infty, \infty)$

35.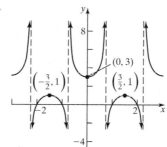

Domain: $\left\{x | x \neq \dfrac{3}{4}k, k \text{ is an odd integer}\right\}$

Range: $\{y | y \leq 1 \text{ or } y \geq 3\}$

37.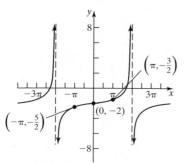

Domain: $\{x | x \neq 2\pi k, k \text{ is an odd integer}\}$

Range: $(-\infty, \infty)$

39.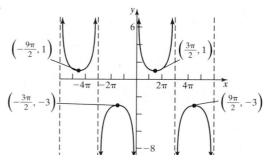

Domain: $\{x | x \neq 3\pi k, k \text{ is an integer}\}$

Range: $\{y | y \leq -3 \text{ or } y \geq 1\}$

41. $\dfrac{2\sqrt{3}}{\pi}$ **43.** $\dfrac{6\sqrt{3}}{\pi}$

45. $(f \circ g)(x) = \tan(4x)$ $(g \circ f)(x) = 4 \tan x$ **47.** $(f \circ g)(x) = -2 \cot x$ $(g \circ f)(x) = \cot(-2x)$

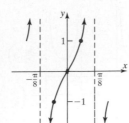

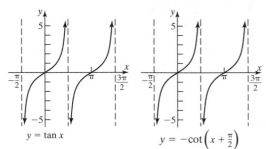

49. (a) $L(\theta) = \dfrac{3}{\cos \theta} + \dfrac{4}{\sin \theta} = 3 \sec \theta + 4 \csc \theta$ **(c)** Approximately 0.83   **51.**

    (b)   **(d)** Approximately 9.86 ft

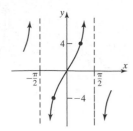

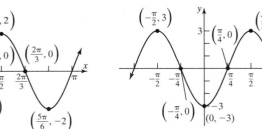

$y = \tan x$ $y = -\cot\left(x + \dfrac{\pi}{2}\right)$

2.8 Assess Your Understanding (page 193)

1. phase shift   **2.** F

3. Amplitude $= 4$ **5.** Amplitude $= 2$ **7.** Amplitude $= 3$

Period $= \pi$ Period $= \dfrac{2\pi}{3}$ Period $= \pi$

Phase shift $= \dfrac{\pi}{2}$ Phase shift $= -\dfrac{\pi}{6}$ Phase shift $= -\dfrac{\pi}{4}$

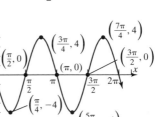

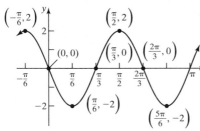

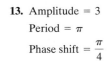

9. Amplitude $= 4$ **11.** Amplitude $= 3$ **13.** Amplitude $= 3$

Period $= 2$ Period $= 2$ Period $= \pi$

Phase shift $= -\dfrac{2}{\pi}$ Phase shift $= \dfrac{2}{\pi}$ Phase shift $= \dfrac{\pi}{4}$

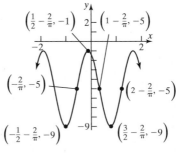

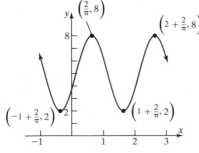

15. $y = 2 \sin\left[2\left(x - \dfrac{1}{2}\right)\right]$ or $y = 2 \sin(2x - 1)$ **19.**

17. $y = 3 \sin\left[\dfrac{2}{3}\left(x + \dfrac{1}{3}\right)\right]$ or $y = 3 \sin\left(\dfrac{2}{3}x + \dfrac{2}{9}\right)$

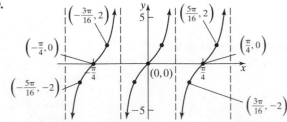

21.

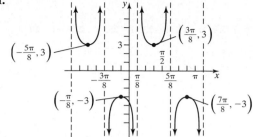

23.

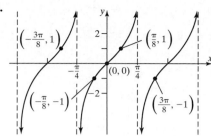

25.

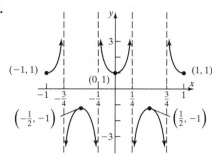

27. Period $= \dfrac{1}{15}$ s; amplitude $= 120$ amp;

amplitude $= 120$ amp; phase shift $= \dfrac{1}{90}$ s

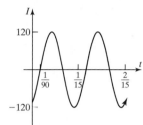

29. (a)

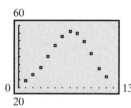

(b) $y = 15.9 \sin\left[\dfrac{\pi}{6}(x - 4)\right] + 40.1$ or $y = 15.9 \sin\left(\dfrac{\pi}{6}x - \dfrac{2\pi}{3}\right) + 40.1$

(c)

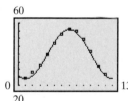

(d) $y = 15.62 \sin(0.517x - 2.096) + 40.377$

(e)

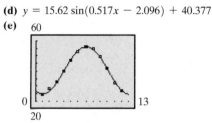

31. (a)

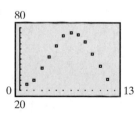

(b) $y - 24.95 \sin\left[\dfrac{\pi}{6}(x - 4)\right] + 50.45$ or $y = 24.95 \sin\left(\dfrac{\pi}{6}x - \dfrac{2\pi}{3}\right) + 50.45$

(c)

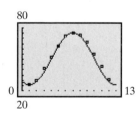

(d) $y = 25.693 \sin(0.476x - 1.814) + 49.854$

(e)

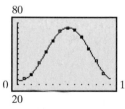

33. (a) 4:08 PM

(b) $y = 4.4 \sin\left[\dfrac{4\pi}{25}(x - 0.5083)\right] + 3.8$ or

$y = 4.4 \sin\left[\dfrac{4\pi}{25}x - 0.2555\right] + 3.8$

(c)

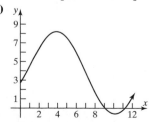

(d) 8.2 ft

35. (a) $y = 1.6 \sin\left(\dfrac{2\pi}{365}x - 1.39\right) + 12.15$

(b) 12.43 h

(c)

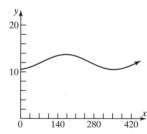

(d) The actual hours of sunlight on April 1, 2007, were 12.43 hours. This is the same as the predicted amount.

37. (a) $y = 6.97 \sin\left(\dfrac{2\pi}{365}x - 1.39\right) + 12.45$

(b) 13.67 h

(c)

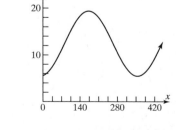

(d) The actual hours of sunlight on April 1, 2007, were 13.38 hours. This is close to the predicted amount of 13.67 hours.

Review Exercises *(page 199)*

1. $\dfrac{3\pi}{4}$ **3.** $\dfrac{\pi}{10}$ **5.** $135°$ **7.** $-450°$ **9.** $\dfrac{1}{2}$ **11.** $\dfrac{3\sqrt{2}}{2} - \dfrac{4\sqrt{3}}{3}$ **13.** $-3\sqrt{2} - 2\sqrt{3}$ **15.** 3 **17.** 0 **19.** 0 **21.** 1 **23.** 1 **25.** 1 **27.** -1 **29.** 1

31. $\cos\theta = \dfrac{3}{5}$; $\tan\theta = \dfrac{4}{3}$; $\csc\theta = \dfrac{5}{4}$; $\sec\theta = \dfrac{5}{3}$; $\cot\theta = \dfrac{3}{4}$

33. $\sin\theta = -\dfrac{12}{13}$; $\cos\theta = -\dfrac{5}{13}$; $\csc\theta = -\dfrac{13}{12}$; $\sec\theta = -\dfrac{13}{5}$; $\cot\theta = \dfrac{5}{12}$

35. $\sin\theta = \dfrac{3}{5}$; $\cos\theta = -\dfrac{4}{5}$; $\tan\theta = -\dfrac{3}{4}$; $\csc\theta = \dfrac{5}{3}$; $\cot\theta = -\dfrac{4}{3}$

37. $\cos\theta = -\dfrac{5}{13}$; $\tan\theta = -\dfrac{12}{5}$; $\csc\theta = \dfrac{13}{12}$; $\sec\theta = -\dfrac{13}{5}$; $\cot\theta = -\dfrac{5}{12}$

39. $\cos\theta = \dfrac{12}{13}$; $\tan\theta = -\dfrac{5}{12}$; $\csc\theta = -\dfrac{13}{5}$; $\sec\theta = \dfrac{13}{12}$; $\cot\theta = -\dfrac{12}{5}$

41. $\sin\theta = -\dfrac{\sqrt{10}}{10}$; $\cos\theta = -\dfrac{3\sqrt{10}}{10}$; $\csc\theta = -\sqrt{10}$; $\sec\theta = -\dfrac{\sqrt{10}}{3}$; $\cot\theta = 3$

43. $\sin\theta = -\dfrac{2\sqrt{2}}{3}$; $\cos\theta = \dfrac{1}{3}$; $\tan\theta = -2\sqrt{2}$; $\csc\theta = -\dfrac{3\sqrt{2}}{4}$; $\cot\theta = -\dfrac{\sqrt{2}}{4}$

45. $\sin\theta = \dfrac{\sqrt{5}}{5}$; $\cos\theta = -\dfrac{2\sqrt{5}}{5}$; $\tan\theta = -\dfrac{1}{2}$; $\csc\theta = \sqrt{5}$; $\sec\theta = -\dfrac{\sqrt{5}}{2}$

47.

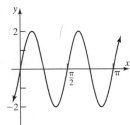

Domain: $(-\infty, \infty)$;

Range: $[-2, 2]$

49.

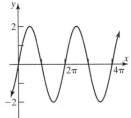

Domain: $(-\infty, \infty)$;

Range: $[-2, 2]$

51.

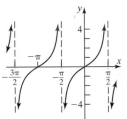

Domain: $\left\{ x \,\middle|\, x \neq \dfrac{k\pi}{2},\, k \text{ is an odd integer} \right\}$;

Range: $(-\infty, \infty)$

53.

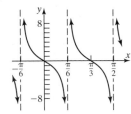

Domain: $\left\{ x \,\middle|\, x \neq \dfrac{k\pi}{6},\, k \text{ is an odd integer} \right\}$;

Range: $(-\infty, \infty)$

55.

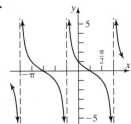

Domain: $\left\{ x \,\middle|\, x \neq \dfrac{(4k-1)\pi}{4},\, k \text{ is an integer} \right\}$;

Range: $(-\infty, \infty)$

57.

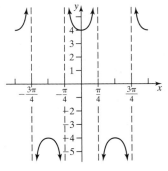

Domain: $\left\{ x \,\middle|\, x \neq \dfrac{k\pi}{4},\, k \text{ is an odd integer} \right\}$;

Range: $\{ y \,|\, y \leq -4 \text{ or } y \geq 4 \}$

59.

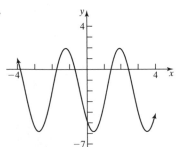

Domain: $(-\infty, \infty)$;

Range: $[-6, 2]$

61.

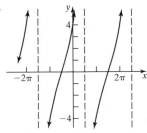

Domain: $\left\{ x \,\middle|\, x \neq \dfrac{(2k-1)\pi}{2},\, k \text{ is an odd integer} \right\}$;

Range: $(-\infty, \infty)$

63. Amplitude $= 4$; period $= 2\pi$

65. Amplitude $= 8$; period $= 4$

67. Amplitude = 4

Period = $\dfrac{2\pi}{3}$

Phase shift = 0

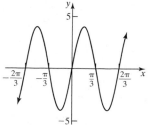

69. Amplitude = 2

Period = π

Phase shift = $\dfrac{\pi}{2}$

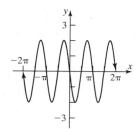

71. Amplitude = $\dfrac{1}{2}$

Period = $\dfrac{4\pi}{3}$

Phase shift = $\dfrac{2\pi}{3}$

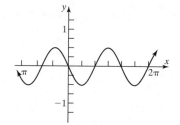

73. Amplitude = $\dfrac{2}{3}$

Period = 2

Phase shift = $\dfrac{6}{\pi}$

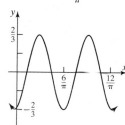

75. $y = 5\cos\left(\dfrac{x}{4}\right)$ **77.** $y = -6\cos\left(\dfrac{\pi}{4}x\right)$

79. $\sin\theta = \dfrac{5}{13}$; $\cos\theta = \dfrac{12}{13}$; $\tan\theta = \dfrac{5}{12}$; $\csc\theta = \dfrac{13}{5}$; $\sec\theta = \dfrac{13}{12}$; $\cot\theta = \dfrac{12}{5}$

81. $\sin\theta = -\dfrac{4}{5}$; $\cos\theta = \dfrac{3}{5}$; $\tan\theta = -\dfrac{4}{3}$; $\csc\theta = -\dfrac{5}{4}$; $\sec\theta = \dfrac{5}{3}$; $\cot\theta = -\dfrac{3}{4}$ **83.** $\dfrac{\pi}{5}$

85. Domain: $\left\{x \mid x \neq \dfrac{k\pi}{2}, k \text{ is an odd integer}\right\}$; range: $\left\{y \mid y \leq -1 \text{ or } y \geq 1\right\}$

87. $\dfrac{\pi}{3} \approx 1.047$ ft; $\dfrac{\pi}{3} \approx 1.047$ ft^2 **89.** Approximately 114.59 rev/h

91. 0.1 rev/s $= \dfrac{\pi}{5}$ radian/s **93.** 839.10 ft **95.** 23.32 ft **97.** 2.15 mi

99. (a) $\dfrac{1}{15}$ s **(b)** 220 amp **(c)** $-\dfrac{1}{180}$ s **(d)**

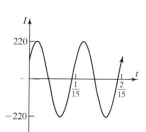

101. (a) $y = 2.465\sin\left(\dfrac{2\pi}{365}x - 1.39\right) + 12.165$ **(b)** 12.60 h **(c)**

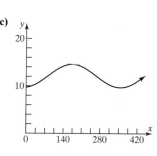

(d) The actual hours of sunlight on April 1, 2007, were 12.6 hours. This is the same as the predicted amount.

Chapter Test *(page 202)*

1. $\dfrac{13\pi}{9}$ **2.** $-\dfrac{20\pi}{9}$ **3.** $\dfrac{13\pi}{180}$ **4.** $-22.5°$ **5.** $810°$ **6.** $135°$ **7.** $\dfrac{1}{2}$ **8.** 0 **9.** $-\dfrac{1}{2}$ **10.** $-\dfrac{\sqrt{3}}{3}$ **11.** 2 **12.** $\dfrac{3(1-\sqrt{2})}{2}$ **13.** 0.292 **14.** 0.309

15. -1.524 **16.** 2.747 **17.**

	$\sin\theta$	$\cos\theta$	$\tan\theta$	$\sec\theta$	$\csc\theta$	$\cot\theta$
θ in QI	+	+	+	+	+	+
θ in QII	+	−	−	−	+	−
θ in QIII	−	−	+	−	−	+
θ in QIV	−	+	−	+	−	−

18. $-\dfrac{3}{5}$

19. $\cos \theta = -\dfrac{2\sqrt{6}}{7}$; $\tan \theta = -\dfrac{5\sqrt{6}}{12}$; $\csc \theta = \dfrac{7}{5}$; $\sec \theta = -\dfrac{7\sqrt{6}}{12}$; $\cot \theta = -\dfrac{2\sqrt{6}}{5}$

20. $\sin \theta = -\dfrac{\sqrt{5}}{3}$; $\tan \theta = -\dfrac{\sqrt{5}}{2}$; $\csc \theta = -\dfrac{3\sqrt{5}}{5}$; $\sec \theta = \dfrac{3}{2}$; $\cot \theta = -\dfrac{2\sqrt{5}}{5}$

21. $\sin \theta = \dfrac{12}{13}$; $\cos \theta = -\dfrac{5}{13}$; $\csc \theta = \dfrac{13}{12}$; $\sec \theta = -\dfrac{13}{5}$; $\cot \theta = -\dfrac{5}{12}$ **22.** $\dfrac{7\sqrt{53}}{53}$ **23.** $-\dfrac{5\sqrt{146}}{146}$ **24.** $-\dfrac{1}{2}$

25.

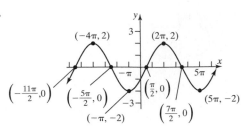

26.

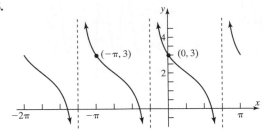

27. $y = -3\sin\left(3x + \dfrac{3\pi}{4}\right)$ **28.** 78.93 ft² **29.** 143.5 rpm **30.** The ship is about 838 ft from the statue. **31.** The building is 122.4 ft tall.

Cumulative Review *(page 202)*

1. $\left\{-1, \dfrac{1}{2}\right\}$ **2.** $y - 5 = -3(x + 2)$ or $y = -3x - 1$ **3.** $x^2 + (y + 2)^2 = 16$

4. A line; slope $\dfrac{2}{3}$; intercepts $(6, 0)$ and $(0, -4)$ **5.** A circle; center $(1, -2)$; radius 3 **6.**

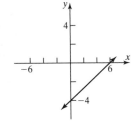

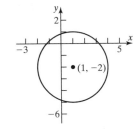

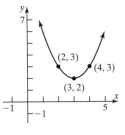

7. (a)

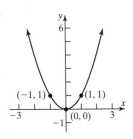

(b)

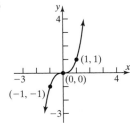

(c)

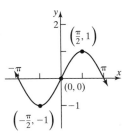

(d)

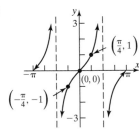

8. $f^{-1}(x) = \dfrac{1}{3}(x + 2)$ **10.**

9. -2

11. $3 - \dfrac{3\sqrt{3}}{2}$

12. $y = 3\cos\left(\dfrac{\pi}{6}x\right)$

13. $f(x) = -3x - 3$;

$m = -3$; $(-1, 0)$, $(0, -3)$

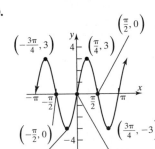

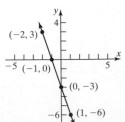

CHAPTER 3 Analytic Trigonometry

3.1 Assess Your Understanding *(page 216)*

7. $x = \sin y$ **8.** $\dfrac{\pi}{2}$ **9.** $\dfrac{\pi}{5}$ **10.** F **11.** T **12.** T **13.** 0 **15.** $-\dfrac{\pi}{2}$ **17.** 0 **19.** $\dfrac{\pi}{4}$ **21.** $\dfrac{\pi}{3}$ **23.** $\dfrac{5\pi}{6}$ **25.** 0.10 **27.** 1.37 **29.** 0.51 **31.** -0.38

33. -0.12 **35.** 1.08 **37.** $\dfrac{4\pi}{5}$ **39.** $-\dfrac{3\pi}{8}$ **41.** $-\dfrac{\pi}{8}$ **43.** $-\dfrac{\pi}{5}$ **45.** $\dfrac{1}{4}$ **47.** 4 **49.** Not defined **51.** π

53. Restricted domain of f: $\left[-\dfrac{\pi}{2}, \dfrac{\pi}{2}\right]$ **55.** Restricted domain of f: $\left[0, \dfrac{\pi}{3}\right]$ **57.** Restricted domain of f: $\left[-\dfrac{\pi+2}{2}, \dfrac{\pi-2}{2}\right]$

$f^{-1}(x) = \sin^{-1}\dfrac{x-2}{5}$ $f^{-1}(x) = \dfrac{1}{3}\cos^{-1}\left(-\dfrac{x}{2}\right)$ $f^{-1}(x) = -\tan^{-1}(x+3) - 1$

Domain of f^{-1}: $[-3, 7]$ Domain of f^{-1}: $[-2, 2]$ Domain of f^{-1}: $(-\infty, \infty)$

59. Restricted domain of f: $\left[-\dfrac{\pi+2}{4}, \dfrac{\pi-2}{4}\right]$ **61.** $\left\{\dfrac{\sqrt{2}}{2}\right\}$ **63.** $\left\{-\dfrac{1}{4}\right\}$ **65.** $\{\sqrt{3}\}$ **67.** $\{-1\}$

$f^{-1}(x) = \dfrac{1}{2}\left[\sin^{-1}\left(\dfrac{x}{3}\right) - 1\right]$ **69. (a)** 13.92 h or 13 h, 55 min **(b)** 12 h **(c)** 13.85 h or 13 h, 51 min

Domain of f^{-1}: $[-3, 3]$ **71. (a)** 13.3 h or 13 h, 18 min **(b)** 12 h **(c)** 13.26 h or 13 h, 15 min

73. (a) 12 h **(b)** 12 h **(c)** 12 h **(d)** It is 12 h. **75.** 3.28 min **77. (a)** $\dfrac{\pi}{3}$ square units **(b)** $\dfrac{5\pi}{12}$ square units **79.** 4250 mi

3.2 Assess Your Understanding *(page 223)*

4. $x = \sec y$; ≥ 1; 0; π **5.** $\dfrac{\sqrt{2}}{2}$ **6.** F **7.** T **8.** T **9.** $\dfrac{\sqrt{2}}{2}$ **11.** $-\dfrac{\sqrt{3}}{3}$ **13.** 2 **15.** $\sqrt{2}$ **17.** $-\dfrac{\sqrt{2}}{2}$ **19.** $\dfrac{2\sqrt{3}}{3}$ **21.** $\dfrac{3\pi}{4}$ **23.** $\dfrac{\pi}{6}$ **25.** $\dfrac{\sqrt{2}}{4}$

27. $\dfrac{\sqrt{5}}{2}$ **29.** $-\dfrac{\sqrt{14}}{2}$ **31.** $-\dfrac{3\sqrt{10}}{10}$ **33.** $\sqrt{5}$ **35.** $-\dfrac{\pi}{4}$ **37.** $\dfrac{\pi}{6}$ **39.** $-\dfrac{\pi}{2}$ **41.** $\dfrac{\pi}{6}$ **43.** $\dfrac{2\pi}{3}$ **45.** 1.32 **47.** 0.46 **49.** -0.34 **51.** 2.72 **53.** -0.73

57. $\dfrac{1}{\sqrt{1+u^2}}$ **59.** $\dfrac{u}{\sqrt{1-u^2}}$ **61.** $\dfrac{\sqrt{u^2-1}}{|u|}$ **63.** $\dfrac{\sqrt{u^2-1}}{u}$ **65.** $\dfrac{1}{u}$ **67.** $\dfrac{5}{13}$ **69.** $\dfrac{3\pi}{4}$ **71.** $-\dfrac{3}{4}$ **73.** $\dfrac{5}{13}$ **75.** $\dfrac{\pi}{6}$ **77.** $-\sqrt{15}$

79. (a) $\theta = 31.89°$ **(b)** 54.64 ft in diameter **(c)** 37.96 ft high **81. (a)** $\theta = 22.3°$ **(b)** $v_0 = 2940.23$ ft/s **83.**

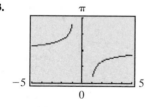

3.3 Assess Your Understanding *(page 229)*

3. identity; conditional **4.** -1 **5.** 0 **6.** T **7.** F **8.** T **9.** $\dfrac{1}{\cos\theta}$ **11.** $\dfrac{1+\sin\theta}{\cos\theta}$ **13.** $\dfrac{1}{\sin\theta\cos\theta}$ **15.** 2 **17.** $\dfrac{3\sin\theta+1}{\sin\theta+1}$

19. $\csc\theta \cdot \cos\theta = \dfrac{1}{\sin\theta} \cdot \cos\theta = \dfrac{\cos\theta}{\sin\theta} = \cot\theta$ **21.** $1 + \tan^2(-\theta) = 1 + (-\tan\theta)^2 = 1 + \tan^2\theta = \sec^2\theta$

23. $\cos\theta(\tan\theta + \cot\theta) = \cos\theta\left(\dfrac{\sin\theta}{\cos\theta} + \dfrac{\cos\theta}{\sin\theta}\right) = \cos\theta\left(\dfrac{\sin^2\theta + \cos^2\theta}{\cos\theta\sin\theta}\right) = \cos\theta\left(\dfrac{1}{\cos\theta\sin\theta}\right) = \dfrac{1}{\sin\theta} = \csc\theta$

25. $\tan u\,\cot u - \cos^2 u = \tan u \cdot \dfrac{1}{\tan u} - \cos^2 u = 1 - \cos^2 u = \sin^2 u$ **27.** $(\sec\theta - 1)(\sec\theta + 1) = \sec^2\theta - 1 = \tan^2\theta$

29. $(\sec\theta + \tan\theta)(\sec\theta - \tan\theta) = \sec^2\theta - \tan^2\theta = 1$ **31.** $\cos^2\theta(1 + \tan^2\theta) = \cos^2\theta\sec^2\theta = \cos^2\theta \cdot \dfrac{1}{\cos^2\theta} = 1$

33. $(\sin\theta + \cos\theta)^2 + (\sin\theta - \cos\theta)^2 = \sin^2\theta + 2\sin\theta\cos\theta + \cos^2\theta + \sin^2\theta - 2\sin\theta\cos\theta + \cos^2\theta$
$= \sin^2\theta + \cos^2\theta + \sin^2\theta + \cos^2\theta = 1 + 1 = 2$

35. $\sec^4\theta - \sec^2\theta = \sec^2\theta(\sec^2\theta - 1) = (1 + \tan^2\theta)\tan^2\theta = \tan^4\theta + \tan^2\theta$

37. $\sec u - \tan u = \dfrac{1}{\cos u} - \dfrac{\sin u}{\cos u} = \dfrac{1 - \sin u}{\cos u} \cdot \dfrac{1 + \sin u}{1 + \sin u} = \dfrac{1 - \sin^2 u}{\cos u(1 + \sin u)} = \dfrac{\cos^2 u}{\cos u(1 + \sin u)} = \dfrac{\cos u}{1 + \sin u}$

39. $3\sin^2\theta + 4\cos^2\theta = 3\sin^2\theta + 3\cos^2\theta + \cos^2\theta = 3(\sin^2\theta + \cos^2\theta) + \cos^2\theta = 3 + \cos^2\theta$

41. $1 - \dfrac{\cos^2\theta}{1+\sin\theta} = 1 - \dfrac{1-\sin^2\theta}{1+\sin\theta} = 1 - \dfrac{(1+\sin\theta)(1-\sin\theta)}{1+\sin\theta} = 1 - (1-\sin\theta) = \sin\theta$

43. $\dfrac{1+\tan v}{1-\tan v} = \dfrac{1+\dfrac{1}{\cot v}}{1-\dfrac{1}{\cot v}} = \dfrac{\dfrac{\cot v+1}{\cot v}}{\dfrac{\cot v-1}{\cot v}} = \dfrac{\cot v+1}{\cot v-1}$ **45.** $\dfrac{\sec\theta}{\csc\theta} + \dfrac{\sin\theta}{\cos\theta} = \dfrac{\dfrac{1}{\cos\theta}}{\dfrac{1}{\sin\theta}} + \tan\theta = \dfrac{\sin\theta}{\cos\theta} + \tan\theta = \tan\theta + \tan\theta = 2\tan\theta$

47. $\dfrac{1+\sin\theta}{1-\sin\theta} = \dfrac{1+\dfrac{1}{\csc\theta}}{1-\dfrac{1}{\csc\theta}} = \dfrac{\dfrac{\csc\theta+1}{\csc\theta}}{\dfrac{\csc\theta-1}{\csc\theta}} = \dfrac{\csc\theta+1}{\csc\theta-1}$

49. $\dfrac{1-\sin v}{\cos v} + \dfrac{\cos v}{1-\sin v} = \dfrac{(1-\sin v)^2 + \cos^2 v}{\cos v(1-\sin v)} = \dfrac{1-2\sin v+\sin^2 v+\cos^2 v}{\cos v(1-\sin v)} = \dfrac{2-2\sin v}{\cos v(1-\sin v)} = \dfrac{2(1-\sin v)}{\cos v(1-\sin v)} = \dfrac{2}{\cos v} = 2\sec v$

51. $\dfrac{\sin\theta}{\sin\theta-\cos\theta} = \dfrac{1}{\dfrac{\sin\theta-\cos\theta}{\sin\theta}} = \dfrac{1}{1-\dfrac{\cos\theta}{\sin\theta}} = \dfrac{1}{1-\cot\theta}$

53. $(\sec\theta-\tan\theta)^2 = \sec^2\theta - 2\sec\theta\tan\theta + \tan^2\theta = \dfrac{1}{\cos^2\theta} - \dfrac{2\sin\theta}{\cos^2\theta} + \dfrac{\sin^2\theta}{\cos^2\theta} = \dfrac{1-2\sin\theta+\sin^2\theta}{\cos^2\theta} = \dfrac{(1-\sin\theta)^2}{1-\sin^2\theta} = \dfrac{(1-\sin\theta)^2}{(1-\sin\theta)(1+\sin\theta)}$

$= \dfrac{1-\sin\theta}{1+\sin\theta}$

55. $\dfrac{\cos\theta}{1-\tan\theta} + \dfrac{\sin\theta}{1-\cot\theta} = \dfrac{\cos\theta}{1-\dfrac{\sin\theta}{\cos\theta}} + \dfrac{\sin\theta}{1-\dfrac{\cos\theta}{\sin\theta}} = \dfrac{\cos\theta}{\dfrac{\cos\theta-\sin\theta}{\cos\theta}} + \dfrac{\sin\theta}{\dfrac{\sin\theta-\cos\theta}{\sin\theta}} = \dfrac{\cos^2\theta}{\cos\theta-\sin\theta} + \dfrac{\sin^2\theta}{\sin\theta-\cos\theta}$

$= \dfrac{\cos^2\theta-\sin^2\theta}{\cos\theta-\sin\theta} = \dfrac{(\cos\theta-\sin\theta)(\cos\theta+\sin\theta)}{\cos\theta-\sin\theta} = \sin\theta + \cos\theta$

57. $\tan\theta + \dfrac{\cos\theta}{1+\sin\theta} = \dfrac{\sin\theta}{\cos\theta} + \dfrac{\cos\theta}{1+\sin\theta} = \dfrac{\sin\theta(1+\sin\theta)+\cos^2\theta}{\cos\theta(1+\sin\theta)} = \dfrac{\sin\theta+\sin^2\theta+\cos^2\theta}{\cos\theta(1+\sin\theta)} = \dfrac{\sin\theta+1}{\cos\theta(1+\sin\theta)} = \dfrac{1}{\cos\theta} = \sec\theta$

59. $\dfrac{\tan\theta+\sec\theta-1}{\tan\theta-\sec\theta+1} = \dfrac{\tan\theta+(\sec\theta-1)}{\tan\theta-(\sec\theta-1)} \cdot \dfrac{\tan\theta+(\sec\theta-1)}{\tan\theta+(\sec\theta-1)} = \dfrac{\tan^2\theta+2\tan\theta(\sec\theta-1)+\sec^2\theta-2\sec\theta+1}{\tan^2\theta-(\sec^2\theta-2\sec\theta+1)}$

$= \dfrac{\sec^2\theta-1+2\tan\theta(\sec\theta-1)+\sec^2\theta-2\sec\theta+1}{\sec^2\theta-1-\sec^2\theta+2\sec\theta-1} = \dfrac{2\sec^2\theta-2\sec\theta+2\tan\theta(\sec\theta-1)}{-2+2\sec\theta}$

$= \dfrac{2\sec\theta(\sec\theta-1)+2\tan\theta(\sec\theta-1)}{2(\sec\theta-1)} = \dfrac{2(\sec\theta-1)(\sec\theta+\tan\theta)}{2(\sec\theta-1)} = \tan\theta + \sec\theta$

61. $\dfrac{\tan\theta-\cot\theta}{\tan\theta+\cot\theta} = \dfrac{\dfrac{\sin\theta}{\cos\theta} - \dfrac{\cos\theta}{\sin\theta}}{\dfrac{\sin\theta}{\cos\theta} + \dfrac{\cos\theta}{\sin\theta}} = \dfrac{\dfrac{\sin^2\theta-\cos^2\theta}{\cos\theta\sin\theta}}{\dfrac{\sin^2\theta+\cos^2\theta}{\cos\theta\sin\theta}} = \dfrac{\sin^2\theta-\cos^2\theta}{1} = \sin^2\theta - \cos^2\theta$

63. $\dfrac{\tan u-\cot u}{\tan u+\cot u} + 1 = \dfrac{\dfrac{\sin u}{\cos u} - \dfrac{\cos u}{\sin u}}{\dfrac{\sin u}{\cos u} + \dfrac{\cos u}{\sin u}} + 1 = \dfrac{\dfrac{\sin^2 u-\cos^2 u}{\cos u\sin u}}{\dfrac{\sin^2 u+\cos^2 u}{\cos u\sin u}} + 1 = \sin^2 u - \cos^2 u + 1 = \sin^2 u + (1-\cos^2 u) = 2\sin^2 u$

65. $\dfrac{\sec\theta+\tan\theta}{\cot\theta+\cos\theta} = \dfrac{\dfrac{1}{\cos\theta} + \dfrac{\sin\theta}{\cos\theta}}{\dfrac{\cos\theta}{\sin\theta} + \cos\theta} = \dfrac{\dfrac{1+\sin\theta}{\cos\theta}}{\dfrac{\cos\theta+\cos\theta\sin\theta}{\sin\theta}} = \dfrac{1+\sin\theta}{\cos\theta} \cdot \dfrac{\sin\theta}{\cos\theta(1+\sin\theta)} = \dfrac{\sin\theta}{\cos\theta} \cdot \dfrac{1}{\cos\theta} = \tan\theta\sec\theta$

67. $\dfrac{1-\tan^2\theta}{1+\tan^2\theta} + 1 = \dfrac{1-\tan^2\theta+1+\tan^2\theta}{1+\tan^2\theta} = \dfrac{2}{1+\tan^2\theta} = \dfrac{2}{\sec^2\theta} = 2\cos^2\theta$

69. $\dfrac{\sec\theta-\csc\theta}{\sec\theta\csc\theta} = \dfrac{\sec\theta}{\sec\theta\csc\theta} - \dfrac{\csc\theta}{\sec\theta\csc\theta} = \dfrac{1}{\csc\theta} - \dfrac{1}{\sec\theta} = \sin\theta - \cos\theta$

71. $\sec\theta - \cos\theta = \dfrac{1}{\cos\theta} - \cos\theta = \dfrac{1-\cos^2\theta}{\cos\theta} = \dfrac{\sin^2\theta}{\cos\theta} = \sin\theta \cdot \dfrac{\sin\theta}{\cos\theta} = \sin\theta\tan\theta$

73. $\dfrac{1}{1-\sin\theta} + \dfrac{1}{1+\sin\theta} = \dfrac{1+\sin\theta+1-\sin\theta}{(1+\sin\theta)(1-\sin\theta)} = \dfrac{2}{1-\sin^2\theta} = \dfrac{2}{\cos^2\theta} = 2\sec^2\theta$

75. $\dfrac{\sec \theta}{1 - \sin \theta} = \dfrac{\sec \theta}{1 - \sin \theta} \cdot \dfrac{1 + \sin \theta}{1 + \sin \theta} = \dfrac{\sec \theta (1 + \sin \theta)}{1 - \sin^2 \theta} = \dfrac{\sec \theta (1 + \sin \theta)}{\cos^2 \theta} = \dfrac{1 + \sin \theta}{\cos^3 \theta}$

77. $\dfrac{(\sec v - \tan v)^2 + 1}{\csc v (\sec v - \tan v)} = \dfrac{\sec^2 v - 2 \sec v \tan v + \tan^2 v + 1}{\dfrac{1}{\sin v}\left(\dfrac{1}{\cos v} - \dfrac{\sin v}{\cos v}\right)} = \dfrac{2 \sec^2 v - 2 \sec v \tan v}{\dfrac{1}{\sin v}\left(\dfrac{1 - \sin v}{\cos v}\right)} = \dfrac{\dfrac{2}{\cos^2 v} - \dfrac{2 \sin v}{\cos^2 v}}{\dfrac{1 - \sin v}{\sin v \cos v}} = \dfrac{2 - 2 \sin v}{\cos^2 v} \cdot \dfrac{\sin v \cos v}{1 - \sin v}$

$= \dfrac{2(1 - \sin v)}{\cos v} \cdot \dfrac{\sin v}{1 - \sin v} = \dfrac{2 \sin v}{\cos v} = 2 \tan v$

79. $\dfrac{\sin \theta + \cos \theta}{\cos \theta} - \dfrac{\sin \theta - \cos \theta}{\sin \theta} = \dfrac{\sin \theta}{\cos \theta} + 1 - 1 + \dfrac{\cos \theta}{\sin \theta} = \dfrac{\sin^2 \theta + \cos^2 \theta}{\cos \theta \sin \theta} = \dfrac{1}{\cos \theta \sin \theta} = \sec \theta \csc \theta$

81. $\dfrac{\sin^3 \theta + \cos^3 \theta}{\sin \theta + \cos \theta} = \dfrac{(\sin \theta + \cos \theta)(\sin^2 \theta - \sin \theta \cos \theta + \cos^2 \theta)}{\sin \theta + \cos \theta} = \sin^2 \theta + \cos^2 \theta - \sin \theta \cos \theta = 1 - \sin \theta \cos \theta$

83. $\dfrac{\cos^2 \theta - \sin^2 \theta}{1 - \tan^2 \theta} = \dfrac{\cos^2 \theta - \sin^2 \theta}{1 - \dfrac{\sin^2 \theta}{\cos^2 \theta}} = \dfrac{\cos^2 \theta - \sin^2 \theta}{\dfrac{\cos^2 \theta - \sin^2 \theta}{\cos^2 \theta}} = \cos^2 \theta$

85. $\dfrac{(2 \cos^2 \theta - 1)^2}{\cos^4 \theta - \sin^4 \theta} = \dfrac{[2 \cos^2 \theta - (\sin^2 \theta + \cos^2 \theta)]^2}{(\cos^2 \theta - \sin^2 \theta)(\cos^2 \theta + \sin^2 \theta)} = \dfrac{(\cos^2 \theta - \sin^2 \theta)^2}{\cos^2 \theta - \sin^2 \theta} = \cos^2 \theta - \sin^2 \theta = (1 - \sin^2 \theta) - \sin^2 \theta = 1 - 2 \sin^2 \theta$

87. $\dfrac{1 + \sin \theta + \cos \theta}{1 + \sin \theta - \cos \theta} = \dfrac{(1 + \sin \theta) + \cos \theta}{(1 + \sin \theta) - \cos \theta} \cdot \dfrac{(1 + \sin \theta) + \cos \theta}{(1 + \sin \theta) + \cos \theta} = \dfrac{1 + 2 \sin \theta + \sin^2 \theta + 2(1 + \sin \theta) \cos \theta + \cos^2 \theta}{1 + 2 \sin \theta + \sin^2 \theta - \cos^2 \theta}$

$= \dfrac{1 + 2 \sin \theta + \sin^2 \theta + 2(1 + \sin \theta)(\cos \theta) + (1 - \sin^2 \theta)}{1 + 2 \sin \theta + \sin^2 \theta - (1 - \sin^2 \theta)} = \dfrac{2 + 2 \sin \theta + 2(1 + \sin \theta)(\cos \theta)}{2 \sin \theta + 2 \sin^2 \theta}$

$= \dfrac{2(1 + \sin \theta) + 2(1 + \sin \theta)(\cos \theta)}{2 \sin \theta(1 + \sin \theta)} = \dfrac{2(1 + \sin \theta)(1 + \cos \theta)}{2 \sin \theta(1 + \sin \theta)} = \dfrac{1 + \cos \theta}{\sin \theta}$

89. $(a \sin \theta + b \cos \theta)^2 + (a \cos \theta - b \sin \theta)^2 = a^2 \sin^2 \theta + 2ab \sin \theta \cos \theta + b^2 \cos^2 \theta + a^2 \cos^2 \theta - 2ab \sin \theta \cos \theta + b^2 \sin^2 \theta$

$= a^2(\sin^2 \theta + \cos^2 \theta) + b^2(\cos^2 \theta + \sin^2 \theta) = a^2 + b^2$

91. $\dfrac{\tan \alpha + \tan \beta}{\cot \alpha + \cot \beta} = \dfrac{\tan \alpha + \tan \beta}{\dfrac{1}{\tan \alpha} + \dfrac{1}{\tan \beta}} = \dfrac{\tan \alpha + \tan \beta}{\dfrac{\tan \beta + \tan \alpha}{\tan \alpha \tan \beta}} = (\tan \alpha + \tan \beta) \cdot \dfrac{\tan \alpha \tan \beta}{\tan \alpha + \tan \beta} = \tan \alpha \tan \beta$

93. $(\sin \alpha + \cos \beta)^2 + (\cos \beta + \sin \alpha)(\cos \beta - \sin \alpha) = (\sin^2 \alpha + 2 \sin \alpha \cos \beta + \cos^2 \beta) + (\cos^2 \beta - \sin^2 \alpha)$

$= 2 \cos^2 \beta + 2 \sin \alpha \cos \beta = 2 \cos \beta(\cos \beta + \sin \alpha) = 2 \cos \beta(\sin \alpha + \cos \beta)$

95. $\ln|\sec \theta| = \ln|\cos \theta|^{-1} = -\ln|\cos \theta|$ **97.** $\ln|1 + \cos \theta| + \ln|1 - \cos \theta| = \ln(|1 + \cos \theta||1 - \cos \theta|) = \ln|1 - \cos^2 \theta| = \ln|\sin^2 \theta| = 2 \ln|\sin \theta|$

99. $g(x) = \sec x - \cos x = \dfrac{1}{\cos x} - \cos x = \dfrac{1}{\cos x} - \dfrac{\cos^2 x}{\cos x} = \dfrac{1 - \cos^2 x}{\cos x} = \dfrac{\sin^2 x}{\cos x} = \sin x \cdot \dfrac{\sin x}{\cos x} = \sin x \cdot \tan x = f(x)$

101. $f(\theta) = \dfrac{1 - \sin \theta}{\cos \theta} - \dfrac{\cos \theta}{1 + \sin \theta} = \dfrac{1 - \sin \theta}{\cos \theta} \cdot \dfrac{1 + \sin \theta}{1 + \sin \theta} - \dfrac{\cos \theta}{1 + \sin \theta} \cdot \dfrac{\cos \theta}{\cos \theta} = \dfrac{1 - \sin^2 \theta}{\cos \theta(1 + \sin \theta)} - \dfrac{\cos^2 \theta}{\cos \theta(1 + \sin \theta)}$

$= \dfrac{\cos^2 \theta}{\cos \theta(1 + \sin \theta)} - \dfrac{\cos^2 \theta}{\cos \theta(1 + \sin \theta)} = 0 = g(\theta)$

103. $1200 \sec \theta (2 \sec^2 \theta - 1) = 1200 \dfrac{1}{\cos \theta}\left(\dfrac{2}{\cos^2 \theta} - 1\right) = 1200 \dfrac{1}{\cos \theta}\left(\dfrac{2}{\cos^2 \theta} - \dfrac{\cos^2 \theta}{\cos^2 \theta}\right) = 1200 \dfrac{1}{\cos \theta}\left(\dfrac{2 - \cos^2 \theta}{\cos^2 \theta}\right) = \dfrac{1200 (1 + 1 - \cos^2 \theta)}{\cos^3 \theta}$

$= \dfrac{1200 (1 + \sin^2 \theta)}{\cos^3 \theta}$

3.4 Assess Your Understanding *(page 239)*

4. − **5.** − **6.** F **7.** F **8.** F **9.** $\dfrac{1}{4}(\sqrt{6} + \sqrt{2})$ **11.** $\dfrac{1}{4}(\sqrt{2} - \sqrt{6})$ **13.** $-\dfrac{1}{4}(\sqrt{2} + \sqrt{6})$ **15.** $2 - \sqrt{3}$ **17.** $-\dfrac{1}{4}(\sqrt{6} + \sqrt{2})$

19. $\sqrt{6} - \sqrt{2}$ **21.** $\dfrac{1}{2}$ **23.** 0 **25.** 1 **27.** −1 **29.** $\dfrac{1}{2}$ **31.** (a) $\dfrac{2\sqrt{5}}{25}$ (b) $\dfrac{11\sqrt{5}}{25}$ (c) $\dfrac{2\sqrt{5}}{5}$ (d) 2

33. (a) $\dfrac{4 - 3\sqrt{3}}{10}$ (b) $\dfrac{-3 - 4\sqrt{3}}{10}$ (c) $\dfrac{4 + 3\sqrt{3}}{10}$ (d) $\dfrac{25\sqrt{3} + 48}{39}$

35. (a) $-\dfrac{5 + 12\sqrt{3}}{26}$ (b) $\dfrac{12 - 5\sqrt{3}}{26}$ (c) $\dfrac{-5 + 12\sqrt{3}}{26}$ (d) $\dfrac{-240 + 169\sqrt{3}}{69}$

37. (a) $-\dfrac{2\sqrt{2}}{3}$ **(b)** $\dfrac{-2\sqrt{2}+\sqrt{3}}{6}$ **(c)** $\dfrac{-2\sqrt{2}+\sqrt{3}}{6}$ **(d)** $\dfrac{9-4\sqrt{2}}{7}$ **39.** $\dfrac{1-2\sqrt{6}}{6}$ **41.** $\dfrac{\sqrt{3}-2\sqrt{2}}{6}$ **43.** $\dfrac{8\sqrt{2}-9\sqrt{3}}{5}$

45. $\sin\left(\dfrac{\pi}{2}+\theta\right)=\sin\dfrac{\pi}{2}\cos\theta+\cos\dfrac{\pi}{2}\sin\theta=1\cdot\cos\theta+0\cdot\sin\theta=\cos\theta$

47. $\sin(\pi-\theta)=\sin\pi\cos\theta-\cos\pi\sin\theta=0\cdot\cos\theta-(-1)\sin\theta=\sin\theta$

49. $\sin(\pi+\theta)=\sin\pi\cos\theta+\cos\pi\sin\theta=0\cdot\cos\theta+(-1)\sin\theta=-\sin\theta$

51. $\tan(\pi-\theta)=\dfrac{\tan\pi-\tan\theta}{1+\tan\pi\tan\theta}=\dfrac{0-\tan\theta}{1+0\cdot\tan\theta}=-\tan\theta$ **53.** $\sin\left(\dfrac{3\pi}{2}+\theta\right)=\sin\dfrac{3\pi}{2}\cos\theta+\cos\dfrac{3\pi}{2}\sin\theta=(-1)\cos\theta+0\cdot\sin\theta=-\cos\theta$

55. $\sin(\alpha+\beta)+\sin(\alpha-\beta)=\sin\alpha\cos\beta+\cos\alpha\sin\beta+\sin\alpha\cos\beta-\cos\alpha\sin\beta=2\sin\alpha\cos\beta$

57. $\dfrac{\sin(\alpha+\beta)}{\sin\alpha\cos\beta}=\dfrac{\sin\alpha\cos\beta+\cos\alpha\sin\beta}{\sin\alpha\cos\beta}=\dfrac{\sin\alpha\cos\beta}{\sin\alpha\cos\beta}+\dfrac{\cos\alpha\sin\beta}{\sin\alpha\cos\beta}=1+\cot\alpha\tan\beta$

59. $\dfrac{\cos(\alpha+\beta)}{\cos\alpha\cos\beta}=\dfrac{\cos\alpha\cos\beta-\sin\alpha\sin\beta}{\cos\alpha\cos\beta}=\dfrac{\cos\alpha\cos\beta}{\cos\alpha\cos\beta}-\dfrac{\sin\alpha\sin\beta}{\cos\alpha\cos\beta}=1-\tan\alpha\tan\beta$

61. $\dfrac{\sin(\alpha+\beta)}{\sin(\alpha-\beta)}=\dfrac{\sin\alpha\cos\beta+\cos\alpha\sin\beta}{\sin\alpha\cos\beta-\cos\alpha\sin\beta}=\dfrac{\frac{\sin\alpha\cos\beta+\cos\alpha\sin\beta}{\cos\alpha\cos\beta}}{\frac{\sin\alpha\cos\beta-\cos\alpha\sin\beta}{\cos\alpha\cos\beta}}=\dfrac{\frac{\sin\alpha\cos\beta}{\cos\alpha\cos\beta}+\frac{\cos\alpha\sin\beta}{\cos\alpha\cos\beta}}{\frac{\sin\alpha\cos\beta}{\cos\alpha\cos\beta}-\frac{\cos\alpha\sin\beta}{\cos\alpha\cos\beta}}=\dfrac{\tan\alpha+\tan\beta}{\tan\alpha-\tan\beta}$

63. $\cot(\alpha+\beta)=\dfrac{\cos(\alpha+\beta)}{\sin(\alpha+\beta)}=\dfrac{\cos\alpha\cos\beta-\sin\alpha\sin\beta}{\sin\alpha\cos\beta+\cos\alpha\sin\beta}=\dfrac{\frac{\cos\alpha\cos\beta-\sin\alpha\sin\beta}{\sin\alpha\sin\beta}}{\frac{\sin\alpha\cos\beta+\cos\alpha\sin\beta}{\sin\alpha\sin\beta}}=\dfrac{\frac{\cos\alpha\cos\beta}{\sin\alpha\sin\beta}-\frac{\sin\alpha\sin\beta}{\sin\alpha\sin\beta}}{\frac{\sin\alpha\cos\beta}{\sin\alpha\sin\beta}+\frac{\cos\alpha\sin\beta}{\sin\alpha\sin\beta}}=\dfrac{\cot\alpha\cot\beta-1}{\cot\beta+\cot\alpha}$

65. $\sec(\alpha+\beta)=\dfrac{1}{\cos(\alpha+\beta)}=\dfrac{1}{\cos\alpha\cos\beta-\sin\alpha\sin\beta}=\dfrac{\frac{1}{\sin\alpha\sin\beta}}{\frac{\cos\alpha\cos\beta-\sin\alpha\sin\beta}{\sin\alpha\sin\beta}}=\dfrac{\frac{1}{\sin\alpha}\cdot\frac{1}{\sin\beta}}{\frac{\cos\alpha\cos\beta}{\sin\alpha\sin\beta}-\frac{\sin\alpha\sin\beta}{\sin\alpha\sin\beta}}=\dfrac{\csc\alpha\csc\beta}{\cot\alpha\cot\beta-1}$

67. $\sin(\alpha-\beta)\sin(\alpha+\beta)=(\sin\alpha\cos\beta-\cos\alpha\sin\beta)(\sin\alpha\cos\beta+\cos\alpha\sin\beta)=\sin^2\alpha\cos^2\beta-\cos^2\alpha\sin^2\beta$
$=(\sin^2\alpha)(1-\sin^2\beta)-(1-\sin^2\alpha)(\sin^2\beta)=\sin^2\alpha-\sin^2\beta$

69. $\sin(\theta+k\pi)=\sin\theta\cos k\pi+\cos\theta\sin k\pi=(\sin\theta)(-1)^k+(\cos\theta)(0)=(-1)^k\sin\theta,\ k$ any integer

71. $\dfrac{\sqrt{3}}{2}$ **73.** $-\dfrac{24}{25}$ **75.** $-\dfrac{33}{65}$ **77.** $\dfrac{63}{65}$ **79.** $\dfrac{48+25\sqrt{3}}{39}$ **81.** $\dfrac{4}{3}$ **83.** $u\sqrt{1-v^2}-v\sqrt{1-u^2}:\ -1\le u\le1;\ -1\le v\le1$

85. $\dfrac{u\sqrt{1-v^2}-v}{\sqrt{1+u^2}}:\ -\infty<u<\infty;\ -1\le v\le1$ **87.** $\dfrac{uv-\sqrt{1-u^2}\sqrt{1-v^2}}{v\sqrt{1-u^2}+u\sqrt{1-v^2}}:\ -1\le u\le1;\ -1\le v\le1$

89. Let $\alpha=\sin^{-1}v$ and $\beta=\cos^{-1}v$. Then $\sin\alpha=\cos\beta=v$, and since $\sin\alpha=\cos\left(\dfrac{\pi}{2}-\alpha\right)$, $\cos\left(\dfrac{\pi}{2}-\alpha\right)=\cos\beta$.

If $v\ge0$, then $0\le\alpha\le\dfrac{\pi}{2}$, so $\left(\dfrac{\pi}{2}-\alpha\right)$ and β both lie on $\left[0,\dfrac{\pi}{2}\right]$. If $v<0$, then $-\dfrac{\pi}{2}\le\alpha<0$, so $\left(\dfrac{\pi}{2}-\alpha\right)$ and β both lie on $\left(\dfrac{\pi}{2},\pi\right]$.

Either way, $\cos\left(\dfrac{\pi}{2}-\alpha\right)=\cos\beta$ implies $\dfrac{\pi}{2}-\alpha=\beta$, or $\alpha+\beta=\dfrac{\pi}{2}$.

91. Let $\alpha=\tan^{-1}\dfrac{1}{v}$ and $\beta=\tan^{-1}v$. Because $v\ne0$, $\alpha,\beta\ne0$. Then $\tan\alpha=\dfrac{1}{v}=\dfrac{1}{\tan\beta}=\cot\beta$, and since

$\tan\alpha=\cot\left(\dfrac{\pi}{2}-\alpha\right)$, $\cot\left(\dfrac{\pi}{2}-\alpha\right)=\cot\beta$. Because $v>0$, $0<\alpha<\dfrac{\pi}{2}$, and so $\left(\dfrac{\pi}{2}-\alpha\right)$ and β both lie on $\left(0,\dfrac{\pi}{2}\right)$.

Then $\cot\left(\dfrac{\pi}{2}-\alpha\right)=\cot\beta$ implies $\dfrac{\pi}{2}-\alpha=\beta$, or $\alpha=\dfrac{\pi}{2}-\beta$.

93. $\sin(\sin^{-1}v+\cos^{-1}v)=\sin(\sin^{-1}v)\cos(\cos^{-1}v)+\cos(\sin^{-1}v)\sin(\cos^{-1}v)=(v)(v)+\sqrt{1-v^2}\sqrt{1-v^2}=v^2+1-v^2=1$

95. $\dfrac{\sin(x+h)-\sin x}{h}=\dfrac{\sin x\cos h+\cos x\sin h-\sin x}{h}=\dfrac{\cos x\sin h-\sin x(1-\cos h)}{h}=\cos x\cdot\dfrac{\sin h}{h}-\sin x\cdot\dfrac{1-\cos h}{h}$

97. (a) $\tan(\tan^{-1}1+\tan^{-1}2+\tan^{-1}3)=\tan((\tan^{-1}1+\tan^{-1}2)+\tan^{-1}3)=\dfrac{\tan(\tan^{-1}1+\tan^{-1}2)+\tan(\tan^{-1}3)}{1-\tan(\tan^{-1}1+\tan^{-1}2)\tan(\tan^{-1}3)}$

$=\dfrac{\frac{\tan(\tan^{-1}1)+\tan(\tan^{-1}2)}{1-\tan(\tan^{-1}1)\tan(\tan^{-1}2)}+3}{1-\dfrac{\tan(\tan^{-1}1)+\tan(\tan^{-1}2)}{1-\tan(\tan^{-1}1)\tan(\tan^{-1}2)}\cdot3}=\dfrac{\frac{1+2}{1-1\cdot2}+3}{1-\frac{1+2}{1-1\cdot2}\cdot3}=\dfrac{\frac{3}{-1}+3}{1-\frac{3}{-1}\cdot3}=\dfrac{-3+3}{1+9}=\dfrac{0}{10}=0$

(b) From the definition of the inverse tangent function $0<\tan^{-1}1<\dfrac{\pi}{2}$, $0<\tan^{-1}2<\dfrac{\pi}{2}$, and $0<\tan^{-1}3<\dfrac{\pi}{2}$, so $0<\tan^{-1}1+\tan^{-1}2+\tan^{-1}3<\dfrac{3\pi}{2}$.

On the interval $\left(0,\dfrac{3\pi}{2}\right)$, $\tan\theta=0$ if and only if $\theta=\pi$. Therefore, from part (a), $\tan^{-1}1+\tan^{-1}2+\tan^{-1}3=\pi$.

99. $\tan\theta = \tan(\theta_2 - \theta_1) = \dfrac{\tan\theta_2 - \tan\theta_1}{1 + \tan\theta_1\tan\theta_2} = \dfrac{m_2 - m_1}{1 + m_1 m_2}$

101. $2\cot(\alpha - \beta) = \dfrac{2}{\tan(\alpha - \beta)} = 2\left(\dfrac{1 + \tan\alpha\tan\beta}{\tan\alpha - \tan\beta}\right) = 2\left(\dfrac{1 + (x+1)(x-1)}{(x+1) - (x-1)}\right) = 2\left(\dfrac{1 + x^2 - 1}{x + 1 - x + 1}\right) = \dfrac{2x^2}{2} = x^2$

103. $\tan\dfrac{\pi}{2}$ is not defined; $\tan\left(\dfrac{\pi}{2} - \theta\right) = \dfrac{\sin\left(\dfrac{\pi}{2} - \theta\right)}{\cos\left(\dfrac{\pi}{2} - \theta\right)} = \dfrac{\cos\theta}{\sin\theta} = \cot\theta.$

3.5 Assess Your Understanding (page 248)

1. $\sin^2\theta; 2\cos^2\theta; 2\sin^2\theta$ **2.** $1 - \cos\theta$ **3.** $\sin\theta$ **4.** T **5.** F **6.** F **7. (a)** $\dfrac{24}{25}$ **(b)** $\dfrac{7}{25}$ **(c)** $\dfrac{\sqrt{10}}{10}$ **(d)** $\dfrac{3\sqrt{10}}{10}$

9. (a) $\dfrac{24}{25}$ **(b)** $-\dfrac{7}{25}$ **(c)** $\dfrac{2\sqrt{5}}{5}$ **(d)** $-\dfrac{\sqrt{5}}{5}$ **11. (a)** $-\dfrac{2\sqrt{2}}{3}$ **(b)** $\dfrac{1}{3}$ **(c)** $\sqrt{\dfrac{3 + \sqrt{6}}{6}}$ **(d)** $\sqrt{\dfrac{3 - \sqrt{6}}{6}}$

13. (a) $\dfrac{4\sqrt{2}}{9}$ **(b)** $-\dfrac{7}{9}$ **(c)** $\dfrac{\sqrt{3}}{3}$ **(d)** $\dfrac{\sqrt{6}}{3}$ **15. (a)** $-\dfrac{4}{5}$ **(b)** $\dfrac{3}{5}$ **(c)** $\sqrt{\dfrac{5 + 2\sqrt{5}}{10}}$ **(d)** $\sqrt{\dfrac{5 - 2\sqrt{5}}{10}}$

17. (a) $-\dfrac{3}{5}$ **(b)** $-\dfrac{4}{5}$ **(c)** $\dfrac{1}{2}\sqrt{\dfrac{10 - \sqrt{10}}{5}}$ **(d)** $-\dfrac{1}{2}\sqrt{\dfrac{10 + \sqrt{10}}{5}}$ **19.** $\dfrac{\sqrt{2 - \sqrt{2}}}{2}$ **21.** $1 - \sqrt{2}$ **23.** $-\dfrac{\sqrt{2 + \sqrt{3}}}{2}$

25. $\dfrac{2}{\sqrt{2 + \sqrt{2}}} = (2 - \sqrt{2})\sqrt{2 + \sqrt{2}}$ **27.** $-\dfrac{\sqrt{2 - \sqrt{2}}}{2}$ **29.** $-\dfrac{4}{5}$ **31.** $\dfrac{\sqrt{10(5 - \sqrt{5})}}{10}$ **33.** $\dfrac{4}{3}$ **35.** $-\dfrac{7}{8}$ **37.** $\dfrac{\sqrt{10}}{4}$ **39.** $-\dfrac{\sqrt{15}}{3}$

41. $\sin^4\theta = (\sin^2\theta)^2 = \left(\dfrac{1 - \cos(2\theta)}{2}\right)^2 = \dfrac{1}{4}[1 - 2\cos(2\theta) + \cos^2(2\theta)] = \dfrac{1}{4} - \dfrac{1}{2}\cos(2\theta) + \dfrac{1}{4}\cos^2(2\theta)$

$= \dfrac{1}{4} - \dfrac{1}{2}\cos(2\theta) + \dfrac{1}{4}\left(\dfrac{1 + \cos(4\theta)}{2}\right) = \dfrac{1}{4} - \dfrac{1}{2}\cos(2\theta) + \dfrac{1}{8} + \dfrac{1}{8}\cos(4\theta) = \dfrac{3}{8} - \dfrac{1}{2}\cos(2\theta) + \dfrac{1}{8}\cos(4\theta)$

43. $\cos(3\theta) = 4\cos^3\theta - 3\cos\theta$ **45.** $\sin(5\theta) = 16\sin^5\theta - 20\sin^3\theta + 5\sin\theta$ **47.** $\cos^4\theta - \sin^4\theta = (\cos^2\theta + \sin^2\theta)(\cos^2\theta - \sin^2\theta) = \cos(2\theta)$

49. $\cot(2\theta) = \dfrac{1}{\tan(2\theta)} = \dfrac{1 - \tan^2\theta}{2\tan\theta} = \dfrac{1 - \dfrac{1}{\cot^2\theta}}{2\left(\dfrac{1}{\cot\theta}\right)} = \dfrac{\dfrac{\cot^2\theta - 1}{\cot^2\theta}}{\dfrac{2}{\cot\theta}} = \dfrac{\cot^2\theta - 1}{\cot^2\theta}\cdot\dfrac{\cot\theta}{2} = \dfrac{\cot^2\theta - 1}{2\cot\theta}$

51. $\sec(2\theta) = \dfrac{1}{\cos(2\theta)} = \dfrac{1}{2\cos^2\theta - 1} = \dfrac{1}{\dfrac{2}{\sec^2\theta} - 1} = \dfrac{1}{\dfrac{2 - \sec^2\theta}{\sec^2\theta}} = \dfrac{\sec^2\theta}{2 - \sec^2\theta}$ **53.** $\cos^2(2u) - \sin^2(2u) = \cos[2(2u)] = \cos(4u)$

55. $\dfrac{\cos(2\theta)}{1 + \sin(2\theta)} = \dfrac{\cos^2\theta - \sin^2\theta}{1 + 2\sin\theta\cos\theta} = \dfrac{(\cos\theta - \sin\theta)(\cos\theta + \sin\theta)}{\sin^2\theta + \cos^2\theta + 2\sin\theta\cos\theta} = \dfrac{(\cos\theta - \sin\theta)(\cos\theta + \sin\theta)}{(\sin\theta + \cos\theta)(\sin\theta + \cos\theta)} = \dfrac{\cos\theta - \sin\theta}{\cos\theta + \sin\theta}$

$= \dfrac{\dfrac{\cos\theta - \sin\theta}{\sin\theta}}{\dfrac{\cos\theta + \sin\theta}{\sin\theta}} = \dfrac{\dfrac{\cos\theta}{\sin\theta} - \dfrac{\sin\theta}{\sin\theta}}{\dfrac{\cos\theta}{\sin\theta} + \dfrac{\sin\theta}{\sin\theta}} = \dfrac{\cot\theta - 1}{\cot\theta + 1}$

57. $\sec^2\dfrac{\theta}{2} = \dfrac{1}{\cos^2\left(\dfrac{\theta}{2}\right)} = \dfrac{1}{\dfrac{1 + \cos\theta}{2}} = \dfrac{2}{1 + \cos\theta}$

59. $\cot^2\dfrac{v}{2} = \dfrac{1}{\tan^2\left(\dfrac{v}{2}\right)} = \dfrac{1}{\dfrac{1 - \cos v}{1 + \cos v}} = \dfrac{1 + \cos v}{1 - \cos v} = \dfrac{1 + \dfrac{1}{\sec v}}{1 - \dfrac{1}{\sec v}} = \dfrac{\dfrac{\sec v + 1}{\sec v}}{\dfrac{\sec v - 1}{\sec v}} = \dfrac{\sec v + 1}{\sec v}\cdot\dfrac{\sec v}{\sec v - 1} = \dfrac{\sec v + 1}{\sec v - 1}$

61. $\dfrac{1 - \tan^2\left(\dfrac{\theta}{2}\right)}{1 + \tan^2\left(\dfrac{\theta}{2}\right)} = \dfrac{1 - \dfrac{1 - \cos\theta}{1 + \cos\theta}}{1 + \dfrac{1 - \cos\theta}{1 + \cos\theta}} = \dfrac{\dfrac{1 + \cos\theta - (1 - \cos\theta)}{1 + \cos\theta}}{\dfrac{1 + \cos\theta + 1 - \cos\theta}{1 + \cos\theta}} = \dfrac{2\cos\theta}{1 + \cos\theta}\cdot\dfrac{1 + \cos\theta}{2} = \cos\theta$

63. $\dfrac{\sin(3\theta)}{\sin\theta} - \dfrac{\cos(3\theta)}{\cos\theta} = \dfrac{\sin(3\theta)\cos\theta - \cos(3\theta)\sin\theta}{\sin\theta\cos\theta} = \dfrac{\sin(3\theta - \theta)}{\dfrac{1}{2}(2\sin\theta\cos\theta)} = \dfrac{2\sin(2\theta)}{\sin(2\theta)} = 2$

65. $\tan(3\theta) = \tan(\theta + 2\theta) = \dfrac{\tan\theta + \tan(2\theta)}{1 - \tan\theta\tan(2\theta)} = \dfrac{\tan\theta + \dfrac{2\tan\theta}{1 - \tan^2\theta}}{1 - \dfrac{\tan\theta(2\tan\theta)}{1 - \tan^2\theta}} = \dfrac{\tan\theta - \tan^3\theta + 2\tan\theta}{1 - \tan^2\theta - 2\tan^2\theta} = \dfrac{3\tan\theta - \tan^3\theta}{1 - 3\tan^2\theta}$

67. $\frac{1}{2}(\ln|1 - \cos(2\theta)| - \ln 2) = \ln\left(\frac{|1 - \cos(2\theta)|}{2}\right)^{1/2} = \ln|\sin^2\theta|^{1/2} = \ln|\sin\theta|$ **69.** $\frac{\sqrt{3}}{2}$ **71.** $\frac{7}{25}$ **73.** $\frac{24}{7}$ **75.** $\frac{24}{25}$ **77.** $\frac{1}{5}$ **79.** $\frac{25}{7}$

81. (a) $W = 2D(\csc\theta - \cot\theta) = 2D\left(\frac{1}{\sin\theta} - \frac{\cos\theta}{\sin\theta}\right) = 2D\frac{1 - \cos\theta}{\sin\theta} = 2D\tan\frac{\theta}{2}$ **(b)** $\theta = 24.45°$

83. (a) $R = \frac{v_0^2\sqrt{2}}{16}\cos\theta(\sin\theta - \cos\theta)$ **(b)** **(c)** $\theta = 67.5°$ makes R largest.

$= \frac{v_0^2\sqrt{2}}{32}(2\cos\theta\sin\theta - 2\cos^2\theta)$

$= \frac{v_0^2\sqrt{2}}{32}[\sin(2\theta) - \cos(2\theta) - 1]$

85. $A = \frac{1}{2}h(\text{base}) = h\left(\frac{1}{2}\text{base}\right) = s\cos\frac{\theta}{2}\cdot s\sin\frac{\theta}{2} = \frac{1}{2}s^2\sin\theta$ **87.** $\sin(2\theta) = \frac{4x}{4 + x^2}$ **89.** $-\frac{1}{4}$

91. $\frac{2z}{1 + z^2} = \frac{2\tan\left(\frac{\alpha}{2}\right)}{1 + \tan^2\left(\frac{\alpha}{2}\right)} = \frac{2\tan\left(\frac{\alpha}{2}\right)}{\sec^2\left(\frac{\alpha}{2}\right)} = \frac{\dfrac{2\sin\left(\frac{\alpha}{2}\right)}{\cos\left(\frac{\alpha}{2}\right)}}{\dfrac{1}{\cos^2\left(\frac{\alpha}{2}\right)}} = 2\sin\left(\frac{\alpha}{2}\right)\cos\left(\frac{\alpha}{2}\right) = \sin\left(2\cdot\frac{\alpha}{2}\right) = \sin\alpha$

93.

95. $\sin\dfrac{\pi}{24} = \dfrac{\sqrt{2}}{4}\sqrt{4 - \sqrt{6} - \sqrt{2}}$

$\cos\dfrac{\pi}{24} = \dfrac{\sqrt{2}}{4}\sqrt{4 + \sqrt{6} + \sqrt{2}}$

97. $\sin^3\theta + \sin^3(\theta + 120°) + \sin^3(\theta + 240°) = \sin^3\theta + (\sin\theta\cos 120° + \cos\theta\sin 120°)^3 + (\sin\theta\cos 240° + \cos\theta\sin 240°)^3$

$= \sin^3\theta + \left(-\frac{1}{2}\sin\theta + \frac{\sqrt{3}}{2}\cos\theta\right)^3 + \left(-\frac{1}{2}\sin\theta - \frac{\sqrt{3}}{2}\cos\theta\right)^3$

$= \sin^3\theta + \frac{1}{8}(3\sqrt{3}\cos^3\theta - 9\cos^2\theta\sin\theta + 3\sqrt{3}\cos\theta\sin^2\theta - \sin^3\theta) - \frac{1}{8}(\sin^3\theta + 3\sqrt{3}\sin^2\theta\cos\theta + 9\sin\theta\cos^2\theta + 3\sqrt{3}\cos^3\theta)$

$= \frac{3}{4}\sin^3\theta - \frac{9}{4}\cos^2\theta\sin\theta = \frac{3}{4}[\sin^3\theta - 3\sin\theta(1 - \sin^2\theta)] = \frac{3}{4}(4\sin^3\theta - 3\sin\theta) = -\frac{3}{4}\sin(3\theta)$ (from Example 2)

3.6 Assess Your Understanding (page 253)

1. $\frac{1}{2}[\cos(2\theta) - \cos(6\theta)]$ **3.** $\frac{1}{2}[\sin(6\theta) + \sin(2\theta)]$ **5.** $\frac{1}{2}[\cos(2\theta) + \cos(8\theta)]$ **7.** $\frac{1}{2}[\cos\theta - \cos(3\theta)]$ **9.** $\frac{1}{2}[\sin(2\theta) + \sin\theta]$ **11.** $2\sin\theta\cos(3\theta)$

13. $2\cos(3\theta)\cos\theta$ **15.** $2\sin(2\theta)\cos\theta$ **17.** $2\sin\theta\sin\frac{\theta}{2}$ **19.** $\dfrac{\sin\theta + \sin(3\theta)}{2\sin(2\theta)} = \dfrac{2\sin(2\theta)\cos\theta}{2\sin(2\theta)} = \cos\theta$

21. $\dfrac{\sin(4\theta) + \sin(2\theta)}{\cos(4\theta) + \cos(2\theta)} = \dfrac{2\sin(3\theta)\cos\theta}{2\cos(3\theta)\cos\theta} = \dfrac{\sin(3\theta)}{\cos(3\theta)} = \tan(3\theta)$ **23.** $\dfrac{\cos\theta - \cos(3\theta)}{\sin\theta + \sin(3\theta)} = \dfrac{2\sin(2\theta)\sin\theta}{2\sin(2\theta)\cos\theta} = \dfrac{\sin\theta}{\cos\theta} = \tan\theta$

25. $\sin\theta[\sin\theta + \sin(3\theta)] = \sin\theta[2\sin(2\theta)\cos\theta] = \cos\theta[2\sin(2\theta)\sin\theta] = \cos\theta\left[2\cdot\frac{1}{2}[\cos\theta - \cos(3\theta)]\right] = \cos\theta[\cos\theta - \cos(3\theta)]$

27. $\dfrac{\sin(4\theta) + \sin(8\theta)}{\cos(4\theta) + \cos(8\theta)} = \dfrac{2\sin(6\theta)\cos(2\theta)}{2\cos(6\theta)\cos(2\theta)} = \dfrac{\sin(6\theta)}{\cos(6\theta)} = \tan(6\theta)$

29. $\dfrac{\sin(4\theta) + \sin(8\theta)}{\sin(4\theta) - \sin(8\theta)} = \dfrac{2\sin(6\theta)\cos(-2\theta)}{2\sin(-2\theta)\cos(6\theta)} = \dfrac{\sin(6\theta)}{\cos(6\theta)}\cdot\dfrac{\cos(2\theta)}{-\sin(2\theta)} = \tan(6\theta)[-\cot(2\theta)] = -\dfrac{\tan(6\theta)}{\tan(2\theta)}$

31. $\dfrac{\sin\alpha + \sin\beta}{\sin\alpha - \sin\beta} = \dfrac{2\sin\dfrac{\alpha + \beta}{2}\cos\dfrac{\alpha - \beta}{2}}{2\sin\dfrac{\alpha - \beta}{2}\cos\dfrac{\alpha + \beta}{2}} = \dfrac{\sin\dfrac{\alpha + \beta}{2}}{\cos\dfrac{\alpha + \beta}{2}}\cdot\dfrac{\cos\dfrac{\alpha - \beta}{2}}{\sin\dfrac{\alpha - \beta}{2}} = \tan\dfrac{\alpha + \beta}{2}\cot\dfrac{\alpha - \beta}{2}$

33. $\dfrac{\sin \alpha + \sin \beta}{\cos \alpha + \cos \beta} = \dfrac{2 \sin \dfrac{\alpha + \beta}{2} \cos \dfrac{\alpha - \beta}{2}}{2 \cos \dfrac{\alpha + \beta}{2} \cos \dfrac{\alpha - \beta}{2}} = \dfrac{\sin \dfrac{\alpha + \beta}{2}}{\cos \dfrac{\alpha + \beta}{2}} = \tan \dfrac{\alpha + \beta}{2}$

35. $1 + \cos(2\theta) + \cos(4\theta) + \cos(6\theta) = [1 + \cos(6\theta)] + [\cos(2\theta) + \cos(4\theta)] = 2 \cos^2(3\theta) + 2 \cos(3\theta) \cos(-\theta)$

$= 2 \cos(3\theta)[\cos(3\theta) + \cos \theta] = 2 \cos(3\theta)[2 \cos(2\theta) \cos \theta] = 4 \cos \theta \cos(2\theta) \cos(3\theta)$

37. (a) $y = 2 \sin(2061\pi t) \cos(357\pi t)$ **(b)** $y_{\max} = 2$ **(c)**

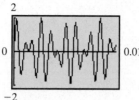

39. $I_u = I_x \cos^2 \theta + I_y \sin^2 \theta - 2I_{xy} \sin \theta \cos \theta = I_x \cos^2 \theta + I_y \sin^2 \theta - I_{xy} \sin 2\theta$

$= I_x \left(\dfrac{\cos 2\theta + 1}{2} \right) + I_y \left(\dfrac{1 - \cos 2\theta}{2} \right) - I_{xy} \sin 2\theta$

$= \dfrac{I_x}{2} \cos 2\theta + \dfrac{I_x}{2} + \dfrac{I_y}{2} - \dfrac{I_y}{2} \cos 2\theta - I_{xy} \sin 2\theta$

$= \dfrac{I_x + I_y}{2} + \dfrac{I_x - I_y}{2} \cos 2\theta - I_{xy} \sin 2\theta$

$I_v = I_x \sin^2 \theta + I_y \cos^2 \theta + 2I_{xy} \sin \theta \cos \theta = I_x \left(\dfrac{1 - \cos 2\theta}{2} \right) + I_y \left(\dfrac{\cos 2\theta + 1}{2} \right) + I_{xy} \sin 2\theta$

$= \dfrac{I_x}{2} - \dfrac{I_x}{2} \cos 2\theta + \dfrac{I_y}{2} \cos 2\theta + \dfrac{I_y}{2} + I_{xy} \sin 2\theta$

$= \dfrac{I_x + I_y}{2} - \dfrac{I_x - I_y}{2} \cos 2\theta + I_{xy} \sin 2\theta$

41. $\sin(2\alpha) + \sin(2\beta) + \sin(2\gamma) = 2 \sin(\alpha + \beta) \cos(\alpha - \beta) + \sin(2\gamma) = 2 \sin(\alpha + \beta) \cos(\alpha - \beta) + 2 \sin \gamma \cos \gamma$

$= 2 \sin(\pi - \gamma) \cos(\alpha - \beta) + 2 \sin \gamma \cos \gamma = 2 \sin \gamma \cos(\alpha - \beta) + 2 \sin \gamma \cos \gamma = 2 \sin \gamma [\cos(\alpha - \beta) + \cos \gamma]$

$= 2 \sin \gamma \left(2 \cos \dfrac{\alpha - \beta + \gamma}{2} \cos \dfrac{\alpha - \beta - \gamma}{2} \right) = 4 \sin \gamma \cos \dfrac{\pi - 2\beta}{2} \cos \dfrac{2\alpha - \pi}{2} = 4 \sin \gamma \cos \left(\dfrac{\pi}{2} - \beta \right) \cos \left(\alpha - \dfrac{\pi}{2} \right)$

$= 4 \sin \gamma \sin \beta \sin \alpha = 4 \sin \alpha \sin \beta \sin \gamma$

43.
$$\sin(\alpha - \beta) = \sin \alpha \cos \beta - \cos \alpha \sin \beta$$
$$\sin(\alpha + \beta) = \sin \alpha \cos \beta + \cos \alpha \sin \beta$$
$$\sin(\alpha - \beta) + \sin(\alpha + \beta) = 2 \sin \alpha \cos \beta$$
$$\sin \alpha \cos \beta = \dfrac{1}{2} [\sin(\alpha + \beta) + \sin(\alpha - \beta)]$$

45. $2 \cos \dfrac{\alpha + \beta}{2} \cos \dfrac{\alpha - \beta}{2} = 2 \cdot \dfrac{1}{2} \left[\cos \left(\dfrac{\alpha + \beta}{2} + \dfrac{\alpha - \beta}{2} \right) + \cos \left(\dfrac{\alpha + \beta}{2} - \dfrac{\alpha - \beta}{2} \right) \right] = \cos \dfrac{2\alpha}{2} + \cos \dfrac{2\beta}{2} = \cos \alpha + \cos \beta$

3.7 Assess Your Understanding *(page 258)*

3. $\dfrac{\pi}{6}, \dfrac{5\pi}{6}$ **4.** $\left\{ \theta \,\middle|\, \theta = \dfrac{\pi}{6} + 2\pi k, \theta = \dfrac{5\pi}{6} + 2\pi k, k \text{ is any integer} \right\}$ **5.** F **6.** F **7.** $\left\{ \dfrac{7\pi}{6}, \dfrac{11\pi}{6} \right\}$ **9.** $\left\{ \dfrac{\pi}{3}, \dfrac{2\pi}{3}, \dfrac{4\pi}{3}, \dfrac{5\pi}{3} \right\}$ **11.** $\left\{ \dfrac{\pi}{4}, \dfrac{3\pi}{4}, \dfrac{5\pi}{4}, \dfrac{7\pi}{4} \right\}$

13. $\left\{ \dfrac{\pi}{2}, \dfrac{7\pi}{6}, \dfrac{11\pi}{6} \right\}$ **15.** $\left\{ \dfrac{\pi}{3}, \dfrac{2\pi}{3}, \dfrac{4\pi}{3}, \dfrac{5\pi}{3} \right\}$ **17.** $\left\{ \dfrac{4\pi}{9}, \dfrac{8\pi}{9}, \dfrac{16\pi}{9} \right\}$ **19.** $\left\{ \dfrac{7\pi}{6}, \dfrac{11\pi}{6} \right\}$ **21.** $\left\{ \dfrac{3\pi}{4}, \dfrac{7\pi}{4} \right\}$ **23.** $\left\{ \dfrac{2\pi}{3}, \dfrac{4\pi}{3} \right\}$ **25.** $\left\{ \dfrac{3\pi}{4}, \dfrac{5\pi}{4} \right\}$

27. $\left\{ \dfrac{3\pi}{4}, \dfrac{7\pi}{4} \right\}$ **29.** $\left\{ \dfrac{11\pi}{6} \right\}$ **31.** $\left\{ \theta \,\middle|\, \theta = \dfrac{\pi}{6} + 2k\pi, \theta = \dfrac{5\pi}{6} + 2k\pi \right\}; \dfrac{\pi}{6}, \dfrac{5\pi}{6}, \dfrac{13\pi}{6}, \dfrac{17\pi}{6}, \dfrac{25\pi}{6}, \dfrac{29\pi}{6}$

33. $\left\{ \theta \,\middle|\, \theta = \dfrac{5\pi}{6} + k\pi \right\}; \dfrac{5\pi}{6}, \dfrac{11\pi}{6}, \dfrac{17\pi}{6}, \dfrac{23\pi}{6}, \dfrac{29\pi}{6}, \dfrac{35\pi}{6}$ **35.** $\left\{ \theta \,\middle|\, \theta = \dfrac{\pi}{2} + 2k\pi, \theta = \dfrac{3\pi}{2} + 2k\pi \right\}; \dfrac{\pi}{2}, \dfrac{3\pi}{2}, \dfrac{5\pi}{2}, \dfrac{7\pi}{2}, \dfrac{9\pi}{2}, \dfrac{11\pi}{2}$

37. $\left\{ \theta \,\middle|\, \theta = \dfrac{\pi}{3} + k\pi, \theta = \dfrac{2\pi}{3} + k\pi \right\}; \dfrac{\pi}{3}, \dfrac{2\pi}{3}, \dfrac{4\pi}{3}, \dfrac{5\pi}{3}, \dfrac{7\pi}{3}, \dfrac{8\pi}{3}$ **39.** $\left\{ \theta \,\middle|\, \theta = \dfrac{8\pi}{3} + 4k\pi, \theta = \dfrac{10\pi}{3} + 4k\pi \right\}; \dfrac{8\pi}{3}, \dfrac{10\pi}{3}, \dfrac{20\pi}{3}, \dfrac{22\pi}{3}, \dfrac{32\pi}{3}, \dfrac{34\pi}{3}$

41. $\{0.41, 2.73\}$ **43.** $\{1.37, 4.51\}$ **45.** $\{2.69, 3.59\}$ **47.** $\{1.82, 4.46\}$ **49.** $\{2.08, 5.22\}$ **51.** $\{0.73, 2.41\}$ **53.** $\dfrac{\pi}{3}, \dfrac{2\pi}{3}, \dfrac{4\pi}{3}, \dfrac{5\pi}{3}$

55. (a) $-2\pi, -\pi, 0, \pi, 2\pi, 3\pi, 4\pi$

(b)

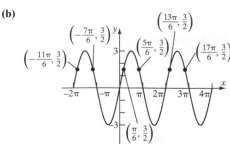

(c) $\left\{-\dfrac{11\pi}{6}, -\dfrac{7\pi}{6}, \dfrac{\pi}{6}, \dfrac{5\pi}{6}, \dfrac{13\pi}{6}, \dfrac{17\pi}{6}\right\}$

(d) $\left\{x \,\middle|\, -\dfrac{11\pi}{6} < x < -\dfrac{7\pi}{6} \text{ or } \dfrac{\pi}{6} < x < \dfrac{5\pi}{6} \text{ or } \dfrac{13\pi}{6} < x < \dfrac{17\pi}{6}\right\}$

57. (a) $\left\{x \,\middle|\, x = -\dfrac{\pi}{4} + k\pi, k \text{ is any integer}\right\}$

(b) $-\dfrac{\pi}{2} < x < -\dfrac{\pi}{4}$ or $\left(-\dfrac{\pi}{2}, -\dfrac{\pi}{4}\right)$

59. (a), (d)

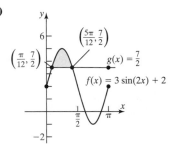

(b) $\left\{\dfrac{\pi}{12}, \dfrac{5\pi}{12}\right\}$

(c) $\left\{x \,\middle|\, \dfrac{\pi}{12} < x < \dfrac{5\pi}{12}\right\}$ or $\left(\dfrac{\pi}{12}, \dfrac{5\pi}{12}\right)$

61. (a), (d)

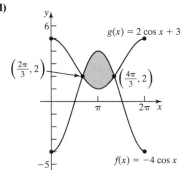

(b) $\left\{\dfrac{2\pi}{3}, \dfrac{4\pi}{3}\right\}$

(c) $\left\{x \,\middle|\, \dfrac{2\pi}{3} < x < \dfrac{4\pi}{3}\right\}$ or $\left(\dfrac{2\pi}{3}, \dfrac{4\pi}{3}\right)$

63. (a) 0 s, 0.43 s, 0.86 s

(b) 0.21 s

(c) $[0, 0.03] \cup [0.39, 0.43] \cup [0.86, 0.89]$

65. (a) 150 mi

(b) 6.06, 8.44, 15.72, 18.11 min

(c) Before 6.06 min, between 8.44 and 15.72 min, and after 18.11 min

(d) No

67. 28.90° **69.** Yes; it varies from 1.25 to 1.34. **71.** 1.47 **73.** If θ is the original angle of incidence and ϕ is the angle of refraction, then $\dfrac{\sin \theta}{\sin \phi} = n_2$.

The angle of incidence of the emerging beam is also ϕ, and the index of refraction is $\dfrac{1}{n_2}$. Thus, θ is the angle of refraction of the emerging beam.

3.8 Assess Your Understanding *(page 266)*

5. $\left\{\dfrac{\pi}{2}, \dfrac{2\pi}{3}, \dfrac{4\pi}{3}, \dfrac{3\pi}{2}\right\}$ **7.** $\left\{\dfrac{\pi}{2}, \dfrac{7\pi}{6}, \dfrac{11\pi}{6}\right\}$ **9.** $\left\{0, \dfrac{\pi}{4}, \dfrac{5\pi}{4}\right\}$ **11.** $\left\{\dfrac{\pi}{2}, \dfrac{2\pi}{3}, \dfrac{4\pi}{3}, \dfrac{3\pi}{2}\right\}$ **13.** $\{\pi\}$ **15.** $\left\{\dfrac{\pi}{3}, \dfrac{2\pi}{3}, \dfrac{4\pi}{3}, \dfrac{5\pi}{3}\right\}$ **17.** $\left\{\dfrac{\pi}{4}, \dfrac{5\pi}{4}\right\}$

19. $\left\{0, \dfrac{\pi}{3}, \pi, \dfrac{5\pi}{3}\right\}$ **21.** $\left\{\dfrac{\pi}{2}, \dfrac{3\pi}{2}\right\}$ **23.** $\left\{0, \dfrac{2\pi}{3}, \dfrac{4\pi}{3}\right\}$ **25.** $\left\{0, \dfrac{\pi}{3}, \dfrac{\pi}{2}, \dfrac{2\pi}{3}, \pi, \dfrac{4\pi}{3}, \dfrac{3\pi}{2}, \dfrac{5\pi}{3}\right\}$ **27.** $\left\{0, \dfrac{\pi}{5}, \dfrac{2\pi}{5}, \dfrac{3\pi}{5}, \dfrac{4\pi}{5}, \pi, \dfrac{6\pi}{5}, \dfrac{7\pi}{5}, \dfrac{8\pi}{5}, \dfrac{9\pi}{5}\right\}$

29. $\left\{\dfrac{\pi}{6}, \dfrac{5\pi}{6}, \dfrac{3\pi}{2}\right\}$ **31.** $\left\{\dfrac{\pi}{2}\right\}$ **33.** $\{0\}$ **35.** $\left\{\dfrac{\pi}{3}, \dfrac{5\pi}{3}\right\}$ **37.** No real solution **39.** No real solution **41.** $\left\{\dfrac{\pi}{2}, \dfrac{7\pi}{6}\right\}$ **43.** $\left\{0, \dfrac{\pi}{3}, \pi, \dfrac{5\pi}{3}\right\}$

45. $\left\{\dfrac{\pi}{4}\right\}$ **47.** $-1.31, 1.98, 3.84$ **49.** 0.52 **51.** 1.26 **53.** $-1.02, 1.02$ **55.** 0, 2.15 **57.** 0.76, 1.35 **59.** $\dfrac{\pi}{3}, \dfrac{2\pi}{3}, \dfrac{4\pi}{3}, \dfrac{5\pi}{3}$ **61.** $0, \dfrac{\pi}{3}, \pi, \dfrac{5\pi}{3}$ **63.** π

65. (a) 60° **(b)** 60°

(c) $A(60°) = 12\sqrt{3}$ in.²

(d)

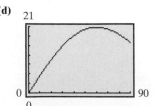

$\theta_{\max} = 60°$
Maximum area = 20.78 in.²

67. 2.03, 4.91

69. (a) 30°, 60° **(b)** 123.6 m

(c)

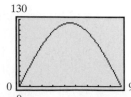

Review Exercises *(page 270)*

1. $\dfrac{\pi}{2}$ **3.** $\dfrac{\pi}{4}$ **5.** $\dfrac{5\pi}{6}$ **7.** $\dfrac{\pi}{4}$ **9.** $\dfrac{3\pi}{8}$ **11.** $-\dfrac{\pi}{3}$ **13.** $\dfrac{\pi}{7}$ **15.** 0.9 **17.** -0.3 **19.** Not defined **21.** $-\dfrac{\pi}{6}$ **23.** $-\dfrac{\pi}{4}$ **25.** $-\sqrt{3}$ **27.** $\dfrac{2\sqrt{3}}{3}$

29. $\dfrac{4}{5}$ **31.** $-\dfrac{4}{3}$ **33.** $f^{-1}(x) = \dfrac{1}{3}\sin^{-1}\left(\dfrac{x}{2}\right)$; Domain of f = range of f^{-1}: $(-\infty, \infty)$; Domain of f^{-1} = range of f: $[-2, 2]$

35. $f^{-1}(x) = \cos^{-1}(3 - x)$; Domain of f = range of f^{-1}: $(-\infty, \infty)$; Domain of f^{-1} = range of f: $[2, 4]$

37. $\sqrt{1 - u^2}$ **39.** $\dfrac{1}{u}$ **41.** $\tan\theta\cot\theta - \sin^2\theta = 1 - \sin^2\theta = \cos^2\theta$ **43.** $\sin^2\theta(1 + \cot^2\theta) = \sin^2\theta\csc^2\theta = 1$

45. $5\cos^2\theta + 3\sin^2\theta = 2\cos^2\theta + 3(\cos^2\theta + \sin^2\theta) = 3 + 2\cos^2\theta$

47. $\dfrac{1 - \cos\theta}{\sin\theta} + \dfrac{\sin\theta}{1 - \cos\theta} = \dfrac{(1 - \cos\theta)^2 + \sin^2\theta}{\sin\theta(1 - \cos\theta)} = \dfrac{1 - 2\cos\theta + \cos^2\theta + \sin^2\theta}{\sin\theta(1 - \cos\theta)} = \dfrac{2(1 - \cos\theta)}{\sin\theta(1 - \cos\theta)} = 2\csc\theta$

49. $\dfrac{\cos\theta}{\cos\theta - \sin\theta} = \dfrac{\dfrac{\cos\theta}{\cos\theta}}{\dfrac{\cos\theta - \sin\theta}{\cos\theta}} = \dfrac{1}{1 - \dfrac{\sin\theta}{\cos\theta}} = \dfrac{1}{1 - \tan\theta}$

51. $\dfrac{\csc\theta}{1 + \csc\theta} = \dfrac{\dfrac{1}{\sin\theta}}{1 + \dfrac{1}{\sin\theta}} = \dfrac{1}{1 + \sin\theta} = \dfrac{1}{1 + \sin\theta}\cdot\dfrac{1 - \sin\theta}{1 - \sin\theta} = \dfrac{1 - \sin\theta}{1 - \sin^2\theta} = \dfrac{1 - \sin\theta}{\cos^2\theta}$

53. $\csc\theta - \sin\theta = \dfrac{1}{\sin\theta} - \sin\theta = \dfrac{1 - \sin^2\theta}{\sin\theta} = \dfrac{\cos^2\theta}{\sin\theta} = \cos\theta\cdot\dfrac{\cos\theta}{\sin\theta} = \cos\theta\cot\theta$

55. $\dfrac{1 - \sin\theta}{\sec\theta} = \cos\theta(1 - \sin\theta)\cdot\dfrac{1 + \sin\theta}{1 + \sin\theta} = \dfrac{\cos\theta(1 - \sin^2\theta)}{1 + \sin\theta} = \dfrac{\cos^3\theta}{1 + \sin\theta}$ **57.** $\cot\theta - \tan\theta = \dfrac{\cos\theta}{\sin\theta} - \dfrac{\sin\theta}{\cos\theta} = \dfrac{\cos^2\theta - \sin^2\theta}{\sin\theta\cos\theta} = \dfrac{1 - 2\sin^2\theta}{\sin\theta\cos\theta}$

59. $\dfrac{\cos(\alpha + \beta)}{\cos\alpha\sin\beta} = \dfrac{\cos\alpha\cos\beta - \sin\alpha\sin\beta}{\cos\alpha\sin\beta} = \dfrac{\cos\alpha\cos\beta}{\cos\alpha\sin\beta} - \dfrac{\sin\alpha\sin\beta}{\cos\alpha\sin\beta} = \cot\beta - \tan\alpha$

61. $\dfrac{\cos(\alpha - \beta)}{\cos\alpha\cos\beta} = \dfrac{\cos\alpha\cos\beta + \sin\alpha\sin\beta}{\cos\alpha\cos\beta} = \dfrac{\cos\alpha\cos\beta}{\cos\alpha\cos\beta} + \dfrac{\sin\alpha\sin\beta}{\cos\alpha\cos\beta} = 1 + \tan\alpha\tan\beta$

63. $(1 + \cos\theta)\left(\tan\dfrac{\theta}{2}\right) = (1 + \cos\theta)\cdot\dfrac{\sin\theta}{1 + \cos\theta} = \sin\theta$

65. $2\cot\theta\cot 2\theta = 2\left(\dfrac{\cos\theta}{\sin\theta}\right)\left(\dfrac{\cos 2\theta}{\sin 2\theta}\right) = \dfrac{2\cos\theta(\cos^2\theta - \sin^2\theta)}{2\sin^2\theta\cos\theta} = \dfrac{\cos^2\theta - \sin^2\theta}{\sin^2\theta} = \cot^2\theta - 1$

67. $1 - 8\sin^2\theta\cos^2\theta = 1 - 2(2\sin\theta\cos\theta)^2 = 1 - 2\sin^2(2\theta) = \cos(4\theta)$ **69.** $\dfrac{\sin(2\theta) + \sin(4\theta)}{\cos(2\theta) + \cos(4\theta)} = \dfrac{2\sin(3\theta)\cos(-\theta)}{2\cos(3\theta)\cos(-\theta)} = \tan(3\theta)$

71. $\dfrac{\cos(2\theta) - \cos(4\theta)}{\cos(2\theta) + \cos(4\theta)} - \tan\theta\tan(3\theta) = \dfrac{-2\sin(3\theta)\sin(-\theta)}{2\cos(3\theta)\cos(-\theta)} - \tan\theta\tan(3\theta) = \tan(3\theta)\tan\theta - \tan\theta\tan(3\theta) = 0$

73. $\dfrac{1}{4}\left(\sqrt{6} - \sqrt{2}\right)$ **75.** $\dfrac{1}{4}\left(\sqrt{6} - \sqrt{2}\right)$ **77.** $\dfrac{1}{2}$ **79.** $\sqrt{2} - 1$ **81. (a)** $-\dfrac{33}{65}$ **(b)** $-\dfrac{56}{65}$ **(c)** $-\dfrac{63}{65}$ **(d)** $\dfrac{33}{56}$ **(e)** $\dfrac{24}{25}$ **(f)** $\dfrac{119}{169}$ **(g)** $\dfrac{5\sqrt{26}}{26}$ **(h)** $\dfrac{2\sqrt{5}}{5}$

83. (a) $-\dfrac{16}{65}$ **(b)** $-\dfrac{63}{65}$ **(c)** $-\dfrac{56}{65}$ **(d)** $\dfrac{16}{63}$ **(e)** $\dfrac{24}{25}$ **(f)** $\dfrac{119}{169}$ **(g)** $\dfrac{\sqrt{26}}{26}$ **(h)** $-\dfrac{\sqrt{10}}{10}$ **85. (a)** $-\dfrac{63}{65}$ **(b)** $\dfrac{16}{65}$ **(c)** $\dfrac{33}{65}$ **(d)** $-\dfrac{63}{16}$ **(e)** $\dfrac{24}{25}$

(f) $-\dfrac{119}{169}$ **(g)** $\dfrac{2\sqrt{13}}{13}$ **(h)** $-\dfrac{\sqrt{10}}{10}$ **87. (a)** $\dfrac{-\sqrt{3} - 2\sqrt{2}}{6}$ **(b)** $\dfrac{1 - 2\sqrt{6}}{6}$ **(c)** $\dfrac{-\sqrt{3} + 2\sqrt{2}}{6}$ **(d)** $\dfrac{8\sqrt{2} + 9\sqrt{3}}{23}$ **(e)** $-\dfrac{\sqrt{3}}{2}$ **(f)** $-\dfrac{7}{9}$

(g) $\dfrac{\sqrt{3}}{3}$ **(h)** $\dfrac{\sqrt{3}}{2}$ **89. (a)** 1 **(b)** 0 **(c)** $-\dfrac{1}{9}$ **(d)** Not defined **(e)** $\dfrac{4\sqrt{5}}{9}$ **(f)** $-\dfrac{1}{9}$ **(g)** $\dfrac{\sqrt{30}}{6}$ **(h)** $-\dfrac{\sqrt{6}\sqrt{3} - \sqrt{5}}{6}$

91. $\dfrac{4 + 3\sqrt{3}}{10}$ **93.** $-\dfrac{48 + 25\sqrt{3}}{39}$ **95.** $-\dfrac{24}{25}$ **97.** $\left\{\dfrac{\pi}{3}, \dfrac{5\pi}{3}\right\}$ **99.** $\left\{\dfrac{3\pi}{4}, \dfrac{5\pi}{4}\right\}$ **101.** $\left\{\dfrac{3\pi}{4}, \dfrac{7\pi}{4}\right\}$ **103.** $\left\{0, \dfrac{\pi}{2}, \pi, \dfrac{3\pi}{2}\right\}$ **105.** $\left\{\dfrac{\pi}{3}, \dfrac{2\pi}{3}, \dfrac{4\pi}{3}, \dfrac{5\pi}{3}\right\}$

107. $\{0, \pi\}$ **109.** $\left\{0, \dfrac{2\pi}{3}, \pi, \dfrac{4\pi}{3}\right\}$ **111.** $\left\{0, \dfrac{\pi}{6}, \dfrac{5\pi}{6}\right\}$ **113.** $\left\{\dfrac{\pi}{6}, \dfrac{\pi}{2}, \dfrac{5\pi}{6}\right\}$ **115.** $\left\{\dfrac{\pi}{3}, \dfrac{5\pi}{3}\right\}$ **117.** $\left\{\dfrac{\pi}{4}, \dfrac{\pi}{2}, \dfrac{3\pi}{4}, \dfrac{3\pi}{2}\right\}$ **119.** $\left\{\dfrac{\pi}{2}, \pi\right\}$

121. 0.78 **123.** -1.11 **125.** 1.23 **127.** $\{1.11\}$ **129.** $\{0.87\}$ **131.** $\{2.22\}$ **133.** $\left\{-\dfrac{\sqrt{3}}{2}\right\}$

135. $\sin 15° = \sqrt{\dfrac{1 - \cos 30°}{2}} = \sqrt{\dfrac{1 - \dfrac{\sqrt{3}}{2}}{2}} = \sqrt{\dfrac{2 - \sqrt{3}}{4}} = \sqrt{\dfrac{2 - \sqrt{3}}{2}}$;

$\sin 15° = \sin(45° - 30°) = \sin 45°\cos 30° - \cos 45°\sin 30° = \dfrac{\sqrt{2}}{2}\cdot\dfrac{\sqrt{3}}{2} - \dfrac{\sqrt{2}}{2}\cdot\dfrac{1}{2} = \dfrac{\sqrt{6}}{4} - \dfrac{\sqrt{2}}{4} = \dfrac{\sqrt{6} - \sqrt{2}}{4}$;

$\left[\dfrac{\sqrt{2 - \sqrt{3}}}{2}\right]^2 = \dfrac{2 - \sqrt{3}}{4} = \dfrac{4(2 - \sqrt{3})}{4\cdot 4} = \dfrac{8 - 4\sqrt{3}}{16} = \dfrac{6 - 2\sqrt{12} + 2}{16} = \left(\dfrac{\sqrt{6} - \sqrt{2}}{4}\right)^2$

Chapter Test *(page 272)*

1. $\dfrac{\pi}{6}$ **2.** $-\dfrac{\pi}{4}$ **3.** $\dfrac{\pi}{5}$ **4.** $\dfrac{7}{3}$ **5.** 3 **6.** $-\dfrac{4}{3}$ **7.** 0.39 **8.** 0.78 **9.** 1.25 **10.** 0.20

11. $\dfrac{\csc\theta + \cot\theta}{\sec\theta + \tan\theta} = \dfrac{\csc\theta + \cot\theta}{\sec\theta + \tan\theta} \cdot \dfrac{\csc\theta - \cot\theta}{\csc\theta - \cot\theta} = \dfrac{\csc^2\theta - \cot^2\theta}{(\sec\theta + \tan\theta)(\csc\theta - \cot\theta)} = \dfrac{1}{(\sec\theta + \tan\theta)(\csc\theta - \cot\theta)}$

$= \dfrac{1}{(\sec\theta + \tan\theta)(\csc\theta - \cot\theta)} \cdot \dfrac{\sec\theta - \tan\theta}{\sec\theta - \tan\theta} = \dfrac{\sec\theta - \tan\theta}{(\sec^2\theta - \tan^2\theta)(\csc\theta - \cot\theta)} = \dfrac{\sec\theta - \tan\theta}{\csc\theta - \cot\theta}$

12. $\sin\theta\tan\theta + \cos\theta = \sin\theta \cdot \dfrac{\sin\theta}{\cos\theta} + \cos\theta = \dfrac{\sin^2\theta}{\cos\theta} + \dfrac{\cos^2\theta}{\cos\theta} = \dfrac{\sin^2\theta + \cos^2\theta}{\cos\theta} = \dfrac{1}{\cos\theta} = \sec\theta$

13. $\tan\theta + \cot\theta = \dfrac{\sin\theta}{\cos\theta} + \dfrac{\cos\theta}{\sin\theta} = \dfrac{\sin^2\theta}{\sin\theta\cos\theta} + \dfrac{\cos^2\theta}{\sin\theta\cos\theta} = \dfrac{\sin^2\theta + \cos^2\theta}{\sin\theta\cos\theta} = \dfrac{1}{\sin\theta\cos\theta} = \dfrac{2}{2\sin\theta\cos\theta} = \dfrac{2}{\sin(2\theta)} = 2\csc(2\theta)$

14. $\dfrac{\sin(\alpha + \beta)}{\tan\alpha + \tan\beta} = \dfrac{\sin\alpha\cos\beta + \cos\alpha\sin\beta}{\dfrac{\sin\alpha}{\cos\alpha} + \dfrac{\sin\beta}{\cos\beta}} = \dfrac{\sin\alpha\cos\beta + \cos\alpha\sin\beta}{\dfrac{\sin\alpha\cos\beta}{\cos\alpha\cos\beta} + \dfrac{\cos\alpha\sin\beta}{\cos\alpha\cos\beta}} = \dfrac{\sin\alpha\cos\beta + \cos\alpha\sin\beta}{\dfrac{\sin\alpha\cos\beta + \cos\alpha\sin\beta}{\cos\alpha\cos\beta}}$

$= \dfrac{\sin\alpha\cos\beta + \cos\alpha\sin\beta}{1} \cdot \dfrac{\cos\alpha\cos\beta}{\sin\alpha\cos\beta + \cos\alpha\sin\beta} = \cos\alpha\cos\beta$

15. $\sin(3\theta) = \sin(\theta + 2\theta) = \sin\theta\cos(2\theta) + \cos\theta\sin(2\theta) = \sin\theta \cdot (\cos^2\theta - \sin^2\theta) + \cos\theta \cdot 2\sin\theta\cos\theta = \sin\theta\cos^2\theta - \sin^3\theta + 2\sin\theta\cos^2\theta$

$= 3\sin\theta\cos^2\theta - \sin^3\theta = 3\sin\theta(1 - \sin^2\theta) - \sin^3\theta = 3\sin\theta - 3\sin^3\theta - \sin^3\theta = 3\sin\theta - 4\sin^3\theta$

16. $\dfrac{\tan\theta - \cot\theta}{\tan\theta + \cot\theta} = \dfrac{\dfrac{\sin\theta}{\cos\theta} - \dfrac{\cos\theta}{\sin\theta}}{\dfrac{\sin\theta}{\cos\theta} + \dfrac{\cos\theta}{\sin\theta}} = \dfrac{\dfrac{\sin^2\theta - \cos^2\theta}{\sin\theta\cos\theta}}{\dfrac{\sin^2\theta + \cos^2\theta}{\sin\theta\cos\theta}} = \dfrac{\sin^2\theta - \cos^2\theta}{\sin^2\theta + \cos^2\theta} = \dfrac{-\cos(2\theta)}{1} = -(2\cos^2\theta - 1) = 1 - 2\cos^2\theta$ **17.** $\dfrac{1}{4}\left(\sqrt{6} + \sqrt{2}\right)$

18. $2 + \sqrt{3}$ **19.** $\dfrac{\sqrt{5}}{5}$ **20.** $\dfrac{12\sqrt{85}}{49}$ **21.** $\dfrac{2\sqrt{13}\left(\sqrt{5} - 3\right)}{39}$ **22.** $\dfrac{2 + \sqrt{3}}{4}$ **23.** $\dfrac{\sqrt{6}}{2}$ **24.** $\dfrac{\sqrt{2}}{2}$ **25.** $\left\{\dfrac{\pi}{3}, \dfrac{2\pi}{3}, \dfrac{4\pi}{3}, \dfrac{5\pi}{3}\right\}$ **26.** $\{0, 1.911, \pi, 4.373\}$

27. $\left\{\dfrac{3\pi}{8}, \dfrac{7\pi}{8}, \dfrac{11\pi}{8}, \dfrac{15\pi}{8}\right\}$ **28.** $\{0.285, 3.427\}$ **29.** $\{0.253, 2.889\}$ **30.** The change in elevation during the time trial was about 1.18 km.

Cumulative Review *(page 272)*

1. $\left\{\dfrac{-1 - \sqrt{13}}{6}, \dfrac{-1 + \sqrt{13}}{6}\right\}$ **2.** $y + 1 = -1(x - 4)$ or $x + y = 3; 6\sqrt{2}; (1, 2)$ **3.** x-axis symmetry; $(0, -3), (0, 3), (3, 0)$

4.

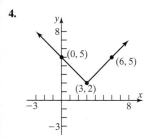

5.

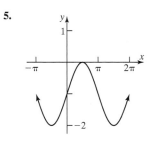

6. (a)

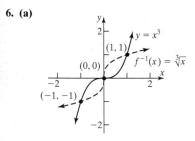

(b)

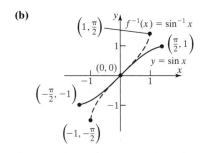

(c)

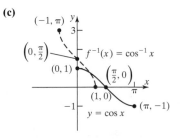

7. (a) $-\dfrac{2\sqrt{2}}{3}$ **(b)** $\dfrac{\sqrt{2}}{4}$ **(c)** $\dfrac{4\sqrt{2}}{9}$ **(d)** $\dfrac{7}{9}$ **(e)** $\sqrt{\dfrac{3 + 2\sqrt{2}}{6}}$ **(f)** $-\sqrt{\dfrac{3 - 2\sqrt{2}}{6}}$ **8.** $\dfrac{\sqrt{5}}{5}$

9. (a) $-\dfrac{2\sqrt{2}}{3}$ **(b)** $-\dfrac{2\sqrt{2}}{3}$ **(c)** $\dfrac{7}{9}$ **(d)** $\dfrac{4\sqrt{2}}{9}$ **(e)** $\dfrac{\sqrt{6}}{3}$

CHAPTER 4 Applications of Trigonometric Functions

4.1 Assess Your Understanding (page 279)

5. T **6.** direction or bearing **7.** T **8.** F **9.** $a \approx 13.74, c \approx 14.62, A = 70°$ **11.** $b \approx 5.03, c \approx 7.83, A = 50°$ **13.** $a \approx 0.71, c \approx 4.06, B = 80°$
15. $b \approx 10.72, c \approx 11.83, B = 65°$ **17.** $b \approx 3.08, a \approx 8.46, A = 70°$ **19.** $c \approx 5.83, A \approx 59.0°, B \approx 31.0°$ **21.** $b \approx 4.58, A \approx 23.6°, B \approx 66.4°$
23. 23.6° and 66.4° **25.** 80.5° **27.** (a) 111.96 ft/s or 76.3 mi/h (b) 82.42 ft/s or 56.2 mi/h (c) Under 18.8° **29.** (a) 2.4898×10^{13} miles
(b) 0.000214° **31.** S76.6°E **33.** The embankment is 30.5 m high. **35.** The buildings are 7984 ft apart. **37.** 69.0° **39.** 38.9°

4.2 Assess Your Understanding (page 288)

4. oblique **5.** $\dfrac{\sin A}{a} = \dfrac{\sin B}{b} = \dfrac{\sin C}{c}$ **6.** F **7.** T **8.** F **9.** $a \approx 3.23, b \approx 3.55, A = 40°$ **11.** $a \approx 3.25, c \approx 4.23, B = 45°$
13. $C = 95°, c \approx 9.86, a \approx 6.36$ **15.** $A = 40°, a = 2, c \approx 3.06$ **17.** $C = 120°, b \approx 1.06, c \approx 2.69$ **19.** $A = 100°, a \approx 5.24, c \approx 0.92$
21. $B = 40°, a \approx 5.64, b \approx 3.86$ **23.** $C = 100°, a \approx 1.31, b \approx 1.31$ **25.** One triangle; $B \approx 30.7°, C \approx 99.3°, c \approx 3.86$ **27.** One triangle;
$C \approx 36.2°, A \approx 43.8°, a \approx 3.51$ **29.** No triangle **31.** Two triangles; $C_1 \approx 30.9°, A_1 \approx 129.1°, a_1 \approx 9.07$ or $C_2 \approx 149.1°, A_2 \approx 10.9°, a_2 \approx 2.20$
33. No triangle **35.** Two triangles; $A_1 \approx 57.7°, B_1 \approx 97.3°, b_1 \approx 2.35$ or $A_2 \approx 122.3°, B_2 \approx 32.7°, b_2 \approx 1.28$ **37.** (a) Station Able is about 143.33 mi
from the ship: Station Baker is about 135.58 mi from the ship. (b) Approximately 41 min **39.** 1490.48 ft **41.** 381.69 ft **43.** The tree is 39.4 ft high.
45. Adam receives 88.3 more frequent flyer miles. **47.** 84.7°; 183.72 ft **49.** 2.64 mi **51.** 38.5 in. **53.** 449.36 ft **55.** 187,600,000 km or 101,440,000 km
57. The diameter is 252 ft.

59. $\dfrac{a-b}{c} = \dfrac{a}{c} - \dfrac{b}{c} = \dfrac{\sin A}{\sin C} - \dfrac{\sin B}{\sin C} = \dfrac{\sin A - \sin B}{\sin C} = \dfrac{2\sin\left(\dfrac{A-B}{2}\right)\cos\left(\dfrac{A+B}{2}\right)}{2\sin\dfrac{C}{2}\cos\dfrac{C}{2}} = \dfrac{\sin\left(\dfrac{A-B}{2}\right)\cos\left(\dfrac{\pi}{2} - \dfrac{C}{2}\right)}{\sin\dfrac{C}{2}\cos\dfrac{C}{2}} = \dfrac{\sin\left(\dfrac{A-B}{2}\right)}{\cos\dfrac{C}{2}}$

61. $\dfrac{a-b}{a+b} = \dfrac{\dfrac{a-b}{c}}{\dfrac{a+b}{c}} = \dfrac{\dfrac{\sin\left[\dfrac{1}{2}(A-B)\right]}{\cos\dfrac{C}{2}}}{\dfrac{\cos\left[\dfrac{1}{2}(A-B)\right]}{\sin\dfrac{C}{2}}} = \dfrac{\tan\left[\dfrac{1}{2}(A-B)\right]}{\cot\dfrac{C}{2}} = \dfrac{\tan\left[\dfrac{1}{2}(A-B)\right]}{\tan\left(\dfrac{\pi}{2} - \dfrac{C}{2}\right)} = \dfrac{\tan\left[\dfrac{1}{2}(A-B)\right]}{\tan\left[\dfrac{1}{2}(A+B)\right]}$

4.3 Assess Your Understanding (page 296)

3. Cosines **4.** Sines **5.** Cosines **6.** F **7.** F **8.** T **9.** $b \approx 2.95, A \approx 28.7°, C \approx 106.3°$ **11.** $c \approx 3.75, A \approx 32.1°, B \approx 52.9°$
13. $A \approx 48.5°, B \approx 38.6°, C \approx 92.9°$ **15.** $A \approx 127.2°, B \approx 32.1°, C \approx 20.7°$ **17.** $c \approx 2.57, A \approx 48.6°, B \approx 91.4°$
19. $a \approx 2.99, B \approx 19.2°, C \approx 80.8°$ **21.** $b \approx 4.14, A \approx 43.0°, C \approx 27.0°$ **23.** $c \approx 1.69, A \approx 65.0°, B \approx 65.0°$ **25.** $A \approx 67.4°, B \approx 90°, C \approx 22.6°$
27. $A = 60°, B = 60°, C = 60°$ **29.** $A \approx 33.6°, B \approx 62.2°, C \approx 84.3°$ **31.** $A \approx 97.9°, B \approx 52.4°, C \approx 29.7°$ **33.** $A = 85°, a = 14.56, c = 14.12$
35. $A = 40.8°, B = 60.6°, C = 78.6°$ **37.** $A = 80°, b = 8.74, c = 13.80$ **39.** Two triangles: $B_1 = 35.4°, C_1 = 134.6°, c_1 = 12.29$;
$B_2 = 144.6°, C_2 = 25.4°, c_2 = 7.40$ **41.** $B = 24.5°, C = 95.5°, a = 10.44$ **43.** 165 yd **45.** (a) 26.4° (b) 30.8 h **47.** (a) 63.7 ft (b) 66.8 ft
(c) 92.8° **49.** (a) 492.6 ft (b) 269.3 ft **51.** 342.33 ft **53.** The footings should be 7.65 ft apart.

55. Suppose $0 < \theta < \pi$. Then by the Law of Cosines, $d^2 = r^2 + r^2 - 2r^2 \cos\theta = 4r^2\left(\dfrac{1 - \cos\theta}{2}\right) \Rightarrow d = 2r\sqrt{\dfrac{1 - \cos\theta}{2}} = 2r\sin\dfrac{\theta}{2}$

Since for any angle in $(0, \pi)$, d is strictly less than the length of the arc subtended by θ, that is, $d < r\theta$, then $2r\sin\dfrac{\theta}{2} < r\theta$, or $2\sin\dfrac{\theta}{2} < \theta$.
Since $\cos\dfrac{\theta}{2} < 1$, then, for $0 < \theta < \pi$, $\sin\theta = 2\sin\dfrac{\theta}{2}\cos\dfrac{\theta}{2} < 2\sin\dfrac{\theta}{2} < \theta$. If $\theta \geq \pi$, then since $\sin\theta \leq 1$, $\sin\theta < \theta$. Thus $\sin\theta < \theta$ for all $\theta > 0$.

57. $\sin\dfrac{C}{2} = \sqrt{\dfrac{1 - \cos C}{2}} = \sqrt{\dfrac{1 - \dfrac{a^2 + b^2 - c^2}{2ab}}{2}} = \sqrt{\dfrac{2ab - a^2 - b^2 + c^2}{4ab}} = \sqrt{\dfrac{c^2 - (a-b)^2}{4ab}} = \sqrt{\dfrac{(c + a - b)(c + b - a)}{4ab}}$
$= \sqrt{\dfrac{(2s - 2b)(2s - 2a)}{4ab}} = \sqrt{\dfrac{(s-a)(s-b)}{ab}}$

4.4 Assess Your Understanding (page 301)

2. Heron's **3.** F **4.** T **5.** 2.83 **7.** 2.99 **9.** 14.98 **11.** 9.56 **13.** 3.86 **15.** 1.48 **17.** 2.82 **19.** 30 **21.** 1.73 **23.** 19.90

25. $K = \dfrac{1}{2}ab\sin C = \dfrac{1}{2}a\sin C\left(\dfrac{a\sin B}{\sin A}\right) = \dfrac{a^2\sin B\sin C}{2\sin A}$ **27.** 0.92 **29.** 2.27 **31.** 5.44 **33.** 9.03 sq ft **35.** $4446.38

37. The area of home plate is about 216.5 in². **39.** $K = \dfrac{1}{2}r^2(\theta + \sin\theta)$ **41.** The ground area is 7517.4 ft².

43. (a) Area $\triangle OAC = \dfrac{1}{2}|OC||AC| = \dfrac{1}{2} \cdot \dfrac{|OC|}{1} \cdot \dfrac{|AC|}{1} = \dfrac{1}{2}\sin \alpha \cos \alpha$

(b) Area $\triangle OCB = \dfrac{1}{2}|BC||OC| = \dfrac{1}{2}|OB|^2 \dfrac{|BC|}{|OB|} \cdot \dfrac{|OC|}{|OB|} = \dfrac{1}{2}|OB|^2 \sin \beta \cos \beta$

(c) Area $\triangle OAB = \dfrac{1}{2}|BD||OA| = \dfrac{1}{2}|OB|\dfrac{|BD|}{|OB|} = \dfrac{1}{2}|OB|\sin(\alpha + \beta)$

(d) $\dfrac{\cos \alpha}{\cos \beta} = \dfrac{\dfrac{|OC|}{1}}{\dfrac{|OC|}{|OB|}} = |OB|$

(e) Area $\triangle OAB = $ Area $\triangle OAC + $ Area $\triangle OCB$

$\dfrac{1}{2}|OB|\sin(\alpha + \beta) = \dfrac{1}{2}\sin \alpha \cos \alpha + \dfrac{1}{2}|OB|^2 \sin \beta \cos \beta$

$\sin(\alpha + \beta) = \dfrac{1}{|OB|}\sin \alpha \cos \alpha + |OB|\sin \beta \cos \beta$

$\sin(\alpha + \beta) = \dfrac{\cos \beta}{\cos \alpha}\sin \alpha \cos \alpha + \dfrac{\cos \alpha}{\cos \beta}\sin \beta \cos \beta$

$\sin(\alpha + \beta) = \sin \alpha \cos \beta + \cos \alpha \sin \beta$

45. $31{,}145$ ft^2 **47. (a)** The perimeter and area are both 36. **(b)** The perimeter and area are both 60.

49. $K = \dfrac{1}{2}ah = \dfrac{1}{2}ab \sin C \Rightarrow h = b \sin C = \dfrac{a \sin B \sin C}{\sin A}$

51. $\cot\dfrac{C}{2} = \dfrac{\cos\dfrac{C}{2}}{\sin\dfrac{C}{2}} = \dfrac{c \sin\dfrac{A}{2}\sin\dfrac{B}{2}}{r\sqrt{\dfrac{(s-a)(s-b)}{ab}}} = \dfrac{c\sqrt{\dfrac{(s-b)(s-c)}{bc}}\sqrt{\dfrac{(s-a)(s-c)}{ac}}}{r\sqrt{\dfrac{(s-a)(s-b)}{ab}}} = \dfrac{c}{r}\sqrt{\dfrac{(s-c)^2}{c^2}} = \dfrac{s-c}{r}$

53. $K = $ area of triangle $QOR + $ area of $ROP + $ area of $POQ = \dfrac{1}{2}ar + \dfrac{1}{2}br + \dfrac{1}{2}cr = r\dfrac{1}{2}(a + b + c) = rs$, so

$r = \dfrac{K}{s} = \dfrac{\sqrt{s(s-a)(s-b)(s-c)}}{s} = \sqrt{\dfrac{(s-a)(s-b)(s-c)}{s}}.$

4.5 Assess Your Understanding *(page 311)*

2. simple harmonic; amplitude **3.** simple harmonic; damped **4.** T **5.** $d = -5\cos(\pi t)$ **7.** $d = -6\cos(2t)$ **9.** $d = -5\sin(\pi t)$ **11.** $d = -6\sin(2t)$

13. (a) simple harmonic **(b)** 5 m **(c)** $\dfrac{2\pi}{3}$ s **(d)** $\dfrac{3}{2\pi}$ oscillation/s **15. (a)** simple harmonic **(b)** 6 m **(c)** 2 s **(d)** $\dfrac{1}{2}$ oscillation/s

17. (a) simple harmonic **(b)** 3 m **(c)** 4π s **(d)** $\dfrac{1}{4\pi}$ oscillation/s **19. (a)** simple harmonic **(b)** 2 m **(c)** 1 s **(d)** 1 oscillation/s

21. (a) $d = -10e^{-0.7t/50}\cos\left(\sqrt{\dfrac{4\pi^2}{25} - \dfrac{0.49}{2500}}\, t\right)$ **23. (a)** $d = -18e^{-0.6t/60}\cos\left(\sqrt{\dfrac{\pi^2}{4} - \dfrac{0.36}{3600}}\, t\right)$ **25. (a)** $d = -5e^{-0.8t/20}\cos\left(\sqrt{\dfrac{4\pi^2}{9} - \dfrac{0.64}{400}}\, t\right)$

(b)
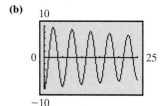

(b)
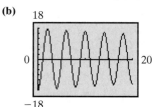

(b)
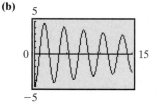

27. (a) The motion is damped.
The bob has mass $m = 20$ kg
with a damping factor of 0.7 kg/s.
(b) 20 m leftward
(c)
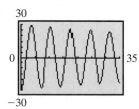
(d) 18.33 m leftward **(e)** $d \to 0$

29. (a) The motion is damped.
The bob has mass $m = 40$ kg
with a damping factor of 0.6 kg/s.
(b) 30 m leftward
(c)
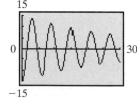
(d) 28.47 m leftward **(e)** $d \to 0$

31. (a) The motion is damped.
The bob has mass $m = 15$ kg
with a damping factor of 0.9 kg/s.
(b) 15 m leftward
(c)
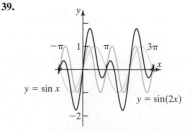
(d) 12.53 m leftward **(e)** $d \to 0$

33.
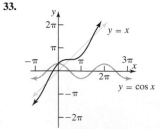

35.
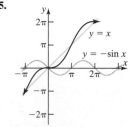

37.
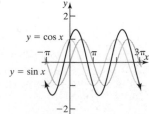

39.
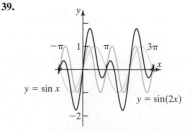

CHAPTER 4 Applications of Trigonometric Functions

4.1 Assess Your Understanding (page 279)

5. T **6.** direction or bearing **7.** T **8.** F **9.** $a \approx 13.74, c \approx 14.62, A = 70°$ **11.** $b \approx 5.03, c \approx 7.83, A = 50°$ **13.** $a \approx 0.71, c \approx 4.06, B = 80°$
15. $b \approx 10.72, c \approx 11.83, B = 65°$ **17.** $b \approx 3.08, a \approx 8.46, A = 70°$ **19.** $c \approx 5.83, A \approx 59.0°, B \approx 31.0°$ **21.** $b \approx 4.58, A \approx 23.6°, B \approx 66.4°$
23. $23.6°$ and $66.4°$ **25.** $80.5°$ **27. (a)** 111.96 ft/s or 76.3 mi/h **(b)** 82.42 ft/s or 56.2 mi/h **(c)** Under $18.8°$ **29. (a)** 2.4898×10^{13} miles
(b) $0.000214°$ **31.** S76.6°E **33.** The embankment is 30.5 m high. **35.** The buildings are 7984 ft apart. **37.** $69.0°$ **39.** $38.9°$

4.2 Assess Your Understanding (page 288)

4. oblique **5.** $\dfrac{\sin A}{a} = \dfrac{\sin B}{b} = \dfrac{\sin C}{c}$ **6.** F **7.** T **8.** F **9.** $a \approx 3.23, b \approx 3.55, A = 40°$ **11.** $a \approx 3.25, c \approx 4.23, B = 45°$
13. $C = 95°, c \approx 9.86, a \approx 6.36$ **15.** $A = 40°, a = 2, c \approx 3.06$ **17.** $C = 120°, b \approx 1.06, c \approx 2.69$ **19.** $A = 100°, a \approx 5.24, c \approx 0.92$
21. $B = 40°, a \approx 5.64, b \approx 3.86$ **23.** $C = 100°, a \approx 1.31, b \approx 1.31$ **25.** One triangle; $B \approx 30.7°, C \approx 99.3°, c \approx 3.86$ **27.** One triangle;
$C \approx 36.2°, A \approx 43.8°, a \approx 3.51$ **29.** No triangle **31.** Two triangles; $C_1 \approx 30.9°, A_1 \approx 129.1°, a_1 \approx 9.07$ or $C_2 \approx 149.1°, A_2 \approx 10.9°, a_2 \approx 2.20$
33. No triangle **35.** Two triangles; $A_1 \approx 57.7°, B_1 \approx 97.3°, b_1 \approx 2.35$ or $A_2 \approx 122.3°, B_2 \approx 32.7°, b_2 \approx 1.28$ **37. (a)** Station Able is about 143.33 mi
from the ship: Station Baker is about 135.58 mi from the ship. **(b)** Approximately 41 min **39.** 1490.48 ft **41.** 381.69 ft **43.** The tree is 39.4 ft high.
45. Adam receives 88.3 more frequent flyer miles. **47.** $84.7°; 183.72$ ft **49.** 2.64 mi **51.** 38.5 in. **53.** 449.36 ft **55.** 187,600,000 km or 101,440,000 km
57. The diameter is 252 ft.

59. $\dfrac{a-b}{c} = \dfrac{a}{c} - \dfrac{b}{c} = \dfrac{\sin A}{\sin C} - \dfrac{\sin B}{\sin C} = \dfrac{\sin A - \sin B}{\sin C} = \dfrac{2 \sin\left(\dfrac{A-B}{2}\right)\cos\left(\dfrac{A+B}{2}\right)}{2 \sin\dfrac{C}{2}\cos\dfrac{C}{2}} = \dfrac{\sin\left(\dfrac{A-B}{2}\right)\cos\left(\dfrac{\pi}{2}-\dfrac{C}{2}\right)}{\sin\dfrac{C}{2}\cos\dfrac{C}{2}} = \dfrac{\sin\left(\dfrac{A-B}{2}\right)}{\cos\dfrac{C}{2}}$

61. $\dfrac{a-b}{a+b} = \dfrac{\dfrac{a-b}{c}}{\dfrac{a+b}{c}} = \dfrac{\dfrac{\sin\left[\frac{1}{2}(A-B)\right]}{\cos\dfrac{C}{2}}}{\dfrac{\cos\left[\frac{1}{2}(A-B)\right]}{\sin\dfrac{C}{2}}} = \dfrac{\tan\left[\frac{1}{2}(A-B)\right]}{\cot\dfrac{C}{2}} = \dfrac{\tan\left[\frac{1}{2}(A-B)\right]}{\tan\left(\dfrac{\pi}{2}-\dfrac{C}{2}\right)} = \dfrac{\tan\left[\frac{1}{2}(A-B)\right]}{\tan\left[\frac{1}{2}(A+B)\right]}$

4.3 Assess Your Understanding (page 296)

3. Cosines **4.** Sines **5.** Cosines **6.** F **7.** F **8.** T **9.** $b \approx 2.95, A \approx 28.7°, C \approx 106.3°$ **11.** $c \approx 3.75, A \approx 32.1°, B \approx 52.9°$
13. $A \approx 48.5°, B \approx 38.6°, C \approx 92.9°$ **15.** $A \approx 127.2°, B \approx 32.1°, C \approx 20.7°$ **17.** $c \approx 2.57, A \approx 48.6°, B \approx 91.4°$
19. $a \approx 2.99, B \approx 19.2°, C \approx 80.8°$ **21.** $b \approx 4.14, A \approx 43.0°, C \approx 27.0°$ **23.** $c \approx 1.69, A = 65.0°, B = 65.0°$ **25.** $A \approx 67.4°, B = 90°, C \approx 22.6°$
27. $A = 60°, B = 60°, C = 60°$ **29.** $A \approx 33.6°, B \approx 62.2°, C \approx 84.3°$ **31.** $A \approx 97.9°, B \approx 52.4°, C \approx 29.7°$ **33.** $A = 85°, a = 14.56, c = 14.12$
35. $A = 40.8°, B = 60.6°, C = 78.6°$ **37.** $A = 80°, b = 8.74, c = 13.80$ **39.** Two triangles: $B_1 = 35.4°, C_1 = 134.6°, c_1 = 12.29$;
$B_2 = 144.6°, C_2 = 25.4°, c_2 = 7.40$ **41.** $B = 24.5°, C = 95.5°, a = 10.44$ **43.** 165 yd **45. (a)** $26.4°$ **(b)** 30.8 h **47. (a)** 63.7 ft **(b)** 66.8 ft
(c) $92.8°$ **49. (a)** 492.6 ft **(b)** 269.3 ft **51.** 342.33 ft **53.** The footings should be 7.65 ft apart.

55. Suppose $0 < \theta < \pi$. Then by the Law of Cosines, $d^2 = r^2 + r^2 - 2r^2 \cos\theta = 4r^2\left(\dfrac{1-\cos\theta}{2}\right) \Rightarrow d = 2r\sqrt{\dfrac{1-\cos\theta}{2}} = 2r\sin\dfrac{\theta}{2}$.

Since for any angle in $(0, \pi)$, d is strictly less than the length of the arc subtended by θ, that is, $d < r\theta$, then $2r\sin\dfrac{\theta}{2} < r\theta$, or $2\sin\dfrac{\theta}{2} < \theta$.

Since $\cos\dfrac{\theta}{2} < 1$, then, for $0 < \theta < \pi$, $\sin\theta = 2\sin\dfrac{\theta}{2}\cos\dfrac{\theta}{2} < 2\sin\dfrac{\theta}{2} < \theta$. If $\theta \geq \pi$, then since $\sin\theta \leq 1$, $\sin\theta < \theta$. Thus $\sin\theta < \theta$ for all $\theta > 0$.

57. $\sin\dfrac{C}{2} = \sqrt{\dfrac{1-\cos C}{2}} = \sqrt{\dfrac{1-\dfrac{a^2+b^2-c^2}{2ab}}{2}} = \sqrt{\dfrac{2ab-a^2-b^2+c^2}{4ab}} = \sqrt{\dfrac{c^2-(a-b)^2}{4ab}} = \sqrt{\dfrac{(c+a-b)(c+b-a)}{4ab}}$
$= \sqrt{\dfrac{(2s-2b)(2s-2a)}{4ab}} = \sqrt{\dfrac{(s-a)(s-b)}{ab}}$

4.4 Assess Your Understanding (page 301)

2. Heron's **3.** F **4.** T **5.** 2.83 **7.** 2.99 **9.** 14.98 **11.** 9.56 **13.** 3.86 **15.** 1.48 **17.** 2.82 **19.** 30 **21.** 1.73 **23.** 19.90
25. $K = \dfrac{1}{2}ab\sin C = \dfrac{1}{2}a\sin C\left(\dfrac{a\sin B}{\sin A}\right) = \dfrac{a^2\sin B\sin C}{2\sin A}$ **27.** 0.92 **29.** 2.27 **31.** 5.44 **33.** 9.03 sq ft **35.** \$5446.38

37. The area of home plate is about 216.5 in². **39.** $K = \dfrac{1}{2}r^2(\theta + \sin\theta)$ **41.** The ground area is 7517.4 ft².

43. (a) Area $\triangle OAC = \frac{1}{2}|OC||AC| = \frac{1}{2} \cdot \frac{|OC|}{1} \cdot \frac{|AC|}{1} = \frac{1}{2}\sin\alpha\cos\alpha$

(b) Area $\triangle OCB = \frac{1}{2}|BC||OC| = \frac{1}{2}|OB|^2\frac{|BC|}{|OB|} \cdot \frac{|OC|}{|OB|} = \frac{1}{2}|OB|^2\sin\beta\cos\beta$

(c) Area $\triangle OAB = \frac{1}{2}|BD||OA| = \frac{1}{2}|OB|\frac{|BD|}{|OB|} = \frac{1}{2}|OB|\sin(\alpha+\beta)$

(d) $\dfrac{\cos\alpha}{\cos\beta} = \dfrac{\dfrac{|OC|}{1}}{\dfrac{|OC|}{|OB|}} = |OB|$

(e) Area $\triangle OAB = $ Area $\triangle OAC + $ Area $\triangle OCB$

$\frac{1}{2}|OB|\sin(\alpha+\beta) = \frac{1}{2}\sin\alpha\cos\alpha + \frac{1}{2}|OB|^2\sin\beta\cos\beta$

$\sin(\alpha+\beta) = \frac{1}{|OB|}\sin\alpha\cos\alpha + |OB|\sin\beta\cos\beta$

$\sin(\alpha+\beta) = \frac{\cos\beta}{\cos\alpha}\sin\alpha\cos\alpha + \frac{\cos\alpha}{\cos\beta}\sin\beta\cos\beta$

$\sin(\alpha+\beta) = \sin\alpha\cos\beta + \cos\alpha\sin\beta$

45. 31,145 ft^2 **47. (a)** The perimeter and area are both 36.   **(b)** The perimeter and area are both 60.

49. $K = \frac{1}{2}ah = \frac{1}{2}ab\sin C \Rightarrow h = b\sin C = \dfrac{a\sin B\sin C}{\sin A}$

51. $\cot\dfrac{C}{2} = \dfrac{\cos\dfrac{C}{2}}{\sin\dfrac{C}{2}} = \dfrac{c\sin\dfrac{A}{2}\sin\dfrac{B}{2}}{r} \Big/ \sqrt{\dfrac{(s-a)(s-b)}{ab}} = \dfrac{c\sqrt{\dfrac{(s-b)(s-c)}{bc}}\sqrt{\dfrac{(s-a)(s-c)}{ac}}}{r\sqrt{\dfrac{(s-a)(s-b)}{ab}}} = \dfrac{c}{r}\sqrt{\dfrac{(s-c)^2}{c^2}} = \dfrac{s-c}{r}$

53. $K = $ area of triangle $QOR + $ area of $ROP + $ area of $POQ = \frac{1}{2}ar + \frac{1}{2}br + \frac{1}{2}cr = r\frac{1}{2}(a+b+c) = rs$, so

$r = \dfrac{K}{s} = \dfrac{\sqrt{s(s-a)(s-b)(s-c)}}{s} = \sqrt{\dfrac{(s-a)(s-b)(s-c)}{s}}$.

4.5 Assess Your Understanding (page 311)

2. simple harmonic; amplitude   **3.** simple harmonic; damped   **4.** T   **5.** $d = -5\cos(\pi t)$ **7.** $d = -6\cos(2t)$ **9.** $d = -5\sin(\pi t)$ **11.** $d = -6\sin(2t)$

13. (a) simple harmonic   **(b)** 5 m   **(c)** $\dfrac{2\pi}{3}$ s   **(d)** $\dfrac{3}{2\pi}$ oscillation/s   **15. (a)** simple harmonic   **(b)** 6 m   **(c)** 2 s   **(d)** $\dfrac{1}{2}$ oscillation/s

17. (a) simple harmonic   **(b)** 3 m   **(c)** 4π s   **(d)** $\dfrac{1}{4\pi}$ oscillation/s   **19. (a)** simple harmonic   **(b)** 2 m   **(c)** 1 s   **(d)** 1 oscillation/s

21. (a) $d = -10e^{-0.7t/50}\cos\left(\sqrt{\dfrac{4\pi^2}{25} - \dfrac{0.49}{2500}}\,t\right)$ **23. (a)** $d = -18e^{-0.6t/60}\cos\left(\sqrt{\dfrac{\pi^2}{4} - \dfrac{0.36}{3600}}\,t\right)$ **25. (a)** $d = -5e^{-0.8t/20}\cos\left(\sqrt{\dfrac{4\pi^2}{9} - \dfrac{0.64}{400}}\,t\right)$

(b)

10	
0	25
−10	

(b)

18	
0	20
−18	

(b)

5	
0	15
−5	

27. (a) The motion is damped. The bob has mass $m = 20$ kg with a damping factor of 0.7 kg/s.
(b) 20 m leftward
(c)

20	
0	25
−20	

(d) 18.33 m leftward   **(e)** $d \to 0$

29. (a) The motion is damped. The bob has mass $m = 40$ kg with a damping factor of 0.6 kg/s.
(b) 30 m leftward
(c)

30	
0	35
−30	

(d) 28.47 m leftward   **(e)** $d \to 0$

31. (a) The motion is damped. The bob has mass $m = 15$ kg with a damping factor of 0.9 kg/s.
(b) 15 m leftward
(c)

15	
0	30
−15	

(d) 12.53 m leftward   **(e)** $d \to 0$

33.

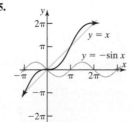

35.

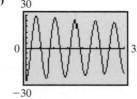

37.

39.

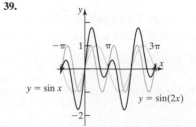

41. $\omega = 1040\pi$;

$d = 0.80 \cos (1040\pi t)$

43. $\omega = 880\pi$;

$d = 0.01 \sin (880\pi t)$

45. **(a)**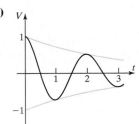

(b) At $t = 0, 2$; at $t = 1, t = 3$

(c) During the approximate intervals $0.35 < t < 0.67$, $1.29 < t < 1.75$, and $2.19 < t \le 3$

47.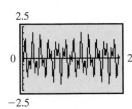

51.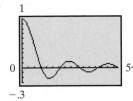

53. $y = \dfrac{1}{x} \sin x$

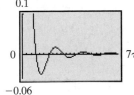

$y = \dfrac{1}{x^2} \sin x$

$y = \dfrac{1}{x^3} \sin x$

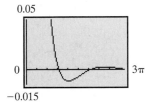

Review Exercises *(page 314)*

1. $A = 70°, b \approx 3.42, a \approx 9.40$ **3.** $u \approx 4.58, A \approx 66.4°, B \approx 23.6°$ **5.** $C = 100°, b \approx 0.65, c \approx 1.29$ **7.** $B \approx 56.8°, C \approx 23.2°, b \approx 4.25$ **9.** No triangle
11. $b \approx 3.32, A \approx 62.8°, C \approx 17.2°$ **13.** No triangle **15.** $c \approx 2.32, A \approx 16.1°, B \approx 123.9°$ **17.** $B \approx 36.2°, C \approx 63.8°, c \approx 4.55$ **19.** $A \approx 39.6°$,
$B \approx 18.6°, C \approx 121.9°$ **21.** Two triangles: $B_1 \approx 13.4°, C_1 \approx 156.6°, c_1 \approx 6.86$ or $B_2 \approx 166.6°, C_2 \approx 3.4°, c_2 \approx 1.02$ **23.** $a \approx 5.23, B \approx 46.0°, C \approx 64.0°$
25. 1.93 **27.** 18.79 **29.** 6 **31.** 3.80 **33.** 0.32 **35.** 12.7° **37.** 29.97 ft **39.** 6.22 mi **41.** 71.12 ft **43.** \$222,983.51 **45.** S4.0°E
47. 79.69 in. **49.** $d = -5 \cos \left(\dfrac{\pi}{3} t\right)$ **51. (a)** simple harmonic **(b)** 2 ft **(c)** $\dfrac{\pi}{2}$ s **(d)** $\dfrac{2}{\pi}$ oscillation/s **53. (a)** simple harmonic **(b)** 3 ft **(c)** 4 s

(d) $\dfrac{1}{4}$ oscillation/s **55. (a)** $d = -13 e^{-0.65t/50} \cos\left(\sqrt{\dfrac{\pi^2}{4} - \dfrac{0.4225}{2500}}\, t\right)$ **(b)**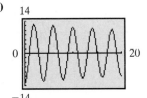

57. (a) The motion is damped. The bob has mass
$m = 30$ kg with a damping factor of 0.5 kg/s

(b) 20 m leftward

(c)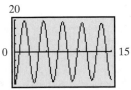

(d) 19.51 m leftward
(e) $d \to 0$

59.

Chapter Test *(page 316)*

1. 61.0° **2.** 1.3° **3.** $a = 15.88, B \approx 57.5°, C \approx 70.5°$ **4.** $b \approx 6.85, C = 117°, c \approx 16.30$ **5.** $A \approx 52.4°, B \approx 29.7°, C \approx 97.9°$
6. $b \approx 4.72, c \approx 1.67, B = 105°$ **7.** No triangle **8.** $c \approx 7.62, A \approx 80.5°, B \approx 29.5°$ **9.** 15.04 square units **10.** 19.81 square units **11.** The area
of the shaded region is 9.26 cm². **12.** 54.15 square units **13.** Madison will have to swim about 2.23 miles. **14.** 12.63 square units

15. The lengths of the sides are 15, 18, and 21. **16.** $d = 5(\sin 42°)\sin\left(\dfrac{\pi t}{3}\right)$ or $d \approx 3.346 \sin\left(\dfrac{\pi t}{3}\right)$

Cumulative Review (page 317)

1. $\left\{\dfrac{1}{3}, 1\right\}$ **2.** $(x + 5)^2 + (y - 1)^2 = 9$ **3.** $\{x \mid x \le -1 \text{ or } x \ge 4\}$ **4.**

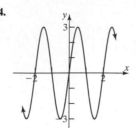

5.

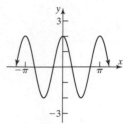

6. (a) $-\dfrac{2\sqrt{5}}{5}$ **(d)** $-\dfrac{3}{5}$

(b) $\dfrac{\sqrt{5}}{5}$ **(e)** $\sqrt{\dfrac{5 - \sqrt{5}}{10}}$

(c) $-\dfrac{4}{5}$ **(f)** $-\sqrt{\dfrac{5 + \sqrt{5}}{10}}$

7. (a)

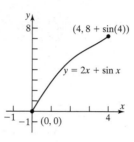

(b)

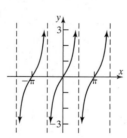

8. (a)

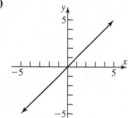

(b)

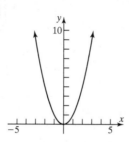

(c)

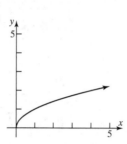

(d)

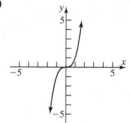

(e)

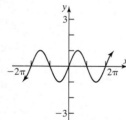

(f)

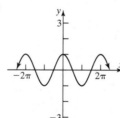

(g)

9. Two triangles: $A_1 \approx 59.0°$, $B_1 \approx 81.0°$, $b_1 \approx 23.05$ or $A_2 \approx 121.0°$, $B_2 \approx 19.0°$, $b_2 \approx 7.59$

CHAPTER 5 Polar Coordinates; Vectors

5.1 Assess Your Understanding (page 327)

5. pole; polar axis   **6.** -2 **7.** $\left(-\sqrt{3}, -1\right)$ **8.** F   **9.** T   **10.** T   **11.** A   **13.** C   **15.** B   **17.** A   **19.**

21.

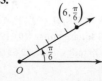

23.

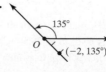

25.

27.

29.

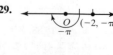

31.

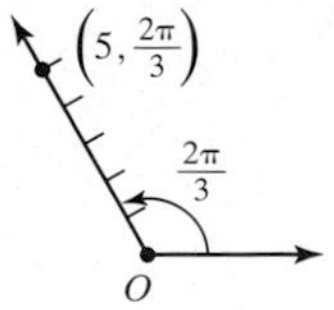

(a) $\left(5, -\frac{4\pi}{3}\right)$
(b) $\left(-5, \frac{5\pi}{3}\right)$
(c) $\left(5, \frac{8\pi}{3}\right)$

33.

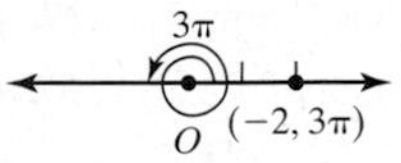

(a) $(2, -2\pi)$
(b) $(-2, \pi)$
(c) $(2, 2\pi)$

35.

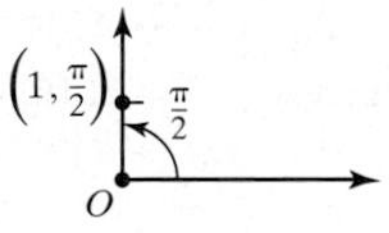

(a) $\left(1, -\frac{3\pi}{2}\right)$
(b) $\left(-1, \frac{3\pi}{2}\right)$
(c) $\left(1, \frac{5\pi}{2}\right)$

37.

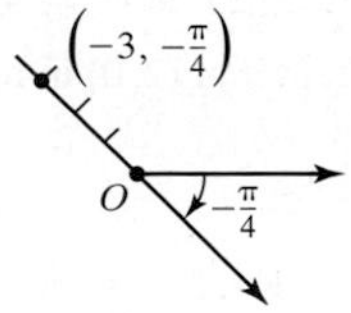

(a) $\left(3, -\frac{5\pi}{4}\right)$
(b) $\left(-3, \frac{7\pi}{4}\right)$
(c) $\left(3, \frac{11\pi}{4}\right)$

39. $(0, 3)$ **41.** $(-2, 0)$ **43.** $(-3\sqrt{3}, 3)$ **45.** $(\sqrt{2}, -\sqrt{2})$ **47.** $\left(-\frac{1}{2}, \frac{\sqrt{3}}{2}\right)$ **49.** $(2, 0)$ **51.** $(-2.57, 7.05)$ **53.** $(-4.98, -3.85)$ **55.** $(3, 0)$

57. $(1, \pi)$ **59.** $\left(\sqrt{2}, -\frac{\pi}{4}\right)$ **61.** $\left(2, \frac{\pi}{6}\right)$ **63.** $(2.47, -1.02)$ **65.** $(9.30, 0.47)$ **67.** $r^2 = \frac{3}{2}$ or $r = \frac{\sqrt{6}}{2}$ **69.** $r^2 \cos^2\theta - 4r\sin\theta = 0$

71. $r^2 \sin 2\theta = 1$ **73.** $r\cos\theta = 4$ **75.** $x^2 + y^2 - x = 0$ or $\left(x - \frac{1}{2}\right)^2 + y^2 = \frac{1}{4}$ **77.** $(x^2 + y^2)^{3/2} - x = 0$ **79.** $x^2 + y^2 = 4$

81. $y^2 = 8(x + 2)$ **83.** **(a)** $(-10, 36)$ **(b)** $\left(2\sqrt{349}, 180° + \tan^{-1}\left(-\frac{18}{5}\right)\right) \approx (37.36, 105.5°)$ **(c)** $(-3, -35)$

(d) $\left(\sqrt{1234}, 180° + \tan^{-1}\left(\frac{35}{3}\right)\right) \approx (35.13, 265.1°)$

5.2 Assess Your Understanding *(page 344)*

7. polar equation **8.** $r = 2\cos\theta$ **9.** $-r$ **10.** F **11.** F **12.** F

13. $x^2 + y^2 = 16$; circle, radius 4, center at pole

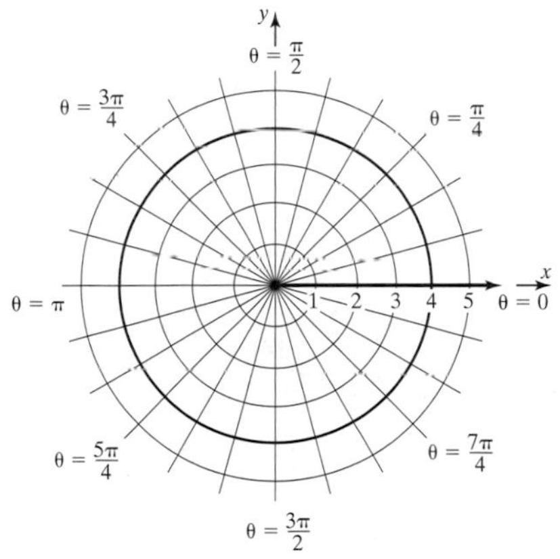

15. $y = \sqrt{3}x$; line through pole, making an angle of $\frac{\pi}{3}$ with polar axis

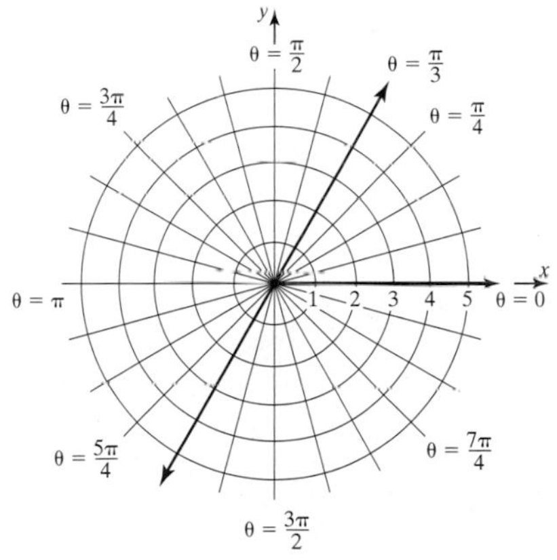

17. $y = 4$; horizontal line 4 units above the pole

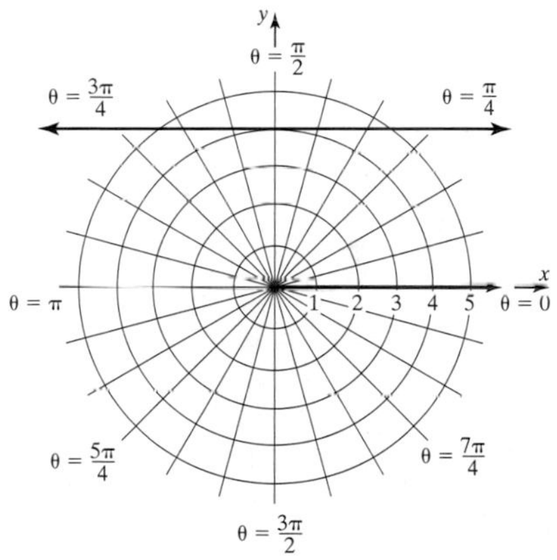

19. $x = -2$; vertical line 2 units to the left of the pole

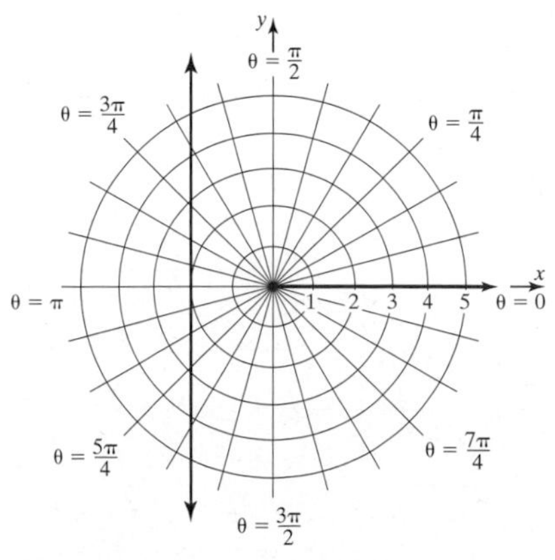

21. $(x - 1)^2 + y^2 = 1$; circle, radius 1, center $(1, 0)$ in rectangular coordinates

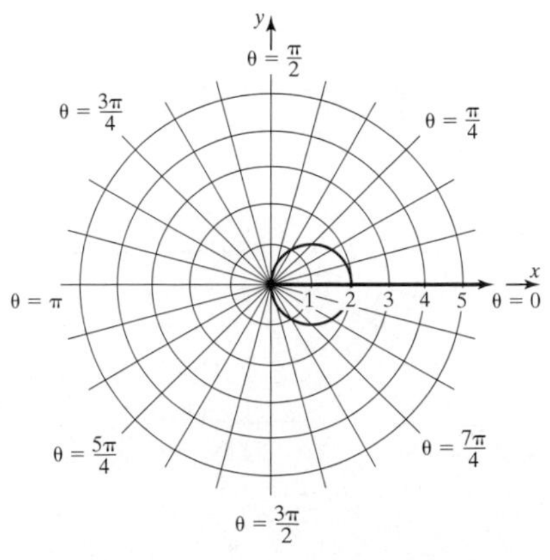

23. $x^2 + (y + 2)^2 = 4$; circle, radius 2, center at $(0, -2)$ in rectangular coordinates

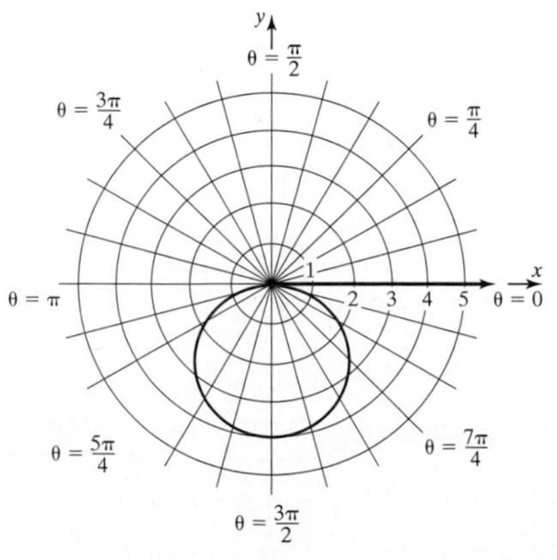

25. $(x - 2)^2 + y^2 = 4$; circle, radius 2, center at $(2, 0)$ in rectangular coordinates

27. $x^2 + (y + 1)^2 = 1$; circle, radius 1, center at $(0, -1)$ in rectangular coordinates

29. E **31.** F **33.** H **35.** D
37. Cardioid

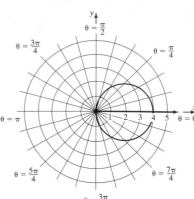

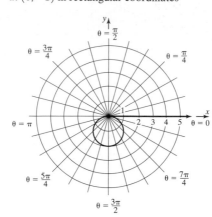

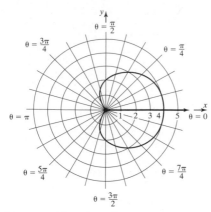

39. Cardioid

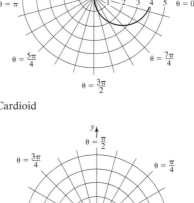

41. Limaçon without inner loop

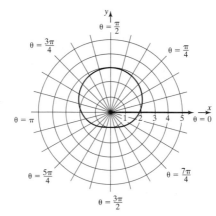

43. Limaçon without inner loop

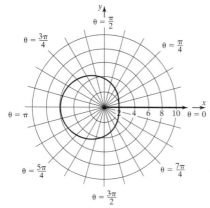

45. Limaçon with inner loop

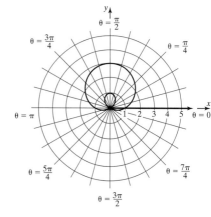

47. Limaçon with inner loop

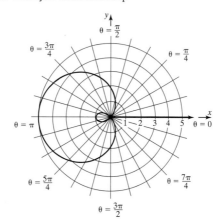

49. Rose

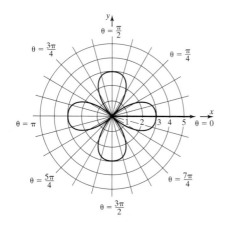

51. Rose

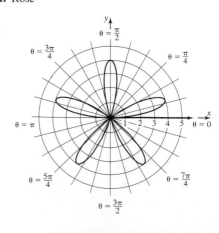

53. Lemniscate

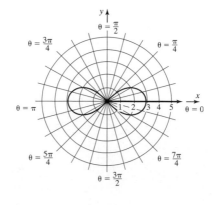

55. Spiral

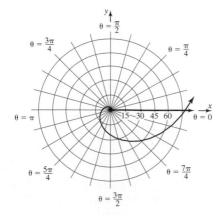

57. Cardioid

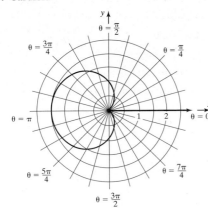

59. Limaçon with inner loop

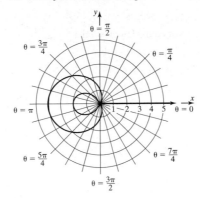

61.

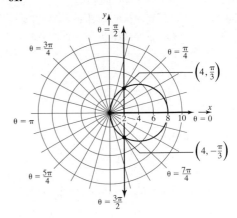

63.

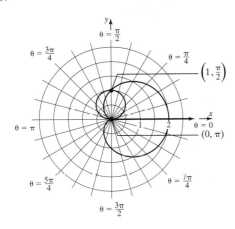

65.

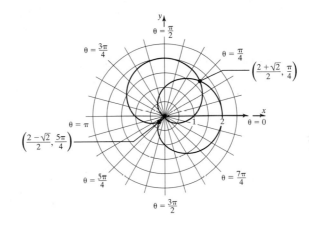

67. $r = 3 + 3\cos\theta$

69. $r = 4 + \sin\theta$

71.

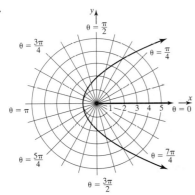

73.

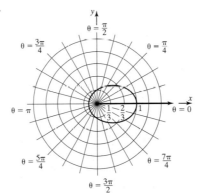

75.

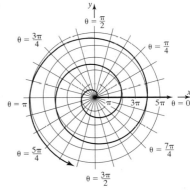

77.

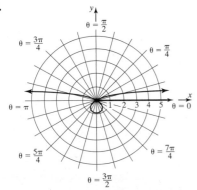

79.

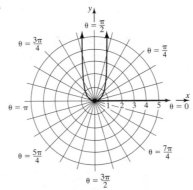

81. $r\sin\theta = a$
$y = a$

83.

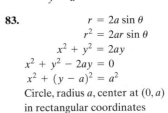

$$r = 2a\sin\theta$$
$$r^2 = 2ar\sin\theta$$
$$x^2 + y^2 = 2ay$$
$$x^2 + y^2 - 2ay = 0$$
$$x^2 + (y - a)^2 = a^2$$

Circle, radius a, center at $(0, a)$
in rectangular coordinates

85.
$$r = 2a \cos \theta$$
$$r^2 = 2ar \cos \theta$$
$$x^2 + y^2 = 2ax$$
$$x^2 - 2ax + y^2 = 0$$
$$(x - a)^2 + y^2 = a^2$$
Circle, radius a, center at $(a, 0)$ in rectangular coordinates

87. (a) $r^2 = \cos \theta$: $r^2 = \cos(\pi - \theta)$
$$r^2 = -\cos \theta$$
Not equivalent; test fails.
$$(-r)^2 = \cos(-\theta)$$
$$r^2 = \cos \theta$$
New test works.

(b) $r^2 = \sin \theta$: $r^2 = \sin(\pi - \theta)$
$$r^2 = \sin \theta$$
Test works.
$$(-r)^2 = \sin(-\theta)$$
$$r^2 = -\sin \theta$$
Not equivalent; new test fails.

Historical Problems *(page 353)*

1. (a) $1 + 4i, 1 + i$ **(b)** $-1, 2 + i$.

5.3 Assess Your Understanding *(page 353)*

5. magnitude; modulus; argument **6.** De Moivre's **7.** three **8.** T **9.** F **10.** T

11.

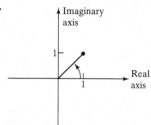

$\sqrt{2}(\cos 45° + i \sin 45°)$

13.

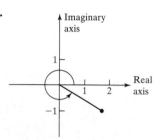

$2(\cos 330° + i \sin 330°)$

15.

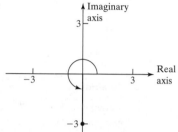

$3(\cos 270° + i \sin 270°)$

17.

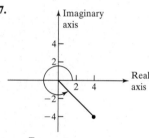

$4\sqrt{2}(\cos 315° + i \sin 315°)$

19.

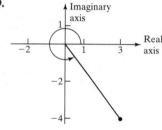

$5(\cos 306.9° + i \sin 306.9°)$

21.

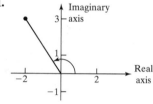

$\sqrt{13}(\cos 123.7° + i \sin 123.7°)$

23. $-1 + \sqrt{3}i$ **25.** $2\sqrt{2} - 2\sqrt{2}i$ **27.** $-3i$ **29.** $-0.035 + 0.197i$ **31.** $1.970 + 0.347i$ **33.** $zw = 8(\cos 60° + i \sin 60°); \dfrac{z}{w} = \dfrac{1}{2}(\cos 20° + i \sin 20°)$

35. $zw = 12(\cos 40° + i \sin 40°); \dfrac{z}{w} = \dfrac{3}{4}(\cos 220° + i \sin 220°)$ **37.** $zw = 4\left(\cos \dfrac{9\pi}{40} + i \sin \dfrac{9\pi}{40}\right); \dfrac{z}{w} = \cos \dfrac{\pi}{40} + i \sin \dfrac{\pi}{40}$

39. $zw = 4\sqrt{2}(\cos 15° + i \sin 15°); \dfrac{z}{w} = \sqrt{2}(\cos 75° + i \sin 75°)$ **41.** $-32 + 32\sqrt{3}i$ **43.** $32i$ **45.** $\dfrac{27}{2} + \dfrac{27\sqrt{3}}{2}i$ **47.** $-\dfrac{25\sqrt{2}}{2} + \dfrac{25\sqrt{2}}{2}i$

49. $-4 + 4i$ **51.** $-23 + 14.142i$ **53.** $\sqrt[6]{2}(\cos 15° + i \sin 15°), \sqrt[6]{2}(\cos 135° + i \sin 135°), \sqrt[6]{2}(\cos 255° + i \sin 255°)$

55. $\sqrt[4]{8}(\cos 75° + i \sin 75°), \sqrt[4]{8}(\cos 165° + i \sin 165°), \sqrt[4]{8}(\cos 255° + i \sin 255°), \sqrt[4]{8}(\cos 345° + i \sin 345°)$

57. $2(\cos 67.5° + i \sin 67.5°), 2(\cos 157.5° + i \sin 157.5°), 2(\cos 247.5° + i \sin 247.5°), 2(\cos 337.5° + i \sin 337.5°)$

59. $\cos 18° + i \sin 18°, \cos 90° + i \sin 90°, \cos 162° + i \sin 162°, \cos 234° + i \sin 234°, \cos 306° + i \sin 306°$

61. $1, i, -1, -i$

63. Look at formula (8); $|z_k| = \sqrt[n]{r}$ for all k.

65. Look at formula (8). The z_k are spaced apart by an angle of $\dfrac{2\pi}{n}$.

67. (a)

$z = a_0$	a_1	a_2	a_3	a_4	a_5	a_6
$0.1 - 0.4i$	$-0.05 - 0.48i$	$-0.13 - 0.35i$	$-0.01 - 0.31i$	$0.01 - 0.35i$	$0.2 - 0.41i$	$0.07 - 0.42i$
$0.5 + 0.8i$	$0.11 + 1.6i$	$-2.05 + 1.15i$	$3.37 - 3.92i$	$-3.52 - 25.6i$	$-641.9 + 181.1i$	$379290 - 232509i$
$-0.9 + 0.7i$	$-0.58 - 0.56i$	$0.87 + 1.35i$	$-1.95 - 1.67i$	$0.13 + 7.21i$	$-52.88 + 2.56i$	$2788.5 - 269.6i$
$-1.1 + 0.1i$	$0.1 - 0.12i$	$-1.10 + 0.76i$	$0.11 - 0.068i$	$-1.09 + 0.085i$	$0.085 - 0.85i$	$-1.10 + 0.086i$
$0 - 1.3i$	$-1.69 - 1.3i$	$1.17 + 3.09i$	$-8.21 + 5.92i$	$32.46 - 98.47i$	$-8643.6 - 6393.7i$	$33833744 + 110529134.4i$
$1 + 1i$	$1 + 3i$	$-7 + 7i$	$1 - 97i$	$-9407 - 193i$	$88454401 + 3631103i$	$7.8 \times 10^{15} + 6.4 \times 10^{14}i$

(b) z_1 and z_4 are in the Mandlebrot set. a_6 for the complex numbers not in the set have very large components.

(c)

z	$|z|$	$|a_6|$
$0.1 - 0.4i$	0.4	0.4
$0.5 + 0.8i$	0.9	444884
$-0.9 + 0.7i$	1.1	2802
$-1.1 + 0.1i$	1.1	1.1
$0 - 1.3i$	1.3	115591573
$1 + 1i$	1.4	7.8×10^{15}

The numbers that are in the Mandlebrot set satisfy the condition $|a_n| \leq 2$.

5.4 Assess Your Understanding *(page 364)*

1. unit **2.** scalar **3.** horizontal; vertical **4.** T **5.** T **6.** F

7.
9. $3\mathbf{v}$

11.
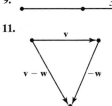

13.
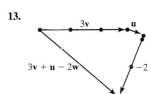

15. T **17.** F **19.** F **21.** T **23.** 12 **25.** $\mathbf{v} = 3\mathbf{i} + 4\mathbf{j}$
27. $\mathbf{v} = 2\mathbf{i} + 4\mathbf{j}$ **29.** $\mathbf{v} = 8\mathbf{i} - \mathbf{j}$ **31.** $\mathbf{v} = -\mathbf{i} + \mathbf{j}$
33. 5 **35.** $\sqrt{2}$ **37.** $\sqrt{13}$ **39.** $-\mathbf{j}$ **41.** $\sqrt{89}$
43. $\sqrt{34} - \sqrt{13}$ **45.** $\mathbf{i}$ **47.** $\dfrac{3}{5}\mathbf{i} - \dfrac{4}{5}\mathbf{j}$ **49.** $\dfrac{\sqrt{2}}{2}\mathbf{i} - \dfrac{\sqrt{2}}{2}\mathbf{j}$
51. $\mathbf{v} = \dfrac{8\sqrt{5}}{5}\mathbf{i} + \dfrac{4\sqrt{5}}{5}\mathbf{j}$ or $\mathbf{v} = -\dfrac{8\sqrt{5}}{5}\mathbf{i} - \dfrac{4\sqrt{5}}{5}\mathbf{j}$

53. $\left\{-2 + \sqrt{21}, -2 - \sqrt{21}\right\}$ **55.** $\mathbf{v} = \dfrac{5}{2}\mathbf{i} + \dfrac{5\sqrt{3}}{2}\mathbf{j}$ **57.** $\mathbf{v} = -7\mathbf{i} + 7\sqrt{3}\mathbf{j}$ **59.** $\mathbf{v} = \dfrac{25\sqrt{3}}{2}\mathbf{i} - \dfrac{25}{2}\mathbf{j}$

61. (a) $(-1, 4)$ **(b)**

63. $\mathbf{F} = 20\sqrt{3}\mathbf{i} + 20\mathbf{j}$

65. $\mathbf{F} = \left(20\sqrt{3} + 30\sqrt{2}\right)\mathbf{i} + \left(20 - 30\sqrt{2}\right)\mathbf{j}$

67. Tension in right cable: 1000 lb; tension in left cable: 845.2 lb

69. Tension in right part: 1088.4 lb; tension in left part: 1089.1 lb

71. The truck must pull with a force of 4635. 2 lb.

73.
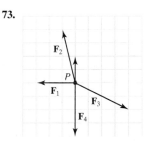

Historical Problem *(page 372)*

$(a\mathbf{i} + b\mathbf{j}) \cdot (c\mathbf{i} + d\mathbf{j}) = ac + bd$
Real part $[(\overline{a + bi})(c + di)] = $ real part$[(a - bi)(c + di)] = $ real part$[ac + adi - bci - bdi^2] = ac + bd$

5.5 Assess Your Understanding *(page 372)*

2. orthogonal **3.** parallel **4.** F **5.** T **6.** F **7. (a)** 0 **(b)** 90° **(c)** orthogonal **9. (a)** 0 **(b)** 90° **(c)** orthogonal **11. (a)** $\sqrt{3} - 1$
(b) 75° **(c)** neither **13. (a)** -50 **(b)** 180° **(c)** parallel **15. (a)** 0 **(b)** 90° **(c)** orthogonal **17.** $\dfrac{2}{3}$ **19.** $\mathbf{v}_1 = \dfrac{5}{2}\mathbf{i} - \dfrac{5}{2}\mathbf{j}, \mathbf{v}_2 = -\dfrac{1}{2}\mathbf{i} - \dfrac{1}{2}\mathbf{j}$
21. $\mathbf{v}_1 = -\dfrac{1}{5}\mathbf{i} - \dfrac{2}{5}\mathbf{j}, \mathbf{v}_2 = \dfrac{6}{5}\mathbf{i} - \dfrac{3}{5}\mathbf{j}$ **23.** $\mathbf{v}_1 = \dfrac{14}{5}\mathbf{i} + \dfrac{7}{5}\mathbf{j}, \mathbf{v}_2 = \dfrac{1}{5}\mathbf{i} - \dfrac{2}{5}\mathbf{j}$ **25.** 9 ft-lb **27. (a)** $\|\mathbf{I}\| = 0.022$; the intensity of the sun's rays is

approximately 0.022 W/cm². $\|\mathbf{A}\| = 500$; the area of the solar panel is 500 cm². **(b)** $W = 10$; ten watts of energy is collected. **(c)** Vectors $\mathbf{I}$ and $\mathbf{A}$ should be parallel with the solar panels facing the sun. **29.** 496.7 mi/h; 38.5° west of south **31.** 8.6° off direct heading across the current, upstream; 1.52 min
33. Force required to keep Sienna from rolling down the hill: 737.6 lb; force perpendicular to the hill: 5248.4 lb
35. $\mathbf{v} = \left(250\sqrt{2} - 30\right)\mathbf{i} + \left(250\sqrt{2} + 30\sqrt{3}\right)\mathbf{j}$; 518.8 km/h; N38.6°E **37.** Timmy must exert 85.5 lb. **39.** 60° **41.** Let $\mathbf{v} = a\mathbf{i} + b\mathbf{j}$. Then
$\mathbf{0} \cdot \mathbf{v} = 0a + 0b = 0$. **43.** $\mathbf{v} = \cos\alpha\mathbf{i} + \sin\alpha\mathbf{j}, 0 \leq \alpha \leq \pi$; $\mathbf{w} = \cos\beta\mathbf{i} + \sin\beta\mathbf{j}, 0 \leq \beta \leq \pi$. If θ is the angle between $\mathbf{v}$ and $\mathbf{w}$, then $\mathbf{v} \cdot \mathbf{w} = \cos\theta$,
since $\|\mathbf{v}\| = 1$ and $\|\mathbf{w}\| = 1$. Now $\theta = \alpha - \beta$ or $\theta = \beta - \alpha$. Since the cosine function is even, $\mathbf{v} \cdot \mathbf{w} = \cos(\alpha - \beta)$.
Also, $\mathbf{v} \cdot \mathbf{w} = \cos\alpha\cos\beta + \sin\alpha\sin\beta$. So $\cos(\alpha - \beta) = \cos\alpha\cos\beta + \sin\alpha\sin\beta$.

45. (a) If $\mathbf{u} = a_1\mathbf{i} + b_1\mathbf{j}$ and $\mathbf{v} = a_2\mathbf{i} + b_2\mathbf{j}$, then, since $\|\mathbf{u}\| = \|\mathbf{v}\|$, $a_1^2 + b_1^2 = \|\mathbf{u}\|^2 = \|\mathbf{v}\|^2 = a_2^2 + b_2^2$, and

$(\mathbf{u} + \mathbf{v})\cdot(\mathbf{u} - \mathbf{v}) = (a_1 + a_2)(a_1 - a_2) + (b_1 + b_2)(b_1 - b_2) = (a_1^2 + b_1^2) - (a_2^2 + b_2^2) = 0$.

(b) The legs of the angle can be made to correspond to vectors $\mathbf{u} + \mathbf{v}$ and $\mathbf{u} - \mathbf{v}$.

47. $(\|\mathbf{w}\|\mathbf{v} + \|\mathbf{v}\|\mathbf{w})\cdot(\|\mathbf{w}\|\mathbf{v} - \|\mathbf{v}\|\mathbf{w}) = \|\mathbf{w}\|^2\mathbf{v}\cdot\mathbf{v} - \|\mathbf{w}\|\|\mathbf{v}\|\mathbf{v}\cdot\mathbf{w} + \|\mathbf{v}\|\|\mathbf{w}\|\mathbf{w}\cdot\mathbf{v} - \|\mathbf{v}\|^2\mathbf{w}\cdot\mathbf{w} = \|\mathbf{w}\|^2\mathbf{v}\cdot\mathbf{v} - \|\mathbf{v}\|^2\mathbf{w}\cdot\mathbf{w} = \|\mathbf{w}\|^2\|\mathbf{v}\|^2 - \|\mathbf{v}\|^2\|\mathbf{w}\|^2 = 0$

49. $\|\mathbf{u} + \mathbf{v}\|^2 - \|\mathbf{u} - \mathbf{v}\|^2 = (\mathbf{u} + \mathbf{v})\cdot(\mathbf{u} + \mathbf{v}) - (\mathbf{u} - \mathbf{v})\cdot(\mathbf{u} - \mathbf{v}) = (\mathbf{u}\cdot\mathbf{u} + \mathbf{u}\cdot\mathbf{v} + \mathbf{v}\cdot\mathbf{u} + \mathbf{v}\cdot\mathbf{v}) - (\mathbf{u}\cdot\mathbf{u} - \mathbf{u}\cdot\mathbf{v} - \mathbf{v}\cdot\mathbf{u} + \mathbf{v}\cdot\mathbf{v})$

$= 2(\mathbf{u}\cdot\mathbf{v}) + 2(\mathbf{v}\cdot\mathbf{u}) = 4(\mathbf{u}\cdot\mathbf{v})$

5.6 Assess Your Understanding (page 382)

2. xy-plane **3.** components **4.** 1 **5.** F **6.** T **7.** All points of the form $(x, 0, z)$ **9.** All points of the form $(x, y, 2)$ **11.** All points of the form $(-4, y, z)$ **13.** All points of the form $(1, 2, z)$ **15.** $\sqrt{21}$ **17.** $\sqrt{33}$ **19.** $\sqrt{26}$ **21.** $(2, 0, 0); (2, 1, 0); (0, 1, 0); (2, 0, 3); (0, 1, 3); (0, 0, 3)$ **23.** $(1, 4, 3); (3, 2, 3); (3, 4, 3); (3, 2, 5); (1, 4, 5); (1, 2, 5)$ **25.** $(-1, 2, 2); (4, 0, 2); (4, 2, 2); (-1, 2, 5); (4, 0, 5); (-1, 0, 5)$ **27.** $\mathbf{v} = 3\mathbf{i} + 4\mathbf{j} - \mathbf{k}$ **29.** $\mathbf{v} = 2\mathbf{i} + 4\mathbf{j} + \mathbf{k}$ **31.** $\mathbf{v} = 8\mathbf{i} - \mathbf{j}$ **33.** 7 **35.** $\sqrt{3}$ **37.** $\sqrt{22}$ **39.** $-\mathbf{j} - 2\mathbf{k}$ **41.** $\sqrt{105}$ **43.** $\sqrt{38} - \sqrt{17}$ **45.** $\mathbf{i}$ **47.** $\frac{3}{7}\mathbf{i} - \frac{6}{7}\mathbf{j} - \frac{2}{7}\mathbf{k}$

49. $\frac{\sqrt{3}}{3}\mathbf{i} + \frac{\sqrt{3}}{3}\mathbf{j} + \frac{\sqrt{3}}{3}\mathbf{k}$ **51.** $\mathbf{v}\cdot\mathbf{w} = 0; \theta = 90°$ **53.** $\mathbf{v}\cdot\mathbf{w} = -2, \theta \approx 100.3°$ **55.** $\mathbf{v}\cdot\mathbf{w} = 0; \theta = 90°$ **57.** $\mathbf{v}\cdot\mathbf{w} = 52; \theta = 0°$

59. $\alpha \approx 64.6°; \beta \approx 149.0°; \gamma \approx 106.6°; \mathbf{v} = 7(\cos 64.6°\mathbf{i} + \cos 149.0°\mathbf{j} + \cos 106.6°\mathbf{k})$

61. $\alpha = \beta = \gamma \approx 54.7°; \mathbf{v} = \sqrt{3}(\cos 54.7°\mathbf{i} + \cos 54.7°\mathbf{j} + \cos 54.7°\mathbf{k})$ **63.** $\alpha = \beta = 45°; \gamma = 90°; \mathbf{v} = \sqrt{2}(\cos 45°\mathbf{i} + \cos 45°\mathbf{j} + \cos 90°\mathbf{k})$

65. $\alpha \approx 60.9°; \beta \approx 144.2°; \gamma \approx 71.1°; \mathbf{v} = \sqrt{38}(\cos 60.9°\mathbf{i} + \cos 144.2°\mathbf{j} + \cos 71.1°\mathbf{k})$ **67. (a)** $\mathbf{d} = \mathbf{a} + \mathbf{b} + \mathbf{c} = \langle 7, 1, 5 \rangle$ **(b)** 8.66 ft

69. $(x - 3)^2 + (y - 1)^2 + (z - 1)^2 = 1$ **71.** Radius = 2, center $(-1, 1, 0)$ **73.** Radius = 3, center $(2, -2, -1)$

75. Radius = $\frac{3\sqrt{2}}{2}$, center $(2, 0, -1)$ **77.** 2 newton-meters = 2 joules **79.** 9 newton-meters = 9 joules

5.7 Assess Your Understanding (page 388)

1. T **2.** T **3.** T **4.** F **5.** F **6.** T **7.** 2 **9.** 4 **11.** $-11A + 2B + 5C$ **13.** $-6A + 23B - 15C$ **15. (a)** $5\mathbf{i} + 5\mathbf{j} + 5\mathbf{k}$ **(b)** $-5\mathbf{i} - 5\mathbf{j} - 5\mathbf{k}$
(c) 0 **(d)** 0 **17. (a)** $\mathbf{i} - \mathbf{j} - \mathbf{k}$ **(b)** $-\mathbf{i} + \mathbf{j} + \mathbf{k}$ **(c)** 0 **(d)** 0 **19. (a)** $-\mathbf{i} + 2\mathbf{j} + 2\mathbf{k}$ **(b)** $\mathbf{i} - 2\mathbf{j} - 2\mathbf{k}$ **(c)** 0 **(d)** 0 **21. (a)** $3\mathbf{i} - \mathbf{j} + 4\mathbf{k}$
(b) $-3\mathbf{i} + \mathbf{j} - 4\mathbf{k}$ **(c)** 0 **(d)** 0 **23.** $-9\mathbf{i} - 7\mathbf{j} - 3\mathbf{k}$ **25.** $9\mathbf{i} + 7\mathbf{j} + 3\mathbf{k}$ **27.** 0 **29.** $-27\mathbf{i} - 21\mathbf{j} - 9\mathbf{k}$ **31.** $-18\mathbf{i} - 14\mathbf{j} - 6\mathbf{k}$ **33.** 0
35. -25 **37.** 25 **39.** 0 **41.** Any vector of the form $c(-9\mathbf{i} - 7\mathbf{j} - 3\mathbf{k})$, where c is a nonzero scalar **43.** Any vector of the form $c(-\mathbf{i} + \mathbf{j} + 5\mathbf{k})$,

where c is a nonzero scalar **45.** $\sqrt{166}$ **47.** $\sqrt{555}$ **49.** $\sqrt{34}$ **51.** $\sqrt{998}$ **53.** $\frac{11\sqrt{19}}{57}\mathbf{i} + \frac{\sqrt{19}}{57}\mathbf{j} + \frac{7\sqrt{19}}{57}\mathbf{k}$ or $-\frac{11\sqrt{19}}{57}\mathbf{i} - \frac{\sqrt{19}}{57}\mathbf{j} - \frac{7\sqrt{19}}{57}\mathbf{k}$
55. 98 cubic units

57. $\mathbf{u} \times \mathbf{v} = \begin{vmatrix} \mathbf{i} & \mathbf{j} & \mathbf{k} \\ a_1 & b_1 & c_1 \\ a_2 & b_2 & c_2 \end{vmatrix} = (b_1c_2 - b_2c_1)\mathbf{i} - (a_1c_2 - a_2c_1)\mathbf{j} + (a_1b_2 - a_2b_1)\mathbf{k}$

$\|\mathbf{u} \times \mathbf{v}\|^2 = \left(\sqrt{(b_1c_2 - b_2c_1)^2 + (a_1c_2 - a_2c_1)^2 + (a_1b_2 - a_2b_1)^2}\right)^2$

$= b_1^2c_2^2 - 2b_1b_2c_1c_2 + b_2^2c_1^2 + a_1^2c_2^2 - 2a_1a_2c_1c_2 + a_2^2c_1^2 + a_1^2b_2^2 - 2a_1a_2b_1b_2 + a_2^2b_1^2$

$\|\mathbf{u}\|^2 = a_1^2 + b_1^2 + c_1^2, \|\mathbf{v}\|^2 = a_2^2 + b_2^2 + c_2^2$

$\|\mathbf{u}\|^2\|\mathbf{v}\|^2 = (a_1^2 + b_1^2 + c_1^2)(a_2^2 + b_2^2 + c_2^2) = a_1^2a_2^2 + a_1^2b_2^2 + a_1^2c_2^2 + b_1^2a_2^2 + b_1^2b_2^2 + b_1^2c_2^2 + a_2^2c_1^2 + b_2^2c_1^2 + c_1^2c_2^2$

$(\mathbf{u}\cdot\mathbf{v})^2 = (a_1a_2 + b_1b_2 + c_1c_2)^2 = (a_1a_2 + b_1b_2 + c_1c_2)(a_1a_2 + b_1b_2 + c_1c_2)$

$= a_1^2a_2^2 + a_1a_2b_1b_2 + a_1a_2c_1c_2 + b_1b_2c_1c_2 + b_1b_2a_1a_2 + b_1^2b_2^2 + b_1b_2c_1c_2 + a_1a_2c_1c_2 + c_1^2c_2^2$

$= a_1^2a_2^2 + b_1^2b_2^2 + c_1^2c_2^2 + 2a_1a_2b_1b_2 + 2b_1b_2c_1c_2 + 2a_1a_2c_1c_2$

$\|\mathbf{u}\|^2\|\mathbf{v}\|^2 - (\mathbf{u}\cdot\mathbf{v})^2 = a_1^2b_2^2 + a_1^2c_2^2 + b_1^2a_2^2 + a_2^2c_1^2 + b_2^2c_1^2 + b_1^2c_2^2 - 2a_1a_2b_1b_2 - 2b_1b_2c_1c_2 - 2a_1a_2c_1c_2$, which equals $\|\mathbf{u} \times \mathbf{v}\|^2$.

59. By Problem 58, since $\mathbf{u}$ and $\mathbf{v}$ are orthogonal, $\|\mathbf{u} \times \mathbf{v}\| = \|\mathbf{u}\|\|\mathbf{v}\|$. If, in addition, $\mathbf{u}$ and $\mathbf{v}$ are unit vectors, $\|\mathbf{u} \times \mathbf{v}\| = 1\cdot 1 = 1$.

61. Assume that $\mathbf{u} = a\mathbf{i} + b\mathbf{j} + c\mathbf{k}, \mathbf{v} = d\mathbf{i} + e\mathbf{j} + f\mathbf{k}$, and $\mathbf{w} = l\mathbf{i} + m\mathbf{j} + n\mathbf{k}$. Then $\mathbf{u} \times \mathbf{v} = (bf - ec)\mathbf{i} - (af - dc)\mathbf{j} + (ae - db)\mathbf{k}$,

$\mathbf{u} \times \mathbf{w} = (bn - mc)\mathbf{i} - (an - lc)\mathbf{j} + (am - lb)\mathbf{k}$, and $\mathbf{v} + \mathbf{w} = (d + l)\mathbf{i} + (e + m)\mathbf{j} + (f + n)\mathbf{k}$. Therefore,

$(\mathbf{u} \times \mathbf{v}) + (\mathbf{u} \times \mathbf{w}) = (bf - ec + bn - mc)\mathbf{i} - (af - dc + an - lc)\mathbf{j} + (ae - db + am - lb)\mathbf{k}$ and

$\mathbf{u} \times (\mathbf{v} + \mathbf{w}) = [b(f + n) - (e + m)c]\mathbf{i} - [a(f + n) - (d + l)c]\mathbf{j} + [a(e + m) - (d + l)b]\mathbf{k}$

$= (bf - ec + bn - mc)\mathbf{i} - (af - dc + an - lc)\mathbf{j} + (ae - db + am - lb)\mathbf{k}$, which equals $(\mathbf{u} \times \mathbf{v}) + (\mathbf{u} \times \mathbf{w})$.

Review Exercises (page 390)

1. $\left(\frac{3\sqrt{3}}{2}, \frac{3}{2}\right)$ **3.** $\left(1, \sqrt{3}\right)$ **5.** $(0, 3)$ **7.** $\left(3\sqrt{2}, \frac{3\pi}{4}\right), \left(-3\sqrt{2}, -\frac{\pi}{4}\right)$

9. $\left(2, -\frac{\pi}{2}\right), \left(-2, \frac{\pi}{2}\right)$

11. $(5, 0.93), (-5, 4.07)$

13. (a) $x^2 + (y - 1)^2 = 1$ (b) circle, radius 1, center $(0, 1)$ in rectangular coordinates

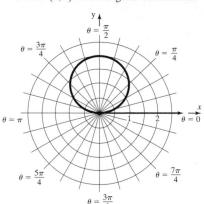

15. (a) $x^2 + y^2 = 25$ (b) circle, radius 5, center at pole

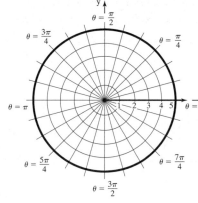

17. (a) $x + 3y = 6$ (b) line through $(6, 0)$ and $(0, 2)$ in rectangular coordinates

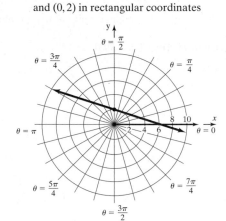

19. Circle; radius 2, center at $(2, 0)$ in rectangular coordinates; symmetric with respect to the polar axis

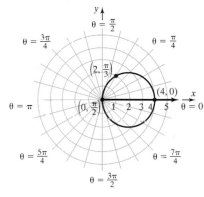

21. Cardioid; symmetric with respect to the line $\theta = \dfrac{\pi}{2}$

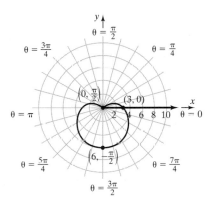

23. Limaçon without inner loop; symmetric with respect to the polar axis

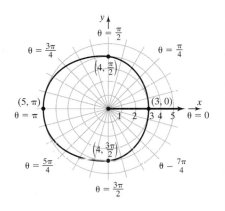

25. $\sqrt{2}(\cos 225° + i \sin 225°)$ **27.** $5(\cos 323.1° + i \sin 323.1°)$

29. $-\sqrt{3} + i$

31. $-\dfrac{3}{2} + \dfrac{3\sqrt{3}}{2}i$

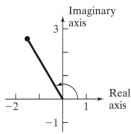

33. $0.10 - 0.02i$

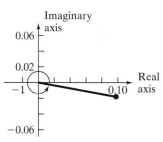

35. $zw = \cos 130° + i \sin 130°$; $\dfrac{z}{w} = \cos 30° + i \sin 30°$ **37.** $zw = 6(\cos 0 + i \sin 0) = 6$; $\dfrac{z}{w} = \dfrac{3}{2}\left(\cos \dfrac{8\pi}{5} + i \sin \dfrac{8\pi}{5}\right)$

39. $zw = 5(\cos 5° + i \sin 5°)$; $\dfrac{z}{w} = 5(\cos 15° + i \sin 15°)$ **41.** $\dfrac{27}{2} + \dfrac{27\sqrt{3}}{2}i$ **43.** $4i$ **45.** 64 **47.** $-527 - 336i$

49. $3, 3(\cos 120° + i \sin 120°), 3(\cos 240° + i \sin 240°)$ or $3, -\dfrac{3}{2} + \dfrac{3\sqrt{3}}{2}i, -\dfrac{3}{2} - \dfrac{3\sqrt{3}}{2}i$

51.

53.

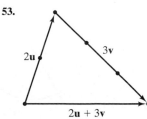

55. $\mathbf{v} = 2\mathbf{i} - 4\mathbf{j}$; $\|\mathbf{v}\| = 2\sqrt{5}$ **57.** $\mathbf{v} = -\mathbf{i} + 3\mathbf{j}$; $\|\mathbf{v}\| = \sqrt{10}$ **59.** $2\mathbf{i} - 2\mathbf{j}$ **61.** $-20\mathbf{i} + 13\mathbf{j}$

63. $\sqrt{5}$ **65.** $\sqrt{5} + 5 \approx 7.24$ **67.** $-\dfrac{2\sqrt{5}}{5}\mathbf{i} + \dfrac{\sqrt{5}}{5}\mathbf{j}$ **69.** $\mathbf{v} = \dfrac{3}{2}\mathbf{i} + \dfrac{3\sqrt{3}}{2}\mathbf{j}$ **71.** $\sqrt{43} \approx 6.56$

73. $\mathbf{v} = 3\mathbf{i} - 5\mathbf{j} + 3\mathbf{k}$ **75.** $21\mathbf{i} - 2\mathbf{j} - 5\mathbf{k}$ **77.** $\sqrt{38}$ **79.** 0 **81.** $3\mathbf{i} + 9\mathbf{j} + 9\mathbf{k}$

83. $\dfrac{3\sqrt{14}}{14}\mathbf{i} + \dfrac{\sqrt{14}}{14}\mathbf{j} - \dfrac{\sqrt{14}}{7}\mathbf{k}$; $-\dfrac{3\sqrt{14}}{14}\mathbf{i} - \dfrac{\sqrt{14}}{14}\mathbf{j} + \dfrac{\sqrt{14}}{7}\mathbf{k}$ **85.** $\mathbf{v} \cdot \mathbf{w} = -11$; $\theta \approx 169.7°$

87. $\mathbf{v} \cdot \mathbf{w} = -4$; $\theta \approx 153.4°$ **89.** $\mathbf{v} \cdot \mathbf{w} = 1$; $\theta \approx 70.5°$ **91.** $\mathbf{v} \cdot \mathbf{w} = 0$; $\theta = 90°$ **93.** Parallel

95. Parallel   **97.** Orthogonal

99. $\mathbf{v}_1 = \frac{4}{5}\mathbf{i} - \frac{3}{5}\mathbf{j}; \mathbf{v}_2 = \frac{6}{5}\mathbf{i} + \frac{8}{5}\mathbf{j}$ **101.** $\mathbf{v}_1 = \frac{9}{10}(3\mathbf{i} + \mathbf{j}); \mathbf{v}_2 = \frac{-7}{10}\mathbf{i} + \frac{21}{10}\mathbf{j}$ **103.** $\alpha \approx 56.1°; \beta \approx 138°; \gamma \approx 68.2°$ **105.** $2\sqrt{83}$ **107.** $-2\mathbf{i} + 3\mathbf{j} - \mathbf{k}$

109. $\sqrt{29} \approx 5.39$ mi/hr; 0.4 mi **111.** Left cable: 1843.21 lb; right cable: 1630.41 lb **113.** 50 foot-pounds

Chapter Test (page 393)

1–3.

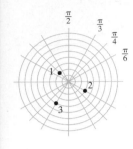

4. $\left(4, \frac{\pi}{3}\right)$

5. $x^2 + y^2 = 49$

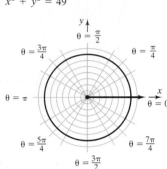

6. $\frac{y}{x} = 3$ or $y = 3x$

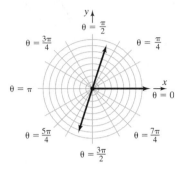

7. $8y = x^2$ or $4(2)y = x^2$

The graph is a parabola with vertex $(0, 0)$ and focus $(0, 2)$.

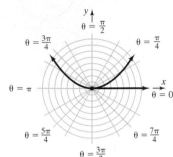

8. $r^2 \cos\theta = 5$ is symmetric about the pole, the polar axis, and the line $\theta = \frac{\pi}{2}$. **9.** $r = 5\sin\theta\cos^2\theta$ is symmetric about the line $\theta = \frac{\pi}{2}$. The tests for

symmetry about the pole and the polar axis fail, so the graph of $r = 5\sin\theta\cos^2\theta$ may or may not be symmetric about the pole or polar axis.

10. $z \cdot w = 6(\cos 107° + i\sin 107°)$ **11.** $\frac{w}{z} = \frac{3}{2}(\cos 297° + i\sin 297°)$ **12.** $w^5 = 243(\cos 110° + i\sin 110°)$

13. $z_0 = 2\sqrt[3]{2}(\cos 40° + i\sin 40°)$, $z_1 = 2\sqrt[3]{2}(\cos 160° + i\sin 160°)$, $z_2 = 2\sqrt[3]{2}(\cos 280° + i\sin 280°)$

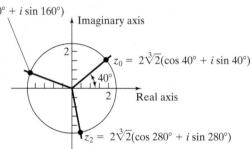

14. $\mathbf{v} = \langle 5\sqrt{2}, -5\sqrt{2}\rangle$ **15.** $\|\mathbf{v}\| = 10$

16. $\mathbf{u} = \frac{\mathbf{v}}{\|\mathbf{v}\|} = \left\langle \frac{\sqrt{2}}{2}, -\frac{\sqrt{2}}{2}\right\rangle$ **17.** 315° off the positive x-axis

18. $\mathbf{v} = 5\sqrt{2}\mathbf{i} - 5\sqrt{2}\mathbf{j}$ **19.** $\mathbf{v}_1 + 2\mathbf{v}_2 - \mathbf{v}_3 = \langle 6, -10\rangle$

20. Vectors $\mathbf{v}_1$ and $\mathbf{v}_4$ are parallel.

21. Vectors $\mathbf{v}_2$ and $\mathbf{v}_3$ are orthogonal.

22. 172.87° **23.** $-9\mathbf{i} - 5\mathbf{j} + 3\mathbf{k}$

24. $\alpha \approx 57.7°, \beta \approx 143.3°, \gamma \approx 74.5°$ **25.** $\sqrt{115}$

26. The cable must be able to endure a tension of approximately 670.82 lb.

Cumulative Review (page 393)

1. $y = \frac{\sqrt{3}}{3}x$ **2.** $x^2 + (y - 1)^2 = 9$ **3.** Symmetry with respect to the y-axis **4.**

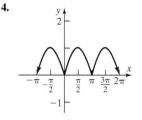

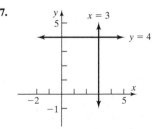

5.

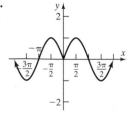

6. $-\frac{\pi}{6}$ **7.**

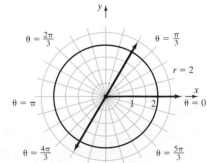

8.

9. Amplitude: 4; period: 2

CHAPTER 6 Analytic Geometry

6.2 Assess Your Understanding *(page 404)*

6. parabola **7.** paraboloid of revolution **8.** T **9.** T **10.** T **11.** *B* **13.** *E* **15.** *H* **17.** *C*

19. $y^2 = 16x$

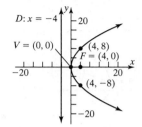

21. $x^2 = -12y$

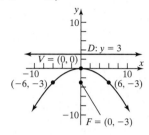

23. $y^2 = -8x$

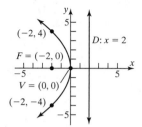

25. $x^2 = 2y$

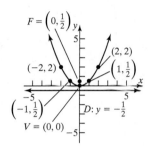

27. $x^2 = \dfrac{4}{3}y$

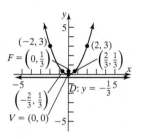

29. $(x - 2)^2 = -8(y + 3)$

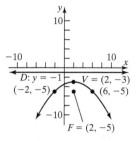

31. $(y + 2)^2 = 4(x + 1)$

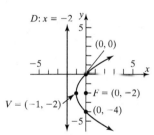

33. $(x + 3)^2 = 4(y - 3)$

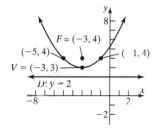
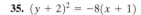

35. $(y + 2)^2 = -8(x + 1)$

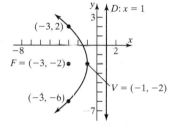

37. Vertex: $(0, 0)$; focus: $(0, 1)$;
   directrix: $y = -1$

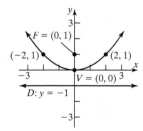

39. Vertex: $(0, 0)$; focus: $(-4, 0)$;
   directrix: $x = 4$

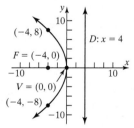

41. Vertex: $(-1, 2)$; focus: $(1, 2)$;
   directrix: $x = -3$

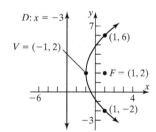

43. Vertex: $(3, -1)$; focus: $\left(3, -\dfrac{5}{4}\right)$;
   directrix: $y = -\dfrac{3}{4}$

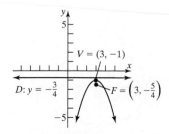

45. Vertex: $(2, -3)$; focus: $(4, -3)$;
   directrix: $x = 0$

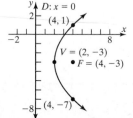

47. Vertex: $(0, 2)$; focus: $(-1, 2)$;
   directrix: $x = 1$

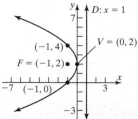

49. Vertex: $(-4, -2)$; focus: $(-4, -1)$;

   directrix: $y = -3$

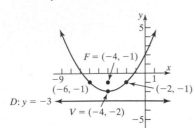

51. Vertex: $(-1, -1)$; focus: $\left(-\frac{3}{4}, -1\right)$;

   directrix: $x = -\frac{5}{4}$

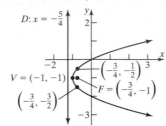

53. Vertex: $(2, -8)$; focus: $\left(2, -\frac{31}{4}\right)$;

   directrix: $y = -\frac{33}{4}$

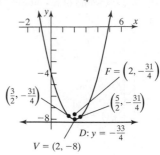

55. $(y - 1)^2 = x$ **57.** $(y - 1)^2 = -(x - 2)$ **59.** $x^2 = 4(y - 1)$ **61.** $y^2 = \frac{1}{2}(x + 2)$ **63.** 1.5625 ft from the base of the dish, along

the axis of symmetry   **65.** 1 in. from the vertex, along the axis of symmetry   **67.** 20 ft   **69.** 0.78125 ft   **71.** 4.17 ft from the base, along the axis of

symmetry   **73.** 24.31 ft, 18.75 ft, 7.64 ft   **75. (a)** $y = -\frac{625}{(299)^2}x^2 + 625$ **(b)** 567 ft: 63.12 ft; 478 ft: 225.67 ft; 308 ft: 459.2 ft   **(c)** No

77. $Cy^2 + Dx = 0, C \neq 0, D \neq 0$ This is the equation of a parabola with vertex at $(0, 0)$ and axis of symmetry the x-axis.

$$Cy^2 = -Dx$$ The focus is $\left(-\frac{D}{4C}, 0\right)$; the directrix is the line $x = \frac{D}{4C}$.

$$y^2 = -\frac{D}{C}x$$ The parabola opens to the right if $-\frac{D}{C} > 0$ and to the left if $-\frac{D}{C} < 0$.

79. $Cy^2 + Dx + Ey + F = 0, C \neq 0$

$$Cy^2 + Ey = -Dx - F$$
$$y^2 + \frac{E}{C}y = -\frac{D}{C}x - \frac{F}{C}$$
$$\left(y + \frac{E}{2C}\right)^2 = -\frac{D}{C}x - \frac{F}{C} + \frac{E^2}{4C^2}$$
$$\left(y + \frac{E}{2C}\right)^2 = -\frac{D}{C}x + \frac{E^2 - 4CF}{4C^2}$$

(a) If $D \neq 0$, then the equation may be written as $\left(y + \frac{E}{2C}\right)^2 = -\frac{D}{C}\left(x - \frac{E^2 - 4CF}{4CD}\right)$.

This is the equation of a parabola with vertex at $\left(\frac{E^2 - 4CF}{4CD}, -\frac{E}{2C}\right)$

and axis of symmetry parallel to the x-axis.

(b)–(d) If $D = 0$, the graph of the equation contains no points if $E^2 - 4CF < 0$, is a single horizontal line if $E^2 - 4CF = 0$, and is two horizontal lines if $E^2 - 4CF > 0$.

6.3 Assess Your Understanding *(page 415)*

7. ellipse   **8.** major   **9.** $(0, -5); (0, 5)$ **10.** F   **11.** T   **12.** T   **13.** C   **15.** B

17. Vertices: $(-5, 0), (5, 0)$

   Foci: $\left(-\sqrt{21}, 0\right), \left(\sqrt{21}, 0\right)$

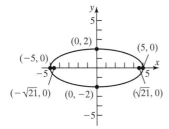

19. Vertices: $(0, -5), (0, 5)$

   Foci: $(0, -4), (0, 4)$

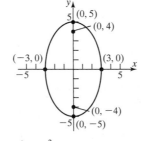

21. $\frac{x^2}{4} + \frac{y^2}{16} = 1$

   Vertices: $(0, -4), (0, 4)$

   Foci: $\left(0, -2\sqrt{3}\right), \left(0, 2\sqrt{3}\right)$

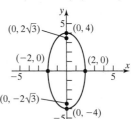

23. $\frac{x^2}{8} + \frac{y^2}{2} = 1$

   Vertices: $\left(-2\sqrt{2}, 0\right), \left(2\sqrt{2}, 0\right)$

   Foci: $\left(-\sqrt{6}, 0\right), \left(\sqrt{6}, 0\right)$

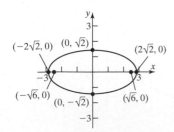

25. $\frac{x^2}{16} + \frac{y^2}{16} = 1$

   Vertices: $(-4, 0), (4, 0), (0, -4), (0, 4)$

   Focus: $(0, 0)$

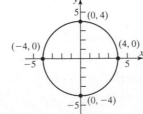

27. $\frac{x^2}{25} + \frac{y^2}{16} = 1$

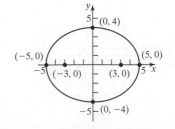

29. $\dfrac{x^2}{9} + \dfrac{y^2}{25} = 1$

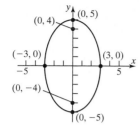

31. $\dfrac{x^2}{9} + \dfrac{y^2}{5} = 1$

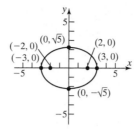

33. $\dfrac{x^2}{25} + \dfrac{y^2}{9} = 1$

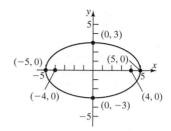

35. $\dfrac{x^2}{4} + \dfrac{y^2}{13} = 1$

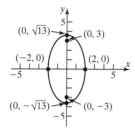

37. $x^2 + \dfrac{y^2}{16} = 1$

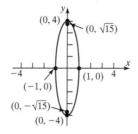

39. $\dfrac{(x+1)^2}{4} + (y-1)^2 = 1$

41. $(x-1)^2 + \dfrac{y^2}{4} = 1$

43. Center: $(3, -1)$; vertices: $(3, -4), (3, 2)$;
foci: $\left(3, -1 - \sqrt{5}\right), \left(3, -1 + \sqrt{5}\right)$

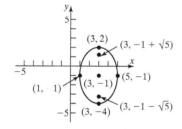

45. $\dfrac{(x+5)^2}{16} + \dfrac{(y-4)^2}{4} = 1$
Center: $(-5, 4)$; vertices: $(-9, 4), (-1, 4)$;
foci: $\left(-5 - 2\sqrt{3}, 4\right), \left(-5 + 2\sqrt{3}, 4\right)$

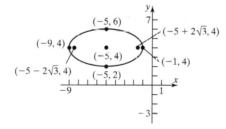

47. $\dfrac{(x+2)^2}{4} + (y-1)^2 = 1$
Center: $(-2, 1)$; vertices: $(-4, 1), (0, 1)$;
foci: $\left(-2 - \sqrt{3}, 1\right), \left(-2 + \sqrt{3}, 1\right)$

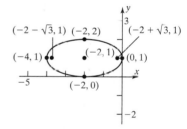

49. $\dfrac{(x-2)^2}{3} + \dfrac{(y+1)^2}{2} = 1$
Center: $(2, -1)$; vertices: $\left(2 - \sqrt{3}, -1\right)$;
$\left(2 + \sqrt{3}, -1\right)$; foci: $(1, -1), (3, -1)$

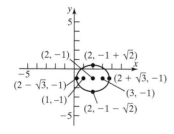

51. $\dfrac{(x-1)^2}{4} + \dfrac{(y+2)^2}{9} = 1$
Center: $(1, -2)$; vertices: $(1, -5), (1, 1)$;
foci: $\left(1, -2 - \sqrt{5}\right), \left(1, -2 + \sqrt{5}\right)$

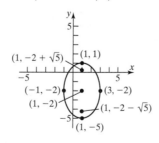

53. $x^2 + \dfrac{(y+2)^2}{4} = 1$
Center: $(0, -2)$; vertices: $(0, -4), (0, 0)$;
foci: $\left(0, -2 - \sqrt{3}\right), \left(0, -2 + \sqrt{3}\right)$

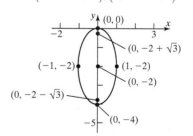

55. $\dfrac{(x-2)^2}{25} + \dfrac{(y+2)^2}{21} = 1$

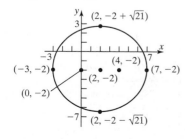

57. $\dfrac{(x-4)^2}{5} + \dfrac{(y-6)^2}{9} = 1$

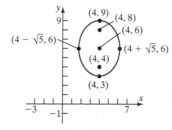

59. $\dfrac{(x-2)^2}{16} + \dfrac{(y-1)^2}{7} = 1$

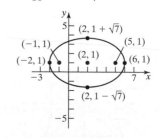

61. $\dfrac{(x-1)^2}{10} + (y-2)^2 = 1$

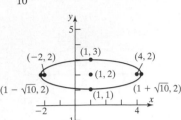

63. $\dfrac{(x-1)^2}{9} + (y-2)^2 = 1$

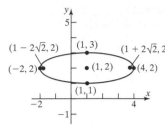

65.

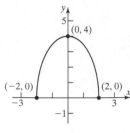

67.

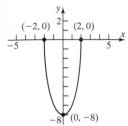

69. $\dfrac{x^2}{100} + \dfrac{y^2}{36} = 1$ **71.** 43.3 ft **73.** 24.65 ft, 21.65 ft, 13.82 ft **75.** 30 ft

77. The elliptical hole will have a major axis of length $2\sqrt{41}$ in. and a minor axis of length 8 in.

79. 91.5 million mi; $\dfrac{x^2}{8649} + \dfrac{y^2}{8646.75} = 1$

81. Perihelion: 460.6 million mi; mean distance: 483.8 million mi; $\dfrac{x^2}{233{,}482} + \dfrac{y^2}{233{,}524.2} = 1$

83. (a) $Ax^2 + Cy^2 + F = 0$ If A and C are of the same sign and F is of opposite sign, then the equation takes the form

$Ax^2 + Cy^2 = -F$ $\dfrac{x^2}{\left(-\dfrac{F}{A}\right)} + \dfrac{y^2}{\left(-\dfrac{F}{C}\right)} = 1$, where $-\dfrac{F}{A}$ and $-\dfrac{F}{C}$ are positive. This is the equation of an ellipse with center at $(0,0)$.

(b) If $A = C$, the equation may be written as $x^2 + y^2 = \dfrac{-F}{A}$. This is the equation of a circle with center at $(0,0)$ and radius equal to $\sqrt{-\dfrac{F}{A}}$.

6.4 Assess Your Understanding *(page 428)*

7. hyperbola **8.** transverse axis **9.** $y = \dfrac{3}{2}x;\ y = -\dfrac{3}{2}x$ **10.** F **11.** T **12.** F **13.** B **15.** A

17. $x^2 - \dfrac{y^2}{8} = 1$

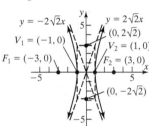

19. $\dfrac{y^2}{16} - \dfrac{x^2}{20} = 1$

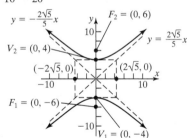

21. $\dfrac{x^2}{9} - \dfrac{y^2}{16} = 1$

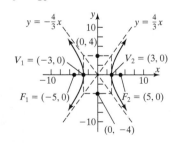

23. $\dfrac{y^2}{36} - \dfrac{x^2}{9} = 1$

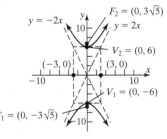

25. $\dfrac{x^2}{8} - \dfrac{y^2}{8} = 1$

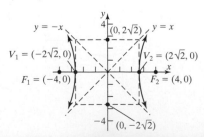

27. $\dfrac{x^2}{25} - \dfrac{y^2}{9} = 1$

Center: $(0,0)$
Transverse axis: x-axis
Vertices: $(-5,0)$, $(5,0)$
Foci: $\left(-\sqrt{34},0\right)$, $\left(\sqrt{34},0\right)$
Asymptotes: $y = \pm\dfrac{3}{5}x$

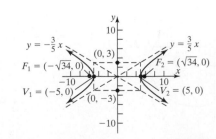

29. $\dfrac{x^2}{4} - \dfrac{y^2}{16} = 1$

Center: $(0, 0)$

Transverse axis: x-axis

Vertices: $(-2, 0), (2, 0)$

Foci: $\left(-2\sqrt{5}, 0\right), \left(2\sqrt{5}, 0\right)$

Asymptotes: $y = \pm 2x$

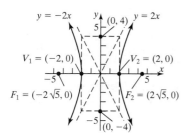

31. $\dfrac{y^2}{9} - x^2 = 1$

Center: $(0, 0)$

Transverse axis: y-axis

Vertices: $(0, -3), (0, 3)$

Foci: $\left(0, -\sqrt{10}\right), \left(0, \sqrt{10}\right)$

Asymptotes: $y = \pm 3x$

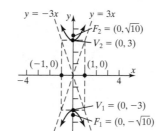

33. $\dfrac{y^2}{25} - \dfrac{x^2}{25} = 1$

Center: $(0, 0)$

Transverse axis: y-axis

Vertices: $(0, -5), (0, 5)$

Foci: $\left(0, -5\sqrt{2}\right), \left(0, 5\sqrt{2}\right)$

Asymptotes: $y = \pm x$

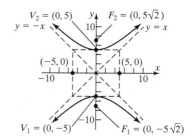

35. $x^2 - y^2 = 1$

37. $\dfrac{y^2}{36} - \dfrac{x^2}{9} = 1$

39. $\dfrac{(x-4)^2}{4} - \dfrac{(y+1)^2}{5} = 1$

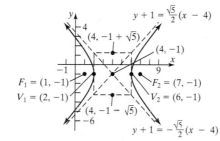

41. $\dfrac{(y+4)^2}{4} - \dfrac{(x+3)^2}{12} = 1$

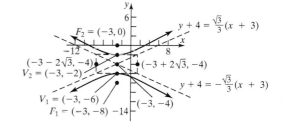

43. $(x-5)^2 - \dfrac{(y-7)^2}{3} = 1$

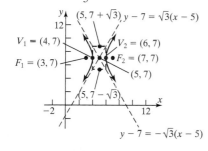

45. $\dfrac{(x-1)^2}{4} - \dfrac{(y+1)^2}{9} = 1$

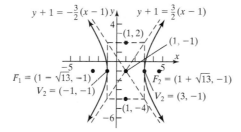

47. $\dfrac{(x-2)^2}{4} - \dfrac{(y+3)^2}{9} = 1$

Center: $(2, -3)$

Transverse axis: parallel to x-axis

Vertices: $(0, -3), (4, -3)$

Foci: $\left(2 - \sqrt{13}, -3\right), \left(2 + \sqrt{13}, -3\right)$

Asymptotes: $y + 3 = \pm \dfrac{3}{2}(x - 2)$

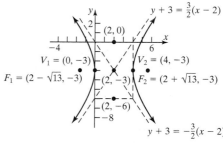

49. $\dfrac{(y-2)^2}{4} - (x+2)^2 = 1$

Center: $(-2, 2)$

Transverse axis: parallel to y-axis

Vertices: $(-2, 0), (-2, 4)$

Foci: $\left(-2, 2 - \sqrt{5}\right), \left(-2, 2 + \sqrt{5}\right)$

Asymptotes: $y - 2 = \pm 2(x + 2)$

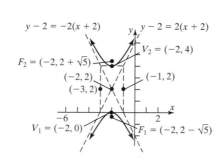

51. $\dfrac{(x+1)^2}{4} - \dfrac{(y+2)^2}{4} = 1$

Center: $(-1, -2)$

Transverse axis: parallel to x-axis

Vertices: $(-3, -2), (1, -2)$

Foci: $\left(-1 - 2\sqrt{2}, -2\right), \left(-1 + 2\sqrt{2}, -2\right)$

Asymptotes: $y + 2 = \pm(x+1)$

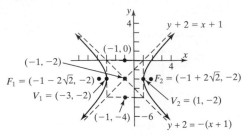

53. $(x-1)^2 - (y+1)^2 = 1$

Center: $(1, -1)$

Transverse axis: parallel to x-axis

Vertices: $(0, -1), (2, -1)$

Foci: $\left(1 - \sqrt{2}, -1\right), \left(1 + \sqrt{2}, -1\right)$

Asymptotes: $y + 1 = \pm(x-1)$

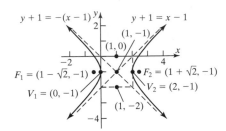

55. $\dfrac{(y-2)^2}{4} - (x+1)^2 = 1$

Center: $(-1, 2)$

Transverse axis: parallel to y-axis

Vertices: $(-1, 0), (-1, 4)$

Foci: $\left(-1, 2 - \sqrt{5}\right), \left(-1, 2 + \sqrt{5}\right)$

Asymptotes: $y - 2 = \pm 2(x+1)$

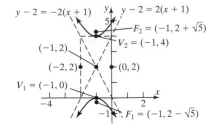

57. $\dfrac{(x-3)^2}{4} - \dfrac{(y+2)^2}{16} = 1$

Center: $(3, -2)$

Transverse axis: parallel to x-axis

Vertices: $(1, -2), (5, -2)$

Foci: $\left(3 - 2\sqrt{5}, -2\right), \left(3 + 2\sqrt{5}, -2\right)$

Asymptotes: $y + 2 = \pm 2(x-3)$

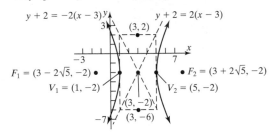

59. $\dfrac{(y-1)^2}{4} - (x+2)^2 = 1$

Center: $(-2, 1)$

Transverse axis: parallel to y-axis

Vertices: $(-2, -1), (-2, 3)$

Foci: $\left(-2, 1 - \sqrt{5}\right), \left(-2, 1 + \sqrt{5}\right)$

Asymptotes: $y - 1 = \pm 2(x+2)$

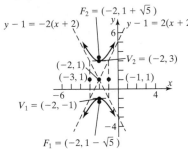

61.

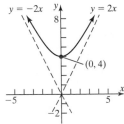

63.

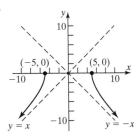

65. Center: $(3, 0)$

Transverse axis: parallel to x-axis

Vertices: $(1, 0), (5, 0)$

Foci: $\left(3 - \sqrt{29}, 0\right), \left(3 + \sqrt{29}, 0\right)$

Asymptotes: $y = \pm \dfrac{5}{2}(x-3)$

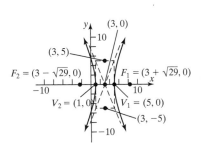

67. Vertex: $(0, 3)$; focus: $(0, 7)$; directrix: $y = -1$

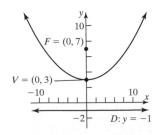

69. $\dfrac{(x-5)^2}{9} + \dfrac{y^2}{25} = 1$

Center: $(5, 0)$; vertices: $(5, 5), (5, -5)$;

foci: $(5, -4), (5, 4)$

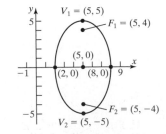

71. $(x - 3)^2 = 8(y + 5)$

Vertex: $(3, -5)$; foci: $(3, -3)$;

directrix: $y = -7$

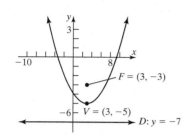

73. The fireworks display is 50,138 ft north of the person at point A.   **75.** The tower is 592.4 ft tall.

77. (a) $y = \pm x$

(b) $\dfrac{x^2}{100} - \dfrac{y^2}{100} = 1, x \geq 0$

81. $\dfrac{x^2}{4} - y^2 = 1$: asymptotes $y = \pm\dfrac{1}{2}x$

$y^2 - \dfrac{x^2}{4} = 1$: asymptotes $y = \pm\dfrac{1}{2}x$

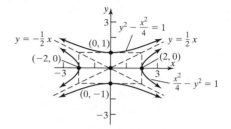

83. $Ax^2 + Cy^2 + F = 0$ If A and C are of opposite sign and $F \neq 0$, this equation may be written as $\dfrac{x^2}{\left(-\dfrac{F}{A}\right)} + \dfrac{y^2}{\left(-\dfrac{F}{C}\right)} = 1$,

$Ax^2 + Cy^2 = -F$ where $-\dfrac{F}{A}$ and $-\dfrac{F}{C}$ are opposite in sign. This is the equation of a hyperbola with center $(0, 0)$.

The transverse axis is the x-axis if $-\dfrac{F}{A} > 0$; the transverse axis is the y-axis if $-\dfrac{F}{A} < 0$.

6.5 Assess Your Understanding *(page 437)*

5. $\cot(2\theta) = \dfrac{A - C}{B}$ **6.** Hyperbola   **7.** Ellipse   **8.** T   **9.** T   **10.** F   **11.** Parabola   **13.** Ellipse   **15.** Hyperbola

17. Hyperbola   **19.** Circle   **21.** $x = \dfrac{\sqrt{2}}{2}(x' - y'), y = \dfrac{\sqrt{2}}{2}(x' + y')$ **23.** $x = \dfrac{\sqrt{2}}{2}(x' - y'), y = \dfrac{\sqrt{2}}{2}(x' + y')$

25. $x = \dfrac{1}{2}(x' - \sqrt{3}y'), y = \dfrac{1}{2}(\sqrt{3}x' + y')$ **27.** $x = \dfrac{\sqrt{5}}{5}(x' - 2y'), y = \dfrac{\sqrt{5}}{5}(2x' + y')$ **29.** $x = \dfrac{\sqrt{13}}{13}(3x' - 2y'), y = \dfrac{\sqrt{13}}{13}(2x' + 3y')$

31. $\theta = 45°$ (see Problem 21)

$x'^2 - \dfrac{y'^2}{3} = 1$

Hyperbola
Center at origin
Transverse axis is the x'-axis.
Vertices at $(\pm 1, 0)$

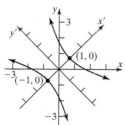

33. $\theta = 45°$ (see Problem 23)

$x'^2 + \dfrac{y'^2}{4} = 1$

Ellipse
Center at $(0, 0)$
Major axis is the y'-axis.
Vertices at $(0, \pm 2)$

35. $\theta = 60°$ (see Problem 25)

$\dfrac{x'^2}{4} + y'^2 = 1$

Ellipse
Center at $(0, 0)$
Major axis is the x'-axis.
Vertices at $(\pm 2, 0)$

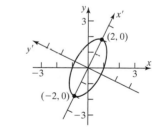

37. $\theta \approx 63°$ (see Problem 27)

$y'^2 = 8x'$

Parabola
Vertex at $(0, 0)$
Focus at $(2, 0)$

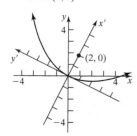

39. $\theta \approx 34°$ (see Problem 29)

$\dfrac{(x' - 2)^2}{4} + y'^2 = 1$

Ellipse
Center at $(2, 0)$
Major axis is the x'-axis.

Vertices at $(4, 0)$ and $(0, 0)$

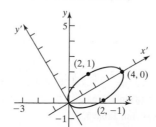

41. $\cot(2\theta) = \dfrac{7}{24}$;

$\theta = \sin^{-1}\left(\dfrac{3}{5}\right) \approx 37°$

$(x' - 1)^2 = -6\left(y' - \dfrac{1}{6}\right)$

Parabola

Vertex at $\left(1, \dfrac{1}{6}\right)$

Focus at $\left(1, -\dfrac{4}{3}\right)$

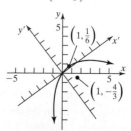

43. Hyperbola **45.** Hyperbola **47.** Parabola **49.** Ellipse **51.** Ellipse

53. Refer to equation (6): $A' = A\cos^2\theta + B\sin\theta\cos\theta + C\sin^2\theta$

$$B' = B(\cos^2\theta - \sin^2\theta) + 2(C - A)(\sin\theta\cos\theta)$$
$$C' = A\sin^2\theta - B\sin\theta\cos\theta + C\cos^2\theta$$
$$D' = D\cos\theta + E\sin\theta$$
$$E' = -D\sin\theta + E\cos\theta$$
$$F' = F$$

55. Use Problem 53 to find $B'^2 - 4A'C'$. After much cancellation, $B'^2 - 4A'C' = B^2 - 4AC$.

57. The distance between P_1 and P_2 in the $x'y'$-plane equals $\sqrt{(x_2' - x_1')^2 + (y_2' - y_1')^2}$.

Assuming that $x' = x\cos\theta - y\sin\theta$ and $y' = x\sin\theta + y\cos\theta$, then

$$(x_2' - x_1')^2 = (x_2\cos\theta - y_2\sin\theta - x_1\cos\theta + y_1\sin\theta)^2$$
$$= \cos^2\theta(x_2 - x_1)^2 - 2\sin\theta\cos\theta(x_2 - x_1)(y_2 - y_1) + \sin^2\theta(y_2 - y_1)^2,\text{ and}$$
$$(y_2' - y_1')^2 = (x_2\sin\theta + y_2\cos\theta - x_1\sin\theta - y_1\cos\theta)^2 = \sin^2\theta(x_2 - x_1)^2 + 2\sin\theta\cos\theta(x_2 - x_1)(y_2 - y_1) + \cos^2\theta(y_2 - y_1)^2.$$
Therefore, $(x_2' - x_1')^2 + (y_2' - y_1')^2 = \cos^2\theta(x_2 - x_1)^2 + \sin^2\theta(x_2 - x_1)^2 + \sin^2\theta(y_2 - y_1)^2 + \cos^2\theta(y_2 - y_1)^2$
$$= (x_2 - x_1)^2(\cos^2\theta + \sin^2\theta) + (y_2 - y_1)^2(\sin^2\theta + \cos^2\theta) = (x_2 - x_1)^2 + (y_2 - y_1)^2.$$

6.6 Assess Your Understanding (page 443)

3. $\frac{1}{2}$; ellipse; parallel; 4; below **4.** 1; <1; >1 **5.** T **6.** T **7.** Parabola; directrix is perpendicular to the polar axis 1 unit to the right of the pole.

9. Hyperbola; directrix is parallel to the polar axis $\frac{4}{3}$ units below the pole.

11. Ellipse; directrix is perpendicular to the polar axis $\frac{3}{2}$ units to the left of the pole.

13. Parabola; directrix is perpendicular to the polar axis 1 unit to the right of the pole; vertex is at $\left(\frac{1}{2}, 0\right)$.

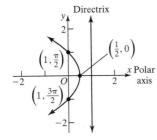

15. Ellipse; directrix is parallel to the polar axis $\frac{8}{3}$ units above the pole; vertices are at $\left(\frac{8}{7}, \frac{\pi}{2}\right)$ and $\left(8, \frac{3\pi}{2}\right)$.

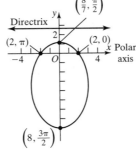

17. Hyperbola; directrix is perpendicular to the polar axis $\frac{3}{2}$ units to the left of the pole; vertices are at $(-3, 0)$ and $(1, \pi)$.

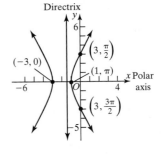

19. Ellipse; directrix is parallel to the polar axis 8 units below the pole; vertices are at $\left(8, \frac{\pi}{2}\right)$ and $\left(\frac{8}{3}, \frac{3\pi}{2}\right)$.

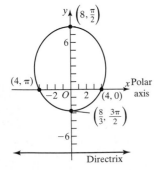

21. Ellipse; directrix is parallel to the polar axis 3 units below the pole; vertices are at $\left(6, \frac{\pi}{2}\right)$ and $\left(\frac{6}{5}, \frac{3\pi}{2}\right)$.

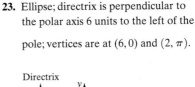

23. Ellipse; directrix is perpendicular to the polar axis 6 units to the left of the pole; vertices are at $(6, 0)$ and $(2, \pi)$.

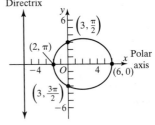

25. $y^2 + 2x - 1 = 0$ **27.** $16x^2 + 7y^2 + 48y - 64 = 0$ **29.** $3x^2 - y^2 + 12x + 9 = 0$ **31.** $4x^2 + 3y^2 - 16y - 64 = 0$

33. $9x^2 + 5y^2 - 24y - 36 = 0$ **35.** $3x^2 + 4y^2 - 12x - 36 = 0$ **37.** $r = \dfrac{1}{1 + \sin\theta}$ **39.** $r = \dfrac{12}{5 - 4\cos\theta}$ **41.** $r = \dfrac{12}{1 - 6\sin\theta}$

43. Use $d(D, P) = p - r\cos\theta$ in the derivation of equation (a) in Table 5.

45. Use $d(D, P) = p + r\sin\theta$ in the derivation of equation (a) in Table 5.

6.7 Assess Your Understanding (page 455)

2. plane curve; parameter **3.** ellipse **4.** cycloid **5.** F **6.** T

7.

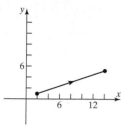

$x - 3y + 1 = 0$

9.

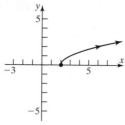

$y = \sqrt{x - 2}$

11.

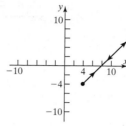

$y = x - 8$

13.

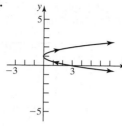

$x = 3(y - 1)^2$

15.

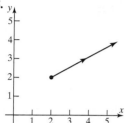

$2y = 2 + x$

17.

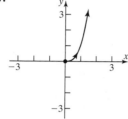

$y = x^3$

19.

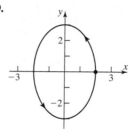

$\dfrac{x^2}{4} + \dfrac{y^2}{9} = 1$

21.

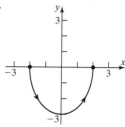

$\dfrac{x^2}{4} + \dfrac{y^2}{9} = 1$

23.

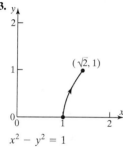

$x^2 - y^2 = 1$

25.

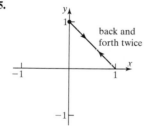

back and forth twice

$x + y = 1$

27. $x = t$ or $x = \dfrac{t + 1}{4}$
  $y = 4t - 1$ $y = t$

29. $x = t$ or $x = t^3$
  $y = t^2 + 1$ $y = t^6 + 1$

31. $x = t$ or $x = \sqrt[3]{t}$
  $y = t^3$ $y = t$

33. $x = t$ or $x = t^3$
  $y = t^{2/3}$ $y = t^2, t \geq 0$

35. $x = t + 2, y = t, 0 \leq t \leq 5$ **37.** $x = 3 \cos t, y = 2 \sin t, 0 \leq t \leq 2\pi$

39. $x = 2 \cos(\pi t), y = -3 \sin(\pi t), 0 \leq t \leq 2$

41. $x = 2 \sin(2\pi t), y = 3 \cos(2\pi t), 0 \leq t \leq 1$

43.

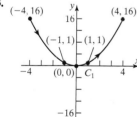

$(-4, 16)$ $(4, 16)$ $(-1, 1)$ $(1, 1)$ $(0, 0)$ C_1

$(-1, 1)$ $(1, 1)$ C_2

$(4, 16)$ $(1, 1)$ C_3

$(4, 16)$ $(1, 1)$ C_4

45.

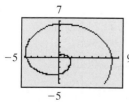

47.

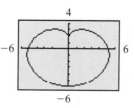

49. (a) $x = 3$
  $y = -16t^2 + 50t + 6$
(b) 3.24 s
(c) 1.56 s; 45.06 ft
(d)

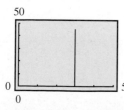

51. (a) Train: $x_1 = t^2, y_1 = 1$;
  Bill: $x_2 = 5(t - 5), y_2 = 3$
(b) Bill won't catch the train.
(c)

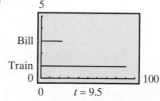

$t = 9.5$

53. (a) $x = (145 \cos 20°)t$
  $y = -16t^2 + (145 \sin 20°)t + 5$
(b) 3.20 s **(c)** 435.61 ft
(d) 1.55 s; 43.43 ft
(e)

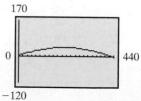

55. (a) $x = (40 \cos 45°)t$

$y = -4.9t^2 + (40 \sin 45°)t + 300$

(b) 11.23 s   **(c)** 317.52 m

(d) 2.89 s; 340.82 m

(e)

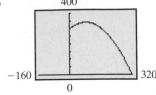

57. (a) Camry: $x = 40t - 5$, $y = 0$; Chevy Impala: $x = 0$, $y = 30t - 4$

(b) $d = \sqrt{(40t - 5)^2 + (30t - 4)^2}$

(c)

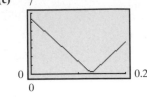

(d) 0.2 mi; 7.68 min

(e) Turn axes off to see the graph:

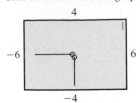

59. (a) $x = \dfrac{\sqrt{2}}{2} v_0 t$, $y = -16t^2 + \dfrac{\sqrt{2}}{2} v_0 t + 3$ **(b)** Maximum height is 139.1 ft.   **(c)** The ball is 272.25 ft from home plate.   **(d)** Yes, the ball will clear the wall by about 99.5 ft.   **61.** The orientation is from (x_1, y_1) to (x_2, y_2).

Review Exercises (page 459)

1. Parabola; vertex $(0, 0)$, focus $(-4, 0)$, directrix $x = 4$ **3.** Hyperbola; center $(0, 0)$, vertices $(5, 0)$ and $(-5, 0)$, foci $\left(\sqrt{26}, 0\right)$ and $\left(-\sqrt{26}, 0\right)$, asymptotes $y = \dfrac{1}{5}x$ and $y = -\dfrac{1}{5}x$ **5.** Ellipse; center $(0, 0)$, vertices $(0, 5)$ and $(0, -5)$, foci $(0, 3)$ and $(0, -3)$ **7.** $x^2 = -4(y - 1)$: Parabola; vertex $(0, 1)$, focus $(0, 0)$, directrix $y = 2$ **9.** $\dfrac{x^2}{2} - \dfrac{y^2}{8} = 1$: Hyperbola; center $(0, 0)$, vertices $\left(\sqrt{2}, 0\right)$ and $\left(-\sqrt{2}, 0\right)$, foci $\left(\sqrt{10}, 0\right)$ and $\left(-\sqrt{10}, 0\right)$, asymptotes $y = 2x$ and $y = -2x$ **11.** $(x - 2)^2 = 2(y + 2)$: Parabola; vertex $(2, -2)$, focus $\left(2, -\dfrac{3}{2}\right)$, directrix $y = -\dfrac{5}{2}$

13. $\dfrac{(y - 2)^2}{4} - (x - 1)^2 = 1$: Hyperbola; center $(1, 2)$, vertices $(1, 4)$ and $(1, 0)$, foci $\left(1, 2 + \sqrt{5}\right)$ and $\left(1, 2 - \sqrt{5}\right)$, asymptotes $y - 2 = \pm2(x - 1)$

15. $\dfrac{(x - 2)^2}{9} + \dfrac{(y - 1)^2}{4} = 1$: Ellipse; center $(2, 1)$, vertices $(5, 1)$ and $(-1, 1)$, foci $\left(2 + \sqrt{5}, 1\right)$ and $\left(2 - \sqrt{5}, 1\right)$

17. $(x - 2)^2 = -4(y + 1)$: Parabola; vertex $(2, -1)$, focus $(2, -2)$, directrix $y = 0$

19. $\dfrac{(x - 1)^2}{4} + \dfrac{(y + 1)^2}{9} = 1$: Ellipse; center $(1, -1)$, vertices $(1, 2)$ and $(1, -4)$, foci $\left(1, -1 + \sqrt{5}\right)$ and $\left(1, -1 - \sqrt{5}\right)$

21. $y^2 = -8x$

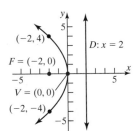

23. $\dfrac{y^2}{4} - \dfrac{x^2}{12} = 1$

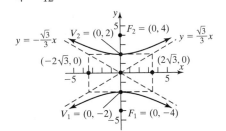

25. $\dfrac{x^2}{16} + \dfrac{y^2}{7} = 1$

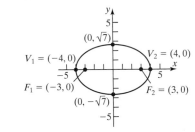

27. $(x - 2)^2 = -4(y + 3)$

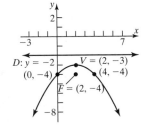

29. $(x + 2)^2 - \dfrac{(y + 3)^2}{3} = 1$

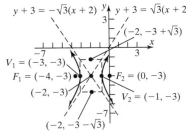

31. $\dfrac{(x + 4)^2}{16} + \dfrac{(y - 5)^2}{25} = 1$

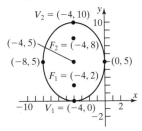

33. $\dfrac{(x + 1)^2}{9} - \dfrac{(y - 2)^2}{7} = 1$

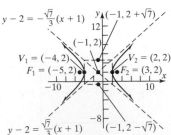

35. $\dfrac{(x - 3)^2}{9} - \dfrac{(y - 1)^2}{4} = 1$

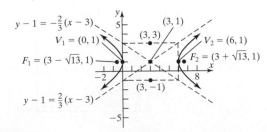

37. Parabola   **39.** Ellipse

41. Parabola   **43.** Hyperbola

45. Ellipse

47. $x'^2 - \dfrac{y'^2}{9} = 1$

Hyperbola
Center at the origin
Transverse axis the x'-axis
Vertices at $(\pm 1, 0)$

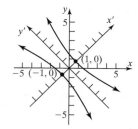

49. $\dfrac{x'^2}{2} + \dfrac{y'^2}{4} = 1$

Ellipse
Center at origin
Major axis the y'-axis
Vertices at $(0, \pm 2)$

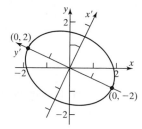

51. $y'^2 = -\dfrac{4\sqrt{13}}{13}x'$

Parabola
Vertex at the origin
Focus on the x'-axis at $\left(-\dfrac{\sqrt{13}}{13}, 0\right)$

53. Parabola; directrix is perpendicular to the polar axis 4 units to the left of the pole; vertex is $(2, \pi)$.

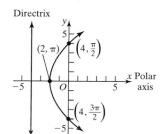

55. Ellipse; directrix is parallel to the polar axis 6 units below the pole; vertices are $\left(6, \dfrac{\pi}{2}\right)$ and $\left(2, \dfrac{3\pi}{2}\right)$.

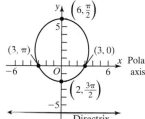

57. Hyperbola; directrix is perpendicular to the polar axis 1 unit to the right of the pole; vertices are $\left(\dfrac{2}{3}, 0\right)$ and $(-2, \pi)$.

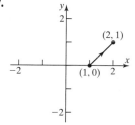

59. $y^2 - 8x - 16 = 0$ **61.** $3x^2 - y^2 - 8x + 4 = 0$

63.

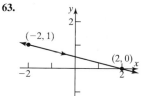

$x + 4y = 2$

65.

$\dfrac{x^2}{9} + \dfrac{(y-2)^2}{16} = 1$

67.

$1 + y = x$

69. $x = t, y = -2t + 4, -\infty < t < \infty$ **71.** $x = 4\cos\left(\dfrac{\pi}{2}t\right), y = 3\sin\left(\dfrac{\pi}{2}t\right), 0 \le t \le 4$ **73.** $\dfrac{x^2}{5} - \dfrac{y^2}{4} = 1$

$x = \dfrac{t-4}{-2}, y = t, -\infty < t < \infty$

75. The ellipse $\dfrac{x^2}{16} + \dfrac{y^2}{7} = 1$ **77.** $\dfrac{1}{4}$ ft or 3 in.   **79.** 19.72 ft, 18.86 ft, 14.91 ft   **81.** 450 ft

83. (a) $x = (80\cos 35°)t$

$y = -16t^2 + (80\sin 35°)t + 6$

(b) 2.9932 s
(c) 1.4339 s; 38.9 ft
(d) 196.15 ft

(e)

Chapter Test (page 461)

1. Hyperbola; center: $(-1, 0)$; vertices: $(-3, 0)$ and $(1, 0)$; foci: $\left(-1 - \sqrt{13}, 0\right)$ and $\left(-1 + \sqrt{13}, 0\right)$; asymptotes: $y = -\dfrac{3}{2}(x + 1)$ and $y = \dfrac{3}{2}(x + 1)$

2. Parabola; vertex: $\left(1, -\dfrac{1}{2}\right)$; focus: $\left(1, \dfrac{3}{2}\right)$; directrix: $y = -\dfrac{5}{2}$

3. Ellipse; center: $(-1, 1)$; foci: $\left(-1 - \sqrt{3}, 1\right)$ and $\left(-1 + \sqrt{3}, 1\right)$; vertices: $(-4, 1)$ and $(2, 1)$

4. $(x + 1)^2 = 6(y - 3)$

5. $\dfrac{x^2}{7} + \dfrac{y^2}{16} = 1$

6. $\dfrac{(y - 2)^2}{4} - \dfrac{(x - 2)^2}{8} = 1$

7. Hyperbola

8. Ellipse

9. Parabola

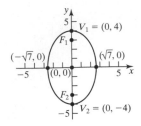

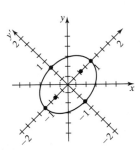

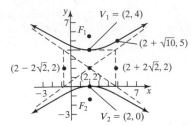

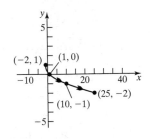

10. $x'^2 + 2y'^2 = 1$. This is the equation of an ellipse with center at $(0, 0)$ in the $x'y'$-plane. The vertices are at $(-1, 0)$ and $(1, 0)$ in the $x'y'$-plane.

11. Hyperbola; $(x + 2)^2 - \dfrac{y^2}{3} = 1$

12. $y = 1 - \sqrt{\dfrac{x + 2}{3}}$

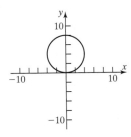

13. The microphone should be located $\dfrac{2}{3}$ ft from the base of the reflector, along its axis of symmetry.

Cumulative Review *(page 461)*

1. (a) $y = 2x - 2$ **(b)** $(x - 2)^2 + y^2 = 4$ **(c)** $\dfrac{x^2}{9} + \dfrac{y^2}{4} = 1$ **(d)** $y = 2(x - 1)^2$ **(e)** $y^2 - \dfrac{x^2}{3} = 1$

2. $\theta = \dfrac{\pi}{12} \pm \pi k$, k is any integer; $\theta = \dfrac{5\pi}{12} \pm \pi k$, k is any integer   **3.** $\theta = \dfrac{\pi}{6}$

4. $r = 8 \sin \theta$ **5.** $\left\{ x \,\middle|\, x \ne \dfrac{3\pi}{4} \pm \pi k, k \text{ is an integer} \right\}$ **6.** $\{22.5°\}$ **7.** $y = \dfrac{x^2}{5} + 5$

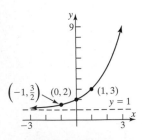

CHAPTER 7 Exponential and Logarithmic Functions

7.1 Assess Your Understanding *(page 474)*

6. $\left(-1, \dfrac{1}{a}\right), (0, 1), (1, a)$ **7.** 1   **8.** 4   **9.** F   **10.** F   **11. (a)** 11.212   **(b)** 11.587   **(c)** 11.664   **(d)** 11.665   **13. (a)** 8.815   **(b)** 8.821   **(c)** 8.824

(d) 8.825   **15. (a)** 21.217   **(b)** 22.217   **(c)** 22.440   **(d)** 22.459   **17.** 3.320   **19.** 0.427   **21.** Neither   **23.** Exponential; $H(x) = 4^x$

25. Exponential; $f(x) = 3(2^x)$ **27.** Linear; $H(x) = 2x + 4$ **29.** B   **31.** D   **33.** F   **35.** E

37.

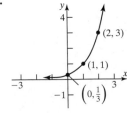

Domain: All real numbers
Range: $\{y \mid y > 1\}$ or $(1, \infty)$
Horizontal asymptote: $y = 1$

39.

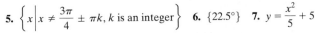

Domain: All real numbers
Range: $\{y \mid y > 0\}$ or $(0, \infty)$
Horizontal asymptote: $y = 0$

41.

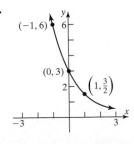

Domain: All real numbers
Range: $\{y \mid y > 0\}$ or $(0, \infty)$
Horizontal asymptote: $y = 0$

43.

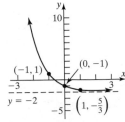

Domain: All real numbers
Range: $\{y \mid y > -2\}$ or $(-2, \infty)$
Horizontal asymptote: $y = -2$

45.

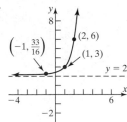

Domain: All real numbers
Range: $\{y|y > 2\}$ or $(2, \infty)$
Horizontal asymptote: $y = 2$

47.

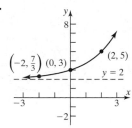

Domain: All real numbers
Range: $\{y|y > 2\}$ or $(2, \infty)$
Horizontal asymptote: $y = 2$

49.

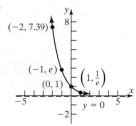

Domain: All real numbers
Range: $\{y|y > 0\}$ or $(0, \infty)$
Horizontal asymptote: $y = 0$

51.

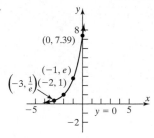

Domain: All real numbers
Range: $\{y|y > 0\}$ or $(0, \infty)$
Horizontal asymptote: $y = 0$

53.

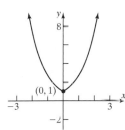

Domain: All real numbers
Range: $\{y|y < 5\}$ or $(-\infty, 5)$
Horizontal asymptote: $y = 5$

55.

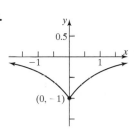

Domain: All real numbers
Range: $\{y|y < 2\}$ or $(-\infty, 2)$
Horizontal asymptote: $y = 2$

57. $\{3\}$ **59.** $\{-4\}$ **61.** $\{2\}$ **63.** $\left\{\dfrac{3}{2}\right\}$

65. $\left\{-\sqrt{2}, 0, \sqrt{2}\right\}$ **67.** $\{6\}$ **69.** $\{-1, 7\}$ **71.** $\{-4, 2\}$

73. $\{-4\}$ **75.** $\{1, 2\}$ **77.** $\dfrac{1}{49}$ **79.** $\dfrac{1}{4}$ **81.** $f(x) = 3^x$

83. $f(x) = -6^x$ **85.** (a) $16; (4, 16)$ (b) $-4; \left(-4, \dfrac{1}{16}\right)$

87. (a) $\dfrac{9}{4}; \left(-1, \dfrac{9}{4}\right)$ (b) $3; (3, 66)$

89. (a) $60; (-6, 60)$ (b) $-4: (-4, 12)$ (c) -2

91.

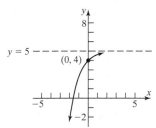

Domain: $(-\infty, \infty)$

Range: $[1, \infty)$

Intercept: $(0, 1)$

93.

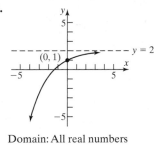

Domain: $(-\infty, \infty)$

Range: $[-1, 0)$

Intercept: $(0, -1)$

95. (a) 74% (b) 47%
97. (a) \$12,123 (b) \$6443
99. 3.35 mg; 0.45 mg

101. (a) 0.632 (b) 0.982 (c) 1
(d)

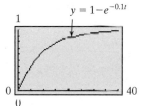

(e) About 7 min

103. (a) 0.0516 (b) 0.0888 **105.** (a) 70.95% (b) 72.62% (c) 100%

107. (a) 5.41 amp, 7.59 amp, 10.38 amp (b) 12 amp (d) 3.34 amp, 5.31 amp, 9.44 amp (e) 24 amp

(c), (f)

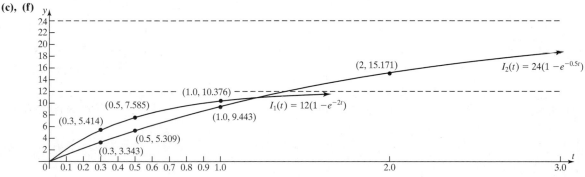

109. $n = 4$: 2.7083; $n = 6$: 2.7181;
$n = 8$: 2.7182788; $n = 10$: 2.7182818

111. $\dfrac{f(x + h) - f(x)}{h} = \dfrac{a^{x+h} - a^x}{h} = \dfrac{a^x a^h - a^x}{h} = \dfrac{a^x(a^h - 1)}{h}$

113. $f(-x) = a^{-x} = \dfrac{1}{a^x} = \dfrac{1}{f(x)}$

115. (a) $f(-x) = \frac{1}{2}(e^{-x} - e^{-(-x)})$

$$= \frac{1}{2}(e^{-x} - e^{x})$$

$$= -\frac{1}{2}(e^{x} - e^{-x})$$

$$= -f(x)$$

(b)

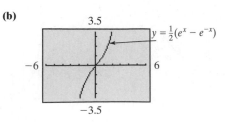

$y = \frac{1}{2}(e^{x} - e^{-x})$

117. $f(1) = 5, f(2) = 17, f(3) = 257, f(4) = 65{,}537; f(5) = 4{,}294{,}967{,}297 = 641 \times 6{,}700{,}417$

7.2 Assess Your Understanding (page 488)

4. $\{x | x > 0\}$ or $(0, \infty)$ **5.** $\left(\frac{1}{a}, -1\right), (1, 0), (a, 1)$ **6.** 1   **7.** F   **8.** T   **9.** $2 = \log_3 9$ **11.** $2 = \log_a 1.6$

13. $x = \log_2 7.2$ **15.** $x = \ln 8$ **17.** $2^3 = 8$ **19.** $a^6 = 3$ **21.** $3^x = 2$ **23.** $e^x = 4$ **25.** 0   **27.** 2   **29.** -4

31. $\frac{1}{2}$ **33.** 4   **35.** $\frac{1}{2}$ **37.** $\{x | x > 3\}; (3, \infty)$ **39.** All real numbers except 0; $\{x | x \neq 0\}$

41. $\{x | x > 10\}; (10, \infty)$ **43.** $\{x | x > -1\}; (-1, \infty)$ **45.** $\{x | x < -1 \text{ or } x > 0\}; (-\infty, -1) \cup (0, \infty)$

47. $\{x | x \geq 1\}; [1, \infty)$ **49.** 0.511   **51.** 30.099   **53.** 2.303   **55.** -53.991 **57.** $\sqrt{2}$

59.

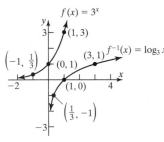

61.

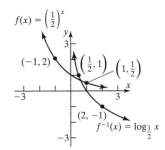

63. B   **65.** D   **67.** A   **69.** E

71. (a) Domain: $(-4, \infty)$

(b)

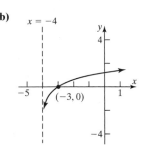

(c) Range: $(-\infty, \infty)$

Vertical asymptote: $x = -4$

(d) $f^{-1}(x) = e^x - 4$

(e) Range of f: $(-\infty, \infty)$

Range of f^{-1}: $(-4, \infty)$

(f)

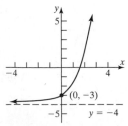

73. (a) Domain: $(0, \infty)$

(b)

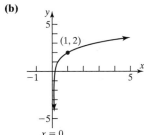

(c) Range: $(-\infty, \infty)$

Vertical asymptote: $x = 0$

(d) $f^{-1}(x) = e^{x-2}$

(e) Range of f: $(-\infty, \infty)$

Range of f^{-1}: $(0, \infty)$

(f)

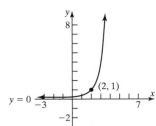

75. (a) Domain: $(0, \infty)$

(b)

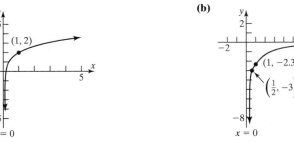

(c) Range: $(-\infty, \infty)$

Vertical asymptote: $x = 0$

(d) $f^{-1}(x) = \frac{1}{2}e^{x+3}$

(e) Range of f: $(-\infty, \infty)$

Range of f^{-1}: $(0, \infty)$

(f)

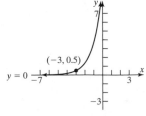

77. (a) Domain: $(4, \infty)$

(b)

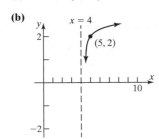

(c) Range: $(-\infty, \infty)$
Vertical asymptote: $x = 4$

(d) $f^{-1}(x) = 10^{x-2} + 4$

(e) Range of f: $(-\infty, \infty)$
Range of f^{-1}: $(4, \infty)$

(f)

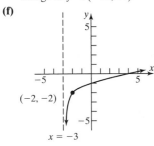

79. (a) Domain: $(0, \infty)$

(b)

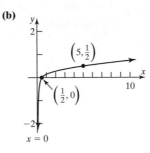

(c) Range: $(-\infty, \infty)$
Vertical asymptote: $x = 0$

(d) $f^{-1}(x) = \dfrac{1}{2} \cdot 10^{2x}$

(e) Range of f: $(-\infty, \infty)$
Range of f^{-1}: $(0, \infty)$

(f)

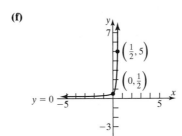

81. (a) Domain: $(-2, \infty)$

(b)

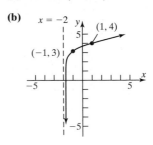

(c) Range: $(-\infty, \infty)$
Vertical asymptote: $x = -2$

(d) $f^{-1}(x) = 3^{x-3} - 2$

(e) Range of f: $(-\infty, \infty)$
Range of f^{-1}: $(-2, \infty)$

(f)

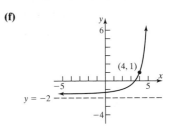

83. (a) Domain: $(-\infty, \infty)$

(b)

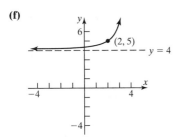

(c) Range: $(-3, \infty)$
Horizontal asymptote: $y = -3$

(d) $f^{-1}(x) = \ln(x + 3) - 2$

(e) Range of f: $(-3, \infty)$
Range of f^{-1}: $(-\infty, \infty)$

(f)

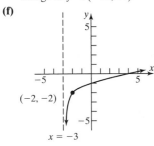

85. (a) Domain: $(-\infty, \infty)$

(b)

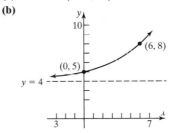

(c) Range: $(4, \infty)$
Horizontal asymptote: $y = 4$

(d) $f^{-1}(x) = 3 \log_2(x - 4)$

(e) Range of f: $(4, \infty)$
Range of f^{-1}: $(-\infty, \infty)$

(f)

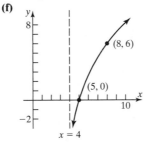

87. $\{9\}$ **89.** $\left\{\dfrac{7}{2}\right\}$ **91.** $\{2\}$ **93.** $\{5\}$

95. $\{3\}$ **97.** $\{2\}$ **99.** $\left\{\dfrac{\ln 10}{3}\right\}$

101. $\left\{\dfrac{\ln 8 - 5}{2}\right\}$

103. $\{-2\sqrt{2}, 2\sqrt{2}\}$ **105.** $\{-1\}$

107. $\left\{5 \ln \dfrac{7}{5}\right\}$ **109.** $\left\{2 - \log \dfrac{5}{2}\right\}$

111. (a) $\left\{x \mid x > -\dfrac{1}{2}\right\}; \left(-\dfrac{1}{2}, \infty\right)$

(b) $2; (40, 2)$ **(c)** $121; (121, 3)$ **(d)** 4

113.

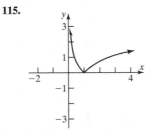

Domain: $\{x \mid x \neq 0\}$
Range: $(-\infty, \infty)$
Intercepts: $(-1, 0), (1, 0)$

115.

Domain: $\{x \mid x > 0\}$
Range: $\{y \mid y \geq 0\}$
Intercept: $(1, 0)$

117. (a) 1 **(b)** 2 **(c)** 3 **(d)** It increases.
(e) 0.000316 **(f)** 3.981×10^{-8}

119. (a) 5.97 km **(b)** 0.90 km

121. (a) 6.93 min **(b)** 16.09 min

123. $h \approx 2.29$, so the time between injections is about 2 h, 17 min.

125. 0.2695 s
0.8959 s

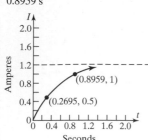

127. 50 decibels (dB) **129.** 90 dB **131.** 8.1

133. (a) $k \approx 11.216$ **(b)** 6.73 **(c)** 0.41% **(d)** 0.14%

135. Because $y = \log_1 x$ means $1^y = 1 = x$, which cannot be true for $x \neq 1$

7.3 Assess Your Understanding *(page 499)*

1. sum **2.** 7 **3.** $r \log_a M$ **4.** F **5.** F **6.** T **7.** 71 **9.** -4 **11.** 7 **13.** 1 **15.** 1 **17.** 2 **19.** $\frac{5}{4}$ **21.** 4 **23.** $a + b$ **25.** $b - a$

27. $3a$ **29.** $\frac{1}{5}(a + b)$ **31.** $2 + \log_5 x$ **33.** $3 \log_2 z$ **35.** $1 + \ln x$ **37.** $\ln x + x$ **39.** $2 \log_a u + 3 \log_a v$ **41.** $2 \ln x + \frac{1}{2}\ln(1 - x)$

43. $3 \log_2 x - \log_2(x - 3)$ **45.** $\log x + \log(x + 2) - 2 \log(x + 3)$ **47.** $\frac{1}{3}\ln(x - 2) + \frac{1}{3}\ln(x + 1) - \frac{2}{3}\ln(x + 4)$

49. $\ln 5 + \ln x + \frac{1}{2}\ln(1 + 3x) - 3 \ln(x - 4)$ **51.** $\log_5 u^3 v^4$ **53.** $\log_3\left(\frac{1}{x^{5/2}}\right)$ **55.** $\log_4\left[\frac{x - 1}{(x + 1)^4}\right]$ **57.** $-2 \ln(x - 1)$ **59.** $\log_2[x(3x - 2)^4]$

61. $\log_a\left(\frac{25x^6}{\sqrt{2x + 3}}\right)$ **63.** $\log_2\left[\frac{(x + 1)^2}{(x + 3)(x - 1)}\right]$ **65.** 2.771 **67.** -3.880 **69.** 5.615 **71.** 0.874

73. $y = \dfrac{\log x}{\log 4}$

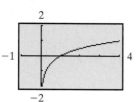

75. $y = \dfrac{\log(x + 2)}{\log 2}$

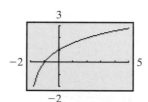

77. $y = \dfrac{\log(x + 1)}{\log(x - 1)}$

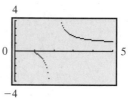

79. (a) $(f \circ g)(x) = x$; $\{x \mid x$ is any real number$\}$ or $(-\infty, \infty)$ **81.** $y = Cx$ **83.** $y = Cx(x + 1)$ **85.** $y = Ce^{3x}$ **87.** $y = Ce^{-4x} + 3$

(b) $(g \circ f)(x) = x$; $\{x \mid x > 0\}$ or $(0, \infty)$ **(c)** 5

(d) $(f \circ h)(x) = \ln x^2$; $\{x \mid x \neq 0\}$ or $(-\infty, 0) \cup (0, \infty)$ **(e)** 2

89. $y = \dfrac{\sqrt[3]{C}(2x + 1)^{1/6}}{(x + 4)^{1/9}}$ **91.** 3 **93.** 1

95. $\log_a\left(x + \sqrt{x^2 - 1}\right) + \log_a\left(x - \sqrt{x^2 - 1}\right) = \log_a\left[\left(x + \sqrt{x^2 - 1}\right)\left(x - \sqrt{x^2 - 1}\right)\right] = \log_a[x^2 - (x^2 - 1)] = \log_a 1 = 0$

97. $\ln(1 + e^{2x}) = \ln[e^{2x}(e^{-2x} + 1)] = \ln e^{2x} + \ln(e^{-2x} + 1) = 2x + \ln(1 + e^{-2x})$

99. $y = f(x) = \log_a x$; $a^y = x$ implies $a^y = \left(\dfrac{1}{a}\right)^{-y} = x$, so $-y = \log_{1/a} x = -f(x)$. **101.** $f(x) = \log_a x$; $f\left(\dfrac{1}{x}\right) = \log_a \dfrac{1}{x} = \log_a 1 - \log_a x = -f(x)$

103. $\log_a \dfrac{M}{N} = \log_a(M \cdot N^{-1}) = \log_a M + \log_a N^{-1} = \log_a M - \log_a N$, since $a^{\log_a N^{-1}} = N^{-1}$ implies $a^{-\log_a N^{-1}} = N$; i.e., $\log_a N = -\log_a N^{-1}$.

7.4 Assess Your Understanding *(page 505)*

5. $\{16\}$ **7.** $\left\{\dfrac{16}{5}\right\}$ **9.** $\{6\}$ **11.** $\{16\}$ **13.** $\left\{\dfrac{1}{3}\right\}$ **15.** $\{3\}$ **17.** $\{5\}$ **19.** $\left\{\dfrac{21}{8}\right\}$ **21.** $\{-6\}$ **23.** $\{-2\}$

25. $\left\{-1 + \sqrt{1 + e^4}\right\} \approx \{6.456\}$ **27.** $\left\{\dfrac{-5 + 3\sqrt{5}}{2}\right\} \approx \{0.854\}$ **29.** $\{2\}$ **31.** $\left\{\dfrac{9}{2}\right\}$ **33.** $\{8\}$ **35.** $\{\log_2 10\} = \left\{\dfrac{\ln 10}{\ln 2}\right\} \approx \{3.322\}$

37. $\{-\log_8 1.2\} = \left\{-\dfrac{\ln 1.2}{\ln 8}\right\} \approx \{-0.088\}$ **39.** $\left\{\dfrac{1}{3}\log_2 \dfrac{8}{5}\right\} = \left\{\dfrac{\ln \frac{8}{5}}{3 \ln 2}\right\} \approx \{0.226\}$ **41.** $\left\{\dfrac{\ln 3}{2 \ln 3 + \ln 4}\right\} \approx \{0.307\}$

43. $\left\{\dfrac{\ln 7}{\ln 0.6 + \ln 7}\right\} \approx \{1.356\}$ **45.** $\{0\}$ **47.** $\left\{\dfrac{\ln \pi}{1 + \ln \pi}\right\} \approx \{0.534\}$ **49.** $\left\{\dfrac{\ln 3}{\ln 2}\right\} \approx \{1.585\}$ **51.** $\{0\}$

53. $\left\{\log_4\left(-2 + \sqrt{7}\right)\right\} \approx \{-0.315\}$ **55.** $\{\log_5 4\} \approx \{0.861\}$ **57.** No real solution **59.** $\{\log_4 5\} \approx \{1.161\}$ **61.** $\{2.79\}$ **63.** $\{-0.57\}$

65. $\{-0.70\}$ **67.** $\{0.57\}$ **69.** $\{0.39, 1.00\}$ **71.** $\{1.32\}$ **73.** $\{1.31\}$ **75.** $\{1\}$ **77.** $\{16\}$ **79.** $\left\{-1, \dfrac{2}{3}\right\}$ **81.** $\{0\}$ **83.** $\left\{\ln\left(2 + \sqrt{5}\right)\right\} \approx \{1.444\}$

85. $\left\{e^{\frac{\ln 5 \cdot \ln 3}{\ln 15}}\right\} \approx \{1.921\}$ **87. (a)** $\{5\}; (5, 3)$ **(b)** $\{5\}; (5, 4)$ **(c)** $\{1\};$ yes, at $(1, 2)$ **(d)** $\{5\}$ **(e)** $\left\{-\dfrac{1}{11}\right\}$

89. (a)

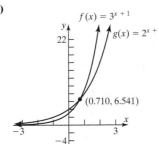

(b) $(0.710, 6.541)$
(c) $\{x \mid x > 0.710\}$ or $(0.710, \infty)$

91. (a), (b)

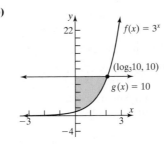

(c) $(\log_3 10, 10)$

93. (a), (b)

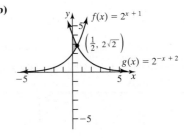

(c) $\left(\dfrac{1}{2}, 2\sqrt{2}\right)$

95. (a)

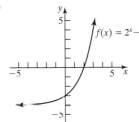

(b) 2
(c) $\{x \mid x < 2\}$ or $(-\infty, 2)$

97. (a) 2010 **(b)** 2027

99. (a) After 2.4 yr
(b) After 6.5 yr
(c) After 10 yr

7.5 Assess Your Understanding *(page 515)*

3. $\$108.29$ **5.** $\$609.50$ **7.** $\$697.09$ **9.** $\$12.46$ **11.** $\$125.23$ **13.** $\$88.72$ **15.** $\$860.72$ **17.** $\$554.09$ **19.** $\$59.71$ **21.** $\$361.93$ **23.** 5.095%
25. 5.127% **27.** $6\dfrac{1}{4}\%$ compounded annually **29.** 9% compounded monthly **31.** 25.992% **33.** 24.573% **35. (a)** About 8.69 yr
(b) About 8.66 yr **37.** 6.823% **39.** 5.09 yr; 5.07 yr **41.** 15.27 yr or 15 yr, 3 mo **43.** $\$104,335$ **45.** $\$12,910.62$ **47.** About $\$30.17$ per share or $\$3017$
49. Not quite. Jim will have $\$1057.60$. The second bank gives a better deal, since Jim will have $\$1060.62$ after 1 yr.
51. Will has $\$11,632.73$; Henry has $\$10,947.89$. **53. (a)** $\$49,581$ **(b)** $\$33,235$ **55.** Approximately $\$602$ billion
57. $\$940.90$ **59.** 2.53% **61.** 34.31 yr **63. (a)** $\$1364.62$ **(b)** $\$1353.35$ **65.** $\$4631.93$

67. (a) 6.12 yr

(b) 18.45 yr

(c) $mP = P\left(1 + \dfrac{r}{n}\right)^{nt}$

$m = \left(1 + \dfrac{r}{n}\right)^{nt}$

$\ln m = \ln\left(1 + \dfrac{r}{n}\right)^{nt} = nt \ln\left(1 + \dfrac{r}{n}\right)$

$t = \dfrac{\ln m}{n \ln\left(1 + \dfrac{r}{n}\right)}$

69. (a) 2.51% **(b)** In 2022 or after 27 yr **71.** 22.7 yr

7.6 Assess Your Understanding *(page 526)*

1. (a) 500 insects
(b) $0.02 = 2\%$
(c) 2500

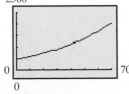

(d) About 611 insects
(e) After about 23.5 days
(f) After about 34.7 days

3. (a) $-0.0244 = -2.44\%$
(b) 500

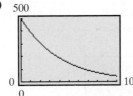

(c) About 391.7 g
(d) After about 9.1 yr
(e) 28.4 yr

5. (a) $N(t) = N_0 e^{kt}$
(b) 5832
(c) 3.9 days
7. (a) $N(t) = N_0 e^{kt}$
(b) 25,198
9. 9.797 g

11. (a) 9727 yr ago

(b)
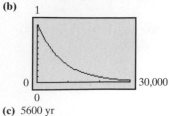

(c) 5600 yr

13. (a) About 5:18 PM

(b)
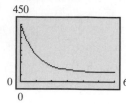

(c) About 14.3 min

(d) The temperature of the pizza approaches 70°F.

15. (a) 18.63°C; 25.07°C

(b)
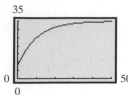

17. 1.7 ppm; 7.17 days or 172 h   **19.** 0.26 mol; 6.58 h or 395 min   **21.** 26.6 days

23. (a) 1000 g   **(b)** 43.9%   **(c)** 30 g

(d)
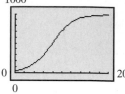

(e) 616.6 g   **(f)** After 9.85 h
(g) About 7.9 h

25. (a) 9.23×10^{-3}, or about 0

(b) 0.81, or about 1

(c) 5.01, or about 5

(d) 57.91°, 43.99°, 30.07°

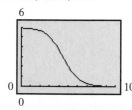

27. (a) In 1984, 91.8% of households
did not own a personal computer.

(b)
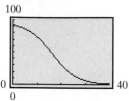

(c) 70.6%   **(d)** During 2011

29. (a)
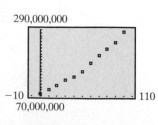

(b) 0.78 or 78%   **(c)** 50 people
(d) As n increases, the probability decreases.

7.7 Assess Your Understanding *(page 534)*

1. (a)
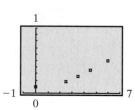

(d)
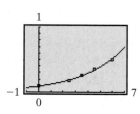

(b) $y = 0.0903(1.3384)^x$
(c) $N(t) = 0.0903e^{0.2915t}$

(e) 0.69
(f) After about 7.26 hr

3. (a)
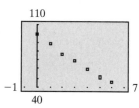

(d)
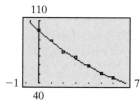

(b) $y = 100.326(0.8769)^x$
(c) $A(t) = 100.326e^{-0.1314t}$

(e) 5.3 weeks   **(f)** 0.14 g
(g) After about 12.3 weeks

5. (a)
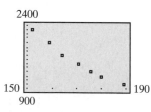

(b) $y = 32{,}741.02 - 6070.96 \ln x$ **(c)**
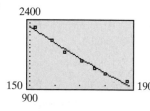

(d) Approximately 168
computers

7. (a)
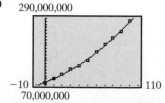

(b) $y = \dfrac{799{,}475{,}916.5}{1 + 9.1968e^{-0.0160x}}$

(c)

(d) 799,475,917

(e) Approximately 291,599,733

(f) 2006

9. (a) 76,100

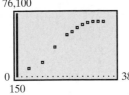

0 · · · · · · · · · · · · 38
150

(b) $y = \dfrac{68{,}684.8}{1 + 18.9416e^{-0.1974x}}$

(c) 76,100

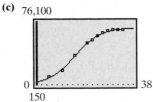

0 · · · · · · · · · · · · 38
150

(d) About 68,685,000 subscribers

(e) About 68,505,000 subscribers

11. (a) 225

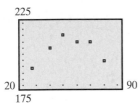

20 · · · · · · · · 90
175

(b) Quadratic with $a < 0$ because of the "upside down U-shape" of the data

(c) $y = -0.0311x^2 + 3.4444x + 118.2493$

(d) 225

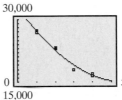

20 · · · · · · · · 90
175

(e) 201

13. (a) 30,000

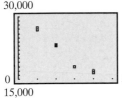

0 · · · · · 5
15,000

(b) Exponential because depreciation of a car is described by exponential models in the theory of finance

(c) $y = 31{,}808.51\,(0.8474)^x$

(d) 30,000

0 · · · · · 5
15,000

(e) $13,899

Review Exercises *(page 539)*

1. (a) 81 **(b)** 2 **(c)** $\dfrac{1}{9}$ **(d)** -3 **3.** $\log_5 z = 2$ **5.** $5^{13} = u$ **7.** $\left\{x \mid x > \dfrac{2}{3}\right\};\left(\dfrac{2}{3}, \infty\right)$ **9.** $\{x \mid x < 1 \text{ or } x > 2\}; (-\infty, 1) \cup (2, \infty)$

11. -3 **13.** $\sqrt{2}$ **15.** 0.4   **17.** $\log_3 u + 2\log_3 v - \log_3 w$ **19.** $2\log x + \dfrac{1}{2}\log(x^3 + 1)$ **21.** $\ln x + \dfrac{1}{3}\ln(x^2 + 1) - \ln(x - 3)$

23. $\dfrac{25}{4}\log_4 x$ **25.** $2\ln(x + 1)$ **27.** $\log\left(\dfrac{4x^3}{[(x + 3)(x - 2)]^{1/2}}\right)$ **29.** 2.124

31.

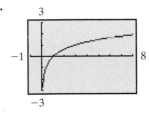

3

−1 ┤ 8

−3

33. (a) Domain of f: $(-\infty, \infty)$

(b)

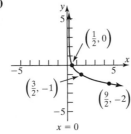

(4, 2)
(3, 1)

(c) Range of f: $(0, \infty)$

Horizontal asymptote: $y = 0$

(d) $f^{-1}(x) = 3 + \log_2 x$

(e) Range of f: $(0, \infty)$

(f)

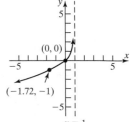

(2, 4)
(1, 3)

$x = 0$

35. (a) Domain of f: $(-\infty, \infty)$

(b)

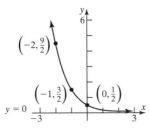

$\left(-2, \dfrac{9}{2}\right)$

$\left(-1, \dfrac{3}{2}\right)$ $\left(0, \dfrac{1}{2}\right)$

$y = 0$
−3   3

(e) Range of f: $(0, \infty)$

(f)

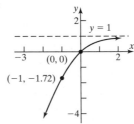

$\left(\dfrac{1}{2}, 0\right)$

$\left(\dfrac{3}{2}, -1\right)$
$\left(\dfrac{9}{2}, -2\right)$

$x = 0$

(c) Range of f: $(0, \infty)$

Horizontal asymptote: $y = 0$

(d) $f^{-1}(x) = -\log_3(2x)$

37. (a) Domain of f: $(-\infty, \infty)$

(b)

$y = 1$

(0, 0)
(−1, −1.72)

(e) Range of f: $(-\infty, 1)$

(f)

(0, 0)

(−1.72, −1)

$x = 1$

(c) Range of f: $(-\infty, 1)$

Horizontal asymptote: $y = 1$

(d) $f^{-1}(x) = -\ln(1 - x)$

39. (a) Domain of f: $(-3, \infty)$ **(c)** Range of f: $(-\infty, \infty)$ **(e)** Range of f: $(-\infty, \infty)$

(b)   Vertical asymptote: $x = -3$ **(f)**

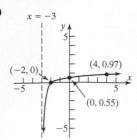

(d) $f^{-1}(x) = e^{2x} - 3$

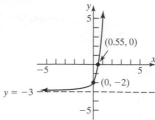

41. $\left\{\dfrac{1}{4}\right\}$ **43.** $\left\{\dfrac{-1 - \sqrt{3}}{2}, \dfrac{-1 + \sqrt{3}}{2}\right\} \approx \{-1.366, 0.366\}$ **45.** $\left\{\dfrac{1}{4}\right\}$ **47.** $\left\{\dfrac{2 \ln 3}{\ln 5 - \ln 3}\right\} \approx \{4.301\}$ **49.** $\left\{\dfrac{12}{5}\right\}$ **51.** $\{83\}$ **53.** $\left\{\dfrac{1}{2}, -3\right\}$

55. $\{-1\}$ **57.** $\{1 - \ln 5\} \approx \{-0.609\}$ **59.** $\left\{\log_3\left(-2 + \sqrt{7}\right)\right\} = \left\{\dfrac{\ln\left(-2 + \sqrt{7}\right)}{\ln 3}\right\} \approx \{-0.398\}$

61. (a), (e)

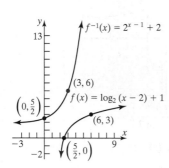

(b) $3; (6, 3)$ **(c)** $10; (10, 4)$

(d) $\left\{x \,\middle|\, x > \dfrac{5}{2}\right\}$ or $\left(\dfrac{5}{2}, \infty\right)$

(e) $f^{-1}(x) = 2^{x-1} + 2$

63. 3229.5 m   **65. (a)** 37.3 W   **(b)** 6.9 dB   **67. (a)** 9.85 yr   **(b)** 4.27 yr   **69.** \$41,668.97   **71.** 24,203 yr ago   **73.** 7,237,271,501   **75.** \$483.67 billion

77. (a)

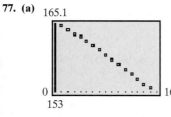

(b) $y = 165.73(0.9951)^x$ **(c)**

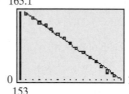

(d) Approximately 83 s

79. (a)

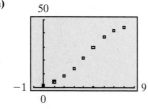

(b) $C = \dfrac{46.93}{1 + 21.273e^{-0.7306t}}$ **(c)**

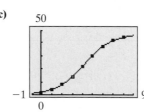

(d) About 47 people; 50 people

(e) 2.4 days; during the tenth hour of day 3

(f) 9.5 days

Chapter Test *(page 542)*

1. $x = 5$ **2.** $b = 4$ **3.** $x = 625$ **4.** $e^3 + 2 \approx 22.086$ **5.** $\log 20 \approx 1.301$ **6.** $\log_3 21 = \dfrac{\ln 21}{\ln 3} \approx 2.771$ **7.** $\ln 133 \approx 4.890$

8. (a) Domain of f: $\{x|-\infty < x < \infty\}$ or $(-\infty, \infty)$

(c) Range of f: $\{y|y > -2\}$ or $(-2, \infty)$; Horizontal asymptote: $y = -2$

(e) Range of f: $\{y|y > -2\}$ or $(-2, \infty)$

(b)

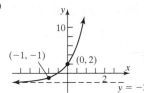

(d) $f^{-1}(x) = \log_4(x + 2) - 1$

(f)

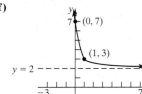

9. (a) Domain of f: $\{x|x > 2\}$ or $(2, \infty)$

(c) Range of f: $\{y|-\infty < y < \infty\}$ or $(-\infty, \infty)$; vertical asymptote: $x = 2$

(e) Range of f: $\{y|-\infty < y < \infty\}$ or $(-\infty, \infty)$

(b)

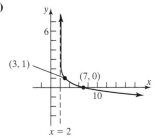

(d) $f^{-1}(x) = 5^{1-x} + 2$

(f)

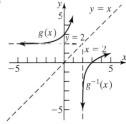

10. $\{1\}$ **11.** $\{91\}$ **12.** $\{-\ln 2\} \approx \{-0.693\}$ **13.** $\left\{\dfrac{1 - \sqrt{13}}{2}, \dfrac{1 + \sqrt{13}}{2}\right\} \approx \{-1.303, 2.303\}$ **14.** $\left\{\dfrac{3\ln 7}{1 - \ln 7}\right\} \approx \{-6.172\}$

15. $\{2\sqrt{6}\} \approx \{4.899\}$ **16.** $2 + 3\log_2 x - \log_2(x - 6) - \log_2(x + 3)$ **17.** About 250.39 days **18. (a)** \$1033.82 **(b)** \$963.42 **(c)** 11.9 yr

19. (a) About 83 dB **(b)** The pain threshold will be exceeded if 31,623 people shouted at the same time.

Cumulative Review *(page 542)*

1. Yes; no **2. (a)** 10 **(b)** $2x^2 + 3x + 1$ **(c)** $2x^2 + 4xh + 2h^2 - 3x - 3h + 1$ **3.** $\left(\dfrac{1}{2}, \dfrac{\sqrt{3}}{2}\right)$ is on the graph. **4.** $\{-26\}$

5.

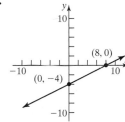

6.

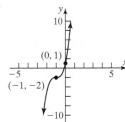

7. (a), (c)

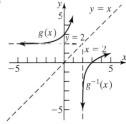

Domain g = range $g^{-1} = (-\infty, \infty)$

Range g = domain $g^{-1} = (2, \infty)$

(b) $g^{-1}(x) = \log_3(x - 2)$,

8. $\left\{-\dfrac{3}{2}\right\}$ **9.** $\{2\}$ **10. (a)** $\{-1\}$ **(b)** $\{x|x > -1\}$ or $(-1, \infty)$ **(c)** $\{25\}$

APPENDIX

A.1 Assess Your Understanding *(page A10)*

1. variable **3.** strict **5.** T **7.** F **9.** $\{1, 2, 3, 4, 5, 6, 7, 8, 9\}$ **11.** $\{4\}$ **13.** $\{1, 3, 4, 6\}$ **15.** $\{0, 2, 6, 7, 8\}$ **17.** $\{0, 1, 2, 3, 5, 6, 7, 8, 9\}$

19. $\{0, 1, 2, 3, 5, 6, 7, 8, 9\}$ **21.** [number line] **23.** $>$ **25.** $>$ **27.** $>$ **29.** $=$ **31.** $<$ **33.** $x > 0$ **35.** $x < 2$

37. $x \le 1$ **39.** ⟶ (number line with point at -2)   **41.** ⟶ (number line with point at -1)   **43.** 1   **45.** 2   **47.** 6   **49.** 4   **51.** -28 **53.** $\dfrac{4}{5}$ **55.** 0   **57.** 1   **59.** 5   **61.** 1

63. 22   **65.** 2   **67.** $x = 0$ **69.** $x = 3$ **71.** None   **73.** $x = 0, x = 1, x = -1$ **75.** $\{x \mid x \ne 5\}$ **77.** $\{x \mid x \ne -4\}$ **79.** 0°C   **81.** 25°C   **83.** 16

85. $\dfrac{1}{16}$ **87.** $\dfrac{1}{9}$ **89.** 9   **91.** 5   **93.** 4   **95.** $64x^6$ **97.** $\dfrac{x^4}{y^2}$ **99.** $\dfrac{x}{y}$ **101.** $-\dfrac{8x^3z}{9y}$ **103.** $\dfrac{16x^2}{9y^2}$ **105.** -4 **107.** 5   **109.** 4   **111.** 2   **113.** $\sqrt{5}$

115. $\dfrac{1}{2}$ **117.** $10; 0$ **119.** 81   **121.** 304,006.671   **123.** 0.004   **125.** 481.890   **127.** 0.000   **129.** $A = lw$ **131.** $C = \pi d$ **133.** $A = \dfrac{\sqrt{3}}{4}x^2$

135. $V = \dfrac{4}{3}\pi r^3$ **137.** $V = x^3$ **139. (a)** $\$6000$ **(b)** $\$8000$ **141.** $|x - 4| \ge 6$ **143. (a)** $2 \le 5$ **(b)** $6 > 5$ **145. (a)** Yes   **(b)** No

147. No; $\dfrac{1}{3}$ is larger; 0.000333...   **149.** No

A.2 Assess Your Understanding *(page A20)*

1. right; hypotenuse   **3.** $C = 2\pi r$ **5.** T   **7.** F   **9.** T   **11.** 13   **13.** 26   **15.** 25   **17.** Right triangle; 5   **19.** Not a right triangle   **21.** Right triangle; 25

23. Not a right triangle   **25.** 8 in.²   **27.** 4 in.²   **29.** $A = 25\pi$ m²; $C = 10\pi$ m   **31.** $V = 224$ ft³; $S = 232$ ft²   **33.** $V = \dfrac{256}{3}\pi$ cm³; $S = 64\pi$ cm²

35. $V = 648\pi$ in.³; $S = 306\pi$ in.²   **37.** π square units   **39.** 2π square units   **41.** $x = 4$ units; $A = 90°$; $B = 60°$; $C = 30°$

43. $x = 67.5$ units; $A = 60°$; $B = 95°$; $C = 25°$ **45.** About 16.8 ft   **47.** 64 ft²   **49.** $24 + 2\pi \approx 30.28$ ft²; $16 + 2\pi \approx 22.28$ ft

51. 160 paces   **53.** About 5.477 mi   **55.** From 100 ft: 12.2 mi; From 150 ft: 15.0 mi

A.3 Assess Your Understanding *(page A24)*

1. $3x(x - 2)(x + 2)$ **2.** prime   **3.** T   **4.** F   **5.** $(x + 6)(x - 6)$ **7.** $2(1 + 2x)(1 - 2x)$ **9.** $(x + 1)(x + 10)$ **11.** $(x - 7)(x - 3)$
13. $4(x^2 - 2x + 8)$ **15.** Prime   **17.** $-(x - 5)(x + 3)$ **19.** $3(x + 2)(x - 6)$ **21.** $y^2(y + 5)(y + 6)$ **23.** $(2x + 3)^2$ **25.** $2(3x + 1)(x + 1)$
27. $(x - 3)(x + 3)(x^2 + 9)$ **29.** $(x - 1)^2(x^2 + x + 1)^2$ **31.** $x^5(x - 1)(x + 1)$ **33.** $(4x + 3)^2$ **35.** $-(4x - 5)(4x + 1)$
37. $(2y - 5)(2y - 3)$ **39.** $-(3x - 1)(3x + 1)(x^2 + 1)$ **41.** $(x + 3)(x - 6)$ **43.** $(x + 2)(x - 3)$ **45.** $(3x - 5)(9x^2 - 3x + 7)$
47. $(x + 5)(3x + 11)$ **49.** $(x - 1)(x + 1)(x + 2)$ **51.** $(x - 1)(x + 1)(x^2 - x + 1)$ **53.** $2(3x + 4)(9x + 13)$ **55.** $2x(3x + 5)$
57. $5(x + 3)(x - 2)^2(x + 1)$ **59.** $3(4x - 3)(4x - 1)$ **61.** $6(3x - 5)(2x + 1)^2(5x - 4)$
63. The possibilities are $(x \pm 1)(x \pm 4) = x^2 \pm 5x + 4$ or $(x \pm 2)(x \pm 2) = x^2 \pm 4x + 4$, none of which equals $x^2 + 4$.

A.4 Assess Your Understanding *(page A37)*

5. equivalent equations   **6.** identity   **7.** F   **8.** T   **9.** add; $\dfrac{25}{4}$ **10.** discriminant; negative   **11.** F   **12.** F   **13.** $\{7\}$ **15.** $\{-3\}$ **17.** $\{4\}$ **19.** $\left\{\dfrac{5}{4}\right\}$

21. $\{-1\}$ **23.** $\{-18\}$ **25.** $\{-3\}$ **27.** $\{-16\}$ **29.** $\{0.5\}$ **31.** $\{2\}$ **33.** $\{2\}$ **35.** $\{3\}$ **37.** $\{0, 9\}$ **39.** $\{0, 9\}$ **41.** $\{21\}$ **43.** $\{-2, 2\}$ **45.** $\{6\}$

47. $\{-3, 3\}$ **49.** $\{-4, 1\}$ **51.** $\left\{-1, \dfrac{3}{2}\right\}$ **53.** $\{-4, 4\}$ **55.** $\{2\}$ **57.** No real solution   **59.** $\{-2, 2\}$ **61.** $\{-1, 3\}$ **63.** $\{-2, -1, 0, 1\}$ **65.** $\{0, 4\}$

67. $\{-6, 2\}$ **69.** $\left\{-\dfrac{1}{2}, 3\right\}$ **71.** $\{3, 4\}$ **73.** $\left\{\dfrac{3}{2}\right\}$ **75.** $\left\{-\dfrac{2}{3}, \dfrac{3}{2}\right\}$ **77.** $\left\{-\dfrac{3}{4}, 2\right\}$ **79.** $\{-6\}$ **81.** $\{-2, -1, 1, 2\}$ **83.** $\{-6, -5\}$ **85.** $\left\{-\dfrac{3}{2}, 2\right\}$

87. $\{-1, 0, 4\}$ **89.** $\{-1, 1\}$ **91.** $\left\{-2, \dfrac{1}{2}, 2\right\}$ **93.** $\{-5, 5\}$ **95.** $\{-1, 3\}$ **97.** $\{-3, 0\}$ **99.** 16   **101.** $\dfrac{1}{16}$ **103.** $\dfrac{1}{9}$ **105.** $\{-7, 3\}$ **107.** $\left\{-\dfrac{1}{4}, \dfrac{3}{4}\right\}$

109. $\left\{\dfrac{-1 - \sqrt{7}}{6}, \dfrac{-1 + \sqrt{7}}{6}\right\}$ **111.** $\{2 - \sqrt{2}, 2 + \sqrt{2}\}$ **113.** $\left\{\dfrac{5 - \sqrt{29}}{2}, \dfrac{5 + \sqrt{29}}{2}\right\}$ **115.** $\left\{1, \dfrac{3}{2}\right\}$ **117.** No real solution

119. $\left\{\dfrac{-1 - \sqrt{5}}{4}, \dfrac{-1 + \sqrt{5}}{4}\right\}$ **121.** $\left\{\dfrac{-\sqrt{3} - \sqrt{15}}{2}, \dfrac{-\sqrt{3} + \sqrt{15}}{2}\right\}$ **123.** No real solution   **125.** Repeated real solution   **127.** Two unequal

real solutions   **129.** $R = \dfrac{R_1 R_2}{R_1 + R_2}$ **131.** $R = \dfrac{mv^2}{F}$ **133.** $r = \dfrac{S - a}{S}$ **135.** $\dfrac{-b + \sqrt{b^2 - 4ac}}{2a} + \dfrac{-b - \sqrt{b^2 - 4ac}}{2a} = \dfrac{-2b}{2a} = \dfrac{-b}{a}$

137. $k = -\dfrac{1}{2}$ or $\dfrac{1}{2}$ **139.** The solutions of $ax^2 - bx + c = 0$ are $\dfrac{b + \sqrt{b^2 - 4ac}}{2a}$ and $\dfrac{b - \sqrt{b^2 - 4ac}}{2a}$.   **141.** (b)

A.5 Assess Your Understanding *(page A41)*

3. ZERO   **4.** F   **5.** $\{-2.21, 0.54, 1.68\}$ **7.** $\{-1.55, 1.15\}$ **9.** $\{-1.12, 0.36\}$ **11.** $\{-2.69, -0.49, 1.51\}$ **13.** $\{-2.86, -1.34, 0.20, 1.00\}$

15. No real solutions   **17.** $\{-18\}$ **19.** $\{-4\}$ **21.** $\left\{\dfrac{46}{5}\right\}$ **23.** $\{3\}$ **25.** $\{2\}$ **27.** $\{-4, 7\}$ **29.** $\left\{-\dfrac{2}{3}, 2\right\}$ **31.** $\{-2, -1, 2\}$ **33.** $\{15\}$ **35.** $\left\{-4, -\dfrac{1}{8}\right\}$

A.6 Assess Your Understanding *(page A49)*

4. real; imaginary; imaginary unit   **5.** $-2i, 2i$ **6.** F   **7.** T   **8.** F   **9.** $8 + 5i$ **11.** $-7 + 6i$ **13.** $-6 - 11i$ **15.** $6 - 18i$ **17.** $6 + 4i$ **19.** $10 - 5i$

21. 37   **23.** $\dfrac{6}{5} + \dfrac{8}{5}i$ **25.** $1 - 2i$ **27.** $\dfrac{5}{2} - \dfrac{7}{2}i$ **29.** $-\dfrac{1}{2} + \dfrac{\sqrt{3}}{2}i$ **31.** $2i$ **33.** $-i$ **35.** i **37.** -6 **39.** $-10i$ **41.** $-2 + 2i$ **43.** 0   **45.** 0   **47.** $2i$

49. $5i$ **51.** $5i$ **53.** $\{-2i, 2i\}$ **55.** $\{-4, 4\}$ **57.** $\{3 - 2i, 3 + 2i\}$ **59.** $\{3 - i, 3 + i\}$ **61.** $\left\{\dfrac{1}{4} - \dfrac{1}{4}i, \dfrac{1}{4} + \dfrac{1}{4}i\right\}$ **63.** $\left\{-\dfrac{1}{5} - \dfrac{2}{5}i, -\dfrac{1}{5} + \dfrac{2}{5}i\right\}$

65. $\left\{-\dfrac{1}{2}-\dfrac{\sqrt{3}}{2}i,-\dfrac{1}{2}+\dfrac{\sqrt{3}}{2}i\right\}$ **67.** $\{2,-1-\sqrt{3}i,-1+\sqrt{3}i\}$ **69.** $\{-2,2,-2i,2i\}$ **71.** $\{-3i,-2i,2i,3i\}$ **73.** Two complex solutions

75. Two unequal real solutions **77.** A repeated real solution **79.** $2-3i$ **81.** 6 **83.** 25 **85.** $2+3i$

87. $z+\bar{z}=(a+bi)+(a-bi)=2a;\ z-\bar{z}=(a+bi)-(a-bi)=2bi$

89. $\overline{z+w}=\overline{(a+bi)+(c+di)}=\overline{(a+c)+(b+d)i}=(a+c)-(b+d)i=(a-bi)+(c-di)=\bar{z}+\bar{w}$

95. $\sqrt{a}\cdot\sqrt{b}=\sqrt{ab}$ only when $\sqrt{a},\sqrt{b}$ are real numbers.

A.7 Assess Your Understanding (page A58)

5. negative **6.** closed interval **7.** $-5,5$ **8.** $-5<x<5$ **9.** T **10.** T **11.** $[0,2];0\le x\le 2$ **13.** $[2,\infty);x\ge 2$ **15.** $[0,3);0\le x<3$
17. (a) $6<8$ **(b)** $-2<0$ **(c)** $9<15$ **(d)** $-6>-10$ **19. (a)** $7>0$ **(b)** $-1>-8$ **(c)** $12>-9$ **(d)** $-8<6$
21. (a) $2x+4<5$ **(b)** $2x-4<-3$ **(c)** $6x+3<6$ **(d)** $-4x-2>-4$

23. $[0,4]$

25. $[4,6)$

27. $[4,\infty)$

29. $(-\infty,-4)$

31. $2\le x\le 5$

33. $-3<x<-2$

35. $x\ge 4$

37. $x<-3$

39. $<$ **41.** $>$ **43.** $\ge$ **45.** $<$ **47.** $\le$ **49.** $>$ **51.** $\ge$

53. $\{x|x<4\}$ or $(-\infty,4)$

55. $\{x|x\ge -1\}$ or $[-1,\infty)$

57. $\{x|x>3\}$ or $(3,\infty)$

59. $\{x|x\ge 2\}$ or $[2,\infty)$

61. $\{x|x>-7\}$ or $(-7,\infty)$

63. $\left\{x\middle|x\le\dfrac{2}{3}\right\}$ or $\left(-\infty,\dfrac{2}{3}\right]$

65. $\{x|x<-20\}$ or $(-\infty,-20)$

67. $\left\{x\middle|x>\dfrac{4}{3}\right\}$ or $\left[\dfrac{4}{3},\infty\right)$

69. $\{x|3\le x\le 5\}$ or $[3,5]$

71. $\left\{x\middle|\dfrac{2}{3}\le x\le 3\right\}$ or $\left[\dfrac{2}{3},3\right]$

73. $\left\{x\middle|-\dfrac{11}{2}<x<\dfrac{1}{2}\right\}$ or $\left(-\dfrac{11}{2},\dfrac{1}{2}\right)$

75. $\{x|-6<x<0\}$ or $(-6,0)$

77. $\{x|x<-5\}$ or $(-\infty,-5)$

79. $\{x|x\ge -1\}$ or $[-1,\infty)$

81. $\left\{x\middle|\dfrac{1}{2}\le x<\dfrac{5}{4}\right\}$ or $\left[\dfrac{1}{2},\dfrac{5}{4}\right)$

83. $\left\{x\middle|x<-\dfrac{1}{2}\right\}$ or $\left(-\infty,-\dfrac{1}{2}\right)$

85. $\left\{x\middle|x>\dfrac{10}{3}\right\}$ or $\left(\dfrac{10}{3},\infty\right)$

87. $\{x|x>3\}$ or $(3,\infty)$

89. $\{x|-4<x<4\};(-4,4)$

91. $\{x|x<-4\text{ or }x>4\};(-\infty,-4)\cup(4,\infty)$

93. $\{x|0\le x\le 1\};[0,1]$

95. $\{x|x<-1\text{ or }x>2\};(-\infty-1)\cup(2,\infty)$

97. $\{x|-1\le x\le 1\};[-1,1]$

99. $\{x|x\le -2\text{ or }x\ge 2\};(-\infty,-2]\cup[2,\infty)$

101. $|x-2|<\dfrac{1}{2};\left\{x\middle|\dfrac{3}{2}<x<\dfrac{5}{2}\right\}$ **103.** $|x+3|>2;\{x|x<-5\text{ or }x>-1\}$

105. $\{x|x\ge -2\}$ **107.** $21<\text{Age}<30$ **109. (a)** Male ≥ 79.66 years **(b)** Female ≥ 83.58 years **(c)** A female can expect to live 3.92 years longer.
111. The agent's commission ranges from \$45,000 to \$95,000, inclusive. As a percent of selling price, the commission ranges from 5% to 8.6%, inclusive.
113. The amount withheld varies from \$98.30 to \$148.30, inclusive. **115.** The usage varies from 675.41 kW · hr to 2500.91 kW · hr, inclusive.
117. The dealer's cost varies from \$15,254.24 to \$16,071.43, inclusive.
119. (a) You need at least a 74 on the fifth test. **(b)** You need at least a 77 on the fifth test.

121. $\dfrac{a+b}{2} - a = \dfrac{a+b-2a}{2} = \dfrac{b-a}{2} > 0;$ therefore, $a < \dfrac{a+b}{2}.$ $b - \dfrac{a+b}{2} = \dfrac{2b-a-b}{2} = \dfrac{b-a}{2} > 0;$ therefore, $b > \dfrac{a+b}{2}.$

123. $(\sqrt{ab})^2 - a^2 = ab - a^2 = a(b-a) > 0;$ thus $(\sqrt{ab})^2 > a^2$ and $\sqrt{ab} > a.$
$b^2 - (\sqrt{ab})^2 = b^2 - ab = b(b-a) > 0;$ thus $b^2 > (\sqrt{ab})^2$ and $b > \sqrt{ab}.$

125. $h - a = \dfrac{2ab}{a+b} - a = \dfrac{ab-a^2}{a+b} = \dfrac{a(b-a)}{a+b} > 0;$ thus $h > a.$ $\quad b - h = b - \dfrac{2ab}{a+b} = \dfrac{b^2-ab}{a+b} = \dfrac{b(b-a)}{a+b} > 0;$ thus $h < b.$

127. Since $0 < a < b,$ then $a - b < 0$ and $\dfrac{a-b}{ab} < 0.$ So $\dfrac{a}{ab} - \dfrac{b}{ab} < 0,$ or $\dfrac{1}{b} - \dfrac{1}{a} < 0.$ Therefore, $\dfrac{1}{b} < \dfrac{1}{a}.$ And $0 < \dfrac{1}{b}$ because $b > 0.$

A.8 Assess Your Understanding *(page A66)*

3. index **4.** T **5.** cube root **6.** F **7.** 3 **9.** -2 **11.** $2\sqrt{2}$ **13.** $-2x\sqrt[3]{x}$ **15.** x^3y^2 **17.** x^2y **19.** $6\sqrt{x}$ **21.** $6x\sqrt{x}$ **23.** $15\sqrt[3]{3}$ **25.** $12\sqrt{3}$

27. $7\sqrt{2}$ **29.** $\sqrt{2}$ **31.** $2\sqrt{3}$ **33.** $-\sqrt[3]{2}$ **35.** $x - 2\sqrt{x} + 1$ **37.** $(2x-1)\sqrt[3]{2x}$ **39.** $(2x-15)\sqrt{2x}$ **41.** $-(x+5y)\sqrt[3]{2xy}$ **43.** $\dfrac{\sqrt{2}}{2}$

45. $-\dfrac{\sqrt{15}}{5}$ **47.** $\dfrac{(5+\sqrt{2})\sqrt{3}}{23}$ **49.** $\dfrac{8\sqrt{5}-19}{41}$ **51.** $\dfrac{5\sqrt[3]{4}}{2}$ **53.** $\dfrac{2x+h-2\sqrt{x^2+xh}}{h}$ **55.** 4 **57.** -3 **59.** 64 **61.** $\dfrac{1}{27}$ **63.** $\dfrac{27\sqrt{2}}{32}$

65. $\dfrac{27\sqrt{2}}{32}$ **67.** $x^{7/12}$ **69.** xy^2 **71.** $x^{2/3}y$ **73.** $\dfrac{8x^{5/4}}{y^{3/4}}$ **75.** $\dfrac{3x+2}{(1+x)^{1/2}}$ **77.** $\dfrac{x(3x^2+2)}{(x^2+1)^{1/2}}$ **79.** $\dfrac{22x+5}{10\sqrt{(x-5)(4x+3)}}$ **81.** $\dfrac{2+x}{2(1+x)^{3/2}}$

83. $\dfrac{4-x}{(x+4)^{3/2}}$ **85.** $\dfrac{1}{x^2(x^2-1)^{1/2}}$ **87.** $\dfrac{1-3x^2}{2\sqrt{x}(1+x^2)^2}$ **89.** $\dfrac{1}{2}(5x+2)(x+1)^{1/2}$ **91.** $2x^{1/2}(3x-4)(x+1)$ **93.** $(x^2+4)^{1/3}(11x^2+12)$

95. $(3x+5)^{1/3}(2x+3)^{1/2}(17x+27)$ **97.** $\dfrac{3(x+2)}{2x^{1/2}}$ **99.** 1.41 **101.** 1.59 **103.** 4.89 **105.** 2.15 **107. (a)** 15,660.4 gal **(b)** 390.7 gal

109. $2\pi\sqrt{2} \approx 8.89$ s **111.** $\dfrac{\pi\sqrt{3}}{6} \approx 0.91$ s

A.9 Assess Your Understanding *(page A80)*

1. undefined; 0 **2.** 3; 2 **3.** $y = b;$ y-intercept **4.** T **5.** F **6.** T **7.** $m_1 = m_2;$ y-intercepts; $m_1 m_2 = -1$ **8.** 2 **9.** $-\dfrac{1}{2}$ **10.** F

11. (a) Slope $= \dfrac{1}{2}$

   (b) If x increases by 2 units, y will increase by 1 unit.

13. (a) Slope $= -\dfrac{1}{3}$

   (b) If x increases by 3 units, y will decrease by 1 unit.

15. Slope $= -\dfrac{3}{2}$

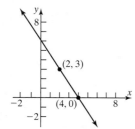

17. Slope $= -\dfrac{1}{2}$

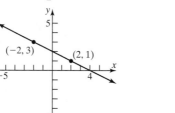

19. Slope $= 0$

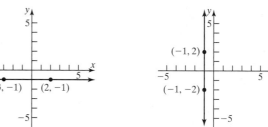

21. Slope undefined

23.

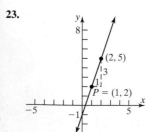

25.

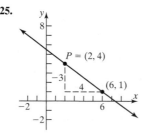

27.

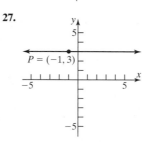

29.

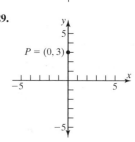

31. $(2,6); (3,10); (4,14)$ **33.** $(4,-7); (6,-10); (8,-13)$ **35.** $(-1,-5); (0,-7); (1,-9)$ **37.** $x - 2y = 0$ or $y = \dfrac{1}{2}x$ **39.** $x + y = 2$ or $y = -x + 2$

41. $2x - y = 3$ or $y = 2x - 3$ **43.** $x + 2y = 5$ or $y = -\dfrac{1}{2}x + \dfrac{5}{2}$ **45.** $3x - y = -9$ or $y = 3x + 9$ **47.** $2x + 3y = -1$ or $y = -\dfrac{2}{3}x - \dfrac{1}{3}$

49. $x - 2y = -5$ or $y = \dfrac{1}{2}x + \dfrac{5}{2}$ **51.** $3x + y = 3$ or $y = -3x + 3$ **53.** $x - 2y = 2$ or $y = \dfrac{1}{2}x - 1$ **55.** $x = 2;$ no slope–intercept form

57. $y = 2$ **59.** $2x - y = -4$ or $y = 2x + 4$ **61.** $2x - y = 0$ or $y = 2x$ **63.** $x = 4;$ no slope–intercept form **65.** $2x + y = 0$ or $y = -2x$

67. $x - 2y = -3$ or $y = \dfrac{1}{2}x + \dfrac{3}{2}$ **69.** $y = 4$

71. Slope = 2; y-intercept = 3

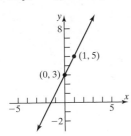

73. Slope = 2; y-intercept = -2

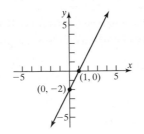

75. Slope = $\frac{1}{2}$; y-intercept = 2

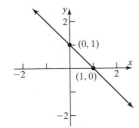

77. Slope = $-\frac{1}{2}$; y-intercept = 2

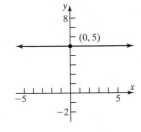

79. Slope = $\frac{2}{3}$; y-intercept = -2

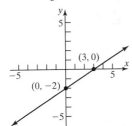

81. Slope = -1; y-intercept = 1

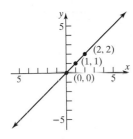

83. Slope undefined; no y-intercept

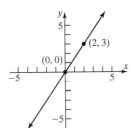

85. Slope = 0; y-intercept = 5

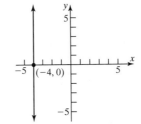

87. Slope = 1; y-intercept = 0

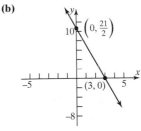

89. Slope = $\frac{3}{2}$; y-intercept = 0

91. (a) x-intercept: 3; y-intercept: 2

(b)

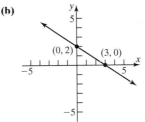

93. (a) x-intercept: -10; y-intercept: 8

(b)

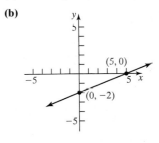

95. (a) x-intercept: 3; y-intercept: $\frac{21}{2}$

(b)

97. (a) x-intercept: 2; y-intercept: 3

(b)

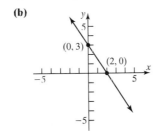

99. (a) x-intercept: 5; y-intercept: -2

(b)

101. $y = 0$ **103.** Parallel   **105.** Neither   **107.** (b)   **109.** (d)

111. $P_1 = (-2, 5)$, $P_2 = (1, 3)$, $m_1 = -\frac{2}{3}$; $P_2 = (1, 3)$, $P_3 = (-1, 0)$, $m_2 = \frac{3}{2}$; because $m_1 m_2 = -1$, the lines are perpendicular and the points $(-2, 5)$, $(1, 3)$, and $(-1, 0)$ are the vertices of a right triangle; thus, the points P_1, P_2, and P_3 are the vertices of a right triangle.

113. $P_1 = (-1, 0)$, $P_2 = (2, 3)$, $m = 1$; $P_3 = (1, -2)$, $P_4 = (4, 1)$, $m = 1$; $P_1 = (-1, 0)$, $P_3 = (1, -2)$, $m = -1$; $P_2 = (2, 3)$, $P_4 = (4, 1)$, $m = -1$; opposite sides are parallel, and adjacent sides are perpendicular; the points are the vertices of a rectangle.

115. $C = 0.20x + 29$; \$51.00; \$75.00   **117.** $C = 0.53x + 1,070,000$

119. (a) $C = 0.08275x + 7.58$, $0 \le x \le 400$ **(b)**

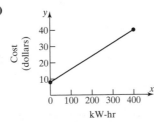

(c) \$15.86
(d) \$32.41
(e) Each additional kW-hr used adds \$0.08275 to the bill.

121. $°C = \frac{5}{9}(°F - 32)$; approximately $21.1°C$

123. (a) $y = -\dfrac{2}{25}x + 30$ **(b)** x-intercept: 375; the ramp meets the floor 375 in. (31.25 ft) from the base of the platform.

(c) The ramp does not meet design requirements. It has a run of 31.25 ft long.

(d) The only slope possible for the ramp to comply with the requirement is for it to drop 1 in. for every 12-in. run.

125. (a) $A = \dfrac{1}{5}x + 20{,}000$ **(b)** \$80,000 **(c)** Each additional box sold requires an additional \$0.20 in advertising.

127. All have the same slope, 2; the lines are parallel.

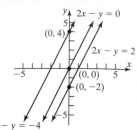

129. (b), (c), (e), (g)

131. (c)

137. No; no

139. They are the same line.

141. Yes, if the y-intercept is 0.

A.10 Assess Your Understanding (page A88)

1. scatter diagram **2.** T **3.** Linear relation, $m > 0$ **5.** Linear relation, $m < 0$ **7.** Nonlinear relation

9. (a)

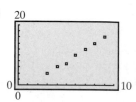

(b) Answers vary. Using $(4, 6)$ and $(8, 14)$, $y = 2x - 2$.

(c)

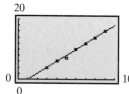

(d) $y = 2.0357x - 2.3571$

(e)

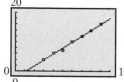

11. (a)

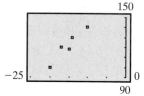

(b) Answers will vary. Using $(-2, -4)$ and $(2, 5)$, $y = \dfrac{9}{4}x + \dfrac{1}{2}$.

(c)

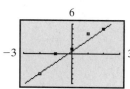

(d) $y = 2.2x + 1.2$

(e)

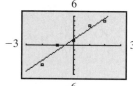

13. (a)

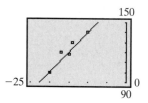

(b) Answers will vary. Using $(-20, 100)$ and $(-10, 140)$, $y = 4x + 180$

(c)

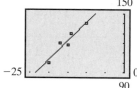

(d) $y = 3.8613x + 180.2920$

(e)

15. (a)

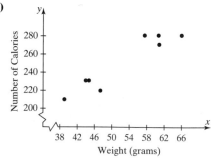

Weight (grams)

(d)

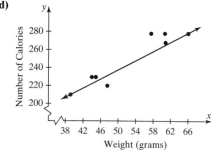

Weight (grams)

(b) Linear

(c) Answers will vary. Using the points $(39.52, 210)$ and $(66.45, 280)$, $y = 2.599x + 107.288$.

(e) 269 calories

(f) If the weight of a candy bar is increased by 1 gram, the number of calories will increase by 2.599.

17. (a)

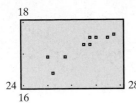

(b) $C(H) = 0.3734H + 7.3268$
(c) If height increases by 1 inch, head circumference increases by about 0.3734 inch.
(d) About 17.0 inches
(e) About 26.98 inches

19. (a) No

(b)

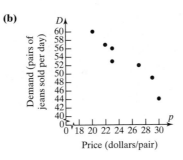

Price (dollars/pair)

(c) $D = -1.3355p + 86.1974$
(d) If the price increases $1, the quantity sold per day decreases by about 1.34 pairs of jeans.
(e) $D(p) = -1.3355p + 86.1974$
(f) $\{p \mid 0 < p \leq 64\}$
(g) About 49 pairs

21.

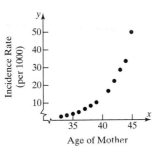

Age of Mother

No, the data do not follow a linear pattern.

23. No linear relation

Index

LICENSING AGREEMENT
MICHAEL SULLIVAN / MICHAEL SULLIVAN, III
Chapter Test Prep Videos CD Trigonometry: A Right Angle Approach, 5e
ISBN: 0-13-602896-9/978-0-13-602896-3
© 2009 Pearson Education, Inc.
Pearson Prentice Hall, Upper Saddle River, NJ 07458
Pearson Prentice Hall™ is a trademark of Pearson Education, Inc.

YOU SHOULD CAREFULLY READ THE TERMS AND CONDITIONS BEFORE USING THE CD-ROM PACKAGE. USING THIS CD-ROM PACKAGE INDICATES YOUR ACCEPTANCE OF THESE TERMS AND CONDITIONS.

Pearson Education, Inc. provides this program and licenses its use. You assume responsibility for the selection of the program to achieve your intended results, and for the installation, use, and results obtained from the program. This license extends only to use of the program in the United States or countries in which the program is marketed by authorized distributors.

LICENSE GRANT: You hereby accept a nonexclusive, nontransferable, permanent license to install and use the program ON A SINGLE COMPUTER at any given time. You may copy the program solely for backup or archival purposes in support of your use of the program on the single computer. You may not modify, translate, disassemble, decompile, or reverse engineer the program, in whole or in part.

TERM: The License is effective until terminated. Pearson Education, Inc. reserves the right to terminate this License automatically if any provision of the License is violated. You may terminate the License at any time. To terminate this License, you must return the program, including documentation, along with a written warranty stating that all copies in your possession have been returned or destroyed.

LIMITED WARRANTY: THE PROGRAM IS PROVIDED "AS IS" WITHOUT WARRANTY OF ANY KIND, EITHER EXPRESSED OR IMPLIED, INCLUDING, BUT NOT LIMITED TO, THE IMPLIED WARRANTIES OF MERCHANTABILITY AND FITNESS FOR A PARTICULAR PURPOSE. THE ENTIRE RISK AS TO THE QUALITY AND PERFORMANCE OF THE PROGRAM IS WITH YOU. SHOULD THE PROGRAM PROVE DEFECTIVE, YOU (AND NOT PEARSON EDUCATION, INC. OR ANY AUTHORIZED DEALER) ASSUME THE ENTIRE COST OF ALL NECESSARY SERVICING, REPAIR, OR CORRECTION. NO ORAL OR WRITTEN INFORMATION OR ADVICE GIVEN BY PEARSON EDUCATION, INC., ITS DEALERS, DISTRIBUTORS, OR AGENTS SHALL CREATE A WARRANTY OR INCREASE THE SCOPE OF THIS WARRANTY. SOME STATES DO NOT ALLOW THE EXCLUSION OF IMPLIED WARRANTIES, SO THE ABOVE EXCLUSION MAY NOT APPLY TO YOU. THIS WARRANTY GIVES YOU SPECIFIC LEGAL RIGHTS AND YOU MAY ALSO HAVE OTHER LEGAL RIGHTS THAT VARY FROM STATE TO STATE.

Pearson Education, Inc. does not warrant that the functions contained in the program will meet your requirements or that the operation of the program will be uninterrupted or error-free. However, Pearson Education, Inc. warrants the CD-ROM(s) on which the program is furnished to be free from defects in material and workmanship under normal use for a period of ninety (90) days from the date of delivery to you as evidenced by a copy of your receipt. The program should not be relied on as the sole basis to solve a problem whose incorrect solution could result in injury to person or property. If the program is employed in such a manner, it is at the user's own risk and Pearson Education, Inc. explicitly disclaims all liability for such misuse.

LIMITATION OF REMEDIES: Pearson Education, Inc.'s entire liability and your exclusive remedy shall be:

1. the replacement of any CD-ROM not meeting Pearson Education, Inc.'s "LIMITED WARRANTY" and that is returned to Pearson Education, or
2. if Pearson Education is unable to deliver a replacement CD-ROM that is free of defects in materials or workmanship, you may terminate this agreement by returning the program.

IN NO EVENT WILL PEARSON EDUCATION, INC. BE LIABLE TO YOU FOR ANY DAMAGES, INCLUDING ANY LOST PROFITS, LOST SAVINGS, OR OTHER INCIDENTAL OR CONSEQUENTIAL DAMAGES ARISING OUT OF THE USE OR INABILITY TO USE SUCH PROGRAM EVEN IF PEARSON EDUCATION, INC. OR AN AUTHORIZED DISTRIBUTOR HAS BEEN ADVISED OF THE POSSIBILITY OF SUCH DAMAGES, OR FOR ANY CLAIM BY ANY OTHER PARTY.

SOME STATES DO NOT ALLOW FOR THE LIMITATION OR EXCLUSION OF LIABILITY FOR INCIDENTAL OR CONSEQUENTIAL DAMAGES, SO THE ABOVE LIMITATION OR EXCLUSION MAY NOT APPLY TO YOU.

GENERAL: You may not sublicense, assign, or transfer the license of the program. Any attempt to sublicense, assign or transfer any of the rights, duties, or obligations hereunder is void.

This Agreement will be governed by the laws of the State of New York. Should you have any questions concerning this Agreement, you may contact Pearson Education, Inc. by writing to:

ESM Media Development
Higher Education Division
Pearson Education, Inc.

1 Lake Street
Upper Saddle River, NJ 07458

Should you have any questions concerning technical support, you may write to:

New Media Production
Higher Education Division
Pearson Education, Inc.
1 Lake Street
Upper Saddle River, NJ 07458

YOU ACKNOWLEDGE THAT YOU HAVE READ THIS AGREEMENT, UNDERSTAND IT, AND AGREE TO BE BOUND BY ITS TERMS AND CONDITIONS. YOU FURTHER AGREE THAT IT IS THE COMPLETE AND EXCLUSIVE STATEMENT OF THE AGREEMENT BETWEEN US THAT SUPERSEDES ANY PROPOSAL OR PRIOR AGREEMENT, ORAL OR WRITTEN, AND ANY OTHER COMMUNICATIONS BETWEEN US RELATING TO THE SUBJECT MATTER OF THIS AGREEMENT.

Windows System Requirements
- Microsoft Windows® XP with Service Pack 2 or Windows Vista
- In addition to the minimum processor requirements for the operating system your computer is running, this CD requires a Pentium III, 600 MHz
- In addition to the RAM required by the operating system your computer is running, this CD requires 128 MB RAM
- 390 MB hard disk space (required if all supplement files are installed to computer)
- VGA display at a screen resolution of 800 x 600
- Color Monitor running "Thousands of Colors"
- CD-ROM or DVD-ROM drive
- Mouse or compatible pointing device
- Internet browser (Internet Explorer 6.x, Netscape Navigator 7, Firefox 2.x) with an Internet connection is required to open web page links.

Macintosh System Requirements
- OS X 10.2.8, 10.3, or 10.4
- PowerPC G4
- 128 MB RAM
- 390 MB hard disk space (required if all supplement files are installed to computer)
- 1,024 × 768 pixel screen resolution with 16-bit video card (24-bit screen display recommended)
- Color monitor running "Thousands of Colors"
- 4x CD-ROM or DVD-ROM drive
- Mouse or compatible pointing device
- Printer is required to print files
- Internet browser (Netscape Navigator 7, Safari 2.x, Firefox 2.x) with an Internet connection is required to open web page links.

Third Party Software
- QuickTime

SUPPORT INFORMATION

If you are having problems with this software, you can get support by filling out the web form located at: http://247.pearsoned.com/
Our technical staff will need to know certain things about your system in order to help us solve your problems more quickly and efficiently. If possible, please be at your computer when you call for support. You should have the following information ready:

- ISBN
- CD-ROM ISBN
- corresponding product and title
- computer make and model
- Operating System (Windows or Macintosh) and Version
- RAM available
- hard disk space available
- Sound card? Yes or No
- printer make and model
- network connection
- detailed description of the problem, including the exact wording of any error messages.

NOTE: Pearson does not support and/or assist with the following:
- third-party software (i.e. Microsoft including Microsoft Office suite, Apple, Borland, etc.)
- homework assistance
- Textbooks and CD-ROMs purchased used are not supported and are non-replaceable. To purchase a new CD-ROM, contact Pearson Individual Order Copies at 1-800-282-0693.

CONICS

Parabola

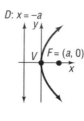
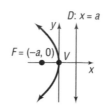
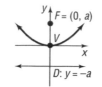

$$y^2 = 4ax \qquad y^2 = -4ax \qquad x^2 = 4ay \qquad x^2 = -4ay$$

Ellipse

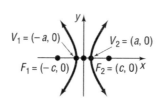

$$\frac{x^2}{a^2} + \frac{y^2}{b^2} = 1, \quad a > b, \quad c^2 = a^2 - b^2 \qquad\qquad \frac{x^2}{b^2} + \frac{y^2}{a^2} = 1, \quad a > b, \quad c^2 = a^2 - b^2$$

Hyperbola

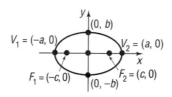
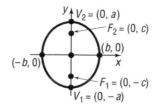

$$\frac{x^2}{a^2} - \frac{y^2}{b^2} = 1, \quad c^2 = a^2 + b^2 \qquad\qquad \frac{y^2}{a^2} - \frac{x^2}{b^2} = 1, \quad c^2 = a^2 + b^2$$

$$\text{Asymptotes:} \quad y = \frac{b}{a}x, \quad y = -\frac{b}{a}x \qquad\qquad \text{Asymptotes:} \quad y = \frac{a}{b}x, \quad y = -\frac{a}{b}x$$

PROPERTIES OF LOGARITHMS

$$\log_a(MN) = \log_a M + \log_a N$$

$$\log_a\left(\frac{M}{N}\right) = \log_a M - \log_a N$$

$$\log_a M^r = r \log_a M$$

$$\log_a M = \frac{\log M}{\log a} = \frac{\ln M}{\ln a}$$

LIBRARY OF FUNCTIONS

Identity Function
$f(x) = x$

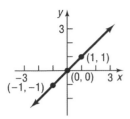

Square Function
$f(x) = x^2$

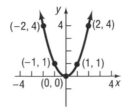

Cube Function
$f(x) = x^3$

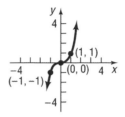

Square Root Function
$f(x) = \sqrt{x}$

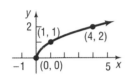

Reciprocal Function
$f(x) = \dfrac{1}{x}$

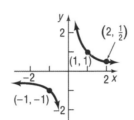

Cube Root Function
$f(x) = \sqrt[3]{x}$

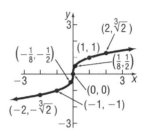

Absolute Value Function
$f(x) = |x|$

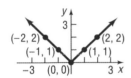

Exponential Function
$f(x) = e^x$

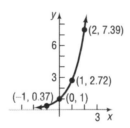

Natural Logarithm Function
$f(x) = \ln x$

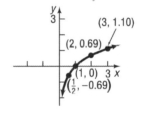

Sine Function
$f(x) = \sin x$

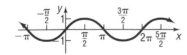

Cosine Function
$f(x) = \cos x$

Tangent Function
$f(x) = \tan x$

Cosecant Function
$f(x) = \csc x$

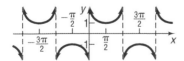

Secant Function
$f(x) = \sec x$

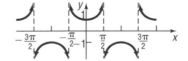

Cotangent Function
$f(x) = \cot x$

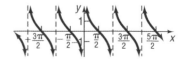

FORMULAS/EQUATIONS

Distance Formula

If $P_1 = (x_1, y_1)$ and $P_2 = (x_2, y_2)$, the distance from P_1 to P_2 is

$$d(P_1, P_2) = \sqrt{(x_2 - x_1)^2 + (y_2 - y_1)^2}$$

Standard Equation of a Circle

The standard equation of a circle of radius r with center at (h, k) is

$$(x - h)^2 + (y - k)^2 = r^2$$

Slope Formula

The slope m of the line containing the points $P_1 = (x_1, y_1)$ and $P_2 = (x_2, y_2)$ is

$$m = \frac{y_2 - y_1}{x_2 - x_1} \qquad \text{if } x_1 \neq x_2$$

$$m \text{ is undefined} \qquad \text{if } x_1 = x_2$$

Point–Slope Equation of a Line

The equation of a line with slope m containing the point (x_1, y_1) is

$$y - y_1 = m(x - x_1)$$

Slope–Intercept Equation of a Line

The equation of a line with slope m and y-intercept b is

$$y = mx + b$$

Quadratic Formula

The solutions of the equation $ax^2 + bx + c = 0, a \neq 0$, are

$$x = \frac{-b \pm \sqrt{b^2 - 4ac}}{2a}$$

If $b^2 - 4ac > 0$, there are two unequal real solutions.
If $b^2 - 4ac = 0$, there is a repeated real solution.
If $b^2 - 4ac < 0$, there are two complex solutions that are not real.

GEOMETRY FORMULAS

Circle

$r = $ Radius, $\;A = $ Area, $\;C = $ Circumference
$$A = \pi r^2 \qquad C = 2\pi r$$

Triangle

$b = $ Base, $\;h = $ Altitude (Height), $\;A = $ area
$$A = \tfrac{1}{2}bh$$

Rectangle

$l = $ Length, $\;w = $ Width, $\;A = $ area, $\;P = $ perimeter
$$A = lw \qquad P = 2l + 2w$$

Rectangular Box

$l = $ Length, $\;w = $ Width, $\;h = $ Height, $\;V = $ Volume, $\;S = $ Surface area
$$V = lwh \qquad S = 2lw + 2lh + 2wh$$

Sphere

$r = $ Radius, $\;V = $ Volume, $\;S = $ Surface area
$$V = \tfrac{4}{3}\pi r^3 \qquad S = 4\pi r^2$$

Right Circular Cylinder

$r = $ Radius, $\;h = $ Height, $\;V = $ Volume, $\;S = $ Surface area
$$V = \pi r^2 h \qquad S = 2\pi r^2 + 2\pi rh$$